PLANTS OF CHINA

A companion to the *Flora of China*

The flora of China is astonishing in its diversity. With 32,500 species of vascular plants, over 50 percent of which are endemic, it has more botanical variety than anywhere else in the world and provides unbroken connections to all its landscapes from tropical to subtropical, temperate and boreal forests.

This book tells the story of the plants of China: from the evolution of the flora through time to the survey of the bioclimatic zones, soundly based on chapters with information on climate, physical geography and soils. The history of botany and its study are also examined, with chapters dedicated to forestry, medicinal plants and ornamentals, with the changing flora, aliens, extinction and conservation also discussed.

An essential read for years to come, *Plants of China* shows that an understanding of the flora of China is crucial to interpreting plant evolution and fossil history elsewhere in the world.

Hong De-Yuan is Professor of the State Key Laboratory of Systematic and Evolutionary Botany at the Institute of Botany, Chinese Academy of Sciences in Beijing. In 2012 he was awarded the Royal Botanic Gardens & Domain Trust Lachlan Macquarie Medal, in recognition of his outstanding achievement in helping protect plant biodiversity. He became a CAS academician in 1991.

Stephen Blackmore CBE FRSE is Queen's Botanist and Honorary Fellow at the Royal Botanic Garden Edinburgh, where he was previously the Regius Keeper. Before that he was Keeper of Botany at the Natural History Museum in London. His research has concentrated on the area of palynology, microscopy and systematics and his achievements have been recognized by three Linnean Society medals.

PLANTS OF CHINA

A companion to the *Flora of China*

EDITED BY

HONG De-Yuan

Institute of Botany, Chinese Academy of Sciences, 20 Nanxincun, Xiangshan, Beijing 100093, China

Stephen Blackmore

Regius Keeper, Royal Botanic Garden Edinburgh, 20A Inverleith Row, Edinburgh EH3 5LR, UK

Shaftesbury Road, Cambridge CB2 8EA, United Kingdom

One Liberty Plaza, 20th Floor, New York, NY 10006, USA

477 Williamstown Road, Port Melbourne, VIC 3207, Australia

314–321, 3rd Floor, Plot 3, Splendor Forum, Jasola District Centre, New Delhi – 110025, India

103 Penang Road, #05–06/07, Visioncrest Commercial, Singapore 238467

Cambridge University Press is part of Cambridge University Press & Assessment, a department of the University of Cambridge.

We share the University's mission to contribute to society through the pursuit of education, learning and research at the highest international levels of excellence.

www.cambridge.org
Information on this title: www.cambridge.org/9781107070172

© Science Press 2015

This publication is in copyright. Subject to statutory exception and to the provisions of relevant collective licensing agreements, no reproduction of any part may take place without the written permission of Cambridge University Press & Assessment.

First published in 2013 by Science Press (Beijing) under ISBN 978-7-030-38574-1

This edition published 2015

A catalogue record for this publication is available from the British Library

ISBN 978-1-107-07017-2 Hardback

Cambridge University Press & Assessment has no responsibility for the persistence or accuracy of URLs for external or third-party internet websites referred to in this publication and does not guarantee that any content on such websites is, or will remain, accurate or appropriate.

CONTENTS

FOREWORD

When the *Flora of China* was first conceived in 1979 and formally agreed upon in 1988, it was always intended that there would be an introductory volume providing an overview, not only of the *Flora of China,* but also of the plants of China generally. *Plants of China* is the realisation of that intention. *Plants of China* is not a formal part of the *Flora,* but summarizes in an accessible way the state of knowledge of the plants of China considered in many dimensions. It will be useful as a guide to China's plants from the various points of view summarized in its chapters.

The *Flora of China* has been completed in 49 volumes published as 45 books. It amounts to a complete English-language revision of the Chinese-language *Flora Reipublicae Popularis Sinicae.* It is a monumental project both in terms of both the taxonomic works produced and the collaboration fostered among Chinese and non-Chinese botanists. The revisions that have been presented have far-reaching implications. During the course of preparing this work, the Chinese and non-Chinese authors published more than 2 200 new names, of which nearly 1 200 represent taxa new to science, among them 14 new genera and almost 1 000 new species. The project involved an amazing 470 authors, providing a unique opportunity for scientists from China, the USA, Canada, Europe, Russia, Australia and Japan, among others, to work together. And, importantly in this electronic age, the entire *Flora* is presented, and searchable, online (http://www.efloras.org/flora_page.aspx?flora_id=2).

In contrast to the floristic volumes, which have been co-authored jointly by Chinese and non-Chinese scientists, *Plants of China* has been written in the main by Chinese authors, all of whom are leaders in their fields. it covers ecology, plant geography, the uses of plants, and many other important features of the nation's plants that could not be treated in detail in the *Flora.* Much of the information in the volume has not previously been published outside China, and it has certainly never been brought together in this way. Thus, *Plants of China* brings knowledge of Chinese plants to a wider international audience than ever before and provides a gateway to the wider literature on this fascinating and valuable array of organisms.

WU Cheng-Yih (WU Zheng-Yi) and Peter H. Raven

Kunming and St. Louis

PREFACE

The *Plants of China* was conceived of, and has been developed as, a companion publication to the *Flora of China*, the remarkable international project which has so successfully completed an English-language revision of the *Flora Republicae Popularis Sinicae* through the efforts of 470 scientists around the world. As anticipated, this task required 25 years of research. With its descriptions of over 31 000 species of plants, the *Flora of China* stands as a high point in modern floristic botany. It was one of the first projects of its kind to use the internet to share information as widely as possible, including to those who might never have access to the hard-copy books themselves.

The completion of the *Flora of China* comes at a pivotal point in history, a time when we have finally come fully to appreciate the fundamental importance of biodiversity in sustaining human lives. Plants have long been taken for granted but we now understand that we neglect them at our peril. They are the base of our food chain and provide the essential ecosystem services that both created and maintain the biosphere. Now that the biodiversity crisis has the attention of governments and people around the world we can fully appreciate the vision and leadership of Peter Raven and WU Cheng-Yih (WU Zheng-Yi) for conceiving the Flora of China project and bringing it to fruition. We thank them for their inspirational leadership and note with regret the recent passing of WU Cheng-Yih (WU Zheng-Yi).

The purpose of this book, as a companion to the *Flora of China*, is to provide a synthesis of the wider knowledge concerning the plants of China. It does so through a series of chapters written by appropriate experts in their respective fields. It also aims to provide a route into the wider botanical literature of the subject, including a great many standard reference works published in China that are less well-known internationally than they deserve to be. In this way we hope that, like the *Flora of China* itself, this book can serve as a bridge between nations. We thank the scientists who have contributed to it for sharing their expertise. We extend a special thanks for the financial support. We are grateful to the National Natural Science Foundation of China (39899400), Chinese Academy of Sciences (KSCX2-SW-122 & KSCX2-YW-Z-0901), and Ministry of Science and Technology of China (2006FY120100).

Extensive though this volume is, it scarcely does justice to either the importance or the vast diversity of the plants of China. It is our hope, nevertheless, that it will serve to excite interest in the subject and we hope that you, the reader, will enjoy the opportunity to visit China and to study its plants. Few opportunities can be more rewarding.

HONG De-Yuan and Stephen Blackmore

Beijing and Edinburgh

CONTRIBUTORS
(arranged alphabetically)

Stephen Blackmore
Regius Keeper, Royal Botanic Garden Edinburgh, 20A Inverleith Row, Edinburgh EH3 5LR, UK

David E. Boufford
Harvard University Herbaria, 22 Divinity Avenue, Cambridge, Massachusetts 02138-2020, USA.

CHEN Ling-Zhi
Institute of Botany, Chinese Academy of Sciences, 20 Nanxincun, Xiangshan, Beijing 100093, China

DAI Er-Fu
Institute of Geographic Sciences and Natural Resources Research, Chinese Academy of Sciences, 11A Datun Road, Chaoyang, Beijing 100101, China

FAN Xiao-Hong
Institute of Plant Quarantine, Chinese Academy of Inspection and Quarantine, Beijing, China

HE Shan-An
Institute of Botany, Jiangsu and Chinese Academy of Sciences, PO Box 1435, 1 Qianhu Houcun, Zhongshanmen Wai, Nanjing 210014, Jiangsu, China

HONG De-Yuan
Institute of Botany, Chinese Academy of Sciences, 20 Nanxincun, Xiangshan, Beijing 100093, China

HU Chi-Ming (HU Qi-Ming)
South China Botanical Garden, Chinese Academy of Sciences, 723 Xingke Road, Tianhe, Guangzhou 510650, China

HU Guang-Wan
Kunming Institute of Botany, Chinese Academy of Sciences, 132 Lanhei Road, Heilongtan, Kunming 650201, Yunnan, China

HU Zong-Gang
Lushan Botanical Garden, Jiangxi and Chinese Academy of Sciences, Jiujiang 332900, Jiangxi, China

HUAI Hu-Yin
Yangzhou University, 88 South University Avenue, Yangzhou, Jiangsu 225009, China

HUANG Hong-Wen
South China Botanical Garden, Chinese Academy of Sciences, 723 Xingke Road, Tianhe, Guangzhou 510650, China

Peter S. Wyse Jackson
President, Missouri Botanical Garden, PO Box 299, St. Louis, Missouri 63166-0299, USA

LI De-Zhu
Kunming Institute of Botany, Chinese Academy of Sciences, 132 Lanhei Road, Heilongtan, Kunming 650201, Yunnan, China

LI Zhen-Yu
Institute of Botany, Chinese Academy of Sciences, 20 Nanxincun, Xiangshan, Beijing 100093, China

MA Hai-Ying
School of Life Sciences, Yunnan University, 2 Cuihu North Road, Kunming 650091, China

MA Jin-Shuang
Chenshan Botanical Garden, Shanghai, 3888 Chenhua Road, Songjiang, Shanghai 201602, China

Peter L. Morrell
Department of Agronomy and Plant Genetics, University of Minnesota, 411 Borlaug Hall, 1991 Upper Buford Circle, St. Paul, Minnesota 55108-6026, USA

Hong QIAN
Curator of Botany, Illinois State Museum Research and Collections Center, 1011 East Ash Street, Springfield, Illinois 62703, USA

Sara Oldfield
Botanic Gardens Conservation International, Descanso House, 199 Kew Road, Richmond, Surrey, TW9 3BW, UK

PEI Sheng-Ji
Kunming Institute of Botany, Chinese Academy of Sciences, 132 Lanhei Road, Heilongtan, Kunming 650201, Yunnan, China

PENG Hua
Kunming Institute of Botany, Chinese Academy of Sciences, 132 Lanhei Road, Heilongtan, Kunming 650201, Yunnan, China

Peter H. Raven
President Emeritus, Missouri Botanical Garden, PO Box 299, St. Louis, Missouri 63166-0299, USA

SUN Hang
Kunming Institute of Botany, Chinese Academy of Sciences, 132 Lanhei Road, Heilongtan, Kunming 650201, Yunnan, China

WANG Juan
Yunnan Academy of Forestry, 2 Lanan Road, Panlong, Kunming 650000, China

WANG Xiu-Hong
Institute of Geographic Sciences and Natural Resources Research, Chinese Academy of Sciences, 11A Datun Road, Chaoyang, Beijing 100101, China

WANG Zhao-Feng
Institute of Geographic Sciences and Natural Resources Research, Chinese Academy of Sciences, 11A Datun Road, Chaoyang, Beijing 100101, China

Mark F. Watson
Royal Botanic Garden Edinburgh, 20A Inverleith Row, Edinburgh EH3 5LR, UK

Alexandra H. Wortley
Royal Botanic Garden Edinburgh, 20A Inverleith Row, Edinburgh EH3 5LR, UK

WU Cheng-Yih (WU Zheng-Yi)
Kunming Institute of Botany, Chinese Academy of Sciences, 132 Lanhei Road, Heilongtan, Kunming 650201, Yunnan, China

XING Fu-Wu
South China Botanical Garden, Chinese Academy of Sciences, 723 Xingke Road, Tianhe, Guangzhou 510650, China

YANG Yu-Ming
Yunnan Academy of Forestry, 2 Lanan Road, Panlong, Kunming 650000, China

YI Ting-Shuang
Kunming Institute of Botany, Chinese Academy of Sciences, 132 Lanhei Road, Heilongtan, Kunming 650201, Yunnan, China

ZHANG Xue-Qin
Institute of Geographic Sciences and Natural Resources Research, Chinese Academy of Sciences, 11A Datun Road, Chaoyang, Beijing 100101, China

ZHANG Yu-Xiao
Kunming Institute of Botany, Chinese Academy of Sciences, 132 Lanhei Road, Heilongtan, Kunming 650201, Yunnan, China

ZHENG Du
Institute of Geographic Sciences and Natural Resources Research, Chinese Academy of Sciences, 11A Datun Road, Chaoyang, Beijing 100101, China

ZHOU Zhe-Kun
Kunming Institute of Botany, Chinese Academy of Sciences, 132 Lanhei Road, Heilongtan, Kunming 650201, Yunnan, China

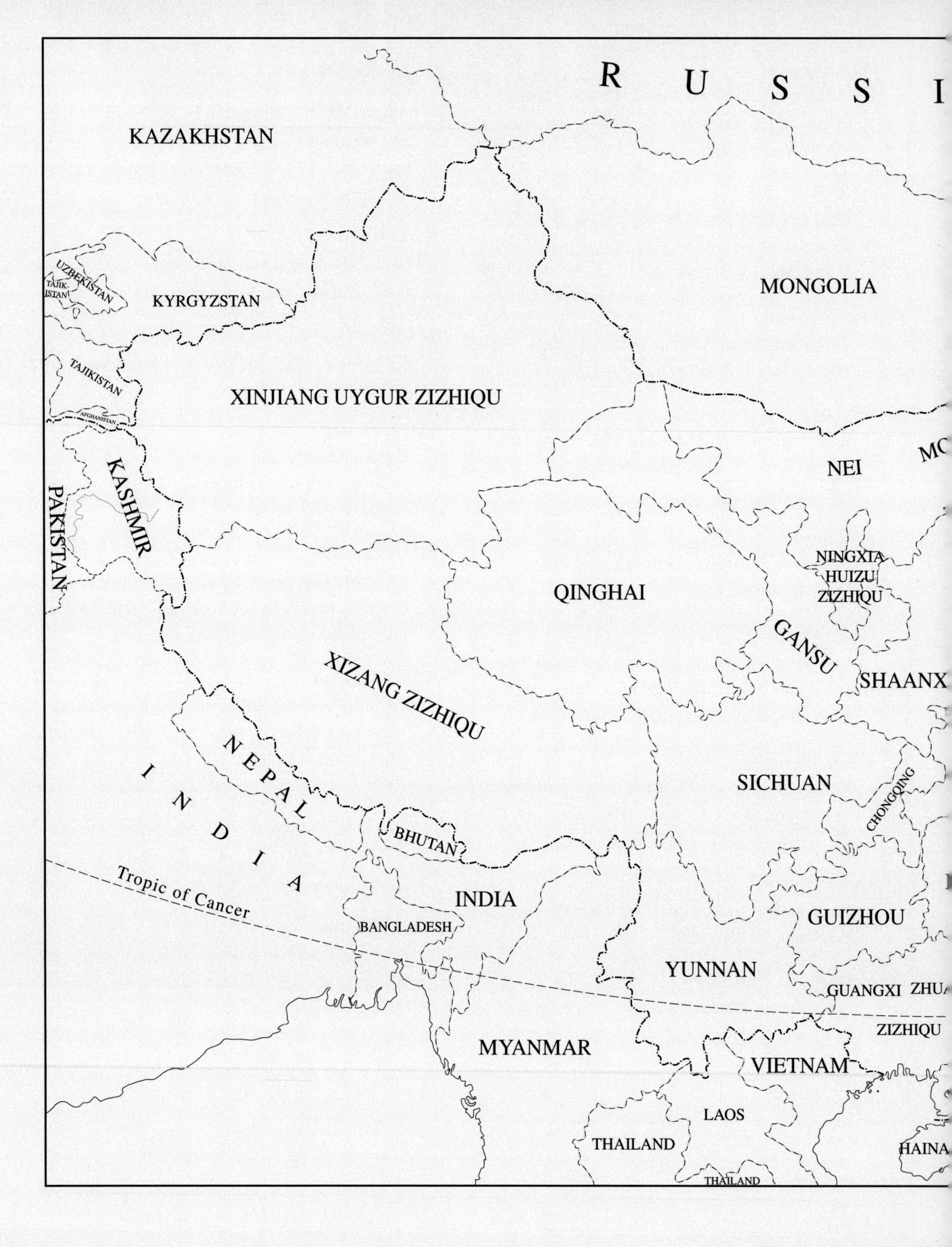

R U S S I
KAZAKHSTAN
MONGOLIA
UZBEKISTAN
TAJIK-ISTAN
KYRGYZSTAN
TAJIKISTAN
XINJIANG UYGUR ZIZHIQU
AFGHANISTAN
KASHMIR
PAKISTAN
NEI
MO
NINGXIA HUIZU ZIZHIQU
QINGHAI
GANSU
SHAANX
XIZANG ZIZHIQU
NEPAL
INDIA
SICHUAN
CHONGQING
BHUTAN
Tropic of Cancer
INDIA
GUIZHOU
BANGLADESH
YUNNAN
GUANGXI ZHU
ZIZHIQU
MYANMAR
VIETNAM
LAOS
THAILAND
HAINA
THAILAND
0 200 400 600 800

The national boundaries of China on this map are drawn after the 1.4M "Relief Map of People's Republic of China" published by China Cartographic Publishing House in 1989

1000 km

CHINA'S ADMINISTRATIVE DIVISIONS

ANHUI–Southeast
BEIJING–Northeast
CHONGQING–North Central
FUJIAN–Southeast
GANSU–North Central
GUANGDONG–Southeast
GUANGXI–South Central
GUIZHOU–South Central
HAINAN–South Central
HEBEI–Northeast
HEILONGJIANG–Northeast
HENAN–Southeast
HONG KONG–Southeast
HUBEI–Southeast
HUNAN–Southeast
JIANGSU–Southeast
JIANGXI–Southeast
JILIN–Northeast
LIAONING–Northeast
MACAO–Southeast
NEI MONGOL–North Central
NINGXIA–North Central
QINGHAI–North Central
SHAANXI–North Central
SHANDONG–Northeast
SHANGHAI–Southeast
SHANXI–Northeast
SICHUAN–North Central
TAIWAN–Southeast
TIANJIN–Northeast
XINJIANG–Northwest
XIZANG–West Central
YUNNAN–South Central
ZHEJIANG–Southeast

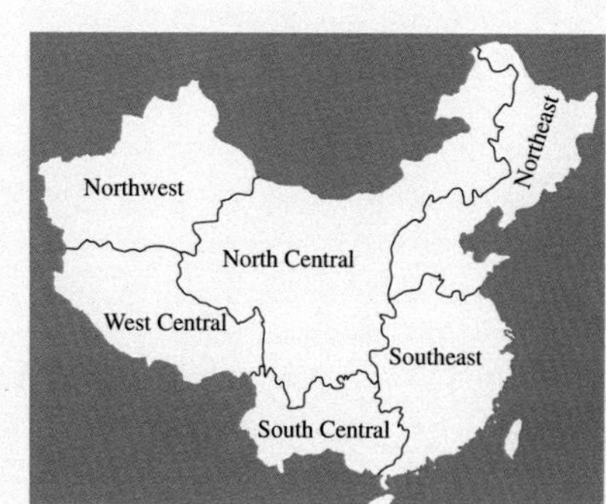

Chapter 1

Introduction

Stephen Blackmore
HONG De-Yuan
Peter H. Raven and Alexandra H. Wortley

1.1 Introduction: China – garden of the world

The flora of China is astonishing in its diversity. With 32 500 species of vascular plants, over 50% of them endemic, it has more plant species and more botanical variety than any other temperate country, and more than all but a few tropical countries.

Just why the flora of China is so diverse is a complex issue: many historical factors can account for the degree of richness of plant life found in different places on Earth, including the changing face of the Earth itself. Some 180 million years ago, before vascular plants had evolved, the continents were gathered together as a gigantic land mass known as Pangaea. The movement of basaltic plates across the surface of the globe caused the continents to separate, at different rates in different regions, sometimes colliding with others. Mountain ranges were forced upwards at regions of continental collision, such as the Himalaya Shan, or separation, such as the Appalachian Mountains of North America. At the same time, oceans changed in outline and extent, and regional climates developed and changed, particularly as a result of changing oceanic currents. For example, the formation of the Circum-Antarctic current followed the separation of Tierra del Fuego from Antarctica about 49 million years ago, which in turn accelerated Antarctic glaciation, eventually leading to the formation of Arctic continental glaciers, and driving the formation of sharpening contrasts between world climatic zones over the past 15 million years. Eventually, in the past three million years or so, the Arctic ice sheets moved to lower latitudes, with the formation of a cyclical series of expansions and contractions of the glaciers (Axelrod *et al.*, 1996).

In general, the flora of China is more numerous and diverse than that of other temperate areas because, firstly, China extends into the tropics, which neither Europe nor the USA (the other major temperate landmasses); secondly, 40% of the landmass of China lies at an elevation of over 2 000 m, including many isolated mountain ranges on which distinct species have developed relatively recently in geological time; and thirdly, from the Mid Miocene Period onwards (the past 15 million years), a time when the climate of the Northern Hemisphere is considered to have become less favorable for plants, and during ice ages of the Pleistocene Epoch, China's land connections to the south provided areas of refuge for many kinds of plants that disappeared from other northern Temperate areas (Axelrod *et al.*, 1996).

At the time the first humans (the genus *Homo*) first appeared on Earth, about 2.3 million years ago, the climates thus cycled between cold and warm, depending on the position of the ice sheets. The vegetation of the planet reflected these climatic and physical factors, with lush equatorial rainforests, prairies and savannas, alpine meadows extending to their vertical limits, boreal forests and arctic tundra. Until around 11 200 years ago, when agriculture was first developed, humans lived in bands of ca. 30–45 people that rarely came into contact with one another; it is estimated that the total global human population on all continents amounted to perhaps three million people. As human numbers increased, at first slowly and then with increasing rapidity, to perhaps 300 million 2 000 years ago, one billion by the early nineteenth century, and 7.1 billion today, we became a major force in shaping the Earth's vegetation. In the plants of China we clearly can see the legacy of this interplay between earth history—mountain-building, glaciation, continental shift and climatic change—and human history, in one of the world's most ancient and continuous civilizations.

China is not only vast, extending over some 9 600 000 km^2, but also has a great latitudinal range, from 18° N in the south, to 53.5° N in the north. Within this territory lie examples of almost every form of landscape, from the shores of the Nan Hai (South China Sea) to tropical and subtropical forests, deserts, the enormous elevated Qinghai-Xizang Plateau, and some of the world's highest mountains. Each of these ecosystems has its own distinctive plant community, some found nowhere else on earth. However, many are surprisingly familiar to those outside China: hundreds of China's more than 31 000 flowering plant species are now grown in gardens around the world. China is, quite literally, the garden of the world—as Ernest H. Wilson put it, "the mother of gardens" (Wilson, 1929).

This book tells the story of the plants of China, a story all the more remarkable in the context of China's recent rapid development into an economic powerhouse and the second largest economy in the world. China is changing rapidly in terms of economics, politics, science, technology, industry, culture and not least demography: by 2025, its urban population is expected almost to double, to one billion people. But underlying these dramatic developments, seen most clearly in the cities, is the natural economy of China: its green resources that are now recognized by the Chinese government as the essential foundation of a sustainable future.

1.2 The flora of China

China has the richest northern temperate flora in the world – containing approximately 31 500 native species of vascular plants, around 8% of the world's estimated total, or 1.5 times as many species as the USA and Canada combined. Around 15,750 species (over 50%) are endemic to the country. China is also notable for the presence of many taxa with origins far back in geological time, once common in North America and Europe but now surviving only in China, particularly gymnosperms such as *Cathaya* and *Pseudolarix* (both Pinaceae), *Ginkgo* (Ginkgoaceae) and *Metasequoia* (Taxodiaceae). This makes an understanding of the flora of China crucial to interpreting plant evolution and fossil history elsewhere in the world.

Furthermore, China is famed for its variety of habitats and for the continuity it provides between them: it is the only country in the world with unbroken connections from tropical to subtropical, temperate and boreal forests. This has contributed to the formation of rich plant assemblages and vegetation types rarely seen elsewhere. Certain regions of China deserve special mention for the richness and uniqueness of their flora, among them Yunnan, with an astonishing 15 000 species, half the total species in the country. Other Chinese "hotspots" of plant diversity are described in Chapter 2. Like elsewhere in the world, some of the China's richest areas of biodiversity are threatened, by various factors including habitat destruction, over-harvesting and global climate change. Chapter 22 to 24 describe some of these threats, and what we might be able to do to protect the flora of China for future generations.

It is not only the size, diversity and uniqueness of the Chinese flora that makes it such a fascinating subject for study, but also the individual features of the plant species themselves. From ancient seed plant lineages to stunning orchids, rapidly-radiating groups to complex taxonomic conundrums and intricate ecological networks, all can be found among the plants of China. To give just one example, China's most speciose genus is *Rhododendron* (Ericaceae), with 571 species, more than half the world total. Over 400 (72%) of these are found nowhere else in the world. Concentrated in the mountainous southwestern of China (Yunnan and Sichuan), many of these species are now cultivated throughout the world for their glossy foliage and showy spring flowers. Natural and artificial hybrids are common, producing more than 25 000 varieties now available to the grower.

1.3 Plants and people in China

China, perhaps more than any other country, has made extensive use of its natural plant resources for food, forestry, horticulture and medicine (see Chapters 15–21). Over 10 000 species have been employed in one or more of these fields, the vast majority in traditional medicinal systems. Many Chinese plant species have also been widely exported and introduced throughout the world. The orange (*Citrus ×aurantium*) and lemon (*C. ×limon*; both Rutaceae), peach (*Amygdalus persica*) and rose (*Rosa* spp.; both Rosaceae), ginger (*Zingiber officinale;* Zingiberaceae), soybean (*Glycine max;* Fabaceae), magnolia (Magnoliaceae), rice (*Oryza sativa;* Poaceae) and, of course, tea (*Camellia sinensis;* Theaceae), all originated in China.

Plants are also highly significant in Chinese culture, being used as symbols in art, architecture and religion. In her beautiful book, *Hidden Meanings in Chinese Art* (Bartholomew, 2006), Terese Tse Bartholomew highlights two main ways in which plants are used symbolically in China: firstly, by association, such as using a long-lived plant to symbolize longevity; secondly (and uniquely to China), through puns based on their Chinese names. Hence, the lotus (*Nelumbo nucifera,* Nelumbonaceae) can symbolize harmony (he, described by a different Chinese character, 和). Most plant symbols can have many meanings depending on their interpretation and the circumstance. *Bambusa* spp. (bamboo, zhu; Poaceae), with their straight, hollow and strong stems, symbolize humility, fidelity and integrity, while a stalk of *Oryza sativa* (rice) with many branches and multiple seeds symbolizes a good harvest. The paeony (mudan; *Paeonia* spp.; Paeoniaceae) is the most popular botanical image in China and, through its beauty, symbolizes wealth, honor and rank. The blossoms of plum (meihua; *Prunus domestica*; Rosaceae), which are said to occur even on withered old branches, symbolize perseverance and vigor in old age; as one of the earliest plants to flower each year, they also symbolize renewal and purity; and the five petals on each flower are used to represent China's traditional five blessings (old age, wealth, health, love of virtue and a peaceful death).

China is justly famed for its knowledge and use of plant species in medicine. As described in Chapter 14, over 10 000 species are reported to be used medicinally in China (He & Gu, 1997; Xiao & Yong, 1998; Pei, 2002; Hamilton, 2004) and, increasingly, the chemical basis and efficacy of these traditional uses is being investigated and the active compounds identified. Perhaps one of the most significant medicinal plants to come out of China in recent years is *Artemisia annua* (sweet wormwood; Asteraceae), used since ancient times in the treatment of fever. In the 1970s, *A. annua* was identified as a possible treatment for malaria, one of the most deadly tropical diseases worldwide, affecting some 300 million people at any one time and killing around 2.5 million each year, based on information in a 1 600-year old herbal. The active chemical component, artemisin (which is stored by the plant in specialized hairs, has now been isolated, described, and shown to have antimalarial activity. In fact it is considered the only effective treatment against many strains of the disease, including drug-resistant *Plasmodium falciparum. Artemisia annua* is now grown for artemisin production in several countries worldwide, across a total of over 11 000 ha; this is still only about half of what would be required to meet world demand (Heemskerk *et al.*, 2006).

1.4 The *Flora of China*: a golden age of Chinese botany

As the remarkable scholar Joseph Needham demonstrated so well in his monumental series *Science and Civilisation in China* (*Volume 6. Biology and Biological Technology*), detailed knowledge of Chinese plants and their uses extends into antiquity; such knowledge always being an integral part of Chinese culture, art and medicine (Various authors, 1984–2000). Nonetheless, the twentieth century in particular goes down as a time of great expansion in our knowledge of the plants of China. From the middle of the nineteenth century, first French missionaries and later British and American plant explorers began to make the botanical riches of China known scientifically and to the horticultural public in Europe and North America (see Chapter 12). It was not until the 1920s, however, that Chinese botanists, at first mostly trained in the west or in one of the colleges in China run by westerners for Chinese students, took on such explorations using international standards for naming plants. From 1949 onwards, the Chinese began to pursue the production of a national flora, the *Flora Reipublicae Popularis Sinicae* (*FRPS*), and between 1959 and 2004 this mammoth project, interrupted by the Cultural Revolution, but involving three generations of Chinese botanists, documented 31 180 species of vascular plants in a series of 80 volumes and 126 books (Ma & Clemants, 2006).

In 1979, the idea of translating the *FRPS* into English, the international language of science, was first mooted at a joint meeting of Chinese and American botanists in Berkeley, California, USA. At this point it was recognized that the Chinese authors of the *FRPS* had not had adequate access to the literature and specimens held in western countries, which was seen as necessary to improve this great work. Thus, in 1989 work officially began on the *Flora of China*, a major international project to produce a fully revised and updated work in which the account of each plant group was to be prepared jointly by at least one Chinese and at least one non-Chinese co-author. The first volume (*Volume 17: Verbenaceae through Solanaceae*) was published in 1994, with the first accompanying illustration volume following in 1998. In 2013, the final text volumes of the *Flora of China* were published. The completion of this entire series in just 24 years makes it one of the most successful Flora projects ever undertaken. Much of this success relates to the friendly, scholarly and cooperative relationships generated through the project, relationships that will, we hope, last for many years to come.

The *Flora of China* covers the vascular plant species of China in 25 volumes. *Volume 1* comprises the indices, summary, classifications, and statistics. *Volumes 2–3* cover the ferns. *Volume 4* includes the gymnosperms and some angiosperms, and *Volumes 5–25* incorporate all the remaining angiosperms. Asteraceae (the largest family) are treated in two volumes (20 and 21). *Plants of China* acts as a companion volume to the series, accessible both to users of the Flora and to a wider audience interested in the flora of China. A complementary project, the *Moss Flora of China*, has also been published in eight volumes by Missouri Botanical Garden Press (Moss Flora of China Editorial Committee, 1999–2011).

Botany in China has seen many changes since the inception of the *Flora of China* project in 1979. As described in Chapter 12, in 1980, Chinese and American botanists joined together for the first joint field expedition involving scientists from a non-Communist country (the Sino-American Sennongjia Expedition). The 1980s and 1990s saw increasing numbers of Chinese and Chinese-International collecting expeditions to many under-explored regions including the Hengduan Shan (Yunnan, Sichuan and Xizang), Wulin Shan (Hubei, Sichuan Hunan and Guizhou), upper reaches of the Hongshui (Guangxi), and Dulongjiang and Gaoligongshan (Yunnan), as well as numerous provincial expeditions.

State support and encouragement for botany, along with other sciences, is today at an all-time high in China. Today, Botanic Gardens Conservation International (BGCI) lists 148 botanical gardens in China, of which three (South China Botanical Garden, Wuhan Botanical Garden and Xishuangbanna Tropical Botanical Garden) are operated under the Chinese Academy of Sciences (CAS). Two Botanical Institutes (Beijing and Kunming) also belong to the CAS, and continue to play a very important role in the development of botany in the country. Botanical research is also conducted in life sciences laboratories, 52 of which hold the prestigious title of State Key Laboratory. Public interest in botany is also growing in China. The Beijing Botanical Garden attracts over 1.4 million visitors per year, almost as many as the Royal Botanic Gardens, Kew, London (1.6 million in 2011). The 2011 International Horticultural Expo at Xi'an Shi, Shaanxi attracted over 15 million visitors during its six-month show.

Botany is an ever-changing science; the *Flora of China* represents only the current state of taxonomic knowledge about the plants of China. Indeed, the rate of taxonomic progress in China is growing at an ever-increasing rate (see Figure 1.1). Take, for example, *Volume 25* (*Orchidaceae*): this volume was published in 2009 and followed a 2003 classification (Chase *et al.*, 2003). However, a detailed new classification of the orchids—*Genera Orchidacearum* (Pridgeon *et al.*, 1999–2009)—is also nearing completion and is expected to provide a revised, contemporary taxonomy of the Chinese species (among others) of this enormous family. The authors of *Volume 25* of the *Flora of China* take care to note in their introduction that "Even when this work [*Genera Orchidacearum*] is completed, such is the speed with which new information and techniques are being developed and published, it will almost certainly require revision" (Chen *et al.*, 2009: 1). Taxonomic interest in the plants of China, from both within and outside the country, shows no sign of slowing down. With the

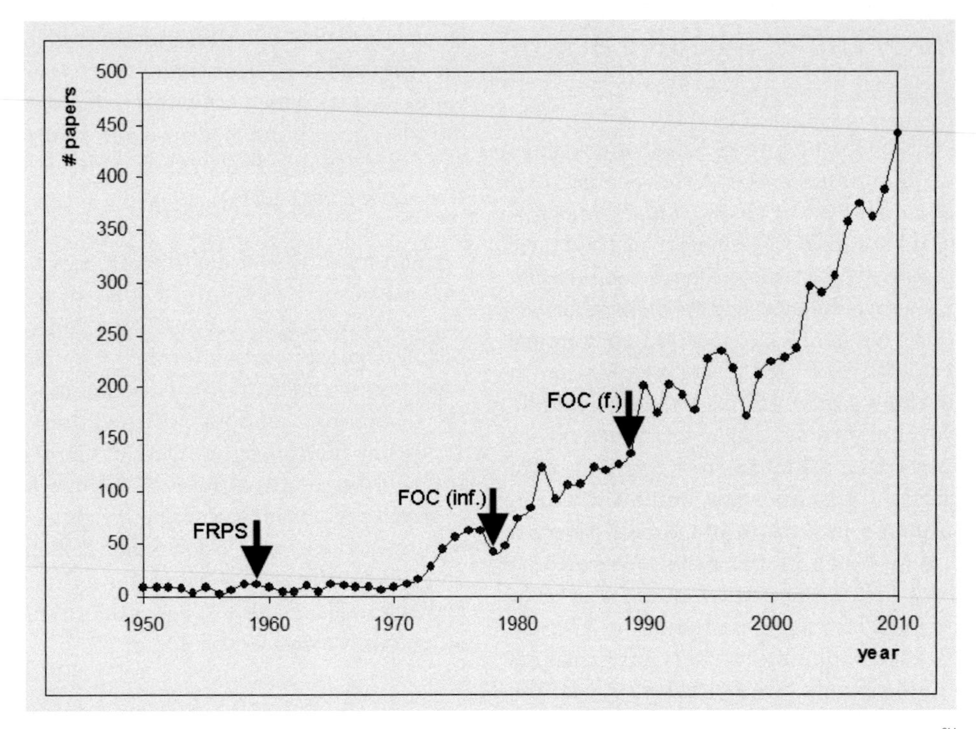

Figure 1.1 Published papers on Chinese plant taxonomy abstracted in Thomson-Reuters Web of Knowledge[SM] since 1950. Arrows indicate the initiation of *FRPS* in 1959 ("FRPS"), the informal inception of the *Flora of China* in 1979 ("FOC (inf.)") and the formal start of the *Flora of China* project in 1989 ("FOC (f.)").

completion of the *Flora of China,* Chinese scientists are now taking the lead in a new, fifty-volume *Flora of the Pan-Himalayas,* working alongside botanists from India, Japan, Nepal, the UK and USA.

1.5 *Plants of China*

Unlike the rest of the *Flora of China, Plants of China* was not formally co-authored by Chinese and non-Chinese authors, but was written largely by Chinese experts. It tells the story of the Chinese flora, beginning with its global significance (Chapter 2, by HUANG Hong-Wen, Director of South China Botanical Garden (SCBG), Sara Oldfield, Secretary General of BGCI and Hong Qian, Curator of Botany and Illinois State Museum). The features underlying the country's characteristic and diverse vegetation are outlined in Chapters 3 to 5: Chapter 3, by ZHENG Du of the Institute of Geographic Sciences and Natural Resources Research, CAS, covers the physical geography of this immense and varied landscape including topography, water features, and of course the enormous impact of the orogeny of the Himalaya Shan and uplift of the Qinghai-Xizang Plateau. It concludes with a new classification of China's geography–a physico-geographical regionalization. Chapter 4, by the same author, details the country's equally diverse climate including the important effect of the monsoon, and also ends with a regionalization of China on the basis of its climate. The soils of China are a product of both its geography and climate, among other factors. Their formation, classification and distribution are explained in Chapter 5, also by ZHENG Du, which concludes with a division of the country's landmass according to soil type.

Chapters 6 and 7 explain the complex and fascinating history of the Chinese flora, while Chapters 8--10 explain the nature of the present-day flora from three different viewpoints. Chapter 6, authored by ZHOU Zhe-Kun, Professor of Botany and Paleobotany at Kunming Institute of Botany (KIB), CAS, explains the origin of the Chinese flora from the very first angiosperms, detailing our knowledge of the history of all major groups of Chinese plants from the fossil record, with an interesting discussion of "living fossils" and endemic taxa. Chapter 7, by the same author, takes a detailed chronological look at the history of China's plants from the early Cretaceous Period (150 million years ago) to the present day. This brings us neatly to Chapter 8, by CHEN Ling-Zhi of the Institute of Botany, CAS (IBCAS), which gives a comprehensive description of the myriad vegetation types found in China today. Chapter 9, by PENG Hua, Professor of Botany and Curator of the Herbarium at KIB, and the late WU Cheng-Yih (WU Zheng-Yi), also of KIB, describes how the flora of China may be partitioned into elements based on distribution and affinities, while Chapter 10 by SUN Hang, Professor of Botany at KIB, outlines a classification for regions of China based on the present flora.

Chapters 11 to 14 delve into the fascinating history of botanical study in China. Chapter 11, again by PENG Hua, charts the development of indigenous knowledge about Chinese plants, over more than two millennia. Chapter 12, by HU Chi-Ming (HU Qi-Ming) of SCBG and Mark F. Watson of the Royal Botanic Garden Edinburgh, describes the history of botanical expeditions and exploration, from the foreign travelers of the eighteenth century, through the era of Chinese exploration, to recent international collaborations. Chapter 13, by HU Zong-Gang of Lushan Botanical Garden, MA Hai-Ying of KIB, MA Jin-Shuang, Professor at Shanghai Chenshan Plant Science Research Center, and HONG De-Yuan, Academician of CAS at IBCAS, charts the particular history of botanical institutions in China, and Chapter 14, by ZHANG Yu-Xiao, KIB, and LI De-Zhu, Director of KIB, explains the growth of floristic, taxonomic and systematic studies of Chinese plants, into today's molecular-phylogenetic era.

Chapters 15 to 21 give a flavor of the amazing range of uses to which the plants of China may be–and have been–put. Chapter 15, by PEI Sheng-Ji, Professor at KIB, provides an overview of the history, diversity and importance of economic plants in China. Chapter 16, by HE Shan-An of Nanjing Botanical Garden, YI Ting-Shuang of KIB, PEI Sheng-Ji and HUANG Hong-Wen, details important crop plants with an emphasis on their wild relatives still found in China, and Chapters 17 to 19 give an interesting insight into the range of plants used in forestry (by PEI Sheng-Ji, YANG Yu-Ming and WANG Juan, both of the Southwest Forestry University, Kunming Shi), medicine (by PEI Sheng-Ji and HUAI Hu-Yin of KIB's department of ethnobotany), and horticulture (by HE Shan-An and SCBG's XING Fu-Wu). Chapter 20, by YI Ting-Shuang, Peter L. Morrell of the University of Minnesota, PEI Sheng-Ji and HE Shan-An, shows us the incredible range of non-native plants that have been adopted for use by the Chinese, including many staple foods and other important species, while Chapter 21, by PEI Sheng-Ji and HU Guang-Wan from KIB, demonstrates just how widely plants are used in China–as beverages, dyes, aromatic oils, rubber, fibers and of course animal feed and forage. This chapter also touches upon novel approaches to using plants as a source of energy, a growing field in today's climate-changing world.

Around 3 000 (almost 10%) of China's plant species are endangered. The final four chapters of *Plants of China* look at some of the problems facing the flora of China, strategies that may be used to solve these, and how the future of the flora might look. Chapter 22, by LI Zhen-Yu of IBCAS, FAN Xiao-Hong of the Institute of Animal and Plant Quarantine, Chinese Academy of Inspection and Quarantine, and David E. Boufford, Senior Research Scientist at the Harvard University Herbaria, looks at the impact of non-native plants in the Chinese flora, many of which have become naturalized and in some cases invasive. Chapter 23, by HUANG Hong-Wen and Sara Oldfield, explores the potential crisis faced by extinction in the flora of China, detailing the threats to a range of plant groups from a variety of human-induced sources. Chapter 24, by HUANG Hong-Wen, Peter S. Wyse Jackson and CHEN Ling-Zhi, looks at the range of national and international strategies that have been implemented to conserve this invaluable resource, and Chapter 25, again by HUANG Hong-Wen, provides a rational but positive glimpse of how the future might pan out, providing case study examples of where the flora of China has been successfully protected for generations to come. This will be facilitated by the new *Red List of China's Plants*, the result of a major national effort headed by QIN Hai-Ning of IBCAS.

The names of the many Chinese botanists featured in this book are written in English notation using both the Hanyu Pinyin and Wade-Giles systems, depending on which has been more commonly used for each individual. Family names are capitalized to avoid confusion. Alternative transliterations for the names of many Chinese botanists are provided in Table 12.1 on pages 213–214. As much of the information in this book is historical, we hope the reader will also find it useful to refer to the Appendix showing Chinese dynasties presented on page 473. To assist with finding the geographic locations mentioned in the text, a map of Chinese provinces may be found on endpaper.

The editors are immensely grateful to all the authors who have contributed to this book. We are also grateful to the following reviewers: H. John B. Birks (Department of Biology, University of Bergen, Norway), David Boufford (Harvard University Herbaria), Peter Del Tredici (Arnold Arboretum of Harvard University), Sandy Knapp (Natural History Museum, London), Yude Pan (United States Department of Agriculture Forest Service), Jan Salick (Missouri Botanical Garden), Dr. Thomas Scholten (Faculty of Science, University of Tübingen, Germany) and Teresa Spicer and Professor Robert Spicer (Centre for Earth, Planetary, Space and Astronomical Research, The Open University), and two anonymous reviewers, as well as to GONG Xiao-Lin (IBCAS), without whose tireless work the volume might never have been brought together. The authors of Chapter 5 are immensely grateful to the authoring committee of *Soil Geography in China*: GONG Zi-Tong, ZHANG Gan-Lin and HUANG Rong-Jin. We thank all those who have provided illustrations and figures–these are credited within the relevant captions.

References

Axelrod, D. I., Al-Shehbaz, I. A. & Raven, P. H. (1996). History of the modern flora of China. In: Zhang, A. L. & Wu, S. G. (eds.) *Floristic Characteristics and Diversity of East Asian Plants*. pp. 43-55. Beijing, China and Berlin, Germany: China Higher Education Press and Springer-Verlag.

Bartholomew, T. T. (2006). *Hidden Meanings in Chinese Art*. San Francisco, CA: Asian Art Museum.

Chase, M. W., Cameron, K. M., Barrett, R. L. & Freudenstein, J. V. (2003). DNA data and Orchidaceae systematics: a new phylogenetic classification. In: Dixon, K. M., Kell, S. P., Barrett, R. L. & Cribb, P. J. (eds.) *Orchid Conservation*. pp. 69-89. Kota Kinabalu: Natural History Publications.

Chen, X. Q., Liu, Z. J., Zhu, G. H., Lang, K. Y., Ji, Z. H., Luo, Y. B., Jin, X. H., Cribb, P. J., Wood, J. J., Gale, S. W., Ormerod, P., Vermeulen, J. J., Wood, H. P., Clayton, D. & Bell, A. (2009). Orchidaceae. In: Wu, Z. Y. & Raven, P. H. (eds.) *Flora of China*. St Louis, MO and Beijing, China: Missouri Botanical Garden Press and Science Press.

Hamilton, A. C. (2004). Medicinal plants, conservation and livelihoods. *Biodiversity and Conservation* 13: 1477-1517.

He, S. A. & Gu, Y. (1997). The challenge for the 21st Century for Chinese botanic gardens. In: Touchell, D. H. & Dixon, K. W. (eds.) *Conservation into the 21st Century. Proceedings of the 4th International Botanic Gardens Conservation Congress (Perth, 1995)*. pp. 21-27. Perth, Australia: Kings Park and Botanic Garden.

Heemskerk, W., Schallig, H. & de Steenhuijsen Piters, B. (2006). *The world of Artemisia in 44 questions*. Amsterdam, The Netherlands: Koninklijk Instituut voor de Tropen.

Ma, J. S. & Clemants, S. (2006). A history and overview of the *Flora Reipublicae Popularis Sinicae* (FRPS, Flora of China, Chinese Edition, 1959-2004). *Taxon* 55(2): 451-460.

Moss Flora of China Editorial Committee (1999-2011). *Moss Flora of China*. St Louis, MO and Beijing, China: Missouri Botanical Garden Press and Science Press.

Pei, S. J. (2002). A brief review of ethnobotany and its curriculum development in China. In: Shinwari, Z. K., Hamilton, A. & Khan, A. A. (eds.) *Proceedings of a Workshop on Curriculum Development in Applied Ethnobotany, Nathiagali, 2-4 May 2002*. p. 41. Lahore, Pakistan: WWF-Pakistan.

Pridgeon, A. M., Cribb, P. J., Chase, M. W. & Rasmussen, F. N. (1999-2009). *Genera Orchidacearum*. New York: Oxford University Press.

Various authors (1984-2000). *Biology and Biological Technology*. Cambridge, UK: Cambridge University Press.

Wilson, E. H. (1929). *China: Mother of Gardens*. Boston, Massachusetts: The Stratford Company.

Xiao, P. G. & Yong, P. (1998). Ethnopharmacology and research on medicinal plants in China. In: Prendergast, H. D. V., Etkin, N. L., Harris, D. R. & Houghton, P. J. (eds.) *Plants for Food and Medicine. Proceedings from a Joint Conference of the Society for Economic Botany and the International Society for Ethnopharmacology, London, 1-6 July 1996*. pp. 31-39. London, UK: Royal Botanic Gardens, Kew.

Chapter 2

Global Significance of Plant Diversity in China

HUANG Hong-Wen
Sara Oldfield and Hong Qian

2.1 Introduction

China is ranked in the top six megadiverse countries of the world (Mittermeier *et al.*), with approximately 32 096 vascular plants (Table 2.1), more than half of these species being endemic (Liu *et al.*, 2003; eFloras, 2008). This tremendous plant diversity comprises pteridophytes (2 278 species), gymnosperms (207 species) and angiosperms (29 611 species), accounting respectively for about 17%, 21% and 10% of the world total (Table 2.1). Amongst the higher plants, 256 endemic genera and 15 000–18 000 endemic species have been recorded (Liu *et al.*, 2003; eFloras, 2008). This rich plant diversity is evolutionarily associated with an ancient geological history, complex and diverse topography and climatic variation, as shall be seen in later chapters.

The flora of China comprises living remnants of the Early Miocene flora of the entire North Temperate Region. As one of the most important centers of diversity of seed plants, China inherited components from ancient northern, Mediterranean and southern continental floras (Ministry of Environmental Protection of the People's Republic of China, 1998). This richness has also given rise to many relict lineages or "living fossils," such as *Metasequoia glyptostroboides* (Taxodiaceae), *Ginkgo biloba* (Ginkgoaceae) and *Cathaya argyrophylla* (Pinaceae).

China's flora is also renowned as the source of numerous food crops, medicinal plants and plants for ornamental horticulture. Furthermore, a long history of agriculture and crop domestication has generated an enormous amount of cultivated germplasm of useful species. This plant diversity has the potential to be of crucial importance to the whole world population, serving as a primary bioresource for livelihoods and sustainable development, as well providing the environmental services on which we depend (China's Strategy Plant Conservation Editorial Committee, 2008). However, the plant diversity of China is increasingly threatened, with an estimated 4 000–5 000 plant species categorized as threatened or endangered (Wang & Xie, 2004), a proportion ranking it as one of the highest priorities for global biodiversity conservation. Conservation of its rich and diverse flora is one of the greatest challenges facing China today.

Table 2.1 The rich diversity of vascular plants in China, dividing according to plant groups. Data for pteridophytes compiled from an updated version of the checklist *Phylum Pteridophyta* in the 2012 edition of the *Species 2000 China Node* (http://www.sp2000.cn); data for seed plants compiled from a database based on the published volumes 4-25 of the *Flora of China* (Wu *et al.*, 1994-2011), plus species that missing from the *Flora of China* or published after the relevant family treatment. Infraspecific taxa were merged into species; exotic species were excluded. Data for the world flora obtained from, but not limited to, Mabberley (2008).

Group	Number of families			Number of genera			Number of species		
	China	World	China %	China	World	China %	China	World	China %
Pteridophytes	64	71	90	221	381	58	2 278	13 025	17
Gymnosperms	10	15	66	36	79	46	207	980	21
Angiosperms	249	ca. 400	62	2 899	ca. 10 000	29	29 611	ca. 300 000	10

2.1.1 Species richness (see Table 2.1)

Pteridophytes.—The pteridophyte flora of China is impressive in terms of the number of both families (64 recorded, ca. 90% of the world total) and genera (221, 58%; Table 2.1). In terms of species, China boasts 2 278, 17% of the world total, the richest diversity of any country (see Table 2.2). In addition, six endemic genera and 500–600 endemic species have been identified in the country (Ministry of Environmental Protection of the People's Republic of China, 1998). Southwestern China is a recognized center of distribution for Asian, and world pteridophytes (Lu, 2004). As many as 2 000 species have been recorded in the four provinces of Sichuan, Guizhou, Yunnan and Guangxi, with approximately 1 500 species in Yunnan alone. This region probably serves as a center for radiative dispersal, with the number of taxa decreasing in all directions from this point. For instance, only 700 species occur on the Indochinese peninsula (Vietnam, Lao PDR and Cambodia), fewer than 640 species in Thailand, 550 species in Malaysia, 1 000 in the Philippines, 452 in Hainan, 639 in Japan, 600 in India, 430 in Australia and 420 in North America (Lu, 2004). Three pteridophyte families in China contain more than 300 species: Dryopteridaceae (13 genera and 700 species), Athyriaceae (20 genera and ca. 400 species) and Thelypteridaceae (20 genera and 300 species; Wu & Chen, 2004).

Table 2.2 Pteridophyte families with more than 50 species in China (Data from Lu, 2004; Wu & Chen, 2004).

Family	Number of species (China/World)	Number of genera (China/World)
Dryopteridaceae	700/1 000	13/14
Athyriaceae	400/500	20/20
Thelypteridaceae	300/600	20/25
Polypodiaceae	250/500	27/40
Aspleniaceae	150/700	8/10
Pteridaceae	100/400	2/11
Aspidiaceae	90/400	8/20
Hymenophyllaceae	80/700	14/34
Dennstaedtiaceae	70/150	4/9
Sinopteridaceae	60/300	9/14
Hemionitidaceae	50/110	5/17
Selaginellaceae	50/700	1/1
Total	2 300/6 060	129/215

Human disturbance, ecosystem degradation and habitat loss caused by the rapid industrialization and urbanization of China have severely threatened natural populations of many pteridophytes species over the past 30 years. Some species are in critical danger or on the brink of extinction. For example *Cystoathyrium chinense* (Cystopteridaceae), *Cyrtomium hemionitis* (Dryopteridaceae; Figure 2.1), *Trichoneuron microlepioides* (Thelypteridaceae) and *Isoetes sinensis* (Isoetaceae; Figure 2.2) are all considered extinct or possibly extinct in the wild.

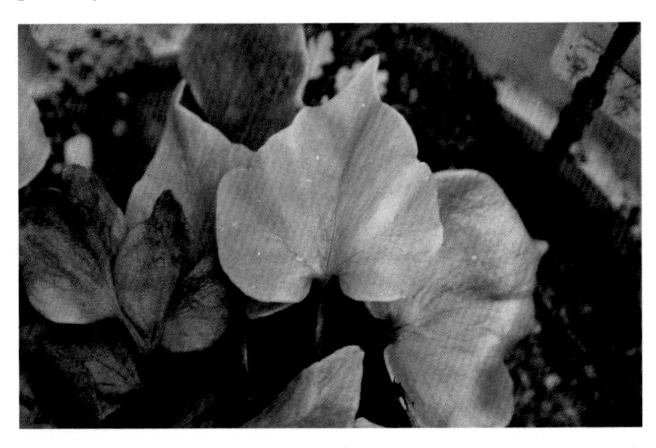

Figure 2.1 *Cyrtomium hemionitis* (image courtesy YANG Ke-Ming, South China Botanical Garden).

Gymnosperms.—As a group, gymnosperms probably emerged in the Paleozoic Era and thrived during the Mesozoic and Cenozoic Eras. Today there are an estimated 17 families, 86 genera and 840 species worldwide, grouped into the four classes Cycadopsida, Ginkgoopsida, Gnetopsida and Pinopsida (Wang, 2004). Although gymnosperms account for less than 1% of seed plants, they have a wide distribution and are major components of boreal and alpine forests in the Northern Hemisphere. The gymnosperm flora of China is one of the richest in the world, including 10 families, 36 genera and 207 species (Table 2.1). Here also they make up less than 1% of the higher plant flora, but are of great ecological importance, with coniferous forests accounting for over 50% of total forest cover in China. The warm climate that persisted from the Mesozoic Era to the Cenozoic Era, and the lower impact of Quaternary glaciations here than elsewhere resulted in China being a favorable refugium, retaining a large number of ancient lineages, relict and endemic taxa that disappeared in other parts of the world. Examples include the monotypic family Ginkgoaceae and monotypic genera *Metasequoia*, *Cathaya*, *Pseudolarix* (Pinaceae), *Pseudotaxus* (Taxaceae), and many relict species of *Cycas* (Cycadaceae; Ministry of Environmental Protection of the People's

Figure 2.2 *Isoetes sinensis*. A, *I. sinensis* in natural habitat (image courtesy KANG Ming, South China Botanical Garden); B, individual *I. sinensis* (image courtesy XU Ke-Xue).

Republic of China, 1998).

Angiosperms.—There are an estimated 400 families, 10 000 genera and 300 000 species of angiosperms worldwide; China has 249 families, 2 899 genera, and 29 611 species, representing about 60%, 31% and 10%, respectively, of the world total (see Table 2.1). In the world's four largest families (Orchidaceae, Asteraceae, Fabaceae and Poaceae), each with more than 10 000 species, 7–10% occur in China. In addition, China has 60 families containing more than 100 species. These 60 largest families are widespread throughout the country, representing basic components of its flora, and encompass 19 700 species, approximately 80% of China's total seed plants (see Table 2.3). Furthermore, many genera with worldwide distributions include a large proportion of species in China (see 2.3.3, below).

Table 2.3 Angiosperm families with more than 400 species in China (eFloras, 2008).

Family	Number of Species					
	>2 000	1 500-2 000	1 000-1 500	800-1 000	500-800	400-500
Asteraceae	Y					
Fabaceae		Y				
Poaceae		Y				
Orchidaceae			Y			
Rosaceae				Y		
Lamiaceae				Y		
Ranunculaceae				Y		
Cyperaceae				Y		
Ericaceae				Y		
Scrophulariaceae					Y	
Apiaceae					Y	
Primulaceae					Y	
Rubiaceae					Y	
Saxifragaceae					Y	
Brassicaceae						Y
Euphorbiaceae						Y
Gentianaceae						Y
Gesneriaceae						Y
Lauraceae						Y

The angiosperm flora in China is renowned for three significant characteristics: the richness of vegetation and forest types, paleofloristic origin, and high endemism. These characteristics are underlined by the tremendous geographical complexity and range of climate zones (see Chapters 3 and 4), in turn responsible for a huge diversity of habitats and ecosystems supporting plant diversity. Thus China's plants range from alpine permafrost species such as *Phyllodoce caerulea* (Ericaceae) to tropical rainforest plants such as *Parashorea chinensis* (Dipterocarpaceae), from extreme xerophytes such as *Reaumuria soongarica* (Tamaricaceae) to marsh and wetland plants such as *Potamogeton* (Potamogetonaceae), from Himalayan cushion plants such as *Androsace* and *Pomatosace* (Primulaceae) to tropical mangroves such as *Bruguiera gymnorhiza* (Rhizophoraceae) from the coast of southern China. The widest spectrum of angiosperm plants is found across China's unique continuous geographic cline from tropical, subtropical and temperate to boreal forests. Each climatic zone harbors its own representative families and genera. For example, deciduous broad-leaved forests are typically represented by Betulaceae; *Quercus* (Fagaceae) mixed with Salicaceae, Caprifoliaceae and Berberidaceae is typically representative of the temperate zone; whereas evergreen forests with Lauraceae, Magnoliaceae, Theaceae and Fagaceae mixed with Hamamelidaceae, Aquifoliaceae, Araliaceae, Nyssaceae and the monotypic Cercidiphyllaceae and Tetracentraceae are typical of the subtropical zone. Similarly, tropical forests in southern China are well represented by many families including Dipterocarpaceae, Annonaceae, Burseraceae, Sapotaceae, Meliaceae, Clusiaceae, Combretaceae, Euphorbiaceae and Datiscaceae.

The richness and phylogenetic significance of the paleofloristic or primitive components of angiosperm groups–such as Magnoliaceae, Ranunculaceae, Tetracentraceae, Cercidiphyllaceae, Saururaceae, Chloranthaceae, Hamamelidaceae and Lardizabalaceae—are well accepted by the botanical community worldwide, particularly the few relict groups that are only found in China.

A high level of endemism is one of the most significant features of the angiosperm flora of China. There are estimated to be approximately 250 genera and 15 000–

18 000 species endemic to China, including many well-known plants such as *Actinidia chinensis* (Actinidiaceae), *Bretschneidera sinensis* (Bretschneideraceae)., *Cercidiphyllum japonicum* (Cercidiphyllaceae), *Helianthemum songaricum* (Cistaceae), *Euptelea pleiosperma* (Eupteleaceae), *Liriodendron chinense* (Magnoliaceae), *Tetracentron sinense* (Tetracentraceae), *Trochodendron aralioides* (Trochodendraceae), *Tetraena mongolica* (Zygophyllaceae) and *Davidia involucrata* (Nyssaceae).

Davidia involucrata (also known as the dove or handkerchief tree; Figure 2.3) is an endemic, monotypic genus, naturally occuring in central and western China. The plant was first discovered by the French missionary Père David, and named after him by the botanist Henri Ernest Baillon. It is probably the most famous of the introductions made by Ernest Wilson to the western world. The nurseryman Henry Veitch expressed an interest in obtaining seeds of the tree and commissioned the young Wilson to go to China and find it. In 1900, at Yichang, Wilson purchased a native houseboat and engaged a crew for an expedition aimed at finding *D. involucrata*. They sailed through the three gorges of the Chang Jiang to arrive in Badong, from where Wilson pursued his journey on foot through the mountains. In May 1900, while collecting in the countryside southwest of Yichang, Wilson suddenly stumbled across a *Davidia* in full flower. With its large snow-white bracts fluttering in the wind, it must have made a dramatic sight. He later wrote that it was, "the most interesting and most beautiful of the trees which grow in the North temperate regions," and described how when the flowers are "stirred by the slightest breeze they resemble huge butterflies or small doves hovering amongst the trees" (Briggs, 1993).

Figure 2.3 Flowers of *Davidia involucrata* (image courtesy HUANG Hong-Wen, South China Botanical Garden).

In China the genus *Actinidia* Lindl. is known by the name "mihoutao" (monkey peach). Throughout the rest of the world it is commonly known as the kiwifruit. *Actinidia* belongs to Actinidiaceae and contains a total of about 75 described taxa (54 species and 21 varieties). *Actinidia* is naturally centered in China but very widespread over eastern Asia, from just south of the equator to cold regions as far as 50° N, making it a constituent of both Holoarctic and Paleotropical floras. Its general pattern of distribution is typical of many Chinese genera: great diversity within China itself with outlier taxa extending to adjoining countries. Thus most *Actinidia* taxa are endemic to China. The two species found in adjoining countries are *A. strigosa* (restricted to Nepal) and *A. hypoleuca* (endemic to Japan).

Currently, commercial cultivation is based on two Chinese endemics, *A. chinensis* var. *deliciosa* and *A. chinensis* var. *chinensis*. The kiwifruit has been domesticated since only the beginning of the last century, when in 1904 *A. chinensis* (Figure 2.4) was introduced to New Zealand, and in 1930 the first kiwifruit orchard was established there. The fruit is widely favored for its unique flavor, rich vitamin C content, high levels of dietary fiber, the variety of minerals it contains, and its alternative medicinal function as an antioxidant and treatment for gastrointestinal complaints. Domestication and commercial cultivation of kiwifruit is considered one of the most successful examples of plant domestication of endemic Chinese plants in the twentieth century.

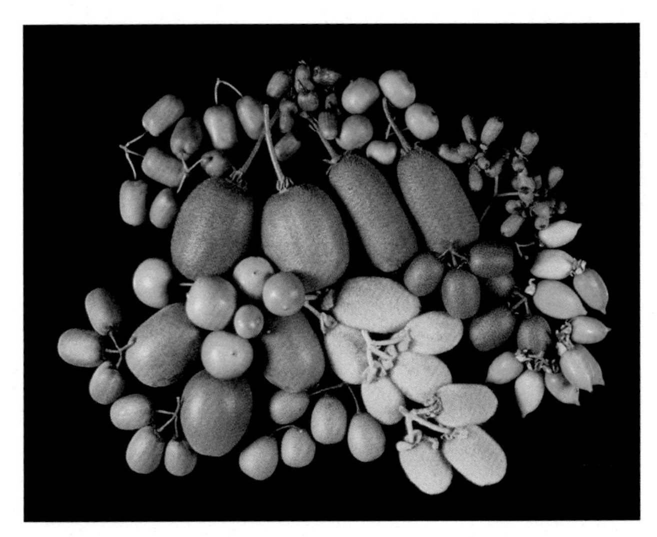

Figure 2.4 Fruits of *Actinidia* (image courtesy New Zealand Institute for Plant and Food Research Ltd).

In China, three regions have been recognized as centers of distribution for endemic genera: eastern Sichuan (including Chongqing Shi), western Hubei and northwestern Hunan; western Sichuan and northwestern Yunnan; and southeastern Yunnan and western Guizhou. The area of eastern Sichuan, western Hubei and northwestern Hunan harbors 82 endemic genera in 46 families, including 39 monotypic genera, 31 oligotypic genera and 12 larger genera. The most impressive center is that of western Sichuan and northwestern Yunnan, with a total of 101 endemic genera in 47 families, including 51 monotypic, 35 oligotypic and 15 larger genera. Southeastern Yunnan and western Guizhou contains 56 endemic genera in 36 families, with 32 monotypic, 17 oligotypic and seven larger genera. Although some of the endemic genera overlap in distribution across two or three of these centers, each center contains 23–30 genera unique only to that area.

2.1.2 Plant diversity patterns and distributions

Diversity patterns.—Within angiosperms, the magnoliid clade is one of the earliest-branching lineages (APGIII, 2009), and contains approximately 9 900 species worldwide (Stevens, 2001 onwards). There are 890 species of Magnoliidae in China (Figure 2.5). The monocot clade includes about 6 000 Chinese species, 3 309 (55%) of which belong to the Commelinidae (Figure 2.5). The eudicots include 22 615 species in China, with 20 600 belonging to the core eudicot clade. The fabid and malvid clades, part of the rosids, contain 4 651 and 1 658 species respectively, in China, while the lamiid and campanulid clades, both part of the asterids, the most derived clade of angiosperms, contain 4 956 and 2 891 species respectively (Figure 2.5).

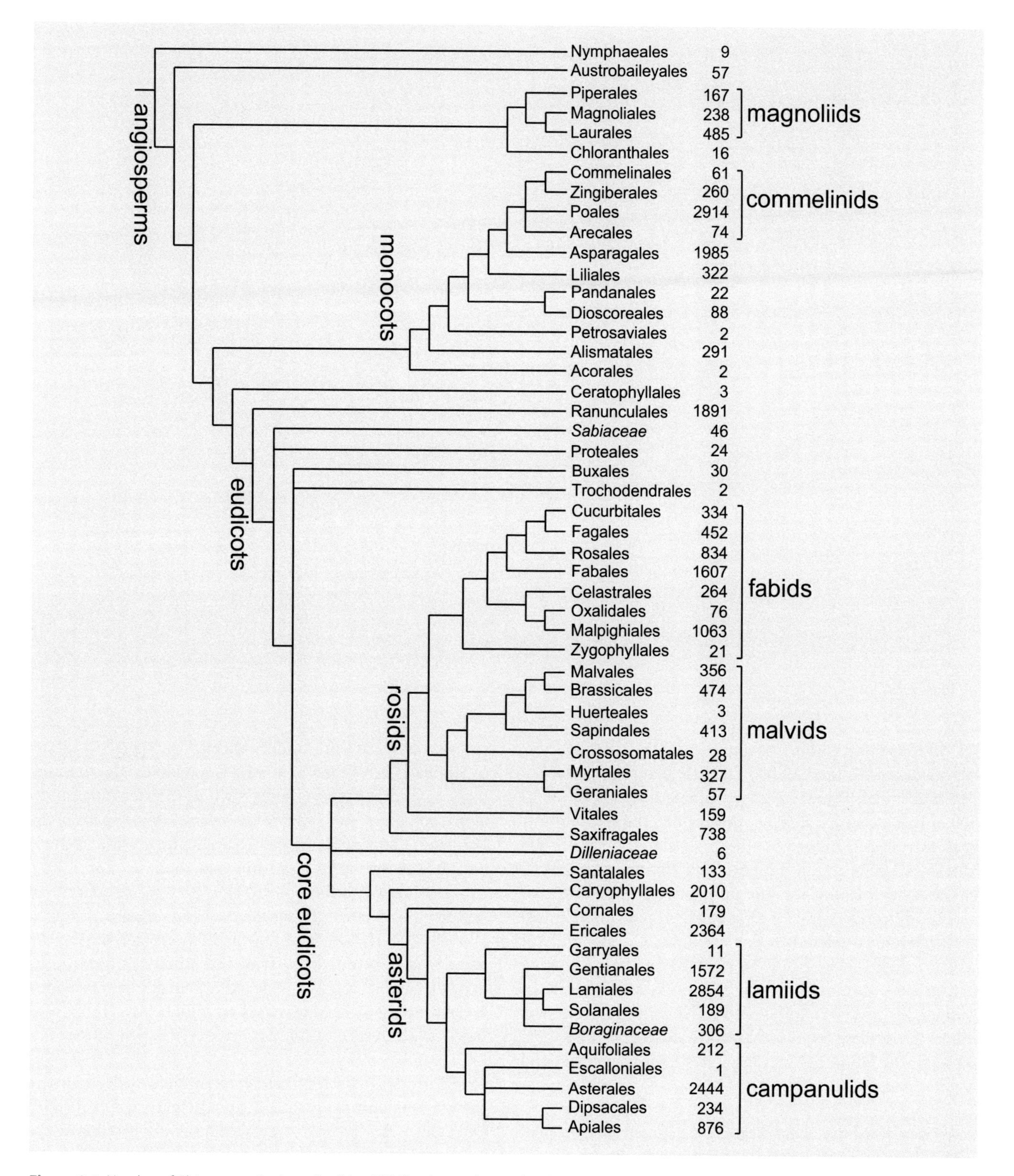

Figure 2.5 Number of Chinese species in each of the APG III orders and some families (indicated by *italics*) that are not assigned to any order. Twenty-five species (one in Cynomoriaceae, 24 in Icacinaceae) are not included in this phylogenetic tree (redrawn from APGIII, 2009).

In this chapter we review the species distributions of vascular plants across the provinces, autonomous regions and municipalities of China based on a large number of sources, primarily Editorial Committee of *Flora Reipublicae Popularis Sinicae* (1959–2004), Wu *et al.* (1994–2011), Wu & Ding (1999), the *Species 2000 China Node* (http://www.sp2000.cn), the *Chinese Virtual Herbarium* (http://www.cvh.org.cn), and the *Flora of China Checklist* at Missouri Botanical Garden (http://www.tropicos.org/Project/FC). In this study, Beijing Shi and Tianjin Shi were included in Hebei, Shanghai Shi in Zhejiang, Chongqing Shi in Sichuan, and Hong Kong, Macao and Shenzhen Shi in Guangdong, thus recognising 28 province-level geographic units in China, grouped into six major regions (Figure 2.6): Northeastern, Northern, Eastern, Southern, Southwestern and Northwestern China.

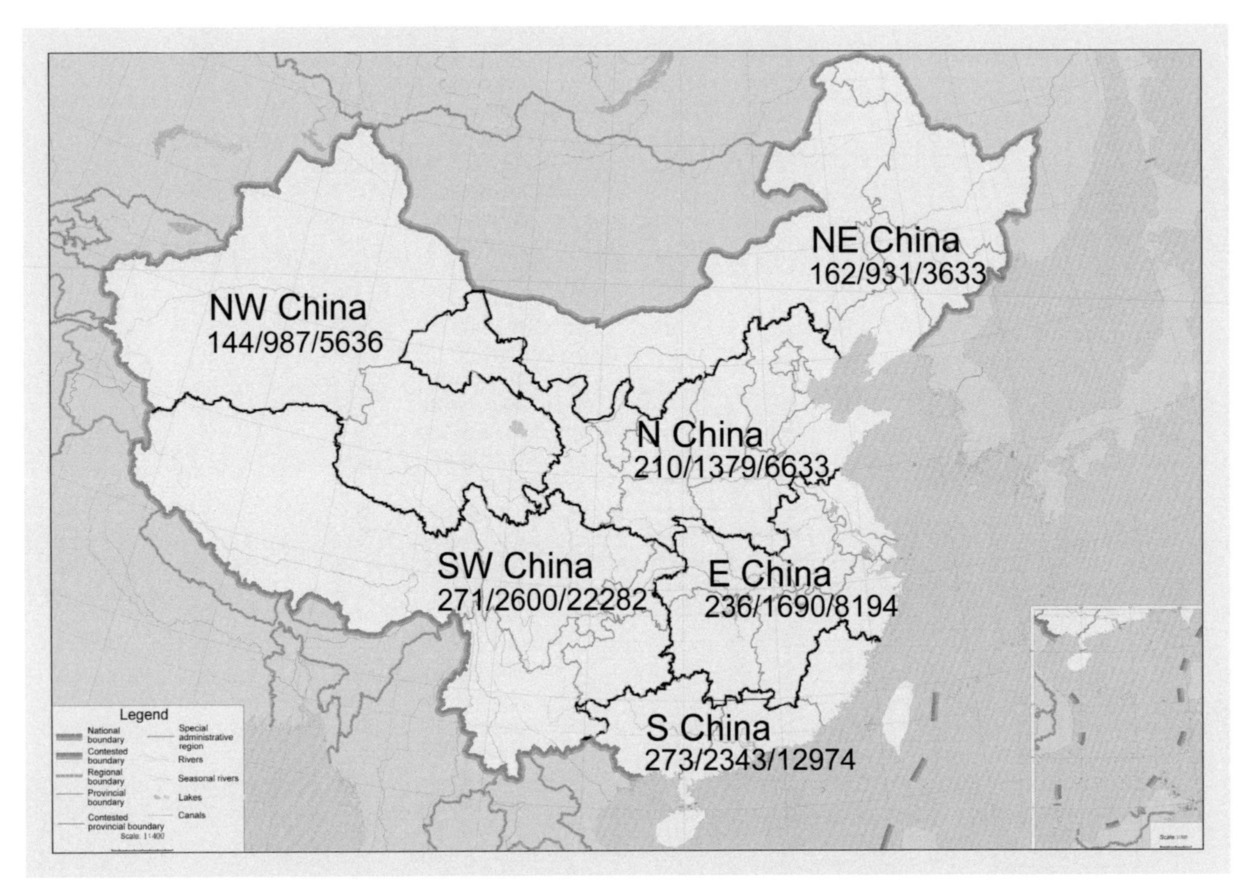

Figure 2.6 Map showing six regions in China with the numbers of families, genera, and species of vascular plants in each region.

Southwestern China was found to be the most species-rich region with over 22 000 vascular plant species (68% of China's total; Figure 2.6); the Chinese distribution of 10 800 of these were restricted to this region. The second most species-rich region was Southern China with ca. 13 000 species, 4 100 of which did not occur elsewhere in China. Northeastern China was the poorest region in terms of species diversity, with 3 633 species. Approximately 690 species were restricted to this region in China, some of which are, however, found in neighboring areas outside China such as North Korea, Mongolia and Russia. In general, the species richness of vascular plants in China was found to decrease from southwest to northeast. The same pattern was found for pteridophytes, gymnosperms, and angiosperms (Table 2.4).

It is believed that the collision of the Indian subcontinent with the Asian continent during the Eocene Epoch played a primary role in generating the high species richness of Southwestern China. As Qian (2002) pointed out, the collision may have profoundly affected the species diversity of Southwestern China in several ways. Firstly, the resulting high, rugged mountain ranges such as the Hengduan Shan and Gaoligong Shan, and large river systems such as the Nu Jiang, Lancang Jiang and Drungjiang, acted as natural barriers preventing species from spreading (Li *et al.*, 1999). A significant number of species became vicariant on separate mountains during the orogenic process, favoring allopatric speciation. The new habitats created during the rapid and continued uplift of the Himalaya Shan, plus existing habitats across a wide altitudinal gradient provided a great array of habitat types in which both relict and newly-evolved species could survive (Wu & Wu, 1996).

Secondly, this tectonic movement resulted in the horizontal compression of southwestern China and adjacent areas (Press & Siever, 1986), which reduced the total area of the region, although some of the horizontal area was transformed into mountain slopes through vertical expansion. This, combined with low rates of extinction and

Table 2.4 The numbers of families, genera, and species of vascular plants in each of the major geographic regions of China. Data compiled from a based on, but not limited to, Editorial Committee of Flora Reipublicae Popularis Sinicae (1959–2004), Wu *et al.* (1994–2011), Wu & Ding (1999), the Species 2000 China Node (http://www.sp2000.cn), the Chinese Virtual Herbarium (http://www.cvh.org.cn), the Flora of China Checklist at Missouri Botanical Garden (http://www.tropicos.org/Project/FC), and numerous provincial floras. Infraspecific taxa were merged into species; exotic species were excluded.

Region	Plant group	Number of families	Number of genera	Number of species
Northeastern China	Pteridophytes	26	49	138
	Gymnosperms	6	14	43
	Angiosperms	130	868	3 452
Northern China	Pteridophytes	40	94	391
	Gymnosperms	9	25	75
	Angiosperms	161	1 260	6 167
Eastern China	Pteridophytes	53	145	832
	Gymnosperms	9	32	83
	Angiosperms	174	1 513	7 279
Southern China	Pteridophytes	61	201	1 297
	Gymnosperms	9	34	101
	Angiosperms	203	2 108	11 576
Southwestern China	Pteridophytes	61	190	1 698
	Gymnosperms	10	34	148
	Angiosperms	200	2 376	20 436
Northwestern China	Pteridophytes	19	29	90
	Gymnosperms	3	9	53
	Angiosperms	122	949	5 493

abundant opportunities for speciation likely resulted in the increased species density seen in Southwestern China. The wide array of habitats across a wide altitudinal range made the region a center not only for survival of relict species but also for speciation and evolution (Chapman & Wang, 2002; Qian, 2002). Thus, Southwestern China is considered both a "cradle" and a "museum" of flora and fauna.

Species distribution.—Each province of China has, on average, 4 523 species of vascular plants, and most vascular plant species are distributed in only few provinces: on average, each species occurs in fewer than four provinces. Approximately 13 440 species of vascular plants are found in only one of the 28 provinces of China (Figure 2.7), and 3 916 of these are restricted to Yunnan alone. In contrast, only about 500 species are distributed across more than 20 provinces, and even fewer across more provinces than this (see Figure 2.7).

2.1.3 Floristic similarity among regions

China spans a wide range of latitudes and longitudes and a great variation in climate. The history of floristic development and evolution also varies greatly across the country. As a result, floristic composition differs substantially among regions. We calculated floristic similarity for all possible pairs of the six major regions of China using the Jaccard (J; Legendre & Legendre). J is equal to one when all taxa are shared between two regions, and zero when none are shared. Calculating the Jaccard index separately for species, genera and families showed that, at the species rank, the lowest similarity was between Southern and Northwestern China (Figure

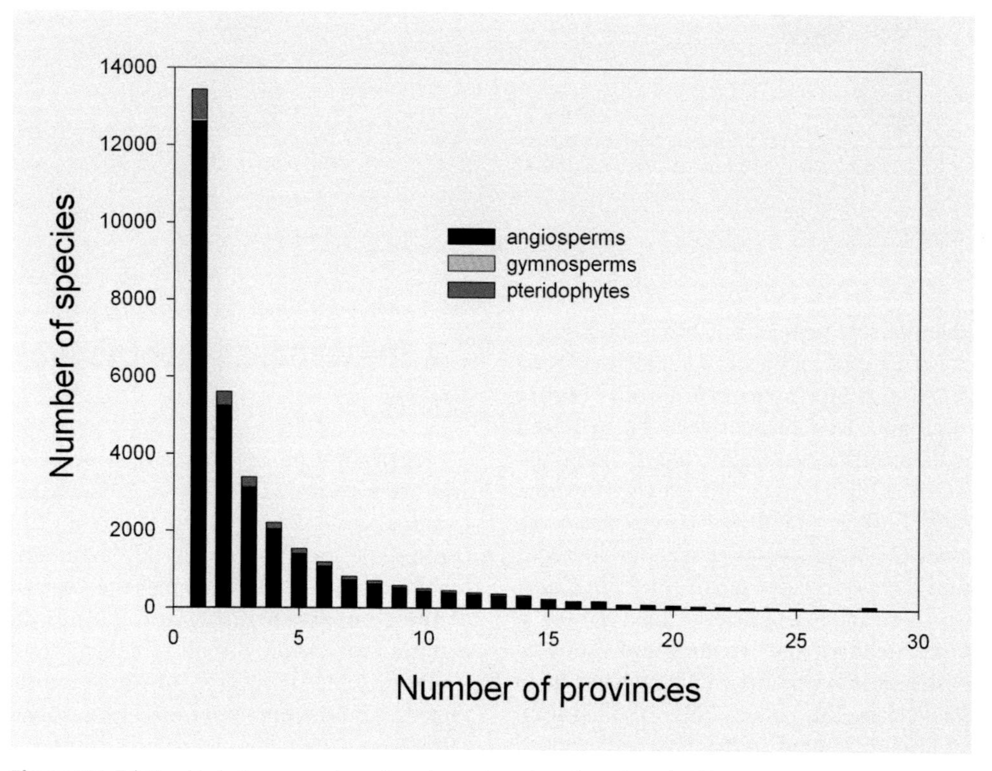

Figure 2.7 Relationship between number of species and number of provinces in China.

2.6; $J = 0.04$), and the highest similarities were between Southern and Eastern China and between Northern and Northeastern China ($J = 0.34$ in both cases). As expected, pairs of regions adjacent to one another were more similar than those far apart (Figure 2.8). For a given pair of regions, similarity increased dramatically between species, genus and family ranks (Figure 2.8). However, the pattern of similarity observed at species rank across pairs of regions was generally retained for genus and family ranks (Figure 2.8).

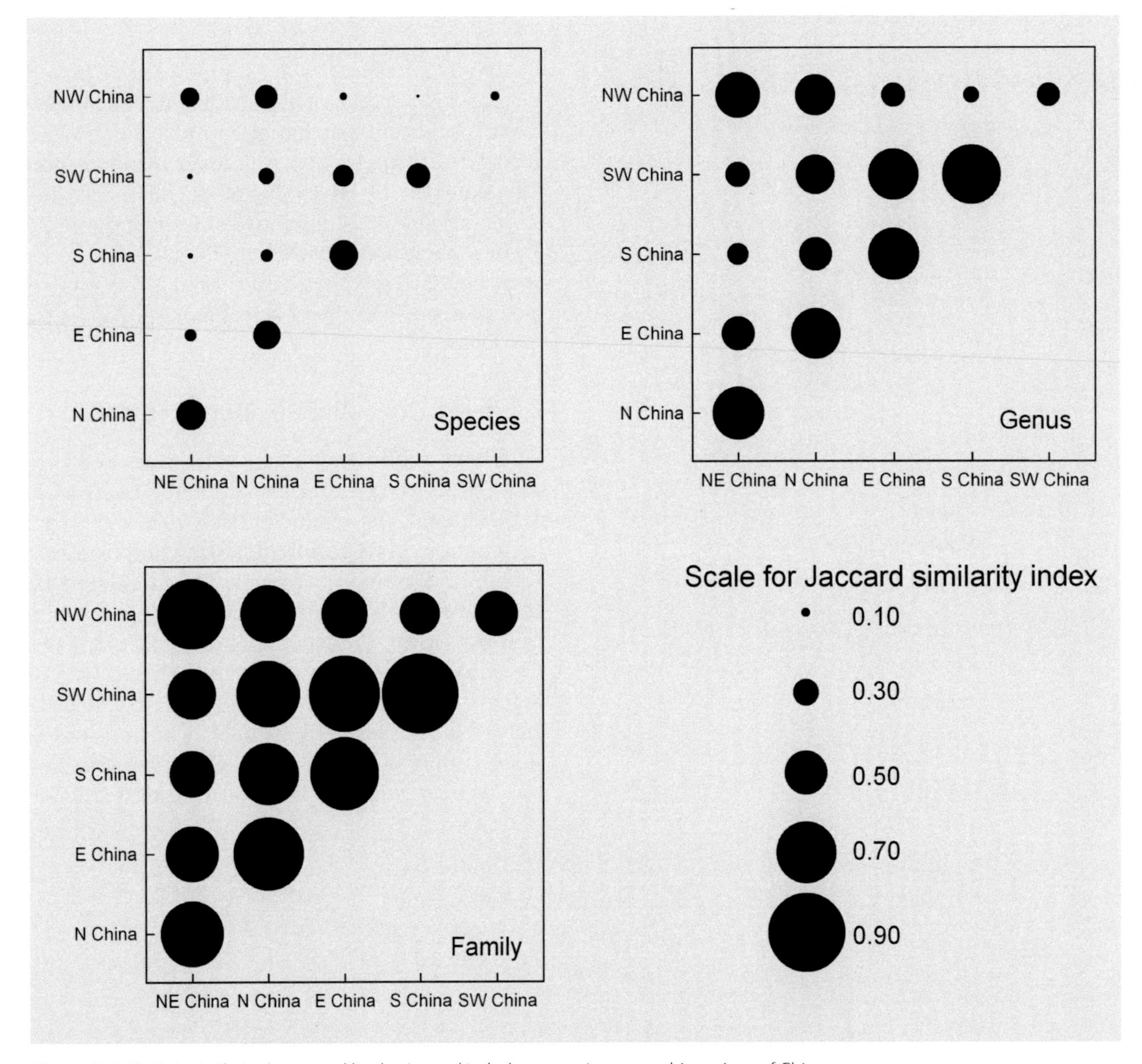

Figure 2.8 Floristic similarity (measured by the Jaccard index) among six geographic regions of China.

Of all possible pairs of regions in China, the highest floristic similarity at familial and generic ranks was found between Southern and Southwestern China (Figure 2.8), with $J = 0.90$ and 0.68, respectively. This pattern suggests that the floras of the two regions share a common evolutionary history at higher taxonomic ranks. However, at the species rank, floristic similarity between these two neighboring regions ($J = 0.28$) was lower than for Southern and Eastern China ($J = 0.34$), Eastern and Northern China ($J = 0.33$), and Northern and Northeastern China ($J = 0.35$), despite the fact that Southern and Southwestern China lie at approximately the same latitudine, while the latter lie in different latitudinal zones. This pattern likely resulted from higher speciation rates in Southwestern China as described above.

2.2 The worldwide contribution of Chinese plants

China is one of the cradles of world civilization and has a recorded history of 5 000 years of civilization (Schirokauer & Brown, 2005). Ancient Chinese civilization contributed greatly to world developments in science and technology with, for example, the celebrated inventions of the compass, printing and gunpowder. The British historian of science Joseph Needham considered that the inventions and discoveries of China transcended those of their European contemporaries, especially around 500 AD (Needham, 1956). Simultaneously, plant domestication and the development of Chinese herbal medicine have made tremendous contributions to global advances in agriculture, food and medicine during China's 5 000-year exploitation of useful plant resources, which will be details in later chapters.

The abundance of cultivated plants in China is unique and unparalleled. During its more than 5 000 years of agricultural development, China has accumulated a rich genetic diversity of cultivated plants, and is one of the eight original centers of farming in the world (Harlan, 1971). Of 1 500 crop plants cultivated worldwide, 300 originated, were domesticated, or underwent differentiation in China, resulting in an approximate 10 000 cultivated taxa. Noteworthy examples include rice (*Oryza*; Poaceae), with 50 000 cultivars and landraces domesticated from just three wild species (*O. rufipogon*, *O. officinalis* and *O. meyeriana*); wheat (*Triticum*; Poaceae) with 30 000 varieties or landraces, millet (*Setaria italica*; Poaceae; 25 000 varieties or landraces), soybean (*Glycine max*; Fabaceae; 20 000 varieties or landraces), and a great number of other major crops or their wild relatives, including buckwheat (*Fagopyrum esculentum*; Polygonaceae), barley (*Hordeum vulgare*; Poaceae), broomcorn millet (*Panicum miliaceum*; Poaceae) and various bean crops (*Vigna* spp.; Fabaceae; Ministry of Environmental Protection of the People's Republic of China, 1998; Zhu, 2004).

China also has a rich genetic pool of vegetables. Around 14 000 varieties or landraces have been documented, belonging to 229 species and 56 families, of which 135 species originated in China. Examples from the 50 most widely used vegetable include Chinese leek (*Allium tuberosum*; Liliaceae), Chinese green onion (*A. fistulosum*), winter melon (*Benincasa hispida*; Cucurbitaceae), stem mustard (*Brassica juncea* var. *tumida*), Chinese cabbage (*Brassica rapa* var. *chinensis*; Brassicaceae) and Chinese water chestnut (*Eleocharis dulcis*; Cyperaceae; Ministry of Environmental Protection of the People's Republic of China, 1998). In addition, there are about 200 wild vegetable species with potential for further exploration for the vegetable industry.

Equally impressively, China is one of the richest countries for fruit and nut plants, with more than 300 fruit crops belonging to 50 families and 81 genera. Of these, more than 50 originated or were domesticated in China; notable examples include peach (*Amygdalus persica*), apricot (*Armeniaca vulgaris*), loquat (*Eriobotrya japonica*) and plum (*Prunus salicina*), all Rosaceae, chestnut (*Castanea mollissima;* Fagaceae), jujube (*Ziziphus jujube*; Rhamnaceae), kumquat (*Citrus japonica*), mandarin (*Citrus reticulata*), pomelo (*Citrus maxima*) and wampee (*Clausena lansium*), all Rutaceae, longan (*Dimocarpus longan*) and litchi (*Litchi chinensis*), both Sapindaceae, persimmon (*Diospyros kaki*; Ebenaceae), walnut (*Juglans*; Juglandaceae), and many other minor fruits (Zhu, 2004).

The wild fruit and nut species naturally occurring in China—of which 1 076 species in 73 families and 173 genera have been documented—have tremendous potential for domestication and commercial cultivation. The largest genera of fruit and nut plants are *Rubus* (Rosaceae) with ca. 200 species, *Ribes* (Saxifragaceae; ca. 60 species), *Actinidia* (ca. 50), *Rosa* (Rosaceae; ca. 40), *Elaeagnus* (Elaeagnaceae; ca. 25), *Vitis* (Vitaceae; ca. 25), *Cerasus* (ca. 20) and *Malus* (ca. 20; both Rosaceae), *Diospyros* (ca. 20), *Quercus* (ca. 20), *Vaccinium* (Ericaceae; ca. 20), *Castanopsis* (Fagaceae; ca. 15), *Viburnum* (Adoxaceae; ca. 15), *Crataegus* (Rosaceae; ca. 15), *Ziziphus* (ca. 15) and *Pyrus* (Rosaceae; ca. 10). These 16 largest genera encompass a total of 628 species, accounting for 60% of the wild fruit and nut plants native to China (Li, 1998). See Chapter 16 for more detail on the crop plants of China.

2.3 Diversity hotspots and endemism

2.3.1 Important areas of plant diversity in China

As described in future chapters, China is a large country with varied climatic zones, topography and soils, reflected in the diversity of its flora. Plant diversity is of course not equally distributed spatially across the Chinese landmass. Rather, patterns of distribution reflect factors such as latitude, altitude, temperature, moisture conditions and degree of human disturbance as described further in Chapters 3–10.

Various different systems have been developed over the past 40 years to characterise priority areas for biodiversity conservation at global, regional and national scales. These have been applied to varying extents in China, and the country's areas of highest plant diversity are generally well-known. The delimitation of areas differs according to different authors as seen in the descriptions below.

Centres of Plant Diversity.—At a global level, the Centres of Plant Diversity project initiated in 1987, undertaken by the World Wide Fund for Nature (WWF) and the International Union for Conservation of Nature (IUCN), aimed to identify which areas around the world, if conserved, would safeguard the greatest number of plant species, document the scientific and economic benefits of conserving these areas, and provide a strategy for their conservation (Davis *et al.*, 1995). Centres of plant diversity were identified primarily on the basis of species richness and endemism, and are also important natural pools of germplasm of potential crop plants, sites with a diverse range of habitat types, or sites threatened or under imminent threat of large-scale devastation. Forty-one sites were identified for China, such as Changbai Mountains region, Taibai Mountain region of Qinling Mountains, Nanling Mountain range, Gaoligong Mountains, Tropical forest of Hainan, Xishuangbanna region, Limestone region of Guangxi, Hengduan Mountains and Minjiang River basin, and other high mountain and deep gorge regions.

Ecoregions.—An "ecoregion" is defined as a large area of land or water that contains a geographically distinct assemblage of natural communities sharing a large majority of their species and ecological dynamics, with similar environmental conditions and interacting ecologically in ways that are critical for their long-term persistence (Hawkins *et al.*, 2008). A number of ecoregions are recognized globally as conservation priorities for the protection of ecosystems and unique habitats for plant species (WWF, 2009). Table 2.5 shows the WWF ecoregions of China.

Table 2.5 WWF ecoregions (from (http://wwf.panda.org/about_our_earth/ecoregions/about/). Note that spellings of geographic names are according to WWF, and do not necessarily match those in the *Flora of China*.

Altai-Sayan Montane Forests
Daurian Steppe
Eastern Himalayan Alpine Meadows
Eastern Himalayan Broadleaf and Conifer Forests
Hengduan Shan Coniferous Forests
Indochina Dry Forests
Mekong River
Middle Asian Montane Steppe and Woodlands
Northern Indochina Subtropical Moist Forests
Russian Far East Rivers and Wetlands
Salween River
Southeast China-Hainan Moist Forests
Southwest China Temperate Forests
Taiwan Montane Forests
Tibetan Plateau Steppe
Xi Jiang Rivers and Streams
Yangtze River and Lakes
Yellow Sea
Yunnan Lakes and Streams

Selected areas of plant diversity in China(WWF, 2009). The Xishuangbanna tropical seasonal rainforests of Yunnan support dipterocarp (Dipterocarpaceae) and fig (*Ficus*; Moraceae) species, while the region's tropical montane forests comprise tree species including *Alstonia scholaris* (Apocynaceae) and *Phoebe puwenensis* (Lauraceae). Oak (*Quercus*), tea (*Camellia*; Theaceae) and evergreen chinquapin (*Castanopsis*) species dominate the evergreen broad-leaved woodlands.

The subtropical evergreen forests of the Yunnan plateau are broad-leaved and adapted to wet summers and cool dry seasons. Here, elevated land has a warming effect on the subtropical forests. Temperate cloud forest communities may be found on the summits of taller mountains, while low hills support seasonally dry subtropical forest. Several species of *Castanopsis*, *Quercus* and Lauraceae dominate this ecoregion. The leaves of these trees are hairy, which helps them conserve water and remain green during the long dry season.

The subtropical evergreen forests of Jiannan grow on the hills that stretch between southern China's coastal plains, the Chang Jiang and Zhu Jiang basins, and the Guizhou and Yunnan plateaux. The climate is mild, but typhoons and monsoons are common. The Nanling Shan are a recognized center of plant diversity and endemism in this area, where protected species such as the subtropical conifer *Cathaya argyrophylla* and the tree fern *Cyathea spinulosa* (Cyatheaceae) may be found. Bamboos are a diverse and important part of the forest here.

The South China-Vietnam subtropical evergreen forests form a transition zone between the tropical forests of Vietnam to the south and the subtropical and mixed forests of southern China, and is rich in species from both areas.

Hainan, the second largest island off the coast of mainland China after Taiwan, harbors monsoon rain forests containing 4 200 species of plants, many of them endemic. There is a surprisingly high diversity of conifer species in the upland interior of Hainan.

The Qin Ling reach elevations as high as ca. 3 700 m, run east-west and are subject to strong, cold winter winds on the northern slopes with wetter, warmer southern slopes. They form an important boundary between two of China's largest watersheds, the Chang Jiang and Huang He. The foothills of the Qin Ling are dominated by deciduous forests of oak (*Quercus*), elm (*Ulmus;* Ulmaceae), walnut (*Juglans*), maple (*Acer*; Aceraceae), and ash (*Fraxinus;* Oleaceae) trees. As elevation increases, broad-leaved evergreen species and conifers begin to dominate. Larch (*Larix;* Pinaceae) and birch (*Betula;* Betulaceae) do well on the highest hills, while *Rhododendron* (Ericaceae) and dwarf bamboos create a thick understory. *Tsuga chinensis* var. *chinensis* (Pinaceae) is endemic to this region, and protected species such as red firs, *Cercidiphyllum japonicum*, and hardy rubber trees are sheltered in nature reserves.

The Daba Shan evergreen forests contain species of trees belonging to ancient lineages, some of them thought to be extinct until recently rediscovered. The foothills of the lower mountains support a mix of evergreen and deciduous trees, mostly species of *Quercus*, while conifers such as the Chinese red pine (*Pinus tabuliformis*; Pinaceae) and *P. armandii* grow on the higher slopes. The Shennongjia Mountain Nature Reserve contains more than 2 600 species of vascular plants including three protected species, and is renowned for the conservation of many such endemic and ancient lineages, as well as wild relatives of crop plants and medicinal plants. The dawn redwood (*Metasequoia glyptostroboides*), a descendent of conifers that arose 100 million years ago, was once thought to be extinct but relict stands were discovered in the 1940s. The dove tree (*Davidia involucrata*) and coffin tree (*Taiwania cryptomerioides*; Taxodiaceae) are also found here. These three species are noteworthy because they represent ancient lineages whose populations were greatly reduced in number during the ice ages of the Pleistocene Epoch 2.5 million to 11 700 years ago, but small numbers survived on the relatively warm southern slopes of the Daba Shan complex.

In the Sichuan Basin, much of the original evergreen broad-leaved forest has been destroyed, so it is hard to know exactly which species used to exist in the ecoregion. Emei Shan is one of the best places to see the remnants of the region's lush subtropical forests. Rare plants that still occur at subtropical elevations include tree ferns (members of

Cyatheales), *Davidia involucrata*, and the conifers *Cathaya argyrophylla* and *Taxus wallichiana* var. *chinensis* (Taxaceae).

The Altay montane forest and forest steppe lie in the region where Kazakhstan, Russia, China and Mongolia meet. Nine-hundred and seventy-four plant species are found here, including 60 endemics. The forests in the Altay region are composed of *Abies sibirica* (Pinaceae), *Pinus* and *Larix*, interspersed with groves of *Betula* and *Populus* (Salicaceae). Spring and summer here bring many beautiful flowers including several types of *Aster* (Asteraceae). Under the forest canopy, *Vaccinium* (including bilberry and cowberry), *Ribes rubrum* (redcurrant), oriental *Spiraea* (Rosaceae) and *Prunus* provide sweet fragrance and tasty fruit for foragers. There are at least 26 species of orchid (Orchidaceae) and 35 species of ferns; the lichen and moss diversity is also unusually high.

The subalpine conifer forests of the Hengduan Shan are a topographically complex ecoregion comprising parallel mountain ranges separated by deep, narrow river valleys. The subalpine zone of the southern Hengduan Shan has abundant conifer forests, with distinct forest types made up of hemlock (*Tsuga*), spruce (*Picea*; Pinaceae), and fir (*Abies*) species. *Acer, Betula*, and mountain ash (*Sorbus*; Rosaceae) form a lower canopy, slender subalpine bamboos and ferns flourish in the understory, with mosses and lichens growing on the bamboo leaves.

The Qionglai Shan and Min Shan, which are among the steepest and tallest mountains on Earth, separate the Qinghai-Xizang Plateau from the Sichuan Basin. Gongga Shan, the highest summit, at 7 556 m, is so steep that the glaciers on its east face tumble down below the tree line before they finally melt. In this ecoregion vegetation is to some extent determined by elevation. *Pinus* and *Picea* can be found in the drier, lower valleys; dense coniferous forests with abundant *Abies* and *Larix* and a thick understory of deciduous *Acer, Betula, Sorbus* and bamboo grow at mid elevations; juniper (*Juniperus;* Cupressaceae) grows at the highest altitudes, where the soil is rocky and well-drained.

The gorges of the Nu Jiang and Langcang Jiang support alpine coniferous and mixed forests. The area remains one of the most intact and biologically diverse parts of China, notably rich in conifer species. A variety of trees grow above the river gorges on summits as high as 3 000–5 000 m. Most of the 20 mid-elevation conifer species reported from Yunnan occur here (Sun *et al.*, 2007). The Gaoligong National Nature Reserve is of great ecological importance, including habitats ranging from subtropical evergreen broad-leaved forest at low elevations to subalpine conifer forests at higher elevations, large tracts of which are relatively undisturbed. The Nu Jiang Nature Reserve contains similar forest habitat and supports red pandas, musk deer, and other forest animals of the Himalaya Shan.

Several different kinds of habitat occur on the Qinghai-Xizang Plateau, from dense scrubland and meadow to steppe and desert, and riparian communities. In the Yarlung Zangbo Jiang valley, the milder climate has made it possible for some coniferous-*Rhododendron* forests to thrive. Depending on the elevation this ecoregion, defined as arid steppe, is dominated by grassland, shrubland, or coniferous trees. Above ca. 5 000 m, cushion plants (low, tightly-massed plants that form dense, cushiony tufts) dominate stable slopes. Alpine shrub and meadows are also found on the Qinghai-Xizang Plateau, and the region also contains the headwaters of China's two largest rivers, the Huang He (Yellow River) and the Chang Jiang (Yangtze).

2.3.2 Hotspots

"Biodiversity hotspot" is a term coined by Norman Myers to delimit and highlight areas of biodiversity richness and threat (Myers, 1988). On a global scale, 25 biodiversity hotspots have been identified and described by Conservation International (Mittermeier *et al.*, 1999), taking into account the earlier results of the WWF-IUCN Centres of Plant Diversity project. The primary criteria for defining global biodiversity hotspots are the number and threatened status of the plant species present. The biodiversity hotspot concept has been taken up by policy makers and is used, for example, in the allocation of financial resources for biodiversity conservation. Two of the 25 global biodiversity hotspots identified by Conservation International (Mittermeier *et al.*, 1999) are located in China: that of "South-Central China" totally within the country, "Indo-Burma" partially.

A subsequent revision of the original 25 identified 34 hotspots (Mittermeier *et al.*, 2004). Of these, the "Mountains of Southwest China" hotspot falls almost entirely within China (with a very small portion in Myanmar), and the "Himalaya," "Indo-Burma" and "Mountains of Central Asia" hotspots fall partly within the country. The Mountains of Southwest China covers the area corresponding to the Hengduan Shan. As such it mainly equates to the South-Central China hotspot of the 1999 report. The Himalaya hotspot has been separated out from the Indo-Burma hotspot in the 2004 revision, and the Mountains of Central Asia includes the Chinese Tian Shan and Muztag Shan massifs.

The Mountains of Southwest China is the largest hotspot in China, stretching over 262 400 km^2 of temperate to alpine mountains between the easternmost edge of the Qinghai-Xizang Plateau and the Central Chinese Plain, to the north of the Indo-Burma hotspot, and the immediate east of the Himalaya hotspot. It is bounded in the northwest by the dry Qinghai-Xizang Plateau, in the north by the Tao He of southern Gansu, in the east by the Sichuan Basin and eastern Yunnan Plateau, and in the south by northeastern Myanmar. The mountains of southwestern China are characterized by extremely complex topography, ranging from below 2 000 m in some valley floors to 7 558 m at the summit of Gongga Shan, and contain the most species-rich temperate and tropical river systems in Asia. The complex topography results in a wide range of climatic conditions: temperatures range from frost-free throughout the year in parts of Yunnan and short, frost-free periods at the northern boundary of the region, to permanent glaciers on the high mountain peaks of Sichuan, Yunnan and Xizang (Tibet). Annual average rainfall in the region exceeds 1 000 mm on southwestern slopes at higher altitudes in Yunnan, while northwestern parts of the region, in the rainshadow of the Qinghai-Xizang Plateau, rarely receive more than 400 mm(Conservation International, 2009).

These climatic and topographic conditions result in a wide variety of vegetation types, including broad-leaved and coniferous forests, bamboo groves, scrub communities, savanna, meadow, prairie, freshwater wetlands, alpine scrub and scree communities. As a result of the dramatic differences in topography, climate and vegetation, and the physical barriers between areas, the Mountains of Southwest China hotspot has evolved a cluster of distinctive mini-hotspots and a huge diversity of habitats within it. The hotspot is probably the most botanically-rich temperate region in the world (Kelley, 2001), although it has not yet been fully documented. Vascular plant diversity is estimated at ca. 12 000 species, representing as much as 40% of all the species in China. Of these, ca. 3 500 species (29%) and at least 20 genera are endemic, including about 100 endemic ferns and 20 endemic gymnosperms. The Hengduan Shan, in particular, are recognized to be an important Cenozoic center of species diversification and a refugium for Laurasian angiosperm groups including several ancient lineages found nowhere else in the world, including representatives of the genera *Circaeaster* (Circaeasteraceae), *Kingdonia* (Ranunculaceae), *Rhodiola* (Crassulaceae; Figure 2.9) and *Rhododendron* (Figure 2.10) (Donoghue *et al.*, 1997; Conservation International, 2004). More than a quarter of the world's *Rhododendron* species, and almost half of Chinese species, are represented in the Hengduan Shan, an astounding 230 different species, many of which are endemic and quite rare.

Figure 2.9 *Rhodiola* spp. (images courtesy GAO Xin-Fen, WU Jiao-Feng and LIU Su, Chinese Virtual Herbarium).

Figure 2.10 Examples of richness of *Rhododendron* diversity in China. A, *R.* spp. B, *R. liliiflorum*. C, *R. irroratum* subsp. *pogonostylum*. D, *R. huangpingense*. E, *R. jinboense*. F, *R. delavayi*. G, *R. yunnanense* (image A courtesy Guizhou Baili Rhododendron National Forest Park; images B-G courtesy CHEN Xiang).

The Hengduan Shan are likewise a centre of diversity for *Primula* (Primulaceae), with 113 of the 500 species in the genus and *Saussurea* (Asteraceae), with 101 of the world's 400 species. Many species in these genera are endemic to the Hengduan Shan, as are conifers such as *Abies forrestii*, *A. ferreana*, *Larix speciosa*, *Pseudotsuga forrestii* (Pinaceae) and *Torreya fargesii* var. *yunnanensis* (Taxaceae), and other woody plants such as *Paeonia delavayi* (Paeoniaceae), orchids including *Calanthe dulongensis*, *Coelogyne taronensis* and *Paphiopedilum tigrinum*, and other herbaceous species including *Acanthochlamys bracteata* (Amaryllidaceae), *Anemoclema glaucifolium* (Ranunculaceae), *Bergenia emeiensis* (Saxifragaceae), *Delphinium smithianum* (Ranunculaceae), *Nomocharis pardanthina* (Liliaceae) and *Salvia evansiana* (Lamiaceae). The endemic medicinal species *Corydalis benecincta* (Papaveraceae), *Incarvillea delavayi* and *I. forrestii* (Bignoniaceae), *Lilium lophophorum* (Liliaceae), *Meconopsis speciosa* (Papaveraceae) and *Neopicrorhiza scrophulariiflora* (Scrophulariaceae) form part of a total list of around 2 000 medicinal plant species recorded for the region.

The Indo-Burma hotspot, as defined by Conservation International (Mittermeier *et al.*, 1999), includes parts of southern China (Yunnan, Guangxi and Guangdong), along the borders with Myanmar, Lao PDR and Vietnam. The Indo-Burma hotspot has been more narrowly redefined as the Indochinese subregion, including, in China, parts of southern and western Yunnan and Hainan. A wide diversity of ecosystems is represented, including mixed wet evergreen, deciduous, and montane forests. There are also patches of shrubland and woodland on karst limestone outcrops and, in some coastal areas, scattered heath forests. In addition, a wide variety of distinctive localized vegetation formations occur in the region, including lowland floodplains, swamps and mangroves, alongside many local centers of endemism, particularly isolated patches of montane and lowland wet evergreen forests. Here there are at least 13 500 vascular plants, of which about 7 000 (52%) are endemic to the hotspot. There is a particularly rich diversity of Orchidaceae and Zingiberaceae, and tropical hardwood trees (Conservation International, 2009).

The Indo-Burma hotspot has a tropical monsoon climate. As a result of its geographical position and climatic conditions, a varied vegetation and complex floristic patterns have developed, reflecting transitions between tropics and subtropics. Biogeographically, the region is located in a transition zone, with tropical Southeast Asia to the south, subtropical east Asia to the north, the Sino-Japanese floristic region to the east and the Sino-Himalayan floristic region to the west (Zhu *et al.*, 2006). The Xishuangbanna region of southern Yunnan, bordering Lao PDR and Myanmar, is particularly rich in plant diversity, as has been widely recognized (e.g. Davis *et al.*, 1995; Zhang & Ma, 2008; Zhu, 2008). The main vegetation types in Xishuangbanna, as described by Zhu *et al.* (2006), are tropical rainforest, tropical seasonal moist forest, tropical montane evergreen broad-leaved forest and monsoon forest. The existence of true tropical rainforest in this area of southern Yunnan was confirmed after long speculation by a China-Russia scientific expedition in the 1950s (Zhu *et al.*, 2006). Tropical rainforest in Xishuangbanna occurs quite locally in limited habitats determined mainly by topography. These rainforests lie at the absolute limit of the moist tropics, further from the Equator than tropical rainforests elsewhere in the world: the climate is barely warm and moist enough for some of the forest types found here, and rainforest can only persist in areas where it creates its own microclimate. Tropical montane evergreen broad-leaved forest is the primary montane vegetation type in Xishuangbanna, occurring over 1 000 m. It forms part of a mosaic with the patches of tropical rainforest. Tropical seasonal moist forest occurs on limestone slopes with shallow soils at altitudes of 650–1300 m (Zhu *et al.*, 2006). Monsoon forest in Xishuangbanna occurs predominantly on the banks of the Lancang Jiang.

Figure 2.11 *Pinus wangii* cone (image courtesy JIN Hong-Gang, Chinese Virtual Herbarium).

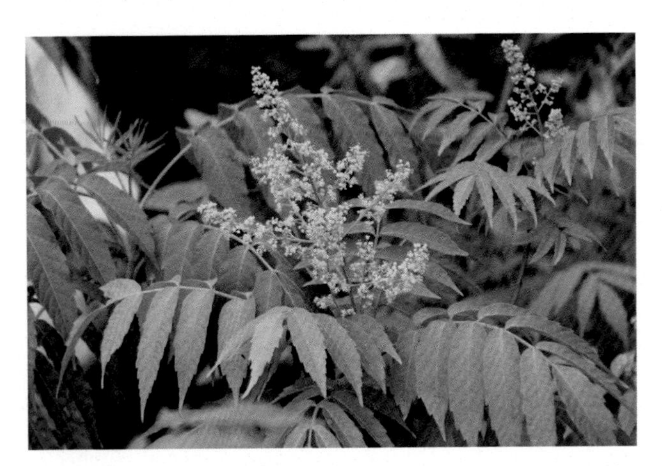

Figure 2.12 Flowers of *Dipteronia dyeriana* (image courtesy YE Yu-Shi, South China Botanical Garden).

Figure 2.13 Leaves of *Pterospermum menglunense* (image courtesy WANG Shao-Ping, South China Botanical Garden).

Figure 2.14 *Cycas balansae* (image courtesy YANG Ke-Ming, South China Botanical Garden).

The flora of Xishuangbanna consists of 3 340 flowering plant species (Zhu, 2008), and there are considered to be over 150 endemics. The endemic species include trees such as *Pinus wangii* (Figure 2.11), *Beilschmiedia brachythyrsa* (Lauraceae), *Dipteronia dyeriana* (Aceraceae; Figure 2.12), *Myristica yunnanensis* (Myristicaceae), *Nyssa yunnanensis* (Nyssaceae) and *Pterospermum menglunense* (Sterculiaceae; Figure 2.13). A number of cycads are also endemic, including *Cycas balansae* (Figure 2.14), *C. micholitzii*, *C. multipinnata* and *C. pectinata* (Figure 2.15).

Figure 2.15 *Cycas pectinata* cone, in cultivation (image courtesy LIAO Li-Fang, South China Botanical Garden).

Hainan is also recognized as part of the Indo-Burma biodiversity hotspot. It has rich plant diversity with about 4 200 species, 630 of which are considered endemic to the island (Carpenter, 2001). Hainan has a tropical humid monsoon climate and generally mountainous topography. The main ecosystem types are tropical montane rainforest, tropical evergreen monsoon forest, tropical semi-deciduous monsoon forest, summit moss forest and savanna. Endemic trees include *Keteleeria hainanensis* (Pinaceae; Figure 2.16), *Sonneratia hainanensis* (Lythraceae), *Chunia bucklandioides* (Hamamelidaceae; Figure 2.17), *Firmiana hainanensis* (Sterculiaceae), *Michelia shiluensis* (Magnoliaceae), *Madhuca hainanensis* (Sapotaceae; Figure 2.18), *Hopea hainanensis* (Figure 2.19) and *Vatica mangachapoi* (Figure 2.20), both Dipterocarpaceae. The endemic cycad *Cycas hainanensis* is restricted to the wetter eastern part of the island.

Figure 2.16 *Keteleeria hainanensis* (image courtesy KONG Ling-Feng, Chinese Virtual Herbarium).

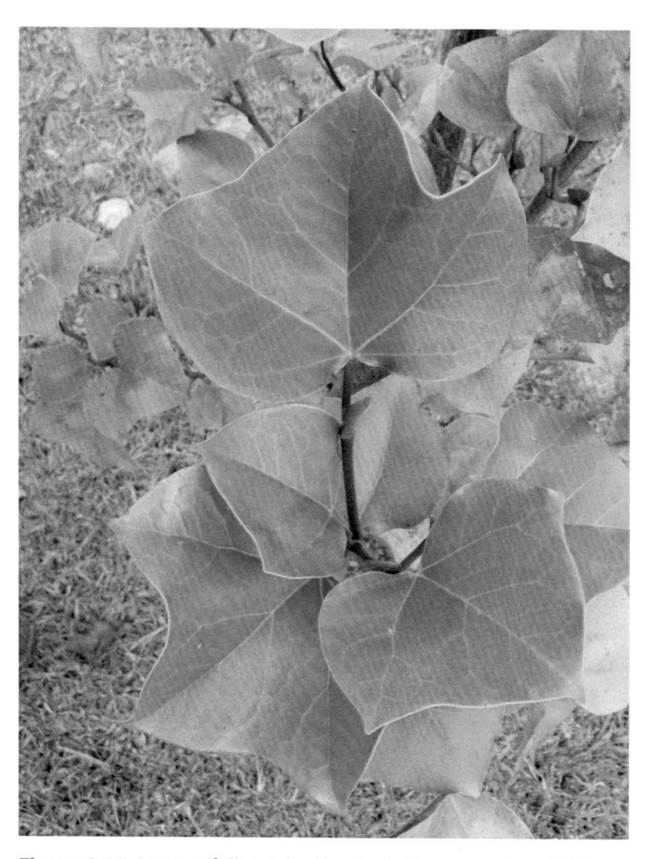

Figure 2.17 Leaves of *Chunia bucklandioides* (image courtesy HU Feng-Qin, Chinese Virtual Herbarium).

Figure 2.18 Fruits of *Madhuca hainanensis* (image courtesy JIN Wen-Chi, Chinese Virtual Herbarium).

Figure 2.19 *Hopea hainanensis*, habit and close-up of fruits (image courtesy YANG Ke-Ming, South China Botanical Garden).

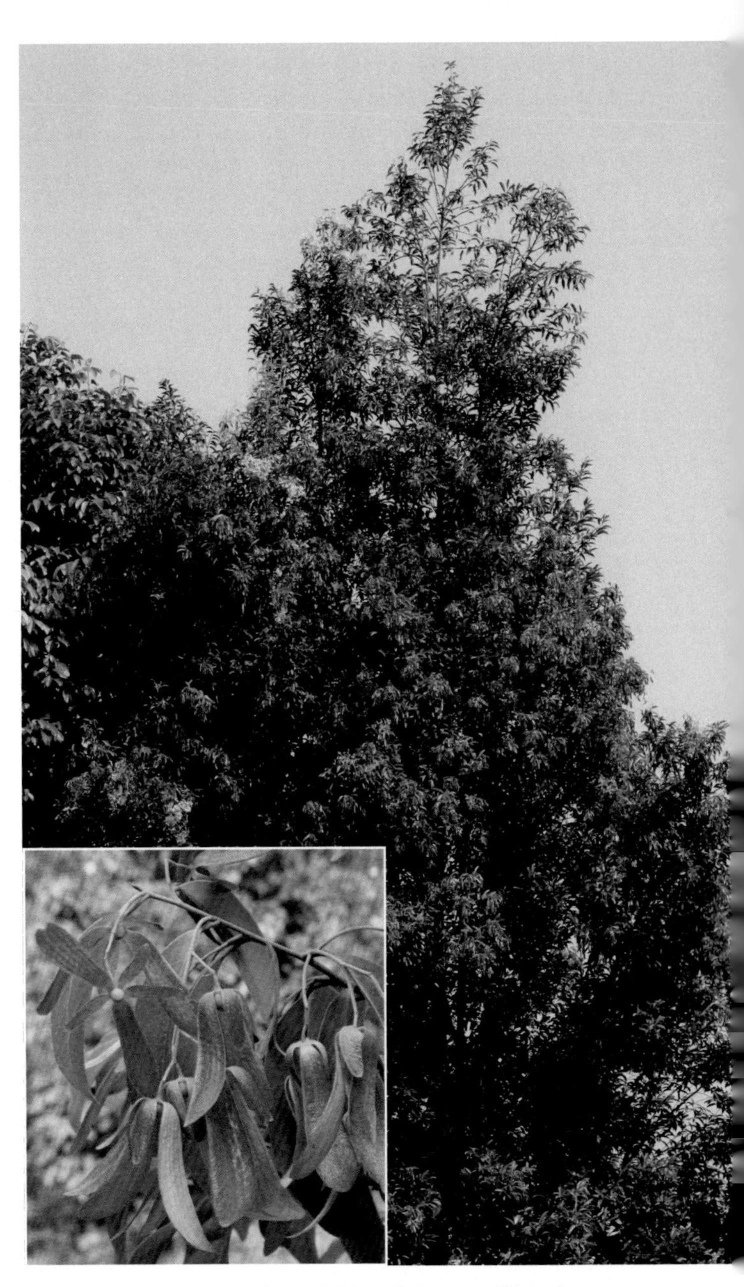

Figure 2.20 *Vatica mangachapoi*, habit and close-up of fruits (image courtesy YANG Ke-Ming, South China Botanical Garden).

Much of Hainan's natural forest vegetation has been cleared for rubber, palm or coffee plantations. Up to 45 plant species are considered to be endangered, and about 450 tree species are harvested for timber. Although there is a high diversity of conifers in the upland interior, most are threatened, such as *Cephalotaxus mannii* (Cephalotaxaceae; Figure 2.21), *Pinus fenzeliana* (Figure 2.22), *P. latteri*, *P. massoniana* var. *hainanensis*, *Keteleeria hainanensis* and *Podocarpus annamiensis* (Podocarpaceae; Figure 2.23). Many angiosperms are also under threat, for example *Chunia bucklandioides* (Figure 2.17), *Hopea hainanensis* (Figure 2.19), *Vatica mangachapoi* (Figure 2.20) and *Dracaena cambodiana* (Liliaceae).

A further hotspot identified consists of Emei Shan, Baoxing and Tianquan in Sichuan as well as Wen Xian and Kang Xian in southern Gansu. The climate in this

Figure 2.21 *Cephalotaxus mannii* (image courtesy KONG Ling-Feng, Chinese Virtual Herbarium).

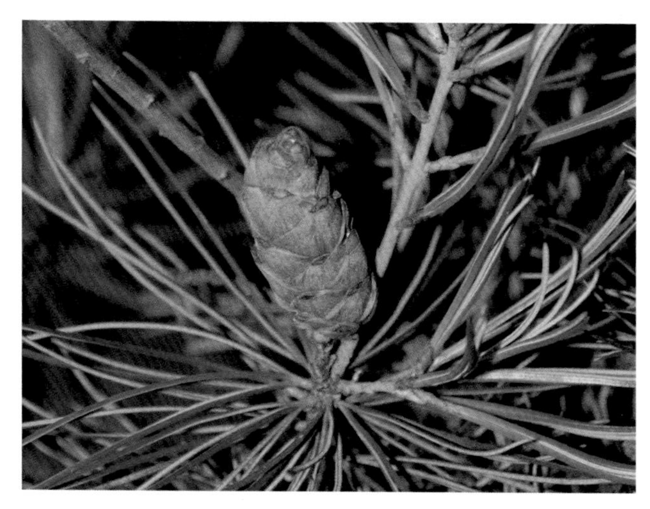

Figure 2.22 *Pinus fenzeliana* cone (image courtesy YE Yu-Shi, South China Botanical Garden).

Figure 2.23 Foliage of *Podocarpus annamiensis* (image courtesy YOU Hui-Xiong, Chinese Virtual Herbarium).

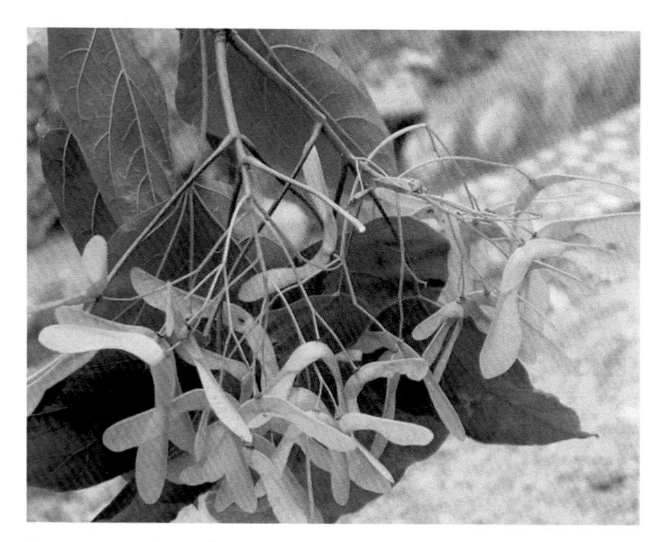

Figure 2.24 Fruits of *Acer amplum* subsp. *catalpifolium* (image courtesy CHEN You-Sheng and GAO Xian-Ming, Chinese Virtual Herbarium).

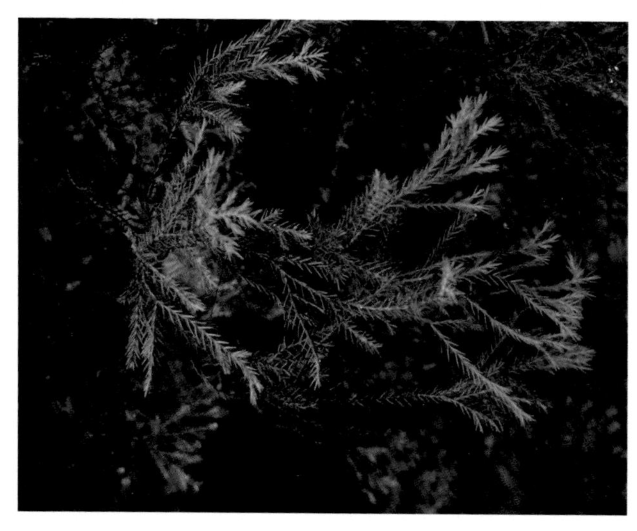

Figure 2.25 *Cupressus chengiana* (image courtesy YANG Ke-Ming, South China Botanical Garden).

Figure 2.26 *Cycas szechuanensis* (image courtesy YANG Ke-Ming, South China Botanical Garden).

area is transitional between the northern subtropical and warm temperate zones. The vegetation includes broad-leaved forest, coniferous forest, alpine shrubland and alpine meadow. Species endemic to this region include *Acer amplum* subsp. *catalpifolium* (Figure 2.24), *Cupressus chengiana* (Cupressaceae; Figure 2.25), *Cycas szechuanensis* (Figure 2.26), *Cystoathyrium chinense*, *Kingdonia uniflora*, *Larix mastersiana* (Figure 2.27), and *Parakmeria omeiensis* (Magnoliaceae).

The western mountains of Guangdong are also considered to be a plant diversity hotspot on the basis of the distribution of threatened species(Zhang & Ma, 2008). The climate in this area is dominated by the subtropical monsoon, being at the boundary between the climates of southern and central China. The land lies above 1 000 m and the major ecosystem types include subtropical evergreen broad-leaved forest, evergreen coniferous and

Figure 2.27 Stand of *Larix mastersiana* (image courtesy CHEN You-Sheng, Chinese Virtual Herbarium).

broad-leaved mixed forest, montane elfin forest, coniferous forest and alpine shrubland. Threatened species endemic to this region include *Dunnia sinensis* (Rubiaceae), *Manglietia pachyphylla* (Magnoliaceae), and *Primulina tabacum* (Gesneriaceae).

An alternative method of defining hotspots within China has been attempted based on endemisms and the number of plant species per unit area, resulting in a set of ten hotspots defined in Table 2.6 (López-Pujol & Zhao, 2004). As Table 2.6 shows, these include five whole provinces (or special regions) plus the areas identified by Conservation International (Mittermeier *et al.*, 1999), with the addition of three more local hotspots. Yunnan as a whole is considered to be a hotspot on the basis of its plant species diversity. Over half the vascular plants of China, nearly 17 000 species (15 000 angiosperms, 100 gymnosperms and 1 500 pteridophytes) are found in this province, which accounts for just 4% of the total landmass, making it the richest Chinese province in terms of plant biodiversity. The Yunnan flora is also noted for its high degree of endemicity. Almost thirty angiosperm genera occur only in Yunnan, including *Chaerophyllopsis* (Apiaceae), *Cyphotheca* (Melastomataceae), *Ferrocalamus* (Poaceae), *Musella* (Musaceae) and *Siliquamomum* (Zingiberaceae). Yunnan is also rich in bamboos, with 250 species, 150 of them being endemic to the province (Yang *et al.*, 2004).

Sichuan is considered the second richest Chinese province in terms of plant biodiversity, with about 9 300 species of vascular plants in over 1 600 genera and 230 families (Sichuan United University Department of Biology, 1994). Of these, 8 453 are angiosperms, 88 are gymnosperms and 708 species are pteridophytes (see Table 2.4, page 13). Due to the ancient origin of the Sichuanese flora, these include plenty of relict and endemic species. Among the angiosperms, 464 species are currently considered as verified endemics (Sichuan United University Department of Biology, 1994). Although the high mountain ridges of western Sichuan were glaciated during the Quaternary Period, this did not affect the survival of many ancient lineages, which could move southwards through the Hengduan Shan to radiate again after the glacial periods (Sichuan United University Department of Biology, 1994).

The other region partially included in the South-Central China hotspot of Mittermeier *et al.* (1999), Xizang (Tibet), is itself another rich area of plant biodiversity, harboring 9 600 species of vascular plants. More significantly, the biogeographical area to which it belongs, the Qinghai-Xizang Plateau, contains over 12 000 species of vascular plants in 1 500 genera. Of these, about 3 500 species and at least 29 genera are endemic to the Plateau, including ca. 100 pteridophytes (Miller, 2003; see Table 2.4). Species endemic to the Qinghai-Xizang Plateau include *Aconitum alpinonepalense*, *A. qinghaiense*, *Delphinium ceratophoroides* and *D. kingianum* (Ranunculaceae), *Clerodendrum tibetanum* (Verbenaceae), *Fritillaria delavayi* (Liliaceae; Figure 2.28), *Paeonia ludlowii* (Paeoniaceae; Figure 2.29), *Primula latisecta* and *P. tsongpenii* (Primulaceae) and *Silene davidii* (Caryophyllaceae).

Table 2.6 Chinese hotpots as defined on basis of endemism and number of plant species per unit area (from López-Pujol & Zhao, 2004).

Hotspot	Area (km^2)	Number of vascular plants (number endemic in parentheses)	Number of vascular plant species per 100 km^2	Total number of bryophyte species
Yunnan	394 000	16 600	4.2	1 500
Sichuan	488 000	9 254 (3 500)	1.9	data not available
Qinghai-Xizang Plateau	2 000 000	12 000 (3 500)	0.6	641
South-Central China	50 000	12 000 (3 500)	2.4	data not available
Northwestern Yunnan	70 000	7 000 (910)	10	data not available
Xishuangbanna	19 200	5 000	26	data not available
Three Gorges Area	53 200	2 800-6 400	5.4-12	data not available
Hainan	15 500	4 200 (630)	27	412
Taiwan	36 000		9.7-12	900-1 404
Hong Kong	1 100	2 145 (25)	195	360

Figure 2.28 *Fritillaria delavayi* (image courtesy CHEN You-Sheng, Chinese Virtual Herbarium).

Despite the geographic isolation and inaccessibility of some areas, increasing human activities, such as logging, livestock grazing and tourism, are substantial threats to the biodiversity of the Qinghai-Xizang Plateau. At present, about 40 species are considered to be seriously threatened, including *Cupressus gigantea*, *Dipentodon sinicus* (Dipentodontaceae), *Mandragora caulescens* (Solanaceae), *Picea smithiana*, *Pinus gerardiana* and *Sophora moorcroftiana* (Fabaceae).

The three additional country-level hotspots identified (López-Pujol & Zhao, 2004) are the Three Gorges area of the Chang Jiang valley, Hong Kong, and Taiwan. Although it is not considered to be a global hotspot, the valley of the Chang Jiang is a highly biodiverse region. The local flora is severely threatened due to rapid economic development of the area, a situation worsened by the Three Gorges Dam (Xie, 2003). Two of the three primary distribution centers for endemic plant groups in China—southwestern Sichuan and northern Yunnan, and eastern Sichuan and western Hubei (the other is southern Yunnan and southwestern Guangxi)—are located in the Chang Jiang valley, and 127 of the 388 species firstly listed in the *China Plant Red Data Book* can be found here (Fu, 1992; Xie, 2003). Eight taxa endemic to the valley are in a very critical situation with only one population: *Abies fanjingshanensis*, *Cystoathyrium chinense*, *Gleditsia japonica* var. *velutina* (Fabaceae), *Picea asperata* var. *aurantiaca*, *Sibbaldia omeiensis* (Rosaceae), *Tangtsinia nanchuanica* (Orchidaceae), *Ulmus gaussenii* and *Yulania zenii* (Magnoliaceae). One of the most interesting areas in the Chang Jiang valley is the "*Metasequoia* area," a region of about 800 km^2 at the junction of Hubei, Sichuan and Hunan where there are still natural populations of *Metasequoia glyptostroboides* (Hu, 1980; Figure 2.30). In this very small, biogeographically-interesting area, at least 550 species of vascular plants occur, many of them considered living fossils and most of them belonging to monotypic or oligotypic genera: in addition to *Metasequoia*, examples include *Cunninghamia lanceolata* (Taxodiaceae), *Eucommia ulmoides* (Eucommiaceae), *Keteleeria davidiana*, *Pseudolarix amabilis*, *Taiwania cryptomerioides*, *Tapiscia sinensis* (Tapisciaceae) and *Tetracentron sinense*.

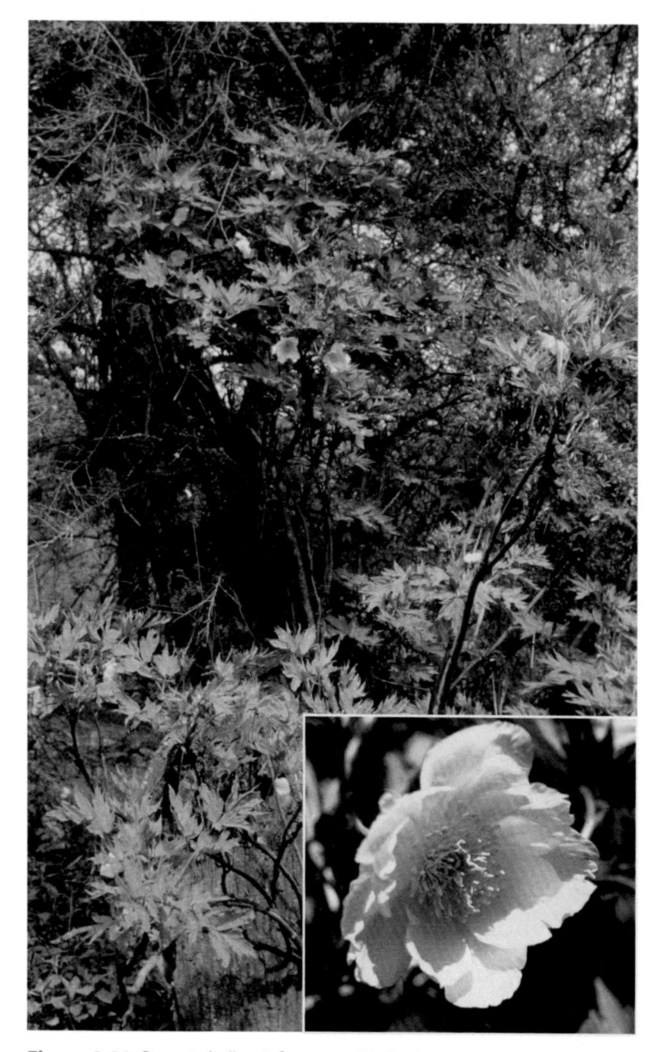

Figure 2.29 *Paeonia ludlowii*, flower and habit (image courtesy HONG De-Yuan, Institute of Botany, Chinese Academy of Sciences).

Hong Kong has a particularly rich flora given its small size. The 2 145 native vascular plant species recorded belong to 973 genera and 224 families. The most species-rich families here are Asteraceae, Cyperaceae, Euphorbiaceae, Fabaceae, Lauraceae, Moraceae, Myrsinaceae, Poaceae, Orchidaceae, Rubiaceae and Theaceae. There are several tree genera with more than ten species: *Camellia*, *Castanopsis*, *Cyclobalanopsis* (Fagaceae), *Ficus*, *Ilex* (Aquifoliaceae), *Lithocarpus* (Fagaceae), *Machilus* (Lauraceae) and *Symplocos* (Symplocaceae). As a part of the flora of Guangdong, Hong Kong's flora is closely related to that of Hainan and Indochina (Xing *et al.*, 1999). There are a number of endemic flowering plant species here including three orchids, *Bulbophyllum bicolor*, *Cheirostylis monteiroi* and *Dendrobium loddigesii*. Other endemic flowering plants include *Croton hancei* (Euphorbiaceae; a Critically Endangered montane shrub restricted to Qingyi Dao), *Illicium angustisepalum* (Illiciaceae; occurring in evergreen broad-leaved forest on Tai Tung Shan and elsewhere), *Zingiber integrilabrum* (Zingiberaceae; Critically Endangered), *Asarum hongkongense* (Aristolochiaceae; occuring in moist shady areas of thickets at 500–800 m on Lantou Dao), and the recently discovered *Balanophora hongkongensis* (Balanophoraceae). In

Figure 2.30 *Metasequoia glyptostroboides* (image courtesy HUANG Hong-Wen, South China Botanical Garden).

addition the bryophytes *Macromitrium brevituberculatum* (Orthotrichaceae) and *Syrrhopodon hongkongensis* (Calymperaceae) are endemic to Hong Kong (Zhang & Corlett, 2003).

The original vegetation of Hong Kong has been extensively modified by centuries of human disturbance; the original climax vegetation is considered to be subtropical evergreen broad-leaved monsoon forest, which is today found only as remnants in steep ravines or close to traditional villages where they have been maintained for cultural reasons. The major types of vegetation are now woodland, shrubland and grassland, with present-day forests comprising secondary vegetation developed after the Second World War. Minor vegetation formations occur in specialised habitats related to freshwater and coastal environments.

The subtropical island of Taiwan has a rich flora with a high degree of endemism. Around 1 000 taxa, about one-quarter of the total flora, are considered to be endemic (Gu, 1998). This diversity is considered noteworthy on a global scale, almost meeting the threshold for inclusion in Conservation International's global hotspots (Mittermeier *et al.*, 2004). Taiwan has a mountainous terrain with rugged uplands covering about two-thirds of the island. A few remnant patches of tropical rainforest occur in the south, with genera such as *Ficus*, *Machilus* and *Cinnamomum* (Lauraceae) being well-represented. This type of rainforest probably once covered much of the island but now almost all land below 700 m and in the foothills is intensively cultivated (Yang & Pan, 1996). Other vegetation types of Taiwan include evergreen broad-leaved forest with Fagaceae and Lauraceae, mixed forest, coniferous forest and alpine grassland. Two plant genera are endemic to Taiwan: *Sinopanax* (Araliaceae) and *Kudoacanthus* (Acanthaceae). Endemic species or subspecific taxa include *Cycas taitungensis*, *Chamaecyparis formosensis* (Cupressaceae), *Amentotaxus formosana* (Taxaceae), *Calocedrus macrolepis* var. *formosana* (Cupressaceae), *Pseudotsuga sinensis* var. *wilsoniana*, *Fagus hayatae* (Fagaceae), *Keteleeria davidiana* var. *formosana*, *Parakmeria kachirachirai*, *Lilium formosanum* (Liliaceae), and *Koelreuteria elegans* subsp. *formosana* (Sapindaceae).

Following on from the Centres of Plant Diversity and global biodiversity hotspots initiatives, the site-based approach to documentation for plant conservation has more recently moved to a finer scale with the development of the Important Plant Areas (IPA) approach. This is based to a large extent on the Important Bird Area model developed by Birdlife International (Oldfield *et al.*, 1998). The IPA selection criteria, which have global application, are threefold and a site qualifies as an IPA if it fulfils one or more of these criteria: it must either hold significant populations of species of global or regional concern; have an exceptionally rich flora in a regional context in relation to its biogeographic zone; or be an outstanding example of a habitat type of global or regional importance. The IPA criteria are being applied with a certain amount of flexibility and adapted to suit national and local priorities. In the Himalaya Shan, for example, IPAs have been described for medicinal plants (Hamilton & Radford, 2007).

In China, six large areas have tentatively been selected as IPAs for medicinal plants, based on their diverse medicinal floras, importance as sources of supply for Chinese traditional medicine and Tibetan medicine, and compatibility with the critical regions for biodiversity conservation identified by the Ministry of Environmental Protection of the People's Republic of China (1998). These initial six areas are Northwestern Yunnan (the southern Hengduan Shan), Western Sichuan (Ming Shan and the northern Hengduan Shan), Southwestern Sichuan (the southeastern Henduan Shan), Southern Gansu, Western Qinghai and Northern Xizang, and Southern Xizang. The Northwestern Yunnan IPA was then used as a case study to identify much smaller sites as a basis for practical conservation interventions (Hamilton & Radford, 2007). The criteria utilized were effectively factors likely to guarantee successful conservation outcomes based, for example, on local knowledge, needs and willingness to undertake externally-stimulated conservation initiatives. Three small sites were identified, all in Ludian, Yulong, Lijiang. Ludian has a rich tradition of medicinal plant use, with 363 medicinal plant species recorded, several of which are locally threatened by over-harvesting, and around 60 of which are in cultivation for medicinal use.

The 41 Centres of Plant Diversity probably remain the most appropriate listing of globally important plant areas in China but, in many cases, very little information has been provided on these and the boundaries of the sites not clearly delimited. They in any case include all 14 sites listed in China's Strategy for Plant Conservation (China's Strategy Plant Conservation Editorial Committee, 2008) as key areas of terrestrial ecosystem biodiversity with global significance for protection,as shown in Table 2.7. Clearly, with China's rich diversity of plants and rapidly increasing threats, identification and protection of IPAs in line with Target 5 of the Global Strategy for Plant Conservation (CBD Secretariat, 2003) is an urgent imperative.

Table 2.7 Key areas of biodiversity of importance for plant conservation (China's Strategy Plant Conservation Editorial Committee, 2008).

	Key biodiversity zone	Area (ha)	Number of nature reserves	Area of nature reserves (ha)	Percentage of area conserved (%)
1	Southern ridge of Hengduan Shan	25 474 000	47	6 193 042	24
2	Min Shan - northern ridge of Hengduan Shan	11 108 000	29	2 995 768	27
3	Junction area of Xinjiang, Qinghai and Xizang	100 925 000	9	59 506 478	59
4	Xishuangbanna (southern Yunnan)	1 675 000	3	317 539	19
5	Mountainous area at the Hunan, Guizhou, Sichuan and Hubei borders	10 204 000	61	1 134 048	11
6	Mountainous areas of central and southern Hainan	1 430 000	21	71 730	5
7	Limestone area of southwestern Guangxi	3 777 000	22	534 448	14
8	Mountainous area at the Zhejiang, Fujian and Jiangxi borders	4 912 000	18	165 467	3
9	Qinling Shan	4 801 000	19	366 560	8
10	Ili-western ridge of Tian Shan	12 361 000	11	1 423 054	12
11	Changbai Shan	14 299 000	42	1 065 471	7
12	Huitu coastal wetland (Liao Jiang estuary, Huang He delta, Yancheng, and the eastern beach of Chongmin, Shanghai)	12 333 000	56	2 624 805	21
13	Songneng-Sanjiang Plain in northeastern China	22 994 000	71	2 310 161	10
14	Lake area of the downstream Chang Jiang	4 141 000	9	85 360	2
	Totals	**230 434 000**	**418**	**79 793 930**	**34**

2.3.3 Endemism

Endemism has been emphasized in discussions of biodiversity in order to assign political responsibility for species in a particular geographical unit, for example at country or state level. China, with its huge land area and diverse of landscapes, ecological conditions and flora, has a predictably significant degree of plant endemism. Table 2.8 shows the degree of endemism for different plant groups in China, and shows that gymnosperms have a particularly high degree of endemism at generic and species level.

Table 2.8 Endemicity in Chinese Plants (Fu, 1992; Li, 2003; Flora of China, 2004).

Taxon	Number of endemic species/ total species	Percent endemic (%)	Number of endemic genera/ total genera	Percent endemic (%)
Bryophytes	225/2 200	10	13/495	2.6
Pteridophytes	-	data not available	6/224	2.6
Gymnosperms	79/232	34	5/35	14
Angiosperms	>9 000/30 000	>30	246/3 128	7.8
Total	**10 000/33 000**	**30**	**270/3 882**	**6.9**

Endemic gymnosperms include *Cathaya argyrophylla*, *Ginkgo biloba* and *Pseudolarix kaempferi*.

Two families, the gymnosperm family Ginkgoaceae and angiosperm family Eucommiaceae are relict and monotypic families endemic to China. Near-endemic monotypic families include Rhoipteleaceae (*Rhoiptelea chiliantha* occurring in Guangxi, Guizhou, Yunnan and northern Vietnam) and Cercidiphyllaceae (*Cercidiphyllum japonicum* occurring in China and Japan). Tetracentraceae has a single species, *Tetracentron sinense,* that occurs in China, Bhutan, India, Myanmar, Nepal and Vietnam, while *Trochodendron aralioides,* the only species in Trochodendraceae, occurs in mainland China, Taiwan, Japan and South Korea. Circaeasteraceae has a single species, *Circaeaster agrestis,* found in China and also in Bhutan, northeastern India, Nepal and Sikkim.

At the generic level, over 250 genera are endemic to mainland China. Table 2.9 lists the Chinese endemic, monotypic or oligotypic genera. At the species level, estimates of the number of higher plants endemic to China varies considerably, from 10 000 to 18 000 depending on the approach used (Fu, 1992; Groombridge, 1994;

Ministry of Environmental Protection of the People's Republic of China, 1998; Pitman & Jørgensen, 2002).

Table 2.9 Monotypic and oligotypic endemic plant genera of China.

Family	Genus	Number of species
Ginkgoaceae	Gingko	1
Pinaceae	Cathaya	1
Pinaceae	Pseudolarix	1
Taxaceae	Pseudotaxus	1
Taxodiaceae	Glyptostrobus	1
Taxodiaceae	Metasequoia	1
Apiaceae	Chuanminshen	1
Asclepiadaceae	Sichuania	1
Asteraceae	Ajaniopsis	1
Campanulaceae	Echinocodon	1
Labiatae	Ombrocharis	1
Poaceae	Ferrocalamus	1
Ranunculaceae	Anemoclema	1
Ranunculaceae	Kingdonia	1
Sapindaceae	Delavaya	1
Tiliaceae	Diplodiscus	1
Ulmaceae	Pteroceltis	1
Zygophyllaceae	Tetraena	1
Asteraceae	Sinoleontopodium	2
Lardizabalaceae	Sinofranchetia	2
Ranunculaceae	Urophysa	2
Sterculiaceae	Craigia	2
Verbenaceae	Schnabelia	2
Hamamelidaceae	Semiliquidambar	3

In addition, there are many genera in which most species occur in China. Notable examples include *Rhododendron* (with 650 out of about 1 000 species), *Saussurea* (320 of 400 species), *Pedicularis* (Scrophulariaceae; 352 of 600), *Camellia* (98 of 119), *Primula* (300 of 500), *Gentiana* (Gentianaceae; 248 of 360), *Aconitum* (211 of 400), *Ligularia* (Asteraceae; 100 of 150), *Corydalis* (300 of 440), *Euonymus* (Celastraceae; 125 of 175) and *Acer* (150 of 200; Flora of China, 2004).

2.3.4 Rare and threatened plants

Assessment of the conservation status of plants at a global level began in the early 1970s. The *Red Data Book* concept was developed by Sir Peter Scott in 1963 during his time as the first Chairman of the IUCN Species Survival Commission (SSC). The first *IUCN Red Data* volumes were published in 1966, for mammals and birds. *Volume Five* of the *Red Data Book* on flowering plants, compiled by Robert Melville of the Royal Botanic Gardens, Kew, was published in 1970. *Volume Five* was the first to acknowledge that, for some groups of species, notably plants and invertebrates, for which information was relatively sparse and in which many species were under threat, a full treatment similar to that achieved for mammals and birds would not be possible (Scott *et al.*, 1987). Building on Melville's work, the IUCN established a Threatened Plants Committee (TPC) of the SSC, with a secretariat based at the Royal Botanic Gardens, Kew. Compiled with assistance from the TPC network of botanists, the first *IUCN Plant Red Data Book* was published (Luces & Synge, 1978). This comprised "Red Data sheets" for 250 selected threatened plant species.

A first national list of Chinese threatened plants was compiled in 1982 under the auspices of the National Environmental Protection Agency and the Botanical Institute of Academia Sinica. It included 354 vascular plant species, each given a numerical conservation rating of one to three (Fu, 1992). In 1987, a revised list of rare and endangered plants protected in China was released by the National Environment Protection Board, again with conservation ratings graded as one to three. Subsequently, the *China Red Data Book* was published (Fu, 1992). This followed the format of the *IUCN Plant Red Data Book,* with data sheets for individual species. It also followed the IUCN categories of "Endangered," "Vulnerable" and "Rare," that were in use at the time. The *1997 IUCN Red List* of plants (Walter & Gillett, 1998) included 312 threatened Chinese plant species, 1% of the national flora. The main source of information for this was the National Environment Protection Board publication of 1987 (Fu, 1992).

In 1994, the IUCN adopted a new *Red List* system with more objective categories of threat status, and criteria for applying them. The 1994 version of the *IUCN Red List Categories and Criteria* were subsequently revised and the *IUCN Red List Categories and Criteria Version 3.1* was published in 2001 (IUCN, 2001). The categories for threatened species now stand at "Critically Endangered," "Endangered" and "Vulnerable." Species that do not meet the thresholds for inclusion in the threat categories are considered to be "Near Threatened," of "Least Concern," "Data Deficient" or "Not Evaluated." The categories "Extinct" and "Extinct in the Wild" complete the *IUCN Red List* system.

As yet, take-up of the post-1994 version of the *IUCN Red List Categories and Criteria* for plants has been relatively patchy and slow on a global scale. The *World List of Threatened Trees* (Oldfield *et al.*, 1998) recorded threatened Chinese woody plants using the *IUCN Red List Categories and Criteria version 2.4.* The current *IUCN Red List* (IUCN, 2011) includes 272 Chinese threatened plant species. There is also one species (*Adiantum lianxianense*; Pteridaceae) classed as extinct. Various processes are currently underway to compile information on the conservation status of Chinese plants using the *IUCN Red List Categories and Criteria Version 3.1.* The IUCN/SSC China Plant Specialist Group has particular responsibility for implementing this process. Other IUCN/SSC Specialist Groups play a role in current "Red Listing" efforts for Chinese plants. The Global Trees Specialist Group has, for example, published global red lists of maples (Gibbs & Chen, 2009), Magnoliaceae (Cicuzza *et al.*, 2007) and oaks (Oldfield & Eastwood, 2007), all groups which have centers of diversity in southern China.

Magnoliaceae provide a case study here: southern China is the world centre of diversity and distribution of the family, with over 40% of species. A significant number are considered globally threatened due to habitat loss and in some cases over-exploitation. In a global evaluation, Cicuzza *et al.* (2007), record 42 Chinese *Magnolia* species as Threatened, Near Threatened or Data Deficient. Of these, 32 are endemic to China. However, the taxonomy and nomenclature used by Cicuzza and colleagues followed that of Figlar & Nooteboom (2004), who consider that the family Magnoliaceae consists of one genus, *Magnolia,* with three subgenera. The treatment followed in the *Flora of China* (Xia *et al.*, 2008) considers Magnoliaceae to comprise 12 genera. There is nevertheless general agreement on delimitation at species level. The Red Listing process for Magnoliaceae involved mapping the species and overlaying forest cover data to show the extent of their potential habitat, adding information on the rate of forest loss, and discussions at an expert workshop. Table 2.10 lists the ten Critically Endangered Chinese Magnoliaceae, all of which are endemic (see also Figure 2.31). Global Red Listing has helped to raise awareness of the conservation needs of Chinese Magnoliaceae, and has boosted international support for measures to survey populations of the threatened species in the wild and to undertake *ex situ* and *in situ* conservation measures.

Table 2.10 Critically Endangered *Magnolia* species of China (Cicuzza *et al.*, 2007).

Species	Distribution	Conservation issues
Michelia coriacea	Southeastern Yunnan	Occurs as scattered individuals in evergreen woods on limestone mountain slopes at 1 200-1 450 m. Field surveys in December 2005 estimated that there are about 300-500 individuals remaining, mainly outside nature reserves. Most of the individuals are sprouted multi-trunks from the base of chopped trees. In Malipo, local people protect large trees as symbols of good luck.
Manglietia grandis	Guangxi, Yunnan	Occurs in forested valleys on limestone mountains between 800 and 1 500 m; this habitat has been extensively cleared and degraded with large areas surrounding nature reserves in Guangxi having been planted with economically-valuable trees.
Manglietia ventii	Yunnan	Distributed in forest fragments and along rivers at 300-1 200 m.
Parakmeria kachirachirai	Taiwan	Fragmented populations in lowland broad-leaved forest. Kenting National Park covers part of the range.
Michelia lacei	Yunnan	This species is known from fewer than five locations with a estimated 50-60 individuals.
Manglietia dandyi	Guangxi, Yunnan	Small populations restricted to broad-leaved evergreen forest between altitudes of 450 and 1 500 m. The forests are unprotected and heavily exploited for firewood and timber. This species provides a construction timber particularly preferred by local inhabitants.
Parakmeria omeiensis	Guizhou, Sichuan	Confined to three locations with fewer than 100 male individuals, found in an area of broad-leaved evergreen forest between 1 000 and 1 200 m. Logging appears to be continuing in the area and no special protection is yet in place for the species.
Manglietia ovoidea	Yunnan	Distributed in evergreen broad-leaved forests at altitudes of 1 700–2 000 m in southeastern Yunnan. It is thought that there are fewer than 50 individuals in four sub-populations.
Lirianthe fistulosa	Yunnan	Major threat is reduction of habitat for banana plantations.
Pachylarnax sinica	Yunnan	Occurs on forested slopes in southeastern Yunnan. Habitat destruction has been the major threat. Conservation action underway involving local nurseries.

China is also an important center of diversity for oaks, with over 100 taxa. This includes species of *Cyclobalanopsis,* which is treated as a distinct genus from *Quercus* in the *Flora of China* (Huang *et al.*, 1999). Oaks are important components of the broad-leaved evergreen forest of the country. Fifteen species of Chinese oaks are recorded by Oldfield & Eastwood (2007) as threatened, with a further 11 species as Data Deficient.

Southwestern China is the global center of diversity for maples (*Acer*). Out of total of 191 taxa evaluated in compiling the *Red List of Maples,* 115 are Chinese (Gibbs & Chen, 2009). Forty of the Chinese *Acer* species are recorded as threatened and 17 as Data Deficient. In addition, the closely related *Dipteronia dyeriana* is considered to be threatened (Gibbs & Chen, 2009). The methodology used in Red Listing the maples followed that for *Magnolia*: a mixture of literature survey, mapping of both species and potential habitat distributions, and expert opinion in applying the *IUCN Red List Categories and Criteria Version 3.1.*

The IUCN/SSC Conifer Specialist Group has compiled Red List assessments for all conifer taxa in mainland China and Taiwan. Very rare conifers included on the *IUCN Red List* include *Abies beshanzuensis,* with only five living plants known in the wild on Baishanzu (Zhejiang), and *Thuja sutchuenensis* (Cupressaceae; Figure 2.32), endemic to a very inaccessible locality in Chengkou, Chongqing, eastern Sichuan, where it is Critically Endangered. This species was first described in 1899 from specimens collected by the French botanist Paul Guillaume Farges in 1892, but was not found again for almost a century, and was listed as Extinct in the Wild, presumably due to over-harvesting for its valuable, scented timber. A small number of specimens were rediscovered in 1999 by a regional botanical team, growing on steep ridges close to where it was first found. This habitat has now been designated a Special Protection Area in order to protect the species (Xiang *et al.*, 2003). The cycads of China have also been assessed using the *IUCN Red List Categories and Criteria Version 3.1.*

Figure 2.31 Examples of endangered Magnoliaceae species in China. A, *Manglietia aromatica*. B, *M. dandyi*. C, *Lirianthe odoratissima*. D, *L. fistulosa*. E, *M. grandis*. (images A, B, C and E courtesy SUN Wei-Bang; image D courtesy YANG Ke-Ming, South China Botanical Garden).

Figure 2.32 Branch of *Thuja sutchuenensis* (image courtesy YANG Quan, PlantPhoto).

In addition to the global and national Red Lists, China has a number of regional Red Lists. The *Red Data Book of Taiwan*, for example, lists 502 species, of which three are already Extinct, 14 Endangered, 62 Vulnerable, and 423 Rare (Taiwan Endemic Species Research Institute, 2004). During the past twenty years, 11 species have been added to the Taiwan Red List. In Hong Kong, about 360 species of vascular plants are considered by the local government to be threatened (Agriculture Fisheries and Conservation Department, 2004).

Agreement and coordination of IUCN Red Listing is a complex process. The IUCN system allows for revision of conservation status as new information becomes available, and specifies that a review should be undertaken every decade. A major impediment for plants may have been the amount of supporting information required to support formal inclusion of species in the *Red Lists*. Whereas further field data is required for many Chinese species assumed to be Rare or Threatened, rules of thumb and proxy data may now be used to assist in Red Listing assessments (Lusty *et al.*, 2007). Herbarium data may also be usefully employed to support the Red Listing process (Brummitt *et al.*, 2008). Recent estimates of the true number of Rare and Endangered plant species in China are ca. 4 000–5 000 species (Fu, 1992; Xu, 1997; Gu, 1998; Wang & Xie, 2004; China's Strategy Plant Conservation Editorial Committee, 2008). Further assessment, as far as possible based on field surveys and taking into account other approaches, is an urgent priority.

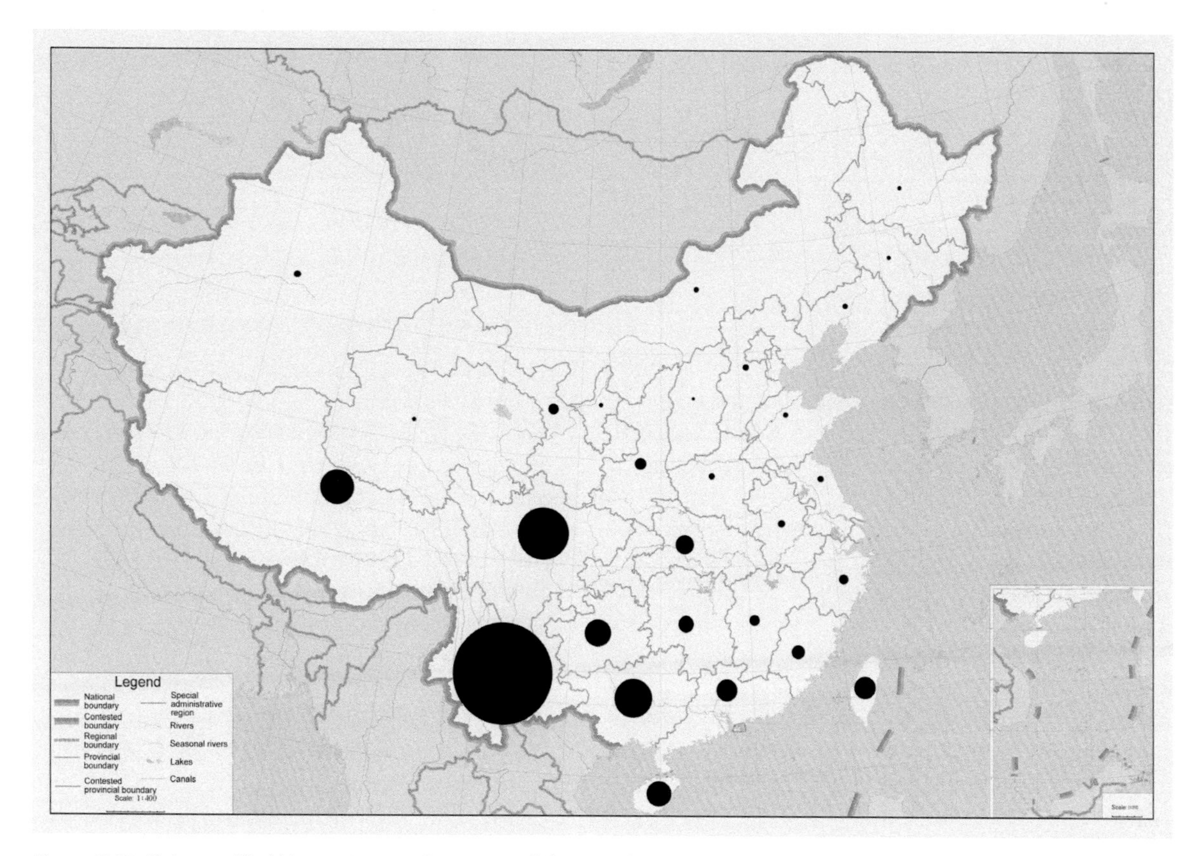

Figure 2.33 Richness of Red List species in each province of China. Circle size represents richness, ranging from 62 to 2 295 species.

Currently, ca. 4 400 plant species are included in the *China Species Red List* (Wang & Xie, 2004). These species are widely distributed across 166 families, but 2 476 (56%) of them belong to only ten families. The Orchidaceae alone includes more than 1 200 Red Listed species, and Ericaceae ca. 500. On average, each province of China has ca. 400 Red Listed species, but the vast majority are restricted to southwestern China (Figure 2.33). Yunnan alone comprises over half (2 295), and Sichuan has 1 077. Six provinces each have 500–1 000 Red Listed species: Guangdong, Guangxi, Guizhou, Hainan, Taiwan and Xixang, while another eight each have 200–500: Anhui, Gansu, Fujian, Hubei, Hunan, Jiangxi, Shaanxi and Zhejiang. Four provinces (Hebei, Heilongjiang, Ningxia and Shangdong) have fewer than 100 Red Listed species, with Ningxia having the fewest (62) of all China's provinces.

References

Agriculture Fisheries and Conservation Department (2004). Agriculture, Fisheries and Conservation Department of Hong Kong SAR. <http://www.afcd.gov.hk/index_e.htm>.

APGIII (2009). An update of the Angiosperm Phylogeny Group classification for the orders and families of flowering plants: APG III. *Botanical Journal of the Linnean Society* 161: 105-121.

Briggs, R. W. (1993). *"Chinese" Wilson*. London, UK: HMSO.

Brummitt, N., Bachman, S. P. & Moat, J. (2008). Applications of the IUCN Red List: towards a global barometer for plant diversity. *Endangered Species Research* 6: 127-135.

Carpenter, C. (2001). *Hainan Island monsoon rain forests (IM0169)*. WWF.

CBD Secretariat (2003). *Global Strategy for Plant Conservation*. Montreal, Canada: CBD Secretariat.

Chapman, G. P. & Wang, Y. Z. (2002). *The Plant Life of China: Diversity and Distribution*. Berlin and Heidelberg, Germany: Springer-Verlag.

China's Strategy Plant Conservation Editorial Committee (2008). *China's Strategy for Plant Conservation*. Guangzhou, China: Guangdong Press Group.

Cicuzza, D., Newton, A. & Oldfield, S. (2007). *The Red List of Magnoliaceae*. Cambridge, UK: Fauna & Flora International.

Conservation International (2004). Mountains of Southwest China. Biodiversity Hotspots. <http://www.biodiversityhotspots.org/xp/Hotspots/china/?showpage=Biodiversity>.

Conservation International (2009). Biodiversity hotspots. <http://www.biodiversityhotspots.org/xp/hotspots/Pages/default.aspx>.

Davis, S. D., Heywood, V. H. & Hamilton, A. C. (1995). *Centres of Plant Diversity. A Guide and Strategy for their Conservation. Volume 2. Asia, Australasia and the Pacific*. Cambridge, UK: WWF and IUCN.

Donoghue, M. J., Boufford, D. E., Tan, B. C. & Pfister, D. H. (1997). Plant and Fungal Diversity of Western Sichuan and Eastern Xizang, China. Biodiversity of the Hengduan Mountains Region, China. <http://hengduan.huh.harvard.edu/fieldnotes/project_description>.

Editorial Committee of Flora Reipublicae Popularis Sinicae (1959-2004). *Flora Reipublicae Popularis Sinicae*. Beijing, China: Sciences Press.

eFloras (2008). eFloras.

Figlar, R. B. & Nooteboom, H. P. (2004). Notes on Magnoliaceae IV. *Blumea* 49: 87-100.

Flora of China (2004). Flora of China Website. <http://flora.huh.harvard.edu/china/mss/treatments.htm>.

Fu, L. K. (ed.) (1992). *China Plant Red Data Book. Rare and Endangered Plants 1*. Beijing, China and New York: Science Press.

Gibbs, D. & Chen, Y. (2009). *The Red List of Maples*. Richmond, UK: BGCI.

Groombridge, B. (ed.) (1994). *Biodiversity Data Sourcebook*. Cambridge, UK: World Conservation Press.

Gu, J. (1998). Conservation of plant diversity in China: achievements, prospects and concerns. *Biological Conservation* 85: 321-327.

Hamilton, A. C. & Radford, E. A. (2007). *Identification and Conservation of Important Plant Areas for Medicinal Plants in the Himalaya*. Salisbury, UK and Kathmandu, Nepal: Plantlife International and Ethnobotanical Society of Nepal.

Harlan, J. R. (1971). Agricultural origins: centers and non-centers. *Science* 174: 468-474.

Hawkins, B., Sharrock, S. & Havens, K. (2008). *Plants and Climate Change: Which Future?* Richmond, UK: BGCI.

Hu, S.Y. (1980). *An Enumeration of Chinese Materia Medica*. Hong Kong: The Chinese University Press.

Huang, C.J., Zhang, Y.T. & Bartholomew, B. (1999). Fagaceae. In: Wu, Z. Y., Raven, P. H. & Hong, D.Y. (eds.) *Flora of China*. pp. 314-400. Beijing, China and St. Louis, MO: Science Press and Missouri Botanical Garden Press.

IUCN (2001). *Red List Categories and Criteria. Version 3.1*. Gland, Switzerland: IUCN Species Survival Commission.

IUCN (2011). IUCN Red List of Threatened Species. Version 2011.2. <http://www.iucnredlist.org/>.

Kelley, S. (2001). Plant hunting on the rooftop of the world. *Arnoldia* 61: 2-13.

Legendre, P. & Legendre, L. (1998). *Numerical Ecology*. Amsterdam, The Netherlands: Elsevier.

Li, H., He, D.M., Bartholomew, B. & Long, C. L. (1999). Re-examination of the biological effect of plate movement: impact of Shan-Malay Plate displacement (the movement of Burma-Malaya Geoblock) on the biota of Gaoligong Mountains. *Acta Botanica Yunnanica* 21: 407-425.

Li, M. J. (1998). *Wild Fruits in China*. Beijing, China: China

Agricultural Press.
Li, Z.Y. (2003). Advances in plant resource conservation in China: sustainable development of plant resources in China. *Acta Botanica Sinica* 45: 124-130.
Liu, J. G., Ouyang, Z. Y., Pimm, S. L., Raven, P. H., Wang, X. K., Miao, H. & Han, N. Y. (2003). Protecting China's biodiversity. *Science* 300: 1240-1241.
López-Pujol, J. & Zhao, A. M. (2004). China: a rich flora needed of urgent attention. *Orsis* 19: 49-89.
Lu, S. G. (2004). Pteridophyte flora of China. In: Wu, C. Y. & Chen, S. C. (eds.) *Flora Reipublicae Popularis Sinicae*. pp. 78-92. Beijing, China: Science Press.
Luces, G. & Synge, H. S. (1978). *The IUCN Plant Red Data Book*. Morges, Switzerland: IUCN.
Lusty, C., Amaral, W. A. N., Hawthorne, W. D., Hong, L. T. & Oldfield, S. (2007). Applying the IUCN Red List Categories in a forest setting. *In:* Chua, L. S. L., Kirton, L. G. & Saw, L. G. (eds.) *Status of Biological Diversity in Malaysia and Threat Assessment of Plant Species in Malaysia, Seminar and Workshop*. Kepong, Malaysia: Forest Research Institute, Malaysia.
Mabberley, D. J. (2008). *Mabberley's Plant Book. A Portable Dictionary of Plants, their Classification and Uses*. Cambridge, UK: Cambridge University Press.
Miller, D. (2003). *Tibet Environmental Analysis. Background Paper in Preparation for USAID's Program*. USAID Bureau for Asia and Near East.
Ministry of Environmental Protection of the People's Republic of China (1998). *Country Status Report of China's Biodiversity*. Beijing, China.
Mittermeier, R. A., Robles Gil, P. & Goettsch Mittermeier, C. (eds.) (1997). *Megadiversity. Earth's Biologically Wealthiest Nations*. Mexico City, Mexico: CEMEX/Agrupaciaon Sierra Madre.
Mittermeier, R. A., Myers, N., Robles Gil, P. & Goettsch Mittermeier, C. (1999). *Hotspots. Earth's Biologically Richest and Most Endangered Terrestrial Ecoregions*. Mexico City, Mexico: CEMEX/Agrupación Sierra Madre.
Mittermeier, R. A., Robles Gil, P., Hoffmann, M., Pilgrim, J., Brooks, T., Goettsch Mittermeier, C., Lamoreux, J. F. & da Fonseca, G. A. B. (2004). *Hotspots Revisited*. Mexico City, Mexico: CEMEX/ Agrupaciaon Sierra Madre.
Myers, N. (1988). Threatened biotas: "hot spots" in tropical forests. *The Environmentalist* 8(3): 187-208.
Needham, J. (1956). *Science and Civilization in China*. Cambridge, UK: Cambridge University Press.
Oldfield, S., Lusty, C. & MacKinven, A. (1998). *The World List of Threatened Trees*. Cambridge: World Conservation Press.
Oldfield, S. & Eastwood, A. (2007). The Red List of Oaks. In: UK: Fauna & Flora International.
Pitman, N. C. A. & Jørgensen, P. M. (2002). Estimating the size of the world's threatened flora. *Science* 298: 989.
Press, F. & Siever, R. (1986). *Earth*. New York: W.H. Freeman.
Qian, H. (2002). A comparison of the taxonomic richness of temperate plants in East Asia and North America. *American Journal of Botany* 89: 1818-1825.
Schirokauer, C. & Brown, M. (2005). *A Brief History of Chinese Civilization*. New York: Houghton Mifflin Press.
Scott, J. M., Csuti, B., Jacobi, J. D. & Estes, J. E. (1987). Species richness-a geographic approach to protecting future biological diversity. *BioScience* 37: 782-788.
Sichuan United University Department of Biology (1994). Plant resources of Sichuan. <http://www.blasum.net/holger/wri/biol/sichuanp.html>.
Stevens, P. F. (2001 onwards). Angiosperm Phylogeny Website. <http://www.mobot.org/MOBOT/research/APweb/>.
Sun, Z. H., Peng, S. J. & Ou, X. K. (2007). Rapid assessment and explanation of tree species abundance along the elevation gradient in Gaoligong Mountains, Yunnan, China. *Chinese Science Bulletin* 52: 225-231.
Taiwan Endemic Species Research Institute (2004). TESRI. <http://www.tesri.gov.tw/english/E_species.asp>.
Walter, K. S. & Gillett, H. J. (1998). *1997 IUCN Red List of Threatened Plants*. Gland: IUCN.
Wang, H. S. (2004). Gymnosperms flora of China. In: Wu, C. Y. & Chen, S. C. (eds.) *Flora Reipublicae Popularis Sinicae*. pp. 95-120. Beijing, China: Science Press.
Wang, S. & Xie, Y. (eds.) (2004). *China Species Red List 1*. Beijing, China: Higher Education Press.
Wu, C. Y. & Ding, T. Y. (1999). Seed Plants of China CD-ROM. *In:* Kunming, China: Yunnan Science & Technology Press.
Wu, Z. Y., Raven, P. H. & Hong, D. Y. (eds.) (1994-2011). *Flora of China*. Beijing, China: Science Press.
Wu, Z. Y. & Wu, S. G. (1996). A proposal for a new floristic kingdom (realm) - the E. Asiatic Kingdom, its delineation and characteristics. In: Zhang, A. L. & Wu, S. G. (eds.) *Floristic Characteristics and Diversity of East Asian Plants, Proceeding of the first international Symposium on Floristic Characteristics and Diversity of East Asian Plants*. pp. 3-42. Beijing, China: China Higher Education Press.
Wu, Z. Y. & Chen, S. C. (eds.) (2004). *Flora Reipublicae Popularis Sinicae*. Beijing, China: Science Press.
WWF (2009). List of Ecoregions. <http://www.panda.org/about_our_earth/ecoregions/ecoregion_list/>.
Xia, N. H., Liu, Y. H. & Nooteboom, H. P. (2008). Magnoliaceae. In: Wu, Z. Y., Raven, P. H. & Hong, D. Y. (eds.) *Flora of China*. pp. 48-90. Beijing, China and St. Louis, MO: Science Press and Missouri Botanical Garden Press.
Xiang, Q. P., Farjon, A., Li, Z. Y., Fu, L. K. & Liu, Z. Y. (2003). *Thuja sutchuenensis*. <http://www.iucnredlist.org/>.
Xie, Z. (2003). Characteristics and conservation priority of threatened plants in the Yangtze Valley. *Biodiversity and Conservation* 12: 65-72.
Xing, F. W., Corlett, R. T. & Chau, K. C. (1999). Study on the Flora of Hong Kong. *Journal of Tropical and Subtropical Botany* 7(4): 295-307.
Xu, Z. F. (1997). The status and strategy for ex situ conservation of plant diversity in Chinese botanic

gardens - discussion of principles and methodologies of ex situ conservation for plant diversity. In: Schei, P. J. & Wang, S. (eds.) *Conserving China's Biodiversity*. pp. 79-95. Beijing, China: China Environmental Science Press.

Yang, J. C. & Pan, F. J. (1996). The current status of native woody vegetation in Taiwan. *In:* Hunt, D. (ed.) *Temperate Trees Under Threat. Proceedings of an IDS Symposium on the Conservation Status of Temperate Trees*. University of Bonn: International Dendrology Society.

Yang, Y., Tan, K., Hao, J., Pei, S. & Yang, Y. (2004). Biodiversity and biodiversity conservation in Yunnan, China. *Biodiversity and Conservation* 13: 813-826.

Zhang, L. & Corlett, R. T. (2003). Phytogeography of Hong Kong bryophytes. *Journal of Biogeography* 30: 1329-1337.

Zhang, Y. B. & Ma, K. P. (2008). Geographic distribution patterns and status assessment of threatened plants in China. *Biodiversity and Conservation* 17: 1783-1798.

Zhu, H., Cao, M. & Hu, H. B. (2006). Geological history, flora, and vegetation of Xishuangbanna, southern Yunnan, China. *Biotropica* 38(3): 310-317.

Zhu, H. (2008). Advances in biogeography of the tropical rain forest in southern Yunnan, southwestern China. *Tropical Conservation Science* 1: 34-42.

Zhu, T. P. (2004). Plant resource in China. In: Wu, Z. Y. & Chen, X. Q. (eds.) *Flora Reipublicae Popularis Sinicae*. pp. 584-657. Beijing, China: Science Press.

Chapter 3 Physical Geography

ZHENG Du and DAI Er-Fu

3.1 Introduction

In this chapter we describe the major factors and processes shaping the topography and physical environment in which the plants of China exist. We also describe the country's resources of surface water and groundwater. We end the chapter by introducing a new method for the regional classification of China's land area, and discussing its implications.

3.2 Major factors affecting topography and landforms

3.2.1 Geographical location

China's territory extends in latitude from its northern boundary at the Heilong Jiang at 53°31' N, to its southernmost point at Zengmu Ansha, with a latitude of about 4° N. The western edge of China lies at the Pamir Shan, longitude 73°40' E, while the most easterly coast lies at 135°5' E. Politically, China is divided into 23 provinces, five autonomous regions and four municipalities directly under the control of the central government.

3.2.2 Land area

China covers, from the confluence of the Heilong Jiang and Wusuli Jiang in the east to the Pamir Plateau, a span of more than 5 200 km. From the Heilong Jiang north of Mohe, southward to the Nansha Qundao near the equator is more than 5 500 km. Thus, when the sun shines high in the south, it is still early morning at Pamir, and while Beijing Shi is still experiencing snow, Hainan's climate is spring-like; that of the Nansha Qundao is hot and humid throughout the year (Zhao, 1986).

The land area of China is ca. 9.6 million km^2, 6.4% of the world's total land area. Only Russia and Canada have a larger land area (17 million km^2 and 10 million km^2 respectively), but these two countries differ from China in being adjacent to the frigid Arctic environment. Only the USA has similar natural conditions in an approximately similar land area.

3.2.3 Marine area

The territorial seas and islands of China are extensive, covering an offshore area of ca. 4.7 million km^2 (The Physical Geography of China Committee, 1985). China's coastline lies on, from north to south, the Bohai Wan, the Huang Hai (Yellow Sea), the Dong Hai (East China Sea), the Nan Hai (South China Sea) and the Taiping Yang (Pacific Ocean), to the east of Taiwan (see Table 3.1).

Table 3.1 Basic information on the four seas of China (modified from The Physical Geography of China Committee, 1985: Table 4.1).

Name	Area (km^2)	Average depth (m)	Maximum depth (m)	Proportion of total sea area taken up by continental shelf (%)
Bohai Wan (Bohai Sea)	77 000	18	70	100
Huang Hai (Yellow Sea)	380 000	44	140	100
Dong Hai (East China Sea)	770 000	370	2 719	67
Nan Hai (South China Sea)	3 500 000	1 212	5 559	50

China's coastline is approximately 18 000 km long, making it one of the longest in the world. On 4 September 1958, the Chinese government issued a statement of territorial waters, designating that the width of China's territorial waters is 12 nautical miles, applicable to both the mainland and coastal islands, as well as Taiwan, Penghu Dao, Dongsha Dao, Xisha Qundao, Zhongsha Qundao, Nansha Qundao, and other islands separated by international waters from the Chinese mainland. China's offshore area contains 6 536 islands larger than 500 m^2—a total land area of nearly 72 800 km^2. Of these, 450 islands are inhabited. The area of coastal beaches is almost 20 800 km^2, and that of continental shelf covered by less than 200 m of water is ca. 1.3–1.5 million km^2.

3.2.4 Distribution of land and sea

China lies between the world's largest continent, Eurasia, and its largest ocean, the Pacific, and is bounded by the largest and highest plateau: the Qinghai-Xizang (Tibetan) Plateau. The uplift of this plateau created a monsoon climate system that has persisted since the Late

Pleistocene Epoch: a dry, cold, northwestern continental monsoon dominates during winter and a moist, warm, southeastern maritime monsoon prevails during summer. The alternation of monsoon systems is an important determinant of China's physical geographic conditions and their regional differences. Moisture in the atmosphere, mainly deriving from the warm and moist maritime monsoon, leads to a distribution of precipitation roughly proportional to distance from the sea (the greater the distance from the sea, the lower the precipitation and the drier the climate). From southeast to northwest, China can be divided into four moisture zones chiefly based on the aridity index adopted by the Physical Regionalization Working Committee of Chinese Academy of Sciences (1959), as described below.

The humid region (nearest to the sea).—This region has an aridity index (Physical Regionalization Working Committee of Chinese Academy of Sciences, 1959) below 1.0 (Huang, 1959), with greater precipitation than evaporation. Forest is the dominant natural vegetation, occupying 32% of the total land area (1% in the cold temperate zone, 3% in the temperate zone, 1% in the warm-temperate zone, 26% in the subtropical zone and 2% in the tropical zone).

The sub-humid region.—This region has an aridity index of 1.0--1.5, signifying a balance between precipitation and evaporation. The dominant natural vegetation is forest-steppe, occupying 15% of the total land area (4% in the cold-temperate zone, 7% in the warm-temperate zone and 3% on the southeastern part of the Qinghai-Xizang Plateau).

The semiarid region.—This region has an aridity index of 1.5–2.0, i.e. less precipitation than evaporation. Steppe is the dominant vegetation, occupying 22% of the total land area (6% in the temperate zone, 3% in the warm-temperate zone and 13% on the Qinghai-Xizang Plateau).

The arid region.—This region has an aridity index of greater than 2.0 and the dominant vegetation is desert-steppe (aridity index 2.0–4.0) or desert (aridity index greater than 4.0). This region occupies 31% of the total land area of China (13% in the temperate zone, 8% in the warm temperate zone occupies and 9% on the Qinghai-Xizang Plateau).

China bears several similarities to the USA in terms of the distribution of land and sea, and in humidity, although the monsoon system is less developed in the USA and the continental climate is not so extreme because there is no equivalent of the Qinghai-Xizang Plateau. Compared to the Indian subcontinent, although both China and India are monsoonal, China suffers from the strong northwestern continental monsoon, which is cold and arid, while an oceanic monsoon system prevails in India, due to the presence of the Qinghai-Xizang Plateau to the north and Indian Ocean to the south.

In addition to its location, another reason for China's monsoon system is that offshore winds and the Coriolis effect deflect warm currents away from the Chinese coast, increasing any heat differences between land and sea. Mainly for these reasons, a continental climate is therefore well developed in China, generally presenting as warmer summers and much colder winters compared to other coastal areas at similar latitudes. For instance, Huma in China has a similar latitude to London, UK (51° N–52° N), but quite different winter temperatures: the mean January temperature in Huma is as low as −28 °C while in London it is ca. 4 °C, similar to Shanghai, which is located at ca. 31° N (The Physical Geography of China Committee, 1985). As it is hardly influenced by westerly marine climates, and due to the fact that the amelioration of the Taiwan Warm Current (also known as the Kuroshio Current) is also limited by the Coriolis effect, this continental climate presents even in eastern coastal locations during winter. Nevertheless, although the coastal current and Taiwan Warm Current exert negligible influence on China's climate system, they do contribute to some extent to water vapor transport in the east of the country. In summer, the southeastern monsoon brings plenty of moisture; in winter, the returning air has a significant impact on Taipei Shi and southeastern coastal parts of the Chinese mainland.

3.2.5 Geological History

Based on the theory of plate tectonics, China is situated on both the Eurasian Plate, the southern extent of which is the Indo-Australian Plate, and on the Pacific and Philippine Plates in the east (Dewey, 1972). For this reason, the geological history of China is a complex and continuous process of interactive propulsion and collision between the Eurasian Plate and other adjacent plates. Moreover, it is also a process of interaction between the Chinese Platform and its surrounding folding zones. These tectonic interactions have varied in direction and intensity with time, and their impact can only be discerned indirectly, for instance by looking at the karst landscapes of southwestern China, related to limestone rocks that appeared in the Paleozoic Era.

More than 2 500 million years ago, especially during the Luliang Movement (1 900–1 700 million years ago) and Jinning Movement (1 000 million years ago), a Presinian Period Platform gradually formed, of which the eastern part comprised mainly platforms and the western part a geosyncline. By the end of the Sinian Periond (around 570 million years ago, towards the end of Precambrian time), the land area had expanded and sea area reduced; however, by the end of Paleozoic Era, this had been reversed. During the Mesozoic Era, dramatic changes to the Chinese mainland occurred, joining the whole country into a continuous and vast land area by the uplift of the ancient Kunlun Shan and Qin Ling at the beginning of this era. These events basically shaped the outline for the tectonic movements of the later Yanshanian Movement, which is deemed to be an important formative stage for China's mainland.

The tectonic influences resulting from the Yanshanian Movement vary across China. It was composed of three orogenic phases, of which the second phase, between the Late Jurassic and Early Cretaceous Periods, was the most active and had the greatest impact (Ren *et al.*, 1992). Although many previous tectonic movements had hardened the basement of the Paleozoic fold belts of western China, the Yanshanian Movement rejuvenated pre-existing structures or, for those involving the basement, bend them into large, broad folds. This phenomenon can be observed today in the Karakoram Shan and Hengduan Shan. The effects of the Yanshanian Movement are so notable that almost all tectonic frameworks, dominating the patterns of the major landforms of China, were established by it. Exceptions are the Himalaya Shan, Taiwan, whose low structures built during the Yanshanian Movement were later turned into high reliefs by the Himalayan Movement, and the Tarim Pendi, which was still under marine transgression. This was followed by a quiet period (Figure 3.1) in both landforms and climate until the beginning of the Paleocene Era and the Himalayan Movement, leaving a red weathering crust across the country.

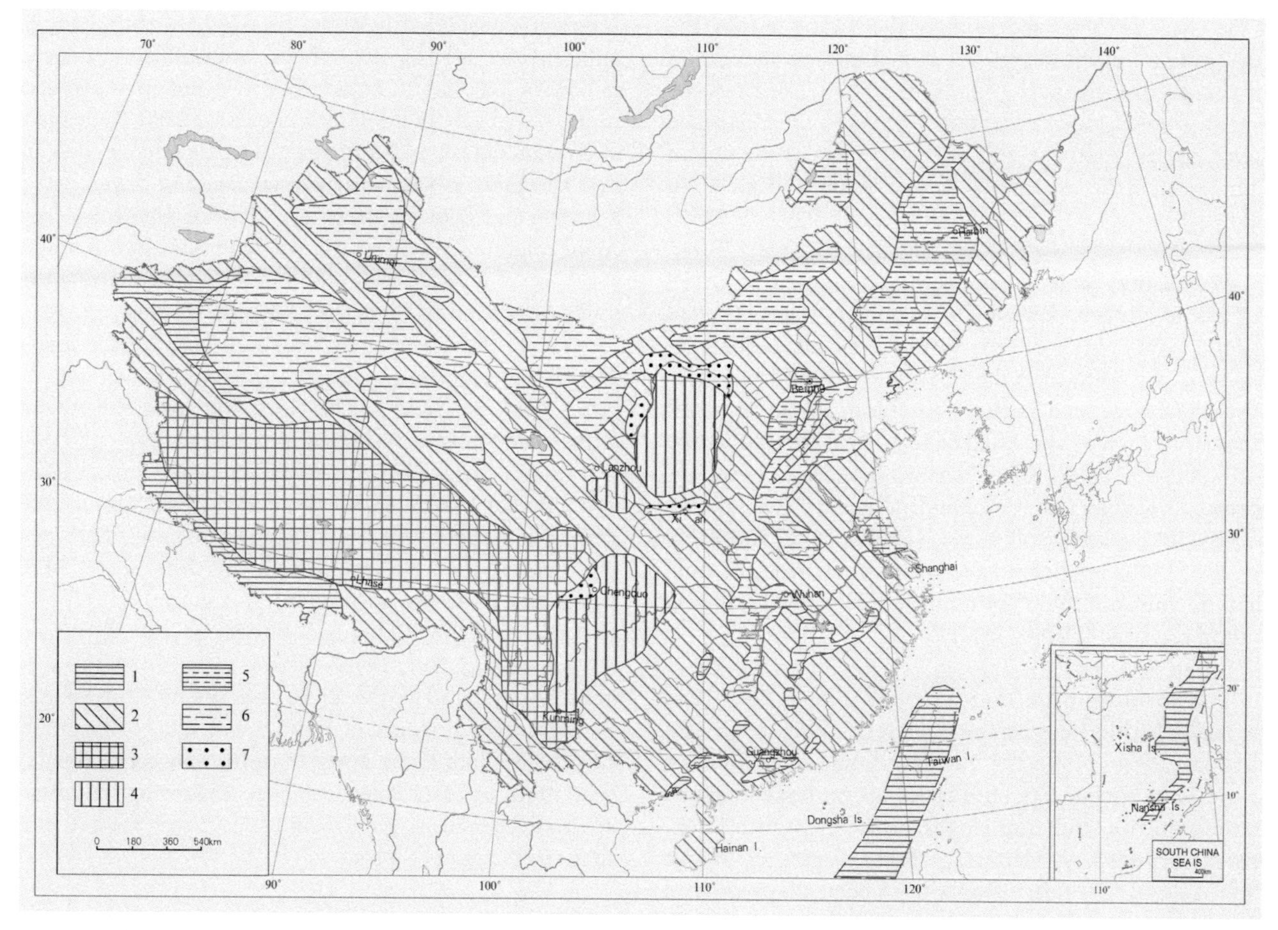

Figure 3.1 A preliminary paleogeographical map of China in the Late Cretaceous and Early Tertiary Periods (Zhao, 1986).
1. Area of Submergence; 2. Uplifted-denundated hill and mountain; 3. Low, rolling peneplain; 4. Mesozoic era depositional basin; 5. Mesozoic-Cenozoic eras depositional basin without saline red beds; 6. Mesozoic-Cenozoic eras depositional basin with saline red beds; 7. Early Tertiary period depositional basin.

Uplift of the Qinghai-Xizang Plateau.—The Qinghai-Xizang Plateau, known as the "roof of the world," with an average altitude of over 4 000 m and surrounded on each side by high mountain ranges descending to the adjacent plains and basins, is the highest, youngest and largest structural landform unit on the Earth. Throughout most of Pliocene Epoch (ca. 5.3–2.6 million years ago), the surface of the Plateau was sustained at an elevation of about 1 000 m. More violent uplift occurred at the end of Pliocene and beginning of the Pleistocene Epoch.

Tectonic evolution of the Qinghai-Xizang Plateau.—The geological history of the Qinghai-Xizang Plateau was actually an evolutionary process from ocean (the Tethys Ocean) to land. According to Pan (1990), Deng (1998) and Shi *et al.* (1998), there are five ophiolite suture zones in the Plateau: the North Qilian, Kunlun, Hongshan-Jinsha Jiang, Bangong-Nu Jiang and Indus-Yarlung Zangbo suture zones. These in turn represent boundaries between six terrains: the North Kunlun, Middle Kunlun, Taxkorgan-Tianshuihai, Karakorum, Gangdisê and Himalayan terrains. All five ophiolite zones are the remains of ancient ocean basins from different geological times. Along with the terrains between them, they decrease in age from north to south (Pan, 1990).

The Tethys Ocean began to appear at the end of Permian Period, and became stable by the end of Triassic Period. This accelerated subduction in the Mid Cretaceous Period, and was complete by the Mid Eocene Epoch, when the Indian subcontinent collided with the Asian continent, leading to the disappearance of the Tethys Ocean from Xizang. The Qinghai-Xizang Plateau then entered a new stage of terrestrial evolution, that of the Himalayan Movement.

The Himalayan Movement in the Cenozoic Era.—The Himalayan Movement is defined as a tectonic movement in the Cenozoic Era, characterized by mountain building, faulting and magmatic activity. The first stage occurred in the Eocene Epoch (ca. 40 million years ago): collision of the Indian and Eurasian Plates resulted in the building of the Gangdisê Shan and rising of the Xizang terrain, the disappearance of the Tethys Ocean and molasse deposition on the piedmont of the Gangdise Shan. Meanwhile, reddish-colored molasse deposits formed in the Oligocene Epoch on the southern and northern sides of the Qilian Shan, indicating an intensive rise in the Qilian Shan following this episode.

The second stage of Himalayan Movement occurred in the Mid Neogene Period (ca. 20 million years ago), and was the main episode of orogeny of the Himalaya Shan. By this time, a system of faulted basins resulting from east-west expansion, such as the Jilong basin, or north-south expansion, such as the Zhada basin, were formed in the north of the Himalaya Shan. Meanwhile, volcanic eruptions occurred, forming a several hundred meter-thick layer of andesite and volcanic ash in the Namling Pendi and Baxoi Pendi of the Gangdise range, while in northern Xizang, alkaline volcanic rocks were formed between 10 and 40 million years ago (Deng, 1998).

The surface of the Plateau in the Early Paleogene Period.—After the collision between the Indian and Eurasian Plates along the Yarlung Zangbo suture zone in the Late Eocene Epoch, the Kailas molasse deposits formed on the southern slopes of the Gangdise Shan. Fossils of *Eucalyptus* (Myrtaceae), *Fagus* (Fagaceae), *Myrica* (Myricaceae) and other plants have been discovered in this formation, indicating that the prevailing climate was warm and humid and the altitude was very low. Pollen data obtained from northern Xizang were rich in *Sequoia* and *Taxodium* (both Taxodiaceae) during the Early Paleogene Period, whilst in the Qaidam Pendi and Gansu Corridor, *Magnolia* (Magnoliaceae), *Ginkgo* (Ginkgoaceae) and Proteaceae were found in Oligocene strata. All these plants grew in relatively warm and low altitude areas, representing a tropical environment (Li, 1995). In addition, numerous remains of the giant rhinoceros (*Elasmotherium*) - the largest terrestrial animal of the Cenozoic Era - alongside Early Paleogene tropical vegetation have been found on the Qinghai-Xizang Plateau and around its margins, such as in Yunnan, Gansu and Xinjiang as well as Pakistan. During this time, the surface of the Plateau is likely to have experienced a warm climate with tropical vegetation.

The Plateau surface in the Neogene Period.—After the rise of the Himalaya Shan, the main Plateau was by this time well-formed as flat or gently sloping ridges or flat-topped uplands. On the eastern margin of the Plateau, the dissected remains of the original surface, incised by large rivers, became separated high above the river gorges. There is evidence that the main surface was formed under a tropical or subtropical, warm and humid climate. Distributed on and around the Plateau are lacustrine and fluvial deposits, like the corresponding sediments adjacent to the main surface, containing a *Hipparion* fauna (Zhang *et al.*, 1981). Such faunas from the Miocene and Pliocene Epochs have been discovered in many places in China and southern Asia (Ji *et al.*, 1981), including one from a karst depression of southern Gansu, dated to between 5.3 and 6.1 million years ago (Li, 1995). Pollen indicative of trees such as cedars, palms and oaks, usually found in subtropical zones, have been found in Pliocene strata from many places in Xizang, Gansu and Yunnan. All this evidence indicates that the main Plateau surface was widely developed in Miocene and Pliocene times, and that the *Hipparion* fauna could migrate freely over the whole of Asia by 5.3 million years ago. The average elevation of the Plateau at that time might have been no more than 1 000 m. (Li *et al.*, 1979).

Violent uplift of the Qinghai-Xizang Plateau over the last 3.4 million years.—The third and final stage of the Himalayan Movement began in the Late Pliocene Epoch (3.4 million years ago). From this time the Himalaya Shan and Qinghai-Xizang Plateau entered a new stage of uplift, periodically very rapid. Due to its great significance in the forming of the Plateau itself, the third stage, which coincides with the geological and geomorphological evolution of the Plateau, is known as the "Qingzang Movement" (Li *et al.*, 1996), (Zhang & Li, 1983). This rapid uplift is evidenced by the Xiyue and Yumen Conglomerate on the northern slopes of Kunlun and Qilian Mountains, and the Boulder Conglomerate in the southern Himalaya Shan, both of Early Pleistocene age. Following the intense uplift of the Plateau, the Chang Jiang, Huang He and other large rivers began to form.

The Qingzang Movement is divided in to three episodes (A, B and C) at 3.4, 2.5 and 1.8 million years ago, respectively (Li, 1995). After the Movement, the Plateau was still elevated, which resulted in fluvial incision and forming of river terraces. Some 2 000 m in height (to what is known as the "first critical altitude") was gained between 2.5 and 1.6 million years ago, inducing the Asian monsoon and subsequent loess deposition in northern China. This suggests that a coupling mechanism existed between tectonic events and climatic changes in Asia during the Quaternary Period. When the Plateau entered the cryosphere at the "second critical altitude" of 3 500 m, an event estimated to have occurred between 1.2 and 0.8 million years ago, the climate and environment of the Plateau were greatly affected. The "third critical altitude" of about 4 000 m was probably reached by 150 000 years ago, when the Plateau, especially its interior and northwestern part (known as its "core"), became extremely cold and dry (Zheng & Li, 1994). The average speed of uplift at the present time is 5.8 per year, with the actual rate being lower in the north and higher in the south. The highest known historical rate of uplift was more than 10 mm per year (Zhang *et al.*, 1991).

The impact of these events on China's geography–not to mention its climate and flora–cannot be underestimated. Under the action of Himalayan Movement, the Tethys Ocean disappeared, the Eurasian Continent became integrated, Taiwan became a mountainous island arising from the ocean floor, the Himalaya Shan were transformed from a sea trench into the highest peaks of the world, and the Qinghai-Xizang Plateau was raised by up to 4 500–5 000 m. Undoubtedly, without the intensive periodic uplift of the Plateau since the early Pleistocene Epoch, the four great topographic steps of major landforms in China would not have formed, the Asian monsoon system would not prevail so strongly, and the huge thickness of loess would not have been deposited in northern China.

3.2.6 Main factors determining China's landforms

The present physical environment of China—its landforms—have undergone a complicated geological history. The pattern and intensity of tectonic movements —particularly the relatively recent movements of the Mesozoic and Cenozoic Eras, property and composition of bedrock, and spatial and temporal changes of climate have all exerted great influence on the formation and development of China's landforms.

Tectonic movements.—According to recent investigations, the landforms of the Chinese mainland, particularly the four great topographic steps, faithfully mirror the major relief forms of the Mohorovičić ("Moho") surface—the boundary between the Earth's mantle and crust. Thus, the mountains and plateaux of China reflect depressed areas of the Moho surface. The boundaries between the four topographic steps (the Qinghai-Xizang Plateau; from the Plateau to the Da Hinggan Ling, Taihang Shan and Wu Shan ranges; from here east to the coast; and the continental shelf) correspond roughly to long-established and deep-rooted fault belts.

The close correlation of China's landforms with geological structure means that the arrangement of all landform units such as mountains, plateaux and plains has a definite orientation, and their boundaries are usually identified with geological structural lines. The longitudinal line formed by the Helan Shan, Liupan Shan and Hengduan Shan, at ca. 106° E, divides China into two major parts with quite different orientations of landforms. To the west, landform units are arranged in a northwest or west-east direction, whereas to the east they lie in northeasterly or north, northeasterly orientation. The landforms of western China comprise mainly inland basins such as the Junggar Pendi and Tarim Pendi, and block mountains, such as the Altay and Tian Shan to the north and, south of these, the Kunlun Shan and Altun Shan. Further south lies the extensive Qinghai-Xizang Plateau with its numerous folding-faulting mountains. Eastern China is essentially composed of a series of parallel, northeastearly-orientated, uplifted and subducted belts. From west to east, the first uplifted belt includes the Da Hinggan Ling, Taihang Shan and Wu Shan systems. The second is represented by the Changbai Shan , Shandong Bandao, and a series of southeastern coastal mountains and hills, and the third, the Taiwan Shan in the east. The first corresponding subduction zone includes the eastern Inner Mongolian Gaoyuan, the Ordos Gaoyuan, the Sichuan Pendi, and the eastern Yunnan Plateau. The second includes the Dongbei Pingyuan, Bohai Wan, the Huabei Pingyuan, and the central Chang Jiang Valley. The third is mainly occupied by the Asian marginal seas – the Dong Hai, Nan Hai, and the sea between Japan and the Korean Peninsula.

Properties and composition of bedrock.—The bedrock of China dates from many different geological periods. The most ancient, crystalline rocks are generally hard and erosion-resistant, so form rugged ridges and high peaks, such as the Tian Shan, Kunlun Shan, Qilian Shan, Qin Ling and Taishan. The diverse sedimentary rocks of the Mesozoic and Cenozoic Eras usually form low mountains and hills, while the loose, unconsolidated deposits of the Quaternary Period are the main materials of the plains of eastern China.

Some of the widely-distributed bedrocks of China include the carbonate bedrocks of southern and southwestern China, which cover a total area of about 1.3 million km^2. Under suitable climatic and structural conditions these may form spectacular karst topography. The near-continuous karst landscape of Guangxi, Guizhou and eastern Yunnan, covering more than 300 000 km^2, is probably the largest and best developed karst topography in the world. The form of the karst landscape varies depending on the climatic and fluvial conditions: the topography is best developed in hot and humid Guangxi, over 50% of which is covered by pure Devonian and Permian Period limestone and dolomite beds 3 000–5 000 m thick. In Guizhou and eastern Yunnan, more than 50% of which are based on limestone and other carbonates of the Carboniferous and Permian Periods, the karst-forming process has also left its mark, but not so clearly as in Guangxi, due to the higher elevation and consequent cooler climate. In Yunnan and eastern Guizhou, the karst landforms were mostly formed during the Paleogene and Neogene Periods when the climate was much warmer. In the northern subtropical zone, including Sichuan, Hubei, Hunan and Zhejiang, most carbonate bedrocks are Paleozoic and the karst topography is characterized by low hills and sinkholes. In the warm-temperate zone of northern China, including Shandong, Shanxi and Hebei, there are a number of Cambrian and Ordovician Period limestone beds, but a karst topography is little developed, characterized only by karst springs and dry valleys.

Continental Mesozoic "red beds" are widely distributed in central and southern China, composed of a Jurassic to

Paleogene reddish conglomerate of sandstone and shale, with a total thickness of several thousand meters. These are mostly fluvial-lacustrine deposits of intermontane basins, formed in hot and dry environments after the Yanshanian Movement. During the Himalayan Movement, these red beds were structurally uplifted and partly tilted, following which, under more humid climatic conditions causing greater erosion, they were dissected into hills and terraces. The Sichuan Pendi, composed mainly of Cretaceous red sandstone and purple shale, is the largest such montane basin in China, but hundreds of smaller basins are found in central and southern China. The landforms within such basins are usually arranged in concentric belts comprising steep-side reddish hills with their summits around the margin of the basin. In southeastern China, where outcrops of conglomerate or sandstone beds have developed, these landforms are known as "Danxia landforms," after Danxia Shan, Guangdong.

North of the Kunlun Shan, Qilian Shan and Qin Ling, Quaternary loess and loess-like deposits are widely distributed on plateaux, mountain slopes, inter-montane basins and piedmont slopes, comprising a total area of more than 600 000 km^2. The Huangtu Gaoyuan (Loess Plateau) itself covers area of about 300 000 km^2 of thick loess and loess-like deposits, the most characteristic loess landform in the world.

Owing to differing configurations of the underlying bedrock, surface landforms differ greatly. In central Gansu, where Paleogene beds had already been dissected into hills before they were mantled with loess deposits, loessic hills dominate the landscape. In eastern Gansu, northern Shaanxi and western Shanxi, where large-scale peneplane landforms were formed prior to the deposition of Pliocene loess, these were turned into high-level loessic plateaux. The areas between the Changcheng (Great Wall of China) and the Wei He, and between the Liupan and Lüliang Shan consisted, before Pliocene loess deposition, of shallow depressions; after the Pliocene deposition they underwent uplift, and finally the two areas were re-deposited with Quaternary loess. This series of events resulted in a characteristic mosaic of flat loess ridges, high loess plains and gentle loess slopes, all prone to erosion, especially the steep slopes with sparse vegetation.

Granite is widely distributed in China, particularly on the southeastern coast where colossal granitic bodies intrude into many folded mountains and usually form the backbones of anticlinic ranges or dome mountains. Sometimes these are eroded and dissected into peaks such as Taibai Ding in the Qin Ling, Huang Shan in Anhui, Tianmu Shan in Zhejiang, and other famous sites. Owing to the hot, humid climate of central and southern China, chemical weathering and other erosional forces are intensive, causing most granitic mountains to be rounded and dome-shaped.

Late Cenozoic basalts appear extensively on the eastern Nei Mongol Gaoyuan, where they erupted and covered the ancient peneplane surface to form a series of large-scale plateaux. Large tracts of basalt are also found in northeastern China and, in southern China, on the southern part of the Leizhou Bandao and the northern part of Hainan.

Climate.—The forces of wind, water, ice and biological agents that work to create different landforms are closely related to, or even determined by, climatic conditions. In particular, temperature and moisture regimes exert strong, direct impact on weathering and erosion, transportation, deposition and other geomorphological processes. Annual precipitation in China is generally determined by distance from the sea, i.e. decreases from southeast to northwest, and the density of river networks follows the same pattern. In the region south of the Huai He and east of the Yunnan-Guizhou Plateau, the length of river networks totals more than 0.5 km per km^2, and more than one km per km^2 in some delta areas, whereas on the Nei Mongol Gaoyuan and in the arid northwest of China the proportion is less than 0.05 km per km^2, with large tracts without any year-round rivers at all. This impacts on the intensity of fluvial erosion, which also decreases from southeast to northwest. In eastern, monsoon-dominated China, the climate is characterized by an unevenness in the seasonal distribution of precipitation, with more than half the annual precipitation concentrated in the summer and a considerable part of this falling as heavy rainstorms, exceeding 100–200 mm per day in some cases. Such an intensity of rain and the subsequent runoff causes intense erosion on slopes and in gullies, sometimes even with flows of rock debris overflow of channels. The impacts of temperature include intense chemical weathering in tropical and subtropical zones, resulting in a reddish weathering crust with a thickness of several dozen meters. On high plateaux and mountains as well as in the northern part of northeastern China, periglacial and permafrost landforms are well developed.

3.3 China's topography

Huge mountain systems and small hills range across China, forming the skeleton of its topography. Mountain topography–counting all mountains, plateaux, mountainous basins and gentler hills—occupies 65% of the total land area. Seven out of the world's 14 mountains above 8 000 m are located in China, and Qomolangma Feng (Mount Everest), at 8 844 m the highest peak in the world, straddles the border between China and Nepal. Besides its absolute relief, China is also notable for its relative relief, which makes its altitudinal zonation more pronounced even than horizontal zonation. The world's second lowest terrestrial point is Aydingkol Hu (−154 m) in the Turpan Pendi, located at the southern foot of the 5 445 m Bogda Feng (The Physical Geography of China Committee, 1985). The mountains of China provide not only its macro-topographic skeleton but also significant geographical boundary lines. According to Zhao (1986) the country has four major mountain systems, described below (Figure 3.2).

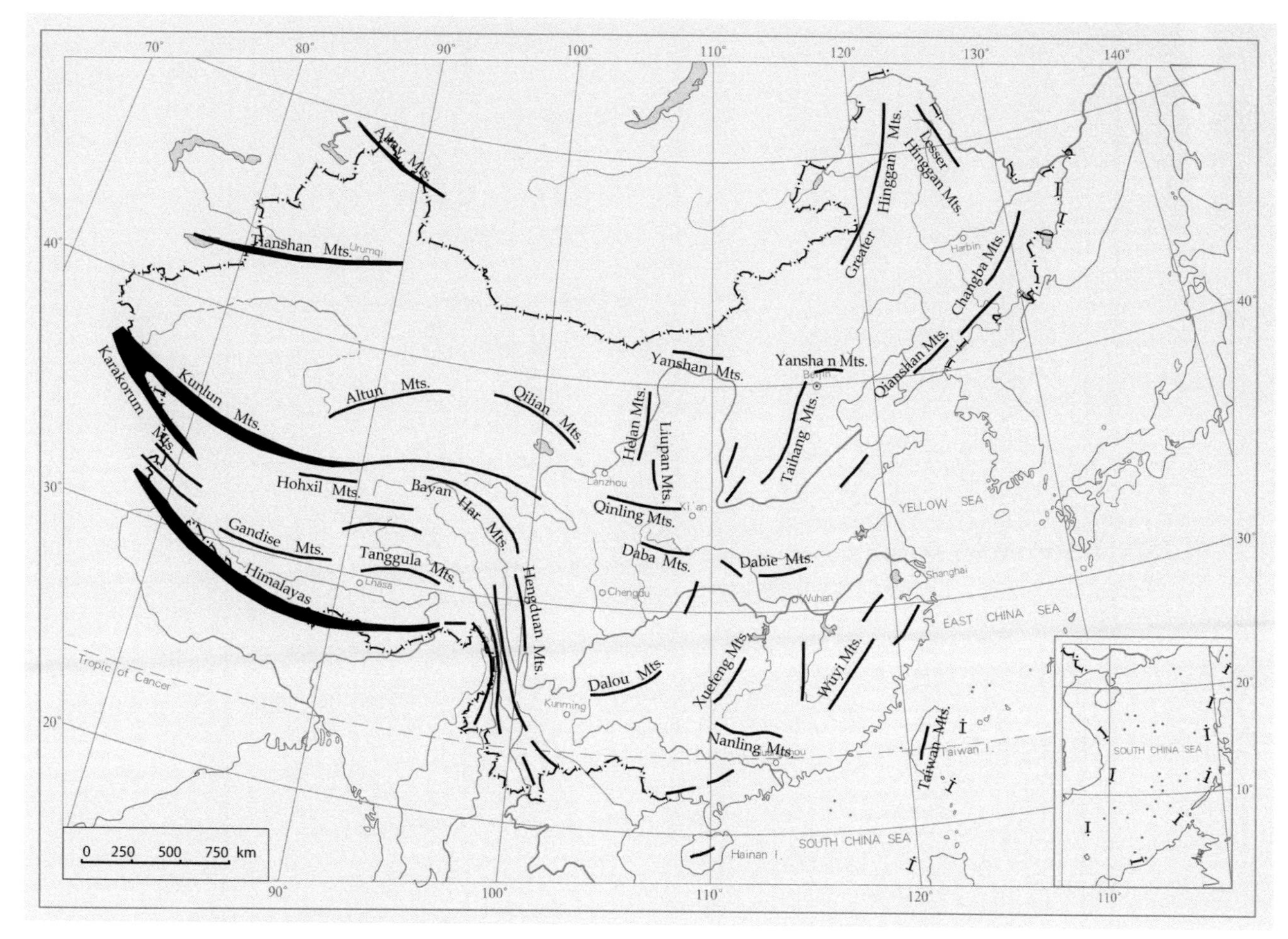

Figure 3.2 The main ranges and mountains of China (Zhao, 1986).

The north-south mountain system.—A north-south-orientated mountain system is located in central China and includes the Helan Shan, Liupan Shan and the famous Hengduan Shan ranges. The Hengduan Shan, located in western Sichuan and northern Yunnan constitutes a series of parallel high ridges, mostly over 4 000 m, separated by deep valleys. These north-south mountains divide China into western and eastern parts: to the west are mainly mountains higher than 3 500 m, some more than 5 000 m, and orientated north-west or northwest-west, while to the east the mountains are mostly lower than 3 500 m, and follow a northeast-east orientation.

The east-west mountain system.—There are three major range of east-west-orientated mountains: from north to south these are the Tian Shan, Yin Shan and Yan Shan, between 40° and 43° N, within which the Tian Shan is 1 500 km long and 250–300 km wide; the Kunlun Shan, Qin Ling and Dabie Shan lie between 33° and 35° N; and the Nanling Shan are located between 25° and 26° N. There are roughly eight degrees of latitude between each of these ranges. The western Kunlun Shan and Tian Shan are 4 000–5 000 m high, while the Yin Shan and Qin Ling are 1 000–2 000 m and Nanling Shan only around 1 000 m. However, because the topography is generally lower in eastern China, all three are important geographical dividing lines.

The northeast-orientated mountain system.—Northeast-orientated ranges are found to the east of the east-west-orientated ranges. From west to east they are the Da Hinggan Ling, Taihang Shan and Xuefeng Shan system, the Changbai Shan-Wuyi Shan system and the Taiwan Shan system.

The northwest-orientated mountain system.—Northwest-orientated mountain ranges are found to the west of the east-west-orientated ranges, and include the Altay Shan, Qilian Shan, Hoh Xil Shan and the Tanggula Shan, all of which are huge ranges with associated glaciers.

3.4 China's terrain

3.4.1 The four great topographical steps

The topography of China is broadly arranged into four great steps, from the Qinghai-Xizang Plateau eastward (Zhao, 1986; Figure 3.3).

The first topographical step.—The Qinghai-Xizang Plateau, with a mean elevation of over 4 000 m, is the first topographical step. It features a range of gradual plateaux, basins and lakes, and its eastern edge contains the headwaters of the Chang Jiang and the Huang He.

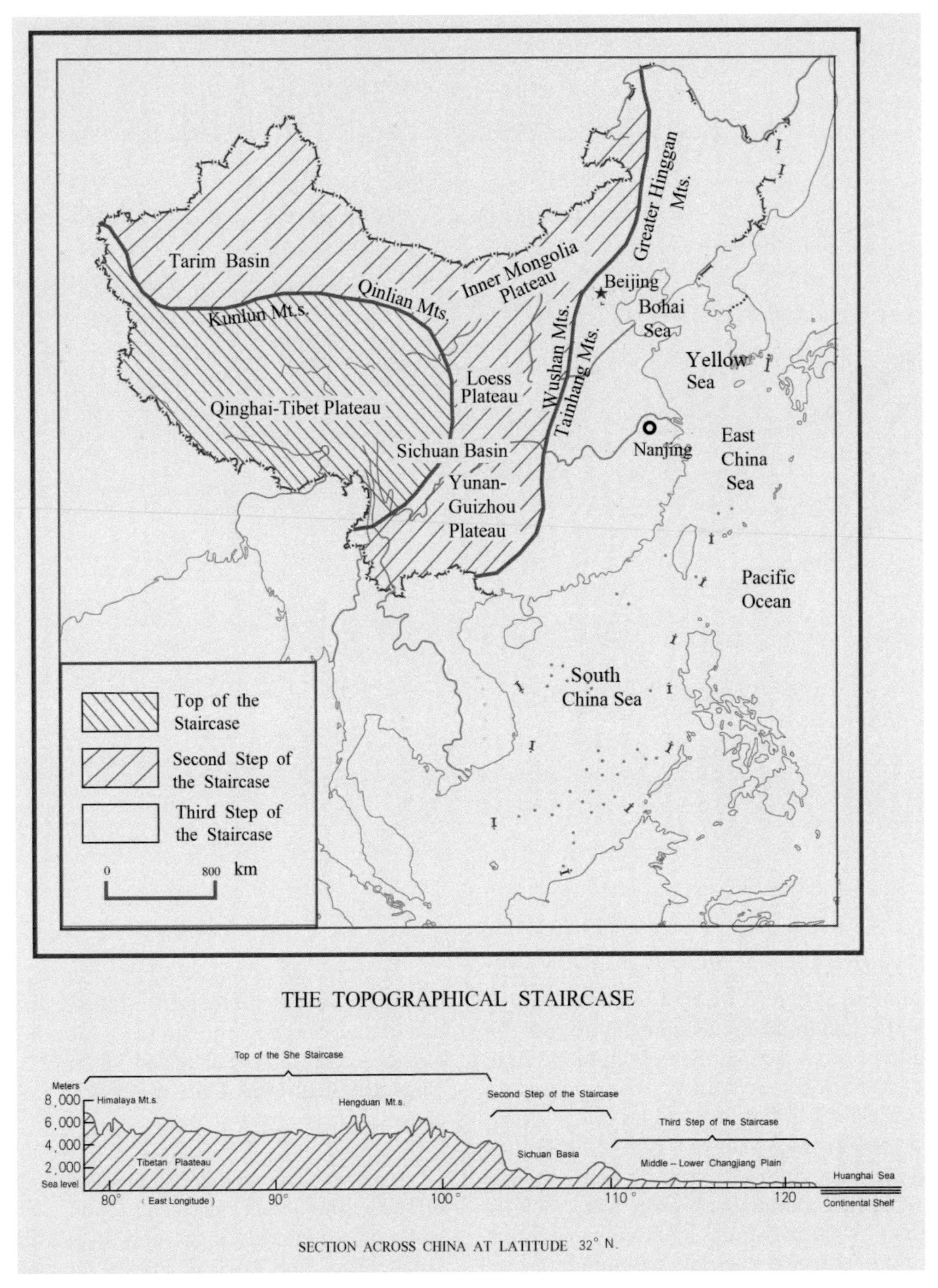

Figure 3.3 The topographical staircase (The China Handbook Editorial Committee, 1983).

The second topographical step.—The second topographical step extends from the eastern margin of the Qinghai-Xizang Plateau eastward to the Da Hinggan Ling, Taihang Shan and Wu Shan, and is composed mainly of plateaux and basins with elevations from 1 000–2 000 m, representing a significant fall in elevation from the first step. Adjacent to the Qinghai-Xizang Plateau is the Tarim Pendi, the largest basin in China, at an elevation of 1 000 m; north of this is the Junggar Pendi at an elevation of ca. 500 m. Between the two basins lie the Tian Shan and fault basins within these mountains. Because of the long distance from the ocean, there is little rainfall and deserts are extensive, although some oases and farmlands are found along rivers. The second topographical step includes four plateaux: the Nei Mongol Gaoyuan, Ordos Gaoyuan, Huangtu Gaoyuan (Loess Plateau) and Yunnan-Guizhou Plateau. Due to various external forces the surfaces of these four plateaux differ greatly, from grassland to dunes, loess ridges, loess hills and karst landforms.

The third topographical step.—The third topographical step stretches from the Da Hinggan Ling, Taihang Shan and Wu Shan eastward to the coast, and contains China's largest plains. At elevations below 500 m, this is the major area of China's agriculture and industry comprising, from north to south, the Dongbei Pingyuan (Northeast China Plain), Huabei Pingyuan (North China Plain), and the plains of the middle and lower Chang Jiang. South of the Chang

Jiang lie extensive hills, mostly comprising red beds. Along the coast and between these plains, there is a series of hills and mountains, generally with elevations of 500--1 500 m, and the step also includes a series of rugged cliffs and rocky inshore islands along the Liaodong Bandao and Shandong Bandao and to the south of the Hangzhou estuary.

The fourth topographical step.—The fourth topographical step comprises the continental shelf, with a water depth generally less than 200 m. There are more than 6 000 islands in China's neighboring seas, many of which are rocky and uplifted. The continental shelf generally has a level or rolling floor, gently dipping towards the southeast, with many drowned deltas and ancient fluvial channels near the coast. According to recent investigations, the sea level has risen by nearly 100 m since the Late Pleistocene and Early Holocene Epochs, reaching present levels ca. 3 000--5 000 years ago. Taiwan, the largest island in China, is located at the conjunction of the Philippine and Eurasian Plates.

3.5 Surface and ground water

Water is an active and mobile element of the physical geographical environment, and is also a natural resource essential to mankind's varied and intense agricultural and industrial activities. Despite natural cycles and renewal, water is a limited resource and China has below world average levels of fresh water *per capita*. In this section we focus only on the surface and groundwater of China.

3.5.1 Surface water

As reported by the 2007 Bulletin of China's Water Resources (Ministry of Water Resources of the People's Republic of China, 2007), the annual amount of total surface water resources is 2,424 billion m^3, or ca. 256 mm of runoff depth. Surface water is chiefly held in rivers and lakes but also includes glaciers and marshes. Glaciers are widely distributed on the Qinghai-Xizang Plateau; marshes are most extensive on the Sanjiang Pingyuan of northeastern China and the Aba area of the northeastern Qinghai-Xizang Plateau. Since glaciers are of limited extent, and support little if any vegetation, we focus here on rivers and lakes.

River systems.—China is essentially a country of great rivers. There are more than 50 000 rivers with a drainage area of over 100 km^2, having a total length of about 420 000 km (The Physical Geography of China Committee, 1985). The chief features of China's major rivers are listed in Table 3.2. The two largest rivers crossing China's territory are the Chang Jiang (Yangtze) and Huang He (Yellow River). The Chang Jiang originates in the Tanggula Shan of Qinghai, and enters the Dong Hai a short distance downstream of Shanghai Shi. The Huang He originates on the northern side of the Bayan Har Shan in Qinghai, forms its "Great Bend" on the western Huangtu Gaoyuan, passes across the Huabei Pingyuan, and enters the Bohai Wan. The other main rivers of China include the Songhua Jiang and Liao He in the northeast of the country, the Hai He in northern China, the Huai He (between the Huang He and Chang Jiang), the Zhu Jiang in southern China and the Yarlung Zangbo Jiang in Xizang.

Table 3.2 Chief features of the major rivers of China (Zhao, 1986). For international rivers, –Heilong Jiang (Amur River), Lancang Jiang (Mekong), Nu Jiang (Salween), Yarlung Zangbo Jiang (Brahmaputra)– data in the table refers only to the sections in China. Note, the Songhua Hu is a tributary of the Heilong Jiang, the Nen Jiang a tributary of the Songhua Hu, the Wei He a tributary of the Huang He (Yellow River), the Jinsha Jiang comprises the upper reaches of the Chang Jiang (Yangtze), and the Han Shui and Jialing are its tributaries.

River	Drainage area (km^2)	Length (km)	Annual discharge	
			Total (10^8 m^3)	Average (m^2/sec)
Heilong Jiang (Amur River)	888 502	3 101	1 181	3 740
Songhua Hu	545 594	1 956	707	2 240
Nen Jiang	283 000	1 379	241	764
Liao He	219 014	1 390	145	459
Hai He	264 617	1 090	233	737
Huang He (Yellow River)	752 443	5 464	574	1 822
Wei He	134 766	818	98	311
Huai He	189 000	1 000	459	1 460
Chang Jiang (Yangtze)	1 808 500	6 300	9 794	31 055
Jinsha Jiang	490 546	—	1 547	4 900
Han Shui	168 851	1 532	574	1 820
Jialing	159 638	1 119	694	2 200
Zhu Jiang	442 585	2 210	3 466	11 000
Lancang Jiang (Mekong)	164 766	2 354	693	2 200
Nu Jiang (Salween)	134 882	2 013	657	2 000
Yarlung Zangbo Jiang (Brahmaputra)	240 280	2 057	1 380	4 370

Table 3.2 shows all rivers with a drainage area greater than 10 000 km^2. Of these, the Songhua Jiang, Liao He, Hai He, Huang He, Huai He, Chang Jiang and Zhu Jiang are the most important. They empty into the Pacific Ocean and their middle and lower reaches are essential regions for China's social and economic development. The combined runoff of these rivers is some 1 500 billion m^3, about 55% of the total for the whole country. However, they are all located in the eastern monsoon area and are therefore prone to serious flooding and drought (Zhao & Chen, 1995).

Based on climatic and geomorphological conditions, China can be divided into two large systems: oceanic and inland. The boundary between them stretches from the northeast to the southwest, ranging from the western piedmont slopes the Da Hinggan Ling in the northeast of China, across the southern fringe of the Nei Mongol

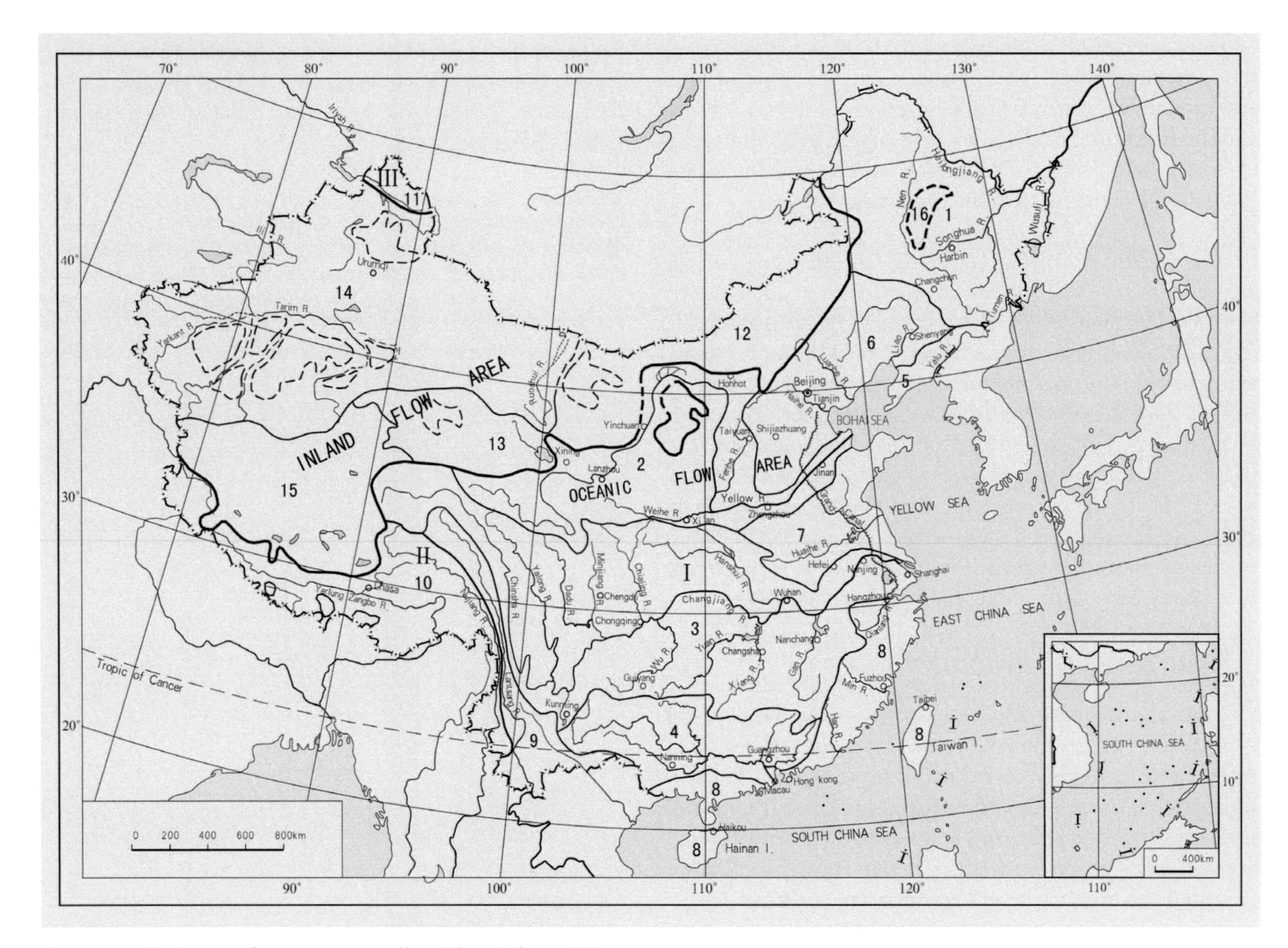

Figure 3.4 Distribution of water systems in China (Zhao & Chen, 1995).

Gaoyuan and the southeast of the Qinghai-Xizang Plateau, to the southwestern border of China (see Figure 3.4). Most of China's rivers are located in the humid southeast (oceanic system) of the country, except for few scattered rivers in the inland northwest, where large areas have no runoff.

The oceanic system, accounting for 64% of the country's land area, can be further subdivided into Pacific, Indian and Arctic drainage basins. Rivers issuing into the Pacific Ocean—in addition to the major rivers listed above—include the Heilong Jiang, Luan He and Lancang Jiang; those flowing into the Indian Ocean include the Nu Jiang and Yarlung Zangbo Jiang; and the Ertix He flows into the Arctic Ocean. The inland system accounts for the remaining 36% of the total land area of China, and includes the Tarim He, Yili He and Hei He, all of which flow into saline inland lakes or peter out in sandy deserts or salt marshes (see Table 3.3).

In the oceanic system, three topographic belts are recognized, corresponding to rivers originating in the southeastern part of the Qinghai-Xizang Plateau; the belt comprising the Da Hinggan Ling, Taihang Shan, Qin Ling and Yunnan Plateau;, and the southeastern coastal belt including the Changbai Shan and Shandong Bandao (Zhao, 1986). These three topographic belts roughly coincide with the eastern uplifted margins of the first, second, and third

Table 3.3 Major drainage basins in China (Zhao, 1986).

Drainage basin	Area of drainage basins (10^3 km^2)	% total land area in China
Oceanic river systems	6 120	64
Pacific Ocean subsystem	5 445	57
Sea of Okhotsk	861	9
Sea of Japan	33	0.3
Bohai Wan (Bohai Sea) and Huang Hai (Yellow Sea)	1 670	17
Dong Hai (East China Sea)	2 045	21
Nan Hai (South China Sea)	825	9
Draining directly into the Pacific	11	0.1
Indian Ocean subsystem	625	7
Bay of Bengal	558	6
Arabian Sea	66	0.7
Arctic Ocean subsystem	51	0.5
Kara Sea	51	0.5
Inland river systems	3 480	36
Nei Mongol	329	3
Desert zone (Gansu, Xinjiang, Qaidam)	2 374	25
Northeastern China	48	0.5
Northwestern Qinghai-Xizang Plateau	729	8
Total	9 600	100

great topographic steps respectively. On the first great topographic step originate the largest rivers in China, such as the Chang Jiang, Huang He, Lancang Jiang and Nu Jiang, which are also major rivers of the world. Rivers such as the Heilong Jiang, Liao He, Huai He and Zhu Jiang, which are not as significant in length or discharge, originate on the second great topographic step. Rivers originating on the third great topographic step, such as the Yalu He and Qiantang Jiang, are smaller than those of the first two steps but are usually abundant in discharge and potential for hydro-electric power generation, because of their location in the rainiest part of China.

In contrast, inland river systems are mostly located in arid and semi-arid northwestern China, the northwest of the Qinghai-Xizang Plateau, and in small "islands" in semi-arid and sub-humid areas of western northeastern China and the central Ordos Gaoyuan. Four drainage areas are recognized among the inland river systems (Zhao, 1986): Nei Mongol, where precipitation is the main source of rivers; Gansu-Xinjiang, in which large rivers including the Shiyang He, Hei He, Shule He, Tarim He, Ili He and Manas He originate in the surrounding high mountains (most of these are not year-round rivers and indeed more than half of the area has no rivers at all); the Qaidam Pendi on the northern Qinghai-Xizang Plateau, which has numerous short rivers (the eastern part has saline lakes and salt marshes and there are no rivers in the west); and the northern Qinghai-Xizang Plateau, comprising a basin with numerous saline lakes, supplemented with small streams (again, more than half of the land area has no year-round rivers).

Based on annual runoff depth (see Figure 3.5, Table 3.4), five major runoff belts have been recognized in China (Zhao, 1986; Zhao & Chen, 1995). The belt of abundant runoff experiences annual precipitation in excess of 1 600 mm, and has an annual runoff depth of at least 900 mm. It includes most of Guangdong, Fujian and Taiwan, mountainous areas of Jiangxi and Hunan, southwestern Yunnan and southeastern Xizang; it coincides roughly with the coverage of tropical and subtropical forests. The second belt, that of adequate runoff, has annual precipitation of 800–1 600 mm and annual runoff depth of 200–900 mm, and includes Guangxi, Yunnan, Guizhou, Sichuan and the middle and lower reaches of the Chang Jiang south of the Qin Ling and Huai He; it is roughly coincident with the zones of subtropical evergreen broad-leaved forest and mixed broad- and needle-leaved forest. The third belt is a transitional belt, with annual precipitation of 400–800 mm and annual runoff depth of 50–200 mm, and includes the Huanghuai Pingyuan, most of Shanxi, Shaanxi and northeastern China, northwestern Sichuan and eastern Xizang; it roughly coincides with the forest-steppe

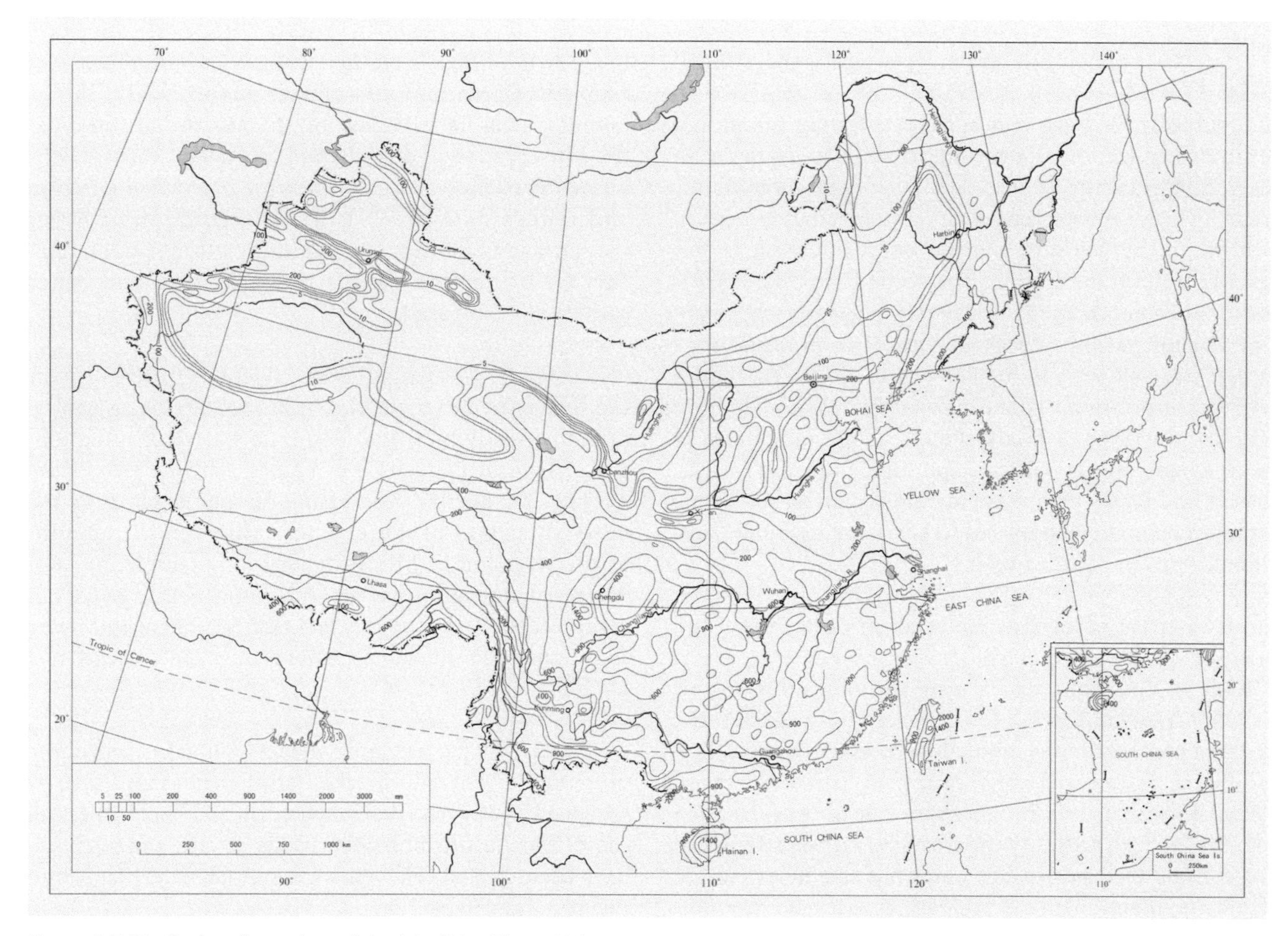

Figure 3.5 Distribution of annual runoff depth in China (Zhao, 1986).

vegetation zone. The final two runoff belts are those of scarce and little runoff. The former has annual precipitation of 200–400 mm and annual runoff depth of 10–50 mm, and includes the west of northeastern China, Nei Mongol, Gansu, Ningxia, western and northern Xinjiang and western Xizang; it is roughly coincident with the semi-arid steppe zone. The latter has annual precipitation below 200 mm and annual runoff depth below 10 mm, and includes the west of Nei Mongol, Junggar Pendi, Tarim Pendi, Qaidam Pendi and the Alxa desert.

Table 3.4 Comparison of annual runoff in the major rivers of China (Zhao & Chen, 1995).

River	Survey station	Catchment area (10^4 km^2)	Survey years	Maximum runoff		Minimum runoff		Ratio of maximum to minimum runoff	Coefficient of variation
				10^8 m^3	Year	10^8 m^3	Year		
Song Hua	Harbin	39	78	84	1932	123	1920	6.9	0.43
Taizi	Liaoyang	1	40	50	1964	11	1978	4.6	0.41
Luan He	Luanxian	4	50	128	1959	16	1936	8.0	0.54
Yongding He	Sanjiadian	5	55	39	1939	8	1930	4.9	0.37
Huang He (Yellow River)	Huayuankou	73	31	1 004	1964	242	1928	3.5	0.25
Huai He	Bengbu	12	54	719	1921	61	1929	11.7	0.62
Chang Jiang (Yangtze)	Datong	171	37	13 727.	1954	7 211	1978	1.9	0.14
Han Jiang	Huangzhuang	14	35	1 072	1964	192	1941	5.6	0.37
Jialing	Beibei	16	37	998	1964	499	1959	2.0	0.21
Qiangtang	Lucifu	3	41	543	1954	173	1979	3.1	0.29
Min Jiang	Zhuqi	5	42	84	1937	276.0	1971	3.1	0.28
Xijiang	Wuzhou	33	37	3 470	1915	1 070	1963	3.2	0.2

Surface runoff reflects both physical geographical factors and human activities. Precipitation is undoubtedly the most crucial factor affecting runoff, but evapotranspiration is also important. For example, in eastern China, the 50 mm runoff isobath corresponds roughly to the 400 mm precipitation isohyet, but in western China. to the 200 mm isohyet, because the latter has greater evapotranspiration. The second most significant factor is geomorphology, which is particularly important for the formation of surface runoff in mountainous areas: steep slopes and high elevation effectively reduce evapotranspiration, increasing surface runoff. The rain shadow causes the leeward slopes of high mountains to have comparatively poor precipitation and runoff while, basins and plains often have low precipitation but strong evapotranspiration compared to surrounding mountains, resulting in less surface runoff.

The effect of surface material on percolation and evapotranspiration is illustrated by karst areas: in Guangxi, Guizhou and eastern Yunnan, most runoff is subterranean, leaving few surface rivers. In deserts, limited precipitation is mostly lost by evapotranspiration and percolation without forming any surface runoff. Vegetation also plays an important role in the formation of surface runoff: evapotranspiration effects of vegetation decrease surface water but also delay runoff. However, the influence of vegetation varies. For example the growth of forests on the Huangtu Gaoyuan increases evapotranspiration but reduces runoff; while in the humid Chang Jiang Valley, forests may enrich runoff by the conservation and regulation of surface water. Lastly, human activities generally increase evapotranspiration and reduce runoff. With the construction of water conservancy facilities and rapid growth of population, the use of surface runoff for irrigation and other activities is increased.

Generally speaking, recent reductions in river runoff can be ascribed to three main reasons: decreasing natural rainfall; greatly increased domestic water consumption; and climate change caused by human activities (see Figure 3.6 for an example). Decreasing rainfall is generally the most important of these factors although the amount of precipitation in Jilin Province, for example, has not decreased during the last 44 years but runoff in rivers has continued to decline, indicating that here increased water consumption for human activities is the main cause (Ren *et al.*, 2002).

Lakes.—China has more than 24 800 lakes, including more than 2 800 natural lakes larger than 1 km^2, such as Qinghai Hu, Poyang Hu, Dongting Hu and Tai Hu. Besides these, many artificial lakes, varying in area and purpose, have been created. The main lakes of China are shown in Figure 3.7 and Table 3.5.

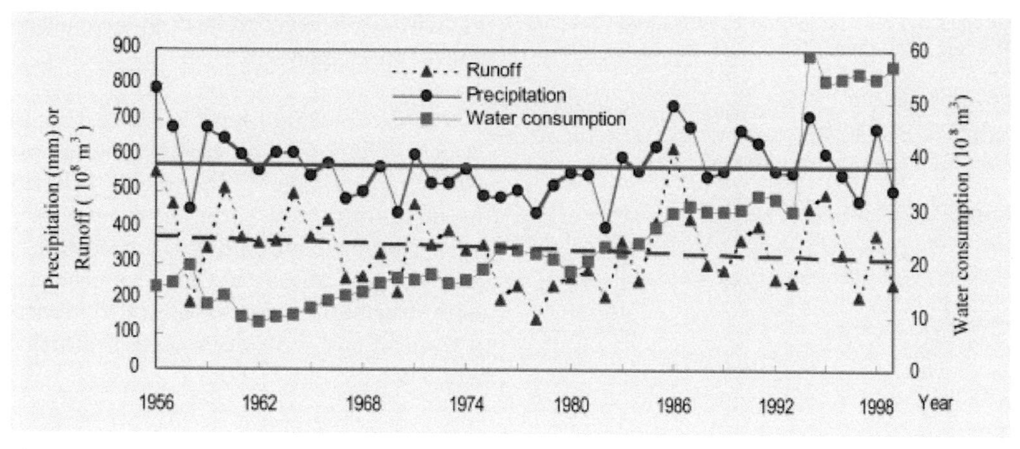

Figure 3.6 Variation in annual precipitation, runoff and water consumption in Jilin, 1956–1999 (Ren et al., 2002).

There are five major lake regions in China (Zhao, 1986). The Northeastern Lake Region in northeastern China comprises large tracts of marsh with small lakes, whose total area is 3 722 km^2, 5% of the total area of lakes in China. A typical example from this region is Wuda Lianchi, located in Dedu, Heilongjiang, the second largest lava-dammed lake in China, with a total area of 18.5 km^2. Tianchi—the largest crater lake in China – lies on the highest peak of the Changbai Shan, has an area of 9.8 km^2 and a maximum depth of 373 m. The Northwestern Lake Region has a total area of lakes of ca. 22 500 km^2, representing ca. 28% of the total lake area of China. Most of the lakes in this region are "wandering" lakes, due to fluctuating and shifting inland river channels. The most famous and largest example is Lop Nur, which once had an area of ca. 3 000 km^2 but is now decreasing due to the growth of a "great ear," of salt deposited by groundwater. One terminal lake of the Hei He has split into two lakes, Gaxun Nur, which has since become entirely dry, and Sogo Nur, now a saline lake. The Bosten Hu, in this region, is at 1 019 km^2 the largest freshwater lake in the inland region of China.

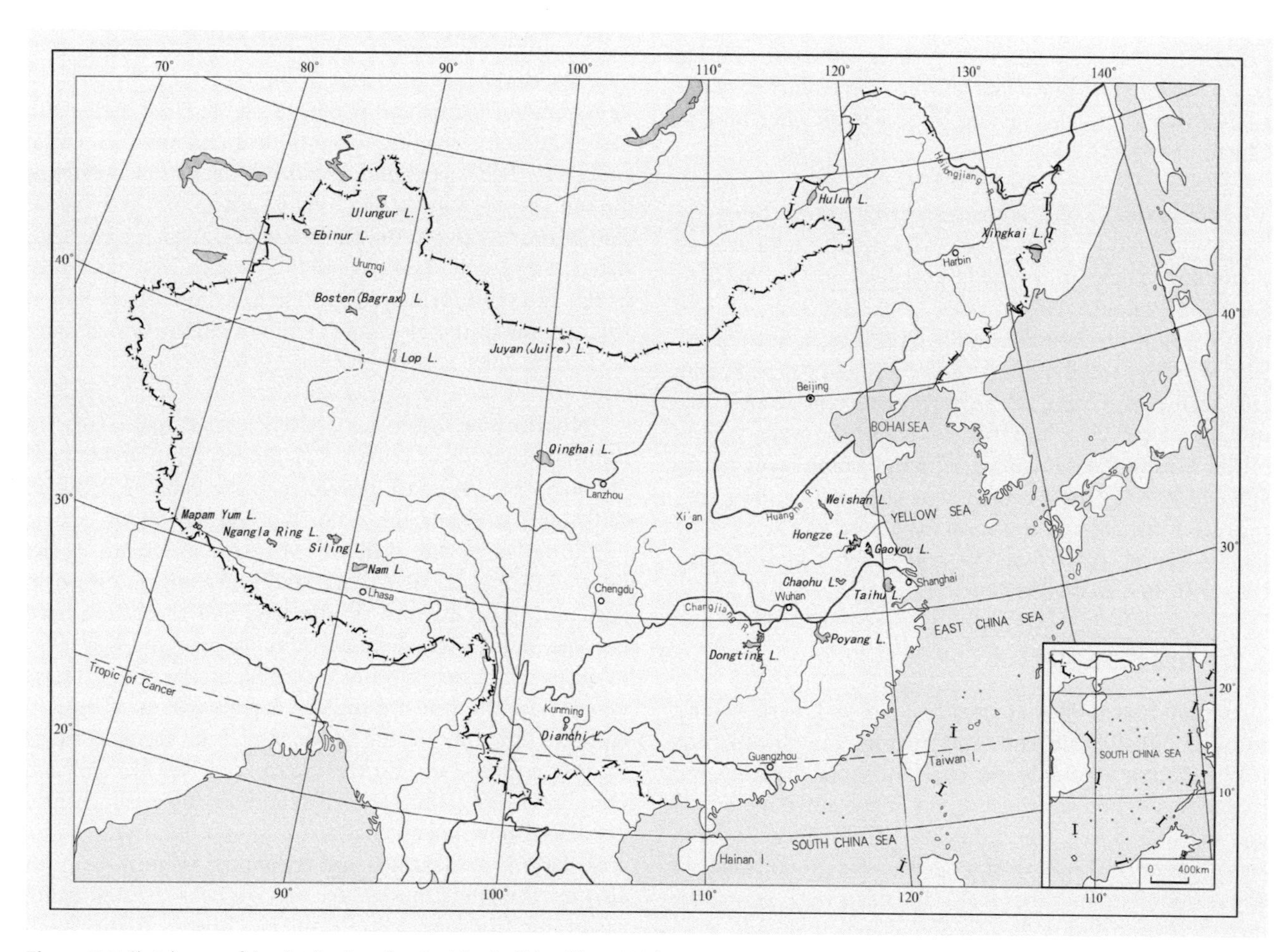

Figure 3.7 Sketch map of the distribution of major lakes in China (Zhao, 1986).

Table 3.5 Major lakes of China (according to http://www.gov.cn/test/2005-07/27/content_17403.htm).

Lake	Province/ autonomous region	Area (km^2)	Elevation (m)
Qinghai Hu	Qinghai	4 583	3 196
Poyang Hu	Jiangxi	3 583	21
Dongting Hu	Hunan	2 740	34
Tai Hu	Jiangsu	2 425	3
Hulun Nur	Nei Mongol	2 315	546
Hongze Hu	Jiangsu	1 960	12
Nam Co	Xizang	1 940	4 718
Siling Co	Xizang	1 640	4 530
Nansi Hu	Shandong	1 266	36–37
Bosten Hu	Xinjiang	1 019	1 048

The Eastern Lake Region, although abundant in precipitation, has only 28% of the country's lake area, similar to the Northwestern Lake Region. However the lakes here are numerous, with abundant inflow and good drainage, comprising mostly freshwater lakes less than 4 m in depth. Poyang Hu, the largest freshwater lake in all of China (3 583 km^2 at high water, 3 050 km^2 at normal water level and just 500 km^2 at low water level), is found here. Dongting Hu, previously the largest freshwater lake is now, due to accelerated silting, only the third largest, having decreased from 4 350 km^2 to 2 625 km^2 between 1949 and 1983, a dramatic change which deserves attention. Tai Hu is the largest freshwater lake on the lower reaches of the Chang Jiang, and the second largest freshwater lake in China.

The lakes of the Southwestern Lake Region are mostly located in karst areas, with a total lake area of 1 188 km^2. Dian Chi is the largest lake on the Yunnan-Guizhou Plateau, with an area of ca. 300 km^2. Due to heavy pollution from Kunming Shi, it is now a "dead" lake with foul-smelling water and poisoned fishes. Fuxian Hu, northeast of Yuxi, at 157 m is the second deepest lake in China after Tianchi. Finally, most of the water bodies of the Xizang Lake Region are inland saline lakes fed by snowmelt. They have a total area of nearly 31 000 km^2, representing 38% of the total lake area of China. The largest lake in this region is Qinghai Hu, a highly saline lake, with an area of 4,580 km^2.

3.5.2 Groundwater

Groundwater, often ignored, is a rare and useful natural resource. In the arid and semi-arid northwest of China it is sometimes the only source of water for drinking and irrigation. Groundwater is also very important in the humid and sub-humid east, where precipitation is concentrated in certain months resulting in a need to use groundwater in dry seasons. Groundwater is also the main water source for metropolises including Beijing Shi, Shanghai Shi and Xi'an Shi. Thus, although there are many advantages to using groundwater, it needs to be carefully managed in order to maintain the supply.

Factors in the formation and distribution of groundwater.—The formation and distribution of groundwater reflects various natural conditions. Of climatic conditions, the most important is the ratio between precipitation and evapotranspiration. For this reason the quantity and quality of groundwater in China corresponds to the north-south distribution of geographical zones from temperate to tropical and with the southeast-northwest trend in humidity.

Geological and geomorphological conditions affect groundwater feeds, movement and drainage. In mountainous areas the main source of groundwater is precipitation and snowmelt, which are drained rapidly by drainage networks on steep slopes, becoming fresh bicarbonate-containing water. There are also close relations and interactions between surface water and groundwater. As surface water infiltrates bedrock it becomes groundwater; generous supplies of groundwater may contribute to an abundance of sub-surface water, as in central and southern China. However, in northern and northeastern China, where there is often a shortage of both surface water and precipitation, the lower reaches of rivers can become important feed sources for groundwater.

3.6 The impact of human activity

Mankind has exerted significant impacts on the physical geography of China and is becoming the most active and essential factor shaping geographical and environmental processes. Tribes of *Homo erectus* lived in China as early as 1.6 million years ago (Zhao, 1986). Agricultural activities at Banpo near Xi'an Shi on the Huangtu Gaoyuan have been dated to ca. 6 000 years ago and in Zhejiang, large quantities of rice and tools for rice cultivation have been discovered in the remains of the Neolithic Hemudu culture which dates back to ca. 7 000 years ago (Zhao, 1986).

Nowadays, although China has successfully industrialized, there remain 121 million ha of arable land (Ministry of Land and Resources of the People's Republic of China, 2008) and the majority of the 1.3 billion people of China, some 890 million (Ministry of Health of the People's Republic of China, 2007), are still engaged in agricultural production. Virgin land and vegetation have become increasingly difficult to find as mankind continues to make significant imprints on the landscape. The Chinese people have transformed the vast expanses of marshland in eastern China into fertile arable land, built terraced fields in hilly regions, and on the plateaux of Nei Mongol and Qinghai-Xizang and western mountainous areas natural grasslands have been converted into vast pastures. Coastal areas have been dammed and reclaimed. Even in the most arid northwestern basins, oases have been created using meltwater from mountain glaciers. Other activities such as afforestation, reservoir construction and the spread of urban and traffic systems mean that most of China's land has been shaped and influenced by human activities. By 2008, 657 million ha of China's land area was occupied by agriculture and 33 million ha by construction; the total land area modified by man was ca. 690 million ha—73% of the total national land area (National Bureau of Statistics of China, 2008).

There have been both successes and failures in the utilization of land and other natural resources in the face of prolonged shortages of food and other socio-economic problems. One outstanding success has been farm irrigation: to guarantee agricultural production, irrigation of farmland was initiated as early as 563 BC by building the Zhengguo Canal from a natural waterway around the Taihang Shan. Later, in the Eastern Zhou Warring States Period (256 BC), the remarkable Dujiang Weir was created on the Chengdu Plain. Other famous successful water conservancy projects of the past include the "karez" water system of the Turpan Pendi in Xinjiang, Hexi oasis, the network of water channels on the Taihu Plain, and the Grand Canal from Beijing Shi to Hangzhou Shi. At present ca. 40% of China's total cropland is under irrigation, about a quarter of the total irrigated area in the world.

Another success is the many wild plants that have been brought into cultivation, contributing to agricultural vegetation including orchards, production forests and pasturelands (see Chapters 16–21 for more on economic plants). Agriculture changes both the natural processes of mineral transfer in the soil and its physicochemical properties, especially through digging, fertilization and irrigation which accelerate the cycling of natural materials. As a result, various kinds of cultivated soils are formed (Ren *et al.*, 1992; see Chapter 5 for more detail).

On the other hand, a lack of a comprehensive understanding of nature has led to mistakes in the utilization of land and other natural resources. For example, for several thousand years and especially during the past century, many regions have been overexploited with heavy losses of cropland, grassland and forest resources as well as severe soil erosion. The Huangtu Gaoyuan (Loess Plateau)—the cradle of Chinese civilization—is the most notorious example. This area is now subject to severe soil erosion and low agricultural productivity, and has become the major source of sediment input the Huang He, following several thousand years of devastation of the natural vegetation, misuse of land, and construction of large dams. In Heilongjiang, extensive use of land and the positioning of most cropland on gentle slopes without conservation measures has caused about half the total cropland to suffer from soil erosion; about a quarter of cropland has lost half of its top stratum of fertile black earth after fewer than 100 years of cultivation. In arid and semi arid northwestern China, the problem of desertification is becoming increasingly serious, chiefly due to poor land use. Here there is also the problem of soil salinization. Overuse of irrigation water in arid areas with inadequate drainage has caused some salinization in about 20% of the total cropland of China.

China's forest area, especially tropical forest, is rapidly decreasing. According to a survey of Hainan in 1956 there were still 839 000 ha of natural forest; after more than 30 years of development, this area has dropped sharply to 660 000 ha (Ding, 1996). Tropical rainforests and monsoon forests have already disappeared, while montane rainforest and ravine forest are now rapidly decreasing as well.

3.7 A physico-geographical regionalization of China

Physical regionalization a classic subject of Chinese geographical study, especially since 1950. Physico-geographical regionalization is an effort to identify differences among areas of the Earth's surface and to document similarities within areas. It considers the total of all physical attributes of the land and hence serves as a guideline for land-use planning. The recognition of physico-geographical regions may also help in the understanding of regional responses to global change. Using techniques of remote sensing and geographical information systems, it is increasingly easy to monitor natural environments and their ecological status. Furthermore, the recognition of physico-geographical regions can provide a scientific basis for the reasonable utilization of natural resources, especially land resources, and the conservation of biodiversity, further promoting sustainable development.

3.7.1 Principles, approach and hierarchical units used

In any hierarchical system, the land may be classified into different types or subdivided into regional units. Since horizontal zonality is characteristic the world over, classification of types at the higher categories is more appropriate for comparative studies. At the lower level, it is more effective for land subdivision to reveal the uniqueness of the areas concerned. Ideally, a system provides a combination of the two approaches (Huang, 1990).

The physico-geographical regions of China should be demarcated based on real differences in the nature of the surface (i.e. geomorphological structure and topographic configuration), and combinations of temperature, moisture conditions, zonal vegetation and soil type (Zheng, 1989). The physico-geographical regionalization system used in China is deduced in a "top-down" way, from high to low levels, dividing by both type and regional demarcation, with the selected hierarchical units being temperature zone, thermal-moisture region and finally natural region. Based on the distribution of natural features in China, the Qinghai-Xizang Plateau is distinctly discordant with the rest of the country and, for this reason, is analyzed separately, although the same general principles are observed (Huang, 1990).

Temperature zone.—The demarcation of land surfaces based on temperature conditions is necessary to understand physical, chemical and biological processes and phenomena as well as agricultural production. In addition to temperature, landforms, vegetation, soils and other physical elements, crops and cultural systems are referenced for demarcating temperature zones. Based on these classification criteria, temperature zones are strictly divided and suitable for comparison.

Thermal-moisture region.—Generally speaking, a combination of thermal conditions and moisture regime is the main factor determining regional differentiation of biodiversity on the terrestrial surface of the Earth. Temperature variation tends to match variation in moisture regime and, in the broad sense of zonality, moisture regime should also reflect changes in the ratio of potential evapotranspiration to precipitation. In global terms, four patterns of regional moisture regime may be recognized: humid, sub-humid, semi-arid and arid, roughly equating to four types of natural vegetation: forest, forest steppe (including meadow), steppe, and desert, respectively. Thermal-moisture regions tend also to match features of land use.

Natural region.—A natural region is recognized by a combination of vegetation and soil type, due to differentiation of landforms or geographical location. Thermal-moisture regions are subdivided into natural regions chiefly on the basis of major differences in landforms and climate. The structures found in each altitudinal belt within a natural region are similar. Only major differences between flat and mountainous land are identified in most cases. In a few cases, climatic differences are taken as the basis of demarcation.

3.7.2 The three natural realms of China

According to the Working Committee of Physical Regionalization of China (Working Committee of "Physical Regionalization of China" - Chinese Academy of Sciences, 1959), China can be divided into three natural realms: the Eastern Monsoon Realm, the Northwest Arid Realm (Nei Mongol and Xinjiang) and the Xizang Alpine Realm. This demarcation has since been recognized and applied by academic circles and government departments in China. Taken as the starting point of a physico-geographical regionalization in China, the country can then be subdivided into further hierarchical units.

The Eastern Monsoon Realm.—Occupying about 45% of the total land area of China, the Eastern Monsoon Realm accounts for 95% of the country's population. Its proximity to the coast, including the Pacific Ocean, means the influence of the summer maritime monsoon is strong, with sharp seasonal variations in both wind direction and precipitation. The climate is humid to sub-humid, and forest is the dominant natural vegetation. Temperature, increasing from north to south, is the chief parameter used for regional differentiation within this realm, but precipitation also decreases conspicuously from east to west in northern and northeastern China. Owing to the absence of Quaternary continental glaciation, both flora and fauna are species-rich and a reddish weathering crust is extensively distributed. With its long history, the impact of human activities is also great. Practically all available arable land in this realm has been cultivated and all natural vegetation modified.

The Northwest Arid Realm.—This Realm, making up the eastern part of the immense Eurasian desert and grassland, occupies about 30% of the total land area of China, but accounts for only 4% of the country's population. Located in the hinterland of the Eurasian continent and surrounded high mountain ranges, the influence of the summer maritime monsoon is rather weak here, and hence the climate is semi-arid to arid. Since the Late Mesozoic Era the area has been subject to gradual desiccation, with some fluctuations. Both flora and fauna are rather poor in species and in number, and zonal vegetation types consist chiefly of desert and steppe. In comparison to the Eastern Monsoon Realm, human impact is less conspicuous. However, numerous fertile oases have been developed along the middle and lower reaches of perennial rivers and large tracts of grassland have been used for pasture for many centuries.

The Xizang Alpine Realm.—The Qinghai-Xizang Plateau occupies about 25% of the total land area of China but contains less than 1% of the country's population. Both flora and fauna are rich in species here and altitudinal zonation is conspicuous. In the northwestern parts of the plateau belts of desert, montane steppe, montane coniferous forest, alpine meadow, sub-nival and nival (i.e. under permanent snow) vegetation can be seen in succession from mountain foot to summit. Because of the influence of strong glaciation and weak chemical weathering, the soil parent materials are usually coarse and thin and soils are generally young, formed only following the last glaciation. Consequently, soil profiles are poorly developed with low fertility. As a whole, the natural conditions of the plateau are unfavorable to human activities.

3.7.3 Criteria for demarcation of selected hierarchical units

Temperature.—The principal temperature criteria used are usually the number of days with mean temperature above 10 °C and accumulated temperature of at least 10 °C, because these factors are related to the duration of the fastest growing season; the two factors are not always concurrent. On the Qinghai-Xizang Plateau, the number of days with mean daily temperature above 5 °C is also related to the growth of alpine pasture, so may be used in conjunction with the above criteria for comparison with lower altitudes.

In an area with fewer days of mean daily temperature above 10 °C and accumulated temperatures below 10 °C, the temperature of the warmest month is proposed as a supplementary criterion. The mean temperature of the warmest month can indicate thermal strength during the growing season, and is highly correlated with the growth and distribution of certain zonal vegetation, as well as crop cultivation. In contrast, the mean temperature of the coldest month, and average minimum temperature, may be taken as supplementary criterion in warmer regions. Based on temperature and its role in natural process and agricultural production, nine temperature zones are recognized in lowland of China (see Table 3.6).

Moisture regime.—Under certain temperature condition, moisture regime becomes a limiting factor that affects plant growth and distribution. Taking annual aridity (the ratio of annual evapotranspiration potential to annual precipitation) as the principal criterion, and annual precipitation as a second criterion, four regional types of moisture regime have been identified: humid, sub-humid, semi-arid and arid (see Table 3.7). A thermal-moisture region, demarcated based only on the annual mean moisture regime, may reflect variation in the ratio of evapotranspiration potential to precipitation or the difference between them. However, this picture is incomplete unless seasonal and long-term variation are taken into account.

Table 3.6 Criteria for demarcating the temperature zones of China.

Temperature zone	Principal criteria		Supplementary criteria	
	Days with mean temperature at least 10 °C	Accumulated temperature over days with at least 10 °C	Mean temperature of warmest month (°C)	Mean temperature of coldest month (°C)
Plateau subpolar	< 50	—	< 12	−18–(−12)
Plateau temperate	50–180	—	12–18	−10–0
Cold-temperate	< 100	< 1 600	< 16	<−30
Mid-temperate	100–170	1 600–3 200	16–24	−30–(−16)
Warm-temperate	171–220	3 200–4 500	—	−16–0
Northern subtropical	> 220	> 4 500	—	0–5
Mid-subtropical	> 230–240	> 5 000	—	5–10
Southern subtropical	> 300	6 000–6 500	—	10–15
Peripheral tropical	365	> 8 000	—	15–18
Mid-tropical	365	> 8 000	—	18–24
Equatorial tropical	365	> 8 000	—	> 24

Table 3.7 Criteria for demarcating regional moisture regimes of China.

Moisture regime	Annual aridity (evaporation/precipitation)	Potential natural vegetation	Related land use issues
Humid	up to 1	Forest	—
Sub-humid	−1.5	Forest-steppe or meadow	Secondary salinization
Semi-arid	1.5–4	Meadow-steppe, steppe of desert-steppe	Rain-fed cultivation
Arid	at least 4	Desert	No cultivation without irrigation

Landforms.—Within any natural zone, landforms can affect natural conditions such as climate, moisture regime, soils and vegetation. Processes such as weathering, erosion and accumulation can reflect differences in bedrock composition and other geological parameters. Furthermore, the same landforms may be characterized by different functions in different thermal-moisture regions. Therefore, physical regions based on climatic factors should be subdivided based on landforms (Huang, 1989). Based on simplified and important differences in plains and mountains, 49 natural regions of China have been recognized (see Table 3.8). Once all factors have been taken into account to determine the eco-geographical regions of China, several key boundaries appear, and are discussed in the following section.

Table 3.8 Physico-geographical regionalization of China.

Temperature zone	Moisture region	Natural regions
I. Cold-temperate zone	IA. Humid region	IA1. Northern Da Hinggan Ling
II. Mid-temperate zone	IIA. Humid region	IIA1. Sanjiang Pingyuan IIA2. Eastern mountains of northeastern China IIA3. Eastern piedmont slopes of northeastern China
	IIB. Sub-humid region	IIB1. Central Dongbei Pingyuan IIB2. Central Da Hinggan Ling IIB3. Piedmont slopes and hills of Sanhe
	IIC. Semi-arid region	IIC1. Western Liaohe Pingyuan IIC2. Southern Da Hinggan Ling IIC3. Eastern Nei Mongol Gaoyuan IIC4. Hulun Buir Gaoyuan
	IID. Arid region	IID1. Western Nei Mongol Gaoyuan and Hetao IID2. Alxa and Hexi corridor IID3. Junggar Pendi IID4. Altay Shan and Tacheng basin IID5. Tian Shan
III. Warm-temperate zone	IIIA. Humid region	IIIA1. Mountains of Liaodong and eastern Shandong
	IIIB. Sub-humid region	IIIB1. Mountains of central Shandong IIIB2. Huabei Pingyuan (North China Plain) IIIB3. Mountains of northern China IIIB4. Plains of southern Shanxi and Wei He valley
	IIIC. Semi-arid region	IIIC1. Loess highlands of central Shanxi, northern Shaanxi and eastern Gansu
	IIID. Arid region	IIID1. Tarim Pendi and Turpan Pendi
IV. Northern subtropical zone	IVA. Humid region	IVA1. Huainan and the middle and lower reaches of the Changjiang IVA2. Qin Ling and Bashan
V. Middle subtropical zone	VA. Humid region	VA1. Hills south of the Chang Jiang VA2. Mountains south of the Chang Jiang and Nanling VA3. Guizhou plateau VA4. Sichuan Pendi VA5. Yunnan plateau VA6. Southern Slopes of the eastern Himalaya Shan
VI. Southern subtropical zone	VIA. Humid region	VIA1. Central and northern. Taiwan VIA2. Hills and plains of Fujian, Guangdong and Guangxi VIA3. Mountains of central Yunnan
VII. Peripheral tropical zone	VIIA. Humid region	VIIA1. Lowlands of southern Taiwan VIIA2. Central and northern Hainan and Leizhou Bandao VIIA3. Hills and valleys of southern Yunnan
VIII. Mid-tropical zone	VIIIA. Humid region	VIIIA1. Southern Hainan, Dongsha Dao, Xisha Qundao and Zhongsha Qundao
IX. Equatorial tropical zone	IXA. Humid region	IXA1. Nansha Qundao
H0. Qinghai-Xizang Plateau polar zone	H0D. Arid region	H0D1. Kunlun Shan
HI. Qinghai-Xizang Plateau sub-polar zone	HIB. Sub-humid region	HIB1. Hilly plateau of Golog-Nagqu
	HIC. Semi-arid region	HIC1. Plateau and broad valleys of southern Qinghai HIC2. Plateau and lake basins of northern Xizang
	HID. Arid region	HID1. Kunlun Shan and plateau
HII. Qinghai-Xizang Plateau temperate zone	HIIA/B. Humid/Sub-humid region	HIIA/B1. High mountains and gorges of western Sichuan and eastern Xizang
	HIIC. Semi-arid region	HIIC1. Plateau and mountains of eastern Qinghai and Qilian HIIC2. Mountains and valleys of southern Xizang
	HIID. Arid region	HIID1. Qaidam Pendi HIID2. Ngari mountains

3.7.4 Discussion of key boundaries

The northern limit of the tropical zone.—Based on Köppen's classification (James, 1922), one of the most widely-used schemes of climatic classification, the limits of the tropical climate zone are defined by the isotherm of 18 °C for the coldest month; other criteria include an "absolute" lack of frost; minimum temperature is not a critical factor for plant growth.

The tropical zone of China is situated at the northern periphery of the tropical realm on Earth, and may be defined on the basis of the mean temperature of the coldest month and absolute minimum temperature. Severe cold spells in winter can do serious harm to rubber and other tropical crops, which are sensitive to low temperatures, making a three-fold division of the tropical realm more valuable on the basis of thermal regimes: the northern limit of the peripheral tropical zone is the northern limit that is frost-free year-round and corresponds with the 15 °C isotherm for the coldest month; the northern limit of the middle tropical zone is the 18 °C isotherm in the coldest month, above which tropical crops are not subjected to damaging cold in winter (Huang, 1990).

The northern limit of the subtropical zone.—China is remarkably for its highly developed subtropicality. An acceptable characterization of subtropical climate is one combining tropical heat in summer with low temperatures in winter. The northern boundary of this realm may be defined either by the polar limit of the distribution of winter wheat to the north of Beijing Shi, or the Qin Ling-Huai He line on the southern boundary of the North China Plain. In the area in between these boundaries, the summer temperature is comparable to that in the tropics, and winter cold is not so severe as to exclude winter wheat. These, together with the cultivation of cotton, rice, grapes and a number of other warm season crops are features of, rather than criteria for, subtropical regions. A total absence of citrus fruits and failure of vegetable cultivation in winter precludes areas from the subtropical zone. In the version of Köppen's classification modified by Trewartha (1980), a temperature of 0 °C in the coldest month has been suggested as marking the boundary between "Group C" (mild-temperate) and "Group D" (continental) climates. In eastern China this isotherm fits nicely with the Qin Ling-Huai He line, the best known geographical divide in China, and is considered to mark the northern boundary of the realm (Huang, 1990).

The boundary between the plateau sub-polar zone and plateau temperate zone.—In discussions of thermal conditions, one problem arises: the existence of the huge Qinghai-Xizang Plateau, with an area of about 2.5 million km^2 and an average elevation exceeding 4 000 m, totally upsets the pattern of temperatures on the Earth's surface. For this reason, the Plateau is considered a separate realm to the rest of China not only in terms of temperature but also in other respects.

The plateau may be divided into the plateau sub-polar and plateau temperate zones, on the basis of temperature. The number of days with a mean daily temperature above 10 °C is regarded as the principal criterion, and mean temperature of the warmest month as a supplementary criterion. The number of days of mean daily temperature above 10 °C are related to the duration of the fast growing season, while the mean temperature of the warmest month indicates thermal strength during the growing season. Both are highly correlated with the growth and distribution of certain plants of the natural vegetation, as well as some agricultural species.

The demarcation line between mountainous and plateau regions is characterized by transitional and gradual features. Several major demarcation lines run along major ridges or across the flanks of the mountains, others may be drawn along the piedmont slopes of the mountains; these depend mainly on regional differentiation of the altitudinal belt.

This physico-geographical regionalization of China, along with the details of physical geography outlined in this chapter, provides a backdrop to understanding the plants of China in their natural habitat.

References

Deng, W. M. (1998). *Cenozoic Intraplate Volcanic Rocks in the Northern Qinghai-Xizang Plateau*. Beijing, China: Geological Publishing House.

Dewey, J. F. (1972). Plate tectonics. *Scientific American* 226: 56-68.

Ding, C. C. (1996). Changes in Hainan tropical natural forest resources and simple analysis. *Central South Forestry Survey Planning* 15(2): 33-35.

Huang, B. W. (1959). Draft of the complex physical geographical division of China. *Chinese Science Bulletin* 18: 594-602.

Huang, B. W. (1989). An outline of the physico-geographical regionalization of China. In: *Geographical Symposium*. pp. 10-20. Beijing, China: Science Press.

Huang, B. W. (1990). Some themes of integrated physical geography. In: *Progress in Geographical Research—The 50th Anniversary of the Establishment of the Institute of Geography, Chinese Academy of Sciences*. Beijing, China: Science Press.

James, P. E. (1922). Köppen's classification of climates: a review. *Monthly Weather Review* 50: 69.

Ji, H. X., Huang, W. B., Chen, J. Y., Xu, Q. Q. & Zheng, S. H. (1981). First discovery of Hipparion fauna in Tibet and its significance in plateau uplifting. In: *The Period, Amplitude and Type of the Uplift of Qinghai-Xizang Plateau*. pp. 19-25. Beijing, China: Science Press.

Li, J. J., Wen, S. X., Zhang, Q. S., Wang, F. B., Zheng, B. X. & Li, B. Y. (1979). A discussion on the period, amplitude and type of the uplift of the Qinghai-Xizang Plateau. *Scientia Sinica* 22(11): 1314-1328.

Li, J. J. (1995). *Uplift of Qinghai - Xizang (Tibet) Plateau and Global Change*. Lanzhou, China: Lanzhou University Press.

Li, J. J., Fang, X. M., Ma, H. Z., Zhu, J. J., Pan, B. T. & Chen, H. L. (1996). Geomorphological and environmental evolution in the upper reaches of the Yellow River during the late Cenozoic. *Science in China Series D: Earth Sciences* 39(4): 380-390.

Ministry of Health of the People's Republic of China (2007). *China Health Statistic Report*.

Ministry of Land and Resources of the People's Republic of China (2008). *National Land Resource Report*.

Ministry of Water Resources of the People's Republic of China (2007). *Bulletin of China's Water Resource*.

National Bureau of Statistics of China (2008). *China Statistic Yearbook*.

Pan, Y. S. (1990). Characteristics and evolution of west Kunlun Mountains. *Geological Science* 3: 224-232.

Physical Regionalization Working Committee of Chinese Academy of Sciences (1959). *Comprehensive Physical Regionalization in China (first draft)*. Beijing, China: Science Press.

Ren, L. L., Wang, M. R., Li, C. H. & Zhang, W. (2002). Impacts of human activity on river runoff in the northern area of China. *Journal of Hydrology* 261(1): 204-217.

Ren, M. E., Bao, H. S. & et al. (1992). *The Development and Improvement of Chinese Natural Regions*. Beijing, China: Scientific Press.

Shi, Y. F., Li, J. J. & Li, B. Y. (eds.) (1998). *Uplift and Environmental Changes of Qinghai-Xizang (Tibetan) Plataeu in Late Cenozoic*. Guangzhou: Guangdong Science and Technology Publishing House.

The Physical Geography of China Committee (1985). *Chinese Physical Geography Pandect*. Beijing, China: Scientific Press.

The China Handbook Editorial Committee (1983). *Geography*. Beijing, China: Foreign Languages Press.

Trewartha, G. L. (1980). Köppen's classification of climates. In: Trewartha, G. L. (ed.) *An Introduction to Climate*. pp. 397-403. New York: McGraw-Hill.

Working Committee of "Physical Regionalization of China" - Chinese Academy of Sciences (1959). *Comprehensive Physical Regionalization of China*. Beijing, China: Science Press.

Zhang, Q. S., Li, B. Y., Yang, Y. C. & Yin, Z. S. (1981). Basic characteristics of neotectonic movement of Qinghai-Xizang Plateau. In: Liu, D. S. (ed.) *Geological and Ecological Studies of Qinghai-Xizang Plateau*. pp. 103-110. Beijing, China: Science Press.

Zhang, Q. S. & Li, B. Y. (1983). Neotectonic movement in Xizang. In: *Quaternary Geology of Xizang*. pp. 110-130. Beijing, China: Science Press.

Zhang, Q. S., Zhou, Y. F., Lu, X. S. & Xu, Q. L. (1991). On the present uplift speed of Tibetan Plateau. *Chinese Science Bulletin* 36(21): 1820-1824.

Zhao, S. Q. (1986). *Physical Geography of China*. Beijing, China and New York: Science Press and John Wiley & Sons.

Zhao, W. J. & Chen, Z. C. (1995). Discussion on the foundation of ferrosols. *Acta Pedologica Sinica* 32(suppl.): 21-33.

Zheng, D. (1989). A study on problems of the physico-geographical regionalization in mountainous and plateaus regions. In: *Geographical Symposium*. pp. 21-28. Beijing, China: Science Press.

Zheng, D. & Li, B. Y. (1994). Evolution and differentiation of the physico-geographical environment of the Qinghai-Xizang (Tibetan) Plateau. *The Journal of Chinese Geography* 4(1-2): 34-47.

Chapter 4

Climate

ZHANG Xue-Qin and ZHENG Du

4.1 Introduction

The geographical features discussed in the previous chapter have significant impacts on the complex and diversified climatic characteristics of China. These were exhibited clearly during the period from 1961 to 2010, on which the following brief introduction focuses, based largely on the up-to-date analysis with the data provided by the China Meteorological Data Sharing Service System (http://cdc.cma.gov.cn/home.do), and on the works of WANG and LI (2007), DING (2013), and ZHENG and colleagues (2010) as well.

4.2 Main features of the monsoon climate

4.2.1 Seasonality

China is subject to a typical monsoon climate, with cool, dry winters and warm, wet summers. The factors controlling climate in China vary significantly from season to season. In winter, polar air masses from Siberia and Mongolia exert control over the whole Eurasian continent and therefore the prevailing winds in China are from the north and west; the climate is cold and dry. In summer, a southerly wind prevails in eastern Asia; the influence of warm, wet ocean airflows extends inland and results in abundant rainfall and high temperatures. Spring and autumn are characterized by a short period of temperate climate.

The extensive influence of the winter and summer monsoon causes significant differences in climate between seasons. Generally, the higher the latitude of a location, the greater the temperature differential from winter to summer. In most parts of China, the rainy and warm seasons occur simultaneously; thus summer is the main rainy season, in which precipitation is abundant and often of high intensity. The main rain belt moves northward across China as the season progresses.

4.2.2 The effect of latitude

China stretches over fifty degrees of latitude, covering climatic zones including temperate, subtropical and tropical. Hence, there are significant regional differences in climate. In winter, the monthly mean temperature can be lower than –24 °C in northern Heilongjiang, while it is above 0 °C in areas to the south of the Huai He, and more than 18 °C on Hainan. The difference in monthly mean temperature between south and north can thus be at least 40 °C in winter.

In northern China, winter is persistently cold. To the north of the Changcheng (Great Wall) and the Tian Shan, winter can last six months; north of the Da Hinggan Ling it may be as long as nine months, with barely any summer at all. In winter, cold air passes through the middle and lower reaches of the Chang Jiang, but in summer this area experiences hot and wet conditions. South of the Nanling Shan, the mean temperature of the coldest month is above 10 °C, which could be considered spring-like. Over the Nan Hai (South China Sea) and its islands, the climate is summery throughout the year. Central Yunnan, where the altitude is ca. 1 500 m, is the only region with continuous spring-like conditions.

4.2.3 The impact of land and oceans

China lies in the southeast of Eurasia, the greatest continent in the world, and to the west of the Pacific, the world's largest ocean. Due to variation in distance from the sea, and its great interior extent, totally different dry continental and wet monsoon climates are formed in China. In winter, due to the thermal difference between the ocean and the continent, the weather phenomena known as the Siberian High (cold, dry air from north of China) and Aleutian Low (which causes cyclones, from the east) are very strongly developed. Because the prevailing winds blow from the continent towards the ocean, and China's geography features high topography in the west with low relief in the east, there arises a situation where frigid polar airflows directly affect southern China. Under such conditions, a dry continental climate prevails throughout the country, with only relatively slight regional differences. From spring, the effects of the ocean begin gradually to become more important.

In summer, the hot, low pressure system in the interior of the mainland, and the Pacific High, develop and strengthen. The impact of the ocean extends northwards and westwards as the summer monsoon intensifies, bringing abundant rainfall to most parts of China. In coastal areas, the surface air temperature isotherm lies roughly parallel to the coastline. The climate of inland areas, on the other hand, is little influenced by the ocean. The high mountains of the west block northward- and westward-moving airflows, resulting in lower precipitation in western China. In arid and semi-arid areas of Gansu and Xinjiang, the climate is still mainly continental, with deserts here and there. Especially in spring, these desert regions become the main sources of sandstorms in China.

4.2.4 The impact of topography

The Qinghai-Xizang (Tibetan) Plateau accounts for a quarter of the whole land area of China, and significantly influences the climate of the country, and even that of the world. In winter, the air temperature of the Qinghai-Xizang Plateau is lower than that of neighboring regions, causing air to flow from the high plateau to the surrounding area, strengthening downward airflows and the winter monsoonal circulation. In summer, the plateau is a relative heat source compared to the colder neighboring free atmosphere. Air rises over the plateau to a greater extent than in neighboring areas, thus intensifying the summer monsoon circulation. The great mass of terrain of the Qinghai-Xizang Plateau also plays a role in blocking winter monsoon flows to the south and extending summer monsoon flows to the north, which leads to hot, dry weather, lack of precipitation and midsummer cloud in Gansu and Xinjiang. Furthermore, when northerly winter airflows blow towards the Qilian Shan, and the southwesterly summer wind progresses northward, these airflows diverge, thus enlarging the region impacted by both the winter and summer monsoons on the Qinghai-Xizang Plateau.

The three types of mountains found in China – east-west-, northeast-southwest- and south-north-orientated – play different roles in influencing the climate. Mountains aligned east-west form important boundaries between the major climatic divisions. For example, the Tian Shan in central Xinjiang is not only the boundary between southern and northern Xinjiang, but also between the mid-temperate and warm-temperate climatic zones. The Yin Shan forms a boundary between regions of pastoralism and cultivation agriculture. The Kunlun Shan and Qin Ling are regarded as the divide between the main climate types of southern and northern China, as well as between the warm-temperate and subtropical zones, while the Nanling Shan forms the main climatic divide in southern China.

The Da Hinggan Ling and Taihang Shan, and the mountains from south of Funiu Shan to the mountains of Guangxi, the Changbai Shan, the hills of the Liaodong Bandao and the Shandong Bandao, as well as the hills in the southeastern coastal region of China are oriented in a northeast-southwest direction. All these mountains act as barriers to the penetration of the southeastern monsoonal flow to inland areas. Rainstorms thus occur frequently on the windward sides of these mountains. In southwestern China, the famous Hengduan Shan is aligned in a north-south direction, with average elevations of 4 000–5 000 m. Air currents appear in the deep valleys between pairs of mountains, and the climate is distinctly different on the mountain summits compared to in deep valleys, making this region one with the most complex climate in China.

4.3 Temperature

4.3.1 Annual mean surface air temperature

The annual mean surface air temperature can reflect the general temperature and heat resources in a given region. The distribution of temperature is strongly influenced by atmospheric circulation, terrain and physiognomy, and the distribution of ocean and land. Mainly influenced by latitude, the annual mean surface air temperature in eastern China increases from north to south. The average annual mean surface air temperature from 1961–2010 was below 0 °C in northern parts of Nei Mongol and Heilongjiang. The lowest average temperature, −7.4 °C, was found at Tianchi Station in the Changbai Shan. The 10 °C isotherm of annual mean temperature generally lies through Anshan in Liaoning, Huailai in Hebei, Taiyuan Shi in Shanxi, and Suizhong in Shaanxi, while the 14 °C isotherm follows the middle to the lower reaches of the Chang Jiang and Huai He valleys. South of 25° N, the warmest part of China, average temperatures are above 20 °C. Annual mean temperature is about 24–26 °C in southern Taiwan and Hainan, and exceeds 26 °C in the Xisha Qundao and on the surface of the Nan Hai (South China Sea; Figure 4.1).

The distribution of temperature is rather complicated in western China, where the isotherms are greatly shaped by both topography and altitude. The Tarim Pendi and Turpan Pendi in Xinjiang are the warmest places in northwestern China, with annual mean temperatures of 10–15 °C. Mean temperature is about 6–8 °C in the Junggar Pendi, and is lowest in the Tian Shan and Kunlun Shan ranges, illustrating the great difference in temperature between basins and mountains. On the Qinghai-Xizang Plateau the temperature declines from southeast to northwest, ranging from 10 °C to 22 °C in the border region of the southeast, through 2–10 °C in the central valley of southern Xizang, to generally below 0 °C on the northern Plateau. In Yunnan, complicated topography with great differences in elevation also causes the annual mean temperature to range from 24 °C in the valleys of southern Yunnan to 4 °C in the northwest. Temperatures in most parts of Guizhou average about 14 °C (Figure 4.1).

4.3.2 Seasonal mean temperature

In winter (classified as December to February), the whole of China is under the control of a polar continental air mass, which results in a severely cold climate in most parts of the country, with the exception of the southern coastal regions and islands. The temperature of the eastern plains is shaped by solar radiation, giving mainly zonal isotherms. The temperature gradient between south and north is the greatest in winter, especially in January, when the mean temperature is about −30 °C at Mohe in far northeastern China (Figure 4.2). The mean January temperatures on the Dongbei Pingyuan (Northeast Plain) are between −10 °C and −20 °C. The 0 °C isotherm lies along the Huai He and the Qin Ling in eastern China, and then extends westward along the eastern side of the Qinghai-Xizang Plateau, and southwest until it reaches southeastern Xizang. Winter temperatures exceed 10 °C over the area to the south of the Nanling Shan, are greater than 18 °C over the coastal regions of southern Hainan and Taiwan, and even reach 26 °C on the Zengmu Ansha at the southernmost tip of China.

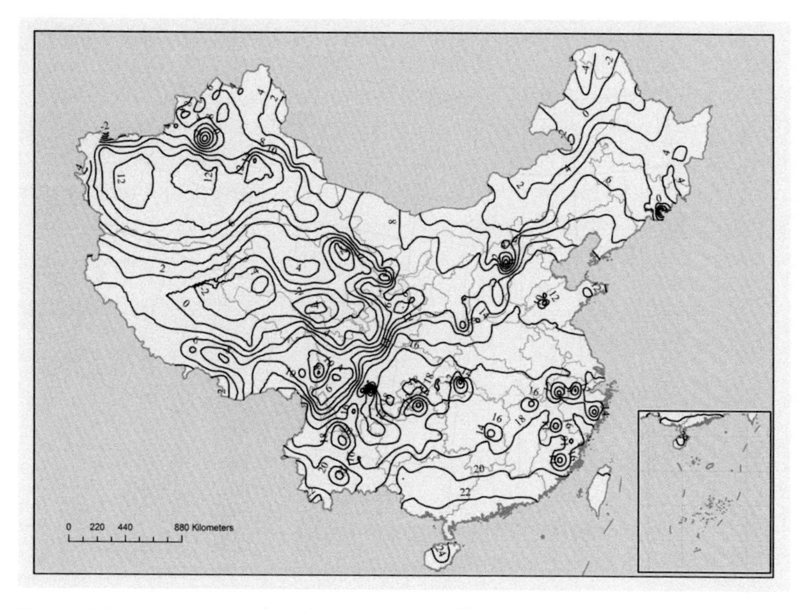

Figure 4.1 Annual mean surface air temperature in China, 1961–2010 (°C).

Of the western regions, the Junggar Pendi in Xinjiang is a cold center where the mean January temperature is lower than −22 °C. The temperature is generally between −8 °C and −10 °C in southern Xinjiang, and between −16 °C and −18 °C on the northern Qinghai-Xizang Plateau, declining with topography and latitude and increasing towards the southeast to as much as −10 °C in the valley of the Yarlung Zangbo Jiang, and 2–10 °C in southeastern Xizang. On the Yunnan Plateau, temperatures range from 16 °C in the south to −4 °C in the northwest, and are around 4–6 °C in Guizhou.

In spring (March to May), temperatures increase throughout the country following the northward retreat of the polar continental air mass. In northeastern China, the weather tends to be sunny with few clouds, so temperatures increase quickly. Influenced by warm moist maritime air flows, the southern part of the country is wet, and temperatures increase more slowly. Thus the temperature in some parts of northern China is higher than that in southern China between late spring and early summer. With the exception of the area around the northern Da Hinggan Ling in northeastern China, the temperature is above 0 °C throughout the country. It is lower than 10 °C in northeastern China and Nei Mongol, between 10 °C and 22 °C from south of the Changcheng to the Zhu Jiang valley, and often over 20 °C south to the Tropic of Cancer. In the west, it is still below 12 °C on the plateaus and hills, with most parts of northern Xizang being only in the range of −8–0 °C. It is, however, higher than 12 °C in southeastern Xizang and the desert regions of Xinjiang.

In summer (June to August), the monthly mean temperature is between 16 °C and 28 °C over most parts of China except for the Qinghai-Xizang Plateau, for which

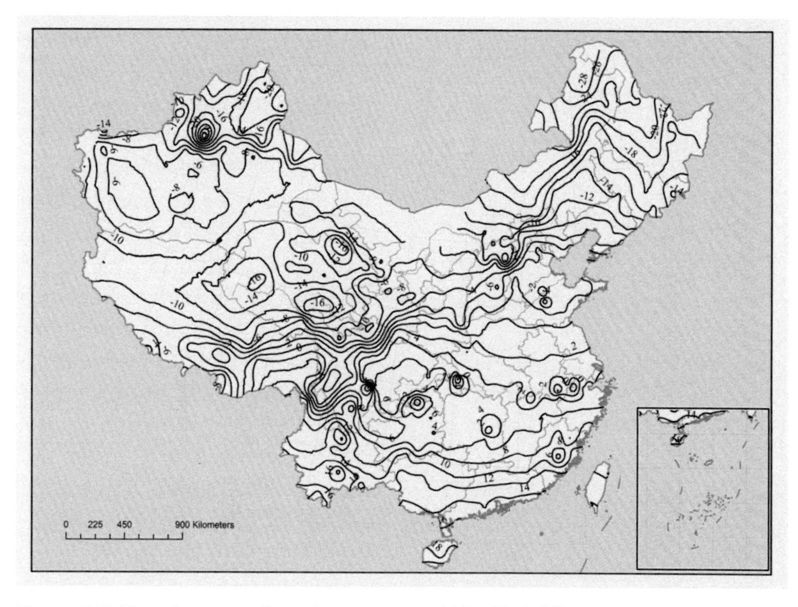

Figure 4.2 Mean January surface air temperature, 1961–2010 (°C).

the difference in temperature between south and north is the smallest of the year. The mean July temperature reaches over 18 °C at Mohe, 22–24 °C on the Dongbei Pingyuan, 26–28 °C on the Huabei Pingyuan (North China Plain) and above 28 °C in almost all regions south to the Huai He and the eastern part of Sichuan. However, in the inshore belts and islands of eastern China it is much cooler than inland, and the 28 °C isotherm nearly parallels the coastline in Zhejiang and Fujian. In the Gobi desert belts of Xinjiang the climate is very dry and the mean July temperature exceeds 24 °C, which is higher than that in wetter regions of eastern China at the same latitude. The Turpan Pendi is the hottest region of China with a monthly mean temperature of up to 33 °C. In contrast, the temperature over large parts of the Qinghai-Xizang Plateau is below 14 °C even in midsummer (Figure 4.3).

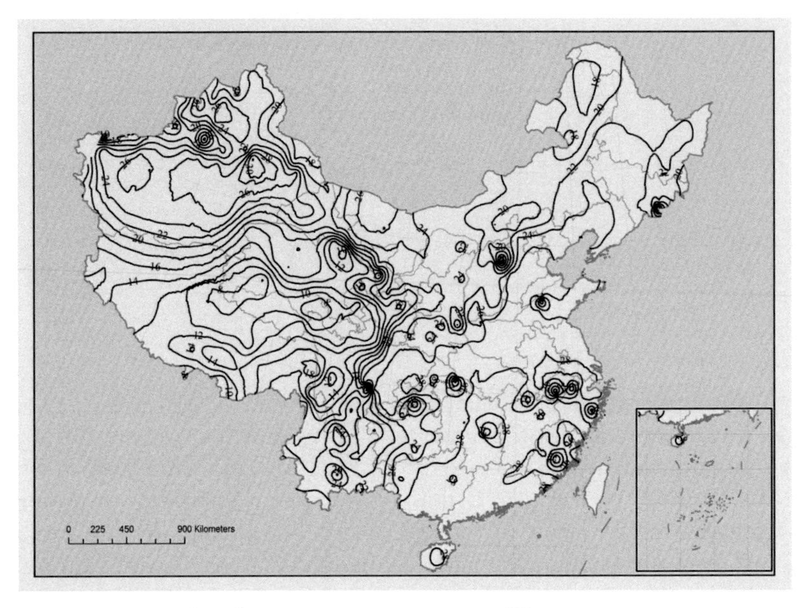

Figure 4.3 Mean July surface air temperature, 1961–2010 (°C).

In autumn (September to November), the winter monsoon gradually intensifies and the temperature difference between north and south increases as temperature decreases rapidly from north to south and drops faster in the north than in the south. In October, the temperature begins to descend to below 0 °C in the northernmost part of Heilongjiang, below 10 °C in northeastern China, Nei Mongol and the Huangtu Gaoyuan, about 12–16 °C on the Huabei Pingyuan and 16–20 °C in the Chang Jiang valley, but remains over 22 °C in southern China. In northwestern China, with the exception of the Tarim Pendi and Turpan Pendi, temperatures are mostly below 10 °C, and below 4 °C for large part of the northern Qinghai-Xizang Plateau.

Accumulated temperatures are presented in Table 4.1 and Figure 4.4.

Table 4.1 Temperature zones of China (Ren & Bao, 1992).

Temperature zone	Proportion of total land area of China (%)	Accumulated temperature during periods of at least 10 °C (°C)
Tropical	4	more than 7 000 (east); more than 6 500 (west)
Subtropical	25	4 500–7 000 (east); 4 500–6 500 (west)
Warm-temperate	18	3 200–4 500
Temperate	34	1 700–3 200
Cold-temperate	3	less than 1 700
Plateau	16	data not available

4.3.3 Annual mean minimum and maximum temperatures

As described above, China is marked by a continental monsoon climate in winter, with January as the coldest month in most areas. Cold air from Siberia frequently invades southward, making China the world's coldest region at the same latitude: the mean January temperature is lower than the world's latitudinal mean by about 14–18 °C in northernmost northeastern China, 10–14 °C in the Huang He valley, 8 °C south to the Chang Jiang and 5 °C in the coastal regions of southern China. The annual mean minimum temperatures in most of northeastern China are below –6 °C, being about –4–4 °C in the plain regions but below –12 °C at higher latitudes north to 50° N, and near –16 °C at Mohe, the northernmost station in China. In northern China, the annual mean minimum temperatures increase from north to south, ranging from below –2 °C on the eastern Nei Mongol Gaoyuan to about –6 °C in the Huang He valley. The annual mean minimum temperatures are from about 4 °C on the Huangtu Gaoyuan, above 10 °C in the region south to the Chang Jiang, above 4 °C in the Sichuan Pendi and above 14 °C in the southeast coastal region. In western China, they are 0–8 °C in the northwest and –10–0 °C in most of the Qinghai-Xizang Plateau, but above 0 °C in southeastern Xizang (Figure 4.5).

In summer, the climate is under the influence of the East Asian summer monsoon which blows from the oceans

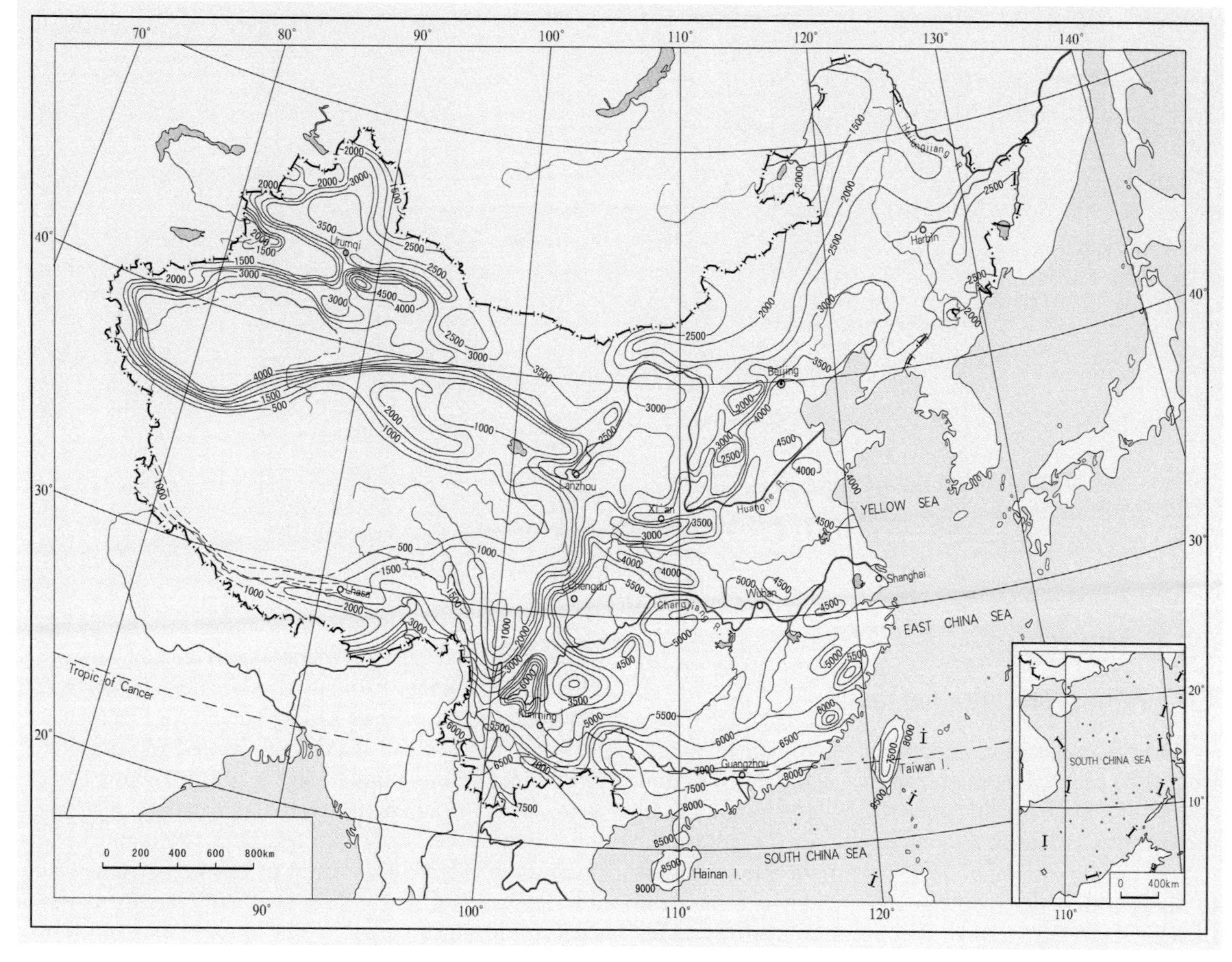

Figure 4.4 Distribution of accumulated temperatures when at least 10 °C in China (°C; Zhao, 1986).

to inland areas. The hottest month in coastal areas is August but inland it is normally July. Exceptionally, on the Yunnan Plateau and the valleys of southern Xizang, the hottest months are found before the rainy season in May and June. Except for the Qinghai-Xizang Plateau, the annual mean maximum temperatures across most of the country exceed 14 °C, being above 20 °C in most of parts of the Huabei Pingyuan, Sichuan, Chongqing Shi and south to the Chang Jiang, and in the desert areas of northwestern China. (Figure 4.6).

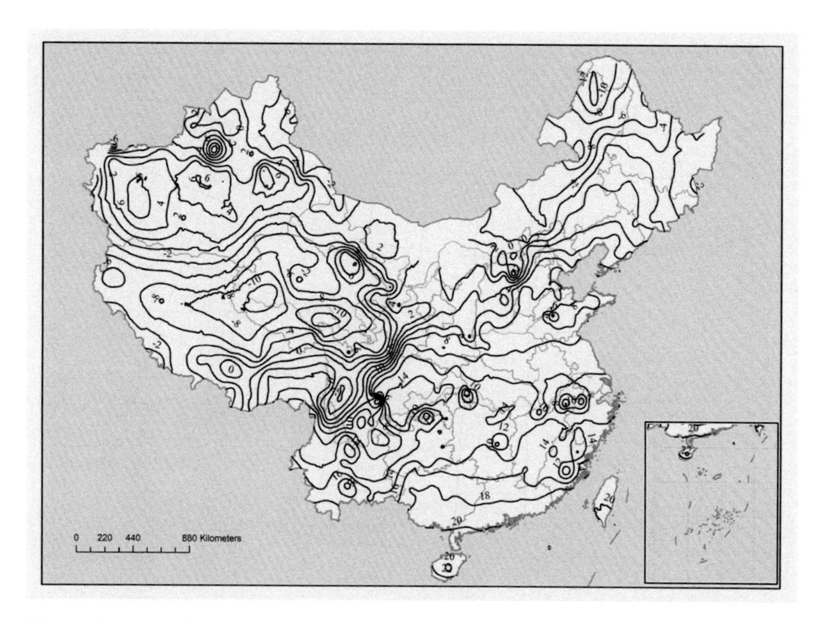

Figure 4.5 Annual mean minimum air temperatures, 1961–2010 (°C).

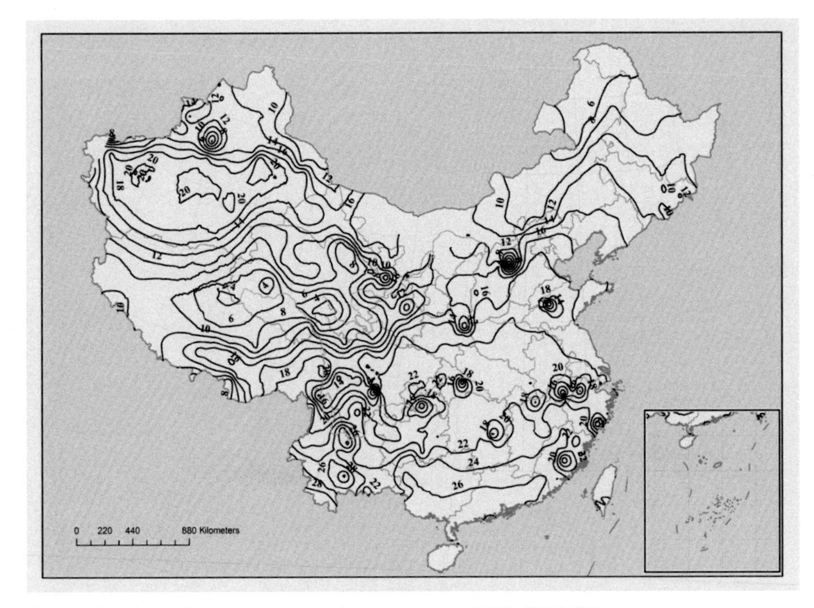

Figure 4.6 Annual mean maximum air temperatures, 1961–2010 (°C).

4.4 Precipitation

4.4.1 Annual precipitation totals

Over China, atmospheric water vapor comes mainly from the oceans. The main features of the precipitation regime of China are characterized by larger amounts in the southeast than in the northwest, in the mountain regions than in the plains, on the windward side compared to the leeward side of upland areas, and over land areas than over nearby islands. Precipitation also increases significantly with the passage of cold fronts, cyclones and typhoons.

The annual total precipitation amount decreases from the south and the east toward inland regions in the northwest. Precipitation is quite abundant in the middle and lower reaches of the Chang Jiang and to its south, where the mean annual precipitation is more than 1 000 mm, with over 1 800 mm on hills in southeastern China, southern Guangdong and Guangxi, mountainous areas of Taiwan and the hills of Hainan. Rainfall totals are above 2 000 mm at the southeastern edge of Xizang but decrease dramatically towards the northwest. Annual rainfall totals are 400–1 000 mm in the middle reaches of the Yarlung Zangbo Jiang, over 500 mm on the eastern Qinghai-Xizang Plateau, and least in the north with less than 100 mm. The total amount is 300–900 mm in most of northeastern China, the Huabei Pingyuan and the southern Huangtu Gaoyuan, below 400 mm in eastern Nei Mongol and 100 mm in northwestern Nei Mongol. An obvious example of rainfall distribution difference associated with terrain is found in Xinjiang, where annual precipitation is about 200 mm across most of the northern Xinjiang, while in southern Xinjiang, precipitation is generally less than 100 mm and is at a minimum in the Tarim Pendi, Qaidam Pendi and Turpan Pendi, each of which has less than 25 mm (Figure 4.7).

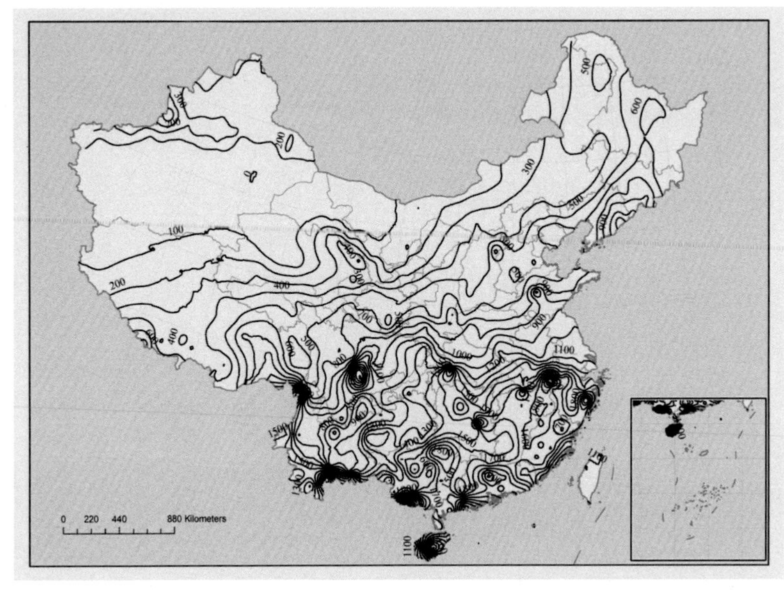

Figure 4.7 Annual precipitation totals, 1961–2010 (mm).

4.4.2 Seasonal precipitation

Because of its vast area and complex terrain, significant differences exist in the timing of the rainy season across China. The country is under the influence of the winter monsoon from October to March, the dry season. The summer monsoon prevails from June to August with plentiful precipitation. April, May and September are transitional months between the dry and wet seasons. The variability in precipitation is remarkable, with summer precipitation accounting for only ~40% of the annual total in regions south of the Qin Ling and Huai He, yet elsewhere making up over 60% of annual precipitation (Figure 4.8). However, only about 10% of annual precipitation occurs in the winter months. Mid-June to mid-July is the period of the well-known plum rain ("meiyu") season, which lasts more than 20 days, in the middle and lower reaches of the Chang Jiang. The meiyu impacts greatly on droughts and floods in eastern China with the hot and dry season followed under the control of the Northwest Pacific Subtropical High.

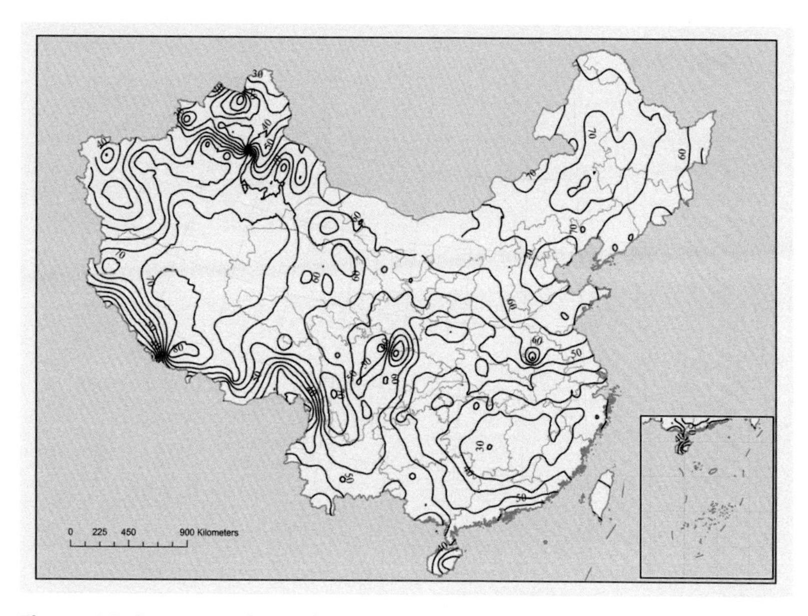

Figure 4.8 Proportion of annual precipitation accounted for by summer precipitation, 1961–2010 (%).

4.4.3 Rain days

The distribution of rain days in China is closely associated with the activities of the winter and summer monsoons and the displacement of the polar front. Hence, seasonal variation in the number of rain days is consistent with the precipitation regime. The season with the most rain days throughout China is the summer, with the fewest in winter. In the area south of the Chang Jiang and in the southeastern foothills where there is more spring rainfall, however, the highest frequency of rain days is observed in May and the lowest in September and October. In the desert regions of Gansu and Xinjiang, the lowest frequency of rain days is also found in autumn. In the Wei He valley, the Sichuan Pendi and Hainan, where there are more autumn rains, however, the highest number of rain days is found in September and October.

The region with the highest frequency of rain days overall, with more than 160 rain days per year, is found at latitudes 25°–30° N in the east of the Qinghai-Xizang Plateau. Among the sub-regions within this zone, Sichuan and Guizhou have maximum rain days, with more than 200 rain days per year. More than 200 rain days also occur in parts of the hilly areas of southeastern China, eastern Taiwan, southwestern Hainan and western Yunnan. The number of annual rain days in Xizang decreases rapidly from southeast to northwest, with more than 180 days in the southern valleys, declining to only about 40 days in the northwest. In northern China, to the Huai He and the Qin Ling, annual rain days are generally fewer than 100, with fewer than 30 days in western Nei Mongol and southern Xinjiang, and fewer than 20 in the Gobi and other desert areas. There are more rain days in far northeastern China, the Tian Shan in Xinjiang and Qilian Shan in Gansu, where there are over 120 rain days per year (Figure 4.9). Figure 4.10, in contrast, shows the annual mean relative humidity during 1961-2010 in China.

4.5 Cloud cover and sunshine

4.5.1 Cloud cover

Sunshine and cloudiness indicate not only the sky conditions of a given region, but also its radiation balance. The annual mean cloud cover over China decreases from south to north. Greatest cloud cover is observed in the Sichuan Pendi and Guizhou, where it exceeds eight tenths (8/10). About 7/10–8/10 cloud cover is found in most of the area to the south of the Qin Ling and the Chang Jiang

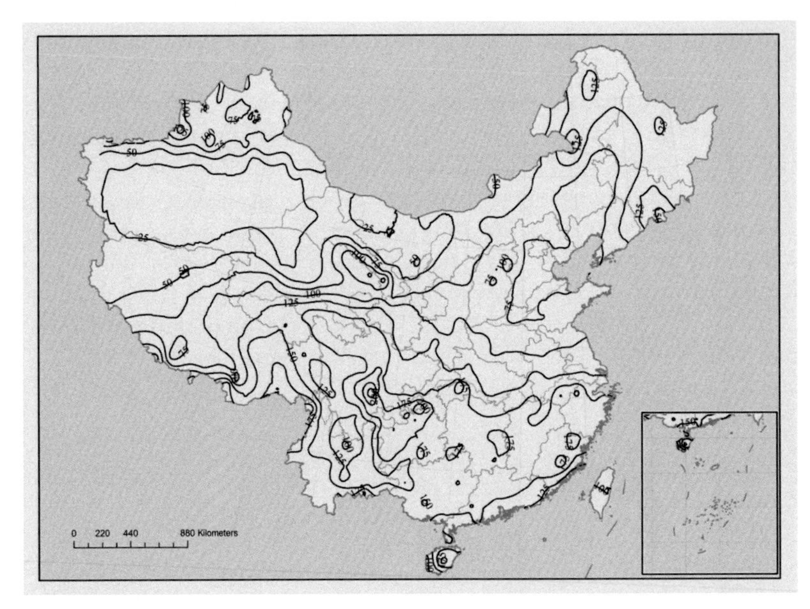

Figure 4.9 Mean number of precipitation days per year, 1961–2010.

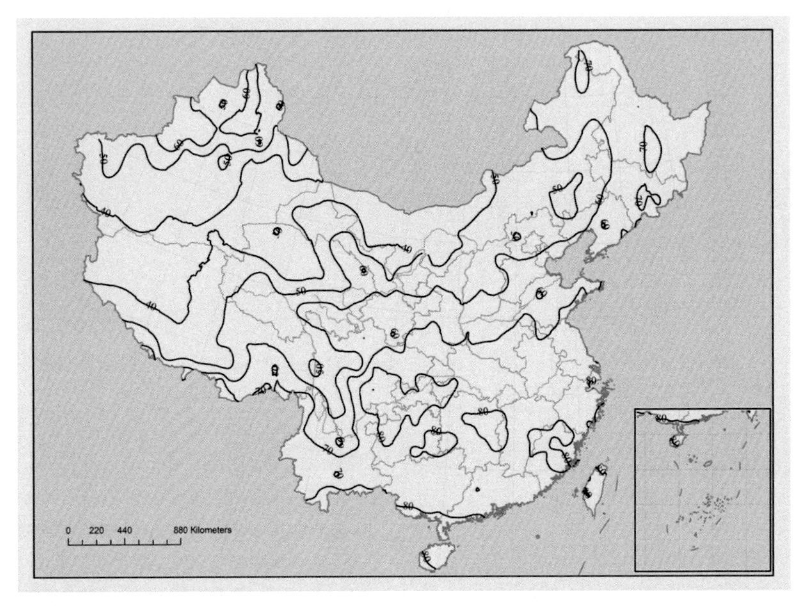

Figure 4.10 Annual mean relative humidity in China, 1961–2010 (%).

as well as in eastern Taiwan. In western Taiwan and coastal parts of Guangdong, 6/10–7/10 cloud cover is observed. About 6/10 cloud cover is found over the Huai He valley, 5/10–6/10 over the southern part of the Huabei Pingyuan, the southern Huangtu Gaoyuan, and hilly regions of northeastern China, 4/10–5/10 in the northern Huabei Pingyuan, the Dingbei Pingyuan and eastern Nei Mongol, and less than 4/10 in western Nei Mongol. In Xinjiang, over 5/10 cloud cover is found in the west while less than 4/10 is observed in the region from the central Tarim Pendi to the east of the region. On the Qinghai-Xizang Plateau, the southeast experiences more than 6/10 cloud cover, with 5/10–6/10 in the north and 3/10–5/10 in the southwest (Figure 4.11).

Due to the onset and retreat of the monsoon, there is a clear alternation between dry and wet seasons, and cloud cover thus displays distinctive seasonal variations. The highest cloud cover over most of China occurs in summer – June in the hilly regions of southern China and August in western Xizang. Other regions with cloudy summers experience greatest cloud cover in July. In the south to the Chang Jiang and in northwestern China, the highest cloud cover is in autumn. The season of least cloud cover also varies across different regions. In most of the south to the Chang Jiang and in Xinjiang, cloud cover is lowest in autumn, while in Hunan and the Sichuan Pendi the lowest cover occurs in August. In most parts of the other regions, the lowest cloud cover is found in winter.

4.5.2 Sunshine duration and percentage

Sunshine duration is an absolute value indicating the hours of sunshine received by a given region, while percentage sunshine reflects the degree of reduction in hours of sun caused by the weather conditions in

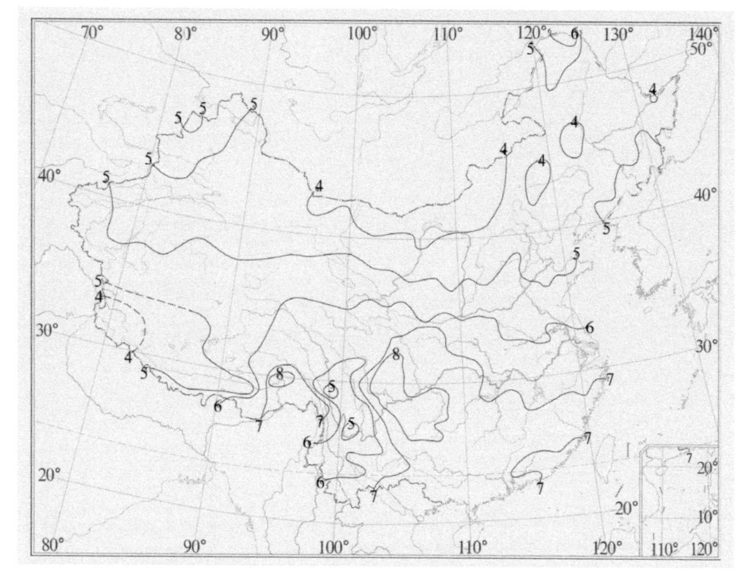

Figure 4.11 Annual mean total cloud cover, 1971–2000 (1-10/10).

that region. The spatial distribution of mean average annual sunshine duration is in direct contrast to cloud cover, i.e. there is greater cloud cover and less sunshine in southeastern China, and less cloud cover and more sunshine for the northwestern and northern parts. Sunshine duration is below 2 000 hours for the region south of the Qin Ling and Huai He, the eastern slopes of the Qinghai-Xizang and Yunnan-Guizhou Plateaus, and the Yarlung Zangbo Jiang valley. In particular, there are less than 1 200 hours of sunshine per year in the Sichuan Pendi and northern Guizhou, the lowest levels in China. In contrast, mean average annual hours of sunshine exceed 2 800 for northwestern inland regions such as Baicheng Shi, Datong, Wuwei, Xining and Lhasa Shi, and top 3 200 hours for the Gobi desert in eastern Xinjiang, the western Qaidam Pendi, Alxa Gaoyuan and southwestern Qinghai-Xizang Plateau, which has the highest annual sunshine duration in China. At Shiquanhe on the southwestern Qinghai-Xizang Plateau, sunshine duration is close to 3 600 hours per year (Figure 4.12).

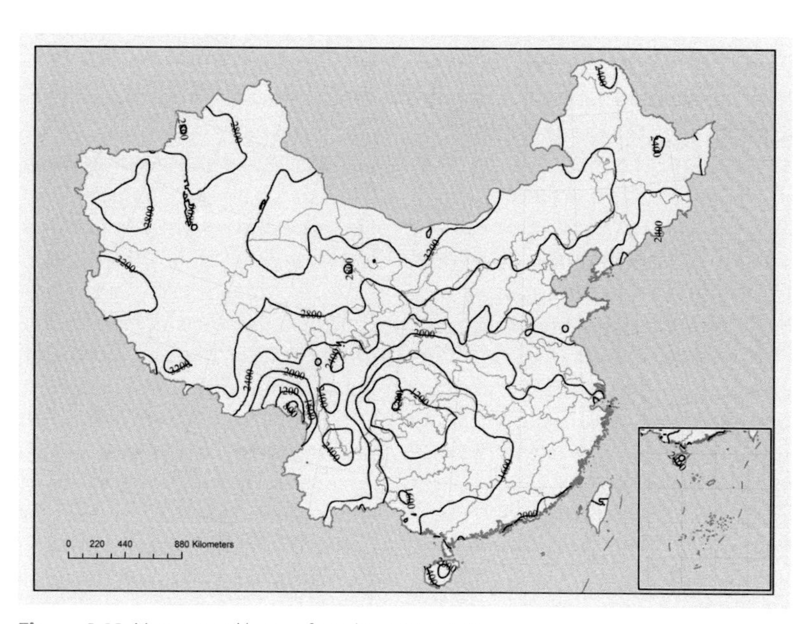

Figure 4.12 Mean annual hours of sunshine, 1961–2010.

The mean annual percentage sunshine is below 50% for regions south of 35° N, the eastern part of the Qinghai-Xizang Plateau and the eastern slopes of the Yunnan-Guizhou Plateau. The lowest values are recorded in the Sichuan Pendi, most of Guizhou, and northwestern Hunan, where the percentage is below 30%. The percentage sunshine is close to 70% for regions north of Harbin Shi, Jinzhou Shi and Yulin. Generally speaking, the percentage sunshine is above 60% for most parts of western China except for the eastern Qinghai-Xizang Plateau. The percentage sunshine exceeds 70% in the Gobi deserts of eastern Xinjiang, the western Qaidam Pendi, Alxa Gaoyuan and southwestern Qinghai-Xizang Plateau, which records the highest percentage of sunshine in China (Figure 4.13).

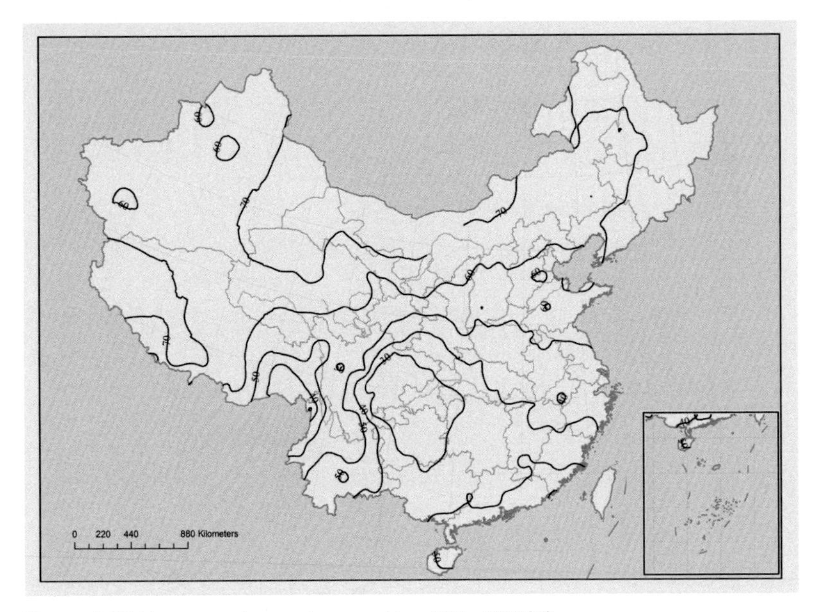

Figure 4.13 Mean annual percentage sunshine, 1961–2010 (%).

4.6 Atmospheric pressure systems and winds

4.6.1 Atmospheric pressure systems

The atmospheric pressure distributions over China exhibit significant seasonal variations. Taking January as an example for winter, the Siberian High dominates the continent of Asia. The pressure at the centre of the Siberian High exceeds 1 040 hPa (hector Pascals, which are equivalent to millibars), 20 hPa higher than that of the North American High. In winter, the cold Siberian High shows the largest extent and the longest duration of any in the Northern Hemisphere. Meanwhile, the Northwest Pacific Subtropical High weakens and retreats southeastward. The Aleutian Low dominates most of the northern Pacific. Together the Siberian High and Aleutian Low are the two main atmospheric systems controlling the climate of eastern Asia in winter. The Siberian High carries strong cold air masses when moving southeastward. It is, however, a comparatively shallow pressure system which always disappears at the 500 hPa pressure level. At 500 hPa, westerly winds prevail over China and the deep East Asia Trough lies over the offshore waters of the Dong Hai (East China Sea) and the Sea of Japan.

In July, a hot low pressure system, the Indian Low, with a central pressure value below 995 hPa, controls the area from the northern India Peninsula to southwestern China. This is the strongest and most extensive hot low pressure influencing almost the entire Asian continent. Meanwhile, the Northwest Pacific Subtropical High extends northwestward and matures, having a central pressure of over 1 025 hPa at sea level. The Indian Low and the Northwest Pacific Subtropical High together dominate the summer climate of China. Spring and autumn are transitional seasons in terms of atmospheric pressure patterns. In spring, at mid to high latitudes, the Siberian High and Aleutian Low weaken while the Indian Low starts to develop. In autumn, in the subtropics, the Indian Low and the Northwest Pacific Subtropical High start to weaken while at mid to high latitudes the Siberian High and the Aleutian Low become active.

4.6.2 Winds

The direction and speed of winds over China show an obvious seasonality as the result of the opposing winter and summer atmospheric pressure patterns described above. The topography also often has an effect on monsoon flow, and localized winds occur in some places. In winter, most of China is located to the east and southeast of a core region of high pressure resulting in northerly prevailing winds except in the northwest of the country and Nei Mongol, where westerly winds prevail. In the northern part of the Huabei Pingyuan the prevailing winds are southwesterly, as a result of topographic troughs. Northerly and northwesterly winds are common in the southern part of the Huabei Pingyuan and the Huangtu Gaoyuan. In the south to the Qin Ling and the lower reaches of the Huang He, northerly and northeasterly winds prevail. Northerly and northwesterly winds dominate the coast from the Liaodong Bandao to the Taiwan Haixia (Taiwan Strait). Southeasterly winds are seen in eastern Yunnan and western Guangxi but southwesterly winds dominate in western Yunnan. Southerly winds prevail in the southeastern part of the Qinghai-Xizang Plateau with northeasterly winds in the northeast and westerly winds in other regions. Northwesterly winds often occur in northwestern Xinjiang, while easterly winds are common in the east. Southeastern Xinjiang is dominated by northeasterly winds, with westerly and southwesterly winds in the western half of the region.

In summer, when the surface air pressure gradient

is from ocean to land, southerly and southeasterly winds prevail in China. Southeasterly winds are often observed from the Liaodong Bandao to the Taiwan Haixia. Southeasterly and southwesterly winds dominate the regions from the Dongbei Pingyuan through the Huabei Pingyuan to southern China and the Yunnan-Guizhou Plateau. A warm and wet air flow from the southwest is present at upper levels over the Sichuan Pendi but northerly winds affect the climate near the ground. Southeasterly winds prevail in southern Xizang but easterly and northeasterly winds are dominant on the northern Qinghai-Xizang Plateau. Northwesterly and westerly winds are observed in most of Xinjiang with northeasterly winds in eastern Xinjiang.

In spring, although northerly winds dominate most of China, southerly and southeasterly winds begin to develop in the south up to the Shandong Bandao including the lower reaches of the Chang Jiang and the coastal regions of Guangdong and Guangxi. In autumn, due to the rapid onset and stabilization of the winter monsoon, steady northerly winds exert their influence over most of China. The general features of wind direction in autumn are similar to those in winter.

Generally speaking, mean annual wind speed is greater in northern than in southern China. It is also higher in coastal areas than inland, and on plateaus than in basins. The mean annual wind speed is generally over 3 m per second (m/s) across most of the Qinghai-Xizang Plateau, north of the Yin Shan, and in coastal regions. It is over 4 m/s in central Nei Mongol, northwestern Xinjiang, and parts of coastal regions. For most areas south of 40° N, mean annual wind speed is below 2 m/s. Overall, mean annual wind speed in China during the period 1971–2004 was 2.4 m/s, with seasonal mean speeds of 2.8 m/s in spring, 2.3 m/s in summer, 2.2 m/s in autumn, and 2.3 m/s in winter.

4.7 Climate classification

4.7.1 Geographical regionalization and climate classification

Geographical regionalization has been widely used by geographers for many years but is still considered a somewhat radical methodology. As is well known, there exist similarities and dissimilarities exist among geographical zones, and zonality is a universal theory governing geographical distributions. Therefore, geographical regionalization exists objectively, and the role of the geographer is merely to develop an image of these geographical realities. At a conceptual level, geographical regionalization can be used to draft regional divisions for practical purposes, and may be considered a process of dividing regions on the basis of certain indices and techniques into interrelated sub-areas (sub-regions), or as a method of revealing the laws hidden in various geographical phenomena.

A map-based format is a good way to describe the distribution of geographical features and phenomena. On the basis of analysis of the various regionalization methods and plans, a scheme for geographical regionalization, composed of five elements, has been proposed (Zheng *et al.*, 2010). Any process of regionalization must start from an understanding of the development of the method: the objectives, aims and actions of that geographical regionalization. The principles and criteria upon which a geographical regionalization are based should be carefully designed and selected, fully to meet the needs of the regionalization. Regionalization models include both those for dividing an area and those for determining the boundaries of sub-areas. A regionalization information system stores the attributes of all sub-areas and provides functions for assessing and simulating a draft geographical regionalization.

As one important factor in geographical features, the classification and division of climate are naturally involved, particularly so in China where the climate is characterized by complexity and diversity due to the country's vast territory and extremely complex landforms. In detail, three major climate-governing factors must be taken into consideration in classifying the climate of China and dividing it over space: geographical latitude, elevation above sea level, and distance from the Pacific Ocean. Of particular importance for climate regionalization in China is the country's location at the southeastern corner of Eurasia, adjacent to the Pacific Ocean, which means that the nature of the surrounding regions varies from open oceans to compact landmasses (Domrös & Peng, 1988).

Chinese climatologists have already conducted a great deal of research (Wang & Li, 2007). ZHU (1931) published the first map of a climate classification of China, in which the country was divided into eight climatic regions based on temperature, precipitation, natural landscape features and weather systems. This pioneering work established, under conditions of insufficient data, the basis for a climate classification of China, and was later modified by TU (1936) with more complete data. Following this, HUANG (1959) used the criterion of accumulative air temperature with a daily mean value of at least 10 °C to denote a heat index, and that of aridity to denote a moisture index. Based on these criteria, as well as sunshine, he divided China into six monsoon climatic zones and one plateau climatic region. The monsoon zones he named were the equatorial, tropical, subtropical, warm-temperate, temperate and cold-temperate monsoon zones. Aridity was separated into four grades. Within this, China was separated into eight first-class regions, 32 second-class regions and 68 third-class regions. In 1966, China Climate Classification charts were published by the China Meteorological Administration (CMA) based on data from 600 stations taken from 1951–1960. This scheme divided China into 22 regions and 45 sub-regions. In this method, the first level division criteria were cumulative daily mean temperature of at least 10 °C, coldest monthly mean temperature, and yearly extreme minimum temperature. Yearly aridity was employed as a second division criterion and seasonal aridity as the third division criterion. Using the same climate classification methods, a new version was published by CMA based on data from 1951–1970, again for about 600 stations. This study included three classes of division: climatic zone, region, and sub-region, which established a solid foundation for the recently-published new scheme for regionalization in China (Zheng *et al.*, 2010).

4.7.2 New scheme for climate regionalization in China

A new scheme for climate regionalization in China has now been established based on daily observations from 609 meteorological stations during the period 1971–2000 (Zheng *et al.*, 2010). Current basic theories, classification methodologies and systems of criteria were used to conduct the regionalization. In addition, five principles were taken into consideration: zonal and azonal integration, genetic unity and relative consistency of regional climate integration, comprehensiveness and integration of leading factors, bottom-up and top-down integration, and spatial continuity and small path omission. The new scheme consists of 12 temperature zones, 24 moisture regions and 56 climatic sub-regions (Figure 4.14). The Qinghai-Xizang Plateau alone includes three temperature zones, nine moisture regions, and 12 climatic sub-regions, with the rest of China covering the remaining nine temperature zones, 15 moisture regions, and 44 climatic sub-regions (see Table 4.2).

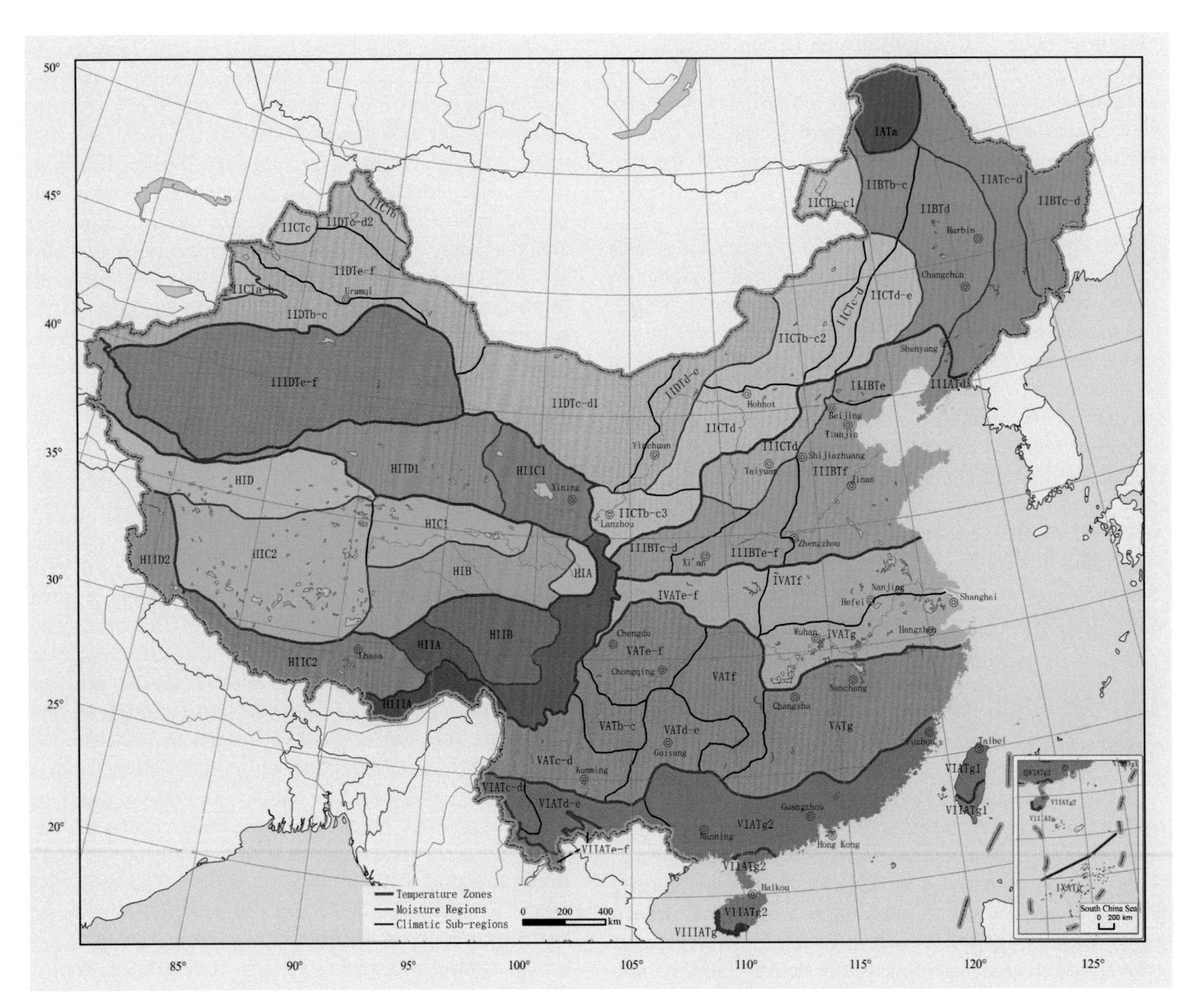

Figure 4.14 New scheme of climate regionalization in China.

Table 4.2 Most recent scheme for climate regionalization in China (Zheng *et al.*, 2010). Note that, in regions where relief varies greatly, two meteorological stations are given, to represent data from both high mountains and river basins or valleys. n/a, data not available.

Temperature zone	Moisture region	Climatic sub-region code	Climatic sub-region	Representative meteorological stations with altitude and main indices							
				Station name	Station code	Station altitude (m)	Days per year with mperature of at least 10°C	Relative aridity	Mean January temperature (°C)	Mean July temperature (°C)	Annual precipitation (mm)
I. cold emperate zone	A. humid region	IATa	Northern Da Hinggan Ling cold-temperate humid region	Tulihe	50434	733	88	0.9	-28.7	16.6	465
II. mid-temperate zone	A. humid region	IIATc-d	Xiaoshan Da Hinggan Ling, Changbai Shan mid-temperate humid region	Yichun	50774	241	124	0.8	-22.5	21.0	627
	B. sub-humid region	IIBTc-d	Sanjiang Pingyuan and its southern hills mid-temperate sub-humid region	Baoqing	50888	83	143	1.3	-17.5	22.3	511
		IIBTd	Song-Liao Plain mid-temperate sub-humid region	Changchun	54161	237	154	1.3	-15.1	23.1	570
		IIBTb-c	Central Da Hinggan Ling mid-temperate sub-humid region	Zhalantun	50639	307	136	1.3	-17.1	21.4	506
	C. semi-arid region	IICTd-e	Western Liao He Pingyuan mid-temperate semi-arid region	Chifeng	54218	571	164	2.2	-10.7	23.6	371
		IICTc-d	Southern Da Hinggan Ling mid-temperate semi-arid region	Linxi	54115	799	142	1.9	-13.6	21.3	385
		IICTb-c1	Hulun Buir Gaoyuan mid-temperate semi-arid region	Hailar	50527	610	115	1.6	-25.1	20.0	367
		IICTb-c2	Eastern Nei Mongol Gaoyuan mid-temperate semi-arid region	Xilinhot	54102	990	129	2.7	-18.8	21.2	287
		IICTd	Ordos Gaoyuan and eastern Hetao Pingyuan mid-temperate semi-arid region	Hohhot	53463	1063	156	1.8	-11.6	22.6	398
		IICTb-c3	Western Huangtu Gaoyuan mid-temperate semi-arid region	Haiyuan	53806	1854	143	2.4	-6.5	19.8	368
		IICTb	Altay Shan mid-temperate semi-arid region	Qinghe	51186	1218	120	3.5	-22.4	18.9	171
		IICTc	Tacheng Pendi mid-temperate semi-arid region	Tacheng	51133	535	158	2.7	-10.4	22.9	282
		IICTa-b	Yili He valley mid-temperate semi-arid region	Zhaosu	51437	1851	97	1.2	-11.3	15.1	492
				Yining	51431	663	180	2.7	-8.8	23.1	269
	D. arid region	IIDTd-e	Western Hetao Pingyuan and western Nei Mongol Gaoyuan mid-temperate arid region	Linhe	53513	1039	164	5.7	-9.9	24.1	146
		IIDTc-d1	Alxa Plateau and Hexi corridor mid-temperate arid region	Zhangye	52652	1483	159	6.1	-9.2	21.5	130
		IIDTe-f	Junggar Pendi mid-temperate arid region	Caijiahu	51365	441	168	5.5	-18.6	25.5	141
		IIDTc-d2	Ergis Valley mid-temperate arid region	Fuhai	51068	501	152	6.2	-18.9	23.1	122
		IIDTb-c	Tian Shan mid-temperate arid region	Dabancheng	51477	1104	147	16.5	-9.9	21.2	70
III. warm-temperate zone	A. humid region	IIIATd	Eastern Liaoning hills warm-temperate humid region	Xiuyan	54486	79	170	0.7	-9.9	23.2	818
	B. sub-humid region	IIIBTe	Yan Shan warm-temperate sub-humid region	Jinzhou	54337	66	184	1.5	-7.9	24.3	568
		IIIBTf	Huabei Pingyuan and central-eastern Shandong hills warm-temperate sub-humid region	Jinan	54823	52	219	1.5	-0.4	27.5	671
		IIIBTe-f	Fen He-Wei He plain and hills warm-temperate sub-humid region	Xi'an	57036	398	211	1.3	-0.1	26.6	553
		IIIBTc-d	Southern Huangtu Gaoyuan warm-temperate sub-humid region	Luochuan	53942	1158	170	1.2	-4.4	21.9	591
	C. semi-arid region	IIICTd	Eastern Huangtu Gaoyuan and Taihang Shan warm-temperate semi-arid region	Taiyuan	53772	778	184	1.8	-5.5	23.4	431
	D. arid region	IIIDTe-f	Tarim and eastern Xinjiang basins warm-temperate arid region	Dunhuang	52418	1139	179	21.6	-8.3	24.6	42

Continued

Temperature zone	Moisture region	Climatic sub-region code	Climatic sub-region	Representative meteorological stations with altitude and main indices							
				Station name	Station code	Station altitude (m)	Days per year with mperature of at least 10°C	Relative aridity	Mean January temperature (°C)	Mean July temperature (°C)	Annual precipitation (mm)
IV. northern subtropical zone	A. humid region	IVATf	Dabie Shan and northern Jiangsu plain northern subtropical humid region	Xinyang	57297	115	227	0.7	2.2	27.4	1106
		IVATg	Mid-lower reaches of the Chang Jiang plain and northern Zhejiang northern subtropical humid region	Wuhan	57494	23	238	0.6	3.7	28.7	1270
		IVATe-f	Qin Ling and Bashan northern subtropical humid region	Ankang	57245	291	235	0.8	3.5	26.9	814
V. mid-subtropical zone	A. humid region	VATg	Southern Chang Jiang hills mid-subtropical humid region	Nanchang	58606	47	246	0.5	5.3	29.2	1624
		VATf	Western Hunan and Jiangxi hills mid-subtropical humid region	Yuanling	57655	152	242	0.5	5.0	27.6	1404
		VATd-e	Guizhou plateau mountains mid-subtropical humid region	Guiyang	57816	1074	232	0.6	5.0	23.9	1129
				Zhijiang	57745	272	239	0.6	4.9	27.1	1231
		VATe-f	Sichuan Pendi mid-subtropical humid region	Chengdu	56294	506	248	0.7	5.6	25.2	871
		VATb-c	Southwestern Sichuan and northern Yunnan mountains mid-subtropical humid region	Yuexi	56475	1659	214	0.6	4.0	21.1	1130
				Xuyong	57608	378	272	0.6	7.9	27.1	1147
		VATc-d	Western Yunnan mountains and central Yunnan plateau mid-subtropical humid region	Huili	56671	1787	261	0.7	7.1	20.8	1149
VI. southern subtropical zone	A. humid region	VIATg1	Northern Taiwan hills and plains southern subtropical humid region	Taipei	58968	7	n/a	n/a	15.8	29.3	2325
		VIATg2	Fujian, Guangdong and Guangxi hills and plains southern subtropical humid region	Guangzhou	59287	7	325	0.5	13.6	28.6	1736
		VIATd-e	Central and southern Yunnan mountains southern subtropical humid region	Jingdong	56856	116	341	0.7	11.3	23.4	1133
		VIATc-d	Southwestern Yunnan mountains southern subtropical humid region	Lincang	56951	1502	342	0.7	11.2	21.4	1163
VII. marginal tropical zone	A. humid region	VIIATg1	Southern Taiwan mountains and plains marginal tropical humid region	Hengchun	59559	24	n/a	n/a	20.6	28.3	2018
		VIIATg2	Hainan and Leizhou hills marginal tropical humid region	Haikou	59758	14	363	0.6	17.7	28.6	1653
		VIIATe-f	Southern Yunnan mountains marginal tropical humid region	Mengla	56969	632	364	0.6	16.0	24.8	1523
VIII. mid-tropical zone	A. humid region	VIIIATg	Southern Hainan lowlands, Dongsha Dao, Zhongsha Dao and Xisha Qundao mid-tropical humid region	Sanya	59948	6	365	0.9	21.6	28.5	1392
IX. equatorial tropical zone	A. humid region	IXATg	Nansha Qundao equatorial tropical humid region	Shanhu Dao	59985	4	365	-	23.5	29.0	1448
HI. plateau sub-cold zone	A. humid region	HIA	Zoigê plateau sub-cold humid region	Zoigê	56079	3440	22	0.9	-10.2	10.8	649
	B. sub-humid region	HIB	Guoluo and Nagqu high mountain and valley plateau sub-cold sub-humid region	Dari	56046	3968	10	1.1	-12.6	9.2	545
	C. semi-arid region	HIC1	Southern Qinghai plateau sub-cold semi-arid region	Wudaoliang	52908	4612	1	2.2	-16.7	5.5	275
		HIC2	northern Xizang lake basin sub-cold semi-arid region	Xainza	55472	4672	11	2.8	-10.1	9.6	299
	D. arid region	HID	Kunlun Shan plateau sub-cold arid region	n/a							

Continued

Temperature zone	Moisture region	Climatic sub-region code	Climatic sub-region	Representative meteorological stations with altitude and main indices							
				Station name	Station code	Station altitude (m)	Days per year with mperature of at least 10 °C	Relative aridity	Mean January temperature (°C)	Mean July temperature (°C)	Annual precipitation (mm)
HII. plateau temperate zone	A. humid region	HIIA	Eastern and southern Hengduan Shan plateau temperate humid region	Kangding	56374	2616	109	0.7	-2.2	15.5	832
	B. sub-humid region	HIIB	Central and northern Hengduan Shan plateau temperate sub-humid region	Qamdo	56137	3306	131	1.5	-2.3	16.0	475
	C. semi-arid region	HIIC1	Qilian Shan and eastern Qinghai high mountain and basin plateau temperate semi-arid region	Xining	52866	2261	140	1.8	-7.4	17.2	374
		HIIC2	Southern Xizang high mountain and valley plateau temperate semi-arid region	Lhasa	55591	3658	148	2.1	-1.6	15.7	426
	D. arid region	HIID1	Qaidam Pendi and northern slopes of Kunlun Shan plateau temperate arid region	Da Qaidam	52713	3173	90	9.4	-13.4	15.5	83
		HIID2	Ngali Mountains plateau temperate arid region	Shiquanhe	55228	4278	80	11.9	-12.4	13.8	75
HIII. plateau subtropical zone	A. humid region	HIIIA	Southern slope of eastern Himalaya Shan plateau subtropical humid region	Chayu	56434	2328	193	0.9	4.3	18.8	808

References

DING, Y. H. (2013). *Climate of China*. Beijing, China: Science Press.

DOMRÖS, M. & PENG, G. B. (1988). *The Climate of China*. Berlin and Heidelberg, Germany: Springer-Verlag.

HUANG, B. W. (1959). Draft of the complex physical geographical division of China. *Chinese Science Bulletin* 18: 594-602.

REN, M. E., BAO, H. S. & ET AL. (1992). *The Development and Improvement of Chinese Natural Regions*. Beijing, China: Scientific Press.

TU, C. W. (1936). Climatic provinces of China. *Acta Geographica Sinica* 3(3): 495-528.

WANG, S. W. & LI, W. J. (eds.) (2007). *Climate of China*. Beijing, China: China Meteorological Press.

ZHENG, J. Y., YIN, Y. H. & LI, B. Y. (2010). A new scheme for climate regionalization in China. *Acta Geographica Sinica* 65(1): 3-12.

ZHU, K. Z. (1931). *On the Regionality of Climate in China. Collected Works of Zhu Kezhen*. Beijing, China: Science Press.

ZHAO, S. Q. (1986). *Physical Geography of China*. Beijing, China and New York: Science Press and John Wiley & Sons.

Chapter 5

Soils

WANG Xiu-Hong
WANG Zhao-Feng and ZHENG Du

5.1 Introduction

Soils provide the basic nutrient substrates for plant growth and anchor for their support. Soil type and moisture levels are key determinants of the species of plants that will be found–or can be cultivated–thereon. Thus, a consideration of soils is a key prerequisite to understanding fully the plants of China.

5.2 Soil formation

The soil-forming process is a function of several variables including the parent material, climate, landform, vegetation and time. In such an ancient and densely-populated country as China, the impact and feedback of past and present human activities are also important. Each type of soil-forming process usually results in more than one soil type and each soil type can result from more than one soil-forming process. For example, within the semi-arid temperate steppe environment, the interplay of calcification and humification produces chernozems; whereas within a wet tropical forest environment, the joint actions of leaching, laterization and humification result in the production of ferralosols (Zhao, 1986).

5.2.1 The major soil-forming processes

Major soil-forming processes include salinization, calcification, argillation, leucinization, laterization, podsolization, humification and gleization (Zhao, 1986; Gong *et al.*, 2007).

Salinization.—Salinization is associated chiefly with steppe and desert environments and poorly-drained sites such as littoral zones, yet it has been estimated that as much as 20% of total Chinese farmland, ca. 20 million ha, is affected to some extent by salinization. There are two types of salinization: saline accumulation and alkaline accumulation. The former includes both modern and residual saline accumulation.

Modern saline accumulation may be brought about by surface water, groundwater or seawater: salinization due to both surface water and groundwater occurs widely on the Chinese northern plain and in areas around lakes and depressions. Salts, such as sulphates, chlorides and sodium salts, accumulate mainly in topsoil, giving it a salt content of around 1%, sometimes reaching 2–3%. Salinization by groundwater (containing similar salts to those listed above) occurs in the low parts or edges of alluvial fans and proluvial fans with fine soils, stagnant groundwater runoff and highly mineralized groundwater. Salinization caused by seawater occurs by mineralization of groundwater, decreasing in intensity from the coast to inland, north to south in China, with chlorides being the main type of salts present.

Residual saline accumulation always occurs in the upper parts of mountainside pediments in desert regions. Saline accumulation results in the formation of a salic horizon, with a thickness of at least 15 cm and soil salt content of more than 20 g per kg (Gong *et al.*, 2007). This typically produces a solonchak soil.

Alkaline accumulation is similar to saline accumulation but with the soils having a higher proportion of sodium. An alkali horizon forms, with a columnar or prismatic structure, an exchangeable sodium percentage of at least 30%, pH greater than nine, and salt content in the topsoil of less than 5 g per kg (Gong *et al.*, 2007). This typically produces a solonetz soil type.

Calcification.—Calcification occurs mainly in semi-arid and arid regions where evaporation on average exceeds precipitation, and rainfall is not sufficient to leach bases and colloids from the soil. Calcium carbonate is precipitated in the "B horizon" (subsoil) in the form of nodules, slabs, and even dense stony layers of caliche, a form of sedimentary rock. The depth of this calcic horizon decreases with level of aridity: in arid regions it may be as close as 20–30 cm below the surface, in semi-arid regions 40–50 cm, and in humid regions 60–80 cm or deeper. Typically the B horizon has a thickness of 15 cm and contains 150–500 g of calcium carbonate per kg. A horizon thicker than 15 cm and with at least 500 g of calcium carbonate per kg is known as a hypercalcic horizon (Gong *et al.*, 2007). Typical end products of this soil-forming process are kastanozems and chernozems.

Argillation.—This process includes the disintegration of primary minerals into secondary minerals and clays in the upper layers of the substrate, which are then deposited in the lower horizons. There are three types of argillation: residual, illuvial and residual-illuvial. Residual argillation occurs mainly in steppe and desert regions where precipitation is low, winters are cold, summers dry and hot, and consequently the products of soil formation are relatively immobile. It is characterized by fine soil particles, no movement or leaching of clays, and no optically-oriented clays in the argic horizon. The thickness of the argic horizon increases from west to east as precipitation and soil moisture increase.

Illuvial argillation occurs mainly in warm and subtropical humid regions, and is characterized only by the mechanical movement of clays and the appearance of optically-oriented clays in the argic horizon. Residual-illuvial argillation generally occurs in the transitional zone between wet and arid regions, and is characterized by clays high in the argic horizon, with optically-oriented clays in the lower part of the horizon. Typically, argillation produces luvisol soils.

Leucinization.—Leucinization is essentially a bleaching of the upper soil horizons under seasonal leaching conditions. Iron and other pigmented materials in the upper soil horizons are removed either by laterally-flowing water or by forming concretions *in situ*, resulting in a whitish layer. Leucinization occurs mainly in humid and sub-humid regions, and is characterized by the occurrence of an albic horizon below the humic and argic horizons. The active iron content of the soil is reduced to ca. 1.4 g per kg, and the amount of manganese is also reduced in the albic horizon. At the same time, the content of iron-manganese concretions in the albic horizon can reach 70–90 g per kg. Sometimes the soil solution contains high silicate levels, up to 20--30 mg per litre (Gong *et al.*, 2007). The end products of leucinization are planosols, the most famous of which is the widespread "baijiang tu" ("whitish soup earth") found in the humid, temperate northeast of China.

Laterization.—Laterization occurs in tropical climates with copious rainfall, permitting sustained bacterial action to break down dead vegetation as fast as it is produced. In the absence of humic acids, insoluble sesquioxides of iron (Fe_2O_3) are formed and accumulate in the soil as red clays, nodules, and rocklike strata (laterites). Silica, on the other hand, is leached out of the soil. The chemical content of plants growing on such soils is reflective of the laterization process, usually having a very low ash content, 500–600 g per kg, lower nitrogen, sulphur, phosphorus, calcium, sodium, potassium and iron content, slightly higher manganese and very high aluminium (generally 0.5 g per kg, sometimes up to 8 g per kg) (Gong *et al.*, 2007). Laterization typically results in ferrosols, soils which tend to be firm and porous rather than sticky and plastic.

Podsolization.—This process, typically producing podsoluvisols, predominates in climates that are cold enough to inhibit bacterial action but with sufficient moisture to permit large green plants to thrive. Extreme podsolization is associated with coniferous forest. Humic acids, produced from the abundant leaf mould and humus in such forests strongly leach the upper soil of bases, colloids, iron and aluminium oxides, leaving a characteristic ash-grey "A2 soil horizon" composed largely of silica. In such soils the ratio of humic acids to fulvic acids (HA/FA) is less than 0.4, and the soil pH is 4–5 in water extraction, 3–4 under salt extraction. The so-called "spodic horizon" has a high cation exchange capacity, large surface area and high moisture-retaining capacity (Gong *et al.*, 2007).

Humification.—The formation of humus is essentially the slow oxidation of organic matter, producing high quantities of organic acids. Humification differs depending on the climatic conditions and vegetation type. In desert regions, the organic matter content of topsoil is generally less than 10 g per kg so humification is weak and the HA/FA usually less than 0.5. In steppe regions, with soil organic matter 10–30 g per kg, HA/FA is greater than one in sub-humid regions, and less than one in arid regions. In meadows, topsoil organic matter can reach 30–80 g per kg, and HA/FA greater than one. In alpine meadows, soil organic matter can exceed 100 g per kg but humification is weak, and in peatlands, the thickness of organic matter is greater than 40 cm (Gong *et al.*, 2007). In forests, the accumulation and influence of organic matter is rather different. Typical products of humification are phaeozems and histosols.

Gleization.—Gleization is characteristic of poorly-drained (but not saline) environments under a moist, cool or cold climate. Low temperatures allow heavy accumulation of organic matters resulting in a surface layer of peaty material; beneath this is the "gleyic" horizon, a thick layer of compacted, sticky, structureless clay of a bluish-grey colour. Under waterlogged conditions, the soil oxidation-reduction potential (Eh) is low, usually below 250 mV (Gong *et al.*, 2007). The gleyic horizon usually contains a small amount of organic matter. Reduction and mechanical leaching result in changes to the soil colloid, giving a low cation exchange and buffering capacity. The typical end products of gleization are gleysols and histosols.

5.2.2 The formation of anthrosols

The strongest impacts of human activity upon soils in China occur in farmland, which is mainly distributed on the eastern plains where conditions are most amenable for farming. The steppe and desert regions of northwestern China and the Qinghai-Xizang Plateau are also susceptible to anthropogenic impacts due to their fragile ecological condition.

Intensity of human activity.—Over the last millennium, the impact of human activity on soils has increased as the Chinese population has doubled. Increased use of fertilizers and the expansion of irrigation have enabled grain yields to increase from 1 000 to 4 500 kg per hectare over the past 50 years. During this process, some land has been made more fertile, while other land has become degraded (Gong *et al.*, 2007).

Human activities have caused the rate of transport of the earth's surface material to increase 17-fold over natural levels. This activity is most intense in China. Of the 136 billion tonnes of surface material transported by man each year, China accounts for over 38 billion tonnes (28%). From a global perspective, the intensity of human activity in China is 3–3.5 times higher than the world average (Gong *et al.*, 2007).

Mechanisms of human impact.—Human activity impacts upon soils both directly and indirectly. Five factors of soil formation that are affected by man are detailed in Table 5.1.

Table 5.1 The effect of human activities on five soil forming factors (from Gong *et al.*, 2007).

Soil forming factor	Human activity		
	Prehistoric	Agricultural civilization	Post Industrial Revolution
Changes in composition of soil-forming material	Slash and burn cultivation; deforestation	Application of organic fertilizers, ash, lime and mud; irrigation, salinization	Application of mineral fertilizers, pesticides and garbage; heavy metal pollution; spread of radioactive substances; acid deposition; dust storms; incorporation of non-biodegradable substances
Changes in surface morphology	Land reclamation	Levelling; construction of terraces; artificial deposition and filling-in of wetlands	Mining; land reclamation; road and urban construction; dam construction; development of tourism
Changes to soil water, temperature and gases	Disturbance and soil fumigation	Irrigation; drainage; ditch-formation; field drainage and coverage	Artificial irrigation and heating; fertilization; industrial emission of greenhouse gases; establishment of artificial climates by agricultural facilities
Changes in biological activities	Fallowing	Replacing of natural with artificial vegetation; application of organic fertilizers through micro-organisms; creation of conditions for biological reproduction through farming	Application of bacterial fertilizers; soil sterilization; artificial stocking with earthworms and other organisms
Changes in timing of soil formation	-	Subsoil exposed by soil erosion; underwater soils replaced by terrestrial soils through drainage and planting	-

Changes in soil composition.—Human-induced changes in soil composition may include desalination or resalination, decalcification or recalcification, dealkalinization or realkalinization, phosphorus accumulation, changes in organic carbon and general soil pollution (Gong *et al.*, 2007).

Soil salt content can be reduced through drainage and washing; by contrast careless irrigation may cause secondary salinization. For example, in the 1960s, the area of soil affected by salinization on the Huanghuai Pingyuan increased by 52% as a result of increased irrigation and reduced drainage; since that time a shift to careful irrigation and farming methods have greatly reduced the secondary salt levels.

Irrigated conditions can cause the continuous leaching of calcium carbonate from soils. In subtropical Guangdong and Guangxi, this leaching of lime from the soil occurs under natural conditions, but long-term recalcification by application of lime, and irrigation using cave water, has increased the amount of lime in the upper meter of soil to 240–1 260 tonnes per ha. Irrigation can also promote the loss of bases from soils (dealkalinization), while in acid soils, cultivation and fertilization can cause not only neutralization of soil acidity, but also realkalinization. Approximately 20% of China's naturally acidic soils, such as iron-rich and red soils, have experienced increased base saturation following cultivation, especially through rice cultivation and addition of lime or ash.

Nearly 80 million tonnes of phosphate fertilisers were applied to soils in China in 1992, and around 60 million tonnes accumulated in the soil, especially in suburban areas, both old and new, under vegetable cultivation. According to a national survey of soil profiles, available phosphorous and total phosphorous levels in topsoil are 16–26 and 4–6 times higher, respectively, than natural levels. Soils account for about twice the carbon storage of the atmosphere; yet since 1945, the level of organic carbon in temperate soils has declined by 20–40%. In some areas, where soils have been created by man, organic carbon has been sequestered.

In China, the area of land deemed to be polluted is already some 629 000 ha. Pollution by solid waste, pesticides and fertilizers mainly comprises heavy metals such as zinc, copper, lead, cadmium, mercury and chromium. These not only change the nature of the soil, but above critical levels can actually by harmful to plants. In some cases, radioactive pollution is also a problem.

Changes in soil moisture.—Changes to natural soil moisture conditions are achieved by irrigation, which has made agriculture possible in arid zones. Soil can be maintained in a moist state (60–70% water) by irrigating seven to nine times per year, resulting in new soil features (Gong *et al.*, 2007). In particular, hydroponic irrigation of paddy fields causes a layer of water to form on the soil surface which slowly infiltrates the soil through the action of the plough, thereby changing the natural soil moisture conditions (Gong *et al.*, 2007).

Formation of new soil by addition of material.—The addition of novel, artificial substances to soils can have very different effects depending upon the original environment, particularly the moisture regime (Gong *et al.*, 2007). On the Huangtu Gaoyuan (Loess Plateau), it is traditional practice to apply manure, a generic term for various substances including organic fertilizer, livestock manure, human waste and spent animal bedding. Long-term application of such material causes a layer to form on the original soil, subjected to tillage and subtended by a plough pan.

In the low-lying Zhu Jiang delta and Tai Hu regions, the former less than a meter above sea level, material taken from local rivers and lake sediments is applied to soils, either as a single application or a repeated process. In a single application, dykes are built to trap sediment, resulting in a raised, dry area that is planted with sugar cane (*Saccharum* spp.), mulberry (*Morus* spp.) or other fruit trees and also stocked with fish ponds. The parent soil becomes anaerobic and, through the impact of groundwater, rapidly oxidised. Sludge from the ponds can be used as a manure which is applied two to three times per year, thickening to form a mat of mud over the soil.

In sites with low vegetation cover and loose soils, sedimentation occurs, most particularly in the arid regions including vast areas of Central Asia and northwestern China. The use of river water for irrigation also generates quantities of sediment. In reports from eight rivers in the Tarim Pendi, Xinjiang, the average annual thickness of sludge accumulated through irrigation was 0.25–0.85 cm. The mineral and chemical composition of such sludge varies from place to place, with those precipitated from topsoils generally being highest in organic matter and mineral nutrients.

Deep soil disturbance.—Mining may be a major cause of soil disturbance, resulting in the surface being stripped away, buried or mixed, and subsoil and parent material layers exposed or partially exposed. Open-pit mining undoubtedly causes great disturbance of the soil surface; deep mining can sometimes cause greater disturbance through ground subsidence.

Disturbance by urban construction.—Processes of urban construction cause great disturbance to soils through mining, transportation, accumulation, mixing and landfill of waste. The humus layer may be removed or buried, some soil layers lost, some mixed, and others inverted so that older soils lie above more recently-formed layers. Thick layers of deposited substances and abnormal distributions of particles also cause soils to differ significantly from their natural state. Typically, soils disturbed by urban construction are also abnormal in colour, usually due to chemical pollutants.

Soil erosion and accumulation.—Around 16.5 million km^2 of soil are subject to natural or artificial erosion worldwide, with the loss of some 75 billion tonnes each year. In present-day China, 38% of the total land area is affected, with 1.8 million km^2 eroded by water and 1.9 million km^2 by wind. Human activity is the main cause of soil erosion, which usually takes the form of the long-term loss of fertile topsoil into rivers or the atmosphere, from which it can re-start the process of soil formation.

Major processes of anthrosol formation.—In China, cultivation should be considered one of the most important soil-forming processes. All cultivated soils have been more or less modified from their original condition (Zhao, 1986). As the population increases, agriculture has spread from river terraces and plains to mountainous areas, marshes and saline beaches; and from humid and sub-humid to arid and semi-arid regions (Gong *et al.*, 2007). Owing to changes in natural conditions, soils have been gradually and persistently ameliorated and cultivated, resulting in the formation of anthrosols, mainly through hydragric, irragric, cumulic and fimic processes (Gong, 1994).

Hydragric processes occur through the levelling of paddy fields for effective rice cultivation and fertilisation, a process which has occurred in China for thousands of years. Terraced fields are constructed in mountainous regions to prevent the loss of water and soil, up to 2 000 m altitude; "polder fields" in lakeland and marshy areas by construction of dams to prevent flooding; and beach fields to prevent invasion by seawater. This reshaping of the terrain and disturbance of soils results in an altered redox state with new leaching and accumulation characteristics (Gong, 1980; Gong, 1983; Gong, 1986). At present, around 2 530 million ha of land are considered to be hydragric anthrosols.

Irrigation of arid areas for agriculture results in large amounts of siltation, which is known as the irragric process (Shi & Gong, 1995). In areas where annual irrigation amounts to about 1 000 mm, the depth of silted materials can reach 0.3--0.5 cm per year (Wang *et al.*, 1996). Year upon year, new soil horizons are formed, whose material constituents, fertility and moisture regime differ from the original soils. In the Hotan region, Xinjiang, along the ancient "Silk Road," irragric horizons in anthrosols can be as deep as several meters. The organic carbon in these irragric horizons formed from silted materials at a depth of 5 m dates from 1 870 years ago. In Ningxia and Hetao, Inner Mongolia, the practice of irrigation has a history of over 2 000 years (Lin, 1996). These kinds of soils are distributed extensively in arid regions, covering a total area of more than 1.5 million ha.

The cumulic process is one of the most important forming process of anthrosols, and can be classified into two types, earth-cumulic and mud-cumulic. The former occurs mainly on the Huangtu Gaoyuan (Loess Plateau), where the application of earth-mixed manure not only adds nutrients to soils, but also constantly thickens them, resulting in a new cultivated horizon covering an old cultivated horizon which itself covers an ancient cultivated horizon in a step-like formation. The latter is found in the Zhu Jiang and Chang Jiang deltas. Low-lying marshes, such as the Zhu Jiang delta, are naturally waterlogged. In order to make full use of natural resources, people here dug out earth from the lower area of the marshes and piled it up on higher areas nearby; the lower parts became fish ponds, the higher, mulberry fields. The mud-cumilic anthrosol mulberry fields started out about one meter higher than the water, increasing by a small amount each year. The leaves of the mulberries feed silkworms, the excrement of which feed the fish, and the fish excrement fertilizes the mulberry fields in a beneficial cycle. Soils formed by this process have hydromorphic features such as snail shells, rusty spots and streaks in their profile (Gong, 1994). The cumulic anthrosols of Shaanxi date from ca. 2 800 years ago, the mud-cumulic anthrosols of the Zhu Jiang delta to the Han Dynasty, more than 2 000 years ago (Lin, 1996). The area covered by these two kinds of soil currently amounts to 1.5 million ha.

Finally, long-term application of domestic garbage and organic manure to soils results in the accumulation of large amounts of phosphorous in their A (topsoil) and B horizons. The available phosphorous content in these fimic anthrosols is even higher than that of guano (Gong *et al.*, 2007). These soils are principally distributed around cities, including ancient cities such as Zibo Shi, Shandong, at around 3 000 years old and, more commonly, modern cities with a long history. These include the vegetable plots of the royal families in Beijing Shi, which were established 600 years ago; those in Hangzhou, dating back more than 1 000 years; and those in Urumqi more than 100 years. The area covered by fimic anthrosols in China is at least 0.7 million ha. In addition, anthroturbation has a positive influence on the formation of anthrosols.

Artificial effects on soil evolution.—Soil is a dynamic natural body constantly subject to many complex chemical, physical and biological (including human) activities, always changing and evolving. In different regions, at different stages, and under the impact of different environmental factors, one soil type may be transformed into another (Zhao, 1986). The relationship between anthrosols and other soil orders is given in figure 5.1. For example, irragric anthrosols are formed from natural soils by accumulated irrigation and ploughing. Hydragric, cumulic and fimic anthrosols are formed by wet cultivation, accumulated transport and intensive fertilization (Gong *et al.*, 1999; Gong *et al.*).

In China, the soils of the eastern plains are the most affected by urbanization and environmental pollution, while those of the west suffer most from over-grazing and the expansion of agriculture onto marginal lands. Thus in the east, protection of prime farmlands and the control of environmental pollution, especially agricultural pollution, are essential to soil protection, while in the west soil protection may be facilitated by sustainable implementation of the government's "Grain for Green" policy, which aims to provides grain subsidies to local communities in exchange for planting trees to restore hillside agricultural land.

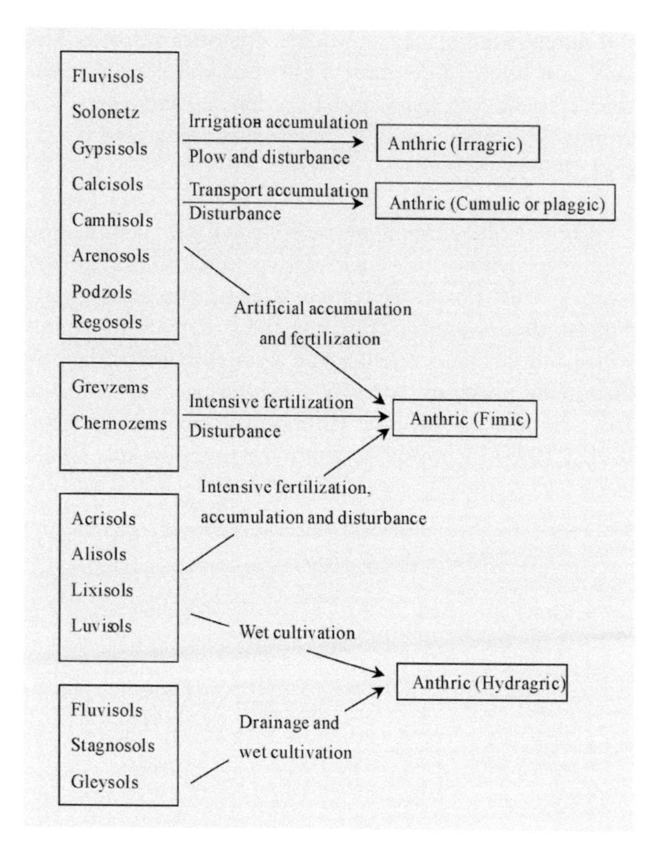

Figure 5.1 The relationships between anthrosols and other anthropogenic soil types (Gong *et al.*, 1999).

5.3 Major soil types and their distribution

This section is based on the work of GONG and colleagues (in press).

5.3.1 Characteristics of Chinese soil taxonomy

Chinese soil taxonomy research was initiated in 1984. The book *Chinese Soil Taxonomy (First Proposal)* was published in 1991 (Cooperative Research Group on Chinese Soil Taxonomy, 1991), a revised proposal was completed in 1995 (Cooperative Research Group on Chinese Soil Taxonomy, 1995), and *Keys to Chinese Soil Taxonomy* (third edition) was released in 2001 (Cooperative Research Group on Chinese Soil Taxonomy, 2001). Chinese soil taxonomy is based on diagnostic horizons and diagnostic characteristics under the guidance of soil genetic theory, and is compatible with international standards.

Chinese soil taxonomy is a six-level classification system, the levels being soil order, suborder, soil group, subgroup, family and series. Names are based upon subsections, with soil order, suborder, soil group and subgroup comprising one section. Families are named after soil characteristics such as particle size, mineral composition or soil temperature conditions. Soil types are determined, in *Keys to Chinese Soil Taxonomy* (Cooperative Research Group on Chinese Soil Taxonomy, 2001), from top to bottom, excluding types one by one on the basis of soil diagnostic horizons and characteristics (Zhao & Shi, 2007). In total, 33 diagnostic horizons and 25 diagnostic characteristics are used, and 14 orders, 39 suborders, 138 groups, 588 subgroups and 750 families recognised (Gong *et al.*, 2007).

Chinese Soil Taxonomy transformed soil classification from a qualitative to a quantitative study, combined soil history with morphogenesis under the guidance of soil genetic theory, and established the order of anthrosols with Chinese characteristics and a series of particular soil diagnostic horizons. In 1995, the International Union of Soil Science also adopted Chinese soil taxonomy (ISSS/ISRIC, 1995) including its approach to anthrosols (ISSS/ISRIC/FAO, 1998). The magnum opus of soil sciences, *Handbook of Soil Science*, has described Chinese soil taxonomy in detail (Sammes, 1999), Russian soil scientists regard Chinese soil taxonomy as a revolutionary advance, and the Japanese Soil Science Society has taken Chinese soil taxonomy as one of the three major soil classification systems of the world. In particular, Chinese anthrosols have become the international standard for classifying anthrosols. Thus *Chinese Soil Taxonomy* has played an important role in the development of international soil classification systems, and was a milestone in the development in Chinese soil classification, alongside the American and World Soil Taxonomy systems (see Table 5.2 for a comparison of the three systems).

Table 5.2 Comparison of Chinese, American and World Soil Taxonomy Systems (from Gong *et al.*, 2007; Shi *et al.*, 2010)

Soil order in Chinese soil taxonomy	Most similar soil type in American system	Most similar soil type in world reference base for soil resources (http://www.fao.org/docrep/W8594E/W8594E00.htm)
Andosols	Andisols	Andosols
Anthrosols	Andisols/oxisols/vertisols/ultisols/mollisols/alfisols/inceptisols	Anthrosols
Argosols	Alfisols/ultisols	Planosols/albeluvisols/alisols/luvisols
Aridosols	Aridisols	Calcisols/solsonchak/gypsisols/luvisols/cambosols/leptosols
Cambosols	Inceptisols/gelisols/mollisols	Cambisols/umbrisols/calcisols/gypsisols/cryosols
Ferralosols	Oxisols	Ferralsols/plinthosols
Ferrosols	Ultisols/alfisols/inceptisols	Acrisols/lixisols/cambisols/plinthisols
Gleyosols	Inceptisols/entisols/gelisols	Gleysols/cryosols/leptosols
Halosols	Aridisols/mollisols/alfisols/inceptisols	Solsonchak/solsonetz
Histosols	Histosols/gelisols	Histosols
Isohumosols	Mollisols	Chernozenm/kastanozem/phaeozem/leptosols
Primosols	Entisols	Regosols/arenosols/fluvisols/leptosols/cryosols
Spodosols	Spodosols	Podzol
Vertosols	Vertisols	Vertisols

5.3.2 Fundamental properties of the soil orders

Argosols.—Argosols are found in almost all versions of Chinese soil taxonomy and soil classifications. In Chinese soil taxonomy, their diagnosis is based on the existence of an argic horizon; whereas elsewhere their zonality is emphasized (Sharma, 1969; Zhang & Shi, 1994; Gong *et al.*, 2007). Diagnostic horizons and diagnostic features of argosols are therefore the argic horizon or certain upper layers, from the soil surface to 125 cm below, containing at least a 0.5 mm thick illuviation clay pan, and in the argillic B horizon, cation exchange capacity being more than 24 centimol per kg. Argosols are equivalent to dark brown, baijiang, brown, yellow brown, cinnamon or limestone soils or yellow earths in general classifications. Soil formation depends mainly on processes of soil clayization and humus accumulation. Argosols are divided, based on soil temperature and water conditions, into boric, ustic, perudic and udic argosols.

In China, argosols are mainly distributed in the area affected by the monsoon, including northeastern, northern and eastern China as well as the northwestern mountains, the southeastern Qinghai-Xizang Plateau and the southwest of southern China. The annual average temperature of these areas ranges from -1 to 17 °C, annual precipitation from 600 to 1 800 mm, soil temperature from frigid to cryic under a mesic to thermic regime, and soil water conditions from ustic through udic to perudic. The natural vegetation includes different types of forest and shrubland, the terrain ranges from mountains to loess ridges, and the soil parent materials are complex and diverse. Argosols are an important resource for forestry and agriculture with great production potential, but differ greatly between subclasses. Therefore, in order to enhance their productivity, measures appropriate to the different types should be adopted.

Andosols.—The parent material of andosols is volcanic eruptive material, including volcanic ash, pumice, cinder and volcanic trash. The soil is fine and very light, with high levels of oxalic-extractable aluminium, amorphous iron and sorbed phosphate. The pedogenesis horizons of andosols are simple: an andic B horizon lies below the A horizon; the near lower horizon comprises cinder or other volcanic trash materials. The pH value of andosols can exceed 9.4, and the sorbed phosphate capacity in the fine soil is generally more than 25%, sometimes as much as 80%. Andosols are diagnosed by their andic properties: volcanic ash, cinder and other volcanic trash materials make up more than 60% of the soil by mass, and its mineral composition comprises mainly short sequence minerals such as allophane, imogolite and hisingerite. The total percentage of amorphous iron, measured by the acidic ammonium oxalate method is at least 2%. The density of the fine soil is less than 0.9 tonnes per m^3 and the sorbed phosphate capacity is more than 85%. Andosols are divided into cryic and udic andosols.

Andosols make up 0.02% of China's soils (Gong *et al.*, 2007), and are mainly distributed in northeastern China, Yunnan and Taiwan, with some deposits on the Qinghai-Xizang Plateau and Hainan. Due to their high organic content, good configuration, low density and abundant phosphorus, they have been used for soil amelioration in Yunnan and Hainan. Andosols tend to be located in volcanic regions, with well-developed tourism, rich deposits of mineral such as peridotite and volcanic lapilli, and resources such as mineral water. These resources and environments need careful protection where andosols are exploited.

Aridosols.—Aridosols are soils with an aridic epipedon comprising a gravel mulch (a surface of gravel and stone eroded by wind), aeolian sand (a layer of sand, sandy gravel or small particles), and a thin crust of organic material which may be absent under conditions of drought or wind erosion. The upper section of the soil profile contains layers of crust with holes, flakes or a foam-like appearance (Huang, 1987; Souiriji, 1991; Lei & Gu, 1996). The soil has a low organic matter content and is strongly mineralized.

Aridosols make up 21% of China's soils, and are mainly distributed in the arid areas of the northwest, including Xinjiang, Gansu, Ningxia, Nei Mongol, Qinghai and parts of Xizang. The climate of these areas is generally continental and arid, with temperature varying according to elevation. Vegetation types include mainly xeric shrubs and small semi-shrubs, with coverage being less than 20%. Aridosols are divided into two suborders on the basis of soil temperature: cryic and orthic aridosols. Due to the arid climate, agriculture cannot occur on aridosols without irrigation. Therefore, most cryic aridosols are used as pastoral land; orthic aridosols at low elevations can be used as cultivated land or pastoral land, but are prone to desertification.

Cambosols.—Cambosols are weakly developed soils, diagnosed by the cambic horizon where there is no material deposition and apparent clayization. Cambosols may be brown, reddish-yellow or purple in colour. The cambic horizon is greater than 5 cm in thickness, fine-textured, and different from argillic, spodic, ferralic, lac-ferric or gleyic horizons, but can feature oxidation and reduction. These soils are also known as cambosols in world soil classifications (ISSS/ISRIC/FAO, 1998), and as inceptisols in the American Soil Taxonomy system (Soil Survey Staff, 2006). Cambosols are divided into gelic, aquic, ustic, perudic and udic cambosols.

In China, cambosols cover 22% of the total land area (Gong *et al.*, 2007), mainly concentrated on the eastern and southeastern Qinghai-Xizang Plateau, the Huabei Pingyuan (North China Plain), and northern Da Hinggan Ling, as well as sparse deposits in the eastern valleys and the plains of northeastern China, the northwestern mountains and irrigated oasis areas, and southeastern mountains, plains, valleys, lakesides and beaches. Due to the great differences in the surrounding conditions and properties of cambosols, their utilization and management must be adapted in accordance with local conditions. Gelic cambosols are used primarily for exploitation of natural vegetation, animal husbandry, tourism and raising wild animals. Aquic cambosols provide the most important base for the production of grain, cotton, oil and fruit in China. Ustic cambosols are the major dry farming soil, being subject to low rainfall, great evaporation, low soil organic matter and poor fertility. Perudic cambosols in general are used for the development of forestry and tourism or stocking wild animals.

Primosols.—Primosols comprise only a thin surface layer, with no diagnostic horizon or features to identify it. The soil is not obviously differentiated into horizons and there is only a very weak degree of profile development. These characteristics are similar to those of primosols in American Soil Taxonomy. In previous soil classifications in China, primosols included mainly wind sands, loess, alluvial soils, and regosols. The properties of primosols largely depend on the characteristics of the parent material. They are diagnosed by weak surface characteristics, and are divided into anthric, sandic, alluvic and orthic primosols.

The area covered by primosols in China is c. 6% of the total land area (Xiong & Li, 1987), or 13% if deserts are included (Gong *et al.*, 2007). Primosols are mainly concentrated on alluvial plains, especially the deltas of major rivers, estuaries and deltas. The centres of primosol distribution are areas with aeolian matter, and those containing newly-formed soils, developed from rock weathering in mountainous regions. In areas with intense human activity, man-made accumulation, irrigation and other activities affect the normal development of the soil causing the formation of anthric primosols (Liu, 1999).

Ferralosols.—Ferralosols contain a clay component dominated mainly by kaolinite minerals and iron and aluminium oxides, with strong allitization; the content of efflorescing minerals is quite low. These soils are similar to some types of latosol (Gong *et al.*, 1978; Chen & Zhao, 1989; Chen & Zhao, 1994). The area covered by ferralosols in China is about 42 200 km^2, ca. 0.5% of the territory, and mainly distributed on the low terraces of Hainan,

Guangdong, Guangxi, Fujian, Taiwan and Yunnan. Ferralosols also coexist with ferrosols, cambosols and anthrosols in the tropical and southern tropical zones (He *et al.*, 1958; Zhao & Zou, 1958; Li & Shi, 1959). This soil type forms under conditions of high temperature and rainfall (Gong *et al.*, 1999), in habitats including tropical rainforest and tropical and subtropical monsoon forest. The parent materials are sediments including Quaternary red earths and neritic sediments. The material of which ferralosols are made up is completely air-slaked; electropositive elements are heavily leached, silicic acid strongly extracted, iron and aluminium oxides are rich and clays significantly reduced. The thickness of the ferralic horizon is at least 30 cm and its clay content at least 80 g per kg; the texture is fine or sandy. The cation exchange capacity of ferralosols is less than 16 centimol per kg and the effective cation exchange capacity less than 12 centimol per kg; the content of efflorescing minerals is less than 10%, and total nitrogen less than 8 g per kg. Ferralosols have no volcanic properties.

The effect of deforestation and cultivation upon ferralosols is to reduce the organic matter content of the surface soil and destroy the soil structure. The cation exchange capacity also decreases, and ultimately fertility is reduced making it necessary to increase the application of phosphate and potassium fertilizers and lime in order to raise the organic and nitrogen contents, adjust the pH value, and restore productivity.

Ferrosols.—Ferrosols include the soils generally known as red, lateritic red, and typical yellow soils, and are similar to the ultisols and acrisols of other soil classification systems (ISSS/ISRIC/FAO, 1998; National Soil Survey Staff, 1998; Soil Survey Staff, 2006). They form mainly under warm climatic conditions, and support a natural vegetation of evergreen broad-leaved or coniferous forest. They are found mainly on hills in subtropical and tropical southern China, including Hainan, Guangdong, Guangxi, Fujian, Taiwan, Jiangxi, Zhejiang, Hunan, Guizhou and Yunnan and some parts of southern Hubei and Anhui (Zhao & Chen, 1999). They cover around 9% of the total land area of China (Gong *et al.*, 2007).

Ferrosols are formed by moderate to rich ferrallitization (Zhao & Chen, 1995), from parent materials which differ from place to place. In cross-section, from top to bottom, they display a humus layer (A), sticky sediment layer (Bt), weathered layer (Bw) or reticulate layer (Bl), passing down gradually to the parent material or parent rock layer. They are diagnosed by the presence of a lac-ferric horizon, at least 30 cm thick, with the texture of a sandy or finer loam. The free ferric oxide content is at least 20 g per kg, or free iron accounts for more than 40% of total iron, and the apparent cation exchange capacity is less than 24 centimol per kg (Institute of Soil - Chinese Academy of Sciences, 2001). Ferrosols are divided according to soil moisture conditions into three subclasses: ustic, perudic and udic ferrosols.

Ferrosols under natural forests or plantations are rich in soil organic matter and nitrogen in the surface layer. Where forests are lost, the content of soil organic matter and soil nitrogen is significantly reduced. Reclaimed ferrosol land has been used to cultivate various tropical and subtropical industrial crops, fruit, grain and oil crops. However, reclaiming such land leads to a reduction in soil organic matter, deterioration of the soil structure and permeability, and a lack of soil nitrogen and phosphorus. Sustainable use of such lands requires application of organic fertilizers including nitrogen and phosphate fertilizers, which can result in acidic conditions where exchangeable potassium is generally replaced by aluminium ions leading to intense leaching. The application of lime, dolomite or limestone powder can neutralize the excessive acid in the soil to reduce the effects of potassium fertilization (Li, 1983).

Gleyosols.—Gleyosols are soils that are frequently saturated with water and subject to deoxidisation reactions, ultimately forming a special soil structure with a typically grey-blue colour. As they are influenced by groundwater or surface water, a ca. 10 cm layer with gleyic features appears in the range from the soil surface to 50 cm below. This layer is grey-blue with a small amount of other staining due to the presence of metal compounds. Gleyosols are divided into permagelic, stagnic and orthic gleyosols.

Gleyosols take up approximately 1% of the country's total land area (Gong *et al.*, 2007), mainly distributed in the bottom of floodplains and lake shores of the Da Hinggan Ling and Xiao Hinggan Ling in northeastern China, the Changbai Shan, Sanjiang Pingyuan and Dongbei Pingyuan. Gleyosols are sparsely distributed in other areas such as the Qinghai-Xizang Plateau, catchments at the northern and southern feet of the Tian Shan, the Huabei Pingyuan, middle and lower Chang Jiang basin, middle and lower Zhu Jiang basin and southeastern coastal regions. Gleyosols are concentrated in wetland areas, important for biological diversity, and are heavily cultivated and therefore require careful management.

Halosols.—Halosols are defined as soils with a salic horizon and an alkalic horizon at certain depths (Gong *et al.*, 1999). The top of the salic horizon ranges from the soil surface to 30 cm below ground, and the alkalic horizon from the surface to 75 cm below. The alkalic horizon is defined as having a solubility in cold water greater than that of gypsum, causing soluble salts to accumulate. In arid areas, the salinity of the alkalic horizon of halosols is at least 20 g per kg; elsewhere it is at least 10 g per kg. The alkalic horizon is a special, argic layer rich in exchangeable sodium–the percentage of which is at least 30%. The pH value of halosols is at least nine and the salinity of the topsoil is less than 5 g per kg.

Halosols, which are divided into alkalic and orthic halosols, take up approximately 4% of China's total area (Gong *et al.*, 2007), mainly in arid and semi-arid zones in northern and coastal areas, roughly to the north of a line following the Huai He, Qin Ling, Bayan Har Shan, Nyainqêntanglha Shan and Gangdisê Shan, as well as on low plains in eastern and southern coastal zones, including parts of Taiwan. They are primarily concentrated in low-lying wet basins, semi-enclosed shallow depressions, river deltas and dry delta areas, where surface and underground runoff is relatively slow. Small areas of halosols are found on riverbanks and in basins and depressions on the west of the Qinghai-Xizang Plateau, and in wide valleys, lakes, and basins of southern Xizang. In most halosol areas, evaporation rates are high, with poor groundwater quality and soil properties, so soil and water are difficult to retain. Careful management, adjusted to the local conditions, is necessary to farm such soils, and includes engineering measures such as irrigation and drainage systems as well as biological ones and action to reduce soil salinity and control the levels of groundwater.

Histosols.—Histosols are an order of soils (similar to those known as "peats" in previous soil classifications) whose main pedogenic process is peatification, and main diagnostic characteristics are their organic soil materials. From the soil surface to 40 cm in depth, the organic carbon content is at least 120 g per kg (excluding live roots). When the top layer of the soil is cultivated turf, the organic content between the soil surface and 25 cm below is at least 80 g per kg; if it comprises litter, usually saturated for no more than one month per year, it is at least 200 g per kg.

Histosols account for only 0.15% of China's soils, mainly in the low-lying areas of the cold temperate and temperate zones, and concentrated in northeastern China and the eastern and northern edges of the Qinghai-Xizang Plateau (Gong *et al.*, 1999). The main climatic characteristics of these regions are low temperature, abundant rainfall and high air moisture levels. The main vegetation is marsh with sedges in most areas, mosses in alpine areas, and few shrubs and trees. Few histosols have been reclaimed as paddy fields. The soil's parent material is usually Quaternary deposits, and the soil texture is very clayey. The top of a histosol profile comprises humic or peat epipedons. The humic epipedon, which is always dark in colour, presents a granular structure containing dense grass roots, with an organic content of over 40%, carbon-nitrogen ratio of 14–20, and total nitrogen content ca. 1–2%. At the bottom of the profile is a gleyic horizon, grey in colour, clayey in texture, with an organic content of ca. 1–2% and total nitrogen less than 0.2%. The total phosphorous content of histosols is ca. 0.3–0.5%; total potassium varies greatly but is only 0.2–0.3% in the peat horizon. Histosols are usually neutral or slightly acid, but may be alkaline when they contain free calcium carbonate. In addition, there are a high number of exchangeable cations, up to 30--50 milliequivalents of hydrogen per 100 g of dry soil, and the base saturation varies from 30–90%.

Histosols are divided into two suborders on the basis of soil temperature conditions: permagelic and orthic histosols. Permagelic histosols have a permagelic soil temperature regime, in which a permafrost layer occurs from the surface to 200 cm below; orthic histosols comprise the remaining histosols. Histosols are mainly used for agriculture in northeastern China, and as pasture in the northwestern mountains and on the Qinghai-Xizang Plateau. Histosol environments are a special ecosystem which deserves protection, particularly at the headstream of rivers (Huang, 1996).

Isohumosols.—Isohumosols are mostly distributed on steppe or wet prairies. This type of soil forms a relatively deep, dark surface due to organic matter accumulation, and has a base saturation of at least 50%. It is diagnosed by a mollic epipedon and isohumic properties. The mollic epipedon contains more than 6 g per kg organic carbon, and its base saturation is at least 50%. The humus reserve ratio from the surface to 20 cm depth, to that in the layer from the top to 100 cm, is no more than 0.4. Isohumosols are divided according to their diagnostic subsurface horizon, lithological characters and soil moisture regimes into bleachic, lithomorphic, haplic, ustic, and udic isohumosols.

In China, isohumosols are mainly distributed in the south of the Nei Mongol Gaoyuan, hills in the south of the Da Hinggan Ling, the Songnen Pingyuan and Sanjiang Pingyuan of Heilongjiang, basins in the Altay Shan and Tian Shan, the Altun Shan, and the northern slopes of the western Kunlun Shan in Xinjiang, Jilin, Gansu and Qinghai. The utilization and management of isohumosols differs greatly between types, due to their very different properties. For example, the fertility of the mollic epipedon in bleachic isohumosols is high, but the productivity of its albic horizon is low (Zhang & Zhang, 1986; Zeng, 1997). The other isohumosols all are important agricultural resources, with lithomorphic isohumosols, rich in phosphorus, being considered a very high-grade phosphate resource.

Spodosols.—Soils containing a spodic horizon are formed by means of cheluviation, and are diagnosed, unsurprisingly, on the basis of containing spodic material from the soil surface to 100 cm in depth. Other diagnostic properties include the absence of andic properties in more than 60% of the soil body from the mineral surface or top of the organic horizon to 60 cm in depth. Spodosols are almost indistinguishable from sombric bleached podzolic soils. Spodosols are usually acidic, with a pH less than 5.9, and iron and aluminium oxides commonly illuviated (Gao *et al.*, 1996; Gong *et al.*, 1999 1999). The diagnostic spodic horizon has a thickness of at least 2.5 cm, and occurs below the albic horizon. Spodic materials take up more than 85% of the soil volume, and organic materials are present at more than 6 g per kg. At least half of the soil body is tightly cemented by a combination of organic material and aluminium or iron. There are two suborders of spodosols, humic and orthic.

Spodosols cover a very small area in China, and are mainly distributed in the north of the Da Hinggan Ling, the northern slopes of the Changbai Shan, an altitudinal zone in the southern and southeastern mountains on the edge of the Qinghai-Xizang Plateau, and the Zhongyang Shan, Yu Shan, Xue Shan and Ali Shan of Taiwan (Xiong *et al.*, 1979; Gao, 1989; Zhao, 1989; Liu & Chen, 1990 1990; Chen & Zhang, 1991; Gong *et al.*, 2007). Spodosols support mainly forestry, and these areas comprise some of China's most important forest production bases.

Vertosols.—Vertosols have a very heavy texture, expand greatly when wet and shrink when dry. The soil profile shows a "smashed" structure and slickensides. The genesis of vertosols is similar to that of China's "Shajiang" soils. Their parent material comprises clay river or lake deposits, basic igneous rocks, calcareous sedimentary rocks or mild loess clay. Vertosols are very high in dilatant minerals such as smectite (Wilding & Tessier, 1988; Zhang & Gong, 1993; Zhang & Liu, 1993; Liu, 1995), and can be recognised by their pattern, evidence of reversal, anthroturbic character and self-swallowing, a process by which the soil is homogenized by pedoturbation. The clay content is more than 300 g per kg and, when split, can contain cracks greater than 5 mm across. Vertosols may also contain ashen-slotted slickensides, wedge-like structures and extrusions. There are three types: aquic, ustic and udic vertosols (Dudal & Eswaran, 1988; Huang & Wu, 1988; Institute of Soil - Chinese Academy of Sciences, 2001).

Vertosols are distributed mainly in lowland areas with frequent, alternating cycles of dry and wet, including the Shandong Bandao, Zhangpu (Fujian), southwestern Guangxi, Leizhou Bandao (Guangdong), northern Hainan and the xerothermic valley of the Jinsha Jiang (Huang & Wu, 1987; Zhu & Tan, 1989; Liu, 1991; Zhang & Gong, 1992; Qiu *et al.*, 1994; He & Huang, 1995; Gong *et al.*, 2007). Most vertosols are used as cultivated land although the soil texture is heavy and it is very low in fertility. Irrigation and drainage, soil exchange and careful fertilization are all necessary to prevent crazing and infertility (Zhang & Gong, 1993; Zhang & Liu, 1993; He *et al.*, 2001 2002; Huang *et al.*, 2001).

Anthrosols.—Anthrosols, as described above, are soils affected greatly, or even created, by human activities, with clearly different characteristics from those of the original soil. Their diagnostic horizons are the anthropic epipedons. Anthrosols are divided into two suborders according to soil water conditions: stagnic anthrosols, with an anthrostagnic soil moisture regime, anthrostagnic epipedon and hydragric horizon, and orthic anthrosols, comprising all other anthrosols.

Anthrosols are found throughout China, concentrated in areas with the greatest farming activity and longest agriculture histories. Therefore in arid regions they are found mostly near water systems, in mountains in the best hydrothermal areas, elsewhere in valleys, on river banks, by lakes, and in easily accessible places. On the whole, anthrosols are more common in the east than the west, in the south than the north, and downstream rather than upstream. Anthrosols are centred on delta areas, especially the Chang Jiang and Zhu Jiang deltas (Gong *et al.*, 1999). Two-thirds of China's cultivated land is classed as medium- to low-yield due to salinization, gleization, drought or low fertility. Therefore farmland should be protected, improved, and rotated effectively, and soil pollution should be eliminated, if it is to continue to support a large population with the limited farmland available.

5.4 Soil geographical regions of China

5.4.1 The regularity of regional distribution

As described in previous chapters, China can be described as forming three topographic "steps," (four including the continental shelf) which give obvious regional characteristics for soil distribution. Due to the monsoon climate, most parts of the country are cold and dry in winter and hot and wet in summer, resulting in a zonal distribution of soil series, from those under moist forest vegetation in the east, through a northeast to southwest region under hapli-udic forest grassland and grassland in central China, to soils under desert steppe vegetation in the west (Gong *et al.*, 1996).

Closest to the ocean (east of a line formed by the Da Hinggan Ling, Taihang Shan and the eastern edge of Qinghai-Xizang Plateau) the climate is humid and temperature decreases from south to north, forming various soils and soil combinations. They include, from south to north, udic ferralosols-udic ferrosols, udic ferrosols-udic ferralosols, udic ferrosols-perudic cambosols, udic argosols-stagnic anthrosols, udic argosols-aquic cambosols, boric argosols-udic isohumosols, and gelic cambosols-orthic spodosols (Figure 5.2).

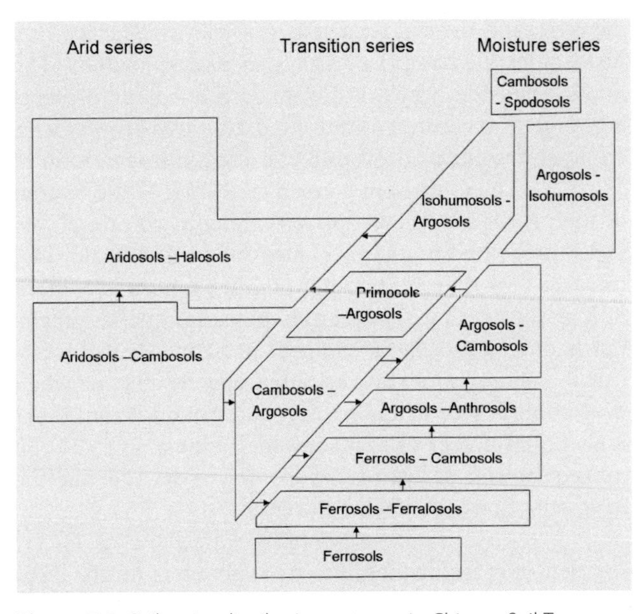

Figure 5.2 Soil series distribution patterns in Chinese Soil Taxonomy (Gong, 1999).

The vast regions to the northwest of a line formed by western Nei Mongol, the Helan Shan and Nyainqêntanglha Shan are weakly affected by the maritime monsoon, and their climate becomes gradually drier towards the west. From east to west, vegetation types here include desertified steppe, steppe-desert and desert; the soil associations include, from south to north, cryic aridosols, gelic cambosols, orthic aridosols and orthic halosols.

In the transition zone between the southeast and the northwest soil series, the climate is sub-humid and semi-arid, and vegetation is mainly grassland. Since this region has a span of more than 20° of latitude, many types of soil occur, based on cambosols and luvisols, and including ustic cambosols-ustic argosols, orthic primosols-ustic argosolsustic cambosols, and ustic isohumosols-boric argosols.

The soil altitudinal spectrum is a reflection of the processes of soil formation, climatic changes and montane bioclimatic changes. As elevation increases, the climate and biodiversity change, while redistribution of material on the slopes causes succession of soil from the foot to the top of mountains. Mountain environments feature a number of basic soil types based on the relative and absolute heights of the mountains, the redistribution of water and heat caused by the terrain, and the slope variation of the constituent units of the altitudinal structure. The most complete soil altitudinal spectrum known is that of Qomolangma Feng (Mount Everest). On the southern slope, from the foothills to the summit, soils appear in order as follows: udic ferrosols/udic cambosols-udic argosols/udic cambosols-gelic cambosols (albic umbri-gelic cambosols-typic matti-gelic cambosols-typic molli-gelic cambosols-typic permi-gelic cambosols). For low altitude mountains the soil altitudinal spectrum is generally relatively simple.

The Qinghai-Xizang Plateau, with an average elevation of over 4 000 m, covers 2.5 million km^2 (Zhang *et al.*, 2002 2002) including high mountains and deep valleys. The soil on the Plateau therefore reflects both altitudinal and horizontal distributional features (Gao *et al.*, 1985 1985; Xiong & Li, 1987 1987). The types of soil altitudinal structure around the plateau are complex and diverse, with ferrosols, argosols, aridosols and cambosols appearing in turn from southeast to northwest (Gao *et al.*, 1985 1985). Soil altitudinal differentiation only occurs in the mountains, but the endemic distribution of soil spectra can also be observed in incised valleys on the Plateau, with soils varying as differing-depth rivers carved out valleys, as well as due to natural environmental conditions (Xiong & Li, 1987).

For small to medium-sized regions, the distribution of soil is affected not only by climate and biology, but also by topography, parent materials and hydrogeological conditions. Various kinds of soil associations are observed in such regions, including the branch-shaped soil association alongside rivers in hills, mountains and highland areas, fan-shaped soil associations in proluvial-fan or alluvial-fan terrains, comb-shaped soil associations in parallel range-gorge regions; and cone-shaped soil associations in areas of volcanic eruptions (Gong *et al.*, 2004).

5.4.2 Principles of soil division

Soil division enables us to categorise soil types in different areas according to their characters, in order to take best advantage of regional natural resources, and utilise soils sustainably in keeping with local conditions. Therefore, soil division has the dual functions of theory and practice, following characteristics of regionality, comprehensiveness, macrocosm and long-term strategy, and in keeping with the forming processes of each soil. Soil formation is a synthetic process such that, in relatively large areas, not only soil regional characteristics are relatively consistent, but also their formation processes.

The formation and distribution of soil is related to several natural and economic factors. Thus soil division cannot be based on unique phenomena, nor can it treat each factor equally, but must be based on analysis of the relationship between all factors including the dominant factor that explains most of the soils' regional differences, to determine the borders of soil division regions. In addition, not all soil division units of the country may be determined by the same key factors, with the factors causing regional variation differing between areas. Therefore different indices are adopted in each region to determine the soil division. The rules of regional soil differentiation are hierarchical, and the internal similarity and external differences of divisional unit systems are related such that any system of soil division should be a multi-level system.

It is common for land-use to combine agricultural production and environmental protection. The national condition of cultivated land resources roughly determines the economic pattern of small-scale peasant economies which, being widely dispersed, cannot be fully integrated with a supermarket economy, and cannot solve the problem of labour transfer from the country to the cities. Therefore current methods of soil division should incorporate both sustainable agricultural development and environmental protection, with their units, systems and indices selected to be meaningful for both. Soil division should emphasize providing basic farmland, preventing soil degradation, and developing advanced agricultural methods and technologies.

5.4.3 Characters of the three soil regions

A three-level soil division system comprising regions, zones and sections has been determined according to soil division principles as outlined above. Chinese soils as a whole are thus divided into three regions, 17 zones and 83 sections. The basic characteristics of the three soil regions are briefly described here.

The eastern moist soil region.—The eastern moist soil region is located southeast of a line running from the Da Hinggan Ling through the Taihang Shan and Qin Ling to the eastern edge of the Qinghai-Xizang Plateau. It covers an area of about 4 million km^2 - about 43% of China's landmass. The region is strongly influenced by the monsoon, and light, heat and water resources are fairly abundant. Annual precipitation ranges from 400 to 2 000 mm. The land comprises mainly low mountains, hills and plains, supplying a variety of soil parent materials (Institute of Geography - Chinese Academy of Sciences, 1959). The main natural vegetation is forest, with some forest-steppe.

The soils in this region are mostly ferrosols and argosols, with anthrosols, cambosols and primosols also present in considerable amounts. Soils to the south of the line of the Qin Ling and Huai He tend to be sticky and acidic; north of the line, except for the cold-temperate zone, they are lime-rich and alkaline, sometimes containing alkalic salts. On the Dongbei Pingyuan (Northeast China Plain), the soils are rich in organic material (Gong *et al.*, 1999; Gong *et al.*, 2007).

The eastern moist soil region provides the main agricultural areas of China, supports 81% of the population and produces 82% of its grain. Water is abundant, especially in the south of the region (Sun & Zhang, 2007), and aquatic resources and scope for tourism are great. Environmental impacts due to human activity are also significant in this region, and include deforestation (Chen & Jiang, 1996), water quality deterioration (Sun & Zhang, 2007) and increasing industrial waste, as well as frequent natural disasters (Zhang, 2000). There is a need, during utilisation of soil resources, for protection of natural forests, construction of artificial forests, protection of cultivated land, control of soil erosion and a comprehensive consideration of the environment.

The central dry and moist soil region.—The dry and moist soils of the central region are located to the east and south of a line of linking Helan Shan, A'nyêmaqên Shan and Nyainqêntanglha Shan, and the southern Xizang watershed, and to the west and north of the eastern moist soil region. The region covers about 2.1 million km^2– about 22% of China. This area is slightly affected by the monsoon, and the climate ranges from moist to semi-moist and partly arid. The natural vegetation is grassland and steppe, with forest-steppe on mountains. The average altitude of the region is around 1 000 m. The natural conditions are complex, giving a variety of soil types, as well as loess and accumulations of loose material. The main soils are cambosols and argosols; there are also isohumosols on the northern Nei Mongol Gaoyuan, primosols on the central Huangtu Gaoyuan and cambosols in the southern Hengduan Shan and the middle reaches of the Yarlung Zangbo Jiang (Gong *et al.*, 2007).

Agriculture, forestry and animal husbandry are the main economic activities in the central dry and moist soil region. The type, structure and function of farming have clear transitional characteristics: dry land is intermixed with scattered paddy fields, and areas of agriculture with those for animal husbandry. Both floods and drought are common problems, and earthquakes and landslides occur frequently due to the location of this region on a boundary between landforms in China. Management of soils in this region requires careful use of water resources, restrained exploitation of bare sandy land, strict protection of natural vegetation, provision of an artificial fodder base, and establishment of a disaster warning and prevention system.

The northwestern arid soil region.—The northwestern arid soil region is located to the west of the central region. It covers ca. 3.4 million km^2, 36% of the country. Located deep inland, the region is very arid, and this feature determines the local economic pattern, focused on animal husbandry with a little irrigated agriculture. The natural vegetation types are desert or desert steppe, and the soils mainly aridosols, halosols, cambosols and primosols. The soil composition is very coarse with very low organic matter content. Especially on the Qinghai-Xizang Plateau, the soil is very thin and simple in profile, with an obvious reaction to lime (Chen *et al.*, 1981; Gao *et al.*, 1981 et al., 1981; Gao *et al.*, 1985 1985).

This region supports only 4% of the Chinese population, and is one of the main centres for ethnic minority groups. The arid climate, lack of water resources and fragile environment allow for limited regional economic development, although the area is rich in natural resources such as oil and natural gas (Gong *et al.*, 1996). In recent years, a series of environmental problems have emerged due to economic development: forest loss, grassland degradation and aggravated desertification (Tang *et al.*, 1992; Liu *et al.*, 2001; Tang & Zhang, 2001; Sun & Zhang, 2007). To protect natural resources in this region would require limiting the population (Wu, 1994), protection of water resources and prohibition of unsustainable exploitation of wasteland (Gong *et al.*, 1996).

Soil is an essential prerequisite for plants, providing material basic to their growth, and soil properties can therefore significantly affect plant condition. Equally, the plant is an important factor affecting soil formation and has a protection role for the soil. Soil without vegetation cover is easily lost, causing soil degradation and desertification. This in turn reduces ecosystem production and crop yields, thus threatening food security. Soil degradation can also alter the energy balance and processes in the soil, affecting global climate and biogeochemical cycling of carbon, nitrogen, sulphur and phosphorus. All these changes can ultimately result in unstable governments and societies (Zhao & Shi, 2007), such that it is in all our interests to protect the environment, plants and soil.

References

Chen, B. M. & Jiang, S. K. (1996). *Water, Land and Climate Resources and Sustainable Development Potential of Farming, Forestry, Animal Husbantry and Fishery in China*. Beijing, China: China Meteorological Press.

Chen, B. R. & Zhang, J. (1991). Preliminary study on chemistry of leacheate and soils under coniferous forest on northern slope of Changbai Mountains. *Acta Pedologica Sinica* 28(4): 372-381.

Chen, H. Z., Gao, Y. X. & Wu, Z. D. (1981). The influence of plateau uplift on the formation of Alpine soil in Qinghai-Xizang region. *Acta Pedologica Sinica* 18(2): 137-147.

Chen, Z. C. & Zhao, W. J. (1989). The classification of ferralsols in China. *Soil* 21(2): 75-79.

Chen, Z. C. & Zhao, W. J. (1994). The red soil in China soil taxonomy (first proposal). In: Taxonomy, T. E. C. o. C. S. (ed.) *New View of China Soil Taxonomy*. pp. 170-178. Beijing, China: Science Press.

Cooperative Research Group on Chinese Soil Taxonomy (1991). *Chinese Soil Taxonomy (First Proposal)*. Beijing, China: Science Press.

Cooperative Research Group on Chinese Soil Taxonomy (1995). *Chinese Soil Taxonomy (Revised Version)*. Beijing, China: Chinese Agricultural Sciences and Technology Press.

Cooperative Research Group on Chinese Soil Taxonomy (2001). *Keys to Chinese Soil Taxonomy*. Hefei, China: Press of University of Science and Technology of China.

Dudal, R. & Eswaran, H. (1988). Distribution, properties and classification of vertisols. In: Wilding, L. P. & Puentes, R. (eds.) *Vertisols: Their Distribution, Properties, Classification and Management*. pp. 1-22. College Station, Texas: Texas A & M University Printing Center.

Gao, Y. X., Chen, H. Z. & Wu, Z. D. (1981). The Relationship of Paleosol and Uplift of Tibetan Plateau. In: The Scientific Expedition Group to Tibetan Plateau (ed.) *The Epoch, Scope and Form of Tibetan Plateau*. pp. 90-101. Beijing, China: Science Press.

Gao, Y. X., Chen, H. C., Wu, Z. D. & et al. (1985). *Tibetan Soil*. Beijing, China: Science Press.

Gao, Y. X. (1989). The classification of spodosols in China. *Soil* 21(2): 71-74.

Gao, Y. X., Guo, X. D. & Zhang, L. D. (1996). The compare study on the diagnostic characteristics of spodic horizon. *Chinese Journal of Soil Science* 27(3): 102-106.

Gong, Z. T., Zhao, Q. G., Zeng, Z. S., Lin, P. & Wang, R. C. (1978). Temporary draft of China soil taxonomy. *Soil* 5: 168-169.

Gong, Z. T. (1980). On the genetic classification of paddy soils in China. *In:* ISSAS (ed.) *Proceeding of Symposium on Paddy Soils*. pp. 129-138. Science Press and Springer-Verlag.

Gong, Z. T. (1983). Pedogenesis of paddy soils and its signifigance in soil classification. *Soil Science* 35: 5-10.

Gong, Z. T. (1986). Origin, evolution, and classification of paddy soils in China. *Advances in Soil Science* 5: 179-200.

Gong, Z. T. (1994). Formation and classification of anthrosols: China's perspective. *Transactions of the 15th World Congress of Soil Science* 6A: 120-128.

Gong, Z .T. (1999). China soil Taxonomy-theory, method and practice. Beijing: Science Press. 1-903

Gong, Z. T., Chen, H. Z. & Wang, H. L. (1996). Distribution of higher categories in Chinese soil taxonomy. *Scientia Geographica Sinica* 16(4): 289-297.

Gong, Z. T., Zhang, G. L. & Luo, G. B. (1999). Diversity of anthrosols in China. *Pedosphere* 9(3): 193-204.

Gong, Z. T., Zhang, G. L. & Qi, Z. P. (2004). *The Soil Series in Hainan Island*. Beijing, China: Science Press.

Gong, Z. T., Zhang, G. L. & Chen, Z. C. (2007). *Pedogenesis and Soil Taxonomy*. Beijing, China: Science Press.

Gong, Z. T., Huang, R. J. & Zhang, G. L. (in press). *Soil Geography in China*. Beijing, China: Science Press.

He, J. H., Shi, H., Lu, Z. X., Gong, Z. T., Liang, K. & Zhang, X. N. (1958). *Survey Report on Hainan Island Soil*. Beijing, China: Science Press.

He, Y. R. & Huang, C. M. (1995). Soil taxonomic classification in Yuanmou dry and hot valley, Yunnan Province. *Mountain Research* 13(2): 73-78.

He, Y. R., Huang, C. M., Gong, G. D., Zhang, D. & Yang, W. Q. (2001). An approach to the mechanism of soil degradation in dry and hot valley region of Jinsha River, Yunnan Province - effect of disparity in features of parent materials on soil degradation. *Southwest China Journal of Agricultural Sciences* 14(suppl.): 9-13.

Huang, C. M., He, Y. R., Zhang, D. & Zhu, H. Y. (2001). Mechanisms involved in soil degradation in dry and hot valleys in Jinshajiang River area, Yunnan Province: II soil water and soil degradation. *Resources and Environment in the Yangtze Basin* 10(6): 578-584.

Huang, R. C. & Wu, S. M. (1987). Geographical distribution of vertisols and vertic soils in China. *Journal of Nanjing Agricultural University* 10(4): 63-68.

HUANG, R. C. & WU, S. M. (1988). A discussion of the classification of vertisols for the second approximation of taxonomy of soils in China. *Journal of Nanjing Agricultural University* 11(1): 65-69.

HUANG, R. J. (1987). The pedogenesis and utilization of desert soil in China. *Collection of Geography* 19: 92-105.

HUANG, X. C. (1996). *The Progress on Swamp Research. Study on Physical Geography and Environment - Analects of Hang Xichou.* Beijing, China: Science Press.

INSTITUTE OF GEOGRAPHY - CHINESE ACADEMY OF SCIENCES (1959). *Integrated Natural Division in China.* Beijing, China: Science Press.

INSTITUTE OF SOIL - CHINESE ACADEMY OF SCIENCES (2001). *The Keys to China Soil Taxonomy.* Hefei, China: University Science and Technology of China Press.

ISSS/ISRIC (1995). *Reference Soil Profile of PRC.* Wageningen, The Netherlands.

ISSS/ISRIC/FAO (1998). *World Reference Base for Soil Resources.* Wageningen, The Netherlands and Rome, Italy.

LEI, W. J. & GU, G. A. (1996). Illuminate on recension on aridosols in China soil taxonomy. *Soil* 28(5): 232-236.

LI, Q. K. & SHI, H. (1959). *Accidence Soil Division of Guangdong, Hunan, Jiangxi and Guangxi in China.* Beijing, China: Science Press.

LI, Q. K. (1983). *Red Soil in China.* Beijing, China: Science Press.

LIN, P. T. (1996). *Soil Classification and Land Utilization in Ancient China.* Beijing, China: Science Press.

LIU, H. J. (1999). Discussion on problems on "China soil taxonomy". *Bulletin in Soil* 30(5): 201-202.

LIU, L. W. (1991). Formation and evolution of vertisols in Huaibei Plain. *Pedosphere* 6(2): 147-153

LIU, L. W. (1995). Study on the ages of vertisols. *Soil* 27(5): 274-278.

LIU, Z. L., ZHU, Z. Y. & HAO, D. Y. (2001). The study of oasis ecosystem damage and conservation in the lower reaches of Black River (Erginar River). *Journal of Arid Land Resources and Environment* 15(3): 1-8.

LIU, Z. Z. & CHEN, Z. X. (1990). The characteristics, genesis and classification of spodic soil in Taman Mountains area. *Records of Committee of Chinese Agricultural Chemistry* 28(2): 148-159.

NATIONAL SOIL SURVEY STAFF (1998). *China Soil Taxonomy.* Beijing, China: China Agriculture Press.

QIU, R. L., XIONG, D. X. & HUANG, R. C. (1994). Genesis and taxonomic classification of vertisols in Yunnan-Guangxi region of China. *Acta Pedologica Sinica* 31(3): 385-395.

SAMMES, M. E. (1999). *Handbook of Soil Science.* Boca Raton, FL: CRC Press.

SHARMA, B. D. (1969). Studies of Indian pollen grains in relation to plant taxonomy Sterculiaceae. *Proceedings of the National Institute of Sciences in India Part B - Biological Sciences* 35(4): 320-359.

SHI, C. H. & GONG, Z. T. (1995). Formation and classification of irragric soils in China. *Acta Pedologica Sinica* S2(4): 437-448.

SHI, X. Z., YU, D. S., XU, S. X., WARNER, E. D., WANG, H. J., SUN, W. X. & ZHAO, Y. C. (2010). Cross-reference for relating Genetic Soil Classification of China with WRB at different scales. *Geoderma* 155: 344-350.

SOIL SURVEY STAFF (2006). *Keys to Soil Taxonomy.* Washington, DC: USDA/NRCS.

SOUIRIJI, A. (1991). Classification of aridisols, past and present, proposal of diagnostic desert epipedon. *In: Sixth International Soil Correlation Meeting.* pp. 175-184. USA.

SUN, H. L. & ZHANG, R. Z. (2007). *Principle and Practice on Regional Rules in Environment Construction in China.* Beijing, China: Science Press.

TANG, C. Q., QU, Y. G. & ZHOU, W. C. (1992). *Utilization on Water Resources in Arid Region in China.* Beijing, China: Science Press.

TANG, C. Q. & ZHANG, J. W. (2001). Water resources and eco environment protection in the arid regions in Northwest of China. *Progress in Geography* 20(3): 227-232.

WANG, J. Z., MA, Y. L. & JIN, G. Z. (1996). *Irragric Soils in China.* Beijing, China: Science Press.

WILDING, L. P. & TESSIER, D. (1988). Genesis of vertisols: shrink-swell phenomena. In: Wilding, L. P. & Puents, R. (eds.) *Vertisols: Their Distribution, Properties, Classification and Management.* pp. 55-81. Texas A & M University Printing Centre.

WU, C. J. (1994). *Loaded Land - Population, Resource, Environment and Economy.* Beijing, China: People's Education Press.

XIONG, G. Y., ZHAO, Q. G. & WANG, M. Z. (1979). Podzolic soils in northern part of Daxinganling. *Acta Pedologica Sinica* 16(2): 110-126.

XIONG, Y. & LI, Q. K. (1987). *China Soil.* Beijing, China: Science Press.

ZENG, Z. S. (1997). *Chinese Planosol.* Beijing, China: Science Press.

ZHANG, F. R. (2000). *The Land Station in China.* Beijing, China: Kaiming Press.

ZHANG, J. M. & SHI, X. Z. (1994). Preliminary study on argosols in China Soil Taxonomy. In: The Editorial Committee of China Soil Taxonomy Books (ed.) *New view of China Soil Taxonomy.* pp. 242-247. Beijing, China: Science Press.

ZHANG, M. & GONG, Z. T. (1992). The distribution, characteristics and taxonomic of vertosols in China. *Acta Pedologica Sinica* 29(1): 1-17.

ZHANG, M. & GONG, Z. T. (1993). Characteristics and potentials productivity of vertisols resources in China. *Resources Science* 3: 1-10.

ZHANG, M. & LIU, L. W. (1993). Age and some genetic characteristics of vertisols in China. *Pedosphere* 3(1): 81-88.

ZHANG, Y. L., LI, B. Y. & ZHENG, D. (2002). A discussion on the boundary and area of the Tibetan Plateau in China. *Geographical Research* 21(1): 1-8.

ZHANG, Z. Y. & ZHANG, Z. F. (1986). Initial report on mechanism of genesis of Beijing soil in three-river-plain of Heilongjiang Province. *Journal of Heilongjiang August First Land Reclamation University* 1: 1-8.

Zhao, Q. G. & Zou, G. C. (1958). *The Soil and its Utilization in Leizhou Peninsula*. Beijing, China: Science Press.

Zhao, Q. G. (1989). The characteristics, taxonomy and utilization of spodosols. *Soil* 21(1): 1-4.

Zhao, Q. G. & Shi, X. Z. (2007). *The General Discussion on Soil Resources*. Beijing, China: Science Press.

Zhao, S. Q. (1986). *Physical Geography of China*. Beijing, China and New York: Science Press and John Wiley & Sons.

Zhao, W. J. & Chen, Z. C. (1995). Discussion on the foundation of ferrosols. *Acta Pedologica Sinica* 32(suppl.): 21-33.

Zhao, W. J. & Chen, Z. C. (1999). Ferrosols. In: Gong, Z. T. (ed.) *China Soil Taxonomy - Theory, Method and Practice*. pp. 488-534. Beijing, China: Science Press.

Zhu, H. J. & Tan, B. H. (1989). Study on the characteristics of vertisol in Fujian Province. *Acta Pedologica Sinica* 26(3): 267-297.

Chapter 6

Origin and Development of the Chinese Flora

ZHOU Zhe-Kun

6.1 Introduction

This chapter focuses on Chinese angiosperms, including early examples of the group and the origin of major angiosperm clades, the history of key angiosperm families with good fossil records in China, biogeography of "living fossil" species, and the origin of endemic elements in the Chinese flora.

6.2 China's earliest angiosperms

China's flora is incredibly diverse, in terms not only of extant species, but also of angiosperm fossil records dating from the Early Cretaceous Period (at least 125 million years ago). In addition to unequivocal angiosperm fossils, there are numerous reports of obscure fossils from this time, which are in poor condition or do not provide sufficient characteristics to confirm their angiosperm affinity. Examples include *Amesoneuron* (Zhou *et al.*, 1990), *Archaeamphora* (Li, 2005), *Archimagnolia* (Tao & Zhang, 1992), *Asiatifolium elegans* (Sun *et al.*, 1992), *Baisia* (Wang, 1984), *Carpolithus* (Zheng & Zhang, 1983; Liu, 1988; Wu, 1999), *Celastrophyllum* (Zheng & Zhang, 1983), *Chaoyangia* (Duan, 1997), *Chengzihella obovata* (Sun *et al.*, 1992), *Lilites*, *Orchidites* (Wu, 1999), *Polygonites*, *Potamogeton* (Yabe & Endo, 1935) and *Typhaera*. These equivocal fossils will not be discussed further in this chapter.

Several fossils discovered at the end of the twentieth century from the Early Cretaceous (125–127 million years old) Yi Xian Formation in northeastern China, are critical to understanding the origin and development of angiosperms. *Archaefructus*, a genus of three species, *A. liaoningensis* (Sun et al., 1998), *A. sinensis* (Sun et al., 2002) and *A. eoflora* (Ji *et al.*, 2004); *Hyrcantha* (also known as *Sinocarpus*; Leng & Friis, 2003; Leng & Friis; Dilcher *et al.*, 2007); and *Leefructus mirus*, a well-preserved fossil with fruit and detailed leaf characters (Sun *et al.*, 2011). *Archaefructus liaoningensis* and *A. eoflora* were found in the lower sediments (estimated at 127 million years old), and *A. sinensis*, *Hyrcantha* and *Leefructus* in the upper layer (125 million years old; Sun *et al.*, 2010). *Archaefructus* has simple determinate axes bearing helical conduplicate carpels enclosing several ovules in each. There appears to be an elongated adaxial region that may have been stigmatic; stamens are often paired or there may be three or four attached to the end of a short shoot. The anthers appear to produce monosulcate pollen, and the leaves, with petioles of varying lengths, terminate in highly-dissected leaves (Sun *et al.*, 2008). Based on these anatomical characters and a cladistic analysis, the new family Archaefructaceae has been established and inferred to lie on the most basal branch of angiosperms (Sun *et al.*, 2002).

The characteristics of the whole *Hyrcantha* fossil, including roots, stems, and branch termination, suggest it was an aquatic plant with emergent reproductive organs (Dilcher *et al.*, 2007). Leng and Friis (2003; 2006) proposed that *Hyrcantha* represents a basally-branching eudicot, whereas Dilcher *et al.* (2007) suggested it was an extinct angiosperm without phylogenetic affinity to any modern taxa. The recently described *Leefructus mirus* is represented by a well-preserved fossil in which many morphological characters are clearly defined, including multi-stranded, upright, herbaceous stems, several leaves clustered at two distinct nodes, three-lobed leaves with the lobe margins further lobed and a unique venation pattern, and a long, simple, auxiliary reproductive pedicle terminating in a collection of five, basally-fused follicles sitting on a flattened receptacle with several small scars around its base. All these characters indicate a close relationship to Ranunculaceae, indicating that *L. mirus* represents the earliest reliable record of eudicots to date (Sun *et al.*, 2010).

From the fossil records described above, it can be concluded that angiosperms were already present in northeastern China 127 million years ago and, by 125 million years ago, at least four different taxa were present. Eudicot pollen is also known to have been present by at least 127 million years ago, and diversified basally-branching eudicot lineages such as *Leefructus* by 122--124 million years ago, indicating that the eudicots had emerged by the Early Cretaceous Period.

6.3 The appearance of the major angiosperm clades in China

In China, Wu *et al.* (1998; 2002) have divided the angiosperms into eight classes: Magnoliopsida, Lauropsida, Piperopsida, Caryophyllopsida, Liliopsida, Ranunculopsida, Hamamelidopsida and Rosopsida. In comparison, the international scientific community recognises eight main monophyletic clades: Amborellales, Nymphaeales, Austrobaileyales, magnoliids, Chloranthales, monocots, Ceratophyllales and eudicots (APG, 1998; APGII, 2003; APGIII, 2009). In this chapter we discuss the history of angiosperms in the context of both systems.

Amborellales is a monotypic order (APGIII, 2009), found within Magnoliopsida in the Wu *et al.* (1998) system. The single species, *Amborella trichopoda,* is a vesselless shrub distributed in the rainforests of New Caledonia, and unfortunately no fossil record of it has yet been found.

Nymphaeales (APGIII, 2009) comprises three families, Cabombaceae, Hydatellaceae and Nymphaeaceae, and again falls within Magnoliopsida in Wu and colleagues' eight class system. Rich fossil records of Nymphaeales are found throughout the world. These include unequivocal waterlily fossils from the Early Cretaceous Period (115–125 million years ago) of Portugal (Friis *et al.*, 2001; Friis *et al.*, 2009) and a *Microvictoria* fossil from the Turonian Age (Late Cretaceous, ca. 90 million years ago) of the USA considered to have affinities to Nymphaeaceae (Gandolfo *et al.*, 2004). Some authors have suggested that *Archaefructus* is an extinct member of Nymphaeaceae (Friis *et al.*, 2003) or sister-group to Hydatellaceae (Endress & Doyle, 2009); under these hypotheses, *Archaefructus* would be included in Nymphaeales and China would therefore be the source of the oldest fossil record of Nymphaeales.

Austrobaileyales include Austrobaileyaceae, Schisandraceae and Trimeniaceae under the APGIII system (APGIII, 2009), all three of which are placed in Magnoliopsida in Wu and colleagues' eight class system. There is no known fossil record of Austrobaileyaceae. The earliest Trimeniaceae record comprises seeds from the Early Cretaceous Period (100 million years ago) from Northern Hemisphere Hokkaido, Japan (Yamada *et al.*, 2008), although Trimeniaceae is now distributed mainly in subtropical to tropical Southeast Asia, eastern Australia, and the Pacific Islands.

The magnoliids (Magnoliidae) comprise four orders under the APG system—Canellales, Piperales, Laurales and Magnoliales; under Wu and colleagues' system Magnoliopsida is divided into five subclasses including 11 orders. This group has an abundant fossil record, including *Cronquistiflora* and *Detrusandra* flowers with probable Magnoliales affinities reported from the Late Cretaceous Period (91–98 million years ago) and Lauraceae fossil flowers and inflorescences from the Cenomanian Age (Late Cretaceous, 91–98 million years ago), both from the USA. Magnoliid fossils are also common in China. The earliest Chinese fossil with affinities to Magnoliidae is a receptacle from the Aptian or Albian Age (Mid Cretaceous) of northeastern China (Tao & Zhang, 1992). *Cinnamomum* and other taxa of magnoliid affinities appeared in the Late Cretaceous Period or Paleocene Epoch, and magnoliid elements are common from this time. A detailed fossil history of Magnoliaceae is provided later in this chapter.

Chloranthales is treated as a monofamilial order in both the APG (APGIII, 2009) and Wu *et al.* (1998; 2002) systems; in the latter it is included in Lauropsida, but is probably sister to Magnoliidae (Moore *et al.*, 2007). The earliest reliable fossil record of Chloranthaceae is a stamen found from the Aptian Age (Early Cretaceous, 93-133 million years ago) strata of the USA (Friis *et al.*, 1986). Another, *Chloranthistemon endressii* is reported from the Santonian Age (Late Cretaceous) of Sweden. There are no macrofossil records of Chloranthaceae from China, although pollen fossils are reported from northeastern China in the Late Cretaceous Period, and become common in the Cenozoic Era (Song *et al.*, 2004).

Monocots are an independent clade in the APG system, referred to as Liliopsida in Wu and colleagues' eight class system. The earliest monocot flower fossils have been found in sediments of the Turonian Age (Mid Cretaceous, 90 million years ago) of New Jersey, USA (Gandolfo *et al.*, 1998)). Many putative monocot fossils have been found in Chinese sediments from before the Late Cretaceous, including *Amesoneuron* (Zhou *et al.*, 1990), *Eragrosites changii* and *Liaoxia chenii* (Cao et al., 1998), *Graminophyllum* (Guo & Li, 1979), *Lilites reheensis* and *Orchidites linearifolius* (Wu, 1999) and *Potamogeton jeholensis* (Zhang, 1980). However, further studies are needed in order to confirm the phylogenetic affinities of these fossils.

Ceratophyllales are small, monogeneric group and sister to the eudicots (APGIII, 2009). Ceratophyllaceae is included in Magnoliopsida in the Wu *et al.* (1998; 2002) system. A single fossil of Ceratophyllales reported has been reported from the Miocene Epoch of Shanwang, Shandong, in eastern China (Writing Group of Cenozoic Plants of China, 1978).

Eudicots are the largest group of angiosperms and include four of the eight classes of the Wu *et al.* (1998; 2002) system: Caryophyllopsida, Ranunculopsida, Hamamelidopsida, and Rosopsida. Both *Leefructus mirus* (Sun *et al.*, 2010) from the Early Cretaceous Period, 124 million years ago, and *Hyrcantha decussata* (Leng & Friis, 2003; Leng & Friis, 2006) also from the Cretaceous Period of northeastern China, are believed to be of eudicot affinity. *Nordenskioldia* (Trochodendraceae) fossils were widely distributed in the Late Cretaceous Period, with fossils reported from northeastern China, Guandong, and the Paleocene sediments at Altay, Xinjiang.

6.4 Fossil histories of major groups in the Chinese flora

Fossil records play a very important role in determining the evolution and diversification of extant angiosperm families. There have been various attempts to review the fossil histories of extant modern families in order to understand their evolution and biogeography. Here we discuss the angiosperm groups that are common in modern Chinese vegetation and have rich fossil records: Fabaceae, Fagaceae, Lauraceae and Magnoliaceae. Some other families very rich in extant diversity, such as Asteraceae, Orchidaceae and Poaceae, are not well represented in the fossil record, and are therefore not discussed here.

6.4.1 Fabaceae

The reported fossil record of Fabaceae in China extends back to the Late Cretaceous Period, but due to poor preservation and rarity of specimens, the earliest records are unreliable. The first reliable legume fossils are fossil fruits of *Acacia aguilona* and leaflets of *Mimosites variabilis*. Both are from the Eocene Fushun coal field, Liaoning, northeastern China. Other Eocene Fabaceae leaflets include *Campylotropis* sp. and *Lespedeza* sp. from Weinan, Shaanxi, central China. Several other legume fossils have been reported from the Eocene Mancer Formation in Gar, Xizang, including *Cassia marshalensis* and *Lespedeza* (Tao, 1965; Writing Group of Cenozoic Plants of China, 1978).

Only two species of Fabaceae have been described from China in the Oligocene Epoch: *Erythrophleum ovatifolium* and *Cercis* sp., both from Jinggu, Yunnan (Writing Group of Cenozoic Plants of China, 1978), the former comprising a well-preserved leaflet impression, the latter two fruit impressions. In the Miocene Epoch, however, Fabaceae are well represented in the fossil record throughout China. At least 15 genera and 28 species have been recorded from about ten localities (Guo & Zhou, 1992). Most of these fossils are well-preserved leaflet impressions and some are impressions with a cuticle. Several fruit impressions have also been recovered.

The fossil record of Fabaceae from the Pliocene Epoch and Quaternary Period is scanty, but it may be that floras from these periods have not comprehensively been studied in China. Only two Pliocene legumes have been described, both from Taigu, Shanxi: *Leguminosites climensis* and *Desmodium* cf. *microphyllum*. *Leguminosites climensis* was determined by Rüffle (1963) as *Podogonium oehningensis*. From the Quaternary Period, a wood fossil from Zhoukoudian, Beijing Shi, has been identified as *Cercis balcki*, and leaflet fossils of three species from Xixabangma, Xizang have also been reported (Guo & Zhou, 1992; see Figure 6.1).

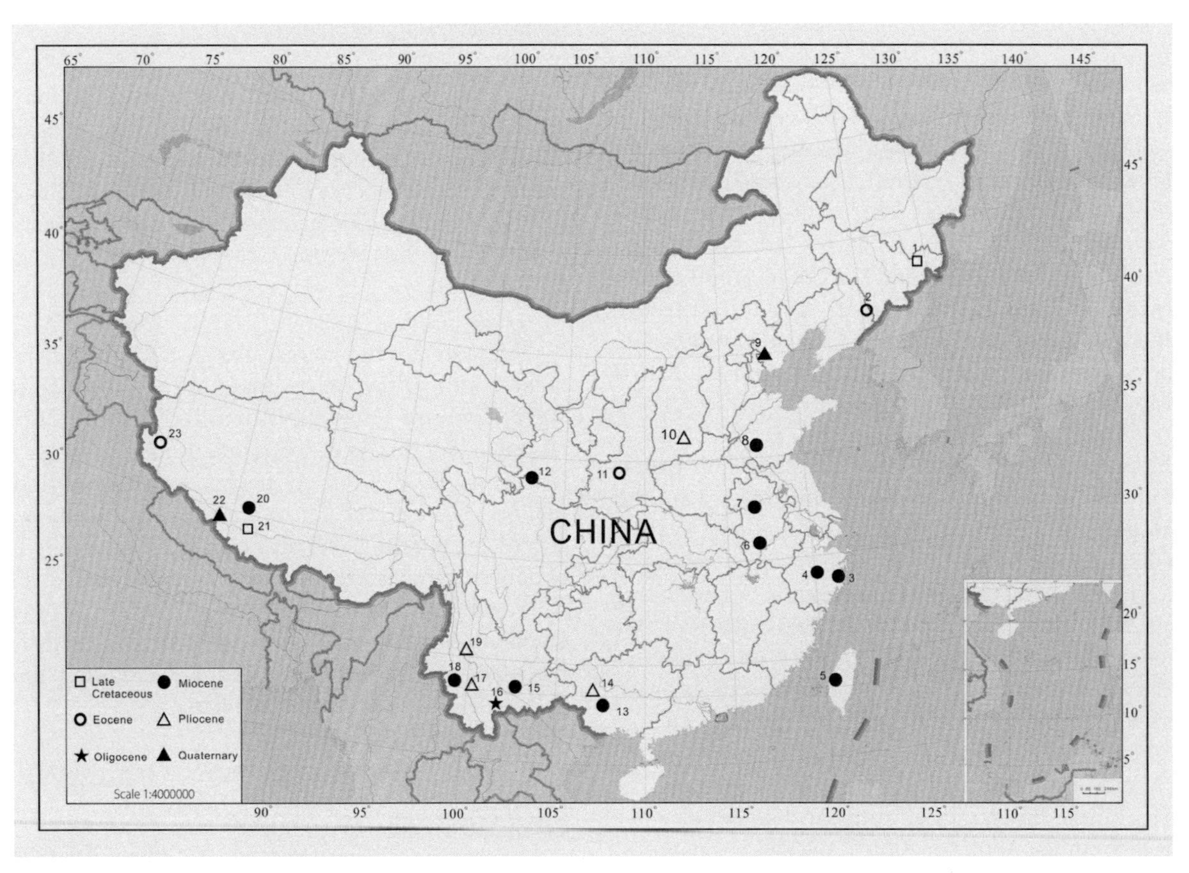

Figure 6.1 Geological distribution of Chinese Fabaceae fossils.

6.4.2 Fagaceae

Fossil records of Fagaceae have been reported from the Late Cretaceous Period and throughout the Cenozoic Era (Figure 6.2). Extant Fagaceae are widely distributed throughout the northern hemisphere, from North America to Europe and Asia. The family's wide distribution, rich fossil record and ecological and economic importance have attracted much research including attempts to use the fossil history of the family to interpret its evolution and biogeography. However, the known record is not yet sufficient to provide a comprehensive history of the family. Two fossil staminate flowers, fruits and cupules of Fagaceae (Herendeen *et al.*, 1995; Sims *et al.*, 1998), along with cupules and leaves of *Fagus* (Manchester & Dillhoff, 2004) have been reported from Late Cretaceous North

America. Providing many detailed reproductive characters, these fossils can provide key insights into understanding evolutionary processes. Here we review the fossil records of Fagaceae which provide a basis for understanding the evolutionary and biogeographical history of the family.

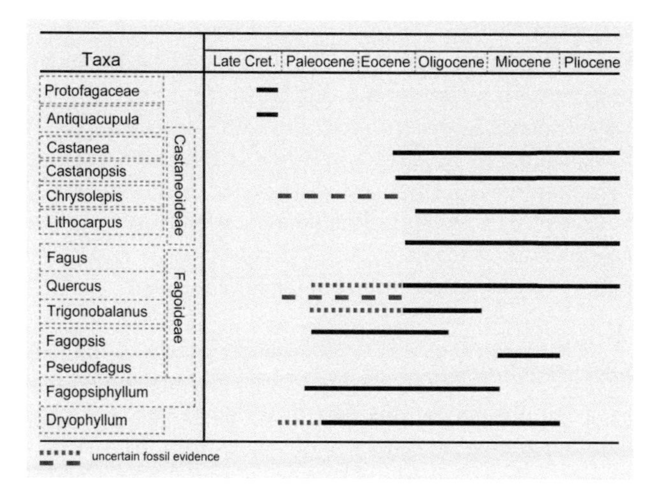

Figure 6.2 Stratigraphic distribution of Fagaceae. Solid lines indicate geological distribution; dashed lines indicate obscure fossil records.

The earliest Fagaceae records known date to the Late Cretaceous Santonian Stage of Georgia, USA, and include two genera of cupulate fruits and flowers: Protofagacea and Antiquacupula (Herendeen *et al.*, 1995; Sims *et al.*, 1998). Protofagacea was established for staminate flowers with associated fruit cupules found from the Buffalo Creek part of the Gaillard formation in central Georgia, and is strongly linked to Fagaceae on the basis of numerous characters including the presence of a cupule, trimerous unisexual flowers in separate dichasial inflorescence units, trigonous and lenticular fruits, and inferior ovaries. *Antiquacupula*, from the same location, has cupules, trimerous flowers and an inferior, trilocular ovary with two apically attached, pendulous, anatropous ovules per locule (Sims *et al.*, 1998). Although they share some characters such as their floral structure, dorsifixed staminate flowers and tricolporate pollen, these two taxa differ in several significant features: *Antiquacupula* has swollen stamen bases, a knoblike extension of the connective and multicellular glands lacking from Protofagacea, and the two genera have differing pollen sizes, exine sculpture and fruit shape. The fact that two distinct genera of Fagaceae occurred in the same deposit suggests that the family originated earlier than the Late Cretaceous Period, in order that it had time to diversify by this date. These Late Cretaceous fossils include some Fagaceae characters but differ from modern types in a number of ways, indicating that they are not members of modern Fagaceae and are assigned to "pre-Fagaceae." Below we consider some Fagaceae records, genus by genus.

Fagaceae records by genus.—Castanea is a small deciduous genus of ten to 12 species. Many fossil leaves and some cupules are attributed to *Castanea*. However, the leaves are very similar to some *Quercus* species and the cupules similar to *Castanopsis*. *Castanea* leaves lack the fimbrial veins normally found in *Quercus* leaves (Zhou *et al.*, 1995), a character which has been used to reassign many of the fossils previously ascribed to *Castanea*; hence there is no reliable fossil record of *Castanea* in Paleogene China. More reliable leaf fossils of *Castanea* have been reported from both northern and southern China from the Miocene Epoch onwards (Writing Group of Cenozoic Plants of China, 1978; Liu *et al.*, 2011; Yao *et al.*, 2011). A partial, immature, pistillate inflorescence found from the Paleocene-Eocene boundary of western Tennessee, USA, has been identified as a new genus and species, *Castanopsoidea columbiana*. The pollen morphology, style and stigma characters of this fossil are castaneoid in appearance. Of the modern castaneoids (*Castanea, Castanopsis, Lithocarpus* and *Chrysolepsis*), it is most similar to *Castanopsis* (Crepet & Nixon, 1989) , and were it to be assigned to this genus it would be the earliest *Castanopsis* record. There are a total of six unequivocal Miocene and Pliocene *Castanopsis* fossil records from Yunnan and Sichuan, making China the richest source of fossils of this genus (Guo, 1978; Writing Group of Cenozoic Plants of China, 1978; Zhou, 1999).

Lithocarpus fossils have been reported from Oligocene Germany: *L. saxonicus* has been assigned to the genus based on cuticle characters, particularly its finger-like appressed hairs (Kvaček & Walther, 1989). Leaf imprints similar in gross morphology to *Lithocarpus* have been reported from China from the Oligocene Epoch to the Pliocene Epoch (Writing Group of Cenozoic Plants of China, 1978; Zhou, 1999).

Fagus has a widespread fossil record in America, Europe and Asia (Iljinskaya, 1982; Kvaček & Walther, 1991; Tanai, 1995; Liu *et al.*, 1996; Denk & Meller, 2001). Manchester and Dillhoff (2004) recently reported the oldest *Fagus* fossils yet known (*F. langevinii*) from the Mid Eocene of western North America. The species was represented by leaves with long petioles, cupules and pollen. The cupule is composed of four ovate valves and spiny appendages with a long peduncle; spiny four-valved cupules completely enclosing two trigonal nuts are diagnostic characters for *Fagus*. Among modern *Fagus* species this fossil seems closest to the New World *F. grandifolia*. In China, *Fagus* leaf fossils have been reported from the Eocene Epoch of Fushun and Yilan in the northeast and Panxian in the southwest (Writing Group of Cenozoic Plants of China, 1978; Zhang, 1983; He & Tao, 1994; Zhou, 1999).

Nixon and Crepet (1989) divided the genus *Trigonobalanus* into three monotypic genera, *Trigonobalanus, Formanodendron* and *Colombobalanus*. Fossils of these three taxa are rare; the first report of a trigonobalanoid fossil was recorded in Upper Eocene Baltic amber, and is now considered to be a species of *Fagus*. The oldest *Trigonobalanus* fossil is now reported from the Paleocene-Eocene of Tennessee, USA. This fossil includes pistillate inflorescences and young fruits, dispersed fruits with and without cupules, and staminate catkins (Crepet & Nixon, 1989). Kvaček and Walther (1989) have reviewed the *Trigonobalanus* fossil records of Europe, and found their affinities of to be uncertain. There are no *Trigonobalanus* fossil records from Asia, despite the presence of modern *Trigonobalanus* species in East Asia.

The *Flora of China* treats *Quercus* and *Cyclobalanopsis* as separate genera (Huang *et al.*, 1999), based on whether the cupule is imbricate-scaled or lamellate (Camus, 1936-1954), although elsewhere these are often treated as subgenera of *Quercus*. *Quercus* is the largest genus of Fagaceae. Reliable fossil records of *Quercus* (see Figure 6.3) from the Eocene Epoch include acorns of *Q. paleocarpa* from America (Manchester, 1994) and leaves (with cuticle) of *Q. subhercynica* from the Mid Eocene of Germany, which are considered to be the earliest European record (Kvaček & Walther, 1989). In China, two leaf remains identified as *Quercus* were found in the Eocene Epoch of Funshun and Huadian in the northeast of the country (Writing Group of Cenozoic Plants of China, 1978), and *Quercus* fossils are common from the Neogene Period onwards: at least six species have been recorded based on leaf fossils throughout China (Writing Group of Cenozoic Plants of China, 1978; Li & Yang, 1984; Zhou, 1999; Liu *et al.*, 2011). The 9--11 *Quercus* species sometimes grouped as section *Heterobalanus* have a distribution centred on the Himalaya Shan and Hengduan Shan, where they remain dominant species today, playing a very important role in both ecosystems. Their fossil records are abundant in Neogene sediments of both regions. When the fossil taxonomy of the section was reviewed (Zhou, 1992), seven species were recognized. The oldest records are *Q. namlingensis* and *Q. wulongensis* from the Mid Miocene (15 million years ago) of Namling, Xizang (Zhou, 1992).

Cyclobalanopsis naitoi is considered an unequivocal representative of *Cyclobalanopsis* from the Eocene Epoch of Japan, and includes both leaves and depressed acorns (Tanai, 1995). The earliest Chinese record of *Cyclobalanopsis* is from the Oligocene Jinggu flora, southern Yunnan. Thirteen species of fagaceous fossils have been reported from the Oligocene Epoch of Yunnan in southwest China (Li & Guo, 1976; Writing Group of Cenozoic Plants of China, 1978; Zhou, 1992; Zhou, 1999). All eight fossil species attributed to *Cyclobalanopsis* were evergreen trees (Writing Group of Cenozoic Plants of China, 1978). Today, one species is an important component of evergreen broad-leaved forests in eastern Asia, indicating that these forests have occurred in Yunnan since the Oligocene Epoch (See Figure 6.3). There are also several extinct genera of Fagaceae. For example *Pseudofagus*, reported from the Clarkia flora of Idaho, USA, is closely related to *Fagus*, and believed to have gone extinct in the Eocene Epoch.

The presence of two fossil Fagaceae genera in the Late Cretaceous Period shows that the family originated and began to diversity at this time or earlier (Figure 6.2, page 89). This timing is roughly synchronous with that of many modern dicot families. The family underwent a major diversification during the Eocene Epoch, by which time all modern subfamilies were present. However, these Late Cretaceous and Eocene fossils differ from modern families and genera, for instance in having very small pollen grains with reticulate to microfoveolate exine ornamentation, well-developed nectary lobes and multicellular glands on the ovary surface. Many Oligocene fossils, however, can be recognized to sectional level. This indicates that modern Fagaceae genera did not evolve until the Eocene Epoch, and the Eocene-Oligocene boundary was likely a period of modernization in the family.

Most extant and fossil Fagaceae occur in the Northern Hemisphere, with a few *Quercus* and *Trigonobalanopsis*

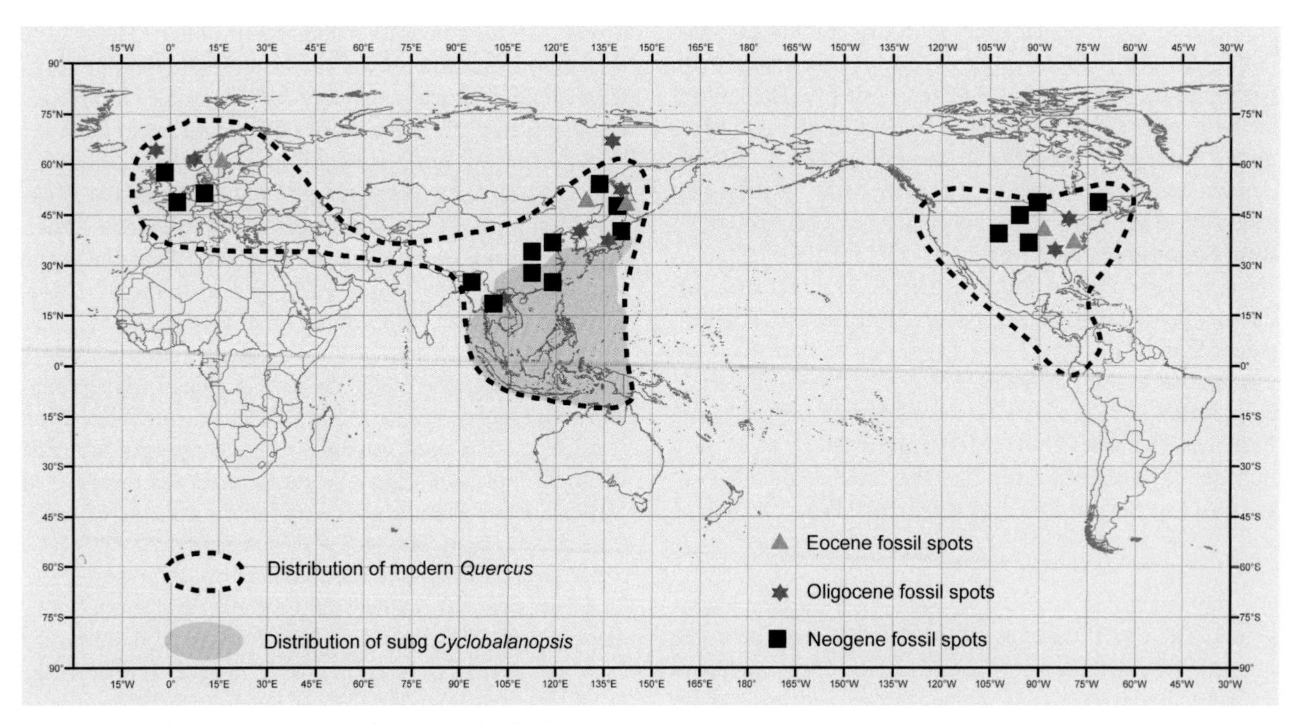

Figure 6.3 Distribution of extant and fossil *Quercus and Cyclobalanopsis.*

species in the Southern Hemisphere. Fagaceae are therefore assumed to have originated in the north. Eastern Asia is the centre of modern generic diversity for the family, with all genera except *Chrysolepis* found here. *Cyclobalanopsis* and *Castanopsis* are found only in eastern and Southeast Asia, as are most species of *Lithocarpus*. Only *Castanea, Fagus* and *Quercus* are today known from Europe. However, the oldest Fagaceae fossils are recorded from North America, and during the Late Cretaceous and Paleogene Periods the family was more diverse in northern America than in eastern Asia, with most genera found in the former. In fact there are no reliable Late Cretaceous Fagaceae fossil records in eastern Asia, although *Quercus* and *Fagus* are reported from Japan and China, and *Castanopsis, Lithocarpus* and *Cyclobalanopsis* from Japan. Eastern Asian Fagaceae fossil records become richer into the Oligocene Epoch. Given the longer history and more intensive study of paleobotany in North America compared to eastern Asia, these discrepancies may be attributable to sampling bias. It is, however, clear that in Europe, Fagaceae diversified later than in either North America or eastern Asia.

Fossil Fagaceae leaves of uncertain generic affinity, with craspedodromous venation and regularly-spaced marginal teeth have often been placed in the genus *Dryophyllum*, a problematic genus in terms of nomenclature, taxonomy and systematics. A thorough re-examination of leaves from the Eocene Epoch of southeastern North America by Jones *et al.* (1988), based on epidermal anatomy as well as leaf architecture, indicated that many species bore closer affinity to Juglandaceae. However, the name *Dryophyllum* continues to be used for Palaeogene fagaceous fossils in Eurasia. The "*Dryophyllum* complex" of fossils from Eocene deposits in Tennessee and Kentucky has been divided into three taxa, two of which were placed in new fagaceous genera: *Berryophyllum,* characterized by leaves with craspedodromous venation, percurrent tertiary veins, and spinose teeth; and *Castaneophyllum* with regularly spaced, slightly spiny, teeth, each fed by a secondary vein. These leaves may well, in fact, represent the extant genus *Castanea*, because there are associated *Castanea*-like cupules and staminate inflorescences with *in situ* pollen (Crepet & Daghlian, 1980; Jones *et al.*, 1988). It is known that the fossils of the *Dryophyllum* complex first occurred in North America in the Late Cretaceous Period, and were widely distributed throughout North America, eastern Asia and Europe in the Eocene Epoch. The complex continued to diversify in eastern Asia and Europe in the Oligocene Epoch, became restricted to eastern Asia in the Miocene Epoch, and became extinct in the Pliocene Epoch.

6.4.3 Lauraceae

Lauraceae is one of the largest plant families, with over 50 genera and 2 500–3 000 species (Rohwer, 1993), comprising mainly trees and shrubs. Lauraceae are usually dominant components of subtropical to tropical broad-leaved evergreen forests, making them important indicators of paleovegetation and paleoclimate (Hu *et al.*, 2007). The family also provides an excellent example of the macrofossil record complementing the palynological record, as the pollen grains are only preserved under exceptional situations due to their highly reduced exine frequently combined with extremely limited production (Erdtman, 1952; MacPhail, 1980; Herendeen *et al.*, 1994). Fossil records of Lauraceae are well represented in the Cretaceous and Tertiary Periods worldwide (LaMotte, 1952; Taylor, 1988), with the earliest record being a charcoalified fossil flower, *Potomacanthus lobatus*, from the Early Cretaceous (Early-Mid Albian Age) of Virginia, USA (Von Balthazar *et al.*, 2007). In China, fossil records of Lauraceae date back to the Late Cretaceous Period (Zhang, 1984), and were widely distributed at this time, being found in both high latitude (Jiayin, Heilongjiang and Dalazi, Jilin; Zhang, 1984; Tao & Zhang, 1990) and low latitude floras (Bali in Guangxi; Guo, 1979).

Only one lauraceous fossil assigned to *Cinnamomum* has been found in the Paleocene Epoch, from the Shanshui flora, Guangdong, southern China (Guo, 1979), but such fossils became quite common and widespread in Eocene floras such as those at Fushun, Liaoning (Li *et al.*, 1995), Changchang, Hainan (Li *et al.*, 2009) and Shinao, Guizhou (Li *et al.*, 1995). Oligocene Lauraceae fossils are found only in southern and southwestern China (Writing Group of Cenozoic Plants of China, 1978), perhaps due to climatic fluctuations following the Eocene Epoch: in the Oligocene Epoch, the climate became cooler and broad-leaved forests correspondingly declined (Zachos *et al.*, 2001).

Lauraceae have an abundant fossil record in the Neogene Period, particularly in the Mid Miocene Epoch when the climate was much warmer than it is at present (Zachos *et al.*, 2001). During this time the distribution of Lauraceae reached mid to high latitude areas (Writing Group of Cenozoic Plants of China, 1978). The Late Miocene climate was also more humid and warmer than at present, and at this time Lauraceae became one of the dominant components of broad-leaved evergreen forests (Zhou, 1985; Tao *et al.*, 2000; Wu, 2009; Xing, 2010). Several Late Miocene floras from southwestern China have been well studied, including Xiaolongtan (Zhou, 1985), Lincang (Tao & Chen, 1983) and Xianfeng (Xing, 2010). Lauraceae and Fagaceae were the dominant families in these subtropical broad-leaved evergreen forest floras. Once the climate began cooling again after the Miocene Epoch into the Pliocene Epoch, lauraceous plants became known only from low latitude floras (Tao *et al.*, 2000; Wu, 2009; Dai *et al.*, 2010).

The Quaternary Period was a cold period, in which only a few records of Lauraceae have been reported, from southern China (Liu, 1990). In the Pleistocene Baisheling flora, dominant elements include *Cinnamomum, Litsea, Quercus, Phoebe* and *Machilus,* indicating a subtropical broad-leaved evergreen forest. In sum, fossil records of Lauraceae in China date back to the Late Cretaceous Period and became abundant after the Paleocene Epoch.

The distribution of Lauraceae corresponds to climatic fluctuations, with a wide distribution during the warm Miocene Epoch, becoming restricted to low latitude areas during the cold Quaternary Period.

6.4.4 Magnoliaceae

Magnoliaceae contains approximately 220–240 species, mainly occurring in temperate, subtropical and tropical regions of the Northern Hemisphere. The *Flora of China* recognises 13 genera. The family has a disjunct distribution with roughly two-thirds of the species currently distributed in eastern and Southeast Asia and the other third in eastern North America and South America (Law, 1984; Nooteboom, 1993; Nie *et al.*, 2008). Such disjunct intercontinental patterns have attracted much interest from botanists over the past century and are considered to be among the most complex biogeographic patterns observed at the global scale (Wen, 2001; Milne & Abbott, 2002; Donoghue & Smith, 2004; Milne, 2006; Nie *et al.*, 2008). Magnoliaceae is a good model for examining these disjunctions. Compared to the abundant records in other regions of the world, Magnoliaceae fossils are relatively rare in China.

There are no credible fossil records for Magnoliaceae from before the Cretaceous Period (Zhang, 2001). Fossil wood, leaves and pollen from as far back as the Jurassic Period (Sahni, 1932; Simpson, 1937; Pan, 2000) have been rejected as Magnoliaceae (Hsu & Bose, 1952; Hughes & Couper, 1958). For example, Sahni (1932) published a record of petrified wood, *Homoxylon rajmahalense*, with affinity to Magnoliaceae, from the Jurassic Period of India, but Hsu & Bose (1952) reassigned this to Bennetitales. To date, the earliest and most reliable fossil of Magnoliaceae is *Archaeanthus* from the Late Albian Age (Early Cretaceous) Dakota Formation at Linnenberger Ranch, Kansas, USA. Dilcher & Crane (1984) interpreted several fossils —impressions of elongate floral axes with numerous multiovulate fruiting follicles (*Archaeanthus linnenbergeri*), bilobed leaves (*Liriophyllum kansense*), isolated perianth parts (*Archaepetala beekeri* and *Archaepetala obscura*) and bud-scales (*Kalymmanthus walkeri*)—as being from the same plant based on the possession of similar resin bodies. Morphological analyses confirmed a relationship to Magnoliaceae (Doyle & Endress, 2010). From the same period, *Lesqueria*, with similar fruits to *Archaeanthus*, may also belong to Magnoliaceae (Crane & Dilcher, 1984).

In China, the earliest fossil record of Magnoliaceae dates back to the Early Cretaceous Period. *Archimagnolia rostrato-stylosa* was uncovered in the last century from the Dalazi Formation (Aptian to Albian Ages) in the northeast of the country (Tao & Zhang, 1992). It comprises the impression of a floral axis (receptacle) with ca. 20 helically-arranged carpels. The base is shallowly cup-shaped and there are two prominent, narrow, elliptical scars at the base, which are thought to mark the position of tepals (Tao & Zhang, 1992). This fossil has been assigned to Magnoliaceae due to the markedly aggregate follicles and spirally arranged tepals, although this identification is uncertain due to the poor preservation of the specimen (Wu *et al.*, 1998), and needs further confirmation.

Compared to the scarce fossil records of the Early Cretaceous Period, those from the Late Cretaceous Period are relatively abundant. Around 20 mega-fossil species are reported from the Northern Hemisphere, including some extinct and two extant genera, *Magnolia* and *Liriodendron* (e.g. Bell, 1957; Bell, 1963; Tanai, 1979; Shilin, 1986). Late Cretaceous Magnoliaceae fossils have been found in North America (Berry, 1916; Bell, 1956; Bell, 1963; Leppik, 1963), northern Asia (Takhtajan, 1974; Tanai, 1979) and Europe (Teixeira, 1948; Teixeira, 1950). The earliest and most reliable fossil record related to *Liriodendron* comprises *Liriodendroidea* seeds from the Cenomanian to Turonian (ca. 93.5 million years ago) sediments of northwestern Kazakhstan, *Liriodendroidea alata*. These are comparable with both extant and fossil *Liriodendron* seeds in several critical characters, including anatropous organization and endotestal seed coat structure with a distinct endotesta of palisade-shaped sclerenchyma cells containing fibrous lignifications and cubic crystals. The fossil differs from modern *Liriodendron* in being much smaller and having a distinct wing to the seeds (Frumin & Friis, 1996; Frumin & Friis, 1999). *Liriodendron* is thought to have been widely distributed in warm-temperate to subtropical areas of North America, Europe and Asia during the Late Cretaceous Period (Tanai, 1979; Shilin, 1986), with more than ten species published from this time. In China, only a few Magnoliaceae pollen records from the north of the country have been reported for the Late Cretaceous Period (Song *et al.*, 2004).

From the first epoch of the Paleogene Period, the Paleocene, nearly ten fossil species of Magnoliaceae have been reported from North America and high latitude parts of Eurasia (LaMotte, 1952). In China, one pollen record is recorded from Yunnan (Song *et al.*, 1999). The small number of Magnoliaceae species occurring in the Paleocene Epoch may be due to environmental changes occurring after the Cretaceous Period. The Early Eocene Epoch was the warmest period of the Cenozoic Era. Magnoliaceae records became more abundant again during this time, with wide distributions in both North America and Europe (LaMotte, 1952; Takhtajan, 1974; Taylor, 1990). However, fossils related to *Liriodendron* are relatively rare from this time, being found only at European and Siberian sites (Berry, 1926; LaMotte, 1952; Mai, 1995). In China, fossils of *Magnolia* have been found from the Eocene Epoch of Xizang and southern China (Tao, 1988; Li *et al.*, 1995).

Compared to the Eocene Epoch, fewer Magnoliaceae species are recorded from the Oligocene Epoch (LaMotte, 1952; Takhtajan, 1974) but the family was still common in North American, European and Asian floras. In China, only two records have been reported: *Magnolia latifolia* from the Oligocene Jinggu flora, a subtropical to tropical evergreen forest in Yunnan (Writing Group of Cenozoic Plants of China, 1978), and another from the Oligocene Shenbei flora (Jin & Shang, 1998). The distribution of Magnoliaceae was at its widest during the warm Miocene Epoch, covering extensive parts of North America, Europe and Asia (Takhtajan, 1974; van der Burgh, 1983; Baghai, 1988). In China, several *Magnolia* species are recorded from the Miocene Shanwang flora, Shandong (Writing Group of Cenozoic Plants of China, 1978), Xiaolongtan, Nanlin and Mangdang floras, Yunnan (Writing Group of Cenozoic Plants of China, 1978; Ge & Li, 1999; Zhao *et al.*, 2004); the latter flora also contains a *Liriodendron* species.

After the Miocene Epoch and into the Pliocene Epoch, the global climate cooled. Few Magnoliaceae fossil records have been found from Europe, Siberia or North America (van der Burgh, 1978; Axelrod, 1980; van der Burgh, 1983). In the Quaternary Period, magnoliaceous records disappeared from high latitude areas such as present-day Greenland, Siberia and Europe, while other species spread to lower latitudes and diversified, possibly as a result of global cooling (Zhang, 2001). In China, *Magnolia* fossils have been found in the Pleistocene Changsheling flora, Guangxi. Thus, on a worldwide scale, the distribution of Magnoliaceae has tracked climatic fluctuations, reaching its widest distributions in warmer periods such as the Eocene and Miocene Epochs.

6.5 History of living fossils and endemic elements in the Chinese flora

The term "living fossil" is used to refer to extant taxa that appear very early in geological history in a similar morphological form to the present day, and whose relatives are mostly extinct, such as *Metasequoia* (Taxodiaceae). Endemism plays an important role in analysing the origin and development of floras. In this section we review the fossil history of both living fossils and taxa endemic to China.

6.5.1 Ginkgoaceae

Ginkgoaceae is a monotypic family; the sole species, *Ginkgo biloba*, is recognised as a flagship species of the Chinese flora. Native populations of *G. biloba* are found only in the provinces of Guizhou (southern China) and Zhejiang (southeastern China). However, in the past, Ginkgoaceae was rich in diversity and wide in distribution, with a fossil record dating back to the Paleozoic Era. Ginkgoalean leaf fossils have been reported from Paleozoic floras of both Angara (present-day Siberia) and Gondwana (Zhou, 2003; Zhou, 2009). Fossils identified to the genus *Ginkgo* also date back to the Paleocene Epoch: *Ginkgo adiantoides*-*Ginkgoites*-type leaves associated with *Ginkgo*-type ovulate organs (without seeds) have been found near Almont, USA.

In China, *Ginkgo yimaensis*-*Ginkgo*-type ovulate organs associated with *Ginkgoites*-type leaves, scale-bearing *Ginkgoitocladus* and adherent *Ginkgocycadophytus*-type pollen have been recorded from the Middle Jurassic Period of Henan, and *Ginkgo apodes*-*Ginkgo*-type ovulate organs associated with *Ginkgoites*-type leaves from the Lower Cretaceous Jehol Formation at Liaoning (Zhou, 2009). Extant *G. biloba* is believed to be most closely related to *G. apodes* (Zhou, 2003). This evidence suggests that *Ginkgo* originated and diversified out of China, before becoming endemic to China in the Quaternary Period (Zhou & Momohara, 2005).

6.5.2 Metasequoia

Metasequoia (Figure 6.4) is perhaps the most famous plant living fossil and is another flagship species of the Chinese flora. The genus *Metasequoia* was established based on fossil conifer leaves and cones from the Late Miocene Tokiguchi Porcelain Clay Formation in Tajimi City, Gifu, Japan, in 1941 (Miki, 1941; Momohara, 2005). Following this, Hu & Cheng (1948) described a new extant species, *M. glyptostroboides*, from central China. The discovery of living *Metasequoia* in China attracted worldwide attention, and the abundant relict and ancient species of this region are now referred to as the "*Metasequoia* flora" (Hu, 1980).

So far, *Metasequoia* fossils have been reported from more than 500 fossil localities (Yang & Jin, 2000; LePage *et al.*, 2005). The fossil record indicates that *Metasequoia* first appeared in the Early Cretaceous (Albian Age; Tao, 1992), and in the Late Cretaceous Period became a dominant component of Northern Hemisphere floras (Yang & Jin, 2000; LePage *et al.*, 2005). During the Paleocene-Eocene thermal maximum, ca. 55 million years ago, *Metasequoia*-dominated forests appeared as far north as the Arctic circle (Leng & Yang, 2009). A major contraction in the northerly extent of the range of *Metasequoia* occurred during the Eocene-Oligocene boundary cooling event and again at the Plio-Pleistocene climatic decay event (Liu *et al.*, 2007). The most recent fossils of the genus were found in the Early Pleistocene strata of Osaka, Japan (Momohara, 1994; LePage *et al.*, 2005; Momohara, 2005). Classification of fossil *Metasequoia* is controversial (Leng, 2005), but all Chinese *Metasequoia* fossils, except those from the Quaternary Period, could be merged into *M. occidentalis*.

As a plant genus with an evolutionary history as long as 100 million years, a wide fossil distribution in the Northern Hemisphere, strong morphological stasis from its first appearance to the present day, a living representative species for comparison, and a fascinating history of discovery, *Metasequoia* is an ideal model for studies of plant evolution, phytogeography, and paleoenvironmental and paleoclimatic reconstruction (Wang, 2010). The global fossil history of *Metasequoia* has been reviewed many times (Florin, 1963; Yang & Jin, 2000; LePage *et al.*, 2005; Zhou & Momohara, 2005; Liu *et al.*, 2007). Here, we summarise its history in China, including new findings from fossils found in Yunnan. There are one Chinese Early Cretaceous

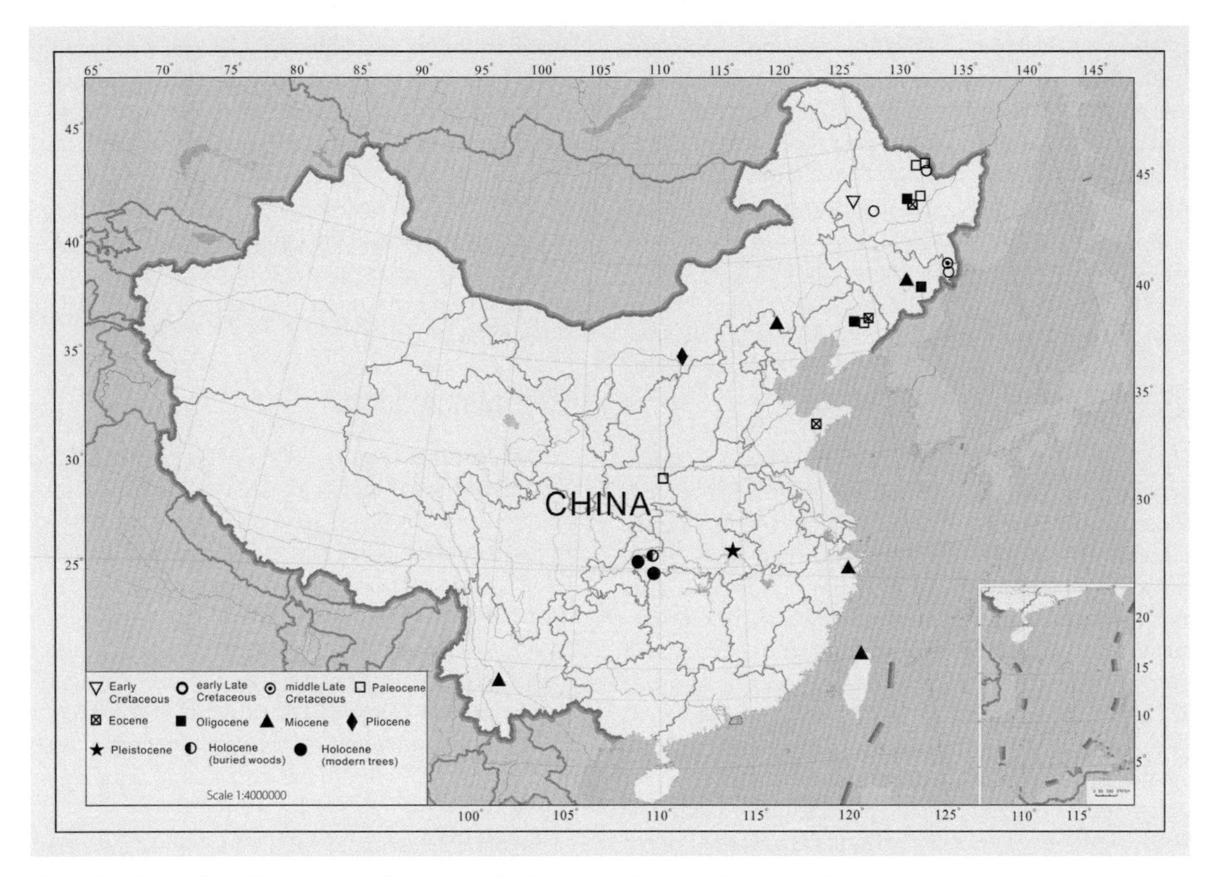

Figure 6.4 Distribution of *Metasequoia* in China during the Mesozoic and Cenozoic Epochs and the present day.

and many Late Cretaceous records of *Metasequoia*, all from the northeast of the country. The Early Cretaceous fossils were reported by Tao (1992) from the Quantou Formation in Lindian, Heilongjiang. Considered likely to be of Albian age, these are the earliest known fossils of *Metasequoia* (Tao & Sun, 1980; Tao, 1992; Yu, 1995).

Metasequoia fossils from the start of the Late Cretaceous Period are known from Heilongjiang, Jilin and the Quantou Formation. They are a common element in the Jiayin flora, found in the Yongantun Formation, Jiayin, Heilongjiang, which may be Cenomanian to Turonian in age (Zhang, 1983). The Quantou fossils might be similar in age (Guo, 1980; Guo, 1983). Guo (1983) reported *Metasequoia* fossils from the lower part of Hunchun the Formation (Jilin), considered to be Cenomanian to Coniacian in age. The Hunchun Formation has also yielded middle Late Cretaceous *Metasequoia* fossils, possibly Turonian to Santonian in age (Guo, 1983), as has Jilin (Guo & Li, 1979), where *Glyptostrobus* and *Metasequoia* make up one-third of all specimens found. *Metasequoia* fossils from the end of the Late Cretaceous Period are known mainly from the Wuyun Formation in Jiayin (Tao & Xiong, 1986a; Tao & Xiong, 1986b) and Tangyuan, Heilongjiang (Zhang *et al.*, 1990). The geological age of the Wuyun Formation has long been disputed, with recent research showing that it is probably of Danian age (Sun *et al.*, 2005).

During the Paleogene Period, *Metasequoia* was also mainly distributed in northeastern China, with a few reports from Henan and Shandong in eastern China. Most Paleocene *Metasequoia* fossils were found in the Wuyun Formation (Guo, 1983; Luo *et al.*, 1983; Tao & Xiong, 1986a; Tao & Xiong, 1986b; Xiong, 1986; Zhang *et al.*, 1990; Liu *et al.*, 1999), as well as the Fushun coal field, Liaoning (Sun *et al.*, 1980; Yang & Jin, 2000), and Xiangcheng Group, Lingbao, western Henan, the latter being a significant distance from northeastern China (Tao *et al.*, 2000). Early Eocene *Metasequoia* fossils have been reported from the strata of the Yilan coal mine, Heilongjiang (He & Tao, 1994) and Pindu, Shandong (Nanjing Institute of Geology and Mineral Resources, 1982). Many more Eocene *Metasequoia* fossils have been found in the Fushun coal field (Endo, 1928; Hu, 1946; Writing Group of Cenozoic Plants of China, 1978; Liu & Li, 2000). Liu & Li (2000) have obtained leaf cuticles from Early Eocene *Metasequoia* fossils from the Guchengzi and Jijuntun Formations of the Fushun coal field, providing the only report on leaf cuticle morphology in Chinese *Metasequoia* fossils. Oligocene reports include those from the Yangliantun Formation in Shenbei coal field, Shenyang, Liaoning (Jin & Shang, 1998), Dalianhe, near the Songhua Jiang, Yilan, Heilongjiang (He & Tao, 1994), and Qingshui near Sanhe in Longjing, Jilin (Guo & Zhang, 2002).

In the Neogene Period, the number of *Metasequoia* fossil records in China declined, becoming almost absent by the end of period. During the Miocene Epoch, the distribution of *Metasequoia* became very restricted and discontinuous. Most Early Miocene fossils are from the north of China, with reports from the Tumenzi Formation at Qiuligou, Jilin (Li & Yang, 1984) and Hannuoba Formation in Weichang, Hebei (Tao *et al.*, 2000). The latter has been dated to ca. 22 million years ago using potassium-argon dating (Li & Xiao, 1980; Liang *et al.*, 2010). From the Mid Miocene Epoch, *Metasequoia* fossils have mainly been found in southern China. Canright (1972) reported Mid Miocene fossils from the Shihti Formation of Taiyang coal mine, 20 km southeast of Taipei. Further fossils have been found in the Xiananshan Formation in Ninghai, Zhejiang (Nanjing Institute of Geology and Mineral Resources, 1982; Guo, 1983; Li, 1984). The latter formation has been dated to ca. 10.5 million years ago by argon-argon dating of the underlying basalt rocks (Ho *et al.*, 2003; Ren *et al.*, 2010). The only Pliocene *Metasequoia* fossil in China was found in Liangcheng, Nei Mongol (Geology Bureau of Nei Mongol Autonomous Region, 1976; Yu, 1995), following contraction of its distribution during this epoch (LePage *et al.*, 2005; Liu *et al.*, 2007).

In the Quaternary Period, *Metasequoia* records encompass fossil wood, buried wood and living populations. Fossil wood has been found in Wuhan, Hubei, and carbon-14 dated to the late Pleistocene Epoch (11280 ± 190 years ago; Qi *et al.*, 1993). Buried *Metasequoia* wood of the modern species, with trunks of up to 2.3 m in diameter, was unearthed from farmland in "*Metasequoia* Valley" in Lichuan, Hubei (Yang *et al.*, 2004). Modern *Metasequoia* is endemic to a narrow region in the Chang Jiang valley on the borders of Hubei, Chongqing and Hunan in southern-central China (Hu & Cheng, 1948; LePage *et al.*, 2005), where the population numbers only 5 396 known individuals (Leng *et al.*, 2007).

Yang & Jin (2000) have described the extirpation of *Metasequoia* from China during the Plio-Pleistocene period, and its subsequent reappearance, as a phytogeographic enigma with two possible explanations. Firstly, that *Metasequoia* survived in central China after the Late Miocene Epoch but was not preserved or detected in the plant fossil record, or secondly that *Metasequoia* became extinct in China in the Late Miocene, and was re-established by migration from Japan before the Early Pleistocene Epoch. Determining the true process will require further paleobotanical and archaeological studies focused on Miocene-Pliocene Epoch and Quaternary Period deposits in central and southeastern China (Yang & Jin, 2000; LePage *et al.*, 2005).

6.5.3 Cathaya *and* Pseudolarix

Cathaya argyrophylla, the sole living species of *Cathaya*, was described in only 1958 by Chun & Kuang (1958). However, reliable fossil records of *Cathaya* occur in the high latitudes of northern America and eastern Asia from as long ago as the Late Cretaceous Period, in Europe from the Paleocene Epoch, and widely in the Northern Hemisphere from the Neogene Period. *Cathaya* records then disappeared from most of the world from the Quaternary Period, except in Guangxi, Sichuan, Guizhou and Hunan (Takahashi, 1988; Liu & Basinger, 2000; Zhou & Momohara, 2005).

The fossil history of *Pseudolarix* is very similar to that of *Cathaya*. The earliest fossils were found from the Late Jurassic and Early Cretaceous Periods, when it was widely distributed in high latitude areas of the Northern Hemisphere, in the Bureya Basin, southeastern Russia and the Fuxin Pendi, northeastern China. During the Late Cretaceous Period its distribution expanded to 70--80° N and the genus became a major element in Arcto-Tertiary forests. However, in China it is known only from the Quaternary Period (Zhou & Momohara, 2005).

6.5.4 Eucommiaceae

The family Eucommiaceae is endemic to China and contains a single species, *Eucommia ulmoides*. The fruit of this species is distinctive and recognisable. The fossil history of Eucommiaceae has been well studied, with at least 18 taxa recorded from 63 fossil localities (Guo, 2000). The oldest fossil *Eucommia* record comprises pollen grains of *Eucommiaceoi-pollites* from Paleocene strata in eastern China (Guo, 2000). Fossil fruits of *Eucommia* from Eocene deposits have been reported by Geng *et al.* (1999) at Fushun, northeastern China, and Guo (1978) from the Sanshui basin. Eocene *Eucommia* fossils are also recorded from Japan and America. Miocene *Eucommia* records span the whole Northern Hemisphere; after this its distribution narrowed. *Eucommia* disappeared from Japan and Europe in the Pliocene Epoch and Quaternary Period, and is now found only in central China. The fossil record does not enable us to determine where *Eucommia* originated, but implies that it had a wide distribution and has been endemic to China since the Quaternary Period (Figure 6.5).

6.5.5 Origin of Chinese endemics

Zhou & Momohara (2005) have summarized three patterns of origin for Chinese endemic plant species based on their fossil histories. The "Arcto-Tertiary origin" pattern (exemplified by *Cathaya* and *Metasequoia*) is characterised by a fossil record showing that related lineages appeared at high latitudes from the Late Cretaceous Period, dispersed into middle and lower latitudes during global cooling events, and then went extinct from most parts of China, becoming endemic to those refugial regions where they remained. The "East Asia indigenous origin" pattern (demonstrated by *Ginkgo* and *Eucommia*) characterises plants that appeared first in China and maintained a wide distribution until after the Quaternary Period. The "boreotropical origin" hypothesis refers to taxa (such as Cephalotaxaceae and *Sargentodoxa*) whose earliest fossil records appeared in the boreotropical regions in the Late Cretaceous Period or Paleocene Epoch and became endemic to China in the Pliocene Epoch or Quaternary Period.

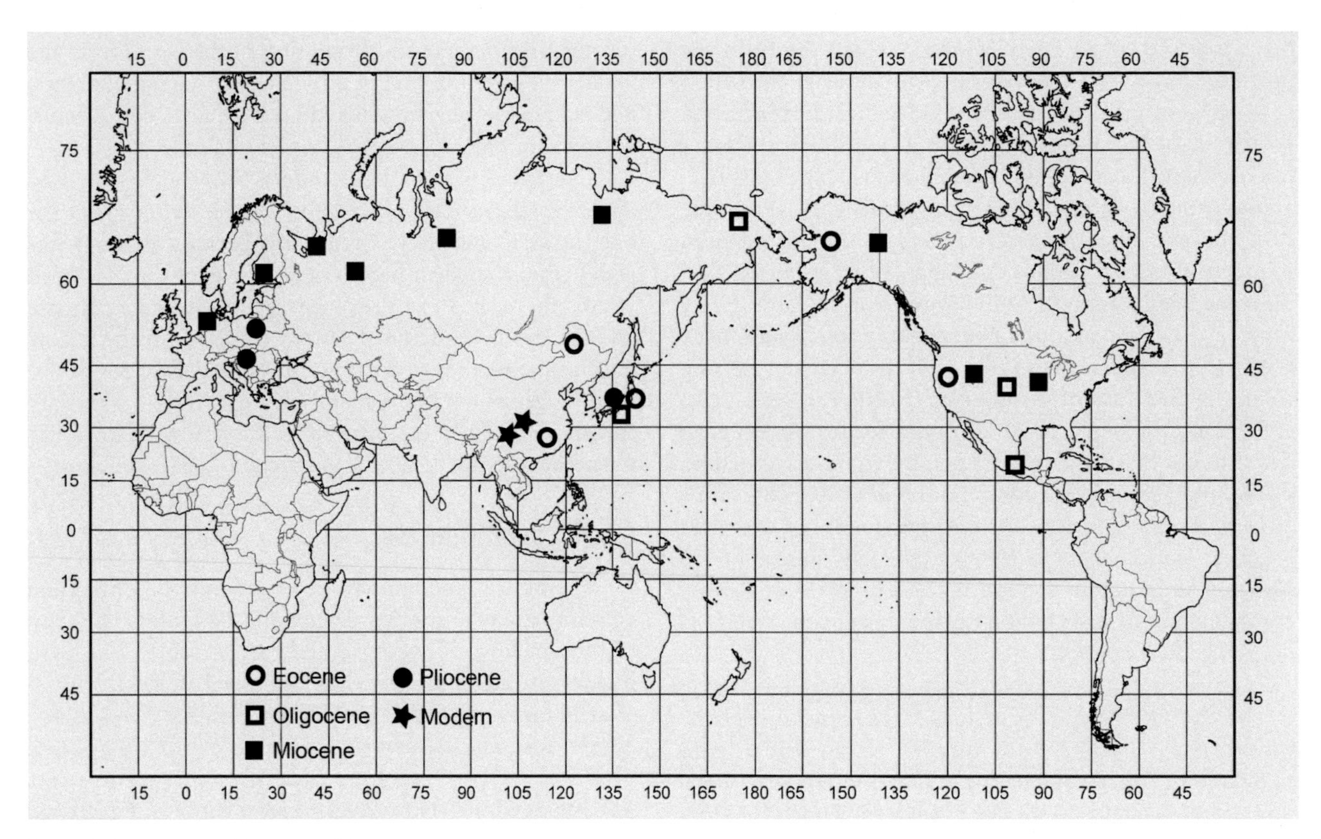

Figure 6.5 Geological distribution of *Eucommia*.

The Chinese seed plant flora has an ancient history. In this chapter we have described how China is the source of the world's oldest angiosperm fossils, and is rich in early angiosperm fossils from the Early to Late Cretaceous Period, since which time the major angiosperm lineages have been present. Significant elements of the Chinese vegetation today, such as Fabaceae, Fagaceae, Lauraceae and Magnoliaceae have been distributed widely in China since the Paleogene Period, and are represented by a continuous fossil record from that time until the present day, although their species representatives may have changed. The floristic composition of eastern Asia is complex in origin, with endemic species having had much wider distributions in the past, their distributions contracting over geological time and becoming endemic to eastern Asia in the Late Pliocene Epoch or Early Quaternary Period. Based on the origin of endemic lineages, the modern flora of eastern Asian therefore came into being around this time.

References

APG (1998). An ordinal classification for the families of flowering plants. *Annals of the Missouri Botanical Garden* 85: 531-553.

APGII (2003). An update of the Angiosperm Phylogeny Group classification for the orders and families of flowering plants: APG II. *Botanical Journal of the Linnean Society* 141: 399-436.

APGIII (2009). An update of the Angiosperm Phylogeny Group classification for the orders and families of flowering plants: APG III. *Botanical Journal of the Linnean Society* 161: 105-121.

Axelrod, D. I. (1980). Contributions to the Neogene paleobotany of Central California. *University of California Publications in Geological Sciences* 121: 3-212.

Baghai, N. L. (1988). *Liriodendron liriodendron* (Magnoliaceae) from the Miocene Clarkia Flora of Idaho. *American Journal of Botany* 75(4): 451-464.

Bell, W. A. (1956). *Lower Cretaceous Floras of Western Canada.*

Bell, W. A. (1957). Flora of the Upper Cretaceous Nanaimo Group of Vancouver Island, British Columbia. *Bulletin of the Geological Survey of Canada* 293: 1-124.

Bell, W. A. (1963). Upper Cretaceous floras of the Dunvegan, Bed Heart, and Milk River Formations of western Canada. *Bulletin of the Geological Survey of Canada* 94: 1-76.

Berry, E. W. (1916). Systematic paleontology, Upper Cretaceous: fossil plants. In: Clarke, W. B. (ed.) *Upper Cretaceous.* pp. 757-901. Baltimore, MD: Maryland Geological Survey.

Berry, E. W. (1926). Tertiary floras from British Columbia. *Canada Department of Mines Geological Survey Bulletin* 42: 91-116.

Camus, A. (1936-1954). *Les Chênes. Monographie du Genre Quercus.* Paris: Lechevalier.

Canright, J. E. (1972). Evidence of the existence of *Metasequoia* in the Miocene of Taiwan. *Taiwania* 17: 222-228.

Cao, Z. Y., Wu, S. Q., Zhang, P. A. & Li, J. R. (1998). Discovery of fossil monocotyledons from Yixian Formation, western Liaoning. *Chinese Science Bulletin* 43: 230-233.

Chun, W. H. & Kuang, K. Z. (1958). Genus novum Pinacearum ex sina australi et occidentali. *Botanicheskii Zhurnal* 43: 461-476.

Crane, P. R. & Dilcher, D. L. (1984). *Lesqueria*: an early angiosperm fruiting axis from the mid-Cretaceous. *Annals of the Missouri Botanical Garden* 71: 384-402.

Crepet, W. L. & Daghlian, C. P. (1980). Castaneoid inflorescences from the Middle Eocene of Tennessee and the diagnostic value of pollen (at the subfamily level) in the Fagaceae. *American Journal of Botany* 67(5): 739-757.

Crepet, W. L. & Nixon, K. C. (1989). Earliest megafossil evidence of Fagaceae: phylogenetic and biogeographic implications. *American Journal of Botany* 76: 842-855.

Dai, J., Sun, B. N., Xie, S. P., Lin, Z. P., Wen, W. W. & Wu, J. Y. (2010). Cuticular microstructure of *Litsea* cf. *chunii* from the Pliocene of Tengchong, Yunnan Province. *Journal of Lanzhou University (Natural Sciences)* 46(1): 22-28.

Denk, T. & Meller, B. (2001). Systematic significance of the cupule/nut complex in living and fossil *Fagus*. *International Journal of Plant Sciences* 162(4): 869-897.

Dilcher, D. L. & Crane, P. R. (1984). *Archaeanthus*: an early angiosperm from the Cenomanian of the western interior of North America. *Annals of the Missouri Botanical Garden* 71: 351-383.

Dilcher, D. L., Sun, G., Ji, Q. & Li, H. Q. (2007). An early infructescence *Hyrcantha decussata* (comb. nov.) from the Yixian Formation in northeastern China. *Proceedings of the National Academy of Sciences of the United States of America* 104: 9370-9374.

Donoghue, M. J. & Smith, S. A. (2004). Patterns in the assembly of temperate forests around the Northern Hemisphere. *Philosophical Transactions of the Royal Society of London Series B, Biological Sciences* 359: 1633-1644.

Doyle, J. A. & Endress, P. K. (2010). Integrating Early Cretaceous fossils into the phylogeny of living angiosperms: Magnoliidae and eudicots. *Journal of Systematics and Evolution* 48: 1-35.

Duan, S. Y. (1997). The oldest angiosperm - a tricarpous female reproductive fossil. *Sciences in China (Series D)* 27(6): 519-524.

Endo, S. (1928). A new Paleogene species of *Sequoia*. *Japanese Journal of Geology and Geography* 6: 27-29.

Endress, P. K. & Doyle, J. A. (2009). Reconstructing the ancestral angiosperm flower and its initial specializations. *American Journal of Botany* 96(1): 22-66.

Erdtman, G. (1952). *Pollen Morphology and Plant Taxonomy: Angiosperms (An Introduction to Palynology. I).* Stockholm: Almqvist & Wiksell.

Florin, R. (1963). The distribution of conifer and taxad genera in time and space. *Acta Horti Bergiani* 20: 121-312.

Friis, E. M., Crane, P. R. & Pedersen, K. R. (1986). Floral evidence for Cretaceous chloranthoid angiosperms. *Nature* 320(6058): 163-164.

Friis, E. M., Pedersen, K. R. & Crane, P. R. (2001). Fossil evidence of water lilies (Nymphaeales) in the Early Cretaceous. *Nature* 410: 357-360.

Friis, E. M., Doyle, J. A., Endress, P. K. & Leng, Q. (2003). *Archaefructus* - angiosperm precursor or specialized early angiosperm? *Trends in Plant Science* 8(8): 369-373.

Friis, E. M., Pedersen, K. R. & Crane, P. R. (2009). Early Cretaceous mesofossils from Portugal and eastern North America related to the Bennettitales-Erdtmanithecales-Gnetales group. *American Journal of Botany* 96(1): 252-283.

Frumin, S. & Friis, E. M. (1996). Liriodendroid seeds from the Late Cretaceous of Kazakhstan and North Carolina, USA. *Review of Palaeobotany and Palynology* 94: 39-55.

Frumin, S. & Friis, E. M. (1999). Magnoliid reproductive organs from the Cenomanian-Turonian of north-western Kazakhstan: Magnoliaceae and Illiciaceae. *Plant Systematics and Evolution* 216: 265-288.

Gandolfo, M. A., Nixon, K. C. & Crepet, W. L. (1998). A new fossil flower from the Turonian of New Jersey: *Dressiantha bicarpellata* gen. et sp. nov. (Capparales). *American Journal of Botany* 85(7): 964-974.

Gandolfo, M. A., Nixon, K. C. & Crepet, W. L. (2004). Cretaceous flowers of Nymphaeaceae and implications for complex insect entrapment pollination mechanisms in early angiosperms. *Proceedings of the National Academy of Sciences of the United States of America* 101(21): 8056-8060.

Ge, H. R. & Li, D. Y. (1999). *Cenozoic Coal-bearing Basins and Coal-forming Regularity in West Yunnan*. Kunming, China: Yunnan Science and Technology Press.

Geng, B. Y., Manchester, S. R. & Lu, A. M. (1999). The first discovery of *Eucommia* fruit fossil in China. *Chinese Science Bulletin* 44(16): 1506-1508.

Geology Bureau of Nei Mongol Autonomous Region (1976). *Palaeontological Atlas of Huabei Region. Mesozoic and Cenozoic*. Beijing, China: Geology Press.

Guo, S. X. (1978). Pliocene flora of Western Sichuan. *Acta Palaeontologica Sinica* 18(6): 547-560.

Guo, S. X. (1979). Late Cretaceous and Early Tertiary floras from southern Guangdong and Guangxi with their stratigraphic significance. In: Institute of Vertebrate Paleontology and Paleoanthropology and Nanjing Institute of Geology and Paleontology - Academia Sinica (ed.) *Mesozoic and Cenozoic Red Beds of South China*. pp. 223-231. Beijing, China: Science Press.

Guo, S. X. & Li, H. M. (1979). Late Cretaceous flora from Hunchun of Jilin. *Acta Palaeontologica Sinica* 18(6): 547-560.

Guo, S. X. (1980). Late Cretaceous and Eocene floral provinces of China. In: Academia Sinica (ed.) *Paper for the 1st Conference of the IOP, Reading, UK*. pp. 1-9.

Guo, S. X. (1983). Note on phytogeographic provinces and ecological environment of Late Cretaceous and Tertiary floras in China. In: Lu, Y. H., Mu, A. Z. & Wang, Y. (eds.) *Palaeobiogeographic Provinces of China*. pp. 164-177. Beijing, China: Science Press.

Guo, S. X. & Zhou, Z. K. (1992). The mega fossil legumes from China. In: Herendeen, P. S. & Dilcher, D. L. (eds.) *Advances in Legume Systematics: Part 4. The Fossil Record*. pp. 207-223.

Guo, S. X. (2000). Evolution, palaeobiogeography and palaeocology of Eucommiaceae. *Palaeobotanist* 49(2): 65-83.

Guo, S. X. & Zhang, G. F. (2002). Oligocene Sanhe flora in Longjing County of Jilin, Northeast China. *Acta Palaeontologica Sinica* 41(2): 193-210.

He, C. X. & Tao, J. R. (1994). Paleoclimatic analysis of Paleogene flora in Yilan County, Heilongjiang. *Acta Botanica Sinica* 36(12): 952-956.

Herendeen, P. S., Crepet, W. L. & Nixon, K. C. (1994). Fossil flowers and pollen of Lauraceae from the Upper Cretaceous of New Jersey. *Plant Systematics and Evolution* 189: 29-40.

Herendeen, P. S., Crane, P. R. & Drinnan, A. N. (1995). Fagaceous flowers, fruits, and cupules from the Campanian (Late Cretaceous) of central Georgia, USA. *International Journal of Plant Sciences* 156: 93-116.

Ho, K. S., Chen, J. C., Lo, C. H. & Zhao, H. L. (2003). 40Ar-39Ar dating and geochemical characteristics of late Cenozoic basaltic rocks from the Zhejiang-Fujian region, SE China: eruption ages, magma evolution and petrogenesis. *Chemical Geology* 197: 287-318.

Hsu, J. & Bose, M. N. (1952). Further information on on *Homoxylon rajmahalense* Sahni. *Journal of the Indian Botanical Society* 31: 1-12.

Hu, H. H. (1946). Notes on a Palaeogene species of *Metasequoia* in China. *Bulletin of the Geological Society of China* 26: 105-107.

Hu, H. H. & Cheng, W. C. (1948). On the new family Metasequoiaceae and on *Metasequoia glyptostroboides*, a living species of the genus *Metasequoia* found in Szechwan and Hupeh. *Bulletin of the Fan Memorial Institute of Biology New Series* 1(2): 199-212.

Hu, S. Y. (1980). The *Metasequoia* flora and its phylogeographical significance. *Journal of the Arnold Arboretum* 61: 41-94.

Hu, Y. Q., Ferguson, D. K., Li, C. S., Xiao, Y. P. & Wang, Y. F. (2007). *Alseodaphne* (Lauraceae) from the Pliocene of China and its paleoclimatic significance. *Review of Palaeobotany and Palynology* 146: 277-285.

Huang, C. J., Zhang, Y. T. & Bartholomew, B. (1999). Fagaceae. In: Wu, Z. Y., Raven, P. H. & Hong, D. Y. (eds.) *Flora of China*. pp. 314-400. Beijing, China and St. Louis, MO: Science Press and Missouri Botanical Garden Press.

Hughes, N. F. & Couper, R. A. (1958). Palynology of the Brora coal of the Scottish Middle Jurassic. *Nature* 181: 1482-1483.

Iljinskaya, I. A. (1982). *Fagus* L. In: Takhtajan, A. L. (ed.) *Magnoliaophyta Fossilia SSSR. Vol. 2. Ulmaceae-Betulaceae*. pp. 60-73. St. Petersburg, Russia.

Ji, Q., Li, H. Q., Bowe, M., Liu, Y. S. & Taylor, D. W. (2004). Early Cretaceous *Archaefructus eoflora* sp. nov.

with bisexual flowers from Beipaio, Western Liaoning, China. *Acta Geologica Sinica* 78: 883-896.

Jin, J. H. & Shang, P. (1998). Discovery of Early Tertiary flora in Shenbei coalfield, Liaoning. *Acta Scientiarum Naturalium Universitatis Sunyatseni* 37(6): 129-130.

Jones, J. H., Manchester, S. R. & Dilcher, D. L. (1988). *Dryophyllum* Debey ex Saporta, Juglandaceous not Fagaceous. *Review of Palaeobotany and Palynology* 56(3-4): 205-211.

Kvaček, Z. & Walther, H. (1989). Paleobotanical studies in Fagaceae of the European Tertiary. *Plant Systematics and Evolution* 162: 213-229.

Kvaček, Z. & Walther, H. (1991). Revision of the Central European Tertiary Fagaceae according to leaf epidermal characteristics. IV. *Fagus* Linnaeus. *Feddes Repertorium* 102(7-8): 471-534.

LaMotte, R. S. (1952). *Catalogue of the Cenozoic Plants of North America Through 1950.*

Law, Y. W. (1984). A preliminary study on the taxonomy of the family Magnoliaceae. *Acta Phytotaxonomica Sinica* 22: 80-109.

Leng, Q. & Friis, E. M. (2003). *Sinocarpus decussatus* gen. et sp. nov., a new angiosperm with basally syncarpous fruits from the Yixian Formation of Northeast China. *Plant Systematics and Evolution* 241: 77-88.

Leng, Q. (2005). Cuticular analysis of living and fossil *Metasequoia*. In: LePage, B. A., Williams, C. & Yang, H. (eds.) *The Geobiology and Ecology of Metasequoia*. pp. 197-217. Dordrecht, Netherlands: Springer.

Leng, Q. & Friis, E. M. (2006). Angiosperm leaves associated with *Sinocarpus* infructescences from the Yixian Formation (mid-Early Cretaceous) of NE China. *Plant Systematics and Evolution* 262(3-4): 173-187.

Leng, Q., Fan, S. H., Wang, L., Yang, H., Lai, X. L., Cheng, D. D., Ge, J. W., Shi, G. L., Jiang, Q. & Liu, X. Q. (2007). Database of native *Metasequoia glyptostroboides* trees in China based on new census surveys and expeditions. *Bulletin of the Peabody Museum of Natural History* 48(2): 185-233.

Leng, Q. & Yang, H. (2009). A new approach to reconstruct Paleogene atmospheric hydrology at high latitudes. *Science Foundation in China* 17(2): 46-49.

LePage, B. A., Yang, H. & Matsumoto, M. (2005). The evolution and biogeographic history of *Metasequoia*. In: LePage, B. A., Williams, C. J. & Yang, H. (eds.) *The Geobiology and Ecology of Metasequoia*. pp. 3-114. Dordrecht, Netherlands: Springer.

Leppik, E. E. (1963). Reconstruction of a Cretaceous *Magnolia* flower. *Advances and Frontiers in Plant Science* 4: 79-94.

Li, H. M. & Guo, S. X. (1976). The Miocene flora from Namling of Xizang. *Acta Palaeontologica Sinica* 15: 7-18.

Li, H. M. (1984). Neogene floras from eastern Zhejiang, China. In: Whyte, R. (ed.) *The Evolution of the East Asia Environment. II. Palaeobotany, Palaeozoology and Palaeanthropology*. pp. 461-466. Hong Kong, China: Hong Kong: Centre of Asian Studies, University of Hong Kong.

Li, H. M. & Yang, G. Y. (1984). A Miocene flora from Qiuligou, Jilin. *Acta Palaeontologica Sinica* 23(2): 204-213.

Li, H. Q. (2005). Early Cretaceous Sarraceniacean-like pitcher plants from China. *Acta Botanica Gallica* 152(2): 227-234.

Li, J. & Xiao, Z. G. (1980). Introduction of Qipanshan section (K-50-16). In: *1/200000 Geological Map of People's Republic of China*. Beijing, China: Geological Map Printing Plant of China.

Li, J., Qiu, J., Liao, W. & Jin, J. (2009). Eocene fossil *Alseodaphne* from Hainan Island of China and its paleoclimatic implications. *Science in China Series D: Earth Sciences* 52(10): 1537-1542.

Li, X. X., Zhou, Z. Y., Cai, C. Y., Sun, G., Ouyang, S. & Deng, L. H. (1995). *Fossil Floras of China Through the Geological Ages*. Guangzhou, China: Guangdong Science and Technology Press.

Liang, X. Q., Wilde, V., Ferguson, D. K., Kvaček, Z., Ablaev, A. G., Wang, Y. F. & Li, C. S. (2010). *Comptonia naumannii* (Myricaceae) from the early Miocene of Weichang, China, and the palaeobiogeographical implication of the genus. *Review of Palaeobotany and Palynology* 163: 52-63.

Liu, Y. J., Li, C. S. & Wang, Y. F. (1999). Studies on fossil *Metasequoia* from North-East China and their taxonomic implications. *Botanical Journal of the Linnean Society* 130(3): 267-297.

Liu, Y. J. & Li, C. S. (2000). On *Metasequoia* in Eocene Age from Liaoning Province of Northeast China. *Acta Botanica Sinica* 42(8): 873-878.

Liu, Y. J., Arens, N. C. & Li, C. S. (2007). Range change in *Metasequoia*: relationship to palaeoclimate. *Botanical Journal of the Linnean Society* 154: 115-127.

Liu, Y. S. (1990). Cuticular studies on two Pleistocene species of Lauraceae in Baise Basin, Guangxi. *Acta Botanica Sinica* 32(10): 805-808.

Liu, Y. S., Momohara, A. & Mei, S. W. (1996). A revision on the Chinese megafossils of *Fagus* (Fagaceae). *Journal of Japanese Botany* 71(3): 168-177.

Liu, Y. S. & Basinger, J. F. (2000). Fossil *Cathaya* (Pinaceae) pollen from the Canadian High Arctic. *International Journal of Plant Sciences* 161(5): 829-847.

Liu, Y. S., Utescher, T., Zhou, Z. K. & Sun, B. N. (2011). The evolution of Miocene climates in North China: preliminary results of quantitative reconstructions from plant fossil records. *Palaeogeography, Palaeoclimatology, Palaeoecology* 304(3-4): 308-317.

Liu, Z. (1988). Plant fossils from the Zhidan Group between Huating and Longxian, southwestern part of Ordos Basin. *Bulletin of Xi'an Institute of Geology and Mineral Resources, Chinese Academy of Geological Sciences* 24: 91-100.

Luo, Y. X., Zhang, Z. C. & Li, W. R. (1983). Late Mesozoic and Tertiary strata in Jiayin Sunke region, Heilongjiang Province. *China Journal of Stratigraphy* 7(3): 169-183.

MacPhail, M. K. (1980). Fossil and modern *Beilschmiedia* (Lauraceae) pollen in New Zealand. *New Zealand Journal of Botany* 18: 453-457.

MAI, D. H. (1995). *Tertiäre Vegetationsgeschichte Europas*. Stuttgart, Germany: Gustav Fischer Verlag, Jena.

MANCHESTER, S. R. (1994). Fruits and seeds of the Middle Eocene Nut Beds Flora, Clarno Formation, Oregon. *Palaeontographica Americana* 58: 1-205.

MANCHESTER, S. R. & DILLHOFF, R. M. (2004). *Fagus* (Fagaceae) fruits, foliage, and pollen from the middle Eocene of Pacific Northwestern North America. *Canadian Journal of Botany* 82: 1509-1517.

MIKI, S. (1941). On the change of flora in eastern Asia since the Tertiary period (I). The caly or liginite bed flora in Japan with special reference to the *Pinus trifolia* beds in central Hondo. *Japanese Journal of Botany* 8: 303-341.

MILNE, R. I. & ABBOTT, R. J. (2002). The origin and evolution of Tertiary relict floras. *Advances in Botanical Research* 38: 281-314.

MILNE, R. I. (2006). Northern hemisphere plant disjunctions: a window on Tertiary land bridges and climate change? *Annals of Botany* 98(3): 465-472.

MOMOHARA, A. (1994). Floral and paleoenvironmental history from the late Pliocene to middle Pleistocene in and around central Japan. *Palaeogeography, Palaeoclimatology, Palaeoecology* 108(3-4): 281-293.

MOMOHARA, A. (2005). Paleoecology and history of *Metasequoia* in Japan, with reference to its extinction and survival in East Asia. In: LePage, B. A. & Williams, C. (eds.) *The Geobiology and Ecology of Metasequoia. Topics in Geobiology*. pp. 115-136. Dordrecht, Netherlands: Springer.

MOORE, M. J., BELL, C. D., SOLTIS, P. S. & SOLTIS, D. E. (2007). Using plastid genome-scale data to resolve enigmatic relationships among basal angiosperms. *Proceedings of the National Academy of Sciences of the United States of America* 104(49): 19363-19368.

NANJING INSTITUTE OF GEOLOGY AND MINERAL RESOURCES (1982). *Paleontological Atlas of East China. III. Mesozoic and Cenozoic Volume*. Beijing, China: Geological Publishing House.

NIE, Z. L., WEN, J., AZUMA, H., QIU, Y. L., SUN, H., MENG, Y., SUN, W. B. & ZIMMER, E. A. (2008). Phylogenetic and biogeographic complexity of Magnoliaceae in the Northern Hemisphere inferred from three nuclear data sets. *Molecular Phylogenetics and Evolution* 48: 1027-1040.

NIXON, K. C. & CREPET, W. L. (1989). *Trigonobalanus* (Fagaceae): taxonomic status and phylogenetic status and phylogenetic relatioinship. *American Journal of Botany* 76(6): 828-841.

NOOTEBOOM, H. P. (1993). Magnoliaceae. In: Kubitzki, K., Rohwer, J. G. & Bittrich, V. (eds.) *The Families and Genera of Vascular Plants*. pp. 391-401. New York: Springer-Verlag.

PAN, K. (2000). Fossil seed of *Magnolia* from middle Jurassic of Yanliao region, North China. In: Liu, Y. H. (ed.) *Proceedings of the International Symposium on the Family Magnoliaceae*. Beijing, China: Science Press.

QI, G. F., YANG, J. J. & SU, J. Z. (1993). Two unearthed ancient woods excavated from Wuhan City. *Acta Botanica Sinica* 35(9): 722-726.

REN, W. X., SUN, B. N. & XIAO, L. (2010). Quantitative reconstruction on paleoelevation and paleoclimate of Miocene Xiananshan Formation in Ninghai, Zhejiang Province. *Acta Micropalaeontologica Sinica* 27(1): 93-98.

ROHWER, J. G. (1993). Lauraceae. In: Kubitzki, K., Rohwer, J. G. & Bittrich, V. (eds.) *The Families and Genera of Vascular Plants*. pp. 366-391. Berlin, Germany: Springer-Verlag.

RÜFFLE, L. (1963). Die obermiozäne (sarmatische) Flora vom Randeckar Maar. *Paläontologische Abhandlungen* 1: 139-298.

SAHNI, B. (1932). *Homoxylon rajmahalense*, gen. et sp. nov., a fossil angiospermous wood, devoid of vessels, from the Rajmahal Hills, Bihar. *Memoirs of the Geological Survey of India, Palaeontologia Indica, New Series* 20(2): 1-19.

SHILIN, P. V. (1986). *Late Cretaceous Flora of Kazakhstan*. Nauka Kazakhstan SSR.

SIMPSON, J. B. (1937). Fossil pollen in Scottish Jurassic coal. *Nature* 139: 673.

SIMS, H. J., HERENDEEN, P. S. & CRANE, P. R. (1998). New genus of fossil Fagaceae from the Santonian (Late Cretaceous) of central Georgia, USA. *International Journal of Plant Sciences* 159: 391-404.

SONG, Z. C., ZHENG, Y., LI, M., ZHANG, Y., WANG, W., ZHAO, C., ZHOU, S., ZHU, Z. & ZHAO, Y. (1999). *Fossil Spores and Pollen of China. 1. Late Cretaceous and Tertiary Spores and Pollen*. Beijing, China: Science Press.

SONG, Z. C., WANG, W. M. & HUANG, F. (2004). Fossil pollen records of extant angiosperms in China. *The Botanical Review* 70(4): 425-458.

SUN, C. L., SUN, G. & WANG, L. X. (2010). New materials of fossil plants from Upper Triassic of western Liaoning, China. *In: EPPC-8 Conference*. Budapest, Hungary.

SUN, G., GUO, S. X., ZHENG, S. L., PIAO, T. Y. & SUN, X. K. (1992). First discovery of the earliest angiospermous megafossil flora in the world. *Science in China Series B* 5: 543-548.

SUN, G., DILCHER, D. L., ZHENG, S. L. & ZHOU, Z. K. (1998). In search of the first flower: a Jurassic angiosperm, *Archaefructus*, from northeast China. *Science* 282: 1692-1695.

SUN, G., JI, Q., DILCHER, D. L., ZHENG, S. L., NIXON, K. C. & WANG, X. F. (2002). Archaefructaceae, a new basal angiosperm family. *Science* 296: 899-904.

SUN, G., QUAN, C., SUN, C. L., SUN, Y. W., LUO, K. L. & LÜ, J. S. (2005). Some new knowledge on subdivisions and age of Wuyun Formation in Jiayin of Heilongjiang, China. *Journal of Jilin University (Earth Science Edition)* 35(2): 137-142.

SUN, G., DILCHER, D. L. & ZHENG, S. L. (2008). A review of recent advances in the study of early angiosperms from northeastern China. *Palaeoworld* 17: 166-171.

SUN, G., DILCHER, D. L., WANG, H. & CHEN, Z. (2011). A eudicot from the Early Cretaceous of China. *Nature* 471: 625-628.

SUN, X. J., DU, N. Q. & SUN, M. R. (1980). Palynological

investigation on Fushun Group (Paleocene) of northeastern China. In: Hong, Y. C. (ed.) *Stratigraphy of Fushun coal field, Liaoning and the fossil assemblages.* Beijing, China: Science Press.

Takahashi, K. (1988). Palynology of the Upper Cretaceous Futaba Group. *Science Bulletin of the Faculty of Liberal Arts of Nagasaki University* 28: 67-183.

Takhtajan, A. (1974). Magnoliophyta Fossilia. *URSS* 1: 1-188.

Tanai, T. (1979). Late Cretaceous floras from the Kuji District, Norheastern Honshu, Japan. *Journal of the Faculty of Science, Hokkaido University. Series 4, Geology and Mineralogy* 19(1-2): 75-136.

Tanai, T. (1995). Fagaceous leaves from the Paleogene of Hokkaido, Japan. *Bulletin of the National Science Museum. Series C, Geology and Paleontology* 21(3-4): 71-101.

Tao, J. R. (1965). A Late Eocene florula from the district Weinan of central Shensi. *Acta Botanica Sinica* 13: 272-282.

Tao, J. R. & Sun, X. J. (1980). The Cretaceous floras of Linding Xian, Heilongjiang Province. *Acta Botanica Sinica* 22(1): 75-79.

Tao, J. R. & Chen, M. H. (1983). Neogene flora of south part of the watershed of Salween-Mekong-Yangtze rivers (the Linczan region) Yunnan. In: *Exploration of Hengduan Mountain Area.* pp. 74-89. Beijing, China: Beijing Publishing House of Sciences and Technology.

Tao, J. R. & Xiong, X. Z. (1986a). The latest Cretaceous flora of Heilongjiang Province and the floristic relationship between East Asia and North America. *Acta Phytotaxonomica Sinica* 24(1): 1-15.

Tao, J. R. & Xiong, X. Z. (1986b). The latest Cretaceous flora of Heilongjiang Province and the floristic relationship between East Asia and North America (cont.). *Acta Phytotaxonomica Sinica* 24(2): 1-15, 121-135.

Tao, J. R. (1988). Plant fossils fossils from the the Lepuqu Formation in Lhaze County, Xizang and their palaeoclimatological significance. *Memoir of the Institute of Geology, Chinese Academy of Sciences* 3: 223-238.

Tao, J. R. & Zhang, C. B. (1990). Early Cretaceous angiosperms of the Yanji basin, Jilin province. *Acta Botanica Sinica* 32(3): 220-229.

Tao, J. R. (1992). The Tertiary vegetation and flora and floristic regions in China. *Acta Phytotaxonomica Sinica* 30(1): 25-43.

Tao, J. R. & Zhang, C. B. (1992). Two angiosperm reproductive organs from the Early Cretaceous of China. *Acta Phytotaxonomica Sinica* 30: 423-426.

Tao, J. R., Zhou, Z. K. & Liu, Y. S. (2000). *The Evolution of the Late Cretaceous-Cenozoic Floras in China.* Beijing, China: Science Press.

Taylor, D. W. (1988). Eocene floral evidence of Lauraceae: corroboration of the North American megafossil record. *American Journal of Botany* 75(7): 948-957.

Taylor, D. W. (1990). Paleobiogeographic relationships of angiosperms from the Cretaceous and early Tertiary of the North American area. *Botanical Review* 56: 279-417.

Teixeira, C. (1948). *Flora Mesozóica Portuguesa, Parte I.* Lisbon, Portugal: Memorias Services Geologicos Portugal.

Teixeira, C. (1950). *Flora Mesozóica Portuguesa, Parte II.* Lisbon, Portugal: Memorias Services Geologicos Portugal.

van der Burgh, J. (1978). The Pliocene flora flora of of Fortuna-Garsdorf Garsdorf I. Fruits and seeds of angiosperms. *Review of Palaeobotany and Palynology* 26(1-4): 173-211.

van der Burgh, J. (1983). Allochthonous seed and fruit floras from the Pliocene of the lower Rhine Basin. *Review of Palaeobotany and Palynology* 40(1-2): 33-90.

Von Balthazar, M., Pedersen, K. R., Crane, P. R., Stampanoni, M. & Friis, E. M. (2007). *Potomacanthus lobatus* gen. et sp. nov., a new flower of probable Lauraceae from the Early Cretaceous (Early to Middle Albian) of eastern North America. *American Journal of Botany* 94(12): 2041-2053.

Wang, L. (2010). *Morphology and anatomy of Metasequoia leaves and their environmental indicative values: evidence from the comparative studies of "living fossil"and fossils.* Beijing, China: Nanjing Institute of Geology and Palaeontology; The Graduate University of Chinese Academy of Sciences.

Wang, Z. (1984). Plant Kingdom. In: Tianjin Institute of Geology and Mineral Resources (ed.) *Palaeontological Atlas of North China.* pp. 223-384. Beijing, China: Geological Publishing House.

Wen, J. (2001). Evolution of eastern Asian-Eastern North American biogeographic disjunctions, a few additional issues. *International Journal of Plant Sciences* 162: S117-S122.

Writing Group of Cenozoic Plants of China (1978). *Cenozoic Plants from China, Fossil Plants of China.* Beijing, China: Science Press.

Wu, J. Y. (2009). The Pliocene Tuantian flora of Tengchong, Yunnan Province and its paleoenvironmental analysis. Lanzhou, China: Lanzhou University.

Wu, S. (1999). A preliminary study of the Jehol flora from western Liaoning. In: Chen, P. & Jin, F. (eds.) *The Jehol Biota,.* pp. 7-57. Hefei, China: Chinese Science and Technology Publishing House.

Wu, Z. Y., Tang, Y. C., Lu, A. M. & Chen, Z. D. (1998). On primary subdivisions of the Magnoliophyta - towards a new scheme for an eight-class system of classification of the angiosperms. *Acta Phytotaxonomica Sinica* 36(5): 385-402.

Wu, Z. Y., Lu, A. M., Tang, Y. C., Chen, Z. D. & Li, D. Z. (2002). Synopsis of a new "polyphyletic-polychronic-polytopic" system of the angiosperms. *Acta Phytotaxonomica Sinica* 40: 289-322.

Xing, Y. W. (2010). The Late Miocene Xianfeng flora, Yunnan, Southwest China and its quantitative palaeoclimatic reconstructions. Ph.D., Kunming, China: Kunming Institute of Botany, Chinese Academy of Sciences.

Xiong, X. Z. (1986). Paleocene flora from the

Wuyun Formation in Jiayin of Heilongjiang. *Acta Palaeontologica Sinica* 25: 571-576.

Yabe, H. & Endo, S. (1935). *Potamogeton* remains from the Lower Cretaceous? Lycoptera bed of Jehol. *Proceedings of the Imperial Academy* 11(7): 274-276.

Yamada, T., Nishida, H., Umebayashi, M., Uemura, K. & Kato, M. (2008). Oldest record of Trimeniaceae from the Early Cretaceous of northern Japan. *BMC Evolutionary Biology* 8.

Yang, H. & Jin, J. H. (2000). Phytogeographic history and evolutionary stasis of *Metasequoia*: geological and genetic information contrasted. *Acta Palaeontologica Sinica* 39(suppl.): 288-307.

Yang, J. M., Yang, X. Y. & Liang, H. (2004). The discovery of buried *Metasequoia* wood in Lichuan, Hubei, China, and its significance. *Acta Palaeontologica Sinica* 43(1): 124-131.

Yao, Y. F., Bruch, A. A., Mosbrugger, V. & Li, C. S. (2011). Quantitative reconstruction of Miocene climate patterns and evolution in Southern China based on plant fossils. *Palaeogeography Palaeoclimatology Palaeoecology* 304: 291-307.

Yu, Y. F. (1995). Origin, evolution and distribution of the Taxodiaceae. *Acta Phytotaxonomica Sinica* 33(4): 362-389.

Zachos, J., Pagani, M., Sloan, L., Thomas, E. & Billups, K. (2001). Trends, rhythms, and abberations in global climate, 65 Ma to present. *Science* 292: 686-693.

Zhang, F. K. (1984). Fossil record of Mesozoic mammals of China. *Vertebrata Palasiatica* 22(1): 29-38.

Zhang, G. F. (2001). Fossil records of Magnoliaceae. *Acta Palaeontologica Sinica* 40: 433-442.

Zhang, J. H. (1983). Discovery of old Tertiary flora from Panxian of Guizhou and its significance. In: Committee of Stratigraphy and Paleontology - Geological Society of Guizhou Province (ed.) *Papers of Stratigraphy and Paleontology of Guizhou*. pp. 133-143. Guiyang, China: People's Publishing House of Guizhou.

Zhang, Q. R. (1980). The Paleocene sporo-pollen assemblage in Nanxiong Basin, Guangdong. *In:* Yichang Institute of Geology and Mineral Resources-Chinese Academy of Geology Sciences (ed.) *Yichang Institute of Geology and Mineral Resources, Chinese Academy of Geology Sciences (Stratigraphy and Palaeontology, Special Issue)*. pp. 106-117. Yichang, China: Yichang Institute of Geology and Mineral Resources, Chinese Academy of Geology Sciences.

Zhang, Y., Zhai, P. M., Zhang, S. L. & Zhang, W. (1990). Late Cretaceous-Paleogene plants from Tangyuan, Heilongjiang. *Acta Palaeontologica Sinica* 29: 237-245.

Zhao, L. C., Wang, Y. F., Liu, C. J. & Li, C. S. (2004). Climatic implications of fruit and seed assemblage from Miocene of Yunnan, southwestern China. *Quaternary International* 117: 81-89.

Zheng, S. L. & Zhang, W. (1983). The Middle and Late Early Cretaceous floras in the Boli Basin of Heilongjiang Province, China. *Bulletin of Shenyang Institute of Geology and Mineral Resources* 7: 68-98.

Zhou, Z. K. (1985). The Miocene Xiaolongtan fossil flora in Kaiyuan, Yunnan, China. M.Sc.: Nanjing Institute of Geology and Palaeontology, Chinese Academy of Sciences.

Zhou, Z. K. (1992). A taxonomical revision of fossil evergreen sclerophyllous oaks from China. *Acta Botanica Sinica* 34(12): 954-609.

Zhou, Z. K., Wilkinson, H. & Wu, Z. Y. (1995). Taxonomical and evolutionary implications of the leaf anatomy and architecture of *Quercus* L. subg. *Quercus* from China. *Cathaya* 7: 1-34.

Zhou, Z. K. (1999). Fossils of the Fagaceae and their implication in systematics and biogeography. *Acta Phytotaxonomica Sinica* 37(4): 369-385.

Zhou, Z. K. & Momohara, A. (2005). Fossil history of some endemic seed plants of east Asia and its phytogeographical significance. *Acta Botanica Yunnanica* 27(5): 449-470.

Zhou, Z. Y., LI, H. M., Cao, Z. Y. & Nau, P. S. (1990). Some Cretaceous plants from Pingzhou (Pingchau) island, Hong Kong. *Acta Palaeontologica Sinica* 29(4): 415-426.

Zhou, Z. Y. (2003). Mesozoic ginkgoaleans: phylogeny, classification and evolutionary trends. *Acta Botanica Yunnanica* 25(4): 377-396.

Zhou, Z. Y. (2009). An overview of fossil Ginkgoales. *Palaeoworld* 18: 1-22.

Chapter 7

History of Vegetation in China

ZHOU Zhe-Kun

7.1 Introduction

This account of the history of the vegetation of China spans from the early Cretaceous Period, 125 million years ago, when angiosperms first appeared in the terrestrial fossil record (Sun *et al.*, 1998; Sun *et al.*, 2002), to the early Pleistocene Epoch, which is considered representative of today's plant assemblages. We describe the paleovegetation and paleoflora of China based on evidence from macrofossils—mainly leaves but also stems, fruits, seeds, wood and flowers—and microfossils (pollen grains and spores).

7.2 Early Cretaceous Period (Berriasian to Albian Ages, 150–70 million years ago)

Although confirmed angiosperm fossils of Archaefructaceae have been reported from the Early Cretaceous Period (ca. 125 million years ago) of northeastern China (Sun *et al.*, 1998; Sun *et al.*, 2002; see Chapter 6), very few angiosperm-like pollen grains are found in sediments from this time, indicating that angiosperms were only rare elements of this gymnosperm-dominated vegetation. In China, the flora of the Early Cretaceous Period has been divided into two floristic regions, northern and southern (Li & Qing, 1994). The northern flora is characterised by the *Disacciatrileti-Cicatricosisporites* assemblage. For example, the Early Cretaceous assemblage of the Fuxin Formation in southern Nei Mongol (Inner Mongolia) is dominated (69% of the pollen and spores found) by gymnosperms such as *Abietineapollenites, Cedripites, Piceaepollenites* and *Pinuspollenites*. Pteridophytes such as Lygodiaceae are the other major component of this flora, making up 30% of the microfossils (Li, 2005). Similarly, in northeastern China the microflora assemblage is dominated by gymnosperms, followed by pteridophytes (Gao *et al.*, 1994).

The southern floristic region is distinguished by the *Classopollis-Schizaeoisporites* assemblage (Li & Qing, 1994). For example the Early Cretaceous Baihedong Formation near Guangzhou Shi is dominated by gymnosperm pollen such as *Classopollis annulatus, Exesipollenites Inaperturopollenites,, Monosulcites* and *Psophosphaera,*which together make up 60–91% of total pollen. Other common components include pteridophyte forms such as *Cicatricosisporites, Lygodiumsporites, Plicatella* and *Toroisporites*. Angiosperm pollen is very rare—ca. 2%—and represented by *Tricolpopollenites* and *Tricolporopollenites* (Song & Zhong, 1985). Other assemblages from southern China are similar, all characterised by a predominance of *Cicatricosisporites, Classopollis, Lygodiumsporites* and *Schizaeoisporites,* with very few angiosperm pollen grains. To summarise, the Early Cretaceous vegetation of both northern and southern China comprised mainly gymnosperms and pteridophytes, with very rare angiosperms; this is very different from the flora of today.

7.3 Late Cretaceous Period (Cenomanian to Maastrichtian Ages, 70–65 million years ago)

One of the most distinctive features of the Late Cretaceous flora is the conspicuous number of angiosperm pollen grains and leaf fossils. Based on assemblages of microfossils, the Late Cretaceous floristic regions of China have been divided into northeastern "mesopalynofloristic" and northwestern-southeastern "xeropalynofloristic" regions; the latter has further been subdivided into two subregions, northwestern and southeastern (see Figure 7.1).

The northeastern, mesopalynofloristic region, as seen in the northern part of the Yin Shan and Yan Shan (Figure 7.1) is characterised by spores of the tree fern family Cyatheaceae and pollen of Pinaceae. However, pollen grains of xerophytes such as Ephedraceae and *Classopollis* appear in several horizons. During later periods, *Triprojectacites* and brevaxonate *Borealipollis* pollen become dominant, indicating a warm, semi-arid climate. *Borealipollis* pollen is typically elongate with colpus-like apertures and a multi-layered, striate exine (Batten & Christopher, 1981; Li *et al.*, 2011), and is believed related to Proteaceae (Song *et al.*, 2005). The Turonian-Santonian palynoflora is dominated by pteridophyte spores of *Cyathidites, Deltoidospora, Gleicheniidites, Polypodiaceaesporites, Schizaeoisporites,* and the gymnosperm *Pinuspollenites*, with pollen of Podocarpaceae and Cupressaceae as subordinate elements, and the genus *Quantouenpollenites* as a distinctive feature. Angiosperms are represented by *Beaupreadites, Borealipollis, Cranwellia, Gothanipollis, Syncolpopollenites* and *Xingjiangpollis*; the phylogentic affinities of these taxa are unclear. Angiosperm pollen grains become the main component in the Campanian Age, with at least 30–50% of total pollen grains. In the Maastrichtian Age, the

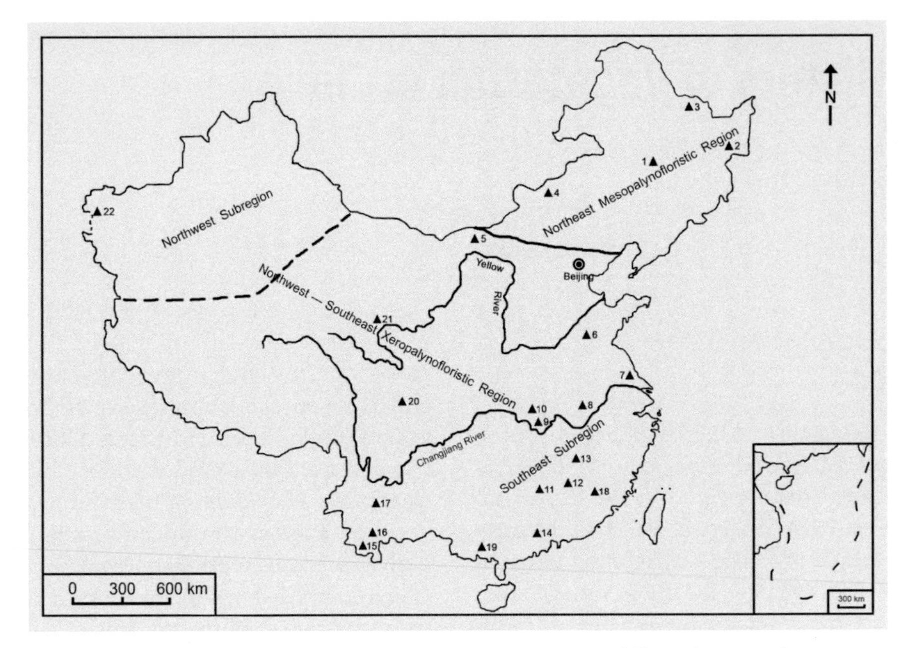

Figure 7.1 Chinese palynofloristic regions of the Late Cretaceous Period (from Zhang, 1993).

northeastern mesopalynofloristic region is characterised by pollen grains comparable to those of modern families, such as Betulaceae, Juglandaceae and Ulmaceae (Zhang, 1993).

The northwestern-southeastern xeropalynofloristic region, as seen in the southern part of the Yin Shan and Yan Shan (Figure 7.1), is distinguished by a large increase in pollen of xerophytic taxa. At Xinjiang, in the northwestern subregion, during the Cenomanian Age, pteridophytes accounted for 75–88%, gymnosperms 9–19%, and angiosperms only 3–9% of pollen and spores found. In the Turonian to Santonian Ages, the percentage of angiosperm pollen increased to 13%, with brevaxonate tricolpate pollen such as *Cranwellia* becoming common and *Lythraites* and *Xinjiangpollis* abundant by the Coniacian to Santonian Ages. In the southeastern subregion, during the Cenomanian Age, pteridophytes and gymnosperms were generally dominant, and the percentage of angiosperm pollen grains only 1–7%. The main pteridophyte representative is *Schizeoisporites*, the main gymnosperms *Classopollis* and *Ephedripites*, and the main angiosperms *Psilatricolpites*, *Retitricolpites* and *Tricolporopollenites*. During the Turonian to Santonian Ages, the pteridophyte and gymnosperm assemblage was similar to that of the Cenomanian Age, but the angiosperm *Cranwellia* becomes more common. Most sediments of the Campanian Age are red in colour, indicating a dry environment, and palynofloras are rare. The gymnosperm genera *Classopollis* and *Exesipollenites* dominate, comprising 20% of the assemblage, pteridophytes 10–20%, and angiosperms about 5%. In the Maastrichtian Age, angiosperm pollen becomes common and the composition more complex. Pollen of modern families such as Betulaceae and Ulmaceae, as well as *Betpakdalina*, *Callistopollenites*, *Cranwellia*, *Jianghanpollis*, *Pentapollenites* and *Proteacidites* occur (Zhang, 1993). At this time, the northeastern corner of present-day China was humid, while areas further south were more or less arid and semi-arid (Figure 7.2, page 109). Pteridophytes and gymnosperms dominated the vegetation, with angiosperms increasing progressively in abundance in the Late Cretaceous Period.

The palynoflora reported from Zhongba, Xizang, which is Santonian to Maastrichtian in age, is dominated by gymnosperm pollen such as *Cedripites*, *Classopollis*, *Cyadopites*, *Exesipollenites*, *Piceapollis* and *Tsugaepollenites*. Angiosperm pollen such as *Betulaepollenites*, *Caryapollenites*, *Clavatricolpites*, *Engelhardtioidites*, *Juglanspollenites*, *Momipites*, *Proteacidites*, *Quercoidites*, *Sporopollis*, *Tricolpites*, *Tricolpopollenites*, *Tricolporopollenites* and *Ulmipollenites* is also reported from this locality (Li *et al.*, 2008).

There is a rich record of plant macrofossils from the Late Cretaceous Period in China (Guo, 1979; Wang, 1984; Tao & Zhang, 1990; Zhou *et al.*, 1990; Li *et al.*, 1995a; Tao *et al.*, 2000; Quan, 2006; Quan & Sun, 2008; Quan *et al.*, 2009), including extant gymnosperms such as *Ginkgo* (Ginkgoaceae) and *Glyptostrobus*, *Metasequoia* and *Sequoia* (Taxodiaceae), from Heilongjiang, Jilin, and the Dongbei Pingyuan (Zhang, 1983b; Li *et al.*, 1995c; Quan, 2006). Many genera and families of angiosperms are also recorded from Late Cretaceous sediments, including *Alangium* (Alangiaceae) from Hebei (Wang, 1984), *Amesoneuron* from Hong Kong (Zhou *et al.*, 1990), *Aralia* (Araliaceae) from Xizang (Guo, 1975), *Arthollia*, *Aspidiophyllum*, *Beringiaphyllum*, *Betuliphyllum*, *Carpolithus*, *Celastrophyllum*, *Ceratophyllum* (Ceratophyllaceae), *Cercidiphyllum* (Cercidiphyllaceae), *Cissites*, *Clematites*, *Corylopsiphyllum*, *Corylus* (Betulaceae), *Dryophyllum*, *Juglandites*, *Leguminosites*, *Monocotylophyllum* and *Platanus* (Platanaceae) from northwestern China (Guo, 1979; Guo & Li, 1979; Guo, 1984; Wang, 1984; Tao & Zhang, 1990; Zhou *et al.*, 1990; Guo, 2000; Quan, 2006) and *Cinnamomum* (Lauraceae) and *Nectandra* from Guangxi (Guo, 1979).

To summarise, gymnosperm forests comprising *Ginkgo*, *Glyptostrobus*, *Metasequoia* and *Sequoia* have existed in northeastern China since the Late Cretaceous Period. Angiosperm elements of the Late Cretaceous differed from their modern relatives in morphology, and most cannot be identified to modern taxa. The Late Cretaceous angiosperm elements of northeastern China resemble those of modern-day deciduous forests, while those of southern China resemble today's evergreen forests.

7.4 Paleogene Period (Paleocene to Oligocene Epochs, 65–22 million years ago)

The history of the Chinese flora from the Late Cretaceous Period to the Mid Eocene Epoch was strongly influenced by geological and climatic events. The collision of the Indian and Asian continental plates led to the uplift of the Himalaya Shan range and the gradual retreat of the Tethys Sea which, along with the global Paleocene-Eocene Thermal Maximum event, caused extensive changes to the climate and vegetative composition of China. Angiosperms, especially woody angiosperms, replaced gymnosperms as the predominant element in the terrestrial vegetation.

7.4.1 Paleocene Epoch

In the Paleocene Epoch, the middle and low latitudes of China were traversed by a widespread arid belt, and it was only coastal areas of the modern Dong Hai (East China Sea) and Nan Hai (South China Sea) that remained moist (Sun & Wang, 2005). Only a few Paleocene macrofloras and palynofloras are reported from China during this time (Li, 1965; Guo, 1979; Guo *et al.*, 1984; Guo, 1986a; Guo, 1986b; Tao & Xiong, 1986; Xiong, 1986). Based on a combination of macrofossils and palynofloras, the vegetation has been divided into two regions: a northern humid temperate region and southern arid region (Sun & Wang, 2005).

The northern humid temperate region.—This region is defined as lying to the north of the ancient Tian Shan-Yin Shan-Yan Shan range (Li *et al.*, 1995a; Tao *et al.*, 2000l; Jin *et al.*, 2003). Representative floras of this region include the Wulungu flora from Xinjiang (Guo *et al.*, 1984) and the Wuyun flora from Heilongjiang, northeastern China (Tao & Xiong, 1986; Hao *et al.*, 2010). From the Paleocene Wulungu flora (Guo *et al.*, 1984), 38 species (four gymnosperms and 34 angiosperms) of macrofossils in 35 genera and 24 families are recorded. The main elements of this flora are *Betula* spp. (Betulaceae), *Cercidiphyllum diversifolium* (Cercidiphyllaceae), *Ditaxocladus planiphyllus*, *Platanus* sp. (Platanaceae), *Sequoia langsdorfii* (Taxodiaceae) and *Tetrastigma caloneurum* (Vitaceae), all of which are deciduous. The percentage of species with toothed leaves is over 60%. The flora thus represents a mixed needle-leaved and deciduous broad-leaved forest and reflects a warm-temperate to subtropical climate.

The geological age of the Wuyun Formation was first considered to be Late Cretaceous or Paleocene-Eocene in age (Liu, 1983; Luo *et al.*, 1983; Tao & Xiong, 1986; Feng *et al.*, 2000; Sun *et al.*, 2003; Chen *et al.*, 2004; Li *et al.*, 2004). However, it is now widely accepted to date from the Early Paleocene Epoch (Sun *et al.*, 2003; Li *et al.*, 2004). More than 60 species have been reported from the Wuyun flora (Tao & Xiong, 1986; Xiong, 1986; Sun *et al.*, 2005; Hao *et al.*, 2010), comprising temperate and subtropical broad-leaved elements with a few coniferous trees and some ferns. Deciduous trees, such as *Betula*, *Juglans* and *Pterocarya* (both Juglandaceae), with some subtropical elements such as *Engelhardia* (Juglandaceae) , *Liquidambar* (Hamamelidaceae) and Magnoliaceae, as well as conifers related to Pinaceae and Taxodiaceae, dominate. This vegetation is characteristic of a warm-temperate to subtropical climate. A subtropical climate has been inferred at Wuyun during the Danian Age (Wu, 1980) from reconstructions of the vegetation and paleoclimate of the region (Hao *et al.*, 2010) using the coexistence approach (Mosbrugger & Utescher, 1997). In summary, during the Paleocene Epoch, vegetation in the northern humid temperate region of China was made up of mixed needle-leaved and deciduous broad-leaved forests, characteristic of a warm-temperate to subtropical climate. Angiosperms were dominant and the morphological characters of macrofossils are comparable with those of their modern relatives.

The southern arid region.—This region is defined to run from the north of the ancient Tian Shan-Yin Shan-Yan Shan range to the continental shelf of the Nan Hai (Jin *et al.*, 2003). Several floras have been described from the region (Guo, 1979; Li *et al.*, 1984), with their most distinctive characteristic being the prevalence of the xerophytic genus *Palibinia*. Most of the floras are considered to be Late Paleocene in age. The composition of vegetation indicates that this region was quite arid in the Paleocene Epoch, as do Paleocene palynofloras from the Nanxiong basin of Guangdong and Hainan, which include *Celtispollenites*, *Ephedripites*, *Pterisisporites*, *Ulmipollenites* and *Ulmoideipites* (Li, 1989). Overall, evidence from Paleocene floras indicates that the Late Paleocene Epoch was more arid than the Early Paleocene Epoch.

7.4.2 Eocene Epoch

The Eocene Epoch is a distinctive geological epoch in which the global climate experienced an extremely warm period followed by a dramatic cooling event (Irving, 2008). From the end of the Paleocene Epoch to the Early Eocene Epoch is termed the Paleocene-Eocene Thermal Maximum. In China, a hot Early to Early-Mid Miocene climate was followed by a cooler Late-Mid to Late Eocene climate. The general trend in temperature change in China during this time matched that elsewhere in the world (Su *et al.*, 2009), and consequently, the vegetation also experienced dramatic changes. Eocene terrestrial fossil records are very rich in China. Three latitudinal vegetation zones have been defined based on vegetation types and floristic elements (Guo,

1986a; Li *et al.*, 1995b; Tao *et al.*, 2000; Sun & Wang, 2005).

The subtropical humid vegetation zone.—This zone encompasses a similar area to during the Paleocene Epoch: northern China, bounded by the southern rim of the Tian Shan-Yin Shan range. Several fossil sites have been found in this zone, including the Mid to Late Eocene Fushun macroflora of Liaoning, Shulan macroflora of Jilin, and Yilan macroflora of Heilongjiang. The Fushun macroflora is well studied, with 70 reported species (Writing Group of Cenozoic Plants of China, 1978). This flora is dominated by Betulaceae (12 species in four genera), Fagaceae (six species in three genera) and Taxodiaceae (four species in four genera), and also contains some typical subtropical elements such as *Cinnamomum, Lindera* (Lauraceae), *Piper* (Piperaceae) and *Sabalites*. This macrofossil assemblage suggests that the Fushun flora represents a deciduous broad-leaved forest with some associated coniferous and evergreen elements. Using leaf margin analysis, SU and colleagues (2009) reconstructed a mean annual temperature for the Fushun flora of ca. 5 °C, suggesting a temperate climate.

The Yilan macroflora is also well studied (He & Tao, 1994; He & Tao, 1997), and comprises two layers. The lower section (Section A) is considered to be Early Eocene in age (He & Tao, 1994; He & Tao, 1997), and is dominated by evergreen subtropical to tropical elements such as *Magnolia* (Magnoliaceae; three species) and Lauraceae (*Litsea* and *Lindera*). The upper layer (Section B) is considered to be Late Eocene in age, and is dominated by temperate elements such as *Alnus* (Betulaceae), *Quercus* (Fagaceae), *Salix* (Salicaceae) and *Sophora* (Fabaceae). Using the same leaf margin method, SU and colleagues (2010) reconstructed a mean annual temperature for the Yilan flora as ca. 17 °C for Section A and ca. 9 °C for Section B. This indicates that the climate cooled from the Early to Late Eocene Epoch, congruent with the changing vegetation assemblage.

KOU (2005) reconstructed the paleoclimate of the Mid to Late Eocene Hunchun flora using the coexistence approach. Mean annual temperature and mean annual precipitation were ca. 15 °C and ca. 1 100 mm respectively, which also indicate a warm-temperate to subtropical climate. In summary, fossil records indicate that the Eocene Epoch of northern China was characterised by mixed broad-leaved evergreen and deciduous forest in the subtropical humid vegetation zone. A large quantity of pollen of the aquatic plants *Potamogeton* (Potamogetonaceae) and *Sparganium* (Typhaceae) and freshwater algae (*Pediastrum*) were also found at this time (Sun & Wang, 2005).

The subtropical, tropical arid and semi-arid vegetation zone of central China.—This zone developed from the subtropical arid vegetation of Paleocene southern China, but its southern boundary advanced northwards to ca. 25° N in the east, and to north of the Yarlung Zangbo Jiang on the Qinghai-Xizang Plateau in the west (Sun & Wang, 2005). Many Eocene fossil sites have been reported from this zone (Li, 1965; Liu & Kong, 1978; Li, 1979; Chen *et al.*, 1983; Guo, 1986a). In the western part of the zone, the vegetation has been studied from fossils reported from the Tarim Pendi, Qaidam Pendi and Xining-Minhe Pendi. These pollen assemblages contain high frequencies of the arid elements *Ephedripites* and *Nitraria* (Nitrariaceae); *Quercoidites, Ulmipollenites* and *Ulmoideipites* were also common. A number of pollen types probably related to tropical and subtropical plants are distributed rather widely but at low frequencies, including *Ilexpollenites, Liquidambarpollenites, Sapindaceoidites* and *Myrtaceidites.* Conifer pollen, assigned to *Picea* and *Pinus* (both Pinaceae), also occurred, and Taxodiaceae was present in trace amounts. These pollen data suggest that desert vegetation, mainly comprising *Ephedra*-like plants and *Nitraria*, was distributed across the basins, with trees present on the mountains (Wang *et al.*, 1990; Sun & Wang, 2005).

Twelve Eocene plant macro-fossil locations have been reported from southeastern and central China, including the genus *Palibinia* (with small, sclerophyllous, narrow-lobed leaves resembling *Banksia* [Proteaceae] or *Comptonia* [Myricaceae]), with a low diversity of fossil types (Li *et al.*, 1995a; Sun & Wang, 2005). CHEN and colleagues (1983) reported 11 plant species from the Relu macroflora of Sichuan: *Albizia* sp. (Fabaceae), cf. *Alstonia* sp. (Apocynaceae), *Banksia puryearensis, Comptonia* sp., *Eucalyptus reluensis* (Myrtaceae), *Hemiptelea paradavidii* (Ulmaceae), *Myrica* sp. (Myricaceae), *Palibinia pinnatifida, Phyllites* sp., *Pistacia* sp. (Anacardiaceae) and *Viburnum* sp. (Adoxaceae). CHEN and colleagues (1983) proposed that this flora was dominated by *Eucalyptus*, in agreement with the idea of a Northern Hemisphere origin for *Eucalyptus* (Geng & Tao, 1982), but their viewpoint is still debated. GUO (1986a) reported further species from the Relu flora, including *Arundo goeppertii* (Poaceae), *Chamaecyparis* sp. (Cupressaceae), *Myrtophyllum* sp., *Palibinia* sp., *Rhus turcomanica* (Anacardiaceae), *Syzygium* sp. (Myrtaceae), *Trapa paulula* (Trapaceae) and *Zelkova ungeri* (Ulmaceae), and considered that the fossils previously ascribed to *Eucalyptus* belonged instead to *Myrtophyllum*. Regardless of this identity, both CHEN and colleagues (1983) and GUO (1986a) agree that the Relu macroflora indicates a dry and hot environment. SU and colleagues (2009) reconstructed the mean annual temperature of the Relu flora as ca. 17 °C, rather higher than the temperature of the area today. Overall, fossil records from this region indicate the presence of a dry vegetation belt across the middle latitudes of China, from east to west, in the Eocene Epoch.

The southern tropical humid zone.—This zone is located to the south of the central zone described above. The main fossil sites of the region are found in Guangdong, Hainan, Guizhou and Xizang (Guo, 1979; Zhang, 1980; Geng & Tao, 1982; Zhang, 1983; Tao, 1988; Lei *et al.*, 1992; Li *et al.*, 1995b; Yao *et al.*, 2009). From the Shinao macroflora in Guizhou, southwestern China, eighteen species have been reported, including *Acer* sp. (Aceraceae), *Actinodaphne* sp. (Lauraceae), *Corylus* sp. (Betulaceae), *Dryophyllum* sp.,

Fagus sp. (Fagaceae), *Lindera* sp., *Meliosma* sp. (Sabiaceae), *Phoebe* sp. (Lauraceae) and *Sequoia* sp. (Zhang, 1983). The dominant families are Lauraceae (with four monospecific genera) and Fagaceae (three monospecific genera). This suggests that the Shinao macroflora represented a mixed evergreen and broad-leaved forest. The estimated mean annual temperature for Eocene Shinao is ca. 9 °C (Su *et al.*, 2010).

Another well studied fossil site is the Liuqu macroflora, Xizang. To date, 39 species in 25 genera and 21 families have been reported (Geng & Tao, 1982; Tao, 1988; Fang *et al.*, 2005), including *Annona* (Annonaceae), *Ficus* (Moraceae), *Grewiopsis* (Tiliaceae), *Magnolia*, *Mallotus* (Euphorbiaceae), *Platanus* and *Sapindus* (Sapindaceae). Most of these genera are predominantly tropical or subtropical, and the assemblage is considered indicative of tropical forest (Tao, 1988; Fang *et al.*, 2005). The Eocene mean annual temperature of Xizang has been reconstructed as ca. 20 °C (Su *et al.*, 2010).

In addition to macrofossils, China bears rich Eocene palynofossil records. The Eocene palynoflora of Changchang basin on Hainan contains pollen and spores from 40 genera in 32 families of spermatophytes and 11 genera in eight families of pteridophytes. At the family level, the Changchang palynoflora is dominated by pantropical elements: 38% of the pollen grains are pantropical, 6% tropical-subtropical and 31% subtropical-temperate. At the genus level, 8% of the grains are pantropical, 18% tropical-subtropical and 68% subtropical-temperate (Yao *et al.*, 2009; Zhao *et al.*, 2009). YAO and colleagues (2009) reconstructed the paleoclimate of Changchang using the coexistence approach (Mosbrugger & Utescher, 1997), and estimated its mean annual Eocene temperature as 10–14 °C and mean annual precipitation as 780–1 110 mm, indicating a subtropical climate, which is supported by the tropical-subtropical forest pollen assemblage (Zhao *et al.*, 2009).

The Paleogene palynoflora of the Buxin group in Guangzhou Shi, Guangdong, has been divided into first, second, and third formations. Two spore and pollen assemblages have been described, the *Ulmipollenites-Pterisporites* assemblage, in the first formation, dated to the Late Paleocene Epoch, and the *Quercoidites-Pentapollenites* assemblage, in the second formation, dated to the Early to Mid Eocene Epoch. Angiosperm pollen grains increase from 63% in the first formation to 77% in the second formation, and include *Cuputiferoipollenites*, *Faguspollenites* and *Quercoidites* (Fagaceae), *Euphorbiacites* (Euphorbiaceae), *Myrtaceidites* (Myrtaceae), *Pentapollenites* and *Santalumidites* (Santalaceae), *Plicapolis* and *Ulmipollenites* (Ulmaceae), *Proteacidites* (Proteaceae), *Rhoipites* (Anacardiaceae), *Rutaceoipollis* (Rutaceae), *Sapindaceidites* (Sapindaceae), *Zonicerapollis* (Caprifoliaceae), *Lygodiumsporites* and *Toroisporis* (Lygodiaceae), and *Pterisporites* (Pteridaceae) These elements indicate that the Buxin palynoflora represented tropical-subtropical vegetation (Li & Qing, 1994).

LIU and YONG (1999) reported a continuous palynological sequence from the Nadu and Baigang Formations in the Bose Basin of Guangxi in southern China. This palynological flora consists of 160 different pollen and spore types grouped into 121 genera and 62 families. From base to top, the four most common pollen and spore assemblages are *Polypodiaceae-Pinuspollenites*, *Quercoidites-Ulmipollenites*, *Alnipollenites-Tricolporopollenites* and *Pinuspollenites-Quercoidites*. These are indicative of a mixed evergreen and deciduous broad-leaved forest. The main elements in such forests today are Betulaceae, Fagaceae, Juglandaceae and Ulmaceae. The presence of Pinaceae and broad-leaved angiosperm fossils together indicates this was a mountainous region, where Pinaceae could grow at higher levels and broad-leaved angiosperms in valley bottoms. This type of paleovegetation is similar to the modern mixed evergreen and deciduous broad-leaved forests found in mountainous areas of southern China today, and indicates a humid subtropical climate in the Eocene Epoch.

The Late Eocene Ningming palynoflora of Guangxi has been studied by WANG and colleagues (2003). It is characterised by pollen grains of angiosperms such as Betulaceae, Fagaceae, Hamamelidaceae, Juglandaceae, Polygonaceae, Ranunculaceae, Ulmaceae and *Echitricolporites* (Asteraceae), and gymnosperms such as *Abietineaepollenites*, *Cedripites*, *Keteleeriaepollenites*, *Piceaepollenites*, *Pinuspollenites*, *Tsugaepollenites* (Pinaceae), some Ephedraceae, Podocarpaceae and Taxodiaceae, and a few Polypodiaceae such as *Polypodiisporites* and *Lygodiumsporites*. The herbaceous pollen grains of this palynoflora render the vegetation similar to that of Baise (Late Eocene Epoch). The Ningming palynoflora is considered to date to at least the Late Eocene to Oligocene Epochs (Wang *et al.*, 2003). Macrofossils from the same locality have been divided into 136 morphotypes and identified as 12 species in nine genera and five families (Shi, 2010), including monospecific genera in Cephalotaxaceae and Hamamelidaceae, two monospecific Cupressaceae genera, and nine species in six Lauraceae genera.

In summary, the Eocene Epoch in northeastern and southern China was characterised by humid forests, broad-leaved evergreen or evergreen deciduous mixed forests, with arid vegetation types in other regions of China. In the early Eocene Epoch, during the Paleocene-Eocene Thermal Maximum, the vegetation of northeastern China (e.g. Hunchun) and southern China (e.g. Hainan) was relatively similar. The macrofossils found are very similar in morphology to their modern relatives. In southern China, the main elements of today's evergreen broad-leaved forests, such Fagaceae, Hamamelidaceae and Lauraceae, had already appeared and become dominant in some regions by the Late Eocene Epoch.

7.4.3 Oligocene Epoch

During the Oligocene Epoch, northwestern China was progressively elevated by the uplift of the Qinghai-Xizang Plateau (Tapponier *et al.*, 2001). The Tethys Sea retreated completely from the west Tarim Pendi. Uplift and subsequent erosion in eastern China led to an absence of Oligocene sediments across extensive areas (Sun *et al.*, 2005). The paleoclimate of the Ningming area (see above) during the Oligocene Epoch has been quantitatively reconstructed using leaf margin analysis and climate leaf analysis (http://clamp.ibcas.ac.cn/Clampset2.html). Estimated mean annual temperatures under the former were ca. 20 °C. and ca. 17 °C, with the growing season precipitation under the latter estimated at ca. 1 700 mm, ca. 810 mm during the three consecutive wettest months and ca. 200 mm during the three driest. These figures indicate that a monsoon climate was present by this time (Shi, 2010).

To date, only five Oligocene macrofossil floras and some palynological assemblages have been found in China (Writing Group of Cenozoic Plants of China, 1978; Li *et al.*, 1995b; Jin & Shang, 1998; Guo & Zhang, 2002). However, these floras indicate that the Oligocene Epoch of China had similar types of vegetation and climate to the Late Eocene Epoch. Three Oligocene vegetation zones can be recognised, although their boundaries are not well defined due to the limited fossil record.

The northeastern temperate and humid vegetation zone.—This zone is represented by the Sanhe flora, Jilin. Twenty-four species in 22 genera and 17 families have been recognised in the Sanhe flora (Guo & Zhang, 2002). Fagaceae is the largest family represented, with five species in three genera, followed by Taxodiaceae (three monospecific genera) and Fabaceae (two monospecific genera). The presence of these families indicates mixed coniferous and deciduous broad-leaved forest, reflecting a humid, warm-temperate or temperate climate, with both temperature and precipitation higher than today.

JIN and SHANG (1998) reported on an another Oligocene flora from north of Shenyang, northeastern China. The main elements in this flora were *Betula dissecta, B. fushunensis, Carpinus* sp. (Betulaceae), *Celtis peracuminata* (Ulmaceae), *Cercidiphyllum arcticum, Cinnamomum naitoanum, Dryophyllum fushunense, Firmiana* sp. (Sterculiaceae), *Hamamelites inaequalis, Magnolia* sp., *Metasequoia disticha, Populus* sp. (Salicaceae) *Quercus* sp., *Salix* sp., *Schisandra fushunensis* (Schisandraceae), *Sequoia chinensis,Taxodium tinajorum* (Taxodiaceae) and *Trochodendron* sp. (Trochodendraceae). Its composition is similar to the Eocene Fushun flora, reflecting a warm and humid climate.

The subtropical arid and semi-arid vegetation zone of central China.—Several palynofloras have been reported from this zone. In northwestern China, pollen assemblages are dominated by arid plants, but Chenopodiaceae are significantly more frequent compared to the Eocene Epoch, in some locations even exceeding *Ephedripites* pollen. During the Oligocene Epoch the vegetation of the northwest is considered to have been xerophytic, and the climate arid (Sun & Wang, 2005).

The tropical and subtropical humid vegetation zone of southern China.—Several floras and palynofloras have been reported from this zone. One of the best studied is the Jinggu flora of Yunnan, southwestern China (Writing Group of Cenozoic Plants of China, 1978). Thirty-four species in 25 genera and18 families have been recognised in this flora. Fagaceae is the most common family with four genera—*Castanopsis, Cyclobalanopsis, Lithocarpus* and *Quercus*—and 13 species recognised, followed by Lauraceae with four species in three genera. Both families are major elements of modern subtropical evergreen broad-leaved forests. Other elements from Jinggu characteristic of extant subtropical regions include *Annona, Carya* (Juglandaceae), *Magnolia* and *Terminalia* (Combretaceae). The Jinggu flora and vegetation are therefore identified as a subtropical evergreen broad-leaved forest, reflecting a humid and warm subtropical or tropical climate.

7.5 Neogene Period (22–2.6 million years ago)

The Neogene was characterised by a transition from the "greenhouse" climate of the Paleogene Epoch to the "icehouse" climate of the Quaternary Period. During this time, China underwent major environmental changes and large tectonic movements. Paleobotanical evidence demonstrates that the broad belt of aridity that stretched across China from west to east during the Paleogene Period became restricted to western China during the Miocene Epoch of the Neogene, as a humid belt extended into eastern China (Sun & Wang, 2005; Yao *et al.*, 2009; see Figure 7.2, page 109).

Miocene deposits are widely distributed throughout China. Those in northeastern China are considered to have developed under humid warm-temperate to subtropical climates, while those of northwestern China represent sub-arid temperate climates, evidenced by the presence of microphyllous and papery leaves and absence of evergreen and warm-adapted plants. This "wet east, arid west" pattern has persisted in China since the Miocene Epoch.

Due to geographical variation, environmental change and evolution, the vegetation of the Neogene Period varied in both time and space in China (Wang, 2006), and an understanding of this vegetation development through geological time is important for interpreting paleoclimatological change. Therefore, a great deal of research has been carried out on the Neogene vegetation of China, based on pollen sequences as well as fossils of leaves, wood, and reproductive organs (Xia *et al.*, 1987; Wang,

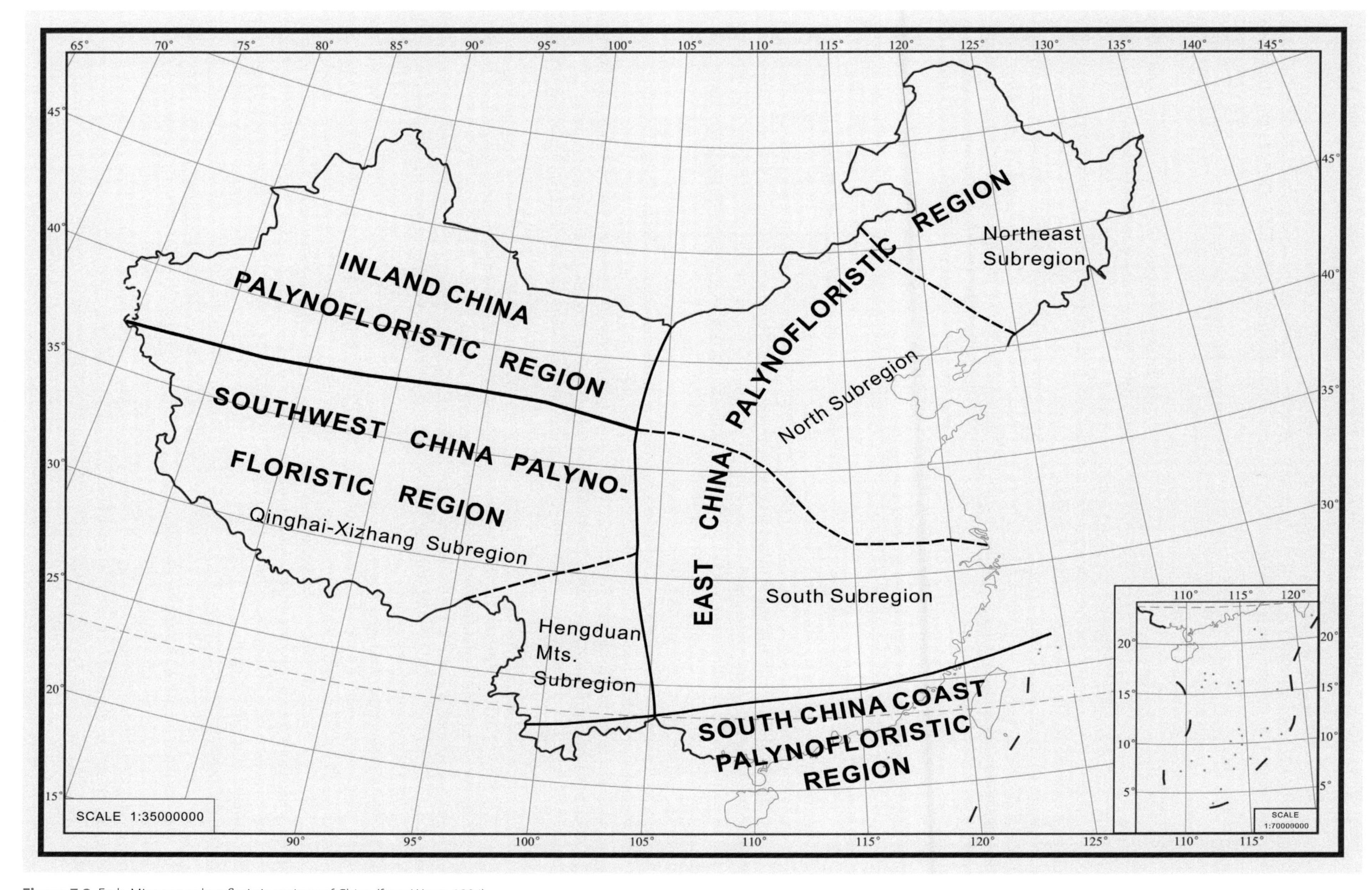

Figure 7.2 Early Miocene palynofloristic regions of China (from Wang, 1994).

1990; Wang, 1994; Wang, 1996b; Wang, 1996a; Wang, 1999; Yi *et al.*, 2005; Wang, 2006). Among others, WANG (1994; 2006) has conducted a comprehensive investigation of palynological data, based on which he divided China in the Neogene Period into four regions (see Figure 7.2): the South China coast, Southwest China, Inland China and East China.

7.5.1 Miocene Epoch

The arid vegetation zone, northwestern China.—In the western part of China, early Miocene palynofloras are found at Dunhuang in Gansu, the southern Junggar Pendi in Xinjiang, and the Xining-Minhe Pendi in Qinghai. The palynoflora of Dunhuang consists of coniferous and broad-leaved deciduous species, the main elements being *Abietineaepollenites*, *Betulaceoipollenites*, *Ephedripites*, *Juglanspollenites*, *Keteleriapollenites*, *Pinuspollenites*, *Quercoidites* and *Salixipollenites* (Ma, 1991). Palynofloras of other areas are similar to that at Dunhuang; for instance the southern Jungar flora is similar in floristic composition and vegetation type but with a higher percentage of *Ephedripites* pollen (Liu *et al.*, 2011).

During the early Miocene Epoch, dense forests are thought to have occupied an extremely large region, with mixed gymnosperm and angiosperm forests in Nei Mongol pre-dating the now widespread steppe and desert (Tao *et al.*, 1994). Results from a study of fossil wood demonstrate that coniferous and broad-leaved mixed forests, indicating a temperate and moist climate, existed here in the early Miocene Epoch (Tao *et al.*, 1994). By the end of the Mid Miocene Epoch, pollen and spore assemblages from the Tongguer Formation, Nei Mongol, indicate a cold, arid climate (Wang, 1990). The development of this cold, dry climate, disadvantageous to the growth of many trees and shrubs, may be a crucial factor in the spread of desert vegetation in northwestern China at this time. Fossil evidence further shows a decline in woody plants and expansion of herbaceous species, including *Artemisia* (Asteraceae), *Chenopodium* (Chenopodiaceae) and many modern-day families (Wang & Zhang, 1990). It is probable that C4 plants appeared and became common at this time. Mid Miocene palynofloras have been found at the same sites as the Early Miocene ones, described above. The vegetation at these sites was similar to that of the Early Miocene Epoch, comprising mixed coniferous and deciduous forests, but with some subtropical elements such as *Engelhardtia* indicating a warm climate, and the appearance of herbs such as *Artemisia* and other Asteraceae.

The Wulong flora, Namling (Xizang) is an important Late Miocene flora. Within this flora, a lower (older) assemblage includes *Betula, Carpinus, Populus* and *Ulmus* (Ulmaceae), common elements of today's deciduous forests. The upper (younger) assemblage contains *Quercus namulinensis, Q. prespathulata, Q. wulongensis, Rhododendron* (Ericaceae), *Salix* and *Thermopsis* (Fabaceae), among others. The three oak species are sometimes ascribed to *Quercus* section *Heterobalanus* (not recognised in the *Flora of China*), which is a dominant element of sclerophyllous evergreen broad-leaved forests in the Himalaya Shan and Hengduan Shan today. The Miocene assemblages of the Wulong flora indicate a transition in the vegetation, from deciduous to sclerophyllous evergreen forest.

The humid monsoon forest vegetation zone.—The vegetation of the humid monsoon zone is much more complex. Two subzones, northern and southern, have been defined. In the Early Miocene Epoch, the vegetation of the northern subzone is represented by palynofloras from the Bohai Wan and Jiyang Basin in Shandong, the Shangdu-Huade Pendi in Nei Mongol, and Wolonggang, Hebei, and a macroflora at Dunhua, Jilin. The main elements of these floras include *Acer, Betula, Castanea* (Fagaceae), *Fagus, Fraxinus* (Oleaceae) *Juglans, Pinus, Quercus, Tilia* (Tiliaceae), *Ulmus* and *Zelkova*, i.e. typical deciduous broad-leaved vegetation. In the southern subzone, the vegetation is represented by floras from Toubei in Jiangxi, Fushan, Beibu Wan, Yingge Hai, and Zhujiang Kou on the north continental shelf of the Nan Hai, and Beibu Wan. The main elements of these microfloras include *Betula, Carpinus, Cornus* (Cornaceae), *Cyclobalanopsis, Engelhardtia, Hamamelis* (Hamamelidaceae), *Liquidambar, Magnolia, Quercus,* Moraceae, Myrtaceae and Nyssaceae, indicating a type of evergreen broad-leaved forest (Yao *et al.*, 2009).

Mid Miocene palynofloras of the northern subzone have been found in the Bohai Wan, Tianchang (Jiangsu), Jiyang (Shandong), Huanan (Heilongjiang) and Hunchun (Jilin), among others. These floras are similar to those of the Early Miocene Epoch, representing deciduous forest. In the southern subzone, the vegetation is represented by the famous Shanwang flora of Shangdong, eastern China. This site contains well-preserved diverse, abundant macrofossils of leaves, fruits, and seeds and even a few flowers. Records from the Shanwang flora to date include 155 species in 104 genera and 55 families, including *Acer, Aconitum* (Ranunculaceae), *Albizia* (Fabaceae), *Alnus, Ampelopsis* (Vitaceae), *Berchemia* (Rhamnaceae), *Betula, Carpinus, Castanopsis, Carya, Castanea, Cercis* (Fabaceae), *Chaneya* (Simaroubaceae), *Cinnamomum, Cornus, Corylus, Cyclobalanopsis, Cyperus* (Cyperaceae), *Eucommia* (Eucommiaceae), *Fagus, Ficus, Fothergilla* (Hamamelidaceae), *Gleditsia* (Fabaceae), *Hamamelis, Indigofera* (Fabaceae), *Juglans, Koelreuteria* (Sapindaceae), *Lindera, Liquidambar, Liriodendron* (Magnoliaceae), *Litsea, Magnolia, Morus* (Moraceae), *Ostrya* (Betulaceae), *Paliurus* (Rhamnaceae), *Platycarya* (Juglandaceae), *Pistacia, Polygonum* (Polygonaceae), *Populus, Quercus, Rhus, Salix, Sapindus, Tetrastigma, Tilia, Toona* (Meliaceae), *Ulmus* and Poaceae. It is clear that the Shanwang flora contains most of the common trees of deciduous forests today, as well as elements of evergreen broad-leaved forest, such as *Castanopsis, Cinnamomum, Cyclobalanopsis, Lindera,*

Litsea and *Magnolia*, along with vines such as *Tetrastigma* and *Ampelopsis* and herbs such as *Aconitum*, *Cyperus* and Poaceae. Most of the fossil leaves here are very well preserved, indicating that they were probably preserved in, or very near, where they grew. From the fossil records of the Mid Miocene Epoch it can be inferred that mixed deciduous and evergreen broad-leaved forests were found at Shanwang.

Overall, the climate and vegetation of the Miocene Epoch in northeastern China were comparable with those of modern times, with a temperate, slightly moist climate and deciduous forests. For instance, the Late Miocene of Heilongjiang, northeastern China, was associated with a temperate, slightly moist climate supporting mixed coniferous and deciduous broad-leaved forests dominated by Betulaceae, Fagaceae and Pinaceae (Liu & Zheng, 1995). The Pliocene vegetation shows little difference in this area, with some increase in *Pinus* and elements such as *Alnus, Fagus, Picea* and *Tilia* (Xia *et al.*, 1987), and the climate remaining temperate and cool.

The Late Miocene southern subzone vegetation is well represented by the Xiaolongtan flora of Kaiyuan, Yunnan. The Xiaolongtan flora is dominated by Fabaceae, including *Albizia, Cassia, Dalbergia, Desmodium, Erythrophleum, Gleditsia, Indigofera, Lespedeza, Ormosia, Robinia* and *Podocarpium*, Fagaceae, including *Castanea, Cyclobalanopsis, Lithocarpus* and *Quercus*, and Lauraceae (*Cinnamomum, Laurus, Litsea, Machilus, Nothophoebe* and *Phoebe*). *Exbucklandia* (Hamamelidaceae), *Ficus, Magnolia* and *Myrica* are also common. More than half (66%) of the genera found are subtropical, for example *Castanopsis, Cyclobalanopsis, Desmos* (Annonaceae), *Indigofera* and *Passiflora* (Passifloraceae; Zhou, 2000; Xia *et al.*, 2009), and it is clear that the Xiaolongtan flora represents an evergreen broad-leaved forest. A second flora of similar age has been located at Xiangfeng in central Yunnan, about 270 km north of Xiaolongtan. The Xiangfeng flora is very similar to that of Xiaolongtan, and dominated by Fagaceae (including *Castanopsis, Cyclobalanopsis, Lithocarpus* and *Quercus*). This floristic composition is very similar to the subtropical evergreen broad-leaved forests found in central Yunnan today. Several other Late Miocene floras are known from Yunnan, including the Shuanghe flora from Jiangchuan, Hunshuitan in Kunming, and the Yuhe flora in Chuxiong. Fagaceae, Lauraceae and Magnoliaceae are again common elements of these floras. Thus in the Late Miocene Epoch we find that evergreen broad-leaved forests, dominated by Fagaceae and Lauraceae, were already widely distributed in central Yunnan and similar to today's evergreen broad-leaved forests. Evergreen broad-leaved forests dominated by Fabaceae, Fagaceae and Lauraceae were widely distributed in southern China in the Miocene Epoch. In general, the Miocene vegetation of the humid monsoon forest vegetation zone was similar to today's evergreen broad-leaved forests in both floristic composition and vegetation structure.

7.5.2 Pliocene Epoch

The vegetation regions defined for the Pliocene Epoch are very similar to those of the Miocene Epoch. Southern China is particularly rich in Pliocene floras, especially the southwest, which includes the Hengduan Shan and Qinghai-Xizang Plateau, the former widely recognised as having a rich Pliocene flora. At least 20 plant fossil assemblages have been studied, mainly distributed in and near the Hengduan Shan, a region in which both climate and plant evolution has been greatly affected by the uplift of the Qinghai-Xizang Plateau. One important example is the Xixabangma flora in Xizang, which is dominated by macrofossils of oaks such as *Quercus namlingensis, Q. preguyavifolia* and *Q. presenescens,* alongside microfossils of *Betula, Cedrus, Picea* and *Tsuga* (Pinaceae). This fossil assemblage contains species indicative of sclerophyllous oak forest (Xu *et al.*, 1973; Zhou *et al.*, 2007), which has also been indicated in Pliocene floras at Eryuan, Yunnan (Tao & Kong, 1973) and Tengchong (Tao & Du, 1982; Li *et al.*, 2004) and in the Lühe Neogene flora (Xu *et al.*, 2000), among others.

The topography of southern China is complex and the vegetation differed from place to place in the past as it does today. In the south of the region, the Late Pliocene Tengchong flora displays high angiosperm diversity, with 19 families of seed plants (including Aceraceae, Betulaceae, Fabaceae, Fagaceae, Juglandaceae, Lauraceae, Magnoliaceae, Moraceae, Pinaceae and Ulmaceae), and one species of fern. The flora was dominated by trees, with a few species of shrubs and herbs (Li *et al.*, 2004), a composition characteristic of evergreen broad-leaved forests (Li *et al.*, 2004). In contrast, in the Hengduan Shan, most Pliocene floras were distinguished by dominance of the same *Quercus* species as at Xixabangma (Tao & Kong, 1973; Tao & Du, 1982; Tao *et al.*, 2000), a flora similar to the sclerophyllous evergreen broad-leaved forests that are widely distributed in the eastern Himalaya Shan and Hengduan Shan today. Thus the vegetation of the Hengduan Shan has remained largely unchanged since the Pliocene Epoch. Pollen sequence analysis of the Late Pliocene Yangyi flora indicates mixed evergreen coniferous and broad-leaved forest, while that of Longling suggests humid evergreen broad-leaved forests (Xu *et al.*, 2003). From this XU and colleagues (2003) concluded that it was much warmer and more humid in southwestern China during the Late Pliocene Epoch than today.

The flora of Lufeng, central Yunnan, differs again from floras of the same geological age in the eastern Himalaya Shan and Hengduan Shan. Lufeng records include the extinct *Dryophyllum* (Fagaceae) together with *Cinnamomum, Paliurus* (Rhamnaceae), *Phoebe, Pistacia, Rhus* and *Sassafras* (Lauraceae), dominant components of evergreen broad-leaved forests or mixed evergreen coniferous and broad-leaved forests, suggesting a subtropical warm, humid climate (Tao & Han, 1990). The pollen and spore composition of the Pliocene Lühe flora shows that angiosperms were the main component, including temperate elements such as *Alnus, Betula, Carpinus, Castanea, Corylus, Quercus* and *Ulmus,* with a few subtropical elements such as *Cyclobalanopsis, Castanopsis, Carya* and *Ilex* (Aquifoliaceae). The dominance of temperate species suggests that Lühe experienced a cool climate during the Pliocene Epoch (Xu *et al.*, 2000), during which time gymnosperms such as *Abies, Pinus* and *Tsuga* were restricted to higher elevations (Xu *et al.*, 2000). Thus we tentatively conclude that here evergreen and deciduous broad-leaved forests grew at lower altitudes while mixed coniferous and broad-leaved forests existed at higher altitudes in the Pliocene Epoch.

In the south of the Qinghai-Xizang Plateau subregion, large areas of broad-leaved forests existed prior to its uplift (Xu *et al.*, 1973; Li & Guo, 1976). Although woody plants remained the dominant elements until the Late Pliocene Epoch, when steppe vegetation began to spread, some widely-distributed tropical and subtropical elements did become extinct, and others became largely restricted to lower latitudes. Conversely, a number of newly-evolved types, such as dicotyledonous herbs, flourished (Wang & Zhang, 1990).

Grasslands formed in regions of northwestern China in the Late Pliocene Epoch (Wang & Zhang, 1990). These were dominated by herbaceous families such as Asteraceae, Chenopodiaceae, Liliaceae and Poaceae, with *Artemisia* becoming the most abundant element in the fossil record in subsequent periods (Li *et al.*, 2003). Deserts also developed to their full extent as a result of the drier climate and gradual local extinction of woody plants (Li *et al.*, 2003). Overall, the vegetation of inland China during the Neogene Period can be summarised as forests of woody plants in the early Mid Miocene Epoch, mixed forest and grassland in the later Mid Miocene and Early Pliocene Epoch, with deserts and regional grasslands in the Late Pliocene Epoch.

7.6 Quaternary Period (2.6 million years ago to present)

The Quaternary Period spans the last 2.6 million years and comprises the Pleistocene and Holocene Epochs. The Quaternary Period was a time of major environmental and climatic change, characterised, in the Northern Hemisphere, by a sequence of roughly 30 glacial-interglacial cycles (Gornitz, 2009). However, China was never covered by an extensive ice-sheet, so many elements survived from earlier periods into the present-day.

The Quaternary Chinese flora is differentiated longitudinally along climatic and topographic lines (Tao, 1992), and may have evolved differently from that in other areas of the world. The main sources of information on Quaternary vegetation history are fossil pollen records extracted from lake, wetland and near-shore marine sediments, supplemented extensively with plant macrofossils and stable carbon isotope records. Pollen data are the most easily obtained and pollen analysis is therefore the most common method of reconstructing Quaternary vegetation patterns and studying climatic change and interactions between the atmosphere, biosphere and human activities (Wang, 1996b; Wang *et al.*, 1997; Tong *et al.*, 1999; Liu *et al.*, 2001; Li *et al.*, 2003; Wang & Sun, 2005; Zhang *et al.*, 2006; Ni *et al.*, 2010). For example, pollen and spore analysis suggests that, during the Pleistocene Epoch, Guangxi was under subtropical evergreen broad-leaved forest with components such as *Cinnamomum, Desmos, Juglans, Litsea, Machilus, Phoebe* and *Quercus* (Liu, 1990).

7.6.1 Pleistocene Epoch

The temperate forest region.—The Pleistocene flora of China has been divided into four regions. The temperate forest region includes northeastern and northern China, and is marked by a cold climate. Hygrophilous species such as *Carya* and *Tsuga* gradually became extinct, broad-leaved deciduous taxa such as *Quercus, Tilia* and *Ulmus* decreased, conifers such as *Picea* and *Pinus* increased, and xerophilic herbs, mainly *Artemisia* and *Chenopodium,* flourished. During the coldest period, the flora was dominated by herbaceous plants, with a few *Pinus* and very few broad-leaved trees. *Picea* was also common at first but rapidly decreased as the climate cooled and dried further, eventually forming a steppe-like habitat. During subsequent warming, conifers and broad-leaved trees reappeared on the steppe. Within the region, woody plants such as *Abies, Betula* and *Picea* were more abundant in the northeast than the north, suggesting that the climate in the northeast was slightly warmer and more humid (Zhao, 1992; Li *et al.*, 1995a; Li, 1998; Tao *et al.*, 2000).

The subtropical forest region.—Subtropical forests occurred in the Pleistocene Epoch to the south of a line formed by the Qin Ling and Huai He. During colder periods, many hygrophilous gymnosperms were found in what are now temperate and subtropical mountainous areas. Some hygrophilous ferns, such as *Gleichenia* (Gleicheniaceae), *Lepisorus* and *Pyrrosia* (Polypodiaceae) and *Osmunda* (Osmundaceae), are also recorded, and the vegetation was somewhat similar to the subtropical montane forests of today. The appearance of many subtropical taxa, such as *Aralia, Carya, Fagus, Ilex, Liquidambar, Magnolia, Pterocarya* and *Symplocos* (Symplocaceae) during warmer periods indicates a warm and humid climate.

The temperate grassland and desert region.—This region includes northwestern China and the Huangtu Gaoyuan (Loess Plateau). From the beginning of the Quaternary Period, this region was dominated by xerophytes, mainly represented by *Artemisia* and *Chenopodium.* Some broad-leaved plants such as *Betula, Juglans, Salix* and *Ulmus* were present in the early Pleistocene Epoch, but these decreased with time and most eventually disappeared. The region alternated between grassland dominated by *Artemisia* and *Chenopodium* and desert-steppe, due to changing precipitation regimes. Strong evaporation during warmer periods resulted in drier conditions and the expansion of desert; desert-steppe developed when temperatures decreased.

Alpine grassland on the Qinghai-Xizang Plateau.—The Pleistocene flora of the Qinghai-Xizang Plateau was distinct from other regions of China, with cold, dry conditions and changing elevation due to uplift. The flora south of the Himalaya Shan differed greatly from that to the west. South of the mountains were found mixed coniferous and broad-leaved forests, with herbaceous *Artemisia*; to the west were *Abies*- and *Picea*-dominated sub-alpine conifer forests with xerophytic *Chenopodium*. After the end of the glacial period, *Artemisia*, Chenopodiaceae and Poaceae are the main elements recorded here, indicating that the vegetation was largely alpine grassland or meadow.

7.6.2 Holocene Epoch

In China, the Holocene Epoch is divided into five climatic stages: the first with a lower temperature than today, but increasing; the second a pre-warm period with similar temperatures to today; the third a high temperature period, 2–3 °C higher than today; the fourth a post-warming period, again with similar temperatures to today; and finally a cooling period with lower temperatures than today. Based on pollen and spore data, the epoch has also been divided into three vegetation stages: corresponding to the Early, Mid, and Late Holocene Epoch. Overall the flora of the Holocene Epoch is considered quite similar to modern-day vegetation. Seven floristic regions have been designated based on vegetation types (Zhao, 1992; Li *et al.*, 1995a; Li, 1998; Tao *et al.*, 2000).

The temperate mixed coniferous and broad-leaved forest region.—This region covered large areas north and east of the Dongbei Pingyuan (Northeast China Plain), with a humid monsoon climate at mid latitudes, and was dominated by *Pinus koraiensis*. During the Early Holocene Epoch, *Betula* was the most prevalent species in this assemblage, with other common species including *Alnus, Corylus, Salix* and *Ulmus*. From this it is inferred that the vegetation consisted of a shrub forest and the climate was cold and humid. Later in the epoch, *Alnus* and *Betula* fossils decreased considerably, while those of mesic plants such as *Acer, Carpinus, Quercus, Tilia* and *Ulmus* gradually increased, suggesting a warmer and wetter climate. The Mid Holocene Epoch was characterised by extensive deciduous broad-leaved forests with *Acer, Alnus, Carpinus, Corylus, Juglans, Quercus, Salix, Tilia* and *Ulmus* as main components, suggesting a warm, moist climate, a little warmer than today. Conifers were also present at higher altitudes. The Late Holocene Epoch witnessed coniferous and broad-leaved forests dominated by *Pinus koraiensis*. Coniferous forests with *Abies, Picea* and *Pinus* began a shift to lower altitudes, while the broad-leaved taxa that flourished in the Mid Holocene Epoch decreased, reflecting the decreasing temperature and humidity (Zhao, 1992; Li *et al.*, 1995a; Li, 1998; Tao *et al.*, 2000).

The temperate grassland region.—This region includes the Dongbei Pingyuan, Nei Mongol and the Huangtu Gaoyuan. Pollen and spore data show that the Early Holocene Epoch here was characterised by large areas of grassland dominated by herbaceous *Artemisia* and *Chenopodium* (Li *et al.*, 2003; Zhao *et al.*, 2008). Woody plants were rare, comprising only a few species such as *Abies, Betula, Fraxinus, Juglans, Quercus, Pinus, Tilia* and *Ulmus*. In the Mid Holocene Epoch, herbaceous vegetation remained dominant, but these woody taxa spread, indicating that both forest and grassland flourished and suggesting a warm, moist climate. During the Late Holocene Epoch, with increasing aridity, the diversity and abundance of herbaceous plants increased and woody plants declined, the region becoming dominated by steppe and grassland communities typical of those seen today (Li *et al.*, 1995a; Li, 1998).

The warm-temperate deciduous broad-leaved forest region.—This region is bounded by the temperate mixed coniferous and broad-leaved forest and temperate grassland regions to the north, subtropical evergreen forest to the south, and the Bohai Wan to the east. During the Early Holocene Epoch, the area around present-day Beijing Shi was characterised by forest-grassland or grassland, the grassland dominated by *Artemisia, Chenopodium*, Asteraceae and Poaceae, and the forest by species such as *Alnus, Betula, Pinus, Quercus, Tilia* and *Ulmus*. Later in the epoch, the proportion of woody plants increased and *Betula*-dominated forest was found in the east of the region, suggesting a warming climate. In the Mid Holocene Epoch, woody plants of warm areas occupied the region, with *Quercus* as the dominant element along with *Betula, Carpinus, Celtis, Corylus, Juglans* and *Pterocarya*. In the south-central part of the region the presence of these species, which now exist in the subtropics, suggests a much warmer climate than present. However in the northern part of the region, herbaceous plants remained a significant feature, and the vegetation type was forest-grassland. In the eastern coastal areas, broad-leaved forests appeared, which may reflect a warm and humid climate. Precipitation decreased westward and the vegetation correspondingly became coniferous and broad-leaved forest or forested-steppe (Zhao, 1992; Li *et al.*, 1995a; Li, 1998; Tao *et al.*, 2000).

The subtropical evergreen broad-leaved forest region.—This is defined as the vast area south of the line formed by the Huai He and Qin Ling. The region is rich in pollen and spore records, especially in the areas around the middle and lower reaches of the Chang Jiang, eastern Guangdong, and mountainous parts of Guangxi. In the delta of the Chang Jiang, evidence for mixed evergreen and deciduous broad-leaved forests appears, with *Alnus, Betula, Carpinus, Carya, Castanea, Castanopsis, Pinus, Quercus, Salix* and *Ulmus* as the main components. Some temperate and cold-resistant elements such as *Abies* and *Picea* coexisted here, suggesting a lower temperature than today. The south subtropical evergreen forest of eastern Guangdong had as its main components *Altingia* (Hamamelidaceae), *Castanopsis, Elaeocarpus* (Elaeocarpaceae) , *Ilex, Lithocarpus, Microtropis* (Celastraceae), *Pinus, Podocarpus, Quercus* and *Randia* (Rubiaceae), suggesting a similar climate to the present day. Subtropical evergreen and deciduous broad-leaved forests were present in the mountainous areas of Guangxi, with species such as *Ilex, Liquidambar* and *Quercus* as the main components and a few coniferous taxa such as *Pinus* and *Tsuga*.

During the Mid Holocene Epoch, broad-leaved evergreen forests and evergreen and deciduous broad-leaved forests developed in the Changjiang Sanjiaozhou and many subtropical elements began to appear, including *Aphanamixis, Melia* and *Dysoxylum* (all Meliaceae), *Castanopsis, Helicia* (Proteaceae), *Juglans, Liquidambar, Lithocarpus, Pinus* and *Quercus*, suggesting a warmer climate than today. As in eastern Guangdong, evergreen broad-leaved elements expanded further, and tropical-subtropical mangrove vegetation also flourished at this time, with *Acanthus* (Acanthaceae), *Avicennia, Bruguiera, Ceriops, Kandelia* and *Rhizophora* (all Rhizophoraceae) and *Excoecaria* (Euphorbiaceae) as the main components (Wang & Sun, 2005); both the number of taxa and their distribution exceeded those of today. The mountainous forests of Guangxi were characterised by subtropical evergreen forest, suggesting a warm or hot, humid climate (Wang & Sun, 2005).

In the Late Holocene Epoch, subtropical elements began gradually to decrease in the middle and lower reaches of the Chang Jiang, while temperate elements and *Pinus* tended to increase, which perhaps indicates a cooling period. Eastern Guangdong underwent a similar transition as evergreen elements decreased, and *Pinus* forests still exist here in some places today (Luo & Sun, 2001). Decreasing temperatures and human activity may both be responsible for the presence of secondary *Pinus* forests. In the mountains of Guangxi, coniferous species began to spread, with *Tsuga* replacing *Castanopsis*, and *Pinus-Quercus* forest developing nearby (Tao *et al.*, 2000).

The temperate desert region.—The temperate desert region generally refers to the northwest of China and the vast areas north of the Qinghai-Xizang Plateau. The Holocene flora of this region has been studied at the Chaiwopu Pendi in Xinjiang and areas around Qinghai Hu. Holocene vegetation types include cold, high scrub, cold alpine meadow and steppe, temperate semi-shrub and desert. In the Early Holocene Epoch, grasslands occurred around Qinghai Hu, dominated by *Artemisia, Chenopodium* and Poaceae. In drier places there is some evidence for the existence of halophytic shrubs such as *Ephedra* (Ephedraceae) and *Nitraria*, and in some more humid places for forests of *Betula, Picea* and *Pinus*. In parts of Xinjiang, *Artemisia, Chenopodium* and Poaceae were dominant, forming 90% of the fossils found, while woody plants (including *Betula, Ephedra, Nitraria, Picea* and *Salix*) were rarely seen. During the Mid Holocene Epoch the number of arboreal species increased to more than 50% of fossils, including most frequently *Picea* and *Pinus* but also *Abies, Betula, Quercus* and *Ulmus*. These taxa characterised a sub-alpine coniferous forest, suggesting a higher temperature than today. In the Late Holocene Epoch, in areas around Qinghai Hu, *Betula* increased and replaced *Picea* as the dominant species. At the same time some xerophytic herbs and shrubs also increased in abundance, suggesting a drying climate. In some parts of Xinjiang *Betula*, together with herbaceous *Chenopodium* and Poaceae, increased, while *Ephedra* and some other woody plants decreased, suggesting a decline in temperature.

The Qinghai-Xizang Plateau region.—Because of its extremely high elevation, this region had more complex vegetation types, including alpine meadow, grassland, alpine shrub and coniferous forest (or mixed coniferous and broad-leaved forest during warm periods). Quantitative paleovegetation data from the Qinghai-Xizang Plateau are extremely rare for the late glacial period and Holocene Epoch (Shen *et al.*, 2006a). Conifer-dominated forests probably covered large areas at elevations between ca. 2 500 and 4 000 m on the northeastern Plateau during the first half of the Holocene Epoch (Shen *et al.*, 2005). Pollen grains of *Betula* and *Picea* are rich from the Early Holocene Epoch, and those of *Pinus* reached a maximum during the Mid Holocene Epoch at Qinghai Hu. Forests also began to decline in the Mid Holocene Epoch (Yan *et al.*, 1999; Herzschuh *et al.*, 2006; Shen *et al.*, 2006b; Herschuh *et al.*, 2010). From about 7 000 years ago there was a decline in arboreal pollen indicating a broad-scale reduction in tree cover on the Qinghai-Xizang Plateau during the Mid Holocene Epoch (Herzschuh & Birks, 2010).

At Hurleg Hu in the Qaidam Pendi, pollen data show that the vegetation changed from desert before the Holocene Epoch, to desert-steppe dominated by *Artemisia* from 11 900 to 9 500 years ago, then to desert dominated by Chenopodiaceae from 9 500 to 5 500 years ago, and finally to steppe-desert dominated by *Artemisia* and Poaceae from 5 500 years ago (Zhao *et al.*, 2007). Paleovegetative reconstructions based on fossil pollen data from the Qilian Shan on the northeastern margin of the Qinghai-Xizang Plateau provide information about changes in vegetation assemblages. A short period towards the end of what is known as the third Marine Isotope Stage, about 50 000–30 000 years ago, was characterised around Hurleg Hu by the expansion of forest dominated by *Abies, Picea* and *Pinus*, suggesting temperatures slightly above those of today. During the last glacial maximum (ca. 18 000 years ago), the vicinity of the lake was covered by sparse alpine vegetation and alpine deserts indicating cold, dry conditions (ca. 4–7 °C colder than today). Higher temperatures are inferred since around 13 000 years ago, with a Holocene temperature optimum characterised by the maximal expansion of mixed forests (*Betula, Picea* and some *Juniperus*) between 9 000 and 7 000 years ago. For the last 7 000 years, the vicinity of the lake has been covered by alpine steppe and meadows, intermixed with sub-alpine shrub vegetation (Herzschuh *et al.*, 2006). The most likely explanation for the widespread forest decline seen on the Qinghai-Xizang Plateau invokes the intensity of the East Asian summer monsoon in the Mid Holocene Epoch (Herzschuh *et al.*, 2011).

The tropical forest region.—This region covered areas of the Leizhou Bandao, Hainan and southern Yunnan. The vegetation types of this region were quite similar to those of today, including tropical montane rainforest, evergreen broad-leaved forest, mangrove forest, evergreen and deciduous broad-leaved forest.

This account of changing vegetation types across all regions of China from the Early Cretaceous Period shows the incredible complexity of vegetation and climate that the country has experienced, as well as the depth of macro- and micro-fossil evidence that has been amassed to interpret it. An understanding of past climatic, geological and vegetation history is essential to understanding the current distribution of plants and vegetation types in China. The following chapters consider in more depth the modern-day vegetation of China from the Quaternary Period onwards.

References

BATTEN, D. J. & CHRISTOPHER, R. A. (1981). Key to the recognition of *Normapolles* and some morphologically similar pollen genera. *Review of Palaeobotany and Palynology* 35(2-4): 359-383.

CHEN, M. H., KONG, Z. C. & CHEN, Y. (1983). On the discovery of Palacogene flora from the western Sichuan plateau and its significance in phytogeography. *Acta Botanica Sinica* 25(5): 482-493.

FANG, A. M., YAN, Z., LIU, X. H., TAO, J. R., LI, J. L. & PAN, Y. S. (2005). The flora of the Liuqu Formation in south Tibet and its climatic implication. *Acta Palaeontologica Sinica* 44(3): 435-445.

GAO, R. Q., ZHAO, C. B., ZHENG, Y. L., SONG, Z. C., HUANG, P. & WANG, X. F. (1994). Palynological study of deep-beds (Lower Cretaceous) in Songliao Basin, China. *Acta Palaeontological Sinica* 33(6): 659-675.

GENG, G. C. & TAO, J. R. (1982). Tertiary plants from Xizang (Tibet). In: Nanjing Institute of Geology and Paleontology & Institute of Botany - Academia Sinica (ed.) *The Series of the Scientific Expedition to the Qinghai-Xizang (Tibet) Plateau, Palaeontology of Xizang*. pp. 110-125. Beijing, China: Science Press.

GORNITZ, V. (2009). *Encyclopedia of Paleoclimatology and Ancient Environments*. Dordrecht, The Netherlands: Springer.

GUO, S. X. (1975). *The Plant Fossils of the Xigaze Group from Mount Jolmo Lungma Region*. Beijng, China: Science Press.

GUO, S. X. (1979). Late Cretaceous and Early Tertiary floras from southern Guangdong and Guangxi with their stratigraphic significance. In: Institute of Vertebrate Paleontology and Paleoanthropology and Nanjing Institute of Geology and Paleontology - Academia Sinica (ed.) *Mesozoic and Cenozoic Red Beds of South China*. pp. 223-231. Beijing, China: Science Press.

GUO, S. X., SUN, Z. H., LI, II. M. & DOU, Y. W. (1984). Paleocene megafossil flora from Altai of Xinjiang. *Bulletin of Nanjing Institute of Geology and Palaeontology, Academica Sinica* 8: 119-146.

GUO, S. X. (1986a). Floral character of Eocene Relu Formation and history of Eucalyptus from Litang of Sichuan. In: *Exploration of Hengduan Mountain Area*. pp. 69-73. Beijing, China: Beijing Publishing House of Sciences and Technology.

GUO, S. X. (1986b). Preliminary interpretation of Tertiary climate by using megafossil floras in China. *Palaeontologica Cathayana* 2: 169-176.

GUO, S. X. & ZHANG, G. F. (2002). Oligocene Sanhe flora in Longjing County of Jilin, Northeast China. *Acta Palaeontologica Sinica* 41(2): 193-210.

HAO, H., FERGUSON, D. K., FENG, G. P., ABLAEV, A., WANG, Y. F. & LI, C. S. (2010). Early Paleocene vegetation and climate in Jiayin, NE China. *Climatic Change* 99: 547-566.

HE, C. X. & TAO, J. R. (1994). Paleoclimatic analysis of Paleogene flora in Yilan County, Heilongjiang. *Acta Botanica Sinica* 36(12): 952-956.

HE, C. X. & TAO, J. R. (1997). A study on the Eocene flora in Yilan County, Heilongjiang. *Acta Phytotaxonomica Sinica* 35(3): 249-256.

HERSCHUH, U., BIRKS, H. J. B., MISCHKE, S., ZHANG, C. J. & BÖHNER, J. (2010). A modern pollen-climate calibration set based on lake sediments from the Tibetan Plateau and its application to a Late Quaternary pollen record from the Qilian Mountains. *Journal of Biogeography* 37: 752-766.

HERZSCHUH, U., KÜRSCHNER, H. & MISCHKE, S. (2006). Temperature variability and vertical vegetation belt shifts during the last ~50,000 yr in the Qilian Mountains (NE margin of the Tibetan Plateau, China). *Quaternary Research* 66: 133-146.

HERZSCHUH, U. & BIRKS, H. J. B. (2010). Evaluating the indicator value of Tibetan pollen taxa for modern vegetation and climate. *Review of Palaeobotany and Palynology* 160: 197-208.

HERZSCHUH, U., BIRKS, H. J. B., LIU, X. Q., KUBATZKI, C. & LOHMANN, G. (2011). What caused the mid-Holocene forest decline on the eastern Tibet-Qinghai Plateau? *Global Ecology and Biogeography* 19: 278-286.

IRVING, E. (2008). Why earth became so hot 50 million years ago and why it then cooled. *Proceedings of the National Academy of Sciences of the United States of America* 105(42): 16061-16062.

JIN, J. H. & SHANG, P. (1998). Discovery of Early Tertiary flora in Shenbei coalfield, Liaoning. *Acta Scientiarum Naturalium Universitatis Sunyatseni* 37(6): 129-130.

JIN, J. H., LIAO, W. B., WANG, B. S. & PENG, S. L. (2003). Global change in Cenozoic and evolution of flora in China. *Guihaia* 23(3): 217-225.

KOU, X. Y. (2005). *Studies on quantitative reconstruction of Cenozoic climates in China by palynological data*. thesis, Beijing, China: Institute of Botany, Chinese Academy of Sciences.

LEI, Y. Z., ZHANG, Q. R., HE, W. & CAO, X. P. (1992). Tertiary. In: Yichang Institute of Geology and Mineral Resources, B. o. G. a. M. R. o. H. p. i. C. (ed.) *Geology of Hainan Island (I): Stratigraphy and Palaeontology*. Beijing, China: Geological Publishing House.

Li, F. X. (2005). A sporopollen assemblage from the early Cretaceous Fuxin Formation in the Kailu Basin Nei Mongol. *Acta Micropalaeontologica Sinica* 22(1): 87-96.

Li, H. M. (1965). Early Tertiary plant fossil from Chashanao, Hengyang, Hunan province, China. *Acta Palaeontologica Sinica* 13(3): 540-547.

Li, H. M. & Guo, S. X. (1976). The Miocene Namlin flora from Tibet. *Acta Palaeontologica Sinica* 15(4): 598-609.

Li, H. M. (1979). Some fossil plants of Myricaceae from China and its stratigraphic significance. In: Institute of Vertebrate Palaeontology and Palaeoanthropology-Nanjing Institute of Geology and Palaeontology-Chinese Academy of Sciences (ed.) *Mesozoic and Cenozoic Red Beds of South China*. pp. 232-239. Beijing, China: Science Press.

Li, J. G., Zhang, Y. Y., Cai, H. W., Guo, Z. Y. & Wan, X. Q. (2008). Cretaceous and Paleogene Palynological successions at Zhongba. *Acta Geologica Sinica* 82(5): 584-593.

Li, M. Y. (1989). Sporo-pollen from Shanghu Formation of Early Paleocene in Nanxiong Basin, Guangdong. *Acta Palaeontologica Sinica* 28(6): 741-749.

Li, M. Y. & Qing, G. R. (1994). Paleocene sporopollen from the Buxin group of the Longgui Basin, Guangdong. *Acta Micropalaeontologica Sinica* 11(1): 55-69.

Li, T. Q., Cao, H., Kang, M. S., Zhang, Z. X., Zhao, N. & Zhang, H. (2011). *Pollen Flora of China Woody Plants by SEM*. Beijing, China: Science Press.

Li, W. Y. (1998). *Vegetation and Environment of the Quaternary China*. Beijing, China: Science Press.

Li, X. H., Li, W. X., Chen, P. J., Wan, X. J., Li, G., Song, B., Jiang, J. H., Liu, J. C., Yin, D. S. & Yan, W. (2004). The age of tuff SHRIMP U-Pb from upper part of Furao Formation, Heilongjiang: an age closest to the boundary of Cretaceous/Tertiary. *Chinese Science Bulletin* 49(8): 816-818.

Li, X. Q., An, Z. S., Zhou, J., Gao, H. J. & Zhao, H. L. (2003). Characteristics of vegetation in the loess plateau area since Holocene. *Marine Geology and Quaternary Geology* 23(3): 109-114.

Li, X. X., Zhou, Z. Y. & Cai, C. Y. (1995a). *Flora in the Geological Time of China*. Guangzhou, China: Guangdong Science and Technology Press.

Li, X. X., Zhou, Z. Y., Cai, C. Y., Sun, G., Ouyang, S. & Deng, L. H. (1995b). *Fossil Floras of China Through the Geological Ages*. Guangzhou, China: Guangdong Science and Technology Press.

Li, Y. T., Sun, X. Y. & Liu, J. Y. (1984). *The Tertiary System of China*. Beijing, China: Geological Regional Press.

Liu, G. W. & Yong, R. Y. (1999). Pollen assemblages of the late Eocen Nadu formation from the Bose basin of Guangxi southern China. *Palynology* 23: 97-114.

Liu, J. Q., Ni, Y. Y. & Chu, G. Q. (2001). Main palaeoclimatic events in the Quaternary. *Quaternary Sciences* 21(3): 239-248.

Liu, Y. A. & Kong, Z. C. (1978). Late Eocene plant fossils from Wucheng, Henan and its botanical and paleoclimatic implications. *Acta Botanica Sinica* 20(1): 59-65.

Liu, Y. S. (1990). Cuticular studies on two Pleistocene species of Lauraceae in Baise Basin, Guangxi. *Acta Botanica Sinica* 32(10): 805-808.

Liu, Y. S. & Zheng, Y. (1995). Neogene floras. In: Li, X. (ed.) *Fossil Floras of China through the Geological Ages*. pp. 506-551. Guangdong, China: Guangdong Science and Technology Press.

Liu, Y. S., Utescher, T., Zhou, Z. K. & Sun, B. N. (2011). The evolution of Miocene climates in North China: preliminary results of quantitative reconstructions from plant fossil records. *Palaeogeography Palaeoclimatology Palaeoecology* 304(3-4): 308-327.

Luo, Y. L. & Sun, X. J. (2001). Vegetation evolution in the northern South China Sea region since 40 ka BP - an attempt to reconstruct palaeovegetation based on biomization. *Acta Botanica Sinica* 43(11): 1202-1206.

Ma, Y. (1991). Tertiary sporo-pollen assemblages from southern Dunhuang Basin, Gansu Province. *Acta Micropalaeontologica Sinica* 8(2): 207-225.

Mosbrugger, V. & Utescher, T. (1997). The coexistence approach - a method for quantitative reconstructions of Tertiary terrestrial palaeoclimate data using plant fossils. *Palaeogeography, Palaeoclimatology, Palaeoecology* 134: 61-86.

Ni, J., Chen, Y., Herzschuh, U. & Dong, D. (2010). Late Quaternary pollen records in China. *Chinese Journal of Plant Ecology* 34(8): 1000-1005.

Shen, C., Liu, K., Tang, L. & Overpeck, J. T. (2006a). Quantitative relationships between modern pollen rain and climate in the Tibetan Plateau. *Review of Palaeobotany and Palynology* 140: 61-77.

Shen, J., Liu, X., Wang, S. & Matsumoto, R. (2005). Palaeoclimatic changes in the Qinghai lake area during the 18,000 years. *Quaternary International* 136: 131-140.

Shen, J., Jones, R. T., Yang, X. D., Dearing, J. A. & Wang, S. (2006b). The Holocene vegetation history of Lake Erhai, Yunnan province southwestern China: the role of climate and human forcings. *The Holocene* 16(2): 265-276.

Shi, G. L. (2010). *Gymnosperms, Lauraceae and Hamamelidaceae of the Oligocene Ningming Flora, Guangxi and its Preliminary Palaeoclimatic Reconstruction*. Ph.D. thesis, Nanjing, China: Nanjing Institute of Geology and Palaeontology, Chinese Academy of Sciences.

Song, Z. S. & Zhong, L. (1985). *Cretaceous and Early Tertiary Sporo-pollen Assemblages from the Sanshui Basin, Guangdong Province*. Beijing, China: Science Press.

Song, Z. S., Wang, W. M. & Huang, F. (2005). Fossil pollen records of extant angiosperms in China. *The Botanical Review* 70(40): 425-458.

Su, T., Xing, Y. W., Yang, Q. S. & Zhou, Z. K. (2009). Reconstruction of mean annual temperature in Chinese Eocene paleofloras based on leaf margin analysis. *Acta Palaeontologica Sinica* 48(1): 65-72.

Su, T., Jacques, F., Chen, W. Y., Huang, Y. J., Liu, Y. S. &

Zhou, Z. K. (2010). Quantitative reconstructions of mean annual temperature for Cenozoic Paleofloras in China based on the Chinese leaf margin analysis model. *Geological Review* 56(5): 638-646.

Sun, G., Dilcher, D. L., Zheng, S. L. & Zhou, Z. K. (1998). In search of the first flower: a Jurassic angiosperm, *Archaefructus*, from northeast China. *Science* 282: 1692-1695.

Sun, G., Ji, Q., Dilcher, D. L., Zheng, S. L., Nixon, K. C. & Wang, X. F. (2002). Archaefructaceae, a new basal angiosperm family. *Science* 296: 899-904.

Sun, G., Sun, C. L., Dong, Z. M., Sun, Y. W., Lü, J. S., Xiong, X. Z., Zhou, Z. L., Yu, F. L., Xing, Y. L., Quan, C., Akhmetiev, M. A., Ashraf, A. R., Bugdaeva, E., Dilcher, D. L., Golovneva, L. B., Johnson, K., Kezina, T., Kodrul, T. & Okada, H. (2003). Preliminary study of the Cretaceous-Tertiary boundary in Jiayin of the Heilongjiang River Area of China. *Global Geology* 22(1): 8-14.

Sun, G., Quan, C., Sun, C. L., Sun, Y. W., Luo, K. L. & Lü, J. S. (2005). Some new knowledge on subdivisions and age of Wuyun Formation in Jiayin of Heilongjiang, China. *Journal of Jilin University (Earth Science Edition)* 35(2): 137-142.

Sun, X. J. & Wang, P. X. (2005). How old is the Asian monsoon system? - Palaeobotanical records from China. *Palaeogeography, Palaeoclimatology, Palaeoecology* 222: 181-222.

Tao, J. R. (1992). The Tertiary vegetation and flora and floristic regions in China. *Acta Phytotaxonomica Sinica* 30(1): 25-43.

Tao, J. R. & Kong, Z. C. (1973). The fossil florule and sporo-pollen assemblage of Shang-in coal series of Eryuan, Yunnan. *Acta Botanica Sinica* 15(1): 120-126.

Tao, J. R. & Du, N. Q. (1982). Neogene flora Tengchong basin in western Yunnan, China. *Acta Botanica Sinica* 24: 273-281.

Tao, J. R. & Xiong, X. Z. (1986). The latest Cretaceous flora of Heilongjiang Province and the floristic relationship between East Asia and North America (cont.). *Acta Phytotaxonomica Sinica* 24(2): 1-15, 121-135.

Tao, J. R. (1988). The Palaeocene flora and Palaeoclimate of Liuqu Formation in Xizang (Tibet). In: Whyte, P., Aigner, J. S. & Jablonski, N. G. (eds.) *Geology, Sea Level Changes, Palaeoclimatology and Palaeobotany*. pp. 520-522. Hong Kong, China: Centre of Asian Studies, University of Hong Kong.

Tao, J. R. & Han, D. F. (1990). A fossil leaf from the fossil ape locality in Lufeng, Yunnan. *Chinese Bulletin of Botany* 7(1): 45-47.

Tao, J. R., Yang, J. J. & Wang, Y. F. (1994). Miocene wood fossils and paleoclimate in Inner Mongolia. *Acta Botanica Yunnanica* 16: 111-116.

Tao, J. R., Zhou, Z. K. & Liu, Y. S. (2000). *The Evolution of the Late Cretaceous-Cenozoic Floras in China*. Beijing, China: Science Press.

Tapponier, P., Xu, Z. Q., Roger, F., Mayer, B., Arnand, N., Wittlinger, G. & Yang, J. S. (2001). Oblique Stepwise Rise and Growth of the Tibet Plateau. *Science* 294: 1671-1677.

Tong, G. H., Chen, Y., Wu, X. H., Li, Z. H., Yang, Z. J., Wang, S. B. & Cao, J. D. (1999). Pleistocene environmental megaevolution as indicated by the sporo-pollen floras in china. *Journal of Geomechanics* 5(4): 11-21.

Wang, D. N., Sun, X. Y. & Zhao, Y. N. (1990). Late Cretaceous to Tertiary palynofloras in Xinjiang and Qinghai, China. *Review of Palaeobotany and Palynology* 65(1): 95-94.

Wang, W. M. (1990). Sporo-pollen assemblages from the Miocene Tongguer Formation of Inner Mongolia and its climate. *Acta Botanica Sinica* 32: 901-904.

Wang, W. M. & Zhang, D. H. (1990). Tertiary sporo-pollen assemblages from the Shangdou-Huade Basin, Inner Mongolia with discussion on the formation of steppe vegetation in China. *Acta Micropalaeontologica Sinica* 7: 239-252.

Wang, W. M. (1994). Paleofloristic and paleoclimatic implications of Neogene palynofloras in China. *Review of Palaeobotany and Palynology* 82: 239-250.

Wang, W. M. (1996a). On the origin and development of steppe vegetation in China. *Palaeobotanist* 45: 447-456.

Wang, W. M. (1996b). A palynological survey of Neogene strata in Xiaolongtan Basin, Yunnan Province of South China. *Acta Botanica Sinica* 38: 743-748.

Wang, W. M., Yu, Z. Y. & Yang, H. (1997). A study on phytoliths and palynomorphs of Quatenary red earth in Xingzi county, Jiangxi province and its significance. *Acta Micropalaeontologica Sinica* 14(1): 41-48.

Wang, W. M. (1999). Neogene palynofloras. In: Song, Z. C., Zheng, Y. H., Li, M. Y., Zhang, Y. Y., Wang, W. M., Wang, D. N., Zhao, C. B., Zhou, S. F., Zhu, Z. H. & Zhao, Y. N. (eds.) *Fossil Spores and Pollen of China: the Late Cretaceous and Tertiary Spores and Pollen*. pp. 763-773. Beijing, China: Science Press.

Wang, W. M., Chen, G. J., Chen, Y. F. & Kuang, G. D. (2003). Tertiary palynostratigraphy of the Ningming basin Guangxi. *Journal of Stratigraphy* 27(4): 324-327.

Wang, W. M. (2006). Correlation of pollen sequences in the Neogene palynofloristic regions of China. *Palaeoworld* 15: 77-99.

Wang, X. M. & Sun, X. J. (2005). Palynological records since the last glacial maximum on the Sunda shelf of the South China Sea. *Advances in Earth Science* 20(8): 833-839.

Wang, Z. (1984). Plant Kingdom. In: Tianjin Institute of Geology and Mineral Resources (ed.) *Palaeontological Atlas of North China*. pp. 223-384. Beijing, China: Geological Publishing House.

Writing Group of Cenozoic Plants of China (1978). *Cenozoic Plants from China, Fossil Plants of China*. Beijing, China: Science Press.

Wu, Z. Y. (ed.) (1980). *Vegetation of China*. Beijing, China: Science Press.

Xia, K., Su, T., Liu, Y. S., Xing, Y. W., Jacques, F. M. B. & Zhou, Z. K. (2009). Quantative climate reconstruction of the late Miocene Xialongtan megaflora from Yunnan, southwest China. *Palaeogeography, Palaeoclimatology,*

Palaeoecology 276: 80-86.
Xia, Y. M., Wang, P. F. & Wang, M. H. (1987). A preliminary discussion on the characteristics of the Neogene and Quaternary palynological assemblages and the palacoclimate in the Three-River Plain. In: Committee for Quaternary Stratigraphy of China (ed.) *Contribution to the Quaternary Glaciology and Geology*. pp. 151-168. Beijing, China: Geological Publishing House.
Xiong, X. Z. (1986). Paleocene flora from the Wuyun Formation in Jiayin of Heilongjiang. *Acta Palaeontologica Sinica* 25: 571-576.
Xu, J. X., Wang, Y. F., Du, N. Q. & Zhang, G. F. (2000). The Neogene pollen/spore flora of Lühe, Yunnan. *Acta Botanica Sinica* 42(5): 526-532.
Xu, J. X., Wang, Y. F. & Du, N. Q. (2003). Late Pliocene vegetation and palaeoclimate of Yangyi and Longling of western Yunnan Province. *Journal of Palaeogeography* 5(2): 217-223.
Xu, R., Tao, J. R. & Sun, X. J. (1973). The discovery of plant fossils in Qomolangma region of Southern Tibet and its meaning. *Acta Botanica Sinica* 15(2): 254-258.
Yan, G., Wang, F. B., Shi, G. R. & LI, S. F. (1999). Palynology and stable isotopic study of palaeoenviroment change on the north-eastern Tibetan Plateau in the last 30,000 years. *Palaeogeogrpahy, Palaeoclimatology, Palaeoecology* 153: 147-159.
Yao, Y. F., Bera, S., Jin, J. H., Ferguson, D. K., Mosbrugger, V., Paudayal, K. N. & Li, C. S. (2009). Reconstruction of paleovegetation and paleoclimate in the early and middle Eocene, Hainan Island, China. *Climatic Change* 92: 169-189.
Yi, T. M., Li, C. S. & Jiang, X. M. (2005). Conifer woods of the Pliocene Age from Yunnan, China. *Journal of Integrative Plant Biology* 47(3): 264-270.
Zhang, J. H. (1983). Discovery of old Tertiary flora from Panxian of Guizhou and its significance. In: Committee of Stratigraphy and Paleontology - Geological Society of Guizhou Province (ed.) *Papers of Stratigraphy and Paleontology of Guizhou*. pp. 133-143. Guiyang, China: People's Publishing House of Guizhou.
Zhang, Q. R. (1980). The Paleocene sporo-pollen assemblage in Nanxiong Basin, Guangdong. *In:* Yichang Institute of Geology and Mineral Resources-Chinese Academy of Geology Sciences (ed.) *Yichang Institute of Geology and Mineral Resources, Chinese Academy of Geology Sciences (Stratigraphy and Palaeontology, Special Issue)*. pp. 106-117. Yichang, China: Yichang Institute of Geology and Mineral Resources, Chinese Academy of Geology Sciences.
Zhang, S. Q., Wang, Y. G., Xin, Y. H., Zhao, Y., LI, Y. G., Wang, Z. C., Shi, W. D. & Shang, X. G. (2006). Discovery of early Pleistocene strata containing plants fossils in the source area of the Yellow River and significance. *Geology in China* 33(1): 78-85.
Zhang, Y. Y. (1993). Late Cretaceous palynofloras of China. *Acta Micropalaeontologica Sinica* 10(2): 131-157.
Zhao, S. J., Cheng, J., Yin, G. M. & Zan, L. H. (2008). Palynological assemblages and palaeoclimatic significance in Beijing plain area since the Middle Pleistocene. *Journal of Palaeogeography* 10(6): 637-646.
Zhao, W., Shen, R. J., Liao, W. B. & Jin, J. H. (2009). Eocene palynoflora from Changchang Basin, Hainan Island. *Journal of Jilin University (Earth Science)* 39(3): 379-396.
Zhao, X. W. (1992). *Introduction to Paleoclimatology*. Beijing, China: Geological Publishing House.
Zhao, Y., Yu, Z. C., Chen, F. H., Ito, E. & Zhao, C. (2007). Holocene vegetation and climate history at Hurleg Lake in the Qaidam Basin, northwest China. *Review of Palaeobotany and Palynology* 145: 275-288.
Zhou, Z. K. (2000). On the Miocene Xiaolongtan flora from Kaiyuan, Yunnan Province. In: Tao, J. R. (ed.) *The Evolution of the Late Cretaceous-Cenozoic Floras in China*. pp. 64-72. Beijing, China: Science Press.
Zhou, Z. K., Yang, Q. S. & Xia, K. (2007). Fossils of *Quercus* sect. *Heterobalanus* can help explain the uplift of the Himalayasx. *Chinese Science Bulletin* 52(2): 238-247.
Zhou, Z. Y., Li, H. M., Cao, Z. Y. & Nau, P. S. (1990). Some Cretaceous plants from Pingzhou (Pingchau) island, Hong Kong. *Acta Palaeontologica Sinica* 29(4): 415-426.

Chapter 8

The Vegetation of China Today

CHEN Ling-Zhi

8.1 Introduction

China, as we have seen, is a vast country incorporating multiple climatic zones, moisture regimes and soil types and therefore diverse natural vegetation types. Furthermore, it harbors an equally exciting range of cultivated crops, which will be detailed in later chapters.

8.2 Principal systems of vegetation classification in China

The principles of vegetation classification are based on both the character of the plant community and the ecological conditions in which it is situated. The main unit of vegetation classification in use in China is the "vegetation type," which roughly equates to the biome. Vegetation types are divided on the basis of ecological conditions, physiognomy and community structure, and there are forty such types in China (Hou, 1960; Wu, 1980; Song, 2001). See Table 8.1 for a full listing.

At a higher level, vegetation types are classified into "groups of vegetation types." Seven such groups are recognized in China: Forest, Scrub, Desert, Steppe, Meadow, Alpine and Wetland (Table 8.1). Within each group, vegetation types are based upon life-form characteristics such as plant size, morphology, tissue structure, longevity, and seasonal dynamics. For example, all vegetation types within the Forest group are constructed of trees, but Forest is divided into ten vegetation types according to differences in stratification, leaf type, leaf quality and seasonal dynamics.

Vegetation types can themselves be divided into "sub-types," depending on differences in ecological conditions. For example, the Evergreen Coniferous Forest vegetation type can be divided into four sub-types: Cold-temperate, Temperate, Warm and Tropical. Not all vegetation types are subdivided; this depends upon their distribution scope and complexity.

Within sub-types are found "formations," based on the actual species present, such as the *Picea* spp. formation, *Abies* spp. formation and *Pinus* spp. formation within the Cold-temperate sub-type of the Evergreen Coniferous Forest vegetation type. Formations are perhaps the most commonly used unit when describing the vegetation of regions in China. You may also see the unit "association" used in classifications. This applies to plant communities having several features in common, but is nowadays rarely used.

8.3 Basic characteristics of the main vegetation types

Forest, desert and steppe are the most plentiful vegetation groups in China and are the only ones covered here. Scrub types are also very rich but, with the exceptions of alpine, sand and saline alkaline habitats, are mostly secondary communities. Meadows are also, with the exception of savanna, usually secondary.

8.3.1 Forest

Twelve types of forest vegetation are recognized in China (see Table 8.1), including many secondary forests. Most are distributed according to altitudinal zones. Some examples are described below.

Deciduous Coniferous Forest (Figures 8.1, 8.2).—Forest dominated by deciduous conifers is distributed in mountainous parts of northen China and the subalpine mountains of subtropical China. A single sub-type, Cold-temperate, and eleven formations are recognized, with the dominant species in all being *Larix* (Pinaceae). The elevation at which this forest type is found increases with latitude from north to south. For example, *Larix gmelinii* (Figure 8.1) forest is found in the cold-temperate zone at elevations of 350—1 200 m; *L. olgensis* forest occurs on the Xiao Hinggan Ling and Changbai Shan in Jilin, on both moist lowlands and as secondary forest in mesic habitats, with an altitudinal range of 500–1 800 m; *L. gmelinii* var. *principis-rupprechtii* forest (Figure 8.2) occurs on mountains in the warm-temperate zone at elevations of 1 500–2 100 m; and *L. sibirica* forest, found on the Tian Shan, reaches elevations of 1 300–2 900 m. *Larix* forests are also common in the mountainous parts of western China. Here they include *L. potaninii* var. *chinensis* forest on the Qin Ling at an elevation of 2 900–3 350 m, *L. potaninii* var. *australis* forest in the Hengduan Shan at 3 200–4 400 m, and *L. griffithii* forest, which is most widely distributed in Xizang and the Himalaya Shan, at an elevation of 2 800–4 000 m. Additional formations include *L. mastersiana*, *L. himalaica* and *L. speciosa* forest, all distributed in the subalpine belt above 2 500 m.

Table 8.1 The vegetation groups, types and sub-types recognized in China (after Hou, 1960; Wu, 1980; Song, 2001).

Vegetation types	Vegetation sub-types
Group I. Forest	
Deciduous Coniferous Forest	Cold-temperate Deciduous Coniferous Forest
Evergreen Coniferous Forest	Cold-temperate Evergreen Coniferous Forest, Temperate Evergreen Coniferous Forest, Warm Evergreen Coniferous Forest, Tropical Evergreen Coniferous Forest
Mixed Coniferous and Broad-leaved Forest	Typical Temperate Mixed Coniferous and Broad-leaved Forest, Subtropical Mixed Coniferous and Broad-leaved Forest (Mixed Tsuga-Broad-leaved Forest)
Deciduous Broad-leaved Forest	Cold-temperate Deciduous Broad-leaved Forest, Temperate Deciduous Broad-leaved Forest, Warm Deciduous Broad-leaved Forest, Riparian Deciduous Broad-leaved Forest in Desert Regions
Mixed Evergreen and Deciduous Broad-leaved Forest	Typical Mixed Evergreen and Deciduous Broad-leaved Forest, Montane Mixed Evergreen and Deciduous Broad-leaved Forest, Montane Mixed Evergreen and Deciduous Broad-leaved Forest on Limestone
Evergreen Broad-leaved Forest	Warm Evergreen Broad-leaved Forest, Warm Humid Evergreen Broad-leaved Forest, Thermophilic Evergreen Broad-leaved Forest (Monsoon Evergreen Broad-leaved Forest), Thermophilic Humid Evergreen Broad-leaved Forest
Sclerophyllous Evergreen Broad-leaved Forest	—
Monsoon Forest	Deciduous Monsoon Forest, Semi-evergreen Monsoon Forest, Monsoon Forest on Limestone
Rainforest	Seasonal Rainforest, Montane Rainforest, Seasonal Rainforest on Limestone
Mangrove	—
Broad-leaved Forest on Coral	—
Bamboo Forest and Scrub	Temperate Bamboo Forest and Bamboo Scrub, Warm Bamboo Forest and Bamboo Scrub, Warm-moist Bamboo Forest and Bamboo Scrub, Tropical Bamboo Forest, Tropical Moist Bamboo Forest
Group II. Scrub	
Evergreen Coniferous Scrub	Alpine-subalpine Evergreen Coniferous Scrub, Montane and Sandy Evergreen Coniferous Scrub
Deciduous Broad-leaved Scrub	Alpine Deciduous Broad-leaved Scrub, Temperate Deciduous Broad-leaved Scrub, Warm Deciduous Broad-leaved Scrub, Deciduous Broad-leaved Scrub on Limestone, Deciduous Broad-leaved Scrub in Hot Dry Valleys, Salty Deciduous Broad-leaved Scrub, Sandy Deciduous Broad-leaved Scrub
Evergreen Broad-leaved Scrub	Warm Evergreen Broad-leaved Scrub, Warm Evergreen Broad-leaved Scrub on Limestone, Thermophilic Evergreen Broad-leaved Scrub, Thermophilic Evergreen Broad-leaved Scrub on Limestone, Tropical Evergreen Broad-leaved Scrub, Hot, Dry Evergreen Broad-leaved Scrub
Evergreen Sclerophyllous Scrub	Alpine-subalpine Evergreen Sclerophyllous Scrub, Montane Evergreen Sclerophyllous Scrub on Limestone
Succulent Thorn Scrub	Succulent Thorn Scrub on Tropical Sandy Beaches, Succulent Thorn Scrub in Hot Dry Valleys
Group III. Desert	
Reduced-Leaved Small Tree Desert	—
Evergreen Sclerophyllous Shrub Desert	—
Reduced-leaved Shrub Desert	—
Succulent Shrub Desert	—
Xerophytic Shrub Desert	—
Succulent Semi-shrub Desert	Succulent Semi-shrub Desert, Salt-Tolerant Succulent Semi-shrub (Small Semi-shrub) Desert
Xerophytic Semi-shrub Desert	—
Dwarf Cushion Semi-shrub Desert	—
Group IV. Steppe	
Clump Grass Steppe	Meadow Clump Grass Steppe, Typical Clump Grass Steppe, Desert Clump Grass Steppe, Alpine Clump Grass Steppe
Rhizomatic Grass Steppe	Meadow Rhizomatic Grass Steppe, Typical Rhizomatic Grass Steppe, Alpine Rhizomatic Grass Steppe
Forb Steppe	Meadow Forb Steppe, Desert Forb Steppe
Semi-shrub and Small Semi-shrub Steppe	Typical Semi-shrub and Small Semi-shrub Steppe, Desert Semi-shrub and Small Semi-shrub Steppe, Alpine Semi-shrub and Small Semi-shrub Steppe
Group V. Meadow	
Clump Grass Meadow	Typical Clump Grass Meadow, Alpine Clump Grass Meadow, Salty Clump Grass Meadow
Rhizomatic Grass Meadow	Typical Rhizomatic Grass Meadow, Alpine Rhizomatic Grass Meadow, Swampy Rhizomatic Grass Meadow, Salty Rhizomatic Grass Meadow
Forb Meadow	Typical Forb Meadow, Alpine Forb Meadow
Grassland	Gramineal Grassland, Fern Grassland
Savanna	Savanna in Hot Dry Valleys, Tropical Savanna
Group VI. Alpine	
Cushion Vegetation	—
Alpine Tundra	—
Rare Alpine Vegetation	—
Group VII. Wetland (swamp and aquatic vegetation)	
Small Tree Swamp	—
Herb Swamp	—
Aquatic Vegetation	Submerged Aquatic Vegetation, Floating Aquatic Vegetation, Emergent Aquatic Vegetation

Figure 8.1 *Larix gmelinii* forest, Da Hinggan Ling (image courtesy SONG Wei-Guo).

Figure 8.2 *Larix gmelinii* var. *principis-rupprechtii* forest, Shanxi (image courtesy CHEN Ling-Zhi).

In general, Larch forests comprise three or four layers: a canopy layer, understory layer, (comprising both sapling and shrub layers), herb layer, and moss (bryophyte) layer, the latter covering the soil surface in cold, moist habitats. Here we use the *Larix gmelinii* formation as an example to describe the community characteristics of *Larix* forests. The dominant species of the canopy layer, which is sometimes divided into two sub-layers, in this formation is *L. gmelinii*. Sometimes the canopy also contains *Betula platyphylla* (Betulaceae), *Picea jezoensis* or *Pinus sylvestris* var. *mongolica* (both Pinaceae) or *Sorbus pohuashanensis* (Rosaceae). At low elevations, *Quercus mongolica* (Fagaceae) may dominate in the lower canopy layer. The understory may be dominated by a variety of species depending on the habitat: *Rhododendron dauricum* (Ericaceae) on sunny or semi-sunny slopes, *Alnus mandshurica* (Betulaceae), *Ledum palustre* and *Vaccinium uliginosum* (both Ericaceae) in moist habitats, and *Betula ovalifolia, Lonicera caerulea* var. *edulis* (Caprifoliaceae), *Salix hsinganica*(Salicaceae), *S. raddeana* and *S. taraikensis* in valleys and on lowland river banks. Other shrubs that may be present include *Cornus alba* (Cornaceae), *Potentilla fruticosa, Rosa luciae var. luciae, Rosa davurica, Sorbaria sorbifolia, Spiraea chamaedryfolia, Spiraea media, Spiraea salicifolia* and *Spiraea sericea* (all Rosaceae), *Ribes nigrum* and *Ribes procumbens* (Saxifragaceae). At low elevations there may also be *Corylus mandshurica* (Betulaceae) and *Lespedeza bicolor* (Fabaceae), while at high elevations *Pinus pumila* may be found. The herb layer can be divided into two sub-layers, with *Aegopodium alpestre* and *Peucedanum terebinthaceum* (Apiaceae), *Artemisia vestita, Atractylodes lancea, Klasea centauroides* and *Saussurea neoserrata* (all Asteraceae), and *Lathyrus humilis* and *Vicia venosa* (Fabaceae) in the first, and *Carex globularis, Carex caespitosa, Carex callitrichos* and *Carex schmidtii* (Cyperaceae), *Convallaria majalis* (Liliaceae), *Fragaria orientalis* (Rosaceae), and *Pyrola asarifolia* subsp. *incarnata* and *Vaccinium vitis-idaea* (Ericaceae) in the second. An obvious moss layer is seen in patches on the soil surface of moist habitats, while moss coverage is low under *L. gmelinii* forest on sunny or semi-sunny slopes. The layered structure of the *L. gmelinii* formation is similar to that of *Larix* formations in temperate and warm-temperate zones, with constituent species always comprising the same genera although different species may be present.

Larix formations on subalpine mountains in western subtropical China may be represented by *L. potaninii* var. *australis* forest. In the canopy layer of this formation may be found *Abies ernestii, A. forrestii* (Pinaceae), *Betula platyphylla, B. chinensis* and *Picea likiangensis* (including *P. likiangensis* var. *rubescens*). Evergreen *Quercus* species are often present. Shrub species include *Rhododendron phaeochrysum* var. *agglutinatum, R. dichroanthum, R. vernicosum, R. yunnanensis* and *Fargesia nitida* (Poaceae), which may be a dominant or common species in the shrub layer. *Cotoneaster microphyllus* (Rosaceae), *Lyonia villosa* (Ericaceae), *Ribes glaciale, Spiraea alpina, Spiraea myrtilloides* and *Sorbus rehderiana* may grow in small amounts in various habitats. The herb layer is sparse, consisting mainly of *Anaphalis bicolor* (Asteraceae), *Anaphalis flavescens, Artemisia roxburghiana, Festuca ovina* (Poaceae) and *Polygonum viviparum* (Polygonaceae; Zhou, 1991; Chen & Chen, 1997).

Evergreen Coniferous Forest.—Forest dominated by evergreen conifers is widely distributed in China, in temperate and warm-temperate zones as well as on hills and mountains in the subtropical zone. It may be primary or secondary. Four sub-types are recognized.

The Cold-temperate Evergreen Coniferous Forest sub-type (Figures 8.3, 8.4) is distributed mainly on mountains in the western part of the cold-temperate, temperate and warm-temperate zones, as well as the subalpine belt in the subtropical zone. Four formations are recognized: *Abies* spp., *Picea* spp., *Pinus* spp. and *Juniperus* spp.

The *Picea* spp. formation is widely distributed in China. *Picea jezoensis* and *P. koraiensis* formations are found on northeastern mountains. The *P. jezoensis* community is adapted to well-drained slopes, and is widely distributed on the Changbai Shan and Xiao Hinggan Ling of Jilin, at elevations of 800–1 800 m. *Picea koraiensis* forest is

Figure 8.3 *Picea jezoensis* forest, Changbai Shan (image courtesy CHEN Ling-Zhi).

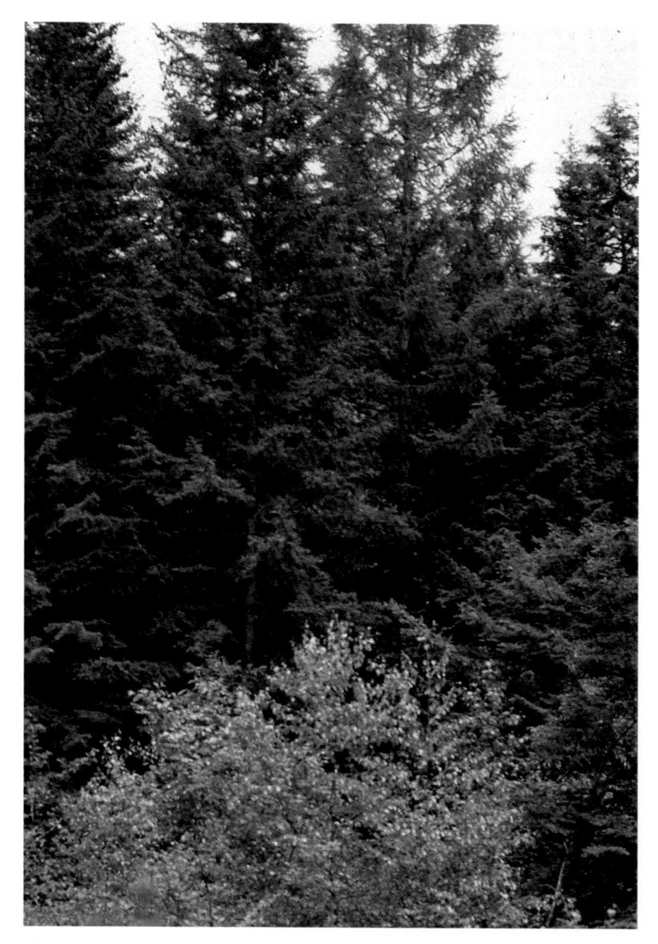

Figure 8.4 *Picea wilsonii* forest, Shanxi (image courtesy CHEN Ling-Zhi).

found in moist lowland habitats beside rivers or streams as well as on mountains in northeastern China. The *Picea meyeri* formation is native to the subalpine belt of the warm-temperate zone, and is found covering large areas of the montane belt on Xiaowutai Shan (Hebei) and Wutai Shan, Guancen Shan and Guandi Shan in Shanxi, at elevations of 1 800–2 700 m. *Picea wilsonii* forest is less widely distributed, and found at elevations of 1 700–2 300 in northern China, and 2 400–2 800 m on subtropical mountains in northwestern Sichuan. *Picea crassifolia* forest is found mainly on the Qilian Shan, with fragmentary populations on the Helan Shan and Luoshan in Ningxia, at elevations of 2 300–3 500 m. *Picea schrenkiana* and *P. obovata* forests are both localized to Xinjiang, the former forming the subalpine forest belt on the northen slopes of the Tian Shan (elevation 1 250–2 800 m), with a few populations in the Junggar Pendi and the montane belt on the western side of the Kunlun Shan (above 3 000 m). The latter grows on the shady southwestern slopes of the Altai (elevation 1 300–2 000 m), sometimes mixed with *Larix sibirica* , but often forming pure formations in river valleys.

The *Picea* formations of montane northeastern China comprise four layers: canopy, shrub, herb and moss layers. Apart from the dominant *Picea* spp., common canopy layer species include *Abies nephrolepis, Betula platyphylla, B. costata, Larix, Sorbus pohuashanensis, Tilia amurensis* (Tiliaceae) and *Ulmus laciniata* (Ulmaceae). Deciduous broad-leaved trees such as *Acer barbinerve* (Aceraceae), *Acer tegmentosum* and *Acer ukurunduense* are more common in the sub-canopy layer. Common shrub layer species include *Corylus mandshurica* and *Viburnum koreanum* (Adoxaceae), with additional shrub layer species incluing *Betula* spp., *Cornus alba, Lonicera* spp., *Ribes* spp., *Rosa* spp. and *Spiraea salicifolia. Ledum palustre* and *Vaccinium vitis-idaea* are found under the shrub layer in different micro-habitats. The main species of the herb layer include *Athyrium multidentatum* (Athyriaceae), *Athyrium spinulosum, Deyeuxia korotkyi* and *Deyeuxia purpurea* (all Poaceae), *Carex callitrichos, Carex pilosa, Dryopteris austriaca* (Dryopteridaceae), *Dryopteris crassirhizoma* and *Dryopteris laeta, Linnaea borealis* (Linnaeaceae), *Pyrola asarifolia* subsp. *incarnate* and *Saussurea amurensis.*

Many *Picea* formations are found in the subalpine belt of western mid-subtropical China (including the mountains of western and northwestern Sichuan, southwestern Yunnan, eastern Xizang and into southern Gansu, at elevations of 2 600–4 200 m). The dominant species of the different formations include *Picea asperata, P. brachytyla, P. brachytyla* var. *complanata, P. likiangensis, P. likiangensis* var. *rubescens* (Figure 8.5) and *P. purpurea. Picea likiangensis* var. *linzhiensis* forest is perhaps the most distinctive coniferous forest formation of Xizang, and is distributed from 2 700 to 4 000 m. *Picea smithiana* and *P. spinulosa* are unique to Xizang, the former only found in Jilong where it is characteristic of lower elevations (2 000–3 000 m). *Picea morrisonicola* formations can be found in the eastern part of the subtropical zone on Yushan and Zhongyang Shan, Taiwan, at elevations of over 3 000 m. In these *Picea* forests of western mid-subtropical China, accompanying canopy species include *Abies* spp., *Juniperus* spp. (Cupressaceae), *Pinus* spp., *Tsuga chinensis* (Pinaceae) and evergreen *Quercus* spp. The shrub layer is dominated by *Chimonocalamus* spp. (Poaceae) and *Rhododendron* spp. with some *Berberis* spp. (Berberidaceae), *Lonicera* spp., *Potentilla* spp., *Ribes* spp.and *Rosa* spp. The herb layer is very diverse, with major taxa including *Carex* spp., *Cystopteris* spp. (Cystopteridaceae), *Deyeuxia* spp., *Dryopteris* spp., *Oxalis* spp. (Oxalidaceae) and *Thalictrum* spp. (Ranunculaceae).

Figure 8.5 *Picea likiangensis* var. *rubescens* forest, Sichuan (image courtesy CHEN Ling-Zhi).

There are two *Abies* spp. formations in northern China: *A. nephrolepis* forest to the northeast and *A. sibirica* forest to the northwest. The former frequently comprises mixed forest with *Picea jezoensis* and *P. koraiensis*, but pure *A. nephrolepis* formations can be seen in moist lowland areas. *Abies sibirica* forest is found only on shady northwestern slopes of the Altai, at elevations of 1 500–2 100 m, and in partial stands on riverside terraces. In contrast to the north, the subalpine zones of subtropical China form a center of distribution for *Abies* forests. *Tsuga chinensis* var. *chinensis* (previously known as *Abies chinensis;* elevation ca. 2 000 m) and *A. fargesii* forest (over 2 400 m) may be seen in the montane belt on the Qin Ling and Bashan ranges. *Abies fargesii* var. *faxoniana* forest (Figure 8.6) is the main subalpine formation of southern Gansu and western Sichuan. The distribution of *A. recurvata* forest overlaps to a large extent that of *A. fargesii* var. *faxoniana* forest. *Abies ernestii* forest (Figure 8.7) is found on the western and northwestern mountains of Sichuan and southeastern Xizang. The vertical distribution of this formation differs depending on the region, and may be 2 100–3 300 m or 3 300–3 700 m. *Abies delavayi, A.fabri* and *A. forrestii* forest are found mainly in western Sichuan and northwestern Yunnan. *Abies georgei* forest is more widely distributed, and can cover large areas at elevations of 3 500–4 200 m; it is found mainly on the southeastern edge of the Qianghai-Xizang Plateau, the mountains alongside the Jinsha Jiang, Lancang Jiang and Nu Jiang in northwestern Yunnan, the mountains west of the Dadu He and the upper reaches of the Min Jiang in Sichuan. By contrast, the distribution of *A. georgei* var. *smithii* forest is small. Both *A. delavayi* and *A. nukiangensis* forests have relatively small distributions, covering northwestern Yunnan, southwestern Sichuan, and the montane belt on the southern side of the eastern Himalaya Shan. *Abies densa* forest is distributed from the Yarlung Zangbo Jiang along the Himalaya Shan to the west, at an elevation of 3 800–4 500 m.

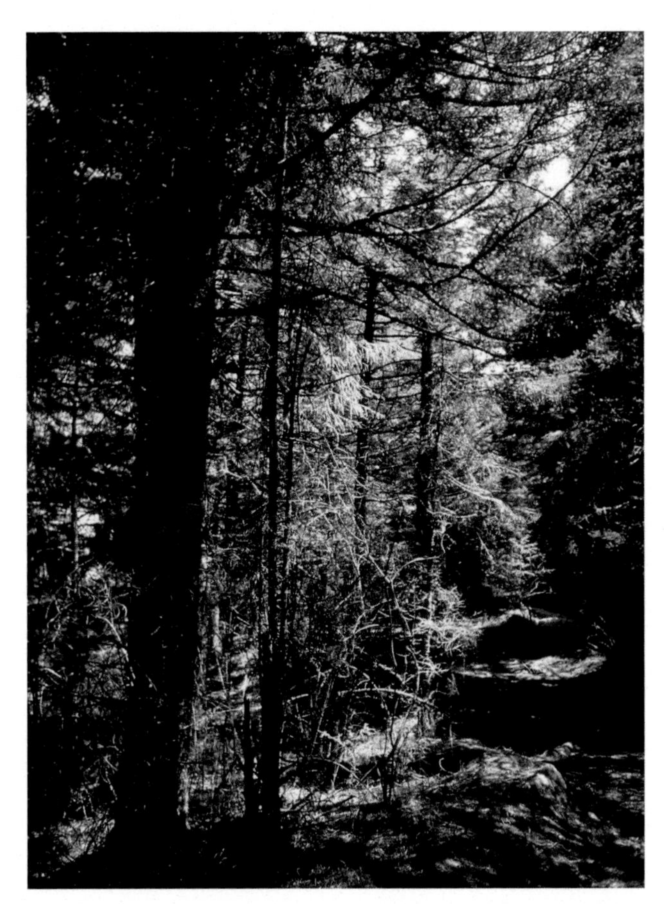

Figure 8.7 *Abies ernestii* forest, Sichuan (image courtesy CHEN Ling-Zhi).

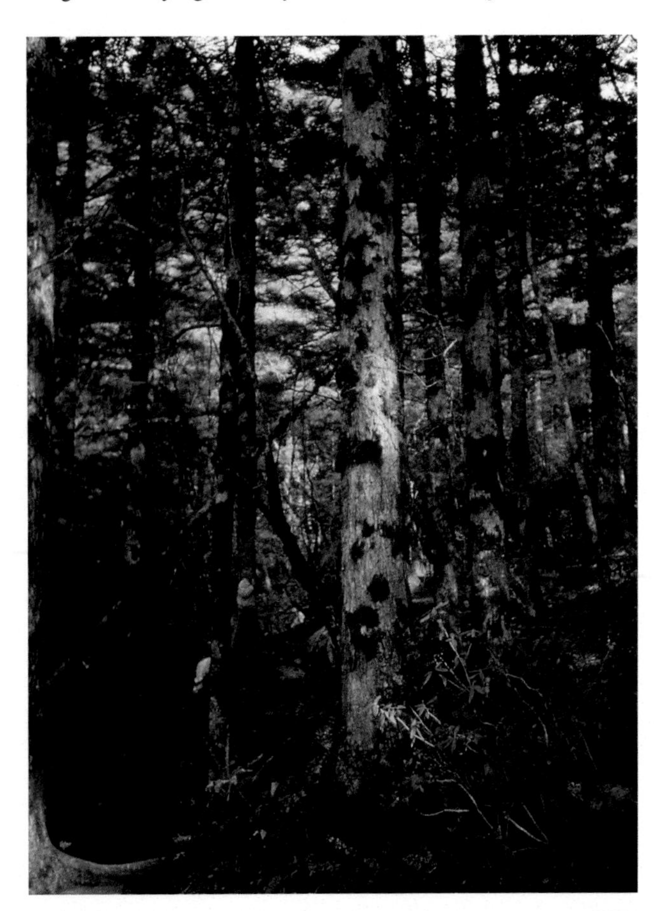

Figure 8.6 *Abies fargesii* forest, Shennongjia, Hubei (image courtesy CHEN Ling-Zhi).

The restricted distributions of the *Abies beshanzuensis* var. *ziyuanensis*, *A. fanjingshanensis* and *A. kawakamii* formations cover a cline from west to east, with *A. fanjingshanensis* forest limited to the montane belt (ca. 2 200 m elevation) on Fanjing Shan in Guizhou, *A.*

beshanzuensis var. *ziyuanensis* forest to the montane belt (1 700–2 900 m) of Ziyuan (Hunan) and *A. kawakamii* forest to Zhongyang Shan and Yushan on Taiwan (ca. 3 000 m).

In these formations of subalpine areas of the subtropical zone, canopy trees may include—in addition to the dominant *Abies*—species of *Betula, Larix, Juniperus, Picea, Populus* (Salicaceae), evergreen *Quercus* and *Tsuga*. The commonest species of the shrub layer are *Rhododendron* spp., alongside species of *Chimonocalamus, Lonicera, Prunus* (Rosaceae), *Ribes, Spiraea* and *Sorbus*. The composition of the herb layer is similar to that of neighboring *Picea* formations. (Wu, 1987; Qi, 1990; Sichuan Forest Editoring Committee, 1992; Zhou, 1994; Lei, 1999).

There are four *Pinus* formations of the Cold-temperate Evergreen Coniferous Forest sub-type: the *P. sylvestris* var. *mongolica* formation is widely distributed on the montane belt (ca. 600–1 200 m) of northeastern China and Nei Mongol, as well as the deserts of Nei Mongol; *P. sibirica* forest is found only on the Altai, Xinjiang, at elevations of 1 900–2 300 m; the *P. pumila* and *P. yunnanensis* var. *pygmaea* formations are both dwarf forests, the former being found on the top of a few mountains in northeastern China, the latter towards the top of mountains (elevation 2 600–3 200 m) in southwestern Sichuan.

We use *Pinus sibirica* to exemplify the characteristics of *Pinus* forests in the cold-temperate zone. The height of this forest is 18–20 m. In addition to *Pinus*, there may be scattered individuals of *Larix sibirica* in the canopy. The shrub layer is dominated by *Vaccinium vitis-idaea* with some *Lonicera* spp., *Rosa acicularis* and *V. myrtillus*. The herb layer is sparse, comprising mostly *Phalaris arundinacea* and *Poa compressa* (both Poaceae), with some *Aquilegia glandulosa* (Ranunculaceae), *Carex* spp. and *Pyrola asarifolia* subsp. *incarnata*. The moss layer is well developed (Xinjiang Comprehansive Expedition Institute of Botany Academia Sinica, 1978; Inner Mongolia and Ningxia Comprehensive Expedition Academia Sinica, 1985).

Finally, *Juniperus* formations are found in the mountains of southwest China, close to the tree line. The main formations are *J. convallium, J. pingii* and *J. tibetica* forest. *Juniperus tibetica* forest is found in the subalpine belt on the eastern edge of the Qinghai-Xizang Plateau and Hengduan Shan, and is characteristically drought- and cold-tolerant, with sparsely-distributed *J. tibetica* trees forming almost pure stands. The shrub layer is also sparse, comprising *Berberis aggregata, Cotoneaster* spp., *Lonicera* spp., *Potentilla fruticosa, Ribes* spp., *Rosa omeiensis, Spiraea mongolica* and *S. alpina*. At the driest and coldest extreme we also find *Caragana* spp. (Fabaceae). Herbs are plentiful, including species characteristic of both cold-temperate coniferous forests and alpine meadows. The main herb species are *Andropogon munroi* (Poaceae), *Deyeuxia scabrescens, Kobresia* spp. (Cyperaceae) and species of *Anaphalis, Artemisia, Leontopodium* (Asteraceae), *Polygonum, Stellera* (Thymelaeaceae) and *Thalictrum*. *Juniperus pingii* forest is distributed on the southen part of the Hengduan Shan around the border between Sichuan and Yunnan. *Juniperus convallium* forest is found in the central-northern part of western Sichuan and eastern Xizang. Both forests are sparse (Qinghai-Tibet Plateau Comprehensive Expedition, 1988).

The Temperate Evergreen Coniferous Forest sub-type is distributed widely in the montane belt of the mid-temperate and subtropical zones. Of the formations within this sub-type, *Pinus* spp. forests are plentiful and widespread, not only as pure stands but also mixed with broad-leaved trees. The most widely distributed forms are *P. tabuliformis* and *P. armandii* forest. The *P. tabuliformis* formation is found on hills and mid-elevation mountains (ca. 1 000–1 600 m) in large parts of the warm-temperate zone as well as in the montane belt (ca. 1 600–2 700 m) of the western subtropical zone. These forests may be natural or planted. The *P. armandii* formation is distributed widely from the middle montane belt of of the southern warm-temperate zone to the montane belt (ca. 1 800–3 400 m) of the western mid-subtropical zone in Guizhou, Sichuan, Yunnan and Xizang. *Pinus bungeana* forest occurs mostly on limestone mountains. It is found in southern and southwestern parts of the warm-temperate zone, and southeastern Sichuan and western Hubei in the central subtropical zone. The *P. sylvestris* var. *sylvestriformis* formation is found in small areas on the Changbai Shan in northeastern China; the *P. densiflora* formation mainly along the eastern coast of the warm-temperate zone; and the *P. taiwanensis* formation on mid-elevation mountains (ca. 800–1 400 m) in the eastern mid-subtropical zone. The center of distribution for *P. tabuliformis* var. *henryi* forest is Daba Shan, and it can also be found in the montane belt (ca. 1 100–2 100 m) on Wu Shan, and on mountains in western Hunan and southern Shanxi. *Pinus densata* forest is found in alpine gorges across a large area of western Sichuan as well as northwestern Yunnan and eastern Xizang. *Pinus wallichiana* forest is distributed widely on the southern slopes of the Himalaya Shan. *Pinus roxburghii* forest occurs in montane valleys in southern Jilong, Xizang. Additional evergreen coniferous forest formations include *Platycladus orientalis* (Cupressaceae) forest on hills and low mountains of the warm-temperate zone, *Chamaecyparis obtusa* var. *formosana* (Cupressaceae) forests on the mountains of Taiwan, and *Cryptomeria japonica* var. *sinensis* (Taxodiaceae) forest on the eastern slopes of mountains in the central subtropical zone.

Pinus tabulifomis forest (Figure 8.8), as found on mid-elevation mountains of the warm-temperate zone, comprises canopy, shrub and herb layers. *Pinus tabuliformis* is completely dominant in the canopy layer with other species, such as *Acer davidii* subsp. *grosseri, A. pictum* subsp. *mono, Malus baccata* (Rosaceae), *Quercus mongolica* and *Toxicodendron vernicifluum* (Anacardiaceae) present only in small amounts. In the shrub layer, *Lespedeza bicolor, Rosa bella* and *Spiraea pubescens* dominate in various habitats, with other common species including *Corylus heterophylla,*

C. mandshurica, *Elaeagnus umbellata* (Elaeagnaceae), *Lonicera maximowiczii*, *Ostryopsis davidiana* (Betulaceae), *Spiraea trilobata* and *Viburnum betulifolium*. The herb layer is mainly composed of *Aster trinervius* subsp. *ageratoides* and *Chrysanthemum chanetii* (both Asteraceae), *Carex siderosticta*, *C. lanceolata*, *Iris ruthenica* (Iridaceae) and *Rubia cordifolia*, with another 20 or so species found occasionally. The structure and constituent species of these layers are similar to those of deciduous broad-leaved forests under similar conditions.

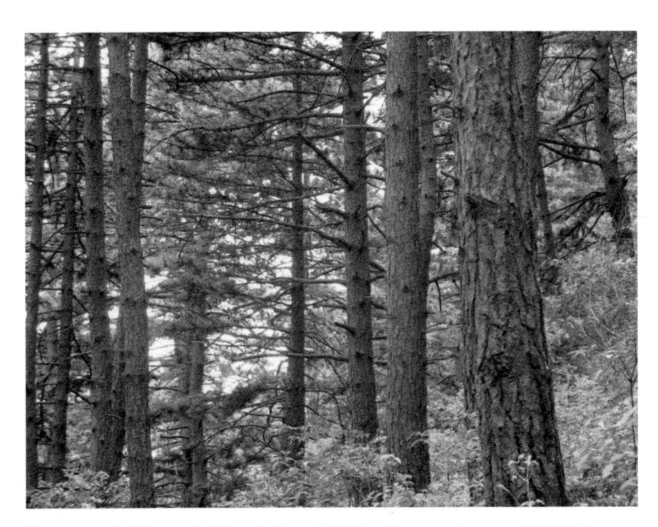

Figure 8.8 *Pinus tabuliformis* forest, Beijing (image courtesy GUO Ke).

Figure 8.9 *Pinus massoniana* forest, Lingchuan, Guangxi (image courtesy CHEN Ling-Zhi).

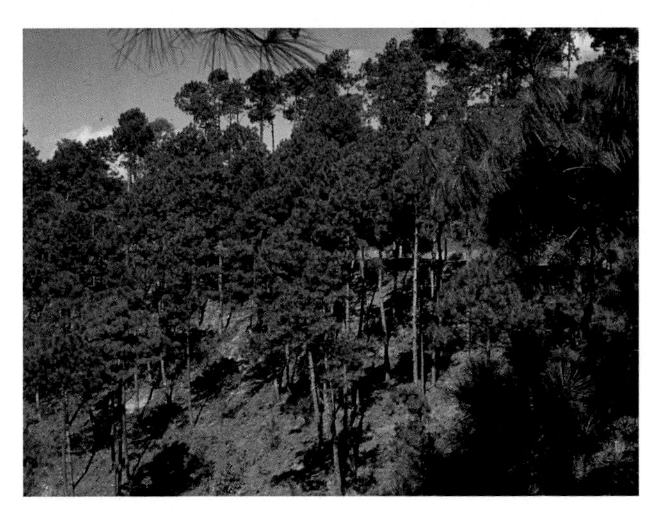

Figure 8.10 *Pinus kesiya* forest, southern Yunnan (image courtesy CHEN Ling-Zhi).

The Warm Evergreen Coniferous Forest sub-type (Figures 8.9, 8.10) is widely distributed in subtropical zones. Here we discuss *Pinus* forest as an example. *Pinus massoniana* forest (Figure 8.9) covers a large area of the eastern part of the subtropical zone in China. Its northern boundary stretches from the Qin Ling to Funiu Shan in Henan and on to the Huai He in Anhui, and the southern boundary from Bose (Guangxi) to the northern Leizhou Bandao (Guangdong). To the west its distribution extends to the Qingyi Jiang in Sichuan. The elevation inhabited by this forest is up to 700 m in the north and 1 500 m in the south. *Pinus yunnanensis* forest is most widespread in the western part of the subtropical zone, in western Guangxi, Guizhou, western Sichuan, Yunnan and eastern Xizang. This formation is found at elevations of 1 500–1 800 m and occasionally up to 2 600 m. *Pinus yunnanensis* var. *tenuifolia* forest is concentrated at the southwestern edge of Guizhou, and *P. kesiya* forest (Figure 8.10) in the montane belt (850–1 850 m elevation) of southern Yunnan. In general, warm *Pinus* forests form the natural secondary forest following destruction of evergreen broad-leaved forests. In addition to *Pinus* forests, warm evergreen coniferous forests include *Cupressus* (Cupressaceae), *Cunninghamia lanceolata* (Taxodiaceae) and *Keteleeria* (Pinaceae), formations, although these are mostly cultivated. *Keteleeria* spp. and *Cupressus* spp. forests are found in the eastern and western parts of the subtropical zone, *Cunninghamia lanceolata* forest mainly in the eastern mid-subtropical zone. Other coniferous forests extend into the eastern part of subtropical zone, including the *Amentotaxus argotaenia* (Taxaceae), *Cathaya argyrophylla* (Pinaceae), *Fokienia hodginsii* (Cupressaceae), *Pseudotsuga sinensis* (Pinaceae) and *Taiwania cryptomerioides* (Taxodiaceae) formations.

Here we describe *Pinus massoniana* forest, the most representative forest of lower mountains in eastern subtropical China (Figure 8.9). Apart from *Pinus massoniana*, the canopy layer contains other broad-leaved and coniferous trees such as *Cunninghamia lanceolata*, *Dalbergia hupeana* (Fabaceae), *Liquidambar formosana* (Hamamelidaceae), *Phyllostachys edulis* (Poaceae), *Pistacia chinensis* (Anacardiaceae), *Pseudotsuga sinensis*, *Quercus acutissima*, *Q. fabri*, *Q. serrata* and *Q. variabilis*. Towards the south of its distribution this forest may also contain *Castanea henryi*, *Castanopsis eyrei*, *Cyclobalanopsis glauca* and *Lithocarpus glaber* (all Fagaceae). The shrub layer may be dominated by *Eurya alata* (Theaceae), *Indosasa crassiflora* (Poaceae), *Loropetalum chinense* (Hamamelidaceae) or *Rhododendron simsii*, depending on the habitat. *Camellia oleifera* (Theaceae), *C. sinensis*, *Eurya brevistyla*, *E. chinensis*, *Glochidion puberum* (Euphorbiaceae), *Gardenia jasminoides* (Rubiaceae), *Helicteres angustifolia* (Sterculiaceae), *Lespedeza* spp., *Rhodomyrtus tomentosa* and *Syzygium grijsii* (both Myrtaceae), *Vaccinium bracteatum* and *V. carlesii* may also be found. Ferns such as *Dicranopteris dichotoma* (Gleicheniaceae), *Dryopteris* spp., *Pteridium aquilinum* var. *latiusculum* (Dennstaedtiaceae) and *Woodwardia japonica* (Blechnaceae) are the commonest plants in the herb layer, with grasses such as *Arundinella hirta, Arthraxon hispidus, Deyeuxia pyramidalis, Imperata cylindrica* var. *major, Miscanthus floridulus* and *Themeda triandra* (all Poaceae) also common, as well as some forbs. In general, *Pinus massoniana* forest is much more species-rich than temperate evergreen coniferous forest.

The Tropical Evergreen Coniferous Forest subtype results from the repeated destruction of natural tropical forest. It comprises mainly *Pinus latteri* forest on the hills and low mountains (elevation less 800 m) of southern Guangdong, southeast Guangxi and Hainan. The community structure is simple, with only canopy and herb layers. The main trees in the canopy layer are conifers, shrubs are sparse, and the herb layer is dominated by tall grasses.

Mixed Coniferous and Broad-leaved Forest.—Widely distributed in China, Mixed Coniferous and Broad-leaved Forests include both natural secondary forests and artificial forests, zonally arranged. There are two vegetation subtypes: Typical Temperate Mixed Coniferous and Broad-leaved Forest, and Subtropical Mixed Coniferous and Broad-leaved Forest (also known as Mixed *Tsuga*-Broad-leaved Forest).

The major representative formation of the Typical Temperate Mixed Coniferous and Broad-leaved Forest subtype is mixed *Pinus koraiensis*-broad-leaved forest (Figures 8.11, 8.12). This formation is distributed widely in the montane belt (elevation 700–1 100 m) of the eastern part of northeastern China. The canopy layer can be divided into two sub-layers, coniferous and broad-leaved. In the coniferous canopy sub-layer, *Pinus koraiensis* is dominant, with other species including *Abies nephrolepis, A. holophylla, Picea jezoensis* and *Larix olgensis*. Common deciduous trees of the broad-leaved canopy sub-layer are *Acer pictum* subsp.

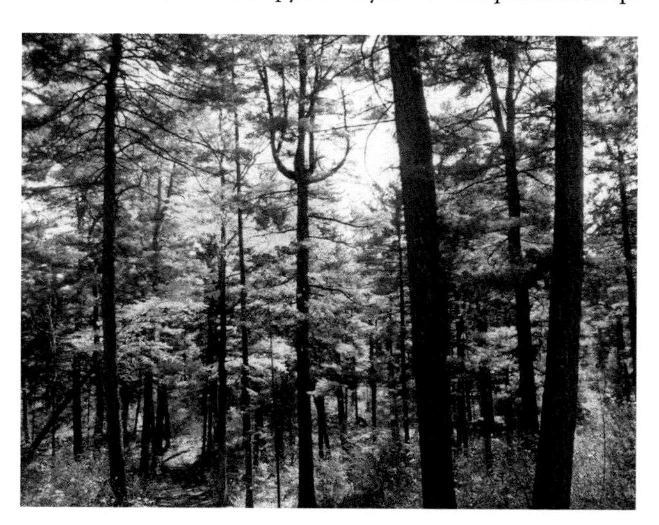

Figure 8.11 *Pinus koraiensis*-broad-leaved mixed forest, Xiao Hinggan Ling (image courtesy SONG Wei-Guo).

Figure 8.12 *Pinus koraiensis*-broad-leaved mixed forest, Changbai Shan (image courtesy SONG Wei-Guo).

mono, Acer tegmentosum, Acer ukurunduense, Betula platyphylla, Fraxinus mandschurica (Oleaceae), *Populus ussuriensis, Quercus mongolica* and *Ulmus davidiana* var. *japonica*. The shrub layer comprises species of *Corylus, Deutzia* and *Philadelphus* (both Saxifragaceae), *Lonicera, Prunus, Ribes, Rosa, Spiraea* and *Viburnum*. The herb layer is very species-rich, with the main taxa being *Athyrium* spp., *Carex* spp. *Deyeuxia* sp. and *Dryopteris* spp., along with additional shade-loving grasses. Lianas such as *Actinidia arguta* (Actinidiaceae), *Actinidia kolomikta* and *Schisandra chinensis* (Schisandraceae) are also found in the herb layer (Chen & Bao, 1964; Chen & Chen, 1997; Li *et al.*, 2001).

Subtropical Mixed Coniferous and Broad-leaved Forest is characteristic of mid-elevation mountains in the subtropical zone, and is represented here by mixed *Tsuga* and broad-leaved forest. *Tsuga* is known to form pure stands, but is usually present in mixed forests. Mixed *T. chinensis* var. *chinensis*-broad-leaved and mixed *T. longibracteata*-broad-leaved forests are found in the montane belt (elevation 1 200–2 000 m) of the eastern mid-subtropical zone. Mixed *T. chinensis* var. *formosana*-broad-leaved forest occurs on the Zhongyang Shan of Taiwan, at elevations of 2 000–3 000 m. Mixed *T. chinensis*-broad-leaved and mixed *T. dumosa*-broad-leaved forests are distributed on mountains (elevation 1 200–3 200 m) in the western subtropical zone. The mixed coniferous and broad-leaved forests found in the eastern part of the subtropical zone contain many evergreen broad-leaved trees such as species of *Castanopsis, Cyclobalanopsis, Illicium* (Illiciaceae), *Lithocarpus, Michelia* (Magnoliaceae) and *Schima* (Theaceae), as well as deciduous trees such as species of *Acer, Betula, Carpinus* (Betulaceae) and *Tilia*. The shrub layer is rich in bamboos such as species of *Chimonocalamus* and *Yushania* (Poaceae), alongside evergreen *Ilex* (Aquifoliaceae), *Rhododendron* and *Symplocos* (Symplocaceae) species, and deciduous shrubs including species of *Lonicera, Ribes, Sorbus* and *Viburnum*. The main plants of the herb layer are ferns, *Carex* spp., and dicots such as *Oxalis* spp. and *Thalictrum* spp.

Deciduous Broad-leaved Forest.—This forest type is characterized by broad-leaved trees that lose their leaves in winter. Deciduous Broad-leaved Forests appear mainly in the warm-temperate zone, but are also widely distributed in temperate and subtropical zones, and along river banks in desert regions. Deciduous Broad-leaved Forest can be divided into four sub-types: Cold-temperate, Temperate, Warm and Riparian in Desert Regions.

Cold-temperate Deciduous Broad-leaved Forest occurs mainly in subalpine areas of the cold-temperate and temperate zones and the southwestern part of the central subtropical zone. It comprises *Betula ermanii* elfin woods, which are found on the Da Hinggan Ling and Changbai Shan, graduating into a *B. ermanii* forest belt at elevations of 1 800–2 000 m under the low temperatures, strong winds and snow found on the Changbai Shan in winter (Figure 8.13). A few *B. ermanii* elfin woods are also found on the top of the Da Hinggan Ling. Intermixed with *B. ermanii* may be a small number of *Abies, Larix* or *Picea* species. *Rhododendron aureum* dominates the shrub layer, with small amounts of *Juniperus sibirica* also present. The herb layer is more species-rich, with main constituent taxa being species of *Cimicifuga* and *Trollius* (both Ranunculaceae), *Deyeuxia, Ligularia* (Asteraceae) and *Sanguisorba* (Rosaceae). Mosses are also found covering the surface of bare rocks. *Betula utilis* forest is also found, as secondary forest following destruction of *Abies* and *Picea* forests in the subtropical subalpine belt; it has a wider distribution than the *B. ermanii* forest but covers only a limited area.

Temperate (typical) Deciduous Broad-leaved Forest (Figures 8.14–8.16) is characteristic of the warm-temperate zone, and may also be found in a mid-level band on

Figure 8.14 *Quercus mongolica* forest, Dongling Shan, Beijing Shi (image courtesy CHEN Ling-Zhi).

Figure 8.13 *Betula ermanii* forest, Changbai Shan (image courtesy CHEN Ling-Zhi).

Figure 8.15 *Betula platyphylla* forest, Baihua Shan, Beijing Shi (image courtesy CHEN Ling-Zhi).

Table 8.2 Main plant species of temperate oak (*Quercus*) forests, showing abundance of the constituent species graded from one (least abundant) to five (most abundant; Chen & Chen, 1997).

Species	*Quercus* formation	
	Q. mongolica	*Q. aliena*
Canopy layer		
Acer pictum subsp. *mono*	2	2–3
A. truncatum	2–3	—
Betula chinensis	2–3	—
B. costata	2	—
B. dahurica	2–3	—
B. platyphylla	2–3	—
B. schmidtii	2	—
Carpinus cordata	2	—
Carpinus turczaninowii	2–3	—
Celtis koraiensis	2	2
Celtis bungeana	2	—
Fraxinus chinensis	2	—
F. mandschurica	2	—
F. chinensis subsp. *rhynchophylla*	2	2–3
Juglans mandshurica	2	3
Juniperus rigida	2–5	—
Kalopanax septemlobus (Araliaceae)	2	—
Maackia amurensis (Fabaceae)	2	2
Phellodendron amurense	2	—
Pinus densiflora	2	2
Pinus tabuliformis	2	2
Populus davidiana	2–3	—
Quercus aliena	2	5
Q. aliena var. *acutiserrata*	—	3
Q. dentata	2	—
Q. mongolica	2–5	2–3
Q. variabilis	3	—
Sorbus alnifolia var. *alnifolia*	2	—
Tilia amurensis	2	—
T. mandshurica	2	2–3
Ulmus davidiana	—	—
U. davidiana var. *japonica*	2	—
U. laciniata	2	—
U. macrocarpa	—	3
Shrub layer		
Alangium platanifolium var. *trilobum* (Alangiaceae)	—	2
Corylus heterophylla	1–4	3
C. mandshurica	1–4	—
Deutzia baroniana	1–4	1
Exochorda serratifolia	1–3	—
Flueggea suffruticosa (Euphorbiaceae)	1	1
Fraxinus bungeana	3	—
Indigofera kirilowii (Fabaceae)	1–4	—
Leptopus chinensis (Euphorbiaceae)	1–2	—
Lespedeza bicolor	1–4	1
Lespedeza davurica	1–2	—
Lespedeza floribunda	1	—
Lindera obtusiloba	4	1
Lonicera praeflorens	1–2	1
Myripnois dioica	1–4	—
Ostryopsis davidiana	2–3	—
Philadelphus pekinensis	1–2	—
Rhamnus arguta (Rhamnaceae)	1–2	—
Rhamnus globosa	2	—
Rhamnus parvifolia	1–2	—
Rhamnus schneideri var. *manshurica*	1–2	1
Rhododendron micranthum	1–4	—
Rhododendron mucronulatum	1–4	—
Rosa xanthina	2	—
Rubus crataegifolius	1–2	1
Rubus saxatilis	1–2	—
Sorbaria sorbifolia	2	—
Spiraea media	1–2	—
Spiraea pubescens	1–4	—
Spiraea trilobata	1–4	—
Symplocos paniculata	1	3
Syringa pubescens subsp. *microphylla* (Oleaceae)	4	—
Syringa pubescens subsp. *patula*	2	—
Viburnum opulus subsp. *calvescens*	1–2	1
Vitex negundo var. *heterophylla*	1–4	—
Weigela florida (Diervillaceae)	2	—
Zabelia biflora (Linnaeaceae)	1–4	—
Herb layer		
Achnatherum pubicalyx	1–3	—
Achnatherum sibiricum	1–3	—
Adenophora trachelioides (Campanulaceae)	1–2	1
Agrimonia pilosa (Rosaceae)	1–2	1
Artemisia latifolia	2–3	
Artemisia stolonifera	1–2	—
Atractylodes lancea	1–2	1
Arundinella hirta	1–3	—
Asparagus oligoclonos (Liliaceae)	1–2	1
Aster scaber	1–2	—
Athyrium crenatum	—	2
Carex callitrichos	2–3	—
Carex humilis	2–3	—
Carex lanceolata	1–3	3
Carex callitrichos var. *nana*	2–4	2
Carex pilosa	1	—
Carex quadriflora	2–3	—
Carex siderosticta	1–3	2–3
Carex ussuriensis	2–3	—
Chrysanthemum lavandulifolium var. *lavandulifolium*	1–2	—
Cleistogenes caespitosa	1–3	—
Convallaria majalis	1–2	—
Deyeuxia pyramidalis	2–3	—
Diarrhena fauriei (Poaceae)	3	—
Dictamnus dasycarpus (Rutaceae)	2	—
Dioscorea nipponica (Dioscoreaceae)	1–2	—
Filifolium sibiricum	3	—
Fragaria orientalis	2	—
Geranium dahuricum	2	—
Isodon japonicus var. *glaucocalyx* (Lamiaceae)	1–3	—
Iris uniflora	1–2	—
Melampyrum roseum (Scrophulariaceae)	1–3	1
Meehania urticifolia (Lamiaceae)	—	3
Phedimus aizoon var. *aizoon* (Crassulaceae)	1–2	—
Platycodon grandiflorus	1–2	—
Polygonatum odoratum	1–3	—
Potentilla fragarioides	1–2	1
Pteridium aquilinum var. *latiusculum*	—	2
Pyrola dahurica	—	2
Sanguisorba tenuifolia var. *alba*	1–3	—
Scutellaria baicalensis	1–2	—
Spodiopogon sibiricus	1–4	—
Syneilesis aconitifolia (Asteraceae)	1–2	—
Thalictrum aquilegiifolium var. *sibiricum*	2	—
Thalictrum ichangense	2–3	—
Themeda triandra	1	1
Vicia cracca	2–3	—
V. unijuga	1–3	1

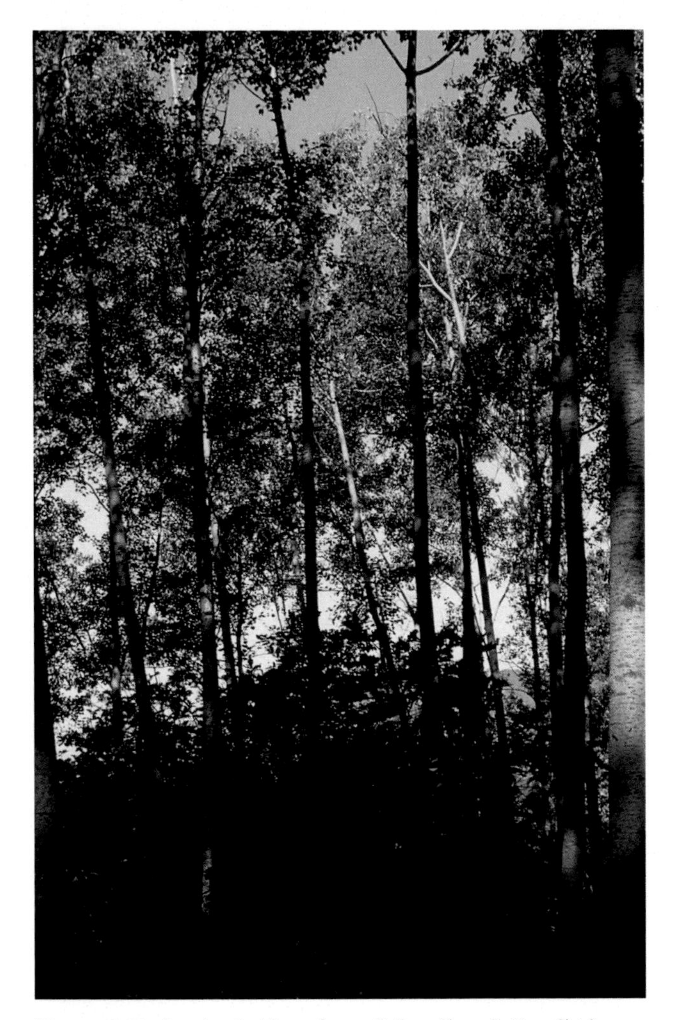

Figure 8.16 *Populus davidiana* forest, Baihua Shan, Beijing Shi (image courtesy CHEN Ling-Zhi).

mountains of the northern subtropical zone. In the temperate zone, most deciduous broad-leaved forests are secondary, developing after the destruction of coniferous or mixed coniferous and broad-leaved forests. The subtype is quite diverse, and the main tree species may include *Betula dahurica, B. platyphylla, Fraxinus chinensis* subsp. *rhynchophylla, Juglans mandshurica, Populus davidiana* (Figure 8.16), *Quercus aliena, Q. mongolica, Tilia mandshurica, T. mongolica* or *Ulmus pumila*. Of these, *Quercus mongolica* (Figure 8.14) and *Betula platyphylla* (Figure 8.15) forests are the most widely distributed, with *Betula* spp. and *Populus* spp. forests the most common secondary forests. In forests dominated by *Quercus*, additional canopy species may include any of those mentioned above. The shrub layer comprises various deciduous species such as *Corylus* spp., *Lespedeza* spp., *Ostryopsis* spp., *Rosa* spp., *Rubus* spp. (Rosaceae) and *Spiraea* spp. The main genera of the herb layer include *Arundinella, Carex, Clematis* (Ranunculaceae), *Deyeuxia, Poa, Polygonatum* and *Veratrum* (both Liliaceae), *Saussurea, Schisandra* and *Spodiopogon* (Poaceae). *Quercus mongolica* forests are found across cold-temperate and temperate zones, and the northern part of the warm-temperate zone, where the species composition is richer than elsewhere due to the more amenable natural conditions (see Table 8.2; Chen & Chen, 1997; Liu & Liu, 2000; Wang & Jiang, 2000; Ma, 2001).

Warm Deciduous Broad-leaved Forest (Figures 8.17–8.19) is distributed mainly on the middle and upper part of mountains in the subtropical zone, in the southern to mid warm-temperate zone. It includes both *Fagus* and *Quercus* forests. There are four *Fagus* formations: *Fagus lucida* forest is the main vegetation found on Fanjing Shan, Guizhou, and is also found in the montane belts of the northern Sichuan Pendi (elevation 1 500–2 000 m) and Sangzhi, Hunan (ca. 1 600 m); *F. engleriana* forest (Figure 8.17) occurs at elevations of 1 000–1 200 m in Shennongjia (Hubei), Huang Shan (Anhui), Anji (Zhejiang) and the northern Sichuan Pendi; *F. longipetiolata* forest (Figure 8.18) is found mainly on Fanjing Shan and Leigong Shan at elevations of 1 400–1 500 m, and may include other evergreen and deciduous broad-leaved trees; and *F. hayatae* forest (Figure 8.19) is native to Taiwan and the montane belt (elevation 1 500–1 800 m) of Chatian Shan (Huang & Tu, 1988; Huang, 1993).

The warm *Quercus* formations are distributed not only in the northern and mid-subtropical zone but also on low mountains in the warm-temperate zone. *Quercus acutissima, Q. aliena, Q. aliena* var. *acutiserrata, Q. baronii, Q. dentata, Q. serrata* and *Q. variabilis* forests are widely

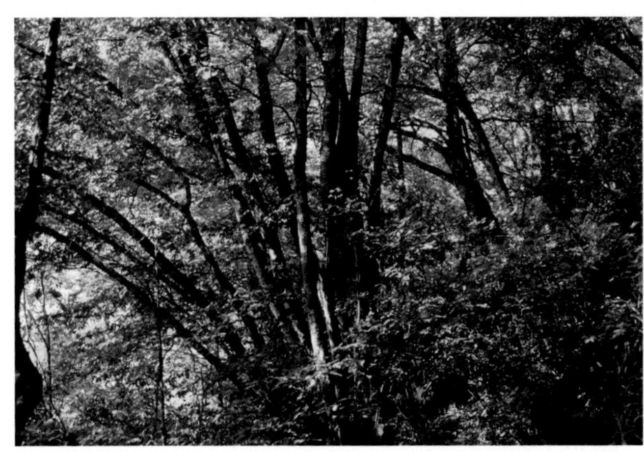

Figure 8.17 *Fagus engleriana* forest, Shennongjia, Hubei (image courtesy CHEN Wei-Lie).

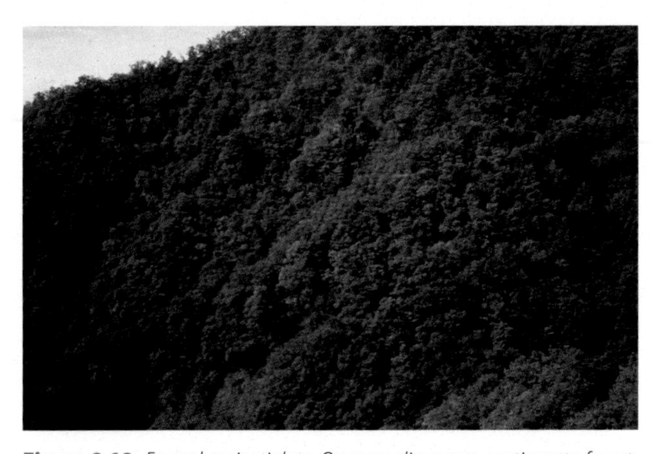

Figure 8.18 *Fagus longipetiolata-Quercus aliena* var. *acutiserrata* forest, Shennongjia, Hubei (image courtesy CHEN Wei-Lie).

Figure 8.19 *Fagus hayatae* forest, Shennongjia, Hubei (image courtesy CHEN Ling-Zhi).

distributed. *Quercus fabri* and *Q. stewardii* forests are distributed mainly in the montane belt of the eastern mid-subtropical zone. *Quercus chenii* forest is found on low mountains of the northern and central subtropical zone. Other forest formations, such as *Liquidambar formosana* and *Platycarya strobilacea* (Juglandaceae) forests are also distributed widely in the subtropical zone. The warm deciduous broad-leaved forests can be divided into canopy, shrub and herb layers. The shrub layer is species-rich and, while the ground-cover of the herb layer is not dense, this also contains many species including *Carex* spp., *Phlomis* spp. (Lamiaceae), *Platycodon* spp. (Campanulaceae), *Polygonatum* spp. and *Rodgersia* spp. (Saxifragaceae). Lianas, including species of *Abelia* (Linnaeaceae), *Celastrus* (Celastraceae), *Clematis, Schisandra* and *Vitis* (Vitaceae), are plentiful.

The main Riparian Deciduous Broad-leaved Forest in Desert Regions formations are *Populus alba, P. euphratica* and *P. pruinosa* forests. These formations include other trees and shrubs including species of *Elaeagnus, Halostachys* (Chenopodiaceae), *Lycium* (Solanaceae), *Salix* and *Tamarix* (Tamaricaceae). The herb layer is very species-poor, mostly comprising salt-tolerant perennials such as *Calamagrostis epigeios* and *Phragmites australis* (both Poaceae), *Alhagi sparsifolia* and *Glycyrrhiza* spp. (both Fabaceae), *Polygonum* spp. and *Schoenoplectus triqueter* (Cyperaceae), plus some ephemeral and annual species.

Mixed Evergreen and Deciduous Broad-leaved Forest.— This vegetation type forms a transition between Evergreen Broad-leaved Forest and Deciduous Broad-leaved Forest. It is mainly distributed on the mountains of the northern subtropical zone and limestone mountains of the central subtropical zone. The forest is divided into three sub-types: Typical Mixed Evergreen and Deciduous Broad-leaved Forest, Montane Mixed Evergreen and Deciduous Broad-leaved Forest, and Montane Mixed Evergreen Broad-leaved and Deciduous Forest on Limestone.

Typical Mixed Evergreen and Deciduous Broad-leaved Forest is found on hills and low mountains in the northern subtropical zone. It comprises deciduous *Fagus* and *Quercus* with evergreen *Castanopsis* and *Cyclobalanopsis*. The formation containing *Castanopsis sclerophylla, Cyclobalanopsis glauca, Quercus serrata* and *Q. variabilis* can be seen on Dabie Shan and other mountains in Henan, the southern slopes of the Qin Ling and the northern side of Jiangnan Qiuling below 500 m elevation, although it is sometimes found on low mountains up to 800 m. The mixed forest formation of *Q. acrodonta* and *Q. variabilis*, and that of *Q. acutissima* and *Q. oxyphylla*, are found on the southern slopes of the Qin Ling and northern slopes of Daba Shan. The *Cyclobalanopsis myrsinaefolia* and *Q. serrata* mixed forest formation is similar to the Deciduous Broad-leaved Forest type, but with many cold-resistant evergreen broad-leaved trees, such as *Castanopsis eyrei, Castanopsis lamontii, Cyclobalanopsis gracilis, Cyclobalanopsis stewardiana, Lithocarpus cleistocarpus* and *L. glaber*, in the canopy sub-layer. Many of the deciduous broad-leaved trees found in these mixed forests are also common in warm deciduous broad-leaved forests. The shrub layer comprises many saplings of the tree species as well as cold-resistant evergreen shrubs such as *Eurya* spp., *Ilex cornuta, Lindera glauca* and *Litsea cubeba* (Lauraceae), *Loropetalum chinense, Rhododendron mariesii, R. simsii, Symplocos sumuntia* and *Vaccinium bracteatum*. The herb layer is species-rich, including *Carex* spp., *Liriope* spp. (Liliaceae), *Lophatherum gracile* and *Miscanthus sinensis* (both Poaceae) and ferns such as *Pteridium aquilinum* and *Woodwardia japonica*.

Fagus lucida is found in Mixed Evergreen and Deciduous Broad-leaved Forests with various evergreen broad-leaved trees such as *Cyclobalanopsis glauca, Cyclobalanopsis multinervis, Cyclobalanopsis myrsinifolia* and *Cyclobalanopsis sessilifolia*, while *F. longipetiolata* forms mixed forests with *Castanopsis eyrei* and *Cyclobalanopsis oxyodon*. These forests are distributed mainly on the mountains of the central subtropical zone in Guizhou, Sichuan, Hubei and Hunan, at elevations of 1 000–1 800 m. Other mixed forests such as the *Davidia involucrata* (Nyssaceae)-*Cyclobalanopsis multinervis* mixed forest are found on the mid-elevation mountains (1 400–1 700

m) of western Hubei and northwestern Hunan. *Davidia involucrata-Schima argentea-Castanopsis platyacantha* mixed forest covers a large area of the montane belt (elevation 1 800–2 200 m) on the western edge of the Sichuan Pendi.

There are many formations of the Montane Mixed Evergreen and Deciduous Broad-leaved Forest sub-type. The main dominant species are *Castanopsis eyrei, Castanopsis lamontii, Cyclobalanopsis gracilis, Cyclobalanopsis multinervis, Cyclobalanopsis myrsinifolia, Cyclobalanopsis oxyodon, Cyclobalanopsis sessilifolia, Cyclobalanopsis stewardiana, Lithocarpus cleistocarpus, L. fenestratus, L. glaber, L. variolosus, Quercus baronii, Q. tarokoensis, Schima argentea* and *S. superba*. Co-dominant species include *Acer caudatifolium, A. cordatum, A. flabellatum, A. sinense, Carpinus viminea, Fagus* spp., *Liquidambar acalycina, Pterostyrax psilophyllus* (Styracaceae) and *Zelkova serrata* (Ulmaceae). These forests are distributed mainly on mid-elevation mountains of the subtropical zone.

The *Fagus longipetiolata* mixed forest formation occurs on the mountains of southern Jiangxi in the central subtropical zone (elevation 1 100–1 400 m), *Schima superba-Acer cordatum-Toxicodendron succedaneum* mixed forest on Wugong Shan (Jiangxi) at elevations of 800–1 000 m, *Cyclobalanopsis multinervis-Liquidanbar acalycina-Acer sinense* mixed forest on the mountains of the southwestern mid-subtropical zone (elevation 1 300–2 000 m), *Cyclobalanopsis myrsinifolia-Carpinus viminea* mixed forest in the montane belt of the eastern mid-subtropical zone (elevation 600–1 000 m), and *Cyclobalanopsis sessilifolia-Trochodendron aralioides* (Trochodendraceae)-*Acer caudatifolium* mixed forest at elevation 2 000–2 300 m on Zhongyang Shan, Taiwan. *Castanopsis lamontii* can form mixed forests with *Fagus lucida, Liquibambar acalycina* or *Pterostyrax psilophyllus*, and such forests are mainly found in the mountains of northern Guangxi. Similarly, *Lithocarpus cleistocarpus* can form mixed forests with either *Fagus longipetiolata* or *Davidia involucrata*, and occurs mainly on mountains at elevations of 1 300–2 400 m on the southwestern edge of the Sichuan Pendi. *Lithocarpus variolosus-Tetracentron sinense* (Tetracentraceae) mixed forest is sparsely distributed on the southwestern mountains of the Sichuan Pendi. *Lithocarpus fenestratus-Liquidambar acalycina* mixed forest and *Schima argentea-Fagus lucida* mixed forest are found in the mountains of western Guangxi. These are not the only formations of Montane Mixed Evergreen and Deciduous Broad-leaved Forest. In general, apart from the dominant and co-dominant tree species, the species composition of mixed forests comprises many more evergreen than deciduous species. In terms of common species, neighboring forests tend to be very similar in composition in terms of both deciduous and evergreen broad-leaved species.

Within the Montane Mixed Evergreen and Deciduous Broad-leaved Forest on Limestone sub-type, *Cyclobalanopsis glauca* mixed deciduous broad-leaved forests are of wide distribution and diverse formations including a variety of evergreen and deciduous broad-leaved species such as *Albizia kalkora* (Fabaceae), *Celtis biondii, Celtis sinensis* (Ulmaceae), *Cyclobalanopsis gracilis, Pistacia chinensis, Platycarya longipes, Platycarya strobilacea, Pteroceltis tatarinowii, Quercus chenii, Triadica rotundifolia* (Euphorbiaceae) and *T. sebifera*. Co-dominant species in these formations can include *Carpinus fargesiana, Celtis bungeana, Fagus longipetiolata, Myrica rubra* (Myricaceae) or *Platycarya strobilacea*. *Cyclobalanopsis glauca* mixed deciduous broad-leaved forests are mainly located on Jinfo Shan and Daba Shan. *Cyclobalanopsis glaucoides* also forms mixed forests with two species of *Platycarya*, which can be seen on the southeastern Yunnan-Guizhou Plateau and western Guangxi. Forests dominated by *Castanopsis sclerophylla* may also include *Liquidambar formosana, Platycarya strobilacea* and *Quercus acutissima, Q. chenii*. Other Mixed Evergreen and Deciduous Broad-leaved Forests, which may feature species of *Cinnamomum* and *Machilus* (both Lauraceae) as the dominant trees, are distributed widely on the limestone mountains of the subtropical zone. Other Mixed Evergreen and Deciduous Broad-leaved Forests on Limestone may be dominated by families such as Magnoliaceae and Rubiaceae (Huang & Tu, 1988; Wang *et al.*, 1997).

Evergreen Broad-leaved Forest.—Evergreen Broad-leaved Forest is the most characteristic forest of the subtropical zone in China, and has been divided into four sub-types: Warm (typical) Evergreen Broad-leaved Forest, Warm Humid Evergreen Broad-leaved Forest, Thermophilic Evergreen Broad-leaved Forest and Thermophilic Humid Evergreen Broad-leaved Forest.

Warm Evergreen Broad-leaved Forests are composed of evergreen broad-leaved tree species of Elaeocarpaceae, Fagaceae, Hamamelidaceae, Lauraceae and Magnoliaceae, with leathery leaves and which remain evergreen throughout the year. The height of the canopy layer in such forests is 15–25 m, and the forest can be divided into three layers: canopy, shrub and herb layers. The canopy layer is further divided into two or three sub-layers. The shrub layer comprises species of the genera *Aidia* (Rubiaceae), *Camellia, Eurya, Gardenia, Lindera, Litsea, Neolitsea* (Lauraceae), *Rhododendron, Symplocos* and *Vaccinium*. The herb layer may also be divided into two or three sub-layers, and comprises species of *Alpinia* (Zingiberaceae), *Carex, Lophatherum* and members of Araceae and Liliaceae; lianas are very rich in this layer.

Castanopsis spp. forests (Figures 8.20–8.22) are mainly distributed on mid- and low-elevation mountains of the central subtropical zone. The formations with the widest distributions are the *Castanopsis eyrei* (Figure 8.21), *Castanopsis fargesii, Castanopsis hystrix, Castanopsis sclerophylla* and *Castanopsis tibetana* (Figure 8.22) forests. *Castanopsis fargesii* forest is one of the main kinds of Evergreen Broad-leaved Forest in the central subtropical zone. It is found at low elevations (300–400 m) in the

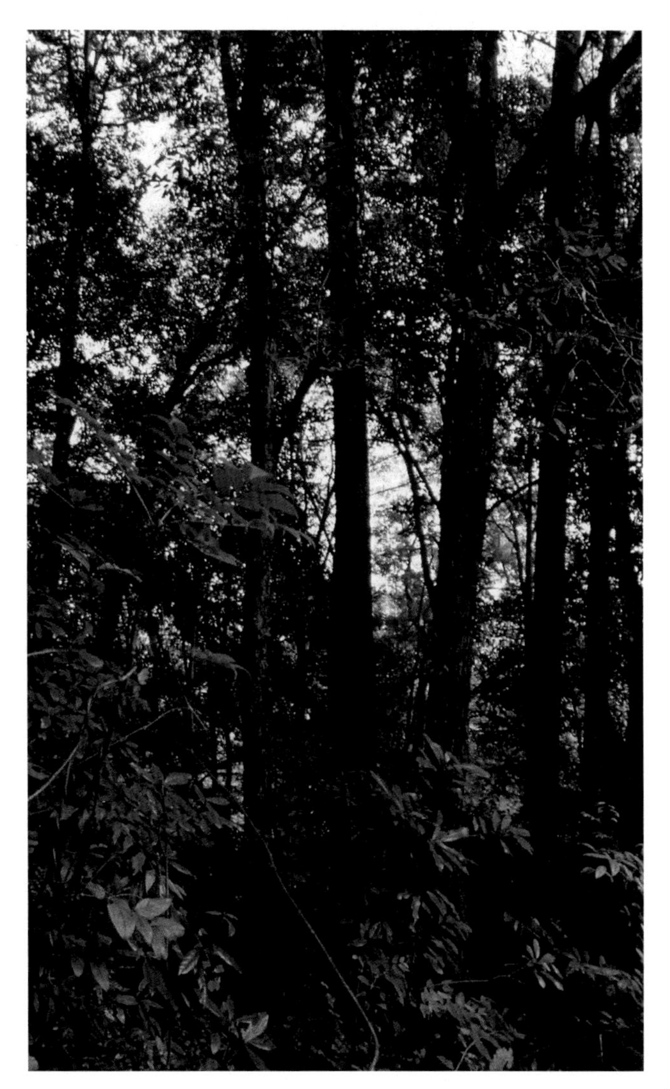

Figure 8.20 *Castanopsis carlesii-C. eyrei* forest, Longxi Shan, Fujian (image courtesy CHEN Wei-Lie).

Figure 8.21 *Castanopsis eyrei* forest, Zhejiang (image courtesy CHEN Bin).

eastern part of the zone. Co-dominant tree species with *C. fargesii* include *Castanopsis carlesii, Castanopsis fissa, Castanopsis tibetana, Cyclobalanopsis glauca, Elaeocarpus chinensis* (Elaeocarpaceae), *E. japonicus, E. sylvestris* and *Machilus pauhoi*. In the central and southern parts of the subtropical zone, *C. fargesii* is co-dominant with a different range of species.

Figure 8.22 *Castanopsis tibetana* forest, Jinfo Shan, Chongqing (image courtesy CHEN Wei-Lie).

Castanopsis eyrei forest (Figure 8.21) is also widely distributed in the eastern and and central parts of the mid-subtropical zone. Co-dominant species and common species in this forest at an elevation of ca. 1 000 m include *Castanopsis fabri, Castanopsis tibetana, Corylopsis multiflora* and *Exbucklandia tonkinensis* (both Hamamelidaceae), *Cunninghamia lanceolata, Cyclobalanopsis gracilis, Elaeocarpus japonicus, Lithocarpus fenestratus, Machilus nanmu, M. thunbergii, Pinus massoniana, Schima argentea* and *S. superba*. *Castanopsis tibetana* forest (Figure 8.22) is mainly distributed from the south of the Chang Jiang to Guangdong and Guangxi, at elevations of 300–900 m. Co-dominant species with *Castanopsis tibetana* include *Castanopsis fargesii, Castanopsis lamontii, Cryptocarya chinensis* (Lauraceae) and *Photinia benthamiana* (Rosaceae). Other common species include *Canarium album* (Burseraceae), *Cryptocarya concinna* and *Syzygium levinei*.

Castanopsis sclerophylla forest is one of the main evergreen broad-leaved formations on low mountains in the central subtropical zone; *Castanopsis carlesii* forest occurs in the central and southern subtropical zone; and *Castanopsis lamontii* forest is one of the main Evergreen Broad-leaved Forest of the central subtropical zone. Co-dominant and common species in the latter formation are similar to those of other Evergreen Broad-leaved Forests described here. *Castanopsis orthacantha* forest is distributed on, and typical of, gentle slopes at elevations of 2 000–2 400 m on the central Yunnan plateau in the western part of the mid-subtropical zone. It comprises Evergreen Broad-leaved Forest with co-dominant species including *Cyclobalanopsis glaucoides, Lithocarpus craibianus, L. dealbatus* and *Schima argentea*.

Cyclobalanopsis spp. forests are mainly represented by the *Cyclobalanopsis glauca* and *Cyclobalanopsis glaucoides* formations. Other formations include *Cyclobalanopsis chungii, Cyclobalanopsis delavayi, Cyclobalanopsis gracilis* and *Cyclobalanopsis kiukiangensis* forests. *Cyclobalanopsis glauca* forest is widely distributed on low mountains (200–600 m elevation) of the eastern and central subtropical zone.

Common species, as well as *Cyclobalanopsis glauca*, include *Castanopsis sclerophylla, Lithocarpus glaber, Litsea coreana* var. *lanuginosa*, *Machilus thunbergii* and *Phoebe sheareri* (Lauraceae). Deciduous trees, such as *Dalbergia hupeana*, *Diospyros kaki* var. *sylvestris* (Ebenaceae), *Quercus fabri*, *Q. serrata*, *Rhus chinensis* (Anacardiaceae) and *Sassafras tzumu* (Lauraceae), are also common in the canopy layer. *Cyclobalanopsis glaucoides* forest instead comprises (in addition to *Cyclobalanopsis glaucoides*) *Keteleeria evelyniana* and *Lithocarpus dealbatus*, and is characteristic of the western mid-subtropical zone. Additional Evergreen Broad-leaved Forest formations, dominated by *Cyclobalanopsis delavayi* and *Lithocarpus confinis*, are found on the Yunnan-Guizhou Plateau and in southern Sichuan.

Machilus spp. formations are distributed mainly in the eastern mid-subtropical zone. *Machilus thunbergii* forest is distributed widely on mountains in central and southern Jiangxi, and southern, eastern and southwestern Hunan, occupying both streamsides and valley slopes below 1 000 m. *Machilus thunbergii* forest is also found in Zhejiang, Fujian and Guangdong, but covering smaller areas. Accompanying taxa include species of *Albizia, Castanopsis, Cinnamomum, Lindera, Phoebe, Ternstroemia* (Theaceae), *Zelkova* and other species of *Machilus*. Other *Machilus* formations include *M. chinensis, M. leptophylla* and *M. pauhoi* forests, all of which are found in small areas on low mountains of the eastern mid-subtropical zone.

There are many other Evergreen Broad-leaved Forest formations, including *Cinnamomum camphora*, *C. japonicum*, C. *subavenium*, *Phoebe bournei* (Figure 8.23), *P. sheareri* and *P. zhennan* forests, all distributed on hills below 1 000 m in the eastern mid-subtropical zone. *Schima superba* forest is more widely distributed, and can be found on hills below 1 300 m in the central and southern subtropical zone. *Schima argentea* forest is found in both eastern and western parts of the central subtropical zone. Further broad-leaved evergreen formations include *Adinandra* spp. (Theaceae), *Altingii* spp. and *Distylium* spp. (both Hamamelidaceae), *Daphniphyllum* spp. (Daphniphyllaceae), *Elaeocarpus* spp., *Manglietia* spp. (Magnoliaceae) and *Michelia* spp forests. These are all of limited distribution, and mostly confined to the eastern mid-subtropical zone (Sichuan Vegetation Cooperative Group, 1980; Huang & Tu, 1988; Anhui Forest Editorial Committee, 1990; Lin, 1990; Qi, 1990).

Figure 8.23 *Phoebe bournei* forest, Longxi Shan, Fujian (image courtesy CHEN Wei-Lie).

Figure 8.24 Warm humid evergreen broad-leaved forest, southern Yunnan (image courtesy ZHU Hua).

The Warm Humid Evergreen Broad-leaved Forest sub-type (Figure 8.24) is mainly distributed on the humid mountains in the eastern mid-subtropical zone. *Lithocarpus* spp. forests are common, and found mainly on the mountains of the central subtropical zone. For instance, the *Lithocarpus glaber* formation is distributed widely on mountains at elevations of ca. 1 300 m in the eastern mid-subtropical zone. There are also many *Lithocarpus* spp. forests in the western mid-subtropical zone, including *L. cleistocarpus, L. craibianus* and *L. variolosus* formations, which are found in the upper montane belt (2000–3 900 m) of northeastern Yunnan. *Lithocarpus corneus* and *L. henryi* forests are found on mountains (at ca. 1 000 m) in the eastern mid-subtropical zone to the southern subtropical zone. Among others, *Lithocarpus chrysocomus* forest is found on Nanling Shan (elevation 1 300–1 800 m), and *L. craibianus* is co-dominant in a community with *Illicium simonsii*. In general, these forests contain small numbers of additional tree species, such as *Castanopsis orthacantha, Cyclobalanopsis glaucoides, Ilex yunnanensis, Phoebe forrestii* and *Schefflera delavayi* (Araliaceae).

Castanopsis platyacantha forest is found in the western mid-subtropical zone, such as in Sichuan and Yunnan. In the *C. platycantha* forests of Yunnan, the dominant species of the canopy layer are *Acer oliverianum, Fagus engleriana, Ilex dunniana, Lithocarpus hancei, Machilus ichangensis, Neolitsea levinei, Schima crenata* and *Symplocos sumuntia*. *Castanopsis neocavaleriei* forest occurs in the montane belt (elevation 1 000–1 800 m) of Jiuwan Dashan in Guangxi. Warm humid evergreen *Cyclobalanopsis* forests include *C. annulata, C. kiukiangensis* and *C. lamellosa* formations. The latter two are distributed sparsely in the montane belt (elevation 1 800–2 200 m) of western Yunnan and Mêdog (Xizang). Many tree species are present in these forests.

Thermophilic Evergreen Broad-leaved Forest (Figure 8.25), also known as Monsoon Evergreen Broad-leaved Forest, is the main Evergreen Broad-leaved forest of the southern subtropical zone. It is the only tropical form of Evergreen Broad-leaved forest, and forms a transition zone between Subtropical Evergreen Broad-leaved Forest, Tropical Monsoon Forest and Seasonal Rainforest. This kind of Thermophilic Evergreen Broad-leaved Forest may have various dominant tree species, including *Castanopsis carlesii* var. *carlesii, Castanopsis chinensis, Castanopsis concinna, Castanopsis hystrix, Castanopsis tonkinensis, Cryptocarya chinensis* and *Cryptocarya concinna*, with other species belonging to Arecaceae, Elaeocarpaceae, Fabaceae, Meliaceae, Moraceae, Myrsinaceae, Myrtaceae and Rubiaceae. The canopy layer can be divided into two or three sub-layers and, in this forest, has many distinctive physical characteristics such as plank-buttresses, cauliflory, drip-tips, stranglers and large woody vines.

Figure 8.25 Thermophilic evergreen broad-leaved forest, Guangdong (image courtesy CHEN Ling-Zhi).

The *Castanopsis hystrix-Cryptocarya chinensis* formation is found on hills and low mountain in the eastern part of the southern subtropical zone, which includes southern Fujian, central-southern Guangdong and Guangxi, and central and northern Taiwan. The main species in the upper canopy layer, in addition to *Castanopsis hystrix* and *Cryptocarya chinensis*, vary depending on the region and may include, for example, *Castanopsis uraiana* and *Syzygium hancei* in southern Fujian, and *Cyclobalanopsis gilva, Cinnamomum* spp., *Cryptocarya concinna* and *Michelia compressa* in central northern Taiwan. Other common species include *Archidendron clypearia* (Fabaceae), *Cinnamomum japonicum, Cinnamomum parthenoxylon, Cryptocarya chingii, Elaeocarpus japonicus, Elaeocarpus sylvestris, Engelhardia roxburghiana* (Juglandaceae), *Pterospermum heterophyllum* (Sterculiaceae) and *Syzygium buxifolium*. Understory species include *Aidia pycnantha, Aidia racemosa, Casearia glomerata* (Flacourtiaceae), *Casearia velutina, Ilex* spp. and *Rhaphiolepis ferruginea* (Rosaceae). The shrub layer contains *Antidesma fordii* (Euphorbiaceae), *Antidesma japonicum, Ardisia* spp. (Myrsinaceae), *Blastus cochinchinensis* (Melastomataceae) and *Psychotria asiatica* (Rubiaceae), as well as seedlings of shade-tolerant trees. The herb layer is tall and dense.

The *Castanopsis chinensis, Cryptocarya chinensis* and *Cryptocarya concinna* formations are the main thermophilic evergreen broad-leaved forests found on hills below 500 m in central-southern Guangdong and Guangxi. *Castanopsis carlesii* var. *carlesii-Cryptocarya chinensis* forest is found in northern Taiwan; *Castanopsis tonkinensis-Cryptocarya concinna* forest in southwestern Guangxi; and *Syzygium hancei-Schefflera heptaphylla-Cryptocarya chinensis* forest in Fujian.

Thermophilic Evergreen Broad-leaved Forest comprising *Castanopsis hystrix* and *Schima* spp. occurs in Simao and Xishuangbanna, Yunnan, and also in Mêdog on the southern slope of the Himalaya Shan. The main species in the upper canopy layer of this formation are *Castanopsis fargesii, C. hystrix, Schima villosa* and *S. wallichii*. Other accompanying species include *C. indica, C. mekongensis, Elaeocarpus prunifolioides, Lithocarpus microspermus, L. xylocarpus, Machilus kurzii, M. robusta* and *Phoebe neurantha*. The second and third forest layers, which comprise tree-like shrubs, include *Helicia* sp. (Proteaceae), *Illicum simonsii, Schefflera heptaphylla, Sloanea sinensis* (Elaeocarpaceae), *Sterculia hainanensis* (Sterculiaceae) and *Symplocos sumuntia*.The main species of the shrub layer are

Ardisia quinquegona and *Psychotria asiatica*. The herb layer contains many large ferns and species of *Alpinia*. A further Thermophilic Evergreen Broad-leaved Forest dominated by *Castanopsis hystrix, C. indica* and *Schima wallichii* forms the main vegetation formation of wide valleys and low mountains (elevation 900–1 300 m) in central-southern Yunnan, and is also common in the montane belt of the tropical zone.

Thermophilic Humid Evergreen Broad-leaved Forest is mainly distributed on humid mountains in the southern subtropical zone. The main formations are *Cyclobalanopsis* spp. and *Lithocarpus* spp. forests, with others including *Exbucklandia tonkinensis, Manglietia fordiana* var. *hainanensis, Rhodoleia parvipetala* (Hamamelidaceae) and *R. championii* forest.

Many kinds of Humid Evergreen Broad-leaved Forest dominated by *Cyclobalanopsis* are found in Taiwan: *Cyclobalanopsis gilva-Lithocarpus brevicaudatus* forest occurs in the montane belt (elevation 600–1 000 m) in the center of the island and also on Zhongyang Shan in the north; *Cyclobalanopsis stenophylloides-Lithocarpus kawakamii* forest in the montane belt (elevation 1 600–2 000 m) of Zhongyang Shan; and *Cyclobalanopsis longinux-Castanopsis carlesii* forest in the montane belt (elevation 1 000–2 000 m) of the central and southern slopes of Zhongyang Shan. Co-dominant species of the latter include *Lithocarpus amgydalifolius* and *L. kawakamii*, and additional species present include *Castanopsis carlesii* var. *carlesii, Cryptocarya concinna, Cyclobalanopsis morii, Cyclobalanopsis pachyloma, Cyclobalanopsis sessilifolia, Cyclobalanopsis stenophylloides, Machilus japonica* var. *kusanoi* and *M. thunbergii.*

Lithocarpus spp. forest is found mainly on the mountains of the western part of the southern subtropical zone. For example, *Lithocarpus xylocarpus* forest is the Humid Evergreen Broad-leaved Formation found on the northern section of Ailao Shan in central Yunnan, and includes co-dominant species such as *Castanopsis wattii* and *Schima noronhae*. Other accompanying species include *Cyclobalanopsis stewardiana, Illicium macranthum, Lithocarpus pachyphyllus* var. *fruticosus, Lyonia ovalifolia, Manglietia insignis, Michelia floribunda, Stewartia sinensis* (Theaceae) and *Styrax perkinsiae* (Styracaceae). *Lithocarpus echinotholus* forest occurs on the mid-elevation mountains of southwestern Yunnan, and is composed of *Lithocarpus echinotholus* with *Schima noronhae* and *Cyclobalanopsis glaucoides*. *Lithocarpus echinotholus* can also form evergreen broad-leaved forest with co-dominant species such as *Acer sikkimense* and *Lirianthe delavayi*. Finally, humid dwarf forests, comprising *Rhododendron* and *Vaccinium* spp., are found at the top of the montane belt (elevation 1 500–3 000 m) in the western part of the central subtropical and tropical zones (Guangdong Institute of Botany, 1976; Hu, 1979; Wu, 1983; Wu, 1987; Huang & Tu, 1988; Peng, 1989; Qi, 1990; Sichuan Forest Editorial Committee, 1992; Wang *et al.*, 1997; Wang & Jiang, 2000).

Monsoon Forest.—Monsoon Forests are found in the tropical zone where dry and moist seasons alternate periodically. In Monsoon Forests, the height of the canopy is lower than in rainforests, the trees shed all or some of their leaves in the dry season and are straight-trunked with lower branches and thick, rough bark. Cauliflory, large woody vines and epiphytes are less frequent than in rainforest. Monsoon Forests are distributed discontinuously across tropical regions of Asia, Africa and America, with perhaps the most typical monsoon rainforests being found in Southeast Asia. In China, Monsoon Forests occur in parts of Guangdong, Guangxi, Hainan, Taiwan, Yunnan and Xizang. They have been divided into three sub-types: Deciduous Monsoon Forest, Semi-Evergreen Monsoon Forest, and Monsoon Forest on Limestone.

Deciduous Monsoon Forest is characterized by the loss of all or a large proportion of the leaves from trees during the dry season. The stratification of the forest is relatively monotonous. It is found mainly in the hot dry valleys and basins of western Hainan and southern Yunnan. The community composition is relatively complicated. The canopy layer is composed of species of *Albizia, Bischofia, Mallotus* and *Phyllanthus* (all three in Euphorbiaceae), *Bombax* (Bombacaceae), *Chukrasia, Dysoxylum, Melia* and *Toona* (Meliaceae), *Cyclobalanopsis, Ficus, Morus* and *Streblus*, (Moraceae), *Lagerstroemia* (Lythraceae), *Lannea* (Anacardiaceae), *Quercus* and *Terminalia* (Combretaceae). The shrub layer is characterized by sun-loving, drought-tolerant species. The herb layer contains many grasses, and few vines. Several formations of Deciduous Monsoon Forest occur in southern China, as described below.

Bombax ceiba-Albizia chinensis forest is found in the lowlands (below 200 m elevation) of western Hainan and southwestern Guangxi, and is more widespread in the valleys (elevation up to 1 200 m) of southern Yunnan. Large trees are sparse in this kind of forest. The main components of the canopy layer are *Albizia chinensis, Bombax malabarica, Erythrina stricta* (Fabaceae) and *Garuga floribunda* (Burseraceae), alongside such species as *Ficus hispida, Lagerstroemia intermedia, Markhamia stipulata* var. *kerrii* (Bignoniaceae) and *Morus macroura*. The shrub layer plants are sparse and clump-forming, including a common *Solanum* sp. and *Woodfordia fruticosa* (Lythraceae). The main species of the herb layer are tall grasses such as *Neyraudia reynaudiana* and *Saccharum arundinaceum* (both Poaceae).

Terminalia nigrovenulosa-Albizia odoratissima-Lannea coromandelica forest is a stable forest formation distributed mainly in the hot dry basins of western and southwestern Hainan. The canopy layer is species-rich and two-layered; in addition to the dominant trees, the upper canopy contains *Aporosa dioica* (Euphorbiaceae), *Buchanania microphylla* (Anacardiaceae), *Ilex* sp. and some evergreen species. The lower canopy layer also contains these species, is dominated by *Aporosa dioica* and *Lannea coromandelica*, and the numerous other species include *Catanopsis jucunda*,

Dimocarpus longan (Sapindaceae), *Diospyros strigosa* and *Vitex pierreana* (Verbenaceae). The shrub layer is relatively species-poor, apart from the many saplings and seedlings of the canopy tree species. The herb layer, however, is species-rich, including species of *Eupatorium* (Asteraceae), *Imperata, Panicum* (Poaceae) and *Saccharum*. Vines are always present.

Terminalia myriocarpa-Erythrina stricta forest is found mainly in the lowland river valleys and plateaus in southern Yunnan, but also appears on the southern slopes of the eastern Himalaya Shan in Xizang. It is usually a secondary forest formed after the destruction of primary Monsoon Forests. This formation is characterized by a rich variety of tall trees forming an irregular canopy, rich in epiphytes and vines. Common canopy-layer species include *Bombax ceiba, Elaeocarpus varunua, Duabanga grandiflora* (Lythraceae), *Falconeria insignis* (Euphorbiaceae), *Tetrameles nudiflora* (Tetramelaceae) and *Toona ciliata*. The shrub layer is relatively species-poor, with common species including *Alchornea tiliifolia* (Euphorbiaceae), *Leea asiatica* (Vitaceae) and *Rhus chinensis*. The main species of the herb layer are tall grasses such as *Imperata cylindrica, Microstegium fasciculatum, Neyraudia reynaudiana* and *Thysanolaena latifolia* (all Poaceae).

Albizia kalkora-Michelia doltsopa forest is frequent in the valleys of western Yunnan below 500 m elevation. Apart from the dominant species, the canopy also features *Dysoxylum excelsum* and *Toona ciliata* among others. The common species of the shrub layer are *Clerodendrum chinense* var. *chinense* (Verbenaceae), *Phyllanthus emblica* and *Wendlandia uvariifolia* (Rubiaceae). The herb layer comprises mainly grasses and ferns. There are many other deciduous monsoon forest formations, including *Bauhinia variegata* (Fabaceae), *Gmelina arborea* (Verbenaceae), *G. hainanensis* and *Protium serratum* (Burseraceae) forests; these are mostly cultivated or develop following the destruction of other monsoon forests.

The Semi-evergreen Monsoon Forest (Figure 8.26) sub-type may be composed of evergreen, or a combination of evergreen and deciduous broad-leaved trees. It occurs in southern southern Guangdong, Guangxi, Hainan, Taiwan and Yunnan, in places with higher levels of moisture. It displays some of the characteristics of rainforest, but the canopy is lower and the forest is poorer in vines and epiphytes than true rainforest.

Figure 8.26 Semi-evergreen monsoon forest, southern Yunnan (image courtesy ZHU Hua).

The *Ficus microcarpa-Gironniera* sp. formation occurs mainly on basalt and granite plateaus and hills of the southern part of the Leizhou Bandao (Guangdong) and northern Hainan. The canopy may be divided into two sub-layers. The upper sub-layer contains, in addition the two dominant species, many specimens of *Aphanamixis polystachya* and *Walsura robusta* (both Meliaceae), *Artocarpus tonkinensis* (Moraceae), *Ficus nervosa, Harpullia cupanioides* (Sapindaceae) and *Radermachera hainaensis* (Bignoniaceae). The lower sub-layer comprises *Baccaurea ramiflora* (Euphorbiaceae), *Garcinia* spp. (Clusiaceae), *Litsea* spp., *Schefflera heptaphylla* and *Sterculia* spp. Except saplings, the shrub layer comprises mainly *Ardisia crenata, Calamus rhabdocladus* (Arecaceae) and *Psychotria asiastica*. The herb layer is dense, but is mainly composed of bamboos and ferns. There are several species of liana, but they are present only at low frequency. *Ficus microcarpa* can also form semi-evergreen monsoon forest with *Syzygium odoratum*, which has a different set of constituent species and is limited in distribution.

Ficus altissima-Chukrasia tabularis forest is widely distributed in open valleys, basins and sunny slopes below 1 000 m elevation in southern and southwestern Yunnan. In this formation the canopy can be divided into two sub-layers: the upper layer contains, in addition to the two dominant species, about ten tree species including *Bischofia javanica, Ficus fistulosa* and *Schima wallichii*. Common species of the lower sub-layer are *Aporosa yunnanensis, Mallotus philippensis, Syzygium cumini* and *Wendlandia scabra*. There are few shrubs, but the herb layer is dense, comprising mainly grasses.

Liquidambar formosana-Schima wallichii forest develops on the low mountains (below 950 m) of southwestern Guangxi following the destruction of other monsoon forests. The canopy comprises two sub-layers, the first containing, in addition to the dominant species, *Albizia* spp., *Betula alnoides, Syzygium cumini* and occasionally *Sterculia lanceolata*. Common species of the lower sub-layer include *Antidesma montanum, Mallotus philippensis* and *Wendlandia* spp. The shrub layer often contains *Aporosa dioica, Psychotria asiatica* and *Ardisia quinquegona*. The main plants of the herb layer are tall ferns, with some grasses. Vine species are rich, but present only as small populations.

Mesua ferrea (Clusiaceae) is an introduced species from Southeast Asia. In China it forms *Mesua ferrea-Mangifera sylvatica* (Anacardiaceae) forest, which can cover large areas, such as in the valley basin of Lincang, Yunnan. Other species present in this forest are *Archidendron clypearia,*

Knema tenuinervia (Myristicaceae), *Mallotus philippensis* and *Markhamia stipulata*. *Senna siamea* (Fabaceae) was also introduced from Southeast Asia, and is commonly planted in tropical southern Yunnan. These artificial *S. siamea* formations contain some of the tree and shrub species found in other Semi-evergreen Monsoon Forest Formations.

Monsoon Forest on Limestone mountains in the tropical zone may be primary or secondary, the latter comprising natural regeneration following the destruction of rainforests on these mountains. The canopy of the primary Monsoon Forest on Limestone may be divided into three sub-layers, with deciduous trees in the upper layer and a mixture of evergreen and deciduous trees in the lower two. Except saplings and seedlings, shrubs are fairly sparse, but vines and epiphytes are found in the shrub layer. Examples of primary Monsoon Forest on Limestone include *Cleistanthus sumatrana* (Euphorbiaceae) forest in southern Yunnan and *Pistacia weinmanniifolia-Cleistanthus sumatrana* forest in valleys covering a small area of southwestern Yunnan at elevations of 600--900 m. Secondary examples include the *Bauhinia variegata-Microstegium fasciculatum* formation of southern Yunnan. *Zenia insignis* (Fabaceae) forest and *Triadica rotundifolia* forest are found in small populations in southwestern Guangxi.

Rainforest.—Tropical rainforest is found only in southern China, in southern Guangdong, southwestern Guangxi, southeastern Hainan, southern Yunnan, southeastern Xizang and the southern part of Taiwan. Most rainforests have only sparse distributions in China, and develop under the influence of the tropical monsoon climate. They therefore differ from equatorial rainforest, but have an affinity with the rainforests of Southeast Asia. The characteristics of Chinese rainforest include tall trees, obvious buttresses, cauliflory, abundant lianas and epiphytes, and a canopy comprising three or four sub-layers. The main constituent families are Annonaceae, Apocynaceae, Burseraceae, Dipterocarpaceae, Moraceae, Myristicaceae, Myrtaceae, Sapindaceae and Tetramelaceae. The shrub and herb layers are composed of Acanthaceae, Araceae, Arecaceae, Euphorbiaceae, Poaceae, Rubiaceae, Rutaceae, Zingiberaceae and tall ferns. Three sub-types of rainforest are recognized in China: Seasonal Rainforest, Montane Rainforest and Seasonal Rainforest on Limestone (Figure 8.27).

Seasonal Rainforest is found mainly in Hainan, southern Taiwan and Yunnan, and the southern side of the Himalaya Shan. There are many formations, most of fairly localized distribution. *Dipterocarpus* spp. (Dipterocarpaceae) forest is the most characteristic rainforest of tropical Southeast Asia. In China, it is limited to the piedmont of the southern side of the eastern Himalaya Shan and southern Yunnan. Two formations are recognized. The first, *Dipterocarpus retusus* var. *retusus-Hopea chinensis* (Dipterocarpaceae) forest is found in Hekou and Jinping, Yunnan, on steep hillsides below 500 m, sometimes up to 1 000 m along river valleys.

Figure 8.27 Seasonal rainforest on limestone, Xishuangbanna, Yunnan (image courtesy ZHU Hua).

The upper canopy layer is mainly composed of *Crypteronia paniculata* (Crypteroniaceae), *Dipterocarpus retusus* var. *retusus*, *Hopea chinensis* and *Tetrameles nudiflora*, with no single clearly dominant species. Other accompanying species include *Amesiodendron chinense* and *Pometia pinnata* (both Sapindaceae), *Aphanamixis polystachya*, *Chukrasia tabularis*, *Dracontomelon duperreanum* (Anacardiaceae), *Duabanga grandiflora*, *Dysoxylum excelsum*, *Lysidice rhodostegia* and *Saraca dives* (both Fabaceae).

Dipterocarpus turbinatus-D. retusus var. *macrocarpus-D. gracilis* forest is mainly distributed on the southern side of the eastern Himalaya Shan (at ca. 100 m with a maximum elevation of 500 m). Again, the dominant species is not obvious, although *D. gracilis* is perhaps the most frequent tree. Other common tree species include *Artocarpus chama*, *Canarium strictum*, *Dimocarpus longan*, *Dysoxylum gotadhora*, *Shorea assamica* (Dipterocarpaceae), *Tetrameles nudiflora* and species of *Cinnamomom*, *Ficus*, *Machilus* and *Phoebe*. The height of the upper canopy layer is at least 30 m. The lower two or three sub-layers of the canopy contain many species. The shrub layer includes *Antidesma* spp., *Maesa montana* (Myrsinaceae), *Pavetta* spp. (Rubiaceae) and species of *Arenga* and *Caryota* (both Arecaceae), *Dendrocalamus* (Poaceae) and *Pandanus* (Pandanaceae). There are many herbs, lianas, epiphytes and ferns.

Shorea and *Parashorea* spp. (Dipterocarpaceae) forests are also widely distributed in the tropical zone but limited in China. Two formations are recognized here: *Parashorea chinensis* and *S. assamica-Dipterocarpus turbinatus* forests. *Parashorea chinensis* forest is sparsely distributed on moist river valleys and lower slopes (elevation 700–950 m) in the eastern part of Mengla, Xishuangbanna, Yunnan, along the middle segment of the Nanshahe, and in a very narrow area of Napo, Guangxi (elevation 800–1 600 m).The canopy can be divided into four sub-layers. *Parashorea chinensis* is dominant, and there are around 125 accompanying canopy species, chief among them being

Baccaurea ramiflora, Barringtonia spp. (Lecythidaceae), *Castanopsis indica, Dysoxylum grande, Ficus langkokensis, Garcinia cowa, Knema tenuinervia , Lithocarpus fohaiensis, Myristica yunnanensis* (Myristicaceae), *Pittosporopsis kerrii* (Icacinaceae), *Pometia pinnata* and *Pseudovaria trimera* (Annonaceae). The main components of the shrub layer are tree saplings, with ca. 50 species. The ca. 20 species of true shrubs include *Drypetes hoaensis* (Euphorbiaceae), *Ixora longshanensis, Lasianthus attenuatus, L. sikkimensis* and *Saprosma ternata* (all Rubiaceae). The formation is rich in liana species, but all are present only in small numbers (Zhu, 2001). The *C. assamica-Dipterocarpus turbinatus* formation occurs mainly in river valleys below 1 000 m on the southern side of the Himalaya Shan in southeastern Xizang, and on mountainsides below 600 m, in western Yingjiang, Yunnan. Additional canopy species in this formation include *Antiaris toxicaria* (Moraceae), *Hydnocarpus anthelminthicus* (Flacourtiaceae), *Knema tenuinervia , Michelia doltsopa* and *Vitex burmensis.*

Vatica spp. (Dipterocarpaceae) and associated species rainforest is also widely distributed in tropical regions, but limited in distribution in China. The *Heritiera parvifolia* (Sterculiaceae)-*Vatica mangachapoi* formation is found on Diaoluo Shan and in Wanling, Hainan (elevation below 500 m). Additional canopy species in this formation include *Baccaurea ramiflora, Carallia brachiata* (Rhizophoraceae), *Cyclobalanopsis blakei, Dillenia turbinata* (Dilleniaceae), *Gironniera subaequalis, Litchi chinensis* (Sapindaceae), *Maclurodendron oligophlebium* (Rutaceae), *Schefflera heptaphylla, Saprosma ternata, Sarcosperma* spp. (Sapotaceae), *Sindora glabra* (Fabaceae) and *Syzygium acuminatissimum* (Huang, 1986; Jiang, 1991). The *Vatica mangachapoi* formation is found in low montane gullies at Bawangling (Hainan). The canopy layer contains co-dominant species *Castanopsis hystrix* and *Hopea hainanensis.* The understory is dominated by *Blastus cochinchinensis, Prismatomeris tetrandra* (Rubiaceae) and *Psychotria asiatica.* A riverine rainforest characterized by *Castanopsis carlesii, Lithocarpus fenzelianus* and *Vatica mangachapoi* is also found on Hainan (Zang *et al.*, 2004). *Vatica* and associated species forests are tall and contain many other tree species, such as *Barringtonia fusicarpa, Chisocheton cumingianus* subsp. *balansae* (Meliaceae), *Lithocarpus fenestratus, Machilus gamblei, Myristica yunnanensis, Nephelium lappaceum* (Sapindaceae), *Dacrycarpus imbricatus* var. *patulus* (Podocarpaceae) and *Pygeum topengii* (Rosaceae; Zhu, 1993).

Hopea spp. forest is a seasonal rainforest, also widely distributed in tropical regions but limited to a small area in China. Two formations are recognized: *Hopea chinensis-Horsfieldia kingii* (Myristicaceae)-*Neolamarckia cadamba* (Rubiaceae) forest and *Hopea chinensis-Canarium pimela* forest. Both are found in montane gullies at 700–900 m on Shiqan Dashan, Guangxi. *Hopea chinensis* is not common, but is characteristic of these formations, of which the dominant species are not clear. Common species in *Hopea chinensis-Horsfieldia kingii-Neolamarckia cadamba* forest include *Dillenia turbinata, Drimycarpus racemosus* (Anacardiaceae), *Schima argentea* and *Xanthophyllum hainanense* (Polygalaceae; Guangxi Forestry Bureau, 1993). Species in *Hopea chinensis-Canarium pimela* forest include *Artocarpus styracifolius, A. tonkinensis, Canarium album, Dysoxylum hongkongense, Eberhardtia aurata* (Sapotaceae), *Elaeocarpus nitentifolius* and *Garcinia multiflora.*

Myristica spp. forest is confined to low mountains and hills (below 600 m) on the northeast of the Hengchun peninsula, Lanyu island, and Lu island of southeastern Taiwan. There is only one *Myristica* rainforest formation in China, comprising *Artocarpus* sp., *Myristica cagayanensis* and *Pterospermum niveum.* Additional constituent species are relatively few, but include *Aphanamixis polystachya, Diospyros philippensis, Neonauclea truncata* (Rubiaceae) and *N. sessilifolia* and (Huang, 1993).

Terminalia spp. formations are characterized by the presence of *Terminalia myriocarpa*, which can co-dominate with *Chukrasia tabularis, Dysoxylum excelsum* or *Pometia pinnata*, among others. *Terminalia myriocarpa-Pometia pinnata* forest occurs in valley bottoms or on piedmont slopes at an elevation of 650–700 m in Mengla, Xishuangbanna. The forest also contains *Baccaurea ramiflora, Celtis philippensis* var. *wightii, Chukrasia tabularis, Epiprinus siletianus* (Euphorbiaceae), *Gmelina arborea, Mitrephora wangii* (Annonaceae) and *Tetrameles nudiflora. Terminalia myriocarpa-Dysoxylum excelsum* seasonal rainforest is found at Mêdog in the eastern Himalaya Shan. The upper canopy height of this formation is 30--40 m; the lower canopy layer contains *Castanopsis* spp., *Ficus* spp., *Lithocarpus pachyphyllus, Syzygium formosum* and *Talauma hodgsonii* (Magnoliaceae), and reaches less than 20 m in height. The shrub layer is sparse and species-poor; the herb layer comprises species of Acanthaceae, Commelinaceae, Urticeae and Zingiberaceae. Additional Seasonal Rainforest formations may be formed by *Garcinia cowa, Knema tenuinerva, Pomatia pinnata* and *Tetrameles nudiflora.*

Montane Rainforest (see Figure 8.28) is effectively a sub-sub-type of Seasonal Rainforest located at elevations

Figure 8.28 Montane rainforest, Jianfenling, Hainan (image courtesy CHEN Bin).

above 700 m on mountains in the tropical zone, yet having all the characteristics of Rainforest. *Alstonia rostrata* (Apocynaceae) forest is distributed widely in the moist montane belt and on mid-elevation slopes on both sides of wet valleys in southern and southeastern Yunnan, centered on the Hong He area and Xishuangbanna. *Alstonia rostrata* dominates the upper canopy layer, which may also contain *Actinodaphne henryi* (Lauraceae), *Balakata baccata* (Euphorbiaceae), *Castanopsis hystrix, Michelia baillonii* and *Schima wallichii*. There are a greater number of species in the lower canopy layer, but all at low frequency.

Dysoxylum pallens-Semecarpus reticulatus (Anacardiaceae)-*Machilus nanmu* forest is found in some montane vallies around 1 300–1 500 m elevation in Xishuangbanna, and in the montane belt (elevation 1 200–1 500 m) on both sides of the lower reaches of the Lancang Jiang. There are different communities within this formation, including *Calophyllum polyanthum* (Clusiaceae)-*Semecarpus retiiculatus* forest, *Dysoxylum pallens-Schima wallichii* forest, and *Semecarpus reticulatus-Xanthophyllum flavescens* forest. Significant accompanying species include *Baccaura ramiflora, Castanopsis hystrix, Cryptocarya densiflora* and *Harpillia cupanioides*.

Madhuca pasquieri (Sapotaceae)-*Altingia yunnanensis* (Hamamelidaceae) forest occurs in the montane belt (elevation 800–1 300 m) of the southern slopes of Wenshan and the Hong He district, Yunnan. The main additional tree species are *Beilschmiedia robusta* (Lauraceae), *Castanopsis* spp., *Ficus* spp., *Lithocarpus* spp., *Semecarpus reticulatus* and *Terminalia myriocarpa*.

Dacrycarpus imbricatus var. *patulus* forests are montane rainforests composed of *Dacrycarpus imbricatus* var. *patulus* mixed with many broad-leaved trees. *Dacrycarpus imbricatus* var. *patulus-Castanopsis kawakamii-Eurya nitida-Ficus vasculosa-Neolitsea ellipsoidea* montane rainforest occurs at elevations of 700–1 000 m on Wuzhi Shan in central Hainan. The upper canopy layer comprises the aforementioned species; the lower two or three canopy layers include *Adinandra hainanensis, Schefflera heptaphylla, Syzygium chunianum* and *Tabernaemontana bufalina* (Apocynaceae). The *Dacrycarpus imbricatus* var. *patulus* montane rainforests on Jianfengling (Hainan) are very species-rich. Here, the first sub-layer of the canopy is composed of *Cinnamomum burmannii, Cyclobalanopsis blakei, Dacrycarpus imbricatus* var. *patulus* and *Elaeocarpus chinensis*. The lower two to three sub-layers comprise *Beilschmiedia longepetiolata, Cryptocarya metcalfiana, Cyclobalanopsis patelliformis, Neolitsea pulchella, Olea hainanensis* (Oleaceae) and *Platea latifolia* (Icacinaceae). The shrub and herb layers are also very species-rich. *Dacrycarpus imbricatus* var. *patulus* formations are also found on Yunkai Dashan and other mountains in western Guangdong.

Dacrydium pectinatum (Podocarpaceae) montane seasonal rainforests are found on mid-elevation mountains (above 1 000 m) on Hainan. For example, the formation comprising *Dacrydium pectinatum, Pentaphylax euryoides* (Pentaphylacaceae), *Symplocos wikstroemiifolia* and *Ternstroemia gymnanthera*, among others, occurs at around 1 360 m on Wuzhi Shan. *Dacrydium pectinatum* is also found at Jianfengling, with an upper canopy layer comprising *Dacrydium pectinatum, Parakmeria nitida* (Magnoliaceae), *Olea brachiata* and *Pentaphylax euryoides*, all in small amounts. The lower canopy layers include numerous *Machilus velutina, Olea brachiata, Pentaphylax euryoides, Symplocos* spp. and *Syzygium araiocladum*.

Aside from the Montane Rainforest formations described here, others of limited distribution occur, such as *Lirianthe henryi* (Magnoliaceae) forest and *Neolitsea phanerophlebia* forest (Guangdong Institute of Botany, 1976; Wu, 1980; Wu, 1987; Zhu, 2001; Zang *et al.*, 2004).

Seasonal Rainforest on Limestone in the tropical zone is dependent on the substrate: in limestone regions it is an important primary forest, found in large parts of Guangxi and Yunnan. *Excentrodendron tonkinense* (Tiliaceae)-*Cephalomappa sinensis* (Euphorbiaceae) forest is found mainly on moist piedmont slopes of limestone mountains below 700 m in southwestern Guangxi. The forest is predominantly evergreen with some semi-evergreen trees. Among the evergreens, *Excentrodendron tonkinense* is dominant; other species include *Cephalomappa sinensis, Cinnamomum pauciflorum, Drypetes congestiflora, D. perreticulata* and *Garcinia paucinervis*. The main species of the shrub layer are *Aglaonema modestum* (Araceae), *Ardisia virens, Aspidistra typica* and *Ophiopogon bodinieri* (both Liliaceae), *Dendrocnide urentissima* (Urticaceae) and *Pyrrosia lingua* (Polypodiaceae) plus some saplings. This forest is rich in lianas.

Parashorea chinensis-Horsfieldia kingii forest is also mainly distributed on limestone in southwestern Guangxi. Where the substrate comprises mostly exposed rock or shallow soil, the organic matter to support the forest may still be found in rock crevices. Common accompanying species of this formation include *Acrocarpus fraxinifolius* (Fabaceae), *Cephalomappa sinensis, Excentrodendron tonkinense, Garcinia paucinervis* and *Saraca dives*, all of which are characteristic of tropical forest on limestone in China. *Canarium bengalense* may also be present. The shrub and herb layers are relatively species-poor. *Horsfieldia kingii* can also form seasonal rainforests with *Cleistanthus petelotii, Cryptocarya densiflora, Deutzianthus tonkinensis* (Euphorbiaceae) or *Ficus glaberrima* (Su, 1988; Guangxi Forestry Bureau, 1993).

Pometia pinnata-Alphonsea monogyna (Annonaceae) forest is found in moist valleys and on lower slopes in Xishuangbanna. *Pometia pinnata* dominates the upper canopy, with a height of 35–40 m. The lower canopy sub-layers are dominated by *Alphonsea monogyna* and *Horsfieldia prainii*. The understory consists mainly of saplings, woody vines and some shrubs, with ferns being important in the herb layer. A second formation, *Pometia pinnata-Celtis philippensis* var. *wightii* forest is found in similar habitats under lower moisture conditions.

Celtis phillippensis var. *wightii-Lasiococca comberi* var. *pseudoverticillata* (Euphorbiaceae) forest is the most common formation on the lower slopes of limestone mountains in China. The upper canopy layer is dominated by *Celtis phillippensis* var. *wightii*, while *Lasiococca comberi* var. *pseudoverticillata* dominates a second sub-layer, and a third sub-layer comprises *Cleidion spiciflorum* and *Sumbaviopsis albicans* (both Euphorbiaceae). Additional tree species include *Chukrasia tabularis, Garuga floribunda* var. *gamblei* (Burseraceae) and *Tetrameles nudiflora*. The shrub and herb layers are similar in character to those of *Pometia pinnata-Alphonsea monogyna* seasonal rainforest (Zhu *et al.*, 1998).

8.3.2 Desert

Desert vegetation is usually sparse and dominated by super-xerophilous semi-shrubs, small semi-shrubs, shrubs or small trees. In general, the leaves of desert plants are reduced or specialized. Desert vegetation types are distributed widely in the dry regions of northwestern China: Gansu, Nei Mongol, Ningxia, Qinghai and Xinjiang. In addition, alpine deserts, dominanted by small, super-xerophyilous, low temperature-tolerant, cushion-forming semi-shrubs, are found in the far north of Xizang and on the Pamir plateau, Altun Shan and Kunlun Shan in Xinjiang. Desert vegetation has been divided into eight vegetation types according to the leaf specializations of the constructive species.

Reduced-leaved Small Tree Desert.—This desert type is dominated by *Haloxylon* spp. (Chenopodiaceae), which are leafless plants something between a small tree and a shrub, reaching up to 6 m in height. The vegetation coverage in these deserts is 30–50%. Only two formations are defined: *Haloxylon ammodendron* desert (Figure 8.29) and *H. persicum* desert. The former is distributed in areas including the Junggar Pendi, the eastern Tarim Pendi (Xinjiang), Qaidam Pendi (Qinghai), and the Alxa Gaoyuan (Nei Mongol). *Haloxylon ammodendron* grows vigorously on loamy soils, and also well on sandy soils on fixed or semi-fixed dunes, between dunes and at desert edges. It can also grow on salty soils. The accompanying species of this formation differ depending on the habitat, in general including *Agriophyllum lateriflorum* and *Horaninovia ulicina* (both Chenopodiaceae), *A. squarrosum, Calligonum* spp., *Haloxylon persicum* and *Psammochloa villosa* (Poaceae) in sandy desert, but salt-tolerant species such as *Lycium* spp., *Nitraria* spp. (Nitrariaceae) and *Tamarix* spp. on more salty soils.

Figure 8.29 *Haloxylon ammodendron* desert, Alashan (image courtesy GUO Ke).

The *Haloxylon persicum* formation occurs in the western part of the central Asian desert. In China, it is found in the western Junggar Pendi of northern Xinjiang. The vegetation coverage of this formation is only 20–30%. Accompanying species include *Calligonum* spp., *Haloxylum ammodendron, Seriphidium terrae-albae* (Asteraceae) and *Stipagrostis pennata* (Poaceae), and the species composition is generally similar to the *H. ammodendron* formation.

Evergreen Sclerophyllous Shrub Desert.—This desert type is divided into two formations, with the dominant species in both being *Ammopiptanthus mongolicus* (Fabaceae), a super-xeromorphic, evergreen, sclerophyllous shrub specific to the deserts of central Asia. The first formation in China is found in the eastern Alxa Shamo and the western Ordos steppe desert, and on piedmont slopes and shallow valleys on Helan Shan in Gansu. The vegetation is usually only 30–90 cm high, with occasional individual shrubs up to 1.6 m, and the vegetation coverage is only 13%. The main accompanying species in the *Ammopiptanthus mongolicus* I formation are *Ajania achilleoides* (Asteraceae), *Amygdalus mongolica* (Rosaceae), *Caragana roborovskyi, Oxytropis aciphylla* (Fabaceae), *Salsola laricifolia* (Chenopodiaceae) and *Zygophyllum xanthoxylon* (Zygophyllaceae). There is a well-developed herb layer beneath the shrubs, mainly comprising grasses and some forbs, such as *Allium mongolicum* (Liliaceae), *Convolvulus tragacanthoides* (Convolvulaceae), *Stipa breviflora, S. bungeana, S. caucasica* subsp. *glareosa, S. tianschanica* var. *gobica* and *Ptilagrostis pelliotii* (all Poaceae). The second formation, *Ammopiptanthus mongolicus* II, is found on the western part of the southern side of the Tian Shan, as well as the region linking the Tian Shan and Kunlun Shan, at elevations of 1 800–2 500 m. This formation is not so species-rich, and includes species such as *Anabasis aphylla* (Chenopodiaceae), *Convolvulus tragacanthoides, Gymnocarpos przewalskii* (Caryophyllaceae) and *Kaschgaria komarovii* (Asteraceae). *Ephedra* (Ephedraceae) species are found within this formation on stony slopes, riversides and ditches (Zhang *et al.*, 2006).

Reduced-leaved Shrub Desert.—This desert type is extremely drought-tolerant, due to the reduction in leaf area of the component shrubs. It is distributed widely in desert regions of Gansu, Nei Mongol, Qinghai and

Xinjiang. Two groups of formations are recognized, characterized by *Ephedra* and *Calligonum* respectively. The main *Ephedra* formations are the *Ephedra przewalskii* and *E. rhytidosperma* formations, while five *Calligonum* formations are defined.

Ephedra przewalskii desert is the main formation of Reduced-leaved Shrub Desert in China, and is distributed widely on the Alxa Gaoyuan, Qaidam Pendi, the edge of Tarim Pendi, the Junggar Pendi and elsewhere. *Ephedra przewalskii* is a switch plant with chlorophyllous branches, reaching 50–100 cm in height. The coverage of vegetation in this formation depends on the local moisture regime, ranging from 1% to 15%. Accompanying species are few, and differ depending on the habitat, with common species including *Agriophyllum squarrosum, Bassia dasyphylla* (Chenopodiaceae), *Calligonum mongolicum, Clematis fruticosa, Convolvulus fruticosus, Gymnocarpos przewalskii, Iljinia regelii* (Chenopodiaceae), *Nitraria sphaerocarpa, Reaumuria soongarica* (Tamaricaceae), *Sympegma regelii* (Chenopodiaceae) and *Zygophyllum xanthoxylon*.

Calligonum mongolicum desert is the most widespread of the *Calligonum* formations, characterized by sandy habitats. *Calligonum mongolicum* desert is found mainly in the eastern Junggar Pendi, northeastern Tarim Pendi, Qaidam Pendi, Alxa and elsewhere in Nei Mongol. The height of this vegetation is 50–150 cm, and its coverage 3–15%. Accompanying species again are few and dependent on habitat, including *Atraphaxis frutescens* (Polygonaceae), *Ephedra przewalskii, Nitraria sphaerocarpa, Reaumuria soongarica* and some annuals such as *Bassia dasyphylla* and *Salsola tragus*. Of the other *Calligonum* formations, for example, *C. alashanicum* desert is found only on mobile sand dunes in western Alxa and Ordos, while the *C. leucocladium* and formation is seen mainly on the sandlands of the northern part and eastern edge of the Gurbantünggüt Shamo in the Junggar Pendi.

Succulent Shrub Desert.—This is perhaps the most important type of desert vegetation. Succulent species have well-developed water storage tissues in their leaves, giving them great drought-resistance. The main species of these deserts are *Nitraria* spp., with others including *Gymnocarpos przewalskii, Tetraena mongolica* (Zygophyllaceae) and *Zygophyllum xanthoxylon. Nitraria sphaerocarpa* desert (Figure 8.30) is found in the Tarim Pendi, Qaidam Pendi, Ejin Qi and Alxa. Vegetation coverage in this species-poor formation is 3–12%. In Xinjiang, species may include *Ephedra przewalskii, Gymnocarpos przewalskii, Reaumuria soongarica* and *Sympegma regelii*; other accompanying species, such as *Calligonum mongolicum, Convolvulus gortschakovii* and *Salsola arbuscula*, occur in other regions, and *Asterothamnus centraliasiaticus* (Asteraceae) is present in Nei Mongol.

Nitraria tangutorum desert (Figure 8.31) is widely distributed in the Junggar Pendi, Tarim Pendi and Qaidam Pendi, Hexi corrior (Gansu), and western Ordos, Alxa and northern Ulanqab, Nei Mongol. This formation is found on fixed and semi-fixed sand dunes and sandlands. *Nitraria tangutorum* builds up into large, dense clumps, limiting the number of other plants in the formation. Those species that are present include *Artemisia scoparia, Atriplex centralasiatica* (Chenopodiaceae), *Atriplex sibirica, Bassia dasyphylla, Salsola collina*, and annual grasses and forbs. Some xerophilous shrubs and semi-shrubs, such as *Ammopiptanthus mongolicus, Krascheninnikovia ceratoides* (Chenopodiaceae), *Reaumuria soongarica* and *Zygophyllum xanthoxylon*, as well as some herbs, are also present on the sandy diluvial fans of piedmont slopes and valley terraces. *Gymnocarpos przewalskii* desert is found in desert regions of Xinjiang, northern Gansu and Nei Mongol, a widespread but patchy distribution. The height of the vegetation is 30-- 50 cm, coverage is low, and the accompanying species differ between habitats.

The *Zygophyllum xanthoxylon* formation is specialized to the steppe-like desert regions of Xinjiang and Nei Mongol, forming a transition between desert and steppe. The height of the vegetation is 50–120 cm, and its coverage ca. 20%. *Zygophyllum xanthoxylon* may form mixed

Figure 8.30 *Nitraria sphaerocarpa* desert, Alashan (image courtesy GUO Ke).

Figure 8.31 *Nitraria tangutorum* desert, Nei Mongol (image courtesy BAI Yong-Fei).

communities with *Ammopiptanthus mongolicus, Nitraria* spp., *Potaninia mongolica* (Rosaceae) and *Reaumuria soongarica*. *Tetraena mongolica* desert is a formation specific to Ordos in Nei Mongol. It is found on sand, gravel and stony substrates in steppe-like desert areas. The vegetation is 20–40 cm high with a coverage of ca. 15%. The main accompanying species are *Caragana brachypoda, Krascheninnikovia ceratoides, Potaninia mongolica, Reaumuria soongarica, Zygophyllum xanthoxylon,* and sometimes species of *Stipa*.

Xerophytic Shrub Desert.—This vegetation type is characterized by shrubby species with rather small leaves having a very thick cutinized layer and, often, dense hairs covering the epidermis. Four formations are recognized. *Potaninia mongolica* desert (Figure 8.32) is a formation specific to eastern Alxa and western Ordos. Its distribution is wide, comprising flat habitats covered with sand. The height of this vegetation is only 8–15 cm, and its coverage 10–13%. Accompanying species include *Brachanthemum fruticulosum* (Asteraceae), *Krascheninnikovia ceratoides, Reaumuria trigyna, Zygophyllum xanthoxylon* and some species of *Allium, Cleistogenes* (Poaceae), *Salsola* and *Stipa*.

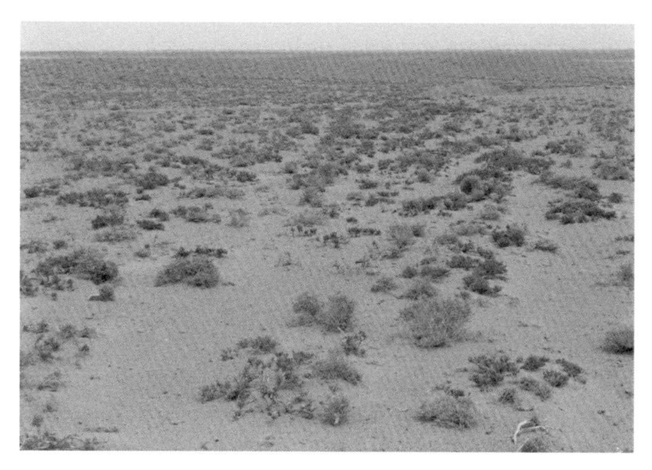

Figure 8.32 *Potaninia mongolica* desert, Alashan (image courtesy GUO Ke).

Helianthemum songaricum (Cistaceae) desert is localized to western Ordos, where it is found on the gravelly piedmont slopes of Arbas Shan. Other dominant species include *Convolvulus tragacanthoides* and *Ptilagrostis mongholica*. The main accompanying species are *Ajania achilleoides, Dracocephalum fruticulosum* (Lamiaceae), *Oxytropis aciphylla, Scorzonera capito* and *Tugarinovia mongolica* (both Asteraceae), *Stipa caucasica* subsp. *glareosa, Tetraena mongolica* and *Zygophyllum xanthoxylon*.

Caragana tibetica desert is formed of cushion-like, xerophytic, dwarf shrubs which are green in summer, and is found in the transition region between desert-steppe and true desert, located mainly in Nei Mongol and in fragments in northern Gansu. This is an area of high plateaus, low mountains and diluvial plains at an elevation of 1 100–1 350 m. The coverage of this desert vegetation is 15–25%. Accompanying species are mainly shrubs and semi-shrubs such as *Ammopiptanthus mongolicus, Caragana korshinskii, Caragana microphylla, Caragana stenophylla, Krascheninnikovia ceratoides, Oxytropis aciphylla* and *Zygophyllum xanthoxylon*. The dominant species of the herb layer are *Cleistogenes songorica, Pennisetum flaccidum* (Poaceae), *Stipa breviflora* and *S. caucasica* subsp. *glareosa*, as well as some forbs.

Amygdalus mongolica desert is specific to eastern Alxa and western Ordos, and can be seen in the low montane belt on Helan Shan, Lang Shan, and Arbas Shan. At low elevations on Helan Shan, the height of mature vegetation may be 20--80 cm, and coverage ca. 25%. The main accompanying species are *Artemisia dracunculus, A. gmelinii* and *Salsola laricifolia*. The herb layer consists mainly of *Stipa bungeana* and *S. caucasica* subsp. *glareosa*.

Succulent Semi-shrub Desert.—This is an important desert vegetation type, which has been divided into two sub-types, Succulent Semi-shrub Desert and Salt-tolerant Succulent Semi-shrub (Small Semi-shrub) Desert. The former includes no less than 14 formations: three *Reamuria* formations, five *Salsola* formations, three *Anabasis* formations, *Sympegma regelii* desert, *Iljinia regelii* desert, and *Nanophyton erinaceum* (Chenopodiaceae) desert. The latter includes *Atriplex cana, Camphorosma monspeliaca* subsp. *lessingii* and *Halocnemum strobilaceum, Halostachys caspica* and *Suaeda physophora* (all Chenopodiaceae) formations, and three *Kalidium* (also Chenopodiaceae) formations.

We use *Reaumuria soongarica* desert to represent the three *Reamuria* formations of Succulent Semi-shrub Desert, because it is the most widespread, extending from the deserts of Xinjiang, Qinghai and Nei Mongol to the steppe region to the east. The community composition differs considerably between regions. Taking a typical community in the desert region of Alxa, the height of the vegetation is 10–30 cm, and its coverage 5–10%. This vegetation is relatively species-poor, with sometimes a few *Calligonum mongolicum, Ephedra przewalskii, Krascheninnikovia ceratoides, Nitraria sphaerocarpa, Salsola passerina, Sympegma regelii, Zygophyllum xanthoxylon* or annual plants present. The other two *Reaumuria* formations are the patchily-distributed *Reaumuria kaschgarica* and *R. trigyna* deserts.

Salsolsa passerina occurs zonally in the eastern part of the desert region of central Asia. In China its distribution is similar to *Reaumuria soongarica* desert, but also occurs in northern Ningxia and Gansu. The distribution is concentrated on the steppe-like desert regions, in valleys between dunes, on diluvial plains and gentle slopes, because *Salsolsa passerina* is best adapted to loamy soil. The height of the community is just 8–10 cm, and its coverage 7–16%. The main accompanying species are herbs: *Allium polyrhizum, Bassia dasyphylla, Cleistogenes songorica, Neopallasia pectinata* (Asteraceae) and *Stipa caucasica* subsp. *glareosa*. *Salsolsa passerina* also forms communities with shrubs and small semi-shrubs such as *Anabasis brevifolia, Kalidium* sp., *Nitraria sphaerocarpa, Potaninia mongolica, Reaumuria soongarica* and *Sympegma regelii.*

Salsola abrotanoides desert is the main montane desert formation of central Asia, forming in habitats similar to *Salsola passerina* desert. *Salsola laricifolia* desert is distributed widely on low and isolated mountains and stony slopes of desert regions in China, often in steppe-like desert regions which have more moist conditions. *Salsola orientalis* desert has a patchy distribution on the piedmont slopes of the diluvial fan of the Junggar Pendi. *Salsola junatovii* desert is found on the southern slopes of the Tian Shan at mid elevations (ca. 2 000 m) between the Ili Pendi and Kashi (Xinjiang).

Anabasis brevifolia desert has a zonal distribution covering a large part of the desert region of Xinjiang, Gansu and Nei Mongol. *Anabasis brevifolia* is a small, succulent, xerophytic semi-shrub, reaching heights of 5–18 cm. It is found on stony substrates on isolated low mountains. In general, accompanying species are rare and found only in habitats with better moisture conditions; they include *Asterothamnus centraliasiaticus, Cancrinia discoidea* (Asteraceae), *Scorzonera divaricata* and *Sympegma regelii.* The coverage of this community is only 2–3%. In habitats with more moist conditions, *Allium mongolicum, Artemisia scoparia, Cleistogenes songorica* and *Stipa caucasica* subsp. *glareosa* can also be found.

Anabasis aphylla desert is distributed in ancient lake basins in the southwest of the Junggar Pendi, on the diluvial fan of low piedmont slopes on the southern side of the Tian Shan, and the northern slopes of the Kunlun Shan. *Anabasis aphylla* forms a monospecific community on the clay takir soils of the Junggar Pendi. Under moister conditions, annual plants may also be found. On the diluvial fan of low piedmont slopes, super-xerophytic shrubs and semi-shrubs, such as *Convolvulus fruticosus, Gymnocarpos przewalskii* and *Iljinia regelii*, are important. The height of this community is 15–20 cm, and its coverage 1–5%.

Anabasis salsa desert is widely distributed on high river terraces, plateaus and high plains in the northern intermontane part of the Junggar Pendi. There are also some communities on the diluvial fan of the northern piedmont slopes of the Tian Shan, Xinjiang. *Anabasis salsa* grows well on salinized or sandy gravels or loamy soil. The height of the vegetation in this formation is 5–20 cm, and its coverage 10–20%, although the coverage and constituent species differ among habitats. Accompanying species may include *Calligonum mongolicum, Ephedra przewalskii, Haloxylum ammodendron, Salsola* spp. or *Seriphidium nitrosum var. gobicum.*

Sympegma regelii is a strongly xerophytic, succulent-leaved semi-shrub, forming a desert community on low mountains, piedmont slopes and alluvial plains in the mountains of Xinjiang, Gansu, Nei Mongol and Ningxia (Figure 8.33). This formation is well adapted to stony habitats, is 20–50 cm high, and covers 3–15% of the land. The main accompanying species are *Ajania achilleoides, Allium polyrhizum, Amygdalus mongolica, Anabasis breviflolia, Asterothamnus centraliasiaticus, Cancrinia discoidea, Gymnocarpos przewalskii, Ptilagrostis pelliotii, P. mongholica, Reaumuria soongarica, Salsola laricifolia, Salsola passerina, Zygophyllum pterocarpum* and *Z. xanthoxylon.*

Figure 8.33 *Sympegma regelii* desert (image courtesy BAI Yong-Fei).

Nanophyton erinaceum desert is mainly distributed on mountains and basin-edges in Xinjiang, and also appears patchily on diluvial fans and old river terraces at higher elevations. *Nanophyton erinaceum* is a small shrub, 5–10 cm high, and this community covers 10–30% of the land surface. Accompanying species include *Allium polyrhizum, Anabasis salsa, Goldbachia laevigata* (Brassicaceae), *Halogeton glomeratus* and *Kochia prostrata* (Chenopodiaceae), *Krascheninnikovia ceratoides, Lappula spinocarpos* (Boraginaceae), *Plantago minuta* (Plantaginaceae), *Salsola brachiata, Seriphidium gracilescens* and *Stipa caucasica* subsp. *glareosa.*

Iljinia regelii desert is found on the mountain diluvial fans, valley bottoms and high plateaus of the extreme desert region of eastern Xinjiang, northwestern Gansu and western Nei Mongol. There are few species, and the height of the vegetation is 20–50 cm, with community coverage 7–10%. Occasional additional species include *Anabasis aphylla, Calligonum calliphysa, Ephedra przewalskii, Haloxylon ammodendron, Limonium gmelinii* (Plumbaginaceae) and *Sympegma regellii.*

Salt-tolerant Succulent Semi-shrub (small semi-shrub) Desert is widely distributed on the highly saline soils of floodplains, lakesides, river banks and depressions in desert regions. *Halostachys caspica* desert is found in Nei Mongol, Xinjiang and northern Gansu, generally on the saline soils of dry valleys and depressions on lakesides in desert regions. The height of *H. caspica* is 80–100 cm, but some species of the formation, such as *Tamarix elongata, T. hispida* and *T. ramosissima*, can reach 1–2 m. Other accompanying salt-tolerant species include *Alhagi sparsifolia* and *Glycyrrhiza inflata* (both Fabaceae), *Kalidium foliatum*

(Chenopodiaceae), *Karelinia caspia* (Asteraceae), *Lycium ruthenicum, Nitraria sibirica* and *Phragmites australis.*

Halocnemum strobilaceum desert is found in saline depressions of the Xinjiang desert region, especially where these are distributed across a large area on the plain before the southern piedment slopes of the Tian Shan, and on the Luobupo Pingyuan. *Halocnemum strobilaceum* is a small, succulent, extremely salt-tolerant semi-shrub, forming monospecific communities on moist salty soils. The height of this community is 30–40 cm, and its coverage is quite dense, at 50% or more. Accompanying plants (where these are present) include *Phragmites australis, Tamarix hispida* and *T. ramosissima*; however these grow only weakly due to the high salinity of the soil. In drier habitats, *Halostachys caspica* and *Karelinia caspia*, among others, may be found.

Kalidium foliatum desert is distributed widely on the moist, loose, salty soils characteristic of desert regions of Nei Mongol, Xinjiang, Qinghai and Gansu. The height of *K. foliatum* is 30–40 cm, and the coverage of this community is ca. 30%. Accompanying species include *Aeluropus* spp. (Poaceae), *Alhagi sparsifolia, Nitraria sibirica, Phragmitis communis, Salsola korshinskyi, Suaeda microphylla* and *Tamarix* spp. The *Kalidium cuspidatum* and *K. gracile* formations are similar to *K. foliatum* desert in terms of habitat and community characteristics. In contrast, *K. schrenkianum* desert (Figure 8.34) is found on the saline, gravelly soils of diluvial fans and sloping deposits of low- to mid-elevation mountains in Xinjiang. The accompanying species are also different here, including *Anabasis aphylla, Halogeton glomeratus, Iljinia regelii, Reaumuria soongarica, Salsola junatovii* and *Sympegma regelii.*

Figure 8.34 *Kalidium schrenkianum* desert (image courtesy GUO Ke).

Atriplex cana desert is always found on alkaline soils in steppe-like deserts and desert-steppe regions. It is distributed mainly in depressions and around salty pools of the Ertix He and Ulungur He, and the southern Junggar Pendi, Xinjiang. The height of *A. cana* in these dry habitats is 25–30 cm, and the vegetation coverage ca. 30%. *Atriplex cana* is highly dominant, but a few *Anabasis salsa* and *Lappula spinocarpos* can be found growing around the salty pools. In addition, *A. cana* can form mixed communities with a variety of herbs, among them *Achnatherum splendens, Apocynum pictum* (Apocynaceae) and *Phragmitis australis*; some shrubs may also be found here. *Suaeda physophora* deserts are found on the saline soils at the edge of the lowland fan of the northern piedmont slopes of the Tian Shan and around salty pools in the lower reaches of the Ulungur He. *Suaeda physophora* may form formations with either *Anabasis salsa* or *Reaumuria soongarica.* Other species may be present in smaller numbers, including *Halocnemum strobilaceum, Kalidium* spp., *Limonium* spp. and *Nitraria sibirica. Camphorosma monspeliaca* subsp. *lessingii* desert is found on slightly less-saline soils around salty pools of the lower reaches the Ulungur He, on the southern piedmont slopes of the Altai and the flood plain of the Manas He, Xinjiang. This formation includes species of salt-tolerant grasses, forbs and shrubs. The communities of *C. monspeliaca* subsp. *lessingi, Atriplex cana* and *Suaeda physophora* deserts intermix in a complex fashion in these areas.

Xerophytic Semi-shrub Desert.—This vegetation type is distributed widely in desert regions of China, in general on sand or sandy gravel. The most widespread formation is that of *Krascheninnikovia ceratoides*; the other three formations are mainly localized to Xinjiang. *Krascheninnikovia ceratoides* desert occurs mainly in steppe-like desert regions of Nei Mongol and Xinjiang, including such habitats as sand, sandy gravel, sandy high plateaus, isolated low mountains and piedmont slopes. The height of the *Krascheninnikovia ceratoides* formation is 30--100 cm depending on the habitat. The level of coverage and nature of accompanying plants differ greatly from site to site; accompanying herbs in Nei Mongol include *Allium mongolicum, Carex ulobasis, Cleistogenes songorica, Stipa caucasica* subsp. *glareosa* and *S. tianschanica* var. *gobica.* Small semi-shrubs such as *Ajania achilleoides* and *Artemisia frigida* are also present, alongside some forbs and annual plants.

Seriphidium gracilescens is widely distributed on southern piedmont slopes of the Altai, the Tacheng valley, and small parts of the Bole valley. In general, it is found on piedmont diluvial fans and the plains between mountains, at elevations of 600–800 m. The height of *Seriphidium gracilescens* is 10–25 cm, and the community coverage 20–30%. Accompanying species tend to be super-xerophytic semi-shrubs such as *Anabasis* spp., *Carex* spp., *Kochia prostrata, Nanophyton erinaceum, Salsola arbuscula, S. laricifolia* and *Stipa* spp.; however the species composition differs among habitats.

Seriphidium borotalense desert is widely distributed in the Bole Valley (Xinjiang) and eastwards along the northern piedmont slopes of the Tian Shan to Dashitou. In general, it is found on piedmont diluvial fans at elevations of 600–1 000 m, where the soil is loamy and not saline. *Seriphidium borotalensis* also forms mono-dominant communities on loams and sandy loams on the northern piedmont slopes of the Tian Shan. Additional species present include *Anabasis salsa, Kochia prostrata, Reaumuria soongarica, Salsola brachiata, Salsola subcrassa, Salsola tragus* and *Suaeda pterantha*, alongside some perennials and ephemeral annuals. *Seriphidium terrae-albae* desert is similar in ecological characteristics to *Seriphidium borotalensis* desert. The main distribution area of this formation is the diluvial fans of the northern piedmont slopes of the Altai, the northern mountains of the Junggar Pendi (Xinjiang), and the edge of the desert belt.

Dwarf Cushion Semi-shrub Desert.—This specialized desert type, dominated by dwarf, cushion-forming semi-shrubs, is adapted to extremely cold continental xerophytic habitats. It occurs mainly in the alpine belt and upper slopes of mountains on the northern and northwestern Qinghai-Xizang Plateau, usually at elevations of 4 200–5 200 m, and never lower than 2 500 m. The various formations of dwarf cushion semi-shrub desert are dominated by *Ajania tibetica, Ephedra intermedia, Krascheninnikovia compacta* or *Seriphidium rhodanthum*, with *Rhodiola tangutica* (Crassulaceae) desert covering only a limited area. The vegetation coverage of these communities is sparse, around 8–20%. Common accompanying species include *Acantholimon hedinii* (Plumbaginaceae), *Carex moorcroftii, Hedinia tibetica* (Brassicaceae), *Thylacospermum caespitosum* (Caryphyllaceae), cold-tolerant *Stipa* spp. and some cushion-forming taxa such as *Astragalus, Oxytropis* and *Saussurea* (Xinjiang Comprehansive Expedition Institute of Botany Academia Sinica, 1978; Inner Mongolia and Ningxia Comprehensive Expedition Academia Sinica, 1985; Zhou *et al.*, 1987; Ningxia Agriculture Exploring Designing Institute, 1988; Qinghai-Tibet Plateau Comprehensive Expedition, 1988).

8.3.3 Steppe

The Steppe vegetation group is characterized by a dominance of perennial xerophilous plants. In China, Steppe communities consist of cold-temperate and mid-temperate herbs, particularly *Stipa*. The main areas of distribution of Steppe are semi-arid areas such as northwestern Nei Mongol, the Huangtu Gaoyuan (Loess Plateau) in the temperate zone, and the Qinghai-Xizang Plateau, a vast area of steppe which is an important constituent of the European-Asian steppe region. Temperate Steppe has been divided into four vegetation types according to plant life form: Clump Grass Steppe, Rhizomatic Grass Steppe, Forb Steppe and Semi-shrub and Small Semi-shrub Steppe.

Clump Grass Steppe.—The main distribution area of Clump Grass Steppe is the semi-arid part of the temperate zone, but it can stretch to both semi-humid and arid areas. Clump Grass Steppe is dominated by perennial, xeromorphic, clump-forming, cold-tolerant hemicryptophytes. It has been divided into four sub-types, according to differences in temperature and humidity regime: Meadow Clump Grass Steppe, Typical Clump Grass Steppe, Desert Clump Grass Steppe and Alpine Clump Grass Steppe.

Meadow Clump Grass Steppe is distributed mainly in the eastern, southern and western parts of the steppe region of the temperate zone, and on mountains in desert regions. We describe three representative formations: *Stipa baicalensis* steppe, *S. kirghisorum* steppe and *Bothriochloa ischaemum* (Poaceae) steppe. *Stipa baicalensis* steppe (Figure 8.35) is a primary meadow-steppe formation specific to the eastern part of the Asian steppe region, distributed widely on hills and plateaus of semi-humid and cool areas west of the Da Hinggan Ling. This is a relatively rich community, with more than 30 species per m^2, mostly in Asteraceae, Fabaceae, Poaceae and Rosaceae. The average height of the meadow-steppe vegetation is 50 cm, and vegetation coverage is 50–70%. Additional co-dominant species include *Leymus chinensis* (Poaceae) and *Hedysarum gmelinii* (Fabaceae), and other common species include *Artemisia rutifolia, Aster alpinus, Carex lanceolata, Helictotrichon hookeri* subsp. *schellianum* (Poaceae), *Hemerocallis minor* (Liliaceae) and *Ligularia mongolica.* Table 8.3 gives an example of the other main components of *Stipa baicalensis* meadow-steppe (Li, 1988).

Figure 8.35 *Stipa baicalensis* meadow-steppe (with *Sanguisorba officinalis*, among others), Nei Mongol (image courtesy BAI Yong-Fei).

Stipa kirghisorum is the dominant species of montane meadow-steppe in desert regions of Central Asia. In China this steppe formation is distributed in scattered patches on the northern slopes of the Tian Shan and the Yili He valley at elevations of 1 600–1 800 m on the western side and 1 800–2 000 m on the eastern. The height of this kind of meadow-steppe is 30–50 cm, and the vegetation coverage more than 50%. It contains mesophytic and xero-mesophytic forbs such as *Fragaria vesca* (Rosaceae), *Galium verum* (Rubiaceae), *Iris ruthenica*, and species of *Aster, Geranium* (Geraniaceae), *Medicago* (Fabaceae), *Phlomis* and *Thalictrum. Stipa kirghisorum* may also form communities with other species, such as *Bromus inermis* (Poaceae), *Festuca ovina* and forbs.

Table 8.3 Main species of *Stipa baicalensis* meadow-steppe, in order of abundance from most to least abundant (Li, 1988).

Stipa baicalensis
Hedysarum gmelinii
Leymus chinensis
Hemerocallis minor
Carex lanceolata
Helictotrichon hookeri subsp. *schellianum*
Artemisia rutifolia
Sanguisorba officinalis
Potentilla longifolia
Ligularia mongolica
Aster alpinus
Medicago ruthenica
Trifolium lupinaster (Fabaceae)
Anthoxanthum glabrum
Klasea centauroides subsp. *centauroides*
Artemisia eriopoda
Iris ensata
Bupleurum scorzonerifolium
Phlomis tuberosa
Nepeta multifida
Carex korshinskii
Stellera chamaejasme
Leontopodium leontopodioides
Galium verum
Koeleria macrantha
Artemisia gmelinii
Potentilla tanacetifolia
Potentilla bifurca
Astragalus melilotoides
Potentilla verticillaris
Tephroseris kirilowii
Iris dichotoma
Pulsatilla ambigua
Scutellaria scordifolia
Oxytropis myriophylla
Scutellaria baicalensis
Thalictrum squarrosum
Saposhnikovia divaricata (Apiaceae)
Allium tenuissimum
Festuca ovina
Phedimus aizoon
Thermopsis lanceolata
Euphorbia esula
Aster altaicus var. *altaicus*
Cleistogenes squarrosa
Linum nutans (Linaceae)
Potentilla acaulis
Vicia cracca
Lathyrus quinquenervius
Saussurea japonica
Erodium stephanianum (Geraniaceae)
Thesium chinense (Santalaceae)
Dianthus chinensis
Adenophora capillaris subsp.*paniculata* (Campanulaceae)
Astragalus laxmannii
Polygonum divaricatum
Agropyron cristatum
Thalictrum petaloideum var. *supradecompositum*
Pedicularis striata
Vicia amoena
Gentianopsis barbata (Gentianaceae)
Agropyron mongolicum var. *villosum*
Filifolium sibiricum
Allium senescens
Thlaspi cochleariforme (Brassicaceae)
Cymbaria daurica
Olgaea leucophylla (Asteraceae)
Androsace longifolia
Orostachys malacophylla (Crassulaceae)
Polygala tenuifolia (Polygalaceae)
Draba nemorosa (Brassicaceae)
Poa attenuata
Bromus inermis
Scabiosa comosa (Dipsacaceae)
Caragana microphylla
Allium anisopodium
Gueldenstaedtia verna
Gentiana squarrosa
Haplophyllum dauricum (Rutaceae)
Saussurea runcinata
Artemisia halodendron

Bothriochloa ischaemum is a perennial clumping grass with short rhizomes; *B. ischaemum* meadow-steppe is a secondary vegetation formation developing after the destruction of warm-temperate forest on the mountains of northern China and the Huangtu Gaoyuan, and by some authorities is classified as a grassland vegetation formation. *Bothriochloa ischaemum* may form a variety of communities, with *Artemisia giraldii, A. gmelinii, Lespedeza davurica, Stipa bungeana* and *Themeda triandra*. A further formation, *Cleistogenes mucronata* steppe, occurs on the Huangtu Gaoyuan and is very similar in species composition to *Bothriochloa ischasmum* steppe.

The dominant species of Typical Clump Grass Steppe are *Stipa* spp., and this vegetation sub-type is the main zonal vegetation of the temperate steppe region. *Stipa grandis* steppe is the major formation found in the central Nei Mongol Gaoyuan, and *S. sareptana* var. *krylovii* steppe takes its place towards the west. *Stipa bungeana* steppe is found on the Huangtu Gaoyuan. *Stipa grandis* is the most characteristic large, dense, xerophytic clump grass. The height of its reproductive branches is 80–100 cm, making it the tallest *Stipa* species in China. *Stipa grandis* steppe (Figure 8.36) is unique to the Nei Mongol Gaoyuan and adjacent regions, with an average vegetation height of 25–50 cm and coverage 25–45%. Over 100 species are found in this formation: *Festuca dahurica* and *Leymus chinensis* are co-dominant species, or sub-dominant to *S. grandis*. Companion species include *Anemarrhena asphodeloides* (Liliaceae), *Artemisia eriopoda, Artemisia frigida, Carex korshinskii, Hedysarum gmelinii, Koeleria macrantha* (Poaceae), *Pulsatilla ambigua* (Ranunculaceae) and *Thalictrum squarrosum* (Li, 1988).

Figure 8.36 Typical *Stipa grandis* steppe, Nei Mongol (image courtesy BAI Yong-Fei).

Stipa sareptana var. *krylovii* steppe is found in drier habitats than *S. grandis* steppe. There are more than 30 major species in the *S. sareptana* var. *krylovii* formation, including abundant *Artemisia frigida, Carex duriuscula* and *Cleistogenes squarrosa*, as well as *Caragana microphylla, Cymbaria daurica* (Scrophulariaceae) and *Potentilla tanacetifolia*. *Stipa bungeana* steppe has a very wide distribution, concentrated on the Huangtu Gaoyuan in the warm-temperate zone—China's main agricultural region on the reaches of the Huang He. In its natural form this steppe is seldom preserved due to long-term agricultural cultivation. The plant species of this formation are diverse and mostly drought-tolerant, often including *Artemisia* spp., *Caragana rosea, Lespedeza davurica* and *Spiraea trilobata.*

Stipa capillata steppe is found mainly on Altay Shan, the northern slopes of Tian Shan, and the mountains of the western Junggar Pendi, Xinjiang. This is the typical montane steppe of elevations between 1 300 and 2 300 m. The height of this steppe vegetation is ca. 20 cm, and its coverage ca. 40%. Grasses, including *Cleistogenes squarrosa, Festuca valesiaca* subsp. *sulcata* and *Stipa sareptana,* are relatively abundant. The main forbs are *Artemisia frigida, Carex duriuscula, Ceratocarpus arenarius* (Chenopodiaceae), *Potentilla acaulis, P. bifurca, Kochia prostrata* and *Saussurea salicifolia.*

Festuca ovina steppe is a typical montane steppe of elevations between 1 000 and 1 500 m in eastern Nei Mongol and on both the eastern and western piedmont slopes of the Da Hinggan Ling. Co-dominant species may include *Chamaerhodos trifida* (Rosaceae), *Filifolium sibiricum* (Asteraceae), *Helictotrichon hookeri* subsp. *schellianum, Koeleria macrantha, Leymus chinensis* and *Stipa baicalensis. Festuca valesiaca* subsp. *sulcata* steppe is the most widespread steppe formation in Xinjiang. This can sometimes be a monospecific formation, but often *F. valesiaca* subsp. *sulcata* is combined with other species such as *Artemisia* spp. and *Koeleria macrantha.* The coverage of the vegetation is 50–60%, and its height 8–10 cm for the leaf layer, 30–35 cm for the reproductive branches. Additional members of this formation include species of *Astragalus, Gentiana* (Gentianaceae), *Leontopodium, Pedicularis* (Scrophulariaceae), *Scorzonera* and *Scutellaria* (Lamiaceae). *Festuca valesiaca* subsp. *sulcata* may also form steppe communities with co-dominant species including *Artemisia frigida, Caragana frutex, C. leucophloea, Potentilla acaulis, Spiraea hypericifolia, Stipa capillata, Stipa kirghisorum* and *Stipa orientalis.*

Cleistogenes squarrosa is found as an accompanying species in many of the above formations, but may also be dominant following a loss of primary *Stipa* steppe caused by over-grazing. *Cleistogenes squarrosa* steppe is found widely on the sandy soils typical of steppe, accompanied by species such as *Agropyron cristatum* (Poaceae), *Artemisia* spp., *Lespedeza* spp. and *Stipa* spp., which may also become abundant or co-dominant.

Agropyron cristatum steppe has a relatively small, scattered distribution in China on sandy areas of montane steppe in the Xinjiang desert region. It is an unstable, successional formation. Vegetation coverage of *A. cristatum* steppe is 25–40%, and the height of the vegetative layer is 10–15 cm. Accompanying species may include *Allium mongolicum, Artemisia* spp., *Cleistogenes squarrosa* and *Stipa* spp. Additional Typical Clump Grass Steppe formations not covered here include *Aristida adscensionis* (Poaceae) steppe and *Poa versicolor* subsp. *stepposa* steppe.

Desert Clump Grass Steppe is also dominated by species of *Stipa*. It is the most drought-resistant sub-type of steppe vegetation, and forms a transition between steppe and desert vegetation. It comprises xerophilous and super-xerophilous perennial dwarf clump grasses and semi-shrubs. The main formations are *Cleistogenes songorica, Stipa breviflora, S. caucasica* subsp. *glareosa, S. tianschanica* var. *gobica* and *S. tianschanica* var. *klemenzii* steppe, with further formations of limited distribution including *S. caucasica* desert-steppe and *S. orientalis* desert-steppe.

Stipa tianschanica var. *gobica* is a strongly drought-tolerant perennial clump-forming dwarf grass reaching 5–15 cm in height. The coverage of *Stipa tianschanica* var. *gobica* steppe vegetation is 20–25%. The main distribution area of this formation is the Ulanqab Gaoyuan and the mid-west of the Ordos Gaoyuan, Nei Mongol, with scattered patches on the mountains of the western desert region. This is a relatively species-poor community, with *Allium polyrrhizum, Artemisia* spp., *Festuca* spp., *Caragana* spp., *Cleistogenes songorica, Cleistogenes squarrosa* and *Reaumuria soongarica* plus some other forbs; the composition differs among regions.

Stipa tianschanica var. *klemenzii* steppe (Figure 8.37) is found mainly on the Nei Mongol Gaoyuan and the mountains to the south, concentrated on the upper slopes where the soil is thin and stony. The plants of this formation are usually rosette-like, densely clump-forming, or cushion-like, often with stolons—all adaptations to extremes of drought, wind and cold. The height of the formation is 10–20 cm, and it may include *Allium mongolicum, Cleistogenes songorica, C. squarrosa, Lagochilus ilicifolius* (Lamiaceae), *Scorzonera divaricata, Stipa sareptana* var. *krylovii,* and small semi-shrubs such as *Artemisia brachyloba, A. frigida* and *Thymus mongolicus* (Lamiaceae).

Figure 8.37 *Stipa tianschanica* var. *klemenzii* desert-steppe, Nei Mongol (image courtesy BAI Yong-Fei).

Stipa breviflora steppe is formed of thermophilic, perennial, clump-forming grasses. It is distributed mainly on the Huangtu Gaoyuan, extending to the low mountains of the Alxa Gaoyuan to the west, central Ningxia to the south, and the high plains of Nei Mongol to the north. This is another species-poor steppe formation, featuring *Allium mongolicum, Allium polyrhizum, Allium tenuissimum, Artemisia capillaris, Artemisia frigida, Astragalus* spp., *Caragana microphylla, Caragana pygmaea, Caragana stenophylla, Cleistogenes songorica* and *Lespedeza potaninii.*

Stipa caucasica subsp. *glareosa* steppe is better-adapted to drought, and distributed widely on the high plateaus of northwestern Nei Mongol and the western Ordos. It also occurs on the Helan Shan, Qilian Shan and Tian Shan, and the Altay and Ngari regions of Xizang. The substrate of these areas comprises coarse, gravelly or sandy soil. Owing to its wide distribution, the accompanying species in the formation differ with locality. In the montane belt of northern Xinjiang, they include *Carex physodes, Festuca valesiaca* subsp. *sulcata, Seriphidium nitrosum* var. *gobicum* and *Stipa capillata,* all of which are specific to the western steppe. On the Qinghai-Xizang Plateau, co-dominant species may include *Ajania fruticulosa, Arenaria bryophylla* (Caryophyllaceae), *Krascheninnikovia ceratoides, Stipa breviflora, S. purpurea* and *S. subsessiliflora.*

The Alpine Clump Grass Steppe formations consist of species of *Stipa* and *Festuca* adapted to cold and drought. This is the main steppe sub-type of the Qinghai-Xizang Plateau and the mountains of northwestern China. Here we discuss the *Stipa purpurea* steppe formation, which is the most widely distributed alpine steppe on the Qinghai-Xizang Plateau, Pamir Plateau, Kunlun Shan, Qilian Shan and Tian Shan. *Stipa purpurea* is specific to the high elevation Qinghai-Xizang and Pamier Plateaus. For example in northern Xizang, *S. purpurea* alpine steppe is found at 4 500–5 100 m elevation. The height of the leaf layer in this formation is ca. 20 cm, that of the reproductive branches 40–50 cm, and the vegetation coverage is 30–50%. Accompanying species include *Allium* spp., *Aster* spp., *Dracocephalum* spp., *Festuca ovina, Kengyilia thoroldiana* var. *thoroldiana* (Poaceae), *Pedicularis* spp., *Pennisetum flaccidum, Poa alpina, Poa araratica, Potentilla* spp., *Saussurea* spp., *Stipa roborowskyi, Stipa subsessiliflora* and *Swertia* spp. (Gentianaceae). Various species of small semi-shrubs, such as *Androsace* (Primulaceae), *Arenaria, Artemisia and Oxytropis*, are common. The generic composition of this formation remains the same across localities, although the species composition differs.

A second formation of Alpine Clump Grass Steppe, *Stipa subsessiliflora* alpine steppe is widely distributed in northern Xizang, but not so widely as the *S. purpurea* steppe. Accompanying species include *Leymus secalinus* var. *secalinus, Kengyilia thoroldiana* var. *thoroldiana, Orinus thoroldii* (Poaceae), *Stipa purpurea,* and species of *Astragalus, Lagotis* (Scrophulariaceae), *Leontopodium, Oxytropis, Potentilla* and *Thermopsis* (Fabaceae). The commonest small semi-shrubs are *Artemisia* spp. The species composition again varies across the distribution of the formation.

Festuca kryloviana alpine steppe is mainly distributed in the central and western parts of the Tian Shan, the western part of the Junggar mountains, and the alpine belt of the Altai, at elevations over 2 300 m. This steppe formation is relatively species-poor, the main species being *Anthoxanthum odoratum* (Poaceae), *Carex* spp., *Festuca kryloviana* and *Trisetum spicatum* (Poaceae). *Festuca kryloviana* can also form various communities with *Kobresia* spp. The growth of clumps of *Festuca kryloviana* is dense and luxuriant, giving coverage of 60–85%, with a leaf layer height of 20–25 cm. Accompanying forbs include species of *Aconitum* (Ranunculaceae), *Allium, Aster, Dracocephalum, Galium, Leontopodium, Polygonum, Potentilla, Scorzonera* and *Taraxacum* (Asteraceae). The commonest grasses are *Helictotrichon* spp. and *Poa* spp.

Rhizomatic Grass Steppe.—This steppe vegetation type is dominated by perennial xerophytic geophytes, and is distributed in temperate semi-humid and semi-arid regions. It has been divided into three sub-types: Meadow Rhizomatic Grass Steppe, Typical Rhizomatic Grass Steppe, and Alpine Rhizomatic Grass Steppe. The Meadow Rhizomatic Grass Steppe formation is specific to the eastern part of the temperate steppe region. Its main distribution area incorporates the Dongbei Pingyuan, Nei Mongol Gaoyuan and Huangtu Gaoyuan. It is also found in the steppe regions of outer Baikal (Russia) and Mongolia. In the semi-humid climatic belt of the eastern and western slopes of the Da Hinggan Ling, the main zonal formation is *Leymus chinensis* meadow-steppe (Figure 8.38), which also appears at lower elevations in humid habitats in the west of the typical steppe region. This formation is very species-rich, with accompanying species including *Artemisia latifolia, Bromus inermis, Carex lanceolata, Limonium bicolor, Stipa* spp., *Thalictrum squarrosum* and *Vicia amoena.* There are also many species of Caryophyllaceae, Liliaceae and Rosaceae present at low frequencies (Li, 1988).

Figure 8.38 *Leymus chinensis* meadow-steppe (with *Limonium bicolor*), Nei Mongol (image courtesy BAI Yong-Fei).

Leymus angustus meadow-steppe is distributed in the mid-montane belt of the Altai and Tian Shan. The height of the grass layer is 20–25 cm. Accompanying species include *Bromus inermis, Elymus gmelinii* var. *gmelinii* and *Phleum phleoides* (Poaceae) and *Helictotrichon hookeri* subsp. *schellianum*, plus many forbs and some shrubs such as *Berberis heteropoda* (Berberidaceae), *Cotoneaster uniflorus, Lonicera tatarica, Rosa spinosissima* and *Spiraea hypericifolia*. These shrubs form an upper layer of vegetation, 50–100 cm high. The final formation of Meadow Rhizomatic Grass Steppe that will be mentioned here, *Pennisetum flaccidum* meadow-steppe, has a wide distribution on sandy soils in the temperate steppe region and on the Qinghai-Xizang Plateau, but covers only a small total area.

The Typical Rhizomatic Grass Steppe sub-type is rare, and represented here by the *Agropyron desertorum* formation. Distributed mainly on sandy soils of the temperate steppe region, this community intermixes with Clump Glass Steppe in various habitats, often with *Stipa caucasica* subsp. *glareosa*. The community coverage is 30–35%, and the vegetation height 25–30 cm. Main accompanying species include *Atraphaxis frutescens* (Polygonaceae), *Cleistogenes squarrosa, Ephedra distachya* and *Seriphidium santolinum*.

The Alpine Rhizomatic Grass Steppe is specific to the Qinghai-Xizang Plateau. The main formations are dominated by *Carex moorcroftii, Orinus kokonorica* (Poaceae) and *O. thoroldii*. *Orinus thoroldii* steppe is found mainly in lake basins of southern Xizang, the valley of the middle and upper reaches of the Yarlung Zangbo Jiang, the south and west of northern Xizang, and southwestern Ngari. The formation occurs in fragmented populations on terraces, plateaus and piedmont slopes with thin, sandy soils. The community is species-poor, the main components being *Artemisia* spp. and *Stipa* spp. *Orinus kokonorica* alpine steppe is found mainly around Qinghai Hu, covering a small area. Common accompanying species include *Aster altaicus* var. *altaicus, Leontopodium ochroleucum, Potentilla* spp. and *Stipa sareptana* var. *krylovii*. *Carex moorcroftii* is the dominant species of alpine steppe, and specific to the Qinghai-Xizang Plateau. It prefers sandy soils and is cold tolerant, but not adapted to saline or alkaline soils, and can form co-dominant communities with *Krascheninnikovia compacta* var. *compacta* or *Stipa* sp.

Forb Steppe.—The species composition of forb-dominated steppe communities is relatively simple. Two sub-types are recognized: Meadow Forb Steppe and Desert Forb Steppe. Meadow Forb Steppe, characterized by *Filifolium sibiricum*, is the characteristic montane steppe formation of central-eastern Asia (see Figure 8.39). In China, *Filifolium sibiricum* steppe is distributed mainly on low mountains of the eastern and western Da Hinggan Ling, the eastern edge of the Hulun Buir Gaoyuan-Xilin Gol Gaoyuan, and the Songnen Pingyuan, Jilin. This is a relatively species-rich formation, 20–40 cm high. *Filifolium sibiricum* is completely dominant but there are some 80 accompanying species, including *Bupleurum scorzonerifolium* (Apiaceae), *Hedysarum gmellinii, Leymus chinensis, Medicago ruthenica* and *Stipa baicalensis* (Li, 1988).

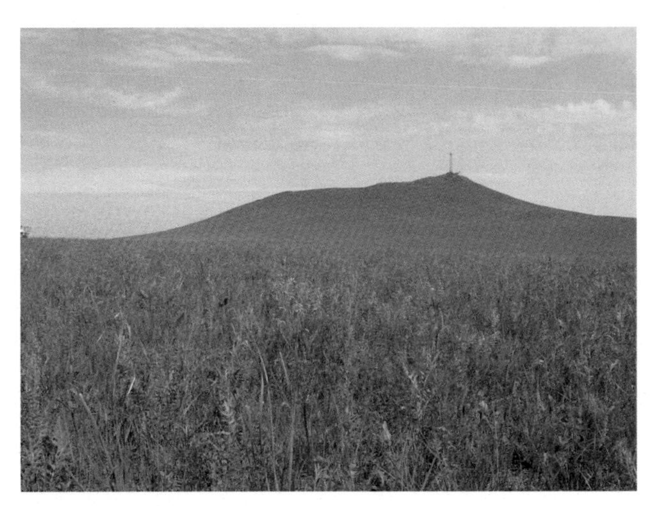

Figure 8.39 *Filifolium sibiricum* meadow-steppe, Nei Mongol (image courtesy BAI Yong-Fei).

Desert Forb Steppe, characterized by *Allium polyrhizum*, is widely distributed but covers only a small total area. It occurs in southwestern Hulun Buir on the Nei Mongol Gaoyuan, the western Junggar Pendi, and sparsely on the mountains of Xinjiang. The constitutent species are rather varied due to this wide distribution. For example, on the Nei Mongol Gaoyuan, common accompanying species include *Artemisia frigida, Aster altaicus var. altaicus* and *Eragrostis minor* (Poaceae); in the mountains of Xinjiang, they include *Anabasis aphylla, Anabasis brevifolia, Suaeda dendroides* and many grasses.

Semi-shrub and Small Semi-shrub Steppe.—This kind of steppe consists of semi-shrubs, adapted to semi-drought or drought conditions. As a primary community, this vegetation type may be adapted to a specific substrate—in which the soil surface layer has been severely eroded—or a particular climate, that of the Qinghai-Xizang Plateau. Semi-shrub and Small Semi-shrub Steppe may also comprise secondary vegetation, following impacts such as livestock grazing. Semi-shrub and Small Semi-shrub Steppe has been divided into three sub-types: Typical Semi-shrub and Small Semi-shrub Steppe, Desert Semi-shrub and Small Semi-shrub Steppe, and Alpine Semi-shrub and Small Semi-shrub Steppe.

Typical Semi-shrub and Small Semi-shrub Steppe is exemplified by the *Thymus mongolicus* formation which develops following erosion of the surface layer of soil. It is distributed mainly on the low mountains and hills of Liaoning and Jilin, eastern Ordos and the central Huangtu Gaoyuan. *Thymus mongolicus* is a small, prostrate, xerophytic semi-shrub, 5–10 cm high, highly branched and with a clumped distribution. The formation is relatively rich in co-occurring species, including *Artemisia frigida, A. scoparia, Cleistogenes squarrosa, Gueldenstaedtia verna*

(Fabaceae), *Lespedeza dahurica, Leymus chinensis* and *Oxytropis microphylla*.

Artemisia frigida steppe is a reduced steppe vegetation, developing from *Stipa* steppe by long-term over-grazing or strong wind erosion. It is found mainly in the typical steppe and desert-steppe regions of the Nei Mongol and Ordos Gaoyuan. The component species of *Artemisia frigida* steppe are relatively abundant, and include members of *Agropyron, Allium, Artemisia, Cleistogenes, Hedysarum, Oxytropis, Potentilla* and *Stipa*. The formation in which *Cleistogenes squarrosa* is co-dominant with *Artemisia frigida* is distributed widely on the Nei Mongol Gaoyuan. The *Artemisia gmelinii* var. *gmelinii* and *Artemisia giraldii* steppe formations are semi-shrub formations distributed on eroded gravel slopes of low mountains and hills in the warm-temperate steppe and meadow-steppe regions. These secondary vegetation formations follow the destruction of deciduous broad-leaved forest or shrubland. *Artemisia gmelinii* var. *gmelinii* can be monodominant or co-dominant with *Artemisia giraldii*. Common components of the community include *Bothriochloa ischaemum, Lespedeza davurica* and *Stipa bungeana*.

The dominant species of Desert Semi-shrub and Small Semi-shrub Steppe are small semi-shrubs such as *Ajania achilleoides, Ajania fruticulosa, Ajania trifida* and *Artemisia dalai-lamae*. *Ajania trifida* desert-steppe is found mainly on the eastern part of the Nei Mongol Gaoyuan, and is the specific steppe of the Gobi of Central Asia. In general, it does not form continuous, large distributions, and *Ajania trifida* may also be co-dominant with *Stipa tianschanica* var. *klemenzii* or *S. caucasica* subsp. *glareosa*. Common species in these formations are *Agropyron desertorum, Allium mongolicum, Artemisia frigida, Astragalus* spp., *Cleistogenes songorica, C. squarrosa, Euphorbia* spp. and *Scorzonera* spp.

Ajania fruticulosa desert-steppe is distributed mainly on the northern part of the Huangtu Gaoyuan in Gansu, southern Ningxia, the mountains of eastern Xinjiang and Tacheng. These areas belong to the arid part of the warm-temperate zone. The component species differ depending on the exact habitat, with common species including *Agropyron desertorum, Artemisia dalai-lamae, Artemisia frigida, Cleistogenes songorica, Stipa breviflora, S. bungeana, S. tianschanica* var. *gobica* and forbs such as *Aster altaicus* var. *altaicus, Cymbaria mongolica* and *Potentilla* spp. *Ajania achilleoides* desert-steppe is found in the western part of the desert-steppe region of the Nei Mongol and Ordos Gaoyuan. The climate is more arid than that of *Ajania trifida* desert-steppe. *Ajania achilleoides* can be co-dominant with *Cleistogenes songorica* or *Krascheninnikovia ceratoides*.

The main small semi-shrubs of the Alpine Semi-shrub and Small Semi-shrub Steppe are *Artemisia* spp. Formations with *A. wellbyi* or *A. younghusbandii* as the dominant species are widely distributed. *Artemisia wellbyi* steppe occurs mainly in the lake basin areas of southern Xizang, the middle and upper reaches of the Yarlung Zangbo Jiang, the south of northern Xizang and southern Ngari, all at elevations of 4 300–4 800 m. The height of *A. wellbyi* steppe is ca. 20 cm, and its coverage 25–40%. The main co-dominant species, each forming different communities with *A. wellbyi*, are *Anaphalis xylorhiza, Orinus thoroldii, Stipa caucasica* subsp. *glareosa* and *S. purpurea*. Common accompanying species include *Astragalus* spp., *Oxytropis* spp., *Pedicularis* spp., *Potentilla* spp., *Rhodiola* spp. and *Saussurea* spp., along with many grasses.

Artemisia younghusbandii high alpine steppe is distributed mainly on the northern piedmont slopes of the central Himalaya Shan and the middle and upper reaches of the Yarlung Zangbo Jiang area. *Artemisia younghushandii* is specific to the Qinghai-Xizang Plateau, and reaches a height of 15–20 cm. The remainder of this community is rather simple, and may include co-dominant species *Orinus thoroldii* or *Potentilla microphylla*, and common species of *Allium, Androsace, Artemisia, Astragalus, Potentilla* and *Stipa*.

Artemisia stracheyi alpine steppe is widely distributed in the steppe regions of Xizang, but with a small total area from the lake basins of southern Xizang to the upper reaches of the Yarlung Zangbo Jiang and northern Xizang, at elevations of 4 400–4 800 m. Usually *Artemisia stracheyi*, a small, cold-tolerant, xerophytic semi-shrub, is an accompanying species in other steppe formations; as a dominant species it is found only in small, scattered patches on the gravel belt of the upper part of the piedmont diluvial fan and lower slopes in this area. The height of this formation is 20–30 cm and its coverage 25–30%. Main accompanying species include *Anaphalis xylorhiza, Androsace tapete, Arenaria bryophylla, Astragalus strictus, Gentiana oreodoxa, Potentilla fruticosa* var. *arbuscula, Senecio* sp., *Stellera chamaejasme* and occasional species of *Carex, Kobresia, Rhodiola* and *Saussurea*.

Artemisia duthreuil-de-rhinsi is a small xerophytic semi-shrub specific to the Qinghai-Xizang Plateau. *Artemisia duthreuil-de-rhinsi* steppe is patchily distributed on arid and piedmont slopes of southern Xizang at elevations of 4 350–4 600 m. The height of this steppe vegetation is ca. 20 cm, and its coverage is 35–50%. Common accompanying species include *Anaphalis xylorhiza, Artemisia moorcroftiana, Artemisia younghusbandii, Carex* spp., *Orinus thoroldii, Oxytropis* spp., *Poa* sp., *Potentilla bifurca* and *Stipa* spp. Other alpine steppe formations not covered here include the *Artemisia minor, A. moorcroftiana* and *Thymus* sp. alpine steppe formations (Inner Mongolia and Ningxia Comprehensive Expedition Academia Sinica, 1985; Qinghai-Tibet Plateau Comprehensive Expedition, 1988).

8.4 Patterns of vegetation distribution in China

As explained in previous chapters, China displays a vast range of topographic and climatic conditions, each supporting different communities of plants. Thus the modern-day vegetation of China can be characterized into specific geographical distribution patterns, particularly in terms of latitudinal zonality, longitudinal zonality and vertical zonality (Hou, 1960). The latitudinal distribution of vegetation in China is influenced most by climatic factors: temperature, precipitation, humidity and day length, whilst longitudinal zonality is determined mostly by hydrological differences.

8.4.1 Latitudinal zonality and montane vertical distributions

The monsoon influences a vast region from east of the Da Hinggan Ling through the Lüliang Shan and Liupan Shan to the eastern edge of the Qinghai-Xizang Plateau. In this region, both precipitation and temperature gradually increase from north to south, resulting in different climatic vegetation zones. Furthermore, specific vertical vegetation distribution patterns develop on mountains.

The Cold-temperate Boreal Coniferous Forest (Taiga Forest) Zone.—This vegetation zone is centered on the Da Hinggan Ling region, north of 48° N. The vegetation is characterized by *Larix gmelinii* forest, with small patches of *Pinus sylvestris* var. *mongolica* forest. Secondary deciduous broad-leaved forest also appears where these coniferous forests have been destroyed, the main formations being *Betula dahurica*, *B. platyphylla* and *Populus davidiana* forest. *Quercus mongolica* forest develops after the destruction of *Larix gmelinii-Quercus mongolica* forest. If additional destruction occurs, the forest can degenerate further into secondary scrub formations, such as *Armeniaca sibirica* (Rosaceae) or *Corylus heterophylla* scrub (see

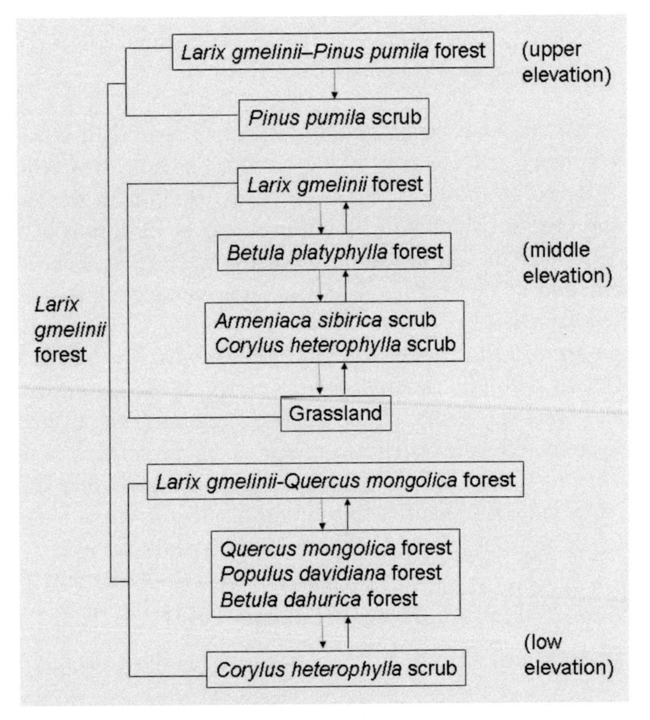

Figure 8.40 Succession profile for natural *Larix gmelinii* forest in the Da Hinggan Ling (after Zhou, 1991). Dotted lines indicate regression; solid lines indicate restoration.

Figure 8.40). There are also many kinds of meadow and swamp vegetation in this zone. Within the zone, different vegetation formations are found at different elevations (vertical zonation; see Table 8.4 for an example from the Da Hinggan Ling).

Table 8.4 Vertical zonation of the natural vegetation of the Da Hinggan Ling (Zhou, 1991). Note that the zones overlap somewhat.

Elevation (m)	Natural vegetation type
above 1 240	*Pinus pumila* elfin forest and scrub, *Betula ermanii* elfin forest, *Juniperus davurica* scrub
1 100–1 450	*Larix gmelinii* forest, *Betula ermanii* woodland
820–1 140	*Larix gmelinii* forest, *Picea jezoensis* forest
450–1 050	*Larix gmelinii* forest, *Pinus sylvestris* var. *mongolica* forest, *Picea koraiensis* forest

The Temperate Mixed Coniferous and Deciduous Broad-leaved Forest Zone.—This region comprises mostly the Xiao Hinggan Ling and Changbai Shan in northeastern China. It occupies a transition zone between taiga forest and deciduous broad-leaved forest. *Pinus koraiensis* normally co-dominates with various species of deciduous tree: *Acer pictum* subsp. *mono*, *Fraxinus mandschurica*, *Phellodendron amurense* (Rutaceae), *Populus ussuriensis*, *Tilia amurensis* and *Ulmus davidiana* var. *japonica*. The forests where *Pinus koraiensis* has been cut, various secondary deciduous broad-leaved forests, such as *Betula*, *Populus* or *Quercus mongolica* forest, developed. *Pinus sylvestris* var. *sylvestriformis* and *Larix olgensis* forests also develop within this mixed forest zone. Secondary scrub found after forest destruction mainly comprises *Corylus mandshurica* and *Lespedeza bicolor*. This zone contains the famous Sanjiang Pingyuan ("Three Rivers Plain") which, being wet, features *Deyeuxia purpurea* meadow and forb meadow formations as well as herbaceous swamp and aquatic vegetation including *Carex* and *Phragmites australis*. Another notable feature of this zone is the Changbai Shan, for which the vertical vegetation characteristics are shown in Table 8.5.

Table 8.5 Vertical zonation of the natural vegetation on Changbai Shan (from Chen & Bao, 1964; Li *et al.*, 2001).

Elevation (m)	Natural vegetation type
above 2100	Small shrub-moss alpine tundra, herbaceous alpine tundra
1 800–2 100	*Betula ermanii* elfin forest
1 400–1 800	*Picea jezoensis* forest, *Abies nephrolepis* forest
1 100–1 400	*Pinus koraiensis* forest, *Abies nephrolepis* forest
700–1 100	*Pinus koraiensis* forest, *Tilia amurensis* forest, *Betula costata* forest, *Pinus koraiensis* forest, *Abies holophylla* forest, *Ulmus* sp. forest

The Warm-temperate Deciduous Broad-leaved Forest Zone.—This forest zone includes the Liaodong peninsula, the lower reaches of the Liao He floodplain, northern China, and the northern slopes of the Qin Ling, and is typified by *Quercus* spp. forests of many kinds: *Q. acutissima*, *Q. dentata*, *Q. variabilis* forests on low mountains; *Q. aliena*,

Q. aliena var. *acutiserrata*, *Q. mongolica* and *Q. serrata* forests at higher elevations. Deciduous broad-leaved forests, dominated by *Betula platyphylla*, *Carpinus cordata*, *Carpinus turczaninowii*, *Juglans mandshurica*, *Populus davidiana*, *Tilia mandshurica* or *T. mongolica*, are frequent. Coniferous forests are also distributed widely in this zone, including *Pinus densiflora* forest on the Liaodong peninsula, *Pinus tabuliformis* forest in much of northern China, and *Pinus armandii* forest at the southwestern edge of this zone. *Platycladus orientalis* forest and *Pinus bungeana* forest are common on limestone mountains. Secondary scrub formations appear after the destruction of forest, usually dominated by *Corylus heterophylla*, *Corylus mandshurica*, *Cotinus coggygria* var. *pubescens* (Anacardiaceae), *Lespedeza bicolor*, *Myripnois dioica* (Asteraceae), *Ostryopsis davidiana*, *Spiraea pubescens*, *S. trilobata* or *Vitex negundo* var. *heterophylla*. Secondary grassland vegetation follows the destruction of these scrub communities.

The Subtropical Evergreen Broad-leaved Forest Zone.— This zone extends from the Qin Ling-Huai He in the north to the Tropic of Cancer. It is conventionally divided into three sub-zones. The Northern Subtropical Deciduous-evergreen Broad-leaved Forest sub-zone covers the southern slopes of the Qin Ling and Funiu Shan. The vegetation is mixed deciduous and evergreen broad-leaved forest, with evergreen broad-leaved forests distributed patchily in valleys. The dominant species of these forests are cold-tolerant evergreen trees such as *Castanopsis sclerophylla*, *Cyclobalanopsis glauca*, *Ilex chinensis* and *Phoebe sheareri*, which can form mixed forests with a variety of deciduous broad-leaved trees. Deciduous broad-leaved forests also occur, dominated by *Quercus acutissima*, *Q. aliena*, *Q. aliena* var. *acutiserrata* or *Q. variabilis*. Of the *Pinus*-dominated forests, *Pinus massoniana* forest is most common on lower slopes, with *P. tabuliformis* var. *henryi* and *P. armandii* forest in the middle montane belt. Secondary deciduous scrubs develop after any forest destruction, dominated by *Coriaria nepalensis* (Coriariaceae), *Cotinus coggygria* var. *cinerea*, *Exochorda racemosa* (Rosaceae) or *Vitex negundo* var. *heterophylla*.

The 3 052 m Shennongjia mountain in Hubei is found in the northern subtropical zone, and extends into the transitional zone between the northern and mid-subtropical zones. We use Shennongjia as an example of a vertical vegetation distribution typical of the northern subtropical zone. At elevations below 800 m (on the northern slope) or 1 000 m (southern slope), the vegetation comprises mixed evergreen and deciduous broad-leaved forest of evergreen *Castanopsis* and *Cyclobalanopsis* with deciduous *Quercus* and *Fagus*; coniferous *Cunninghamia lanceolata* and *Pinus massoniana* forests are also present. Between 800 (or 1 000 on southern slopes) and 1 400 (or 1 600) m the forests are mainly of the deciduous broad-leaved type, such as *Quercus serrata*, *Q. variabilis* or *Castanea seguinii-Platycarya strobilacea* forest; coniferous forests are represented by the *Pinus tabuliformis* var. *henryi* formation. From 1 600 to 2 300 m, deciduous broad-leaved forests include various *Fagus* formations, *Quercus aliena* var. *acutiserrata* or *Betula albosinensis* forest; the coniferous vegetation comprises *Pinus armandii* forest. Above 2 300 m, the main forests are of *Abies fargesii*, with small patches of *Picea brachytyla* forest and mixed forests of *Abies*, *Acer* and *Betula*.

The Mid-subtropical Evergreen Broad-leaved Forest sub-zone could also be called the "typical subtropical zone," since it includes a wide region: the hills of the southern side of the Chang Jiang, mountains of Zhejiang and Fujian, northern Taiwan, the Sichuan Pendi, Guizhou plateau in Yunnan, and the Nanling Shan. The sub-zone is characterized by diverse evergreen broad-leaved forests, dominated by species of *Castanopsis*, *Cinnamomum*, *Cyclobalanopsis*, *Lithocarpus*, *Machilus*, *Phoebe* and *Schima*. The region is also rich in coniferous forests: *Pinus kwangtungensis*, *Pinus massoniana* and *Pinus taiwanensis* forest in the east, and *Pinus armandii* and *Pinus yunnanensis* forest in the west. The influence of Quaternary glaciations has led to the presence of many "relict" species, such as *Cathaya argyrophylla*, *Davidia involucrata*, *Eucommia ulmoides* (Eucommiaceae), *Metasequoia glyptostroboides* (Taxodiaceae) and *Taiwania cryptomerioides*, in this sub-zone. Bamboos and bamboo thickets are also common. The presence of a rich shrub layer in the evergreen broad-leaved forests allows for rapid development of secondary scrub where forests are cut. These scrub formations are dominated by *Lindera glauca*, *Loropetalum chinense*, *Quercus fabri*, *Rhododendron simsii*, *Vaccinium bracteatum* or, on montane limestone, *Bauhinia championii*, *Pyracantha fortuneana* (Rosaceae) or *Zanthoxylum armatum* var. *armatum* (Rutaceae). Where the scrub itself is lost, grasslands or fern-dominated formations may develop.

An example of the vertical vegetation distribution on mountains in this sub-zone is taken from Wuyi Shan, Fujian. The elevation of its highest peak, Huanggang Shan, is 2 158 m. On the southeastern slope of the Huanggang Shan, low elevation scrub (below 350 m elevation) gives way to an evergreen broad-leaved forest belt (350–1 400 m), followed by a mixed coniferous and broad-leaved forest belt (1 200–1 700 m), a montane dwarf moss forest belt (1 700–1 970 m) and finally a montane meadow belt (1 700–2 158 m). The dominant species of the evergreen broad-leaved forests on gentle slopes between 500 and 1 100 m are *Castanopsis fabri*, *Castanopsis fargesii* and *Cyclobalanopsis glauca*. On sunny slopes and ridges *Castanopsis eyrei-Schima superba* forest can be found, and in valleys we find *Castanopsis fordii* and *Castanopsis tibetana* forests. On gentle slopes at elevations below 500 m, small patches of *Castanopsis carlesii* forest can be found. Coniferous forests within the evergreen broad-leaved belt include *Cryptomeria japonica* var. *sinensis* (Taxodiaceae), *Cunninghamia lanceolata*, *Phyllostachys edulis* and *Pinus massoniana* forest. Where evergreen broad-leaved forest is destroyed, further scrub formations develop.

The mixed coniferous and broad-leaved forest belt comprises mainly mixed forests of *Pinus taiwanensis* and broad-leaved trees such as *Castanopsis eyrei, Cyclobalanopsis glauca* and *Schima superba,* on sunny slopes and ridges between 1 100 and 1 700 m. Between 1 500 and 1 800 m we find mixed coniferous and broad-leaved forests dominanted by *Tsuga chinensis* var. *chinensis*, with dominant broad-leaved trees including *Betula luminifera, Castanopsis eyrei* and *Cyclobalanopsis multinervis*. The montane dwarf moss forest consists of *Acer* spp., *Cornus controversa, Cyclobalanopsis multinervis, Eurya* spp., *Rhododendron* spp. and *Symplocos* spp. The montane meadow belt, on gentle terrain, comprises mainly grasses (He & Li, 1994). Figure 8.41 shows the succession of evergreen broad-leaved forest in Tiantong, Zhejiang.

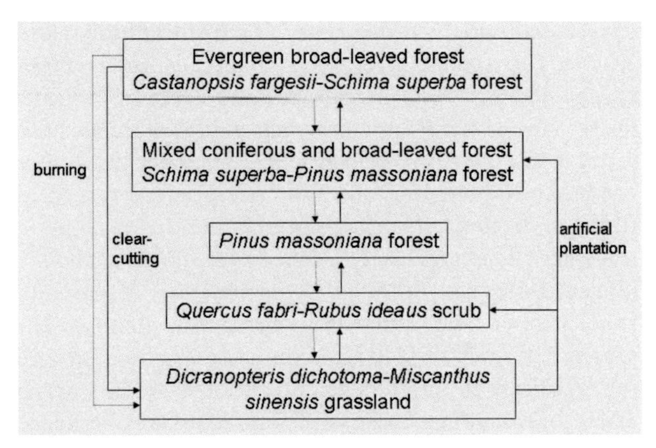

Figure 8.41 Succession of evergreen broad-leaved forest in Tiantong, Zhejiang (from Song, 2001). Dotted lines indicate regression; solid lines indicate restoration.

The range of the Southern Subtropical Evergreen Broad-leaved Forest sub-zone extends from south of the Nanling Shan to near the Tropic of Cancer, and includes most of Taiwan. The vegetation is typically evergreen broad-leaved forest, with many *Castanopsis* and *Lithocarpus* formations, including dominant and common species of the same genera as in evergreen broad-leaved forests of the mid-subtropical sub-zone, but different species. Several families—Elaeocarpaceae, Hamamelidaceae, Lauraceae and Symplocaceae—are more visible in this sub-zone than elsewhere, and the forests often contain tropical elements. The main secondary scrub formation in this sub-zone is *Rhodomyrtus tomentosa* (Myrtaceae) scrub; others include *Aporusa* spp., *Cratoxylum cochinchinense* (Clusiaceae) and *Wendlandia uvariifolia* scrub formations.

The Tropical Monsoon Forest and Rainforest Zone.—This zone is found at the northern edge of the tropical zone, and includes Guangdong, Guangxi, Hainan, southern Yunnan, and the southern slopes of the eastern Himalaya Shan. These tropical forests are composed of species of Combretaceae, Dipterocarpaceae, Euphorbiaceae, Fabaceae, Meliaceae, Moraceae, Myristicaceae, Myrtaceae, Sapotaceae and Sterculiaceae. Common scrub formations include *Baeckea frutescens* (Myrtaceae)-*Helicteres angustifolia* scrub and *Microcos paniculata* (Tiliaceae)-*Cratoxylum cochinchinense* scrub. The vertical zonation of montane vegetation in the tropical monsoon forest and rainforest zone is exemplified by the northwestern slope of Yushan (3 997 m high), Taiwan, in Table 8.6.

Table 8.6 Vertical zonation of the natural vegetation on Yushan (from Huang, 1993).

Elevation (m)	Natural vegetation type
3 600–3 950	Alpine scrub dominated by *Rhododendron morii, R. pseudochrysanthum* and *R. rubropilosum*, alpine meadow consisting of *Phleum alpinum, Brachypodium kawakamii* and *Festuca ovina*
3 400–3 600	Alpine coniferous elfin wood dominated by *Juniperus squamata* var. *squamata*
2 400–3 400	Subalpine coniferous forest dominated by *Abies kawakamii* and *Picea morrisonicola*, with patches of *Tsuga chinensis* var. *formosana* forest
1 800–2 400	Coniferous forest dominated by *Chamaecyparis formosensis* or *Chamaecyparis obtusa* var. *formosana*, mixed evergreen and deciduous broad-leaved forest consisting of *Acer caudatifolium, A. morrisonense, Castanopsis carlesii, Cyclobalanopsis morii, Cyclobalanopsis stenophylloides, Phellodendron amurense, Sassafras randaiense* and *Trochodendron aralioides*, mixed coniferous and broad-leaved forests with *Cunninghamia lanceolata* var. *konishii, Cyclobalanopsis morii, Illicium arborescens, Lithocarpus amygdalifolius, Pinus armandii* var. *mastersiana, Pseudotsuga sinensis* var. *wilsoniana, Sorbus randaiensis* and *Ulmus uyematsui*, deciduous broad-leaved forest dominated by *Fagus hayatae*
900–1 800	Evergreen broad-leaved forest consisting of *Castanopsis kawakamii, Cinnamomum camphora, Cinnamomum micranthum, Cyclobalanopsis gilva, Cyclobalanopsis longinux, Lithocarpus kawakamii, Machilus japonica* var. *kusanoi* and *Machilus zuihoensis*
600–900	Montane seasonal rainforest including *Cryptocarya chinensis, Elaeocarpus sylvestris, Litsea hypophaea, Machilus japonica* var. *kusanoi, Machilus thunbergii, Phoebe formosana* and *Schefflera heptaphylla*
130–600	Tropical seasonal rainforest including *Dendrocnide meyeniana, Ficus formosana, F. nervosa, F. variegata, Helicia formosana, Macaranga tanarius* (Euphorbiaceae) and *Syzygium formosanum*

The Tropical Rainforest Zone.—In contrast to some authors (e.g. Zhu, 1994), we consider that there is no true rainforest zone in continental China; only the outlying Xisha Qundao and Nansha Qundao are located within the rainforest region, and these bear only coral island broad-leaved forest and scrub formations.

8.4.2 Longitudinal features of vegetation distribution

In China, variation in vegetation with longitude is most noticeable in the temperate zone, where there is a clear trend from the moist and semi-moist eastern coast to the semi-arid and arid zones of the continental west. In contrast, tropical and subtropical regions exhibit evergreen broad-leaved forests and thermophilic coniferous forests at all longitudes, due to the presence of plentiful precipitation throughout.

The Forest Region.—Forests are found in the moist monsoon region of temperate northern China, where the annual precipitation is greater than 550 mm. The main forest types are mixed coniferous and broad-leaved, coniferous, and deciduous broad-leaved forests.

The Steppe Region.—This region extends from the western part of the northeast plain of the Nei Mongol Gaoyuan to the western Huangtu Gaoyuan and the northwestern edge of the Xizang Plateau, and has an average annual precipitation of 200–500 mm. The region has been divided into three zones according to precipitation, which decreases gradually from east to west: the Meadow-Steppe Zone, Typical Steppe Zone and Desert-steppe Zone. The steppe region can also be divided according to temperature, into Mid-temperate and Warm-temperate Zones. Meadow-steppe is widely distributed in the temperate zone, near to the forest region. Formations of meadow-steppe here include *Filifolium sibiricum* (Asteraceae), *Leymus chinensis, Stipa baicalensis* and *S. kirghisorum* steppes. On hills, *Quercus mongolica* forest may be found. In the narrow Warm-temperate Zone, the main meadow-steppe formations are *Bothriochloa ischaemum, Cleistogenes mucronata* and *Stipa bungeana* steppe. Deciduous broad-leaved forest and scrub occur in patches on shaded slopes. The largest area of typical steppe is found in the Mid-temperate Zone, with major formations including *Stipa capillata, S. grandis* and *S. sareptana* var. *krylovii* steppe. In the Warm-temperate Zone, typical steppe formations are the same as the meadow-steppe formations: *Bothriochloa ischaemum* and *Stipa bungeana* steppe, with the addition of *Stipa breviflora* steppe. Desert-steppe occupies large areas in the Mid-temperate Steppe Zone. The main formations are *Stipa caucasica* subsp. *glareosa, S. tianschanica* var. *gobica* and *S. tianschanica* var. *klemenzii* steppe. Common desert-steppe formations of the Warm-temperate Zone include *Stipa breviflora* and *S. caucasica* subsp. *glareosa* steppe.

The Desert Region.—To the west of the steppe region of China lies desert. This vast region may be divided into Temperate and Warm-temperate Deserts. The Temperate Desert Region includes the Alxa Gaoyuan, Hexi corridor, Beishan Gobi and the Junggar Pendi; the Warm-temperate Region includes the Dongjiang basin, on the south of the Tian Shan, and the Tarim Pendi. In the Temperate Desert Region, as precipitation decreases from east to west (to less than 50 mm per year), the vegetation changes from steppe-desert to extremely dry shrub-desert or even bare rock. From the lower reaches of the Hei He to the west and north-west, precipitation increases dramatically in winter and spring; here the main vegetation types are Small Tree Desert, Shrub Desert and Small Semi-shrub Desert. On sandlands, *Artemisia, Calligonum, Haloxylon* and *Tamarix* desert formations may be found. On saline soils, deserts are dominated by species of *Halocnemum, Halostachys* or *Kalidium*. At the bottom and in the southern part of the Qaidam Pendi, and on the wide diluvial fan of the piedmont slopes of the Kunlun Shan, the soil is almost bare with the few desert plants growing only in ditches. The Warm-temperate Desert Region forms the extremely dry core of the Central Asian desert. Here, the sparse vegetation comprises super-xerophytic shrubs and semi-shrubs such as species of *Calligonum, Ephedra, Illinia, Nitraria, Reaumuria, Salsola, Sympegma* and *Zygophyllum*. Large areas of mobile sand, gravel and stony slopes are devoid of vegetation. Forest, shrub and meadow formations are found only on lowlands suppled by latent water, or alongside rivers.

8.4.3 Vegetation zonation on the Qinghai-Xizang Plateau

The Qinghai-Xizang Plateau is located in the subtropical zone. Here, strong thermodynamic effects are produced by elevation (which averages an incredible 4 500 m) and the vast area of the Plateau. These greatly affect the atmospheric circulation and natural environment of all of China, and even most of Asia. The Qinghai-Xizang Plateau has been divided into five vegetation zones as described below.

The Alpine Scrub and Meadow Zone.—This zone is located in the central-eastern Qinghai-Xizang Plateau, an area which is relatively low to the southeast (elevation 3 300–3 500 m) and higher in the northwest (4 000–4 500 m). The zonal vegetation types present are alpine scrub and alpine meadow. Alpine scrub formations include alpine evergreen scrub and deciduous scrub, the dominant genera of which are *Potentilla, Rhododendron* and *Salix*. The main species of alpine meadow are *Kobresia* spp. Among them, *K. pygmaea* meadow has the widest distribution. Forb-dominated meadows include *Polygonum macrophyllum* var. *macrophyllum* and *P. viviparum* formations. In addition, alpine swampy meadow and swamp formations may be seen, both dominated by *Carex* spp.

The Alpine Steppe Zone.—This zone is located in the center of the Plateau, an area which includes northern Xizang, the Kekexili Gaoyuan and its eastern fringes. This is the largest vegetation zone on the plateau, with an average elevation of 4 500–4 800 m. The zonal vegetation is alpine steppe, with the most widely distributed formation being *Stipa purpurea* alpine steppe. Other formations include *Artemisia wellbyi, Carex moorcroftii* and *S. subsessiliflora* steppes. *Kobresia pygmaea* alpine meadow is found only on the moist shaded slopes of the southern mountains.

The Alpine Desert Zone.—The Alpine Desert Zone is found on the northwestern Qinghai-Xizang Plateau, and includes the mountains between the Kunlun Shan and Karakoram Shan, and the northern edge of northern Xizang. The average elevation is 4 800–5 300 m, average annual temperature is 8–10 °C below zero, and the average annual precipitation is 20–100 mm. The main, widely distributed, zonal vegetation is *Krascheninnikovia compacta* var. *compacta* desert, sometimes found as a co-dominant formation with *Carex moorcroftii*.

The Temperate Scrub Steppe Zone.—This zone is located in southern Xizang between the central Himalaya Shan, Nyainqêntanglha Shan and Gangdisê Shan. It is a zone of great topographical variation. The vegetation is largely temperate montane scrub and steppe. The scrub formations are composed of species of *Caragana* and *Potentilla*. Steppe formations include *Aristida triseta* (Poaceae), *Artemisia wellbyi*, *Orinus thoroldii* and *Pennisetum flaccidum* steppe.

The Temperate Desert Zone.—The Temperate Desert Zone is found in the wide montane valleys of western Ngari (Xizang), at elevations of ca. 4 200 m. Here, the average annual temperature is 3–8 °C below zero, and the average precipitation 50–170 mm per year. Representative vegetation types include montane desert and steppe-desert, the dominant species of which are *Krascheninnikovia ceratoides*, *Ajania fruticulosa* and *Stipa caucasica* subsp. *glareosa*.

References

ANHUI FOREST EDITORIAL COMMITTEE (1990). *Forest of Anhui*. Beijing, China: Beijing Forestry Press.

CHEN, L. Z. & BAO, X. C. (1964). The main plant communities in various vertical belts on N slope of Changbaishan Mountain, Jilin. *Acta Phytoecologica et Geobotanica Sinica* 2(2): 207-225.

CHEN, L. Z. & CHEN, Q. L. (1997). *Forest Diversity and its Geographical Distribution in China*. Beijing, China: Science Press.

GUANGDONG INSTITUTE OF BOTANY (1976). *Vegetation of Guangdong*. Beijing, China: Science Press.

GUANGXI FORESTRY BUREAU (1993). *The Nature Reserves of Guangxi*. Beijing, China: Forestry Press.

HE, J. Y. & LI, L. H. (1994). Vegetation of Wuyishan Nature Reserve. In: He, J. Y. (ed.) *Studies on Wuyishan Mountain*. pp. 39-117. Xiamen, China: Xiamen University Press.

HOU, X. Y. (1960). *Vegetation of China*. Beijing, China: People Education Press.

HU, S. S. (1979). The community characteristics of evergreen broadleaved forest of Guangxi. *Acta Botanica Sinica* 21(4): 362-370.

HUANG, Q. (1986). Studies on tropical vegetation ecology in Jianfengling Hannan. *Acta Phytoecologica et Geobotanica Sinica* 14(2): 90-105.

HUANG, W. L. & TU, Y. L. (1988). *Vegetation of Guizhou*. Guiyang, China: Guizhou People Press.

HUANG, W. L. (1993). *Vegetation of Taiwan*. Beijing, China: Beijing Environment Press.

INNER MONGOLIA AND NINGXIA COMPREHENSIVE EXPEDITION ACADEMIA SINICA (1985). *Vegetation of Inner Mongolia*. Beijing, China: Science Press.

JIANG, Y. X. (1991). *Tropical Forest Ecosystem on Jianfengling Mountain, Hannan*. Beijing, China: Science Press.

LI, B. (1988). Vegetation and its usage in Xilinhe River Reaches. In: *Research on Steppe Ecosystem*. Beijing, China: Science Press.

LI, J. D., WU, B. H. & SHENG, L. X. (eds.) (2001). *Vegetation of Jilin*. Chungchun, China: Jilin Science and Technology Press

LIN, P. (ed.) (1990). *Vegetation of Fujian*. Fuzhou, China: Fujian Science and Technology Press.

LIU, Y. B. & LIU, Y. C. (2000). *Vegetation of Henan*. Beijing, China: China Forestry Press.

MA, Z. Q. (ed.) (2001). *Vegetation of Shanxi*. Beijing, China: Science Press.

PENG, S. L. (1989). Component, structure, quantitative character of forest community. *Acta Phytoecological et Geobotanica Sinica* 13(1): 10-17.

QI, C. J. (1990). *Vegetation of Hunan*. Chengsha, China: Hunan Science and Technology Press.

QINGHAI-TIBET PLATEAU COMPREHENSIVE EXPEDITION (1988). *Vegetation of Tibet*. Beijing, China: Science Press.

SICHUAN FOREST EDITORIAL COMMITTEE (1992). *Forests of Sichuan*. Beijing, China: China Forestry Press.

SICHUAN VEGETATION COOPERATIVE GROUP (1980). *Vegetation of Sichuan*. Chengdu, China: Sichuan People Press.

SONG, Y. C. (2001). *Vegetation Ecology*. Shanghai, China: East China Normal University Press.

SU, D. M. (1988). Vegetation Survey in Nonggang Nature Reserve. *Guihaia* supplement: 185-212.

WANG, X. P., SUN, S. Z. & LI, X. X. (1997). Studies on the classification of evergreen deciduous broadleaved mixed forest on limestone mountain, Guangxi. *Botanical Research* 17: 235-273.

WANG, X. P. & JIANG, G. M. (2000). Studies on classification and geographical distribution of evergreen broadleaved forest, Guangxi. *Wuhan Botanical Research* 18(3): 195-205.

WU, Z. Y. (ed.) (1980). *Vegetation of China*. Beijing, China: Science Press.

WU, Z. Y. (ed.) (1983). *The Studies on Forest Ecosystem on Ailaoshan Mountain*. Yunnan, China: Yunnan, Science and Technology Press.

WU, Z. Y. (ed.) (1987). *Vegetation of Yunnan*. Beijing, China: Science Press.

XINJIANG COMPREHANSIVE EXPEDITION INSTITUTE OF BOTANY ACADEMIA SINICA (1978). *Vegetation and its usage of Xinjiang*. Beijing, China: Science Press.

ZANG, R. G., AN, S. Q. & TAO, J. P. (2004). *Tropical Biodiversity Maintain Mechnism, Hannan Island*. Beijing, China: Science Press.

ZHANG, Y. Z., PAN, B. R., YIN, K. L. & DUAN, S. M. (2006). The characteristic of floristic composition and structure of *Ammopitanthus* community, Xinjiang. *Arid Zone Research* 23(2): 320-326.

ZHOU, Y. L. (1991). *Vegetation of Daxinganling Mountain*. Beijing, China: Science Press.

ZHU, H. (1993). The floristic characteristics of the tropical rainforest in Xishuangbanna. *Tropical Geography* 13: 149-155.

ZHU, H. (1994). The floristic characteristics of the tropical rainforest in Xishuangbanna. *Chinese Geographical Science* 4(2): 174-185.

ZHU, H., WANG, H. & LI, B. G. (1998). The structure, species composition and diversity of the limestone vegetation in Xishuangbana, SW China. *Gardens' Bulletin (Singapore)* 50: 5-30.

ZHU, H. (2001). *Rainforest in Xishuangbana*. Kunming, China: Yunnan Science and Technology Press.

Chapter 9

Floristic Elements of the Chinese Flora

PENG Hua and WU Cheng-Yih(WU Zheng-Yi)

9.1 Introduction

If the distribution of a large number of plant species, such as all those found in a given area such as China or Africa, is analysed, certain geographical patterns are found to recur consistently. Each of these patterns, together with the groups of taxa which exhibit them, are known as floristic elements, and their recognition may be a valuable method of floristic analysis. Five kinds of floristic element are recognised: geographic elements (based on the modern-day distribution of taxa), generative elements (based on their places of origin), migratory elements (based their migratory routes), historical elements (based on the time taxa occurred in a floristic area) and ecological elements (based on their habitats; Wulff, 1943; Szafer, 1975; Wang, 1992). All floristic areas usually provide examples of all five kinds of floristic elements, but geographic, generative and historical floristic elements are the most important and most frequently discussed. In the most major analysis of Chinese seed plants to date, WU Cheng-Yih (WU Zheng-Yi) and colleagues recognised 15 types (Wu & Wang, 1983; Wu, 1991; Wu *et al.*, 2003a; Wu *et al.*, 2003b; Wu *et al.*, 2006; Wu *et al.*, 2010) based mainly on geographic elements but also including some historical elements.

Research into the patterns (or areal-types, as they were known) of Chinese seed plants at generic level began in 1952 when the joint efforts of a number of Chinese botanists led to the preparation of a key to the genera of seed plants in China (Hou, 1958). Although this key was rather preliminary, it provided a basic and clear framework for research into the patterns of distribution of seed plant genera. Subsequently, a preliminary analysis was made based on all families and genera endemic to China and on every natural geographic region of China, and this was presented during a visit to the then USSR by a delegation of Chinese scientists. In 1959, WU returned to the subject in his teaching at the Biological Department of Yunnan University and wrote the *Outline of Chinese Floristic Geography* (Wu, 1959). In 1964, a comprehensive analysis of the distribution of 2 980 genera of Chinese seed plants known at the time was made (Wu, 1965). Fifteen types and 31 subtypes were detailed by WU at the international *Asia, Africa and Latin America Scientific Conference* in Beijing. His original data and discussions were not published in full, but were outlined in *Chinese Physical Geography, Phytogeography – Part I*, published in 1983 (Wu & Wang, 1983), in which he observed that, "After repeated tests, it has been proved that the characteristics and relationships of each areal-type can be shown by such classification. Furthermore, the research into geographical elements in connection with generative elements could be the solid foundation for regionalization of Chinese flora." (Wu & Wang, 1983: 106). Since 1983, WU and his collaborators have written many papers based on the 1983 studies (e.g. Wu, 1991; Wu *et al.*, 2003b; Wu *et al.*, 2006; Wu *et al.*, 2010) with a few changes being made in later studies according to the most recent literature.

Today, 3 169 genera of seed plants have been recorded in China, according to the *Flora of China,* including 256 exotics, which are not analysed here as they do not help to illuminate the natural floristic types of the country, but are known as Type 16. More continue to be discovered and described, through the *Flora of China* project and beyond. The 2 913 native genera may be classified into WU's 15 areal-types and 31 of his subtypes (Table 9.1; Wu *et al.*, 2003b; Wu *et al.*, 2006; Wu *et al.*, 2010) which are described briefly here. A fuller account is available in *The Areal-Types of Seed Plants and Their Origin and Differentiation* (Wu *et al.*, 2006).

9.2 Widespread (Type 1)

Widespread taxa are defined as genera that are widely distributed across every continent without a specific centre of diversity, or genera which have one, or even more than one, centre of diversity but contain some cosmopolitan species. Taxa recognised as exhibiting this pattern are generally widespread aquatics and mesophytes. Around 90 – ca. 3% – of China's genera fall into this category, of which two are monotypic and eight oligotypic. Many are herbaceous and sub-shrubby mesophytes such as *Anemone* and *Ranunculus* (Ranunculaceae), *Carex* (Cyperaceae), *Cyperus* (Cyperaceae), *Galium* (Rubiaceae), *Geranium* (Geraniaceae), *Gnaphalium* and *Senecio* (Asteraceae) and *Polygonum* (Polygonaceae), forming the dominant species of the herbaceous layer in forests and montane grasslands. Because they are so widespread it is impossible to draw any conclusions about the geographical affinities of the flora of China from these plants so they are excluded from the statistical data outlined below.

9.3 Pantropical (Type 2)

This type consists of those genera which are distributed throughout much of the tropics or which have at least one centre of diversity in the tropics but with some species

Table 9.1 The areal-types and subtypes of Chinese seed plant genera.

Type number	Type name	Total genera	Proportion of genera in China (excluding widespread and exotic genera; %)	Subtype number	Subtype name	Number of genera	Proportion of genera in China (excluding widespread and exotic genera; %)
1	Widespread	90	—	—		—	—
2	Pantropical	326	12	2-1	Disjunct between tropical Asia, Australasia (to New Zealand) and Central to South America	16	1
				2-2	Disjunct between tropical Asia, Africa and Central to South America	43	2
3	Disjunct between tropical Asia and tropical America	32	1	—	—	—	—
4	Old World tropics	181	6	4-1	Disjunct between tropical Asia, Africa (or East Africa and Madagascar) and Australasia	18	1
5	Tropical Asia and tropical Australasia	231	8	5-1	Disjunct between subtropical China and New Zealand	2	<1
6	Tropical Asia to tropical Africa	127	4	6-1	Disjunct between southern and southwestern China, India and tropical Africa	3	<1
				6-2	Disjunct between tropical Asia and eastern Africa or Madagascar	5	<1
7	Tropical Asia	512	18	7-1	Disjunct between Java/Sumatra and the Himalaya Shan to southern and southwestern China	51	2
				7-2	Tropical India to southern China	59	2
				7-3	Myanmar and Thailand to southwestern China	30	1
				7-4	Indochinese peninsula to southern or southwestern China	56	2
8	Northern temperate	145	5	8-1	Circumpolar	5	<1
				8-2	Arctic-alpine	15	1
				8-3	Disjunct between north-temperate and south-temperate regions	136	5
				8-4	Disjunct between temperate Eurasia and South America	2	<1
				8-5	Disjunct between the Mediterranean, eastern Asia, New Zealand and Central and South America	1	<1
9	Disjunct between eastern Asia and North America	111	4	9-1	Disjunct between eastern Asia and Mexico	2	<1
10	Old World temperate	179	6	10-1	Disjunct between the Mediterranean, western or central Asia and eastern Asia	15	1
				10-2	Disjunct between the Mediterranean and Himalaya Shan	8	<1
				10-3	Disjunct between Eurasia and South Africa	34	1
11	Temperate Asia	66	2	—	—	—	—

Continued

Type number	Type name	Total genera	Proportion of genera in China (excluding widespread and exotic genera; %)	Subtype number	Subtype name	Number of genera	Proportion of genera in China (excluding widespread and exotic genera; %)
12	Mediterranean and western to central Asia	114	4	12-1	Disjunct between the Mediterranean to central Asia and South Africa and Australasia	3	<1
				12-2	Disjunct between the Mediterranean to central Asia and Mexico to South America	2	<1
				12-3	Disjunct between the Mediterranean to temperate and tropical Asia and Australasia and South America	3	<1
				12-4	Disjunct between the Mediterranean to tropical Africa and the Himalaya Shan	5	<1
				12-5	Disjunct between the Mediterranean to northern Africa, central Asia, southwestern and North America, South Africa, Chile and Australasia	3	<1
				12-6	Disjunct between the Mediterranean to central Asia, tropical Africa, northern and eastern China and the Jinsha Jiang valley	1	<1
13	Central Asia	97	3	13-1	Eastern central Asia: Xinjiang, Gansu and Qinghai to Nei Mongol	22	1
				13-2	Central Asia to the Himalaya Shan and southwestern China	21	1
				13-3	Western Asia to the Himalaya Shan and Xizang	2	<1
				13-4	Disjunct between central Asia to the Himalaya Shan and Altai, and Pacific North America	3	<1
14	Eastern Asia	305	11	14-1	Sino-Himalayan	136	5
				14-2	Sino-Japanese	93	3
15	Chinese endemic	235	8	—	—	—	—
16	Exotic	256	—	—	—	—	—

distributed in other regions. Widely-distributed pantropical genera, found across the tropics, subtropics and even temperate regions may be categorised further by reference to their phylogeny or distribution; smaller monotypic or oligotypic genera by closely-related genera in the same family (the same approach may be used within many of the elements described below). In China, 326 genera, or ca. 12% of the total (excluding widespread taxa), may be classified as pantropical and subtypes thereof. Up to two-thirds of pantropical taxa, more than 250 genera, contain only one to ten species in China, due to the fact that China lies at the northern geographical limit of such tropical taxa. As an example, *Zornia* (Fabaceae) contains about 75 species distributed in tropical and temperate regions world-wide; yet only two species of the genus occur in China, and these are restricted to the south of the country.

There are 267 fairly typical pantropical genera in China (exemplified by *Connarus*, fig. 9.1), with six large families accounting for almost half of this number: Poaceae (42 genera), Fabaceae (23), Cyperaceae (12), Asteraceae (11), Euphorbiaceae (10) and Rubiaceae (7). About 50 of these genera are limited to tropical areas of Guangdong, Guangxi, Taiwan and Yunnan, especially in Hainan, particularly in families such as Connaraceae, Ochnaceae, Olacaceae, Sphenocleaceae and Surianaceae. For example, *Manilkara* (Sapotaceae), with a single Chinese species out of a world total of 65, is distributed only in the southern areas of Guangxi and Hainan. One species of *Xylopia* (Annonaceae), a genus of 160 species mainly distributed in tropical Africa, occurs in southern Guangxi. *Peltophorum* (Fabaceae; one native Chinese species out of 12) and *Ximenia* (Olacaceae; one of eight) are found in Hainan. *Scaevola* (Goodeniaceae, two out of 80 species) mainly limited to Oceania, is found in China on the southern coast, Taiwan and islands of the Nan Hai (South China Sea). In addition, *Guettarda* (Rubiaceae; one Chinese species out of 60–80), *Remirea* (Cyperaceae), the only species of which is Chinese, and *Sesuvium* (Aizoaceae; one species out of 17) display limited Chinese distributions on the tropical coasts of shallow seas. In all, exclusively tropical genera can be found from far inland to the seas of China. Most are evergreen shrubs or herbs of tropical forests or shallow seas, some are rare, and some are of economic value, such as *Ximenia* and *Cocos* (Arecaceae; one species worldwide including China).

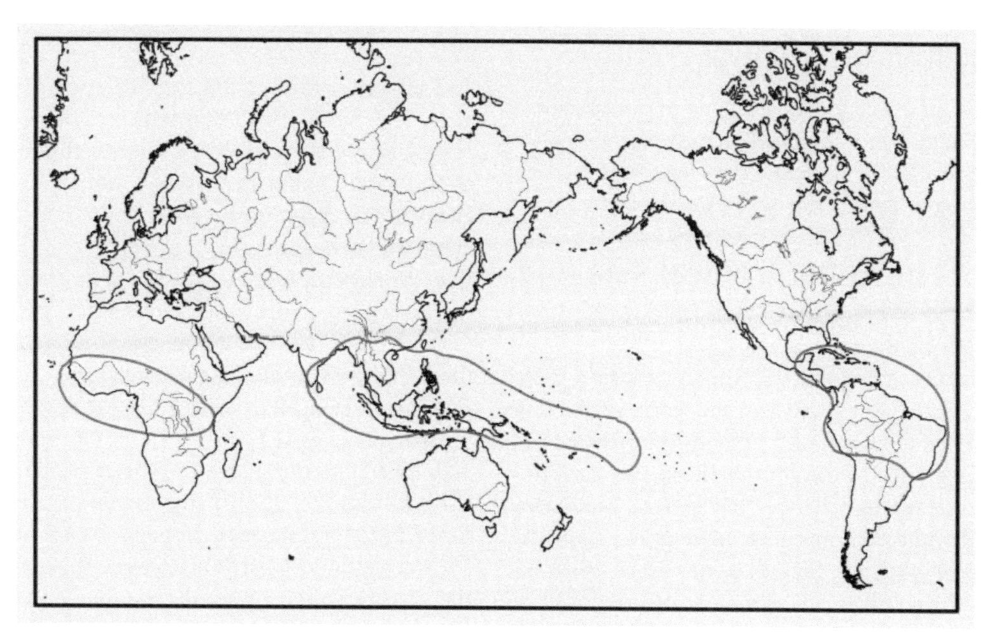

Figure 9.1 The distribution of *Connarus* (Type 2; taken from Wu *et al.*, 2010).

More than 100 typical pantropical genera occur in the subtropics of China, at altitudes of up to 3 000 m in southwestern mountainous areas. Most of these are evergreen trees, shrubs or lianas, which make up an important and sometimes dominant part of the vegetation. These include trees such as *Beilschmiedia* and *Cryptocarya* (Lauraceae), *Chionanthus* (Oleaceae), *Erythroxylum* (Erythroxylaceae), *Sterculia* (Sterculiaceae) and *Terminalia* (Combretaceae), which mainly occur in southwestern, southern and southeastern China, rarely as far north as the Chang Jiang (Yangtze River). Smaller trees and shrubs, such as *Alstonia* (Apocynaceae), *Glochidion* (Euphorbiaceae), *Schefflera* (Araliaceae) and *Symplocos* (Symplocaceae), are also common in this element. Genera of lianas include *Cissampelos* and *Cocculus* (Menispermaceae), and *Uncaria* (Rubiaceae).

Over 80 genera extend as far as northern temperate areas, among which *Ficus* (Moraceae), *Ilex* (Aquifoliaceae) and *Impatiens* (Balsaminaceae) each have more than 100 species in China. Most of these genera are herbaceous, especially the monocots. *Achyranthes* (Amaranthaceae), *Ipomoea* (Convolvulaceae), *Ocimum* (Lamiaceae), *Rotala* (Lythraceae), *Wahlenbergia* (Campanulaceae) and monocots such as *Commelina* and *Murdannia* (Commelinaceae), *Dioscorea* (Dioscoreaceae) and *Eriocaulon* (Eriocaulaceae) as well as Poaceae and Cyperaceae, are very common. *Andropogon*, *Aristida*, *Bothriochloa*, *Heteropogon* and *Imperata* (all Poaceae) are dominant components, often forming large populations, of the tropical and subtropical savanna grasslands of northern and northwestern China.

A few genera of trees extend into northern temperate areas, including *Celtis* (Ulmaceae), *Diospyros* (Ebenaceae) and *Ziziphus* (Rhamnaceae), as do shrubs and lianas including *Bauhinia* (Fabaceae), *Vitex* (Verbenaceae) and *Zanthoxylum* (Rutaceae). Both *Ziziphus jujuba* and *Vitex negundo* var. *heterophylla* form large populations around Beijing Shi. *Celastrus flagellaris* (Celastraceae) can even be found in mixed forests with broad and coniferous leaf trees as far as 50° N in northwestern China, while *Capparis spinosa* (Capparaceae) can extend into the deserts of Xinjiang and Xizang, suggesting that the Chinese temperate flora may have originated in the tropics.

9.3.1 Disjunct between tropical Asia, Australasia (to New Zealand) and Central to South America (Subtype 2-1)

There are two subtypes of the pantropical type. The first includes some 16 genera in China, most of them prominent components of Southern Hemisphere floras, with their northern boundary of distribution lying in the tropics of Asia and America. In China these mainly occur from southern Yunnan to the tropics of Taiwan and the southern subtropics, but some, such as *Ardisia* (Myrsinaceae), *Desmodium* (Fabaceae), *Passiflora* (Passifloraceae), *Perrottetia* (Dipentodontaceae) and *Planchonella* (Sapotaceae), extend to mid-subtropical and temperate parts of the country.

9.3.2 Disjunct between tropical Asia, Africa and Central to South America (Subtype 2-2)

Forty-three Chinese genera may belong to this category, and include *Chaetocarpus* (Euphorbiaceae), *Laurocerasus* (Rosaceae), *Oligomeris* (Resedacea), *Pouzolzia* (Urticaceae) and *Rotula* (Boraginaceae).

9.4 Disjunct between tropical Asia and tropical America (Type 3)

Included in this category are those genera which show a disjunct distribution between the warm regions of America and Asia. They can also extend to northeastern Australia or the islands of the southwestern Pacific. This element is represented by relatively few genera, especially when compared to the pantropical element: only 32 genera, ca. 1% of China's total, are included. In China, most of these occur to the south of the Chang Jiang, where they are important components of tropical and subtropical evergreen forests. These include *Clethra* (Clethraceae), *Meliosma* (Sabiaceae) *Nothaphoebe* (Lauraceae), *Schismatoglottis* (Araceae), *Talauma* (Magnoliaceae), *Turpinia* (Staphyleaceae) and *Waltheria* (Sterculiaceae). The distribution of some members of such genera, such as *Litsea pungens* and *Sageretia thea* (Rhamnaceae), can extend to northern and northeastern China. The reason this category is so poor in species perhaps relates to the fact that tropical South America, located in western Gondwana, became completely separated from other continents by the end of the Cretaceous Period (modern connections between South America and Africa indicate that their floras had shared, pre-Paleocene origins; Takhtajan, 1969; Thorne, 1975; Thorne, 1978).

9.5 Old World tropics (Type 4)

The Old World tropics is taken to mean tropical areas of Asia, Africa, Australasia and their adjacent islands, and is sometimes called the "Paleotropics," to contrast with the American "Neotropics." One hundred and eighty-one genera are classified in this element, ca. 6% of the total genera in China. Monotypic and oligotypic genera are more numerous than in the pantropical element, there are many "primitive" components, and the Old World tropical element is also more concentrated in the tropics, with only around ten genera extending to temperate regions. About 90 of its genera are limited to the tropics, including *Barringtonia* (Lecythidaceae), *Flagellaria* (Flagellariaceae), *Leea* (Leeaceae), *Macaranga* (Euphorbiaceae), *Pandanus* (Pandanaceae) and *Scolopia* (Flacourtiaceae). About 60 genera extend to the subtropics, including *Canarium* (Burseraceae), *Embelia* (Myrsinaceae), *Hypoestes* (Acanthaceae), *Musa* (Musaceae) and *Olax* (Olacaceae); these mainly occur in the Nanling Shan or southern Chang Jiang area, and as far west as Zayü and Mêdog in southeastern Xizang. The number of genera of this element extending to temperate regions falls off sharply, in contrast to the pantropical element; these include *Alangium* (Alangiaceae), *Emilia* (Asteraceae), *Mallotus* (Euphorbiaceae), *Momordica* (Cucurbitaceae) and *Monochoria* (Pontederiaceae). Type 4 is exemplified by *Pittosporum* (Pittosporaceae), in figure 9.2.

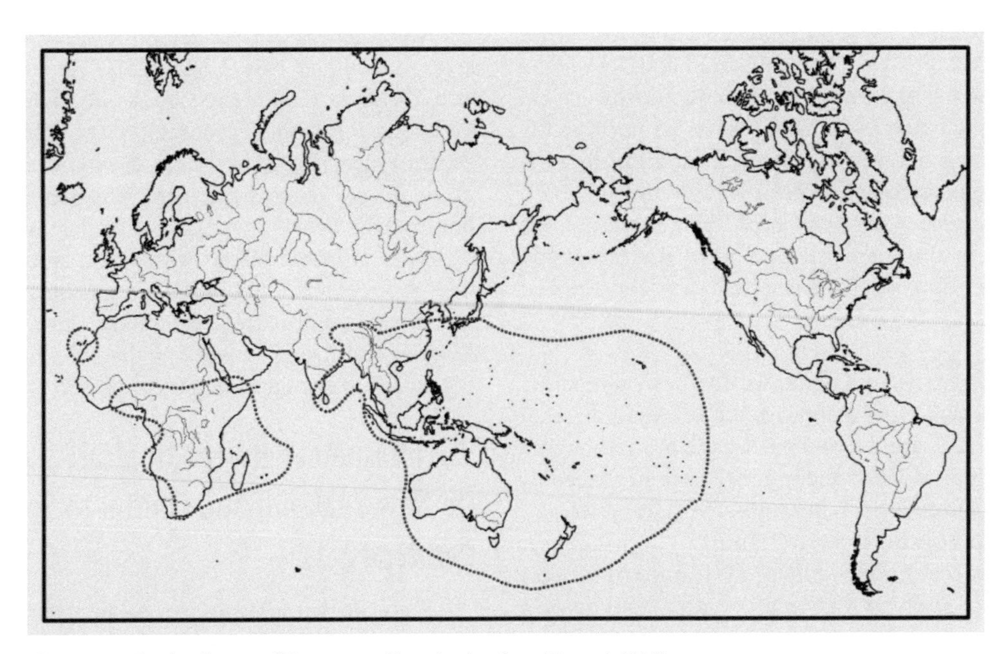

Figure 9.2 The distribution of *Pittosporum* (Type 4; taken from Wu *et al.*, 2010).

9.5.1 Disjunct between tropical Asia, Africa (or East Africa and Madagascar) and Australasia (Subtype 4-1)

About 18 genera may belong to this sub-element. For example, one species of the monotypic family Aponogetonaceae, *Aponogeton lakhonensis*, native to the Old World tropics and South Africa, also occurs in lakes and rivers of the southern Nanling Shan.

9.6 Tropical Asia and tropical Australasia (Type 5)

This element represents the eastern branch of the Old World tropical distribution. Its western boundary sometimes extends as far as Madagascar but never to continental Africa. An example is found in *Corybas* (Orchidaceae), of which 100 species occur in New Guinea, Australia, the Pacific islands, and southeast Asia to the Himalaya Shan, but none on the African continent. In China, 234 genera (8%) are placed in this category. The largest family is Orchidaceae (55 genera), followed by Poaceae (14), Rubiaceae (11), Euphorbiaceae (10), Fabaceae (8), Gesneriaceae (7), Rutaceae (7) and Urticaceae (7), a taxonomic distribution notably different from the pantropical and Old World tropical elements. Several genera in this element, such as members of Cardiopteridaceae, Centrolepidaceae, Cycadaceae, Myoporaceae, Nepenthaceae, Philydraceae, Polygalaceae and Saxifragaceae, are endemic to tropical Asia and tropical Australia (or at least to the Southern Hemisphere). This emphasises the difference between this and the other tropical elements.

Genera belonging to this type and limited to tropical regions include *Horsfieldia* (Myristicaceae), *Nypa* (Arecaceae), *Parsonsia* (Apocynaceae) and *Semecarpus* (Anacardiaceae). About 40 genera extend to the subtropics, including *Baeckea* (Myrtaceae), *Cinnamomum* (Lauraceae), *Cycas* (Cycadaceae), *Garuga* (Burseraceae), *Helicia* (Proteaceae) and *Melastoma* (Melastomataceae). These taxa are distributed as far north as the southern Chang Jiang, and are important components of tropical and subtropical evergreen broad-leaved forests. About ten genera of this element can be found in temperate regions, including *Ailanthus* (Simaroubaceae), *Toona* (Meliaceae), *Wikstroemia* (Thymelaeaceae) and *Zoysia* (Poaceae). From this we conclude that there are some connections between the floras of China and tropical Oceania. In particular, some genera or families originating from, or concentrated in, the Southern Hemisphere can be found in China south of the Chang Jiang. For example one widespread species, *Baeckea frutescens*, in a genus of seventy species, forms large populations in both the north and south Nanling Shan in China, while all other species occur in Australia and New Caledonia, with one in Kalimantan, Indonesia.

9.6.1 Disjunct between subtropical (southwestern) China and New Zealand (Subtype 5-1)

Two genera belong to this subtype, *Dacrycarpus* and *Dacrydium*, both Podocarpaceae. The former occurs in Indonesia, while some species of the latter extend to the Indochinese peninsula and others occur in the Southern Hemisphere.

9.7 Tropical Asia to tropical Africa (Type 6)

This element is described to include those genera distributed in the western part of the Old World tropics, from tropical Africa to India and Malaysia. Although some genera can extend to Fiji or other South Pacific islands, they are never present in Australia. A good example of this pattern is seen in *Lecanthus* (Urticaceae), of which three species occur in tropical Asia, one in tropical Africa, and one in Fiji, but none in Australia. A similar example is found in *Girardinia* (also Urticaceae), which also extends into subtropical to temperate China. One hundred and twenty-seven genera (ca. 4% of China's total) belong to this distribution type. Asteraceae contains the most genera of the type (12), followed by Fabaceae (11), Rubiaceae (8) and Asclepiadaceae (6). Ancistrocladaceae, Pandaceae and Salvadoraceae are endemic to this type of distribution; each is represented in China by a single species. In China, half of the genera belonging to this pattern are restricted to tropical regions, including *Ancistrocladus* (Ancistrocladaceae), *Lannea* (Anacardiaceae), *Pterygota* (Sterculiaceae) and *Sindora* (Fabaceae). These are important components of tropical rainforests and tropical monsoon forests.

About 60 genera extend to subtropical regions, mostly to the Nanling Shan and southern Chang Jiang area. *Pterolobium* (Fabaceae) is a typical example: it contains about ten species, distributed in tropical and subtropical Africa and Asia as far as Indonesia and the Philippines. There are two species in China, distributed across Fujian, Guangdong, Guangxi, Guizhou, Hainan, Hubei, Hunan, Jiangsu, Jiangxi, Sichuan, Yunnan, Zhejiang and the southern Nanling Shan. *Osyris* (Santalaceae), with six or seven species, is found from the Mediterranean to Africa and India; one species occurs from southwestern Xizang to the Nanling Shan. *Trachyspermum* (Apiaceae), has 12 species distributed from tropical to northeastern Africa, central Asia and India, two of which occur naturally in subtropical regions of southwestern China. As these examples show, the tropical Asia to tropical Africa distribution also has some connections with the flora of the Mediterranean.

About ten genera in this category extend to northern or northwestern China and temperate regions of northeastern China, among which only *Ceratostigma* (Plumbaginaceae), with five species, and *Periploca* (Asclepiadaceae), also with five species, are woody. The former is distributed from Xizang to the plateaus of Yunnan, Guizhou and Sichuan and northern China, and the latter occurs from southwestern and southern to northeastern and northwestern China. Others are herbaceous. For example, *Microstegium* and *Miscanthus* (both Poaceae) are main components of thickets to the south of the Chang Jiang. However, they also can be dominant in thickets of northern to northeastern China.

9.7.1 Disjunct between southern and southwestern China, India and tropical Africa (Subtype 6-1)

Three genera belong to this subtype: *Dregea* (Asclepiadaceae), *Rothmannia* (Rubiaceae) and *Terminthia* (Anacardiaceae).

9.7.2 Disjunct between tropical Asia and eastern Africa or Madagascar (Subtype 6-2)

This subtype comprises five genera: *Galeola* (Orchidaceae), *Hancea* (Euphorbiaceae), *Hedychium* (Zingiberaceae; Figure 9.3), *Rhopalocnemis* (Balanophoraceae) and *Woodfordia* (Lythraceae).

9.8 Tropical Asia (Type 7)

Tropical Asia, including India, Sri Lanka, the Indochinese peninsula, Indonesia, the Philippines and New Guinea, is in the centre of the Old World tropics. Its eastern boundary extends to Fiji and other South Pacific islands, but never to Australia, and its northern boundary encompasses southern and southwestern China and Taiwan. The region has one of the richest floras in the world, including many Paleotropical relict species. Five hundred and twelve genera (18% of China's total genera) are included in this category, making this the richest geographical element of the Chinese flora. This type is exemplified in figure 9.3 by *Kaempferia* (Zingiberaceae) and also by *Exbucklandia* (Hamamelidaceae; see figure 9.10).

Within this element, taxa limited to the tropics include around ten species in five genera of Dipterocarpaceae, characteristic trees of both Asian and African rainforests. *Dipterocarpus*, *Hopea* and *Vatica* occur in Yunnan and southern China, where they are the dominant species of tropical and monsoon rainforests. *Parashorea chinensis*, found in Yunnan and Guangxi, is the tallest species, growing to more than 60 m in tropical forests. The two species of *Shorea* are rare in China and restricted to southeastern Xizang and western Yunnan. Non-dipterocarp genera restricted to the tropics include *Dracontomelon* and *Mangifera* (both Anacardiaceae), *Goniothalamus* (Annonaceae), *Heliciopsis* (Proteaceae), *Knema* (Myristicaceae) and *Kopsia* (Apocynaceae).

About 100 genera extend into subtropical regions, including *Actinodaphne* (Lauraceae), *Brassaiopsis* (Araliaceae), *Distyliopsis* (Hamamelidaceae), *Engelhardia* (Juglandaceae), *Ixonanthes* (Erythroxylaceae), *Kadsura* (Schisandraceae), *Nothapodytes* (Icacinaceae), *Sarcandra* (Chloranthaceae) and *Sarcosperma* (Sapotaceae). These are mainly distributed in southwestern China south of the Chang Jiang and Qin Ling. Except for those of tropical families, these also include several basally-branching angiosperm lineages occurring in eastern Asia and North America, mainly in the northern temperate zone.

Some species extending to the subtropics, such as in *Gomphostemma* (Lamiaceae) and *Sonerila* (Melastomataceae), were in the past limited to tropical rainforests. *Petrosavia* (Liliaceae), a parasite of Southeast Asian tropical rainforests, has been found in Guangxi, subtropical forests of southern Yunnan, and Taiwan. These two facts indicate that Chinese subtropical forests, especially those in the south of the country, have a long history, and might have been derived from tropical forests.

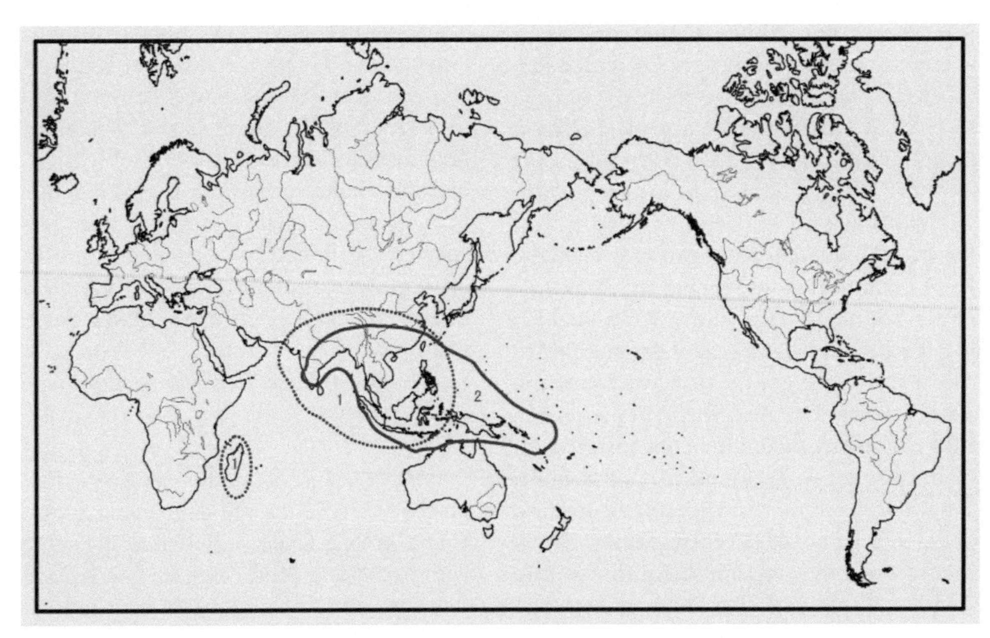

Figure 9.3 The distribution of *Hedychium* (1; Subtype 6-2) and *Kaempferia* (2; Type 7; taken from Wu *et al.*, 2010).

Only about ten, mainly herbaceous, genera of this element extend to temperate zones. For example, *Ellisiophyllum* (Scrophulariaceae) occurs in southwestern China, Gansu, Hebei, southern Xizang and Taiwan. Some species of *Chloranthus* (Chloranthaceae) and *Typhonium* (Araceae) are also found in deciduous forests in northeastern China. These examples suggest that the floras of such regions may have originated in the tropics during the Cenozoic Era (Axelrod, 1952; Axelrod, 1960; van Steenis, 1962; Wu, 1965; Axelrod, 1972; Thorne, 1975; Thorne, 1978). Five subtypes of the tropical Asia element have been described, four of which are discussed here.

9.8.1 Disjunct between Java/Sumatra and the Himalaya Shan to southern and southwestern China (Subtype 7-1)

About 51 genera are present in both Java or Sumatra and the Himalaya Shan, southern or southwestern China. These include *Disepalum* (Annonaceae), *Deutzianthus* (Euphorbiaceae), *Rhodoleia* (Hamamelidaceae), *Pentaphylax* (Pentaphylacaceae) and *Wightia* (Scrophulariaceae).

9.8.2 Tropical India to southern China (particularly southern Yunnan; Subtype 7-2)

About 59 genera fall into this subtype, including *Eleutharrhena* (Menispermaceae), *Gynocardia* (Flacourtiaceae), *Heteropanax* (Araliaceae), *Petrocosmea* (Gesneriaceae) and *Syndiclis* (Lauraceae).

9.8.3 Myanmar and Thailand to southwestern China (Subtype 7-3)

This subtype contains 30 genera, including *Afgekia* (Fabaceae), *Blinkworthia* (Convolvulaceae), *Formanodendron* (Fagaceae), *Pittosporopsis* (Icacinaceae) and *Sladenia* (Sladeniaceae).

9.8.4 Indochinese peninsula to southern or southwestern China (Subtype 7-4)

About 56 genera belong to this subtype, including *Alniphyllum* (Styracaceae), *Arcangelisia* (Menispermaceae), *Carrierea* (Flacourtiaceae), *Craigia* (Tiliaceae), *Fokienia* (Cupressaceae) and *Woonyoungia* (Magnoliaceae). Southern and southwestern China are closely connected to the Indochinese peninsula (in particular Vietnam), which in the past was even considered part of southern China. Such close connections mean the floras of the two areas may be considered parts of the same. This region may have provided natural refugia for many Cenozoic Palaeotropical floristic elements, and a centre of diversity for thermophytes in eastern Asia (Axelrod, 1972; Schuster, 1972; Thorne, 1975).

9.9 Northern temperate (Type 8)

The northern temperate element is represented by genera which are widely distributed in temperate regions of Europe, Asia and North America. It includes 304 genera in China (ca. 11% of the total), with a distinctive high proportion of medium-sized genera (10–100 species). More than half (31) of the 59 large genera, with more than 100 species in China, also belong to this type. In contrast, there are fewer than 30 monotypic and oligotypic genera. This suggests that China may be the main centre of diversity for many relatively young temperate genera. The other characteristic of this element is the number of woody plant taxa—about 50 genera—a level of diversity incomparable with other parts of the world. The main genera of trees in this element include *Abies, Larix* and *Picea* (Pinaceae), *Acer* (Aceraceae), *Aesculus* (Hippocastanaceae), *Alnus, Betula, Carpinus* and *Ostrya* (Betulaceae), *Castanea, Fagus* and *Quercus* (Fagaceae), *Cerasus, Malus* and *Sorbus* (Rosaceae), *Cupressus* and *Juniperus* (Cupressaceae), *Fraxinus* (Oleaceae), *Juglans* (Juglandaceae), *Morus* (Moraceae), *Pinus* (Pinaceae), *Populus* and *Salix* (Salicaceae), *Taxus* (Taxaceae), *Tilia* (Tiliaceae) and *Ulmus* (Ulmaceae). These genera are distributed from southwestern to northeastern China and form the dominant species or important components within temperate broad-leaved deciduous, coniferous, subtropical and tropical montane forests. Areal-type 8 is exemplified by *Actaea* and *Cimicifuga* (both Ranunculaceae; Figure 9.4).

In addition to the tree genera are shrubs including *Berberis* (Berberidaceae), *Cotinus* (Anacardiaceae), *Cornus* (Cornaceae), *Corylus* (Betulaceae), *Elaeagnus* (Elaeagnaceae), *Lonicera* (Caprifoliaceae), *Rhododendron* (Ericaceae), *Rhus* (Anacardiaceae), *Ribes* (Saxifragaceae) and *Viburnum* (Adoxaceae). Herb genera are also abundant and many of these are typical and important components of temperate forests or meadows, such as *Androsace* (Primulaceae), *Cirsium* (Asteraceae), *Lithospermum* and *Mertensia* (both Boraginaceae), *Mentha* and *Prunella* (Lamiaceae). In fact, Xizang and the mountainous areas of southwestern China are centres of diversity for *Primula* (Primulaceae), *Rhodiola* and *Sedum* (Crassulaceae) and *Saussurea* (Asteraceae). Xerophilous subshrubs, such as *Artemisia* (Asteraceae), *Oxytropis* (Fabaceae) and *Silene* (Caryophyllaceae), which are mainly distributed in the arid regions of northwestern China, are the dominant or representative plants of desert or desert-steppe formations within this element.

9.9.1 Circumpolar (circumarctic; Subtype 8-1)

Genera assigned to the circumpolar subtype are distributed in the northern part of the north-temperate and polar region, with a few extending to lower latitudes including amphipolar genera. Five genera belong to this subtype: *Calla* (Araceae), *Chamaedaphne* and *Ledum* (both Ericaceae), *Linnaea* (Linnaeaceae) and *Scheuchzeria* (Scheuchzeriaceae).

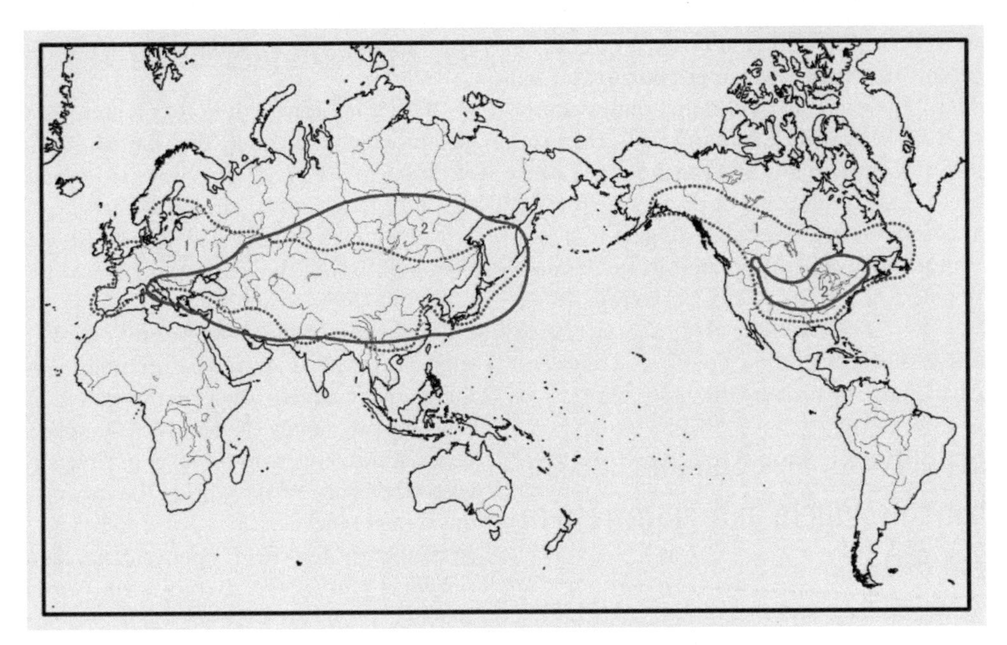

Figure 9.4 The distribution of *Actaea* (1) and *Cimicifuga* (2; both Type 8; taken from Wu *et al.*, 2010).

9.9.2 Arctic-alpine (Subtype 8-2)

The Arctic-alpine subtype includes those genera that are distributed in both arctic and alpine regions of higher latitudes together with those of subtropical and tropical alpine regions. About 14 genera belong to this subtype: *Arctous*, *Cassiope* and *Orthilia* (all Ericaceae), *Braya* and *Eutrema* (both Brassicaceae), *Diapensia* (Diapensiaceae), *Dryas* (Rosaceae), *Koenigia* and *Oxyria* (both Polygonaceae), *Lagotis* (Scrophulariaceae) and *Sibbaldia* (Rosaceae).

9.9.3 Disjunct between north-temperate and south-temperate regions (Subtype 8-3)

Included in this subtype are all those genera with disjunct distributions between north-temperate areas and Australia, southern Africa or South America (or any combination thereof), as well as "pan-temperate" disjunct distributions in both northern and southern temperate regions. About 136 genera belong to the subtype, none of them monotypic, but including several oligotypic genera including *Cinna* and *Scolochloa* (both Poaceae), *Hippuris* (Hippuridaceae) and *Turritis* (Brassicaceae). Woody plants, such as *Ephedra* (Ephedraceae), *Lycium* (Solanaceae), *Sambucus* (Adoxaceae) and *Vaccinium* (Ericaceae), are fewer in this subtype.

9.9.4 Disjunct between temperate Eurasia and South America (Subtype 8-4)

Just two genera are found to be disjunct between Eurasia and South America: *Briza* (Poaceae) and *Hypochaeris* (Asteraceae). It is suggested that such genera previously inhabited a continuous continent prior to the separation of tropical America from Eurasia, after which the diversity of each became centred on one of these regions.

9.9.5 Disjunct between the Mediterranean, eastern Asia, New Zealand and Central and South America (Subtype 8-5)

A single genus, *Coriaria* (Coriariaceae; Figure 9.5), shows a complex disjunct distribution incorporating the Mediterranean region, the Himalaya Shan to temperate and subtropical regions of eastern Asia and the Philippines, New Guinea, New Zealand, South Pacific islands, and the Andes of Mexico and Chile.

9.10 Disjunct between eastern Asia and North America (Type 9)

This element comprises 111 genera (4% of China's total) whose distributions are discontinuous between the temperate and subtropical regions of both eastern Asia and North America. They include *Calycanthus* (Calycanthaceae), *Glehnia* (Apiaceae), *Hamamelis* and *Liquidambar* (Hamamelidaceae), *Illicium* (Illiciaceae), *Liriodendron* (Magnoliaceae), *Nyssa* (Nyssaceae), *Phryma* (Phrymaceae), *Schisandra* (Schisandraceae) and *Shortia* (Diapensiaceae). The distribution of these genera shows that the Asian flora has close historical connections with that of North America and that the element has relations with both the tropics and Gondwana. However the modern-day centres of diversity for these genera tend to lie in eastern Asia or North America, mostly the former, with representatives including *Aralia* and *Panax* (Araliaceae), *Gleditsia* (Fabaceae), *Itea* (Saxifragaceae), *Osmanthus* (Oleaceae), *Pseudotsuga* and *Tsuga* (Pinaceae) and *Torreya*

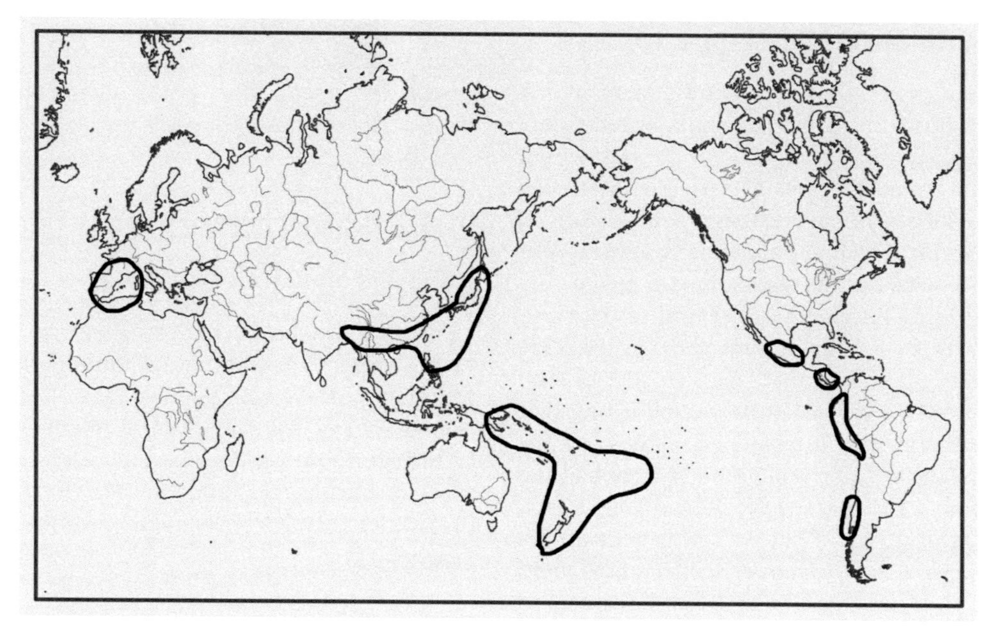

Figure 9.5 The disjunct distribution of *Coriaria* (Subtype 8-5; taken from Wu *et al.*, 2010).

(Taxaceae). Genera centred on North America include *Carya* (Juglandaceae), *Chamaecyparis* (Cupressaceae), *Lyonia* (Ericaceae), *Symphoricarpos* (Caprifoliaceae; Figure 9.6), *Thuja* (Cupressaceae) and *Trillium* (Liliaceae). *Mahonia* (Berberidaceae) is rich in both eastern Asia and North America.

China has more genera in common with the east than the west of North America, because of their similar historical and natural conditions, and due to intense historical changes in geology and climate and the existence of desert in the southwest of North America. Apart from the polytypic genera mentioned above (such as *Illicium*, *Itea* and *Osmanthus*), many oligotypic genera fit this pattern, with 31 out of 50 of these being found in both China and the east of North America (Gray, 1856; Gray, 1957; Good, 1974; Boufford & Spongberg, 1983; Elbert & Little, 1983; Boufford, 1992; Takhtajan, 1997). A few woody examples are shared with the west of North America, including *Pseudotsuga* and *Mahonia*. The herbaceous *Kelloggia* (Rubiaceae) comprises two species, one in southern Sichuan and Yunnan, the other occurring in western North America.

9.10.1 Disjunct between eastern Asia and Mexico (Subtype 9-1)

Only two genera fit this pattern: *Cleyera* (Theaceae) and *Torreya*.

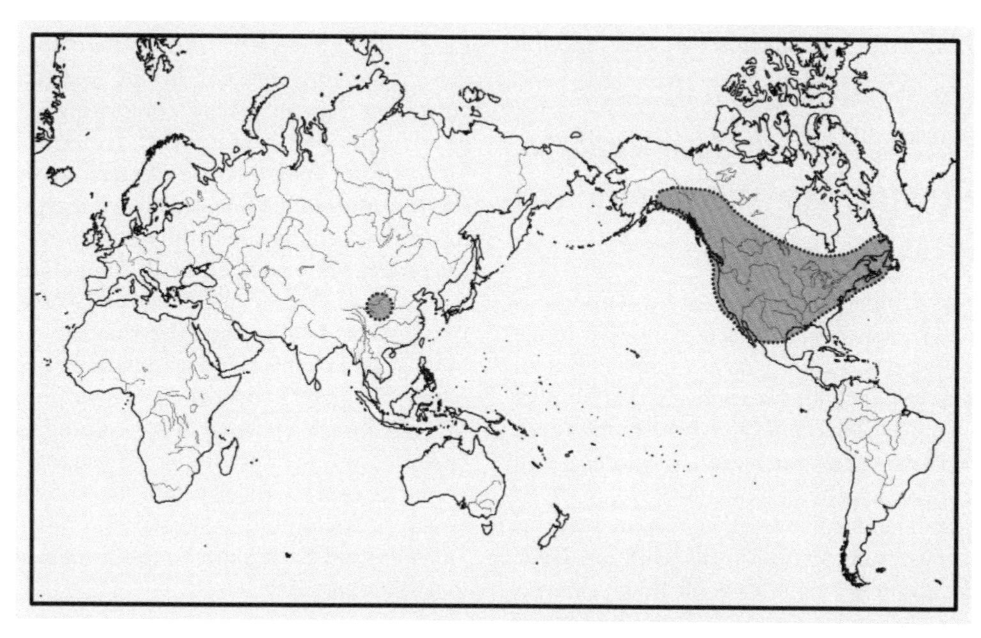

Figure 9.6 The distribution of *Symphoricarpos* (Type 9; taken from Wu *et al.*, 2010).

9.11 Old World temperate (Type 10)

This element generally includes all genera which are widely distributed in temperate and cold-temperate regions from high to middle latitudes of Eurasia, with a few extending to tropical regions of Asia, Africa or even Australia. It comprises 179 genera (ca. 6% of all China's genera). Lamiaceae, which contains many genera with this pattern, is abundant in the Mediterranean region. The Mediterranean (or Mediterranean to central Asia) is also the centre of diversity for Dipsacaceae and Tamaricaceae in this type. Thus, the element has characteristics of both Mediterranean and central Asian floras, indicating that the floras of both Old World temperate regions and the Mediterranean-central Asian region might have had the same origin around the coast of the Tethys Sea. In fact, the modern-day diversity of quite a few genera of this type, such as *Ajuga*, *Calamintha*, *Marrubium* and *Phlomis* (Lamiaceae), *Isatis* (Brassicaceae) and *Tulipa* (Liliaceae), is centred on the Mediterranean, western or central Asia.

The distribution of some genera in this element extends to northern Africa or mountainous regions of tropical Africa (almost to eastern Africa). To the former type belong *Doronicum* (Asteraceae), *Gagea* (Liliaceae), *Lamium* and *Thymus* (Lamiaceae), *Melilotus* (Fabaceae) and *Pyrus* (Rosaceae); the latter includes *Dipsacus* (Dipsacaceae), *Nepeta* (Lamiaceae) and *Verbascum* (Scrophulariaceae). *Trapa* (Trapaceae) can extend to mountainous regions of tropical Asia and some species of *Ligustrum* (Oleaceae) occur in Australia. All these examples suggest that the Eurasian temperate flora might have historical connections with Gondwana. Other genera of the element are mainly distributed in temperate or eastern Asia, including *Chrysanthemum* (Asteraceae), *Daphne* (Thymelaeaceae), *Epimedium* (Berberidaceae), *Herminium* (Orchidaceae), *Hippophae* (Elaeagnaceae), *Meconopsis* (Papaveraceae), *Myricaria* (Tamaricaceae), *Paris* (Liliaceae) and *Syringa* (Oleaceae). Genera centred on temperate Eurasia include *Adonis* and *Callianthemum* (Ranunculaceae), and *Ligularia* (Asteraceae).

9.11.1 Disjunct between the Mediterranean, western or central Asia and eastern Asia (Subtype 10-1)

The centre of diversity for genera of this subtype is often close to eastern Asia; for only a few is it nearer to the Mediterranean region or western Asia. About 15 genera belong to the subtype, including a relatively rich number of woody taxa such as *Amygdalus* and *Pyracantha* (both Rosaceae), *Atraphaxis* (Polygonaceae), *Fontanesia* and *Forsythia* (both Oleaceae), *Paliurus* (Rhamnaceae), *Parrotia* (Hamamelidaceae) and *Zelkova* (Ulmaceae). Most of these are deciduous or evergreen shrubs, with few trees. *Theligonum* (Rubiaceae) is a prime example of the subtype, with four species, distributed disjunctly from the Canary Islands to the Mediterranean, China and Japan. The three species found in China are dispersed in Anhui, Hubei, Shaanxi, Sichuan, Taiwan, Xizang and Zhejiang. The centres of diversity of some other genera show a tendency to lie to the west, for example *Hesperis* (Brassicaceae) and *Origanum* (Lamiaceae).

9.11.2 Disjunct between the Mediterranean and Himalaya Shan (Subtype 10-2)

Eight genera belong to this subtype, including the relict taxa *Cedrus* (Pinaceae) and *Mandragora* (Solanaceae). Even today, native wild populations of *Cedrus*, thought to exist in Xizang, have not been identified. Other examples of this distribution type include *Melissa* (Lamiaceae), *Morina* (Morinaceae) and *Onosma* (Boraginaceae). The distributions of a few genera of this subtype extend as far as Indonesia.

9.11.3 Disjunct between Eurasia and South Africa (Subtype 10-3)

There are 34 genera of this subtype, which occasionally extends in distribution as far as Australia. They include *Peucedanum* (Apiaceae), *Lotus*, *Medicago* and *Trigonella* (Fabaceae), *Matthiola* (Brassicaceae) and *Scabiosa* (Dipsacaceae).

9.12 Temperate Asia (Type 11)

In this element are included genera confined to the temperate regions of Asia, with a range from central Asia (Russia) to eastern Russia and northeastern Asia, and southward to northern China, Korea and northern Japan. Its southernmost boundary lies in the mountainous regions of the Himalaya Shan and southwestern China. Some examples can also be found in subtropical regions while a few extend to tropical Asia and even New Guinea. This distribution type is represented by 66 genera, or ca. 2% of all genera in China. Monotypic and oligotypic genera are more common in this type compared with the temperate elements above, comprising 51, 76% of genera of this type. This indicates that, rather than having a long history, most genera of this element are relatively recently evolved. Most of the monotypic or oligotypic genera belong to phylogenetic groups with some polytypic genera and are recent segregates of north-temperate or widespread genera. Examples include *Filifolium* (Asteraceae), related to *Artemisia*, and *Orostachys* (Crassulaceae), related to *Sedum*. Most genera of this element, especially the monotypic and oligotypic genera such as *Cymbaria* (Scrophulariaceae), *Macropodium* (Brassicaceae) and *Symphyllocarpus* (Asteraceae), are distributed in the north of temperate Asia, i.e. central Asia to Siberia or northeastern Asia. However, *Arctogeron* (Asteraceae), *Chosenia* (Salicaceae), *Filifolium* and *Saposhnikovia* (both Apiaceae) are confined to northeastern Asia.

Some genera have further diversified in the Himalaya

Shan, a second centre of diversity. For example *Bergenia* (Saxifragaceae) has a disjunct distribution between Siberia, northeastern and central Asia and the Himalaya Shan. The seven species in China are mainly dispersed in Xizang, mountainous regions of southwestern China, the Altai in Xinjiang and the Qin Ling. *Crucihimalaya* (Brassicaceae) also occurs in the Himalaya Shan and through southwestern and central Asia to Mongolia and Russia. From these distributions we conclude that the temperate Asian element comprises mostly recently-derived taxa of Laurasian origin. Most genera belonging to the type may have evolved and differentiated from large, north-temperate or worldwide genera, as the temperate climate in Asia, especially central Asia, gradually dried out; further examples diversified during the uplift of the Himalaya Shan.

9.13 Mediterranean and western to central Asia (Type 12)

This element contains genera that are distributed around the Mediterranean, through western or southwestern Asia to central Asia (Russia and Xinjiang in China) and the plateaus of Qinghai-Xizang and Nei Mongol. The category includes 114 genera (ca. 4% of China's total genera) with Brassicaceae (21 genera), Asteraceae (17), Chenopodiaceae (14), Poaceae (8), Boraginaceae (7), Fabaceae (7), Apiaceae (5), Caryophyllaceae (3) and Papaveraceae (3) represented by the most genera and with Lamiaceae, Malvaceae and Plumbaginaceae each including two genera. In China, genera with this distribution pattern are mostly found in Xinjiang, Xizang, northwestern and southwestern China, and are dominant in dry deserts, grassland, alpine and subalpine vegetation. Major examples include *Aeluropus* (Poaceae), *Anabasis* (Figure 9.7), *Camphorosma*, *Ceratocarpus*, *Halocnemum*, *Halostachys*, *Haloxylon*, *Kalidium* and *Nanophyton* (all Chenopodiaceae) and *Reaumuria* (Tamaricaceae).

9.13.1 Disjunct between the Mediterranean to central Asia and South Africa and Australasia (Subtype 12-1)

Three genera belong to this subtype. Xerophilous species of *Zygophyllum* (a genus of ca. 100 species; Zygophyllaceae) are typical, with 19 species in China occurring commonly from Xinjiang to desert and desert-steppe areas of the Ordos Gaoyuan. *Zygophyllum* is thought to have originated in Gondwana before the separation of India and Madagascar and is distributed from the Mediterranean to central Asia, with disjunct populations in South Africa and Australia. *Rochelia* (Boraginaceae) shows a disjunction only with Australia.

9.13.2 Disjunct between the Mediterranean to central Asia and Mexico to South America (Subtype 12-2)

This subtype comprises two genera including *Peganum* (Peganaceae), which contains three species in China, found in the north and northwest of the country. Such taxa imply that the coasts of the Tethys Sea must once have extended westwards to present day Central America and the West Indies and that the character of the floras around these coasts was dry and tropical.

9.13.3 Disjunct between the Mediterranean to temperate and tropical Asia and Australasia and South America (Subtype 12-3)

Three genera are have a disjunct distribution comprising the Mediterranean to temperate and tropical Asia, and Australia and South American. Evergreen shrubs

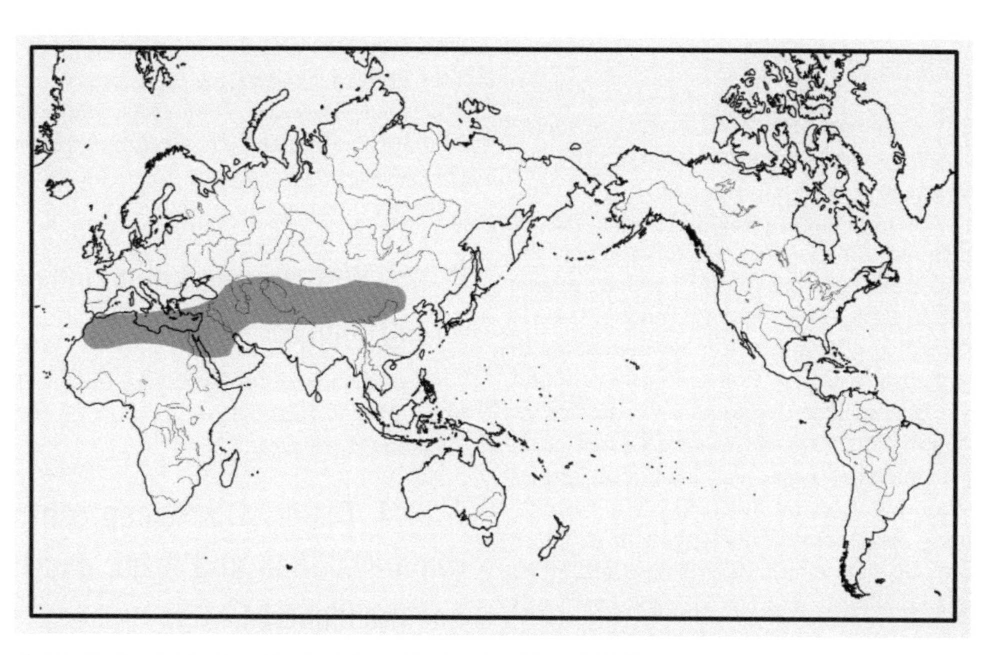

Figure 9.7 The distribution of *Anabasis* (Type 12; taken from Wu *et al.*, 2010).

or small trees of the genus *Olea* (Oleaceae) are distributed from the Mediterranean to tropical Asia and Polynesia and is also found in New Zealand. *Olea* is highly typical of a Mediterranean climate, sometimes known as the "*Olea* climate". Thirteen species of *Olea* occur in China, mainly in the southwest of the country and Taiwan.

9.13.4 Disjunct between the Mediterranean, tropical Africa and the Himalaya Shan (Subtype 12-4)

Five genera in China have a disjunct distribution between the Mediterranean, tropical Africa and the Himalaya Shan, including *Pterocephalus* (Dipsacaceae) which occurs in northwestern China, Xizang and mountainous regions of the southwest.

9.13.5 Disjunct between the Mediterranean to northern Africa, central Asia, southwestern and North America, South Africa, Chile and Australasia (Subtype 12-5)

Three genera belong to this "pan-Mediterranean" subtype, with a distribution encompassing the Mediterranean to northern Africa, central Asia, southwestern and North America, South Africa, Chile and Australasia: *Frankenia* (Frankeniaceae), *Hornungia* (Brassicaceae) and Lavatera (Malvaceae).

9.13.6 Disjunct between the Mediterranean to central Asia, tropical Africa, northern and eastern China and the Jinsha Jiang valley (Subtype 12-6)

Only *Ferula* (Apiaceae) belongs to this subtype.

9.14 Central Asia (Type 13)

The central Asian element includes all genera distributed in central Asia (particularly mountainous regions) but not in western Asia and the Mediterranean region, and represents plants that originated around the eastern part of the ancient Tethys Sea. Ninety-seven genera (ca. 3% of the total in China) are included in this category, and are mainly distributed in mountainous regions of Xinjiang, Gansu, Qinghai, Nei Mongol and Xizang, with a few extending to northern or northeastern China. Common genera found in northwestern China include *Cryptospora* (Brassicaceae), *Ikonnikovia* (Plumbaginaceae), *Kaschgaria* (Asteraceae) and *Lagochilus* (Lamiaceae). *Lagochilus* is a typical species of mountainous regions of central Asia, and is also the only large genus among the genera of this type, with about 35 species. It shows a disjunct distribution from central Asia (Russia) to Xinjiang, Ningxia and Nei Mongol, westwards through Afghanistan to Iran, but no further west. Its centre of diversity lies in the Tian Shan and Pamir-Altai region. The 11 species in China are mainly distributed in northern Xinjiang and the southern Tian Shan, with one extending from Helan Shan in the Alxa Shamo to Xilin Gol, Nei Mongol. Those in Chenopodiaceae are all distributed on desert plains, while *Stephanachne* (Poaceae) extends to Qinghai, Gansu, Nei Mongol and Shanxi, and *Cancrinia* (Asteraceae) and *Lophanthus* (Lamiaceae) are found in Xizang.

9.14.1 Eastern central Asia: Xinjiang (especially Kashi), Gansu and Qinghai to Nei Mongol (Subtype 13-1)

Twenty-two genera belong to this subtype. Among these, *Iljinia* and *Sympegma* (both Chenopodiaceae), distributed from Xinjiang to Gansu and western Nei Mongol, are typical of the deserts of eastern central Asia. *Psammochloa* (Poaccae) and *Pugionium* (Brassicaceae) are typical of the east of this area, forming large populations in Gansu and the on gravels or dunes of the deserts of western Nei Mongol. *Agriophyllum* (Chenopodiaceae), *Neopallasia* (Asteraceae) and *Panzerina* (Lamiaceae) are relatively widespread, with the pioneer *Agriophyllum* extending to northern and northeastern China. The single Chinese species of *Ammopiptanthus* (Fabaceae), the only extant evergreen shrub species in eastern central Asia, has an interesting distribution comprising the western part of southern Xinjiang and Ningxia, and western to southern Nei Mongol.

9.14.2 Central Asia to the Himalaya Shan and southwestern China (Subtype 13-2)

Twenty-one genera have distributions covering central Asia to the Himalaya Shan and the southwest of China. They are mainly distributed in Xinjiang, the high mountains of Xizang and Yunnan, and Sichuan, and include *Lasiocaryum* and *Lindelofia* (Boraginaceae), *Megacarpaea* (Brassicaceae) and *Thylacospermum* (Caryophyllaceae). The genera extending to Yunnan and Sichuan are *Littledalea* (Poaceae), *Notholirion* (Liliaceae), *Paraquilegia* (Ranunculaceae) and *Trachydium* (Apiaceae). *Atelanthera* (Brassicaceae), *Cortia* and *Cortiella* (both Apiaceae) and *Trikeraia* (Poaceae) occur only in Xizang.

9.14.3 Western Asia to the Himalaya Shan and Xizang (Subtype 13-3)

Two genera occur from western Asia to the Himalaya Shan and Xizang: *Mattiastrum* (Boraginaceae) and *Perovskia* (Lamiaceae).

9.14.4 Disjunct between central Asia, the Himalaya Shan and Altai, and Pacific North America (Subtype 13-4)

Three genera belong to the first of these subtypes,

Christolea, *Hedinia* and *Stroganowia* (all Brassicaceae), and suggest past connections between the flora of central Asia and that of North America across the Bering Strait. For example *Christolea,* a genus considered to have originated in central Asia or the western Himalaya Shan, has disjunct populations in Kamchatka (Russia) and Alaska (USA).

9.15 Eastern Asia (Type 14)

This element includes genera distributed from the Himalaya Shan to Japan, excluding the far east of Russia, and are never found as far south as Vietnam, the Philippines or Java. Its northwestern boundary coincides with the limit of the forests of northwestern China. This element might be confused with that of Temperate Asia, but has distribution ranges that are generally less extensive and genera that are nearly all confined to forest habitats, with centres of diversity lying within the area defined by the Himalaya Shan and Japan. The category is the third largest element, after the tropical Asian and pantropical elements, with about 305 genera (ca. 11% of China's total). About 76 genera are typical of the area, mainly woody taxa such as *Actinidia* (Actinidiaceae), *Aucuba* (Aucubaceae), *Cephalotaxus* (Cephalotaxaceae), *Corylopsis* and *Loropetalum* (both Hamamelidaceae), *Eleutherococcus* (Araliaceae), *Enkianthus* (Ericaceae), *Hovenia* (Rhamnaceae), *Stachyurus* (Stachyuraceae; Figure 9.8) and *Vernicia* (Euphorbiaceae). Herbaceous genera include *Ainsliaea* (Asteraceae), *Cardiocrinum* (Liliaceae), *Codonopsis* (Campanulaceae), *Dichocarpum* (Ranunculaceae), *Euryale* (Nymphaeaceae) and *Houttuynia* (Saururaceae).

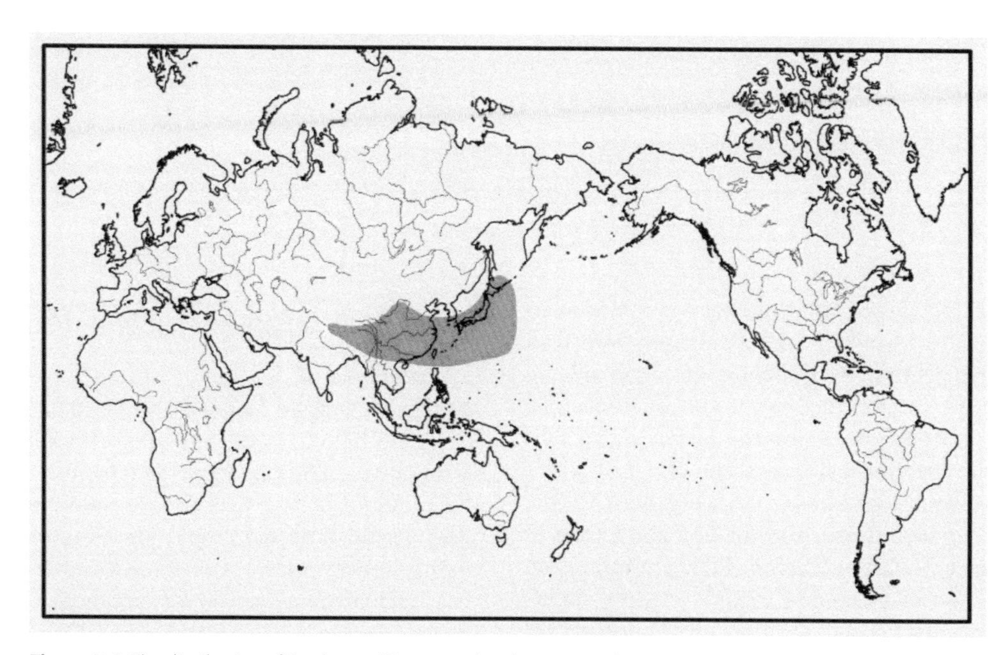

Figure 9.8 The distribution of *Stachyurus* (Type 14; taken from Wu *et al.*, 2010).

9.15.1 Sino-Himalayan (Subtype 14-1)

This subtype includes genera occurring from the Himalaya Shan to southwestern China, sometimes to Shaanxi, Gansu, eastern China and even Taiwan. Outside China, some genera may extend to the Indochinese peninsula. This subtype may be confused with the tropical Asian element, but is distinguished by its taxa, which are never found in Japan. It comprises 136 genera, of which ca. 80% are monotypic or oligotypic. Most of the monotypic genera are herbs of Xizang and the high mountains of southwestern China, such as *Chionocharis* (Boraginaceae), *Hemiphragma* and *Oreosolen* (both Scrophulariaceae), *Leptocodon* (Campanulaceae), *Milula* (Liliaceae) and *Rubiteucris* (Lamiaceae). In addition to these, more recently-evolved, herbs there are eight genera of woody plants considered to represent older lineages, including *Decaisnea* (Lardizabalaceae), *Hemiptelea* (Ulmaceae) and *Tetracentron* (Tetracentraceae). Of the 71 oligotypic genera, 18 (a higher proportion than for the monotypics) are woody, including *Dobinea* (Anacardiaceae), *Prinsepia* (Rosaceae) and *Toricellia* (Toricelliaceae), endemic to eastern Asia. The Subtype is exemplified by *Hemsleya* (Cucurbitaceae; Figure 9.9).

Older, relict taxa often have distinct distributions. For example, *Prinsepia* (Rosaceae) includes about five species, of which one is found from the Himalaya Shan to Yunnan, Guizhou, the Huangtu Gaoyuan (Loess Plateau), Liaodong Bandao and Taiwan. Relict genera also occur in predominantly tropical families, for example *Luculia* and *Neohymenopogon* (Rubiaceae) and *Merilliopanax* (Araliaceae). The element also includes genera in families of a predominantly eastern Asian-North American distribution, such as *Colquhounia* and *Leucosceptrum* (Lamiaceae), *Leycesteria* (Caprifoliaceae) and *Thamnocalamus* (Poaceae). Some oligotypic herbaceous genera within this element have very interesting distributions. For example, *Cryptochilus* (Orchidaceae)

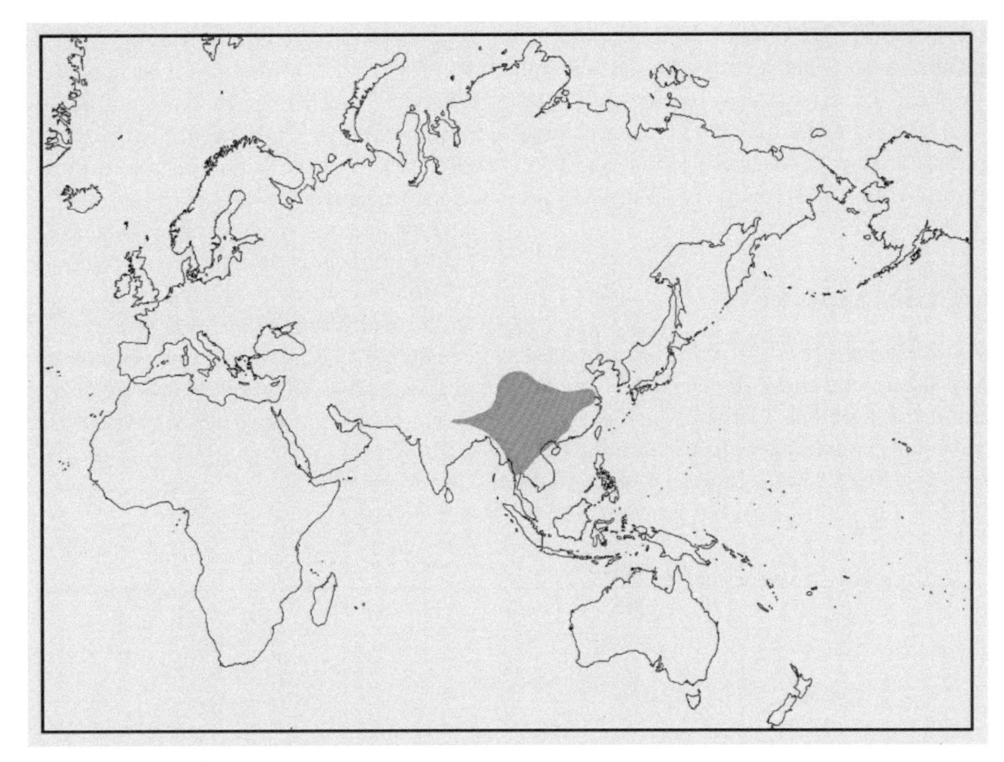

Figure 9.9 The distribution of *Hemsleya* (Subtype 14-1; taken from Wu *et al.*, 2010).

and *Neopicrorhiza* (Scrophulariaceae) are distributed from the Himalaya Shan to, respectively, northwestern or southeastern China. One species of *Tricarpelema* (Commelinaceae) occurs in Bhutan and northeastern India; one is endemic to Sichuan. Three species of *Dicranostigma* (Papaveraceae) have disjunct distributions covering the valley of the Chang Jiang, northern China, the Huangtu Gaoyuan, Nepal, and the Himalaya Shan in Xizang. *Anisadenia* (Linaceae), *Beesia* (Ranunculaceae) and *Gomphogyne* (Cucurbitaceae) may extend to central China, Shaanxi or southern Gansu. *Triplostegia* (Dipsacaceae) extends to Taiwan.

These examples show the relationships between the Sino-Himalayan flora and other parts of China. A common origin for floristic elements of the Chang Jiang valley in Yunnan and the Huangtu Gaoyuan in northern China is indicated by the disjunct distributions of *Prinsepia* and *Dicranostigma*. The division between the two subtypes of the eastern Asian element, Sino-Himalayan and Sino-Japanese, may lie in the area from northwestern Yunnan to the valley of the Chang Jiang in Sichuan, a region which may also represent a real boundary of the ancient Tethys Sea, based on the distributions of the flora. Many Sino-Himalayan elements extend to Taiwan or the Philippines, but never to Hainan. This matches the fact that the latter has a closer relationship with the tropical Asian flora, while Taiwan and the Himalaya Shan have contemporary geological histories, both having formed as a result of the orogeny of the Himalaya Shan themselves. As recently as 10 000 years ago, Taiwan was directly connected with parts of mainland China due to falling sea levels during Quaternary glaciations (Zhao, 1982).

9.15.2 Sino-Japanese (Subtype 14-2)

The range of this subtype extends from the valley of the Chang Jiang in Sichuan and Yunnan to Japan. About 93 genera belong to the subtype, with a high proportion of monotypic and oligotypic genera and, in contrast with the previous subtype, there are many relict deciduous trees, shrubs and bamboos. The monotypic genera are mainly representatives of families endemic to eastern Asia, such as *Cercidiphyllum* (Cercidiphyllaceae), *Nandina* (Berberidaceae) and *Trochodendron* (Trochodendraceae). Also included here are *Disanthus* (Hamamelidaceae; Figure 9.10), *Idesia* (Flacourtiaceae), *Kalopanax* (Araliaceae), *Kerria* and *Rhodotypos* (Rosaceae), *Kirengeshoma* and *Platycrater* (both Saxifragaceae) and *Sinomenium* (Menispermaceae). There are 46 oligotypic genera, most of which are herbs such as *Conandron* (Gesneriaceae), *Deinanthe* (Saxifragaceae), *Discocleidion* (Euphorbiaceae), *Farfugium* (Asteraceae), *Fatsia* (Araliaceae), *Hosiea* (Icacinaceae), *Macleaya* (Papaveraceae), *Metaplexis* (Asclepiadaceae) and *Trapella* (Pedaliaceae). There are 12 polytypic genera, including many bamboos such as *Pleioblastus*, *Pseudosasa*, *Sasa* and *Semiarundinaria* (all Poaceae). Woody genera include *Firmiana* (Sterculiaceae) and *Weigela* (Diervillaceae) and amongst the herbaceous genera are *Crepidiastrum* and *Sinosenecio* (both Asteraceae), *Hemiboea* (Gesneriaceae) and *Hosta* (Liliaceae).

9.16 Chinese endemic (Type 15)

The diversity of many Chinese endemic genera is centred on Yunnan or elsewhere in southwestern China, and the number of endemic genera gradually reduces towards

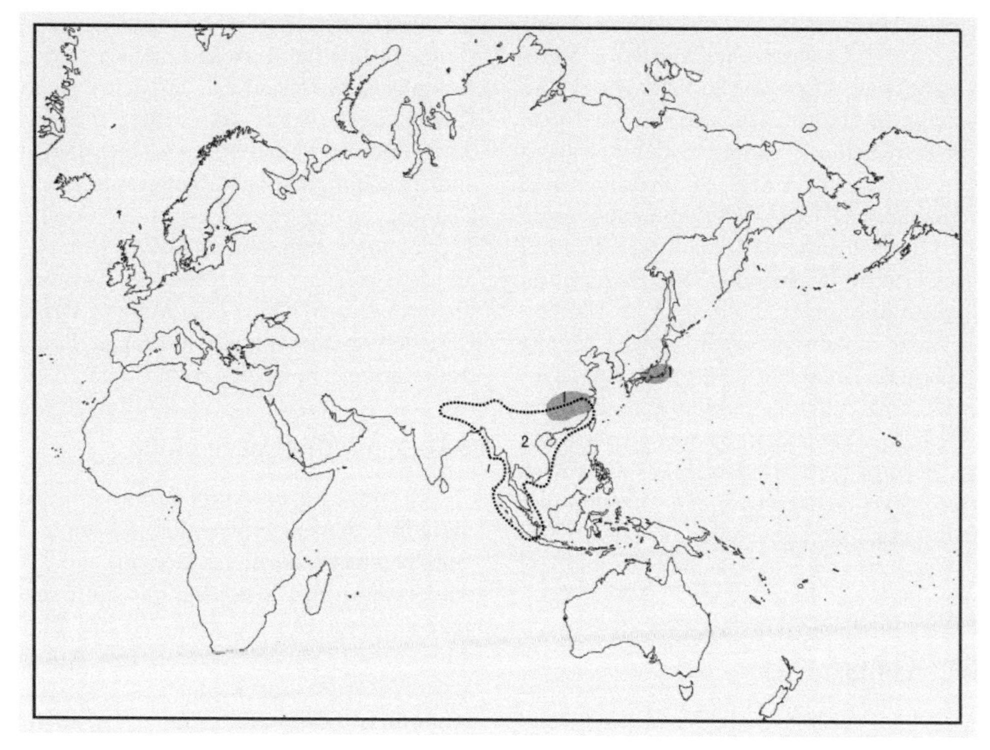

Figure 9.10 The distribution of *Disanthus* (1; Subtype 14-2) and *Exbucklandia* (2; Type 7; taken from Wu *et al.*, 2010).

the northeastern, eastern and northwestern provinces. However many endemics are also found in tropical and subtropical regions south of a line between the Qin Ling and Shandong. A few genera treated here do extend into neighbouring countries such as Myanmar, Laos, Vietnam, Thailand, Korea, the far east of Russia, and Mongolia, but these are centred in the natural floristic regions of China and have their limits not far from the national boundaries. Two hundred and thirty-five genera (ca. 8% of China's total genera) are included in this category, and can be classified into five groups corresponding to southwestern, eastern and central, southern, northern, and northwestern China.

9.16.1 Southwestern China

Included here are those genera which are mainly distributed in Yunnan, Sichuan, Guizhou and eastern Xizang, or are centred in these areas, radiating towards northwestern, northern, eastern or southern China. Among them, several are endemic to Yunnan, including *Holocheila* (Lamiaceae), *Shangrilaia* (Brassicaceae) and *Smithorchis* (Orchidaceae). Others are distributed from Yunnan, Sichuan and Guizhou to Xizang, or in two or three of the southwestern provinces, and include *Diplazoptilon* and *Nouelia* (Asteraceae), *Psammosilene* (Caryophyllaceae) and *Pterygiella* (Scrophulariaceae). More than 20 genera are distributed from southwestern to central or eastern China, including *Chimonanthus* (Calycanthaceae), *Clematoclethra* (Actinidiaceae), *Dipteronia* (Aceraceae) and *Sinofranchetia* (Lardizabalaceae). In addition, many genera extend from southwestern to southern and southeastern or northwestern and northeastern China. The former include *Delavaya* and *Eurycorymbus* (both Sapindaceae), *Handeliodendron* (Hippocastanaceae), *Tetrapanax* (Araliaceae) and *Whytockia* (Gesneriaceae).

9.16.2 Eastern and central China

This group includes genera distributed mainly in eastern Sichuan, the lower and middle reaches of the Chang Jiang in the east, from western Hubei, with some extending to the Qin Ling or Nanling Shan. They share southern and northern boundaries with those genera distributed from southwestern to central and eastern China, and have the same floristic characteristics. About 20 genera belong to this group, including *Fortunearia* (Hamamelidaceae), *Heptacodium* (Caprifoliaceae), *Omphalotrigonotis* (Boraginaceae) and *Sinojackia* (Styracaceae). *Monimopetalum* (Celastraceae), *Ombrocharis* (Lamiaceae), *Psilopeganum* (Rutaceae), *Sinowilsonia* (Hamamelidaceae) and *Triaenophora* (Scrophulariaceae) also occur in central or eastern China. The "living fossil" *Metasequoia glyptostroboides* (Taxodiaceae) is distributed across eastern Sichuan and western Hubei, and more recently has been found at Longshan in northwestern Hunan and Shizhu in eastern Sichuan. *Pseudolarix amabilis* (Pinaceae) and *Pseudotaxus chienii* (Taxaceae) are also relict species dispersed in central to eastern China. The latter is also found in Guangxi and mountainous regions of northern Guangdong.

9.16.3 Southern China

Those genera mainly distributed in the south or the tropical and subtropical regions of southeastern China, Taiwan and neighboring islands, with a few extending to central or eastern China (Zhejiang), are recognized as a group. Most are herbs or lianas of tropical families, and their distributions in these areas are rarely continuous. For example, of five genera in Gesneriaceae, *Bournea* is found in Fujian and Guangdong, *Primulina* in Guangdong and Guangxi, *Cathayanthe* in Hainan, *Titanotrichum* in Fujian and Taiwan, and *Petrocodon* is widely distributed from southern China to Hunan and Hubei. Similarly, of three genera in Asclepiadaceae, *Dolichopetalum* occurs from Longlin (Guangxi) to southeastern Yunnan, *Merrillanthus* is disjunct between Gaoyao (Guangdong) and Hainan, and *Pentastelma* is found only on Diaoluo Shan on Hainan. The other families only contain one or two genera. Among them *Chunia* (Hamamelidaceae), *Setiacis* (Poaceae) and *Wenchengia* (Lamiaceae) are endemic to Hainan. *Kudoacanthus* (Acanthaceae) and *Sinopanax* (Araliaceae) occur in Taiwan, *Barthea* (Melastomataceae) in southern China, and *Heteroplexis* (Asteraceae) and *Dayaoshania* (Gesneriaceae) in Guangxi. The islands of Taiwan and Hainan in particular have characteristic floras including a number of endemic genera and species.

9.16.4 Northern China

This group covers genera distributed in the humid to semi-arid warm-temperate zone of northern to northwestern China (Shaanxi, Gansu, Qinghai, Ningxia and Nei Mongol) and northeastern China (Liaoning). All nine genera are monotypic. Typical examples include *Xanthoceras* (Sapindaceae) and *Myripnois* (Asteraceae). The former is a deciduous tree, the northernmost monotypic genus in Sapindaceae, occurring from northern China to Shaanxi, Gansu, Nei Mongol and Liaoning. The latter is a rare genus of woody, deciduous shrub distributed on dry slopes in northeastern to northwestern China. Both are thought to represent relict species. *Kolkwitzia amabilis* (Linnaeaceae) may be another relict species, mainly distributed in Shaanxi, the eastern part of the Qin Ling and the Zhongtiao Shan between Henan and Shanxi, and also dispersed in eastern Gansu and southern Nei Mongol. *Porolabium* (Orchidaceae), *Pteroxygonum* (Polygonaceae) and *Stilpnolepis* (Asteraceae) are distributed in northern or northwestern China (Nei Mongol, Qinghai and Gansu). *Carlesia* (Apiaceae) and *Oresitrophe* (Saxifragaceae) occur from northern to northeastern China.

9.16.5 Northwestern China

This group consists of those genera which are distributed in northwestern China (Gansu, Ningxia, Qinghai and western Nei Mongol) and Xizang; some taxa may extend into Xinjiang, but their main distribution areas are on the Qinghai-Xizang Plateau. *Tetraena mongolica* (Zygophyllaceae) is a relict species of the western Ordos and eastern Alxa Shamo between Nei Mongol and Ningxia. Other taxa, such as *Pomatosace* (Primulaceae), *Sinochasea* (Poaceae) and *Synstemon* (Brassicaceae), are more derived evolutionary elements, presumably alpine plateau taxa originating with the uplift of the Qinghai-Xizang Plateau. *Przewalskia* (Solanaceae) is more widespread in Qinghai, Gansu, Sichuan and Xizang, and is considered to be derived under the alpine climate of the Qinghai-Xizang Plateau, from components with a temperate Eurasian distribution.

References

AXELROD, D. I. (1952). A theory of angiosperm evolution. *Evolution* 6: 29-60.

AXELROD, D. I. (1960). *The Evolution of Flowering Plants.* Chicago, IL: Chicago University Press.

AXELROD, D. I. (1972). Ocean-floor spreading in relation to ecosystematic problem. *In:* Allen, R. T. & James, F. C. (eds.) *A Symposium on Ecosystematics*. pp. 15-76. Fayetteville, AR: University of Arkansas Museum.

BOUFFORD, D. E. & SPONGBERG, S. A. (1983). Eastern Asian - eastern North American phytogeographical relationships: a history from the time of Linnaeus to the twentieth century. *Annals of the Missouri Botanical Garden* 70: 423-439.

BOUFFORD, D. E. (1992). Affinities in the floras of Taiwan and eastern North America. In: *Phytogeography and Botanical Inventory of Taiwan*. Taipei, Taiwan: Institute of Botany, Academia Sinica.

ELBERT, L. & LITTLE, J. (1983). North American trees with relationships in eastern Asia. *Annals of the Missouri Botanical Garden* 70(4): 605-615.

GOOD, R. (1974). *The Geography of the Flowering Plants.* London.

GRAY, A. (1856). Statistics of the flora of the northern United States. *American Journal of Science and Arts, Series 2* 22: 204-232.

GRAY, A. (1957). Statistics of the flora of the northern United States. *American Journal of Science and Arts, Series 2* 23: 62-84, 369-403.

HOU, K. Z. (1958). *A Dictionary of the Families and Genera of Chinese Seed Plants.* Beijing, China: Science Press.

SCHUSTER, R. M. (1972). Continental movements, "Wallace's line" and Indomalayan-Australasian dispersal of land plants. *Botanical Review* 38(1): 35-50.

SZAFER, W. (1975). *General Plants Geography.* Warsaw, Poland: Panstwowe Wydawnictwo Naukowe.

TAKHTAJAN, A. (1969). *Flowering Plants: Origin and Dispersal.* Edinburgh, UK: Oliver & Boyd.

TAKHTAJAN, A. (1997). *Diversity and Classification of Flowering Plants.* New York: Columbia University Press.

THORNE, R. F. (1975). Angiosperm phylogeny and geography. *Annals of the Missouri Botanical Garden* 62(2): 362-367.

THORNE, R. F. (1978). Plate tectonics and angiosperm distribution. *Notes from the Royal Botanic Garden Edinburgh* 36(2): 297-315.

VAN STEENIS, G. G. G. J. (1962). The land-bridge theory in botany, with particular reference to tropical plants. *Blumea* 11: 235-372.

WANG, H. S. (1992). *Floristic Plant Geography.* Beijing, China: Sciences Press.

WU, C. Y. (1959). *Outline of Chinese Floristic Geography.* Kunming, Yunnan: Botanical Society of Kunming.

WU, C. Y. (1965). The tropical floristic affinity of the flora of China. *Chinese Science Bulletin* 1: 25-33.

WU, C. Y. & WANG, H. S. (1983). The areal-type and floristic origin of Chinese seed plants. In: *Chinese Physical Geography, Phytogeography.* Beijing, China: Sciences Press.

WU, C. Y. (1991). The areal-types of Chinese genera of seed plants. *Acta Botanica Yunnanica Plus* IV: 1-139.

WU, C. Y., LU, A. M., TAN, Y. C., CHEN, Z. D. & LI, D. Z. (2003a). *The Families and Genera of Angiosperms in China - a Comprehensive Analysis.* Beijing, China: Sciences Press.

WU, C. Y., ZHOU, Z. K., LI, D. Z., PENG, H. & SUN, H. (2003b). The areal-types of the world families of seed plants. *Acta Botanica Yunnanica* 25(3): 245-257.

WU, C. Y., ZHOU, Z. K., SUN, H., LI, D. Z. & PENG, H. (2006). *The Areal-Types of Seed Plants and Their Origin and Differentiation.* Beijing, China: Sciences Press.

WU, C. Y., SUN, H., ZHOU, Z. K. & ET AL. (2010). *Floristics of Seed Plants from China.* Sciences Press.

WULFF, E. V. (1943). *An Introduction to Historical Plant Geography.* Waltham, MA: Chronica Botanica.

ZHAO, Z. B. (1982). Primary research into evolvement of Taiwan Channel. *Taiwan Channel* 1(1): 20-24.

Chapter 10

Phytogeographical Regions of China

SUN Hang and WU Cheng-Yih (WU Zheng-Yi)

10.1 Introduction

Our knowledge of the flora of China, the richest and most varied in the Northern Hemisphere, has grown through the completion of the *Flora of China* (Wu *et al.*, 1994–2011), *Flora Reipublicae Popularis Sinicae* (Editorial Committee of Flora Reipublicae Popularis Sinicae, 1959–2004) and many regional floras, as well as a large numbers of research papers on floristics and regional phytogeography (Wu *et al.*, 2010). The formation of the flora in a certain region is an integration of plant evolution in time and space: it reflects not only the results of the interaction between plants and environment, but also the processes of floristic development over geological history. Understanding a flora in these terms can provide important insights into the scientific basis of biodiversity protection, sustainable exploitation of plant resources, and future strategies for agriculture, forestry and land use.

In 1823, the Danish botanist Joakim Frederik Schouw proposed that for the recognition of a particular geographical area as a "floristic kingdom," at least half of the species and a quarter of the genera present should be endemic (Schouw, 1823). He also considered that the region should house endemic families or be the center of diversity for particular families. Kingdoms can be divided into "regions," established on the basis of high levels of endemism at generic, species or sometimes higher levels (Takhtajan, 1986). In addition, floristic regions are characterised by the presence of a particular series of families that predominate in the region, the quantitative correlations between these dominant families being relatively stable (Tolmatchev, 1974). On the basis of these principles, floristic analyses of China use four ranks: kingdom, subkingdom, region and subregion. Subkingdoms are recognised on the basis of their distribution of endemics, centers of diversity and similarity of geological history and floristic evolution. In the same way, subregions recognise the distribution range of endemic species or dominant or characteristic species in the vegetation of the region. The boundaries between kingdoms and subkingdoms are drawn to maximize the overlapping distribution limits of endemic, dominant or characteristic families in the vegetation. For regions and subregions, boundaries are drawn mainly on the basis of overlapping limits of distribution for endemic genera or species.

Among earlier workers, Diels (1901; 1905) discussed the geographic distribution of Chinese plants in general, limiting his studies to the center and west of the country. Wilson (1913) described in detail the plants of western China, and analysed the distribution of its vegetative types in altitudinal zones. Sargent (1913) summarized in the same work a comparison of the Chinese flora with that of eastern North America, indicating the affinities between these two widely separated regions and building on the earlier work of Asa Gray (1846; 1859; 1878). A year earlier, Ward (1912) reported on the plant formations of the arid regions of western China, and was also the first to define a "Sino-Himalayan flora" (Ward, 1916; Ward, 1927). Later he made detailed studies on the phytogeography of Xizang (Tibet; Ward, 1935; Ward, 1936). In his early studies Handel-Mazzetti (1916; 1921) reviewed the vegetation and flora of parts of western China and later suggested dividing the whole of China into eight "phytogeographic provinces" (Handel-Mazzetti, 1930; Handel-Mazzetti, 1931). Engler (1926) mentioned China in connection with his discussion of the geographic distribution of the conifers of the world and recognized five regions within the country. Hu (1926; 1929b; 1929a; 1940) described and analysed the forest flora of southeastern China and Yunnan, and made a comparative study between the woody floras of China and eastern North America. In the 1930s he further discussed the characteristics and affinities of the Chinese flora (Hu, 1935; Hu, 1936), recognizing eight major phytogeographic components. Lee (1934) summarized the work of Wilson and Handel-Mazzetti and suggested the addition of another province, the "Eastern-central Hardwood Type" as distinct from the "Middle-Chinese-Japanese Laurel Province" originally defined by Handel-Mazzetti. Further important contributions to the recognition of phytogeographical regions were made by LIU (1934; 1936), HAO (1938), LI (1944) and Walker (1944).

Two significant analyses of the world's floristic regions were provided by Good (1974), in which China formed part of the "Boreal" and "Palaeotropical Kingdoms," and Takhtajan (1986), who recognised fifteen Chinese phytogeographic regions belonging to the "Holarctic" and "Palaeotropic Kingdoms." WU (1979) and WU and WANG (1983) used Good and Takhtajan's kingdoms but developed the analysis further, recognizing seven subkingdoms and 22 regions. More recently, WU and WU (1996) have suggested a new "East Asia Kingdom," including three of the subkingdoms recognised by WU (1979) and WU and WANG (1983) namely the "Sino-Japan," "Sino-

Himalaya" and "Qinghai-Tibetan Plateau Subkingdoms"[1]. In addition the "Asia Desert Subkingdom" was upgraded and recognised as the "Tethys Kingdom," consisting of one subkingdom, two regions and five subregions within China.

10.2 A scheme for the floristic regions of China

Based the principles and methods of division outlined above, the flora of China can be divided into four kingdoms, seven subkingdoms, 24 regions and 49 subregions as shown in Table 10.1 and Figure 10.1, and described below. Kingdom, subkingdom, region and subregion codes correspond the those on the map in Figure 10.1.

10.2.1 The Holarctic Kingdom (I)

The Holarctic Kingdom, as initially proposed by Diels (1929) and accepted by many others including Good (1974), Schaefer (1958) and Takhtajan (1986), encompassed Europe, most of North America and all of Asia outside the tropics, including almost all of China north of the Tropic of Cancer. This proposal was influenced by the prevailing view that the East Asia Kingdom derived from the Cenozoic flora of the Arctic. Proponents of this view believed that the Arctic flora spread to the Himalaya Shan via Japan and then further east and south, gradually forming the Chinese flora. The present study agrees that the eastern Asian flora was affected by the mid-latitude Cenozoic flora of the Arctic, but considers its cradle to have been the southern wing of Laurasia, making it an ancient flora that should be recognised as independent from the Holarctic Kingdom. Therefore, the Holarctic Kingdom as defined here primarily includes Europe, Siberia, Canada and North America, especially east of the Rocky Mountains. In China, the Holarctic Kingdom includes only the Da Hinggan Ling, the Nei Mongol-Xinjiang steppe, Altai, and the forest and desert-steppe flora of the Tian Shan. These areas largely correspond to the "Boreal Subkingdom" of the Holarctic Kingdom according to the system of Takhtajan (1986).

The flora of the Holarctic Kingdom is relatively depauperate, with only a few endemic families. Leitneriaceae and Garryaceae, found in the northern USA, together with Eucommiaceae, found in China, are deemed to have originated as some of the earliest-branching

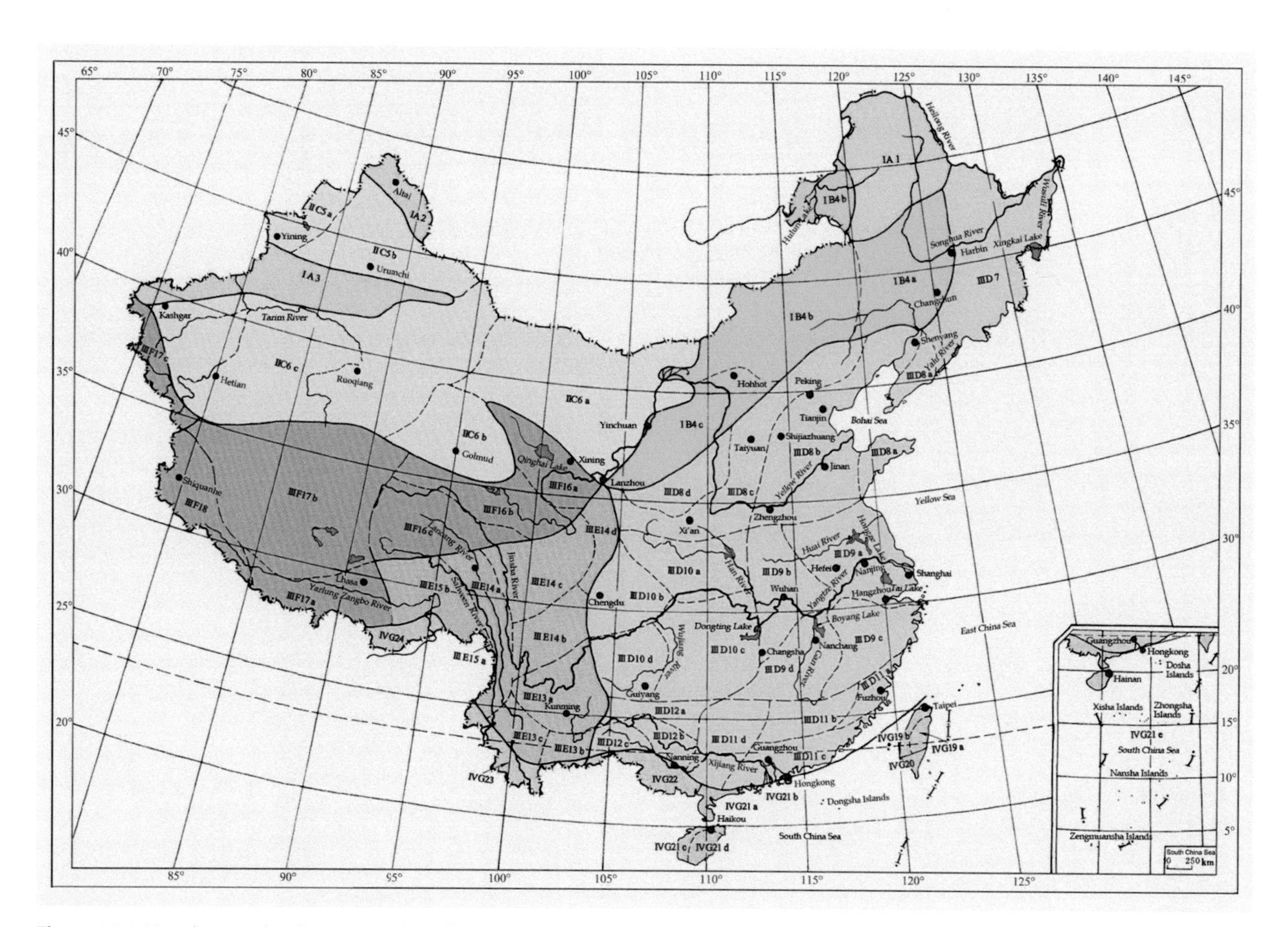

Figure 10.1 Map showing the phytogeographical floristic regions of China as discussed in this chapter. Regional codes are indicated in the text and in Table 10.1.

[1] Note that, while in general this volume, like the *Flora of China*, uses "Xizang," for the region sometimes known as Tibet, we retain the original geographic terms used in the named regions of other authors, to avoid confusion. Other geographic terms are treated similarly.

Table 10.1 The phytogeographical floristic regions of China. Note that, for ease of comparison, here and in the main text, names of kingdoms, subkingdoms, regions and subregions reflect widespread usage and are not necessarily those used in the *Flora of China*. Elsewhere in the text (aside from the names of kingdoms, subkingdoms, regions and subregions), geographic names follow the *Flora of China*.

Kingdom	Subkingdoms	Regions	Subregions
I. Holarctic	IA. Eurasian Forest	IA1. Daxinganling	
		IA2. Altai	
		IA3. Tian Shan	
	IB. Eurasian Steppe	IB4. Mongolia Steppe	IB4 a. Northeast China Plains Forest and Steppe
			IB4 b. East Mongolian Steppe
			IB4 c. Erdos, Shaanxi, Gansu and Ningxia Desert-steppe
II. Tethyan	IIC. Central Asiatic Desert	IIC5. Dzungaria	IIC5 a. Tacheng-Ili
			IIC5 b. Dzungaria
		IIC6. Kashgar	IIC6 a. Southwest Mongolia
			IIC6 b. Qaidam Basin
			IIC6 c. Kashgar
III. Eastern Asiatic	IIID. Sino-Japanese Forest	IIID7. Northeast China	
		IIID8. North China	IIID8 a. Liaoning-Shandong Peninsular
			IIID8 b. North China Plains
			IIID8 c. North China Mountains
			IIID8 d. Loess Plateau
		IIID9. East China	IIID9 a. Yellow River and Huaihe River Plains
			IIID9 b. Jianghan Plain
			IIID9 c. South Zhejiang Mountains
			IIID9 d. South Jiangxi and East Hunan Hills
		IIID10. Central China	IIID10 a. Qinling-Bashan Mountains
			IIID10 b. Sichuan Basin
			IIID10 c. Sichuan-Hubei-Hunan Border
			IIID10 d. Guizhou Plateau
		IIID11. Lingnan Mountains	IIID11 a. North Fujian Mountains
			IIID11 b. North Guangdong
			IIID11 c. East Nanling
			IIID11 d. Guangdong and Guangxi Mountains
		IIID12. Yunnan, Guizhou and Guangxi	IIID12 a. Guizhou-Guangxi Border
			IIID12 b. Hongshui River
			IIID12 c. Southeast Yunnan Limestone
	IIIE. Sino-Himalayan	IIIE13. Yunnan Plateau	IIIE13 a. Central Yunnan Plateau
			IIIE13 b. East Yunnan
			IIIE13 c. Southwest Yunnan
		IIIE14. Hengduan Mountains	IIIE14 a. Three Rivers Gorge
			IIIE14 b. South Hengduan Mountains
			IIIE14 c. North Hengduan Mountains
			IIIE14 d. Taohe-Minshan
		IIIE15. East Himalayan	IIIE15 a. Dulong Jiang-Northern Myanmar
			IIIE15 b. Southeast Xizang
	IIIF. Qinghai-Xizang Plateau	IIIF16. Tangut	IIIF16 a. Qilianshan
			IIIF16 b. A'nyemaqen
			IIIF16 c. Tangut
		IIIF17. Xizang, Pamir and Kunlun	IIIF17 a. Middle and Upper Reaches of the Yalu Tsangpo River
			IIIF17 b. Changthang Plateau
			IIIF17 c. Pamir-Karakorum-Kunlun
		IIIF18. West Himalayan	
IV. Paleotropical	IVG. Malesian	IVG19. Taiwan	IVG19 a. Taiwan High Mountains
			IVG19 b. North Taiwan
		IVG20. South Taiwan	
		IVG21. South China Sea	IVG21 a. West Guangdong-North Hainan
			IVG21 b. East Guangdong Islands
			IVG21 c. South Hainan
			IVG21 d. Central Hainan
			IVG21 e. South China Sea Islands
		IVG22. Bei Buwan (Tonkin Bay)	
		IVG23. Yunnan, Myanmar and Thailand	
		IVG24. Southern Fringe of the East Himalayan	

groups in the flora of eastern Asia. The heterotrophic Sarraceniaceae occur in the southern USA, and three monotypic families, Fouquieriaceae, Simmondsiaceae and Pterostemonaceae, are restricted to the southwestern USA and Mexico. Takhtajan (1986) listed many families in his Holarctic Kingdom, most of which are endemic to eastern Asia. He also listed 300 endemic genera, but few of them were members of ancient lineages. These facts indicate that the flora of the Holarctic Kingdom is a relict of the Cretaceous-Paleogene flora. However, some components of the flora of the Holarctic Kingdom of the were able to survive in the Himalaya Shan and Hengduan Shan of northwestern Yunnan (Sun, 2002).

The Eurasian Forest Subkingdom (IA).—The Eurasian Forest Subkingdom corresponds to the "Boreal Subkingdom" of Takhtajan (1986), excluding the "Arctic Province" and "Subarctic Province" of North America. It stretches from the northern edge of the European and Siberian forests to the steppe, desert and forests of eastern Asia in the south. This is the largest subkingdom of the Holarctic Kingdom and comprises coniferous forest of a comparatively simple composition. In China it encompasses three regions: the Da Hinggan Ling, Altai and Tian Shan. The number of endemic families and genera is comparatively small and there are many dominant or characteristic genera shared with the Eastern Asiatic Kingdom. There are extensive forests of *Abies, Larix* and *Picea* (all Pinaceae) which, if they are destroyed, are first succeeded by *Betula* (Betulaceae) or *Populus* (Salicaceae). Towards the southern border of the subkingdom can be found some deciduous evergreen broad-leaved forests consisting of *Acer* (Aceraceae), *Carpinus* (Betulaceae), *Fagus* and *Quercus* (both Fagaceae), *Tilia* (Tiliaceae) and *Ulmus* (Ulmaceae). The flora of the Eurasian Forest Subkingdom originated from similar eastern Asian subalpine floras, which gradually spread downwards and northwards after Quaternary glacial cycles, and in doing so, became less diverse (Wu *et al.*, 2010).

The Daxinganling Region (IA1) extends from the Heilong Jiang in the north to Ertix He in the west, the Hailar plateau in the southwest, Wuchagou (Arxan) and Horqin Youyi Qianqi in the southeast and Sunwu and Wuda Lianchi in the east. As a result of harsh climatic conditions, the flora of this region is very poor, and consists of 1 019 species in 391 genera and 89 families (Fu *et al.*, 1995). No endemic genera are found here, but eight genera occur nowhere else in China. These include the herbaceous *Anemarrhena* (Liliaceae) and *Linnaea* (Linnaeaceae) and four small evergreen shrubs, *Chamaedaphne, Empetrum, Ledum* and *Phyllodoce* (all Ericaceae). All of these are well adapted to snow cover. *Calla* (Araceae) and *Scheuchzeria* (Scheuchzeriaceae) are circumboreal elements. Endemic species are few, numbering only 11. Siberian elements tend to dominate, especially those of eastern Siberia. For instance, *Larix gmelinii* var. *gmelinii* replaces *Larix olgensis*, an element of the flora of northeastern China and *Larix gmelinii* var. *principis-rupprechtii* of northern China. *Betula platyphylla* and *Populus davidiana* replace *B. pendula* and *Populus tremula* of the flora of the Altai. *Pinus sylvestris* var. *mongolica* and *Pinus sibirica* are dominant species in the forests of the region, while *Quercus mongolica* forests abut the mixed broad-leaved deciduous and *Pinus koraiensis* forests of northeastern China, which some researchers have considered to be part of a "Manchurian Province" (Krasnov, 1899; Good, 1974; Takhtajan, 1987).

According to paleobotanical evidence (Tao, 1992; Tao, 2002), many subtropical montane elements of eastern Asia, such as *Carpinus, Fagus, Liquidambar* (Hamamelidaceae), *Quercus, Tilia* and *Zelkova* (Ulmaceae), were present in this area during the late Neogene and early Quaternary Periods. Some tropical elements, such as *Chloris virgata* and *Setaria viridis* (both Poaceae), *Commelina communis* (Commelinaceae) and *Cuscuta chinensis* (Convolvulaceae), still remain, indicating that this region is transitional between the Eastern Asiatic Kingdom and the Holarctic Kingdom, and is a young flora formed since the Quaternary Period.

The Altai Region (IA2) is a naturally well-defined floristic region which was called the "Altai-Sayan Province" by Kuznetsov (1912) and Takhtajan (1986), or "Altai-Trans-Baikalia Province" by Good (1974). Only a part of the Altai-Sayan Province extends into the most northeastern part of Xinjiang, China. The Altai Shan form a chain that stretches from northwest to southeast, only the southern tip of which lies in China. The dominant species of forests in this region are *Abies sibirica, Larix sibirica, Picea obovata* and *Pinus sibirica*, which are similar to the species of boreal coniferous forest; the lower forest tree line is at about 1 100 m. The alpine and subalpine zones feature shrubs such as *Betula rotundifolia, Caragana* spp. (Fabaceae) and *Spiraea* spp. (Rosaceae). The desert-steppe flora comprises mainly Asteraceae, Caryophyllaceae, Chenopodiaceae, Ephedraceae and Poaceae, which form the base of the altitudinal vegetation zones.

In total the flora of the Altai Region consists of 1 584 species in 467 genera and 80 families. The four families with greatest number of species are Asteraceae, Ranunculaceae, Rosaceae and Fabaceae. The dominant distribution pattern is north-temperate, followed by the Siberia-Altai pattern of WU and colleagues (2010). This region has 120 endemic species (Takhtajan, 1986) including *Androsace ovczinnikovii* (Primulaceae), *Cynoglossum viridiflorum* (Boraginaceae), *Galium paniculatum* (Rubiaceae), *Patrinia intermedia* (Valerianaceae), *Swertia obtusa* (Gentianaceae) and many others (Wu *et al.*, 2010). Several genera of the Altai Region are rare elsewhere, including *Limnocharis* (Alismataceae), *Macropodium* (Brassicaceae), which extends to Japan and *Stenocoelium* (Apiaceae). Chinese endemic species found in the region include *Arenaria aksayqingensis* (Caryophyllaceae), *Betula halophila, Elymus sylvaticus* (Poaceae) and *Tanacetum petraeum* (Asteraceae), *Linaria longicalcarata* (Scrophulariaceae), *Ranunculus chinghoensis* (Ranunculaceae), *Salix metaglauca, Salix neolapponum* and *Salix paraphylicifolia* (Feng *et al.*, 2003).

Some Holarctic elements found also in the Himalaya Shan and southwestern China, such as *Bergenia* (Saxifragaceae), *Coluria* and *Sibiraea* (both Rosaceae), are found in the alpine and subalpine zones, showing that the region has a certain floristic relationship to that of the Himalaya Shan (Wu *et al.*, 2010).

The western boundary of the Tian Shan Region (IA3) lies outside China; its northern boundary runs from Qapqal in the west, following the Tian Shan, to Barkol in the east. The southern boundary runs from Wuqia eastward along the border of the Tarim Pendi to the northern edge of the Hami basin where the southern and northern boundaries meet. The region is thus wide in the west and narrow in the east. Deserts or semi-deserts occur at lower elevations and, at higher elevations, montane conifer forests consisting of *Abies, Larix* or *Picea* occur on shaded, northward-facing slopes, with grassland, semi-desert or desert facing south.

The flora of the Tian Shan Region consists of 2 134 species in 546 genera and 93 families. The dominant families are Asteraceae (367 species), Fabaceae (267), Poaceae (185), Brassicaceae (147) and Ranunculaceae (103); with the largest genera being *Astragalus* (Fabaceae; 67 species), *Oxytropis* (Fabaceae; 47), *Carex* (Cyperaceae; 46) and *Allium* (Liliaceae; 40). One genus, *Synstemon* (Brassicaceae), is endemic to the southern Alashan (Nei Mongol) and Barkol (Xinjiang); there are 57, mostly herbaceous, genera endemic to the Tian Shan Region (Wu *et al.*, 2010). One hundred and forty-two species are endemic to the region, including *Aconitum sinchiangense, Delphinium ellipticovatum, D. tianshanicum, D. wentsaii* (all Ranunculaceae), *Pyrola xinjiangensis, Ranunculus balikunensis, R. hamiensis* and *Salsola junatovii* (Chenopodiaceae; Feng *et al.*, 2003). Interestingly, there are some typical eastern Asia elements, such as *Circaeaster agrestis* (Circeasteraceae), in this region, together with some genera, such as *Aconitum, Androsace, Chrysosplenium* (Saxifragaceae), *Primula* (Primulaceae) and *Saxifraga* (Saxifragaceae), whose modern center of diversity is in eastern Asia. So, the flora of the Tian Shan Region is considered a descendant of the ancient Tethys Kingdom, which was influenced by elements of the Eurasian Forest Subkingdom during the uplift of the Tian Shan, and is a young flora originating between the Holarctic and Tethys Kingdoms. It also has a connection with the Eastern Asiatic Kingdom. For instance *Eremurus* (Liliaceae), a typical Tethyan element mainly distributed in Europe to western and central Asia, has three species in the region and a disjunct species, *E. chinensis*, in the Hengduan Shan. *Colutea* (Fabaceae) has one species, *C. delavayi*, occurring in the valleys of the Hengduan Shan and all other species in the Mediterranean to northwestern China (Wu *et al.*, 2003).

The Eurasian Steppe Subkingdom (IB).—The second subkingdom extends from the lower Danube in Hungary eastward to the Altai Shan, central Mongolia and Nei Mongol in China. It forms a belt surrounding the north of the Asian desert and connecting forest to steppe at this northern boundary (Alexinh, 1957). The subkingdom is almost equivalent to the "Mongolian Province" of Grubov (1959; Grubov & Fedorov, 1964) and Takhtajan (1986). However the genera listed by Takhtajan (1986) as being characteristic are those of the deserts of eastern Asia, rather than those of steppe. Within the borders of China, the steppe is a prolongation of the Eastern European Province in the Circumboreal Region via the Altai Sayan and Transbaikalian Provinces, and its vegetation is dominated by *Stipa* (Poaceae) and *Artemisia* (Asteraceae; Wu *et al.*, 2010). At its eastern boundary the vegetation comprises *Larix* forests in the Da Hinggan Ling and *Pinus koraiensis* forests in the Xiao Hinggan Ling and Changbai Shan. Steppe dominated by *Leymus* (Poaceae) and *Filifolium* (Asteraceae) occurs alongside mixed deciduous broad-leaved forest, and in the east are sparse woodlands of *Corylus heterophylla* (Betulaceae) and *C. mandshurica* (which are replaced by *C. avellana* in eastern Europe), *Quercus mongolica* (in place of *Quercus robur* in eastern Europe), *Salix* and *Ulmus* in sandy places or valleys (Wu *et al.*, 2010). This vegetation has been impoverished by the influence of Pleistocene glaciations (Takhtajan, 1986). *Cymbaria* and *Dodartia* (both Scrophulariaceae) are characteristic genera of this steppe, while *Anemarrhena* and *Filifolium* are characteristic of the transitional zone from steppe to the forests of northeastern Asia. Heavily disturbed steppe communities feature summer ephemeral plants and noxious weeds such as *Agriophyllum, Corispermum* and *Halogeton* (all Chenopodiaceae), *Artemisia* and *Pugionium* (Brassicaceae), while *Euphorbia* (Euphorbiaceae), *Stellera* (Thymelaeaceae) and *Thymus* (Lamiaceae) can be found in secondary vegetation. Only one region of this subkingdom lies in China.

The Mongolian Steppe Region (IB4) is located in middle latitudes of Asia and extends from the eastern edge of the Da Hinggan Ling to the southeastern edge of the Yan Shan, the southwestern edge of the Huangtu Gaoyuan (Loess Plateau), the western edge of the Alxa desert and the northern edge of the Nei Mongol steppe. It forms a wide, flat plateau in Nei Mongol, between the Da Hinggan Ling and Yin Shan at elevations between 700 and 1 000 m. The Da Hinggan Ling and Yin Shan form natural barriers around an arc-like strip of vegetation. The vegetation of the region is predominantly steppe and meadow-steppe, with patches of forest, mountain thickets and meadow in the Da Hinggan Ling and on the shaded side of the Yin Shan. The flora of this region includes about 1 600 species in 541 genera and 110 families (Fu *et al.*, 1995). The richest families are Asteraceae (315 species), Poaceae (216), Fabaceae (153) and Cyperaceae (126) with the richest genera being *Carex* (110 species), *Artemisia* (56) and *Astragalus* (40). The dominant species in meadow-steppe include *Filifolium sibiricum, Leymus chinensis* and *Stipa baicalensis*. Steppe vegetation comprises perennial herbs adapted to cold, xeric conditions, with dominant and characteristic species including *Artemisia frigida, Cleistogenes squarrosa* (Poaceae), *Stipa bungeana, S. grandis* and *S. sareptana* var. *krylovii*. The flora is mainly composed of north-temperate elements (Zhang *et al.*, 2006). No endemic genera are found in this region, but there are many endemic species shared with the steppe of Mongolia, such as *Allium leucocephalum, A. mongolicum, Amygdalus pedunculata* and *Sibbaldia sericea* (both Rosaceae) and *Gypsophila desertorum* (Caryophyllaceae). There also are species endemic to China, including *Oxytropis ramosissima, Rheum racemiferum* (Polygonaceae), *Scrophularia alaschanica* (Scrophulariaceae), *Seriphidium finitum* (Asteraceae) and *Thesium brevibracteatum* (Santalaceae). Two genera, *Anemarrhena* and *Ostryopsis* (Betulaceae), which are wider endemics in China are also found in this region (Zhang *et al.*, 2006). Three subregions are recognised: the Northeastern China Plains Forest and Steppe Subregion, the East Mongolian Steppe Subregion and the Erdos, Shaanxi, Gansu and Ningxia Desert-steppe Subregion.

The Northeast China Plains Forest and Steppe Subregion (IB4 a) extends from its northern boundary at Horqin Youyi Qianqi; to the east at Suijiang, Xifeng where it meets the flora of northeastern China. The southern boundary falls at Kaiyuan and Zhangwu, among others, where it meets the flora of northern China, and the western boundary connects with the East Mongolian Steppe Subregion. The main vegetation types of the subregion are meadow-steppe composed of *Filifolium sibiricum, Leymus chinensis* and *Stipa baicalensis,* with salinized meadows in sandy areas, dominated by *Leymus chinensis* and *Puccinellia* spp. (Poaceae). The flora is composed of north-temperate and temperate Asian elements and, according to Cao *et al.* (1995), there are 1 047 species in 429 genera and 98 families in this subregion. The largest families are Asteraceae (149 species), Poaceae (102), Cyperaceae (66) and Fabaceae (56), while the most speciose genera are *Carex* (39 species), *Artemisia* (33), *Polygonum* (Polygonaceae; 21), *Potentilla* (Rosaceae; 20) and *Viola* (Violaceae; 17). Seven species are endemic to this subregion. Some species replace those of the same or similar genera in neighboring subregions; for instance, *Vicia geminiflora* (Fabaceae) occurs in this subregion instead of *V. bungei* as found in northern and eastern China, *Calamagrostis kengii* and *Deyeuxia purpurea* (Poaceae) replace *D. macilenta* in the East Mongolian Steppe Subregion, and *Cleistogenes hackelii* var. *hackelii* occurs instead of *Cleistogenes festucacea* and *Limonium sinense* (Plumbaginaceae) instead of *L. aureum, L. flexuosum* and *L. tenellum* on the East Mongolian Steppe.

The East Mongolian Steppe Subregion (IB4 b) lies to the west of the Northeastern China Plains Forest and Steppe Subregion. Its western boundary is at Wuyuan and Linhe; to the southwest it stretches to Yinshan at the northeastern edge of the Loess Plateau Subregion. The East Mongolian Steppe Subregion includes foothills of the western side of the Da Hinggan Ling in the northeast; at its center are the wide, flat prairies of Hulun Buir and Xilin Gol. The southwestern boundary is an arid plain at an elevation of 1 000—1 500 m, while salinized swamps, meadows and desert lie at the border between China and Mongolia. The flora contains about 1 200 species in 430 genera and 90 families. Typical communities are composed of *Stipa*, especially *S. bungeana, S. capillata* and *S. grandis*. Although to the west of the Da Hinggan Ling the main vegetation is steppe, in the northeast are coniferous forests with Siberian elements such as *Larix gmelinii* and *Pinus sylvestris* var. *mongolica*. The west of the subregion lies in a transitional zone between steppe and the Alxa desert, with clump grass steppe formed of *Stipa sareptana* var. *krylovii* and *Artemisia frigida* instead of *S. grandis. Allium polyrhizum, Kalidium foliatum* (Chenopodiaceae) and *Reaumuria soongarica* (Tamaricaceae) are found in salinized fields, while shrubs and sparse *Ulmus pumila* trees flourish on stable sand (Wu *et al.*, 2010).

The Erdos, Shaanxi, Gansu and Ningxia Desert-steppe Subregion (IB4 c) is located to the south of Yinshan; its southern edge is the western side of the Guancen Shan and the south of the Mu Us Shadi. It includes the Ordos Gaoyuan and Mu Us Shadi. *Stipa bungeana* is the dominant species of the most representative community in the area, but is reduced to a remnant in some hills as the result of intensive disturbance. Other taxa, such as *Thymus mongolicus* and *T. quinquecostatus* var. *asiaticus,* dominate in heavily-grazed places. The *Artemisia ordosica* formation, which is endemic to this subregion, is well developed and includes *Agropyron desertorum* and *Pennisetum flaccidum* (both Poaceae), *Corethrodendron fruticosum* var. *mongolicum* (Fabaceae) and *Sophora alopecuroides* (Fabaceae). The flora of this subregion includes 1 060 species in 440 genera and 99 families, with the richest families being Poaceae, Asteraceae, Fabaceae and Rosaceae. The dominant vegetation is typical of Eurasian steppe formations, but is strongly influenced by the flora of eastern Asia, especially that of northern China. Thus, the flora also contains species such as *Ailanthus altissima* (Simaroubaceae), *Vitex negundo* var. *heterophylla* (Verbenaceae) and *Xanthoceras sorbifolium* (Sapindaceae). The subregion marks the northernmost point of the range of *Pinus tabuliformis* and westernmost of *Picea crassifolia,* reflecting the close relationship between this area and the northern China flora. *Kolkwitzia amabilis* (Linnaeaceae), an endemic species to China, is distributed in the Huang He gorges of this subregion. Two species of *Ostryopsis* are found in the subregion: *O. davidiana* and *O. nobilis.* The former is found in the valley of the Jinsha Jiang in northwestern Yunnan, the latter in northern China where it is a dominant shrub species. Both *Kolkwitzia* and *Ostryopsis* are palaeoendemics and *Ostryopsis* is considered related to the Tethyan Kingdom discussed below (Wu *et al.*, 2010).

10.2.2 The Tethyan Kingdom (II)

The boundaries of this kingdom are disputed, some researchers treating it as very small and others very large (Wulff, 1944). Here, we consider the Tethyan Kingdom to include the "Macronesian Region" of Takhtajan (1986) together with his "Mediterranean" and "Irano-Turanian" Regions excluding the Qinghai-Xizang Plateau. This is a temperate desert flora, differing from the tropical deserts and savannas of the Sahara-Arabian Region to the south, which many researchers have considered to be part of the paleotropics (Engler, 1899; Engler, 1912). In China, the Tethyan Kingdom includes a single subkingdom, with two regions and five subregions.

The Central Asiatic Desert Subkingdom (IIC).—Compared to the floras of other parts of China, the Asiatic Desert Subkingdom flora is depauperate, with about 1 700 species in 484 genera and 82 families (Pan *et al.*, 2001). There are eight characteristic genera in this subkingdom, of which Chenopodiaceae contains the most (Yin, 1997). Of the major families, Chenopodiaceae contains 20 endemic genera, Asteraceae 16, Brassicaceae 12, and Apiaceae, Boraginaceae, Fabaceae, Lamiaceae and Papaveraceae all five each (Pan *et al.*, 2001). Seven genera—*Ammopiptanthus* (Fabaceae), *Potaninia* (Rosaceae), *Pugionium*, *Stilpnolepis* and *Tugarinovia* (both Asteraceae), *Synstemon* and *Tetraena* (Zygophyllaceae) are endemic to Nei Mongol and part of southern Mongolia. Of these, *Stilpnolepis* and *Synstemon* are recently-evolved endemic species; *Tetraena* (Zygophyllaceae), a monotypic genus, is considered to be a paleoendemic or relict taxon (Wu *et al.*, 2010).

The features of these families and genera indicate that this flora is completely different from those of the Holarctic, Eastern Asiatic and Paleotropical Kingdoms. The Mediterranean, western and central Asia are centers of diversity for no fewer than six endemic families: Cistaceae, Cynomoriaceae, Frankeniaceae, Ixioliriaceae, Nitrariaceae and Peganaceae. These regions also are the center of diversity for Apiaceae, Brassicaceae, Chenopodiaceae, Lamiaceae, Papaveraceae, Plumbaginaceae and Zygophyllaceae. Zhu (1996) considers the area to be the center of origin for Chenopodiaceae. Most of the 135 characteristic genera are endemic to central Asia or the entire Tethyan Kingdom; only four are endemic to China. This flora is dominated by shrubs and herbs. Only *Ammodendron* (Fabaceae), *Haloxylon* (Chenopodiaceae), *Populus*, *Tamarix* (Tamaricaceae) and *Ulmus* are trees. The natural vegetation is dominated by desert plants, salt-tolerant shrubs and subshrubs, meadow and marsh plants. The flora of halophytic areas comprises 38 families, 124 genera and 305 species, with common genera including *Achnatherum* (Poaceae), *Alhagi* (Fabaceae), *Halocnemum* and *Suaeda* (both Chenopodiaceae), *Kalidium*, *Karelinia* (Asteraceae), *Lycium* (Solanaceae), *Nitraria* (Nitrariaceae), *Reaumuria*, *Salsola* and *Tamarix*. Sixty-three percent of these species belong to the four families Chenopodiaceae, Asteraceae, Poaceae and Fabaceae (Xi *et al.*, 2006).

The Dzungaria Region (IIC5) extends from the southern edge of the Altai to the northern edge of the Tian Shan. Its western edge is the border between China and Kazakhstan and its eastern edge runs to Barkol. The annual precipitation in this desert region is 250—300 mm at Yili and Tacheng, with just 200 mm around the Junggar (Dzungar) Pendi. The region is more humid and warmer than other places in east-central Asia. At least ten genera in this region do not occur in the eastern part of central Asia, including *Conringia*, *Diptychocarpus*, *Euclidium*, *Lachnoloma*, *Litwinowia*, *Matthiola* and *Pachypterygium* (all Brassicaceae), *Consolida* (Ranunculaceae), *Russowia* and *Schischkinia* (both Asteraceae). About 203 ephemeral plants in 104 genera and 24 families are found in this region (Mao, 1991), including *Alyssum* spp., *Litwinowia* spp., *Tauscheria* spp.and *Tetracme* spp. (Brassicaceae), *Astragalus commixtus*, *Astragalus contortuplicatus*, *Medicago monantha*, *Trigonella arcuata* and *Trigonella cancellata* (Fabaceae), *Bromus tectorum* and *Poa bulbosa* (Poaceae), *Ferula* (Apiaceae), *Gagea* and *Tulipa* (Liliaceae), *Garhadiolus papposus* (Asteraceae) and *Nonea caspica* (Boraginaceae).

This flora likely originated relatively recently in central Asia (such as Kazakhstan) after the retreat of the Tethys Sea. Typical Tethyian elements include *Artemisia absinthium*, *Artemisia macrantha*, *Artemisia marschalliana*, *Atraphaxis decipiens* (Polygonaceae), *Atraphaxis frutescens*, *Calligonum klementzii*, *Calligonum ebinuricum*, *Calligonum calliphysa* (treated under *Calligonum junceum* in many Chinese works), *Calligonum leucocladum*, *Chamaesphacos ilicifolius* (Lamiaceae), *Ferula dubjanskyi*, *F. fukanensis*, *F. sinkiangensis*, *F. songarica*, *Heliotropium acutiflorum* (Boraginaceae), *Lachnoloma lehmannii*, *Lepidium perfoliatum*, *Leptaleum filifolium* and *Tauscheria lasiocarpa* (Brassicaceae), *Nitraria roborowskii*, *N. sphaerocarpa*, *N. tangutorum*, *Peganum harmala* (Peganaceae), *Reaumuria soongarica*, *Rochelia leiocarpa*, *Salsola dshungarica*, *Salsola orientalis*, *Silene songarica* (Caryophyllaceae), *Tamarix androssowii*, *Tetracme quadricornis*, *Tribulus terrestris* (Zygophyllaceae), *Tulipa biflora*, *Tulipa iliensis*, *Tulipa kolpakovskiana*, *Zygophyllum brachypterum*, *Z. fabago* and *Z. obliquum*. Other, ephemeral plants of the region include *Alyssum desertorum*, *Eremopyrum orientale* (Poaceae), *Eremostachys moluccelloides* (Lamiaceae), *Eremurus inderiensis*, *Euphorbia* spp., *Gagea filiformis*, *Ixiolirion tataricum* (Amaryllidaceae), *Leontice incerta* (Berberidaceae), *Leptaleum filifolium* and *Malcolmia africana* (Brassicaceae) and *Trigonella arcuata* (Pan *et al.*, 2001). This region includes two subregions.

The Tacheng-Ili Subregion (IIC5 a) is located in westernmost China with its eastern edge at the mountains of the western Junggar, its northern edge at the Kazakhstan border and its southern edge at the Tian Shan. This subregion embraces the Tacheng basin, Emin valley and Yili He valley among others. In the west, the vegetation is continuous with the eastern Lake Balkhash area, Kazakhstan. The flora is dominated by *Seriphidium* (Asteraceae) and this subregion represents the eastern limit of many Kazakhstan desert plants, such as *Ammodendron bifolium*, *Astragalus iliensis*, *Astragalus lanuginosus*, *Astragalus sphaerophysa* and *Astragalus stenoceras*. There are many endemic species, such as *Alyssum dasycarpum*, *Calligonum trifarium*, *Caragana tekesiensis*, *Ferula sinkiangensis*, *Halimocnemis karelinii* (Chenopodiaceae), *H. villosa*, *Ikonnikovia kaufmanniana* (Plumbaginaceae) and *Saussurea pulviniformis* (Asteraceae). More than 160 spring ephemeral or ephemeroid plants are found in the Yili He valley. Of 35 species of *Tulipa* occurring in China, six out of the seven found here are endemic; *Ferula sinkiangensis* is also endemic to the area.

The Dzungaria Subregion (IIC5 b) lies between the Altai and Tian Shan. Its western edge is at the Junggar mountains and its eastern edge is at the edge of the region, at Barkol. This subregion is wide in the west and narrow in the east and includes the second largest desert basin, the Gurbantünggüt Shamo. The dominant vegetation comprises small shrubs and sub-trees such as *Anabasis brevifolia* (Chenopodiaceae), *Anabasis salsa, Artemisia songarica, Calligonum rubicundum, Ephedra distachya* (Ephedraceae), *Seriphidium santonilum, S. terrae-albae, Tamarix elongata, T. leptostachya* and *T. ramosissima*. These grow together with leafless shrubs such as *Haloxylon persicum* and *Calligonum* species. The floristic elements are mainly those of the Tethys Kingdom and central Asia. Subregionally-endemic species include *Astragalus qingheensis, Atraphaxis irtyschensis* (Polygonaceae), *Calligonum ebinuricum, C. jeminaicum, C. trifarium, Salix burqinensis* and *Tulipa sinkiangensis* (Dang & Pan, 2001). The number of spring ephemeral plants is fewer than in the first subregion, with only 40–50 species.

The Kashgar Region (IIC6) extends from the eastern edge of Alxa Qi to the western edge of the Tarim Pendi; its northern edge lies at the Tian Shan and its southern runs between the Karakorum Shan, Kunlun Shan and Qilian Shan. The region lies in the center of Asia and the climate is extremely arid with precipitation of about 20 mm a year. It includes deserts such as the Taklimakan Shamo, Tengger Shamo and Badain Jaran Shamo. Representative vegetation types include gobi, desert and desert-steppe. *Populus euphratica* forms riverine forests, *Artemisia* (especially sections *Absinthum* and *Dracunculus*) are common, and species of Chenopodiaceae are fewer than in the flora of western central Asia. Here, ephemeral plants are summer-, rather than spring-flowering. The flora is relatively depauperate although there are still some endemic genera as well as characteristic genera and species.

The flora of this region is considered to be older than the flora of western central Asia, and many genera have relationships with the Tethys flora. For instance, *Ammopiptanthus* contains two evergreen species which show a disjunct distribution between the eastern and western sides of this region and are regarded as a relict and a phylogenetically-isolated clade (Wang *et al.*, 2006). *Cornulaca alaschanica* (Chenopodiaceae) shows a disjunct distribution between Africa and the edge of the Alxa Shamo. *Cynomorium* (Cynomoriaceae), also containing two species, is distributed in the Sahara (*Cynomorium coccineum*) and desert areas of China (*Cynomorium songaricum*), extending into northern and northeastern China. *Pugionium*, another phylogenetically-isolated genus, is found in this region and in southwestern Mongolia. *Capparis* (Capparaceae), has about 210 pantropical species of which one, *Capparis himalayensis,* is a remnant in this region; *Capparis spinosa* is a corresponding species of the Mediterranean, also found in Xinjiang and Xizang. *Frankenia* (Frankeniaceae), with about 35–40 species, is an archaic genus distributed in the Mediterranean (six species), Africa (five), southwestern America, Australia and the Kashgar Region (one species, *F. pulverulenta*). Endemic species found here include *Myricaria pulcherrima* (Tamaricaceae), *Reaumuria kaschgarica* and *Tamarix tarimensis*, some of which originated before the retreat of the Tethys Sea, others after. This region includes three subregions.

The Southwest Mongolia Subregion (IIC6 a) stretches from the north of the Qilian Shan to the border with Mongolia in the north, to the Kashgar Subregion in the west and to east of Urad Zhongqi in the east. It embraces the Alxa Gaoyuan, Tengger Shamo, Badain Jaran Shamo, Ulanqab desert and the Hexi corridor. It lies at an elevation of 1 400—1 600m and the climate is cold with about 20 mm of precipitation annually. The dominant families include Asteraceae, Brassicaceae, Caryophyllaceae, Chenopodiaceae, Fabaceae, Poaceae, Polygonaceae, Ranunculaceae, Tamaricaceae and Zygophyllaceae, and characteristic families also include Ephedraceae and Elaeagnaceae (Liu *et al.*, 2007). The dominant species of desert vegetation in this subregion are *Potaninia mongolica, Reaumuria soongarica, Tetraena mongolica* and *Zygophyllum* spp. To the east can be found *Picea crassifolia* forests, alpine thickets and alpine meadows. The plains areas of the subregion are home to 543 species in 228 genera and 55 families (Liu *et al.*, 2007), while probably over 1 000 species occur in the mountains and Hexi corridor areas.

Common elements between the Southwest Mongolia and Dzungaria Subregions include *Ammopiptanthus, Cynomorium* and *Gymnocarpos*. Endemic and near-endemic genera of the Southwest Mongolia Subregion include *Potaninia, Synstemon* and *Tetraena*, and the subregion is a center of diversity for *Stilpnolepis* and *Tugarinovia*. The eastern part of the subregion is influenced by the floras of northern China and the Qinghai-Xizang Plateau, including species such as *Picea crassifolia*, which is the only true tree in this region in addition to shrubby *Juniperus* (Cupressaceae). *Kobresia humilis, K. myosuroides* subsp. *myosuroides, K. pygmaea* and *Stipa purpurea* are common species in meadows. It is extremely arid in the west of this subregion, where rainfall is less than 30 mm per year and no more than 150 species, dominated by Chenopodiaceae, can be found. The flora resembles that of central Asia deserts, including *Anabasis brevifolia, Ephedra przewalskii, Frankenia pulverulenta, Halogeton glomeratus, Iljinia regelii* (Chenopodiaceae), *Nitraria sphaerocarpa, Salsola abrotanoides* and *Zygophyllum xanthoxylon*. This is considered an ancient flora, located at the intersection between three subkingdoms, the Eurasian Steppe, Sino-Japanese Forest and Qinghai-Xizang Plateau Subkingdoms.

The Qaidam Basin Subregion (IIC6 b) is located in the Qimantag Shan and Burhan Budai Shan, which are bounded by the northern parts of the Kunlun Shan to the south and to the north by Danghe Nanshan, with the eastern edge extending to Delingha. The Qaidam Pendi (Basin) is an inland basin surrounded by high mountains. The annual rainfall here is only 15–20 mm, with a maximum of 160–170 mm in Delingha. The mean annual temperature is around 0–4°C, falling to 0 °C even in midsummer (Wu *et al.*, 2010); the soil is extremely saline. The vegetation here is mainly woody desert and desert grassland, only in the east and southeast, near the border with the Tangut Region, are found fragments of *Picea crassifolia* and *Juniperus przewalskii* communities (Du & Sun, 1990). The flora is very depauperate with only 255 species in 130 genera and 41 families (Dang & Pan, 2001). Most species are members of Asteraceae (63), Poaceae (54), Chenopodiaceae (39), Fabaceae (38) and Tamaricaceae (16). The dominant genera (those with at least six species) are *Artemisia, Astragalus, Carex, Kobresia, Potentilla, Saussurea* and *Tamarix*. Elements of the Mediterranean and central Asian floras are dominant, including *Asterothamnus, Brachanthemum* and *Cancrinia* (Asteraceae), *Alyssum, Anabasis, Cynomorium, Dilophia* and *Hedinia*, (Brassicaceae), *Haloxylon, Incarvillea* (Bignoniaceae), *Krascheninnikovia* and *Sympegma* (Chenopodiaceae), *Malcolmia, Nitraria, Psammochloa* (Poaceae), *Reaumuria* and *Zygophyllum*. Common or dominant species include *Ephedra intermedia, E. przewalskii, Kalidium foliatum, Krascheninnikovia ceratoides, Haloxylon ammodendron, Reaumuria kaschgarica, R. trigyna, Suaeda corniculata* and *Suaeda prostrata*. Four species of *Nitraria* are distributed in the deserts of China, all of which can be found in this subregion, while only 13 species of *Artemisia* occur here.

Due to the large areas of saline soils, this subregion is characterised by many halophytes, such as *Salicornia europaea* (Chenopodiaceae), *Saussurea salsa* and *Youngia paleacea* (Asteraceae). There are also many elements of the Himalaya Shan and Qinghai-Xizang Plateau in the surrounding mountains, especially meadow plants such as *Carex moorcroftii, Cremanthodium discoideum* (Asteraceae), *Kobresia pygmaea, K. tibetica, Lancea tibetica* (Scrophulariaceae) and *Stipa purpurea*. The distribution of *Calligonum zaidamense* is also centered on this subregion but extends to Linze in the Hexi corridor, while *Salsola zaidamica* extends to the Hexi corridor and Turpan Pendi. Species endemic to this subregion and neighboring areas include *Artemisia ordosica, Juniperus przewalskii, Oxytropis ochrocephala, Seriphidium mongolorum* and *Reaumuria kaschgarica* (Dang & Pan, 2001).

The Kashgar Subregion (IIC6 c) is located to the south of the Tian Shan, north of the Kunlun Shan. Its western edge is at Pamir and its eastern edge at the Shule He at the foot of the Dongshan and Qilian Shan. This subregion contains the Tarim Pendi, the largest basin in China, and surrounding areas. The climate is extremely dry with an annual rainfall of only 25–40 mm in most of the subregion, annual evaporation of 2 100–3 400 mm and annual average temperatures of ca. 11 °C. The vegetation is mainly shrubby desert with the dominant structural species being *Anabasis brevifolia, Calligonum roborowskii, Ephedra przewalskii* and *Nitraria sphaerocarpa*, the *Calligonum* species replacing *C. aphyllum, Calligonum calliphysa* and *Calligonum rubicundum* from the Junggar Pendi (Liu, 1995). Other species, such as *Reaumuria kaschgarica, Sympegma regelii* and *Zygophyllum kaschgaricum*, are found in this subregion but not in western central Asia. This flora is very depauperate with only about 200 species, of which 38 are endemic to China (Dang & Pan, 2001) including *Astragalus hamiensis, Calligonum korlaense, Calligonum roborowskii, Calligonum yengisaricum, Glycyrrhiza inflata* (Fabaceae), *Hippolytia kaschgarica* (Asteraceae), *Limonium lacostei, Nitraria tangutorum, Reaumuria trigyna, Tamarix chinensis, T. karelinii, T. taklamakanensis* and *Zygophyllum loczyi*. Species endemic to the subregion include *Astragalus hotianensis, Astragalus toksunensis, Calligonum korlaense, Calligonum pumilum, Calligonum yengisaricum, Caragana dasyphylla, Caragana polourensis, Galitzkya potaninii* (Brassicaceae), *Hippolytia kaschgarica, Myricaria pulcherrima, Salsola junatovii, Suaeda rigida, Suaeda stellatiflora, Tamarix sachensis* and *T. tarimensis* (Dang & Pan, 2001).

10.2.3 The Eastern Asiatic Kingdom (III)

Most of the Eastern Asiatic Kingdom has previously been treated as part of the Holarctic Kingdom, although its southern part has been placed in the Paleotropical Kingdom by many authors (Takhtajan, 1986; Takhtajan, 1987). However, WU and WU (1996) proposed and described the Eastern Asiatic Kingdom as a new Kingdom and their treatment is follwed here. The Eastern Asiatic Kingdom has China at its core, and is the richest area for gymnosperms, with 12 families (63% of the 19 families known in the world). Only one of these is not native to China (Sciadopityaceae, found in Japan), and the other 11 are restricted to China. The Eastern Asiatic Kingdom contains most of the families known from the Northern Hemisphere as well as Podocarpaceae, which is mainly distributed in the Southern Hemisphere. At generic level, the Eastern Asiatic Kingdom, especially its forest areas in China, is home to ten of the 12 genera of Pinaceae, four of the five of Taxaceae and six of the ten of Taxodiaceae. It includes genera such as *Cathaya* (Pinaceae) and *Metasequoia* (Taxodiaceae), famous as "living fossils."

Seventy-one genera (with 820 species) are distributed across all parts of the Eastern Asiatic Kingdom. There are about 143 genera in the Sino-Himalayan Subkingdom, with approximately 810 species occurring in mainland China, and 101 genera in the Sino-Japanese Forest Subkingdom (322 species in mainland China). The Eastern Asiatic Kingdom is characterized by a high level of endemism, with 31 families found only in this area (Wu & Wu, 1996). The majority of these endemic families can be labeled as relict taxa, and tend to be phylogenetically isolated. The Eastern Asiatic Kingdom has some relationships with tropical Southeast Asia, with which it shares five families: Daphniphyllaceae, Lowiaceae, Mastixiaceae (including the living fossil *Diplopanax*), Pentaphragmataceae and

Pentaphylacaceae. Some of these families show disjunct distributions; their southern border can extend to Sumatra, Sulawesi or New Guinea. The number of endemic genera in the Eastern Asiatic Kingdom exceeds 600 (Wu *et al.*, 1996). Genera endemic to China represent a relatively high level of diversity with a total of 249 (covering ca. 541 species). Finally, 42 genera are endemic to Korea and Japan (including the Bonin and Ryukyu Islands). About 50% of all species in China are endemic and these are distributed across the mainland from the northeast to the Hengduan Shan. While the ratio of endemics to non-endemics in the Eastern Asiatic Kingdom is lower than in the Cape Floristic Region (68%) or Australia (75%), it is approximately equal to that of Paleotropical and Neotropical regions and significantly higher than the Holarctic and Holantarctic Kingdoms. At family level, the ratio of endemics to non-endemics in the Eastern Asiatic Kingdom is significantly higher than in these other regions (Wu *et al.*, 2010).

At least 28 of the 40 families of basally-branching angiosperms (Magnoliidae *sensu* Wu *et al.*, 2002) occur in this kingdom, including the monotypic Glaucidiaceae and Pteridophyllaceae. Eastern Asia, especially China, is considered to be the center of diversity for 75% of these early-branching families. Other families, such as Annonaceae and Lauraceae, which are mainly distributed in the tropics and Southern Hemisphere also reach Eastern Asia and especially China. Basally-branching taxa of Rosideae such as Actinidiaceae and Theaceae also exist as ancient components of the flora of the Eastern Asiatic Kingdom. The ancient character of the kingdom is also shown in its concentration of early-branching families once ascribed to the "Hamamelididae:" Betulaceae, Buxaceae, Cercidiphyllaceae, Daphniphyllaceae, Eucommiaceae, Eupteleaceae, Fagaceae, Hamamelidaceae, Juglandaceae, Myricaceae, Platanaceae, Tetracentraceae and Trochodendraceae. All except five or six families of this group – found in Madagascar, tropical South Africa, California, Mexico and South America – are present in the Eastern Asiatic Kingdom. The kingdom also contains some derived taxa, such as in Myrtanae (*Pellacalyx*; Rhizophoraceae), Rutanae (*Tapiscia*; Tapisciaceae) and Cornanae (Aucubaceae and *Diplopanax*; Mastixiaceae). China also houses early-branching lineages of several highly derived sub-classes such as Lamiidae and Asteridae. In the monocots, the Eastern Asiatic Kingdom, especially China, harbors all the three clades of Liliopsida as published by Takhtajan (1997). Remarkably, all early-branching genera of Alismatidae and many of those of Liliidae and Arecidae can be found in the Eastern Asiatic Kingdom; for example *Croomia* (Stemonaceae), *Dioscorea* (Dioscoreaceae), *Paris* (Liliaceae), *Schizocapsa* and *Tacca* (both Taccaceae). In summary, if the Eastern Asiatic Kingdom, especially its south, to the Chang Jiang, is not the exclusive place of origin of gymnosperms and angiosperms, it is certainly one of their centers of origin or key centers of speciation and differentiation (Wu & Wu, 1996).

The Sino-Japanese Forest Subkingdom (IIID).—It seems that there are many similarities between the floras of China and Japan, making it possible to regard them as a single floristic region as Grisebach (1872) first proposed. Drude (1890), Engler (1899) and Diels (1901) also recognised and studied this area, giving it different names. The "Sino-Japanese Subkingdom" was treated as a subkingdom of the Holarctic Kingdom by Alexinh (1957). It also contains part of the Himalaya Shan, making it similar to Ward's (1927) concept of the "Sino-Himalaya."

The northeastern border of the Sino-Japanese Forest Subkingdom lies in the taiga forests of Siberia and the *Pinus-Quercus* forests of Korea. The northwestern border of the subkingdom is represented by a transition between grassland and subalpine shrub and meadow. The southwestern and southeastern borders comprise *Pinus-Quercus* forest and evergreen-deciduous broad-leaved forest. Most parts of Japan lie in this subkingdom, while in China its border extends along the southern border of the grasslands of northern China to Lanzhou, then makes a southward turn to the western margin of the Sichuan Pendi and the eastern margin of the Yunnan plateau. This is the core area of the Eastern Asiatic Kingdom, which has experienced almost no tectonic change since the Cretaceous Period and no large-scale continental ice sheet coverage during the Quaternary Period. This stable tectonic and environmental background provides ideal refugia for plant species persisting since the Cenozoic Era or even earlier. This subkingdom shows a great diversity of floral components with a total of more than 20 000 species and more than 90 endemic genera.

Within the subkingdom, thanks to variations in moisture and temperature, there are several natural climatic zones, including temperate, warm-temperate and subtropical. Within each zone is deciduous forest and evergreen broad-leaved forest dominated by species of *Betula, Populus, Quercus, Tilia* and genera of Fagaceae, Hamamelidaceae, Lauraceae and Magnoliaceae. Coniferous forests are formed of thermophilous species of *Calocedrus, Chamaecyparis, Cupressus* and *Fokienia* (all Cupressaceae), *Cryptomeria* and *Cunninghamia* (both Taxodiaceae), *Keteleeria, Pinus, Pseudolarix, Pseudotsuga* and *Tsuga* (all Pinaceae). This subkingdom can be divided into six regions and 19 subregions.

The Northeast China Region (IIID7) covers the east of northeastern China, far eastern Russia (Khabarovsk) and northern Korea, and has been treated as the "Manchurian Province" by Good (1974) and Takhtajan (1986). According to Takhtajan's thinking, the Manchurian Province also included eastern and northeastern Mongolia and the Da Hinggan Ling, but here we recognize a more restricted region. Nevertheless, Takhtajan described many of the characteristic species of the region, with *Pinus koraiensis* as a dominant and representative tree species mixed with deciduous broad-leaved forest. Both basally-branching taxa, such as *Astilboides* and *Mukdenia* (both Saxifragaceae, both extending to Korea) and *Plagiorhegma* (Berberidaceae), and derived taxa such as the monotypic genera *Brachybotrys* (Boraginaceae), *Diplandrorchis* (Orchidaceae) and *Symphyllocarpus* (Asteraceae), are present. The saprophytic *Diplandrorchis* seems on the basis of morphology to be the basalmost-branching taxon of Orchidaceae, and is found only in China. *Omphalotrix*, a monotypic genus of Scrophulariaceae, occurs around swamps and extends into Hebei. This region contains 1 776 species of seed plants in 575 genera and 116 families. It has temperate characteristics and a close relationship to Japan. One hundred and nineteen Chinese endemic species are found in the region, some of them dominant or founder species of certain communities (Fu *et al.*, 1995).

The North China Region (IIID8) was first described as the "Seventh Province" by Takhtajan (1986). Its northwestern border was defined by WU (1979) as the natural boundary between the Qin Ling and the Central China Region. The vegetation comprises mixed coniferous and deciduous forest characterised by *Pinus densiflora, P. tabuliformis* and *Quercus,* and coastal vegetation of the Shandong Bandao. There are 15 characteristic genera, of which five are strictly endemic. Eight of the characteristic genera are monotypic; the other seven are oligospecific, including two tree genera, *Pteroceltis* (Ulmaceae), which occurs in limestone areas, and *Xanthoceras* (Sapindaceae), which is distributed from northern China to northwestern Nei Mongol and Ningxia in Gansu. Five genera are considered to be endemic or semi-endemic: *Carlesia* (Apiaceae), *Myripnois* and *Opisthopappus* (Asteraceae), *Oresitrophe* (Saxifragaceae) and *Taihangia* (Rosaceae). Of these, *Myripnois* and *Oresitrophe* can also be considered relict genera. *Bolbostemma* (Cucurbitaceae) is a genus of two species, one distributed from northern to central China, the other with two distinct varieties occurring in the middle Chang Jiang and the middle Hong He. The same distribution pattern is also seen in the two varieties of the shrub *Ostryopsis*. This repeated pattern provides strong evidence that these plants are relict components of Laurasian origin. Another example is provided by *Macleaya cordata* (Papaveraceae) and *M. microcarpa,* which share the same distribution pattern. WANG and colleagues (1995) regarded the North China Region as the richest flora in northern China, with 3 829 species in 919 genera and 151 families. About 1 600 of these species are endemic to China, with around 200 species and varieties endemic to this particular region, including *Larix gmelinii* var. *principis-rupprechtii* and *Picea neoveitchii*. The region is divided into four subregions.

The Liaoning-Shandong Peninsular Subregion (IID8 a) lies in the northeastern part of the North China Region, and covers the Liaodong Bandao and Shandong Bandao, excluding the hills of southern Shandong and the area to the south of the Huai He. The two peninsulas share close communication and therefore similar floristic components. Thanks to the presence of the Huang Hai and Bohai Wan, the climate has a distinctly oceanic character with abundant precipitation supporting a rich plant diversity. There are 2 000 species (Wang *et al.*, 1995), of which Sino-Japanese components play an important role. *Pinus densiflora* is a diagnostic species of the subregion. There are no endemic genera at subregional level but there are around 70 endemic species, including *Salix donggouxianica, S. koreensis* var. *shandongensis, Tilia jiaodongensis* and *Ulmus pseudopropinqua*.

The North China Plains Subregion (IIID8 b) covers the Liaohe Pingyuan and a broad area east to the Yan Shan and Taihang Shan. The landscape is dominated by alluvial plains, coastal plains and low hills, no higher than 200 m, except in the mountainous areas of southern Shandong. Because of agricultural activity and climate change no natural forest survives, leaving only fragmentary natural vegetation of secondary bushes and grasses. The lack of diversity of the area is shown by the presence of only 610 species, of which 30 occur only in this subregion, including *Corispermum huanghoense* and *Tournefortia sibirica* var. *angustior* (Boraginaceae; Wang, 1997).

The North China Mountains Subregion (IIID8 c) is located in the northern Qin Ling and covers the large area of the Zhongtiao Shan, Taihang Shan, Wutai Shan and Shanxi highlands. Wutai Shan, rising to 3 058 m, has vertical vegetation zones of temperate deciduous broad-leaved montane forest, taiga and subalpine meadow thicket. There are 3 000 seed plants in this subregion, half of which are endemic to China. The subregion itself contains five genera endemic to the North China Region, and ca. 300 subregionally-endemic species such as *Myripnois dioica, Opisthopappus taihangensis, Oresitrophe rupifraga, Ostryopsis davidiana, Saxifraga unguiculata* var. *limprichtii* and *Taihangia rupestris* (Wang, 1997). There are also some Sino-Himalayan and tropical components—*Mallotus repandus* and *Phyllanthus urinaria* (both Euphorbiaceae), *Sapindus saponaria* (Sapindiaceae) and *Stephania japonica* (Menispermaceae)—in this subregion, which represents the intersection of the Sino-Japanese Forest and Sino-Himalaya Subkingdoms.

The Loess Plateau Subregion (IIID8 d) covers a large extent of the Huangtu Gaoyuan (Loess Plateau) including the area west to the Lüliang Shan, northern Shaanxi and the plateau of central Gansu. Hardly any natural vegetation remains because of human activities over an extended period of time, with only sparse riverine forest and scrub. The subregion houses 864 seed plant genera in 147 families (Zhang *et al.*, 2002a). Components of the floras of northwestern China and central Asia occur in the area, including *Ajania fruticulosa, Artemisia sphaerocephala, Jurinea mongolica* and *Stilpnolepis centiflora* (all Asteraceae) and *Kalidium foliatum* (Wang, 1997). Two subregionally-endemic genera have been recognised in the subregion, *Synstemon* and *Tetraena,* and 164 endemic subgeneric taxa are found here, including *Astragalus taiyuanensis, Oxytropis ganningensis* and *Salix shihtsuanensis* var. *globosa* (Zhang *et al.*, 2002a).

The East China Region (IIID9) was treated by Diels

(1901) and Takhtajan (1986), together with central China, as the "Central China Province," and is regarded as the center of distribution of *Ginkgo* (Ginkgoaceae), *Pseudolarix* and *Pseudotaxus* (Taxaceae). There are 22 endemic, characteristic genera occurring in this region, at least nine of which are strictly endemic to it. These include *Changium* (Apiaceae), *Changnienia* (Orchidaceae), *Cyclocarya* (Juglandaceae), *Fortunearia* (Hamamelidaceae), *Ginkgo, Heptacodium* (Caprifoliaceae), *Monimopetalum* (Celastraceae), *Pseudolarix, Pseudotaxus, Sheareria* (Asteraceae), *Sinojackia* (Styracaceae) and *Speirantha* (Liliaceae). *Cunninghamia* occurs across a large area south to the Chang Jiang. The centers of genetic diversity of wild *Gingko* are thought to lie in southwestern China, in Guizhou and probably also in Tianmu Shan, Zhejiang (Gong *et al.*, 2008). The East China Region shows a high degree of floristic similarity to Japan, with which it shares 86% of genera, including *Berchemiella* (Rhamnaceae), *Chikusichloa* (Poaceae), *Cryptomeria, Disanthus* (Hamamelidaceae), *Kirengeshoma, Peltoboykinia* and *Platycrater* (all Saxifragaceae) and *Theligonum* (Rubiaceae; Zheng, 1984). At species level, the East China Region shares 130 species with the Eastern Asiatic Kingdom and 700 with the Sino-Japanese Forest Subkingdom (Hao & Yang, 1996). LIU and colleagues (1995) concluded that the East China Region is a natural floristic region, with 4 259 plant species belonging to 1 180 genera and 174 families including several basally-branching angiosperm families such as Fagaceae, Juglandaceae and Magnoliaceae. Gymnosperms are strongly represented here by 22 genera in eight families. There are 425 species endemic to the region, and 1 722 Chinese endemics found in the region. It is divided into four subregions.

The Yellow River and Huaihe River Plains Subregion (IIID9 a) includes most parts of Anhui, Jiangsu and part of southeastern Shandong. Two rivers, the Huang He and Huai He, drain the area. Its altitude is ca. 100–200 m, with Dabie Shan (1 774 m) at the western margin. Historically, the area's natural forests have been destroyed by human activities, leaving only small patches of mixed forest in the hills. Only in a few places have evergreen forests, dominated by *Castanopsis sclerophylla, Cyclobalanopsis glauca* and *Lithocarpus* spp. (all Fagaceae), *Ilex chinensis* (Aquifoliaceae) and *Phoebe sheareri* (Lauraceae), remained. In the deciduous broad-leaved forests, which merge gradually into the North China Region, *Albizia kalkora* (Fabaceae), *Castanea seguinii* (Fagaceae), *Celtis sinensis* (Ulmaceae), *Platycarya strobilacea* (Juglandaceae), *Quercus acutissima* and *Quercus variabilis* are dominant. There are 20 species endemic to the subregion, such as *Ulmus chenmoui.*

The Jianghan Plain Subregion (IIID9 b) covers the plains of the Chang Jiang and Hanjiang, an area with numerous rivers, streams and lakes. Twenty subregionally-endemic species are found here including *Gueldenstaedtia verna* (Fabaceae) and *Pinus fenzeliana* var. *dabeshanensis,* with evergreen broad-leaved forest fragments formed of *Castanopsis sclerophylla, Cinnamomum camphora* (Lauraceae), *Cyclobalanopsis glauca, Lithocarpus glaber* and *Schima superba* (Theaceae). Because of the numerous rivers and streams, hydrophytic vegetation develops in many places, with taxa including *Najas marina* (Hydrocharitaceae), *Nymphoides peltata* (Menyanthaceae), *Potamogeton* spp. (Potamogetonaceae) and *Spirodela polyrhiza* (Lemnaceae).

The South Zhejiang Mountains Subregion (IIID9 c) covers most parts of Zhejiang, northern Fujiang and southern Jiangxi. This is a mountainous area including Wuyi Shan, Kuocang Shan, Donggong Shan and Huang Shan, with an average altitude of ca. 1 500 m and highest peak at Huanggang Shan (2 158 m). Below 1 200 m, *Pinus massoniana* forests and evergreen broad-leaved forests occur, including *Altingia gracilipes* (Hamamelidaceae), *Castanopsis eyrei, Castanopsis fargesii, Choerospondias axillaris* (Anacardiaceae), *Cyclobalanopsis sessilifolia, Diospyros japonica* (Ebenaceae), *Meliosma myriantha* (Sabiaceae), *Michelia maudiae* (Magnoliaceae) and *Schima superba.* Mixed evergreen and deciduous forests occur above 1 200 m. There are no subregionally endemic genera, but 63 endemic species, including *Clerodendrum kiangsiense* (Verbenaceae), *Photinia zhejiangensis* (Rosaceae) and *Sinosenecio wuyiensis* (Asteraceae).

The South Jiangxi and East Hunan Hills Subregion (IIID9 d) covers most of Jiangxi and the southeastern part of Hunan. The key landscape forms are hills and basins and the natural vegetation has largely been destroyed by agricultural activities. Coniferous forests and bamboo are the dominant vegetation, with *Cunninghamia lanceolata, Keteleeria fortunei* var. *cyclolepis, Phyllostachys edulis* (Poaceae), *Phyllostachys makinoi, Phyllostachys nigra* var. *henonis* and *Pinus massoniana.* Above 1 200 m, *Pinus taiwanensis, Pseudotsuga sinensis* var. *sinensis* and *Tsuga chinensis* occur. Fragments of evergreen and deciduous broad-leaved forests consisting of *Castanopsis, Cyclobalanopsis* and *Liriodendron chinense* (Magnoliaceae) remain around Jiuling Shan, Luoxiao Shan and Mufu Shan. No subregionally-endemic genera are found, but there are approximately 20 endemic species including *Cyclobalanopsis ningangensis, Mazus saltuarius* (Scrophulariaceae) and *Tilia membranacea.*

The Central China Region (IIID10) lies in the central part of China where the natural vegetation in former times would have been evergreen broad-leaved forest. Now, besides farmland, characteristic species include *Pinus massoniana* and other thermophilous species of Cupressaceae, Pinaceae and Taxaceae. These are replaced to the north and at higher altitudes by *Fagus* or *Quercus.* Woody genera with Sino-Japanese and eastern Asia-North America disjunct distributions occur in this area. There are 5 600 plant species in 1 279 genera and 207 families. Four thousand and thirty-five of the species are Chinese endemics, and 1 548 of these are endemic to the Central China Region (Qi *et al.*, 1995). Thus, endemism is an important characteristic of this region and endemic species can be described as an autochthonous type here. Species from the Eastern Asiatic Kingdom number 1 045 and, together with the Chinese endemic components, total 5 080 species (over 90% of the plant species of the region). Some species are relicts, rarely found in other regions, which could be the result of the area's long tectonic history with no ocean inundation since the Triassic Period, the limited influence of glaciation, with many mountain refugia, and the mountainous landscape providing opportunities for old species to survive and new species to arise.

There are 51 genera endemic or characteristic to the Central China Region, 28 of which are strict endemics. Among these, 31 are monotypic, 17 oligospecific and two polytypic: *Ancylostemon* and *Isometrum* (both Gesneriaceae). No fewer than 30 genera are regarded as "paleo endemic" components, including *Bretschneidera* (Bretschneideraceae, which is also found in northern Thailand and Vietnam), *Cathaya, Davidia* (Nyssaceae), *Eucommia* (Eucommiaceae) and *Metasequoia*, with Eucommiaceae being a Chinese endemic family. Around nine of the endemic genera are trees - including *Camptotheca* (Nyssaceae), *Chimonanthus* (Calycanthaceae), *Dipteronia* (Aceraceae), *Poliothyrsis* (Flacourtiaceae), *Sinowilsonia* (Hamamelidaceae) and *Tetrapanax* (Araliaceae) - six are shrubs, one bamboo, four lianas - *Archakebia, Sargentodoxa* and *Sinofranchetia* (all Lardizabalaceae) and *Clematoclethra* (Actinidiaceae)- and ca. 30 herbaceous plants.

The Qinling-Bashan Mountains Subregion (IIID10 a) lies in the northern part of the North China Region, at the interface of Gansu, Shaanxi, Sichuan and Hubei. The subregion incorporates the Daba Shan, Wudang Shan and Shennongjia mountains, the latter rising to 3 105 m, and has an average altitude of 1 000–2 000 m. The annual average temperate is above 15 °C and rainfall is more than 800 mm per year. The vegetation of this subregion is mainly deciduous broad-leaved forest at the base of the mountains, dominated by *Quercus aliena* var. *acutiserrata* and *Q. engleriana,* replaced at higher elevations by forests of *Fagus engleriana* and, above 2 000 m, by coniferous forests in which the dominant species are *Abies fargesii* and *Tsuga chinesis.* The species-level diversity is considerable (Ying, 1994) with more than 3 000 seed plant species in 963 genera and 178 families. There is a close relationship between the Qin Ling and northwestern China as well as the North China Region (Ying, 1994). Forty-nine genera endemic to China occur in the subregion, and *Echinocodon* (Campanulaceae) and *Kungia* (Crassulaceae) are endemic to it. In addition, there are more than 190 subregionally-endemic species.

The Sichuan Basin Subregion (IIID10 b) lies at an altitudinal range of 200–700 m and is surrounded by high mountains. In the northwest of the subregion lies the main peak of Longmen Shan, rising to 4 982 m, while to the west is Emei Shan (3 099 m), to the east Huaying Shan (1 704 m), and to the south Da'anshan (2 251 m). Due to its low topography, the temperature of this subregion is always higher than elsewhere, and it retains some tropical floristic elements such as *Archidendron clypearia* (Fabaceae), *Combretum wallichii* and *Quisqualis indica* (both Combretaceae), *Cryptocarya* spp. (Lauraceae), *Lasianthus henryi* (Rubiaceae), *Melastoma* sp. (Melastomataceae) and *Meliosma thorelii.* There are no endemic genera but the number of endemic species is relatively large (208). It is noteworthy that about 400 type specimens have been collected from Emei Shan alone (Wu & Peng, 1993).

The Sichuan-Hubei-Hunan Border Subregion (IIID10 c) includes Fanjing Shan (2 571 m) and Leigong Shan (2 179 m). Here the vegetation type varies with elevation, including, from low to high, evergreen broad-leaved forest, mixed evergreen and deciduous broad-leaved forest, subalpine coniferous forest, and subalpine meadow and thickets. The subregion has a relatively high species diversity, with more than 300 subregionally-endemic species and at least five endemic genera: *Cathaya, Ombrocharis* (Lamiaceae), *Psilopeganum* (Rutaceae), *Tangtsinia* (Orchidaceae) and *Triaenophora* (Scrophulariaceae). Famous relict taxa found in this area include *Cathaya, Dipteronia, Liriodendron, Metasequoia, Taiwania* (Taxodiaceae) and *Tetracentron* (Tetracentraceae), as well as species of *Abies*, all of which indicate that this was an important refugium during glacial periods. Most of the genera found here are of a north-temperate distribution type, however there are some tropical genera such as *Derris, Ormosia* and *Zenia* (all Fabaceae), *Eurycorymbus* (Sapindaceae) and *Passiflora* (Passifloraceae), which indicate links with the South China Mountains Region. Some Sino-Himalayan elements, such as *Decaisnea* (Lardizabalaceae) and *Dipentodon sinicus* (Dipentodontaceae) can also be found here.

The Guizhou Plateau Subregion (IIID10 d) covers the broad karst landforms of the Guizhou plateau located to the south of a line linking Nanchuan, Hejiang, Yibin and Pingshan, east of Hexi corridor, west of the Fanjing Shan and Yunwu Shan and north of Miao Ling, including northern Guizhou and northeastern Yunnan. The altitude of valleys in the subregion is about 300–500 m with the summits of the mountains ranging from 1 500 to 2 000 m. The vegetation comprises subtropical coniferous forest and evergreen broad-leaved forest. There are two genera, *Dicercoclados* (Asteraceae) and *Tengia* (Gesneriaceae), and 120 species endemic to this subregion. The vegetation differs from the Yunnan Plateau Region in that *Pinus yunnanensis* is replaced by *P. massoniana* and the evergreen broad-leaved forests are dominated by *Castanopsis fargesii, Castanopsis platyacantha, Fagus engleriana, Lithocarpus cleistocarpus, Quercus engleriana* and *Sycopsis triplinervia* (Hamamelidaceae). There are some remnant fragments of rainforest in valleys, which contain many tropical elements such as *Altingia chinensis, A. multinervis, Beilschmiedia kweichowensis* (Lauraceae) and *Cryptocarya chinensis.*

The Lingnan Mountains Region (IIID11) corresponds to the "South China Province" of Fedorov (1959) and to part of the "India-Malaysia Subkingdom" of Takhtajan

(1986). It is mainly located south of the Nanling Shan, covering western Guangxi, Hunan, Guizhou, Guangdong, Jiangxi, Fujian and as far as north as Wenzhou in Zhejiang. The southern border lies in the north of southeastern Guangxi and on the Leizhou Bandao. The vegetation here mainly comprises *Pinus* and *Quercus* forests mixed with some Indian and Malaysian elements. The southern border of the region lies at the northernmost extent of the range of some tropical families such as Ancistrocladaceae, Dipterocarpaceae, Flagellariaceae, Myristicaceae and Nepenthaceae. There are 25 endemic or typical genera, 14 of which are strictly endemic and all of which are monotypic, with the exception of *Indocalamus* (Poaceae). Endemic genera include *Apterosperma* and *Euryodendron* (both Theaceae), *Semiliquidambar* (Hamamelidaceae), two genera of Melastomataceae, two of Rubiaceae and five of Gesneriaceae, which indicates the influence of Paleotropical floristic elements. *Glyptostrobus* (Taxodiaceae), which is sister to North American *Taxodium*, is considered a relict genus of this area. According to Zhang (1994), the region represents the southern extent of *Abies* and *Tsuga* and the northern extent of tropical gymnosperms such as *Amentotaxus* and *Torreya* (both Taxaceae), *Cephalotaxus* (Cephalotaxaceae) and *Gnetum* (Gnetaceae). Almost no deciduous species of Betulaceae or Fagaceae can be found here. However *Bombax* (Bombacaceae), a deciduous tropical element, can be found at about 24° N. The region holds 102 Chinese endemic genera.

The North Fujian Mountians Subregion (IIID11 a) lies in northern and eastern Fujian, extending north from the Daiyun Shan (Zeng, 1983), and is deeply influenced by the surrounding ocean. Hills and plateaus are the main landscape features of the area, which has an average altitude of about 100–500 m. Yandang Shan in the north has an elevation of 1 237 m and, at 1 856 m, Daiyun Shan Mountain is its highest point. The typical vegetation is evergreen monsoon forest dominated by *Castanea henryi, Cleyera pachyphylla* and *Stewartia sinensis* (both Theaceae), *Cornus elliptica* (Cornaceae), *Cyclobalanopsis myrsinifolia, Fagus longipetiolata, Lithocarpus cleistocarpus,* and *Oyama sieboldii* and *Yulania cylindrica* (both Magnoliaceae), together with some deciduous taxa such as *Carpinus, Fraxinus* (Oleaceae) and *Sorbus* (Rosaceae). The gymnosperms *Cephalotaxus latifolia* and *Tsuga chinensis* var. *chinensis* appear in the subregion (Zeng, 1983). There are some remnant fragments of tropical rainforest in valleys. *Kandelia obovata* (Rhizophoraceae) is found at the northern limit of its distribution on the coasts of the subregion, while *Altingia chinensis, Castanopsis carlesii, C. fargesii, Machilus chinensis* (Lauraceae) and *Meliosma rigida* are the dominant species of evergreen broad-leaved forest. There are no subregionally-endemic genera.

The North Guangdong Subregion (IIID11 b) includes Hunan, southern Jiangxi, southwestern Fujian and northern Guangdong, and has an average altitude of ca. 500–1 000 m. The vegetation is evergreen broad-leaved forest mixed with some tropical elements. Dominant species are largely members of Elaeocarpaceae, Fagaceae, Hamamelidaceae, Lauraceae, Magnoliaceae and Theaceae, sometimes with species of *Carpinus*. This is a rich flora, mostly comprising subtropical taxa and widely-distributed tropical elements. There are no subregionally-endemic genera, but the Chinese endemics *Disanthus* and *Semiliquidambar* are present.

The East Nanling Subregion (IIID11 c) includes the eastern Nanling Shan (in eastern and southeastern Guangdong) and the southern Daiyun Shan in Fujian (Zheng & Zhang, 1983). The vegetation is evergreen broad-leaved forest dominated by Fagaceae, Hamamelidaceae, Lauraceae, Magnoliaceae, Moraceae, Myrsinaceae and Theaceae (Wang & Wu, 1997). Characteristic taxa include *Amentotaxus, Aucuba* (Aucubaceae), *Cephalotaxus, Cyclocarya, Fokienia, Helwingia* (Helwingiaceae), *Michelia, Pteroceltis, Pyrenaria* (Theaceae), *Stachyurus* (Stachyuraceae) and *Taiwania* plus some famous "living fossils" or relict taxa such as *Glyptostrobus pensilis, Pseudolarix amabilis* and *Pseudotaxus chienii* . In addition some tropical elements, *Bowringia callicarpa* (Fabaceae) and *Mappianthus iodoides* (Icacinaceae), occur in eastern Guangdong, and mangrove forest, with *Acanthus ilicifolius* (Acanthaceae), *Aegiceras corniculatum* (Myrsinaceae) and *Kandelia obovata,* is found along the coast (Zheng & Zhang, 1983).

The Guangdong and Guangxi Mountains Subregion (IIID11 d) lies in western Guangdong, southern Guangxi and southeastern Hunan, connecting to the Central China Region. Within this subregion lie large mountains, with the highest peak at Dayao Shan (1 979 m). The topology of the landscape is complex, with deep valleys extending over the entire area. The vegetation exhibits significant latitudinal zonation: from northern montane rainforests to subtropical forests at southern and mid latitudes. In the south the forests include *Artocarpus* and *Ficus* (Moraceae), *Cryptocarya, Endospermum* (Euphorboacedae) and *Syzygium* (Myrtaceae). In central and northern areas there are evergreen broad-leaved forests dominated by *Castanopsis lamontii, Castanopsis tibetana, Castanopsis tonkinensis, Lithocarpus chrysocomus, Machilus versicolora* and *Michelia mediocris*. The diversity of species is very rich, with 2 335 species belonging to 870 genera and 213 families on Dayao Shan alone. These include Chinese endemic genera such as *Cathaya, Diplopanax, Eurycorymbus* and *Fokienia*. The many subregionally-endemic species include *Ternstroemia insignis* (Theaceae) and *Vaccinium yaoshanicum* (Ericaceae).

The Yunnan, Guizhou and Guangxi Region (IIID12), which includes southeastern Yunnan, Guangxi and southern and southwestern Guizhou, comprises mainly limestone landscapes. The natural vegetation is largely evergreen broad-leaved forest dominated by *Castanopsis, Cyclobalanopsis, Lithocarpus, Keteleeria* and *Pinus*, and there are many endemics and monotypic taxa. Six thousand, two hundred and seventy-six species in 1 454 genera and 248 families are recorded in the region. Of a total of 29 endemic or characteristic genera, 21 are strictly endemic to the region, 22 are monotypic and four are oligospecific. There are no regionally-endemic families, although Rhoipteleaceae (which occurs only in China and Vietnam around the Beibu Wan) occurs here. The region is also a center of diversity for Chinese endemic genera and species. The presence of *Pachylarnax* (Magnoliaceae) and *Delavaya* (Sapindaceae) indicates a relationship with the Yunnan Plateau Region; that of *Eurycorymbus* shows links to Taiwan. The presence of tropical and subtropical elements—including 14 genera of Gesneriaceae—in this region indicates that the Eastern Asiatic Kingdom had connections with tropical regions. Although this region shows transitional characteristics between the tropics and subtropics it can still be regarded as a part of the Eastern Asiatic Kingdom, modified by the underlying limestone substrate. The region can be further divided into three subregions.

The Guizhou-Guangxi Border Subregion (IIID12 a) covers the drainage basin of the Nanpan Jiang and the area north to a line between Nandan, Hechi, Yishan and Liujiang, mainly on limestone landforms. The natural vegetation has been almost destroyed by human activities; however, fragments of evergreen broad-leaved forest and limestone-based thickets remain. The limestone forests are dominated by *Boniodendron minus* (Sapindaceae), *Carpinus pubescens, Celtis sinensis, Platycarya strobilacea* and other species. Many Chinese endemic genera are also found here. In addition there are some mixed coniferous and broad-leaved forests, dominated by *Calocedrus* spp., *Pinus kwangtungensis* and *Pseudotsuga brevifolia*, which are particular and characteristic to limestone areas. Thus, the geographical background of this subregion has enriched the local flora. The subregion houses more than 40 Chinese endemic genera, of which *Thamnocharis* (Gesneriaceae) is endemic to the subregion. There are 172 subregionally-endemic species including *Acer sycopseoides, Machilus miaoshanensis, Michelia angustioblonga* and *Photinia lochengensis.*

The Hongshui River Subregion (IIID12 b) extends from the southern subtropics to include southern Guizhou, northern Guangxi and the north of the You Shui valley. The Hongshui He meanders through the area from northwest to southeast, providing a passage for the southeastern monsoon. Because of this, tropical vegetation types including seasonal tropical rainforest and evergreen broad-leaved monsoon forests can be found in this valley, and typically tropical species such as *Horsfieldia kingii* (Myristicaceae) and *Parashorea chinensis* (Dipterocarpaceae) can be found at Du'an and Tianlin, among other places. The range of specialised microhabitats is relatively low in this subregion and the diversity of plant species is also relatively low, with few endemic genera and species. There are ca. 100 subregionally-endemic species, including *Craigia kwangsiensis* (Tiliaceae), *Lycoris guangxiensis* (Amaryllidaceae) and *Machilus glaucifolia*, and they are non-dominant in the community. The Hongshui He valley and other special geographical conditions make this subregion a transitional zone between tropical and subtropical floristic elements. In the east of the subregion *Decaspermum gracilentum* (Myrtaceae), *Pinus massoniana* and *Schefflera pauciflora* (Araliaceae) are common in forests on non-limestone substrates. In the west, *Cyclobalanopsis delavayi, Lithocarpus confinis* and *Keteleeria evelyniana* indicate a mix with floristic elements of the Yunnan-Guizhou Plateau, in particular, and a flora closely related to that of southeastern Yunnan. Some major floristic elements belonging to central and southern China, such as *Keteleeria fortunei, Pinus kwangtungensis* and *Pinus massoniana,* reach their westernmost extent in this subregion.

The Southeast Yunnan Limestone Subregion (IIID12 c) ranges from Qiubei and Wenshan east of longitude 103°30' E to the south of Jinzhong Shan in Longlin and north of Funing and Maguan in the "Golden Triangle". This area is the smallest subregion of Yunnan, Guizhou and Guangxi and the richest in endemic species. It is located on the edge of the Yunnan Plateau with a well developed karst landscape. The vegetation comprises evergreen subtropical broad-leaved monsoon forest, with tropical and subtropical species as the main elements. The genus *Lagarosolen* (Gesneriaceae) is endemic to the subregion, which also shares six endemic genera, including *Malania* (Olacaceae), *Parepigynum* (Apocynaceae) and *Saniculiphyllum* (Saxifragaceae), with northern Guizhou and northwestern and northern Guangxi. There are 263 subregionally-endemic species, so the area is known as an important center of species diversity and endemism. The significance of this subregion lies in the fact that it is an area of interchange between the Sino-Himalayan and Sino-Japanese elements of the Asian flora. The western delimitation of the subregion is close to the "Tanaka Line" proposed by Tanaka (1954) based on the distribution of *Citrus* (Rutaceae), which was subsequently shown to be an important division between these two major floristic elements of eastern Asia (Li & Li, 1992). WU and LI (1965) proposed that it represents the westernmost boundary of the Sino-Japanese flora (Wu & Li, 1965; Li & Li, 1992). The flora of southeastern Yunnan is in fact very similar to that of Vietnam, with which it shares about 68 species found nowhere else, including *Amentotaxus yunnanensis, Elaeocarpus balansae, Fissistigma poilanei* (Annonaceae), *F. tonkinense, Helicia grandis* (Proteaceae), *Illicium petelotii* (Illiciaceae), *Neolitsea polycarpa* (Lauraceae), *Phoebe macrocarpa, Pittosporum merrillianum* (Pittosporaceae) and *Syzygium vestitum.*

The Sino-Himalayan Subkingdom (IIIE).—The concept of the "Sino-Himalayan Flora" was put forward by Ward (1927) and redefined by WU (1979) as a parallel subregion to the Sino-Japanese Subregion. WU (1979) pointed out that the area, situated between 20° N and 40° N, has one of the world's richest floras, encompassing temperate, subtropical and the northern margins of the tropical flora, with more than 20 000 species in total. The Yunnan plateau and parts of the Hengduan Shan alone are home to more than 12 000 species, with alpine plants being especially abundant. This is in part because the complex terrain, climate and large number of refugia have allowed a number of relict plants to survive from the Cenozoic Era, and in part because the great differences in vertical elevation and deeply dissected mountain ranges mean that many new plant species continue to evolve. Thus, the floristic elements of the subregion are both old and new. The vertical zonation is obvious, with sometimes as many as eight distinct vegetation zones at different altitudes. At lower elevations, the forest vegetation is similar to that of the Sino-Japanese flora and is made up of similar genera, although the warmth-loving conifers and fagaceous species are entirely different. Closely-related species often display a vicariant distribution between the two floras. *Tsuga, Picea, Abies, Larix* and *Juniperus* appear sequentially in the subalpine and alpine zones. At lower elevations *Calocedrus, Keteleeria* and *Pinus* are usually dominant in coniferous forests.

The Yunnan Plateau Region (IIIE13) was included in the "Sikang-Yunnan Province" by Takkhtajan (1986), but we suggest separation of the high-altitude areas north of Lijiang and Zhongdian as part of the Hengduan Mountains Region. Defined in this way, the Yunnan Plateau Region includes central to eastern Yunnan (except northeastern Yunnan) and part of southern Sichuan. There are about 5 545, basically subtropical, species belonging to 1 491 genera and 249 families (Li, 1995), of which 18 genera, including 15 monotypics, are endemic to China. *Dichotomanthes* (Rosaceae) is dominant in both shrub and forest vegetation. *Bolbostemma* (Cucurbitaceae), *Holocheila* (Lamiaceae), *Musella* (Musaceae), *Psammosilene* (Caryophyllaceae), *Rhabdothamnopsis* (Gesneriaceae) and many other endemic genera are also common in the region, which is divided into three subregions.

The Central Yunnan Plateau Subregion (IIIE13 a) includes the Yi minority autonomous prefecture of Liangshan in southern Sichuan, is located to the east of Cangshan in Dali and extends east to Wumeng in the Lupanshui area of Guizhou. The average elevation is about 2 000 m while the highest is over 4 000 m. The Jinsha Jiang and its tributaries form deep valleys where many different kinds of vegetation occur including savanna-like dry shrubland, sparse and dense coniferous forests, evergreen broad-leaved forests, alpine shrub and meadow. The sparse coniferous forests are composed of *Keteleeria evelyniana* and *Pinus yunnanensis* mixed with broad-leaved forests of *Castanopsis delavayi* and *C. orthacantha*. There are about 3 000 species in the subregion, including strict endemics such as *Rhabdothamnopsis*. The area is the center of diversity for genera including *Musella, Nouelia* (Asteraceae) and *Trailliaedoxa* (Rubiaceae). Numerous species are endemic to both China and this subregion, including *Cycas panzhihuaensis* (Cycadaceae) and *Lirianthe delavayi* (Magnoliaceae). Similarities exist between species of this region, such as *Quercus cocciferoides, Q. franchetii* and *Olea europaea* subsp. *cuspidata*, and those of the modern-day Mediterranean (*Q. coccifera, Q. ilex* and *O. europaea* subsp. *europaea* respectively), which seem to indicate that some Tethys remnants occur in the subregion.

The East Yunnan Subregion (IIIE13 b) is located in southeastern Yunnan, west of Wenshan and Maguan and east of Mojiang; its southern extent lies close the border with neighboring countries or that of the Paleotropical Kingdom. Within the subregion, the Ailao Shan extend and branch southward, with the highest mountain peaks up to 3 100 m in altitude. Montane rainforest below 1 000 m is dominated by *Altingia yunnanensis, Crypteronia paniculata* (Crypteroniaceae), *Dipterocarpus retusus* (Dipterocarpaceae), *Madhuca pasquieri* (Sapotaceae) and *Terminalia myriocarpa* (Combretaceae) and is gradually replaced with increasing elevation above 1 000 m, by the main montane vegetation, evergreen broad-leaved forest composed of Fagaceae, Hamamelidaceae, Lauraceae and Magnoliaceae. Near mountain summits is found dwarf forest dominated by Aquifoliaceae, Ericaceae, Fagaceae and Lauraceae, with many epiphytic bryophyte. *Anisadenia* (Linaceae), *Choerospondias, Decaisnea, Dipentodon, Neohymenopogon* (Rubiaceae) and *Tetracentron* can also be found above 2 400 m, especially in the high elevation dwarf forests dominated by Ericaceae. Since this area was almost unaffected by Quaternary glaciations, a number of Cenozoic relicts are present, including *Diplopanax, Liriodendron, Mastixia* (Mastixiaceae), *Rhoiptelea* (Rhoipteleaceae), and seven species of *Cycas*. There are no endemic genera or higher taxa in the subregion, but seven Chinese endemic families are found here.

The Southwest Yunnan Subregion (IIIE13 c) is located in the west of the Ailao Shan and around the middle part of the Lancang Jiang, including Simao and most of the Lincang and Baoshan areas. The terrain is diverse, with mountains, valleys, plateaus and basins. The vegetation of the subregion is similar to that of the Yunnan Plateau Region but the species composition is quite different: *Pinus kesiya* replaces *P. yunnanensis*, and Fagaceae species are even richer in the evergreen broad-leaved forest. Dominant species are *Castanopsis ferox, C. fleuryi, C. hystrix* and *C. orthacantha*. The areas surrounding the Lancang Jiang, related rivers and their tributaries below 1 100 m have a hot, dry climate and vegetation consisting of sparse shrubs and grasses. Dominant tree species are *Buchanania latifolia* and *Lannea coromandelica* (both Anacardiaceae), *Chukrasia tabularis* (Meliaceae) and *Garuga forrestii* (Burseraceae). The peak of Wuliang Shan in Jingdong, at 3 370 m is the highest mountain in the subregion, and bears mixed coniferous and broad-leaved forest, characterized by *Acer* spp. and *Tsuga dumosa*, and dwarf moss forest with Ericaceae and abundant bamboos near the summit. Overall the tree flora is poorer than in the Eastern Yunnan Subregion.

The Hengduan Mountains Region (IIIE14) has long been recognized as a natural floristic region (Handel-Mazzetti, 1931; Ward; Wu, 1979; Wu, 1987), but with many differences in its boundaries and scope. Here we define it as extending west to the Hengduan Shan, east to the Erlang Shan and Emei Shan, northeast to the Tao He and southwest to Baoshan; it therefore includes western Sichuan and most parts of northwestern Yunnan. The most remarkable landscape features of this region are the tall, parallel north-south mountain ranges and deep river valleys. For much of the region the relative difference in relief between valleys and peaks is 1 500–2 500 m, while for Gongga Shan and Moirigkawagarbo the difference is more than 4 000 m. Some of the valleys, in rain shadows, experience drought conditions. For example Benzilan in Dêqên, at an altitude of 2 025 m, has a maximum temperature of 36 °C and only 286 mm of rainfall per year, whereas the pass of the Baima Xue Shan, at 4 292 m, receives 807 mm per year. LI and LI (1993) confirmed that there are 226 families and 1 325 genera with 7 954 species in this area (not including the Tao He basin on the border of Gansu). There are at least 12 large genera with more than one hundred species, including *Aconitum, Astragalus, Carex, Gentiana* (Gentianaceae), *Pedicularis* (Scrophulariaceae), *Primula, Rhododendron* (Ericaceae), *Salix, Saussurea* and *Saxifraga*. These taxa have diversified relatively recently in the region. *Rhododendron*, as an example, has been divided by FANG and MIN (1999) into eight subgenera, 15 sections and 58 subsections at a global level, and in the Hengduan Mountains Region there are four sub-genera, nine sections and 37 subsections, with a total of 241 species. The region thus holds 26% of the *Rhododendron* species in the world and 43% of the species occurring in China. The level of endemism is very significant, with 26 Chinese endemic or sub-endemic genera, of which 12 are strictly endemic to the region. It is divided into four subregions.

The Three Rivers Gorge Subregion (IIIE14 a) extends north to Dêngqên (Xizang) and Nangqên (Qinghai) and south to Tengchong, with the Gaoligong Shan in the west and the Yun Ling in the east. The Gaoligong Shan, Nu Shan and Biluo Xueshan are separated by the primary mountain gorges of the Nu Jiang, Lancang Jiang and Jinsha Jiang. Between the parallel mountain systems is a strongly undulating terrain with a relative difference of up to 4 700 m between the river gorges and mountain peaks. The vertical zonation of the vegetation from valleys to mountains is distinct and ranges from tropical and subtropical vegetation to alpine scree. *Metanemone* (Ranunculaceae) and *Smithorchis* (Orchidaceae) are the only genera strictly endemic to the subregion; however, species-level endemism is much greater as a result of recent radiations. The parallel river valleys form a north-south passage for plant dispersal, creating conditions for tropical plants to extend northwards and temperate plants southwards. Examples include *Acrocarpus* (Fabaceae), *Aspidopterys* and *Hiptage* (both Malpighiaceae), *Gnetum, Sterculia* (Sterculiaceae) and *Terminalia*, which are found along the Nu Jiang and Jinsha Jiang at the northernmost extent of their ranges.

The South Hengduan Mountains Subregion (IIIE14 b) extends north to Lijiang, Yanyuan and Xichang, west to the Erlang Shan and south to Kangding, forming the border between the Qinghai-Xizang and Yunnan plateaus. The landscape is composed of ridges, broad plateaus and deep valleys with vegetation similar to the previous subregion but with very different species composition. There are more strictly endemic genera in this subregion, including *Anemoclema* (Ranunculaceae), *Formania* (Asteraceae), *Hemilophia* (Brassicaceae), *Skapanthus* (Lamiaceae) and *Tremacron* (Gesneriaceae), together with ca. 20 genera endemic to China—such as *Antiotrema* (Boraginaceae), *Diuranthera* (Liliaceae), *Kingdonia* (Ranunculaceae) and *Musella*—which are shared with the eastern Himalaya Shan or the Yunnan plateau but are centered on the Southern Hengduan Mountains. The subregion is very rich in endemic species, which number 1 171 (Li & Li, 1993) making it the core of the Hengduan Mountains Region flora.

The North Hengduan Mountains Subregion (IIIE14 c) is located north of Kangding, upstream of the Dadu He. Here several mountains, including Siguniang Shan and Zhegu Shan, are more than 5 000 m high, while valleys mostly lie at 1 000–2 000 m. The upper reaches of the Yalong Jiang and Xianshui He comprise basins with an undulating, domed plateau at 4 500–4 700 m. The differences in altitude between valleys and summits are smaller than in the previous subregion, and warm-temperate mixed coniferous and broad-leaved forests are well developed. Species diversity is high, with 232 subregionally-endemic species and the subregionally-endemic genera *Acanthochlamys* (Amaryllidaceae; treated outside of *Flora of China* in Acanthochlamydaceae or Velloziaceae), *Platycraspedum* (Brassicaceae) and *Salweenia* (Fabaceae; Li & Li, 1993). The species composition of the subregion gradually merges into that of Tangut and northern China, with clear substitutions of species. For example, *Abies fargesii* var. *faxoniana* replaces *A. forrestii*, *Corylus heterophylla* replaces *C. yunnanensis*, *Picea likiangensis* var. *rubescens* replaces *P. likiangensis* var. *likiangensis*, and *Pinus densata* replaces *P. yunnanensis*.

The Taohe-Minshan Subregion (IIIE14 d) includes the Tao He in Gansu and the Min Jiang in northern Sichuan. The upper reaches of the Tao He form a wide valley, cutting only gently into the landscape, with an undulating surface and steep mountains. The middle reaches of the Tao He, and middle to upper reaches of the Min Jiang comprise mountains and canyons. Downstream, the Tao Ha comes close to the Huangtu Gaoyuan, with loess hills and valleys. The vegetation type is relatively simple, with a large area of evergreen broad-leaved forest. The floristic composition is typically temperate, with a total of 448 genera and 1 346 species (Zhang *et al.*, 1997). Transitions from the Sino-Japanese flora to the Sino-Himalayan flora are very obvious and there are many elements from northern China, such as *Ajania potaninii*, *Caryopteris mongholica* (Verbenaceae), *Convolvulus tragacanthoides* (Convolvulaceae)—which appears on saline land in northern China and in the valleys of this subregion—and *Sophora albescens*. *Betula platyphylla*, *Populus davidiana* and *Quercus mongolica* form deciduous broad-leaved forest, *Pinus tabuliformis* replaces *P. densata* here, and *Abies fargesii* var. *faxoniana*,

A. fargesii var. *fargesii*, *Picea asperata*, *Picea crassifolia* and *Picea purpurea* are major components of dense coniferous forests. There are no subregionally-endemic genera but 14 genera endemic to China are present, including *Clematoclethra*, *Pomatosace* (Primulaceae), *Przewalskia* (Solanaceae), *Pteroxygonum* (Polygonaceae) and *Sinojohnstonia* (Boraginaceae). About 66 Chinese endemic species are present, including *Angelica nitida* (Apiaceae) and *Rheum sublanceolatum*. The flora of this subregion is not only close to that of northern Sichuan but also that of the Huangtu Gaoyuan, making it a point of intersection between the Sino-Japanese, Sino-Himalayan and Qinghai-Xizang Plateau floras.

The East Himalayan Region (IIIE15) extends from the Hengduan Shan west of the Nu Jiang and includes the drainage basin of the Jiujiang in northwestern Yunnan. The eastern boundary lies at Lhünzhub and Baqên, the north is formed by the Himalaya Shan and the south by the forest at Bianbar and Dêngqên. Evergreen broad-leaved forest vegetation occurs in the south but its distribution is limited to a small area concentrated in the "Big Bend" of the Yarlung Zangbo Jiang at Mêdog, southeastern Xizang and connecting with the montane evergreen broad-leaved forests of the southern wing of the eastern Himalaya Shan, dominated by *Cyclobalanopsis kiukiangensis*, *Cyclobalanopsis lamellosa*, *Michelia doltsopa* and *Quercus lanata*. At high altitudes, *Abies delavayi* var. *motuoensis*, *Picea likiangensis* var. *linzhiensis* and *Tsuga dumosa* form dense coniferous forests, together with sclerophyllous forests of *Quercus aquifolioides* and *Rhododendron* shrubs. In the north, xerophytic shrubs such as *Sophora moorcroftiana* occur in the valleys, with the huge endemic tree species *Cupressus gigantea* in the Yarlung Zangbo Jiang valley. The dominant species in this vegetation are similar to those of the Hengduan Mountains Region. The region is divided into two subregions.

The Dulong Jiang-Northern Myanmar Subregion (IIIE15 a) is located to the west of the Gaoligong Shan, extends north to the Xizang border, and has western and southern boundaries in Myanmar. This subregion includes the upper reaches of the Nmai Hka and mostly lies within in Myanmar, with only a narrow strip along the Jiujiang basin in China. The flora of the subregion is not yet completely known. LI (1994) has studied the flora of the Jiujiang basin, which includes 158 families, 673 genera and 1 920 species of seed plants. Most species belong to Orchidaceae (141), Asteraceae (107), Ericaceae (105), Rosaceae (98) and families endemic to eastern Asia such as Eupteleaceae. At the generic level, 351 genera are considered tropical and 184 temperate, with 88 endemic to eastern Asia and ca. 12 endemic to China—including *Berneuxia* (Diapensiaceae), *Davidia* and *Syncalathium* (Asteraceae)—but none endemic to the subregion itself. At species level, temperate taxa account for 84% of the total in the region. There are 671 species belonging to eastern Asia and 169 species endemic to the Jiujiang or the wider subregion, including *Albizia sherriffii*, *Ampelopsis gongshanensis* (Vitaceae), *Cephalotaxus lanceolata*, *Gentiana qiujiangensis*, *Litsea taronensis* (Lauraceae), *Rehderodendron gongshanense* (Styracaceae) and *Rhododendron gongshanense*.

The Southeast Xizang Subregion (IIIE15 b) is located west of the Hengduan Shan, east of Lhünzhub and Baqên, north of the Himalaya Shan and south of the forest at Bianbar and Dêngqên. The relative altitudinal difference in the mountains is 2 000 m in general. The central and lower reaches of the Yarlung Zangbo Jiang pass through the middle of the subregion and act as a passage for the southwesterly monsoon into the interior of Xizang, bringing large amounts of water vapour and causing a warm, humid climate with an average annual temperature 8–12 °C and annual precipitation 600–800 mm. Because of the influence of the terrain there are dramatic differences in climate, with the east being wet and the mountains and valleys of the west being very dry. There are 276 endemic species in the subregion (Li & Wu, 1984). Evergreen broad-leaved forest vegetation has *Quercus lanata* as the dominant species at lower elevations, mixed with *Cupressus torulosa* and *Pinus bhutanica*, but restricted to small areas. Larger areas of vegetation comprise sclerophyllous forest, dominated by *Quercus aquifolioides*, and sparse coniferous forest dominated by *Pinus densata*, mixed with *Pinus armandii* in the Palong Jiang and Yarlung Zangbo Jiang valleys and elsewhere. *Larix speciosa* occurs in high altitude, dense coniferous forest dominated by *Abies delavayi*, *Abies forrestii* and *Picea likiangensis* var. *linzhiensis* between 3 000 and 4 000 m. In some dry valleys, *Cotoneaster microphyllus* (Rosaceae) and *Sophora moorcroftiana* are common shrubs. *Cupressus gigantea* is endemic to the Yarlung Zangbo Jiang valley in this subregion and elsewhere.

The Qinghai-Xizang Plateau Subkingdom (IIIF).—The Qinghai-Xizang Plateau is the highest, largest and most recently-formed plateau in the world. The plateau in the broad sense has the Himalaya Shan at its southern boundary, the Kunlun Shan, Altun Shan and Qilian Shan to the north, with 3 000–4 000 m altitude difference, and is separated from the deserts of central Asia by the Tarim Pendi, Hexi corridor and the Qaidam Pendi. Its western border lies along the Karakorum Shan, Pamir, Kashmir, Pakistan and Afghanistan; the southeast and east join to the Sino-Himalayan Subkingdom. The plateau can be described as a closed system surrounded by mountain ranges (Zhang *et al.*, 2002b), open only in the northeast via the Huang He, the southeast via the Yarlung Zangbo Jiang through its "Grand Canyon," and the southwest through the Shiquanghe-Sênggê Zangbo valley. It is thus a unique natural geographical unit with a young flora formed during the uplift of the mountains and plateau (Wu, 1987; Wu *et al.*, 1995). Alpine shrubland with *Caragana*, *Chesneya* (Fabaceae), three species of *Hippophae* (Elaeagnaceae), many species of *Potentilla* and small, leathery-leaved species of *Rhododendron*, and alpine meadows with *Festuca* (Poaceae), *Kobresia* and *Poa*, both occupy the extent of the plateau from northeast to southwest. In the northeast and southeast more complex landscapes, including lower elevation canyons, contain small fragments of subalpine coniferous forest or scattered trees. In these areas it is formed of *Pinus* only; elsewhere valley forest can include *Cupressus, Juniperus, Larix, Picea* and, rarely, *Abies*. Towards the northwest of the plateau and the upper reaches of the Yarlung Zangbo Jiang, the terrain is higher with a colder and drier climate. Here, shrub or alpine meadows gradually give way to meadow-steppe or shrubby steppe, the former composed of *Littledalea, Orinus, Ptilagrostis* and *Trikeraia* (all Poaceae), often mixed with *Leymus* and *Pennisetum*; the latter mainly *Artemisia* and *Stipa*, sometimes with *Ajania* or *Hippolytia*. The drier the area the more species of *Artemisia* are present. *Krascheninnikovia* is commonly found at the edge of alpine vegetation and desert, and in the meadow-steppe are species of *Acantholimon* (Plumbaginaceae), *Arenaria, Astragalus, Oxytropis* and *Thylacospermum* (Caryphyllaceae). There are nine strictly endemic genera to the subkingdom, and no endemic families. This subkingdom include three regions and six subregions.

The Tangut Region (IIIF16) is China's northernmost region, at the northern border of the Qilian Shan, forming a transition zone between this area and the Mediterranean/Tethys-Central Asian Subregion. Along the Datong He there are some coniferous forests of *Picea crassifolia, Picea wilsonii* and *Pinus tabulaeformis*. Shrub and shrub-meadows are formed by a variety of deciduous shrubs and small-leaved, evergreen *Rhododendron*. Sunny slopes are inhabited mostly by *Stipa* steppe, with the dominant species above 3 200 m being *S. purpurea* and the endemic *S. aliena* and, below this altitude, *S. bungeana* and endemic *S. przewalskyi*. Common *Artemisia* species above 3 200 m include *A. dalai-lamae, A. demissa, A. hedinii, A. nanschanica* and *A. tangutica*. The region houses 2 050 species in 520 genera and 90 families of seed plants. The only genera endemic to the area are *Coelonema* (Brassicaceae), *Pomatosace* (Primulaceae; distributed as far as the Tao He), *Sinadoxa* (Adoxaceae; under *Picea* forest) and *Xanthopappus* (Asteraceae).

The Qilianshan Subregion (IIIF16 a) has a northern and eastern boundary at the Dangjin Shankou of the Qilian Shan and extends south to the Xiqing Shan and the west to the Qaidam Pendi. *Picea crassifolia* dominates the cold temperate coniferous forests which are widely distributed in the subregion. *Juniperus przewalskii, Picea wilsonii* and *Pinus tabuliformis* are also common, and on shaded slopes and in river valleys is found deciduous broad-leaved forest dominated by *Betula albosinensis, B. platyphylla, B. utilis, Populus davidiana* and *P. simonii*. The dominant shrubby species are *Rhododendron capitatum, R. thymifolium* and *Salix oritrepha*, with *Kobresia humilis, K. pygmaea, Stipa bungeana* and *Stipa purpurea* common in meadow and steppe, as on the Qinghai-Xizang Plateau. The Qilian Shan form a huge geographical barrier, blocking dispersal from the deserts of central Asia to the Tangut area, while the eastern part of the subregion contains low-elevation valleys with a warm, humid climate. There are many floristic elements from northern China, such as *Corylus mandshurica, Ostryopsis davidiana, Quercus mongolica, Sorbaria* spp. (Rosaceae) and *Koelreuteria paniculata* (Sapindaceae). There are 1 747 species and 390 genera of seed plants, with half of these genera being widely distributed in the temperate zone. Ten Chinese endemic genera are found here, but only *Coelonema* is strictly endemic to the subregion.

The A'nyemaqen Subregion (IIIF16 b) is located around the headwaters and upper reaches of the Huang He, adjacent to the Bayan Har Shan, southward to the south of the Xiqing Shan. The climate of the subregion is dry with few broad-leaved forests, but some coniferous forest can be found in valleys. There is a total of 922 species of seed plants, of which only four are strictly endemic to the subregion. There are no strictly endemic genera although *Scrofella* (Scrophulariaceae) is endemic to here and the northern Hengduan Shan. The types of species found here are similar to the Qilianshan Subregion, but *Picea purpurea* occurs in small forest patches or mixed with *P. crassifolia*. Shrubs comprise a variety of *Salix* species and five species of *Rhododendron*. *Stipa grandis, Stipa przewalskyi* and *S. purpurea* are the dominant herbaceous species in the extensive areas of grassland.

The Tangut Subregion (IIIF16 c) ranges from the Bayan Har Shan, south of the Golog Shan, to the southern slope of the Tanggula Shan, with its western boundary lying roughly along the Qinghai-Xizang Highway, including most parts of the southern Qinghai Plateau and northern Qamdo. The subregion also contains the headwaters of the Chang Jiang, Lancang Jiang and Nu Jiang, and the climate is relatively humid because of the influence of the Indian Ocean monsoon. The shaded slopes of the valley bottoms contain deciduous forest, dominated by *Betula* and *Populus*, and *Picea* forests, and the sunny slopes *Cupressus* woodland. The vegetation composition is similar to that of the northern Hengduan Shan, with alpine meadow and grassland being better developed than in the A'nyemaqen Subregion. It is also richer in species, with 1 249, including 34 endemics and the subregionally-endemic genus *Sinadoxa*.

The Xizang, Pamir and Kunlun Region (IIIF17) extends from the Pamir in western Xinjiang to the middle and upper reaches of the Yarlung Zangbo Jiang by way of northern Xizang. The main types of vegetation are alpine meadow, grassland and desert-steppe (Li, 1960), and common and dominant species include *Alyssum canescens, Artemisia nanschanica, Carex moorcroftii, Hedinia tibetica, Kobresia pygmaea, Oxytropis microphylla, Phyllolobium heydei* (Fabaceae) and *Stipa purpurea*. Although the elevation is quite high and the climate dry, there are 18 regionally-endemic species in seven families and 12 genera, the most prominent of which are four species of *Oxytropis*, three of *Puccinellia*, and three species each of Brassicaceae and Caryophyllaceae. This region can be divided into three subregions.

The Middle and Upper Reaches of the Yalu Tsangpo

River Subregion (IIIF17 a) lies mainly on the northern slopes of the Himalaya Shan and the middle and upper reaches of the Yarlung Zangbo Jiang valley. The main vegetation formations are *Sophora moorcroftiana* shrub, alpine meadow and alpine grassland, and dominant families are Brassicaceae, Caryophyllaceae, Fabaceae and Poaceae. The genera *Ajaniopsis* (Asteraceae) and *Parapteropyrum* (Polygonaceae) are endemic to the subregion, and there are about 159 endemic species in 84 genera and 33 families, concentrated in Asteraceae (32 species), Poaceae (12), Fabaceae (11 species), Ranunculaceae (10), Gentianaceae (9 species), Papaveraceae (7 species) and Saxifragaceae (7 species). Key endemic-rich genera are *Delphinium* with seven species, *Gentiana* (5), *Corydalis* (Papaveraceae, 6), and *Saxifraga* (6).

The Changthang Plateau Subregion (IIIF17 b) lies next to the Tangut Subregion, surrounded by the Gangdisê Shan, the western Nyainqêntanglha Shan and the Kunlun Shan-Karakoram Shan range. The area lies at ca. 4 200—4 700 m and the climate is extremely cold. It is controlled by the Indian monsoon, so is wetter in the southeast than the northwest. Cushion plants are well-developed in alpine and nival meadows in the subregion. There are no forests, only shrub communities formed of *Ephedra monosperma, Hippophae tibetana, Krascheninnikovia compacta* var. *compacta* and *Myricaria prostrata*. There are 300 species in 135 genera, which are almost all Himalayan and Qinghai-Xizang Plateau elements. Four of the 21 founder or dominant species of grassland or meadow are also found in the Himalaya Shan-Qinghai-Xizang Plateau, and ca. eight species are Qinghai-Xizang Plateau endemics, including *Androsace tapete, Kobresia pygmaea, Kobresia robusta, Littledalea racemosa* and *Saussurea wellbyi*. Temperate and eastern Asian elements tend to dominate, while 38 genera of central Asian and Mediterranean elements are also found. In the late Miocene and Early Pliocene Epochs trees, including *Carya, Liquidambar* and *Tsuga*, were present and in the Mid to Late Pliocene Epoch broad-leaved forests of Betulaceae, Fagaceae and Ulmaceae were also found here (Tao, 2002). The homogenous vegetation of this subregion is probably the result of the uplift of the plateau and influence of glacial periods.

The Pamir-Karakorum-Kunlun Subregion (IIIF17 c) lies mainly in Xinjiang, with its northern margin extending to Wuta in Xinjiang. The southeastern border stretches to the Pi Shan, the eastern border falls at the Tarim Pendi, and the western border lies outside China. The vegetation comprises montane and alpine desert with unusual occurrences of alpine meadow in the east of the subregion. Coniferous forest is found from 3 400–3 600 m, with *Berberis* spp., *Isopyrum anemonoides* (Ranunculaceae), *Lonicera* spp. (Caprifoliaceae), *Picea schrenkiana* and *Sorbus* spp. In total the subregion houses ca. 700 species in 240 genera. Asteraceae, Brassicaceae, Fabaceae, Poaceae and Ranunculaceae contribute the majority of species. Ninety-three Northern Hemisphere genera are found in this area, 28 of which are Mediterranean-central Asia elements. Characteristic taxa include species of *Acantholimon, Arnebia* and *Lithospermum* (both Boraginaceae), *Cynomorium* and *Orobanche* (Orobanchaceae).

The West Himalayan Region (IIIF18) is more or less the same as described by Hooker (1907), Turrill (1953) and Takhtajan (1986), and only small parts of it extend into China, in the western part of the Himalaya Shan. Although we consider *Cedrus deodara* to be absent from China (in contrast to the treatment in the *Flora of China*), the forests contain other typical representatives of the region including *Abies spectabilis, Picea smithiana, Pinus gerardiana* and *Pinus roxburghii*. *Biebersteinia odora* (Biebersteiniaceae) and *Ferula* spp. (Apiaceae) are important elements shared between this region and the Irano-Turanian flora. *Arenaria compressa* and *Cicer microphyllum* (Fabaceae) are also found in the region, with *Capparis himalayensis, Koelpinia linearis* and *Tragopogon gracilis* (both Asteraceae) and *Tauscheria lasiocarpa* (Brassicaceae) in warm dry valleys. There are few strictly endemic genera, but characteristic genera include *Eurycarpus, Hornungia* and *Pycnoplinthus* (all Brassicaceae) and *Hedysarum*.

10.2.4 The Paleotropical Kingdom (IV)

According to the definitions of WU (1979) and Takhtajan (1986), the "Paleotropic Region" includes tropical areas of Africa and Asia, extending from Southeast Asia into northern Australia (Wu, 1979) and some Pacific Ocean islands (including Fiji, Polynesia and New Caledonia). According to WU (1979), the eastern African part of the Palaeotropic Region includes not only Madagascar but also the "Saharo-Arabian Region," which Takhtajan (1986) had included in his "Tethyan Subkingdom." The vegetation of the area includes tropical rainforest, monsoon forest, savanna and tropical desert. The 40 endemic families that Takhtajan (1986) listed include two families of pteridophytes and two families, Mastixiaceae and Lowiaceae, found in eastern and Southeast Asia with their earliest-branching taxa in eastern Asia. Therefore, the true number of endemic seed plant families is now considered to be 36–37. Although it seems equivalent to eastern Asia in terms of the number of endemic families, the Paleotropical Kingdom contains only four basally-branching families, Degeneriaceae (endemic to Fiji), Rafflesiaceae, Nepenthaceae and Welwitschiaceae (a relict family in southwest Africa). Most of the 40 families discussed by Takhtajan (1986) are specialized and relict groups, some of which can also be found in tropical areas of China. Although one genus of Dipterocarpaceae appears in the Guyana highlands, the remainder of the family are endemic to this kingdom.

The Malesian Subkingdom (IVG).—Takhtajan (1986) divided the Palaeotropic Kingdom into five subkingdoms: "Africa," "Madagascar," "India-Malaysia," "Polynesia" and "New Caledonia." WU (1979) revived Engler's (1899) delimitation, which suggests that the "Saharo-Arabian Region," as a subkingdom, should be included in the Paleotropical Kingdom. The tropical area of China extends from the eastern wing of the Himalaya Shan (at ca. 90° E, south of Cona in Xizang) to Taiwan (ca. 124° E, on Chiwei Yu) and from the Leizhou Bandao and Beibu Wan to the Nansha Qundao at 3°49' N. The northern border extends to Mêdog (29°30' N) in Xizang but in Guangxi and Guangdong reaches 22° N and to 26° N over the sea around Taiwan. The area thus lies within the north of world's tropical zone, but the two ends extend beyond the Tropic of Cancer. Although the Nansha Qundao and Zengmu Ansha are close to the equator, very few plants live in these areas, so this part of China cannot be included in the "Equatorial Region." Accordingly, all of tropical China should be contained in the traditional "India-Malaysia Subkingdom," here called the Malesian Subkingdom, with the majority of the flora being similar to that of Malaysia. This subkingdom can be divided into five regions and seven subregions.

The Taiwan Region (IVG19) includes Taiwan, Ludao and volcanic islands such as Lanyu (75 km north of the Philippines and 144 km west of mainland of China, from which it is separated by the Taiwan Haixia). Two-thirds of the region is mountainous. The Zhongyang Shan extend from the north and northeast to the south and southwest, and include Taidong Shan, Xueshan-Yushan and Ali Shan. Most of these are 3 000–3 500 m high, with the highest peak being Yushan at 3 997 m. Only Ali Shan in the west is below 3 000 m. The north and west of the region comprises low coastal flatlands strongly influenced by the monsoon. The mean annual temperature of Taiwan is above 21 °C (except in alpine areas) and the mean annual precipitation is 2 500 mm. The lowest-lying vegetation is mangrove forest on the west coast, above which are found, with increasing elevation, coastal forest, rainforest, monsoon rainforest, lowland evergreen broad-leaved forest, montane *Machilus-Castanopsis* forest, subalpine coniferous forest and alpine scrub. According to ZHANG (1995) there are 25 species of gymnosperm (in 17 genera and eight families), 3 346 species of dicot (in 1 194 genera and 167 families) and ca. 1 000 species of monocot in Taiwan. In contrast, in the *Flora of Taiwan*, HSIEH and colleagues (1994) suggest that there are only ca. 2 400 dicots (and, as the previous author, ca. 1 000 monocots). Both sources indicate that a very abundant seed plant flora exists in Taiwan, similar to that of Hainan, with 3 584 species (Wu *et al.*, 1996).

There are no endemic families in the flora of Taiwan. Trochodendraceae, for example, which cannot be found in mainland China, is however distributed in Japan and Korea. Most of the genera once thought to be endemic to Taiwan have been sunk into other genera, with only *Sinopanax* remaining as a strictly endemic genus. However, the region holds several characteristic genera such as *Astronia* (Melastomataceae), *Crepidiastrum* (Asteraceae), *Cypholophus*, *Leucosyke* and *Pipturus* (all Urticaceae), *Cyrtandra* (Gesneriaceae), *Melanolepis* (Euphorbiaceae), *Osmoxylon* (Araliaceae), *Palaquium* (Sapotaceae), *Ryssopterys* and *Tristellateia* (both Malpighiaceae). All of these are Malaysian floristic elements which are distributed in the Philippines, Malaysia (*Pipturus* extends to Australia) and Taiwan, but not in mainland China. Of its 1 106 genera of angiosperms, Taiwan shares 996 genera (90%) with mainland China, and the only gymnosperm of the island that cannot be found on the mainland is *Chamaecyparis* (Cupressaceae). In contrast four gymnosperm genera, *Amentotaxus, Cunninghamia, Keteleeria* and *Taiwania,* are found both on Taiwan and in mainland China but nowhere else in the world. This suggests that the flora of Taiwan has a close relationship with that of mainland China.

At the species level, most non-endemic taxa in Taiwan are very similar, or identical, to those of the southern mainland. There are 900 to 950 endemic species in Taiwan (including pteridophytes), which account for ca. 20% of the total of ca. 4 200 vascular plants. The flora of Taiwan also contains many subspecies and varieties also found on the Chinese mainland and Japan, however. If these are also taken into account, the percentage of endemics increases to ca. 40% of taxa. There has been a longstanding difference of opinion regarding the affinities of the Taiwan flora: Wulff (1944), LI (1950), Takhtajan (1986) and ZHANG (1995) suggested that it should be included in the Holarctic Kingdom, as a province of eastern Asia in the "Northern Subkingdom." In contrast, WU (1979) and WU and WU (1996) suggested it should be included in the Paleotropical Kingdom because its low-elevation vegetation is entirely tropical. Here we follow the latter authors who based their theory on increased detail and evidence. This region can be divided into two subregions.

The Taiwan High Mountains Subregion (IVG19 a) covers the series of high mountains south of the Yilan plain and east of Ali Shan, including Yushan, Xueshan, Zhongyang Shan and Taidong Shan, most of which have peaks over 3 500 m with distinct vertical zonation in vegetation types. Influenced by the landform, there is no mangrove forest on the east coast of Taiwan. Instead the vegetation types, from low to high elevations, are tropical rainforest, montane rainforest, evergreen broad-leaved forest, mixed broad-leaved and coniferous forest, coniferous forest, and alpine *Rhododendron* shrub. The taxa above the low-elevation vegetation are very similar to those in the south of the Hengduan Shan and eastern Himalaya Shan, but with differences at the species level. This subregion holds most of the taxa endemic to Taiwan, including *Abies kawakamii, Chamaecyparis formosensis, C. obtusa* var. *formosana, Picea morrisonicola, Pinus morrisonicola, Pseudotsuga sinensis* var. *wilsoniana* and *Tsuga chinensis* var. *formosana*. All nine *Salix* species found here, as well as all seven *Actinidia* (Actinidiaceae) and all five *Carpinus* species, are endemic to Taiwan (Zhang, 1995). Many families—including Aceraceae, Aristolochiaceae, Asclepiadaceae, Begoniaceae, Ericaceae, Gentianaceae, Gesneriaceae, Lauraceae, Liliaceae, Loganiaceae and Menispermaceae—have more than 50% of species found on Taiwan as endemics. For a further 21 families, including Euphorbiaceae, Fagaceae, Hamamelidaceae, Rosaceae and Symplocaceae, more than 30% of species are endemic.

The North Taiwan Subregion (IVG19 b) includes the flat uplands north of the Hengchun Peninsula and west to Ali Shan. Here, primary forest has been destroyed and only remnants of mangrove forests remain along the coast, of which *Avicennia marina* (Verbenaceae) and *Kandelia obovata* are common components. In addition, some elements of the tropical coastal flora of eastern Asia, such as *Barringtonia asiatica* (Lecythidaceae), *Calophyllum inophyllum* (Clusiaceae), *Excoecaria agallocha* (Euphorbiaceae), *Hernandia nymphaeifolia* (Hernandiaceae), *Pongamia pinnata* (Fabaceae), *Scaevola hainanensis* (Goodeniaceae), *S. taccada* and *Tournefortia argentea* (Boraginaceae), are also present. The hills are covered with evergreen broad-leaved forest dominated by *Castanopsis*, *Cinnamomum* (Lauraceae), *Ficus*, *Lithocarpus*, *Machilus*, *Michelia* (Magnoliaceae) and many others, and monsoon rainforest with *Chamaecyparis formosensis*, *Keteleeria davidiana* var. *formosana*, *Trochodendron aralioides* (Trochodendraceae) and many species of *Actinodaphne* (Lauraceae), *Cryptocarya*, *Elaeocarpus*, *Ficus* and *Lagerstroemia* (Lythraceae).

The South Taiwan Region (IVG20) includes the Hengchun Peninsula and some small islands such as Lanyu, Xiaolan Yu and Ludao. The Hengchun Peninsula is located at the southern end of Taiwan, and includes Zhongyang Shan (less than 2 000 m high). The islands of Lanyu and Ludao rise to 548 m and 2 823 m, respectively, and therefore the regional landform can be described as hilly and mountainous. The climate is hot and wet, with an mean annual temperature of 25 °C and mean annual precipitation of 3 400 mm. The vegetation is mainly composed of mangrove forest, tropical rainforest and monsoon rainforest. The mangrove forest is well protected, whereas the coastal forests of *Allophylus timoriensis* (Sapindaceae), *Barringtonia asiatica*, *Hernandia nymphaeifolia*, *Morinda citrifolia* (Rubiaceae) and *Pongamia pinnata* have largely been destroyed. The main components of the tropical rainforest include *Dysoxylum cumingianum* (Meliaceae), *Ficus* spp. and *Pometia pinnata* (Sapindaceae). The flora of this region is very rich, mainly in tropical elements. There are 590 genera in the region, none of them endemic. There have been different opinions on the Hengchun and Lanyu floras, especially that of Lanyu. According to Kano (1933), Lanyu and Ludao should be regarded as distinct from Taiwan. However, LI (1950) and GENG (1956) suggested they should be combined together as the area in which the floras of southern China, the Phillipines and Malaysia converge. An analysis by LIU and YANG (1974) found that 146 of the 658 vascular plant species in Lanyu are not found on Taiwan, and 78% of the remainder occur only in southern Taiwan. Sixty-five percent of these species show a close relationship between Lanyu and Taiwan. However, Zhang (1967) pointed out that 46 out of 445 Lanyu genera studied are distributed in the Phillipines but not in Taiwan, suggesting the flora of Lanyu is closer to that of the Phillipines.

The South China Sea Region (IVG21) includes the southern boundary of the Chinese mainland and more than 1 130 islands including Hainan and other small islands off Hainan and Guangdong: Xisha Qundao, Zhongsha Qundao, Dongsha Dao and Nansha Qundao (Wu *et al.*, 1996). This region may also include the Leizhou Bandao and the south of Yunkai Dashan. Of 961 species (in 565 genera and 145 families) found on the Leizhou Bandao, 91% are shared with Hainan. The islands of the Zhu Jiang estuary alone contain 704 species in 484 genera and 156 families. The Xisha Qundao houses 312 species of angiosperms in 208 genera and 78 families, with 70% of genera being pantropical. The Nansha Qundao has 78 species (including cultivated species) in 20 genera and 46 families. Hainan has about ten characteristic families but none endemic. According to WU (1996), at generic level there are nine strict endemics and another dozen characteristic genera. Some genera found here are also distributed in southern China, Guangxi, Guizhou and Yunnan, including *Ampelocalamus* (Poaceae), *Chuniophoenix* (Arecaceae) and *Semiliquidambar*; others extend to the Qin Ling and Vietnam. *Sargentodoxa*, found here, is endemic to China and there are six strict regionally-endemic genera: *Cathayanthe* and *Metapetrocosmea* (both Gesneriaceae), *Chunia* (Hamamelidaceae), *Merrillanthus* and *Pentastelma* (both Asclepiadaceae) and *Wenchengia* (Lamiaceae). The region can be divided into five subregions.

The West Guangdong-North Hainan Subregion (IVG21 a) includes the Leizhou Bandao and northern Hainan. The primary vegetation has long since been destroyed but small patches of evergreen monsoon rainforest can still be found, dominated by *Acronychia pedunculata* (Rutaceae), *Ficus microcarpa*, *Antiaris toxicaria* (Moraceae), *Garcinia oblongifolia* (Clusiaceae), *Homalium cochinchinense* (Flacourtiaceae), *Schefflera heptaphylla* and *Sterculia lanceolata*. Mixed forest is composed of *Engelhardia roxburghiana* (Juglandaceae), *Liquidambar formosana* and *Pinus fenzeliana*, and mangrove forest of *Aegiceras corniculatum* (Myrsinaceae), *Avicennia marina*, *Kandelia obovata* and *Rhizophora stylosa* (Rhizophoraceae). There are 961 species of seed plants on the Leizhou Bandao, in 565 genera and 145 families (Xing *et al.*, 1993). Among them, tropical genera number 464, 82% of the total, and temperate genera only 39. In terms of family distribution type, the dominant elements are tropical Asian and Chinese endemics, the former accounting for 54% and the latter 12%. There are no subregionally-endemic genera but 23 endemic species. It is worth noting that some species of this subregion occur mainly in Australia, including *Bruguiera sexangula* (Rhizophoraceae), *Centrolepis banksii* (Centrolepidaceae), *Dapsilanthus disjunctus* (Restionaceae), *Schoenus calostachyus* and *Tricostularia undulata* (both Cyperaceae), *Sonneratia ×hainanensis* (Lythraceae) and *Sonneratia ×gulngai*.

The East Guangdong Islands Subregion (IVG21 b) includes the coast of the Zhu Jiang estuary and a series of islands from Hailing Dao to Nan'ao Dao, excepting Hong Kong. This area did not become separate from mainland China until the Quaternary Period and therefore most of the plants are similar to those of the mainland. On the hills, subtropical evergreen coniferous forest is dominated by *Pinus massoniana* and shrub vegetation by *Baeckea frutescens* (Myrtaceae) and *Loropetalum chinense* (Hamamelidaceae). Some species of subtropical and temperate areas, such as *Acer laevigatum* var. *laevigatum*, *A. palmatum*, *A. sino-oblongum* and *Michelia maudiae*, can also be found. The coastal mangrove forest is dominated by *Aegiceras corniculatum* and *Kandelia obovata*. Ten endemic species are shared with Hong Kong, including *Callicarpa longibracteata* (Verbenaceae) and *Camellia granthamiana* (Theaceae).

The South Hainan Subregion (IVG21 c) is located west of Wuzhi Shan and south of the hills of northern Hainan, and has a hot, dry climate. Its vegetation includes widespread psammophytic, deciduous monsoon rainforest dominated by *Albizia odoratissima*, *A. procera*, *Lagerstroemia balansae*, *Lannea coromandelica*, *Streblus ilicifolius* (Moraceae) and *Terminalia nigrovenulosa*. Above 1 000 m, the montane rainforest is characterized by *Castanopsis echinocarpa*, *Dacrycarpus imbricatus* and *Dacrydium pectinatum* (both Podocarpaceae), *Vatica mangachapoi* (Dipterocarpaceae) and *Xanthophyllum hainanense* (Polygalaceae). Dwarf forest, found on mountain summits, consists mainly of *Altingia obovata* and *Rhododendron* spp. The flora of this subregion is more diverse than that of the Eastern Guangdong Islands Subregion, but less so than that of the Central Hainan Subregion. There are no subregionally-endemic genera but 118 endemic species adapted to the dry climate, including *Croton laui* (Euphorbiaceae).

The Central Hainan Subregion (IVG21 d) includes the mountainous area of central Hainan and the hills and valleys of its southeastern coast. The elevation of this subregion is mostly 300–1 000 m, with the highest peak of Wuzhi Shan at 1 867 meters. The climate is hot and wet. The vegetation types, in sequence from coast to mountain summit, are mangrove forest, coastal forest, lowland rainforest, montane rainforest and mountain-top dwarf forest. The mangrove forest contains some species common to the last three subregions, such as *Bruguiera gymnorhiza*, *B. sexangula*, *Ceriops tagal* (Rhizophoraceae), *Excoecaria agallocha*, *Lumnitzera racemosa* (Combretaceae) and *Rhizophora apiculata*. Most of the coastal forest has been destroyed, with only *Arytera littoralis* (Sapindaceae), *Heritiera parvifolia* (Sterculiaceae) and *Pandanus austrosinensis* (Pandanaceae) remaining. The lowland rainforest is dominated by *Alphonsea monogyna* (Annonaceae), *Amesiodendron chinense* and *Litchi chinensis* (both Sapindaceae), *Dillenia turbinata* (Dilleniaceae), *Hancea hookeriana* and *Koilodepas hainanense* (both Euphorbiaceae), *Hopea reticulata* (Dipterocarpaceae), *Madhuca hainanensis*, *Ormosia pinnata*, *Scleropyrum wallichianum* (Santalaceae), *Sindora glabra* (Fabaceae), *Vatica mangachapoi* and *Xanthophyllum hainanense*. Between 700 and 1 380 m, where the humidity is high, montane rainforest is dominated by *Adinandra hainanensis* (Theaceae), *Alseodaphne hainanensis* (Lauraceae), *Canarium album* (Burseraceae), *Cyclobalanopsis championii*, *C. patelliformis*, *Dacrycarpus imbricatus*, *Dacrydium pectinatum*, *Schima superba*, *Syzygium araiocladum* and *Syzygium chunianum*. Dwarf forest appears above 1 300 m, dominated by tropical species such as *Lithocarpus hancei*, *Lyonia ovalifolia* var. *rubrovenia* (Ericaceae), *Myrsine seguinii* (Myrsinaceae), *Pinus kwangtungensis*, *Rhododendron simiarum* and *Ternstroemia gymnanthera*. Mosses are abundant because of the high level of humidity. This subregion is the core area of the islands of the Nan Hai, its rich flora including almost all genera endemic to the entire region. Some genera are restricted to this subregion, including *Cathayanthe*, *Chunia*, *Pentastelma*, *Setiacis* (Poaceae) and *Wenchengia*. Genera centered on this area but not strictly endemic to it include *Scorpiothyrsus* (Melastomataceae). The number of species endemic to the subregion is 122, 10% of those endemic to the region.

The South China Sea Islands Subregion (IVG21 e) includes the Dongsha Dao, Xisha Qundao, Zhongsha Qundao, Nansha Qundao and other small islands nearby. The climate is dry and thus the vegetation is low and simple, mainly shrubs and grassland. The dominant species are *Erythrina variegata* (Fabaceae), *Guettarda speciosa* (Rubiaceae), *Morinda citrifolia*, *Pisonia grandis* (Nyctaginaceae), *Premna serratifolia*, *Scaevola taccada* and *Tournefortia argentea*. There are only 226 species of seed plants in the Xisha Qundao (Chen *et al.*, 1983; Xing *et al.*, 1993) and 28 species in the Nansha Qundao, most of which are pantropical (Xing *et al.*, 1996). No endemic species occur in the subregion. Among the limited species in this subregion are two species of *Tournefortia*: *T. sibirica*, considered a relict species of ancient coasts distributed from Romania to Japan, and the pantropical *T. argentea*.

The Bei Buwan Region (IVG22) is almost equivalent to "Paleotropics Kingdom C," "Indomalesian 17," "Indochinese Region 5" and the "North Indo-Chinese Province" in Takhtajan (1986). In fact, the area forms a bridge between northern Vietnam, Guangxi (the area south of Bose, Nanning and Yulin) and southeastern Yunan (areas below 1 000 m, such as Maguan, Malipo and Funing). In the Chinese part of the region alone there are 4 303 species of seed plants in 1 294 genera and 255 families. This complex flora has the character of a northern tropical flora (Fang *et al.*, 1995), rich in endemic genera and species. The area below 700 m is covered by monsoon rainforest, characterized by typical tropical families such as Dipterocarpaceae and Myristicaceae. Common canopy species include *Antiaris toxicaria*, *Canarium* spp., *Garcinia oblongifolia*, *Hopea chinensis*, *Saraca dives* (Fabaceae) and *Vatica guangxiensis* and, in Hekou, southern Yunnan, *Dipterocarpus*. This flora is very similar to the evergreen monsoon rainforest of Hainan, but with mostly deciduous trees, such as *Garuga pinnata* and *Spondias lakonensis* (Anacardiaceae). Other dominant families include Anacardiaceae, Annonaceae, Arecaceae, Bignoniaceae, Clusiaceae, Lauraceae, Meliaceae, Moraceae, Sapindaceae, Sapotaceae and Tiliaceae. In the formation dominated by *Cephalomappa sinensis* (Euphorbiaceae) and *Garcinia paucinervis, Aglaia* spp. (Meliaceae), *Cleistanthus* spp., *Deutzianthus tonkinensis* and *Drypetes* spp. (all Euphorbiaceae), *Ficus glaberrima, Horsfieldia* spp., and *Parashorea chinensis* can also be found. There is no significant vertical zonation in the montane monsoon rainforest here. Where secondary forest is found, it is dominated by *Pistacia weinmanniifolia* (Anacardiaceae), *Platycarya strobilacea, Triadica rotundifolia* (Euphorbiaceae) and *Sinosideroxylon* spp. (Sapotaceae). The vegetation of the mountains is similar to subtropical evergreen broad-leaved monsoon forest, with Fagaceae, Hamamelidaceae, Lauraceae, Magnoliaceae and Theaceae dominating. Analysis of floristic components has revealed that there are significantly more tropical and subtropical families, genera and species than in the Yunnan, Guizhou and Guangxi Region (Fang *et al.*, 1995).

Of the species endemic to the region, only 40% are temperate, and this region represents the northernmost distribution of such important tropical families as Dichapetalaceae, Dilleniaceae, Dipterocarpaceae, Myristicaceae, Nepenthaceae and Plagiopteraceae. At generic level, this region contains 34 genera endemic to China, five of which (all in Gesneriaceae) are endemic to the region: *Allostigma, Calcareoboea, Dolicholoma, Gyrogyne* and *Metabriggsia* (Li, 1999; Li & Wang, 2004). There are also 66 characteristic genera shared with Vietnam. Many genera seem to extend to the northern boundary of the Yunnan, Guizhou and Guangxi Region, including *Annamocarya* (Juglandaceae), *Boniodendron* and *Paranephelium* (both Sapindaceae), *Diplopanax, Eustigma* and *Mytilaria* (both Hamamelidaceae), *Woonyoungia* (Magnoliaceae), *Melliodendron* (Styracaceae), *Neocinnamomum* (Lauraceae), *Pavieasia* (Sapindaceae), *Rehderodendron* and *Rhoiptelea*. At the species level, there are 2 171 species endemic to China in this region, ca. 50% of the total flora. Three hundred and two species are endemic to the region (including parts of Vietnam).

Southwestern Guangxi has a mainly karst landscape and houses 210 endemic species whereas the non-karst area in southern Guangxi has only half that number. If Guangxi, Guizhou and Yunnan are included, there are 636 species endemic to this karst area, in 247 genera and 104 families. Of these, 54% are woody plants and 40% herbs. Some noteworthy woody taxa include *Camellia* (section *Archecamellia*), *Carpinus*, and species of Pittosporaceae and Euphorbiaceae; herbs include species of *Aspidistra* (Liliaceae), *Ophiorrhiza* (Rubiaceae) and some Gesneriaceae. In southwestern Guangxi, where montane forest is widespread, monsoon rainforest consists mainly of tropical elements such as Dipterocarpaceae and Lecythidaceae. Characteristic taxa of this region include *Archidendron* spp., *Cephalomappa* spp., *Cleistanthus petelotii, Deutzianthus* spp., *Excentrodendron* spp. (Tiliaceae), *Garcinia paucinervis, Hydnocarpus hainanensis* (Flacourtiaceae), *Parashorea chinensis* and *Streblus tonkinensis*. The distribution of these taxa extends to northern Vietnam. Some tropical taxa of this region show a close relationship to tropical parts of Yunnan, including *Acrocarpus fraxinifolius, Caryota obtusa* (Arecaceae), *Dracaena cochinchinensis* (Liliaceae), *Excentrodendron* spp., *Garcinia paucinervis, Knema globularia* (Myristicaceae), *Saraca dives, Parashorea chinensis* and *Xerospermum bonii* (Sapindaceae). Some other taxa show a connection with southwestern Guangxi and Hainan, including *Actephila merrilliana* (Euphorbiaceae), *Capparis micracantha, Diospyros hainanensis, Drypetes hainanensis, Diplodiscus trichospermus* (Tiliaceae), *Horsfieldia kingii* and *Symplocos poilanei* (Symplocaceae). Four genera and 210 species are endemic to this area which, combined with southeastern Yunnan, forms a core area of endemism for the region.

Southern Guangxi has ca. 1 380 km of coastline, sloping into the ocean. The dominant landforms are hills and mountains with acidic soil, with scattered residual karst hills south of Yulin and isolated karst peaks between Qinzhou and Shangsi. The vegetation comprises montane monsoon forest with mangrove forest on the coast, the latter having more in common with the coastal forests of Guangdong than Hainan. Several typical tropical taxa such as *Antiaris toxicaria, Diplocyclos palmatus* (Cucurbitaceae), *Hopea chinensis, Manglietia fordiana* var. *hainanensis* (Magnoliaceae), *Michelia macclurei, Nepenthes mirabilis* (Nepenthaceae), *Orchidantha chinensis* (Lowiaceae), *Pinus latteri* and the genus *Calamus* (Arecaceae), have the southern extent of their distribution in southern Guangxi. Compared with other parts of Guangxi, Guizhou and Yunnan, this area has fewer endemic genera and species (71 endemic species). It shares many endemic taxa with the floras of southern Guangxi and Hainan, such as *Chieniodendron hainanense, Orophea hainanensis* and *Polyalthia laui* (all Annonaceae) and *Homalium sabiifolium*. Some common taxa in this area, such as *Antiaris toxicaria, Beilschmiedia intermedia, Erythrophleum fordii* (Fabaceae), *Hopea* spp., *Litsea glutinosa* and *Syzygium* spp., are also important elements in the flora of the islands of the Nan Hai.

The Yunnan, Myanmar and Thailand Region (IVG23) includes northern Thailand (Chiang Mai), northern Lao PDR (Savannakhet and Sam Neua) and parts of Myanmar (Shan and Banman). In Yunnan, it extends from a northern boundary at Simao and Lincang to Dehong in the west and the Hong He River in the east. According to LI (1995) the area is occupied by tropical rainforest and tropical monsoon rainforest. The flora includes 4 915 species in 1 447 genera and 248 families. In general, it has a north-tropical character and forms a transition between tropical and subtropical floras. In the east of the region, such as Hekou and Jinping, it is characterized by *Crypteronia paniculata, Dipterocarpus retusus* and *Hopea chinesis*, in the center of the region (e.g. Xishuangbanna) by *Hopea chinensis, Parashorea chinensis* and *Vatica guangxiensis* and in the west—around the Yingjiang He and nearby rivers, below 250 m—by *Dipterocarpus turbinatus* and *Shorea assamica* (Dipterocarpaceae). Most of these taxa are widespread across northern tropical Southeast Asia and have only a fragmentary distribution in the valleys of Yunnan. In addition, tree species such as *Antiaris toxicaria, Lasiococca comberi* (Euphorbiaceae), *Pometia pinnata, Pouteria grandifolia* (Sapotaceae), *Terminalia myriocarpa* and *Tetrameles nudiflora* (Tetramelaceae) are common in tropical forest.

The semi-evergreen monsoon forest is dominated by *Chukrasia tabularis* and *Ficus altissima* (the former typical of southern Dehong). In this area, most mixed monsoon and deciduous rainforests are secondary, and are dominated by *Albizia chinensis, Bombax ceiba, Erythrina* spp. and *Melia azedarach* (Meliaceae), mixed with *Bauhinia variegata* (Fabaceae), *Eriolaena spectabilis* (Sterculiaceae), *Garuga forrestii, Haldina cordifolia* and *Mitragyna rotundifolia* (both Rubiaceae) and *Sterculia pexa*. Because there are many canyons and mountains, the area is mainly covered by montane rainforest (at elevations of 700–1 300 m), characterized by *Altingia yunnanensis*, among others. *Dacrycarpus imbricatus* (Podocarpaceae) occurs to the east of Ailao Shan, *Diploknema yunnanensis* (Sapotaceae) can be found in western Yunnan, and *Michelia baillonii, Phoebe glaucifolia* and *Semecarpus reticulatus* (Anacardiaceae) are distributed in southern Yunnan. The land between 1 000 and 1 750 m elevation is covered by montane monsoon evergreen broad-leaved forest, characterized by *Castanopsis, Cyclobalanopsis* and *Lithocarpus*. Above this is found mossy evergreen broad-leaved forest (to 2 900 m) or, on the mountain summits and exposed slopes (2 700–2 900 m), dwarf forest with mosses. This flora is similar to that of Hainan, Taiwan, the Beibu Wan and the eastern Himalaya Shan, and to tropical montane floras further afield in Malaysia and New Guinea. Orchidaceae is the largest family here, with 82 genera and 253 species, followed by Fabaceae (74 genera and 203 species), Lauraceae (eight genera and 155 species) and Rubiaceae (44 genera and 139 species). More than ten tropical families that never occur in the Holarctic Kingdom are present, most of them also found in Taiwan and Hainan. Some families found here are not found in other regions of China, including Crypteroniaceae and Rafflesiaceae. There are ca. 40 genera endemic to the region, of which three are endemic to the Chinese part of the region: *Cyphotheca* (Melastomataceae), *Paralamium* (Lamiaceae) and *Sinosassafras* (Lauraceae).

The Southern Fringe of the East Himalaya Region (IVG24) is almost equivalent to the "Bengal Province" of Chatterjee (1940) and Turrill (1953), which includes Mêdog in southeastern Xizang, the area south of the big bend of the Yarlung Zangbo Jiang and extends to Bhutan, low-altitude eastern Nepal and northeastern India. The flora is slightly less diverse than neighboring regions, but its vertical distribution is more concentrated, ranging from tropical to holarctic (from Paleotropical India and Malaysia to the northern wing of the eastern Himalaya Shan) within relatively few km. According to SUN and ZHOU (1996; 1997b), tropical and subtropical vegetation occurs from 580 to 2 500 m. Rainforest and monsoon rainforest, found below 500 m in the tropical part of the region, is composed of three species of *Dipterocarpus* and two of *Parashorea*. This flora has the character of a Malaysian flora and is abundant in elements from tropical India and Myanmar, such as *Altingia, Canarium, Garcinia, Lagerstroemia, Talauma* (Magnoliaceae), *Terminalia* and *Tetrameles* (Wu, 1979). Montane rainforest is distributed from 600 to 1 200 m and can be divided into four formations: *Altingia excelsa, Altingia excelsa-Lagerstroemia minuticarpa, Altingia excelsa-Terminalia myriocarpa* and *Castanopsis hystrix-Altingia excelsa. Chloranthus* spp. (Chloranthaceae), *Exbucklandia* spp. (Hamamelidaceae), *Gynocardia odorata* (Flacourtiaceae), *Talauma hodgsonii* and many epiphytes can be found in these forests. Subtropical montane rainforest is found between 1 400 and 2 000 m, while above a zone of cloud forest is found a community dominated by *Castanopsis*. These forests also contain species of *Agapetes* (Ericaceae; in which they are very diverse, with as many as 19 species), *Alcimandra* (Magnoliaceae), *Craigia, Exbucklandia, Michelia, Schima* and various lianas and mosses. Subtropical montane rainforest with *Quercus lanata*, and montane decidous forest with *Cyclobalanopsis lamellosa*, are distributed between elevations of 1 500 and 2 000 m. Coniferous forests are dominated by *Pinus bhutanica* and *Pinus kesiya*.

There are 1 679 species in the Chinese part of this region, in 726 genera and 180 families. The main families in the region are Aucubaceae, Carlemanniaceae, Daphniphyllaceae, Dipentodontaceae, Fabaceae, Lamiaceae, Lauraceae, Orchidaceae, Poaceae, Rosaceae and Rubiaceae. There are also some families characteristic of India and Malaysia, such as Myristicaceae and Pandanaceae. At generic level, tropical components (which include pantropical and tropical Asian components) have a slight majority in the flora (58%), and temperate components (dominated by north-temperate and eastern Asian components) account for 40% (Sun & Zhou, 1997a; Sun & Zhou, 1997b). There are no genera strictly endemic to the region. Near-endemic or characteristic genera, such as *Agapetes, Carlemannia* (Carlemanniaceae), *Dobinea* (Anacardiaceae), *Gamblea* and *Merrilliopanax* (both Araliaceae) and *Indofevillea* (Cucurbitaceae), are rare in other areas of the northern tropics in China. Besides the dominant species *Cyclobalanopsis lamellosa, Cyclobalanopsis kiukiangensis, Michelia doltsopa* and *Quercus lanata, Albizia sherriffii* is also characteristic of the region. The region is abundant in regionally-endemic species, with ca. 177, some of which are also dominant or constructive species of forests. These include *Cyclobalanopsis motuoensis, Lithocarpus obscurus, L. pseudoxizangensis, L. xizangensis, Manglietia caveana* and *Phoebe motuonan*. Other common endemic species in forests include *Agapetes medogensis, Agapetes xizangensis, Beesia deltophylla* (Ranunculaceae), *Chimonobambusa metuoensis* (Poaceae), *Cypripedium subtropicum* (Orchidaceae), *Hornstedtia tibetica* (Zingiberaceae), *Maesa cavinervis* (Myrsinaceae), *Saurauia sinohirsuta* (Actinidiaceae), *Ternstroemia biangulipes* and *Zanthoxylum motuoense* (Sun & Zhou, 2002).

References

ALEXINH, B. B. (1957). *Phytogeography*. Beijing, China: High Education Press.

CAO, W., FU, P. Y., LIU, S. Z. & ET AL. (1995). Studies on the flora of the seed plants from the flora subregion of NE. China plain. *Acta Botanica Yunnanica* Supplement VII: 22-31.

CHATTERJEE, D. (1940). Studies on the endemic flora of Indian and Burma. *Journal of the Royal Asiatic Society, Bengal* 5(1): 19-67.

CIIEN, B. Y., CHEN, W. Q. & WU, H. M. (1983). Notes on plants from Xisha islands, China. *Collected papers from South China Botanical Garden, Chinese Academy of Science* 1: 129-157.

DANG, R. L. & PAN, X. L. (2001). The Chinese endemic plant analysis in northwest desert of China. *Bulletin of Botanical Research* 21(4): 521-526.

DIELS, L. (1901). Die Flora von Central-China. Nach der verh (ndenon Literatur und neu mitgeteilten original Materiale). *Botanische Jahrbücher* 29: 169-659.

DIELS, L. (1905). Uber die Pflanzengeographie von Inner-China nach den Ergebnissen neuerer Sammlungen. *Zeitschrift der Gesellschaft für Erdkunde zu Berlin* 1905: 748-756.

DIELS, L. (1929). *Pflanzengeographie*. Berlin/Leipzig, Germany: Grunter.

DRUDE, O. (1890). *Handbuch der Pflanzengeographie*. Stuttgart: Engelhorn.

DU, Q. & SUN, S. Z. (1990). *The Vegetation in Chaidamu Area and Their Application*. Beijing, China: Science Press.

EDITORIAL COMMITTEE OF FLORA REIPUBLICAE POPULARIS SINICAE (1959-2004). *Flora Reipublicae Popularis Sinicae*. Beijing, China: Sciences Press.

ENGLER, A. (1899). *Die Entwickelung der Pflanzengeographie in den Letzten Hundert Jahrenm*. Berlin, Germany: A. V. Humboldt-Gentenarschrieft.

ENGLER, A. (1912). *Syllabus der Pflanzenfamilien*. Berlin, Germany.

ENGLER, A. (1926). *Die Naturlichen Pflanzenfamilien Nebst ihren Gattungen un Wichtigsen Arten Insbesondre der Nutpflanzen*. Leipzig, Germany: W. Engelmann.

FANG, R. Z., BAI, P. Y., HUANG, G. B. & WEI, Y. G. (1995). A floristic study on the seed plants from tropics and subtropics of Dian-Qian-Gui. *Acta Botanica Yunnanica* Supplement VII: 111-150.

FANG, R. Z. & MIN, T. L. (1999). The floristic studies on the *Rhododendron*. In: Lu, A. M. (ed.) *The Geography of Spermatophytic Families and Genera*. Beijing, China: Science Press.

FEDOROV, A. H. A. (1959). Flora of SW. China and its significance to Euro-Asia Flora Kingdom. *Bulletin of Botanical Research* 8(2): 161-176.

FENG, Y., YAN, C. & YIN, L. K. (2003). The endemic species and their distribution in Xinjiang. *Acta Botanica Boreali-Occidentalia Sinica* 23(2): 263-273.

FU, P. Y., LIU, S. Z., LI, J. Y., CAO, W., QIN, Z. S., DING, X. H. & DING, T. N. (1995). Studies on the flora of the seed plants from the Daxinganling flora region. *Acta Botanica Yunnanica* Supplement VII: 1-10.

GENG, X. (1956). Papers collection on plants taxonomy and phytogeography. *Forestry Papers Collection* 4: 101-106.

GONG, W., ZENG, Z., CHEN, Y. Y., CHEN, C., QOIU, Y. X. & FU, C. X. (2008). Glacial refugia of *Gingko biloba* and human impact on its genetic diversity: evidence from chloroplast DNA. *Journal of Integrative Plant Biology* 50(3): 368-374.

GOOD, R. (1974). *The Geography of the Flowering Plants*. London.

GRAY, A. (1846). Analogy between the flora of Japan and that of the United States. *American Journal of Science and Arts* 2: 135-136.

GRAY, A. (1859). Diagnostic characters of phanerogamous plants, collected in Japan by Charles Wright, botanist of the U S North Pacific Exploring Expedition, with observations upon the relationship of the Japanese flora to that of North America and of other parts of the northern temperate zone. *Memoirs of the American Academy of Arts* 6: 377-453.

GRAY, A. (1878). Forest geography and archeology. *American Journal of Science* 16: 85-94.

GRISEBACH, A. (1872). *Die Begetation der Erde Nach Ihrer Klimatischen Anordnung*. Leipzig, Germany.

GRUBOV, V. I. (1959). *A Contribution to the Botanico-Geographic Subdivision of Central Asia*. Leningrad, Russia: Academy of Science Union des Républiques Socialistes Soviétiques and Society of Botany Union des Républiques Socialistes Soviétiques.

GRUBOV, V. I. & FEDOROV, A. (1964). Flora and vegetation of China. In: Zaichikov, V. T. (ed.) *Physical Geography of China*. pp. 324-428. Moscow, Russia.

HANDEL-MAZZETTI, H. (1916). Vorlaufige Ubersicht uber die Vegetationsstufen und Formationenvon Juennan und SW-Setschuan. *Anzeiger der Kaiserlichen Akademie der Wissenschaften, Mathematisch-Naturwissenschaftliche Classe* 53: 195-215.

HANDEL-MAZZETTI, H. (1921). Ubefrsicht uber die

wichtigsten Vegetationsstufen und Formationen von Yunnan und SW-Setschuan. *Botanische Jahrbücher für Systematik, Pflanzengeschichte und Pflanzengeographie* 56: 578-597.

Handel-Mazzetti, H. (1930). The phytogeographic structure and affinities of China. *In: Eighth International Botanical Congress, Cambridge*. pp. 315-319.

Handel-Mazzetti, H. (1931). Die Pflanzengeographsche Gliederung und Stllung Chines. *Botanische Jahrbücher für Systematik, Pflanzengeschichte und Pflanzengeographie* 64: 309-323.

Hao, K. S. (1938). Pflanzengeographische Studien uber den Kokonor-See und uber das angrenzende Gebiet. *Botanische Jahrbücher für Systematik, Pflanzengeschichte und Pflanzengeographie* 68: 515-668.

Hao, R. M. & Yang, Z. B. (1996). The floristic similarity between East China and Japan. *Acta Botanica Yunnanica* 18(3): 269-276.

Hooker, J. D. (1907). A sketch of the flora of British India. *Imperial Gazetteer of India, Oxford* 3(1, 4): 157-212.

Hsieh, C. F., Shen, C. F. & Yang, K. C. (1994). Introduction to the flora of Taiwan 3: floristics, phytogeography and vegetation. In: Editorial COmmittee of the Flora of Taiwan (ed.) *Flora of Taiwan*. pp. 7-18. Taipei, Taiwan: ROC.

Hu, H. H. (1926). A preliminary survery of the forest of southeastern China. *Contributions from the Biological Laboratory of the Science Society of China* 2: 1-20.

Hu, H. H. (1929a). Further observartion of the forest flora of southeastern China. *Bulletin of the Fan Memorial Institute of Biology* 1: 51-52.

Hu, H. H. (1929b). The nature of the forest flora of southeastern China. *Peking Natural History Bulletin* 4: 47-56.

Hu, H. H. (1935). A comparison of the ligneous flora of China and eastern North America. *Bulletin of the Chinese Botanical Society* 1: 79-97.

Hu, H. H. (1936). The characteristics and affinities of Chinese flora. *Bulletin of the Chinese Botanical Society* 2: 67-84.

Hu, H. H. (1940). Constituents of the flora of Yunnan. *Proceedings of the Sixth Pacific Science Congress* 4: 641-654.

Kano, T. (1933). Zoogeography of Botal Tobago Island (Kotosho) with a consideration on the northern portion of Wallaces line. *Bulletin of the Biological Society of Japan* 9: 381-399, 475-491, 591-613, 675-721.

Krasnov, A. H. (1899). *Geography of Plants*. Kharkov, Russia: Kharkov University.

Kuznetsov, N. I. (1912). A contribution to the division of Siberia into botanico-geographic provinces. *Memoirs of the St. Petersburg Academy of Sciences* 24(1): 1-174.

Lee, J. S. (1934). The ecological distribution of Chinese trees. *Peking Natural History Bulletin* 9: 1-6.

Li, H. & Wu, S. G. (1984). The flora structure of E. Tibet area. *Geographical Research* 8(2): 64-70.

Li, H. (1994). The nature and characteristics of seed plants from Du Long Jiang area. *Acta Botanica Yunnanica* Supplement VI: 1-100.

Li, H. L. (1944). The phytogeographical divisions of China, with special reference to the Araliaceae. *Proceedings of the Academy of Natural Sciences of Philadelphia* 96: 249-277.

Li, H. L. (1950). Phytogeographical affinities of southern Taiwan. *Taiwania* 1: 103-128.

Li, S. Y. (1960). The relationship between characteristics and formation of vegetation and desiccation in North slope of Kunlun Mountain. *Acta Botanica Sinica* 8(2): 87-105.

Li, X. W. & Li, J. (1992). On the validity of Tanaka Line and its significance viewed from the distribution of Eastern Asiatic genera in Yunnan. *Acta Botanica Yunnanica* 14(1): 1-12.

Li, X. W. & Li, J. (1993). A preliminary floristic study on the seed plants from the region of Hengduan Mountain. *Acta Botanica Yunnanica* 15(3): 217-231.

Li, X. W. (1995). A floristic study on the seed plants from the region of Yunnan Plateau. *Acta Botanica Yunnanica* 17(11): 1-14.

Li, Z. Y. (1999). The distribution of Gesneriaceae subfamily. In: Lu, A. M. (ed.) *The Geography of Spermatophytic Families and Genera*. pp. 497-515. Beijing, China: Science Press.

Li, Z. Y. & Wang, Y. Z. (2004). *Plants of Gesneriaceae in China*. Zhengzhou, China: Henan Science and Technology Press.

Liu, F. X., Liu, S. L., Yang, Z. B., Hao, R. M., Yao, G., Hang, Z. Y. & Li, N. (1995). A floristic study on the seed plants from the region of East China. *Acta Botanica Yunnanica* Supplement VII: 93-110.

Liu, J. & Yang, Y. B. (1974). The floristic relationship between Taiwan Island and the islands nearby. *Chinese Forestry Quarterly* 7(4): 69-114.

Liu, S. E. (1934). An introduction to phytogeography of North and West China. *Botanical Bulletin of Peking Institute of Botany* 2(9): 423-451.

Liu, S. E. (1936). An introduction to phytogeography of South and Southwest China. *Beijing Biology Journal* 1(1).

Liu, Y. J., Wang, J. H., Ma, L. Q. & Zhang, D. K. (2007). Floristic analysis of desert spermatophytes in Gansu Province. *Chinese Journal of Ecology* 26(10): 1521-1527.

Liu, Y. X. (1995). A study on origin and formation of the Chinese desert floras. *Acta Phytotaxonomica Sinica* 33(2): 131-143.

Mao, Z. M. (1991). The characteristic of spring ephemerals flora. *Arid Zone Research* 9(1): 11-12.

Pan, X. L., Dang, R. L. & Wu, G. H. (2001). *Floristic and Unitlization from Arid and Desert Region from NW China*. Beijing, China: Science Press.

Qi, C. J., Yu, X. L. & Xiao, Y. T. (1995). A study on the flora of the seed plants from the floristic region of Central China. *Acta Botanica Yunnanica* Supplement VII: 55-92.

Sargent, S. C. (1913). Introduction. In: Wilson, E. H. (ed.) *A Naturalist in Western China*. pp. xvii-xxxvii. London: Methuen.

Schaefer, W. (1958). *Common Phytogeography Principle*.

Beijing, China: High Education Press.

Schouw, F. (1823). *Grunzuge einer Algemeinen Pflanzengeopgraphie*. Berlin, Germany: G. Reimer.

Sun, H. & Zhou, Z. K. (1996). The character and origin of the flora from the Big Bend Gorge of Yalutsangpu (Brahmabutra) River, Eastern Himalayas. *Acta Botanica Yunnanica* 18(2): 185-204.

Sun, H. & Zhou, Z. K. (1997a). The phytogeographical affinities and nature of the Big Bend Gorge of the Yalu Tsangpo River, S.E. Tibet, E. Himalayas. *Chinese Journal of Applied and Environmental Biology* 3(2): 184-190.

Sun, H. & Zhou, Z. K. (1997b). The vegetation of the Big Bend Gorge of Yalu Tsangpo River, S.E. Tibet, E. Himalayas. *Acta Botanica Yunnanica* 19(1): 57-66.

Sun, H. (2002). Evolution of Arctic-Tertiary flora in Himalayan-Hengduan Mountains. *Acta Botanica Yunnanica* 24(6): 671-688.

Sun, H. & Zhou, Z. K. (2002). *Seed Plants of the Big Bend Gorge of Yalu Tsangpo in S.E. Tibet, E. Himalayas*. Kunming, China: Yunnan Science and Technology Press.

Takhtajan, A. (1986). *Floristic Regions of the World*. Berkeley, CA, London, UK, and Los Angeles, CA: California Press.

Takhtajan, A. (1987). Flowering plant origin and dispersal, the cradle of the angiosperms revisited. In: Whitmore, T. C. (ed.) *Biogeographical Evolution of the Malay Archipelago*. pp. 26-31. Oxford, UK: Clarendon Press.

Takhtajan, A. (1997). *Diversity and Classification of Flowering Plants*. New York: Columbia University Press.

Tanaka, T. (1954). *Species Problem in Citrus*. Tokyo, Japan: Japanese Society for the Promotion of Science.

Tao, J. R. (1992). The Tertiary vegetation and flora and floristic regions in China. *Acta Phytotaxonomica Sinica* 30(1): 25-43.

Tao, J. R. (2002). *The development and evolution of flora from Upper Cretaceous to Cainozoic era in China*. Beijing, China: Science Press.

Tolmatchev, A. I. (1974). *Introduction to the Geography of Plants*. Leningrad, Russia: University Press.

Turrill, W. B. (1953). *Pioneer Plant Geography*. The Hague: Martinus Nijhoff.

Walker, E. H. (1944). *The Plants of China and Their Usefulness to Man*. Washington, DC.

Wang, H. C., Sun, H., Compton, J. A. & Yang, J. B. (2006). A phylogeny of Thermopsideae (Leguminosae: Papilionaceae) inferred from nuclear ribosomal internal transcribed spacer (ITS) sequences. *Botanical Journal of the Linnean Society* 151: 365-373.

Wang, H. S., Zhang, Y. L. & Huang, J. S. (1995). A floristic study on the seed plants in the North China region. *Acta Botanica Yunnanica* Supplement VII: 32-54.

Wang, H. S. (1997). *Floristic Geography in North China*. Beijing, China: Science Press.

Wang, Y. H. & Wu, W. C. (1997). A study on the flora of Raoping, Guangdong Province. *Journal of South China Agricultural University* 18(4): 64-68.

Ward, F. K. (1912). Some plant formations from the arid regions of western China. *Annals of Botany* 26: 1105-1110.

Ward, F. K. (1916). On the Sino-Himalayan flora. *Transactions of the Botanical Society of Edinburgh* 27: 12-53.

Ward, F. K. (1927). The Sino-Himalayan flora. *Proceedings of the Linnean Society of London* 139: 67-74.

Ward, F. K. (1935). A sketch of the geography and botany of Tibet, being materials for a flora of the country. *Journal of the Linnean Society. Botany* 50: 239-265.

Ward, F. K. (1936). A sketch of the vegetation and geography of Tibet. *Proceedings of the Linnean Society of London* 148: 133-160.

Wilson, E. H. (1913). *A Naturalist in Western China*. London, UK: Methuen.

Wu, C. Y. (1979). The regionalization of Chinese flora. *Acta Botanica Yunnanica* 1: 1-20.

Wu, C. Y. & Wang, H. S. (1983). *Chinese Physical Geography: Phytogeography*. Beijing, China: Science Press.

Wu, C. Y. (1987). The origin and evolution of Tibetan flora. In: Wu, C. Y. (ed.) *Flora of Tibet*. pp. 874-902. Beijing, China: Science Press.

Wu, C. Y., Lu, A. M., Tang, Y. C., CHen, Z. D. & Li, D. Z. (2002). Synopsis of a new "polyphyletic - polychromic - polytopic" system of the angiosperms. *Acta Phytotaxonomica Sinica* 40(4): 289-322.

Wu, C. Y., Lu, A. M., Tang, Y. C., Chen, Z. D. & Li, D. Z. (2003). *The Families and Genera of Angiosperms in China*. Beijing, China: Science Press.

Wu, D. L., Xing, F. W., Ye, H. G., Li, Z. X. & Chen, B. H. (1996). Study on the spermatophytic flora of South China Sea islands. *Journal of Tropical and Subtropical Botany* 4(1): 1-22.

Wu, J. L. & Peng, F. X. (1993). Plants from E' Mei Mountain that has been cited as type specimens. In: *Papers collection from annual meeting in 60th anniversary of Botanical Society of China*. Beijing, China: Chinese Technology Press.

Wu, S. G., Yang, Y. P. & Fei, Y. (1995). On the flora of the alpine region in the Qinghai-Xizang (Tibet) Plateau. *Acta Botanica Yunnanica* 17(3): 235-246.

Wu, Z. Y., Sun, H., Zhou, Z. K., Li, D. Z. & Peng, H. (2010). *Floristics of Seed Plants from China*. Beijing, China: Science Press.

Wu, Z. Y. & Li, X. W. (1965). *The First Volume of Reports of Tropical and Subtropical Flora in Yunnan*. Beijing, China: Science Press.

Wu, Z. Y., Raven, P. H. & Hong, D. Y. (eds.) (1994-2011). *Flora of China*. Beijing, China: Science Press.

Wu, Z. Y. & Wu, S. G. (1996). A proposal for a new floristic kingdom (realm) - the E. Asiatic Kingdom, its delineation and characteristics. In: Zhang, A. L. & Wu, S. G. (eds.) *Floristic Characteristics and Diversity of East Asian Plants, Proceeding of the first international Symposium on Floristic Characteristics and Diversity of East Asian Plants*. pp. 3-42. Beijing, China: China Higher Education Press.

Wulff, E. V. (1944). *Historical Phytogeography*. Beijing,

China: Science Press.

Xi, J. B., Tian, C. Y., Yan, P. & Dong, Z. C. (2006). Primary research on the halophyte flora in Xinjiang. *Chinese Journal of Eco-Agriculture* 14(1): 7-10.

Xing, F. W., Li, Z. X., Ye, H. G., Chen, B. H. & Wu, D. L. (1993). A research on phytogeography of Xisha islands, China. *Tropical Geography* 13: 250-257.

Xing, F. W., Wu, D. L., Ye, H. G., Li, Z. X., Chen, B. H. & Li, T. H. (1996). *Flora of Nansha Islands and Nearby*. Beijing, China: Ocean Press.

Yin, L. K. (1997). Diversity and ex-situ conservation of plamts in the desert region of temperate zone in China. *Chinese Biodiversity* 5(1): 40-48.

Ying, J. S. (1994). An analysis of the flora of Qinling Mountain range: its nature, characteristics and origins. *Acta Phytotaxonomica Sinica* 32(5): 389-410.

Zeng, W. B. (1983). The flora and phytogeographical subdivision of Fujian. *Journal of Xiamen University (Natural Science)* 22(2): 217-226.

Zhang, C. H., Liu, G. H. & Zhao, X. H. (2006). An analysis on geographical components of seed plants in Hunshandake desert at genus level. *Journal of Desert Research* 26(6): 1024-1031.

Zhang, H. D. (1994). On the origin of Cathaysian flora. *Acta Scientiarum Naturalium Universitatis Sunyatseni* 33(2): 1-9.

Zhang, H. H. (1995). The regions of Taiwan Flora. In: *Zhang Hong-da's Paper Collection*. pp. 131-147. Guangzhou, China: Zhongshan University Press.

Zhang, Q. E. (1967). Forest plants in Lanyu Island. *Taiwan Forestry Quarterly* 3(2): 1-42.

Zhang, W. H., Li, D. W., Liu, G. B. & Xu, X. H. (2002a). Floristic characters of seed plants in Loess Plateau area. *Bulletin of Botany* 22(3): 373-379.

Zhang, Y. J., Pu, X. & Sun, J. Z. (1997). A preliminary study on the spermatophytic flora from Taohe River in Gansu. *Acta Botanica Yunnanica* 19(1): 15-22.

Zhang, Y. L., Li, B. Y. & Zheng, D. (2002b). A discussion on the boundary and area of the Tibetan Plateau in China. *Geographical Research* 21(1): 1-8.

Zheng, M. (1984). The floristic relationship between E. China and Japan. *Acta Phytotaxonomica Sinica* 22(1): 1 -5.

Zheng, S. L. & Zhang, W. (1983). The Middle and Late Early Cretaceous floras in the Boli Basin of Heilongjiang Province, China. *Bulletin of Shenyang Institute of Geology and Mineral Resources* 7: 68-98.

Zhu, G. L. (1996). Origin, differentiaion, and geographic distribution of the Chenopodiaceae. *Acta Phytotaxonomica Sinica* 34(5): 486-504.

Chapter 11

Development of Chinese Botany

PENG Hua

11.1 Introduction

Like any other scientific subject, the history and development of botany reflects a progressing relationship between man and nature. This began with the collection and utilization of plants and evolved alongside our expanding and deepening reliance upon them.

China, as we have seen in the preceding chapters, is abundant in natural resources including plants. Its flora is much richer than that of Europe, for instance, because it did not experience the widespread ice ages of the Pleistocene Epoch which had a heavy impact on the western part of the Eurasian land-mass (Needham, 1986). China is not only one of the four great ancient civilizations (along with Mesopotamia, Egypt and Greece), but is also among the first places where plants were studied (Lu *et al.*, 1991). However, Chinese knowledge of plants, which began with a greater focus on their practical utilization than on academic study, differs greatly from other systems. There are three major differences between ancient Chinese and western botany.

Firstly, rather than beginning as an independent science, the earliest appearance of botany in ancient China was in literature shared with other subjects. Thus Chinese botanical knowledge has mostly been documented within ancient works on medicine, agriculture or horticulture, or in encyclopedias. This tradition lasted for centuries before the fusion with modern western botanical science in the nineteenth century (Needham, 1986). In the west, specific botanical research is considered to have begun in Greece in the third century BC (Needham, 1986) by Theophrastus, known as the "father of botany." Theophrastus wrote exhaustive works on most aspects of botany including description, classification, distribution, propagation, germination and cultivation. He also distinguished between two main groups of flowering plants—now known as dicots and monocots—and between flowering plants and conifers (roughly equivalent to today's angiosperms and gymnosperms). Nonetheless the absence of specialized botanical works from ancient China does not imply that Chinese natural historians overlooked any important aspects of botany, as proven by the existence of unique taxonomic and botanical terms in the Chinese language, for example in the *Erya*, the oldest surviving Chinese dictionary, written by anonymous authors during the time from the Warring States Period (475–221 BC) to the Western Han Dynasty (202 BC—8AD).

Secondly, there are great differences in language—including terminology and nomenclature—between the botany of China and that of Europe. Chinese scholars have been using ideographic words and systems to name, describe and classify plants since long before the introduction of Latin binomials. The introduction of the latter, scientifically-defined names, in the west enabled them to be separated permanently from common names. Thus, in the west, traditional Chinese botanical nomenclature has sometimes been considered "unscientific" when in fact Chinese characters may be just as precise as Latin binomials and may indicate precise facts about each plant's characteristics (Needham, 1986). Generally, today it is accepted that Latin binomials are to be used in order to facilitate international communication.

Finally, the process of development of botany itself has differed between China and elsewhere. In the west, botanical study can be divided into four phases: early history (before the third century AD), the "Dark Ages" from about 200 to 1400, a transition period until about 1800 when great progress was made (such as the phase known as the Renaissance), and modern times from the nineteenth century onwards (Needham, 1986). During this time, however, Chinese botany followed the same traditional methods for more than two millennia, almost wholly uninfluenced by the botany of the outside world, with which there was little traffic or communication.

In this chapter we recount the development of botany and the plant sciences in China, introducing the main literature in chronological order. Since Chinese botanical works have been rather evenly spaced in history and do not fall into distinct stages we have arbitrarily divided Chinese botanical history into stages corresponding to those of the west for ease of comparison. The first comprises the time up to and including the Han Dynasty (to 220 AD), the second from the Wei to the end of the Yuan Dynasty (220–1368), and the third occurred from 1368–1840, during the Ming and Qing Dynasties. More recent times are covered in Chapter 14. In the second part of the chapter we consider a case study: the development of botanical knowledge in one particular culture, Tibetan medicine. The plant knowledge of ethnic minorities is a crucial component of Chinese botany as a whole and this provides an instructive example.

11.2 Historical development of Chinese botanical knowledge

11.2.1 Botanical history through the Han Dynasty

The earliest record regarding plants in ancient China is the cultivation of rice (*Oryza sativa*; Poaceae) ca. 7 000 years ago (The Committee of Protecting the Historic Cultural Relics in Zhejiang Province, 1976). An archeological study of the Banpo historic cultural site in Shaanxi shows that, by ca. 6 000 years ago, the Chinese had also started to cultivate millet (*Setaria italica*; Poaceae) and rapeseed (*Brassica rapa* var. *oleifera*; Brassicaceae; The Institute of Archaeology Chinese Academy of Social Sciences, 1963). Four to five thousand years ago the nutritious soybean (*Glycine max*; Fabaceae) was first cultivated from wild plants native to China; this was later to become one of the most important contributions of the Chinese people to the food and agriculture of the world. The soybean was considered significant enough to be recorded in the *Book of Poetry,* collating folk poetry from the Western Zhou Dynasty (ca. 1100–771 BC) to the middle of the Spring and Autumn Period (around 600 BC). Inscriptions on oracle bones from the Shang Dynasty (ca. 1600–1100 BC) indicate that Chinese names for these cultivated crop plants were already in existence (Chen, 1956).

Silk – products of which are now used worldwide – has been found at the Qianshanyang Neolithic cultural site in Zhejiang, which suggests that the ancient Chinese had started collecting leaves of mulberry (*Morus alba*; Moraceae) for feeding silkworms as many as 7 000 years ago (Li, 1984). Many more, similar, archeological discoveries and records evidence a very long history of cultivating, using and naming plants (for their identification) in China. Early naturalization by the ancient Chinese gradually developed into the agriculture which has made China the center of origin of more crops than anywhere else (The Writing Group of the School of Physics at Peking University, 1978).

The first mention of plants in the Chinese literature comes in the *Zhouli* (*Rites of the Zhou*) written in the fourth century BC. A simple classification of plants into herbs and trees was described in the *Erya,* establishing the botanical and taxonomic basis for other books and records to follow (Chen & Huang, 2005). The *Erya* also named more than 1 000 plants and animals, and described morphological characters of about 600 of these (The Writing Group of the School of Physics at Peking University, 1978). The *Book of Poetry* further records 132 kinds of plant with reference to agriculture and their usage (Chinese Botanical Society, 1983). The *Guanzi,* written during the Warring States Period, discusses the adaptation of different plants to different terrains, groundwater levels and soil types, concluding that plant growth is closely related to abiotic conditions. This work, which laid the foundations of Chinese soil science, is also the earliest document on ecological geobotany in the world.

The *Lüshi Chunqiu* (*Master Lü's Spring and Autumn Annals*), which records the systematic and theoretical agricultural knowledge developed during the Warring States Period, is the earliest agricultural monograph preserved in China (The Writing Group of the School of Physics at Peking University, 1978). The *Shan Hai Jing* (*Classic of the Mountains and Seas*), which may also have been written during the Warring States Period, documents 124 medicines and the corresponding diseases they were used to treat, among which are 51 herbs. This represents the first documentation of the use of medicinal herbs, which had already been collected and studied for a very long time in ancient China (Chen & Huang, 2005). Also by the Warring States Period, sugarcane (*Saccharum officinarum*; Poaceae) was already being planted in China for sugar manufacture, the specific methods of which are recorded in the *Yiwuzhi* (*Records of Rarities*), written by YANG Fu during the Eastern Han Dynasty (25–220 AD; The Writing Group of the School of Physics at Peking University, 1978). During both the Western and Eastern Han Dynasties, many books on agriculture and medicinal herbs were published and became known to the public. Experiences of farming work, including techniques for cultivating crops, were summarized in the *Fan Sheng-chih Shu,* written in the first century BC by FAN Sheng-Chih. The records in the *Qian Hanshu* (*History of the Former Han Dynasty*), completed in 111 AD, indicate that the ancient Chinese had already used hothouse cultivation for vegetables at that time (The Writing Group of the School of Physics at Peking University, 1978).

At this point we should introduce one important economic plant, *Cannabis sativa* (Cannabaceae; hemp), which has been cultivated and used for ca. 7 000 years (The Institute of Archaeology Chinese Academy of Social Sciences, 1963). In 105 AD, Cai Lun successfully made high-quality paper from cannabis based on ancient techniques (Pan, 1973; Pan, 1974; Liu & Hu, 1976). This papermaking technique spread to the Middle East by the eighth century and to Europe by the twelfth century. The papermaking technique is one of the great inventions of the ancient Chinese and a significant contribution to world culture (Xu, 1975). In 166 AD, CUI Shi, a politician of the Eastern Han Dynasty, wrote the *Si Min Yue Ling,* recording the agricultural activities across the 12 months of a year and discussing the technique of transplanting male cannabis plants away from the females prior to flowering to avoid fruit production. This was also the first book in China to explain the relationship between plant gender and plant reproduction (The Writing Group of the School of Physics at Peking University, 1978).

SHEN Nong's *Bencao Jing* (*Herbal Classic*), written around 100–200 AD, is the earliest preserved Chinese monograph on medicinal plants, recording and describing the characters, habitats, taste and medicinal effects of 365 medicines, 252 of which are herbs. Some of the medicines mentioned are still widely used today. The *Herbal Classic* marks the beginning of the "herbal period" in the historical development of botanical study in China (Pan, 1973; The Writing Group of the School of Physics at Peking University, 1978; Chinese Botanical Society, 1983; Chen & Huang, 2005). At the same time, in the west, botanical study produced famous works such as Theophrastus's *Historia Plantarum,* written around 300 BC, and Dioscorides' *Materia Medica,* around 64 AD (Needham, 1986). These books continued to be used and copied for centuries. During this latter period, botanical study in China made rapid and progress, especially in terms of agriculture and medicinal herbs (Chinese Botanical Society, 1983).

11.2.2 Botanical history from the Wei Dynasty to the Yuan Dynasty

In 304 AD, JI Han wrote *Grasses and Trees in South China,* which described 80 plants and is the earliest documentation of a local flora (The Writing Group of the School of Physics at Peking University, 1978). Then came the most important work on agricultural botany and the first encyclopedia of agriculture in ancient China, the *Qi Min Yao Shu* (*Main Techniques for the Welfare of the People*), written by JIA Si-Xie from 533 to 544 AD. This contained systematic agricultural science knowledge including cultivation techniques for different crops, vegetables, fruit and timber trees, animal husbandry, and processing agricultural products (Chinese Botanical Society, 1983; Liu, 2006).

There were no further works on agricultural botany until the twelfth century, when publications such as the *Nong Shu* (*On Farming*), written by CHEN Fu in 1149, the *Nong Sang Ji Yao* (*Agricultural Series*), by the government of the Yuan Dynasty in 1273, another *Nong Shu* by WANG Zhen in 1313, and the *Nong Sang Yi Shi Cuo Yao* (*Summary of Agriculture Food and Clothing*) by LU Ming-Shan in 1314, emerged. The summaries of the rich experience and techniques of farming cultivation and silk production provided in these books not only boosted agriculture in ancient China but also represent the accumulation of knowledge on plants of that time (The Writing Group of the School of Physics at Peking University, 1978).

During the period from the eighth to the middle of the fourteenth century many books on specific plants were also written. The *Chajing* (*Tea Classics*) written by LU Yu in 758 was the first monograph on tea in the world and introduced methods for collecting and making high-quality tea. The *Lizhipu* (*Lychee Book*) by CAI Xiang, completed in 1059, summarizes techniques for cultivating the lychee (*Litchi chinensis;* Sapindaceae) and records 32 varieties of the species. Further monographs on *Paeonia* (Paeoniaceae; the *Luo Yang Mu Dan Ji* by OU Yang-Xiu in 1031), *Chrysanthemum* (Asteraceae; the *Ju Pu* by LIU Meng in 1104), *Saccharum officinarum* (WANG Zhuo's 1154 *Tang Shuang Pu*), orange (*Citrus ×aurantium;* Rutaceae; HAN Yan-Zhi's 1178 *Ju Lu*), bamboo (Poaceae; DAI Kai-Zhi's fifth century *Zhu Pu*), plum (*Prunus domestica*; Rosaceae; FAN Cheng-Da I's 1186 *Fan Cun Mei Pu*) and jujube (*Ziziphus jujuba*; Rhamnaceae; LIU Guan's thirteenth century *Da Zao Pu*), followed. These books, and the knowledge they contain, are valuable cultural treasures in the botanical history of China (The Writing Group of the School of Physics at Peking University, 1978; Chinese Botanical Society, 1983; Lu *et al.*, 1991; Lu, 1998).

The development of herbal botany in China owes a lot to TAO Hong-Jing (ca. 450–536 AD). Years of war and repeated copying by hand had resulted in the accumulation of errors in SHEN Nong's *Herbal Classic*. TAO, who was very interested in herbs, rectified as much as possible from surviving literature and information. In later years he wrote his own book, the *Ben Cao Jing Ji Zhu* (*Annotations on Materia Medica*), based on years of studying and practicing, completed in 502 AD. He added hundreds of medicines, recording a total of 730, making a relatively complete and very precise, ground-breaking resource that continued to be used by students of herbalism in much later dynasties (Chinese Botanical Society, 1983; Chen & Huang, 2005).

During the Tang Dynasty (618–907 AD), communication with western countries became more frequent and many drugs and medicines were imported from abroad which led the government urgently to incorporate the new and challenge the old herbal medicinal doctrine. In 659, at the request of the government, SU Jing and a team of 20 men worked together to write the *Xin Xiu Ben Cao* (*Newly Compiled Tang Materia Medica*). This work is in three parts (text, illustrations and captions) and records 850 medicines. It represents the first government-sponsored pharmacopeia in the world, produced over 800 years earlier than the first European pharmacopeia, the 1489 *Florentine Codex* (The Writing Group of the School of Physics at Peking University, 1978).

During the Northern Song Dynasty (960–1127 AD), a folk doctor named TANG Shen-Wei made a careful study of past herbal literature and a wide collection of useful medicines with popular instructions for their use, which he synthesized into his *Jing Shi Zheng Lei Bei Ji Ben Cao* (*Classified Materia Medica for Emergencies*), finally finished in 1082. It recorded 1 746 medicines and at the same time preserved the content of many ancient herbal books and documents such as the *Herbal Classic* and the *Ben Cao Jing Ji Zhu*. TANG's opus has been translated into foreign languages, spread to Korea and Japan, and is regarded as the work with the greatest influence in the field of herbalism prior to the *Compendium of Materia Medica* 500 years later (The Writing Group of the School of Physics at Peking University, 1978; Chen & Huang, 2005).

The *Quan Fang Bei Zu* (*The Foliage*), which is considered to be the first Chinese botanical dictionary (Xu & Xing, 1982), was written by CHEN Jing-Yi in 1253. This book included a collection of poems about plants, including those describing their morphological characters and usage. Although the *Quan Fang Bei Zu* was only a slim volume, it was the earliest book on the ancient flora of China, and similar works such as the *Qun Fang Pu*, composed by WANG Xiang-Jin in 1621, and the *Guang Qun Fang Pu* by WANG Hao in 1708, were based upon it (Chinese Botanical Society, 1983). While this level of progress was continuing in China, botanical study in the west was still primarily limited to copying of previous works (Needham, 1986).

11.2.3 Botanical history from 1368—1840

This was a period of some important works on agriculture and herbalism in China. Firstly, in 1376, YU Zong-Ben, under the pen name GUO Tuo-Tuo, wrote the *Zhong Shu Shu*, recording in detail grafting techniques for fruit trees (The Writing Group of the School of Physics at Peking University, 1978). A book about edible plants—the *Jiu Huang Ben Cao* (*Materia Medica for the Relief of Famine*)—was finished in 1406 by ZHU Su, aimed at those facing periods of famine caused by war or natural disaster. Four hundred and fourteen edible plants were recorded, including 245 herbs, 80 trees, 20 cereals, 23 fruits and 46 vegetables. Among them, 270 plants were new to the public, all cultivated by ZHU himself in his garden. He recorded the characters of each plant with illustrations and even methods for cooking them. The contents of this valuable book were later incorporated by LI Shi-Zhen into his *Ben Cao Gang Mu* (*Compendium of Materia Medica*) in 1596 and even by Japanese botanists in twentieth century works (The Writing Group of the School of Physics at Peking University, 1978; Chinese Botanical Society, 1983; Chen & Huang, 2005). In 1547, MA Yi-Long dedicated his *Nong Shuo* to details of the cultivation of rice, being the first to advance a philosophical view to explain the principles of cultivation (The Writing Group of the School of Physics at Peking University, 1978).

1596 saw the publication of LI Shi-Zhen's masterpiece, the *Compendium of Materia Medica*. Li was famed as the "prince of pharmacists" and Joseph Needham even translated the title of his book as *The Great Pharmacopoeia* (Needham, 1986). Thirty years in the making, this work recorded 1 892 medicines, of which 1 173 were plants (Lu *et al.*) and 1 109 were illustrated by his sons. LI sorted the medicines into 16 kinds, of which five were plants (Chinese Botanical Society, 1983; Chen & Huang, 2005), in a classification similar to that of Linnaeus, the "father of taxonomy," 130 years later. The *Compendium of Materia Medica* also provided brief introductions to 41 historical herbals and a list of synonyms for medicines with more than one name. These inclusions imply that LI was consciously creating nomenclatural rules and criteria, something that did not occur in Europe for another 300 years (Needham, 1986). The *Compendium of Materia Medica* was translated into Japanese, Latin, French, Polish and English, and was dispersed almost world-wide. It is regarded as being at the highest level of all the herbal works, home and abroad, at that time, far more significant than any other herbal work in Chinese history, and is still considered important today for its direct promotion of botany across the world (The Writing Group of the School of Physics at Peking University, 1978; Chinese Botanical Society, 1983; Needham, 1986; Chen & Huang, 2005).

During the seventeenth century and until the middle of the nineteenth century, while western botany was in a process of transition to the modern science, botanical study in China remained in a period of herbals. Nonetheless many works of this period are worthy of mention (Chinese Botanical Society, 1983). At the start of the seventeenth century, GENG Yin-Lou wrote the *Guo Mai Min Tian*, about farming and management techniques including initiatives for the artificial selection of seeds. Published in 1637, SONG Ying-Xing's *Tian Gong Kai Wu* (*The Exploitation of the Works of Nature*) was praised by Needham as an encyclopedia of the arts and crafts in seventeenth century China (Needham, 1988). This volume refers to techniques for almost every aspect of agriculture and the handicraft industry including folk knowledge on dyeing plants (Wu & Chen, 2004). At roughly the same time, XU Guang-Qi published the *Nong Zheng Quan Shu*, in which he summarized all the techniques recorded in past agriculture works and developed his own creative opinions on, and modifications of, these techniques (The Writing Group of the School of Physics at Peking University, 1978).

The earliest preserved example of horticultural literature in China is the 1688 *Hua Jing*, in which more than 300 plant varieties and cultivation methods were contained. The author, CHEN Hao-Zi, pointed out that cultivation could change some of the morphological and physiological characters of plants and emphasized the importance of human action in this respect (The Writing Group of the School of Physics at Peking University, 1978). Also during the Qing Dynasty (1644–1911), a number of herbal works emerged based on the *Compendium of Materia Medica*. Two outstanding and typical medical achievements of this dynasty – the *Ben Cao Gang Mu Shi Yi* (*Supplementary Amplifications of the "Compendium of Materia Medica"*) by ZHAO Xue-Min and the *Zhi Wu Ming Shi Tu Kao* by WU Qi-Jun – are particularly worthy of mention.

ZHAO Xue-Min's first draft of the *Supplementary Amplifications of the "Compendium of Materia Medica"* was completed in 1765. However, it took almost 100 years of revision and enlargement by ZHANG Ying-Chang before it was finally published in 1864. This work provided additions and corrections to LI's *Compendium*, including 716 new medicines, taking in not only folk medicines but also foreign ones. It followed the classification of the *Compendium* with the addition of two plant categories: rattans and flowers. ZHAO and ZHANG also considered the problem of the names of medicines and described the morphological characters of most medicinal plants, providing a valuable reference for future botanical study.

In all, the *Supplementary Amplifications* contained 921 medicines. Although they were not illustrated and some mistakes were made, the significance and world-wide influence of this book were agreed on by botanists both at home and abroad (The Writing Group of the School of Physics at Peking University, 1978).

The *Zhi Wu Ming Shi Tu Kao* was published in 1848 and was entirely about plants. The taxonomic method of the book was based on that of the *Compendium of Materia Medica*. There were 38 volumes on 12 types of plants, covering a total of 1 714 plants. Among them, 519 were new compared to the *Compendium*. Careful and detailed descriptions were provided of morphological characters, the colors and flavors of the medicinal parts, habitat and distribution. These were accompanied by quite delicate and beautiful illustrations drawn from fresh material, better than those in any previous medical books. In fact, the *Zhi Wu Ming Shi Tu Kao* covered the flora of a relatively large area of China. The author, WU Qi-Jun, was an official doctor and was allowed by the emperor of the Qing Dynasty to travel widely within the country. WU's book covered plants from 19 provinces of China, most of them distributed in Jiangxi, Hubei, Yunnan, Shanxi, Henan, Guizhou and Hunan. This was the first time that plants from border areas such as Yunnan had been recorded, and the work represented greater progress than made by any past herbalist, particularly given the difficulties of collecting and documenting plants alone in remote areas with the restricted materials and equipment of the time (The Writing Group of the School of Physics at Peking University, 1978; Chinese Botanical Society, 1983). WU's achievements did not stop here. Prior to publication of the *Zhi Wu Ming Shi Tu Kao* he also compiled 800 medical and herbal works from Chinese history into the *Zhi Wu Ming Shi Tu Kao Chang Bian,* a valuable resource for studying the history of Chinese herbalism and botany (The Writing Group of the School of Physics at Peking University, 1978). As with other works, the *Zhi Wu Ming Shi Tu Kao* had its limits and errors, including some confusion between plants with the same name and deficiencies in recording plants' medical uses. However its importance in the historical heritage of botanical science and its contribution to botanical development is widely acknowledged. This work is considered to represent the highest level of Chinese botanical work in history, and to mark the end of the period of ancient Chinese herbal botany (The Writing Group of the School of Physics at Peking University, 1978; Chinese Botanical Society, 1983).

After the middle of the nineteenth century, China experienced a series of wars, and little progress was made in botanical study until the twentieth century, with the exception of the 1858 publication, by LI Shan-Lan and Alexander Williamson, of *Botany*, a translation of John Lindley's *Elements of Botany*. This, and later works on botany, were based on modern botanical systems, indicating the efforts of Chinese botanists to work in accordance with western botany (Chinese Botanical Society).

11.3 A case study of botanical development in China: ancient Tibetan medicinal works

Many ethnic minorities, whether inhabiting forest, desert, island, polar lands or mountainous regions, have been in some way isolated from the outside world. Under these circumstances, they harvested from each unique environment not only their food but also medicinal herbs. The gradual accumulation of knowledge and experience surrounding such natural resources may develop into a botanical lore that is passed down from generation to generation. Sadly, under the influence of modern civilization and with a lack of formal records, some such knowledge has today often been partly or completely lost (Verma, 1997; Yang, 2000).

China is a multi-racial country with 56 ethnic groups, of which 55 are officially recognized as ethnic minorities. These minorities are found across more than 18 provinces, covering 65% of the total territory of China. The various and abundant traditional ethnic medicinal knowledge systems of these groups together make up the great Chinese medical system (Huai & Pei, 2001). Tibetan medicine is one of the most important ethnic medicines based on systematic principles.

Traditional Tibetan medicine began more than 2 000 years ago, and elements of it have since been incorporated into Han Chinese, Indian, ancient Arabic and Greek medicinal cultures, among others; thus it has become uniquely well-known as a minority medicinal system throughout the world (Beckwith, 1979; Shang *et al.*, 2006). During the long process of development of Tibetan medicine, several records and documents were written. The earliest record of Tibetan medical knowledge of plants refers to highland barley (*Hordeum vulgare* var. *vulgare*; Poaceae), following the unification of written characters in Tibet in the seventh century AD under Songzan Ganbu. Highland barley was used to make wine, and the residue was applied to the body to cure injuries (Chen & Huang, 2005).

In 720, a Buddhist priest and a Tibetan translator worked together to complete the *Yue Wang Yao Zhen*, the earliest preserved work on Tibetan medicine, which played an important role in the growth and promoting the development of the discipline. Its contents included the causes of illness, pathology, diagnosis, cures and medication (i.e. almost every aspect of medicine). The theories in the book were based on the practical experiences of the Tibetan people, whose distinct ethnic characteristics differ from the medicinal practices of the dominant Han group. About 780 kinds of medicines were listed, 440 of which were plants, and more than 300 of which are endemic to the Qinghai-Xizang Plateau. These include *Phlomis younghusbandii* (Lamiaceae, known as pang xie jia), *Delphinium brunonianum* (Ranunculaceae; nang ju cui que hua) and *Aconitum naviculare* (Ranunculaceae; chuan kui wu tou), all of limited distributed on the Plateau (Zhen & Cai, 2002; Chen & Huang, 2005).

NING Tuo, a member of a longstanding, distinguished family of Tibetan doctors, made a detailed study of the medicinal literature and records of different nationalities, and collected together Tibetan folk medicinal practices over many years. He combined this knowledge into a famous Tibetan medicinal book, the *Si Bu Yi Dian*, written over 20 years to 765. This work included 443 prescriptions and 1 002 herbs, classified into eight types. NING believed that medicines, like everything else, were derived from the five elements, ground, water, fire, wind and air. This was later summarized as the "theory of the assumption of five elements" and the "hypothesis of three factors." NING also claimed that combining different materials of different origins produced six tastes of medicines (Duo, 1998; Zhen & Cai, 2002; Chen & Huang, 2005)—sweet, sour, salty, bitter, pungent and acerbic. He also defined eight traits and 17 curative effects of Tibetan medicine (Ding *et al.*, 2001; Chen & Huang, 2005). This medical work is taken as a sign that Tibetan medicine was becoming an independent and mature medicinal subject at this time. Unfortunately, unlike other medical works of the time, it was ordered to be concealed by the ruler of Xizang due to religious reasons, and was not widely distributed (Ding *et al.*, 2001; Chen & Huang, 2005).

Between the thirteenth and fourteenth centuries, Des Sangjiejiacuo, a famous Tibetan scientist of the Qing dynasty, revised and annotated the *Si Bu Yi Dian*, and wrote a new medical work—the *Si Bu Yi Dian Lan Liu Li*—in a more popular language. Nowadays, this is the most popular form of the *Si Bu Yi Dian*. Around this time the Dalai Lama also built a school of medicine, to provide opportunities for civilians to learn the *Si Bu Yi Dian* and, with his encouragement and support, ancient Tibetan medicine reached its first peak of prosperity (Zhen & Cai, 2002).

The Tibetan pharmacist Dimaer Danzengpengcuo conducted almost 20 years of fieldwork in western and eastern Qinghai, western Sichuan and eastern Xizang, verifying the medicines in over 130 documents, and wrote the classic Tibetan medical work the *Jing Zhu Ben Cao*, completed in 1736, which is considered to be the Tibetan equivalent of the Han *Compendium of Medica Materia* (Zhen & Cai, 2002; Chen & Huang, 2005). The *Jing Zhu Ben Cao* contained chapters on 13 types of medicines, covering a total of 2 294 medicines, about one-third of which are only distributed on the Qinghai-Xizang Plateau (Chen & Huang, 2005). It became a widespread monograph on medicinal herbs, covering their morphological traits, medicinal functions and properties, places of origin, usage, habitat and tastes (Li *et al.*, 2002). Dimaer Danzengpengcuo also paid attention to – and attempted to correct – instances of synonymy and shared names. He also provided methods for identifying true medicines from fake ones (Chen & Huang, 2005).

The *Jing Zhu Ben Cao* embodies the unique features of the traditional Tibetan medicinal system. The distinctive ethnic characteristics of this work lie in the abundant Tibetan experiences of practicing with endemic medicines, special prescriptions and preparation methods (Li *et al.*, 2002). About 75% of the medicines recorded in it are still used in the medical field, and the work has been translated into Sanskrit and English and spread throughout the world. It promoted the progression of Tibetan medicine and attracted a lot of attention in the medical field across the world. It deserves its position as one of the brightest pearls of traditional Chinese medicine (Chen & Huang, 2005).

As we have seen, great achievements and progress were made in Chinese traditional botanical knowledge by both the Han and other ethnic groups during their long history. From the 1850s to 1960s, modern botany began to spread into China and was gradually accepted by Chinese botanists. This reduced the attention that was paid to traditional botanical knowledge relating to agriculture and medicines. In more recent years, however, traditional botanical knowledge has been gaining more attention from Chinese botanical scientists, as a valuable part of Chinese botany. Using the study methods of modern botanical systems, it is hoped that more information can be retrieved from these traditional disciplines (Chinese Botanical Society, 1983).

References

BECKWITH, C. I. (1979). The introduction of Greek medicine into Tibet in the seventh and eighth centuries. *Journal of the American Oriental Society* 99: 297-313.

CHEN, M. J. (1956). *Yin Xu Pu Ci Zong Shu*. Beijing, China: Science Press.

CHEN, Z. M. & HUANG, S. B. (2005). *Herbalism*. Nanjing, China: Southeast University Press.

CHINESE BOTANICAL SOCIETY (1983). *Bibliography of Chinese Botany*. Beijing, China: Science Press.

DING, Y. F., LI, F. & YANG, H. (2001). A brief exposition on study of different nationalities' traditional drugs of China. *Journal of Medicine and Pharmacy of Chinese Minorities* 7: 20-22.

DUO, J. (1998). A summary of "assumption of five elements" of Tibetan medicine. *Journal of Medicine and Pharmacy of Chinese Minorities* 1: 3-4.

HUAI, H. Y. & PEI, S. J. (2001). Medicinal ethnobotany and its advances. *Chinese Bulletin of Botany* 2: 129-136.

LI, F. (1984). *The History of Cultivating Plants in China*. Beijing, China: Science Press.

LI, L. Y., ZHANG, D., WEI, Y. F., ZHONG, G. Y., QIN, S. Y., CIREWN, B. Z. & GESANG, B. Z. (2002). Conservation of endangered species resources of Tibetan medicine in China. *Journal of Chinese Materia Medica* 8: 561-564.

LIU, P. (2006). "Qi Min Yao Shu": an ancient work on agriculture. *Journal of Library and Information Sciences in Agriculture* 18(12): 42-44.

LIU, R. Q. & HU, Y. X. (1976). The Preliminary Study on the Ancient Papers in China. *Cultural Relics* 5: 74-79, 101-102.

LU, J. X. (1998). *The Compendium of the Science History in Ancient China*. Shijiazhuang, China: Science and Technology Press of Hebei Province.

LU, S. W., XU, X. S. & SHEN, M. J. (1991). *Botany*. Beijing, China: Higher Education Press.

NEEDHAM, J. (1986). *Botany*. Cambridge, UK: Cambridge University Press.

NEEDHAM, J. (1988). *Agriculture*. Cambridge, UK: Cambridge University Press.

PAN, J. X. (1973). About the origin of papermaking - one of the studies on special topics of the history of papermaking technique in ancient China. *Cultural Relics* 9: 45-51.

PAN, J. X. (1974). The earliest fiber paper in the world. *Chemistry Online* 5: 44-47.

SHANG, Y. H., LIU, C., PENG, L. X., MENG, Q. Y. & LIU, Y. (2006). The status and prospects of research on Tibetan medicine. *Journal of Southwest University for Nationalities* 1: 140-144.

THE COMMITTEE OF PROTECTING THE HISTORIC CULTURAL RELICS IN ZHEJIANG PROVINCE (1976). The found of the primitive society cultural site in He Mudu. *Cultural Relics* 8.

THE INSTITUTE OF ARCHAEOLOGY CHINESE ACADEMY OF SOCIAL SCIENCES (1963). *The Historical Cultural Site in Banpo Xi'an*. Beijing, China: Cultural Relics Press.

THE WRITING GROUP OF THE SCHOOL OF PHYSICS AT PEKING UNIVERSITY (1978). *The Events of Technology in Ancient China*. Beijing, China: People's Education Press.

VERMA, S. K. (1997). Comparative studies on folk drug of tribals of Chontanagpur and Santhal Pargana of Bihar India. *Ethnobotany* 9: 20.

WU, C. Y. & CHEN, S. C. (2004). *Flora Reipublicae Popularis Sinicae*. Beijing, China: Science Press.

XU, M. Q. (1975). The papermaking technique is the great invention made by ancient Chinese. *Report on the Papermaking Technique* 1: 1-2.

XU, W. X. & XING, R. L. (1982). The "Quan Fang Bei Zu" a first dictionary for Chinese plants. *Acta Phytotaxonomica Sinica* 20(3): 380-384.

YANG, J. S. (2000). Tibetan medicine. In: Xi, J. C. (ed.) *Ethnobotany and Sustainable Utilization of Plant Resources*. pp. 65-75. Kunming, China: Science and Technology Press of Yunnan.

ZHEN, Y. & CAI, J. F. (2002). A summary on Tibetan medicine. *Tibetan Studies* 2: 79-85.

Chapter 12

Plant Exploration in China

HU Chi-Ming (HU Qi-Ming) and Mark F. Watson

12.1 Introduction

China's long and illustrious history as one of the world's great civilizations is mirrored by an extensive tradition of the study and use of its plant resources stretching back thousands of years. This ancient botanical history is characterized by the accumulation of vast information stores on plants, plant products and their uses within the pages of herbals, encyclopedias and dictionaries. Many of these are scholarly works showing a deep knowledge and profound understanding of plants and their classification according to their properties and effects. Some, such as WU Qi-Jun's *Zhi Wu Ming Shi Tu Kao*, are exceedingly well-illustrated, and the plants identifiable with modern species, and so continue to be both usable and useful. These early books have been discussed in detail in Chapter 11.

Modern botanical methods, taking a scientific approach to systematic studies, began to become established in China at the end of the nineteenth century, when western science gradually permeated the country, spread primarily by missionaries and foreign explorers. In the following pages we explore the development of plant exploration in China, dividing it into three sections: exploration driven by professional and amateur naturalists from foreign countries; exploration led by Chinese botanists themselves; and the post-1980 era where Chinese and foreign scientists joined forces and worked together. The main locations of the herbarium specimens they collected are indicated using the standard international herbarium acronyms (Holmgren *et al.*, 1990), with the primary set given first. For example, PE indicates Beijing, KUN Kunming, K the Royal Botanic Gardens, Kew, E the Royal Botanic Garden Edinburgh, LE St. Petersburg, and A Harvard University. As in previous chapters, Chinese personal names have been transliterated using the system by which each person was most commonly known (usually either Hanyu Pinyin or the earlier Wade-Giles). However, since many of the collectors discussed here have been widely known by two different transliterations, both are provided for cross-reference where applicable in Table 12.1, along with the Chinese characters of the name.

12.2 Exploration by collectors from foreign countries, 1700–1949

This period embraces the early European collectors, and although previous authors have documented it in detail, the following paragraphs give a brief review and provide some context. Bretschneider (1898) comprehensively covers the years before 1898 in his classic *History of European Botanical Discoveries in China*, and Lancaster (1989) extends this into the modern period in his *Travels in China—A Plantsman's Paradise*. Other useful general references include Lyte's *The Plant Hunters* (Lyte, 1983) and McClean's *A Pioneering Plantsman* (McLean, 1997). Lancaster (1989) and Lyte (1983) are particularly rich sources of portraits for the botanists mentioned below, and see also Cox (1945), Tyler-Whittle (1997) and Fan (2004).

12.2.1 The pre-1860 era: early contacts with western botanists

As China was strictly closed to outsiders until 1860, its botanical richness and incredible plant diversity was virtually unknown to western countries. The Swedish explorer and student of Carl Linnaeus, Pehr Osbeck (1723–1805), was among the first of the early European explorers to venture into China, spending four months studying the flora, fauna and people of Guangzhou (then known as Canton) between 1750 and 1752. Osbeck sent many herbarium specimens to Linnaeus back in Sweden (LINN) and was just in time for Linnaeus to include them in the first edition of his monumental *Species Plantarum*, published in 1753. Osbeck was responsible for most of the Chinese plants in *Species Plantarum*, and Linnaeus named the showy genus *Osbeckia* (Melastomataceae) in his honour. In 1757, the Qing court confined European trade to Guangzhou Shi, and for nearly a century that city was the sole Chinese center for European natural history. Traders and envoys explored the city's gardens and markets for attractive flowers and ornamental trees and shrubs, and its shops and homes for botanical paintings (Mueggler, 2011). The Opium War (1840–1842) and the Treaty of Nanjing (1842) opened up new territories around the five newly created "treaty ports" of Hong Kong-Guangzhou Shi, Fuzhou Shi (known to the British as Foochow), Xiamen (Amoy), Shanghai Shi and Ningbo Shi. However, apart from diplomats in Beijing Shi, no foreigners were allowed to wander more than a few kilometers outside the city walls and so, throughout the period from 1700 to 1860, it was largely only Chinese ornamental plants for sale in markets and nurseries around the treaty ports that were collected and sent back by foreign collectors.

Table 12.1 Chinese botanists mentioned in Chapter 12, with their names in Chinese characters, commonly used alternative transliterations, and dates. b, born; d, died. Note that, where birth and death dates are unknown a floruit (fl.) statement is given denoting the period of time when the person was botanically active.

Name in Hanyu Pinyin	Name in Chinese characters	Name in other transliterations	Dates	Figure reference
CAI Xi-Tao	蔡希陶	TSAI Hse-Tao	1911–1981	Figure 12.7C
CHEN Feng-Huai	陈封怀	CHEN Feng-Hwai	1900–1993	Figure 12.7B
CHEN Huan-Yong	陈焕镛	CHUN Woon-Young	1890–1971	Figure 12.8D
CHEN Mou	陈 谋		1903–1935	
CHEN Nian-Qu	陈念劬	CHUN Nim-Keoi	fl. 1929–1932	
CHEN Shao-Qing	陈少卿	CHUN Shao-Hing	1911–1997	Figure 12.8E
CHEN Shu-Kun	陈书坤		b. 1936	
CHEN Wei-Lie	陈伟烈		b. 1939	Figure 12.12E
DAI Lun-Kai	戴伦凯		b. 1930	Figure 12.12B
FANG Rui-Zheng	方瑞征	FANG Rhui-Cheng	b. 1932	
FANG Wen-Pei	方文培		1899–1983	Figure 12.6D
FENG Guo-Mei	冯国楣	FENG Kuo-Mei	1917–2007	Figure 12.8C
FENG Qin	冯 钦	FUNG Hom	1896–1989	
FU Guo-Xun	傅国勋		b. 1932	
FU Pei-Yun	傅沛云		b. 1926	
GAO Xi-Peng	高锡朋	KO Sik-Pang	fl. 1930–1940	
GUO Ben-Zhao	郭本兆	KUO Pung-Chao	b. 1920	
HAO Jing-Sheng	郝景盛	HAO Kin-Shen	1903–1957	Figure 12.9D
HE Ting-Nong	何廷农	HO Ting-Nung	1938–2011	Figure 12.11C
HE Zhu	何 铸		1924–1991	
HONG De-Yuan	洪德元		b. 1937	Figure 12.10F
HOU Kuan-Zhao	侯宽昭	HOW Foon-Chew	1908–1959	Figure 12.9A
HU Qi-Ming	胡启明	HU Chi-Ming	b. 1935	Figure 12.11H
HU Xian-Su	胡先骕	HU Hsen-Hsu	1894–1968	Figure 12.6B
HUANG Ren-Huang	黄仁煌		fl. 1960–1988	
HUANG Rong-Fu	黄荣富		b. 1940	
HUANG Zeng-Quan	黄增泉	HUANG Tseng-Chieng	b. 1931	
HUANG Zhi	黄 志	WANG Chi	fl. 1929–1940	
JIANG Shu	姜 恕		b. 1925	
JIANG Xing-Lin	蒋兴麐	TSIANG Hsing-Ling	b. 1927	
JIANG Ying	蒋 英	TSIANG Ying	1898–1982	Figure 12.7A
JIAO Qi-Yuan	焦启源	CHIAO Chi-Yuan	1901–1968	
KONG Xian-Wu	孔宪武	KUNG Hsien-Wu	1897–1984	Figure 12.9E
LAI Shu-Shen	赖书绅		b. 1931	
LANG Kai-Yong	郎楷永	LANG Kai-Yung	b. 1937	Figure 12.10D
LI An-Ren	李安仁	LI An-Jen	b. 1927	
LI Bo-Sheng	李勃生		b. 1946	
LI Guo-Feng	李国凤		b. 1935	Figure 12.10G
LI Heng	李 恒		b. 1929	Figure 12.12H
LI Hong-Jun	李洪均		b. 1929	
LI Liang-Qian	李良千	LI Liang-Ching	b. 1952	
LI Pei-Qiong	李沛琼	LI Pei-Chun	b. 1936	
LI Shu-Xin	李书心		b. 1926	
LI Xin	李 欣		b. 1928	Figure 12.11A
LI Ze-Xian	李泽贤		b. 1937	Figure 12.11E
LI Zhen-Yu	李振宇		b. 1952	
LIANG Chou-Fen	梁畴芬		b. 1921	
LIANG Xiang-Ri	梁向日	LIANG Hoeng-Yip	fl. 1931–1954	
LIAO Ri-Jing	廖日京	LIAO Jih-Ching	b. 1929	
LIU Guo-Jun	刘国均		b. 1933	
LIU Ji-Meng	刘继孟	LIOU Ki-Meng, LIU Chi-Mon	1909–1993	
LIU Shang-Wu	刘尚武		b. 1934	Figure 12.11B
LIU Shen-E	刘慎谔	LIOU Tchen-Ngo	1897–1975	Figure 12.9C
LIU Tang-Shui	刘棠瑞		1910–1997	

Continued

Name in Hanyu Pinyin	Name in Chinese characters	Name in other transliterations	Dates	Figure reference
LU An-Ming	路安民		b. 1939	
MA Yu-Quan	马毓泉	MA Yu-Chuan	1916–2008	
NI Zhi-Cheng	倪志诚	NI Chi-Cheng	b. 1942	
NIE Min-Xiang	聂敏祥		b. 1932	
PENG Jing-Yi	彭镜毅	PENG Ching-I	b. 1950	
PU Fa-Ding	溥发鼎	PU Fa-Ting	b. 1936	
QIAN Chong-Shu	钱崇澍	CHIEN Sung-Shu	1883–1965	Figure 12.6C
QIN Hai-Ning	覃海宁		b. 1960	
QIN Ren-Chang	秦仁昌	CHING Ren-Chang	1898–1986	Figure 12.6F
QIU Shao-Ting	邱少婷		b. 1961	
SHAN Ren-Hua	单人骅		1909–1986	
SHEN Guan-Mian	沈观冕		b. 1935	
SONG Zi-Pu	宋滋圃	SOONG Tze-Pu	b. 1925	
SU Zhi-Yun	苏志云		b. 1936	
SUN Hang	孙 航		b. 1963	Figure 12.12G
TANG Jin	唐 进	TANG Tsin	1897–1984	Figure 12.7F
TAO De-Ding	陶德定		b. 1937	
WANG Bing	王 兵		b. 1957	
WANG Fa-Zan	汪发缵	WANG Fa-Tsuan	1899–1985	Figure 12.8B
WANG Jin-Ting	王金亭	WANG Chin-Ting	b. 1932	Figure 12.12F
WANG Qi-Wu	王启无	WANG Chi-Wu	1913–1987	Figure 12.7E
WANG Wen-Cai	王文采	WANG Wen-Tsai	b. 1926	Figure 12.12A
WANG Xue-Wen	王学文		b. 1935	
WANG Ying-Ming	王映明		b. 1938	
WANG Zhan	王 战		b. 1911	
WANG Zuo-Bin	王作宾	WANG Tso-Ping	1908–1989	Figure 12.10A
WEI Fa-Nan	韦发南		b. 1941	
WU Su-Gong	武素功	WU Su-Kung	1935–2013	Figure 12.12D
WU Yu-Hu	吴玉虎		b. 1951	
WU Zheng-Yi	吴征镒	WU Cheng-Yih	1916–2013	Figure 12.12C
XIA Wei-Ying	夏纬瑛	HSIA Wei-Ying	1895–1987	Figure 12.8A
XIN Shu-Zhi	辛树帜	SIN, S.S.	1894–1977	Figure 12.9B
XING Fu-Wu	邢福武		b. 1956	Figure 12.11F
XIONG Ji-Hua	熊济华		b. 1923	
XIONG Yao-Guo	熊耀国	HSIUNG Yao-Kuo	1910–2004	Figure 12.11G
YANG Chang-You	杨昌友		b. 1928	
YANG Yong-Chang	杨永昌	YANG Yung-Chang	b. 1927	
YAO Gan	姚 淦		b. 1937	
YE Hua-Gu	叶华谷		b. 1956	
YIN Wen-Qing	尹文清		b. 1927	
YING Jun-Sheng	应俊生	YING Tsun-Shen	b. 1934	Figure 12.10E
YU De-Jun	俞德浚	YÜ Te-Tsun	1908–1986	Figure 12.7D
YUE Jun-San	岳俊三	YUE Jun-Shan	b. 1934	
ZANG Mu	臧 穆		1928–2011	
ZENG Huai-De	曾怀德	TSANG Wai-Tak	fl. 1921–1947	
ZHANG Gui-Cai	张桂才		b. 1931	Figure 12.11D
ZHANG Qing-En	张庆恩		1920–2005	
ZHANG Yong-Tian	张永田	CHANG Yong-Tian	b. 1936	Figure 12.10C
ZHANG Zhi-Song	张志松		fl. 1952–1966	
ZHAO Zuo-Cheng	赵佐成		b. 1941	
ZHENG Wan-Jun	郑万钧	CHENG Wan-Chun	1904–1983	Figure 12.6E
ZHENG Zhong	郑 重		b. 1936	
ZHONG Bu-Qiu	钟补求	TSOONG Pu-Chiu	1900–1981	Figure 12.10B
ZHONG Guan-Guang	钟观光	TSOONG Kuan-Kwang	1868–1940	Figure 12.6A
ZHOU Li-Hua	周立华		b. 1934	
ZHOU Zhe-Kun	周浙昆		b. 1956	
ZUO Jing-Lie	左景烈	TSO Ching-Lieh	fl. 1929–1935	

British collectors.—It is generally accepted that first foreign botanical collector in China was James Cunningham (whose date of birth is not known but who died around 1709), a surgeon of the British East India Company sent to Xiamen in 1698. He moved to Zhoushan (transliterated by the British as Chusan) in 1701, where he stayed for at least a further two years. Cunningham collected dried herbarium specimens, sending back to the UK more than 600 preserved collections (BM, OXF). Nearly a century later the Irish amateur botanist Sir George Leonard Staunton (1737–1801) collected herbarium specimens (BM, K, AWH, G, P) and living plants in the Beijing Shi area on behalf of Sir Joseph Banks, president of the Royal Society, while engaged as a diplomatic secretary to Lord Macartney's Embassy to the QIAN Long Emperor in 1793.

Over the following years the British East India Company extended its trade connections with China, sending growing numbers of staff to serve in its factories (commercial stations) in the treaty ports. Between 1800 and 1860, several pioneering botanists slowly edged their way from the confines of these coastal areas towards the interior of China. Scottish gardener William Kerr (date of birth not known, died 1814) was sent to China by Banks as a professional plant collector, to learn horticultural techniques and send back living material of new plants for the Royal Botanic Gardens, Kew. He lived in Guangzhou Shi from 1804 to 1813, and purchased many plants from nurseries in the surrounding area, over 230 of which were successfully grown for the first time in Europe at Kew. *Kerria* (Rosaceae), a familiar shrub of British gardens, is named after him.

Also stationed in Ghangzhou Shi at about the same time was the keen English amateur naturalist John Reeves Senior (1774–1856). Reeves lived in China from 1812 to 1831, becoming an inspector of tea for the British East India Company. Although he too was restricted in his travels, Reeves built a network of local contacts to supply him with specimens from further afield, using which he amassed a large collection of plants, mammals, reptiles, birds, molluscs and fishes. Reeves understood the importance of illustrations in showing what living plants looked like, and commissioned local Chinese artists to prepare colored drawings of a wide variety of the native plants and animals he encountered. These drawings, now at the Natural History Museum, London, are among the first Chinese examples of scientific botanical art in the western tradition. As well as wild species, Reeves also collected cultivated plants from nearby nurseries and gardens, sending them back to London (K, BM, CGE). Reeves was responsible for the first introduction into the UK of many Chinese ornamentals including azaleas (certain *Rhododendron;* Ericaceae), *Camellia* (Theaceae), tree peonies (certain *Paeonia;* Paeoniaceae) and *Chrysanthemum* (Asteraceae). Reeves's son, John Russel Reeves Junior (1804–1877), followed in his father's footsteps and worked for 30 years in China, also with the British East India Company. Like his father, the younger Reeves was a keen amateur naturalist and also sent many herbarium specimens (K) and some living plants back to England.

Hong Kong was an important British trading port and several amateur botanists found themselves working there and used the opportunity to collect plants from nearby islands. The first was the physician Abel Clarke (1789–1826), who was appointed by Joseph Banks to accompany Lord Amhurst's Embassy to the Chinese Court in 1816–1817. He collected plants in Hong Kong, northern China (especially Beijing Shi), Nanjing and Guangzhou. Unfortunately most of his specimens were lost at sea on his return to England, but a small duplicate set was given to George Staunton (discussed above) and so were saved. Richard Brinsley Hinds (1812–1847), surgeon of HMS Sulphur, was also a very able naturalist and collected specimens of 140 species from Hong Kong in 1841 (BM, K). Hinds was closely followed by Captain (later Lieutenant Colonel) John George Champion (1815–1854) of the Ninety-Fifth Regiment of Foot, who made numerous collecting trips around the islands between 1847 and 1850. Champion was a keen naturalist with a good eye, and so his impressive Hong Kong collections consist of almost 600 species of vascular plants and ferns (K, OXF, CGE). Four years later the American botanist Charles Wright (1811–1885) visited Hong Kong twice in 1854 and 1855, collecting over 500 species (GH). Wright sent his specimens to George Bentham in England (K), adding to the collections that formed the basis for Bentham's *Flora Hongkongiensis* (Bentham, 1861).

The most outstanding British plant collector during these years was the Scottish horticulturist Robert Fortune (1812–1880), first employed by the Royal Horticultural Society and later by the British East India Company. Fortune was a traveller and professional plant collector who visited China four times between 1843 and 1861 (1843–1845, 1848–1851, 1853–1856 and 1861; Fortune, 1847). He was sent to buy plants from nurseries in southern and southeastern China, visiting Fuzhou Shi, Guangzhou Shi, Hong Kong, Ningbo Shi, Shanghai Shi and Xiamen, and even managed to explore the Wuyi Shan range where the finest-quality black tea was made. During his four visits he never went more than 50 km from a treaty port, yet he was responsible for the first introductions from China to British gardens of nearly 190 species or varieties of plants, of which more than 120 were entirely new to science. He is particularly famous for the successful introduction of thousands of tea (*Camellia sinensis*) plants to the Darjeeling area, which formed the foundation of the extensive tea industry in India. For this, and other living material, Fortune pioneered the use of the "Wardian case," with which he successfully transported living plants on long sea voyages. His herbarium collections (K, BM, LE, G, P, W, and others) provided important early records of many Chinese plant species, and from these about 25 were described as new species. Fortune is commemorated in the name of the Chusan palm, *Trachycarpus fortunei* (Arecaceae), a hardy palm introduced by him in 1844, and a familiar sight in

British gardens.

Henry Fletcher Hance (1827–1886) was a distinguished self-trained botanist who made a significant contribution to the knowledge of plants in the early botanical history of southern China. Hance lived in Hong Kong and Huangpu (transliterated by the British as Whampoa, ca. 30 km south of Guangzhou Shi), as a customs officer, between 1844 and 1887. Although his own collecting was of no great importance, over the 43 years he was in China he persuaded anyone who was at all interested in plants, wherever they might be going, to collect specimens and send their material to him. His herbarium contained over 20 000 specimens, all of which he sent back to England, and are now primarily held in London (BM, K, LE, P, B, W, and others).

French collectors.—The golden age of French botanists in China would come later in the nineteenth century, but it is worth highlighting that the second botanical collector to arrive in China, after Cunningham, was Pierre Nicholas le Chéron d'Incarville (1706–1757), a French Jesuit priest and amateur botanist stationed in Beijing Shi. Arriving in 1740, Incarville collected plants from the mountains around China's northern capital, sending herbarium specimens and seed back to Paris (P). He was responsible for the introduction of many ornamental plants into European gardens, notably *Ailanthus altissima* (Simaroubaceae) and *Lamprocapnos spectabilis* (Papaveraceae). The spectacular genus *Incarvillea* (Bignoniaceae) is named in his honour. More than a century later Paul-Herbert Perny (1818–1907) became the second French botanist to collect in China. He was also a missionary, but stationed further south, in Guizhou, from 1850 to 1860, and in Sichuan from 1862 to1868. Perny was the first naturalist to explore Guizhou and compiled a sizeable herbarium (P) and numerous zoological collections which included many new species.

Russians collectors.—During this period, Russian botanists were becoming increasingly active in northern provinces of China. Pre-eminent amongst these was Alexander Georg (Alexander Andrejewitsch) von Bunge (1803–1890), Professor at the University of Tartu (Estonia) and later director of the botanic garden there. During his career, Bunge travelled widely in Siberia, Mongolia and northern China. He joined an ecclesiastical mission to Beijing Shi in 1829, and collected herbarium specimens in the nearby regions until his return home in 1831 (LE, BR, F, P, OXF). Four years later he published his *Enumeratio Plantarum Quas in China Boreali Collegit A. Bunge* (Bunge, 1835), in which he documented 240 species, including 17 new genera and 189 new species. About the same time the physician Porphyri Yevdokimovich Kirlov (1801 to ca. 1864) spent more than ten years in China, and collected herbarium specimens mainly around Beijing Shi and in Mongolia (LE). One of Bunge's successors at the Russian Ecclesiastical Mission in Beijing Shi was Alexander Alexeyewich Tatarinov (1817–1886). Tatarinov was a medic based at the mission between 1840 and 1850 who collected many plants in the vicinity of Beijing Shi and from Mongolia during his journey to and from China (LE). Like John Reeves Senior, Tatarinov commission a set of colored illustrations of Chinese plants by a native artist, and he compiled an illustrated atlas of Chinese medicinal plants. To this illustrious group of Russian botanical explorers should be added the distinguished Ukrainian botanist Nicolai Stephanovich Turczaninov (1796–1864), a civil servant whose posting to Irkutsk gave him opportunity to travel extensively in Siberia, and was the first botanist to collect plants in the Heilong Jiang valley, in 1833 (KW, C, G, P, GOET, OXF).

Carl Johann (Karl Ivanovich) Maximowicz (1827–1891) was a great authority on the floras of central and eastern Asia, and a world-renowned Russian botanist of the nineteenth century. He spent most of his life studying the floras of the countries he visited in the Far East, initially collecting in the neighborhood of the Heilong Jiang and Wusuli Jiang for three years from 1853 with Leopold von Schrenck, and then in 1859 undertaking a five-year expedition to northeastern China and Japan. Maximowicz was not only a prolific collector but also a great author on the botany of the Far East. His publications (e.g. Maximowicz, 1839; Maximowicz, 1877; Maximowicz, 1889b; Maximowicz, 1889a) are among the most important contributions ever made to the understanding of the flora of eastern Asi,a and his collections are held mainly in St. Petersburg (LE, K, US).

12.2.2 1860-1949: botanical exploration by western scientists gains momentum

When the "Convention of Peking" was signed in 1860, western allied powers forced China completely to open its borders to foreigners. Many visitors jumped at the opportunity to venture into these newly-opened lands and the treaty ushered in a new era of plant exploration in China. Since then a large number of amateur naturalists, traveller-explorers, customs officers, missionaries and professional botanists have visited virtually every corner of China, penetrating deeply into remote areas of the mainland interior.

French collectors.—Apart from the notable exceptions of Incarville and Perny, French involvement in the botanical exploration of China did not really start until 1860, with the diplomat and agronomist Gabriel Eugène Simon (1829–1896). Simon was a talented botanist and plant collector, who travelled widely in China between 1860 and 1871, both in his official capacity and in pursuit of his botanical interests. He formed an extensive herbarium and seed collection that he later sent to Paris (P).

The vast extent of the true richness of the flora of China was first revealed to western botanists by French missionaries, particularly those stationed in the network of Catholic missions in western China (Fournier, 1932). Many French priests resided in Yunnan for long periods, and several became prolific plant collectors. Jean Pierre Armand David (1826–1900), commonly known as Père David, was the first of the French missionaries to collect plants. David went to China as a Vincentian priest in 1862, aiming to convert people to Roman Catholicism, but his biggest impact was in natural history as he travelled widely in northern, central and western China as well as Mongolia, avidly collecting plants and zoological specimens, many of which turned out to be new species. David was a hugely talented all-round naturalist, with extensive knowledge of geology, mineralogy, ornithology, zoology and botany. He also had a high regard and respect for the local people and their culture and it was this, and his gentle manner, which enabled him to travel safely in regions where people were hostile to, and distrustful of, foreigners. Like the other French missionaries, the herbarium specimens he collected were first sent to Paris (P) for identification, and duplicates later sent on to other institutes to help with this process (including K, LE, E, G, B).

A little later, Jean Marie Delavay (1834–1895) caused a botanical sensation when he sent the first collections of plants from Yunnan back to Europe. The great diversity of temperate plants found there would later prompt Ernest Wilson to call this area the "mother of gardens." Delavay worked for Missions Etrangères de Paris (Foreign Missions of Paris) and was first posted to China in 1867, initially to Guangdong and then to northwestern Yunnan. He returned briefly to Paris in 1881 where he met Père David and agreed to collect for Adrian Franchet (see below) who worked for the Muséum National d'Histoire Naturelle, Paris. Delavay returned to China in 1882, basing himself in Wenshan (then known as Tapintze) from where he explored the spectacularly biodiverse mountains of northwestern Yunnan. Unlike many of the other foreign plant collectors in China, Delavay preferred to work alone, not even taking local people with him to help carry equipment or collections. He was a great explorer and methodical plant collector, gathering one of the largest personal collections ever made, supported by detailed collecting notes. He continued his explorations of the mountain flora of northwestern Yunnan until 1888 when he contracted bubonic plague and was eventually forced to travel to Hong Kong to recuperate, still collecting plants along the way! By 1891, Delavay still had not fully recovered his health, and decided to return to France for more comprehensive treatment. However, he was restless and returned to China in 1893 to continue his collecting. He added a further 1 550 plant collections to his already impressive total, but in 1895 finally succumbed to his illness and died in Yunnan. Delavay's incredible collections totalled over 200 000 herbarium specimens, representing over 4 000 species (P, LE, E, G, K, PC, A, B) and some seed. Nearly half the species that he collected proved new to science, and over 1 500 species and genera are based on his discoveries (Franchet & Delavay, 1889–1890). It soon became clear that his work had opened up a unique floral region, richer perhaps than any area of comparable extent elsewhere in the world. The UK, France, Germany, Russia, Austria and Sweden vied with each other to send their professional collectors to China to capitalise on this opportunity.

Several other French missionaries made significant collections of plants from China. Jean André Soulié (1858--1905) was stationed on the Xizang border with Sichuan from 1886 and sent back over 7 000 specimens (P, PC, A, B, E, BR, G, K) from the surrounding mountains before he was killed by a band of Tibetan warrior-monks. Paul Guillaume Farges (1844–1912) was sent to northeastern Sichuan in 1867, and between 1892 and 1903 collected over 4 000 herbarium specimens (P, F, G, L), and some seed, before moving south to Chongqing Shi where he stayed until his death. Many of these were species of ornamental value, which Ernest Wilson would later re-collect and introduce into western gardens. In 1897, a shipment of seed Farges sent to Maurice de Vilmorin in France included a few seeds of *Davidia involucrata* (Nyssaceae). Nothing grew from these for more than a year, and they appeared dead, but finally germinated in 1889 in the same month that Earnest Wilson landed in Hong Kong on his mission to rediscover the very same species.

Émile-Marie Bodinier (1842–1901) collected plants around Beijing Shi before moving south to spend many years in Hong Kong and Guizhou (P, A). Edouard-Ernest Maire (1848–1932) followed in the footsteps of Delavay, spending 11 years (1905–1916) in Yunnan, and collecting a large number of herbarium specimens (E, DAO) and some seed, particularly in the Kunming Shi area. Back in Paris, Augustin Abel Hector Léveillé (1863–1918), Adrien René Franchet (1834–1900) and other French taxonomists worked on these collections and played an important role in the description and classification of the plants of the area. Léveillé was in charge of several of the Jesuit priests who were sent out to spread knowledge of their religion across China, Korea and Japan (including Bodinier and Maire) and, although he himself never visited Asia, he collated the specimens his priests sent back into his private herbarium (later sold to E). Unfortunately he lived far from Paris, and so had few opportunities to compare these new collections with those already documented. Instead he relied upon literature and memory, which led to many inaccuracies in the identification of specimens in his herbarium (E, P), and mistakes in the lithographic reproduction of his handwritten *Flore de Kouy-Tcheou* (Léveillé, 1914–1915) and *Flore du Pekin, Chang-hai, Tche-li and Kiangsu* (Léveillé, 1916–1917).

Russian collectors.—Emil Bretschneider (1833–1901) was physician to the Russian Legation in Beijing Shi between 1866 and 1883. During his time in China he explored the regions around the capital, collecting many herbarium specimens, seed and other living material which he sent to European and North American institutes (such as K, LE, A, BM, B, GH and MO). Bretschneider was one of the greatest historians of botanical exploration in Asia and the author of a classic work on the subject (Bretschneider, 1898). The most intense period of foreign exploration followed the publication of Bretschneider's book in 1898 and continued up until 1949 when China closed its doors to foreign visitors. In this period many prominent professional botanical explorers and collectors not only gathered valuable materials for herbaria of leading botanical institutions in the west, but also greatly enriched the gardens of the world through their discoveries in China.

In 1874–1875, whilst Bretschneider was in Beijing Shi, three other great Russian explorers paid visits to China. The first was the soldier, explorer and naturalist Nikolai Mikhaylovich Przhevalsky (or Mikhailovich Prezwalski; 1839–1888), who undertook several extensive journeys throughout central and eastern Asia between 1867 and 1885, although he never reached his final goal, Lhasa Shi. His herbarium specimens and living collections were sent to St. Petersburg (LE) and included many new species. They are especially important as early records from modern-day Qinghai. He died of typhus soon after setting out on his fifth journey in present-day Kyrgyzstan, and has been commemorated in the names of many plant species, notably the Solanaceous genus *Przewalskia*, and the only extant species of wild horse, *Equus ferus przewalskii* (Przewalski's horse). In 1874–1875, Pavel Jakovlevich Piassetsky (1843–1919) accompanied Captain Sosnovski's expedition exploring a trade route from Junggar into northwestern China and south to the Chang Jiang. He was an army surgeon, but also a very competent naturalist who made herbarium (LE) and zoological collections throughout his journey. A year later Grigory Nikolayaevich (Grigori Nikolaevich) Potanin (1835–1920) made the first of four great journeys into eastern Asia, including Mongolia and China (Delmar Morgan, 1877). Potanin was an eminent explorer and naturalist who made vast collections of herbarium (LE, E, B, BR), seed and zoological specimens over an eight-year period: he collected 12 000 herbarium specimens on his third expedition alone. In China he explored mainly the provinces of Guansu, Shanxi and Sichuan, and many of the plants he collected were new species.

The last great Russian collector of note in this era was the explorer and botanist Vladimir Leontjevich Komarov (1869–1945), senior editor of the thirty-volume *Flora SSSR* (*Flora of the USSR*) and president of the Academy of Sciences of the USSR from 1936 until his death. Komarov travelled widely in southeastern Siberia, the Heilong Jiang valley and northwestern China, reaching the Korean border in 1896. He revisited Korea the following year, returning to Russia via Jilin to the Siberian coast. Between 1895 and 1897 Komarov collected over 6 000 herbarium specimens (LE, K, GH, B, MO, W), including many new species.

British and Irish collectors.—Charles Maries (1851–1902) was the first of several eminent British professional plant hunters of this period employed by nursery companies to collect plants in Asia on their behalf. Between 1877 and 1879 James Veitch and Sons of Chelsea sponsored Maries to collect in Japan, Taiwan and mainland China on three journeys before he went to India to take up a permanent position. Although his gathering of plant material was most successful in Japan, he made significant collections around Ningbo Shi in Zhejiang, where Robert Fortune had collected some twenty years earlier (K, MO). Maries managed to penetrate further inland than Fortune, reaching up the Chang Jiang as far as Yichang in Hubei. Amongst his numerous introductions were over 500 new species, many of which were named in his honour.

By 1880 the British Consular Service had opened consulates in more than twenty Chinese cities, employing over 200 officers, and the Maritime Customs Service employed more than 600, two-thirds of them British (Mueggler, 2011). Many of these officials were passionate natural historians, using their spare time to explore their environs and hiring locals to collect for them when they were confined to office work. Among these were Augustus Raymond Margary (1846–1875), who was murdered whilst collecting orchids in southwestern Yunnan, and Irishman Augustine Henry (1857–1930), who in 1882 was stationed in Yichang Shi as a medical official and assistant for the Chinese Maritime Customs (O'Brian, 2011). Henry was following in Maries' footsteps, initially collecting plants as part of his official duties—he was required to report on vegetable products for import and export and on Chinese medicinal drugs. He developed a strong interest in botany, and in the following seven years avidly collected herbarium specimens in Yichang and the nearby mountains of western Hubei and Sichuan. Henry continued collecting after he was transferred to Taiwan and then to southern Yunnan, and during his 20 years of service in China sent back an enormous herbarium of more than 15 000 specimens to London (K, now widespread at BM, E, LE, MO, G, W), of which at least 500 represented new species. Henry became a leading authority on the Chinese upland flora, and during his time in China he became increasingly worried about the large-scale deforestation he witnessed. Henry was an early conservationist and in letters accompanying his specimens he repeatedly urged the Royal Botanic Gardens, Kew, to send a trained botanist properly to explore and inventory the region. Although Henry himself collected some seed, his best horticultural discoveries were introduced into cultivation by Ernest Wilson, the man sent following Henry's requests.

Amateur naturalist Antwerp E. Pratt (active around 1880–1915; Figure 12.1) travelled in China between 1887 and 1890, exploring the upper Chang Jiang, Sichuan and the mountains around Kangding. He was mainly interested in entomology but, when he passed through Yichang, Henry persuaded him to take with him a trained native collector who gathered some 500 herbarium specimens, many of which were new species (BM, E, G, K, LE, US).

Around 1897 the Qing court signed at treaty allowing the British access to the trading port of Tengyue, a centre of trade between Yunnan and Myanmar, which was then controlled by the British and known as Burma. A British

Figure 12.1 Antwerp E. Pratt alongside *Rhododendron faberi* subsp. *prattii*, ca. 1890 (image from the archive at the Royal Botanic Garden Edinburgh).

consulate and Chinese maritime customs office were soon established, and the way was paved for British explorers to travel into southwestern China via Myanmar. Ernest Henry Wilson (1876–1930) was the first collector to make use of this opportunity, travelling to China in 1899 as a consequence of Augustine Henry's plea. Wilson became hooked on China and would return three more times (1903–1905, 1906–1908 and 1909–1911) over the next 12 years in search of hardy ornamentals (Briggs, 1993). Having trained as a junior gardener at the Royal Botanic Gardens, Kew, and studied at the Royal Horticultural Society, Wilson was about to embark on full-time botanical studies at the Royal College of Science when he was nominated by the Royal Botanic Gardens, Kew, and offered a position with James Veitch and Sons to work as their plant collector in China. He jumped at the chance and began a career that would see him become one of the most successful of all professional plant collectors in China. Wilson's extensive botanical collections speak for themselves, but he was also a gifted author and speaker on plants, and his fame soon earned him the nickname "Chinese Wilson."

On his first journey Wilson travelled west via the Arnold Arboretum, Boston, USA, where he learnt the latest techniques for collecting, processing and shipping seeds and other living material safely. While in Boston he met Charles Sargent, a man who would play a significant role in his future expeditions and later career. Soon after his arrival in China, Wilson met up with Augustine Henry in Simao, southern Yunnan, and was briefed on the location of the legendary dove or handkerchief tree, *Davidia involucrata*, which he sought. In later years he would travel in Sichuan and Hubei collecting plants on behalf, first, of James Veitch and Sons and, on his later two trips, the Arnold Arboretum. Although he had one main aim on his first expedition—to find and collect seed of *D. involucrata*—Wilson was a prolific collector of herbs, trees and shrubs, introducing a huge number (possibly more than 2 000 species) of living plants and seed into cultivation in the west, and gathering more than 50 000 herbarium specimens (A, K, E, BM, CAS, F, P, MO). Wilson was the first of a generation of plant hunters bringing scientific rigor to their work. He was also an accomplished photographer, and his collection of over 700 glass plate photographs forms an important record of plants and their habitats at the turn of the nineteenth century. After further collecting trips in Taiwan, Korea, Japan and elsewhere, Wilson returned to the USA and took up the post of keeper of the Arnold Arboretum. Wilson had a strong English persona and, unlike his contemporaries, refused to learn the Chinese language or to adopt local customs or dress—much against the advice of Harry Veitch of James Veitch and Sons. Like Augustine Henry he was appalled by the deforestation he witnessed and horrified by man's destruction of his environment.

George Forrest (1873—1932; Figure 12.2) worked a the herbarium of the Royal Botanic Garden Edinburgh and was undoubtedly the most successful of all plant hunters in China. He amassed vast collections of living material and herbarium specimens on the seven long expeditions he undertook, mainly in western Yunnan (1904–1907, 1910–1911, 1912–1914, 1917–1919, 1921–1922, 1924–1925 and 1930–1932; McLean, 1997; 2004; Mueggler, 2011). Forrest was a methodical, dedicated and energetic collector with a keen eye for plants. Unlike his contemporaries, Forrest and his team of local collectors returned to the same areas on many occasions in an attempt to gather comprehensively all the plants from each area (an example of one of there camps is shown in Figure 12.3). This makes his herbarium collections particularly useful as they provide a detailed inventory of the areas in which he worked. Forrest's initiative in training local people to collect was handsomely rewarded, as he was able to accumulate specimens on a scale never seen before. Forrest built up a strong team with his local partner ZHAO Cheng-Zhang, and at one point had around 40 Naxi people engaged in small groups bringing plant material to him (Mueggler, 2011). Even after his death from heart failure in the field, his native collectors carried on and Forrest's final sponsor Henry McLaren (later Lord Aberconway) continued to pay them for the seed and specimens they sent to Bodnant Garden, Wales.

Like his contemporaries, Forrest was sponsored by commercial horticultural companies, in his case mostly the Liverpool cotton-broker Arthur Kiplin Bulley, founder of the nursery and seedsman's firm Bees Limited of Chester. As well as discovering numerous new species himself, he introduced many of the species that had earlier been collected as herbarium specimens by French missionaries. In total, Forrest collected over 31 000 herbarium specimens (E, PE, K, BM, MO, P, W, B, A) and introduced an unprecedented number of living plants into cultivation. These included 300 new species of *Rhododendron* and numerous species of *Camellia, Gentiana* (Gentianaceae), *Lilium* (Liliaceae), *Magnolia* and allied genera (Magnoliaceae) and *Primula* (Primulaceae), most of which were cultivated for the first time at the Royal Botanic Garden Edinburgh, and captured the interest of western gardeners. Unlike previous collectors, Forrest had a second set of his herbarium specimens set aside to be returned to China when there was a suitable herbarium to take them. His wish was fulfilled in the early 1980s when this duplicate set was sent to Beijing (PE).

Figure 12.2 George Forrest and his pet dog in Yunnan, pictured in front of his tent, with bird specimens hanging from a pole and piles of herbarium collections in the background (image from the Forrest collection in the archive at the Royal Botanic Garden Edinburgh).

Figure 12.3 "Camp Chopadola," one of Forrest's camps, on the eastern flank of the Yulong Xueshan, taken looking westwards from an altitude of 3 350 m in June 1906 (image from the Forrest collection in the archive at the Royal Botanic Garden Edinburgh). Note the herbarium specimens drying under rocks.

Francis (Frank) Kingdon Ward (1885–1958) was a multi-talented botanist, explorer, geographer and prolific author. Between 1909 and 1956 he travelled widely and extensively in eastern Asia on around 25 expeditions, especially in western China, southeastern Xizang, Myanmar and northeastern India, and probably collected plants in this area over a longer period than any other plant hunter (McLean, 1997). Kingdon Ward collected seeds, living plants and herbarium material (BM, NY, E, A, CGE, F, K, MO, PE), and wrote 25 books, including many popular works about his travels and the flora and geography of the lands he visited. His first plant-collecting expedition to China was in 1911, where, like Forrest, he was sponsored by Bulley. At times Wilson, Forrest and Kingdon Ward were in the field at the same time, rivaling each other for the most interesting discoveries, and jealously guarding their collecting territories.

At about the same time, but at the opposite end of the country, William Purdom (1880–1921) collected in northwestern China between 1909 and 1911. Purdom was another professional plant hunter sponsored by James Veitch and Sons and the Arnold Arboretum. He spent most of his time in Gansu, Nei Mongol, Shanxi and the Xizang border regions collecting seeds, living plants and herbarium specimens (B, E, K, GH, C, US). In 1914 he returned to collect in Gansu with Reginald John Farrer (1880–1920; Figure 12.4), and in 1915 was appointed to the Chinese Government Forestry Commission. Farrer was an impassioned plant hunter and accomplished botanical artist with a particular interest in alpine gardening. He is famed for his writings on plant hunting and rock gardening, although he made only one expedition to China, travelling by the Trans-Siberian railway via Beijing Shi (Illingworth & Routh, 1991). His expedition was cut short by the outbreak of the First World War, and many of his collections were lost in transit during his hurried return. Farrer also undertook one fateful expedition to northern Myanmar with Euan Hillhouse Methven Cox (1893–1977) in 1919--1920, during which they fell out and went their separate ways, with Farrer spending a miserable, lonely time on the Yunnan border where he died of dysentery. Cox introduced many ornamental plants from the Myanmar-China border from this trip, and wrote many books on plant collecting, including the classic *Plant Hunting in China* (Cox, 1945).

Figure 12.4 Reginald John Farrer, wearing Buddhist robes (image from the Farrer family collection in the archive at the Royal Botanic Garden Edinburgh).

The last two British botanical collectors in this section are Frank Ludlow (1885–1972) and Major George Sherriff (1898–1967), who made a unique and highly effective exploration team on several major expeditions to southeastern Xizang and the eastern Himalaya Shan, especially in Bhutan (Fletcher, 1975). Ludlow was an academic, schoolmaster and the naturalist of the team, whilst the dedicated amateur botanist Sherriff brought his disciplined military experience to the expedition logistics. They first met in 1929 in Kashi, when Ludlow was visiting from Srinagar, India, and between 1933 and 1949 travelled together throughout the Himalaya Shan from Kashmir to southwestern China, covering Bhutan and southeastern Xizang. Ludlow and Sherriff made thorough collections in the areas they visited, including seed, living plants and herbarium specimens (BM, E, MO). They were particularly interested in *Meconopsis* (Papaveraceae), *Primula* and *Rhododendron*, introducing several new species of these into cultivation. The standard of their collecting notes, herbarium specimens and photographic record (including both still images and ciné film) is outstanding, and they also pioneered the use of air transport for bringing live plants back to the UK. When China closed its doors to foreigners Ludlow and Sherriff, who at the time were working in Xizang, returned to the UK, where Ludlow devoted the rest of his life to working on their collections at the Natural History Museum, London.

Other European and American collectors.—Germany arrived relatively late on the botanical scene in China, with missionary Ernst Faber (1839–1899) being the first to collect plants, in 1865. He travelled widely in China and had the distinction of being the first botanist to climb the 3 099 m Emei Shan, Sichuan, one of China's four sacred Buddhist mountains. Faber collected many herbarium specimens which he sent to London (K, now also at LE, MO, B, C, W) and to Henry Hance in Huangpu. Fifty years later, in 1914, the German-born Austrian Camillo Karl Schneider (1876–1951) travelled to Yunnan and western Sichuan with his compatriot Heinrich von Handel-Mazzetti (1882–1940), sponsored by the Austro-Hungarian Dendrological Society. Whist they were away the world was thrown into turmoil by the outbreak of the First World War, and Schneider opted to leave China in 1915 and head for the USA. Schneider was a professional botanist specializing in trees, and in later years he worked at the Arnold Arboretum, studying his own collections (A, B, E, G, W, GH, K) and those of Ernest Wilson, and specializing in the genera *Berberis* (Berberidaceae) and *Syringa* (Oleaceae). Tragically his study on *Berberis* was destroyed in a British bombing raid on Berlin in 1943. Handel-Mazzetti was also a professional botanist but had wider interests extending to mosses and liverworts, and was an accomplished phytogeographer. He decided to stay in China and made extensive collections of herbarium specimens (W, H, E, A, G, MO, and others) in the Yunnan Sanjiangbingliu ("Three Parallel Rivers") area of Yunnan, Sichuan and southeastern Xizang. His collections of mosses and liverworts are particularly important as the first specialist bryophyte collections made in China, and he is considered a pioneer of research into Chinese cryptogams. Handel-Mazzetti recorded his travels in *Naturbilder aus Südwest China* (Handel-Mazzetti, 1927), which has been translated into English by Winstanley (1996) with an added biography.

Guiseppe Giraldi (1848–1901) became Italy's first botanical collector in China when he travelled to Shaanxi as a missionary in 1888. He was a keen amateur botanist and gathered over 5 000 herbarium specimens (FI, H, K. LE, W, BM, A) before his untimely death in the field. Giraldi also collected some seed and plants that he sent to the botanical garden in Florence.

The French continued their long involvement in the botanical exploration of China, again through missionaries with interests in natural history. François Ducloux (1864–1945) was sent to China in 1889, becoming head of the missionary school in Kunming Shi, Yunnan. In 1897 he began collecting herbarium specimens and took a special interest in ferns. Over the next fifteen years, before his departure in 1912, Ducloux sent back over 6 000 herbarium specimens to Paris (P, B, MO, NY). At the same time, missionary Jean-Théodore Monbeig (1875–1914) was also sent to China and stationed in Batang, southeastern Xizang. He too collected widely in the surrounding area, sending back herbarium specimens to Paris (P, K, E), before he was brutally murdered by Tibetan warrior-monks. The last French collector of this period was the colorful Prince Henri d'Orléans (1867–1901). Although born in England, Prince Henri was son of Prince Robert, Duke of Chartres, and grandson of the deposed "king of the French," Louis-Philippe. Sponsored by his father, Prince Henri left France in 1889 on a journey with Gabriel Bonvalot and Constant de Deken through Siberia to Thailand. From Siberia the party headed south to Xizang through Sichuan and Yunnan, and into Vietnam, arriving in Hanoi in 1890 (Bonvalot, 1982). The prince collected many herbarium specimens as well as geological and zoological specimens, all of which were sent to Paris (P). In 1895 he returned to China to explore the upper reaches of the Lancang Jiang, Nu Jiang and Ayeyarwady, with most of his collections being made in Yunnan.

Europe's "low countries" were represented at this time by the Dutchman Frank (Frans) Nicholas Meyer (1875–1918) and Belgian Joseph Hers (1884–1965). Meyer, an American of Dutch birth, was a professional plant hunter employed in 1905 by the USA Department of Agriculture to collect economically-useful plants. He made four major expeditions to central Asia including China, and collected vast quantities of seed and living plants which he sent back to the USA before he was drowned in the Chang Jiang in 1918 (Fairchild, 1921). Hers first travelled to China in 1905 as an interpreter to the Belgian Consul in Shanghai Shi, before becoming general secretary of the General Company of Chinese Railways and Tramways. He was primarily an administrator but also an amateur naturalist with an interest in woody plants. Although based in Henan, Hers travelled widely in remote areas of Gansu, Shaanxi, Henan and Jiangsu before leaving China in 1924. During his time in China he amassed a large herbarium collection (A, K, F, GH, NY), as well as collecting seed of many trees, which he sent to the National Botanic Garden of Belgium in Brussels and the Arnold Arboretum.

The Swedish professional botanist and plant collector Karl August Harald (Harry) Smith (1889–1971) had a huge impact on the exploration of China. Between 1921 and 1935 he undertook three major, wide-ranging expeditions in China, visiting the Beijing Xishan, Yunnan, Sichuan and Xizang, making large collections of herbarium specimens (UPS, GB, S, E, K, MO, W) and gathering seed, which he sent to Magnus Johnson Nurseries, Stockholm, and Gothenburg Botanic Garden. Also from Sweden came David Hummel (1893–1984), a medical doctor with the Sino-Swedish expedition of 1927–1931 under the leadership of Sven Heden and involving HAO Kin-Shen (HAO Jing-Sheng) from the Institute of Botany, Beijing. The expedition explored the Gobi of northern Sichuan, southern Gansu and Qinghai, with Hummel collecting many herbarium specimens and seed, as well as reptiles, fish and insects. In 1930 Hummel also visited Sichuan and Gansu.

The last but certainly not the least botanical explorer in this section is the Austrian-born Joseph Francis Charles Rock (1884–1962; Figure 12.5), who became a naturalized citizen of the USA in 1913. This eccentric, hot tempered, multi-talented man was a renowned explorer, geographer, plant hunter, linguist and photographer, even attending to the medical needs of his fellow villagers in remote northwestern Yunnan (Mueggler, 2011). Between 1922 and 1949 Rock spent the majority of his time in China, mainly in Gansu, Sichuan, Yunnan and southeastern Xizang. He was supported by various American institutes and collected vast quantities of herbarium specimens and seed (A, BM, E, K, P, US, PH, HPH), especially of *Rhododendron* spp. In 1922, Rock took up residence in Lijiang (northwestern Yunnan), using it as a base to explore nearby mountain ranges (including the Yulong Xueshan and Daxue Shan) and further north into the Muli Zangzu Zizhixian of Sichuan. In two years he collected some 80 000 specimens and seed of many ornamental plants, as well as pursuing his interests in researching the language and history of the indigenous people. After a brief trip to the USA, Rock returned to China in 1925–1927, this time to explore southeastern Xizang, Qinghai and Sichuan, again collecting thousands of herbarium specimens and hundreds of batches of seed. He returned to the USA only to find that he had to go straight back to China leading the National Geographical Society's Southwest Expedition to the Muli Zangzu Zizhixian and northwestern Yunnan (1928–1930). Over the next three years Rock returned twice to northwestern Yunnan (1930–1931 and 1932–1933), again making thousands of collections. The second of these was his last collecting trip, as he dedicated his final stay in Lijiang (1934–1938) to the study of the literature and language of the local Naxi

Figure 12.5 Joseph Francis Charles Rock with his field team and guards in southwestern China, 1923—1924 (image from the Joseph Rock collection in the archive at the Royal Botanic Garden Edinburgh).

people and their Dongba culture. Like Wilson and Forrest, his photographic glass plates, exposed and developed in the field, provide a rich archive of the people, plants and landscapes of China at that time (Goodman, 2006). Other Americans such as Floyd McClure and Albert Steward also feature in this period as members of Chinese botanical institutes, which are discussed later in the chapter.

Japanese collectors and the exploration of Taiwan.— While Robert Fortune was the first botanist to collect plants in Taiwan, in 1854, he only had a few days on the island while his ship was in port. Fortune was followed by Charles Wilford (who died in 1893) from the Herbarium of the Royal Botanic Gardens, Kew, in 1858, Robert Swinhoe (1836–1877, K) British Consul in Formosa from 1863-1864, and the aforementioned Augustine Henry between 1892 and 1895. Henry made extensive collections in southern Taiwan and in 1896 published a list of indigenous plants of Taiwan which included over 1 300 species recorded since Fortune's time. The French missionary Père Urbain Jean Faurie (1847–1915) was stationed in Japan and, although better known for his collections from Korea, visited Taiwan between 1882 and 1894, and again in 1896 (P, E). The first collections from Taiwan by a Japanese botanist were made in 1874 by Manjiro Kurita (ca. 1833–1889), a general naturalist who visited and collected plants in the southern part of the island. In his later publications he recorded 101 species of plants from Taiwan, sending his specimens to the Imperial Museum in Tokyo (TNS, TI).

The Japanese occupied Taiwan from after the end of the 1894–1895 Sino-Japanese War until the end of the Second World War in 1945. During this period many Japanese botanists collected extensively on the island, among them Bunzo Hayata (1874–1934), Takiya Kawakami (1871–1915), Syuniti (Shun'ichi) Sasaki (1888–1960) , Yoshimatsu Yamamoto (1893–1947), Genkei Masamune (1899–1993), Ryozo Kanehira (1882–1948) and Yushun Kudo (1887–1932; see Ohashi, 2009 for a detailed discussion of the history of botanical exploration of Taiwan). In addition,

Masao Kitagawa (1909–1995) undertook a botanical expedition to northwestern China between 1933 and 1936 (TI), and visited Hebei in 1940. Hisao Migo (1900–1985) made several collecting trips to southern Jiangsu and northern Zhejiang between 1933 and 1945 whilst employed by the Shanghai Science Institute during the Japanese occupation of Shanghai (TI).

12.3 Exploration by Chinese botanists

This period may be designated as the modern period of botanical research in China, in which Chinese researchers themselves started working on their own flora using scientific methodologies. In the following paragraphs we recognise two major stages: initiation and creation (ca. 1920–1949), and rapid development (1950–1990s).

12.3.1 Initiation and creation, ca. 1920–1949

The first Chinese botanist to collect herbarium material systematically on a large scale was TSOONG Kuan-Kwang (ZHONG Guan-Guang; 1868–1940; Figure 12.6A), a professor at Beijing University (then the National Peking University), China's first modern university. ZHONG was a pioneer of modern botanical research in China, beginning his collecting in 1918, just six years after SUN Yat-Sen founded the Republic of China. Initally, his fieldwork focused on Fujian and Guangdong, but he then extended his exploration into Anhui, Jiangxi, Hubei, Hunan, Yunnan and Zhejiang, and continued collecting up until 1936. ZHONG amassed over 25 000 specimens (PE, NAS) and founded China's first herbarium, at Beijing University. He is commemorated in the generic names *Tsoongia* (Verbenaceae) and *Tsoongiodendron* (now synonymized under *Michelia*; Magnoliaceae). However, the period of modern botany in China was really established a few years later, in the 1920s, when four students—CHIEN Sung-Shu (QIAN Chong-Shu), HU Hsen-Hsu (HU Xian-Su), CHUN Woon-Young (CHEN Huan-Yung) and ZHONG Xin-Xuan—returned to China after studying abroad. These four "founders of plant taxonomy" in China were all trained at botanical institutes across the USA between 1916 and1925, notably at Harvard University, Boston. Although the main interest of the latter, ZHONG Xin-Xuan, was fungi, he organised and supervised many general botanical expeditions in Fujian and Hubei while teaching at Xiamen and Wuhan Universities.

As opportunities for systematic botanical research in China were very limited at that time, these pioneering scientists began their careers by teaching botany at various universities around China. This gave them the chance to inspire, educate and train the next generation, passing on the knowledge and skills gained during their studies abroad. In this way they were responsible for nurturing a host of prominent students, many of them becoming specialists in plant taxonomy. Working in university departments also gave these early Chinese botanists the opportunity for botanical exploration and investigation. The collections they made, and the people they trained, layed down a solid foundation for the development of modern botanical science in China. It took a few more years before the conditions were right for the foundation of a number of herbaria and scientific institutes which would become the leading centers of botanical research in China (Chen *et al.*, 1993; see also Chapter 13). Here, we describe the botanical exploration of the time under the major institutes that supported it, even during the difficult and restrictive times of the Second World War.

The Biological Laboratory of the Science Society of China, Nanjing.—The first research institution specializing in botanical inventory in China was the Biological Laboratory of the Science Society of China, established at Nanjing in 1922. Its staff, many working concurrently for the Biology Department of the National Southeastern University, carried out considerable botanical exploration in Anhui, Fujian, Jiangsu, Sichuan, Yunnan and Zhejiang. Among the more important collectors were HU Hsen-Hsu (HU Xian-Su), CHIEN Sung-Shu (QIAN Chong-Shu), CHING Ren-Chang (QIN Ren-Chang), FANG Wen-Pei, CHENG Wan-Chun (ZHENG Wan-Jun), TSO Ching-Lieh (ZUO Jing-Lie) and CHEN Mou.

HU Hsen-Hsu (HU Xian-Su; 1894–1968; Figure 12.6B) studied first at the University of California at Berkeley, USA, from 1912 to 1916, before graduating with a doctorate from Harvard University in 1925. Along with his colleagues at the Science Society of China, HU was a key leader of this first biological research institute in China and played an important role in the founding of the Botanical Society of China. During his career he held professorships and directorships at the Southeastern University, Fan Memorial Institute of Biology, Zhongzheng University and the Institute of Botany, Beijing. Hu initiated large-scale inventories and surveys of the flora of China and personally collected extensively during two expeditions to Fujian, Jianxi and Zhejiang, in 1919–1920. He authored of an enormous number of scientific papers on the flora of China, and played a key part in identifying and naming living specimens in the historic discovery of the "living fossil" *Metasequoia glyptostroboides* (dawn redwood; Taxodiaceae) in China in 1948. The genus *Huodendron* (Styracaceae) is named in his honour.

CHIEN Sung-Shu (QIAN Chong-Shu; 1883–1965; Figure 12.6C) studied for six years at Chicago and Harvard Universities (1910–1916), returning to teach in colleges and universities in China before his appointment in 1927 as director of the Botany Department in the Biology Institute of the Science Society of China. In 1950 CHIEN became the first director of the newly-formed Institute of Botany of the Chinese Academy of Sciences (CAS). Between 1906 and 1938 he collected plants with CHUN Woon-Young (CHEN Huan-Yung) and CHING Ren-Chang (QIN Ren-Chang) in Hubei, and with CHENG Wan-Chun (ZHENG Wan-Jun) in Sichuan. Unfortunately CHIEN's ca. 10 000 specimens were all destroyed during the Second World War.

FANG Wen-Pei (1899–1983; Figure 12.6D) graduated from the Southeastern University before travelling to the UK for postgradute research on a revision of the genus *Acer* (Aceraceae) for his doctorate at the University of Edinburgh. During his time in Edinburgh he also studied *Rhododendron* taxonomy at the Royal Botanic Garden, and became an authority on the genus. In 1937 he returned to take up a professorship at Sichuan University, and collected widely in Sichuan, making ca. 10 000 collections as well as organizing several important expeditions in the province.

CHENG Wan-Chun (ZHENG Wan-Jun; 1904–1983; Figure 12.6E) was a famous dendrologist and academician of the CAS. From 1926 to 1939 he worked at the Biological Laboratory of the Science Society of China at Nanjing and made several botanical expeditions to Zhejiang (1926), Sichuan (1930–1931, 1936 and 1938) and Yunnan (1944; KUN, YUKU).

TSO Ching-Lieh (ZUO Jing-Lie; active from 1929 to 1935) also graduated from the Southeastern University and was a visiting scholar at the Royal Botanic Garden Edinburgh from 1935 to 1936, at the same time as FANG Wen-Pei. In 1926, under the auspices of the Biological Laboratory of the Science Society of China, TSO and others made extensive collections around Nanjing and in other parts of Jiangsu. From 1929 to 1933 he was employed at the Agriculture and Forestry Institute, Sun Yat-sen University, during which time he collected in Guangdong, Guangxi, Hainan and Hong Kong, making ca. 3 000 collections from Nanjing and Jiangsu, and 3 300 collections from Guangdong, Guangxi, Hainan and Hong Kong (IBSC).

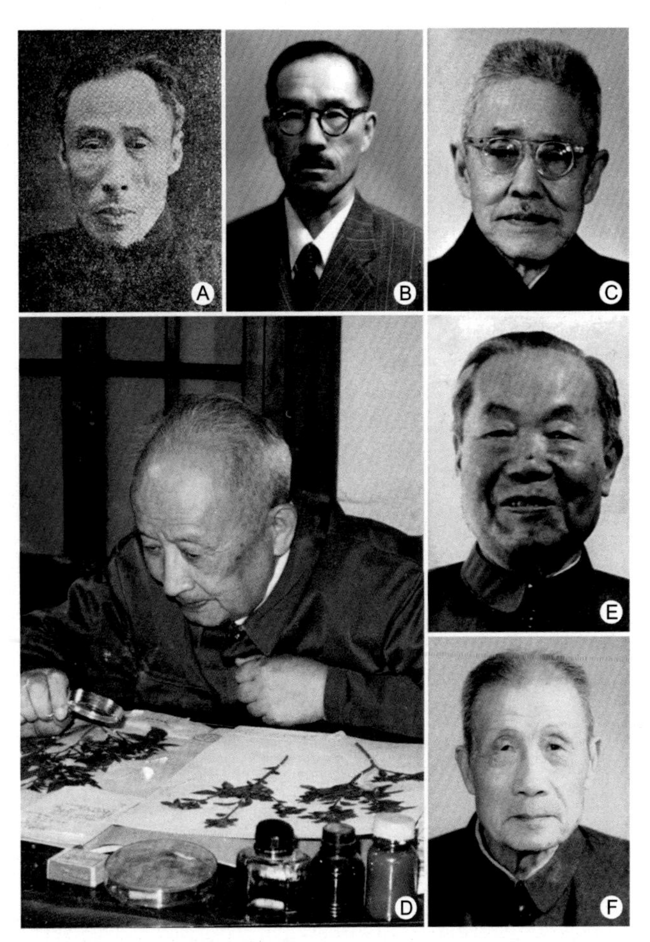

Figure 12.6 A, TSOONG Kuan-Kwang (ZHONG Guan-Guang). B, HU Hsen-Hsu (HU Xian-Su). C, CHIEN Sung-Shu (QIAN Chong-Shu). D, FANG Wen-Pei. E, CHENG Wan-Chun (ZHENG Wan-Jun). F, CHING Ren-Chang (QIN Ren-Chang).

The last of the important collectors associated with Nanjing is CHEN Mou (1903–1935), who was an assistant lecturer at the Forestry Department of Agriculture College, National Central University. In 1934–1935 he participated in a joint expedition to Yunnan and, although more than 3 000 collections of vascular plants were taken, CHEN died on the way back to Kunming Shi and his collections were all destroyed during the Second World War.

The Natural History Museum, Academia Sinica, Nanjing.—The Botanical Section of the Natural History Museum of the Academia Sinica, founded in 1928 at Nanjing, was directed successively by CHING Ren-Chang (QIN Ren-Chang; Figure 12.6F) and TSIANG Ying (JIANG Ying; Figure 12.7A). The museum sponsored several expeditions to Guangxi, Guizhou, Jiangxi and Sichuan, and of special note are the expeditions CHING made to the mysterious Shiwan Dashan, a mountain range extending from east to west along the Guangxi-Guangdong border, and the extensive collections TSIANG made in Guizhou and Yunnan.

CHING Ren-Chang (QIN Ren-Chang; 1898–1986; Figure 12.6F) is considered to be the "father" of Chinese pteridology. He graduated from the University of Nanjing in 1925 before taking a junior lecturship at the Southeastern University. Between 1929 and 1932 he studied in the herbaria at the University of Copenhagen, Denmark, and the Royal Botanic Gardens, Kew. On his return to China he was appointed curator of the herbarium at the Fan Memorial Institute of Biology, before becoming a founder and the first director of the Lushan Botanical Garden in 1934. In 1954 he was appointed a professor at the Institute of Botany, Beijing. CHING was a prolific author of scientific papers on the classification of ferns, and between 1923 and 1939 collected extensively in many provinces, including Gansu, Ningxia and Zhejiang (1923), Fujian (1924), southern Anhui, Jiangsu and Zhejiang (1925), Guangxi (1928), Jiangxi (1928), Sichuan (1937) and Yunnan (1939–1940).

TSIANG Ying (JIANG Ying; 1898–1982; Figure 12.7A) graduated from the Forestry Department of the University of Nanjing in 1925, and taught in the Department of Biology, Sun Yat-sen University, Guangzhou, between 1928 and 1930. After three years as herbarium curator at the Natural History Museum, Nanjing, he joined the Agriculture and Forestry Institute. After the Second World War, TSIANG moved to the Forestry Experimental Institute of Taiwan and was curator of Taiwan Botanical Garden, and in 1949 he was given concurrent professorships at the South China Agriculture University and the South China Institute of Botany. Between 1928 and 1948 he collected extensively in Guangdong, Guangxi and Yunnan, and made some short collecting tours to Hong Kong and elsewhere (making ca. 5 600 collections in total). TSIANG is the author of many taxonomic papers, especially on Apocynaceae and Asclepiadaceae.

The Fan Memorial Institute of Biology, Beijing Shi.— The botanical division of the Fan Memorial Institute of Biology was initially headed by HU Hsen-Hsu (HU Xian-Su), who was promoted to director in 1932. Funded by the China Foundation for the Promotion of Education and Culture and the Shang Chih Society, HU was able to build the institute's herbarium into the largest in China. In 1934 HU established the Lushan Arboretum and Lushan Botanical Garden, Jiangxi, firstly in co-operation with the local government, and then with the Yunnan Botanical Institute at Kunming in 1939. HU made a plan for botanical research across the whole of China and organized a series of successful expeditions to Jiangxi, Jilin, Shanxi, Sichuan and Yunnan, including some previously unknown areas. He was an inspirational and energetic teacher whose students included a number of prominent botanists, among them CHEN Feng-Hwai (CHEN Feng-Huai), FANG Wen-Pei, TSAI Hse-Tao (CAI Xi-Tao), YÜ Te-Tsun (YU De-Jun), WANG Chi-Wu (WANG Qi-Wu), TANG Tsin (TANG Jin), WANG Fa-Tsuan (WANG Fa-Zan) and FENG Kuo-Mei (FENG Guo-Mei). In 1950 the Fan Memorial Intitute of Biology was merged with the Botanical Insitute of Beijing to form the Institute of Plant Taxonomy, Academia Sinica (now the Institute of Botany, CAS).

CHEN Feng-Hwai (CHEN Feng-Huai; 1900–1993; Figure 12.7B), educated at the Southeastern University, joined the Fan Memorial Institute of Biology in 1931. Between 1929 and 1931 he made three botanical excursions to northeastern Jilin, especially Jingpo Hu and its vicinity. CHEN was a visiting scholar at the Royal Botanic Garden Edinburgh, 1934–1936, working on Primulaceae under the guidance of William Wright Smith. After returning to China in 1936 he worked actively in several botanical gardens and institutions until his retirement in 1983. He was a founder of Lushan Botanical Garden, Nanjing Botanical Garden, Wuhan Botanical Garden and South China Botanical Garden, and was the first director of the latter three gardens.

TSAI Hse-Tao (CAI Xi-Tao; 1911–1981; Figure 12.7C) was a largely self-taught botanist who joined the Fan Memorial Institute of Biology in 1930, and from 1931 to 1959 made extensive explorations in Yunnan collecting vast numbers of specimens (13 627 collections; PE, KUN, IBSC, GH). He founded the Xishuangbanna Tropical Botanical Garden in 1959 and was its first director. His achievements in tropical agriculture and forestry are particularly remarkable, and he was one of the chief promoters of the introduction of improved varieties of tobacco (*Nicotiana tabacum;* Solanaceae) and the pará rubber tree (*Hevea brasiliensis;* Euphorbiaceae) to Yunnan.

Figure 12.7 A, TSIANG Ying (JIANG Ying). B, CHEN Feng-Hwai (CHEN Feng-Huai). C, TSAI Hse-Tao (CAI Xi-Tao). D, YÜ Te-Tsun (YU De-Jun). E, WANG Chi-Wu (WANG Qi-Wu). F, TANG Tsin (TANG Jin).

YÜ Te-Tsun (YU De-Jun; 1908–1986; Figure 12.7D) was a well-known and influential botanist and horticulturist, and academician of the CAS. YÜ graduated from the National Peiping (now known as Beijing) Normal University and joined the Fan Memorial Institute of Biology in 1931. In 1932 he was sent to Sichuan to take charge of the botany section of the Academy of Science of West China, and collected in western Sichuan for three years. Between 1937 and 1939 he collected extensively in Yunnan, especially in mountainous zones, collecting not only specimens, but also seeds and bulbs, some of which were sent to the Arnold Arboretum and the Royal Botanic Garden Edinburgh (7 085 colletions were made from Sichuan, 11 952 from Yunnan; PE, KUN, IBSC, A, E, K). In 1947 YÜ travelled to the UK to study Rosaceae at the Royal Botanic Garden Edinburgh, returning to China in 1950. He was the founder and first director of the Beijing Botanical Garden, editor-in-chief of *Flora Reipublicae Popularis Sinicae* (see Chapter 14) and author of many systematic papers, especially on Rosaceae.

WANG Chi-Wu (WANG Qi-Wu; 1913–1987; Figure 12.7E) graduated from Qinghua University and joined the Fan Memorial Institute of Biology in 1933. In 1934 he made several botanical excursions to Dongling, Xiling and other places in Hebei. From 1935 to 1937 he collected extensively in the alpine regions of northwestern Yunnan and adjacent southwestern Sichuan. He extended his collecting area to tropical Yunnan between 1936 and 1937, exploring the botanically-unknown forests and jungles of the malaria-ridden south (making 1 600 collections from Hebei and 24 923 from Yunnan; PE, IBSC, KUN, A, K). In 1945 he emigrated to the USA, and later was appointed a professor at the Universities of Florida and Idaho.

TANG Tsin (TANG Jin; 1897–1984; Figure 12.7F) graduated from Peiping (later known as Beijing) Agriculture University in 1926, and joined the Fan Memorial Institute of Biology two years later. He was a visiting scholar at the

Royal Botanic Gardens, Kew, from 1935 to 1938, before returning to teach at the Yixing School of Fine Art. He returned to the Fan Memorial Institute of Biology as a professor and expert on the Liliaceae and Cyperacae in 1947. Between 1927 and 1928 TANG made collecting trips in Hebei, Jiangsu and Zhejiang (Tianmu Shan), and in 1929 he worked with HSIA Wei-Ying (XIA Wei-Ying; Figure 12.8A) on an intensive survey in Shanxi covering about three-fifths of the province in a period of five months (including at least 8 000 collections; PE). HSIA (1895–1987) was a botanical historian, firstly at the Institute of Botany of the National Academy of Peiping, and later at the Fan Memorial Institute of Biology. Between 1927 and 1933 he made several collecting trips to Gansu, Hebei, Nei Mongol, Shaanxi, Shanxi and Zhejiang, making ca. 800 collections.

WANG Fa-Tsuan (WANG Fa-Zan; 1899–1985; Figure 12.8B) graduated from the Southeastern University in 1927 and joined the Fan Memorial Institute of Biology in 1929. Like several of his contemporaries he travelled to Europe, where he was a visiting scholar at the Royal Botanic Gardens, Kew, and other herbaria between 1935 and 1938. Whilst abroad he developed his expertise on the families Cyperaceae, Liliaceae and Orchidaceae. On his return to China he worked at the Yunnan Agriculture and Forestry Institute and the Agriculture Department of Fudan University, before returning to the Fan Memorial Institute of Biology in 1949. In 1929 he started collecting plant specimens in Hebei. In 1930, he collected in Sichuan with TANG Tsin (TANG Jin), and between 1939 and 1941 he worked with LIU Ying to make several collecting trips around Kunming Shi (ca. 1 000 collections from Sichuan and 1 893 from Yunnan, PE, KUN).

The last alumnus of the Fan Memorial Insitute mentioned here is FENG Kuo-Mei (FENG Guo-Mei; 1917–2007; Figure 12.8C), who became a professor at the Kunming Institute of Botany. He started his career at Lushan Botanical Garden in 1936, and two years later moved with CHING Ren-Chang (QIN Ren-Chang) to Lijiang, Yunnan, because of the disruption caused by the Second World War. FENG collected continuously and extensively in Yunnan until 1962, developing a broad range of knowledge on the flora of the province, especially the genus *Rhododendron* and alpine plants (making at least 7 600 collections; KUN, PE).

The Institute of Botany, Sun Yat-sen University, Guangzhou.—The founder and first directory of the Agriculture and Forestry Institute (now South China Institute of Botany) and Institute of Botany at Sun Yat-Sen University, CHUN Woon-Young (CHEN Huan-Yung; 1890–1971; Figure 12.8D), had a definite purpose: to survey the flora of Guangdong and Hainan, and to publish a Flora of Guangdong. Botanical exploration work towards this had already started in 1927 when CHUN was appointed a professor at the university's Biology Department, but after the founding of the Institute a systematic survey of the flora of Guangdong, Hainan and part of Guangxi was initated and lasted for five years until the Japanese invaded Guangzhou in 1938. No less than 109 expeditions were undertaken, resulting in a total of 31 832 herbarium collections from Guangdong and Hainan. In addition, at almost the same time (1928–1934), SIN, S.S. (XIN Shu-Zhi), then a professor in the Biology Department of Sun Yat-Sen University, made four expeditions to Guangxi and collected extensively in the Yao Shan region along the Guangdong-Guangxi border.

CHUN Woon-Young (CHEN Huan-Yung) graduated from Harvard University in 1919 and on returning to China undertook a nine-month expedition to Hainan. Unfortunately all the collections from this botanically-unexplored island were destroyed by a fire in a warehouse in Shanghai Shi before they could be studied. Between 1927 and 1944 he collected in Guangdong, Guangxi and Hong Kong (making 7 567 collections; IBSC). CHUN was the author of many systematic papers, especially on Fagaceae, Gesneriaceae, Hydrangeaceae and Magnoliaceae, and was one of the first editors-in-chief of *Flora Reipublicae Popularis Sinicae*. He is commemorated in the genera *Chunechites* (now a synonym of *Urceola*; Apocynaceae), *Chunia* (Hamamelidaceae), *Chuniophoenix* (Arecaceae), and *Woonyoungia* (Magnoliaceae).

The Agriculture and Forestry Institute at Sun Yat-Sen University recruited an effective team of professional plant collectors and technicians to help with the extensive surveys and collections that were needed for the Flora of Guangdong project. This team included CHEN Nian-Qu (CHEN Nian-Ju; active from 1929 to 1932), who collected extensively in Guangdong, Hainan and Hong Kong, amassing ca. 4 800 collections (SCIB, PE), LIANG Xiang-Ri (LIANG Hoeng-Yip; active 1931–1954), who made large collections in Guangdong, Guangxi and Hainan (ca. 10 000

Figure 12.8 A, HSIA Wei-Ying (XIA Wei-Ying). B, WANG Fa-Tsuan (WANG Fa-Zan). C, FENG Kuo-Mei (FENG Guo-Mei). D, CHUN Woon-Young (CHEN Huan-Yung). E, CHEN, S.H. (CHEN Shao-Qing).

collections; IBSC), WANG Chi (HUANG Zhi; active 1929–1958), who made extensive collections in Guangdong, Guangxi and Hong Kong, gathering over 14 000 herbarium collecitons (IBSC), and KO Sik-Pang (GAO Xi-Peng; active 1930–1940), who collected widely in Guangdong, Guangxi and Hainan, and also undertook a short collecting tour to Yunnan (totalling ca. 6 300 collections; IBSC). Many new genera, including *Pseudochirita* (Gesneriaceae), *Wenchengia* (Lamiaceae) and *Zenia* (Fabaceae), were collected by the latter. Lastly, CHEN, S.H. (CHEN Shao-Qing; 1911–1997; Figure 12.8E) also started as a plant collector and was later appointed a senior engineer at the South China Institute of Botany. From 1933 to 1934 he accompanied TSIANG Ying (JIANG Ying) collecting in Yunnan, and after joining the Agriculture and Forestry Institute of Sun Yat-Sen University in 1934 he collected continuously in Guangdong, Guangxi, Hainan and Hunan until his retirement in 1880 (making 18 727 collections; IBSC, PE). He was very experienced in field botany and had the reputation of being a "walking encyclopedia" of plant identification.

HOW Foon-Chew (HOU Kuan-Zhao; 1908–1959; Figure 12.8F) graduated from Sun Yat-Sen University before joining the South China Institute of Botany. Between 1933 and 1957 he collected extensively in Guangdong, Hainan and Hong Kong, and made a short collecting tour to Yunnan in 1940 (taking at least 4 700 collections in total; IBSC, PE). HOW is the editor of *Flora of Guangzhou* (Hou, 1956) and author of *A Dictionary of The Families and Genera of Chinese Seed Plants* (Hou, 1958) and many papers, especially on Fabaceae, Rhizophoraceae and Rubiaceae. SIN, S.S. (XIN Shu-Zhi; 1894–1977; Figure 12.9A) studied at Berlin University, Germany, before his appointment as a professor and director of the Department of Biology of Sun Yat-Sen University, 1928–1932. From 1950 until his retirement this well-known biologist and agricultural historian was a professor and dean of the Northwest Agriculture College. Between 1928 and 1929 he made four expeditions to Dayao Shan and Daming Shan along the Guangxi-Guangdong border (making ca. 2 700 collections; IBSC, SYS).

The Institute of Botany, National Academy of Beijing.— The Institute of Botany of the National Academy of Beijing (founded as the National Academy of Peiping) supported an extensive programme of fieldwork in northwestern and northern China in the 1930s, including Gansu, Hubei, Jilin, Ningxia, Qinghai and Shanxi. The more important collectors on these expeditions were LIOU Tchen-Ngo (LIU Shen-E), HAO Kin-Shen (HAO Jing-Sheng), HSIA Wei-Ying (XIA Wei-Ying; described above), KUNG Hsien-Wu (KONG Xian-Wu), WANG Zuo-Bin (WANG Tso-Ping) and LIU Chi-Mon (LIU Ji-Meng).

LIOU Tchen-Ngo (LIU Shen-E; 1897–1975; Figure 12.9B) was a plant geographer who studied in France from 1911 to 1929, eventually graduating with a doctorate. On his return he was appointed director and professor of the Institute of Botany, National Academy of Peiping, and from 1949 onwards was professor at the Institute of Forestry and Soil, CAS. In 1930 LIOU participated in a Sino-French Academic Expedition to Xinjiang, and in 1931 he undertook a long expedition from Hebei via Gansu, Xinjiang, Kashmir and Pakistan, finally arriving in northeastern India in 1933. LIOU collected over 4 500 plant specimens, but all were lost in transit when sent back from India. In 1938 he joined TSOONG Kuan-Kwang (ZHONG Guan-Guang) collecting on Emei Shan and in many other localities in Sichuan, and between 1941 and 1949 he collected widely in Yunnan (6 238 collections).

HAO Kin-Shen (HAO Jing-Sheng; 1903–1957; Figure 12.9C) was an undergraduate student at Beijing University when he joined David Hummel on Sven Hedin's Sino-Swedish Expedition (1927–1931) and conducted an extensive botanical tour in northern Sichuan, southern Gansu and Qinghai, as far northwest as Qinghai Hu. After graduating in 1931 he continued his field exploration, collecting in Henan and Shaanxi in 1932. Between 1933 and 1937, HAO studied forestry in Germany and was awarded his doctorate. KUNG Hsien-Wu (KONG Xian-Wu; 1897–1984; Figure 12.9D) worked as a botanist at the Institute of Botany of the National Academy of Peiping from 1931 to 1937, during which time he made botanical excursions to northern Hebei and eastern Jilin (1931) and along the Qin Ling in Shaanxi (1933–1937; totaling ca. 6 000 collections).

WANG Zuo-Bin (WANG Tso-Ping; 1908–1989; Figure 12.9E) and LIU Chi-Mon (LIU Ji-Meng; 1909–1993) were professional plant collectors at the Institute of Botany, National Academy of Peiping. Between 1933 and 1934 WANG made several collecting expeditions in Hebei, Nei Mongol, Shaanxi and northern Shanxi, extending his collecting in 1935 into southern Gansu, Qinghai and northern Sichuan, and venturing into western Hubei in 1939 (PE). From 1934 LIU collected extensively in western Hebei, and on Taihang Shan and Heng Shan in Shanxi, and in 1939 he made a collecting trip to Funiu Shan, Henan. In 1949 LIU joined the Beijing Botanical Garden as a technician.

Figure 12.9 A, HOW Foon-Chew (HOU Kuan-Zhao). B, SIN, S.S. (XIN Shu-Zhi. C, LIOU Tchen-Ngo (LIU Shen-E). D, HAO Kin-Shen (HAO Jing-Sheng). E, KUNG Hsien-Wu (KONG Xian-Wu).

The herbarium at Lingnan University, Guangzhou.—In the first five years following its establishment in 1916, the herbarium of Lingnan University, Guangzhou organised just a few short collecting expeditions in the Guangzhou area. It was only after the American agriculturalist Floyd Alonzo McClure (1897–1970) joined the herbarium in 1921 that more extensive explorations were made. McClure learnt Chinese and then travelled to China in 1920 for botanical studies with a particular interest in bamboos. From 1921 to 1927 McClure, with Lingnan's professional collector TSANG Wai-Tak (ZENG Huai-De; active 1928–1937) made three expeditions to Hainan, and for the next twenty years TSANG went on to collect extensively in Guangdong, Guangxi, Hainan and Hunan, also visiting Vietnam many times (taking at least 31 000 collections; SYS, IBSC). In 1924 McClure and his assistant FENG Qin (FUNG Hom; 1896–1989) started systematically to collect living bamboo speciemens, and established a special bamboo garden in the university, which had over 600 varieties in cultivation. McClure was responsible for importing many bamboo species into the USA, from where they have been distributed across the world. From 1930 to 1940 FENG made many botanical excursions in Guangdong and Hong Kong (taking ca. 2 900 collections; SYS).

The herbarium of the University of Nanjing.—The herbarium at the University of Nanjing was established in 1922, and in the early years fieldwork was confined to the vicinities of Nanjing and other parts of Jiangsu. Albert Newton Steward (1897–1959), an American graduate in agriculture, joined the university in 1921, and over the next decade made several collecting trips to Huangshan in Anhui and Lushan in Jiangxi, with CHIAO Chi-Yuan (JIAO Qi-Yuan; 1901–1968). In 1930, CHIAO explored extensively the hill districts of central and western Shandong. Steward returned to the USA in the late 1920s to complete his doctorate at Harvard University, but returned to China and in the early 1930s explored Guangxi, Guizhou and Sichuan with CHIAO and their assistants.

12.3.2 Rapid development, 1950-1990s

The founding of the People's Republic of China in 1949 ended the long chaos caused by the Second World War and Chinese Civil War. Under a climate of political unification, and steady improvement in national economic growth, developments in science and technology in China greatly accelerated. Institutions, universities and colleges with biological science departments were established in every province, and many nationwide programs on the study of natural resources were undertaken. All the previously politically inaccessible areas were now open to exploration and Chinese botanical fieldwork entered a "golden age."

From 1952 to 1966 the Institute of Botany (Beijing Shi) organized several collecting trips to Guizhou, Sichuan and Xizang. In Xizang, TSOONG Pu Chiu (ZHONG Bu-Qiu; 1900–1981; Figure 12.9F), CHANG Yong-Tian (ZHANG Yong-Tian; born in 1936; Figure 12.10A), LANG Kai-

Figure 12.10 A, WANG Zuo-Bin (WANG Tso-Ping). B, TSOONG Pu-Chiu (ZHONG Bu-Qiu). C, CHANG Yong-Tian (ZHANG Yong-Tian). D, LANG Kai-Yong (LANG Kai-Yung). E, YING Jun-Sheng (YING Tsun-Shen). F, HONG De-Yuan. G, LI Guo Feng.

Yong (LANG Kai-Yung; born 1937; Figure 12.10B), YING Jun-Sheng (YING Tsun-Shen; born 1934; Figure 12.10C), HONG De-Yuan (born 1936; Figure 12.10D) and FU Guo-Xun (born 1932) collected in Bomi, Cona, Lhasa Shi, Nyalam, Rongpu Si, Yadong, Yigong, Zhangmu and other sites near Qomolangma Feng (Mount Everest), gathering about 13 000 collections (PE). In Guizhou, CHANG Yong-Tian (ZHANG Yong-Tian) and ZHANG Zhi-Song (born 1932) collected on Leigong Shan, Fanjing Shan and various other places, gathering 22 000 collections (PE, HGAS).

FANG Wen-Pei, professor of the Biology Department of Sichuan University, supervised a series of botanical expeditions to various parts of Sichuan, making more than 36 000 collections (PE, SZ, BISC). The most important collectors on these expeditions included HE Zhu (1924–1991), LI Guo Feng (born 1935; Figure 12.10E),TSIANG Hsing-Ling (JIANG Xing-Lin; born 1927), LI Xin (born 1928; Figure 12.10F), SONG Zi-Pu (SOONG Tze-Pu; born 1925) and XIONG Ji-Hua (born 1923).

In northwestern China, the foundation of the Xinjiang Institute of Biology, Pedology and Desert Research in 1958 and the Northwest Plateau Institute of Biology in 1962 at Xining, both affiliated to the Academia Sinica, enhanced the investigation of this vast area. SHEN Kuan-Mien (SHEN Guan-Mian; born 1935), LIU Guo-Jun (born 1933), YANG Chang-You (born 1928) and WANG Bing (born 1957) between them collected about 47 000 numbered specimens in Xinjiang (XJBI, XJA, SHI). KUO Pung-Chao (GUO Ben-Zhao; born 1920), LIU Shang-Wu (born 1934; Figure 12.11A), HO Ting-Nung (HE Ting-Nong; 1938–2011; Figure 12.11B), YANG Yong-Chang (born 1927) and HUANG Rong-Fu (born 1940) collected extensively on the Qinghai-Xizang plateau, gathering more than 17 000 specimens (HNWP).

In northeastern China, many botanical explorations in Heilongiang, Jilin, Liaoning and eastern Nei Mongol were organized and executed by the Institute of Applied Ecology,

Figure 12.11 A, LI Xin. B, LIU Shang-Wu. C, HO Ting-Nung (HE Ting-Nong). D, ZHANG Gui-Cai. E, LI Ze-Xian. F, XING Fu-Wu. G, HSIUNG Yao-Kuo (XIONG Yao-Guo). H, HU Chi-Ming (HU Qi-Ming).

Academia Sinica, and in total more than 80 000 collections were accumulated (IFP). More important collectors here included FU Pei-Yun (born 1926), LI Shu-Xin (born 1926) and WANG Chan (WANG Zhan; born 1911).

At the same time, the South China Institute of Botany (now South China Botanical Garden) organized a series of botanical expeditions to various parts of Guangdong, Guangxi, Hainan, Hunan, Hong Kong, Macao and the South Sea Islands, resulting in more than 66 000 numbered specimens (IBSC, IBK). The most important collectors here included ZHANG Gui-Cai (born 1932; Figure 12.11C), LI Ze-Xian (born 1937; Figure 12.11D), WANG Xue-Wen (born 1935), YE Hua-Gu (born 1956), XING Fu-Wu (born 1956; Figure 12.11E), LIANG Chou-Fen (born 1921) and WEI Fa-Nan (born 1941).

The Jiangsu Institute of Botany, Lushan Botanical Garden and Wuhan Botanical Garden were responsible for surveying eastern and central China. SHAN Ren-Hwa (SHAN Ren-Hua; 1909–1986), YAO Gan (born 1937) and YUE Jun-San (YUE Jun-Shan; born 1934) collected in Anhui, Jiangsu and Jiangxi, gathering about 12 000 specimens (NAS). HSIUNG Yao-Kuo (XIONG Yao-Guo; 1910–2004; Figure 12.11F), HU Chi-Ming (HU Qi-Ming; born 1935; Figure 12.12A), LAI Shu-Shen (born 1931) and NIE Ming-Xiang (born 1932) collected extensively in various parts of Jiangxi, making more than 22 000 collections (LBG, PE). Finally LI Hong-Jun (born 1931), HUANG Ren-Huang (no birth date available) and WANG Ying-Ming (born 1938) collected more than 15 000 specimens in Hubei.

On the other side of the Taiwan Haixia, the staff of Taiwan National University, the Institute of Botany, Academia Sinica (Taiwan), the Natural Science Museum of Taiwan at Taichung, and the Department of Forest Resources Management and Technology, National Pingtung Polytechnic Institute organized a series of botanical expeditions throughout the island and a large number of specimens were accumulated. Important collectors included LIAO Jih-Ching (LIAO Ri-Jing; born 1929), LIU Tang-Shui (1910–1997), HUANG Tseng-Chieng (HUANG Zeng-Quan; born 1931), CHANG Ching-En (1920–2005), PENG Ching-I (PENG Jing-Yi; born 1950) and CHIU Shau-Ting (born 1961).

During this period the flourishing of several major collaborative scientific explorations, especially involving Chinese institutes, deserves special mention. One or another of the Academia Sinica's Insitute of Botany (Beijing), Kunming Institute of Botany, South China Institute of Botany (now South China Botanical Garden) and the Northwest Plateau Institute of Biology usually took the lead as the botanical partner in many of these major nationwide, trans-regional explorations, some of which are described below.

The "Yellow River Expedition," organized by the Academia Sinica (CAS), was a groundbreaking multidisciplinary research project involving 12 professional partners. Members of the botany team included TSOONG Pu-Chiu (ZHONG Bu-Qiu), WANG Wen-Tsai (WANG Wen-Cai; born in 1926; Figure 12.12B), LI An-Jen (LI An-Ren; born 1927), JIANG Shu (born 1925) and DAI Lun-Kai (born 1930; Figure 12.12C) from the Institute of Botany, and MA Yu-Chuan (MA Yu-Quan; 1916–2008) and LI Bo-Sheng (born 1946) of Nei Mongol University. Between 1954 and 1957 they investigated the middle reaches of the Huang He (Yellow River) in Gansu, Nei Mongol, Shaanxi and Shanxi (collecting at least 10 000 specimens; PE). At about the same time the Institute of Botany also began collaborating with newly-formed institutes in other provinces, helping to explore these botanically poorly-known regions and build capacity for biodiversity inventory and botanical research throughout China. Between 1955 and 1958 several expeditions were undertaken to various locations in Gansu and the Qaidam area of Qinghai (producing over 3 000 collections; PE). From 1956 to 1959 the Institute of Botany joined with the Xinjiang Institute of Biology, Pedology and Desert Research to investigate the plants of the Altai, Tian Shan and Kunlun Shan (ca. 24 000 collections made; PE).

A few years later (1959–1961) JIANG Shu, YING Jun-Sheng (YING Tsun-Shen) and CHANG Yong-Tian (ZHANG Yong-Tian) took part in another large-scale multidisiplinary study, the "South to North Water Diverse Expedition," and about 10 000 numbered specimens were collected (PE). At almost the same time (1958–1961) the Academia Sinica and the Commercial Department of the Central Government ran a nationwide programme researching wild plants of economic value. This program involved many botanical institutes, universities and commercial departments of every province, collected about 200 000 numbered specimens and published the results in the book *Economic Plants of China* (Wang *et al.*, 2004).

Also notable during this period was the "Sino-USSR Yunnan Expedition," a joint venture under the auspices of the Institute of Botany, Academia Sinica, and the botanical

Figure 12.12 A, WANG Wen-Tsai (WANG Wen-Cai). B, DAI Lun-Kai. C, WU Cheng-Yih (WU Zheng-Yi). D, WU Su-Kung (WU Su-Gong). E, CHEN Wei-Lie. F, WANG Chin-Ting (WANG Jin Ting). G, SUN Hang. H, LI Hen (LI Heng).

institutes of the Academy of Sciences of Russia. From 1955 to 1957 the team collected extensively in southeastern Yunnan, southern Sichuan and Emei Shan (ca. 10 000 collections; PE KN, LE). Leading Chinese collectors during these times included WU Cheng-Yih (WU Zheng-Yi; born 1916; Figure 12.12D), WANG Wen-Tsai (WANG Wen-Cai), FENG Kuo-Mei (FENG Guo-Mei) and WU Su-Kung (WU Su-Gong; 1935–2013; Figure 12.12E). Russian participants included Andrey Aleksandrovich Fedorov, Igorj Alexandrovich Lincevski and Moïsseï Kirpichnikov.

During the Cultural Revolution (1966–1976), political turmoil greatly disrupted scientific botanical research in China. All theoretical studies stopped, and only a few field explorations were carried out in the name of investigating Chinese herbal medicine in Guangdong (collections held at IBSC), Jiangxi (LBG), Hubei, Qinghai (HNWP) and Xizang (PE). After this temporary lull, further major exporations and large-scale inventory projects were launched to investigate poorly-known areas. One such area was the Qinghai-Xizang Plateau and surrounding regions, so the Institute of Botany (Beijing), Kunming Institute of Botany and the Northwest Plateau Institute of Biology teamed up for an extensive programme of expeditions here. This comprehensive programme of exploration lasted four years (1973–1976) and covered most parts of Xizang, resulting in 15 000 numbered specimens. Important collectors in this programme included, in 1973, WU Su-Kung (WU Su-Gong) and NI Chi-Cheng (NI Zhi Cheng; born 1942). In 1974 WU and NI were joined by LANG Kai-Yong (LANG Kai-Yung), CHEN Shu-Kun (born 1936), CHENG Shu-Zhi, YANG Yong-Chang, HUANG Rong-Fu, TAO De-Ding (born 1937) and ZANG Mu (1928–2011); in 1975 NI, WU and LANG continued this work alone, but were joined in 1976 by HUANG, TAO, YIN Wen-Qing (born 1927) and SU Zhi-Yun (born 1936). At the same time (1975–1976) WU Cheng-Yih (WU Zheng-Yi) led a "taxonomy" group into Xizang on two separate occasions, collecting about 4 000 numbered specimens. A second, "botany" group, under the leadership of CHEN Wei-Lie (born 1939; Figure 12.12F), LI Bo-Sheng (born 1946) and WANG Chin-Ting (WANG Jin-Ting; born 1932; Figure 12.12F), collected about 14 000 specimens. A third "forestry" group collected 4 500 specimens and a final "pasture and meadow" group around 2 000. These ca. 39 500 collections in total (PE, KUN, HNWP) formed the basis of the five-volume *Flora of Xizang* (Wu, 1988), and the institutes continued their separate programmes of exploration in these areas.

In 1987–1989, three expeditions to the Kunlun Shan-Karakoram Shan area of southern Xinjiang and western Xizang were conducted by WU Su-Kung (WU Su-Gong) and WU Yu-Hu (born 1951; making at least 3 700 collections; PE, HNWP, KUN). In 1979, ZHOU Li-Hua (born 1934), a professor of the Northwest Plateau Institute of Biology, made two collecting trips in Qinghai (1 612 collections; HNWP), in 1980 CHEN Wei-Lie and LI Bo-Sheng collected in Mêdog, southeastern Xizang (ca. 3 000 collections; PE), from 1979 to 1999 HUANG Rong-Fu collected extensively in various parts of Qinghai and Xizang (ca. 3 000 collections; HNWP), in 1982–1983 LI Bo-Sheng participated in the "Mountaineering Scientific Expedition" organized by the CAS, and collected in Mêdog and Namjagbarwa, southeastern Xizang (ca. 8 000 collections; PE), from 1989 to 1990, YAO Gan (born 1937), a senior technician of the Nanjing Institute of Botany, collected in Nyingchi, Mêdog and southeastern Xizang (ca. 3 000 collections; NAS), and finally in 1992–1993, SUN Hang (born 1963; Figure 12.12G) and ZHOU Zhe-Kun (born 1956), professors at the Kunming Institute of Botany, made a botanical expedition to Mêdog, an area previously almost unknown to botanists. SUN and ZHOU stayed there from summer to the spring of the following year in order to observe seasonal changes and collect plants flowering in early spring (ca. 7 000 specimens were taken; KUN).

As well as these poorly-known areas, familiar biodiversity hotspots were also the focal point of exploration during these times. In 1976–1977, the Insitute of Botany, Beijing, teamed up with the Wuhan Institute of Botany to continue botanical exploration to Shennongjia Linqu. This well-known and biodiverse area is found at the western end of the Daba Shan, on the Sichuan border with Hubei, north of the Chang Jiang. Leading botanists on this project were LU An-Ming (born in 1939), YING Jun-Sheng (YING Tsun-Shen), ZHENG Zhong (born 1936) and WANG Ying-Ming (born 1938). Between them they collected 8 000 specimens (PE, HIB).

At the far eastern extent of the Himalaya Shan, the complex mountain systems and deep river gorges of the Hengduan Shan are spectacularly rich in biodiversity. In 1981 the "Hengduan Shan Expedition" took a multidisciplinary approach to survey this region comprehensively. Over the next two years, botanical exploration was mainly carried out by the Institute of Botany, Beijing, led by WU Su-Kung (WU Su-Gong), LI Pei-Chun (LI Pei-Qiong; born 1936), LANG Kai-Yong (LANG Kai-Yung), WANG Ji-Ning and LI Liang-Ching (LI Liang-Qian; born 1952). From 1981 to 1983, they collected extensively in northwestern Yunnan, western Sichuan and eastern Xizang, gathering more than 42 000 collections (PE, KUN) and publishing the results in *Vascular Plants of Hengduanshan Mountain* (Wang, 1993). At the same time other staff at the Institute of Botany worked with botanists from the South China Institute of Botany in Guangzhou Shi, the Kunming Institute of Botany, Chengdu Institute of Botany and Wuhan Institute of Botany on a major collaborative programme—the "Wulin Shan Expedition." This programme explored the botanically poorly-known mountain ranges along the borders of southwestern Hubei and southeastern Sichuan down to western Hunan and eastern Guizhou, and the upper reaches of the Hongshui He. This included large areas of northwestern Guangxi, a limestone area with a rich calcicolous flora. Between 1987 and 1990 ca. 32 000 collections were made (CDBI, IBSC, PE, WUHAN) on expeditions led by LU An-Ming, LI Zhen-Yu (born 1952) and QIN Hai-Ning (born 1960) of the Institute of Botany, Beijing, ZHANG Gui-Cai (born 1934) and YE Hua-Gu (born 1956) of the South China Institute of Botany, FANG Rhui-Cheng (FANG Rui-Zheng; born 1932) and SU Zhi-Yun (born 1936) of the Kunming Institute of Botany, ZHAO Zuo-Cheng (born 1941) and PU Fa-Ting (PU Fa-Ding; born 1936) of Chengdu Institute of Botany, and WANG Ying-Ming of Wuhan Institute of Botany.

"The Drungjiang Expedition and Gaoligong Shan Expedition" focused on the remote and inaccessible Drungjiang area, located in northwestern Yunnan on the border with northern Myanmar and southeastern Xizang, and home to an incredibly diverse flora. With extreme topography featuring steep rugged mountains, high peaks, deep river valleys and very high rainfall, botanical exploration in this area is both arduous and demanding. LI Hen (LI Heng; born 1929; Figure 12.12H), a professor at the Kunming Institute of Botany, led the first expedition to Drungjiang. Cut off from returning to Kunming Shi, the team spent the whole winter and spring of 1990–1991 in the field and collected 7 075 numbered specimens (KUN). The results were published in *Flora of Dulongjiang Region* (Li, 1993), and laid the foundations for further exploratory expeditions in the Gaoligong Shan range. Between 1996 and 1998 a further eight botanical expeditions were made to the Gaoligong Shan, which runs south along the China-Myanmar border from the Drungjiang area. Under the leadership of LI Hen (LI Heng), about 7 000 numbered specimens were collected (KUN), and the results recorded in the detailed and comprehensive *Plants of Gaoligongshan* (Li, 2000). More recently, an international team of botanists, zoologists and anthropologists has undertaken a "Biotic Survey of the Gaoligong Shan," one of the biggest international all-organism exploratory surveys ever undertaken in China (Figure 12.13). This involved many Chinese and non-Chinese institutes engaged in numerous multidisciplinary expeditions between 2002 and 2007, and resulted in the collection of over 25 000 plant specimens (KIB, CAS, E) as well as insects and other invertebrates, mammals, fish and amphibians. The botanical component of the survey was carried out by staff from the Kunming Institute of Botany, the California Academy of Sciences (USA) and the Royal Botanic Garden Edinburgh (UK).

Figure 12.13 The Biotic Survey of the Gaoligong Shan: botanists struggle to cross a river in spate in the Gaoligong Shan region of western Yunnan, 2006 (image courtesy JIN Xiao-Hua).

Except for the large-scale explorations organized by the CAS and national botanical institutions, university biology departments and provincial institutions have all been active in exploring their local provinces. Together they have accumulated collections of local flora, of which the more important are held at Anhui Normal University (ANUB), Xiamen University (AU), the South Forestry College (CSFI), Fujian Institute of Subtropical Botany (FJSI), Fujian Normal University (FNU), Fudan University (FUS), Guizhou Academy of Sciences (HGAS), Hangzhou Botanical Garden (HHBG), Nei Mongol University (HIMC), The Hong Kong Herbarium (HK), Hunan Normal University (HNNU), Hangzhou University (HZU), Guangxi Institute of Botany (IBK), Shandong University (JSPC), Lanzhou University (LZU), Nanjing University (N), the Northeast Forestry University (NEFI), Nanjing Forestry University (NF), the Southwest Forestry College (SWFC), Sun Yat-Sen (Zhongshan) University (SYS), Sichuan University (SZ), Wuhan University (WH), the Northwestern Institute of Botany (WUK) and Xinjiang August First Agricultural College (XJA); see FU (1993) for further details of these collections.

12.4 International collaboration (1980–present)

China effectively closed its borders to foreign scientists in 1949. Their re-opening in the 1980s ushered in a new age of botanical exploration. Unlike the preceding years, when only a few Russian scientists were allowed to work in China, botanists from many foreign countries joined forces with Chinese collaborators on expeditions, shared investigations on the materials they collected, and co-authored papers documenting their discoveries. The first to re-connect were those American and European institutes with historic links to China, but soon there was a blossoming of activity, with Chinese botanists developing collaborations with scientists in all parts of the world, and the current distribution of herbarium collections well illustrates this. Earlier expeditions by foreigners and the Chinese had tended to be conducted independently, with the specimens taken back to the institutes of the foreign collectors. This practice limited access to historic collections and so stifled joint research projects. It also had the unfortunate result that the type specimens (vouchers for newly-described plant names) of Chinese species collected by foreign botanists were not available for consultation in China. Nowadays, expeditions routinely collect specimens in multiple sets that are shared between participating institutes. This enables scientists in different countries to study the same collections, and ensures that type specimens of any newly-described species are always present in China. Due to the great number and diversity of field exploration projects that have been carried out in China since 1980 it is impossible to give a comprehensive account here; the following are selected examples.

Sponsored jointly by the Academia Sinica (CAS) and the Botanical Society of America, the "Sino-America Shennongjia Expedition" (1980; Figures 12.14, 12.15) was the first joint field expedition in the People's Republic of China to involve foreign scientists from a non-Communist country. It represented a breakthrough in China-USA relations following the establishment of embassies in Beijing and Washington, and the signing of a bilateral trade agreement between the two countries. The team consisted of TANG Yen-Chen, YING Jun-Sheng (YING Tsun-Shen), CHENG Chong, HE Shan-An and ZHANG Ao-Luo from China, and Stephen A. Spongberg, David E. Boufford, Bruce Bartholomew, Theodore R. Dudley and James L. Luteyn from the USA. In October 1980, following in the footsteps of Judson Linsley Gressitt's 1948 expedition (Gressitt, 1953; Figures 12.16, 12.17), the team spent about six weeks in the Shennongjia Linqu, and visited eastern Sichuan, the type locality of *Metasequoia glyptostroboides*, and Lichuan, the site of the largest, extant, wild population of *M. glyptostroboides* (2 085 collections were taken; A, CM, HIB, KUN, NAS, PE). The following year, the Academia Sinica teamed up with the UK's Royal Society in arranging the "Sino-British Expedition" to northwestern Yunnan, including FENG Kuo-Mei (FENG Guo-Mei), MING Tien-Lu, FANG Rhui-Cheng (FANG Rhui-Zheng), TAO De-Ding, GUAN Kai-Yun, LI Chun-Chao, PAN Hu-Gen, LU Zhen-Wei, LU Ren-Fu and HE Qin-Gen from the Kunming Institute of Botany, and David F. Chamberlain, Peter A. Cox, Peter Hutchinson, Charles Roy Lancaster and Robert J. Mitchell from the UK (Figures 12.18, 12.19). The team focussed on the Diancang Shan above Dali, and gathered 1

Figure 12.14 The Sino-America Shennongjia Expedition: the entire team of botanists, interpreters, guides and drivers, pictured in 1980 in Lichuan, Hubei (image courtesy James L. Luteyn).

Figure 12.15 The Sino-America Shennonjia Expedition: the American botanists of the team, pictured with locals in front of *Metasequoia glyptostroboides* in Shennongjia, Hubei. From left to right, David E. Boufford, James L. Luteyn, Bruce Bartholomew, Stephen A. Spongberg, Theodore R. Dudley (image courtesy James L. Luteyn).

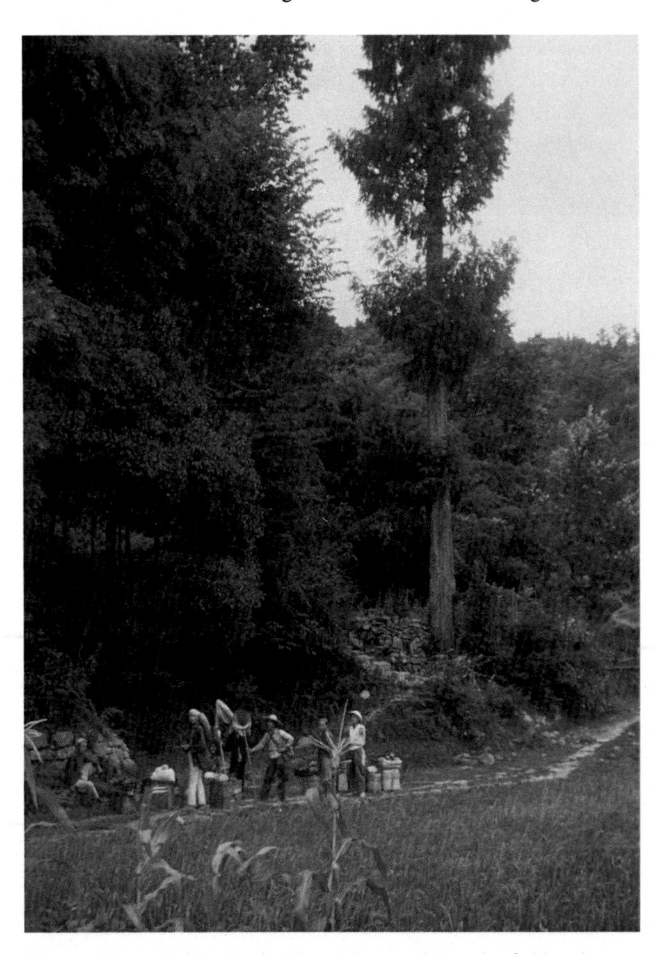

Figure 12.16 Judson Linsley Gressitt's expedition: the fieldwork team moving along a trail in Shuishanba, Lichuan, in 1948 (image courtesy Bruce Bartholomew from an original by Linsley Gressitt).

Figure 12.17 Judson Linsley Gressitt's expedition: collecting in the field near Xiaohe, 1948 (image courtesy Bruce Bartholomew from an original by Linsley Gressitt).

Figure 12.18 The Sino-British Expedition: resting in the field on the Diancang Shan, near Yangbi, Yunnan, 1981. From left to right, Peter Hutchinson, FANG Rhui-Cheng (FANG Rhui-Zheng), Peter A. Cox, MING Tien-Lu, TAO De-Ding, C. Roy Lancaster, David F. Chamberlain, Robert J. Mitchell, FENG Kuo-Mei (FENG Guo-Mei), Gaby Lock, and two local guides (image courtesy Robert J. Mitchell from a photograph by GUAN Kai-Yun).

Figure 12.19 The Sino-British Expedition: recording and pressing specimens after a day collecting on the Diancang Shan near Dali, Yunnan, 1981 (image courtesy Robert J. Mitchell).

500 herbarium collections (E, KUN), plus living plants and seed. Lancaster would later write his acclaimed *Travels in China—a Plantsman's Paradise* (Lancaster, 1989), based on this and his later experiences in China.

In the decades that followed, botanists from the Royal Botanic Garden Edinburgh were involved in several expeditions to southwestern China, usually in collaboration with their sister institute the Kunming Institute of Botany. Notable amongst these were the "Sino-British Lijiang Expedition," 1987, the "Chungtien, Lijiang, Dali Expedition," 1990 (Figure 12.20), the "Sino-Scottish Expedition," 1992, the "Kunming, Edinburgh, Göteborg Expedition," 1993 (Figure 12.21), the Alpine Garden Society's "Chinese Expeditions" (Figure 12.22), all to northwestern Yunnan, and the "Chengdu-Edinburgh Expedition" to Sichuan in 1991. Many botanists and horticulturalists from the Royal Botanic Garden Edinburgh and other western institutes have been involved in these expeditions, with David F. Chamberlain, David G. Knott, David G. Long and Ronald J. D. McBeath playing prominent roles. On the Chinese side, the Kunming Institute of Botany's GUAN Kai-Yun, FEI Yong and ZHOU Zhe-Kun, with PU Fa-Ting (PU Fa-Ding) from the Chengdu Institute of Botany, were notable participants. There were also Sino-American expeditions to Yunnan (including the Diancang Shan and Dali) in 1984 and northeastern Guizhou in 1986. In the 1990s the California Academy of Sciences and Northwest Plateau Institute of Botany also ran joint expeditions exploring the remote regions of Qinghai (Figure 12.23).

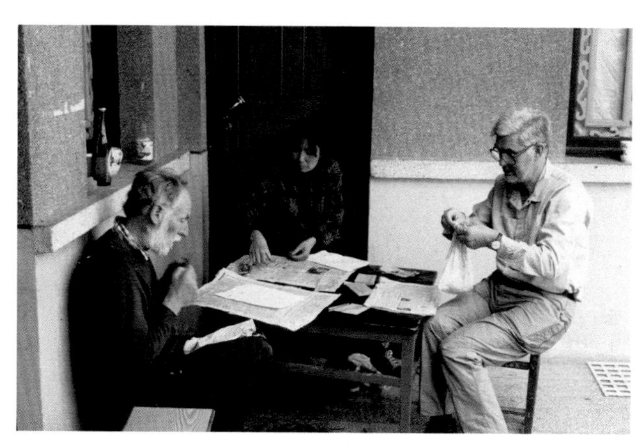

Figure 12.20 The Chungtien, Lijiang, Dali Expedition: Ronald J.D. McBeath, ZHANG Chan-Qin (Chinese expedition leader) and Christopher Brickell cleaning seed collections outside their hotel room in Lijiang, Yunnan, 1990 (image courtesy David G. Long).

Figure 12.21 The Kunming, Edinburgh, Göteborg Expedition: Crinan Alexander, GUAN Kai-Yun, Björn Aldén and Ronald J.D. McBeath examining *Rhododendron* in Yunnan, 1993 (image courtesy Mark F. Watson).

Figure 12.22 An Alpine Garden Society Chinese Expedition: Henry Noltie and local children on the Zhongdian plateau, Yunnan, 1995 (image courtesy Mark F. Watson).

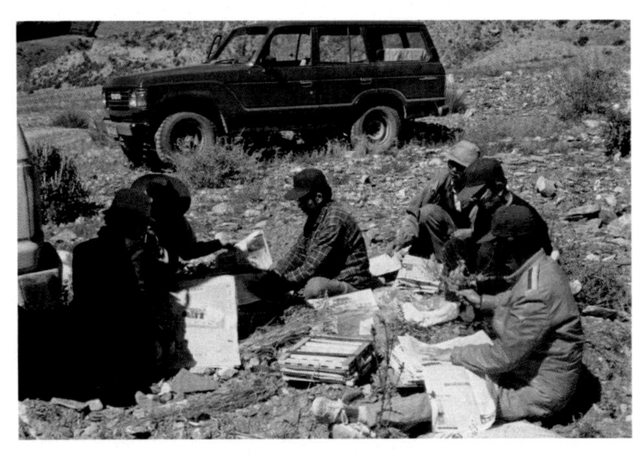

Figure 12.23 A California Academy of Sciences and Northwest Plateau Institute of Botany joint expedition to Qinghai: pressing specimens in the field on the Qinghai-Xizang Plateau, 1996. From left to right, CHEN Shi-Long, Bruce Bartholomew, LIU Jian-Quan, LU Xue-Feng, LIU Shang-Wu, HO Ting-Nung (HE Ting-Nong; image courtesy Mark F. Watson).

In 1988, Lord Charles Howick of Howick Arboretum, UK, teamed up with Charles M. Erskine and Hans J. Fliegner of the Royal Botanic Gardens, Kew, and William A. McNamara of Quarryhill Botanical Garden, USA, on a seed-collecting expedition to Sichuan. This was the beginning of a major series of 25 expeditions that continues to this day, focused on Sichuan and adjacent areas in Yunnan, Xizang and Hubei, aiming to explore poorly-known areas and introduce new stock of living material into western horticultural collections. Many other western and Chinese botanists and horticulturists have been involved in this series of expeditions, notably Mark Flanagan of Windsor Great Park, UK, Tony Kirkham of the Royal Botanic Gardens, Kew, and ZHANG Shou-Zhou from Shenzhen Fairylake Botanical Garden, Guangdong.

The desire to grow Chinese plants in Europe and North America, which was a focus for the landmark expeditions of the early twentieth century, was re-established between 1991 and 2008 by the "North American-China Plant Exploration Consortium." In this series of 11 expeditions, staff members from several arboreta in North America collaborated with many botanical institutes in China to visit temperate areas across the whole of China for field research, training and collection activities. A total of 1 350 accessions of living material were gathered, most with herbarium vouchers (see Del Tredici, 2010 and other articles in *Arnoldia* volume 68, for details).

Finally, like the Biotic Survey of the Gaoligong Shan discussed earlier, the "Biodiversity of the Hengduan Mountains" project is a recent, large-scale inventory program, and is coordinated by Harvard University, USA, the Kunming Insitute of Botany, and the Institute of Botany, Beijing (Figures 12.24, 12.25). This project undertook collaborative expeditions in 1997–1998, 2000, and 2004–2009, to little-known areas of this extremely biodiverse mountainous region in southwestern China. As well as the large number of new collections made, the project also

Figure 12.24 The Biodiversity of the Hengduan Mountains project 2006 expedition team, Zhongdian, Yunnan. From left to right, Richard H. Ree, YUE Ji-Pei, YUE Liang-Liang, XU Bo, CHEN Jia-Hui, ZHANG Da-Cai, HUI Long, LI Hui-Xin, ZHU Wei-Dong, LI Ying-Hui, GE Zai-Wei, David E. Boufford, SUN Hang (image courtesy Susan Kelley).

Figure 12.25 The Biodiversity of the Hengduan Mountains project 2007 expedition team, Xiaojin, Sichuan. Front row, from left to right, Richard H. Ree, JIA Yu, Susan Kelley, Kazumi Fujikawa; back row, from left to right, HUO Shun-Jun, HUI Hong, ZHANG Jian-Wen, David E. Boufford, XU Bo, LI Hui-Xin, local guide, GE Zai-Wei.

digitized and brought together electronically the research collections of many of the previous expeditions discussed above (see http://hengduan.huh.harvard.edu/fieldnotes for more details).

Despite the completion of the *Flora of China* project, botanical exploration has of course not finished. There are still under-studied areas in China and so work continues to complete the inventory of species, search for new species and understand biodiversity at the species, habitat and ecosystem levels. Chinese botanists are also now looking to surrounding countries and working with their neighbours to help document the biodiversity of adjacent lands and to extend their own knowledge in placing Chinese plants into a broader perspective. A good example is the recently started *Flora of the Pan Himalayas* (www.flph.org), an international project led by the Institute of Botany, Beijing, and involving all countries in the wider Himalaya Shan region, from Afghanistan's Wakhan Corridor to Myanmar and the western parts of Yunnan, Sichuan and Gansu. With an estimated 21 000 species in this global biodiversity hotspot, and many botanically very poorly-known areas (such as northern Myanmar), this ambitious project will provide a focus for fieldwork for several years to come.

References

BENTHAM, G. (1861). *Flora Hongkongiensis*. London, UK: Reeve and Co.

BONVALOT, G. (1982). *De Paris au Tonkin à Travers le Tibet Inconnu*. Paris, France: Paris.

BRETSCHNEIDER, E. V. (1898). *History of European Botanical Discoveries in China*. London, UK: Sampson Low, Marston & Co.

BRIGGS, R. W. (1993). *"Chinese" Wilson*. London, UK: HMSO.

BUNGE, A. (1835). *Enumeratio plantarum quas in China boreali collegit Dr. Al. Bunge. Anno 1831. Petropoli*. Petropoli, Russia.

CHEN, S. C., LI, J. L., ZHU, X. Y. & ZHANG, Z. Y. (1993). *Bibliography of Chinese Systematic Botany*. Guangzhou, China: Guangdong Science and Technology Press.

COX, E. H. M. (1945). *Plant Hunting in China*. London, UK: Scientific Book Guild.

DEL TREDICI, P. (2010). The case for plant exploration. *Arnoldi* 68(2): 1.

DELMAR MORGAN, E. (1877). Brief notice of M. Prejevalsky's recent journey to Lob-Nor and Tibet, and other Russian explorations. *Proceedings of the Royal Geographical Society of London* 22(1): 51-53.

FAIRCHILD, D. (1921). An agricultural explorer in Asia. *Asia, the American Magazine on the Orient* January 1921.

FAN, F. T. (2004). *British Naturalists in Qing China: Science, Empire, and Cultural Encounter*. Cambridge, MA: Harvard University Press.

FLETCHER, H. R. (1975). *A Quest for Flowers*. Edinburgh, UK: Edinburgh University Press.

FORTUNE, R. (1847). *Three Years' Wanderings in the Northern Provinces of China*. London, UK: John Murray.

FOURNIER, P. (1932). *Voyages et Découvertes Scientifiques des Missionnaires Naturalistes Français*. London, UK: Paul Lechevalier & Fils.

FRANCHET, A. R. & DELAVAY, J. M. (1889-1890). *Plantae Delavayanae*. Paris, France: P. Klinksieck.

FU, L. (1993). *Index Herbariorum Sinicorum*. Beijing, China: China Science and Technology Press.

GOODMAN, J. (2006). *Joseph F. Rock and His Shangri-La*. Hong Kong: Caravan.

GRESSITT, J. L. (1953). The California Academy-Lingnan dawn redwood expedition. *Proceedings of the California Academy of Sciences* 28(2): 25-58.

HANDEL-MAZZETTI, H. (1927). *Naturbilder aus Südwest China*. Vienna, Austria: Österreichscher Bundesverlag.

HOLMGREN, P. K., HOLMGREN, N. H. & BARNETT, L. C. (1990). *Index Herbariorum. Part 1: The Herbaria of the World*. New York: New York Botanic Garden.

HOU, K. Z. (ed.) (1956). *Flora of Guangzhou*. Guangdong, China: South China Botanical Research Institute.

HOU, K. Z. (1958). *A Dictionary of the Families and Genera of Chinese Seed Plants*. Beijing, China: Science Press.

ILLINGWORTH, J. & ROUTH, J. (1991). *Reginald Farrer: Dalesman, Planthunter, Gardener*. Lancaster, UK: Centre for North-West Regional Studies Lancaster University.

LANCASTER, R. (1989). *Travels in China - A Plantsman's Paradise*. Woodbridge, UK: Antique Collectors Club.

LÉVEILLÉ, A. A. H. (1914-1915). *Flore de Kouy-Tcheou*. Le Mans, France.

LÉVEILLÉ, A. A. H. (1916-1917). *Flore du Pekin, Chang-hai, Tche-li and Kiangsu*. Le Mans, France.

LI, H. (1993). *Flora of Dulongjian Region*. Kunming, China: Yunnan Science and Technology Press.

LI, H. (ed.) (2000). *Flora Gaoligong Mountains*. Beijing, China: Science Press.

LYTE, C. (1983). *The Plant Hunters*. London, UK: Orbis Publishing.

MCLEAN, B. (1997). *A Pioneering Plantsman: AK Bulley and the Great Plant Hunters*. London, UK: The Stationery Office.

MCLEAN, B. (2004). *George Forest, Plant Hunter*. Woodbridge, UK: Antique Collectors Club.

MUEGGLER, E. (2011). *The Paper Road: Archive and Experience in the Botanical Exploration of West China and Tibet*. Berkeley and Los Angeles, CA, and London, UK: University of California Press.

O'BRIAN, S. (2011). *In the Footsteps of Augustine Henry and his Chinese Plant Collectors*. Woodbridge, UK: Antique Collectors Club.

OHASHI, H. (2009). Bunzo Hayata and his contributions to the Flora of Taiwan. *Taiwania* 54: 1-27.

TYLER-WHITTLE, M. S. (1997). *The Plant Hunters*. New York: The Lyons Press.

WANG, W. C. (ed.) (1993). *Vascular Plants of the Hengduan Mountains*. Beijing, China: Science Press.

WANG, Y. Z., QIU, H. N. & FU, D. Z. (2004). In: Wu, C. Y. & Chen, S. C. (eds.) *Flora Reipublicae Popularis Sinicae*. pp. 658-732. Beijing, China: Science Press.

WINSTANLEY, D. (1996). *A Botanical Pioneer in South West China: Experiences and Impressions of an Austrian Botanist During the First World War*. Brentwood, UK: Winstanley.

WU, C. Y. (ed.) (1988). *Flora Xizangica*. Beijing, China: Science Press.

Chapter 13

History of Chinese Botanical Institutions

HU Zong-Gang
MA Hai-Ying
MA Jin-Shuang and HONG De-Yuan

13.1 Introduction

As discussed in Chapter 11, western science was introduced to China during the late Qing Dynasty and early years of the Republic of China, a period of great change including the abolition of the imperial examination system and the introduction of western-style schools, during which countless Chinese students travelled to Europe, the USA and Japan. Natural history, including botany, was introduced to new schools in China at that time but because most of the botanical teachers were Japanese only the plants of Japan were taught. Initially such courses were at a basic level and there was no botanical research in China—genuine scientific studies only came to China with the return of the overseas students from Europe and the USA to join the activity of "saving the country with science." Thus, biological study was introduced by PING Chi (BING Zhi), HU Hsen-Hsu (HU Xian-Su), and TSOU Ping-Wen (ZOU Bing-Wen), who had studied in the USA and who established the first biology department at the Southeast University in 1921. The following year they established the first biological institution in China, the Biological Laboratory of the Science Society of China. Other biological institutions followed sequentially, including the Fan Memorial Institute of Biology, the Natural History Museum of Academia Sinica, and the Institute of Agricultural and Forestry Botany of Sun Yat-Sen University. From those early institutes were derived others including Lushan Forestry Botanical Gardens, Yunnan Institute of Agricultural and Forestry Botany and the Institute of Botany of Guangxi University. Another source of botanical institutions in China was the Institute of Botany of the National Academy of Peiping (NAP; Peiping is now known as Beijing Shi), founded by the French-Chinese student LIOU Tchen-Ngo (LIU Shen-E) who also set up the Northwest Institute of Agricultural and Forestry Botany. In addition to these two categories, the Sun Yat-Sen Memorial Botanical Garden in Nanjing was established in memory of SUN Yat-Sen. In this chapter we present the histories of these primary professional institutions; this is not an exhaustive history of all the universities that also carried out botanical studies in departments of biology or agriculture. For transliterations of some of the Chinese names mentioned in this chapter, see Table 12.1 in Chapter 12.

13.2 The Biological Laboratory of the Science Society of China

The Science Society of China, founded in the USA in 1914 by Chinese students, aimed to connect Chinese students in the USA and to promote scientific research in China. Most of its early members returned to China after completing their studies and many came to teach at the Southeast University in Nanjing, including many biologists famous in later life, such as PING Chi (BING Zhi), HU Hsen-Hsu (HU Xian-Su), CHUN Woon-Young (CHEN Huan-yung), CHEN Zhen and CHIEN Sung-Shu (QIAN Chong-Shu). On 18 August 1922, they established the Biological Laboratory in Nanjing, modeled on the Wistar Institute of Anatomy and Biology in the USA. The Science Society of China considered biological research to be useful for national welfare and livelihoods, and laid the foundations for biological research. Its founders had high moral principles and a serious scientific spirit, working for scientific progress rather than individual gain. They gave careful and patient guidance to their students, providing them with the best materials, sometimes at personal expense. They also financed and helped to establish other institutes including the Fan Memorial Institute of Biology, the Institute of Biology of West China Academy of Sciences and the Henan Provincial Museum, and provided significant help including the exchange of specimens and literature and cooperation in collection and research, to the Natural History Museum of Academia Sinica in Nanjing.

After returning from the USA, the Science Society of China was allocated two small houses in Nanjing by the Jiangsu provincial government. In 1930, as their scientific work expanded, a two-storey building with a total of 36 rooms was sponsored by the China Foundation for the Promotion of Education and Culture (hereafter abbreviated as the "China Foundation"). This was a non-governmental organization set up in 1924 by ordinary citizens from China and the USA to provide funds for cultural and educational development. The funding of scientific research, with biology as a highest priority, was further enhanced after REN Hong-Jun, chairman of the Science Society of China, was appointed secretary of the China Foundation.

In 1931, the Mingfu Library of the Science Society of China was built in Shanghai Shi, to where the Science Society also moved, with the buildings in Nanjing opening as a public exhibition. In 1937, when war with Japan broke out, the Biological Laboratory moved to Beibei in Chongqing Shi, where it borrowed accommodation from the West China Academy of Sciences and built a laboratory in the spring of 1940. During the evacuation of the Shanghai Shi site, only some of the most important specimens and books could be relocated—most were left behind and eventually looted by the Japanese.

When it was first established, the Biological Laboratory was directed by PING Chi (BING Zhi), head of the zoology section, with HU Hsen-Hsu (HU Xian-Su) heading up the botany section from 1922 to 1928, except between 1923 and 1925 when he was again studying in the USA and his role was taken up by CHUN Woon-Young (CHEN Huan-yung). In 1928, HU moved north to create the Fan Memorial Institute of Biology and CHIEN Sung-Shu (QIAN Chong-Shu) led the botany section of the Biological Laboratory. When the Biological Laboratory moved to Chongqing Shi, PING stayed in Nanjing because his wife was seriously ill, so CHIEN became acting director. After the war, PING became director again, but only the zoology section remained active; the botany department was suspended as CHIEN was teaching at Fudan University. The entomologist YANG Wei-Yi took the directorship briefly in June 1946, but resigned soon afterwards; PING was then restored as director and remained until 1949.

Initially, as a non-governmental academic institution, the Science Society of China relied financially upon membership fees, donations and income from periodicals but these were not sufficient to support the work of the laboratory. When the Biological Laboratory was established, funds were still inadequate although the Society provided 240 Yuan per month for operating expenses. Most of the researchers taught at the Southeast University, working only part-time and unpaid at the laboratory. When the Fuming Library, another public institution of the Science Society of China, began to prosper it provided a budget to purchase books for the Biological Laboratory. From 1923, the Science Society of China received financial support from the Jiangsu provincial government and the Biological Laboratory was allotted 300 Yuan each month, enabling paid researchers to be hired. However, the appropriation was stopped in 1931. From 1926 the Biological Laboratory received an annual subvention of 15 000 Yuan each year from the China Foundation, with a supplementary 5 000 Yuan for construction and equipment. From 1929, another 9 000 Yuan was allotted to the laboratory by the Science Society of China for regular expenditure, and the support from the China Foundation was increased to 40 000 Yuan per year. The laboratory began to solicit funds from the public in 1933 which were put into shares in the Chinese Publishing Company. In this way, from 1933 to 1936, more than 40 000 Yuan was collected for the laboratory. When war broke out in the Pacific in 1942, the China Foundation stopped its financial support and the Biological Laboratory could not carry on. The few staff who stayed on had to work elsewhere to maintain a minimum standard of living. After the war, the botany section was closed, with only PING Chi (BING Zhi) and a few staff members continuing some research.

Initially, the major focus of the botany section was the collection of higher plant specimens for taxonomic study together with some morphological and ecological research. Before the war with Japan the laboratory collected not only in and around Nanjing, but also in Anhui, Jiangsu, Jiangxi, Sichuan, Xizang and Zhejiang. In 1929 FANG Wen-Pei collected in southwestern Sichuan and northern Yunnan, and in 1930 CHENG Wan-Chun (ZHENG Wan-Jun) went to eastern Xizang. In 1934, a joint expedition to borders of Yunnan and Myanmar was organized by the botany section of the Biological Laboratory and the Agriculture School of National Central University. In the same year, a one-year survey of the plants of Gansu, Qinghai and Xinjiang was conducted for the National Defense Committee. In 1935 the laboratory joined the agricultural-forestry investigation team of the Ministry of Industries to collect specimens in Fujian, Jiangxi and Zhejiang. A great number of specimens were collected and provided a basis for taxonomic study, herbarium collection, museum exhibition and exchange with other botanical institutions inside and outside of China. This work included taxonomic studies on specific groups, such as CHIEN Sung-Shu (QIAN Chong-Shu)'s study of Orchidaceae and Urticaceae, HU Hsen-Hsu (HU Xian-Su)'s study of *Torreya* (Taxaceae) and Styracaceae, CHUN Woon-Young (CHEN Huan-Yung)'s work on Lauraceae, CHENG's study on gymnosperms, PEI Chien (PEI Jian)'s research on Verbenaceae, SUN Hsiong-Tsai (SUN Xiong-Cai)'s work on Lamiaceae, FANG's study of Aceraceae, KENG Yi-Li (GENG Yi-Li)'s work on Poaceae and WANG Chu-Chia (WANG Zhi-Jia)'s study on the algae of the Chang Jiang.

An anatomical study of fern stem structure conducted by CHANG Chin-Yueh (ZHANG Jing-Yue), and a study on sexual differentiation in higher plants by YEN Tsu-Kiang (YAN Chu-Jiang), were the first plant morphological studies in China. CHANG also studied the anatomical structure and flower anatomy of *Firmiana simplex* (Sterculiaceae), and his paper on the anatomical structure of Pteridophytes (Chang, 1926) was the first paper on plant morphology published independently by a Chinese scientist. Major ecological studies at this time included CHIEN Sung-Shu (QIAN Chong-Shu)'s study on the Huang Shan, Anhui, and the forests and rock plants of Zhongshan, PEI Chien (PEI Jian)'s study on plants and their populations in Nanjing, and WANG Chen-Ju (WANG Zhen-Ru)'s study of aquatic plant populations in Xuanwu Hu and elsewhere. CHIEN's paper *A Preliminary Study on Flora of the Yellow Mountain* (Chien, 1927) was the first published ecological study by a Chinese scientist.

During the war with Japan, plant collection was carried out only in limited areas including Kangding, Taining and on the Huo Ju Shan in western Sichuan, while most botanical work comprised specialized studies with particular aims, such as the investigation of pasture grassland by the Resource Committee, the search for wood for sleepers by the Construction Department of Sichuan, the investigation of forest condition and search for paper resources for the Ministry of Economy, and the study of the degeneration of bamboos in Nanchuan. Scientific studies were mainly focused on the compilation of Floras for Nanjing, Sichuan and Zhejiang. The first volumes of the *Forest Flora of China* (Chien, 1937) were published during this time. Studies on conifers and Chinese Lamiaceae were also carried out during the war.

Most scientific papers written in the laboratory were published in its own English-language journal, *Contributions from the Biological Laboratory of the Science Society of China*, which was initiated in 1925. From 1925 to 1929, five volumes were published, each including five issues and usually only one paper in each issue. From 1930, i.e. the sixth volume of the journal, papers on botany and zoology were published in different issues. Until 1942, 12 volumes were issued by the botany section. The laboratory exchanged the journal with more than 500 institutions worldwide, so that the Biological Laboratory obtained more than 400 journals from foreign countries such as Belgium, France, Germany, Japan, the UK and USA. Thus, through distribution of its publications, the Biological Laboratory obtained journals that were difficult to purchase and, more importantly, gained worldwide recognition, soon becoming known internationally for its outstanding scientific studies.

13.3 The Fan Memorial Institute of Biology

Soon after the establishment of the Biological Laboratory, the Science Society of China planned to establish an institute in Beijing Shi to extend its area of study to northern China. In July 1927 this proposal was approved by the China Foundation but unfortunately, the general secretary, FAN Lian-Yuan (also known as FAN Ching-Sheng), died in December of the same year. To fulfill FAN's ambition, his successor REN Hong-Jun conducted many negotiations with the China Foundation of Culture and Education and the Shangzhi Society, finally securing agreement between them to set up an institute of biology. The Shangzhi Society, mainly composed of former Chinese students in Japan, with the goals of science and social improvement, was initiated by FAN Lian-Yuan who donated 300 000 Yuan in 1919 to set up several enterprises, but its results were not promising. FAN then donated the remainder of his estate, 150 000 Yuan, to the new institute, with overhead costs being covered by the China Foundation. At this time, the Kuomintang had just succeeded in its "Northern Expedition" and the government of the Republic of China had been moved to Nanjing Shi. As a result, Beijing Shi, the old capital, was renamed Peiping and remained the cultural center of northern China. Thus, the full name of the institute was the "Peiping Jing-Sheng Memorial Institute of Biology," translated in English as "The Fan Memorial Institute of Biology," in recognition of FAN Lian-Yuan. In October 1928, a large ceremony was held in Beijing Shi to mark the establishment of the institute.

Entrusted by the Shangzhi Society, the China Foundation organized the committee of the Fan Memorial Institute of Biology under three sections: one nominated by the Shangzhi Society, another by the China Foundation, and the third by the committees of the former two sections. According to the constitution, committee membership was honorary and unpaid, with succeeding committee members elected by their predecessors. Two committee sessions were held each year prior to 1934 and one a year from 1934 to 1937, with only occasional, poorly-attended sessions following the war with Japan.

PING Chi (BING Zhi) was appointed as the first director, responsible for implementing the decisions of the committee. He and HU Hsen-Hsu (HU Xian-Su) were, respectively, heads of the zoology and botany sections, but because PING was also the director in Nanjing, he came to the institute for only two months each year. HU acted as director in PING's absence and became director when, in1932, PING resigned because he could not be director of two institutes. HU served as director until 1949, except for a time in 1940 when he travelled to southern China and YANG Wei-Yi acted as director.

The Fan Memorial Institute of Biology was a non-governmental institute for which the Shangzhi Society provided 150 000 Yuan, to be managed by the China Foundation, with the objective of doubling these funds to 300 000 Yuan, after which the Institute would be supported by interest earned on this capital investment. However, the funds grew slowly and by 1936 had not achieved this target.

Initially, the former residence of FAN Ching-Sheng, at 83 Shifuma Street was donated by his younger brother FAN Xu-Dong as a small, temporary base until, six months later, PING Chi (BING Zhi) successfully submitted a proposal which led to a new building, completed in April 1934, in Wenjing Street on the vacant lot of the Beijing library. Late in 1941 the Fan Memorial Institute of Biology was occupied by the Japanese. After the war, the botany section was restored to Wenjin Street, but not at its pre-war scale. When the Chinese Academy of Sciences (CAS) was founded in 1949, the Fan Memorial Institute of Biology was displaced and its site used as the Academy's general office. Later the site belonged to the State Council of the People's Republic of China and it was finally demolished during the cultural revolution.

In 1928 the Institute boasted 11 staff at, most of them students of PING Chi (BING Zhi) and HU Hsen-Hsu (HU Xian-Su) at the Southeast University or from the Biological Laboratory of the Science Society of China, including SHOU Zhen-Huang, SHEN Jia-Rui, HE Qi, TANG Tsin (TANG Jin), WANG Fa-Tsuan (WANG Fa-Zan), FENG Cheng-Ru, and ZHANG Dong-Yin. The Institute expanded rapidly so that by 1937 it had more than 50 staff and was the largest biological institution in China. The staff at that time included HU Hsen-Hsu (HU Xian-Su), CHING Ren-Chang (QIN Ren-Chang), LI Liang-Ching (LI Liang-Qing), SHOU Zong-Huang, WANG Zong-Qing, TANG Tsin (TANG Jin), CHEN Feng-Hwai (CHEN Feng-Huai), WANG Fa-Tsuan (WANG Fa-Zan), HU Zan-Zhong, TANG

Yao, WANG Chi-Wu (WANG Qi-Wu), YÜ Te-Tsun (YU De-Jun), TSAI Hse-Tao (CAI Xi-Tao), FENG Cheng-Ru, HSIA Wei-Kun (XIA Wei-Kun), LIU Ying, HUANG Fu-Zhen, LÜ Lie-Ying, ZHANG Ying-Bo, FENG Zhong-Yuan and DENG Xiang-Kun. Plant collecting was carried out in many areas including Hebei, Jilin, Shanxi and Yunnan. The latter was the most intensively-collected area and turned out to be the richest in species numbers. In addition the Institute also conducted research on seed plants, ferns, fungi and algae with the ultimate goal being to complete a Flora of China along with other institutes. Some research on wood science and cytology was also undertaken, and from 1929 the *Bulletin of the Fan Memorial Institute of Biology* was published.

After northeastern China was occupied by the Japanese in 1931, northern China also came under threat. Considering that the Institute might move to southern China when necessary, HU Hsen-Hsu (HU Xian-Su) established branches there through cooperation with local governments. These included the Lushan Forestry Botanical Garden and the Yunnan Provincial Institute of Agricultural and Forestry Botany. Although the Fan Memorial Institute of Biology supported the establishment of branches in southern China, the China Foundation hesitated over evacuating the whole institute from Beijing Shi, which later turned out to be a mistake. When Beijing Shi was occupied, the institute was allowed to continue its scientific research under the control of the Japanese, reflecting the good relations between the China Foundation and the USA. However, when war broke out in the Pacific in December 1941, the Japanese army took over the institute and drove away the staff, suspending publication of the *Bulletin*. Shortly after the war with Japan, the Lushan Forestry Botanical Garden also became endangered, and therefore most staff members of the Fan Memorial Institute of Biology gathered at the Yunnan Provincial Institute of Agricultural and Forestry Botany. At this time, income from the China Foundation decreased as did the budget of the Institute. Its staff were widely scattered and experienced all kinds of hardships. In 1940, the National Chongcheng University was created in Taihe, Jiangxi, and HU was nominated as its president. He arranged for the staff evacuated from the Fan Memorial Institute of Biology to work in the new university, employing them as professors or lecturers in the Agriculture College. This group included YANG Wei-Yi, PENG Hong-Shou, TANG Tsin (TANG Jin), SHOU Zong-Huang and HE Qi. Even in such challenging times, some papers were prepared for publication and an issue of the *Bulletin* was finally published in July 1943.

In 1945, after the war, HU Hsen-Hsu (HU Xian-Su) strived to restore the Fan Memorial Institute of Biology but in spite of his endeavors it could not regain its earlier standing. Support from the China Foundation was far less than before the war so HU proposed to transfer the Fan Institute to the Education Ministry of the Republic of China. However, this was not achieved even after several years' effort. Eventually the Fan Memorial Institute of Biology was only partly restored, with some of the botany section maintained and a few staff members returning. These included HU, TANG Tsin (TANG Jin), CHANG Chao-Chien (ZHANG Zhao-Qian), FU Shu-Hsia (FU Shu-Xia), HSIA Wei-Kun (XIA Wei-Kun), LÜ Lie-Ying and FENG Cheng-Ru. Despite working under poor conditions, the Institute made some excellent achievements. In cooperation with the National Forestry Research Bureau of the Ministry of Agriculture and Forestry, it completed an investigation of the forest plants of southern Yunnan and northern Jiangxi through its affiliated institutes, the Yunnan Provincial Institute of Agricultural and Forestry Botany and Lushan Forestry Botanical Garden, respectively. It also edited and published a second volume of *The Forest Flora of China* (Hu, 1948), including Betulaceae and Corylaceae. Two issues of the *Bulletin* were published, with the most famous paper being the description of the rediscovery of the "living fossil" *Metasequoia glyptostroboides* (Taxodiaceae; Hu & Cheng, 1948).

When the CAS was founded, in November 1949, the Fan Memorial Institute of Biology and the Institute of Botany of the NAP were combined as the Institute of Plant Taxonomy, CAS. The former Lushan Forestry Botanical Garden and Yunnan Provincial Institute of Agricultural and Forestry Botany were combined into the Institute of Plant Taxonomy as two work stations of the new institute. Soon afterwards, in November 1950, the Ministry of Culture and the CAS signed an agreement transferring the biological books of the Beijing Library to the Institute of Plant Taxonomy on indefinite loan.

When botanical studies started in China, most of the type specimens of Chinese plants were stored in herbaria overseas. In 1930, when CHING Ren-Chang (QIN Ren-Chang) went study in Europe, he applied for funds to photograph type specimens collected from China and, with the support of HU Hsen-Hsu (HU Xian-Su), eventually secured funds from the China Foundation on condition that the photographs were kept by the Fan Memorial Institute of Biology. CHING immediately set to work and, in a period of over a year, took more than 18 000 pictures, mostly at the Royal Botanical Gardens, Kew, the British Museum (Natural History) and the National Institutes of France and Sweden. At the request of many institutions in China, the pictures were subsequently printed many times and provided important basic materials for plant taxonomy in China.

13.3.1 Lushan Forestry Botanical Garden

When, in 1917, HU Hsen-Hsu (HU Xian-Su) returned from his first period of study in the USA to be the deputy director of Lushan Forestry Bureau he focused his research on the natural environment and plant species of Lushan. Returning from his second study trip, in 1925, to work at the Biological Laboratory of the Science Society of China in Nanjing, he began to conceive a plan to establish a botanic garden. When he was invited to be an editor of the *Lushan Chronicles* in May 1931, HU again visited Lushan for his research, and carried out a thorough investigation of the vegetation and plants of Lushan which, at that time, was the "Summer Palace" of the government of the Republic of China and therefore an ideal place for building a botanic garden.

In January 1933, XIONG Shi-Hui, the chairman of Jiangxi, invited scientists from Jiangxi, including HU Hsen-Hsu (HU Xian-Su), to return home to discuss a construction strategy for the province. At the meeting HU successfully proposed the setting up of the Jiangxi Provincial Academy of Agricultural Sciences, which began immediately. In December of the same year, whilst attending the first directors' meeting of the Jiangxi Provincial Academy of Agricultural Sciences, he proposed and gained approval for the establishment of the Lushan Forestry Botanical Garden by the Jiangxi Provincial Academy of Agricultural Sciences and the Fan Memorial Institute of Botany. On hearing about thc plan, CHING Ren-Chang (QIN Ren-Chang) was enthusiastic and willing to take responsibility for building the garden, and became its first director. The site chosen for the botanic garden was the area around Sanyi at Lushan, where the School of Agriculture and Forestry, affiliated to the Jiangxi Provincial Academy of Agricultural Sciences, was located. CHING visited to survey the site in the spring of 1934 and concluded that it was an ideal place for a botanic garden because of its complex and varied terrain, rich soil, abundant water and beautiful environment. The opening ceremony of the botanic garden was on 20 August 1934, just as the nineteenth annual session of the Science Society of China was taking place.

The management of the botanic garden followed that of the Fan Memorial Institute of Biology, with a committee of seven members; the directors of the Fan Memorial Institute of Biology, the Jiangxi Provincial Academy of Agricultural Sciences, and the Lushan Forestry Botanical Garden were *ex officio* members, each of whom recommended two others. At its establishment, many research assistants, including LEI Zhen and WANG Ju-Yuan, and some trainees, such as FENG Kuo-Mei (FENG Guo-Mei) and HSIUNG Yao-Kuo (XIONG Yao-Guo), were hired. Aiming to make it the best botanic garden in eastern Asia, HU Hsen-Hsu (HU Xian-Su) even sent CHEN Feng-Hwai (CHEN Feng-Huai) to the Royal Botanic Garden Edinburgh to learn landscape architecture for botanic gardens, with the support of the China Foundation. He returned in the summer of 1936, and was employed as a horticultural craftsman.

The garden focused its research on forest plants and horticultural plants, mainly characterized by gymnosperms and alpine plants, such as *Rhododendron* (Ericaceae) and *Primula* (Primulaceae), respectively. Collection of seeds was emphasized from the very beginning, some for cultivation in the garden to increase the number of species grown, and some for exchange with other botanic gardens or agricultural and forestry institutions to supplement its species resources. The collection areas were mainly around Lushan and other mountains in southeastern China, extending to Taibai Shan and Zhongnan Shan in Shaanxi, Wutai Shan in Shanxi, and Emei Shan in Sichuan. Seeds collected in Yunnan by WANG Chi-Wu (WANG Qi-Wu) and YÜ Te-Tsun (YU De-Jun) were also sent to Lushan Botanical Garden by post. Seeds were exchanged with 68 botanic gardens, agricultural and forestry institutions around the world. By 1938, more than 3 000 species of living plants were cultivated in the garden, and some specific gardens had been built including glasshouses, nurseries, a conifer garden, herbal flower garden and alpine garden.

As Jiujiang, where Lushan is located, was occupied by the Japanese in October 1938, most of the books and specimens of the garden were moved to American schools at Lushan, and some staff retreated to Kunming Shi. Along with the staff of the Fan Memorial Institute of Biology, they joined the Yunnan Provincial Institute of Agricultural and Forestry Botany. However, the institute had insufficient accommodation for so many people, so CHING Ren-Chang (QIN Ren-Chang) led the staff from Lushan Forestry Botanical Garden to Lijiang where he set up a work station of the Lushan Forestry Botanical Garden, continuing research on alpine plants and specimen collection. In October 1945, CHEN Feng-Hwai (CHEN Feng-Huai) returned to take charge of the restoration of Lushan Forestry Botanical Garden, which made limited progress until 1949.

13.3.2 The Yunnan Provincial Institute of Agricultural and Forestry Botany

At the inception of the Fan Memorial Institute of Biology laboratory, TSAI Hse-Tao (CAI Xi-Tao), WANG Chi-Wu (WANG Qi-Wu) and YÜ Te-Tsun (YU De-Jun) were sent to Yunnan where, over a period of seven years, they collected more than 40 000 plant specimens, wood samples, seedlings and seeds, surpassing any former works. The high diversity of plant species in Yunnan deeply impressed the Fan Memorial Institute of Biology, who recognized the need for a permanent research institute for long-term study. In May 1937, HU Hsen-Hsu (HU Xian-Su) addressed a letter to the director of the Education Department of Yunnan, GONG Zi-Zhi, to discuss a plan for establishing a research institute in Yunnan following the model of Lushan Forestry Botanical Garden. The proposal was approved and the new institute was required to work to support its own development, the national economy and the livelihoods of the people of Yunnan, according to the tenets of the Fan Memorial Institute of Biology.

In the spring of 1938, TSAI Hse-Tao (CAI Xi-Tao), who had been collecting in Yunnan for a long time, was sent to Kunming Shi to select a site and begin preparation of the new Yunnan Provincial Institute of Agricultural and

Forestry Botany. The Education Department of Yunnan lent him the Zunjing Temple of the Kunhua People's Education Center as a preparation site, and instructed him to select a site for the institute from neighboring urban areas of Kunming Shi. After a wide survey, Heilongtan Park was considered the best site, and the institute was built there. As funds from the Yunnan Education Department were scant, HU Hsen-Hsu (HU Xian-Su) went to Chongqing Shi to seek support from the contemporary government of the Republic of China. A contract with the Yunnan Provincial Education Department to set up the Yunnan Provincial Institute of Agricultural and Forestry Botany was signed on 1 July 1938, and the institute was founded. HU held the directorship, and appointed WANG Fa-Tsuan (WANG Fa-Zan), who was returning from abroad, as vice director. After that HU returned to Beijing while TSAI, YÜ Te-Tsun (YU De-Jun), CHEN Feng-Hwai (CHEN Feng-Huai), WANG Chi-Wu (WANG Qi-Wu), ZHANG Bo-Ying and other researchers stayed in Kunming Shi. In March 1939, WANG Fa-Tsuan (WANG Fa-Zan) applied for a prerequisite of 10 000 Yuan from the Yunnan Provincial Education Department to purchase land around Heilongtan Park to build a nursery and a garden, as well as a herbarium and some laboratories. The application was submitted to, and approved by, LONG Yun, chairman of Yunnan. The institute purchased ca. 5.5 ha of land and the buildings were completed in November of the same year. HU returned to take charge of the institute in February 1940 and to lead appeals for more funds. However, he was appointed the first president of Chongcheng University later in the same year and succeeded as director by CHENG Wan-Chun (ZHENG Wan-Jun). YÜ was director for a short period after the war with Japan, before he went to study in Britain and was replaced by TSAI. At that time, the institute was short of both funds and manpower, and the employees had to support themselves by production, particularly of tobacco which became the main pillar of industry in Yunnan.

13.4 The Institute of Agricultural and Forestry Botany of Sun Yat-Sen University

Around the time that the Fan Memorial Institute of Biology began in Beijing, CHEN Woon-Young (CHEN Huan-Yung) went to Guangzhou Shi to establish the Institute of Agricultural and Forestry Botany at Sun Yat-Sen University. This meant that botanical institutes were distributed in different geographic areas from south to north which provided a good basis for plant resource investigation in China.

Arriving in southern China, CHEN Woon-Young (CHEN Huan-Yung) first went to Hong Kong to collect plant specimens, and then to Sun Yat-Sen University, teaching botany. In the fall of 1928, he turned to the College of Agriculture and established a botanical laboratory which was expanded into an institute after one year. The Institute planned to investigate the plants of Guangdong and publish a Flora of Guangzhou, to provide knowledge about the plants of the province as the foundation for future crop improvement and the development of agricultural and forestry industries to attract investment from overseas Chinese. The Institute was founded in December 1929 with CHEN as director and TSIANG Ying (JIANG Ying) and TSO Ching-Lieh (ZUO Jing-Lie) as major researchers. In spring 1935, the Institute of Agricultural and Forestry Botany was transferred to the Institute of Research as the Section of Agricultural and Forestry Botany. The work of the institute continued as before but graduate education was also begun. When war with Japan broke out and Sun Yat-Sen University moved to Chengjiang in Yunnan, the section of Agricultural and Forestry Botany alone was moved to Hong Kong.

During this time, CHEN Woon-Young (CHEN Huan-Yung) was director of the College of Science, so that the section of Agricultural and Forestry Botany was closely associated with the College and staff members were sent to Chengjiang to work with the Department of Biology. When Hong Kong was occupied by the Japanese, CHEN persevered to preserve the collections, and set up an institute of botany within the puppet Guangdong University. Around the same time, after the Sun Yat-Sen University moved from Yunnan to northern Guangdong, TSIANG Ying (JIANG Ying) set up the Institute of Botany in the Agriculture College of Sun Yat-Sen University in Guangdong. So CHEN and TSIANG each headed an institute of botany during the war. After the war, Sun Yat-Sen University returned to Guangzhou Shi and the Institute of Botany was moved to Fazheng road, still as part of the College of Agriculture. When CHEN was accused of being a traitor to the Chinese nation, TSIANG became the acting director. In 1946, the Institute of Research was dissolved by the Ministry of Education, and each institute belonged to a college. The section of Agricultural and Forestry Botany was restored as the Institute of Botany again and belonged to the College of Science. In fact, the Institute of Research had existed in name only during the war and each institute had already been affiliated to its own college. TSIANG went to Taiwan in July 1946, and WU Yin-Chan worked as acting director. After the People's Republic of China was founded in 1949, CHEN was reappointed as director.

When CHEN Woon-Young (CHEN Huan-Yung) established the Institute of Agricultural and Forestry Botany in the School of Agriculture, the government invited him to join the preparatory committee for the Guangzhou Botanical Garden and Guangzhou Zoo, which had been conceived since the time when Guangzhou University was renamed Sun Yat-Sen University, and to be the director of the former. However, the Guangzhou Botanical Garden was not aimed at preserving plant resources and supporting plant taxonomic research, so CHEN set up another garden for specimens on the Shipai campus at the same time, in 1931. This garden had an area of about six hectares, and included about 16 000 numbers of plants collected from the field. Ferns and Orchidaceae, as well as all the plants collected in Beijing Shi and Hainan, were grown in sheds.

Each plant was identified and labeled with its scientific name, and those unidentified were tied with a number or locality. It was CHEN, S.H. (CHEN Shao-Qing) who really took charge of everything to do with the garden, and many visitors were attracted from the very beginning. Sometimes schools in the city had their students visit the garden, taking botany courses there. By 1935, the collection had developed to a considerable size, with most subtropical genera being represented. As this Specimens Garden was also responsible for the landscaping of the new Shipai campus, a plant nursery with an area more than 50 ha was created, to propagate seedlings. As more and more living plants were brought back from the field, a glasshouse was built to hold the plants, and the whole exhibition area was gradually completed. Unfortunately, the Specimens Garden was destroyed after Guangzhou was occupied by Japanese troops in October 1938. Staff of the Institute tried to restore it, but this was eventually fruitless.

The Institute of Agricultural and Forestry Botany was undoubtedly the first to start floristic research. Because of the area's very high plant diversity it was not practicable for the Flora of Guangdong to be completed in a short time and in this chapter we describe only those studies that were completed. From 1930, the Institute started to publish its own journal, *Sunyatsenia*. Firstly, after coming to Guangzhou, CHEN Woon-Young (CHEN Huan-Yung) began his research on conifers, Betulaceae, Fabaceae, Juglandaceae and Lauraceae with the results published in *Arnoldia* and the scientific journal of Sun Yat-Sen University. From 1936, CHEN continued his study on the plants of Guangdong, Guangxi and Hainan, as well as the families Gesneriaceae and Fagaceae.

TSIANG Ying (JIANG Ying) was another famous researcher who in 1928 came to work as an assistant in the Department of Biology of the College of Science, but went to Nanjing in 1930 to work as an assistant researcher in the Museum of Natural History, Academia Sinica. Invited by CHEN Woon-Young (CHEN Huan-Yung), he returned to teach at Sun Yat-Sen University and continued his research at the Institute, publishing many important papers, such as studies on Asian Apocynales (Tsiang, 1932; Tsiang, 1934; Tsiang, 1936; Tsiang, 1939; Tsiang, 1941), Annonaceae in Guangdong (Tsiang, 1935) and the publication of a new genus of Apocynaceae in China (Tsiang, 1937). From 1936 he carried out studies on dendrology in China, and the families Apocynaceae and Cucurbitaceae.

TSO Ching-Lieh (ZUO Jing-Lie) was invited to the Institute in spring 1929 as a research professor, to take charge of plant collection in Guangdong. Soon after this, he went to Hainan to make collections, but his work was set back by the political situation there and was only carried out in spring 1932, when southern China became more stable. The expedition team included himself and four others. TSO returned first, while the collector CHEN Nian-Qu (CHEN Nian-Ju) and technicians LI Yao, CHEN Wen and LIN Chuan (LIN Quan) remained in Hainan for seven months, even going deep into Wuzhi Shan and making many collections there. In 1933 TSO went to Shiwan Dashan in Guangxi to collect plants, after which he went to lecture at the Biological Department of Shandong Province.

Other researchers included WANG Xian-Zhi, studying Euphorbiaceae and Tiliaceae, HOW Foon-Chew (HOU Kuan-Zhao) on Rubiaceae, CHEN Shu-Zhen on Styracaeae, Symplocaceae and Begoniaceae, LI Ri-Guang studying the medicinal plants of Guangdong, WANG Xiao on Papilionaceae, LIANG Shi-Han on toxic members of Fabaceae, CHEN Lu-Si continuing his study on the ecology of *Glyptostrobus pensilis* (Taxodiaceae) and taxonomic studies on the genus *Ilex* (Aquifoliaceae), and MA Xin-Yi from the College of Agriculture studying *Lonicera* (Caprifoliaceae).

The collections of the Institute of Agricultural and Forestry Botany were started soon after CHEN Woon-Young (CHEN Huan-Yung) went to Guangzhou, focusing at first on Hong Kong, while HUANG Ji-Zhuang and TSIANG Ying (JIANG Ying) were sent to collect at sites including Guangzhou Shi and Yingde. Later, when the plant research laboratory was set up, TSIANG and TSO Ching-Lieh (ZUO Jing-Lie) were in charge of plant collection, focused on areas including Gaozhou, Luofu Shan and Lechang. After the laboratory expanded into an institute, more staff were employed, amongst whom the main collectors included CHEN Nian-Qu (CHEN Nian-Ju), TSO, KO, S.P. (GAO Xi-Peng), WANG Chi (HUANG Zhi), and LIANG Xiang-Ri (LIANG Hoeng-Yip). Collecting was carried out on a large scale and, to ensure that high quality specimens were obtained, CHEN Woon-Young (CHEN Huan-Yung) trained the collectors personally in the methods he had learnt at the Arnold Arboretum. To collect plants thoroughly, he divided Guangdong into four areas and organized four teams, each with a four-year plan and each staying in the same area collecting specimens all year. By the end of 1930 a total of 15 514 dried specimens, 385 specimens in spirit, 249 seed specimens, 75 wood specimens and 232 slides of wood had been registered, and there were still many unregistered specimens. Later, the Institute received funds from the China Foundation and put even more effort into collecting. In 1933 alone, 8 930 numbers were collected, making a total of 31 836 specimens by the end of that year. The localities visited already accounted for three-fifths of the area of Guangdong, with the most intensively collected areas being around the Bei Jiang and Hong Kong, followed by southern Guangdong and the Xi Jiang. The Dong Jiang area was not socially stable, so only Luofu Shan was collected intensively, with its upper reaches and Fenghuang Shan only visited once. For Hainan, four long-term expeditions and three short-term collecting visits were conducted.

In July 1935, TSOONG Chi-Hsin (ZHONG Ji-Xin) and LI Yao collected in Ruyuan and Yaoshan in the Bei Jiang region. LAU, S.K. (LIU Xin-Qi) visited Wengyuan, and HOW Foon-Chew (HOU Kuan-Zhao) Hainan. In 1936, collecting trips were made to Lechang, Luofu

Shan, Zhaoqing, Yishan, Hepu and Ruyuan, as well as Hainan by LAU and Guizhou by TENG Shi-Wei (DENG Shi-Wei). A three-year plan was made for collection in Guizhou with, in the first year, collection centering on Zhenfeng and neighboring counties. Work began in the heat of midsummer, but collecting nonetheless had very good results. Unfortunately, the team met misfortune in Guizhou, with TENG and other six staff dying from serious infections. The monotypic genus *Tengia* (Gesneriaceae) was named in memory of TENG by CHUN Woon-Young (CHEN Huan-Yung).

During the war with Japan, a plan was made to evacuate Sun Yat-Sen University. With the permission of the university, CHUN Woon-Young (CHEN Huan-Yung) transferred important specimens, books and equipment into temporary storage in Jiulong (Hong Kong) using six batches of trucks. Later he bought a wasteland site with his own money and built a two-storey house, completed in September 1938, and named the Hong Kong Office of the Institute of Agricultural and Forestry Botany of Sun Yat-Sen University. Nine other staff worked there, together with the director, although financial support from the China Foundation had stopped and funds from Sun Yat-Sen University were scant. CHUN borrowed money from his family-owned newspaper, the *Chinese Daily*, to maintain the institute. During the time in Hong Kong, they mainly studied certain families and genera and identified the specimens transferred from the College of Science. The journal *Sunyatsenia* continued in Hong Kong, publishing ten issues and nearly 30 papers written by researchers from the institute and elsewhere.

When the Japanese army occupied Hong Kong the Institute was again placed in a dangerous situation. LIN Ru-Heng, the director of the Education Department of Guangdong in the puppet government of WANG Jing-Wei, said that he could help, so CHUN Woon-Young (CHEN Huan-Yung) decided to transfer the specimens and books back to Guangzhou again. In January 1942, the puppet government arranged a ship to "rescue" the staff of the institute, and most of them left Hong Kong apart from HOW Foon-Chew (HOU Kuan-Zhao) and a few others. In March, LIN asked a Japanese secret agent in southern China to send liaison officers to Hong Kong, together with three important officials in the government of Guangdong and a staff member of the Institute, LI Zhong-Luo, to help complete the transfer of all items from the Institute. After many negotiations with a chief officer of the Japanese army, permission was granted for all items from the institute, but not private possessions, to be shipped back in two ships which left Hong Kong on 25 April and arrived in Guangdong three days later.

Before the evacuation to Hong Kong, CHUN Woon-Young (CHEN Huan-Yung) asked TSIANG Ying (JIANG Ying) to lead some staff to combine with the College of Agriculture of Sun Yat-Sen University, and establish a new institute of botany there for the future restoration of the Institute. In February 1942, the Agriculture College of Sun Yat-Sen University was set up in Yizhang, Hunan, and the Institute was established there too. The herbarium was a spacious and bright building with four floors, in which TSIANG and a dozen staff worked until the war ended. During this time, the institute obtained more than 40 000 specimens from various places: 28 200 of these were collected from Yangming Shan, Mangshan and Yizhang in Hunan and Lechang, Pingshan and, Jiufeng mountain in Guangdong, 3 800 specimens were received through exchange with the Institute of Botany of Guangxi University, Sichuan University, Zhejiang University, and Guangdong Provincial School of Science and Art, and 934 specimens were collected from Chengjiang in Yunnan Province and moved to Hunan, among which were 119 were medicinal plant specimens. For identification, in September and October of 1942, TSIANG carried 4 500 specimens with him to Liuzhou, Guangxi. Later all the specimens in northern Guangdong were returned to Guangzhou and combined with the herbarium of the Guangdong Institute of Botany, the *de facto* Institute of Agricultural and Forestry Botany of Sun Yet-Sen University.

13.4.1 The Institute of Botany, Guangxi University

As CHUN Woon-Young (CHEN Huan-Yung) did an excellent job of creating and developing the Institute of Agricultural and Forestry Botany at Sun Yat-Sen University, the China Foundation altered his position from teaching professor to research professor. Since the plants of Guangdong and Guangxi are very similar, in 1934, the China Foundation thought it would be more efficient to study the plants from the both provinces together, and CHUN was therefore invited to be a research professor at both Sun Yat-Sen University and Guangxi University. Considering that a similar institute of botany was necessary at Guangxi University, CHUN discussed with MA Jun-Wu, the president of Guangxi University, and a former British Consulate on Baihe mountain in the Wuzhou delta was chosen for the site. CHUN himself took on the directorship of the Institute of Botany of Guangxi University, and sent WANG Xian-Zhi to Wuzhou to establish it, which was achieved in March 1935.

The most important work of this Institute was to set up a herbarium, initially with 6 000 duplicate specimens from Guangdong, and to start plant collection. Guangxi was well known for its rich diversity in plant species, but little studied, so the province was divided into different areas for collection. Two expedition teams were sent out in the year, one, led by KO, S.P. (GAO Xi-Peng), to Longzhou and neighboring areas in May, and the other, led by LIANG Xiang-Ri (LIANG Hoeng-Yip), to the mountain of Shiqan Dashan in September. Within one year, more than 4 700 dried specimens were collected, as well as more than 200 spirit collections, more than 30 bags of seeds and more than 100 seedlings. In the second year, collection was carried out on a larger scale, focused at Yao Shan and Bose on the borders of Yunnan, and Quanzhou, neighboring Hunan. The Yaoshan expedition team was led by WANG Chi (HUANG Zhi), and the Quanzhou team by TSOONG Chi-Hsin (ZHONG Ji-Xin).

In the summer of 1938, the Japanese army began to attack southern China, putting Guangzhou and Wuzhou in imminent danger. CHUN Woon-Young (CHEN Huan-Yung) resigned the directorship of the Institute of Botany of Guangxi University. The contemporary president of the university, BAI Peng-Fei, transferred the Institute of Botany to the College of Agriculture and invited CHANG Chao-Chien (ZHANG Zhao-Qian) to be director. Before Zhang arrived, WANG Yi-Tao, director of the Agriculture College, appointed ZHOU Bai-Jia as acting director of the Institute of Botany, and called him to Wuzhou to discuss transactions with KO, S.P. (GAO Xi-Peng), the technician designated by CHUN to deal with the matter. At that time Guangxi University had already been bombed several times and the Institute of Botany had stopped almost all work. In late September, Guangzhou had nearly fallen into enemy hands, and then WANG Yi-Tao led CHANG and WANG Jue-Ming to Wuzhou for investigation. Afterwards, the Institute of Botany was moved to Shatang in Liuzhou, and WANG Chen-Ju (WANG Zhen-Ru) became director. In 1944, the Institute was moved to Rongjiang in Guizhou with the rest of the university. However, during this transfer many specimens were damaged by a raging flood on the Rongjiang, while those left in Shatang encountered a fire, and those transferred to Nandan fell into enemy hands. This series of unexpected disasters caused enormous losses to the institute and all work stopped abruptly. After the war, the Institute was reconstructed at Yanshan, Guilin, and renamed the Institute of Economic Botany of the National University of Guangxi. The Institute still belonged to the College of Agriculture and CHUN was still director and dean of the Department of Forestry.

13.5 The Institute of Botany of the National Academy of Peiping

The NAP was founded as an independent institution affiliated to the Ministry of Education by the national government of the Republic of China on in September 1929, with LI Yu-Ying as president and LI Shu-Hua as vice president. As both LI Yu-Ying and LI Shu-Hua had experience of studying in France, and LI Yu-Ying had even led a work-study program in France, the NAP was strongly colored with French culture from the very beginning. Most of the employees had a background of study in France, and the China-France Education Foundation was also closely associated with the NAP. LIOU Tchen-Ngo (LIU Shen-E), who had almost finished his study in France, was invited by LI Yu-Ying to return to China to take charge of the Institute of Botany, one of three institutes together with the Institute of Biology (later renamed the Institute of Physiology) and the Institute of Zoology all located in the Natural Museum, site of the present day Beijing Zoo. At first an 18-room building was provided, with eight rooms allocated to the herbarium. In the following year, as the number of specimens increased, ten rooms used for zoological specimens were rebuilt as a herbarium, a space which was soon full again. Finally, in 1993, a 2 000 m^2, three-storey building was constructed on land provided by the Natural Museum, with funds from the China-France Education Foundation and the National Academy. The first floor hosted the Institute of Biology and part of the Institute of Zoology, while the Institute of Botany alone occupied part of the second floor and had the whole third floor for the herbarium. The building, named the Lamarck Hall in honor of the great French biologist, was occupied by Japanese troops during the war and was only returned to the Institute of Botany in 1945. In 1949, after the CAS was founded, the Hall was given to the Institute of Plant Taxonomy of the CAS. It was listed as a protected archive by the West City District of Beijing in August 1989, and in 1998 the Institute of Botany of the CAS moved to Xiangshan, where it remains to this day.

In October 1929, after becoming the director of the Institute of Botany, LIOU Tchen-Ngo (LIU Shen-E) hired HSIA Wei-Ying (XIA Wei-Ying) and KUNG Hsien-Wu (KONG Xian-Wu) as research assistants, WANG Zuo-Bin (WANG Tso-Ping) and LIU Chi-Mon (LIU Ji-Meng) as trainees, WU Xi-Zhou as garden clerk, and XIA Wei-Zhan as plant illustrator. His colleague LIN Yong (LIN Rong) in France was invited as a part-time research professor; and in return, LIOU himself worked as part-time professor in the university where LIN worked. His students in the university also came to work at the institute, including HAO Kin-Shen (HAO Jing-Sheng), WANG Yun-Chang (WANG Yun-Zhang) and TCHOU Yen-Tcheng (ZHU Yan-Cheng). They carried out extensive and careful investigations and collection in northern and northwestern China, and published a series of five volumes of a Flora of northern China (Liou, 1931-1936). Their papers were published in *Contributions from the Institute of Botany, National Academy of Peiping*. In 1934, the Institute invited the famous herbalist TSOONG Kuan-Kwang (ZHONG Guan-Guang) to work as professor on the *Compendium of Materia Medica* (unpublished) and his son TSOONG Pu-Chiu (ZHONG Bu-Qiu) also came to the Institute. Later, FU Kun-Tsun (FU Kun-Jun), WANG Zhen-Hua, and WANG Zong-Xun were also employed here, and the first illustrator was replaced by JIANG Xing-Qiang. During the war, the Institute of Botany established branches in Wugong, Shaanxi (see below), and in Kunming Shi, Yunnan. After the war, in 1946, LIN returned to Beijing and restored the Institute at its original site, Lamark Hall. LIOU also returned to Beijing in 1947. The specimens and books conserved in Wugong and Kunming Shi were gradually transferred back to Beijing Shi. At that time, the northwestern branch of the Institute was headed by WANG Zhen-Hua and Kunming Shi branch by TCHOU Yen-Tcheng (ZHU Yan-Cheng). The journal *Contributions from the Institute of Botany National Academy of Peiping* was issued from March 1931, with one volume of four series issued each year. The journal was suspended after the first series of the fifth volume in January 1937 and resumed in 1949 with the publication of the first issue of Volume Six. However, it was permanently stopped in December 1949, after the fourth series of Volume Six.

Plant collection had already begun before the Institute of Botany was set up. As early as September 1929, when LIOU Tchen-Ngo (LIU Shen-E) was preparing to establish the Institute, he collected specimens on the Beijing Xishan and Dongling Shan, the first collecting activity of the new Institute. Each year after this the Institute sent out some staff to collect specimens in Hebei and neighboring provinces. Before the war broke out, they had extended their activities to northwestern China. Major expeditions included those of LIOU and WANG Zuo-Bin (WANG Tso-Ping) to Fujian, Guangdong, Jiangsu and Zhejiang in 1930, KUNG Hsien-Wu (KONG Xian-Wu) and LIU Chi-Mon (LIU Ji-Meng) to Tieling and Qianshan in Liaoning in 1930, LIOU to Xinjiang and Xizang (and later to India) in 1931, HAO Kin-Shen (HAO Jing-Sheng) and WANG to Hebei, Henan and Shaanxi in 1932, KUNG and WANG to Shaanxi in 1933, HSIA Wei-Ying (XIA Wei-Ying) and BAI Yin-Yuan to Gansu and Ningxia in 1933, LIU to Shanxi in 1934, and LIOU and TSOONG Pu-Chiu (ZHONG Bu-Qiu) to the Huangshan, Anhui, in 1935. Of these, the most difficult was the trip LIOU made to Xinjiang and Xizang and then on to India in 1931. With HAO, he started from Beijing Shi in May 1931, travelling through Nei Mongol to Xinjiang as part of the "China-France Northwest Expedition Team." This part of the expedition was actually was a test drive supported by Citroën to assess a new type of jeep called a "climbing car," specially developed for driving on sand. For this reason, there was little time for specimen collection, which could take place only when the car stopped due to technical problems. The expedition team was dismissed after arriving in Xinjiang (HAO had already departed from Nei Mongol). As Xinjiang is a vast area with great, but difficult-to-reach plant resources, LIOU and a Swedish botanist decided to stay here and continue collecting, later travelling further, into Xizang, and finally returned via India. The journey took a total of two years and resulted in more than 2 000 specimens; its scientific significance is momentous as the basis of LIOU's "floristic theory of China" (Liou *et al.*, 1959).

Major researchers of the Institute of Botany included LIOU Tchen-Ngo (LIU Shen-E), who studied the impact of ice ages on the floristics of central and eastern Asia and also street trees and ferns. LIN Yong (LIN Rong) studied the family Asteraceae, and WANG Fa-Tsuan (WANG Fa-Zan) the taxonomy of the monocots. HSIA Wei-Ying (XIA Wei-Ying) conducted archival research and revision of plants in ancient Chinese books, KUANG Ko-Zen (KUANG Ke-Ren) compiled a dictionary of plant terminology and also studied Juglandaceae, TSIEN Cho-Po (JIAN Zhuo-Po) worked on the family Saxifragaceae and TSUI Yu-Wen (CUI You-Wen) Caryophyllaceae. In the fall of 1930, LIN returned to China as a professor in the College of Agriculture, Beijing University, dean of the Department of Biology and professor at the Institute of Botany. He specialized in the taxonomy of seed plants and edited the *Flora of Illustrata Plantarum Nordicum Sinicarum,* of which the first volume (Convolvulaceae) was published in 1931 (Ling & Liou, 1931), and the second (Gentianaceae) in 1933 (Ling & Liou, 1933). Later, LIN chose the complex and difficult families Gentianaceae and Asteraceae as research topics, and published on the taxonomy of Gentianaceae and other higher plants, gaining respect from colleagues and counterparts worldwide. In Asteraceae, he first identified the specimens held at the Institute, collated relevant literature, published new species and made a primary classification of the family into sections. In 1942 he went to Yong'an, Fujian, to set up the Institute of Botany and Zoology of the Academy of Fanjian. After the war, he returned to Beijing Shi and published several papers on the plants of Fujian.

When the Institute of Botany was set up in 1929, HAO Kin-Shen (HAO Jing-Sheng) was still a student at Beijing University, but studied under and worked with LIOU Tchen-Ngo (LIU Shen-E). In April 1930 he joined the "China-Sweden Expedition to Northwest China" with LIOU. Following rivers including the Bailong Jiang, they collected specimens from Chongqing Shi to southern Gansu, Qinghai Hu and further afield. He studied in Germany from 1933 and received a doctorate, working on the phytogeography of Qinghai. HAO returned to China in 1939 and continued his research, later completing the third volume of *Flora of Illustrata Plantarum Nordicum Sinicarum,* Caprifoliaceae (Hao, 1934) and a gymnosperm Flora of China (Hao, 1945).

TSOONG Kuan-Kwang (ZHONG Guan-Guang) was a pioneer of modern science in China and the first person to collect plant specimens in large numbers, who in later years devoted himself to the study of herbalism. In spring 1933, he was invited to work at the Institute of Botany as a special research professor, and began to work on herbal plants, comparing the Tang edition of *Classified Materia Medica*, LI Shi-Zhen's *Compendium of Materia Medica* and other classical books in order to determine the correct name for each of the traditional Chinese medicinal plants. Based on specimens, he determined the localities, fruiting season and ideal collecting time for each species. When Beijing Shi and Tianjin Shi were occupied by the Japanese, he took specimens, books and manuscripts to his home town of Ningbo Shi in Zhejiang, to continue his research. He passed away on 30 September 1940, his work uncompleted. TSOONG was a prolific researcher and, in addition to working on the *Compendium of Materia Medica,* wrote *Plants in Classic of Mountains and River,* an unpublished work of about 1 500 000 Chinese characters.

Among LIOU Tchen-Ngo (LIU Shen-E)'s wider interests, he published an outline of phytogeography in western and northern China (Liou, 1934), a ground-breaking study based on a thorough survey of plants in western and northern China analyzed with modern scientific methods. During his time in Wugong, LIOU studied the plants of Taibai Shan and published an important paper on their division (Liou, 1939). At the same time, he led colleagues in the Institute in compiling a survey of the flora of Taibai Shan. The most important paper from his time in Kunming Shi covered the phytogeography of Yunnan (Liou, 1944).

LIOU Tchen-Ngo (LIU Shen-E) had always regarded the construction of a botanic garden as pivotal to the work of the Institute. In cooperation with the National Natural Museum, a botanic garden with an area of ca. 24 ha was set up in April 1930 at the previous vegetable garden of the agricultural experimental site, and named the Botanical Garden of the National Natural Museum; this was one of the earliest botanic gardens in China. The garden's collections came from those of the National Natural Museum, or were transplanted from nearby Beijing Xishan, Dongling Shan or other mountains, collected from elsewhere in the wild, or exchanged with other botanic gardens. The collections were arranged systematically according to Engler's classification. Existing glasshouses were used to cultivate plants, so in just the first year a total of more than 500 species were planted into the garden. In the second year, planting focused on alpine plants from farther afield including, in the next spring, seedlings and rhizomes of about 100 species from Wuling Shan such as *Abelia* sp. (Linnaeaceae), *Carya cathayensis* (Juglandaceae), *Hydrangea macrophylla* var. *normalis* (Saxifragaceae), *Larix* spp. (Pinaceae), *Rhododendron* spp. and *Syringa tomentella* (Oleaceae). In the same year, HSIA Wei-Ying (XIA Wei-Ying) compiled a species list of the cultivated plants in the Botanical Garden of the National Natural Museum (Hsia, 1931). By 1934, the number of species cultivated in the garden had reached nearly 2 000. However, in 1935, the land on which the garden was sited was withdrawn by the municipal government and used to grow cabbages, and the botanic garden was therefore sadly destroyed.

After the institute moved to Wugong in 1936, LIOU Tchen-Ngo (LIU Shen-E) negotiated with the School of Northwest Agricultural and Forestry to set up a new botanic garden there. The School agreed to provide some land and a small amount of money so that an area of ca. 60 ha could be set aside for research and to supply teaching materials. At first it was only a plant nursery, growing living plants collected from field which, after a few years, contained about 300 species and over 7 000 individual plants. In 1940, the School agreed to expand this and provide additional the funds to build a full botanic garden. Half of the garden was designated to plant classification and the other half to ecology. After a year of hard work, the garden began to take shape, housing more than 1 000 species of plants. Those in the ecological area were cultivated in different zones according to their natural habitats, so as to study their ecology and their relationship with the environment. In 1942, the Institute encountered a serious shortage of funds and in order to sustain the botanic garden had to rent out vacant land within it to local farmers. However, this strategy only enabled the garden to survive for one more year, and it was eventually closed in 1943, as the land was withdrawn by the School.

When LIOU Tchen-Ngo (LIU Shen-E) was in Kunming Shi in 1942, although not able to build a botanic garden, he provided some land on his own nursery to cultivate seedlings and to plant seeds of ornamental plants collected in the field, including more than 100 species and varieties of *Camellia* (Theaceae) from Yunnan, more than 50 species of *Rhododendron*, dozens of Orchidaceae and more than 2 000 other herbs and trees, all new to cultivation. He also attempted to grow fruit trees. After returning to Beijing Shi, LIOU tried hard to restore the botanic garden, from 1948 onwards in cooperation with the National Natural Museum. Firstly, two dilapidated glasshouses were repaired, the existing land was prepared and leveled, and some seedlings and flowers collected in Beijing Shi were planted in the garden. From the spring of that year, wild collections were also introduced. In June, LI Shu-Hua, vice president of the NAP, brought back some precious Orchidaceae from a trip to Taiwan, including 52 individuals of more than 40 species. After just one year of the restoration project, more than 500 species of herbs and trees were cultivated in the botanic garden; more than 3 000 potted plants of various sizes were stored in cellars and glasshouses, and the nursery housed dozens of fruit trees, more than 100 grafted fruit trees, and hundreds of street trees.

13.5.1 Kunming Institute of Botany of the National Academy of Peiping

In the fall of 1938, the NAP set up an office in Kunming Shi, and later some affiliated institutes followed. In 1939, in order to study the plants of Yunnan and to join the NAP in Kunming, LIOU Tchen-Ngo (LIU Shen-E) asked HAO Kin-Shen (HAO Jing-Sheng), who had just returned from Germany, to re-establish the Institute of Botany of the NAP in Kunming Shi. In 1941, HAO moved to teach at the National Central University at Chongqing Shi, so LIOU himself moved to Kunming Shi to continue restoring the Institute. While in Kunming Shi, HAO collaborated with the Department of Construction of Yunnan Province to collect some specimens, but work proceeded very slowly. The NAP provided 300–500 Yuan per month for the project and continued cooperation with the Department of Construction for plant collection. They recruited a local person in Dali, WANG Han-Chen, to collect specimens there. WANG had previously been hired by British collector George Forrest to collect specimens in Yunnan and had much experience. LIOU also went west to Dali, Lijiang and Luxi, and south to Yuanjiang, to collect herbarium specimens, wood samples, seedlings and seeds. Later he opened a nursery in Kunming Shi to cultivate alpine plants and to write books. TSIEN Cho-Po (JIAN Zhuo-Po) and KUANG Ko-Zen (KUANG Ke-Ren) were invited by LIOU to work at the Institute of Botany in Kunming at different times. In 1946 LIOU led some staff members back north, and the Kunming Institute of Botany was downsized to a station, headed by TCHOU Yen-Tcheng (ZHU Yan-Cheng).

13.6 The Northwest Investigation Institute of Botany

In the fall of 1936, the National Northwest School of Agriculture and Forestry was established in Wugong, Shaanxi, with SIN, S.S. (XIN Shu-Zhi) as its principal. Considering that the development of agriculture and forestry in northwestern China must be based on a complete investigation of the plant resources in the region, SIN consulted with LIOU Tchen-Ngo (LIU Shen-E), and invited him to take charge of the creation of the Northwest Investigation Institute of Botany, which opened in Wugong in November 1936. The School provided the fourth floor of its teaching building as the Institute's site. Soon after this, they signed an agreement on the joint establishment of the Northwest Investigation Institute of Botany, and the cooperation between them lasted for thirteen years, until 1949.

When the Investigation Institute was established, KUNG Hsien-Wu (KONG Xian-Wu), WANG Zhen-Hua, SHEN Jia-Kang and FU Kun-Tsun (FU Kun-Jun) came with LIOU Tchen-Ngo (LIU Shen-E) to Wugong. After Beijing Shi was occupied by the Japanese, most members of the Institute of Botany in Beijing Shi also came to Wugong, including HSIA Wei-Ying (XIA Wei-Ying), TSOONG Pu-Chiu (ZHONG Bu-Qiu), WANG Zuo-Bin (WANG Tso-Ping), LIU Chi-Mon (LIU Ji-Meng), WANG Zong-Xun and JIANG Xing-Qiang, together with necessary books and equipment. For protection from air raids, specimens and books were moved from the Institute to Wuhou Temple in Mian, where FU and WANG Zong-Xun were responsible for them, until they could be returned to Beijing Shi after the war.

In its early years the Investigation Institute focused on collecting the plants of northwestern China, especially those of Taibai Shan in the Qin Ling where more than 20 visits were made to compile an illustrated Flora (Liou, 1939). Investigations to other places were also carried out from time to time, but mostly before the war. Once the war broke out, funds were very limited and the staff were only paid 60% of their salaries. The Institute had to seek cooperation with other institutions to continue plant collection, including Lanzhou Science and Education Center and the Department of Agricultural Improvement of Shaanxi Province in 1939, Xikang Technical School in 1940, and the Cultivation Area of Huanglong Mountain in Shaanxi in 1941. LIOU Tchen-Ngo (LIU Shen-E) transferred to Kunming Shi in early 1941, and WANG Yun-Chang (WANG Yun-Zhang) succeeded him in the directorship of the Investigation Institute. The situation after this time was even more difficult, and the Institute struggled to survive.

13.7 The Academia Sinica

The Academia Sinica, founded in 1928 by the nationalist government with a mission to develop modern western science and to engage China with international academia, was the principal academic institute during the time of the Republic of China. It made a great contribution to science in the 21 years of its operation in mainland China and its botanical and zoological studies can be divided into three periods, reflecting changes over those years.

13.7.1 The period of the Natural History Museum (1929–1934)

Soon after the Academia Sinica was established, an expedition to Guangxi was organized to investigate geology, biology, agriculture and forestry and to collect specimens. The biological work was undertaken by the Biological Laboratory of the Science Society of China, with CHING Ren-Chang (QIN Ren-Chang) leading the botanical group while FANG Bing-Wen and CHANG Lin-Ding led the zoological group. The expedition lasted eight months, traveled through 15 counties and collected 3 400 numbers and over 30 000 herbarium sheets of woody plants, herbs and ferns.

The Academia Sinica considered it necessary to invite specialists to conduct research in Guangxi so that the results of the expedition could be fully exploited, and to set up a museum for research and exhibition. In January 1929, TSAI Yuan-Pei, president of the Academia Sinica, invited LI Si-Guang, PING CHI (BING ZHI), CHIEN Sung-Shu (QIAN Chong-Shu), YAN Fu-Li, LI Ji, GUO Tan-Xian and QIAN Tian-He to form a committee to develop the museum. They established the National Natural History Museum of the Academia Sinica in January 1930 with QIAN Tian-He as its director. In 1933, QIAN resigned and XU Wei-Man succeeded him. An estate of more than five hectares was acquired in Nanjing for the Museum and, later, another two-storey building was built for offices and a herbarium; in 1931 the display house was rebuilt as a two-storey concrete building. By the end of 1933 the Museum held more than 30 000 specimens, mainly from Guangxi, Guizhou and Jiangxi.

At the Museum were based a botany group and a zoology group, most of whom had taken part in the expedition to Guangxi. The botany group was headed by CHING Ren-Chang (QIN Ren-Chang), with principal staff members including the collector CHEN Chang-Nian and illustrator FENG Zhan-Ru. The botany group held the editorship of the journal *Sinensia: Contributions from the Metropolitan Museum of Natural History, Academia Sinica*. In 1930, CHING received support from the China Foundation to study ferns at Copenhagen University, Denmark. During this period, he represented Academia Sinica at the Fifth International Botanical Congress in Cambridge, and in spring 1932 he visited many museums and botanic gardens in Europe before returning to China to work at the Fan Memorial Institute of Biology.

Before he departed for abroad, CHING Ren-Chang (QIN Ren-Chang) recommended TSIANG Ying (JIANG Ying) from the Institute of Agricultural and Forestry Botany of Sun Yat-Sen University to work as an assistant at the Fan Memorial Institute of Biology, dealing with the business of the plant section. TSIANG joined the Institute and went immediately to Guizhou for 11 months, from April 1930, collecting more than 7 000 numbers and 100 000 sheets of specimens. After returning to Nanjing, he began to organize the herbarium and started taxonomic research, identifying the specimens. Having studied Apocynaceae for years he continued his research on the family and the closely-related Asclepiadaceae. TSIANG also collected in Jiangxi in 1932 but had to stop because of social unrest. In 1933 he was invited by CHUN Woon-Young (CHEN Huan-Yung) to teach at Sun Yat-Sen University and therefore moved to Guangzhou.

From October 1931, TSOONG Kuan-Kwang (ZHONG Guan-Guang), who was already in his sixties, came to work at the Museum as an editor. With this position, he could finally settle down to work on his books and completed three papers (Tsoong, 1932-1933b; Tsoong, 1932-1933a; Tsoong, 1933); the latter being recorded and broadcast on the radio.

13.7.2 The period of the Institute of Zoology and Botany

DING Wen-Jiang became chief secretary of the Academia Sinica in 1934 and wanted to combine its Biological Laboratory with the Natural History Museum. This was not approved by PING Chi (BING Zhi) but he agreed to send his student, the protozoa specialist WANG Jia-Yi, to head the Museum and promised to send more staff to work there. After WANG's arrival, according to the principles of the Academia Sinica, the Natural History Museum was renamed the Institute of Zoology and Botany, and further staff were recruited. DENG Shu-Qun was invited as a full-time research professor in mycology, PEI Chien (PEI Jian) and KENG Yi-Li (GENG Yi-Li) as part-time research professors, and SHAN Ren-Hwa (SHAN Ren-Hua) as a research assistant. After reorganization, the Institute of Zoology and Botany focused mainly on marine biology, with only a few higher plant researchers including PEI, KENG and SHAN. When PEI came to the Institute of Zoology and Botany, he studied the classification of Verbenaceae and other higher plants. In 1934 he completed several papers including studies on *Clematis* (Ranunculaceae; Pei, 1935a), and *Chloranthus* (Chloranthaceae; Pei, 1935b). Following his supervisor, Elmer Drew Merrill, he organized his references in a system of typed cards arranged by family, genus and species according to the plant classification system. He made duplicate sets of cards, one for the Biological Laboratory and the other for the Institute of Zoology and Botany, where they were freely available to the public. Later, SHAN followed the same system, and these cards have become a precious legacy.

KENG Yi-Li (GENG Yi-Li) returned from the USA in 1934 to be a professor and researcher at the National Central University and the Institute of Zoology and Botany, Academia Sinica. He had studied the classification of Poaceae in China since 1927, including when he was in the USA and completed two papers in 1934 on the new genus *Cleistogenes* (Poaceae; Keng, 1934) and new species of Chinese bamboo (Keng, 1935). Based on the specimens of the Institute of Zoology and Botany, KENG completed a study of *Pleioblastus* (Poaceae) in southwestern China in 1935, describing six new species. In the same year he joined Nicholas Roerich's team from the USA Ministry of Agriculture to Bailing Temple, Nei Mongol, to collect seeds of pasture grasses. Later he published a paper on the new of Poaceae he found there (Keng, 1936).

In 1937, after war with Japan broke out, the collections were moved to Nanyue, Hunan for protection and later that year to Yangshuo, Guangxi. They following year they were moved again, to Beibei in Sichuan. The books, equipment and type specimens were transferred without major loss in all three moves. However, the items evacuated with the staff accounted for only one-fifth of the total and most remained at the Institute in Nanjing. When Nanjing was occupied, the specimens were transferred by the Japanese to the Shanghai Institute of Natural Science.

As higher plant classification was not the major interest of the Institute, usually only part-time researchers were hired. During the evacuation, half of the staff were dismissed including all part-time researchers. Though not totally suspended, plant taxonomy was only continued by SHAN Ren-Hwa (SHAN Ren-Hua), who joined the Institute immediately after graduating from the National Central University in 1934. His job was mainly to curate specimens and research the Apiaceae of China under the supervision of PEI Chien (PEI Jian). He published two papers on Apiaceae (Shan, 1936; Shan, 1937). In 1942, the institute employed WU Yin-Chan as a full-time research professor but he left soon afterwards. During the period of the Institute of Zoology and Botany there were no major collecting activities, only some minor collections made along the road during the transfer of the collections.

13.7.3 The period of the Institute of Botany

In 1944, ZHU Jia-Hua, President of the Academia Sinica, successfully proposed to separate the Institutes of Zoology and Botany. Plant physiologist LUO Zong-Luo came to the Institute of Botany, which had only around 500 books and four microscopes. After separation, the Institute was badly short of funds and had to rent a house on Wuzhi Shan, Beibei, as an office. Separate offices were created for higher plant taxonomy, fungi and forestry, with others added later including algae, plant physiology, plant morphology, plant pathology and cytogenetics. After moving to Jingang Temple, PEI Chien (PEI Jian) was invited to be research professor, and headed the higher plant taxonomy research laboratory. Besides the associate professor SHAN Ren-Hwa (SHAN Ren-Hua), the taxonomy laboratory employed assistant researchers and technicians CHEO Tai-Yien (ZHOU Tai-Yan), LAW Yuh-Wu (LIU Yu-Hu), WEI Guang-Zhou and WANG Ke-Hui.

Two years after the war ended, the Academia Sinica was restored and most institutes returned to the Zaijun Experiment Center in Shanghai Shi, which had previously belonged to the Institute of Natural Science. A dozen rooms on the ground floor were allotted to the Institute of Botany, with a herbarium upstairs. The staff increased to 30, including six full-time professors. The equipment in Beibei was shipped back to Shanghai Shi and the Institute took over the specimens and books left by Japanese, many of which had been plundered during the war. In 1945, the Institute purchased books costing 4 000 US dollars from the USA, and in 1946 Academia Sinica allotted 1.5 trillion (10^{12}) Yuan for repairs and equipment. Such significant funding was rare in that time of economic crisis, especially compared with institutes outside of the Academia Sinica.

During this period, SHAN Ren-Hwa (SHAN Ren-Hua), PEI Chien (PEI Jian), CHEO Tai-Yien (ZHOU Tai-Yan), and LAW Yuh-Wu (LIU Yu-Hu) continued some plant taxonomic studies. SHAN remained focused on Apiaceae, identifying and classifying the specimens collected from Gansu, Sichuan, Yunnan and Xizang. In 1945 alone, he described four new species and two new varieties of *Pternopetalum* (Apiaceae). He also conducted palynological research on Apiaceae and Asteraceae to determine their systematic relationships. PEI concentrated on the plants of western China. From 1940 to 1945, the Institute and schools in China organized four investigations to northwestern China. The National Institute of Geology carried out two investigations to the northwest and, in addition to their regular work, collected more than 600 numbers of plant specimens from Xinjiang and more than 700 from Qinghai. A third investigation to northwestern China was carried out by DENG Shu-Qun from the Institute of Zoology and Botany, who went to Gansu and Qinghai to investigate forest plants, but did not collect many specimens. A fourth investigation was carried out by the Gansu provincial government, for forest products, and collected more than 700 numbers. The specimens collected during these investigations were deposited at the Institute of Botany and collated by PEI, who published the Flora of northwestern China (Pei, 1948a; Pei, 1948b; Pei, 1949) and later a list of economic plants of Sichuan and Xizang (Pei, 1947a), the latter recording ten species in four genera of Onagraceae, ten species in seven genera of Verbenaceae and five species of Caprifoliaceae commonly used as medicinal plants. In Shanghai Shi, PEI led the junior staff in the compilation of a Flora of eastern China (Pei, 1947b), himself compiling the Juglandaceae and Ulmaceae treatments. CHEO completed the Brassicaceae treatment (Cheo, 1948), recording 27 species and one variety from 16 genera and LAW the gymnosperm treatment (Liu, 1947), based on the herbarium specimens of the Institute of Botany, in which seven families, 17 genera, 25 species and ten varieties of gymnosperms were recorded. CHEO was re-employed at the Institute of Botany from the National Medicine School of Sichuan Province in 1946 and completed a study of the medicinal plants of Emei Shan, Sichuan (Cheo, 1951), recording 207 species and varieties of medicinal plants in Sichuan. CHING Ren-Chang (QIN Ren-Chang) and PEI helped with the identification of specimens during his research.

13.8 The Premier Memorial Botanical Garden of the Sun Yat-Sen Mausoleum

After SUN Yat-Sen died in 1925, the Chinese Nationalist Party set up a funeral committee which decided to bury his remains at Zijin Shan, Nanjing. The construction of a mausoleum started in January 1926 and the following year, the committee approved the establishment of the Sun Yat-Sen Botanical Garden at the mausoleum, which had been proposed by Professor CHEN Yung (CHEN Rong) from the Department of Agriculture, Jinling University. The funeral committee addressed a letter to the national government in February 1928 stating that "a botanical garden in front of the Xiaoling Mausoleum of the Ming Dynasty is to be built for the purpose of collecting and growing plants of China and of the world for scientific research." The committee decided to invite FU Huan-Guang, former director of the Jiangsu First Forestation Farm, to be its chief technician, and to employ ZHANG Jun-Yu, TANG Di-Xian and WANG Tai-Yi as technicians plus another six staff. In early 1929, FU and CHEN discussed the design of the botanic garden, which was intended to include the whole area of the Xiangling Mausoleum of the Ming Dynasty. Later they invited CHIEN Sung-Shu (QIAN Chong-Shu) and CHING Ren-Chang (QIN Ren-Chang) from the Biological Laboratory of Science Society of China to visit the site and prepare an overall design for the garden. Thirteen feature gardens and houses were planned, over 164 ha, including a rose garden, economic timber garden, plant taxonomy garden, maple garden, fruit tree garden, Rosaceae garden, *Paeonia* (Paeoniaceae) garden, shrub garden, conifer garden, and aquatic plant garden, plus glasshouses for tropical plants. Finally, ZHANG of the Sun Yat-Sen Mausoleum designed the layout of the garden.

The botanic garden was at first formally named the "Botanical Garden Dr. Sun Yat-Sen Memorial Park." The tenets of the garden were to collect and preserve the herbal and woody plants of China, introduce valuable plants from foreign countries, and provide materials for scientific research on taxonomy, morphology, anatomy, physiology and reproductive biology, to serve as a field site for school students and to inspire the interest of the people in the natural beauty of plants and their importance to life. Simply speaking, it was a place to nourish the human spirit and to mould people's temperament. After the burial of Sun Yat-Sen on 1 June 1929, the funeral committee became the management committee of the Sun Yat-Sen Mausoleum, directly affiliated to the national government of China. In July of the same year, the committee clarified the necessity and urgency of setting up the botanic garden, and in particular boosted the construction of a glasshouse for the flowers and trees presented by foreign countries in memory of Sun Yat-Sen. At their fifth session the committee again decided to invite CHEN Yung (CHEN Rong) to be preparation director of the botanic garden, and YE Pei-Zhong as assistant to the preparation committee. However, this was not implemented immediately because of a shortage of funds. The preparation committee of the Premier Memorial Botanical Garden was constituted of the most eminent botanists in Nanjing, including CHIEN Sung-Shu (QIAN Chong-Shu), PING Chi (BING Zhi), QIAN Tian-He, CHING Ren-Chang (QIN Ren-Chang), CHANG Chin-Yueh (ZHANG Jing-Yue), XU Xiang, HU Chang-Chi, CHEN and FU Huan-Guang.

According to the plan, about one million Yuan were needed for the construction of the botanic garden. In 1930, 50 000 Yuan were granted to the Mausoleum as a creation fund for the botanic garden. From July 1930, 500 Yuan per month was allotted to the botanic garden for running expenses, and the salaries of the employees were included in this budget. In this year, an additional 10 000 Yuan was distributed to the garden to level the land, purchase equipment and collect plants. With these funds, the Premier Memorial Botanical Garden developed rapidly. The Rosaceae zone was built in 1931, with a 6 000 Yuan donation from the Jiangsu provincial government and 2 000 Yuan from the local noble school. An exhibition room was built with 20 000 Yuan donated by overseas Chinese. In 1936, an office and herbarium were set up in the garden at a cost of 80 000 Yuan.

When the botanic garden was first established, in May 1930, in a session of the Mausoleum Management Committee, FU Huan-Guang successfully asked for funds to send YE Pei-Zhong abroad to study horticulture at the Royal Botanic Garden Edinburgh. TANG Di-Xian and ZHANG Jun-Yu were appointed to work as gardening technicians in his place. In June 1931, the botanic garden provided further funds for YE to continue his studies in Edinburgh. YE returned to China in early 1932 and worked on the reproduction of introduced plants until war broke out. In July 1933 he was the head of the plant research office, and in September he was promoted to garden technician. SHEN Jun was employed as garden technician in 1934 and, when he left to study in the USA in 1937, SHEN Bao-Zhong was hired, by which time there were already more than 20 staff working in the botanic garden.

Initially the garden comprised a nursery of about 12 ha in area, one hectare for potted plants and a small glasshouse. Later the botanic garden was expanded by 12 ha on wasteland at the eastern corner of Wuwang Tomb. As most of the seeds received were difficult to germinate, the staff had to experiment to find ways of making them germinate. For this reason they also studied asexual reproduction in many species, including *Cedrus deodara* (Pinaceae), *Rhododendron* spp. and *Juniperus chinensis* (Cupressaceae). They achieved good results and by 1937 more than 100 000 individual seedlings, representing over 1 000 species, were in cultivation. In addition to the nursery were the exhibition gardens. The Rosaceae garden occupied 120 ha and was started in 1930. A plant taxonomy garden just inside the wall of the Mingling Mausoleum occupied an area of four hectares and contained 480 species of plants from 130 families. The 36-ha arboretum was located on roadsides outside the Mingling Mausoleum, with trees and shrubs adapted to the climate of Nanjing. The garden of economic plants, started in 1935, had an area of six hectares, focusing on seven species. The conifer garden was located on the eastern slope of the Mingling Bridge, the six-hectare bamboo garden included over 30 valuable species of bamboos and the medicinal plants garden grew more than 200 species in an area of six hectares. Because there was no glasshouse, tropical plants were collected but not cultivated and studies of their reproduction were emphasized. Those collected included more than ten species of Arecaceae, some 300 succulent plants and over 20 Orchidaceae. By 1937, there were more than 3 000 species of plants in cultivation at the Premier Memorial Botanical Garden of the Sun Yat-Sen Mausoleum.

Collecting specimens was an important task for the botanic garden, which carried out many joint expeditions with the Biological Laboratory of the Science Society of China and the Natural History Museum of Academia Sinica, exploring the great mountains of Anhui, Henan, Jiangsu, Jiangxi, Sichuan and Zhejiang. Further seedlings and specimens were procured through exchange with other institutions of agriculture or botany. By 1937 the collections numbered 10 000 herbarium specimens, 1 500 seed specimens, 200 wood specimens and 500 medicinal plants. After the war with Japan broke out, the management committee of the Sun Yat-Sen Mausoleum retreated to Chongqing Shi and its staff dispersed to various places. Under the puppet government established in Nanjing by WANG Jing-Wei, a new management committee was established, but undertook no work and the garden was abandoned. After the war, the display gardens and buildings were razed to the ground and the previously flourishing and prosperous garden was destroyed.

In 1945, after the surrender of Japan and the return of the capital to Nanjing, a renamed management committee of the National Father Mausoleum was established, with SUN Ke as chairman. SHEN Peng-Fei was appointed as the director responsible for the restoration of the garden. However, the financial support available from the government was far less than before the war, so the restoration was not satisfactory. Late in 1946, the committee started fundraising activities but with less success than was needed. Even after three years the restoration of the botanic garden was not complete and there had been numerous changes in leadership of the work. At first ZHANG Jun-Yu returned as head of the horticulture office and botanic garden, but he soon asked for leave of absence and SHENG Cheng-Gui was appointed director of the botanic garden and technician of the horticulture office. After more than a year of hard work he cleared up the damage, rebuilt the nursery, collected specimens and established relationships with other botanic gardens in China and abroad. With the shortage of funds, many things could not be achieved and so he resigned in February 1947. In October of that year CHENG Wan-Chun (ZHENG Wan-Jun) was appointed director of the botanic garden and dean of the Forestry Department of National Central University. Plans were made to rebuild the *Rhododendron* garden, which he began the following spring, but resigned in December 1948. Although the lack of funds caused frequent changes of personnel, the management committee continued to try to appoint learned people; ZHANG, SHENG and CHENG all later became famous botanists.

13.9 The reorganization and establish-ment of new institutions after 1949

13.9.1 The Institute of Botany, Chinese Academy of Sciences

The CAS was founded on 1 November 1949. Under the direction of ZHU Ke-Zhen, vice president of the CAS, the plant taxonomy institutions of China were reorganized under a new plan for plant science. The first step was to set up an institute of plant taxonomy in Beijing Shi. An institute of integrated plant science would be set up only after other disciplines of plant science had been developed. Outside Beijing Shi, because of many restricting factors, only plant taxonomy work stations were set up until conditions improved sufficiently to create separate institutes. Accordingly, the Institute of Plant Taxonomy was founded mainly through the combination and reorganization of the Institute of Botany and the Fan Memorial Institute of Biology. It was located on the site of the previous Institute of Botany of Peiping National Academy of Science. The former locality of the Fan Memorial Institute of Biology became the general office of the CAS.

Upon its formation, there were over 30 staff in the Institute, mostly famous botanists, with CHIEN Sung-Shu (QIAN Chong-Shu) as director. The Institute of Plant Taxonomy was renamed as the Institute of Botany in 1953 and in 1972 as the Beijing Institute of Botany with its leadership transferring to the Beijing municipal government and managed jointly with the CAS. Party affairs and administration were under the leadership of the Beijing municipal government, with scientific affairs being managed jointly. In 1978 the situation changed again, the Institute becoming the Institute of Botany again and attached only to the CAS. Initially following the merger there was a shortage of rooms, a problem that became increasingly serious as the Institute of Botany developed and space for expansion was restricted. Another site was sought for development and in 1982 a new herbarium and office building for the Taxonomy Department was built near Xiangshan. Later, laboratory and ecology buildings were built and finally the whole institute moved to its present site at Xiangshan, Beijing Shi.

Many societies are now associated with the Institute of Botany, including the Botanical Society of China, Ecological Society of Beijing, Beijing Society of Plant Physiology, Chinese Association of Botanic Gardens, Fern Committee of China Flower Association and the editorial committee of the *Flora Reipublicae Popularis Sinicae*. Academic journals issued by the institute today include the *Journal of Integrative Plant Biology*, *Journal of Systematics and Evolution*, *Journal of Plant Ecology*, *Acta Botanica Sinica*, *Biodiversity* and *Life World*.

In January 1950, the plant taxonomy section of the Institute of Botany, Academia Sinica, Nanjing, was reorganized as the East China Work Station of the Institute of Plant Taxonomy. It was at first set up in Shanghai Shi, but moved to Nanjing later that year. In September of the same year, the Yunnan Provincial Institute of Agricultural and Forestry Botany, which was created in cooperation between the Fan Memorial Institute of Biology and the Yunnan Province Department of Education, was combined with the Yunnan Work Station of the Institute of Botany of the NAP to create the Kunming Work Station of the Institute of Plant Taxonomy. In October, the Northwest Investigation Institute of Botany, which was co-founded by the NAP and the Northwest College of Agriculture, was reorganized as the Northwest Work Station of the Institute of Plant Taxonomy. About the same time, the Lushan Forestry Botanical Garden created by the Fan Memorial Institute of Biology and the Jiangxi Provincial Academy of Agriculture were reorganized as the Lushan Work Station of the Institute of Plant Taxonomy. In 1953 the Sun Yat-Sen Mausoleum Botanical Garden was taken over by the Institute of Plant Taxonomy, combined with the East China Work Station of the Institute of Plant Taxonomy and renamed as the Nanjing Yat-Sen Botanical Garden, affiliated with the Institute of Botany, CAS. Among the four work stations, three were created at the scale of institutes and each later developed into an independent institute, except the Lushan Work Station which was downsized and became an alpine botanic garden.

When reorganizing the Institute of Plant Taxonomy, the small botanic garden of the former Institute of Botany of the NAP was maintained, and there was a plan to build a large botanic garden in Beijing Shi. In 1956 the Horticulture Bureau of Beijing began to construct a Beijing Botanical Garden in the area around the Sleeping Buddha Temple in Xiangshan. The garden was divided into two, a north garden, which was the exhibition area, and a south garden, which was the experimental area. The garden was demolished in July 1970 but restored in April 1972, and was completed and opened to public in 1979.

13.9.2 Other institutions

South China Botanical Garden, Chinese Academy of Sciences.—During the reorganization of China's universities in 1952, it was proposed that the Institute of Agricultural and Forestry Botany of Sun Yat-Sen University be affiliated to the CAS. In December 1953, the standing committee decided that the CAS should take over the Institute of Agricultural and Forestry Botany of Sun Yat-Sen University and rename it the South China Institute of Botany of the CAS. CHUN Woon-Young (CHEN Huan-Yung) remained the director of the Institute, and the Institute of Economic Plants of Guangxi University was reorganized as a work station of the South China Institute of Botany. At that time, there were over 120 000 specimens and 12 researchers in the Institute of Agricultural and Forestry Botany of Sun Yat-Sen University, six of whom were above the level of associate professor. In 1956 the South China Institute of Botany was built in Guangzhou, and the Dinghushan Arboretum (now the Dinghushan Protected Reserve) was established in Zhaoqing Shi, establishing a pattern of one institute with two gardens. However, the institute moved to Wu Shan in 1958, which remains its present site. Under the name of Guangdong Institute of Botany, management was undertaken by the Guangdong provincial government from 1970 until 1978, when the institute was re-affiliated with the CAS. Its present name dates from 2003, reflecting the development strategy of the CAS.

The Institute of Botany, Northwest University of Agriculture and Forestry.—The forerunner of the Institute of Botany of the Northwest University of Agriculture and Forestry was the Northwest Institute of Agricultural Biology of the CAS, founded in 1954 as the Northwest Work Station of the Institute of Plant Taxonomy and the Institute of Water and Soil Conservation of the Ministry of Water Resources. Situated in Wugong, Shaanxi, it was later renamed the Northwest Institute of Biological Soil and the Northwest Institute of Water and Soil Conservation of the CAS. In 1965, the botany section was separated to become an integrated plant science institute named the Northwest Institute of Botany of the CAS, while the other part remains today as the Northwest Institute of Water and Soil Conservation of the CAS. In July 1970, the institute was affiliated to Shaanxi and managed by the provincial government. From 1991, it was renamed the Northwest Institute of Botany, Shaanxi Province and the CAS, co-administrated by the provincial government and the CAS. In September 1999, the institute was combined into the Northwest University of Agriculture and Forestry. The journal *Acta Botanica Boreali-Occidentalia Sinica* is issued by this institute.

Jiangsu Institute of Botany, Chinese Academy of Sciences and Jiangsu Provincial Government.—The Premier Memorial Botanical Garden of the Sun Yat-Sen Mausoleum was taken over by the Jiangsu provincial government in 1958 and returned to the CAS two years later as the Nanjing Institute of Botany of the CAS, with the name and function of the Sun Yat-Sen Botanical Garden also being retained. The Institute was greatly affected by the cultural revolution which began in 1966. In 1969 a military unit borrowed the houses of the Institute and in 1972 the Institute was forced to move, first to the Nanjing Normal College, then to the countryside at Liuhe, then to Jiangpu and finally to the previously-suspended Nanjing Agricultural College. In July 1974, the Institute returned to its original site, and since December 1993 it has been affiliated to both the CAS and the Jiangsu provincial government.

Kunming Institute of Botany, Chinese Academy of Sciences.—The Kunming Work Station of the Institute of Plant Taxonomy was upgraded to Kunming Institute of Botany in 1959, under the leadership of the Kunming branch of the CAS. In the same year, the Institute began to build the Xishuangbanna Tropical Botanical Garden. The Kunming branch of the CAS folded in 1962 and the Kunming Institute of Botany and the Xishuangbanna Tropical Botanical Garden became the Kunming branch of the Institute of Botany of the CAS. In 1970, the Kunming branch of the Institute of Botany and the Xishuangbanna Tropical Botanical Garden were managed by the Yunnan provincial government, under the names of the Yunnan Provincial Institute of Botany and Yunnan Provincial Tropical Institute of Botany respectively. They were returned to the CAS in 1978 and renamed the Kunming Institute of Botany of the CAS and Yunnan Tropical Institute of Botany of the CAS, respectively. In 1996, the Yunnan Tropical Institute of Botany was renamed the Xishuangbanna Tropical Botanical Garden of the CAS. The Institute publishes the academic journal *Acta Botanica Yunnanica.*

Guangxi Institute of Botany, Chinese Academy of Sciences and Guangxi Zhuang Minority Autonomous Region.—The forerunner of the Guangxi Institute of Botany was the Guangxi Work Station of the South China Institute of Botany. It was renamed the Guangxi branch of the Institute of Botany of the South China Institute of Botany in 1954 and began to establish the Guilin Botanical Garden in 1958. The Institute was renamed the Guangxi Institute of Botany of CAS in 1958, managed by the CAS. In July 1961 the leadership of the Institute was transferred to the Guangxi Zhuang Minority Autonomous Region. The Institute was under the leadership of the Guangxi Provincial Academy of Sciences from 1982 to 1996. Since then, it has been under co-leadership of the government of the Guangxi Autonomous Region and the CAS. The institute publishes the academic journal *Guihaia.*

The Institute of Applied Ecology, Chinese Academy of Sciences.—The Institute of Applied Ecology, formerly the Institute of Forestry and Soil Science, CAS, was founded in

1954. Initially it was a biological and geological institute, carrying out a wide range of studies covering forestry, soil, plants, microorganisms and environmental sciences. In March 1956, the Institute cooperated with the Shenyang Construction Bureau to build Shenyang Botanical Garden. From 1962 to 1970 the Institute was administrated by the Northeast Branch of the CAS, and in July 1970 it was renamed the Liaoning Provincial Institute of Forestry and Soil, managed by the Liaoning provincial government, first by the Bureau of Science and Technology and later by the Bureau of Agriculture. From 1978 the Institute was under the leadership of both the CAS and the Liaoning Provincial Government. In 1987 the Institute adopted its present name, the Institute of Applied Ecology of the CAS. It is now mainly engaged in research on forest ecology and forest eco-engineering, soil ecology and agricultural eco-engineering, pollution ecology and environmental eco-engineering, with a focus on the rational utilization and protection of natural resources, the improvement of the eco-environment, and the sustainable development of forestry and agriculture. It has ten laboratories, including forest ecology, forestry engineering, eco-climate and an analysis center. The editorship of two journals—*Chinese Journal of Applied Ecology* and *Chinese Journal of Ecology*—has been authorized to the Institute.

Wuhan Botanical Garden, Chinese Academy of Sciences.—Wuhan Botanical Garden of the CAS was formally established and named in 1958, two years after work commenced on the site. In 1963 it was renamed Wuhan Botanical Garden under the Central-South Branch of the CAS. In 1972 it was transferred to the Hubei provincial government, as the Hubei Provincial Institute of Botany. The Garden returned to the CAS in 1978 as the Wuhan Institute of Botany of the CAS, under joint leadership with the Hubei provincial government. In October 2003 it became the Wuhan Botanical Garden of the CAS. The Institute is located on the Moshan in Wuchang and occupies 70 ha. The Botanical Garden is mainly engaged in the study of plant conservation biology, plant ecology, plant resources and bioinformatics, with many projects on plant resources and plant conservation projects in central China. The garden is engaged in a wide range of research, including plant biotechnology, applications of medicinal plants, phytochemistry, restoration ecology, plant systematics and phylogenetics. The Institute is responsible for editing *Acta Botanica Wuhanica.*

Chengdu Institute of Biology, the Chinese Academy of Sciences.—Chengdu Institute of Biology was the successor of the Institute of Agricultural Biology, Sichuan branch of the CAS, founded in 1958. The institute was renamed the Southwest Institute of Biology of the CAS in 1962. From 1972 the Institute was transferred to the provincial government and renamed the Sichuan Institute of Biology. In 1978 it was renamed the Chengdu Institute of Biology of the CAS, governed by both the CAS and the Sichuan provincial government. It currently has ten research laboratories: biochemistry, microorganisms, bio-power, environmental microbes, special microorganisms, plants, phytochemistry, plant cells, plant breeding, and amphibians and reptiles, and has experimental workshops, farms and a drug factory. The institute publishes two academic journals: *Asian Herpetological Research* and *Chinese Journal of Applied and Environmental Biology.*

The Northwest Institute of Plateau Biology, the Chinese Academy of Sciences.—This Institute was founded in 1962, as the Institute of Biology, Qinghai branch of the CAS. In 1970, as affiliations changed, it was renamed the Institute of Biology of the Science and Technology Committee, Revolutionary Committee of Qinghai Province. In July 1977, it assumed its present name, under the leadership of both the CAS and Qinghai provincial government. The institute is mainly engaged in research into plateau plants, and is the editorial institute of *Acta Biological Plateau Sinica.*

References

Chang, C. Y. (1926). A preliminary report on the origin and development of tissues in the rhizome of *Pteris aquilina* L. *Contributions from the Biological Laboratory of the Science Society of China* 2(4): 1-8.

Cheo, T. Y. (1948). The Cruciferae of eastern China. *Botanical Bulletin of Academia Sinica* 2(3): 178-194.

Cheo, T. Y. (1951). Medicinal Plants in Mt. Omei. *Acta Phytotaxonomica Sinica* 1: 369-383.

Chien, S. S. (1927). Preliminary notes on the vegetation and flora of Hwangshan. *Contributions from the Biological Laboratory of the Science Society of China* 8(1): 1-85.

Chien, S. S. (1937). *Forest Flora of China*. Shanghai, China: Science Society of China.

Hao, K. S. (1934). *Flore Illustree du Nord de la Chine, Hopei (Chihli) et Ses Provinces Viosines, III*. Beijing, China: National Academy of Peiping.

Hao, K. S. (1945). *Gymnospermae Sinicae, Illustrations of Chinese Gymnosperms*. Chongqing, China: Zhengzhong Press.

Hsia, W. Y. (1931). A list of cultivated and wild plants from the Botanical Garden of the National Museum of Natural History, Peiping. *Contributions from the Institute of Botany, National Academy of Peiping* 1(3): 39-69.

Hu, H. H. (1948). *Betulaceae - Corylaceae*. Shanghai, China: Science Society of China.

Hu, H. H. & Cheng, W. C. (1948). On the new family Metasequoiaceae and on the *Metasequoia glyptostroboides*, a living species of the genus *Metasequoia* found in Szechuan and Hupeh. *Bulletin of the Fan Memorial Institute of Biology, New Series* 1: 153-161.

Keng, Y. L. (1934). A new generic name of *Cleistogenes* in the grasses of Eurasia. *Sinensia* 5(1-2): 147-157.

Keng, Y. L. (1935). New bamboos and grasses from Chekiang and Kiangsi Provinces. *Sinensia* 6(2): 147-157.

Keng, Y. L. (1936). An expedition in Inner Mongolia. *Science in China* 20: 49-56.

Ling, Y. & Liou, T. N. (1931). *Flore Illustree du Nord de la Chine, Hopei (Chihli) et Ses Provinces Viosines, I* Beijing, China: National Academy of Peiping.

Ling, Y. & Liou, T. N. (1933). *Flore Illustree du Nord de la Chine, Hopei (Chihli) et Ses Provinces Viosines, II*. Beijing, China: National Academy of Peiping.

Liou, T. N. (1931-1936). *Flore Illustree du Nord de la Chine, Hopei (Chihli) et Ses Provinces Viosines, I-V*. Beijing, China: National Academy of Peiping.

Liou, T. N. (1934). Essai sur la geographie botanique du nord et de l'ouest de la China. *Contributions from the Institute of Botany, National Academy of Peiping* 2(9): 423-451.

Liou, T. N. (1939). Taibaishan Forest Plants Division. *In: Collected Works of Prof. Tchenngo Liou*. pp. 74-85. Beijing, China: Science Press.

Liou, T. N. (1944). Plant Geography of Yunnan. *In: Collected Works of Prof. Shizeng Li on his 60th Birthday. Collected Works of Prof. Tchenngo Liou*. pp. 86-111. Beijing, China: Science Press.

Liou, T. N., Feng, Z. W. & Zhao, D. C. (1959). On questions of the classification of the Chinese vegetation. *Acta Botanica Sinica* 8(2): 87-105.

Liu, Y. H. (1947). Gymnosperms of eastern China. *Botanical Bulletin of Academia Sinica* 1: 141-147.

Pei, C. (1935a). A discussion on the allied species of *Clematis armadii* Fr. *Sinensia* 6(3): 38-390.

Pei, C. (1935b). *Chloranthus* of China. *Sinensia* 6(6): 665-688.

Pei, C. (1947a). Noteworthy plants of Szechuan and Sikang. *Botanical Bulletin of Academia Sinica* 1: 1-8.

Pei, C. (1947b). The Juglandaceae, Ulmaceae, Saururaceae and Chloranthaceae of eastern China. *Botanical Bulletin of Academia Sinica* 1: 111-117, 203-208, 283-297.

Pei, C. (1948a). Flowering Plants of Northwestern China I. *Botanical Bulletin of Academia Sinica* 2: 96-106.

Pei, C. (1948b). Flowering Plants of Northwestern China II. *Botanical Bulletin of Academia Sinica* 2: 215-227.

Pei, C. (1949). Flowering Plants of Northwestern China III. *Botanical Bulletin of Academia Sinica* 3: 28-36.

Shan, R. H. (1936). Studies on Umbelliferae of China I. *Sinensia* 7(4): 477-489.

Shan, R. H. (1937). Studies on Umbelliferae of China II. *Sinensia* 8(1): 79-92.

Tsiang, Y. (1935). Annonaceae of Kwangtung. *Journal of Botanical Society of China* 2: 673-708.

Tsiang, Y. (1937). *Chunechites*, a new genus of Apocynaceae from Southern China. *Sunyatsenia* 3: 305-308.

Tsoong, K. K. (1932-1933a). On the importance of Chinese names for plants with suggestions for a proper system of nomenclature. *Sinensia* 3(1-12): 1-8.

Tsoong, K. K. (1932-1933b). Criticisms and correctons of the botanic terms as appeared in the third report of the general committee on scientific terminology, 1914. *Sinensia* 3(1-12): 9-52.

Tsoong, K. K. (1933). Special interests of Chinese botanists and Japanese botanists. *Science of China* 1(4): 5-7.

Chapter 14

History of and Recent Advances in Plant Taxonomy

ZHANG Yu-Xiao and LI De-Zhu

14.1 Introduction

Taxonomy, or systematics, is regarded as the science of organismal diversity, and entails the discovery, description and interpretation of biological diversity, as well as the synthesis of information about it in the form of predictable systems of classification (Judd *et al.*, 2007). As described in Chapter 11, China has a long history of using and studying plants, not only crops and medicinal plants but also bamboos, ornamentals and beverages (for more on the economic uses of plants see Chapters 15–21). The early history of Chinese botany is described in Chapter 11; this chapter deals with the last century or so—a time in which, since the first plant specimen was collected in China and transferred to a European herbarium, western amateurs and professional botanists have made great efforts to gather Chinese specimens and study Chinese plants, because of the richness and importance of plant diversity in China.

14.2 European botanical discoveries in China

Modern taxonomic studies on Chinese plants were first performed by European botanists more than 300 years ago. With their great influence upon, and eventual colonization of, Asia, it became much easier for Europeans to travel and to study plants in China. At first, European botanists (many of them amateurs) explored primarily eastern and southern China, along the coastline. One of the first was James Cunningham, a surgeon of the East India Company and keen botanist, who in the 1690s made several trips to China and collected some 600 herbarium specimens in Xiamen (Fujian) and the Zhoushan Qundao (Zhejiang), which were deposited in the British Museum (now the Natural History Museum), London (BM). The genus *Cunninghamia* (Taxodiaceae) is named after him. Henry F. Hance was a plant taxonomist who worked as a British consular officer in China from 1844 to 1886. During that time, he collected approximately 22 437 plant specimens in Hong Kong, Xiamen and other parts of southern China, all of which were transferred from Hong Kong to the British Museum after his death. Hance also published more than 200 papers, including descriptions of a large number of new species, such as *Chirita anachoreta* (Gesneriaceae) and *Loranthus sampsonii* (now synonymized as *Helixanthera sampsonii*; Loranthaceae). Richard Oldham, a gardener sent out by the Royal Botanic Gardens, Kew, as a botanical collector in eastern Asia, collected 13 000 specimens in Japan, Korea and adjacent islands, and 700 specimens in Taiwan; these were later studied by esteemed botanists such as Daniel Oliver, Joseph D. Hooker and Friedrich A.W. Miquel.

Another British botanist, Augustine Henry also contributed much to early modern Chinese plant taxonomy. He studied medicine in college, but abandoned the profession when, in 1881, he arrived in Shanghai Shi working for the Chinese Imperial Maritime Customs Service. He became interested in plants, and travelled to Hubei, Hainan, Taiwan and Yunnan collecting more than 150 000 plant specimens, most of which were retained at the Royal Botanic Gardens, Kew. He published *A List of Plants from Formosa* in Tokyo in 1896, which listed 1 328 plants native to Taiwan plus 81 cultivated and 20 naturalized taxa. Several other British collectors collected in Taiwan, including William Hancock, and William R. Price. Price explored in Taiwan for 11 months, collecting a variety of plant specimens retained at the Royal Botanic Gardens, Kew. He published his collections in *Plant Collecting in Formosa* (Price, 1982).

The Japanese botanists Tomitaro Makino and Bunzo Hayata also visited Taiwan, the latter more than ten times between 1905 and 1924. Subsequently, he published ten volumes of *Icones Plantarum Formosanarum*, documenting more than 1 000 plant taxa collected by both European and Japanese botanists from 1854 to the 1920s.

Russian botanists focused their explorations on northeastern and northwestern China. Carl J. Maximowicz collected more than 2 000 plant specimens around the Heilong Jiang, Wusuli Jiang and adjacent areas between 1854 and 1860. Nikolai M. Przewalski, a general interested in travel and natural history, collected nearly 1 000 specimens in Nei Mongol, Gansu, Qinghai and Xinjiang, and sent them to Maximowicz for study. Grigori N. Potanin, a geographer, investigated Chinese plants from 1875 to 1895, travelling to Gansu, Nei Mongol, Shaanxi, Sichuan and Xinjiang, and collecting more than 20 000 plant specimens, which were deposited at St. Petersburg Botanical Garden (LE) and studied by Maximowicz. Mikhail V. Pevtsov, Vladimir L. Komarov, and Petr K. Kozlov also dedicated much of their study to the flora of northeastern and northwestern China.

Southwestern China has the most diverse flora in China, and as such has attracted the attention of European botanists since the seventeenth century. Until the nineteenth century, this was also the easiest entry point from Europe into China, and many western missionaries entered the country through Guizhou, Sichuan, and Yunnan. As well as spreading Catholicism, many of them were interested in plant taxonomy and collected tremendous numbers of specimens in southwestern China. Among them, Jean

Pierre Armand David (Père David) collected in northern China in 1862 and in southwestern China in 1868 and 1869, discovering a large number of new species of both animals and plants. Pierre J.M. Delavay conducted many expeditions in Yunnan from 1882 to 1895, making more than 200 000 plant specimens. The collections of both David and Delavay are retained at the Natural History Museum in Paris (P). Another French missionary, François Ducloux, collected ca. 6 000 specimens in western Sichuan and Yunnan between 1897 and 1945. Other European collectors in southwestern China, particularly Sichuan and Yunnan, included Ernest H. Wilson, a British horticulturalist, George Forrest, a plant collector of the Royal Botanic Garden Edinburgh, Francis (Frank) Kingdon-Ward, a British botanist and explorer and Heinrich Handel-Mazzetti, an Austrian botanist. Handel-Mazzetti wrote *Symbolae Sinicae* (published between 1929 and 1937), the first comprehensive flora of China. The Austrian scholar Joseph F. Rock, sent by the United States National Geographic Society, and Swedish taxonomist Karl August Harald (Harry) Smith also collected thousands of specimens in southwestern China. The history of European botanical discoveries in China has been described in greater detail by Bretschneider (1898), and more recently by Cox (1945) and Lancaster (1989), as well as in Chapter 12 of this volume.

14.3 The Flora Reipublicae Popularis Sinicae

With the accumulation of plant specimens in European herbaria, it became feasible by the end of nineteenth century to compile a checklist to the plants of China. Forbes and Hemsley (1886; 1889; 1905) made a first attempt, the *Index Florae Sinensis*, "an enumeration of all the plants known from China proper, Formosa, Hainan, the Corea, the Luchu archipelago, and the island of Hongkong, together with their distribution and synonymy," which was published sequentially in the *Journal of the Linnean Society of London (Botany)*. This is regarded as the earliest attempt to write a national Flora for China.

With changes in regime and social development occurring in China at the beginning of the twentieth century, Chinese botanists started to realize the importance of plant taxonomy and made their own efforts to investigate the flora of the country. One of these pioneers was HU Hsen-Hsu (HU Xian-Su), considered to be a founder of modern botany in China. HU received his doctoral degree from Harvard University in 1925, with the dissertation *Synopsis of Chinese Genera of Phanerogams with Descriptions of Representative Species* (Ma & Barringer, 2005). From 1919 to 1920, HU organized several botanical expeditions to Fujian, Jiangxi and Zhejiang, and collected a great number of herbarium specimens. At the same time, other botanists made collections all over the country, including CHUN Woon-Young (CHEN Huan-Yong), TSOONG Kuan-Kwang (ZHONG Guan-Guang), CHIEN Sung-Shu (QIAN Chong-Shu), CHING Ren-Chang (QIN Ren-Chang) (Wu & Chen, 2004) and YÜ Te-Tsun (YU De-Jun) (Figure 14.1).

After being elected as president of the Botanical Society of China in 1934, HU Hsen-Hsu (HU Xian-Su)

Figure 14.1 YÜ Te-Tsun (YU De-Jun) and his team collecting plant specimens in the Dulongjiang area, northwestern Yunnan, 1938 (image courtesy Kunming Institute of Botany archive).

presented a proposal for the compilation of a national Flora of China. Although Chinese herbarium collections gradually accumulated, they were far from enough for the compilation of the Flora, and the specimens collected by western botanists were preserved in European and American herbaria where they could not easily be accessed by Chinese taxonomists. The project was also hampered by insufficient library facilities and limited specialists. Despite those disadvantages, taxonomic studies were established in the 1930s and 1940s on Chinese pteridophytes (under CHING Ren-Chang [QIN Ren-Chang]), gymnosperms (CHENG Wan-Chun [ZHENG Wan-Jun]) and major families of angiosperms (see Table 14.1). Several monographs were published in this period, including the *Illustrated Manual of Chinese Trees and Shrubs* (Chen, 1937) and *Icones Plantarum Sinicarum* (Hu & Chun, 1927-1937), and great progress was made by LI Hui-Lin, HUANG Tseng-Chieng (HUANG Zeng-Quan) and LIU Tang-Shui (LIU Tang-Rui) on the Flora of Taiwan. HU Shiu-Ying (HU Xiu-Ying) at Harvard University also conducted detailed research on *Ilex* (Aquifoliaceae) and Malvaceae, among others, which provided important starting points for the national flora. During this time, CHING also photographed more than 20 000 Chinese specimens (mostly types) at the herbaria of the Royal Botanic Gardens, Kew (K) and other major herbaria in Sweden and Austria. With the support of his mentor WU Wen-Cheng (WU Wen-Zhen), WU Cheng-Yih (WU Zheng-Yi) made approximately 30 000 cards recording original references to Chinese plant names, linked to CHING's specimen photos and the *Symbolae Sinicae* (Handel- Mazzetti, 1929–1937). All these endeavors by the first and second generations of Chinese plant taxonomists laid the foundations for the national Flora.

Although a few of the early western and Chinese botanists made individual efforts to write catalogs and incomplete Floras of China, it was not until around the middle of the twentieth century that Chinese botanists began to make a collective effort to publish a national Flora. In 1950, the newly organized Chinese Academy of Sciences (CAS, originally known as "Academia Sinica") convened a meeting on plant taxonomy and formally called for compilation of a national Flora of China. Subsequently, *Claves Familiarum Generumque Plantarum Sinicarum*—a

Table 14.1 Examples of taxonomic studies established in the 1930s and 1940s on major families of angiosperms.

Taxonomic group	Lead researchers
Aceraceae and Ericaceae	FANG Wen-Pei
Annonaceae, Apocynaceae and Asclepiadaceae	TSIANG Ying (JIANG Ying)
Apiaceae	SHAN Ren-Hwa (SHAN Ren-Hua)
Araliaceae	LI Hui-Lin and HO Ching (HE Jing)
Begoniaceae and Rosaceae	YÜ Te-Tsun (YU De-Jun)
Betulaceae, Proteaceae, Styracaceae, Theaceae and Ulmaceae	HU Hsen-Hsu (HU Xian-Su)
Campanulaceae	TSOONG Pu-Chiu (ZHONG Bu-Qiu)
Chenopodiaceae and Polygonaceae	KUNG Hsien-Wu (KONG Xian-Wu) and TSIEN Cho-Po (JIAN Zhuo-Bo)
Asteraceae	LIN Yong (LIN Rong) and CHANG Chao-Chien (ZHANG Zhao-Qian)
Brassicaceae	CHEO Tai-Yien (ZHOU Tai-Yan)
Dioscoreaceae and Verbenaceae	PEI Chien (PEI Jian)
Fagaceae, Gesneriaceae and Lauraceae	CHUN Woon-Young (CHEN Huan-Yong)
Lamiaceae and Papaveraceae	WU Cheng-Yih (WU Zheng-Yi)
Juglandaceae and Myricaceae	KUAN Ke-Chien (GUAN Ke-Jian)
Orchidaceae and Liliaceae	WANG Fa-Tsuan (WANG Fa-Zan) and TANG Tsin (TANG Jin)
Poaceae	KENG Yi-Li (GENG Yi-Li)
Primulaceae	CHEN Feng-Hwai (CHEN Feng-Huai)
Rubiaceae	HOW Foon-Chew (HOU Kuan-Zhao)
Scrophulariaceae	LI Hui-Lin and TSOONG Pu-Chiu (ZHONG Bu-Qiu)

systematic listing of families and genera—was published as a series in *Acta Phytotaxonomica Sinica* (e.g. Hu *et al.*, 1953; Tang *et al.*, 1954), and simplified and illustrated treatments of some major families were also published. These included the *Illustrated Treatment of the Principal Plants of China, Leguminosae* (Wang & Tang, 1955), *Illustrations of Important Chinese Plants, Pteridophytes* (Fu, 1957) and *Flora Illustralis Plantarum Primarum Sinicarum, Gramineae* (Keng, 1959). The first local Chinese Flora written in-country was the *Flora of Guangzhou*, compiled by HOW Foon-Chew (HOU Kuan-Zhou; How, 1956); HOW also published the *Dictionary of the Families and Genera of Chinese Seed Plants*, in which Latin names, descriptions and notes were arranged alphabetically (How, 1958). Through the publication of these books, a great many plant specimens were identified and preliminary work for a Flora of China was well established.

In May 1959, CHIEN Sung-Shu (QIAN Chong-Shu), HU Hsen-Hsu (HU Xian-Su) and other 24 botanists made a formal proposal, and the CAS initiated the *Flora Reipublicae Popularis Sinicae* (*FRPS*) project to produce a Chinese-language Flora of China. The work was organized by an editorial committee of 25 members, mostly first-generation botanists of China. The first conference of the committee was held in Beijing in November 1959, at which general rules and guidelines for compilation were discussed and established. CHIEN and CHUN Woon-Young (CHEN Huan-Yong) were appointed as editors-in-chief, with CHING Ren-Chang (QIN Ren-Chang; a fern taxonomist) as general secretary. Volume 2 (*Pteridophytes, Part 1*) was the first published volume of the *FRPS* (Ching, 1959). Subsequently, Volume 11 (*Cyperaceae, Part 1*; Tang & Wang, 1961) and Volume 68 (*Scrophulariaceae, Part 2*; Tsoong, 1963) were also published. The publication of those three volumes symbolized the launch of the *FRPS* project. From 1961 to 1966, the editorial committee made a few modifications to the rules of compilation, and decided to deal with families of important economic value first. However, due to the "Cultural Revolution," the *FRPS* project was adjourned from 1966 to 1973.

The CAS held another meeting on the compilation of a national Flora and Fauna in 1973, and the *FRPS* project began to recover. By this time, both of the editors-in-chief had passed away. LIN Yong (LIN Rong) of the Institute of Botany, CAS, was designated as new editor-in-chief, with WU Cheng-yih (WU Zheng-Yi), TSUI Hung-Pin (CUI Hong-Bin), TSIEN Cho-Po (JIAN Zhuo-Bo) and HONG De-Yuan as associate editors-in-chief. Some middle-aged botanists were also admitted to the committee. Volume 36 (*Rosaceae, Part 1*) was published in 1974 (Yu, 1974), with another three volumes coming three years later: Volumes 63 (Tsiang & Li, 1977), 65 (Part 2; Wu & Li, 1977b) and 66 (Wu & Li, 1977a). From then on, the *FRPS* entered a "golden era." In 1977, YÜ Te-Tsun (YU De-Jun; Institute of Botany, CAS) was appointed editor-in-chief with WU and TSUI associate editors-in-chief. The editorial committee was enlarged to include some 40 members, representing 27 botanical research institutions or university departments. At the same time, for the sake of wide education, another series of books—*Iconographia Cormophytorum Sinicorum* (Institute of Botany, Chinese Academy of Sciences, 1972–1976)—was compiled, and two supplementary volumes published in 1982 and 1983 (Institute of Botany, Chinese Academy of Sciences, 1982; Institute of Botany, Chinese Academy of Sciences, 1983). This was the largest iconograph of plants in the world at that time, and the first comprehensive monograph on the Chinese flora.

Between 1978 and 1980 another 17 volumes of *FRPS* were published, and by 1986 41 books in 36 volumes had been produced, covering ca. 10 960 species. After the death of YÜ Te-Tsun (YU De-Jun) in 1986, WU Cheng-Yih (WU Zheng-Yi) of the Kunming Institute of Botany, CAS, was promoted to editor-in-chief, with TSUI Hung-Pin (CUI Hong-Bin) as associate editor-in-chief, along with another 21 members. At that time, volumes on many major families such as Asteraceae, Poaceae, Orchidaceae, Fabaceae, Ericaceae and Rubiaceae were still under compilation (Li, 2008). In 1996, the editorial committee was reorganized to include only 18 members with seven senior consultants, all members of whom were aged at least 60; a shortage of young, qualified taxonomists threatened the continuation of the project, with 44 books in 35 volumes remaining unpublished. To ensure the completion of the project, four younger botanists were elected to the committee in 1997. Eventually, in September 2004, all 126 books of the 80 volumes of this *magnum opus* were complete (Yang *et al.*, 2005). The monumental work was very much the fruit of long-term teamwork, over more than 45 years, involving

312 botanists across three generations and 164 artists producing 9 081 line drawings. Some 31 228 species in 3 444 genera and 302 families were covered by the *FRPS*. Excluding 48 species that were treated repeatedly, the number of vascular plants in the *FRPS* stands at 31 180 species in 3 434 genera and 300 families (Ma & Clemants, 2006).

Perhaps the most important role in, and contribution to, the *FRPS* project, was played by the late Professor WU Cheng-Yih (WU Zheng-Yi). A founding member of the first editorial committee of 1959–1972, associate editor-in-chief from 1973 to 1986, and editor-in-chief from 1987 to 2004 (Wu & Chen, 2004), he witnessed all stages of the project and led a major part of it to completion. Under his editorship, 82 books in 54 volumes of the *FRPS* were published, covering some 20 197 species in 2 019 genera and 166 families. His knowledge of Chinese plant diversity was widely appreciated in botanical circles across the world (Li, 2008).

14.4 The Flora of China

Because the *FRPS* was written in Chinese, and was not readily accessible to people outside the country, the need for an English-language Flora of China was recognized. Furthermore, at the initiation of the *FRPS* in 1959, most of its authors were unable to compare accounts of species described prior to 1950s against type specimens deposited in western herbaria. This, to some extent, made it difficult accurately to apply published names to specimens collected after the 1950s. In 1979, a Chinese botanical delegation, the first of its kind since the Cultural Revolution, visited the USA. Delegation members YÜ Te-Tsun (YU De-Jun) and WU Cheng-Yih (WU Zheng-Yi) proposed that a project be started to revise the *FRPS* through international collaboration. During the 1987 International Botanical Congress in Berlin, with the approval of the CAS, WU discussed details of the collaboration with Peter H. Raven, director of Missouri Botanical Garden, and agreed in principle to a joint international project between the CAS and Missouri Botanical Garden. This eventually became the *Flora of China* (*FOC*) project which officially commenced on 7 October 1988 when WU and Raven became co-chairs of the joint editorial committee, signing a formal agreement beneath a grove of *Metasequoia glyptostroboides* (Taxodiaceae) in Missouri Botanical Garden.

The *FOC* project has provided unique opportunities for collaboration between hundreds of Chinese and non-Chinese botanists and has been an important step in introducing the flora of China to the world. The *FOC* was designed to be published in 25 text volumes (one being an index and statistics), with 24 volumes of illustrations to accompany the floristic volumes, plus a companion introduction and overview volume. It follows approximately the same sequence of families as the *FRPS*, which is a modified Englerian system, treating the monocots after the dicots. Volume 17 was the first in the series to be published, covering Lamiaceae, Solanaceae and Verbenaceae (Wu & Raven, 1994), followed by Volumes 15 and 16 in 1995 (Wu *et al.*, 1995) and 1996 (Wu *et al.*, 1996), respectively. The first accompanying illustration volume (*Flora of China Illustrations, Volume 17*) was published in 1998 (Wu & Raven, 1998). During a joint editorial committee meeting in August 2001 in Kunming, CHEN Yi-Yu, then vice president of the CAS, and Raven signed a new contract on the project for the next five years. After the Kunming meeting, HONG De-Yuan was nominated to associate co-chair of the joint editorial committee, which in 2002 was expanded to include some middle-aged Chinese botanists. Several families have been revised in the *FOC* relative to the *FRPS*. For instance Poaceae, which in *FRPS* included some 1 942 species in 247 genera across multiple volumes, was treated in a single *FOC* volume (Chen *et al.*, 2006) containing 1 790 species in 227 genera—the latter relegating a large number of species to synonyms but also including many species not recorded in the *FRPS*.

The *FOC* series was completed in 2013. Because of the significance of the Chinese flora in a world-wide context, the publication of the *FOC* contributes greatly to our botanical knowledge of the world (Wu, 1989). The *FOC* provides botanists around the world not only with an English-language Flora of China, but also with an update of our knowledge of plant taxonomy and phytogeography in China.

14.5 Chinese regional and provincial Floras

The first Chinese regional Flora is considered to be Berthold C. Seemann's *Flora of the Island of Hongkong* (Seemann, 1857), which included 585 dicots, 119 monocots, three conifers, 64 ferns and fern allies, seven mosses and three lichens. In 1861, George Bentham published another Flora of Hong Kong, *Flora Hongkongensis: a Description of the Flowering Plants and Ferns of the Island of Hongkong*, covering over 1 000 species. Other regions for which Floras were written by European, Russian or Japanese botanists around the end of the nineteenth and beginning of the twentieth century include Shanghai Shi (Debeaux, 1875), Taiwan (Henry, 1896; Hayata, 1911; Hayata, 1911–1921), central China (Diels, 1900–1901), Yunnan (Franchet, 1889–1890), northeastern China (in part; Maximowicz, 1866–1877; Komarov, 1901–1907; Yabe, 1912), northwestern China (Maximowicz, 1889) and Xizang (Hemsley, 1902). For more detail on these regional Floras see Ma (2011).

In the 1920s, Chinese botanists began to collect plant specimens across the country. TSOONG Kuan-Kwang (ZHONG Guan-Guang) was the first to use scientific methods to collect plant specimens in China, in the late 1910s, although Père Siméon Ten, a follower of the missionary-collector Père François Ducloux, collected specimens for him as early as 1906. TSOONG collected more than 1.5 million herbarium specimens, and many genera and species have been named after him, including *Tsoongia* (Verbenaceae), *Tsoongiodendron* (Magnoliaceae, now a synonym of *Michelia*) and *Herminium tsoongii* (Orchidaceae, now a synonym of *Peristylus forceps*). TSOONG's collections provided abundant materials for the compilation of national and local Floras.

Other local and regional floras published by Chinese

botanists included HU Hsen-Hsu (HU Xian-Su)'s *A List of Chekiang Plants* (Hu, 1921a) and *A list of Kiangsi Plants including the Tsung An District of Fukien* (Hu, 1921b), based on a two-year field exploration. CHIEN Sung-Shu (QIAN Chong-Shu), CHENG Wan-Chun (ZHENG Wan-Jun) and PEI Chien (PEI Jian) published a series of papers on the vascular plants of Zhejiang (Cheng, 1933a; Cheng, 1933b; Chien & Cheng, 1934; Chien *et al.*, 1936), and CHIEN also published *A list of Kiangsu Plants* (Chien, 1919; Chien, 1920; Chien, 1921), while PEI conducted a brief study on the flowering plants of northwestern China (Pei, 1948a; Pei, 1948b; Pei, 1949). Another founder of modern Chinese botany, CHUN Woon-young (CHEN Huan-Yong), focused mainly on the plants of southern China, especially Guangdong and Hainan (Chun, 1934a; Chun, 1934b; Chun, 1940). WU Cheng-Yih (WU Zheng-Yi) made a brief survey of the vegetation of part of southwestern China in the 1940s (Kunming Institute of Botany, Chinese Academy of Sciences, 2006). At the same time, some western botanists continued collecting in China and published a few regional Floras, including *A List of Flowering Plants from Inner Asia* (Ostenfeld & Paulsen, 1922), *Plantae Sinensis* (Smith, 1924-1947), *Lineamenta Florae Manshuricae* (Kitagawa, 1939), *Flora Kainantensis* (Masamune, 1943) and a *Manual of Vascular Plants of the Lower Yangtze Valley, China* (Steward, 1958). See MA (2011) for more details.

After the establishment of the People's Republic of China in 1949, with the launching of the *FRPS* project, Chinese botanists also began compiling regional Floras, for which the earlier works provided some references and material. The first illustrated regional Flora in China was the *Illustrated Flora of Ligneous Plants of Northeast China* edited by LIOU Tchen-Ngo (LIU Shen-E; Liou, 1955). This was followed by the *Flora of Guangzhou* (How, 1956). During the 1960s and 1970s only three provincial floras were completed: the first edition of the *Flora of Beijing* (He, 1961; He, 1962; He, 1975), *Flora Hainanica* (Chun, 1964; Chun, 1965; Chun, 1974; Chun, 1977) and the first edition of the *Flora of Taiwan* (Li *et al.*, 1975–1979). Another five local floras were begun or partially published: *Flora Hupehensis,* the first edition of the *Flora Intramongolica, Flora of Jiangsu, Flora Yunnanica* and *Flora Plantarum Herbacearum Chinae Boreali-Orientalis.* From the 1980s to 1990s, the compilation of local Floras advanced rapidly, with ca. 20 local Floras completed and published including *Flora Xizangica* (see Liu *et al.*, 2007b). However, due to lack of funding and local expertise, a few local floras remain incomplete, including the *Flora of Gansu, Flora of Jiangxi* and *Flora Sichuanica.* Second editions of the *Flora of Beijing* (He *et al.*, 1984; He *et al.*, 1987; He *et al.*, 1992), *Flora Intramongolica* (Commissione Redactorum Florae Intramongolicae, 1989–1998) and *Flora of Taiwan* (Editorial Committee of the Flora of Taiwan Second Edition, 1993–2003) have also been published. To date, 32 of the 34 administrative divisions in China have local floras, although some of these are still being compiled (Ma & Liu, 2008). The flora of Jilin was covered by the *Illustrated Flora of Ligneous Plants of Northeast China* (Liou, 1955) and *Flora Plantarum Herbacearum Chinae Boreali-Orientalis* (Liou, 1958–2005), and that of Shaanxi by the *Flora Loess-Plateaus Sinicae* (Instituto Botanico Boreali-Occidentali Academiae Sinicae Edita, 1989–2000) and *Flora Tsinlingensis* (Instituto Botanico Boreali-Occidentali Academiae Sinicae Edita, 1974–1983).

Most of these local Floras were arranged according to CHING Ren-Chang (QIN Ren-Chang)'s system for pteridophytes, CHENG Wan-Chun (ZHENG Wan-Jun)'s system for gymnosperms and Engler's classification for angiosperms, with the exception of the *Flora of Guangdong, Flora of Guangxi, Flora Hainanica* and *Flora Yunnanica* which followed Hutchinson's system for angiosperms. All are in Chinese except the *Floras of Hong Kong* and *Taiwan*, which are in English. The latter two also include not only species descriptions and distribution information but also voucher details, making it easier to check and trace specimens. The *Flora of Hong Kong* also features, in addition to line drawings, photographs of most plants at the end of each volume. The *Flora of Macau* demonstrates another style of local Flora. It was initiated in 2003 based on herbarium collections and, within five years, three volumes, covering lycophytes and ferns, gymnosperms and angiosperms, as well as an index volume, were published. The most distinctive feature of this Flora is that every species is represented by a photograph in addition to the text, reinforcing the practical utility of the work. The *Flora of Macau* also features an introduction, index and species names in three languages: traditional Chinese, English, and Portuguese. The *Flora of Jiangxi, Flora Intramongolica* and *Flora of Zhejiang* all include a volume or chapters focusing on physical geography, the history of plant collecting, local floristic features and the use of plant resources, while the *Flora of Anhui, Flora Hebeiensis, Flora Shandongica, Flora Shanxiensis* and *Flora Xizangica* include general introductions to geological history, vegetation and phytogeography. The *Plants of Shanghai* is divided into two volumes, the first covering a brief history of botanical studies, physical geography, vegetation, keys and descriptions of native and naturalized plants in Shanghai, and the second on economic plants; this format is unique among local Floras in China. Other local Floras have been edited approximately following the style of *FRPS.*

In terms of species richness, *Flora Yunnanica* contains the highest species diversity among all published local Floras. *Flora Yunnanica* consists of 21 volumes covering bryophytes, pteridophytes, gymnosperms and angiosperms. Nearly 15 000 vascular plants are found in Yunnan, which constitutes some 48% of the entire flora of China. The editor in chief of this flora, WU Cheng-Yih (WU Zheng-Yi), together with other plant taxonomists, completed this masterpiece over 33 years of work (1973–2006).

In addition to the local Floras defined according to administrative divisions, several regional floras have been compiled based on natural geographic areas, such as the *Cryptogamic Flora of the Yangtze Delta and Adjacent Regions* (Hsu, 1989a), *Encyclopedia of Plants in Three Gorges of Yangtze River of China* (Peng, 2005), *Flora in Desertis Reipublicae Populorum Sinarum* (Liu, 1985–1992), *Flora Loess-Plateaus Sinicae* (Instituto Botanico Boreali-Occidentali Academiae Sinicae Edita, 1989–2000), *Flora*

Sinensis in Area Tang-yang (an area covering central Ningxia, northeastern Gansu, the Alxa Zouqi of Nei Mongol, and northern Shaanxi; Office of Agriculture Modernization of Ningxia Hui Autonomous Region—Animal Husbandry Bureau of Ningxia Hui Autonomous Region and Instituto Botanico Boreali-Occidentali Academiae Sinicae, 1988, 1996), *Flora Tsinlingensis* (Instituto Botanico Boreali-Occidentali Academiae Sinicae Edita, 1974-1983) and supplement (Li & Li, 2013), *The Shaanxi-Gansu-Ningxia Basin Flora* (Lo & Hsu, 1957), and *Vascular Plants of the Hengduan Mountains* (The Comprehensive Scientific Expedition to the Qinghai-Xizang Plateau, Chinese Academy of Sciences, 1993–1994). See Ma (2011) for more details. Such regional Floras facilitate further research, such as ecological studies, at a regional level. At the same time, monographs of certain plant groups, such as ferns, were also produced, including the *Preliminary Study on Pteridophyte Flora from Guangxi, China* (Zhou, 2000), *Pteridophyte Flora of Guizhou* (Wang & Wang, 2001), and *Yunnan Ferns of China* (Jiao & Li, 2001).

Overall, most of the Chinese local Floras are consistent with *FRPS* in general information, style of compilation and systems of classification, but tend to be more detailed in terms of distribution information, resources and utility, and local names of taxa. New genera and species were described during compilation of some local Floras, and some taxonomic revisions were made. For example, *Taihangia* (Rosaceae) was described in the progress of compiling *Flora Hebeiensis* (Yu & Li, 1980); *Yinshania* (Brassicaceae) in *Flora Intramongolica* (Ma & Zhao, 1979), and *Gaoligongshania* (Poaceae) in *Flora Yunnanica* (Sun, Li & Hsueh, 2003), published after the bamboo volume of *FRPS*. Thus, local floras have played a crucial role in improving our knowledge of the flora of China.

14.6 Taxonomic and monographic studies of important Chinese plants

The development of plant taxonomy in China has been shaped significantly by changes in Chinese society and regimes over the past three centuries or so. It is only since the 1920s that modern botany has taken root in the country. Since then, Chinese botanists have described a number of new species, genera, and families, some discoveries with great influence on botany world-wide. Such taxonomic work has provided materials for the compilation of *FRPS*, *FOC* and local Floras while, at the same time, the compilation of Floras has itself boosted taxonomic study and the recognition of plant diversity in China, the richest temperate flora in the world.

In the first 50 years of the twentieth century, Chinese taxonomy focused mainly on collecting specimens and describing new plants following the traditions of western botanists. During this period, approximately 400 scientific papers were published by Chinese botanists, including two new fern families (Christenseniaceae, now Marattiaceae, and Monachosoraceae, now Dennstaedtiaceae), 33 new genera, and hundreds of new species, according statistics compiled by Wang (1994). Perhaps the most significant discovery was that of the "living fossil" *Metasequoia glyptostroboides* (Taxodiaceae) in western Hubei (Hu & Cheng, 1948), as described in previous chapters, which made news around the world—including in the journal *Science* (Merrill, 1948)—and was subsequently the subject of many papers (see Ma, 2003) and planted in botanical gardens world-wide. Despite a lack of specimens and literature, Chinese botanists conducted studies on bryophytes, lycophytes and ferns, gymnosperms and angiosperms, particularly in relation to the production of *FRPS*. CHING Ren-Chang (QIN Ren-Chang), for instance, made tremendous contributions to Chinese pteridophyte taxonomy, proposing a new classification system for Polypodiaceae despite the harsh conditions of the Second World War (Ching, 1940).

After the founding of the People's Republic of China, a great deal of work was done on Chinese families and genera. Many more new taxa were discovered, such as a second living fossil, *Cathaya argyrophylla* (Pinaceae). New Chinese records were made for four families, Corsiaceae, Crypteroniaceae, Datiscaceae and Resedaceae (Wu & Wang, 1957a; Wu & Wang, 1957b; Zhang *et al.*, 1999). From the 1950s to the 1980s some 2 000 taxonomic papers were published, including descriptions of nine new families, more than 100 new genera and ca. 3 000 new species (Wang, 1994). *Parashorea chinensis* (Dipterocarpaceae; Cooperation Group of "*Parashorea chinensis*", 1977) and *Sapria himalayana* (Rafflesiaceae; Li, 1975) were both discovered in southern Yunnan, confirming the presence of tropical elements here. In the late 1970s, chemotaxonomy, cytotaxonomy, numerical taxonomy and palynology were gradually brought to bear, and these new methods and technologies brought plant taxonomy in China into a new era.

Chemotaxonomy is a method of biological classification based on similarities and differences in the structure of certain chemical compounds among species, and was introduced to China in the 1960s. For example, comparative studies of alkaloids, steroids, triterpenoids, cyanogenic glycosides and fatty acids in the monocots, Nymphaeales and Ranunculaceae suggested to YANG Tsung-Ren (YANG Chong-Ren) and CHOW Jun (ZHOU Jun) that the Liliales had a close affinity with Ranunculaceae and probably originated from within this dicot family (Yang & Chow, 1978). Gradually, chemotaxonomy was exploited in more plant groups, including pteridophytes, gymnosperms and angiosperms. In two books on the subject (Chen, 1990; Zhou & Duan, 2005), the authors introduce the development of plant chemotaxonomy in China, the chemical composition of different plant groups, and discuss the utility of chemotaxonomy in the identification of species and hybrids.

Cytotaxonomy is an approach to classification based on cellular structure and function, especially the number and structure of chromosomes. The first cytological study in China was on chromosome numbers in *Dioscorea* section *Stenophora* (Dioscoreaceae; The Research Group of *Dioscorea*, Kiangsu Institute of Botany, 1976), examining 16 species and two varieties. This revealed that the basic chromosome number of section *Stenophora* was x=10 and enabled taxonomic treatments of species and varieties to be made based on cytology, morphological characters and geographical distributions. Subsequently, with the advent of new techniques, plant cytotaxonomy developed very rapidly in China. In the 1980s about 100 papers were published on the subject (Wan & Chi, 1987), covering lycophytes and ferns, gymnosperms and angiosperms. For lycophytes and ferns, studies between 1981 and 2007 recorded chromosome numbers for 395 species from China, 15% of the country's total (Wang *et al.*, 2007). About 190 of these (ca. 48%) were found to be polyploid. Through more than 20 years of cytotaxonomic study in lycophytes and ferns, it has been proved that chromosome numbers, basic numbers, karyotypes, aneuploidy, ploidy levels and reproductive modes are useful in understanding the evolution, speciation and taxonomy of the group.

Cytological data for most gymnosperm families are now known, and they have certain taxonomic importance at generic and species levels. For angiosperms, studies remain to be carried out in several families. *Plant Cytotaxonomy* (Hong, 1990) details the importance of the number and structure of chromosomes in plant taxonomy, speciation and evolution, and summarizes data on chromosome number and polyploidy in algae, fungi, bryophytes, lycophytes, ferns, gymnosperms and angiosperms. HSU Bing-Sheng (XU Bing-Sheng) and colleagues (Hsu & Huang, 1985; Hsu & Yang, 1988; Hsu, 1989b) published three series of indices dealing with plant chromosome numbers reported in the Chinese literature from the 1960s to the 1980s, and in 2006 published the fourth series as a book, an *Index to Plant Chromosome Numbers Reported in Chinese Literature for 1989–1999* (Hsu, 2006), which included 4 230 chromosome counts for algae, bryophytes, lycophytes, ferns and seed plants from 811 articles and two monographs, corresponding to 220 species and 324 infraspecific taxa in 98 genera and 183 families. Although it does not include observations made after 2000, or published in western journals, the *Index* is still important for Chinese plant cytotaxonomy. A second monograph series, *Chromosome Atlas of Major Economic Plants Genome in China* (Chen *et al.*, 1993–2009) is also a valuable reference for cytological study. Its five volumes cover Chinese fruit trees and their wild relatives, crops and their wild relatives, ornamental plants, bamboos and medicinal plants.

Numerical taxonomy—grouping using numerical algorithms—began to develop in the late 1950s and early 1960s (Sokal & Sneath, 1963), in parallel with the growth of computers. The method was introduced to China in the late 1970s, but was not widely exploited in plant taxonomy until the 1980s (Wan & Chi, 1987). Most studies in China were performed on angiosperms and were usually to resolve taxonomic problems at generic or species level, rarely at higher levels. Study-subjects included monocots such as *Alisma* (Alismataceae; Zhang & Chen, 1994), Bambusoideae (note that subfamilial groups within Poaceae are not recognized in Flora of China; Chen *et al.*, 1983; Li & Hsueh, 1988) and *Elymus* (Chen & Xu, 1989; both Poaceae), *Carex* (Cyperaceae; Zhang *et al.*, 2010), Dioscoreaceae (Xi *et al.*, 1990), *Hemerocallis* (Liliaceae; Xiong *et al.*, 1997), and dicots such as *Aconitum* (Ranunculaceae; Li *et al.*, 1992), Caprifoliaceae (Hsu, 1983), *Cinnamomum* (Lauraceae; Tao & Zhong, 1988), *Cyclobalanopsis* (Liu *et al.*, 2008b) and *Quercus* (Peng *et al.*, 2007; both Fagaceae), Euphorbiaceae (Xu *et al.*, 2004), *Jasminum* (Oleaceae; Miao & Hu, 1983), *Panax* (Araliaceae; Xu & Li, 1983), *Rehmannia* and *Triaenophora* (Scrophulariaceae; Li *et al.*, 2008), *Schizopepon* (Cucurbitaceae; Lu & Zhang, 1985) and Styracaceae (Fan, 1996).

With regard to palynology, the spores of some 1 000 species across 174 genera and 52 families of lycophytes and ferns in China were described in *Sporae Pteridophytorum Sinicorum* (Palynology Research Group (Morphology Laboratory, Institute of Botany, Chinese Academy of Science, Beijing), 1960b). More recently, *Spores of Polypodiales (Filicales) from China* utilized light, transmission electron and scanning electron microscopes to observe morphological features of spores and the structure of their walls (Wang & Dai, 2010). The authors also demonstrated relationships among species and genera, and the evolution of spore characters. For angiosperms and gymnosperms, there are plenty of pollen monographs. The *Pollen Flora of China* (Palynology Research Group (Morphology Laboratory, Institute of Botany, Chinese Academy of Science, Beijing), 1960a; updated in Wang *et al.*, 1997a) included images and descriptions of pollen morphology for more than 1 400 species in 912 genera and 121 families of gymnosperms and angiosperms. *Pollen Flora of Seed Plants* (Wei *et al.*, 2003) is distinguished by its use of scanning electron micrographs of pollen grains. It comprises more than 1 500 micrographs of ca. 700 species in 350 genera and 113 families. Most recently, LI and colleagues (2011d) published a monograph on pollen morphology of Chinese woody plants. Some 1 097 species in 434 genera and 123 families were examined, involving 111 gymnosperms, 963 dicots and 23 monocots. Those monographs provide some taxonomic information at generic or familial levels. Additional works focused on specific genera or families include *A Study on Pollen Morphology and its Phylogeny of Polygonaceae in China* (Zhang & Zhou, 1998), in which the authors not only observed the morphology of pollen grains, but also discussed the distribution of Polygonaceae in China and speculated on the possible timing and locality of its origin.

Some monographs of specific genera and families have used comprehensive combinations of all these kinds of taxonomic disciplines. LIU (1971) utilized evidence from cytotaxonomy, fossils, geography, morphology and palynology in *Abies* (Pinaceae). On this basis he divided the genus into two subgenera, eleven sections and four central distribution regions. In addition, the systematic position of the genus and evolution within it were discussed. HONG De-Yuan (1984) took 49 representative taxa in 19 genera of the tribe Veroniceae (Scrophulariaceae; note that tribes are not generally recognized in the *Flora of China*), and used cladistic analysis together with previous results from anatomy, embryology, chemotaxonomy, cytotaxonomy and other studies, to establish a new classification system for the tribe. He also discovered an evolutionary reversal in pollen morphology from tricolporate to triporate, and hypothesized that the Veroniceae originated before the complete break-up of Pangaea, diversifying subsequently in both Laurasia and Gondwana. LI De-Zhu (1993) published a monograph of *Hemsleya* (Cucurbitaceae) by integrating information from morphology, anatomy, embryology, palynology, cytology and phytochemistry to study the phylogeny and taxonomy of 24 species of this Sino-Himalayan genus. He also used phenetic and cladistic methods to infer a phylogeny and proposed a new classification system for *Hemsleya*, which included six subsections, three sections, and two subgenera (Li, 1993). LI Heng (1998) published a monograph of *Paris* with twenty-four species in two subgenera and eight sections based on gross morphology, cytology, embryology, phytochemistry and immunoserology. MING Tian-Lu (2000) monographed *Camellia* world-wide from the aspects of anatomy, cytology, morphology and systematics, using a cladistic philosophy to explore phylogeny which, combined with other data, resulted in a division of the genus into 14 sections and two subgenera. *Gentiana* is the largest genus in Gentianaceae, consisting of approximately 500 species across the world. HO Ting-Nung (HE Ting-Nong) and LIU Shang-Wu studied huge numbers of specimens of this genus at both Chinese and western herbaria, and carried out studies in anatomy, biogeography, cytology, embryology, morphology and palynology. They summarized their results in *A Worldwide Monograph of* Gentiana (Ho & Liu, 2001). Their detailed taxonomic treatment of *Gentiana*, using cladistics and evolutionary taxonomy, proposed that the genus be divided into 22 series and 15 sections. HONG is subsequently working on a second monograph, *Peonies of the World* with three volumes, two of which have been published so far (Hong, 2010; Hong, 2011). In the first volume, the author makes an updated taxonomic revision of *Paeonia* (Paeoniaceae), based on extensive field observations, population sampling, examination of a more than 5 000 specimens (including types) and statistical analysis of characters, recognizing 32 species (Hong, 2010). The second volume includes 356 color photographs from the field, illustrating polymorphism and diversity in *Paeonia* (Hong, 2011). The third volume will focus on phylogeny and evolution of *Paeonia*.

In addition to these studies, iconographic monographs have also been compiled. For instance, *Native Orchids of China in Colour* (Chen *et al.*, 1999a) includes 791 photos of 403 species in 117 genera of Chinese wild orchids, with a description of each species in both Chinese and English, making the work accessible to western readers.

Chinese botanists have made a number of attempts at theoretical studies and systems of classification. For the lycophytes and ferns, CHING Ren-Chang (QIN Ren-Chang) proposed a systematic arrangement of Chinese species at familial and generic levels which included 11 orders, 63 families and 222 genera (Ching, 1978a; Ching, 1978b). For gymnosperms, CHENG Wan-Chun (ZHENG Wan-Jun) proposed a system of classification alongside the publication of *FRPS Volume 7* (Delectis Florae Reipublicae Popularis Sinicae Agendae Academiae Sinicae Edita, 1978), and this system has been recognized by some international botanists and adopted in most Chinese local Floras. HU Hsen-Hsu (HU Xian-Su; Hu, 1950) proposed a polyphyletic system of classification for angiosperms based on the Wieland system (Wieland, 1929), which may be the first system of classification of world-wide angiosperms proposed by a Chinese researcher. TSOONG Pu-Chiu (ZHONG Bu-Qiu; 1955; 1956) revised *Pedicularis* (Scrophulariaceae), based on a previous study by LI (1948; 1949). TSOONG discussed speciation, center of origin and relationships among species of *Pedicularis* according to floral morphology, phyllotaxis and patterns of distribution, and put forward a new classification of this large genus with a distinct hypothesis for its origin and dispersal.

Rhoipteleaceae is a monotypic family described by Handel-Mazzetti in 1932, which is distributed in southern Guizhou and Yunnan as well as Vietnam (Chang, 1981). The systematic position of this family had been controversial since its establishment. CHANG (1981) showed that the fundamental pattern of morphological and anatomical features—wood anatomy, resinous peltate scales, apetaly, bicarpellate pistils, one-seeded fruits and exalbuminous seeds—of Rhoipteleaceae was similar to that of Juglandaceae, and not closely related to Betulaceae or Ulmaceae. This relationship has since been supported by molecular phylogeny (Chen *et al.*, 1998). TANG and colleagues (1983) considered phytogeographical, cytological and palynological studies of Chinese Stachyuraceae, resulting in changing the status of some taxa, and discussed the sexuality of flowers in the group. CHEN Sing-Chi (CHEN Xin-Qi) discussed the origin and early differentiation of Orchidaceae based on fossil records, morphological evolution and the relationships among main groups (Chen, 1982). He suggested that the family could be divided into four subfamilies and five tribes, and that it originated in eastern Asia as early as the Cretaceous Period. This was the first hypothesis for Orchidaceae proposed by a Chinese botanist. YÜ Te-Tsun (YU De-Jun) illustrated the systematic position of Rosaceae among angiosperms, the subdivisions of the family and the relationships among genera (Yu, 1984).

The delimitation of genera within Magnoliaceae has been debated for more than a century. LAW Yuh-Wu (LIU Yu-Hu) has proposed a new classification for the family, mainly at subfamilial and tribal level, on the basis of external morphology, wood anatomy and palynology (Law, 1984). Although there remains much debate on generic relationships and recognition of genera, his subfamilial division has been widely adopted. Finally, the Fabaceae is the largest family of flowering plants in China after Asteraceae and Poaceae, and the taxonomy of this family at tribal and lower levels has been controversial for many years. ZHU Xiang-Yun has put forward an outline for the classification of Chinese Fabaceae with some new taxonomic treatments, based on previous systems presented by western botanists (Zhu, 2004).

With this massive accumulation of morphological, molecular and phytogeographical data, it became more and more evident that the division of angiosperms into dicots and monocots is artificial. In a summary paper for a major National Natural Science Foundation of China project, *Floristic Geography of the Chinese Seed Plants*, the late WU Cheng-Yih (WU Zheng-Yi) and colleagues (1998; 2002) proposed a new "polyphyletic-polychronic-polytopic" system of angiosperms based on the genealogical relationships of major groups, and other classification systems. The angiosperms (Magnoliophyta) were divided into eight classes, 40 subclasses, 202 orders, and 572 families. The eight classes are Magnoliopsida, Lauropsida, Piperopsida, Caryophyllopsida, Liliopsida, Ranuculopsida, Hamamelidopsida and Rosopsida. Later, WU and colleagues (2003) conducted a comprehensive analysis of families and genera (over 3 100 genera and 346 families) of angiosperms in China in light of the polyphyletic-polychronic-polytopic system, including information on morphology, molecular phylogenetics, fossils and geographical distribution. Since 1998, the "Angiosperm Phylogeny Group" (APG) classification (APG, 1998; APGII, 2003; APGIII, 2009) has dominated angiosperm classification elsewhere in the world, so the polyphyletic-polychronic-polytopic system may not be a dominant one for long. However, it remains a very important monographic study of Chinese plants based on a life-time of observation and a synthesis of botanical knowledge by one of the most knowledgeable botanists of our era (Figure 14.2).

Figure 14.2 HU Jin-Tao presenting the National Supreme Science and Technology Award of China to WU Cheng-Yih (WU Zheng-Yi) in Beijing, 2008 (image courtesy YANG Yun-Shan).

14.7 Molecular phylogenetics, biogeography and phylogeography

14.7.1 Molecular phylogenetics

Molecular biological techniques have been exploited in reconstructing plant phylogenies since the 1980s. Until 1991, the number of molecular-phylogenetic studies increased gradually, and about half were based on restriction fragment variation, especially in chloroplast DNA; after 1991 the number of studies based on nucleotide sequences rose dramatically (Donoghue, 1994). Taking seed plants as an example, Chase and colleagues (1993) first applied *rbc*L sequences to reconstruct phylogeny using two data sets, one containing 475 species, the other 499. Their results indicated that the *rbc*L gene contains evidence appropriate for phylogenetic analysis at higher taxonomic levels. Several of the major lineages recovered corresponded well with existing taxonomic schemes: although the morphological division into monocots and dicots was not supported, the major lineages corresponded well with pollen morphology (particularly aperture number). This analysis provided the first framework for future studies using both molecular and morphological data. Although *rbc*L provides useful information for phylogeny reconstruction at various levels, many important questions of plant evolution could not be resolved by this gene alone, leading subsequent studies to exploit additional genes and DNA regions from both the chloroplast and nuclear genomes (Soltis & Soltis, 1995).

The potential resolution obtained with eight loci from different genomes was explored at different hierarchical levels during a 1993 symposium of the American Institute of Biological Sciences (Soltis & Soltis, 1995). Due to the limited information provided by any single locus, molecular phylogenetics has since tended to use multiple regions from different genomes for phylogenetic reconstruction. For instance, Soltis and colleagues (1999) used the two plastid genes *rbc*L and *atp*B, and nuclear 18S rDNA, from 560 angiosperms and seven non-flowering seed plants to reconstruct the phylogeny of angiosperms; in the same year QIU and colleagues (1999) reported a phylogenetic analysis of DNA sequences from five mitochondrial, plastid and nuclear genes across all basally-branching angiosperm and gymnosperm lineages. Subsequently, increasing numbers of loci, and even genome-scale data have been employed to resolve the phylogenetic relationships of the seed plants (Qiu *et al.*, 2006; Moore *et al.*, 2007; Burleigh *et al.*, 2011; Soltis *et al.*, 2011). The new APG classification system for angiosperms was put forward at familial and ordinal levels based on these studies (APG, 1998; APGII, 2003; APGIII, 2009). Since plant classifications were previously constructed based primarily on morphological characters, a further trend in molecular-phylogenetic research has been to analyze molecular data sets in conjunction with morphological and other character data, so that the most natural taxonomic system could be built. Nandi and colleagues (1998) reconstructed the phylogeny of 162 extant angiosperms based on 252 non-molecular characters plus *rbc*L sequences. Their results indicated that new groupings found in the non-molecular analysis were congruent with those in the *rbc*L trees. Palaeobotanical discoveries and fossil data have also provided new evidence into the origin and evolution of angiosperms (Crane *et al.*, 1995; Moore *et al.*, 2007). The practice of phylogenetic analysis in plants thus now comprises a comprehensive science involving many different types of data.

In China, as in the rest of the world, plant molecular phylogenetics gradually developed from the 1990s.

At first, research was based primarily on molecular marker techniques, such as restriction fragment length polymorphisms (e.g. Huang *et al.*, 1996; Wang *et al.*, 1997b) and random amplified polymorphic DNA (e.g. Zhou *et al.*, 1999; Zou *et al.*, 1999). CHEN and FENG(1998) translated a number of important papers on molecular-phylogenetic studies in angiosperms (including several of those discussed above) into Chinese in the book *Advances in Plant Phylogenetics*. These led several Chinese research teams to start to use sequence data to reconstruct phylogenies. CHAW and colleagues (1997) analyzed a set of 65 nuclear 18S rRNA sequences from ferns, gymnosperms and angiosperms to infer the evolution and phylogeny of the seed plants; their results indicated that seed plants as a whole were monophyletic, with gymnosperms and angiosperms also both being monophyletic groups. In one of the earliest lower-level studies, GE and colleagues (1997) employed the nuclear ribosomal DNA internal transcribed spacer (ITS) to resolve relationships among species of *Adenophora* (Campanulaceae), finding that sequence divergence was relatively low among species but nonetheless allowing some conclusions about sister relationships within the genus to be drawn. As elsewhere, over the following years, Chinese researchers utilized increasing numbers of molecular loci and combined these with morphological, including fossil, data.

For example, WANG and colleagues (2000) conducted molecular-phylogenetic studies in Pinaceae using the chloroplast *mat*K gene, mitochondrial *nad*5 and low-copy nuclear 4CL genes. Pinaceae is the largest extant family of gymnosperms, comprising more than 200 species, and forms the dominant component of boreal, coastal and montane forests in the northern hemisphere. However, due to morphological convergence within the family, generic relationships and subdivisions have long been disputed. WANG and colleagues' combined analysis of the three gene sequences generated a well-resolved and strongly-supported phylogeny, which agreed to a certain extent with previous phylogenetic hypotheses based on morphological, anatomical and immunological data. In addition, the team estimated divergence times between genera based on sequence divergence in the *mat*K gene, which were found to correspond well with the fossil record (Wang *et al.*, 2000).

Around the same time, the phylogeny of Betulaceae was reconstructed using *rbc*L, ITS and morphological data (Chen *et al.*, 1999b). The results confirmed the monophyly of the family, and two sister clades were recovered, corresponding to subfamilies Betuloideae and Coryloideae (not recognized in the *Flora of China*). Again the pattern of divergence in the family agreed with the fossil record. A third prominent contribution to early Chinese molecular phylogenetics was made by GE and colleagues (1999), who used two nuclear genes (*Adh*1 and *Adh*2) and one chloroplast gene (*mat*K) to infer phylogenetic relationships in *Oryza* (rice; Poaceae). Their reconstructed phylogeny largely supported previously-recognized rice taxa, and the origins of allopolyploid species, which account for more than one-third of rice diversity, were illuminated. Since 2000, molecular-phylogenetic studies in China have accelerated significantly, covering a wide range of taxa at different hierarchical levels, facilitated by increasing numbers of available molecular regions and innovative sequencing techniques.

Lycophytes and ferns.—Molecular-phylogenetic studies in the lycophytes and ferns started much later than in the seed plants. However, already Chinese botanists have investigated Athyriaceae, Cheilanthoideae and Cryptogrammoideae of Pteridaceae, Cyatheaceae, Dryopteridaceae and Tectariaceae, at generic and familial levels. WANG and colleagues (2003) examined 34 representative species of Athyriaceae using the chloroplast *trn*L-F region to infer relationships among genera and subfamilies. They then re-divided the family into five subfamilies based on molecular and morphological data: Cystopterioideae, Athryoideae, Deparioideae, Diplazioideae and Rhachidosoroideae (Wang *et al.*, 2004).

Dryopteridoid ferns are distributed all over the world, with China and adjacent regions as their center of diversity and therefore a key region for studying the phylogeny of Dryopteridaceae, the delimitation of which, as well as infrafamilial relationships within the family, has long been controversial. LI and LU (2006) reconstructed the phylogeny of 105 species of Dryopteridaceae based on *rbc*L sequence data: the delineation of the family was redefined and relationships among genera, the affinity between Peranemaceae and Dryopteridaceae, and other relative relationships were discussed.

Tectariaceae has been considered a tribe of Polypodiaceae, a subfamily of Dryopteridaceae, or an independent family (Liu *et al.*, 2007a). Analyses of *rbc*L and *atp*B sequences from eight Chinese genera of Tectariaceae (*sensu* Ching, 1978a; Ching, 1978b) have indicated that the family as circumscribed was polyphyletic: *Ctenitis, Dryopsis, Lastreopsis* and *Pleocnemia* were better transferred to Dryopteridaceae, a taxonomy largely followed in the *Flora of China*. Tectariaceae has thus been newly delimited based on molecular phylogeny. LIU and colleagues (2008a) reconstructed the phylogeny of 184 species in 179 genera of pteridophytes as a whole, and presented a tentative classification of Chinese lycophytes and ferns at the familial level. Most recently, WANG and colleagues (2010a; 2010b) utilized four chloroplast regions to investigate the molecular phylogeny of Lepisoreae (a tribe not recognized in *Flora of China* but *sensu* Ching, 1978a; Polypodiaceae; Ching, 1978b; Hennipman *et al.*, 1990), resulting in a new classification of the tribe alongside a generic key. *Lepisorus* was revealed to be paraphyletic, with *Drymotaenium miyoshianum* and *Belvisia* nested in it (these have since been transferred to *Lepisorus* in the *Flora of China*). Nine well-supported major clades were recovered and the evolution of three morphological characters plus karyotypes were inferred. The utility of sequences from the three plant genomes (chloroplast, mitochondrial and nuclear) in current pteridophyte molecular phylogenetics has been summarized by LIU and colleagues (2009a).

Gymnosperms.—One of the most disputable issues in

the molecular phylogenetics of gymnosperms has been the relationship between Gnetales and the other groups (cycads, conifers and *Ginkgo*). CHAW and colleagues (2000) analyzed sequences of mitochondrial small subunit rRNA, nuclear small subunit rRNA and chloroplast *rbc*L gene. Although they found different relationships in the phylogenies drawn from different genes, the results indicated that Gnetales might be extremely divergent conifers, and that the morphological similarities between angiosperms and Gnetales had independent origins. As molecular technology improved and it becomes easier to obtain genomes, WU and colleagues (2007) used the complete chloroplast genome of *Cycas taitungensis* (Cycadaceae) and the 56 protein-coding genes in the chloroplast genome of *Gnetum parvifolium* (Gnetaceae), to conduct phylogenetic analyses on major seed plant lineages; their results were congruent with recent hypotheses for Gnetales—sister to the conifers or sister to the Pinaceae (known as the "gnepines" hypothesis). Given the likelihood of phylogenies being misled through "long branch attraction" in the case of Gnetales, its position was still considered unresolved. In 2010, ZHONG and colleagues (2010) analyzed several chloroplast protein data sets of the major seed plant lineages using the maximum likelihood techniques to ascertain the position of Gnetales. Their results supported the gnepines hypothesis and indicated that both long branch attraction and parallel substitutions had seriously biased previous inferences of the phylogenetic position of Gnetales using chloroplast genome data. Other notable Chinese phylogenetic studies on gymnosperms include those on Araucariaceae (Liu *et al.*, 2009b), cycads (Chaw *et al.*, 2005), Cupressaceae (Peng & Wang, 2008; Chen *et al.*, 2009; Mao *et al.*, 2012) and Pinaceae (Lin *et al.*, 2010).

Angiosperms.—Nearly 30 000 angiosperm species are found in China. Accordingly, Chinese molecular-phylogenetic studies have focused mainly on this group. The international APG effort (APG, 1998) included a number of Chinese botanists, and around the same time QIU and colleagues (1999) were the first to identify the "ANITA" group (*Amborella*, Nymphaeaceae, Illiciales, Trimeniaceae and *Austrobaileya*) as the basalmost-branching grade of angiosperms. ZHANG and GE (2007) have summarized some of the interesting and important advances in molecular-evolutionary study in China, including developments in plant molecular phylogenetics of angiosperms, such as studies on *Nannoglottis* (Asteraceae), *Oryza*, Betulaceae, Lythraceae,and Fagales. In another study, SHI and colleagues (2005) sequenced three genes (18S rRNA, *rbc*L and *mat*R) from the nuclear, chloroplast and mitochondrial genomes of mangroves to reconstruct phylogeny, obtaining a high-resolution tree and inferring that both vivipary and salt secretion experienced multiple independent origins.

The circumscription of Ranunculales has long been disputed. Until recently there had been consensus that the order comprises three main constituent clades: two earlier-diverging ranunculids, Eupteleaceae and Papaveraceae, and a well-supported clade of core Ranunculales comprising five families (Berberidaceae, Circaeasteraceae, Lardizabalaceae, Menispermaceae and Ranunculaceae; Kim *et al.*, 2004). However, the relationships between these three clades remained unresolved. WANG and colleagues (2009b) used an enlarged sample of taxa and three plastid regions (*rbc*L, *mat*K and *trn*L-F) plus nuclear ribosomal 26S rDNA, in combination with 65 morphological characters, to reconstruct the phylogenetic relationships within Ranunculales. Their results were consistent with those of Kim and colleagues (2004) but received stronger support and better resolution of relationships among the three main clades. Based on both morphological and molecular evidence, an outline classification for Ranunculales was proposed, including two new subfamilies, Menispermoideae and Tinosporoideae, in Menispermaceae and the new tribe Callianthemeae (Ranunculaceae). Working on Saxifragales, an early-branching eudicot lineage sister to Vitales and the rosids, as part of the continuous efforts of the group led by Douglas and Pam Soltis at the Florida Museum of Natural History, USA, JIAN and colleagues (2008) identified an ancient, rapid radiation during the Cretaceous Period.

The rosids are a major clade of angiosperms containing 13 orders and about one-third of all angiosperm species. Molecular analyses have recognized two major groups within the rosids, fabids and malvids, but relationships within these clades remained to be resolved with more data and taxon sampling. ZHU and colleagues (2007) analyzed two data sets for rosids, one composed of the mitochondrial *mat*R gene for 174 species representing 72 families, the other a combined data set for 91 taxa and four genes (*mat*R, *rbc*L, *atp*B and 18S rDNA). Their results showed that *mat*R appeared useful in higher-level angiosperm phylogenetics and alone identified a novel deep relationship within rosids, that of the COM clade (Celastrales, Oxalidales, Malpighiales and Huaceae). The four-gene analysis supported the placements of Geraniales and Myrtales as basally-branching lineages of the rosids and placed Crossosomatales as sister to malvids, while the core rosids included fabids, malvids and Crossosomatales (Zhu *et al.*, 2007). Two years later, based on analyses of sequences of two nuclear and ten plastid genes plus plastid inverted repeats from more than100 rosid species, WANG and colleagues (2009a) provided a further improved rosid phylogeny, identifying a rapid radiation over a period of fewer than 15 million years, and possibly only four to five million years. ZHANG and colleagues (2011b) investigated evolutionary relationships among families of Rosales using two nuclear genes and ten plastid loci. Rosales were strongly supported as monophyletic, Rosaceae was sister to all other members of the order, the remaining families were nested into two subclades, and a sister relationship between Dirachmaceae and Barbeyaceae was recovered for the first time with robust support.

Magnoliaceae is one of the oldest extant lineages of flowering plants and has played a crucial role in understanding the origin and diversification of angiosperms. This family is usually divided into two subfamilies, Magnolioideae and Liriodendroideae, but controversy remains over more detailed subdivisions. NIE and colleagues (2008) employed three nuclear genes to infer the phylogeny and biogeography of Magnoliaceae. They retrieved twelve well-supported groups within the family, largely consistent with recent taxonomic revisions at sectional and subsectional levels; however, the deeper relationships of subfamily Magnolioideae remain unresolved. In addition to these studies on higher taxonomic levels, a great many papers focusing on tribes and genera have been published, including Apioideae (*sensu* Pimenov & Leonov, 1993; Zhou *et al.*, 2009), Bambuseae (Yang *et al.*, 2008; Zeng *et al.*, 2010; Yang *et al.*, 2013), Oryzeae (Tang *et al.*, 2010), *Gaultheria* (Ericaceae; Lu *et al.*, 2010), *Musa* (Musaceae; Li *et al.*, 2010), *Soroseris* and *Syncalathium* (both Asteraceae; Zhang *et al.*, 2011a).

14.7.2 Biogeography

Advances in molecular phylogenetics have not only reinforced our understanding of plant evolutionary relationships but also promoted the study of plant biogeography in China. China has the most diverse vegetation of any country in the North Temperate Region, with very high levels of endemism (Ying, 2001), providing a fascinating base for plant biogeographic studies.

One of the most controversial topics in Chinese biogeography is the disjunct distribution of plants between eastern Asia and North America. Chinese botanists, particularly those who studied in the USA, have made great contributions to the study of intercontinental disjunctions. For example, one early paper by WEN and Zimmer (1996) on the biogeography of *Panax* (Araliaceae), has been cited ca. 150 times. WEN (1999) has summarized the earlier molecular-biogeographic advances based on this paper. XIANG and colleagues (1998; 2000; 2004) made further analyses of biogeographic patterns and evolution in a number of genera representing diverse angiosperm lineages, while Qian (2001; 2002b; 2002a) made a large-scale comparison of generic endemism and species richness of vascular plants between eastern Asia (particularly China) and North America. Since the 2000s, a younger generation of botanists has joined this effort.

Kelloggia (Rubiaceae) consists of only two, highly morphologically distinct, species, *K. chinensis* from the Hengduan Mountains of southwestern China and Bhutan, and *K. galioides* occurring in the western USA. To test the monophyly of the genus and infer its biogeographic history, NIE and colleagues (2005) estimated the molecular phylogeny of *Kelloggia* and relatives based on three chloroplast regions. The monophyly of the genus was strongly supported, but the inferred divergence time between the two species indicated that the genus might have originated in Asia and migrated to western North America through long-distance dispersal. Other oligotypic genera with disjunct eastern Asia-North America distributions have also been investigated, including *Hamamelis* (Hamamelidaceae; Xie *et al.*, 2010), *Phryma* (Phrymaceae; Nie *et al.*, 2006), *Sassafras* (Lauraceae; Nie *et al.*, 2007), and the *Adiantum pedatum* complex (Adiantaceae; Lu *et al.*, 2011). Other disjunct distribution patterns also occur. For example, the *Persea* group (Lauraceae) has a tropical and subtropical amphi-pacific disjunct distribution, and includes a total of 400 to 450 species in seven currently-recognized genera. Several questions are raised by this distribution and by differing opinions about generic delimitation, including whether recognized genera are monophyletic, whether American *Persea* is sister to the Asian genera, and when the disjunctions within the group and within *Persea* itself originated. To answer these questions, LI and colleagues (2011b) utilized two nuclear regions to reconstruct the phylogeny of the *Persea* group and estimate its divergence time. Their results indicated that the delimitations of *Persea*, *Alseodaphne*, *Dehaasia* and *Nothaphoebe* should be revised, that the group originated from a radiation in Early Eocene Laurasia, and that its amphi-pacific disjunction results from the disruption of the boreotropical flora by climatic cooling during the Mid to Late Eocene Epoch. Further examples of the amphi-pacific disjunct distribution pattern include *Toxicodendron* (Anacardiaceae; Nie *et al.*, 2009) and *Dendropanax* (Araliaceae; Li & Wen, 2013). Recently, MAO and colleagues (2012) sampled 162 species of all 32 genera of Cupressaceae, using ca. 10 000 nucleotides of plastid, mitochondrial and nuclear sequences per species and 16 fossil calibration points, and revealed that Cupressaceae originated during the Triassic Period when the supercontinent Pangea was still intact.

14.7.3 Phylogeography

Tremendous numbers of DNA-based phylogeographic studies have been reported in Europe, North America and Japan; however it was not until 2001 that similar investigations began to thrive in China. CHIANG and colleagues (2001) have studied the phylogeographic pattern of *Kandelia obovata* (Rhizophoraceae) in the mangroves of eastern China based on nucleotide variation in chloroplast and mitochondrial DNA. HUANG and colleagues (2001) have studied the molecular phylogeography of *Cycas taitungensis*, a relict species in Taiwan using organelle DNA haplotypes. QIU and colleagues (2011) have summarized phylogeographic studies from China and adjacent areas over the past decade, including the Qinghai-Xizang Plateau, Sino-Himalayan and Sino-Japanese Subkingdoms (*sensu* Wu & Wu, 1996). They found that three recurrent phylogeographic scenarios were identified by different case studies, which broadly agreed with long-standing biogeographic or paleoecological hypotheses, and also discovered an additional four instances that conflicted with the *a priori* predictions of paleo-data. Changes in plant distribution patterns are today a key subject in molecular phylogeography in China (e.g. Gao *et al.*, 2007; Jia *et al.*, 2012). In a speciel issue of *Journal of Systematics and Evolution* in 2012, some 11 phylogeographic studies on Chinese plants were collected (Liu et al., 2012), which covered Qinghai–Tibetan Plateau and southwestern China, western China, northern and northeast China, and southern and southeastern China. Those studies improved our knowledge of glacial refugia and postglacial range expansion of some representative species and their close relatives in China.

14.8 DNA barcoding

Identification is a keystone of biology. However, the total number of species in existence far outnumbers the names commonly used in English, Chinese or other living languages. In addition diagnostic characters vary greatly between different taxonomic groups. It takes time and effort to train a taxonomist, at a time when fewer and fewer young students are interested in what is seen as a traditional and unfashionable, albeit extremely important, discipline. Many students elect to learn the more modern biological techniques of molecular biology and biochemistry. Thus, in China and the rest of the world, taxonomists are themselves decreasing in number. The rise of DNA barcoding is expected to mitigate, at least in part, this problem (Li *et al.*, 2011a).

The concept of DNA barcoding was proposed to identify species rapidly and accurately by using short and standardized DNA markers (Floyd *et al.*, 2002; Hebert *et al.*, 2003). In fact, the idea of species identification using molecular evidence dates back to 1982, when it was proposed to determine the origin of fresh meat (Kang'ethe *et al.*, 1982). Since 2003, the DNA barcoding approach has been heavily promoted, mainly by zoologists, to provide tools for the recognition of species as an improvement on, or supplement to, traditional morphology-based taxonomy (Hebert *et al.*, 2003; Hebert & Gregory, 2005; Packer *et al.*, 2009). It seems that a portion of the mitochondrial cytochrome oxidase C subunit 1 (CO1) fulfills almost all animal barcoding requirements. However, although great efforts have been made in plant barcoding (Chase *et al.*, 2005; Kress *et al.*, 2005; Cowan *et al.*, 2006), progress has been hampered by three factors. Firstly, it is difficult to design universal primers for the targeted homologous markers across all plants. Secondly, the proposed DNA markers can easily be amplified and sequenced in some families or genera but not in others. Finally, for a given DNA marker, the genetic gaps between species are distinct in some plant groups but lacking in others. Despite these problems, DNA barcoding can be made applicable in plants by combining two or three DNA markers to make a standardized plant barcode. The CBOL (Consortium for the Barcode of Life) Plant Working Group (2009) proposed a combination of *rbc*L and *mat*K as the core plant barcode. In November 2009, during the Third International Barcoding of Life Conference in Mexico City, both the plastid *trn*H-*psb*A and ITS (or ITS2) regions were suggested as complementary markers to this proposed core barcode, to be evaluated further within 18 months. The candidate DNA barcoding markers have proven useful in diverse applications, such as uncovering cryptic species and delimiting closely-related species, establishing community phylogenies for diverse ecosystems, and discerning counterfeit plant materials in medicine and other fields (Li *et al.*, 2011a).

Plant DNA barcoding research in China has grown rapidly during recent years (Ren & Chen, 2010). In August 2009, the Barcoding Chinese Plants Project, based at the Germplasm Bank of Wild Species (GBOWS) at the Kunming Institute of Botany, was initiated with support from the CAS. A coordinated effort has been made among 60 research groups (together known as the China Plant BOL Group) from 22 research institutes and universities to evaluate the candidate barcode markers, and to barcode around 6 000 species of plants in China, including the seed collection held at the GBOWS facility (Li *et al.*, 2011a). So far the group has sampled 6 286 individuals representing 1 757 species in 141 genera, 75 families and 42 orders of seed plants, using four different methods of data analysis to assess the effectiveness and universality of the core barcode, *rbc*L + *mat*K, and the two complementary markers, *trn*H-*psb*A and ITS. The results so far imply that the three plastid markers show high levels of universality (87–93%), while ITS was relatively universal in angiosperms (79%) but comparatively worse in gymnosperms. ITS has shown the highest discriminatory power of the four markers, and a combination of ITS and any plastid DNA marker is able to discriminate many more species than the *rbc*L + *mat*K combination. The group therefore proposed that ITS or ITS2 should be incorporated into the core barcode for seed plants (China Plant BOL Group, 2011). Another study conducted by Chinese scientists, in which 50 790 plant and 12 221 animal ITS2 sequences were downloaded from GenBank (http://www.ncbi.nlm.nih.gov/genbank/), proved that the ITS2 locus should be used not only as a universal DNA barcode for plant discrimination but also as a complementary locus to CO1 to identify animal species (Yao *et al.*, 2010).

Prior to 2011 there were few studies applying plant DNA barcoding to ecology, and elucidating the ecological mechanisms underlying community assembly in subtropical forests remains a central challenge for ecologists. PEI and colleagues (2011) reconstructed a molecular phylogeny of 183 woody plant species in the 20 ha "Dinghushan forest dynamics plot" in Guangdong, from three DNA barcode loci (*rbc*L, *mat*K and *trn*H-*psb*A), to examine whether niche-based or neutral processes determine the assembly of species in this seasonal subtropical forest. They also compared the results of analyses utilizing different methods. Five habitat types and three spatial scales were used to quantify the community phylogenetic structure in the selected region. Their analyses demonstrated that there was a phylogenetic signal in plant-habitat associations (i.e. closely-related species tended to prefer similar habitats) and that patterns of co-occurrence within habitats were typically non-random with respect to phylogeny. The assembly of local-scale communities was explained by non-neutral niche-based processes acting upon evolutionarily-conserved habitat preferences (Pei *et al.*, 2011). This research was perhaps the first major application of DNA barcoding in ecological studies in China.

In 2011, a special issue of the *Journal of Systematics and Evolution* was published focusing on plant DNA barcoding in China (Li *et al.*, 2011a). It included 13 papers estimating the universality and efficiency of the core plant barcode, *mat*K + *rbc*L, and other candidate barcodes in gymnosperms and angiosperms. Among these, LI and colleagues (2011e) examined the universality of the candidate *mat*K primers in gymnosperms and proposed one pair of primers exhibiting highly successful amplification and sequencing in the group, with a specific new primer for Ephedraceae. In the same volume YU and colleagues (2011) developed a new pair of *mat*K primers for barcoding angiosperm species, which resulted in successful amplification and sequence analysis. Several barcodes were tested in *Hedyotis* (Rubiaceae; Guo *et al.*, 2011), *Ligustrum* (Oleaceae; Gu *et al.*, 2011), *Pugionium* (Brassicaceae; Wang *et al.*, 2011), *Primula* (Primulaceae; Yan *et al.*, 2011), *Pterygiella* (Scrophulariaceae; Dong *et al.*, 2011), *Tetrastigma* (Vitaceae; Fu *et al.*, 2011), *Gentianopsis paludosa* (Gentianaceae) and species often marketed in its place in Tibetan medicine (Xue & Li, 2011), and in Juglandaceae (Xiang *et al.*, 2011), Potamogetonaceae (Du *et al.*, 2011) and Zingiberaceae (Shi *et al.*, 2011). LI and colleagues (2011c) also reviewed the applications and limitations of *rbc*L, *mat*K, *trn*H-*psb*A and ITS in identifying herbal medicinal species. We are convinced that studies such as these will stimulate many future DNA barcoding studies on the great diversity of plants found in China, and enhance the creation of a reference database for plant barcoding, towards the eventual goal of a digitized encyclopedia of all Chinese plants.

14.9 Conclusions

Modern plant taxonomy has been developing in China for less than a century, since Chinese botanists started collecting and identifying plants systematically in the late 1910s. Developments in plant taxonomy in China were closely linked to the Chinese social revolution, but advancements in the field could also not be realized without innovative new botanical methods and techniques. Chemotaxonomy, comparative anatomy and embryology, cytotaxonomy, numerical taxonomy, palynology and molecular phylogenetics all provide the data for plant taxonomic and phylogenetic studies. The compilation of the *Flora Reipublicae Popularis Sinicae*, *Flora of China* and local Floras, by several generations of Chinese plant taxonomists, has significantly improved plant taxonomy in China. Now nearly 32 000 species—almost all higher plants (ferns and lycophytes, gymnosperms and angiosperms)—in China can be identified. These works provide the basis for future in-depth floristic analysis, molecular-phylogenetic, biogeographic and phylogeographic studies, as well as supporting evidence-based conservation of plant species.

Over the past decades, progress in genomics—for instance in the model species *Arabidopsis thaliana* (Brassicaceae), *Carica papaya* (Caricaceae), *Oryza sativa*, *Populus trichocarpa* (Salicaceae), *Vitis vinifera* (Vitaceae), *Zea mays* (Poaceae) and, more recently, *Picea abies* (Pinaceae)—has provided great opportunities for systematists and biogeographers. Rapid improvements in sequencing techniques (including "next" and "third generation" sequencers) have made it much easier to attain genomic information compared to the traditional Sanger sequencing method, and DNA barcoding technology also promises to make plant taxonomy more convenient and accessible to end-users. By combining the floristic data accumulated over the past century with information and methodologies developed in the genomic era, it is hoped that the study of plant systematics and biogeography in China and elsewhere will provide a solid evolutionary basis for plant diversity conservation and sustainable use of plant resources in the next decade and beyond (Li, 2008).

From the late sixteenth to the end of the twentieth century, a major focus in plant taxonomy was the writing of floristic works (Frodin, 2001). Traditional hard-copy floras are typically bulky in volume, often difficult to use, and take many years to compile, often leading to much of the information in the Flora being out of date by the time of publication. The rapid development of computer technology and the convenient availability of the internet, with its tools for information access and exchange, have led to the initiation of "eFloras" (http://www.efloras.org), a web-based program developed to enable access to online "electronic" Floras. Through a web interface, users can browse online floristic treatments by volume, family and genus, and can search by name, distribution data or text. Using web-based forms, editors and authors can also correct and update data in real-time (Brach & Song, 2006).

Recently, we put forward the idea of "next-generation Floras" or "iFloras" (Li *et al.*, 2012), which would be hyperlinked, dynamic, and available in a variety of formats (including "smart-phone" platforms), providing access to interactive keys as in all eFloras, but also fully incorporating DNA sequence data and molecular phylogenies as well as "cloud" and computer-assisted storage and identification tools. These iFloras would be outputted in three dimensions: taxon delimitation on the first axis, phylogenetic history on the second axis, and environmental factors on the third axis. Such integration would combine the diagnostic power of DNA sequence data with the traditional knowledge-base of taxonomy, providing powerful new tools accurately and rapidly to identify, monitor, conserve and utilize biodiversity (Li *et al.*, 2011a). If user-friendly iFloras can be achieved in the near future, both professional botanists and the general public would benefit, towards a fully bio-literate world.

References

APG (1998). An ordinal classification for the families of flowering plants. *Annals of the Missouri Botanical Garden* 85: 531-553.

APGII (2003). An update of the Angiosperm Phylogeny Group classification for the orders and families of flowering plants: APG II. *Botanical Journal of the Linnean Society* 141: 399-436.

APGIII (2009). An update of the Angiosperm Phylogeny Group classification for the orders and families of flowering plants: APG III. *Botanical Journal of the Linnean Society* 161: 105-121.

Brach, A. R. & Song, H. (2006). eFloras: new directions for online floras exemplified by the Flora of China Project. *Taxon* 55: 188-192.

Bretschneider, E. V. (1898). *History of European Botanical Discoveries in China*. London, UK: Sampson Low, Marston & Co.

CBOL Plant Working Group (2009). A DNA barcode for land plants. *Proceedings of the National Academy of Sciences of the United States of America* 106: 12794-12797.

Chang, C. Y. (1981). Morphology of the family Rhoipteleaceae in relation to its systematic position. *Acta Phytotaxonomica Sinica* 19: 168-178.

Chaw, S. M., Zharkikh, A., Sung, H. M., Lau, T. C. & Li, W. H. (1997). Molecular phylogeny of extant gymnosperms and seed plant evolution: analysis of nuclear 18s rRNA sequences. *Molecular Biology and Evolution* 14: 56-68.

Chaw, S. M., Parkinson, C. L., Cheng, Y. C., Vincent, T. M. & Palmer, J. D. (2000). Seed plant phylogeny inferred from all three plant genomes: monophyly of extant gymnosperms and origin of Gnetales from conifers. *Proceedings of the National Academy of Sciences of the United States of America* 97: 4086-4091.

Chaw, S. M., Walters, T. W., Chang, C. C., Hu, S. H. & Chen, S. H. (2005). A phylogeny of cycads (Cycadales) inferred from chloroplast *mat*K gene, *trn*K intron, and nuclear rDNA ITS region. *Molecular Phylogenetics and Evolution* 37: 214-234.

Chen, C. H., Huang, J. P., Tsai, C. C. & Chaw, S. M. (2009). Phylogeny of *Calocedrus* (Cupressaceae), an eastern Asian and western North American disjunct gymnosperm genus, inferred from nuclear ribosomal nrITS sequences. *Botanical Studies* 50: 425-433.

Chen, R. Y., Li, X. L., Song, W. Q., Liang, G. L., Zhang, P. X., Lin, R. S., Zong, W. X., Chen, C. B. & Fung, H. (1993-2009). *Chromosome Atlas of Major Economic Plants Genome in China*. Beijing, China: Science Press.

Chen, S. C. (1982). The origin and early differentiation of the Orchidaceae. *Acta Phytotaxonomica Sinica* 20: 1-22.

Chen, S. C., Tsi, Z. H. & Luo, Y. B. (1999a). *Native Orchids of China in Colour*. Beijing, China: Science Press.

Chen, S. L., Xu, K. X. & Sheng, G. Y. (1983). On the numerical classification and determination of taxa of Chinese bamboos with leptomorph rhizomes. *Acta Phytotaxonomica Sinica* 21: 113-120.

Chen, S. L. & Xu, K. X. (1989). Discussion on the validity of *Roegneria* C. Koch (Gramineae) as genus by numerical taxonomic method. *Acta Phytotaxonomica Sinica* 27: 190-196.

Chen, S. L., Li, D. Z., Zhu, G. H., Wu, Z. L., Lu, S. L., Liu, L., Wang, Z. P., Sun, B. X., Zhu, Z. D., Xia, N. H., Jia, L. Z., Guo, Z. H., Chen, W. L., Chen, X., Yang, G. Y., Phillips, S. M., Stapleton, C. M. A., Soreng, R. J., Aiken, S. G., Tzvelev, N. N., Peterson, P. M., Renvoize, S. A., Olonova, M. V. & Ammann, K. (2006). Poaceae. In: Wu, Z. Y., Raven, P. H. & Hong, D. Y. (eds.) *Flora of China*. Beijing, China and St. Louis, MO: Science Press and Missouri Botanical Garden Press.

Chen, X. Q. (1990). *Plant Chemotaxonomy*. Beijing, China: Higher Education Press.

Chen, Y. (1937). *Illustrated Manual of Chinese Trees and Shrubs*. Nanjing, China: Agricultural Association of China.

Chen, Z. D. & Feng, M. (1998). *Advances in Plant Phylogenetics*. Beijing, China: Science Press.

Chen, Z. D., Wang, X. Q., Sun, H. Y., Han, Y., Zhang, Z. X., Zou, Y. P. & Lu, A. M. (1998). Systematic position of the Rhoipteleaceae: evidence from nucleotide sequences of *rbc*L gene. *Acta Phytotaxonomica Sinica* 36: 1-7.

Chen, Z. D., Manchester, S. R. & Sun, H. Y. (1999b). Phylogeny and evolution of the Betulaceae as inferred from DNA sequences, morphology, and paleobotany. *American Journal of Botany* 86: 1168-1181.

Cheng, W. C. (1933a). An enumeration of vascular plants from Chekiang I. *Contributions from the Biological Laboratory of the Science Society of China, Botanical Series* 8: 298-306.

Cheng, W. C. (1933b). An enumeration of vascular plants from Chekiang II. *Contributions from the Biological Laboratory of the Science Society of China, Botanical Series* 9: 58-91.

Chiang, T. Y., Chiang, Y. C., Chen, Y. J., Chou, C. H., Havanond, S., Hong, T. N. & Huang, S. (2001).

Phylogeography of *Kandelia candel* in East Asiatic mangroves based on nucleotide variation of chloroplast and mitochondrial DNAs. *Molecular Ecology* 10: 2697-2710.

Chien, S. S. (1919). A list of Kiangsu plants. *Science (Shanghai)* 4: 117-124.

Chien, S. S. (1920). A list of Kiangsu plants. *Science (Shanghai)* 5: 207-212, 603-622, 729-748, 800-814, 1147-1165.

Chien, S. S. (1921). A list of Kiangsu plants. *Science (Shanghai)* 6: 211-219, 318-335, 417-424, 622-637, 720-733.

Chien, S. S. & Cheng, W. C. (1934). An enumeration of vascular plants from Chekiang III. *Contributions from the Biological Laboratory of the Science Society of China, Botanical Series* 9: 240-304.

Chien, S. S., Cheng, W. C. & Pei, C. (1936). An enumeration of vascular plants from Chekiang IV. *Contributions from the Biological Laboratory of the Science Society of China, Botanical Series* 10: 93-155.

China Plant BOL Group (2011). Comparative analysis of a large dataset indicates that internal transcribed spacer (ITS) should be incorporated into the core barcode for seed plants. *Proceedings of the National Academy of Sciences of the United States of America* 108: 19641-19646.

Ching, R. C. (1940). On natural classification of the family Polypodiaceae. *Sunyatsenia* 5(201-268).

Ching, R. C. (1959). Pteridophytes (1). In: Chien, S. S. & Chun, W. Y. (eds.) *Flora Reipublicae Popularis Sinicae*. Beijing, China: Science Press.

Ching, R. C. (1978a). The Chinese fern families and genera: systematic arrangement and historical origin. *Acta Phytotaxonomica Sinica* 16: 1-19.

Ching, R. C. (1978b). The Chinese fern families and genera: systematic arrangement and historical origin (cont.). *Acta Phytotaxonomica Sinica* 16: 16-37.

Chun, W. Y. (1934a). Additions to our Knowledge of the Hainan Flora. *Sunyatsenia* 2(23-48, 203-344).

Chun, W. Y. (1934b). Contributions to the Flora of Kwangtung and South-eastern China. *Sunyatsenia* 1: 209-316.

Chun, W. Y. (1940). Additions to the flora of Kwangtung and South-eastern China. *Sunyatsenia* 4: 169-271.

Chun, W. Y. (1964). *Flora Hainanica*. Beijing, China: Science Press.

Chun, W. Y. (1965). *Flora Hainanica*. Beijing, China: Science Press.

Chun, W. Y. (1974). *Flora Hainanica*. Beijing, China: Science Press.

Chun, W. Y. (1977). *Flora Hainanica*. Beijing, China: Science Press.

Commissione Redactorum Florae Intramongolicae (1989-1998). *Flora Intramongolica*. Huhhot, China: Typis Intramongolica Popularis.

Cooperation Group of "*Parashorea chinensis*" (1977). A rare and valuable new tree discovered in Yunnan - *Parashorea chinensis*. *Acta Phytotaxonomica Sinica* 15: 10-21.

Cox, E. H. M. (1945). *Plant Hunting in China*. London, UK: Scientific Book Guild.

Crane, P. R., Friis, E. M. & Pedersen, K. R. (1995). The origin and early diversification of angiosperms. *Nature* 374: 27-33.

Debeaux, J. O. (1875). Florule de Shang-hai (Province de Kiang-sou). *Actes de la Société Linnéenne de Bordeaux* 30: 57-130.

Delectis Florae Reipublicae Popularis Sinicae Agendae Academiae Sinicae Edita (1978). *Gymnospermae*. Beijing, China: Science Press.

Diels, L. (1900-1901). Die Flora von Central-China. *Botanische Jahrbücher für Systematik, Pflanzengeschichte und Pflanzengeographie* 29: 169-659.

Dong, L. N., Wortley, A. H., Wang, H., Li, D. Z. & Lu, L. (2011). Efficiency of DNA barcodes for species delimitation: A case in Pterygiella Oliv. (Orobanchaceae). *Journal of Systematics and Evolution* 49(3): 189-202.

Donoghue, M. J. (1994). Progress and prospects in reconstructing plant phylogeny. *Annals of the Missouri Botanical Garden* 81: 405-418.

Du, Z. Y., Qimike, A., Yang, C. F., Chen, J. M. & Wang, Q. F. (2011). Testing four barcoding markers for species identification of Potamogetonaceae. *Journal of Systematics and Evolution* 49(3): 246-251.

Editorial Committee of the Flora of Taiwan Second Edition (1993-2003). *Flora of Taiwan*. Taipei, Taiwan: Department of Botany, National Taiwan University.

Fan, G. S. (1996). Generic numerical taxonomy of Styracaceae from Asia. *Guihaia* 16: 305-307.

Floyd, R., Abebe, E., Papert, A. & Blaxter, M. L. (2002). Molecular barcodes for soil nematode identification. *Molecular Ecology* 11: 839-850.

Franchet, A. R. (1889-1890). *Plantae Delavayanae sive Enumeratio plantarum quas in provincia chinensi Yunnan, collegit J. M. Delavay*. Paris, France: P. Klincksieck.

Frodin, D. G. (2001). *Guide to Standard Floras of the World*. Cambridge, Uk: Cambridge University Press.

Fu, S. H. (1957). *Illustrations of Important Chinese Plants, Pteridophytes*. Beijing, China: Science Press.

Fu, Y. M., Jiang, W. M. & Fu, C. X. (2011). Identification of species within *Tetrastigma* (Miq.) Planch. (Vitaceae) based on DNA barcoding techniques. *Journal of Systematics and Evolution* 49(3): 237-245.

Gao, L. M., Möller, M., Zhang, X. M., Hollingsworth, M. L., Liu, J., Mill, R. R., Gibby M., Li, D. Z. (2007). High variation and strong phylogeographic pattern among cpDNA haplotypes in *Taxus wallichiana* (Taxaceae) in China and North Vietnam. *Molecular Ecology* 16: 4684–4698.

Ge, S., Schaal, B. A. & Hong, D. Y. (1997). A reevaluation of the status of *A. lobophylla* based on ITS sequence, with reference to the utility of ITS sequence in *Adenophora*. *Acta Phytotaxonomica Sinica* 35: 385-395.

Ge, S., Sang, T., Lu, B. R. & Hong, D. Y. (1999). Phylogeny of rice genomes with emphasis on origins of allotetraploid species. *Proceedings of the National*

Academy of Sciences of the United States of America 96: 14400-14405.

Gu, J., Su, J. X., Lin, R. Z., Li, R. Q. & Xiao, P. G. (2011). Testing four proposed barcoding markers for the identification of species within *Ligustrum* L. (Oleaceae). *Journal of Systematics and Evolution* 49(3): 213-224.

Guo, X., Simmons, M. P., But, P. P. H., Shaw, P. C. & Wang, R. J. (2011). Application of DNA barcodes in *Hedyotis* L. (Spermacoceae, Rubiaceae). *Journal of Systematics and Evolution* 49(3): 203-212.

Hayata, B. (1911). Materials for a flora of Formosa. *Journal of the College of Sciences, Imperial University, Tokyo* 30: 1-471.

Hayata, B. (1911-1921). *Icones Plantarum Formosanarum*. Taihoku, Taiwan: Bureau of Productive Industry, Government of Formosa.

He, S. Y. (1961). *Flora of Beijing*. Beijing, China: Beijing Publishing House.

He, S. Y. (1962). *Flora of Beijing*. Beijing, China: Beijing Publishing House.

He, S. Y. (1975). *Flora of Beijing*. Beijing, China: Beijing Publishing House.

He, S. Y., Xing, Q. H., Yin, Z. T. & Jiang, X. F. (1984). *Flora of Beijing*. Beijing, China: Beijing Publishing House.

He, S. Y., Xing, Q. H., Yin, Z. T. & Jiang, X. F. (1987). *Flora of Beijing*. Beijing, China: Beijing Publishing House.

He, S. Y., Xing, Q. H., Yin, Z. T. & Jiang, X. F. (1992). *Flora of Beijing (Revised)*. Beijing, China: Beijing Publishing House.

Hebert, P. D. N., Cwinska, A., Ball, S. L. & De Waard, J. R. (2003). Biological identifications through DNA barcodes. *Proceedings of the Royal Society of London, Series B Biological Sciences* 270: 313-321.

Hemsley, W. B. (1902). The flora of Tibet or High Asia; being a consolidated account of the various Tibetan botanical collections in the herbarium of the Royal Gardens, Kew, together with an exposition of what is known of the flora of Tibet. *Journal of the Linnean Society. Botany* 35: 124-265.

Hennipman, E., Veldhoen, P. & Kramer, K. U. (1990). Polypodiaceae. In: Kubitzki, K. (ed.) *Families and Genera of Vascular Plants I. Pteridophytes and gymnosperms*. pp. 203-230. Berlin, Germany: Springer-Verlag.

Henry, A. (1896). A list of plants from Formosa with some preliminary remarks on the geography, nature of the flora and econcmic botany of the island. *Transactions Asiatic Society of Japan* 24(Supplement): 1-118.

Ho, T. N. & Liu, S. W. (2001). *A Worldwide Monograph of Gentiana*. Beijing, China: Science Press.

Hong, D. Y. (1984). Taxonomy and evolution of the Veroniceae (Scrophulariaceae) with special reference to palynology. *Opera Botanica* 75: 5-60.

Hong, D. Y. (1990). *Plant Cytotaxonomy*. Beijing, China: Science Press.

Hong, D. Y. (2010). *Peonies of the World: Taxonomy and Phytogeography*. London, UK and St. Louis, MO: Royal Botanic Gardens, Kew and Missouri Botanical Garden.

Hong, D. Y. (2011). *Peonies of the World: Polymorphism and Diversity*. London, UK and St. Louis, MO: Royal Botanic Gardens, Kew and Missouri Botanical Garden.

How, F. C. (1956). *Flora of Guangzhou*. Beijing, China: Science Press.

How, F. C. (1958). *A Dictionary of the Families and Genera of Chinese Seed Plants*. Beijing, China: Science Press.

Hsu, P. S. (1983). A preliminary numerical taxonomy of the family Caprifoliaceae. *Acta Phytotaxonomica Sinica* 21: 26-33.

Hsu, P. S. & Huang, S. F. (1985). Index to plant chromosome numbers reported in Chinese literature. *Investigatio et Studium Naturae* 5: 1-116.

Hsu, P. S. & Yang, D. Q. (1988). Index to plant chromosome numbers reported in Chinese literature II. *Investigatio et Studium Naturae* 8: 1-82.

Hsu, P. S. (1989a). *Cryptogamic Flora of the Yangtze Delta and Adjacent Regions*. Shanghai, China: Shanghai Scientific and Technical Publishers.

Hsu, P. S. (1989b). Index to plant chromosome numbers reported in Chinese literature III. *Investigatio et Studium Naturae* 9: 1-87.

Hsu, P. S. (2006). *Index to Plant Chromosome Numbers Reported in Chinese Literature for 1989-1999*. Beijing, China: China Science and Technology Press.

Hu, H. H. (1921a). A list of Chekiang plants. *Science (Shanghai)* 6: 70-101.

Hu, H. H. (1921b). A list of Kiangsi plants (including the Tsung An district of Fukien). *Science (Shanghai)* 6: 1144-1171, 1232-1247.

Hu, H. H. & Chun, W. Y. (1927-1937). *Icones Plantarum Sinicarum*. Beijing, China: Fan Memorial Institute of Biology.

Hu, H. H. & Cheng, W. C. (1948). On the new family Metasequoiaceae and on *Metasequoia glyptostroboides*, a living species of the genus *Metasequoia* found in Szechwan and Hupeh. *Bulletin of the Fan Memorial Institute of Biology New Series* 1(2): 199-212.

Hu, H. H. (1950). A new classification system of the angiosperms based on multiple origins. *Science of China* 1: 243-253.

Hu, H. H., Fu, S. H. & Chien, S. S. (1953). Claves familiarum generumque plantarum sinicarum. *Acta Phytotaxonomica Sinica* 2: 173-337.

Huang, J. L., Zou, Y. P., Zhang, Z. X. & Hong, D. Y. (1996). PCR-RFLP analysis of *rbc*L gene in the tribe Convallarieae. *Acta Phytotaxonomica Sinica* 34: 531-537.

Institute of Botany - Chinese Academy of Sciences (1972-1976). *Iconographia Cormophytorum Sinicorum*. Beijing, China: Institute of Botany.

Institute of Botany - Chinese Academy of Sciences (1982). *Iconographia Cormophytorum Sinicorum. Supplementary Volume I*. Beijing, China: Institute of Botany.

Institute of Botany - Chinese Academy of Sciences (1983). *Iconographia Cormophytorum Sinicorum. Supplementary Volume II*. Beijing, China: Institute of Botany.

Instituto Botanico Boreali-Occidentali Academiae Sinicae Edita (1974-1983). *Flora Tsinlingensis*. Beijing,

China: Science Press.

Instituto Botanico Boreali-Occidentali Academiae Sinicae Edita (1989-2000). *Flora Loess-Plateaus Sinicae*. Beijing, China: Science Press, China Forestry Press, Scientific and Technical Documentation Press.

Jia, D. R., Abbott, R. J., Liu, T. L., Mao, K. S., Bartish, I. V. & Liu, J. Q. (2013). Out of the Qinghai–Tibet Plateau: evidence for the origin and dispersal of Eurasian temperate plants from a phylogeographic study of *Hippophae rhamnoides* (Elaeagnaceae). *New Phytologist* 194: 1123–1133.

Jian, S., Soltis, P. S., Gitzendanner, M. A., Moore, M. J., Li, R., Hendry, T. A., Qiu, Y. L., Dhingra, A., Bell, C. D. & Soltis, D. E. (2008). Resolving an ancient, rapid radiation in Saxifragales. *Systematic Biology* 57: 38-57.

Jiao, Y. & Li, C. S. (2001). *Yunnan Ferns of China*. Beijing, China: Science Press.

Judd, W. S., Campbell, C. S., Kellogg, E. A., Stevens, P. F. & Donoghue, M. J. (2007). *Plant Systematics: A Phylogenetic Approach*. Sunderland, Massachussetts: Sinauer Associates, Inc.

Kang'ethe, E. K., Jones, S. J. & Patterson, R. L. S. (1982). Identification of the species of origin of fresh meat using as enzyme-linked immunoabsorbent assay procedure. *Meat Science* 7: 229-240.

Keng, Y. L. (1959). *Flora Illustralis Plantarum Primarum Sinicarum, Gramineae*. Beijing, China: Science Press.

Kim, S., Soltis, D. E., Soltis, P. S., Zanis, M. J. & Suh, Y. (2004). Phylogenetic relationships among early-diverging eudicots based on four genes: were the eudicots ancestrally woody? *Molecular Phylogenetics and Evolution* 31: 16–30.

Kitagawa, M. (1939). Lineamenta Florae Manshuricae. *Report of the Institute of Scientific Research, Manchoukuo* 3(Appendix 1): 1-488.

Komarov, V. L. (1901-1907). Flora Manchuriae. *Acta Horti Petropolitani* 20-25: 1-559, 551-452, 551-334.

Kunming Institute of Botany - Chinese Academy of Sciences (2006). Wu Chengyih Corpus. Beijing, China: Science Press

Lancaster, R. (1989). *Travels in China - A Plantsman's Paradise*. Woodbridge, UK: Antique Collectors Club.

Law, Y. W. (1984). A preliminary study on the taxonomy of the family Magnoliaceae. *Acta Phytotaxonomica Sinica* 22: 80-109.

Li, C. X. & Lu, S. G. (2006). Phylogenetic analysis of Dryopteridaceae based on chloroplast *rbc*L sequences. *Acta Phytotaxonomica Sinica* 44: 503-515.

Li, D. Z. & Hsueh, C. J. (1988). A study on the genus *Dendrocalamus* Nees from China (I). *Journal of Bamboo Research* 7(3): 1-19.

Li, D. Z., Xu, K. X. & Xiao, P. G. (1992). A preliminary numerical classification on infrageneric taxa of the Chinese *Aconitum*. *Journal of Wuhan Botanical Research* 10: 226-234.

Li, D. Z. (1993). *Systematics and Evolution of Hemsleya (Cucurbitaceae)*. Kunming, China: Yunnan Science and Technology Press.

Li, D. Z. (2008). Floristics and plant biogeography in China. *Journal of Integrative Plant Biology* 50(7): 771-777.

Li, D. Z., Liu, J. Q., Chen, Z. D., Wang, H., Ge, X. J., Zhou, S. L., Gao, L. M., Fu, C. X. & Chen, S. L. (2011a). Plant DNA barcoding in China. *Journal of Systematics and Evolution* 49: 165-168.

Li, D. Z., Wang, Y. H., Yi, T. S., Wang, H., Gao, L. M. & Yang, J. B. (2012). The next generation flora, iFlora. *Plant Diversity and Resources* 34: 525-531.

Li, H. (1998). The Genus *Paris* (Trilliaceae). Beijing, China: Science Press.

Li, H. L. (1948). A revision of the genus *Pedicularis* in China I. *Proceedings of the Academy of Natural Sciences of Philadelphia* 100: 205-378.

Li, H. L. (1949). A revision of the genus *Pedicularis* in China II. *Proceedings of the Academy of Natural Sciences of Philadelphia* 101: 1-214.

Li, H. L., Liu, T. S., Huang, T. C., Koyama, T. & DeVol, C. E. (1975-1979). *Flora of Taiwan*. Taipei, Taiwan: Epoch Publishing Co., Ltd.

Li, H. W. (1975). The Rafflesiaceae in China. *Acta Phytotaxonomica Sinica* 13: 79-82.

Li, L., Li, J., Rohwer, J. G., van der Werff, H., Wang, Z. H. & Li, H. W. (2011b). Molecular phylogenetic analysis of the *Persea* group (Lauraceae) and its biogeographic implications on the evolution of tropical and subtropical Amphi-Pacific disjunctions. *American Journal of Botany* 98: 1520-1536.

Li, L. F., Häkkinen, M., Yuan, Y. M., Hao, G. & Ge, X. J. (2010). Molecular phylogeny and systematics of the banana family (Musaceae) inferred from multiple nuclear and chloroplast DNA fragments, with a special reference to the genus *Musa*. *Molecular Phylogenetics and Evolution* 57: 1-10.

Li, M., Cao, H., But, P. P. H. & Shaw, P. C. (2011c). Identification of herbal medicinal materials using DNA barcodes. *Journal of Systematics and Evolution* 49: 271-283.

Li, R. & Wen, J. (2013). Phylogeny and biogeography of *Dendropanax* (Araliaceae), an amphi-Pacific disjunct genus between tropical/subtropical Asia and the neotropics. *Systematic Botany* 38: 536-551.

Li, S. F. & Li, B. (2013). *Flora Tsinlingensis (Supplementum) Spermatophyte*. Beijing, China: Science Press.

Li, T. Q., Cao, H., Kang, M. S., Zhang, Z. X., Zhao, N. & Zhang, H. (2011d). *Pollen Flora of China Woody Plants by SEM*. Beijing, China: Science Press.

Li, X. D., Zan, Y. Y., Li, J. Q. & Yang, S. Z. (2008). A numerical taxonomy of the genera *Rehmannia* and *Triaenophora*. *Acta Phytotaxonomica Sinica* 46: 730-737.

Li, Y., Gao, L. M., Poudel, R. C., Li, D. Z. & Forrest, A. (2011e). High universality of *mat*K primers for barcoding gymnosperms. *Journal of Systematics and Evolution* 49: 169-175.

Lin, C. P., Huang, J. P., Wu, C. S., Hsu, C. Y. & Chaw, S. M. (2010). Comparative chloroplast genomics reveals the evolution of Pinaceae genera and subfamilies. *Genome Biology and Evolution* 2: 504-517.

Liou, T. N. (1955). *Illustrated Flora of Ligneous Plants of*

Northeast China. Beijing, China: Science Press.

Liou, T. N. (1958-2005). *Flora Plantarum Herbacearum Chinae Boreali-Orientalis*. Beijing, China: Science Press.

Liu, H. M., Zhang, X. C., Chen, Z. D., Dong, S. Y. & Qiu, Y. L. (2007a). Tectariaceae is polyphyletic : evidence from the chloroplast *rbc*L and *atp*B genes. *Science China Life Sciences* 37: 575-584.

Liu, H. M., Wang, L., Zhang, X. C. & Zeng, H. (2008a). Advances in the studies of lycophytes and monilophytes with reference to systematic arrangement of families distributed in China. *Journal of Systematics and Evolution* 46: 808-829.

Liu, H. M., Zhang, X. C. & Zeng, H. (2009a). Applicaiton of DNA sequences in pteridophyte phylogenetic study. *Chinese Bulletin of Botany* 44: 143-158.

Liu, J. Q., Sun, Y. S., Ge, X. J., Gao, L. M., Qiu, Y. X. (2012). Phylogeographic studies of plants in China: Advances in the past and directions in the future. *Journal of Systematics and Evolution* 50: 267-275.

Liu, L. Y., Zhang, M. L., Li, J. Q. & Peng, Y. S. (2008b). A numerical taxonomic study of the genus *Cyclobalanopsis* Oersted from China. *Journal of Wuhan Botanical Research* 26: 466-475.

Liu, N., Zhu, Y. J., Wei, Z. X., Chen, J., Wang, Q. B., Jian, S. G., Zhou, D. W., Shi, J., Yang, Y. & Zhong, Y. (2009b). Phylogenetic relationships and divergence times of the family Araucariaceae based on the DNA sequences of eight genes. *Chinese Science Bulletin* 54: 2648-2655.

Liu, Q. R., Yu, M. & Ma, J. S. (2007b). Review on the Chinese local floras. *Guihaia* 27: 844-849.

Liu, T. S. (1971). *A Monograph of the Genus Abies*. Taipei, Taiwan: The Department of Forestry College of Agriculture, National Taiwan University.

Liu, Y. H. (1985-1992). *Flora in Desertis Reipublicae Populorum Sinarum*. Beijing, China: Science Press.

Lo, T. Y. & Hsu, W. Y. (1957). *The Shaanxi-Gansu-Ningxia Basin Flora*. Beijing, China: Beijing Forestry Press.

Lu, A. M. & Zhang, Z. Y. (1985). Studies on the genus *Schizopepon* Maxim. (Cucurbitaceae). *Acta Phytotaxonomica Sinica* 23: 106-120.

Lu, J. M., Li, D. Z., Lutz, S., Soejima, A., Yi, T. S. & Wen, J. (2011). Biogeographic disjunction between eastern Asia and North America in the *Adiantum pedatum* complex (Pteridaceae). *American Journal of Botany* 98: 1680-1693.

Lu, L., Fritsch, P. W., Cruz, B. C. & et al. (2010). Reticulate evolution, cryptic species, and character convergence in the core East Asian clade of *Gaultheria* (Ericaceae). *Molecular Phylogenetics and Evolution* 57: 364-379.

Ma, J. S. & Clemants, S. (2006). A history and overview of the *Flora Reipublicae Popularis Sinicae* (FRPS, Flora of China, Chinese Edition, 1959-2004). *Taxon* 55(2): 451-460.

Ma, J. S. (2003). Notes on discovery of *Metasequoia glyptostroboides*. *Plants (Beijing)* 3: 37-40.

Ma, J. S. & Barringer, K. (2005). Dr. Hsen-Hsu Hu (1894-1968): a founder of modern plant taxonomy in China. *Taxon* 54: 559-566.

Ma, J. S. & Liu, Q. R. (2008). The Local Floras of China. <http://www.metasequoia.org/local.htm>.

Ma, J. S. (2011). *The outline of taxonomic literature of Eastern Asian higher plants*. Beijing, China: Higher Education Press.

Ma, Y. C. & Zhao, Y. Z. (1979). *Yinshania*, a new genus of Chinese Cruciferae. *Acta Phytotaxonomica Sinica* 17: 113-114.

Mao, K. S., Milne, R. I., Zhang, L. B., Peng, Y. L., Liu, J. Q., Thomas, P., Mill, R. R. & Renner, S. S. (2012). Distribution of living Cupressaceae reflects the breakup of Pangea. *Proceedings of the National Academy of Sciences of the United States of America* 109: 7793-7798.

Masamune, G. (1943). *Flora Kainantensis*. Taihoku, Taiwan: Taiwan Sotokufu Gaijabu.

Maximowicz, C. J. (1866-1877). Diagnoses Plantarum Novarum Japoniae et Mandshuriae. *Bulletin de l'Académie Impériale des Sciences de Saint-Petersbourg* 0-22: 485-490, 429-439, 460-473, 225-231, 225-232, 373-381, 212-226, 142-180, 417-456, 435-472, 275-296, 371-402, 158-186, 247-287, 475-540, 430-472, 209-264.

Maximowicz, C. J. (1889). *Enumeratio Plantarum Hucusque in Mongolia Nec Non Adjacente Parte Turkestaniae Sinensis Lectarum. Fascicule 1, Thalamiflorae et Disciflorae*. Leningrad, Russia: Petropoli, Typis Academiae Imperialis Scientiarum Petropolitanae.

Merrill, E. D. (1948). A living *Metasequoia* in China. *Science* 107: 140.

Miao, B. M. & Hu, J. Q. (1983). A numerical taxonomy of Chinese *Jasminum*. *Acta Phytotaxonomica Sinica* 21: 297-305.

Ming, T. L. (2000). *Monograph of the Genus Camellia*. Kunming, China: Yunnan Science and Technology Press.

Moore, M. J., Bell, C. D., Soltis, P. S. & Soltis, D. E. (2007). Using plastid genome-scale data to resolve enigmatic relationships among basal angiosperms. *Proceedings of the National Academy of Sciences of the United States of America* 104(49): 19363-19368.

Nandi, O. W., Chase, M. W. & Endress, P. K. (1998). A combined cladistic analysis of angiosperms using *rbcL* and non-molecular data sets. *Annals of the Missouri Botanical Garden* 85: 137-212.

Nie, Z. L., Wen, J., Sun, H. & Bartholomew, B. (2005). Monophyly of *Kelloggia* Torrey ex Benth. (Rubiaceae) and evolution of its intercontinental disjunction between western North America and eastern Asia. *American Journal of Botany* 92: 642-652.

Nie, Z. L., Sun, H., Beardsley, P. M., Olmstead, R. G. & Wen, J. (2006). Evolution of biogeographic disjunction between eastern Asia and eastern North America in *Phryma* (Phrymaceae). *American Journal of Botany* 93: 1343-1356.

Nie, Z. L., Wen, J. & Sun, H. (2007). Phylogeny and biogeography of *Sassafras* (Lauraceae) disjunct between eastern Asia and eastern North America. *Plant Systematics and Evolution* 267: 191-203.

NIE, Z. L., WEN, J., AZUMA, H., QIU, Y. L., SUN, H., MENG, Y., SUN, W. B. & ZIMMER, E. A. (2008). Phylogenetic and biogeographic complexity of Magnoliaceae in the Northern Hemisphere inferred from three nuclear data sets. *Molecular Phylogenetics and Evolution* 48: 1027-1040.

NIE, Z. L., SUN, H., MENG, Y. & WEN, J. (2009). Phylogenetic analysis of *Toxicodendron* (Anacardiaceae) and its biogeographic implications on the evolution of north temperate and tropical intercontinental disjunctions. *Journal of Systematics and Evolution* 47: 416-430.

OFFICE OF AGRICULTURE MODERNIZATION OF NINGXIA HUI AUTONOMOUS REGION - ANIMAL HUSBANDRY BUREAU OF NINGXIA HUI AUTONOMOUS REGION AND INSTITUTO BOTANICO BOREALI-OCCIDENTALI ACADEMIAE SINICAE. (1998, 1996). (*Flora Sinensis in Area Tang-yang*). Yinchuan, China: Ningxia People's Press.

OSTENFELD, C. E. H. & PAULSEN, O. W. (1922). A list of flowering plants from inner Asia collected by Dr. Sven A. Hedin determined by various authors and compiled by Carl E.H.Ostenfeld and Ove W. Paulsen. In: Hedin, S. A. (ed.) *Southern TIber Discoveries in Former TImes Compared with My Own Researches in 1906-1908*. pp. 1-193. Stockholm, Sweden: Lithographic Institute of the General Staff of the Swedish Army.

PALYNOLOGY RESEARCH GROUP (MORPHOLOGY LABORATORY - INSTITUTE OF BOTANY - CHINESE ACADEMY OF SCIENCE - BEIJING) (1960a). *Pollen Flora of China*. Beijing, China: Science Press.

PALYNOLOGY RESEARCH GROUP (MORPHOLOGY LABORATORY - INSTITUTE OF BOTANY - CHINESE ACADEMY OF SCIENCE - BEIJING) (1960b). *Sporae Pteridophytorum Sinicorum*. Beijing, China: Science Press.

PEI, C. (1948a). Flowering Plants of Northwestern China I. *Botanical Bulletin of Academia Sinica* 2: 96-106.

PEI, C. (1948b). Flowering Plants of Northwestern China II. *Botanical Bulletin of Academia Sinica* 2: 215-227.

PEI, C. (1949). Flowering Plants of Northwestern China III. *Botanical Bulletin of Academia Sinica* 3: 28-36.

PEI, N. C., LIAN, J. Y., ERICKSON, D. L., SWENSON, N. G., KRESS, W. J., YE, W. H. & GE, X. J. (2011). Exploring tree-habitat associations in a Chinese subtropical forest plot using a molecular phylogeny generated from DNA barcode loci. *PLoS One* 6: e21273.

PENG, D. & WANG, X. Q. (2008). Reticulate evolution in *Thuja* inferred from multiple gene sequences: implications for the study of biogeographical disjunction between eastern Asia and North America. *Molecular Phylogenetics and Evolution* 47: 1190-1202.

PENG, Y. S., CHEN, L. & LI, J. Q. (2007). Study on numerical taxonomy of *Quercus* L. (Fagaceae) in China. *Journal of Wuhan Botanical Research* 25: 149-157.

PENG, Z. H. (2005). *Encyclopedia of Plants in Three Gorges of Yangtze River of China*. Beijing, China: Science Press.

PIMENOV, M. G. & LEONOV, M. V. (1993). *The Genera of the Umbelliferae: A Nomenclator*. London, UK: Royal Botanic Gardens, Kew.

PRICE, W. R. (1982). *Plant Collecting in Formosa*. Taiwan: The Chinese Forestry Association.

QIAN, H. (2001). A comparison of generic endemism of vascular plants between East Asia and North America. *International Journal of Plant Sciences* 162: 191-199.

QIAN, H. (2002a). Floristic relationships between eastern Asia and North America: test of Gray's hypothesis. *The American Naturalist* 160: 317-332.

QIAN, H. (2002b). A comparison of the taxonomic richness of temperate plants in East Asia and North America. *American Journal of Botany* 89: 1818-1825.

QIU, Y. L., LEE, J. H., BERNASCONI-QUADRONI, F., SOLTIS, D. E., SOLTIS, P. S., ZANIS, M. J., ZIMMER, E. A., CHEN, Z. D., SAVOLAINEN, V. & CHASE, M. W. (1999). The earliest angiosperms: evidence from mitochondrial, plastid and nuclear genomes. *Nature* 402: 404-407.

QIU, Y. X., FU, C. X. & COMES, H. P. (2011). Plant molecular phylogeography in China and adjacent regions: tracing the genetic imprints of Quaternary climate and environmental change in the world's most diverse temperate flora. *Molecular Phylogenetics and Evolution* 59: 225-244.

REN, B. Q. & CHEN, Z. D. (2010). DNA barcoding plant life. *Chinese Bulletin of Botany* 45: 1-12.

SEEMANN, B. C. (1857). *Flora of the Island of Hongkong: The Botany of the Voyage of H.M.S.Herald, under the Command of the Captain Henry Kellett, R.N., C.B. during the year 1845-51*. London, UK: Lovell Reeve.

SHI, L. C., ZHANG, J., HAN, J. P., SONG, J. Y., YAO, H., ZHU, Y. J., LI, J. C., WANG, Z. Z., XIAO, W., LIN, Y. L., XIE, C. X., QIAN, Z. Z. & CHEN, S. L. (2011). Testing the potential of proposed DNA barcodes for species identification of Zingiberaceae. *Journal of Systematics and Evolution* 49(3): 261-266.

SHI, S., HUANG, Y., ZENG, K., TAN, F., HE, H., HUANG, J. & FU, Y. (2005). Molecular phylogenetic analysis of mangroves: independent evolutionary origins of vivipary and salt secretion. *Molecular Phylogenetics and Evolution* 34: 159-166.

SMITH, H. (1924-1947). Plantae Sinenses. *Acta Horti Gothoburgensis* 1-17: multiple entries.

SOKAL, R. R. & SNEATH, P. H. A. (1963). *Principles of Numerical Taxonomy*. San Francisco, CA: W.H. Freeman.

SOLTIS, D. E., SOLTIS, P. S. & CHASE, M. W. (1999). Angiosperm phylogeny inferred from multiple genes as a tool for comparative biology. *Nature* 402(25 November 1999): 402-404.

SOLTIS, P. S. & SOLTIS, D. E. (1995). Alternative genes for phylogenetic reconstruction in plants: introduction. *Annals of the Missouri Botanical Garden* 82: 147-148.

STEWARD, A. N. (1958). *Manual of Vascular Plants of the Lower Yangtze, China*. Corvallis, OR: Oregon State College.

SUN, B. X., LI, D. Z. & HSUEH, C. J. (2003). *Poaceae (Gramineae) Flora Yunnanica*. Beijing, China: Science Press.

TANG, L., ZOU, X. H., ACHOUNDONG, G., POTGIETER, C., SECOND, G., ZHANG, D. Y. & GE, S. (2010). Phylogeny and biogeography of the rice tribe (Oryzeae): Evidence from combined analysis of 20 chloroplast fragments. *Molecular Phylogenetics and Evolution* 54: 266-277.

Tang, T. & Wang, F. T. (1961). Cyperaceae (1). In: Chien, S. S. & Chun, W. Y. (eds.) *Flora Reipublicae Popularis Sinicae*. Beijing, China: Science Press.

Tang, Y. C., Wu, C. Y., Ling, Y., Chen, F. H., Fu, S. H., Hu, H. H., Kuang, K. Z., Chen, Y., Feng, C. W., Ma, Y. C., Tsiang, Y., Wang, W. T., Pei, C., Sun, Y. Z., Chun, W. Y., How, F. C., Hao, K. S., Tsoong, P. C., Chang, C. C., Tang, T., Wang, F. T., Keng, Y. L. & Keng, P. C. (1954). Claves familiarum generumque plantarum sinicarum. *Acta Phytotaxonomica Sinica* 2: 339-470.

Tang, Y. C., Cao, Y. L., Xi, Y. Z. & He, J. (1983). Systematic Studies on Chinese Stachyuraceae (1) - phytogeographical, cytological, palynological. *Acta Phytotaxonomica Sinica* 21: 236-253.

Tao, G. F. & Zhong, Y. (1988). Study on numerical chemotaxonomy of *Cinnamomum* in Hubei Province. *Acta Phytotaxonomica Sinica* 26: 409-417.

The Comprehensive Scientific Expedition to the Qinghai-Xizang Plateau - Chinese Academy of Sciences (1993-1994). *Vascular Plants of the Hengduan Mountains*. Beijing, China: Science Press.

The Research Group of *Dioscorea* - Kiangsu Institute of Botany (1976). Studies on Chinese *Dioscorea* sect. *Stenophora* PR. et Burk. and their chromosome numbers. *Acta Phytotaxonomica Sinica* 14: 65-72.

Tsiang, Y. & Li, P. T. (1977). *Asclepiadaceae and Apocynaceae*. Beijing, China: Science Press.

Tsoong, P. C. (1955). A new system for the genus *Pedicularis*. *Acta Phytotaxonomica Sinica* 4: 71-147.

Tsoong, P. C. (1956). A new system for the genus *Pedicularis*. *Acta Phytotaxonomica Sinica* 5: 19-73, 205-278.

Tsoong, P. C. (1963). Scrophulariaceae (2). In: Chien, S. S. & Chun, W. Y. (eds.) *Flora Reipublicae Popularis Sinicae*. Beijing, China: Science Press.

Wan, B. & Chi, H. (1987). Literature metrological analysis of phytotaxonomy method. *Guihaia* 7: 189-192.

Wang, F. S., Chien, N. F., Zhang, Y. L. & Yang, H. Q. (1997a). *Pollen Flora of China*. Beijing, China: Science Press.

Wang, F. T. & Tang, T. (1955). *Illustrated Treatment of the Principal Plants of China, Leguminosae*. Beijing, China: Science Press.

Wang, H. C., Moore, M. J., Soltis, P. S., Bell, C. D., Brockington, S. F., Alexandre, R., Davis, C. C., Latvis, M., Manchester, S. R. & Soltis, D. E. (2009a). Rosid radiation and the rapid rise of angiosperm-dominated forests. *Proceedings of the National Academy of Sciences of the United States of America* 106(10): 3853-3858.

Wang, L., Qi, X. P., Xiang, Q. P., Heinrichs, J., Schneider, H. & Zhang, X. C. (2010a). Phylogeny of the paleotropical fern genus *Lepisorus* (Polypodiaceae, Polypodiopsida) inferred from four chloroplast DNA regions. *Molecular Phylogenetics and Evolution* 54: 211-225.

Wang, L., Wu, Z. Q., Xiang, Q. P., Heinrichs, J., Schneider, H. & Zhang, X. C. (2010b). A molecular phylogeny and a revised classification of tribe Lepisoreae (Polypodiaceae) based on an analysis of four plastid DNA regions. *Botanical Journal of the Linnean Society* 162: 28-38.

Wang, M. L., Chen, Z. D., Zhang, X. C., Lu, S. G. & Zhao, G. F. (2003). Phylogeny of the Athyriaceae: evidence from chloroplast *trn*L-F region sequences. *Acta Phytotaxonomica Sinica* 41: 416-426.

Wang, M. L., Hsieh, Y. T. & Zhao, G. F. (2004). A revised subdivision of the Athyriaceae. *Acta Phytotaxonomica Sinica* 42: 524-527.

Wang, P. S. & Wang, X. Y. (2001). *Pteridophyte Flora of Guizhou*. Guiyang, China: Guizhou Scientific and Technical Publisher.

Wang, Q., Yu, Q. S. & Liu, J. Q. (2011). Are nuclear loci ideal for barcoding plants? A case study of genetic delimitation of two sister species using multiple loci and multiple intraspecific individuals. *Journal of Systematics and Evolution* 49(3): 182-188.

Wang, Q. X. & Dai, X. L. (2010). *Spores of Polypodiales (Filicales) from China*. Beijing, China: Science Press.

Wang, R. X., Lu, S. G. & Deng, X. C. (2007). Cytotaxonomic studies of the Chinese pteridophytes: A review. *Acta Phytotaxonomica Sinica* 45: 98-111.

Wang, W., Lu, A. M., Ren, Y., Endress, M. E. & Chen, Z. D. (2009b). Phylogeny and classification of Ranunculales: evidence from four molecular loci and morphological data. *Molecular Phylogenetics and Evolution* 11: 81-110.

Wang, X. Q., Han, Y., Deng, Z. R. & Hong, D. Y. (1997b). Phylogeny of the Pinaceae evidenced by molecular biology. *Acta Phytotaxonomica Sinica* 35: 97-106.

Wang, X. Q., Tank, D. C. & Sang, T. (2000). Phylogeny and divergence times in Pinaceae: evidence from three genomes. *Molecular Biology and Evolution* 17: 773-781.

Wang, Z. R. (1994). *History of the Chinese Botany*. Beijing, China: Science Press.

Wei, Z. X., Wang, H., Gao, L. M., Zhang, X. L., Zhou, L. H. & Feng, Y. X. (2003). *Pollen Flora of Seed Plants*. Kunming: Yunnan Science and Technology Press.

Wen, J. & Zimmer, E. A. (1996). Phylogeny and biogeography of *Panax* L. (the ginseng genus, Araliaceae): inferences from ITS sequences of nuclear ribosomal DNA. *Molecular Phylogenetics and Evolution* 6: 167-177.

Wen, J. (1999). Evolution of the eastern Asian and eastern North American disjunct distributions in flowering plants. *Annual Review of Ecology and Systematics* 30: 421-455.

Wieland, G. R. (1929). Antiquity of the angiosperms. *Proceedings of the International Congress of Plant Sciences* 1: 429-456.

Wu, C. S., Wang, Y. N., Liu, S. M. & Chaw, S. M. (2007). Chloroplast genome (cpDNA) of *Cycas taitungensis* and 56 cp protein-coding genes of *Gnetum parvifolium*: insights into cpDNA evolution and phylogeny of extant seed plants. *Molecular Biology and Evolution* 24: 1366-1379.

Wu, C. Y., Lu, A. M., Tang, Y. C., CHen, Z. D. & Li, D. Z.

(2002). Synopsis of a new "polyphyletic - polychromic - polytopic" system of the angiosperms. *Acta Phytotaxonomica Sinica* 40(4): 289-322.

WU, C. Y. & CHEN, S. C. (2004). *Flora Reipublicae Popularis Sinicae*. Beijing, China: Science Press.

WU, C. Y. & WANG, W. T. (1957a). Preliminary reports on studies of the plants of tropical and subtropical regions of Yunnan I. *Acta Phytotaxonomica Sinica* 6: 183-254.

WU, C. Y. & WANG, W. T. (1957b). Preliminary reports on studies of the plants of tropical and subtropical regions of Yunnan (continued). *Acta Phytotaxonomica Sinica* 6: 267-300.

WU, C. Y. & LI, H. W. (1977a). Labiatae (2). In: Yu, T. T. (ed.) *Flora Reipublicae Popularis Sinicae*. Beijing, China: Science Press.

WU, C. Y. & LI, H. W. (1977b). Labiatae (1). In: Yu, T. T. (ed.) *Flora Reipublicae Popularis Sinicae*. Beijing, China: Science Press.

WU, C. Y., LU, A. M., TANG, Y. C., CHEN, Z. D. & LI, D. Z. (2003). *The Families and Genera of Angiosperms in China - a Comprehensive Analysis*. Beijing, China: Sciences Press.

WU, Z. Y., RAVEN, P. H. & HONG, D. Y. (eds.) (1995). *Flora of China*. Beijing, China and St. Louis, MO: Science Press and Missouri Botanical Garden Press.

WU, Z. Y., RAVEN, P. H. & HONG, D. Y. (eds.) (1996). *Flora of China*. Beijing, China and St. Louis, MO: Science Press and Missouri Botanical Garden Press.

WU, Z. Y. & RAVEN, P. H. (eds.) (1998). *Flora of China Illustrations*. Beijing, China and St. Louis, MO: Science Press and Missouri Botanical Garden Press.

WU, Z. Y. (1989). The publication of Flora of China will greatly contribute to the scientific circle of the world. *Annals of the Missouri Botanical Garden* 77: 427-429.

WU, Z. Y. & RAVEN, P. H. (1994). Verbenaceae, Lamiaceae and Solanaceae. In: Wu, Z. Y., Raven, P. H. & Hong, D. Y. (eds.) *Flora of China*. Beijing, China: Science Press.

WU, Z. Y. & WU, S. G. (1996). A proposal for a new floristic kingdom (realm) - the E. Asiatic Kingdom, its delineation and characteristics. In: Zhang, A. L. & Wu, S. G. (eds.) *Floristic Characteristics and Diversity of East Asian Plants, Proceeding of the first international Symposium on Floristic Characteristics and Diversity of East Asian Plants*. pp. 3-42. Beijing, China: China Higher Education Press.

WU, Z. Y., TANG, Y. C., LU, A. M. & CHEN, Z. D. (1998). On primary subdivisions of the Magnoliophyta - towards a new scheme for an eight-class system of classification of the angiosperms. *Acta Phytotaxonomica Sinica* 36(5): 385-402.

XI, J. Q., MAO, Z. X., YANG, J. W. & XI, K. C. (1990). A study on numerical taxonomy of Dioscoreaceae plants in Zhejiang Province. *Acta Phytotaxonomica Sinica* 28: 442-451.

XIANG, X. G., ZHANG, J. B., LI, A. M. & LI, R. Q. (2011). Molecular identification of species in Juglandaceae: a tiered method. *Journal of Systematics and Evolution* 49(3): 252-260.

XIE, L., YI, T. S., LI, R., LI, D. Z. & WEN, J. (2010). Evolution and biogeographic diversification of the witch-hazel genus (*Hamamelis* L., Hamamelidaceae) in the Northern Hemisphere. *Molecular Phylogenetics and Evolution* 56: 675-689.

XING, F. W. (2005, 2006, 2007). *Flora of Macau* (vols. 1-3). Macau & Guangzhou: Department of Gardens and Green Areas, Civic and Municipal Affairs Bureau of Macau Special Administrative Region & South China Botanical Garden, Chinese Academy of Sciences.

XIONG, Z. T., CHEN, S. C. & HONG, D. Y. (1997). Numerical taxonomic studies of *Hemerocallis* (Liliaceae) from China. *Acta Phytotaxonomica Sinica* 35: 311-316.

XU, K. X. & LI, D. Z. (1983). A preliminary research on the numerical classification of the genus *Panax* from China. *Acta Phytotaxonomica Sinica* 21: 34-43.

XU, Z. L., YU, B. Y. & XU, L. S. (2004). Numerical analysis of the Euphorbiaceae. *Journal of Tropical and Subtropical Botany* 12: 399-404.

XUE, C. Y. & LI, D. Z. (2011). Use of DNA barcode *sensu lato* to identify traditional Tibetan medicinal plant *Gentianopsis paludosa* (Gentianaceae). *Journal of Systematics and Evolution* 49(3): 267-270.

YABE, Y. (1912). *An Enumeration of Plants Hitherto Known from South Manchuria Dairen (Dalian)*. Central Laboratory of the South Manchuria Railway Company.

YAN, H. F., HAO, G., HU, C. M. & GE, X. J. (2011). DNA barcoding in closely related species: A case study of *Primula* L. sect. *Proliferae* Pax (Primulaceae) in China. *Journal of Systematics and Evolution* 49(3): 225-236.

YANG, T. R. & CHOW, J. (1978). On the origin of monocotyledonae from comparative studies of the chemical constituents. *Acta Phytotaxonomica Sinica* 16 (1): 1-9.

YANG, H. M., ZHANG, Y. X., YANG, J. B. & LI, D. Z. (2013). The monophyly of *Chimonocalamus* and conflicting gene trees in Arundinarieae (Poaceae: Bambusoideae) inferred from four plastid and two nuclear markers. *Molecular Phylogenetics and Evolution* 68: 340-356.

YANG, H. Q., YANG, J. B., PENG, Z. H., GAO, J., YANG, Y. M., PENG, S. & LI, D. Z. (2008). A molecular phylogenetic and fruit evolutionary analysis of the major groups of the paleotropical woody bamboos (Gramineae: Bambusoideae) based on nuclear ITS, GBSSI gene and plastid *trn*L-F DNA sequences. *Molecular Phylogenetics and Evolution* 48: 809-824.

YANG, Q. E., ZHU, G. H., HONG, D. Y., WU, Z. Y. & RAVEN, P. H. (2005). World's largest flora completed. *Science* 309: 2163.

YANG, T. R. & CHOW, J. (1978). On the origin of Monocotyledonae from comparative studies of the chemical constituents. *Acta Phytotaxonomica Sinica* 16: 1-9.

YAO, H., SONG, J. Y., LIU, C., LUO, K., HAN, J. P., LI, Y., PANG, X. H., XU, H. X., ZHU, Y. J., XIAO, P. G. & CHEN, S. L. (2010). Use of ITS2 region as the universal DNA barcode for plants and animals. *PLoS One* 5: e13102.

YING, T. S. (2001). Species diversity and distribution pattern of seed plants in China. *Biodiversity Science* 9: 393-398.

YU, J., XUE, J. H. & ZHOU, S. L. (2011). New universal *mat*K primers for DNA barcoding angiosperms. *Journal of Systematics and Evolution* 48: 176-181.

Yu, T. T. (1974). Rosaceae (1). In: Ling, Y. (ed.) *Flora Reipublicae Popularis Sinicae*. Beijing, China: Science Press.

Yu, T. T. & Li, C. L. (1980). *Taihangia* Yu et Li: a new genus of Rosaceae from China. *Acta Phytotaxonomica et Geobotanica* 18: 469-472.

Yu, T. T. (1984). Origin and evolution of Rosaceae. *Acta Phytotaxonomica Sinica* 22: 431-444.

Zeng, C. X., Zhang, Y. X., Triplett, J. K., Yang, J. B. & Li, D. Z. (2010). Large multi-locus plastid phylogeny of the tribe Arundinarieae (Poaceae: Bambusoideae) reveals ten major lineages and low rate of molecular divergence. *Molecular Phylogenetics and Evolution* 56: 821-839.

Zhang, D. X., Saunders, R. M. K. & Hu, C. M. (1999). *Corsiopsis chinensis* gen. et sp. nov. (Corsiaceae): first record of the family in Asia. *Systematic Botany* 24: 311-314.

Zhang, J. B., Li, X. D. & Li, J. Q. (2010). A numerical taxonomic study of the *Carex* section *Racemosae* (Cyperaceae) in China. *Journal of Wuhan Botanical Research* 28: 279-285.

Zhang, J. W., Nie, Z. L., Wen, J. & Sun, H. (2011a). Molecular phylogeny and biogeography of three closely related genera, *Soroseris, Stebbinsia*, and *Syncalathium* (Asteraceae, Cichorieae), endemic to the Tibetan Plateau, SW China. *Taxon* 60: 15-26.

Zhang, M. L. & Chen, J. K. (1994). A numerical taxonomic study on the genus *Alisma* in China. *Journal of Wuhan Botanical Research* 12: 18-24.

Zhang, S. D., Soltis, D. E., Yang, Y., Li, D. Z. & Yi, T. S. (2011b). Multi-gene analysis provides a well-supported phylogeny of Rosales. *Molecular Phylogenetics and Evolution* 60: 21-28.

Zhang, X. P. & Zhou, Z. Z. (1998). *A Study on Pollen Morphology and its Phylogeny of Polygonaceae in China*. Hefei, China: Press of University of Science and Technology of China.

Zhang, Y. P. & Ge, S. (2007). Molecular evolution study in China: progress and future promise. *Philosophical Transactions of the Royal Society of London Series B, Biological Sciences* 362: 973-986.

Zhong, B., Yonezawa, T., Zhong, Y. & Hasegawa, M. (2010). The position of Gnetales among seed plants: overcoming pitfalls of chloroplast phylogenomics. *Molecular Biology and Evolution* 27: 2855-2863.

Zhou, H. G. (2000). *The Preliminary Study on Pteridophyte Flora from Guangxi, China*. Beijing, China: China Meteorological Press.

Zhou, J., Gong, X., Downie, S. R. & Peng, H. (2009). Towards a more robust molecular phylogeny of Chinese Apiaceae subfamily Apioideae: Additional evidence from nrDNA ITS and cpDNA intron (*rpl*16 and *rps*16) sequences. *Molecular Phylogenetics and Evolution* 53: 56-68.

Zhou, R. H. & Duan, J. A. (2005). *Plant Chemotaxonomy*. Shanghai, China: Shanghai Scientific and Technical Publishers.

Zhou, Y. H., Zheng, Y. L., Yang, J. L., Yan, J. & Jia, J. Z. (1999). Phylogenetic relationships among ten *Elymus* species based on random amplified polymorphic DNA. *Acta Phytotaxonomica Sinica* 37: 425-432.

Zhu, X. Y., Chase, M. W., Qiu, Y. L., Kong, H. Z., Dilcher, D. L., Li, J. H. & Chen, Z. D. (2007). Mitochondrial *matR* sequences help to resolve deep phylogenetic relationships in rosids. *BMC Evolutionary Biology* 7(217): 1-15.

Zhu, X. Y. (2004). Outline of the classified system of the Chinese Leguminosae. *Bulletin of Botanical Research* 24: 20-27.

Zou, Y. P., Cai, M. L. & Wang, Z. P. (1999). Systematic studies on *Paeonia* sect. *Moutan* DC. based on RAPD analysis. *Acta Phytotaxonomica Sinica* 37: 220-227.

Chapter 15

Introduction to Economic Plants

PEI Sheng-Ji

15.1 Introduction: the history of plant utilization

The utilization of plants in China dates back as long as Chinese civilization itself. Archeological and paleobotanical studies show that, as early as 5 000–7 000 or even 12 000 years ago, crop seeds and agricultural tools were found in both northern and southern China. Harvested seeds of trees such as *Corylus* (Betulaceae), *Quercus* (Fagaceae) and *Ziziphus* (Rhamnaceae), and pine nuts, have also been found at archeological sites of Hunan's Yangshao and Longshan cultures. This makes China one of the world's major centers of plant domestication (Vavilov, 1926; Diamond, 2002; Doebley *et al.*, 2006), with about 300 commonly-cultivated plant species thought to have been first domesticated in the country (Zheng, 2000). Within China, there are two major centers where multiple plant species were domesticated: a northern center along the Huang He (Yellow River), where foxtail millet (*Setaria italica;* Poaceae; 8 700 years ago) and common or broomcorn millet (*Panicum miliaceum;* Poaceae; 10 300 years ago) were domesticated (Nasu *et al.*, 2007; Hunt *et al.*, 2008; Jones & Liu, 2009; Lu *et al.*, 2009), and a southern center along the Chang Jiang where rice (*Oryza sativa;* Poaceae) was likely domesticated, 7 000—8 000 years ago (Liu *et al.*, 2007; Fuller *et al.*, 2009; Jones & Liu, 2009).

Written records of plant utilization can be traced back to 2700 BC, during period of emperor SHEN Nong (Department of Native and Waste Products, Ministry of Commerce of China and Institute of Botany, Chinese Academy of Sciences, 1961; Pei, 1987), at which time five cereal crops were cultivated in China: rice (*Oryza sativa*), wheat (*Triticum* spp.), foxtail millet, broomcorn millet and soybean (*Glycine max*; Fabaceae). According to SHEN Nong's records, "medicine and food were of the same origin," and began with the "period of tree arts and five cereals." SHEN Nong's book, the *Shen Nong Ben Cao Jing,* is suggested to be one of the earliest sources of folk knowledge on the use of plants—one of the first herbals. It comprises 365 plants, animals and minerals useful in medicine at that time (Pei, 2001). Ancient written records by SHI Jing, ZHOU Li and in the *Erya* show that both the domestication of animals and the cultivation of plants were established and developed in the Xia, Shang and Western Zhou dynasties (2100–771 BC). At this time, in addition to the five staples, five fruits - peach (*Amygdalus persica*), plum (*Prunus salicina*), apricot (*Armeniaca vulgaris*; all Rosaceae), Chinese date (*Ziziphus jujuba*; Rhamnaceae) and chestnut (*Castanea mollissima;* Fagaceae)—were also cultivated. At the same time, technologies for wine-making, paper-making, production of fibers and silks, vegetable oils and dyes were developed. For example, five colorant plants are recorded in ancient writings: green (*Arthraxon hispidus*; Poaceae), yellow (*Maclura cochinchinensis*, Moraceae, and *Curcuma longa,* Zingiberaceae), black (*Quercus* spp. Fagaceae) and blue (*Polygonum tinctorium*; Polygonaceae). In the fourth century BC *Shan Hai Jing* (*Book of the Mountain and Seas*), more than 20 useful aromatic plants are recorded including *Citrus* (Rutaceae), *Cymbopogon* (Poaceae), *Elsholtzia fruticosa* var. *fruticosa* (Lamiaceae) and *Valeriana jatamansi* (Valerianaceae). The use of tea also has at least 2 000 years of history in China, while the use of plant lacquers can be traced back 2 200 years. Recent studies suggest that the earliest literature on Tibetan medicine dates from the eighth century AD (Yang, 1988).

The introduction and domestication of both native and exotic, wild and cultivated plants into the gardens of China has a very long history. Perhaps the earliest garden in the world was that of Chinese medicinal plants garden established by SHEN Nong in 2800 BC (Johri & Babu, 1974; Pei, 1984). In ancient China, most gardens and parks were attached to, and formed part of the grounds of, the wealthy, or were royal estates. Many medicinal, aromatic and ornamental plants, fruits, vegetables and other useful plants were cultivated, such as the flowering plum (*Armeniaca mume*; Rosaceae), *Chrysanthemum* spp. (Asteraceae), orange (*Citrus×aurantium*; Rutaceae), *Cymbidium* spp. (Orchidaceae), *Ginkgo biloba* (Ginkgoaceae), *Litchi chinensis* (Sapindaceae), *Lonicera* (Caprifoliaceae), *Mentha* (Lamiaceae), banana (*Musa* spp.; Musaceae), waterlily (*Nymphaea* spp.; Nymphaeaceae), *Osmanthus* (Oleaceae), *Paeonia* spp. (Paeoniaceae), bamboos (Poaceae) and palms (Arecaceae). These plants, native to and domesticated in China, are now known around the world for their economic importance and ornamental value (Pei, 1984).

China is one of several sites in the Old and New Worlds that from an early time possessed sophisticated agriculture, horticulture, medical and economic plant-related knowledge. Through its long history, China has accumulated a rich knowledge of botany, agronomy, horticulture, forestry technology and medical science. Incomplete estimation suggests that the extensive ancient literature on plants comprises more than 20 000 documents.

The American botanist Walter Swingle has said that "the Chinese literature is undoubtedly the best in the world as source material for the history of utilization of wild plants and their domestication as cultivated crops" (Li, 1961; Pei, 1984), while the "father of economic botany," Richard Evans Schultes, stated in his foreword to HU Shiu-Ying's *An Enumeration of Chinese Meteria Medica* that "I believe, that China does indeed have one of the world's most extensive, most elaborate and most complex bodies of folk medicine or ethnopharmacology. The home of an ancient system of writing, the region where paper was invented, a country long famous for conservation of tradition and inhabited by a people known for their dedication to learning, China by all arguments had to possess a rich and carefully preserved heritage of the healing arts" (Hu, 1980).

Although botany as a folk and empirical science is highly developed in China, botany as a field of modern scientific enterprise was introduced relatively recently (see Chapters 11 and 12), by a combination of western priests and pioneer Chinese botanists who were trained in foreign countries. Between 1921 and 2008 a number of important works on the economic plants of China were published, including *Economic Plants of North China* (Tui, 1953), *Forest-Plant Taxonomy of China* (Chen, 1937), *Manual of Economic Plants of China* (Hu, 1955), *Economic Botany* (Hu, 1953) and *Chinese Economic Trees* (Chun, 1921). Between 1956 and 1958, a country-wide field investigation and inventory of wild useful plants, known as the "Small Autumn Harvesting," was carried out, involving more than 800 professional botanists and thousands of local experts and rural para-taxonomists. This resulted, in 1959, in the publication of the comprehensive *Economic Plants of China* (Department of Native and Waste Products - Ministry of Commerce of China and Institute of Botany - Chinese Academy of Sciences, 1961), recording 2 411 species of economic plants arranged in ten categories.

Soon after this, many regional and local records of economic plants were published, such as *Economic Plants of Yunnan* (Yunnan Institute of Botany, 1972), which recorded 349 common economic plants in Yunnan under six categories, and *Development of Economic Plants and Animals in Sichuan Province* (Editorial Board of Development of Economic Plants and Animals of Sichuan Province, 1988), which recorded some 8 000 species of plants and animals including 4 100 species of medicinal plants, 300 species of oil and fat-bearing plants, 200 aromatic plants, 100 wild fruits, 200 fiber-producing plants, 1 000 forage plants and more. More recently, in 2008, a more complete monographic publication on the economic plants of China was published by Science Press. *Plant Resources of China* (Zhu *et al.*, 2007) comprises 2 800 useful plants of China in 15 economic use categories. Finally, the publication of *Ethnobotany* (Pei & Huai, 2007), in Chinese, provides a scientific method and approach for the study of the economic plants traditionally used in China. Ethnobotany as a scientific subject—the study of traditional useful plants and plant resources, their conservation and diversity—today plays an important role in such a bio-culturally diverse country as China.

15.2 Diversity of economic plants

China is one of the most diverse countries in the world in terms of plant resources and associations, as described in previous chapters. On this foundation is built a rich diversity of economic plants (undoubtedly one of the richest in the world) and human cultural interactions, developed through thousands of years of history. According to the *Economic Plants of China* (Department of Native and Waste Products, Ministry of Commerce of China and Institute of Botany, Chinese Academy of Sciences, 1961), China has 2 411 economic plant species: 468 species used for fibers, 466 medicinal plants, 430 species used for fats or oils, 278 species used for starches or sugars, 320 species producing tannins, 320 producing aromatic or essential oils, 50 producing pesticides, 30 producing resins and gums, 30 producing rubber or gutta and 43 with other uses.

In 1983, a broader classification system for plant resources was proposed at a meeting commemorating the fiftieth anniversary of the Botanical Society of China (Wu *et al.*, 1983), comprising five categories: edible plant resources, medicinal plant resources, plant resources with industrial uses, plant resources with eco-protective and environmental restoration uses, and plant germplasm resources; it was further indicated that the total number of plants fitting into these categories might number as many as 20 000 species. This was confirmed by an inventory of plants commonly used in Traditional Chinese Medicine, listing 11 146 plant species (Chinese Natural Corporation of Traditional and Herbal Medicine, 1995). The number plants used medicinally in China makes up around one-third of the total flora and half of the total number of plants used commercially or locally at present.

However, the plant resources of China are diverse not only in terms of taxa but also in the variety of usages and technologies employed. The definition of economic plants in China is no longer restricted to market-traded plants, as almost all plants and plant products have commercial value in the market-oriented economy of China today. With rapid economic globalization, many locally-useful plants and native plant products, as well as home-consumed plants, have been introduced into regional, national and international markets. This has resulted in great pressure on wild plants and habitats, creating challenges to the conservation of plant resources all over China. On the other hand it creates greater opportunities for the domestication of wild economic plants and the development of cultivation technologies. Since the 1990s, hundreds of new wild medicinal plants, vegetables, fruits, nuts and beverage plants have been cultivated on a large scale, resulting in products worth billions of US dollars per year.

15.3 The contribution of economic plants in China

The contribution of economic plants to China and its people may be viewed in four dimensions: contributions to the national economy, contributions to healthcare, contributions to community livelihoods, and contributions to the global economy, agriculture, horticulture and forestry.

15.3.1 Contributions to the national economy

In past millennia, the economy of China was based on agriculture, and plants therefore played an important role. The recent development of the Chinese economy has accelerated its industrialization development, but the role of plants remains as important as ever. The contribution of economic plants to the national economy comprises in large part the impact of natural products and by-products ranging from industrial materials to medicines, foods, beverages and forest products. With the great progress of science and technology in the last fifty years, many local plant products have been developed into market-oriented, new natural products, of which "special economic plants" are a good example.

Special economic plants refer to new products developed on the basis of the traditional uses of plants, including aromatic plant products, wild fruits and nuts, functional foods, beverages and herbal teas, mostly harvested from the wild or cultivated locally as minor crops but under-developed for very long time. According to incomplete statistical data, the total productive value of special economic plants in 2009 was around 46 billion US dollars, of which wild vegetables made very significant contribution: of the 700 species of wild vegetable used throughout China, 209 have been developed into large-scale production, with an annual production value of more than 1.4 billion US dollars. Furthermore, in 2008, the production value of aromatic plant products was 12.4 billion US dollars (Zhang, 2010). Bamboos alone, including wild, semi-wild and those traditionally cultivated in south China, comprise over 250 species, of which 25 are economically important. According to the Chinese Academy of Forestry Sciences, the total production value of bamboo in 2009 was 6.4 billion US dollars. In the single county of Anji (Zhejiang), bamboo production contributed 1.5 billion US dollars, mainly from *Phyllostachys* spp. (Poaceae; Yang & Hui, 2010).

15.3.2 Contributions to healthcare

The contribution of medicinal plants to the health of the 1.3 billion Chinese people is reflected in the use of traditional medicines among all ethnic groups. The large number of medicinal plants used in China is not only important in traditional medical systems but also makes a great contribution to plant-based pharmaceuticals. For example, the total production of plant-based medicine in the mountainous, southwestern province of Guizhou was, in 2009, 1.4 billion US dollars. Today, more than 70% of the Chinese population still use traditional medicine. Of the 11 146 medical plants recognized, 492 species are cultivated and the remaining 10 654 are harvested from the wild (Pei, 2009).

15.3.3 Contributions to community livelihoods

Today, 65% of the Chinese population are rural people, many dependent on wild plants and small-scale cultivation for their livelihood in terms of both subsistence and income generation. This is particularly true in mountainous, forested and pastureland areas of China. Economic plants are very important resources with great potential for poverty reduction in remote areas. For example, they contribute greatly to the economic development of western China, an area with a population of 405 million, many of which are ethnic minorities. Economic plants play a significant role in the rural life of China today in terms of both economy and culture (Pei, 2001).

15.3.4 Contributions to the global economy, agriculture, horticulture and forestry

The economic plants of China have historically made a great contribution to world agriculture and horticulture development. Among 666 important cultivated plants of the world, 136 species originated in China. Many economic plants from China, such as soybean, tea, rice, *Citrus* and chestnut, are of worldwide importance and are important economic products of many other countries today (Pei, 1984). Economic globalization increases the demand for industrial plant materials and natural products such as foods, beverages, fruits and nuts, cosmetics, natural dyes, gums and medicines, of which China is one of the largest suppliers in the world.

The selection of important economic plants from the more than 20 000 economic plants of China for inclusion in Chapters 16–21 was very difficult. We adopted a strategy of finding the top 100–150 species in each category based on three main criteria: plants that are important to the economy at national and local levels; plants that are endemic to China, whose cultivation originated in China, or for which China is home to wild relatives with promising germplasm; and plants with strong potential to develop the economy, environment and cultural life of the Chinese people. Over 900 species (not including subspecies, varieties and cultivars) are described in the following chapters under the broad groupings of crops, forest plants, medicinal, ornamental and other native and exotic plants, with subcategories within these. There is a certain degree of arbitrariness in assigning certain species to such categories, and also some overlap, because many plants have multiple uses.

References

Chen, Y. (1937). *Forest-Plant Taxonomy of China*. Nanjing, China: The Agronomy Society of China.

Chinese Natural Corporation of Traditional and Herbal Medicine (1995). *Chinese Medicinal Resources*. Beijing, China: Science Press.

Chun, W. Y. (1921). *Chinese Economic Trees*. Shanghai, China: Commercial Publishing House.

Department of Native and Waste Products - Ministry of Commerce of China and Institute of Botany - Chinese Academy of Sciences (1961). *Economic Plants of China*. Beijing, China: Science Press.

Diamond, J. M. (2002). Evolution, consequences and future of plant and animal domestication. *Nature* 418: 700-708.

Doebley, J. F., Gaut, B. S. & Smith, B. D. (2006). The molecular genetics of crop domestication. *Cell* 127: 1309-1321.

Editorial Board of Development of Economic Plants and Animals of Sichuan Province (1988). *Development of Economic Plants and Animals of Sichuan Province*. Chengdu, Sichuan: Social Academy of Sciences Press.

Fuller, D. Q., Qin, L., Zheng, Y. & Zhao, Z. (2009). The domestication process and domestication rate in rice: spikelet bases from the Lower Yangtze. *Science* 323: 1607-1610.

Hu, H. H. (1953). *Economic Botany*. Shanghai, China: Zhonghua Publishing House.

Hu, H. H. (1955). *Manual of Economic Plants of China*. Beijing, China: Science Press.

Hu, S. Y. (1980). *An Enumeration of Chinese Materia Medica*. Hong Kong: The Chinese University Press.

Hunt, H. V., Linden, M. V., Liu, X. & G., M. M. (2008). Millets across Eurasia: chronology and context of early records of the genera *Panicum* and *Setaria* from archaeological sites in the Old World. *Vegetation History and Archaeobotany* 17(suppl. 1): 5-18.

Johri, B. M. & Babu, C. R. (1974). *Botanical Garden in India: Past, Present and Future*. Kuala Lumpur, Malaysia.

Jones, M. K. & Liu, X. (2009). Origins of agriculture in East Asia. *Science* 324: 730-731.

Li, H. L. (1961). Botanical Sciences. *In:* Gould, S. H. (ed.) *Science in China: New York Meeting of the American Association for the Advancement of Sciences*. Washington, DC.

Liu, L., Lee, G. A., Jiang, L. & Zhang, J. (2007). Evidence for the early beginning (c. 9,000 cal BP) of rice domestication in China: a response. *The Holocene* 17: 1059-1068.

Lu, H., Zhang, J., Liu, K. B. & Wu, N. (2009). Earliest domestication of common millet (*Panicum miliaceum*) in East Asia extended to 10,000 years ago. *Proceedings of the National Academy of Sciences of the United States of America* 106: 7367-7372.

Nasu, H., Momohara, A., Yasuda, Y. & He, J. J. (2007). The occurrence and identification of *Setaria italica* (L.) P. Beauv. (foxtail millet) grains from the Chengtoushan site (ca. 5800 cal B.P.) in central China, with reference to the domestication centre in Asia. *Vegetation History and Archaeobotany* 16: 481-494.

Pei, S. J. (1984). *Botanical Gardens in China*. Honolulu, Hawaii: Universtiy of Hawaii Press.

Pei, S. J. (1987). Medicinal plants in tropical area of China. *In: Symposium 5-35, 14th Botanical Congress*. Berlin, Germany: PUDOC.

Pei, S. J. (2001). Ethnobotanical approach of Traditional medicine studies. *Journal of the Graduate School of the Chinese Academy of Sciences* 19(2).

Pei, S. J. & Huai, H. Y. (2007). *Ethnobotany*. Shanghai, China: Shanghai Science and Technology Press.

Pei, S. J. (2009). Application of traditional knowledge in forest management: ethnobotanical indicators of sustainable forest use. *Forest Ecology and Management* 257: 2017-2021.

Tui, Y. W. (1953). *Economic Plants of North China*. Beijing, China: Chinese Academy of Sciences.

Vavilov, N. I. (1926). *Studies on the Origin of Cultivated Plants*. Leningrad, Russia: Institute of Applied Botany and Plant Breeding.

Wu, C. Y., Zhou, J. & Pei, S. J. (1983). Rational utilization and conservation of natural plant resources. In: Beijing Botanical Society of China (ed.) *Reports and Abstracts Presented at a Meeting Commemorating the 50th Anniversary of the Botanical Society of China*. Beijing, China: Botanical Society of China.

Yang, J. S. (1988). A review on the derivation of Xizhang (Tibetan) drugs and advances of the research. *Acta Botanica Yunnanica* Supplement I: 13.

Yang, Y. M. & Hui, C. M. (2010). *China's Bamboo Culture / Resources / Cultivation / Utilization*. Beijing, China: INBAR.

Yunnan Institute of Botany (ed.) (1972). *Economic Plants of Yunnan*. Kunming, China: Yunnan People's Publishing House.

Zhang, W. M. (2010). Applications of ethnobotany in the researches of wild plant species. *In: Fifth National Symposium on Ethnobotany and Fourth Asia-Pacific Forum on Ethnobotany, September 10-13, 2010*. Beijing, China: Minzu University of China.

Zheng, D. (2000). Diversity of crop genetic resources in China. *Review of China Agricultural Science and Technology* 2: 45-49.

Zhu, T. P., Liu, L. & Zhu, M. (2007). *Plant Resources of China*. Beijing, China: Science Press.

Chapter 16

Crop Plants and Their Wild Relatives

HE Shan-An
YI Ting-Shuang
PEI Sheng-Ji and HUANG Hong-wen

16.1 Introduction

The history of agricultural development in China dates back over 10 000 years, making the introduction and domestication of both native and exotic edible plants into farming and cultivation as old as Chinese civilization itself (Chen, 2002). China is one of the world's important centers where the cultivation of plants began (Diamond, 2002; Doebley *et al.*, 2006). At present, of the 666 species of cultivated plants recognized around the world for their economic importance, at least 136 (20%) originated in China (Sheng & Zhang, 1979). These include many food plants such as cereals, fruits and vegetables, which have been cultivated and developed by the Chinese people throughout their history, as evidenced by many ancient classical writings on agriculture and horticulture as well as local records, archaeological and molecular data (Chen, 2002; Hunt *et al.*, 2008; Fuller *et al.*, 2009; Jones & Liu, 2009).

Today, the food plants of China remain highly diverse in terms of the species, varieties, genes and ecosystems distributed over all the country, providing a basic, life-supporting natural resource for the Chinese population. Such food plant resources may be divided into primary resources (wild plants and original taxa) and secondary resources (crops and cultivars), both of which are of equal importance to the economic development of China. In this chapter, we describe cultivated food plants and their wild relatives in three categories: fruits, vegetables, and crops and edible oil-bearing plants. These delimitations follow traditional Chinese usage, i.e. some species (such as many Cucurbitaceae), for which it is actually the fruit that is used, are treated under vegetables.

16.2 Cultivated fruits and their wild relatives

Worldwide, around 2 800 species of trees are used by humans for their fruit (Tanaka, 1951); of these about 300 species are cultivated. Since the nineteenth century, it has been understood that northern temperate fruit trees were derived from two original centers, east and west; China was the core of the eastern center (Kikuchi, 1953). With its vast territory and range of altitude and climate, China is abundant in fruit tree resources, and is the world's most important and largest original center for fruit trees. In total, ca. 1 300 species and 160 lower taxa of fruit trees are in cultivation or harvested from the wild in China (Liu, 1998). Eighty-five percent of these are still not commercially cultivated. Importantly, more than half of them are endemic to China. A quarter of the world's cultivated fruit trees originated in China; a similarly large proportion of the world's wild fruit trees are presumed to be found in the country.

The cultivation of fruit trees in China has a history of more than 6 000 years, with some ancient *Ginkgo biloba* (Ginkgoaceae) trees themselves being over 3 000 years old. Evidence of Siberian filberts (*Corylus heterophylla*; Betulaceae) and Chinese chestnuts (*Castanea mollissima*; Fagaceae) has been found in archeological remains at Banpo, Shaanxi, dating to around 6 000 years ago. The earliest written taxonomic reference to fruit trees, the *Qi Min Yao Shu* (*Essential Techniques for the Peasantry*) by JIA Si-Xie, dates from around 534–550 AD in the Northern Dynasty, and records the jujube and 42 other species of fruit trees. WANG Xiang-Jin's *Manual of Flowers Growing*, published in 1621 during the Ming Dynasty, represented a taxonomic advance, separating fruit trees into those with drupaceous fruits, nuts and others. There are still large wild fruit populations, some of them having several thousand individuals, of species such as Chinese gooseberry (also known as kiwifruit; *Actinidia* spp.; Actinidiaceae), almond (*Amygdalus communis*), apricot (*Armeniaca vulgaris*), Chinese hawthorn (*Crataegus pinnatifida*), strawberry (*Fragaria* ×*ananassa*), apple (*Malus* spp.) and raspberry (*Rubus* spp.; all Rosaceae), seguin chinkapin (*Castanea seguinii*), orange (*Citrus* ×*aurantium*; Rutaceae), persimmon (*Diospyros kaki*; Ebenaceae), walnut (*Juglans regia*; Juglandaceae), pomegranate (*Punica granatum*; Lythraceae), blueberry (*Vaccinium* spp.; Ericaceae), grape (*Vitis vinifera*; Vitaceae) and jujube (*Ziziphus jujuba*; Rhamnaceae). These populations are concentrated in Gansu, Guizhou, Hunan, Jiangxi, Sichuan, Yunnan, Xinjiang and northeastern China.

Chinese wild fruit trees provide many special and useful genetic resources, for example, the sand-stabilizing *Elaeagnus angustifolia* (Elaeagnaceae) and *Nitraria sibirica* (Nitrariaceae), cold-resistant *Vaccinium uliginosum*, drought- and alkali-resistant *Ziziphus jujuba* var. *spinosa*, and *Vitis amurensis*, an excellent source of wine. *Citrus trifoliata*, *Malus baccata* and *Pyrus betulifolia* (Rosaceae) provide outstanding resistant rootstocks for grafted oranges, apples and pears respectively, while *Cerasus tomentosa* (Rosaceae) and *Malus sikkimensis* are famous dwarfing rootstocks for peach and apple. Fruits rich in vitamin C include *Actinidia* spp., *Hippophae rhamnoides* (Elaeagnaceae), *Rosa roxburghii* and *R. davidii*, while

Elaeagnus fruits are rich in potassium.

The largest family of wild fruit trees in China is Rosaceae, with over 400 wild species. Other large families providing edible fruits in China include Actinidiaceae, Elaeagnaceae, Fagaceae, Moraceae, Rutaceae and Saxifragaceae. At the genus level, the largest taxa are *Rubus* (with 208 wild species), *Rosa* (95 species), *Vaccinium* (92), *Elaeagnus* (67), *Diospyros* (60), *Cotoneaster* (Rosaceae; 59), *Ribes* (Saxifragaceae; 59), *Actinidia* (52), *Cerasus* (43), *Vitis* (37), *Malus* (25), *Crataegus* (18), *Pyrus* (14), *Ziziphus* (12), *Citrus* (11 species and hybrids) and *Morus* (Moraceae; 11 species). Between them these 16 genera total 863 species in China, including 514 endemic species (60%).

Wild fruit tree resources are distributed mainly in mountainous areas. They are generally little exploited due to several factors: lack of knowledge of basic information about the taxa, lack of knowledge of the species' characters, and lack of a scientific basis for a pathway of development and utilization. For these reasons, the production and quality of fruits from most wild resources cannot be sustained. Furthermore, overuse of resources, destruction of forests and even the risk of extinction can lead to wild resources becoming exhausted shortly after their potential is explored and at the same time causing damage to the wider ecological environment. Such scenarios have already been played out in relation to Chinese hawthorn, *Rosa roxburghii* and sea buckthorn (*Hippophae rhamnoides*). At present, facing an upsurge in blueberry consumption, *Vaccinium uliginosum* resources in northeastern China are experiencing a similar challenge. To summarize, humans have a need to use plant resources, but once wild resources are recognized and utilized, they are often quickly exhausted. Past experience with cereals and other crops over thousands of years of history suggests that cultivation is the best and only effective measure to resolve this problem while satisfying human requirements and eliminating threats to wild resources.

The taxa discussed below, many recognized as rare or endangered, have been selected based on criteria of economic importance, endemism, and future development potential. Information regarding the current growing situation, usage and production levels is based largely upon DONG and LIU (2006) and LIU (1998).

16.2.1 Important cultivated fruits and their wild relatives

Ginkgo biloba.—Ginkgo biloba (yin xing) is a large, deciduous, very long-lived tree, with at least 80 plants recorded at more than 1 000 years old. The oldest individuals are over 3 000 years old. Two hundred and twenty-four ancient trees are found at 300–1 200 m elevation in the western Tianmu Shan nature reserve, but after around half a century of discussion it is still not clear whether this represents a true wild population. Many cultivars and varieties are known and are classified based on the shape of their seeds. Distributed in China from northern Guangdong in the south to southern Liaoning, and from Taiwan in the east to Yunnan and Guizhou in the west, the species has also been cultivated for over 3 000 years. The use of *Ginkgo* nuts has a history of at least 600 years, mainly in Anhui, Guangxi, Jiangsu, Shandong Sichuan and Zhejiang. Although the outer part of the fruit is poisonous and strongly corrosive, the cooked seeds are edible and are considered to be a tonic. Annual nut production in China is now around 20 000 tons. The leaves are also used for medicine and the wood for furniture, and the species is recognized as a fine garden tree.

Pinus koraiensis.—Pinus koraiensis (hong song; Pinaceae) is an evergreen coniferous tree with large seeds each weighing more than a gram, and is one of the longest-cultivated plants in China. The species is found in Jilin, Heilongjiang and Liaoning, and is cultivated in large quantities. The seeds are edible and the timber considered to be of good quality. The related species *P. armandii* (hua shan song), from Gansu, Ningxia, Qinghai, Yunnan, Xizang and elsewhere, also has edible seeds.

Torreya.—Torreya (fei shu; Taxaceae), is a genus of six species and two varieties worldwide, with three species and two varieties endemic to China. Two species, *T. grandis* (fei shu) and *T. nucifera* (you fei), have edible seeds which mature in the second year after pollination; only those of the former are used as nuts. *Torreya grandis* is distributed mainly in Anhui and Zhejiang, but also in Fujian, Guizhou, Hunan, Jiangsu and Jiangxi. Varieties of *T. grandis* are divided mainly on fruit shape: the best cultivar, generated from natural variation and propagated by vegetative methods, is found only in Zhejiang, where it is cultivated over about 3 000 ha and produces 600–1 000 tons of nuts annually. Zhejiang is the natural germplasm and growing center of the variety with, in Zhuji, 1 376 trees aged more than 500 years and 27 trees more than 1 000 years. Through seed propagation the genetic diversity of the cultivars is thought to be rich, with 85% polymorphic bands found in microsatellite inter-simple sequence repeat (ISSR) analysis (Dai *et al.*, 2008). Wild relatives of *T. grandis* include *T. grandis* var. *jiulongshanensis* (jiu long shan fei shu), a rare variety from the Jiulong Shan in Jiangxi, *T. fargesii* var. *yunnanensis* (yunnan fei), distributed only in northwestern Yunnan and northern Myanmar and *T. jackii* (chang ye fei shu) from southern Zhejiang, northern Fujian and eastern Jiangxi.

Schisandra chinensis.—Schisandra chinensis (wu wei zi; Schisandraceae) is a deciduous woody vine distributed in Hebei, Heilongjiang, Jilin, Liaoning and Nei Mongol. The edible fruits are also used as medicine and in wine-making; dried, they are used as a portable food by hunters. Additional widely-used species include *S. henryi* (yi geng wu wei zi), distributed around southern China, and *S. sphenanthera* (hua zhong wu wei zi), from Guangdong and Taiwan.

Ficus carica.—Ficus carica (common fig; wu hua guo; Moraceae) is a shrub with pear-shaped berries, which has been cultivated for over 2 000 years, mainly in Xinjiang and southern China. The berries are used fresh and processed. Its relative, *F. pumila* (pi li), of southern China and Shaanxi, in particular, *F. pumila* var. *awkeotsang* (ai yu zi) is also edible and distributed Fujian, Taiwan and Zhejiang. The fruits are used in making jellies, by extraction with water. These taxa are all also used as medicines; when extracted in water they have been found to reduce blood glucose and lipids (Zhong & Xu, 2000).

Morus alba.—Morus alba (white mulberry; sang; Moraceae) is a deciduous shrub or small tree, native to central and northern China and cultivated in the country for more than 3 000 years. The fruits are eaten fresh, made into wine and used in traditional Chinese medicine. There are many varieties, including one in Yunnan with berries several centimeters long. Related species, also with edible fruits, include *M. australis* (ji sang) from the provinces south of Hebei, *M. cathayana* (hua sang) from northern and southern China, *M. mongolica* (meng sang) found from Heilongjiang to Xizang and *M. nigra* (black mulberry; hei sang) from Hebei, Shandong and Xinjiang, which is potentially cultivated.

Carya cathayensis.—Carya cathayensis (Chinese hickory; shan he tao; Juglandaceae) is a deciduous tree found in Anhui, Guizhou, Jiangxi and Zhejiang. It is mainly used in forestry, with edible nuts as a by-product. These have a special flavor and are also used for oil. Related edible species include *C. hunanensis* (hu nan shan he tao) distributed in Guangxi, Guizhou and Hunan, *C. kweichowensis* (gui zhou shan he tao) from mountainous regions over 1 000 m in Guizhou, and *C. tonkinensis* (yue nan shan he tao) from mountainous regions of Guangxi and Yunnan (the most important cultivated species, *C. illinoinensis* (mei guo shan he tao), was introduced from the USA and is cultivated mainly in eastern and central China).

Juglans.—Juglans (he tao shu) comprises 23 species worldwide, five of which are native to China. Archeological evidence comprising spores and pollen more than 6 000 years old, as well as the distribution of a wild population in Yili, Xinjiang, imply that China is one of the origins of cultivation of *Juglans*. The cultivated walnut, *J. regia* (he tao) is a deciduous tree living up to 500 years, found growing in northwestern, northern and southwestern China. There are many varieties, divided into early- and late-ripening groups; the most precocious flower and fruit just two or three years after sowing. The edible kernels are also oil-rich and, with a total output of over 300 000 tons per year, this is the most productive nut species in the world. *Juglans sigillata* (pao he tao), commonly called the "iron walnut" due to its particularly thick shell and hard wood, has been cultivated since the Ming Dynasty (1368–1644). Wild relatives include *J. mandshurica* (he tao qiu), also with edible nuts, distributed in northeastern China, Gansu, Hebei, Henan, Hubei, Hunan, Shaanxi, Shanxi, Sichuan and Taiwan. This species may provide a strongly stress-resistant rootstock for other walnut species.

Myrica.—Myrica rubra (yang mei; Myricaceae) is a broad-leaved evergreen tree from Jiangsu, Taiwan, Zhejiang and southern China. Archeological evidence suggests it has been used for more than 7 000 years. The flesh of the fruits is juicy, edible, sweet and sour. It is a slow-growing species, tolerant to poor soils and fire. Related species include *M. esculenta* (mao yang mei), with each inflorescence producing multiple drupes, found in Sichuan and Yunnan up to altitudes as high as 2 800 m. This widely-used species has two types of cultivar: sharp-leaved and big-leaved. *Myrica nana* (yun nan yang mei), from the high mountains of central Yunnan and western Guizhou has edible, sour fruits; *M. adenophora* (qing yang mei), is from Guangdong and Guangxi, and there may also be a variety in Taiwan.

Castanea.—Castanea (li shu) comprises seven species worldwide, with four in Asia and three in China. Based on genetic diversity studies, it is hypothesized that the Chinese chestnut, *C. mollissima* (Chinese chestnut), is the first-branching species of the genus and that the center of its distribution is the Chinese mainland, possibly around the reaches of the Chang Jiang and Shennongjia region. The species then spread westwards to form the European chestnut, *C. sativa* (Huang, 1998). Archeological records indicate that *Castanea* has been utilized in China for more than 6 000 years and in Europe for about 2 000—3 000 years. *Castanea mollissima* (ban li) is a deciduous tree cultivated widely in northern China to Liaoning and Jilin and southern China to Guangdong and Guangxi. Individual trees of this species are known to be more than 700–800 years old. *Castanea mollissima* is the only species in the genus resistant to chestnut blight (*Cryphonectria parasitica*) and has a higher genetic diversity than other species of *Castanea*. The nuts have a high quality taste and easily-removable inner pericarp. The area under cultivation in China is about one-third of the world total and the yield is ca. 600 000 tons, which is around two-thirds of the world total. There are around 300 cultivars of *C. mollissima*, with ca. 50 in commercial use. The nuts may be eaten raw or cooked, but are usually stir-fried or processed. *Castanea henryi* (zhui li) is distributed south of the Chang Jiang. A natural hybrid of *C. mollissima* and *C. henryi* has also been found in the wild, selections of which are used in orchards. *Castanea seguinii* (seguin chinkapin; mao li), has a similar distribution to *C. mollissima*, mainly in the Huang He and Chang Jiang valleys. The nuts are small but the species is precocious and strongly tolerant of poor soils.

Corylus.—Corylus (zhen shu) comprises 20 species worldwide, eight of which are native to China. The wild nuts have been collected and used for more than 6 000 years and cultivated for more than 1 000 years. *Corylus heterophylla* (Siberian filbert; zhen) is a small tree distributed mainly in Jilin, Heilongjiang and Liaoning, but also in provinces such as Guangdong, Hebei, Jiangsu, Shanxi and Sichuan. It is widely cultivated and there are many varieties. Related species include *C. chinensis* (hua zhen), large trees up to 40 m tall from the Bashan and Qin Ling as well as southwestern China, and *C. yunnanensis* (dian zhen), small trees with edible nuts from Guizhou, Hubei, Sichuan and western Yunnan. Microsatellite ISSR analysis, palynological and cytological studies suggest that the Chinese species of *Corylus* can be divided into northeastern and southwestern groups (Wang *et al.*, 2008).

Actinidia.—Actinidia (mi hou tao shu) comprises 54 species worldwide, with 52 species in China, ca. 40 of them endemic. Research conducted at Wuhan Botanical Garden (Chinese Academy of Sciences) indicates that there have been three ancient whole-genome duplication events in the evolution of *Actinidia*. Its distribution is centered on an area between the Chang Jiang and Zhu Jiang, including the Yunnan-Guizhou Plateau and hilly regions south of the Chang Jiang and Nanling Shan (Liang, 1983). *Actinidia* is one of the oldest fruit trees in China, wild-harvested or cultivated for more than 3 000 years. Current annual production in the country is ca. 500 000 tons, accounting for 25% of the world total. The fruits, which are rich in vitamin C, are used fresh or processed. The most important and largest-fruited cultivated species, *A. chinensis* (Chinese gooseberry or kiwifruit; zhong hua mi hou tao), comprising three varieties, *A. chinensis* var. *chinensis*, *A. chinensis* var. *deliciosa* and *A. chinensis* var. *setosa*, is a climbing vine found in eastern, northwestern, southern and southwestern China. Great achievements have been made in domestication and breeding of this species, with many cultivars, such as the female "Jiantao" and male "Moshan 4" released by Wuhan Botanical Garden, patented and commercially cultivated around the world. The large, green-fleshed variety *A. chinensis* var. *deliciosa* (mei wei mi hou tao) is one of the originators of the widespread "Hayward" cultivar, selected from a seed introduced to New Zealand in 1904. *Actinidia chinensis* var. *setosa* (ci mao mi hou tao) is distributed in Taiwan.

Other economically important species of *Actinidia* in terms of direct use or breeding potential include *A. arguta* (ruan zao mi hou tao), *A. arguta* var. *giraldii* (shaan xi mi hou tao), *A. callosa* (ying chi mi hou tao), *A. eriantha* (mao hua mi hou tao), *A. kolomikta* (gou zao mi hou tao), *A. latifolia* var. *mollis* (chang rong mi hou tao), *A. macrosperma* (da zi mi hou tao), *A. melanandra* (hei rui mi hou tao), *A. rufa* (shan li mi hou tao) and *A. zhejiangensis* (zhe jiang mi hou tao), all of which are widely distributed in the wild in China. Taxa important for their germplasm but not yet recognized in the *Flora of China* also include *A. arguta* var. *purpurea* (zi guo mi hou tao), *A. chinensis* var. *chinensis* forma *rufopulpa* (hong rou mi hou tao) and *A. deliciosa* var. *coloris* (cai se mi hou tao; Huang, 2013). Although China is the origin and center of diversity for *Actinidia*, at least ten species are presently endangered here.

Vaccinium.—Vaccinium (yue ju shu;) comprises about 450 species, with 92 species in China, of which 51 are endemic. *Vaccinium bracteatum* (nan zhu) is a small evergreen tree distributed from southern Jiangsu and Anhui to Hainan, Taiwan and Yunnan. The edible fruits are sweet but not very juicy when fresh. They are used in the preparation of black rice, a traditional, healthy, seasonal food of early summer in China. *Vaccinium uliginosum* (bog bilberry; du si yue ju; Figure 16.1) is a small deciduous shrub of northeastern China and Nei Mongol. The fruits again are edible but cannot be used fresh; they are good for processing. This species is not yet cultivated but is harvested from its wild habitat—marshes or sparse forests. As many as tens of thousands of tons may be harvested annually, meaning the wild resource is in urgent need of protection. A third species, *V. vitis-idaea* (lingonberry; yue ju; Figure 16.2) is a dwarf evergreen shrub also from northeastern China and Nei Mongol. These fruits can be eaten fresh or used in sauces and wine-making.

Figure 16.1 Bog bilberry (*Vaccinium uliginosum*) fruits. A, at Mohe, Heilongjiang. B, at Huma, Heilongjiang (images courtesy WANG, L.S.).

Diospyros.—Diospyros (shi shu) comprises about 500 species worldwide, with 57 species, six varieties and one form in China, mainly distributed in tropical and subtropical regions. According to the FAO, in 2004 China was the world's largest producer of *Diospyros* in terms of both area under cultivation (91% of the world total) and total production (37%; Jia *et al.*, 2006). Eight of China's species have edible fruits. *Diospyros kaki* (persimmon; shi) is a deciduous tree distributed in Fujian, Guangdong, Guizhou, Hainan Hubei, Hunan, Jiangsu, Sichuan and Zhejiang, the fruits of which are eaten both fresh and processed. It has been cultivated for more than 2 000 years and there are as many as 2 000 varieties known worldwide,

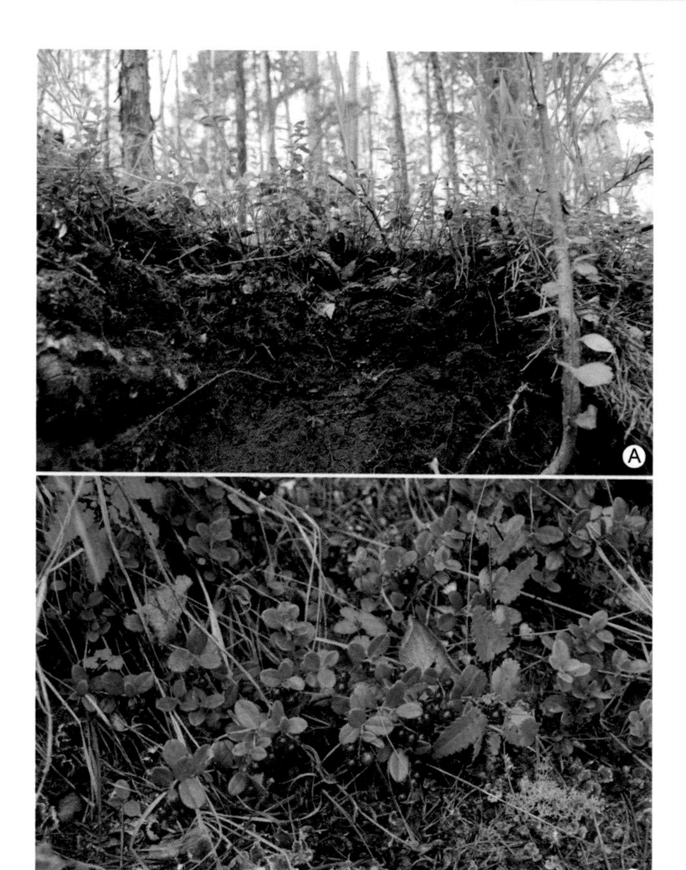

Figure 16.2 Lingonberry (*Vaccinium vitis-idaea*). A, plants in natural habitat, Tahe, Heilongjiang. B, fruits (images courtesy WANG, L.S.).

800 of them in China and Japan. The annual production in China in 2004 was about 1.7 million tons, 70% of the world total. The small-fruited *D. japonica* (shan shi) is native to western Zhejiang and Jiangxi and is also found in Jiangsu. *Diospyros morrisiana* (luo fu shi) is found in Guangdong, Hainan and Taiwan, and *D. rhombifolia* (lao ya shi) in Fujian, Jiangsu and Zhejiang. Immature fruits of the latter are used as persimmon lacquer and the species is also used a rootstock. *Diospyros philippensis* (yi se shi) has large, edible fruits, with an average weight of at least 100 g; found in Guangdong, Hainan and Taiwan, the species also produces good timber.

Ribes.—*Ribes* (cha biao zi shu) has 160 species worldwide and 59 species in China, of which 25 are endemic and five introduced. Although there are many wild species in China, all the cultivated varieties (including *R. nigrum*, blackcurrant, and *R. rubrum*, redcurrant) are introduced. The wild species *R. burejense* (ci guo cha biao zi) is distributed in the Changbai Shan and Xiao Hinggan Ling as well as Hebei and Shanxi and also has edible fruits. The strongly cold-resistant *R. mandshuricum* (dong bei cha biao zi) has two varieties, *R. mandshuricum* var. *subglabrum* (guang yao cha biao zi) and *R. mandshuricum* var. *villosum* (nei meng cha biao zi).

Amygdalus.—*Amygdalus* (tao shu) comprises 40 species worldwide, with 11 species in China, of which four are endemic. *Amygdalus persica* (peach; tao) is a deciduous tree native to northern China, originating in the mountainous areas of Shaanxi, Sichuan, Yunnan, Xizang and southern Henan. It has been cultivated throughout the country for 4 000 years and is one of the four major deciduous fruit trees in the world. Worldwide, 1.4 million ha are cultivated, with an annual production of over 15 million tons, 36% of which is in China. The fruits are delicious fresh or canned, dried or juiced. Related species include *A. davidiana* (shan tao), distributed in Gansu, Hebei, Henan, Shaanxi, Shanxi and Sichuan. The flesh of this species is bitter but the kernels are used for oil and the species can also provide a rootstock for grafted peach or plum trees, being resistant to cold and drought as well as tolerant of salty and alkaline soils. *Amygdalus mira* (guang he tao) is the highest-altitude species, distributed in Sichuan and Xizang. Large tracts of trees over 1 000 years old still exist. *Amygdalus kansuensis* (gan su tao) is found in Gansu and has robust roots with strong drought- and cold-resistance. Red-rooted types are immune to the root nematode *Meloidogyne incognita*, while white rooted types have strong resistance to it. *Amygdalus ferganensis* (xin jiang tao) has previously been classified either as a species or as a subspecies of *A. persica*. Recent random amplification of polymorphic DNA (RAPD) analysis confirms the latter hypothesis. It is cultivated widely in Xinjiang. *Amygdalus communis* (bian tao) is also cultivated in Xinjiang for its edible kernels, as well as in Gansu, Shaanxi and Shanxi; there are also wild escaped populations in Xinjiang. Related species include *A. tangutica* (xi kang bian tao) from southern Gansu, Qinghai and northwestern Sichuan, *A. nana* (ai bian tao) from Xinjiang, *A. mongolica* (meng gu bian tao) from Gansu, Nei Mongol and Ningxia, and *A. pedunculata* (chang geng bian tao), distributed in Nei Mongol, Ningxia and Shaanxi, of which populations are known to be decreasing.

Armeniaca.—*Armeniaca* (xing shu) comprises 11 species worldwide, ten of which are found in China and five of which are endemic, making China the likely center of origin of the cultivated apricot. *Armeniaca vulgaris* (apricot; xing) can still be found wild in the Tian Shan and Qin Ling and in Xizang. It has been cultivated widely in northern, northwestern and eastern China for over 3 000 years. Resistant to cold, drought and poor soils, it is a pioneer tree of bare mountainsides. Both the fruits—which are used fresh, dried, canned and as jam—and the kernels are edible; the kernels are also used for oil and medicine. Worldwide, three million tons are produced each year, 700 000 tons of which are produced in China. There are over 2 000 cultivars in China, including *A. vulgaris* var. *ansu* (ye xing), which has only low resistance to cold and drought, *A. vulgaris* var. *meixianensis* (shan mei xing) and *A. vulgaris* var. *zhidanensis* (zhi dan xing), native to northern Shaanxi. Related species include *A. sibirica* (shan xing, xi bei li ya xing) from northern China, of which the fruits are used for processing and the kernels are edible. Morphological and molecular-genetic evidence has revealed that the cultivar known as "da bian xing" is a natural hybrid of *A. vulgaris* and *A. sibirica*. *Armeniaca sibirica* varieties include *A. sibirica* var. *pubescens* (mao xing), *A. sibirica* var. *multipetala* (zhong ban shan xing) and *A. sibirica* var. *pleniflora* (laio mei xing). *Armeniaca mandshurica* (dong bei xing) is native to northeastern China and cultivated in northern China; the fruits are glacéed and used for jams and baking. *Armeniaca holosericea* is found at high elevations in Qinghai, Shaanxi, Sichuan and Xizang. It is strongly resistant to cold and drought. *Armeniaca dasycarpa* (zi xing) is native to Xinjiang and is believed by some authors (e.g. Jia *et al.*, 2006) to be a hybrid of *A. vulgaris* and *Prunus cerasifera* (Rosaceae). *Armeniaca mume* (mei) is mainly used as an ornamental and for its edible fruits. Native to western Sichuan and western Yunnan, it is cultivated in Guangdong, Hunan, Jiangsu, Zhejiang and south of the Chang Jiang. The fruits are small and sour, mostly used after processing, as preserved plums, syrup or wine. *Armeniaca hongpingensis* (hong ping mei) also produces edible fruits for processing.

Cerasus.—Cerasus (cherry; ying shu) includes 150 species, 43 of which are found in China, where 29 are endemic and five introduced. Cultivated species include the native *C. pseudocerasus* (ying tao), a small deciduous tree found in Gansu, Hebei, Henan, Jiangsu, Jiangxi, Liaoning, Shaanxi, Shandong Sichuan and Zhejiang. This species has been cultivated for more than 3 000 years. Modern cherry cultivation began in the 1880s, when *C. avium* (ou zhou tian ying tao) was introduced to China from its native Europe and western Asia. Related species include *C. tomentosa* (mao ying tao), a shrub with small fruits, the kernels rich in oil (over 42%), from northeastern, northern and northwestern China, Nei Mongol, Sichuan, Yunnan, Xizang and elsewhere. The fruits are edible and the species is also a potential dwarfing rootstock. *Cerasus humilis* (ou li) is a small shrub distributed in northeastern China, Hebei, Henan, Nei Mongol and Shandong, and is resistant to cold and wet conditions. *Cerasus japonica* (Japanese bush cherry; yu li) is another small shrub found in Hebei, Heilongjiang, Jilin, Liaoning, Shandong and Zhejiang. It is strongly resistant to cold, drought and wet.

Chaenomeles.—Chaenoleles (mu gua shu; Rosaceae) are deciduous or evergreen shrubs, subshrubs or small trees, with five species, all found in China, where three are endemic. *Chaenomeles sinensis* (mu gua) has large, sour, pyriform fruits with a unique aroma. It is both distributed and cultivated in central, eastern and southwestern China. The fruits are eaten cooked, or soaked and processed, and are also used in medicine. The species has been cultivated for over 3 000 years in China, yet has limited application in modern fruit cultivation. *Chaenomeles cathayensis* (mao ye mu gua) is widely cultivated in Fujian, Gansu, Guangxi, Guizhou, Hubei, Hunan, Jiangsu, Jiangxi, Shaanxi, Sichuan, Xizang, Yunnan and Zhejiang, *C. speciosa* (zhou pi mu gua) is from Fujian, Gansu, Guangdong, Guizhou, Hubei, Jiangsu, Shaanxi, Sichuan, Xizang and Yunnan, *C. thibetica* (xi zang mu gua) is from western Sichuan and eastern Xizang, and *C. japonica* (ri ben mu gua), native to Japan, is cultivated as an ornamental in Fujian, Hubei, Jiangsu, Shaanxi and Zhejiang.

Cotoneaster.—Cotoneaster (xun zi shu; Rosaceae) comprises 90 species worldwide, of which 59 are found in China, where 37 are endemic. Although their main use is as ornamentals, *C. submultiflorus* (mao ye shui xun zi), native to western China and central Asia, has edible fruits and is commonly used as a rootstock for other fruit taxa. *Cotoneaster multiflorus* (shui xun zi), distributed in northeastern, northwestern and southwestern China and Nei Mongol is used as a dwarfing rootstock for apple, pear and hawthorn species. *Cotoneaster acuminatus* (jian ye xun zi), from Sichuan, Yunnan and Xizang, also has edible fruits and is used as a rootstock. *Cotoneaster acutifolius* (hui xun zi) is distributed in Gansu, Hebei, Henan, Hubei, Nei Mongol, Qinghai, Shaanxi, Shanxi and Xinjiang, *C. melanocarpus* (hei guo xun zi) in northeastern and northern China, Nei Mongol and Xinjiang and *C. pannosus* (zhan mao xun zi) in Sichuan and Yunnan; the latter also has edible fruits and is used as a rootstock and ornamental.

Crataegus.—Crataegus (shan zha shu) is a large genus of around 1 000 species worldwide, only 18 of which are found in China, where ten are endemic. In China, the total area under *Crataegus* cultivation is 240 000 ha and annual fruit production is 450 000 tons. The main cultivated species of northern China is *C. pinnatifida* var. *major* (Chinese hawthorn; shan zha), a small tree from northern, northeastern, northwestern and eastern China. *Crataegus pinnatifida* has been cultivated for more than 2 000 years for its large fruits, up to 12 g in weight. The fruits are sour, usually used in jelly, wine, juice, or snacks, or canned or processed. Several hundred cultivars exist, and are divided into three groups on fruit color: red, orange and yellow. Other varieties include *C. pinnatifida* var. *psilosa* (wu mao shan zha). *Crataegus scabrifolia* (yun nan shan zha) is distributed in Guangxi, Guizhou, Sichuan and Yunnan. Cultivation of this species is concentrated in southwestern China; again there are three cultivar groups having different fruit colors. Related wild species include *C. cuneata* (ye shan zha) of central, eastern and southwestern China and the variety *C. cuneata* var. *tangchungchangii* (xiao ye ye shan zha). *Crataegus altaica* (a er tai shan zha) is found in Xinjiang, *C. hupehensis* (hu bei shan zha) in Jiangsu, Jiangxi, Shaanxi and Sichuan, and *C. dahurica* (guang ye shan zha) in Heilongjiang and Nei Mongol. *Crataegus dahurica* var. *laevicalyx* (guang shan zha) is found in Hebei.

Eriobotrya.—Eriobotrya (pi pa shu; Rosaceae) comprises 30 species, distributed mainly in subtropical and tropical Asia. Thirteen species are native to China, found in Hubei and Sichuan, three of them endemic. The fruits are edible and tasty; the leaves are used in medicine. *Eriobotrya* has been cultivated widely in China for more than 2 000 years, where the annual production is now over 200 000 tons, more than half the world total. *Eriobotrya japonica* (loquat; pi pa) is native to the upper reaches of the Chang Jiang in Hubei and Sichuan and extended to the east. Both Jiangsu and Zhejiang are famous growing regions for this species. Related species include *E. obovata* (dao luan ye pi pa), *E. malipoensis* (ma li po pi pa) and the small-fruited *E. seguinii* (xiao ye pi pa). *Eriobotrya prinoides* (li ye pi pa) has fruit with a diameter of 6–7 cm, while a variaty of *E. prinoides* collected in Sichuan and known as da du he pi pa, with fruits only 1.5–3 cm in diameter was once considered to be the ancestor of *E. japonica*. In fact it is mostly likely a hybrid of *E. japonica* and *E. prinoides* (Zhang *et al.*, 1990). Microsatellite ISSR analysis suggests that traditional classifications based upon flowering time do not reflect natural relationships in the genus (Wang *et al.*, 2010).

Fragaria.—Fragaria (strawberry; cao mei shu) is a perennial herb with aggregate fruits formed from enlarged receptacles, distributed widely across China. There are about 20 species, with 11 of them, and many varieties, in China - the richest wild resource of *Fragaria* in the world. Total world production is 3.6 million tons, of which 75 000 tons are produced in China – the second most productive country in the world. *Fragaria ×ananassa* (cao mei) is native to South America but cultivated widely in China. Related, native species include *F. vesca* (ye cao mei) distributed in Gansu, Guizhou, Heilongjiang, Henan, Jilin, Nei Mongol, Qinghai, Shaanxi, Shandong, Shanxi, Sichuan, Xinjiang and Yunnan. This species is the dominant source of germplasm providing cold-, drought- and disease-resistance as well as tolerance of poor soils and also has an outstanding aroma. *Fragaria orientalis* (dong fang cao mei), distributed in Gansu, Hebei, Heilongjiang, Hubei, Jilin, Liaoning, Nei Mongol, Qinghai, Shaanxi, Shandong and Shanxi, is also strongly resistant to cold. *Fragaria pentaphylla* (wu ye cao mei), from Gansu, Henan, Qinghai, Shanxi and Sichuan, has a strong dormancy trait while *F. nilgerrensis* (huang mao cao mei), distributed in Guizhou, Hubei, Hunan, Shaanxi, Sichuan, Taiwan and Yunnan, is strongly resistant to aphids and foliage diseases. *Fragaria mandshurica* (dong bei cao mei), found in Heilongjiang, Jilin and Nei Mongol, displays extreme resistance to cold.

Malus.—Malus (apple; ping guo shu) comprises 55 species, widely distributed in northern temperate zones. There are 25 species in China, 15 of them endemic. *Malus pumila* (ping guo) is the most important fruit tree in this genus. In 2001, the total growing area of *M. pumila* in China was 2.4 million ha, 43% of the world total, producing 22 million tons of apples, 36% of the world total. There are four major growing regions: the Bohai Wan, central China, the northwestern plateaus, and the southwestern mountains. The other two cultivated or semi-cultivated species in China are *M. asiatica* (hua hong), cultivated in northern and northeastern China and *M. prunifolia* (Chinese crab apple; qiu zi), used mainly to provide fresh fruit and also rootstocks.

Related species include *M. sieversii* (xin jiang ye ping guo; Figure 16.3), of which large wild populations are still distributed in Xinjiang, belonging to relict populations of ancient temperate broad-leaved forests. Its rich diversity has attracted the most attention, both from within China and abroad, and it is considered to be the species most closely related to the ancestor of cultivated species. Due to excessive grazing, over-harvesting, deforestation and pests, mainly the apple borer (*Agrilus mali*), wild populations have sharply decreased, from 9 300 ha to 1 800 ha in the past 50 years, with consequent effects on genetic diversity.

Figure 16.3 Wild apple (*Malus sieversii*) A, trees in Hamasi, Xinjiang. B, tree. C, fruits (images courtesy HE, S.A.).

The species is now ranked as Vulnerable on the IUCN *Red List of Threatened Species* (IUCN, 2011). *Malus sieversii* can also be found in Gansu and Shaanxi, where it has been used to supply rootstocks for cultivated apples. *Malus baccata* (Siberian crab apple; shan jing zi) is distributed in northeastern, northern and northwestern China and is used as a rootstock to produce apples of high adaptability. *Malus hupehensis* (hu bei hai tang), from central, eastern, northwestern, southern and southwestern China is used as a dwarf rootstock for apples. *Malus doumeri* (tai wan hai tang) from Guangdong, Guizhou, Hunan, Jiangxi, Taiwan, Yunnan and Zhejiang is used as a rootstock for apples in subtropical regions, while its fruits are used in pickles. *Malus kansuensis* (long dong hai tang), distributed in Gansu, Henan, Hubei, Shaanxi and Sichuan is also used as a dwarf rootstock. *Malus mandshurica* (Manchurian crab apple; mao shan jing zi), from northeastern, northern and northwestern China, is extremely resistant to freezing and adaptive to shade and is used as a frost-resistant apple rootstock. *Malus sikkimensis* (xi jin hai tang), found in western Sichuan, southern and southeastern Xizang, and northwestern Yunnan, provides a rootstock particularly compatible with numerous apple cultivars. Using RAPD analysis, 17 wild *Malus* species have been divided into five natural groups; *M. sikkimensis* was found to be the most distant from all other species (Zhang, 2008).

Prunus L.—*Prunus* (li shu) comprises about 30 species distributed in Asia, Europe and northern America; seven are found in China, two of which are endemic and three introduced. The cultivated species *P. salicina* (Chinese plum; li) is a deciduous tree native to the reaches of the Chang Jiang and now distributed all over China, where it has been cultivated for more than 3 000 years, making it one of the longest-cultivated plants in China. In fact archeological evidence for its utilization dates back more than 7 000 years. It is grown over an area of more than 1.5 million ha, about 60% of the world total. China's annual fruit production is more than 4.5 million tons, about 50% of the world total. The variety *P. salicina* var. *pubipes* (mao geng li) is found in Gansu, Sichuan and Yunnan. *Prunus domestica* (common plum; ou zhou li) is native to Europe and southwestern Asia and large wild populations are found in Xinjiang. Molecular evidence suggests that it is a natural hybrid of *P. cerasifera* and *P. spinosa* (Ruan *et al.*, 2002). *Prunus ussuriensis* (dong bei li) has smaller fruits but is resistant to extreme cold and cultivated in northern China as a frost-hardy fruit tree. *Prunus simonii* (xing li) is native to Hebei and widely cultivated in northern China for its edible fruits. There are many cultivars of this species, which seems morphologically intermediate between *P. salicina* and *Armeniaca vulgaris*. Finally, the many cultivars of *P. cerasifera* (ying tao li) are widely grown throughout much of China as a garden plant for fruits and flowers.

Pyracantha fortuneana.—*Pyracantha fortuneana* (huo ji; Rosaceae) is an evergreen shrub with fruits that, although small, are abundant. It is highly resistant to drought and poor soils, tolerates a wide pH range and is adapted to altitudes of 600–2 000 m. The fruits are rich in nutrients and used for wine-making and as a food substitutes in times of famine in remote regions. *Pyracantha fortuneana* is found in Fujian, Guangxi, Guizhou, Henan, Hubei, Hunan, Jiangsu, Shaanxi, Sichuan, Xizang, Yunnan and Zhejiang.

Pyrus.—*Pyrus* (pear; li shu) comprises about 30 species worldwide in two, eastern and western, groups. The former is mostly distributed in China, the Korean peninsula and Japan and the latter mainly in Europe, northern Africa and central Asia. Thirteen species are found in China, eight of them endemic. The history of pear cultivation dates back more than 2 500 years and today the total area under cultivation in China is about one million ha, with annual production more than nine million tons, about half of the world total. In 2009, these statistics were 1.3 million ha and 1.4 million tons of fruits, indicating a dramatic increase in fruit production per unit area over the past few years.

Pyrus calleryana (dou li) is recognized on the basis of pollen morphology as the earliest-branching species of the genus (Li & Yang, 2002) and is commonly used as a rootstock for other pear species. It is distributed widely, in Anhui, Fujian, Guangdong, Guangxi, Henan, Hubei, Hunan, Jiangsu, Jiangxi, southern Shaanxi, Shandong, Taiwan and Zhejiang, with a possible center of distribution in the Qin Ling of Shaanxi. There are three varieties. *Pyrus betulifolia* (du li) is found in Anhui, Gansu, Guizhou, Hebei, Henan, Hubei, Jiangsu, Jiangxi, Liaoning, Nei Mongol, Shaanxi, Shandong, Shanxi, Xizang and Zhejiang and is the main rootstock used to provide precocity to pears. *Pyrus phaeocarpa* (he li) is distributed in Gansu, Hebei, Shaanxi, Shandong, Shanxi and Xinjiang and is also used as a rootstock. *Pyrus xerophila* (mu li) is found in Gansu, Henan, Shaanxi, Shanxi, Xinjiang and Xizang. In northwestern China, it is often long-lived and is used as a rootstock to provide high resistance to drought and diseases. *Pyrus pyrifolia* (Chinese pear; sha li) is distributed in Anhui, Fujian, Guangdong, Guangxi, Guizhou, Hubei, Hunan, Jiangsu, Jiangxi, Sichuan, Yunnan and Zhejiang. There are many varieties, cultivated in warm and wet regions, and this is an important fruit tree in the Chang Jiang and Zhu Jiang regions. *Pyrus ussuriensis* (qiu zi li) is found in Gansu, Hebei, Heilongjiang, Jilin, Liaoning, Nei Mongol, Shaanxi, Shandong and Shanxi and is often used as a rootstock.

Pyrus sinkiangensis (xin jiang li), native to Xinjiang, includes many varieties including triploids and possible natural hybrids; they are cultivated in Gansu, Qinghai and Shaanxi. *Pyrus bretschneideri* (bai li) is found in Gansu, Hebei, Henan, Shaanxi, Shandong, Shanxi and Xinjiang and is commonly cultivated in northern China, preferring a sunny, relatively dry climate and loamy soils. There are many forms with excellent fruits. Data from pollen morphology, karyotype studies and quantitative taxonomy suggest that *P. bretschneideri* is one of the most derived species in the genus (Li & Yang, 2002). Previously Yu (1979) has suggested that *P. sinkiangensis* is a natural hybrid of *P. bretschneideri* and *P. communis* (European pear). Recent molecular-genetic evidence suggests it does contain elements from species including *P. bretschneideri* or *P. pyrifolia*, *P. communis* and *P. armeniacifolia* and is probably a hybrid of these species (Teng *et al.*, 2004). In addition, *P. communis* has been shown to be genetically distant from most of China's native species, and related more closely to *P. armeniacifolia* (Jia *et al.*, 2006).

Rosa.—Rosa (rose; qiang wei shu) includes about 200 species widely distributed from subtropical to cold-temperate regions. There are 95 species in China, 65 of them endemic. Perhaps the most important ornamental plant in the world, the hips of many wild species are also used as fruits. Species used for their fruits include *R. bella* (mei qiang wei) from Hebei, Henan, Jilin, Nei Mongol and Shanxi, the hips of which are used in tea, jam and wine and have a high vitamin content, and *R. laevigata* (jin ying zi), distributed widely in Anhui, Fujian, Guangdong, Guangxi, Guizhou, Hainan, Hubei, Hunan, Jiangsu, Jiangxi, southern Shaanxi, Sichuan, Taiwan, Yunnan and Zhejiang and used in tea, sugar and for fermenting wine. *Rosa roxburghii* (ci li, sao si hua) is distributed in southern Gansu, Guangxi, Guizhou, Hubei, Hunan, Jiangxi, southern Shaanxi, Sichuan, Xizang, Yunnan and Zhejiang. The edible fruits are very rich in vitamins and have a sour-sweet taste; they are used in juices and wines. *Rosa rugosa* (mei gui) is native to eastern Jilin, Liaoning and northeastern Shandong and is widely cultivated in China including many horticultural forms. In the wild it is endangered by over-harvesting. The fruits are rich in vitamins and used in juices, jams, wine and tea.

Rubus.—Rubus (raspberry, blackberry and other berries; xuan gou zi shu) are herbs distributed widely throughout China, concentrated in four provinces of the south-west: Guizhou, Sichuan, Yunnan and Xizang. There are around 700 species worldwide and 208 in China, of which 139 are endemic. Most species have high adaptability and survival ability. They are used mainly for their small fruit, in traditional Chinese medicine and as ground-cover plants. The native wild resource is abundant, yet few cultivars have been developed in China. Species with suitable small fruits include *R. chingii* (zhang ye fu pen zi), a strongly adaptive species suitable for low elevations with tasty fruits and used in traditional Chinese medicine, *R. corchorifolius* (shan mei), and *R. ellipticus* var. *obcordatus* (zai yang pao), a high fertility species with rich resources in southwestern China, commonly sold at local markets. *Rubus coreanus* (cha tian pao) has many varieties with different fruit sizes and colors; a form with tasty orange fruits has been reported from Guizhou. *Rubus hirsutus* (peng lei) is an outstanding species with large, colorful, tasty fruits. *Rubus lambertianus* (gao liang pao) fruits have the most stable pigment. This species is distributed widely and abundantly, making it a promising resource for juice-processing. *Rubus niveus* (hing pao ci teng) is used as ground-cover; other ornamental species include *R. lineatus* (juan mao xuan gou zi) and *R. rolfei* (gao shan xuan gou zi), a dwarf, creeping shrub. The latter is ideal ground-cover in high-latitude regions and the variety "emerald carpet" has been released in Canada.

Chinese *Rubus* have been used as breeding material for raspberries for over 100 years; the main species used for this purpose include *Rubus biflorus* (fen zhi mei), famed for its vigor and resistance to leaf spot and anthracnose and released as the variety "Dixie." *Rubus innominatus* var. *kuntzeanus* (wu xian bai ye mei), released the variety "Van Fleet" in the USA in 1911 and extended the cultivation possibilities of raspberries southwards. *Rubus parvifolius* (mao mei) is a species of high adaptability, especially characterized for its disease- and drought-resistance; in the 1980s a new variety was released in the USA from material collected in Guizhou. Other important wild resources include *R. peltatus* (dun ye mei), with the largest fruit of all wild species, and *R. pluribracteatus* (da wu pao).

Sorbus.—Sorbus (hua qiu shu; Rosaceae) comprises about 100 species in the world and 67 in China, of which 43 are endemic. *Sorbus alnifolia* (Korean whitebeam; shui yu hua qiu) is a deciduous tree distributed in Anhui, Fujian, Gansu, Hebei, Heilongjiang, Henan, Hubei, Hunan, Jiangsu, Jiangxi, Jilin, Liaoning, Shaanxi, Shandong, Shanxi, Sichuan, Taiwan and Zhejiang, the fruits of which are used fresh and in wine-making. Other fruit-bearing *Sorbus* include *S. caloneura* (mei mai hua qiu), *S. coronata* (guan e hua qiu), *S. corymbifera* you guo hua qiu) and *S. hemsleyi* (jiang nan hua qiu).

Elaeagnus.—Elaeagnus (oleaster; hu tui zi shu) comprises about 90 species, of which 67 are found in China and 55 are endemic to the country. *Elaeagnus angustifolia* (sha zao) is excellent at fixing desert soils, but the fruits are also used fresh, for jam and wine-making. Its more than 20 cultivars are divided into four groups: those with free stones, those with large fruits and sticky stones, a common sweet group and a common astringent group. Among them "xin jiang da sha zao" is considered the best for fresh fruit, these being sweet and weighing on average 1.7 g. It is distributed in Gansu, Hebei, Henan, Liaoning, Nei Mongol, Ningxia, Qinghai, Shaanxi, Shanxi and Xinjiang, over more than 200 000 ha and with an annual production of 1.5 million tons of fruit. *Elaeagnus conferta* (mi hua hu tui zi, yang nai guo) is found in Guangxi, Hainan and Yunnan; the fruits are used fresh, preserved and in wine-making. *Elaeagnus umbellata* (niu nai zi) has sweet fruits, very high in potassium and vitamin C and is distributed in Gansu, Hubei, Jiangsu, Liaoning, Shaanxi, Shandong, Shanxi, Sichuan, Xizang, Yunnan and Zhejiang. *Elaeagnus pungens* (hu tui zi) is found in Anhui, Fujian, Guangdong, Guangxi, Guizhou, Hubei, Hunan, Jiangsu, Jiangxi and Zhejiang.

Hippophae.—Hippophae (sea buckthorn; sha ji shu) includes seven species distributed from northwestern Europe to eastern Asia, with a center of diversity on the Qinghai-Xizang Plateau. All seven species are found in China, where four are endemic. *Hippophae rhamnoides* (common sea buckthorn; sha ji) has succulent, aromatic fruits, in which the endocarp is particularly easy to separate from the dark brown seed. It is distributed in Gansu, Hebei, Nei Mongol, Qinghai, Shaanxi, Shanxi, Sichuan, Xinjiang, northern and southeastern Xizang and Yunnan. The edible fruits are high in vitamins and nutrients and are used in juices, jams, wines and vinegar; the seeds are used for oil. The plants have a deep root system and high resistance to drought, cold and salinity. There are five subspecies: *H. rhamnoides* subsp. *wolongensis* (wo long sha ji); *H. rhamnoides* subsp. *turkestanica* (zhong ya sha ji), distributed in Xinjiang at elevations of 800–3 000 m; *H. rhamnoides* subsp. *sinensis* (zhong guo sha ji); *H. rhamnoides* subsp. *yunnanensis* (yun nan sha ji), distributed in Sichuan, Yunnan and Xizang at elevations of 2 200–3 700 m; and *H. rhamnoides* subsp. *mongolica* (meng gu sha ji), found mainly in Xizang. Related species include *H. neurocarpa* (lei guo sha ji), mainly distributed in Gansu, Qinghai, Sichuan and Xizang at elevations of 3 400–4 300 m, and *H. gyantsensis* (jiang zi sha ji), distributed in Xizang and endemic to the Himalaya Shan region. Its leaves are used to make tea and its fruits for polishing gold and silver. *Hippophae salicifolia* (liu ye sha ji), distributed in southern Xizang, has similar uses, while *H. tibetana* (xi zang sha ji), from Gansu, Qinghai and Xizang at elevations of 3 300–5 200 m, has large fruits and is used in breeding programs.

Syzygium.—Syzygium (pu tao shu; Myrtaceae) comprises around 1 200 species, with 80 species in China and 45 endemic to the country; two are introduced. *Syzygium jambos* (Malabar plam; pu tao) is an evergreen tree, probably native to western Malesia and Southeast Asia and widely cultivated for its fruits throughout the tropics. It is introduced, cultivated and sometimes escaped, in Fujian, Guangdong, Guangxi, Guizhou, Hainan, Sichuan and Yunnan, where it now covers 10 000 ha and produces 150 000 tons of fruit annually. However the variety *S. jambos var. tripinnatum* (da hua chi nan) is believed to be native to Taiwan and *S. jambos* var. *linearilimbum* (xian ye pu tao) is believed native to Yunnan.

Punica granatum.—Punica granatum (shi liu) is native to central Asia, Iran, Afghanistan and northwestern India. It was introduced into China along the Silk Road by the envoy Zhang Qian in the second century AD. Wild populations have been found in eastern Xizang at elevations of 1 700—3 000 m and it is possible the species is native here. It is widely cultivated in China, with two cultivar groups being prized for their fruits and flowers respectively. More than 200 cultivars are known for their fruits, which are eaten fresh or processed.

Terminalia catappa.—Terminalia catappa (lan ren shu; Combretaceae) is a tree up to 20 m tall with a trunk diameter of up to 2 m, planted throughout the tropics as a shade tree. Distributed in Guangdong, Hainan, Taiwan and southeastern Yunnan, the edible fruits are used for their nuts, sometimes known as "tropical almonds."

Phyllanthus emblica.—Phyllanthus emblica (Indian gooseberry; yu gan zi; Euphorbiaceae) is a deciduous tree distributed in Fujian, Guangdong, Guangxi, Guizhou, Hainan, Jiangxi, Sichuan, Taiwan and Yunnan. The sour fruits are rich in vitamin C, and are used fresh and dried. The seeds, roots and leaves are used as medicine and the dried leaves sometimes as pillow-stuffing. Of the ca. 100 cultivars, about ten are commonly grown, for example "shi tou," with fruits weighing 8–13 g.

Hovenia acerba.—Hovenia acerba (oriental raisin; zhi ju; Rhamnaceae) is a deciduous tree distributed in Anhui, Fujian, Gansu, Guangdong, Guangxi, Guizhou, Hainan, Henan, Hubei, Hunan, Jiangsu, Jiangxi, Shaanxi, Sichuan, southeastern Xizang and Yunnan. The fleshy, sweet-fruiting rachis is edible and is used for sugar and wine. Related species include *H. dulcis* (bei zhi ju) and *H. trichocarpa* (mao guo zhi ju).

Ziziphus jujuba.—There are two hypotheses as to the origin of *Ziziphus jujuba* (jujube; zao): some consider it to be native to China, others that it has multiple origins. Historical records and archeological evidence suggest that the species is native to the middle and lower reaches of the Huang He, China. Cultivated widely in most parts of China, the main growing regions are the five provinces of Henan, Hubei, Shaanxi Shandong and Shanxi. Archeological evidence indicates that *Z. jujuba* has been used in China for more than 7 000 years and cultivated for at least 3 000 years. Today, the total area under cultivation is more than a million hectares, with annual fruit production more than a million tons. The fruits are used fresh or dried and preserved, the dried ones being a traditional tonic food in China. *Ziziphus* is also an important honey plant in China, due to its fragrant, nectar-rich flowers. The species is characterized by a notable precocity and there are several hundred cultivars including seedless and thornless varieties. *Ziziphus jujuba* var. *spinosa* (suan zao) is distributed in Anhui, Gansu, Hebei, Henan, Jiangsu, Liaoning, Nei Mongol, Ningxia, Shaanxi, Shanxi and Xinjiang, usually on dry, sunny slopes. It is used as a rootstock and the kernel is a traditional and very effective Chinese medicine for insomnia. Fossil evidence shows that this variety has been present in China for at least 12 million years and it is suggested to be the earliest-branching of *Z. jujuba* (Jia *et al.*, 2006). The related species *Z. mauritiana* (dian ci zao) has less juicy fruits and is native to Guangdong, Guangxi, Sichuan and Yunnan, as well as being cultivated in Fujian and Taiwan.

Vitis.—Vitis (grapevine; pu tao shu) comprises about 60 species, with 37 in China, of which 30 are endemic and one introduced. These occur naturally in all provinces except Nei Mongol, Ningxia, Qinghai and Xinjiang. *Vitis vinifera* (pu tao), the best-known source of edible grapes and wine, has a long history of cultivation all over the world. The seeds are also used for oil. Newly-introduced cultivars of *V. vinifera* provide the major resource for Chinese grape cultivation, which in 2001 covered 253 000 ha, seventh in the world. China's production is about 3.5 million tons of fruits per year, fifth in the world. In China, the grape industry focuses on both fresh fruits and processing, with rapid development in the use of grapes for red wine since the 1990s.

Many native Chinese *Vitis* species are important germplasm resources: *V. romanetii* (qiu pu tao) fossils have been found in Shandong dating to the Miocene Epoch. *Vitis amurensis* (shan pu tao) is native to Anhui, Hebei, Heilongjiang, Jilin, Liaoning, Shandong, Shanxi and Zhejiang. It is the most cold-resistant species in the genus and therefore provides very important germplasm for grape breeding in China; it is also used as a rootstock. *Vitis davidii* (ci pu tao) is found in Anhui, Chongqing, Fujian Yunnan, Gansu, Guangdong, Guangxi, Guizhou, Hubei, Hunan, Jiangsu, Jiangxi, Shaanxi, Sichuan and Zhejiang, and is adapted to warm and wet weather and resistant to diseases and pests. It bears fruit up to 2.5 cm in diameter, the largest among all wild species and provides another valuable germplasm source for *Vitis* breeding. *Vitis pseudoreticulata* (hua dong pu tao) bears fruits with a high sugar content and is also an important resource in breeding for the wetter southern regions of China, having high resistance to wet and hot conditions as well as diseases. Early in the twenty-first century, genes providing resistance to powdery mildew, *Uncinula necator*, were discovered from the *V. pseudoreticulata* cultivar "Baihe 35-1," and a cDNA gene bank for resistance to powdery mildew was constructed. *V. heyneana* (mao pu tao), distributed in Anhui, Chongqing Shi, Fujian, Gansu, Guangdong, Guangxi, Guizhou, Hebei, Henan, Hubei, Hunan, Jiangsu, Jiangxi, Shaanxi, Shandong, Shanxi, Sichuan, Xizang, Yunnan and Zhejiang, is highly resistant to freezing and diseases and is used both for fresh fruits and wine-making. A study of evolutionary relationships in 37 wild species of *Vitis* has indicated that *V. davidii* and *V. heyneana* are the earliest-branching species of the genus (Chao & Yuan, 1990).

Dimocarpus longan.—Dimocarpus longan (longan; long yan; Sapindaceae) is an evergreen tree native to Guangdong, Guangxi, Hainan and Yunnan. There is a large wild population of more than 70 000 individuals in Yunnan as well as smaller populations in Hainan and Guangxi. The tree is known in China for its long life, and the fruits are used fresh, dried and canned. *Dimocarpus* has been cultivated for more than 2 000 years in southern China, where it covers an area of more than one million ha, producing around one million tons of fruit per year. Related species include *D. fumatus* subsp. *calcicola* (hui yan xiao shao zi) and *D. confinis,* (long li), which differs from *D. longan* only in its rough fruit surface.

Litchi chinensis.—Litchi chinensis (lychee; li zhi; Figure 16.4) is an evergreen tree bearing drupes that are 18–86% edible, and are eaten either fresh or dried. It is native to China, where old trees are several hundred, often 1 000 years old. The most ancient trees are distributed on Wuzhi Shan and its wild populations are in Guangxi and Yunnan too.

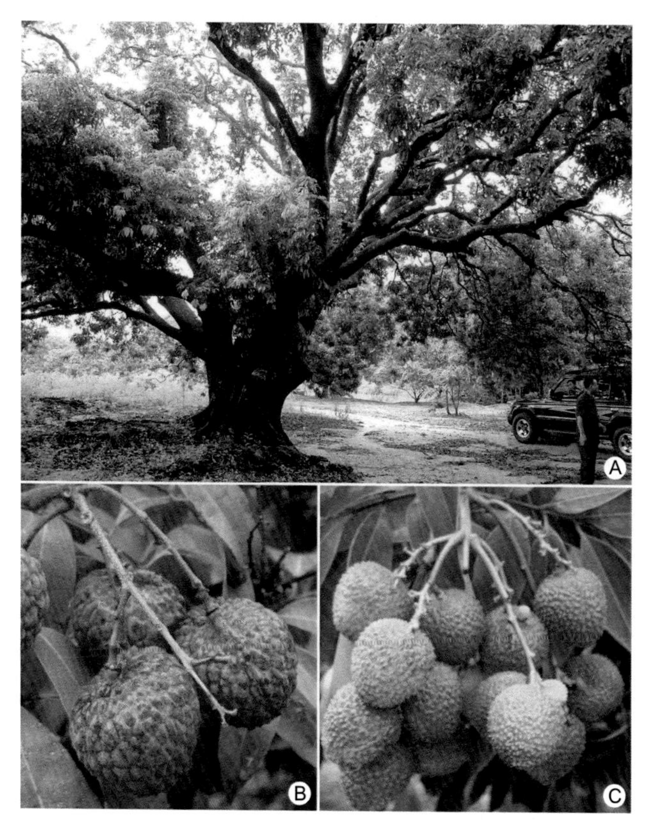

Figure 16.4 Cultivated litchi (*Litchi chinensis*). A, a tree more than 720 years old in Tangping, Yangdong, Guangdong. The diameter of the trunk of this specimen is more than 1.5 m, its height more than 15 m and its crown ca. 300 m^2. B, fruits of cultivar "Nuomici." C, fruits of cultivar "Feizixiao."(images courtesy LI, J. G.)

The species is also found in Fujian, Guangdong and Taiwan. The history of lychee cultivation dates back more than 2 000 years, and traditional cultivation methods can yield as many as 100–200 kg of fruit per tree. China leads the world in both growing area (588 000 ha) and production (128 000 tons per year). Famous cultivars include those known as "fei zi xiao" and "nuo mi zi".

Canarium album.—Canarium album (Chinese olive; gan lan; Burseraceae) is a tall tree distributed in Fujian, Guangdong, Guangxi, Guizhou, Hainan, Sichuan, Taiwan, Yunnan and Zhejiang. It is the most important wild fruit tree resource in Hainan, where it is also under cultivation. The fruits, used fresh, are considered healthy, and have an astringent flavor but a pleasant, sweet after-taste; they are favored as a traditional snack, good for digestion and quenching thirst. The related *C. pimela* (wu lan) is found in Guangdong, Guangxi, Hainan and southern Yunnan, and is used in pickles.

Choerospondias axillaris.—Choerospondias axillaris (Nepali hog plum; nan suan zao; Anacardiaceae) is a deciduous tree distributed in Anhui, Fujian, southern Gansu, Guangdong, Guangxi, Guizhou, Hubei, Hunan, Jiangxi, Sichuan, Taiwan, southeastern Xizang, Yunnan and Zhejiang. The fruits are used fresh and in wine-making.

Dracontomelon duperreanum.—Dracontomelon duperreanum (ren mian zi; Anacardiaceae) is an evergreen tree more than 20 m tall distributed in Guangdong, Guangxi and southeastern Yunnan. The fruits are used fresh.

Pistacia.—Pistacia (pistachio; huang lian mu shu; Anacardiaceae) comprises about ten species with two in China, one of which is endemic. *Pistacia vera* (a yue hun zi) is native to central Asia and cultivated in Xinjiang. It has an edible kernel used as a nut or for oil. Related species include *P. chinensis* (huang lian mu), distributed in Anhui, Fujian, Gansu, Guangdong, Guangxi, Guizhou, Hainan, Hebei, Henan, Hubei, Hunan, Jiangsu, Jiangxi, Shaanxi, Shandong, Shanxi, Sichuan, Taiwan, southeastern Xizang, Yunnan and Zhejiang, the seeds of which are used as a biofuel resource while the wood is used for furniture and yields a yellow dye. *Pistacia weinmanniifolia* (qing xiang mu), found in southwestern Guangxi, southwestern Guizhou, southwestern Sichuan, southeastern Xizang and Yunnan has leaves rich in aromatic compounds which, when extracted, are used in the manufacture of incense and candles. Its resin is also used in medicine.

Citrus.—Citrus (gan ju shu) includes between 20 and 25 species in eastern, southern and Southeast Asia, Australia and the southwestern Pacific islands, with many cultivated taxa widely naturalized in warm countries. Eleven species and hybrids are native, naturalized or extensively cultivated in China, three of these are endemic and five are hybrids cultivated to a limited extent. The genus is one of the three most important for fruits worldwide. In 2000, the world total production of citrus fruits was over 100 million tons, of which China accounted for 11%, third after Brazil and the US.

Citrus maxima (pomelo; you) is cultivated and naturalized in southern China, which is the most important growing area in the world for this species, with over 120 cultivars, the best-known being "sha tian you". No truly wild plants (which presumably would have much smaller fruit) have been found. *Citrus maxima* is a parent species of *C.* ×*aurantium* (sweet orange and grapefruit), and provides a valuable germplasm resource for the future *Citrus* industry. *Citrus* ×*aurantium* (suan cheng) is cultivated and sometimes naturalized in most areas south of the Qin Ling. Related to this species are three groups originating from hybrids of *C. maxima*, *C. reticulata* and backcrosses with the parental species, many of which arose in China. These are the sour orange group (with the bitter orange being most like the original cross); the sweet orange group (the commercially most important being backcrosses with *C. reticulata*); and the grapefruit group (representing backcrosses with *C. maxima* and first produced in the Caribbean).

Citrus reticulata (tangerine or mandarin orange; gan ju) is commonly cultivated south of the Qin Ling, and wild populations have been found in Hunan. *Citrus cavaleriei* (yi chang cheng) is native to southern Gansu, northern Guangxi, Guizhou, western Hubei, northwestern and western Hunan, southern Shaanxi, Sichuan and Yunnan. Wild populations have been found in Yunnan. *Citrus medica* (xiang yuan) is cultivated and sometimes naturalized in Guangxi, southwestern Guizhou, Hainan, Sichuan, eastern Xizang and Yunnan. This species is a parent, with *C.* ×*aurantium*, of *C.* ×*limon* (lemon). An interesting cultivar of this species is *C. medica* "fingered" (fo shou), known as the buddha-hand citron. *Citrus* ×*junos* (xiang cheng) is cultivated and sometimes naturalized in southern Anhui, southern Gansu, Guizhou, Hubei, Hunan, southern Jiangsu, southern Shaanxi, Sichuan, Yunnan and Zhejiang. The parents of this species are possibly *C. cavaleriei* and *C. reticulata*. *Citrus* ×*aurantifolia* (lime; lai meng) is occasionally naturalized in Yunnan.

Citrus japonica (kumquat; jin gan) is found in southern Anhui, Fujian, Guangdong, southeastern Guangxi, Hainan, Hunan, Jiangxi and Zhejiang. There is a considerable amount of variation within some wild populations, and cultivars are placed into two groups based on fruit shape, the round and the oval kumquats. *Citrus trifoliata* (zhi) is distributed in Anhui, Chongqing Shi, southern Gansu, northern Guangdong, northern Guangxi, Guizhou, Henan, Hubei, Hunan, Jiangsu, northwestern Jiangxi, southern Shaanxi, Shandong and Zhejiang. An important rootstock, especially for *C. reticulata*, providing for cold-hardiness. The fruits are not edible, but are commonly used in medicine.

Clausena lansium.—Clausena lansium (wampee; huang pi; Rutaceae), found in Fujian, Guangdong, Guangxi, southern Guizhou, Hainan, Sichuan and southeastern Yunnan, produces edible berries, used fresh or processed.

Olea europaea subsp. *cuspidata.—Olea europaea* subsp. *cuspidata* (xiu lin mu xi lan; Oleaceae) is a small evergreen tree distributed in Yunnan from 600–2 800 m, and also found in Afghanistan, India, Kashmir, Nepal, Pakistan, eastern and southern Africa and southwestern Asia. It is used as a rootstock for other olive taxa.

Lonicera caerulea.—Lonicera caerulea (honeyberry; lan dian guo; Caprifoliaceae) is a deciduous shrub found in Gansu, Hebei, Heilongjiang, Jilin, Nei Mongol, Ningxia, Shandong, Sichuan and Xinjiang. The fruits are rich in healthy "functional" nutrients and have good potential for use fresh, processed, and in wine-making.

Musa.—Musa (banana; ba jiao shu; Musaceae) comprises about 30 species, mainly found in Southeast Asia. Eleven species are found in China, of which two are endemic and three introduced. Banana is the world's third most important fruit, after grapes and citrus, in the world, and is grown mainly in Africa. It is, however, also one of the four major fruits grown in southern China, occupying 250 000 ha with an annual production, in 2002, of more than five million tons. *Musa acuminata* (xiao guo ye jiao) is native to western Guangxi and Yunnan, and cultivated in Fujian, Guangdong, Guangxi, Taiwan and Yunnan.

The wild plants are diploid and bear fruits containing numerous seeds, making them inedible. Cultivated plants are triploid with seedless, edible fruits. *Musa acuminata* var. *chinensis*(xiang jiao) is native to southern China. Most cultivars of short bananas in the world belong to this species.

Related species include *M. balbisiana* (ye jiao), from Guangdong, Guangxi, Hainan, Xizang and western Yunnan, the berries of which are used to feed pigs, *M. ×paradisiaca* (da jiao), cultivated in Fujian, Guangdong, Guangxi, Hainan, Taiwan and Yunnan for its edible fruits, and *M. basjoo* (ba jiao), cultivated in Fujian, Guangdong, Guangxi, Guizhou, Hubei, Hunan, Jiangsu, Jiangxi, Sichuan, Yunnan and Zhejiang for medicinal and ornamental purposes. *Musa formosana* (tai wan ba jiao) is native to Taiwan, *M. itinerans* (a kuan jiao) and *M. rubra* (a xi jiao) to Yunnan. *Musa coccinea* (hong jiao) is distributed in Guangdong, Guangxi and southeastern Yunnan and *M. sanguinea* (xue hong jiao) in Xizang. Both *M. coccinea* and *M. sanguinea* are cultivated as ornamentals.

16.3 Cultivated vegetables and their wild relatives

China has a history of vegetable cultivation of over 3 000 years, covering a variety of cultivation patterns, sophisticated and distinctive cultural techniques, long-term optimization of vegetable selection and breeding, as well as a distinguished culinary knowledge. Archeological evidence suggests that the use of vegetable species dates back more than 5 000 years. This has provided China with perfect conditions to enrich and diversify the genetic resources of its vegetables. As early as 1 800 years ago, vegetables were described in China's first dictionary, the *Shuo Wen Jie Zi* (100–121 AD), as "edible herbaceous plants." However, the tender stems or buds of some woody plants, as well as some fungi (not treated in this book) and algae, can also be used as vegetables. Many parts of vegetables may be edible, including the roots, stems, leaves, tender flowers, fruits and seeds as well as a variety of abnormal organs, such as fleshy roots, tubers, corms, bulbs or leaf balls (Editorial Board of Chinese Agriculture Encyclopedia, 1990).

In China, vegetable crops and vegetable-growing have a close relationship with people's daily life, and play an important role in the economic development of both urban and rural areas. Because of their unique nutritional value, vegetables are essential in the human diet, giving them a high economic value compared to other crops. Also, due to their relatively short growing season and higher multiple cropping index, they can be highly economic to grow. According to Ministry of Agriculture statistics, the area under vegetable cultivation in China is about 15% of the total under crops, but production, in 2003, was more than 37% of the country's total. Compared to grains, vegetables take up only 22% of the sown area but contribute 85% in terms of production. Thus the gross value of vegetables per unit land is about four times that of grains. Net production values are also higher: around 304 billion Yuan for grains; more than 338 billion Yuan for vegetables. Overall, the net income from vegetable cultivation is 376 Yuan for each of the 900 million farmers across the country (Zhang, 2005).

In recent years, with improving living standards and an expanding diet, there has been a significant increase in the cultivation of different types of vegetable. Including wild and semi-wild species, there are about 200 types of vegetables in the world, only 56 of which are widely cultivated (Editorial Board of Chinese Agriculture Encyclopedia, 1990). According to *Chinese Vegetable Cultivation* (Institute of Vegetables and Flowers - Chinese Academy of Agricultural Science, 2010), almost 300 species, subspecies and varieties of vegetable, in ca. 50 families, are used in China. Those cultivated today in China represent 45 families and 158 species and lower taxa. Among the great number of vegetable species, it seems there is a developing trend towards focusing on traditional and indigenous Chinese vegetables, such as green onion (*Allium fistulosum*; Liliaceae), cabbage (*Brassica oleracea;* Brassicaceae), Chinese cabbage (*B. rapa* var. *chinensis*), wild rice stem (*Zizania latifolia*; Poaceae), lotus stem (*Nelumbo nucifera*; Nelumbonaceae) and bamboo shoots (various Poaceae). At the same time, especially since the 1980s, there has been a significant increase in interest in wild vegetables such as purslane (*Portulaca oleracea;* Portulacaceae), dandelion (*Taraxacum* spp.; Asteraceae) and the leaves of wolfberry (*Lycium* spp.; Solanaceae) due to their health benefits. There are at least 200 vegetable species in the wild with high development potential in China.

16.3.1 Important cultivated vegetables and their wild relatives

Houttuynia cordata.—Houttuynia cordata (ji cai; Saururaceae) is a Chinese native, distributed from near sea level to 2 500 m in the northwest and north of the country, mostly in Guizhou, Sichuan and Yunnan. The fleshy rhizomes, tender stems and leaves are all used as a vegetable, and have stomachic and anti-inflammatory functions. It is becoming popular as a health-food.

Nelumbo nucifera.—Nelumbo nucifera (lotus; lian) is cultivated throughout China except in Nei Mongol, Qinghai and Xizang, with cultivation concentrated in the lakes, ponds and paddies of Anhui, Fujian, Hubei, Hunan, Jiangsu, Shandong and Zhejiang. Both the rhizomes and seeds are edible and considered healthy. Archeological evidence indicates that the history of utilization of *N. nucifera* dates back more than 7 000 years. Indeed, lotus seeds more than 1 300 years old, discovered in Pulandian, Liaoning, in the 1950s, were able to germinate and successfully form full-grown lotus plants. Today, the total area cultivated in China is 270 000 ha, with an annual production of edible rhizomes totaling over seven million tons.

Brasenia schreberi.—Brasenia schreberi (chun cai;

Cabombaceae) is native to Anhui, Jiangsu and Zhejiang, and today found in Hunan, Jiangxi, Sichuan, Taiwan and Yunnan. The young underwater leaves are collected as a vegetable and are characterized by a coating of slippery gel with high nutritional and medical value. This unique vegetable is famous worldwide, particularly that produced in Xi Hu, Hangzhou Shi. *Brasenia schreberi* has been cultivated in China for over 2 000 years.

Amaranthus tricolor.—Amaranthus tricolor (Chinese spinach; xian cai; Amaranthaceae) may be divided into three types on the basis of leaf color: green, red and multicolored. A very adaptable species, it is widely distributed in central and southern China. It is also one of the longest-cultivated plants in China, recorded since the second century AD. The tender stems and leaves are used as a highly nutritious vegetable; the whole plants are also used medicinally.

Portulaca.—Portulaca (purslane; ma chi xian shu) comprises about 150–200 species, found mostly in arid tropical and subtropical regions, particularly in Africa and South America, with a few species extending into temperate regions. Five species are found in China, two of which are endemic and one introduced. *Portulaca oleracea* (ma chi xian) is distributed widely throughout the country, especially in Fujian, Guangdong, Guangxi, Hainan (including the Xisha Qundao), Taiwan and southern Yunnan. It is cultivated but also found as a weed. A relatively new and increasingly popular vegetable, the tender stems and leaves are considered to be very beneficial to health, and are eaten as a salad, preserved, dried or pickled. It is also used as a medicine.

Benincasa hispida.—Benincasa hispida (winter melon; dong gua; Cucurbitaceae) is believed to have originated in China and southern Asia; wild populations with small fruits are found in Xishuangbanna, Yunnan. Its history of cultivation in China dates back at least 2 000 years. The gourds are eaten as a vegetable, fresh or processed, and are commonly used in Chinese soups; the seeds are used medicinally.

Cucumis.—Cucumis (cucumbers and melons; huang gua shu; Cucurbitaceae) includes about 32 species distributed in tropical and temperate regions, mostly in Africa, with four in China. Perin *et al.* (2002) have produced a genetic map of the genus. *Cucumis melo* (muskmelon; tian gua), with many different cultivar groups, is found all over the country, where it has been cultivated for more than 4 000 years. Annual production is over 14 million tons, more than half of the total world production. *Cucumis sativus* (cucumber; huang gua), has been cultivated in India for over 3 000 years and was introduced to China in the seventeenth century. Worldwide it is cultivated over 2.4 million ha, with an annual output of more than 40 millions tons. China is the most significant contributor to this, with 1.5 million ha producing 26 million tons. The fruits are used as a vegetable, fresh, cooked, processed, or juiced; the stems are also used as a medicine. *Cucumis sativus* var. *hardwickii* (xi nan ye huang gua) is found in Guangxi, Guizhou and Yunnan, as well as elsewhere in southern Asia. A wild variation found in Xishuangbanna, southern Yunnan, has fruits weighing as much as 5 kg. The related species *C. hystrix* (ye huang gua), native to western Yunnan and southern Asia, is resistant to nematodes and tolerant of low light and low temperatures (Zhuang & Chen, 2003).

Momordica.—Momordica (ku gua shu; Cucurbitaceae) comprises 45 species, mostly native to tropical Africa with some cultivated in other tropical regions. Three species are found in China, one of which is cultivated. Although the main cultivated species, *Momordica charantia* (bitter melon; ku gua) was introduced to China from tropical Asia during the thirteenth century, the related taxon *M. cochinchinensis* is distributed in Anhui, Fujian, Guangdong, Guangxi, Guizhou, Hunan, Jiangsu, Jiangxi, Sichuan, Taiwan, Xizang, Yunnan and Zhejiang. The seeds, roots and leaves of the latter are used in traditional Chinese medicine.

Brassica.—Brassica (yun tai shu) includes about 40 species, primarily native to the Mediterranean region, particularly southwestern Europe and northwestern Africa. There are six species in China, where the genus is extremely important as vegetables. Archeological evidence has shown that some of them have been utilized for over 6 000 years. The total annual yield of 1 200 million tons is made up of several species and numerous cultivars created during this long history of cultivation. *Brassica juncea* (Chinese mustard; jie cai) has been used for the full 6 000 year history, and cultivated widely in southwestern and northwestern China since the sixth century BC. At least two varieties are considered to have been generated in China. Two different types of *B. juncea* are consumed: "leaf mustard" and "oil mustard" types. Varieties include *B. juncea* var. *napiformis* (jie cai ge da), widely cultivated as a vegetable and for forage, and *B. juncea* var. *tumida* (zha cai), a vegetable and very famous pickle of Sichuan, distributed mainly there and in Yunnan. *Brassica oleracea* is an introduced species, covered in Chapter 20.

Brassica rapa (man jing) has been naturalized and cultivated in China for more than 2 600 years, and is widely distributed in the northwestern and southwestern alpine zones, Fujian and southern Zhejiang. The fleshy roots are used as a vegetable (the turnip). Varieties include *B. rapa* var. *chinensis* (bok choy; qing cai), cultivated throughout China since the third century, and including many cultivars. This is one of the most common vegetables on the everyday dining table, especially in eastern, central and southern China. *Brassica rapa* var. *glabra* (Chinese cabbage; bai cai) is naturalized and has a long history of cultivation as a domestic vegetable throughout China, over 2 000 years. It is a popular and important dietary component in northern China, with an annual production of 160 million tons, occupying an amazing 17% of the total vegetable consumption in China. Several genetic maps of this variety have been established and quantitative trait loci (QTLs) for cold-, heat-and disease-resistance as well as male sterility and bolting tolerance have been located (Sun *et al.*, 2007). *Brassica rapa* var. *oleifera* is treated as an oil crop in Section 16.4.

Glycine.—Glycine (da dou shu; Fabaceae) includes about nine species distributed across the tropical, subtropical and temperate regions of the eastern hemisphere. There are six species in China, two of which are endemic and one introduced. *Glycine max* (soybean; da dou), cultivated throughout temperate and tropical regions, is also cultivated throughout China, where it is the most important source of oil and also provides about 35% of human protein. The beans are highly nutritious—with a protein content of ca. 40% and fatty acid content of ca. 20%, both higher than that of a cereal. The tender seeds are treated as a vegetable, while mature seeds are used to make a variety of tofu (bean curd) products, a very popular food on the Chinese dinner table. Currently, more than 2 000 kinds of food are based on soybeans, which are also widely used in light industry and pharmaceuticals.

China has a history of soybean cultivation spanning 4 000–5 000 years. According to the *Book of Songs* (written around six century BC), soybean domestication and cultivation date back as early as prehistoric matriarchal society in China, and soybean remains have been discovered at a number of prehistoric sites in China. The ancient Chinese understood the effects of soybean cultivation in improving soil fertility and developed intercropping and rotation systems using soybean for improving the yield of cereal crops as early as the twenty-first century BC.

It is believed that the cultivated *Glycine max*, which is not known in the wild, has been developed from *G. soja* (wild soybean; ye da dou), which occurs widely throughout China in Anhui, Fujian, Gansu, Guangdong, Guangxi, Guizhou, Hebei, Heilongjiang, Hubei, Hunan, Jiangsu, Jiangxi, Jilin, Liaoning, Nei Mongol, Ningxia, Shaanxi, Shandong, Shanxi, Sichuan, Xizang, Yunnan and Zhejiang, as well as in Japan, Korea and Russia. *Glycine max* and *G. soja* hybridize easily together. *Glycine soja* has many excellent characters including a high protein content, numerous flowers and pods, high salt tolerance, high linolenic acid content, cold resistance and disease resistance, making it a valuable breeding material for cultivated soybeans. With its high protein content, the wild soybean is also used as fodder in some regions of China. *Glycine soja* is protected in China through multiple special reserves, but its populations continue to decrease through habitat loss. *Glycine gracilis* (kuan ye man dou), found in villages and along roadsides in Heilongjiang, Jilin and Liaoning, has sometimes been treated as a variety of *G. soja*. Because many of the morphological characters separating *G. soja* and the cultivated soybean (*G. max*) are found in hybrids, more studies are required to infer the phylogenetic relatioships of *Glycine* species.

Vavilov (1951) provided convincing evidence that China is the origin and center of cultivated soybeans, but the exact area of origin remains controversial. Some researchers believe it was most likely northeastern China, based on the natural distribution of wild soybean species, wild relatives and highly diverse cultivated varieties in the region, plus local written records. Others, however, favor the Huang He basin, one of the places of origin of Chinese agriculture, where wild soybeans are also widely distributed. A few researchers suggest that soybean first originated around the Chang Jiang and southern region, based on extensive sampling of landraces and wild soybeans across the entire distribution range in eastern Asia (Guo *et al.*, 2010). Soybean is a typical "short-day plant," sensitive to daylight duration. Through its long history of domestication in different regions and under different selection preferences, the wild type—characterized by short plants with small pods and seed—has been replaced by larger, cultivated types with an upright habit and large pods and seeds, but traditional landraces are extremely diverse. For example, the growth habit of northern landraces is generally indeterminate, while that of southern landraces is determinate. Seed, pod and flower characters vary widely. The variety of natural soybean resources in China have been inventoried and documented; current collections of soybean germplasm total more than 25 000 landraces or accessions in China (Wang *et al.*, 2012).

China has made significant contributions to the world soybean industry in the centuries since domestication, with the soybean first introduced from China to Korea about 2 500 years ago and then to Japan, India, Indonesia and Vietnam. More recently, soybeans were introduced to the west in 1712 via Germany and in 1739 to Paris by French missionaries. In 1790 they were introduced to the Royal Botanic Gardens, Kew, UK, where extensive studies and cultivation trials were conducted, which led to extended soybean cultivation across Europe by the late nineteenth century. Finally, the introduction of soybean into the USA may have occurred via China, Japan or Europe. In 1890 the US Department of Agriculture organized a large-scale trial of commercial soybean production, and the country has now become the world's largest soybean producer.

Lablab purpureus.—The young pods of *Lablab purpureus* (lablab or hyacinth bean; bian dou; Fabaceae) are cooked as a vegetable.

Vicia faba.—Vicia faba (broad bean; can dou; Fabaceae) is cultivated throughout China, where the tender seeds are used as a vegetable; the mature seeds can also be eaten or processed. The related *Vicia sativa* (common vetch; jiu huang ye wan dou) has been divided into two subspecies based on corolla size, calyx teeth, the color and morphology of the legume and morphology of the leaflets. *Vicia sativa* subsp. *sativa* is cultivated and naturalized across China; *V. sativa* subsp. *nigra* is the wild form, found in damp grassy places, valleys, riverbanks and beds, farms, fields, open wastelands and on roadsides in Fujian, Guizhou, Hunan, Sichuan, Taiwan, Xinjiang and Xizang, as well as elsewhere in Eurasia and Africa.

Vigna.—Vigna (jiang dou shu; Fabaceae) comprises about 100 species distributed in tropical regions; fourteen are found in China, one of which is introduced. *Vigna angularis* (red or adzuki bean) is treated as a staple crop in Section 16.4, below. *Vigna radiata* (mung bean; lü dou) has a 2 000-year history of cultivation in China, documented in many ancient Chinese agricultural books, including the *Qi Min Yao Shu*, *Therapeutic Materia Medica* from the seventh century AD and *Kai Bao Materia Medica* (973–974 AD). The origin of the mung bean has, however, been controversial, with India, Africa, the Nile Basin, central Asia and China all being proposed as possible centers of origin. Large-scale surveys in China indicating the wide occurrence of wild mung beans suggest that China may be at least one of the centers of origin. *Vigna radiata* var. *sublobata* (san lie ye lü dou; Fabaceae) is thought to be the wild ancestor of the cultivated variety *Vigna radiata* var. *radiata*, and is found on open wastelands, roadsides and thicket margins in Taiwan, as well as in East Asia, Cambodia, India, Indonesia, Laos, Sri Lanka, Thailand, Vietnam and Africa. In 1979, a field expedition organized by the Chinese Academy of Agricultural Sciences and other institutes, conducted in Guangxi, Guizhou and Yunnan, discovered large tracts of wild and variant types of mung beans in hot, dry valleys at altitudes of 1 000–1 800 m. It is now considered that mung bean is naturally distributed almost all over the country, and more than 3 000 landraces or accessions have been collected and preserved in China.

Trapa.—Trapa (water caltrop; ling shu; Trapaceae) comprises two aquatic species distributed in subtropical and temperate regions of Africa, Asia and Europe; both are found in China. An ancient carbonized *Trapa* fruit from Zhejiang is thought to be 20 000–30 000 years old, and the genus is known to have been used in China for more than 7 000 years. *Trapa natans* (ou ling) is widely cultivated in Anhui, Fujian, Guangdong, Guangxi, Guizhou, Hainan, Hebei, Heilongjiang, Henan, Hubei, Hunan, Jiangsu, Jiangxi, Jilin, Liaoning, Nei Mongol, Shaanxi, Shandong, Sichuan, Taiwan, Xinjiang, Xizang, Yunnan and Zhejiang for its fruits, which contain abundant starch and are eaten raw or cooked, as well as used as a medicine, in starch production and wine-making. *Trapa incisa* (xi guo ye ling) is distributed worldwide including, in China, Anhui, Fujian, Guangdong, Guizhou, Hainan, Hebei, Heilongjiang, Henan, Hubei, Hunan, Jiangsu, Jiangxi, Jilin, Liaoning, Shaanxi, Sichuan, Taiwan, Yunnan and Zhejiang.

Toona sinensis.—Toona sinensis (Chinese mahogany; xiang chun; Meliaceae; see also Chapter 17 for its use as a timber tree) originated in China and is widely distributed, from Liaoning to Yunnan. The tender leaves are used as a vegetable in China and Malaysia, and as animal fodder in India. The trees are widely used medicinally, the bark as an astringent and depurative, the powdered root as a refreshment and diuretic, and the young leaves as a carminative.

Oenanthe javanica.—Oenanthe javanica (Chinese celery; shui qin; Apiaceae) is cultivated widely in China, where the tender stems are used as a vegetable. *Oenanthe javanica* subsp. *rosthornii* (luan ye shui qin) is found in Chongqing Shi, Fujian, Guangdong, Guangxi, Guizhou, Hunan, Sichuan, Taiwan and Yunnan, and is reputed to have medicinal value. Related wild species include *O. linearis* (xian ye shui qin), distributed in Guizhou, Hubei, Sichuan, Taiwan, Xizang and Yunnan and *O. thomsonii* (duo lie ye shui qin) from Chongqing Shi, Guangdong, Guizhou, Hubei, Jiangxi, Sichuan, Xizang and Yunnan. *O. thomsonii* subsp. *stenophylla* (zhai ye shui qin) is used in Guizhou and Yunnan as a substitute for *O. javanica*.

Lycium.—Lycium (gou qi shu; Solanaceae) comprises about 80 species distributed in South America and South Africa, with a few in temperate Europe and Asia. Seven species are found in China. Two species are known as the wolfberry or goji berry: *Lycium chinense* (gou qi) originated in China and is widely distributed in Anhui, Fujian, Gansu, Guangdong, Guangxi, Guizhou, Hainan, Hebei, Heilongjiang, Henan, Hubei, Hunan, Jiangsu, Jiangxi, Jilin, Liaoning, Nei Mongol, Ningxia, Qinghai, Shaanxi, Shanxi, Sichuan, Taiwan, Yunnan and Zhejiang. It is found in historical records dating back 3 000 years, and is known to have been utilized for more than 2 000 years. The young leaves are used as a vegetable, the dried fruits in medicine, and it is also a popular tonic. *Lycium barbarum* (ning xia gou qi), native to Ningxia, is very famous as a tonic food and medicine, while the young leaves are also used as a vegetable. Related wild species include *L. yunnanense* (yun nan gou qi), endemic to Yunnan, and *L. ruthenicum* (hei guo gou qi), a drought-resistant shrub used to control erosion, found in the deserts of Gansu, Nei Mongol, Ningxia, Qinghai, northern Shaanxi, Xinjiang and Xizang.

Solanum.—Solanum (qie shu; Solanaceae) includes about 1 200 species, mostly found in tropical and subtropical regions, especially in the Americas. Forty-one species are found in China, about half of which are introduced. *Solanum violaceum* (ci tian qie), from southern China, has fruits used for medicine. *Solanum mammosum* (nipplefruit; ru qie) is found in Guangdong, Guangxi and Yunnan, and is occasionally cultivated for its beautiful fruit. *Solanum torvum* (turkey berry; shui qie) is distributed in Fujian, Guangdong, Guangxi, Guizhou, Hainan, Taiwan, southern Xizang and Yunnan. The young fruits are cooked and used medicinally for improving the eyesight; the leaves for treating skin diseases. *Solanum capsicoides* (cockroach berry; niu qie zi), distributed in Fujian, Guangdong, Guangxi, Guizhou, Hainan, Hunan, Jiangsu, Jiangxi, Sichuan, Taiwan, Yunnan and Zhejiang, is also cultivated in Henan and Liaoning for its bright, orange-red, poisonous, ornamental fruits.

Ipomoea.—Ipomoea (morning glories; fan shu shu; Convolvulaceae) comprises approximately 500 species, widely distributed from tropical to warm temperate regions, especially in North and South America; 29 species are found in China. *Ipomoea aquatica* (water spinach; weng cai) is native to southern China, where it has been used for over 1 600 years and cultivated since the seventeenth century. The young stems and leaves are used as a vegetable and for fodder. There are many cultivars, varying in leaf shape. *Ipomoea batatas* (sweet potato; fan shu) is well known as a food crop for its tubers, but the young leaves and tender petioles are also used as a vegetable, becoming more popular in recent years.

Chrysanthemum indicum.—Chrysanthemum indicum (ye ju; Asteraceae) is native to northeastern, northern, central, southern and southwestern China. It is heterogenous species, with various ecological and geographical types. The young shoot tips are used as a vegetable, particularly in southern Jiangsu and Zhejiang, where it is recognised as a special wild vegetable, In Nanjing, it is one of three famous wild vegetables. The leaves, flowers and whole plants are also used medicinally.

Glebionis coronaria.—Glebionis coronaria (garland chrysanthemum; tong hao; Asteraceae) originated in China and has been cultivated widely for more than 900 years in the south and center of the country. The tender shoot tips and young leaves are used as a vegetable. *Glebionis carinata* (hao zi gan) and *G. segetum* (nan tong hao) have also been used as a vegetable under the same common name in China.

Taraxacum mongolicum.—Taraxacum mongolicum (dandelion; pu gong ying; Asteraceae) is distributed all over China. The young plants are used as a vegetable, particularly the flower heads in soups. It is becoming popular for its health benefits, and is now grown under glass to ensure production for the Spring Festival market. Many related wild species have also been trialled for use as a vegetable, including *T. coreanum* (chao xian pu gong ying, bai hua pu gong ying) and *T. formosanum* (tai wan pu gong ying), both widely distributed across the country.

Sagittaria.—Sagittaria (ci gu shu; Alismataceae) comprises about 30 species distributed in tropical and temperate regions with seven species in China, two of which are endemic. *Sagittaria trifolia* subsp. *leucopetala* (threeleaf arrowhead; hua xia ci gu) is cultivated south of the Chang Jiang, mainly in Anhui, Fujian, Guangxi, Guizhou, Hainan, Henan, Shaanxi, Yunnan and Zhejiang, for its tubers, which are used as a vegetable. Related wild taxa include *S. trifolia* subsp. *trifolia* (ye ci gu), which has been recorded in China since the third century AD and is found wild in Anhui, Beijing, Fujian, Gansu, Guangdong, Guangxi, Guizhou, Hainan, Henan, Hubei, Jiangsu, Liaoning, Qinghai, Shaanxi, Shandong, Sichuan, Taiwan, Xinjiang, Yunnan and Zhejiang, *S. lichuanensis* (li chuan ci gu), from Fujian, Guangdong, Guizhou, Hubei , Jiangsu, Jiangxi and Zhejiang, *S. potamogetonifolia* (xiao ci gu) from Anhui, Fujian, Guangdong, Guangxi, Hainan, Hubei, Hunan, Jiangxi, Yunnan and Zhejiang, *S. pygmaea* (ai ci gu), from Anhui, Fujian, Guangdong, Guangxi, Guizhou, Hainan, Henan, Hubei, Hunan, Jiangsu, Jiangxi, Shaanxi, Shandong, Sichuan, Taiwan, Yunnan and Zhejiang, and *S. tengtsungensis* (teng chong ci gu), from Xizang and Yunnan.

Amorphophallus.—Amorphophallus (mo yu shu; Araceae) comprises about 200 species, mainly found in Southeast to eastern Asia, Africa, northern Australia and the Pacific islands; 16 species are found in China, with seven endemic. Seven species are cultivated for their edible tubers. The main cultivated species, *A. konjac* (konjac; hua mo yu), native to Yunnan, has been recorded in China since the fifth century BC and is now found in Gansu, Guangdong, Guangxi, Guizhou, Fujian, Shanxi, Sichuan and Taiwan. The edible tubers are used as a vegetable and a health-food in weight loss, due to their low calorie content. The powdered tuber, comprising mainly polysaccharides, is used to make various kinds of food. *Amorphophallus albus* (bai mo yu) is endemic to China and found in southern Sichuan, northeastern Yunnan and cultivated elsewhere in Yunnan. This species produces the highest quality powdered tuber, containing up to 60% of the hemicellulose glucomannan.

Eleocharis.—Eleocharis (bi qi shu; Cyperaceae) includes about 250 species worldwide, with 35 species in China, nine of which are endemic. *Eleocharis dulcis* (Chinese water chestnut; bi qi) has edible tubers, with a high nutrient content and claimed health benefits. It is distributed in Guangxi, Hainan, Hubei, western Hunan, Jiangsu and Taiwan, and cultivated across most of China, as it has been for more than 2 000 years.

Bamboo shoots.—Bamboo shoots (sun) are obtained from a group of Poaceae, mainly in the genera *Phyllostachys* (gang zhu shu) and *Bambusa* (le zhu shu). China is the center of diversity for bamboos, with more than 500 species, the tender shoots of about 20 of which are harvested as a vegetable, used in both hot and cold dishes, as well as dried. The history of this utilization is over 3 000 years. China is one of three regions of the world in which bamboo shoots are grown, and produces about three million tons annually. The export market for processed bamboo shoots is about 800 000 tons per year.

Chinese bamboo shoots are comonly divided into three seasonal groups for marketing: winter (dong sun), spring (chun sun) and summer bamboo shoots (xia sun). Sources of winter bamboo shoots include *Phyllostachys edulis* (mao zhu), the most important cultivated species, widely distributed in southern China and famous in Jiangsu and Zhejaing. Sources of spring bamboo shoots include *P. violascens* (zao zhu), cultivated in Anhui, Fujian, Hunan, Jiangsu, Jiangxi, Yunnan and Zhejiang and again famous in Jiangsu and Zhejiang, *P. glauca* (dan zhu), distributed in Anhui, Henan, Hunan, Jiangsu, Shaanxi, Shandong, Shanxi, Yunnan and Zhejiang, *P. dulcis* (bai bu ji zhu), cultivated in Fujian, Jiangsu and Zhejiang, *P. glabrata* (hua bu ji zhu), cultivated in Fujian and Zhejiang, and *P.*

circumpilis (mao ke hua bu ji zhu), cultivated in Zhejiang. Summer sources of bamboo shoots include *Dendrocalamus latiflorus* (ma zhu), distributed in Fujian, Guangdong, Guangxi, Guizhou, Hainan, Sichuan, Taiwan and Yunnan, and cultivated in southwestern Jiangsu and southern Zhejiang, *Bambusa beecheyana* var. *beecheyana* (diao si qiu zhu), found in Guangdong, Guangxi, Hainan and Taiwan, *B. beecheyana* var. *pubescens* (da tou dian zhu) from Guangdong and Taiwan, *B. variostriata* (diao si dan zhu), fromm Guangdong, and *B. oldhamii* (lü zhu), distributed in Fujian, Guangdong, Guangxi, Hainan, Taiwan and southern Zhejiang.

Zizania latifolia.—*Zizania latifolia* (gu; Poaceae) is native to China, distributed in the south, center and north of the country, especially Anhui, Hubei, Hunan, Jiangsu and Zhejiang. It is a very special and traditional vegetable in China, and has been used for 3 000 years. In ancient China the grains were used for food; today it is mainly cultivated as an aquatic vegetable, covering over 50 000 ha. The young shoots and rhizomes are edible when infected, swollen and softened by the fungus *Ustilago esculenta* (hei fen jun), which also prevents flowering. There are dozens of cultivars, falling into two main types, single-cropping (in autumn) and double-cropping (in spring and autumn); the former requires short days while the latter is not sensitive to day length. Wild ecotypes are also found along lakes, ponds and swamps. The tender shoots and seeds of non-infected plants are also used.

Zingiber officinale.—*Zingiber officinale* (ginger; jiang; Zingiberaceae) has been cultivated for more than 2 500 years, and it is believed that southern China is, at least, a secondary origin of the species. Cultivated in Anhui, Fujian, Guangdong, Guangxi, Guizhou, Hainan, Henan, Hubei, Hunan, Jiangxi, Shaanxi, Shandong, Sichuan, Taiwan, Yunnan and Zhejiang, it is a very popular condiment in China. The related *Z. mioga* (Japanese giner; rang he), found in Anhui, Guangdong, Guangxi, Guizhou, Hunan, Jiangsu, Jiangxi, Yunnan and Zhejiang, is occasionally cultivated, for use as a medicine and vegetable.

Allium.—Several species of *Allium* (cong shu) are native to Xinjiang. *Allium chinense* (Chinese onion; jiao tou) is a traditional Chinese vegetable found in Anhui, Fujian, Guangdong, Guangxi, Guizhou, Hainan, Henan, Hubei, Hunan, Jiangxi and Zhejiang. The fragrant bulbs are used in pickles. *Allium fistulosum* (green onion; cong) is native to northwestern Xinjiang and has been cultivated widely in China since 681 BC. Today, it covers more than 500 000 ha. Both the leaves and bulbs are used as a vegetable. There are many cultivars, of two types: the large ones used both as a vegetable and for seasoning, the smaller ones used mainly for seasoning. Related wild species include *A. altaicum* (a er tai cong), distributed in western Heilongjiang, Nei Mongol and northern Xinjiang, which also has edible leaves. *Allium tuberosum* (garlic chives; jiu; Alliaceae) is native to southwestern Shanxi, naturalized in southern China, and has been widely cultivated as a vegetable for more than 2 500 years. The edible leaves and scapes are used in various cooked dishes, particularly mixed with meat for stuffing dumplings. The related species *A. ramosum* (Chinese chives; ye jiu) is also cultivated and grows in Gansu, Hebei, Heilongjiang, Jilin, Liaoning, Nei Mongol, Ningxia, Qinghai, Shaanxi, Shandong, Shanxi and Xinjiang. *Allium hookeri* (kuan ye jiu), from southwestern Sichuan, southeastern Xizang and northwestern Yunnan is also used as a vegetable and has edible wild relatives including *A. funckiifolium* (yu zan ye shan cong) from western Hubei and eastern Sichuan. *Allium ovalifolium* (luan ye shan cong), distributed in southeastern Gansu, northeastern Guizhou, western Hubei, eastern Qinghai, southern Shaanxi, Sichuan and northwestern Yunnan, also has edible leaves, as does *A. macrostemon* (xie bai) from Xizang and Yunnan.

Hemerocallis citrina.—*Hemerocallis citrina* (huang hua cai; Liliaceae) is found in Anhui, Hebei, Henan, Hubei, Hunan, Jiangsu, Jiangxi, Nei Mongol, Shaanxi, Shandong, Sichuan and Zhejiang. This is one of the most ancient and traditional vegetables in China, its edible flowers having been used for over 2 500 years. It is widely cultivated, especially in Hunan. It is, however, slightly poisonous and needs careful treatment, usually steaming and drying the flower buds before use. Three other, less important, species are also used as vegetables: *H. minor* (xiao huang hua cai) from Gansu, Hebei, Heilongjiang, Jilin, Liaoning, Nei Mongol, Shaanxi, Shandong and Shanxi, *H. lilioasphodelus* (bei huang hua cai) from Gansu, Hebei, Heilongjiang, Henan, Jiangsu, Jiangxi, Jilin, Liaoning, Shaanxi, Shandong and Shanxi, and *H. fulva* (xuan cao) from Anhui, Fujian, Guangdong, Guangxi, Guizhou, Hebei, Henan, Hubei, Hunan, Jiangsu, Jiangxi, Shaanxi, Shandong, Shanxi, Sichuan, Taiwan, Xizang, Yunnan and Zhejiang.

Lilium.—*Lilium* (bai he shu; Liliaceae) is native to eastern Asia and North America, with most of its diversity found in China, Japan, temperate North America and Europe. It comprises about 115 species, with 55 in China, of which 35 are endemic and one introduced. Of these, 12 species and five varieties are known to have edible bulbs. In China, *Lilium* is found all over the country and has been cultivated for more than 2 200 years. At present about 700 000 tons of bulbs are produced annually. *Lilium brownii* (ye bai he), distributed in eastern, southern and northern China, Yunnan and Sichuan, is cultivated both for its edible bulbs and for medicinal uses. *Lilium brownii* var. *viridulum* (bai he), found in Anhui, Fujian, Gansu, Guangxi, Guizhou, Hebei, Henan, Hubei, Hunan, Jiangsu, Jiangxi, Shaanxi, Shanxi, Sichuan, Yunnan and Zhejiang, is the most popular variety for bulb production. *Lilium davidii* (chuan bai he), from Gansu, Guizhou, Henan, western Hubei, southern Shaanxi, Shanxi, Sichuan and Yunnan, is also widely cultivated for its edible bulbs. *Lilium tigrinum* (juan dan) includes many outstanding cultivars widely grown for both their bulbs and medicinal uses. *Lilium sulphureum* (dan huang hua bai he), distributed in Guangxi, Guizhou,

Sichuan and Yunnan, is cultivated mainly for medicinal use. The bulbs of *L. speciosum* var. *gloriosoides* (yao bai he) are edible and medicinal, and distributed in Anhui, Guangxi, Hunan, Jiangxi, Taiwan and Zhejiang.

Dioscorea.—*Dioscorea* (shu yu shu; Dioscoreaceae) includes more than 600 species, widely distributed in tropical and temperate regions. There are 52 species in China, 21 of which are endemic and two introduced. The genus is of great economic value and has been used for more than 2 500 years. Many wild species are valuable famine foods and sources of both traditional Chinese and western medicines; several species are also widely cultivated. *Dioscorea polystachya* (Chinese yam; shu yu) is distributed in southern Anhui, Fujian, eastern Gansu, northern Guangdong, Guangxi, Guizhou, Hebei, Henan, Hubei, Hunan, Jiangsu, Jiangxi, Jilin, Liaoning, southern Shaanxi, Shandong, Sichuan, Taiwan, northern Yunnan and Zhejiang. The rhizomes are considered to be a tonic, and are used as a vegetable valued for their health-giving properties. The most well-known cultivar is "huai shan yao," which is indigenous to Henan. *Dioscorea japonica* (Japanese yam; ri ben shu yu) is found in Anhui, Fujian, Guangdong, Guangxi, eastern and southern Guizhou, Hubei, Hunan, Jiangsu, Jiangxi, Sichuan, Taiwan and Zhejiang. The rhizomes are edible and commonly used as a vegetable, mainly in cooked dishes but also in salads. *Dioscorea alata* (purple yam; shen shu) is cultivated widely in Fujian, Guangdong, Guangxi, Hubei and Taiwan. The edible tubers are used as a vegetable. *Dioscorea esculenta* (lesser yam; gan shu) has been cultivated in southern China (Guangxi, Hainan and Taiwan) for at least 1 700 years.

16.3.2 Pteridophytes

The use of the tips of young pteridophyte fronds as vegetables is becoming increasingly popular due to their nutrient content and wild nature. A few examples are described below.

Osmundastrum cinnamomeum.—The tender fronds of *Osmundastrum cinnamomeum* (cinnamon fern; wei cai; Osmundaceae), found in northeastern China are used both fresh and dried as a vegetable.

Pteridium aquilinum var. *latiusculum.*—*Pteridium aquilinum* var. *latiusculum* (bracken; jue cai; Dennstaedtiaceae) is distributed widely in the Chang Jiang reaches of northern China. The tips of the young fronds are used as a vegetable in both fresh and cooked dishes. Starch is also extracted from the rhizomes for food and good quality food additives. *Pteridium esculentum* (shi jue), from southwestern China and Hainan, is also used as a vegetable.

Athyrium brevifrons.—*Athyrium brevifrons* (lady fern; hou tui ti gai jue; Athyriaceae) is distributed in Hebei, Heilongjiang, Jilin, Liaoning, Nei Mongol, Shanxi and Shandong. The young leaves are used as a vegetable, fresh, cooked or salted.

16.4 Staple foods and edible oil bearing plants and their wild relatives

Staple foods are among the earliest domesticated crops and were essential to the initial origins of agriculture and the development of modern human society (Diamond, 2002). Staple foods are especially important to countries like China, which has only 10% of the world's arable land to sustain 20% of its human population (Dong & Zheng, 2006). Staple crops in China can largely be divided into three categories: cereals, legumes and tubers (see Table 16.1). At least ten major cereals, 22 legume crops and ten tuber crops have been cultivated in China (Table 16.1). China is the center of distribution of the wild ancestors and wild relatives of cultivated rice (*Oryza sativa*), foxtail millet (*Setaria italica*; Poaceae), broomcorn millet (*Panicum miliaceum*; Poaceae), red (adzuki) bean (*Vigna angularis*) and buckwheat (*Fagopyrum esculentum;* Polygonaceae; Table 16.1). Several species, including rice, foxtail millet, adzuki bean and buckwheat were first domesticated in China (Zohary & Hopf, 2000), and rice remains one of most important food crops worldwide.

Edible oil crops are the second most important class of crops and are widely cultivated across five continents, with the USA, India, China, Brazil, Argentina, Indonesia and Malaysia being the major producers. The four biggest edible oil crops are soybean (also a major vegetable crop, treated under Section 16.3), rapeseed, peanut (*Arachis hypogaea;* Fabaceae) and sunflower (Helianthus annuus; Asteraceae; Fang & Chang, 2007). Twelve major oil crop species are cultivated in China (see Table 16.1). Of these, the soybean was first domesticated in China (Guo *et al.*, 2010; see Section 16.3 for details), and perilla (*Perilla frutescens;* Lamiaceae) possibly so (Nitta *et al.*, 2005); the other ten species are introduced. Five wild *Glycine* species and one wild *Perilla* species are found in China and, whilst the wild ancestor of tigernut (*Cyperus esculentus*; Cyperaceae) is not found here, ca. 60 wild co-genetic species are distributed in China (Table 16.1).

16.4.1 Important staple foods and edible oil bearing plants and their wild relatives

Fagopyrum esculentum.—*Fagopyrum esculentum* (buckwheat; qiao mai), has a long history of domestication and cultivation in China. It was described in the *Book of Songs*, suggesting buckwheat cultivation in China dates back as long as 2 500 years ago to the Chun Qiu period (770–476 BC) of the Western Zhou Dynasty. Archeological findings and tomb artifacts of buckwheat remains at Yangjiawan, Shaanxi, dated to the Western Han Dynasty (202 BC–8 AD) further demonstrate buckwheat cultivation in China more than 2 000 years ago. During the Sui and Tang Dynasties (581–907 AD), buckwheat cultivation gradually expanded in response to demand for food security against natural disasters and limited availability of farmland, due to its very short growing

Table 16.1 Major native staple food and oil crops, or crops with congeneric wild species native to China. Asterisks indicate species not recognized in *Flora of China*. Note that exotic crops are dealt with in Chapter 20.

Scientific name	Chinese name	English name	Number of congeneric wild species native to China
Cereals (all Poaceae)			
Avena sativa	燕麦 (yan mai)	common oat	1
Coix lacryma-jobi	薏苡 (yi yi)	Job's tears	2
Eleusine coracana	穇 (can)	finger millet	1
Hordeum vulgare	大麦 (da mai)	barley	6
Oryza sativa	水稻 (shui dao)	rice	3
Panicum miliaceum	稷 (ji)	broomcorn millet	17
Pennisetum glaucum	御谷 (yu gu)	pearl millet	7
Setaria italica	粱 (liang)	foxtail millet	12
Sorghum bicolor	高粱 (gao liang)	sorghum	2
Triticum aestivum	小麦 (xiao mai)	common wheat	1 (*Aegilops*), 2 (*Triticum*)
Triticum turgidum	圆锥小麦 (yuan zhui xiao mai)	duckbill wheat, cone wheat, rivet wheat	2
Legumes (all Fabaceae)			
Cajanus cajan	木豆 (mu dou)	pigeon pea	6
Canavalia ensiformis	直生刀豆 (zhi sheng dao dou)	jack bean	3
Canavalia gladiata	刀豆 (dao dou)	sword bean	3
Cicer arietinum	鹰嘴豆 (ying zui dou)	chickpea	1
Lathyrus sativus	家山黧豆 (jia shan li dou)	Indian pea	15
Mucuna pruriens var. *utilis*	黧豆 (li dou)	velvet bean	18
Trigonella foenum-graecum	葫芦巴 (hu lu ba)	fenugreek	6
Vicia faba	蚕豆 (can dou)	broad bean	38
Vigna angularis	赤豆 (chi dou)	red bean, adzuki bean	13
Vigna mungo var. *mungo**	黑绿豆 (hei lü dou)	black gram	13
Vigna radiata	绿豆 (lü dou)	mung bean	13
Vigna umbellata	赤小豆 (chi xiao dou)	rice bean	13
Vigna unguiculata	豇豆 (jiang dou)	cowpea	13
Tubers			
Colocasia esculenta (Araceae)	芋 (yu)	taro	6
Dioscorea alata (Dioscoreaceae)	参薯 (shen shu)	purple yam, greater yam	50
D. hispida	白薯茛 (bai shu liang)	white yam	50
D. polystachya	薯蓣 (shu yu)	Chinese yam	50
Ipomoea batatas (Convolvulaceae)	番薯 (fan shu)	sweet potato	27
Solanum tuberosum	阳芋 (yang yu)	potato	37
Other food crops			
Chenopodium quinoa (Chenopodiaceae)	藜谷 (li gu)	quinoa	14
Fagopyrum esculentum (Polygonaceae)	荞麦 (qiao mai)	buckwheat	9
Oil crops			
Brassica rapa (Brassicaceae)	白菜型油菜 (man jing)	turnip	—
B. juncea	芥菜 (jie cai)	Chinese mustard	—
Carthamus tinctorius (Asteraceae)	红花 (hong hua)	safflower	—
Cyperus esculentus var. *sativus* (Cyperaceae)	油莎草 (you sha cao)	tigernut	ca. 60
Glycine max (Fabaceae)	大豆 (da dou)	soybean	5
Perilla frutescens (Lamiaceae)	紫苏 (zi su)	perilla	1

season (historically, it was regarded as a disaster relief or catch crop). There are a large number of records of buckwheat cultivation in Chinese ancient materia medica, poetry and miscellaneous notes. For example, a Tang Dynasty poem by BAI Ju-Yi features the lines "look out front of the field, spectacular buckwheat flowers like snow," while WEN Ting-Yun wrote "crow noisy in sunset, buckwheat flowering mountain." These vivid descriptions reveal the prosperous scenes of buckwheat cultivation. Buckwheat is well known for containing the glycoside rutin, beneficial to health, and polysaccharide glucomannan, of value to elderly diabetes patients.

Two species of buckwheat have been domesticated for cultivation. Common or sweet buckwheat, *Fagopyrum esculentum* occurs widely throughout China and the mountains of northern India. Bitter or Tatar buckwheat, *F. tataricum*, is cold-hardy and cultivated in alpine regions of Asia and Europe, and is found in China in Gansu, Guangxi, Guizhou, Hebei, Heilongjiang, Hubei, Hunan, Jilin, Liaoning, Nei Mongol, Qinghai, Shaanxi, Shanxi, Sichuan, Xinjiang, Xizang and Yunnan. Both cultivated species of buckwheat originated in China and were introduced into Korea and Japan about 2 500 years ago and into Europe, by the fifteenth century, via Mongolia and Russia during the Middle Ages. However, the domestication history of cultivated buckwheat is not clear and its closest wild relatives are not known. China has rich buckwheat germplasm resources: a current inventory of buckwheat germplasm listed 1 814 landraces or accessions of sweet buckwheat and 879 of bitter buckwheat (Institute of Crop Germplasm Resources, Chinese Academy of Agricultural Sciences, 1996). *Fagopyrum dibotrys* (jin qiao) grows in moist valleys and on grassy slopes in the provinces of Anhui, Fujian, Gansu, Guangdong, Guangxi, Guizhou, Henan, Hubei, Hunan, Jiangsu, Jiangxi, Shaanxi, Sichuan, Xizang, Yunnan and Zhejiang, and is also distributed in Bhutan, India, Kashmir, Myanmar, Nepal, Sikkim and Vietnam. This species is not itself a staple but is used medicinally. It has been considered a possible wild ancestor of cultivated buckwheat.

Brassica rapa var. *oleifera* (oilseed rape; yun tai) is considered to be native and endemic to southern China. Green forms are grown mainly in Guangdong and Guangxi; purplish forms in Hubei, Hunan and Sichuan. It is widely cultivated for seed oil, and the tender shoots are also used as a vegetable.

Vigna.—*Vigna angularis* (red or adzuki bean; chi dou) was domesticated and cultivated in China as a food, fodder, green manure and medicinal plant in ancient times. The earliest written record of *V. angularis* is found in the *Shi Sheng Zhi Shu*, the most ancient agricultural book in the world, written during the Western Han Dynasty, which documented red bean planting, sowing, field management, harvesting and yields. SHEN Nong's *Materia Medica* (225–202 BC) also recorded the medicinal value of the red bean, and the *Qi Min Yao Shu* (531–550) recorded detailed information of its cultivation methods. Archeological remains unearthed in the Mawangdui tomb, Changsha, Hunan, from the Han Dynasty, included carbonized seeds of the red bean, providing evidence that it has been cultivated in China for at least 2 000 years. A cake made with red bean paste remains one of the favorite snacks of the Chinese people today. The beans are also used in dietotherapy to strengthen the spleen and remove toxins, among other effects.

Vigna angularis originated in China, its natural range comprising northern and northeastern China, the Huang He and Chiang Jiang regions and Taiwan. Wild *Vigna angularis* is not treated as a separate taxon from the cultivated form. Vavilov (1951) believed China also to be the center of origin of the cultivated red bean. Field surveys have demonstrated that wild and semi-wild red beans occur in the Himalaya Shan, Hubei, Shaanxi, Shandong and Yunnan, and other provinces. Many different variants and landraces have been identified including a wide spectrum of bean colors. Today, most cultivation and production is in China and Southeast Asia. The red bean was introduced into Korea and Japan in the third to eighth century AD, where a secondary cultivation center developed. Later introductions to the USA, Canada, Brazil, New Zealand and the Democratic Republic of Congo, have led to more than 20 countries now growing the bean commercially, with the largest producers remaining China and Japan, followed by South Korea.

Vigna umbellata.—*Vigna umbellata* (ricebean; chi xiao dou, fan dou) was possibly domesticated in China and is now wildly cultivated in Anhui, Beijing Shi, Guangdong, Guangxi, Hainan, Hebei, Jiangsu, Jiangxi, Jilin, Shaanxi, Shandong, Taiwan, Tianjin, Yunnan and Zhejiang, as well as in Japan, Korea and Southeast Asia. Wild *Vigna umbellata* is also found in China, but is not treated as a separate taxon.

Perilla frutescens var. *purpurascens.*—*Perilla frutescens* var. *purpurascens* (perilla; ye sheng zi su; Lamiaceae) grows on wastelands in Fujian, Guangdong, Guangxi, Guizhou, Hebei, Hubei, Jiangsu, Jiangxi, Shanxi, Sichuan, Taiwan, Xizang, Yunnan and Zhejiang, as well as in Japan. *Perilla frutescens* var. *purpurascens* is considered to be the wild ancestor of two cultivated varieties, *P. frutescens* var. *frutescens* and *P. frutescens* var. *crispa*.

Colocasia.—*Colocasia esculenta* (taro; yu tou; Araceae) is a variable species including escapes, naturalizations, re-domestications and selections. It contains multiple intraspecific taxa, some of them even raised to separate species; here, as in the *Flora of China*, we treat them all as a single species. Originating in China and India, it has been known as a crop in China since the third century BC. It is distributed in Anhui, Fujian, Guangdong, Guangxi, Guizhou, Hainan, Hubei, Hunan, Jiangsu, Jiangxi, Sichuan, Taiwan, Yunnan and Zhejiang. The best known Chinese cultivar, "li pu yu," originated in Lipu, Guangxi. The large, fragrant rhizomes were used as tributes to the emperors in ancient times. *Colocasia esculenta* is widely cultivated in Asia, Africa, Europe and America, where both wild and naturalized forms are found in wet places in forests, valleys, swamps and on wastelands. It is also widely distributed on Southeast Asian and Pacific islands. This species is one of most important tuber crops; its petioles and inflorescences are also used as a vegetable. The rhizomes are used medicinally for treating swellings, abscesses, snake and insect bites and swollen lymph nodes. *Colocasia gigantea* (Indian taro; da ye yu; Araceae) is also widely cultivated in Southeast Asia and grows wild in the moist forests of Fujian, Guangdong, Guangxi, Jiangxi, Yunnan; it is cultivated in Anhui, Guizhou, Hunan, Sichuan and Zhejiang as well as Cambodia, Laos, Peninsula Malaysia, Myanmar, Thailand and Vietnam.

Oryza sativa.—The genus *Oryza* (rice) has 22 species and is widely distributed in tropical Asia, Africa and central South America. There are three cultivated species, *O. glaberrima* in Africa, *O. latifolia* in the Americas, and the Asian *O. sativa* (shui dao) which is the most widely cultivated throughout China, Japan, India and Southeast Asia. Production of *O. sativa* extends around the world. China is the original center of cultivated rice with four taxa occurring naturally. In addition to cultivated rice, *O. sativa*, the other three taxa are medicinal wild rice, *O. officinalis*, warts-grain wild rice, *O. meyeriana* subsp. *granulata*, and common wild rice, *O. rufipogon* (ye sheng dao), which is widely distributed in southern China below 700 m, and also from Southeast Asia to Australia. *Oryza rufipogon* bears the closest wild germplasm to cultivated rice and has been widely used in rice breeding. For example the variety "zhong shan shi hao" is a hybrid of Asian cultivated rice and *O. rufipogon* formed in Guandong, which has now been widely cultivated for more than half a century. *Oryza rufipogon* is protected in China, where multiple special reserves have been set up to conserve it; however its populations have decreased greatly through habitat loss.

China's long history of rice domestication and cultivation dates from primitive tribal societies. The earliest written record of rice is found in *Records of the Historian: Summer of the Mind* by QIAN Si-Ma (145–90 BC). Archeological excavations at many Neolithic sites south of the Chang Jiang have discovered the remains of cultivated rice dating back 8 000 years (at 18 Kan, Hunan) to ca. 7 000 years (at Kawado Dom and Luo Gu in Zhejiang). It is evident that rice was cultivated in the Huang He region 5 000 years ago, and in Henan rice remains excavated at Yangshao, determined by carbon-14 analysis, also date back 5 000 years. China's large area and altitudinal range led to the development of two major types of cultivated rice: the heat- and moisture-tolerant "indica" type, suitable for southern low-altitude and tropical regions, and cold-hardy "japonica" type for northern and high-altitude regions. Literature prior to the Eastern Han Dynasty (25–220 AD) has demonstrated that both indica and japonica rice had been widely cultivated all over China since the Western Zhou Dynasty (ca. 1100 BC onwards). The *Qi Min Yao Shu* by JIA Si-Xie (386–557 AD) already documented 13 rice varieties, including four japonica and nine indica varieties. After Chinese reunification in the Tang Dynasty (618–907), rice cultivation and production was rapidly developed in the Chang Jiang region and southern China, with many varieties. From the Song Dynasty (960–1279) onwards, a substantial increase in rice varieties occurred, with more than 300 varieties documented in agricultural and local history books. Many more varieties are recorded in the Ming and Qing Dynasties. The *Shou Shi Tong Kao* (1729 AD), an inventory of 223 counties of 16 provinces, documents more than 3 400 varieties, demonstrating great innovations in rice variety improvement by this time.

This long history of domestication and selection by Chinese people has created thousands of varieties of cultivated rice, making rice the world's most extensively-planted and most important food crop. In the 1950s, use of dwarf varieties contributed to the first great leap in rice production and substantially increased yield; in the 1960s, the international rice breeding program generated the first high-yielding variety, "IR8." Even today, many new dwarf rice varieties and breeding materials continue to be developed by the International Rice Research Institute and many rice research institutes in Southeast Asian countries, directly or indirectly originating from the old Chinese rice varieties. Hybrid rice created in China since the 1970s have also greatly revolutionized world rice production, not only by their high yields, but also to improve heterosis in self-pollinated crops.

Setaria.—*Setaria italica* (foxtail millet; su) is one of the oldest cultivated crops in China, widely used for food in arid regions and has a higher nutritional (protein) value than maize or sorghum, and higher oil content than rice or wheat. It is also a source of high-quality animal feed. Archeological excavations have discovered a large number of carbonized remains in many Neolithic ruins of Chinese primitive society (Dou Ji He and Banpo in Shaanxi, Huang Lian Shu in Henan, Ci Shan in Hebei and sites in Gansu and Qinghai) and carbon-14 analyses have dated these to 6 000–7 000 years ago. Millet was regarded as an important grain from the Shang Dynasty to the Qin Dynasty (from ca. 1600 BC). Until the early twentieth century, millet remained the first arid autumn crop grown in northern China. Millet is a highly nutritious, short growing season crop, cold-hardy, tolerant of poor soils, and better adapted to arid regions than maize, sorghum or wheat. During domestication, the Chinese people have selected and developed many valuable ecotypes and varieties, such as cold-hardy, wet-tolerant, early-maturing and forage types. As early as the fifth century AD, more than 80 varieties were documented in the *Qi Min Yao Shu*.

All cultivated millets around the world were originally introduced from China. Since China has the longest history of millet domestication and is the original center of cultivated millet, more than 25 000 accessions of germplasm are currently conserved in seed banks. Chinese millet varieties are well known for their high yields and quality and have been introduced into the USA, Japan, Korea, India, Russia and other countries. There are currently about 27 countries producing millet throughout the world. The world's largest millet growing areas are in Asia, accounting for about 97% of the world total planting and 96% of world total production. China is the largest producer and accounts for 90% of production. The largest growing areas are in northern China: Heilongjiang is the largest producer province.

Setaria viridis (green foxtail; gou wei cao) grows on slopes, roadsides and grassy wastelands in most Chinese provinces including Anhui, Fujian, Gansu, Guangdong, Guizhou, Hebei, Heilongjiang, Henan, Hubei, Hunan, Jiangxi, Jiangsu, Jilin, Nei Mongol, Ningxia, Qinghai, Shaanxi, Shandong, Shanxi, Sichuan, Taiwan, Xinjiang, Xizang, Yunnan and Zhejiang; it is also distributed in temperate and subtropical regions of Eurasia, and naturalised on other continents. The species has been divided into three subspecies. *Setaria viridis* subsp. *viridis* is the wild ancestor of cultivated foxtail millet (*S. italica*), for which it provides important breeding germplasm; however it is also a cosmopolitan weed of temperate to tropical regions. *Setaria viridis* subsp. *pycnocoma* is widely distributed in temperate and subtropical Eurasia and naturalized in Europe; it may be a hybrid between *S. viridis* subsp. *viridis* and cultivated foxtail millet. *Setaria viridis* subsp. *pachystachys* is distributed on the coasts of Guangdong and Taiwan as well as in Korea and Japan.

Panicum is the largest genus of Poaceae and comprises about 470 species which are mostly distributed in tropical regions. *Panicum miliaceum* (broomcorn millet; ji) occurs naturally at the northern and high-latitude edges of the natural range of the genus and has been domesticated as a specialized annual crop with important agricultural significance. *Panicum maximum* and *P. sumatrense* have also been used in cultivation. Broomcorn millet is one of the most ancient cultivated crops in arid regions of China and was regarded as the top food crop in ancient China. Chinese names for the crop have evolved over different dynasties, from "mi" in the *Cang Jie* articles of the Qin Dynasty, 221–206 BC, to "zhi" in the *Compendium of Materia Medica*, written in 1578 in the Ming Dynasty and "ji" today. The term "shu ji" is also sometimes used, but actually refers to two variants of the same species – "shu" being glutinous broomcorn millet and "ji" a non-glutinous form.

Archeological discoveries of broomcorn millet remains in Neolithic sites at Pei Li Gang, Henan province, suggested cultivation dates back to at least 7 000 years ago determined by carbon-14 analysis. Later discoveries at other sites of Heilongjiang, Liaoning and Nei Mongol have also dated its use back more than 4 000 years. In the *Lu Shi Chun Qiu* (239 BC), broomcorn millet from Yangshan is described as excellent. In ancient times the cultivation of broomcorn millet was regarded as a matter of social stability and livelihood importance. The very concept of the Chinese state evolved via the term "sajk," from the meaning of broomcorn. Chinese society is founded on agriculture, with the principles of agricultural management, known as "hou ji," indicating a very long domestication and cultivation history of broomcorn millet as an important crop.

There are more than 7 500 accessions of millet germplasm, including many with high protein contents, preserved in the seed banks of Chinese Academy of Agricultural Sciences. The wild form of cultivated broomcorn millet, *Panicum miliaceum,* is not separated in the *Flora of China*, because the ancestry of the crop species is unclear, but is sometimes treated as *P. miliaceum* subsp. *ruderale* (ye sheng ji). Wild *P. miliaceum* grows on damp meadows and pastures, lowland grassy bogs, wet riverbanks and lakesides, fallow fields, and damp, nitrogen-rich wastelands in northern China. It is also found in Europe, temperate Asia and naturalized in North America.

Sorghum bicolor.—The genus *Sorghum* (Poaceae) comprises about 24 species distributed in tropical and subtropical regions. *Sorghum bicolor* (sorghum; gao liang) is a high-yielding crop and has 5 000 years history of cultivation in China. It is tolerant of both drought and waterlogging and is particularly adapted to barren and saline soils, in arid regions. In addition to its use as a food, sorghum has also been widely used for wine, sugar, soy sauce and animal feed. Sorghum has been documented in many Chinese ancient books such as the *Guang Zi* (725–645 BC), *Zhuang Zi* (475–221 BC), *Shi Ben* (234–228 BC), *Guang Ya* (123–227 AD), *Bo Wu Zhi* (232–300 AD) and *Qi Min Yao Shu* (533–534 AD). In 1931, carbonized sorghum was first discovered in a late Neolithic ruin from the Jing village in Shanxi. Since then, a large number of carbonized sorghum remains have been discovered from the Warring States period (475–221 BC) in Shijiazhuang, Hebei, and Neolithic Era from Dahe villege, Henan. The remains of sorghum stem bundles have also been found in wall construction sites from the Western Han Dynasty (202 BC–8 AD) in Xi'an, Shaanxi. In addition, carbonized sorghum grains were found in a jar at a Western Zhou Dynasty historic site in Xinyi, Jiangsu, and determined by carbon-14 dating to originate 5 000 years ago.

According to the collections and data of the World Sorghum Genetic Resources Center, China, India and Africa are sorghum germplasm centers of diversity as well as likely places of origin of sorghum. Current collections of Chinese sorghum amount to 16 000 accessions, preserved from more than 34 provinces. Many have been used for hybridization breeding resulting in the release of a number of hybrids into cultivation. Many features of Chinese sorghum are significantly different from those of African and Indian sorghum, suggesting cultivated sorghum could originate from several different wild species. Vavilov (Vavilov, 1935) suggested that China is the the oldest and largest independent center of origin for sorghum domestication and cultivation. The Huang He basin, with its warm and humid climate in ancient times, together with the rich diversity of cultivated sorghum varieties on the Huangtu Gaoyuan (Loess Plateau), suggest these regions as the most likely center of origin of cultivated sorghum in China.

Avena.—The genus *Avena* (Poaceae) has about 30 species worldwide, widely distributed in the Mediterranean region, North Africa, western and eastern Asia. Two species have been domesticated and cultivated for crop production as oats, one of the world's major cereal crops. The common oat, *A. sativa* (yan mai), is widely cultivated in Europe. The Chinese naked oat, *A. chinensis,* was domesticated in China, originally on the Huangtu Gaoyuan in the northwest, where it has been cultivated as a major cereal crop. It has a history of at least 2 400 years of cultivation in China. *A. chinensis* is native to China, naturally occuring across the Taihang Shan and Lüliang Shan in northwestern Shanxi, Yan Shan in northern Hebei and several sites in Nei Mongol, as well as northeastern and southwestern China at elevations of 2 000–3 200 m. It is highly cold-hardy and drought tolerant.

Avena chinensis was widely documented in Chinese ancient historic literature, with the *Qi Min Yao Shu* recording that it was extensively cultivated in China across a wide area from 543–386 BC or even earlier. The cold-climate high mountain ranges in northern China are its center of origin, spreading northward to Nei Mongol, westward to Xinjiang, southward to the mountains of Sichuan and Yunnan, and eastward to Shandong. During the Northern and Southern Dynasties (420–581 AD), *A. chinensis* was extensively cultivated within the country and introduced into Sweden, Russia, the USA and Chile, among others. Today, the Nei Mongol area is its largest producer, followed by Hebei and Shanxi, with widespread cultivation also in northern and southwestern China. Today it is a minor oat crop covering a declining area. However, its high nutritional value with higher protein and unsaturated fatty acids than a variety of other cereal crops gives *A. chinensis* considerable potential for agricultural development in high mountain areas with cold, dry weather conditions. There are very rich germplasm resources of both *A. chinensis* and *A. sativa* in China. As many as 1 663 landraces or accessions have been collected and preserved by the Chinese Academy of Agricultural Sciences, including many accessions of *A. chinensis* varieties with high protein and high linoleic acid content, disease resistance, drought tolerance, high-quality and early-maturing characteristics. In the past *A. chinensis* has been mistaken as a species of European naked oat, *A. nuda,* which is a European weed with small seeds and chromosomes. For a long time, this confusion resulted in the germplasm of *A. chinensis* not receiving attention in international germplasm inventories. China will play an important role in contributing to the further development of *A. chinensis* for global production.

Aegilops tauschii.—*Aegilops tauschii* (goatgrass; shan yang cao; Poaceae) is found on stony slopes, in wheatfields and weedy places in Henan, Shaanxi and Xinjiang as well as from central to western Asia. *Aegilops tauschii* is one of three ancestors of hexaploid wheat and the only one found wild in China.

Coix lacryma-jobi var. *puellarum*.—*Coix lacryma-jobi* var. *puellarum* (Job's tears; xiao zhu yi yi; Poaceae) is one of four varieties of *C. lacryma-jobi* alongside three cultivated forms, *C. lacryma-jobi* var. *lacryma-jobi*, *C. lacryma-jobi* var. *stenocarpa* and *C. lacryma-jobi* var. *ma-yuen*. *Coix lacryma-jobi* var. *puellarum* is found in the moist forests of Hainan, Xizang and Yunnan, and is also distributed in northeastern India, Myanmar, Thailand and Vietnam. As yet, no studies have been conducted to determine whether it is the wild ancestor of the three cultivated varieties.

Hordeum spontaneum.—*Hordeum spontaneum* (dun fu ye da mai; Poaceae), the wild ancestor of cultivated barley (*H. vulgare;* da mai) and sometimes treated as a subspecies of it, is distributed in Sichuan, Xizang and Yunnan as well as central to western Asia and northeastern Africa, and typically grows in disturbed places and on roadsides, and often as a weed in fields of cultivated *Hordeum* and other crops. Based on the morphology of the lemma and awn, *H. spontaneum* has been divided into three varieties, *H. spontaneum* var. *spontaneum*, *H. spontaneum* var. *proskowetzii* and *H. spontaneum* var. *ischnatherum*. The species has been widely used in barley breeding.

References

Chao, W. J. & Yuan, Z. F. (1990). Discussion on classification and relationship of Chinese wild species of grape. *Acta University Agriculture Boreali-Occidentalis* 18(2): 7-12.

Chen, W. (2002). *Nong Ye Kao Gu*. Beijing, China: Cultural Relics Publishing House.

Dai, Z., Chen, L. G. & Tong, P. Z. (2008). Genetic variation and fingerprinting of *Torreya grandis* cultivars detected by ISSR markers. *Acta Horticulturae Sinica* 35(8): 1125-1130.

Diamond, J. M. (2002). Evolution, consequences and future of plant and animal domestication. *Nature* 418: 700-708.

Doebley, J. F., Gaut, B. S. & Smith, B. D. (2006). The molecular genetics of crop domestication. *Cell* 127: 1309-1321.

Dong, Y. & Zheng, D. (2006). Food Crops. In: Dong, Y. & Liu, X. (eds.) *Crops and their Wild Relatives in China*. p. 605. Beijing, China: China Agriculture Press.

Editorial Board of Chinese Agriculture Encyclopedia (1990). *Chinese Agriculture Encyclopedia, Volume of Vegetables*. Beijing, China: Beijing: China Agriculture Press.

Fang, J. & Chang, R. (2007). Economic Crops. In: Dong, Y. & Liu, X. (eds.) *Crops and their Wild Relatives in China*. p. 602. Beijing, China: China Agriculture Press.

Fuller, D. Q., Qin, L., Zheng, Y. & Zhao, Z. (2009). The domestication process and domestication rate in rice: spikelet bases from the Lower Yangtze. *Science* 323: 1607-1610.

Guo, J. A., Wang, Y. S., Song, C., Zhou, J. F., Qiu, L. J., Huang, H. W. & Wang, Y. (2010). A single origin and moderate bottleneck during domestication of soybean (*Glycine max*): implications from microsatellites and nucleotide sequences. *Annals of Botany* 106: 505-514.

Huang, H. W. (1998). Review of current research of the world *Castanea* species and importance of germplasm conservation of China native *Castanea* species. *Journal of Wuhan Botanical Research* 16: 171-176.

Huang, H. W. (2013). *Genus Actinidia: Taxonomy, Germplasm, Domestication and Cultivation*. Beijing, China: Science Press.

Hunt, H. V., Linden, M. V., Liu, X. & Motuzaite-Matuzeviciute, G. (2008). Millets across Eurasia: chronology and context of early records of the genera *Panicum* and *Setaria* from archaeological sites in the Old World. *Vegetation History and Archaeobotany* 17(suppl. 1): 5-18.

Institute of Crop Germplasm Resources - Chinese Academy of Agricultural Sciences (1996). *A Catalog of Chinese Buckwheat Germplasm Resources (Continuation 2)*. Beijing, China: Chinese Agricultural Press.

Institute of Vegetables and Flowers - Chinese Academy of Agricultural Science (2010). *Chinese Vegetable Cultivation*. Beijing, China: China Agriculture Press.

IUCN (2011). IUCN Red List of Threatened Species. Version 2011.2. <http://www.iucnredlist.org/>.

Jia, J. X., Jia, D. X. & Ren, Q. M. (2006). Fruit crops. In: Dong, Y. & Liu, X. (eds.) *Crops and their Wild Relatives in China*. Beijing, China: China Agriculture Press.

Jones, M. K. & Liu, X. (2009). Origins of agriculture in East Asia. *Science* 324: 730-731.

Kikuchi, A. (1953). *Pomology*. Tokyo, Japan: Yokendo.

Li, X. G. & Yang, J. (2002). Application of numerical taxonomy of pollen morphology on origination, evolution and classification of *Pyrus* L. in China. *Journal of Fruit Science* 19(3): 145-148.

Liang, Z. F. (1983). On the distribution of genus *Actinidia*. *Guangxi Botanical Research* 3(4): 229-248.

Liu, M. J. (1998). *Chinese Wild Fruit Trees*. Beijing, China: China Agricultural Press.

Nitta, M., Lee, J. K., Kobayashi, H., Liu, D. J. & Nagamine, T. (2005). Diversification of multipurpose plant, *Perilla frutescens*. *Genetic Resources and Crop Evolution* 52: 663-670.

Perin, C., Hagen, L. S. & de Conto, V. (2002). A reference map of *Cucumis melo* based on two recombinant inbred line populations. *TAG* 104(6-7): 1003-1010.

Ruan, Y., Zhou, P. H. & Lin, C. L. (2002). Phylogenetic relationship among nine *Prunus* species based on random amplified polymorphic DNA. *Acta Horticulturae Sinica* 29(3): 218-223.

Sheng, C. Q. & Zhang, Y. H. (1979). *Acclimatization of Plants*. Shanghai, China: Shanghai Science and Technology Press.

Sun, S. R., Zhao, X. M. & Yang, Z. Q. (2007). Application of molecular markers in the study of Chinese cabbage. *Henan Science* 25(6): 925-929.

Tanaka, T. (1951). *Taxonomy of Fruit Trees*. Tokyo, Japan: Kawade Shobo.

Teng, Y. W., Chai, M. L. & Li, X. G. (2004). A historic retrospect and the progress of taxonomy in the genus *Pyrus*. *Journal of Fruit Science* 21(3): 252-257.

Vavilov, N. I. (1935). The phytogeographical basis for

plant breeding. In: Vavilov, N. I. (ed.) *Theoretical Basis for Plant Breeding*. pp. 17-75. Moscow and Leningrad, Russia: Sel'chozgiz.

Vavilov, N. I. (1951). The origin, variation, immunity and breeding of cultivated plants. *Chronica Botanica* 13: 1-366.

Wang, S. M., Lu, X. X., Li, L. H., Li, Y. & Gao, W. D. (2012). Progress review of germplasm resources conservation and used in past 10 years. In: Institute of Crop Science of Chinese Academy of Agricultural Sciences (ed.) *The Ten Years Progresses of Chinese Crop Germplasm Conservation and Uses*. pp. 1-17. Beijing, China: China Agriculture Press.

Wang, Y. M., Su, S. C., Zhai, M. P., Cheng, L. L., Huang, W. G. & Ma, T. X. (2008). Genetic analysis of *Corylus* in China by SSR. *Journal of Northeast Forestry University* 11: 48-51.

Wang, Y. Q., Fu, Y., Yang, Q., Luo, N., Deng, Q. X., Yan, J., Zeng, J. G. & Ruan, G. L. (2010). Genetic diversity of *Eriobotrya* analyzed by ISSR markers. *Scientia Silvae Sinica* 46(4): 50-57.

Yu, T. T. (1979). *Taxonomy of Fruit Trees in China*. Beijing, China: China Agriculture Press.

Zhang, B. B. (2008). RAPD analysis on the relation of 17 wild *Malus* species. *China Fruits* 2: 24-26.

Zhang, H. Z., Peng, S. A., Cai, L. H. & Fang, D. Q. (1990). The germplasm resources of the genus *Eriobotrya* with special reference on the origin of *E. japonica* Lindl. *Acta Horticulturae Sinica* 17(1): 64-67.

Zhang, Z. H. (2005). A review of current vegetable industry and its prospects (First part). *Journal of Changjiang Vegetables* 5: 1-5.

Zhong, X. Q. & Xu, H. H. (2000). Researches on medicinal plants in genus *Ficus*. *Chinese Traditional and Herbal Drugs* 31(9): 3-4.

Zhuang, F. Y. & Chen, J. F. (2003). RAPD analysis on cultivated cucumber, wild cucumber species, interspecific hybrid and its progenies from backcrossing. *Acta Horticulturae Sinica* 30(1): 47-50.

Zohary, D. & Hopf, M. (2000). *Domestication of Plants in the Old World*. Oxford, UK and New York: Oxford University Press.

Chapter 17

Economic Forest Plants

PEI Sheng-Ji
YANG Yu-Ming and WANG Juan

17.1 Introduction

China is one of the world's richest areas in terms of woody plant resources, with 115 families, 302 genera and around 8 000 species of trees and shrubs. Ninety-five percent of the world's woody plant genera, especially temperate genera, are represented by species in China. Of conifers, particularly, there are 32 genera and 396 Chinese species of Pinaceae, Taxodiaceae and Cupressaceae, i.e. 75% of genera and 33% of all species are represented (Zheng, 1983; Zheng, 1985; Zheng, 1997; Zheng, 2004).

According to official data from the State Forest Agency (State Forestry Administration, 2009), 195 million ha (20%) of China is covered by forest. Within these forests, more than 3 000 plant taxa have economic uses (Wu *et al.*, 2005). China has a long history of utilizing such wild plant resources, with the earliest records of harvesting, cultivating and utilizing wild plants dating back 7 000 years (Zheng, 1997). Perhaps the earliest uses were of bark for clothing, followed by vegetable fibers in weaving and pigment-containing herbs in dyeing. The planting of ornamental plants was initiated around the same time. A number of books have recorded the historical use of forest plants. For instance, the 2 000 year old *Book of Songs* records over 200 plants used as fabrics, fuels and medicines, among others, along with their distribution, classification, cultural significance and ecology. In the Eastern Zhou Dynasty (770–221 BC), oil plants were recorded in the *Rites of the Zhou*, and by the Han and Wei Dynasties (202 BC–265 AD), the cultivation of oil plants was well developed. Since 220 BC, raw lacquer from plants has been used as a coating material. Also during the Han Dynasty, CAI Lun invented technology for making paper from plants. The *Illustrated Catalogue of Plants* and the *Lengthy Compilation of Plants with Illustrations,* both written by WU Qi-Jun in 1848 AD, list 1 714 kinds of wild or cultivated plants (Jiang & Dang, 2002). The range of ethnic cultures found in China have a long history of using forest plants and have accumulated a great deal of indigenous knowledge, which now provides a precious source for plant-based industrial development (Pei & Huai, 2007).

Both harvesting from the wild and cultivation of plants are important in the Chinese traditional economy. The collection of wild plant resources from forests has long provided basic necessities for life including supplemental foods, medicines, clothing and building materials. For instance, houses could be thatched and made of wood or bamboo, furniture also of wood or bamboo, and food comprised mainly wild vegetables, fruits, herbs and food alongside fish and wild animals. With the development of modern science and technology, even more wild forest plants have become utilized, and a major market in forest products has developed. Furthermore, with rapid improvements in living standards and quality of life, the pursuit of natural products such as "green," healthy and organic foods, natural fabrics and cosmetics has become increasingly popular, in turn increasing the demand on forest resources. The extent of use of wild forest plants continues to expand, especially industrial materials, medicines and horticulture. In this chapter we summarize the important forest economic plants of China, including commercial timber trees, industrial oil- and lacquer-producing plants, edible forest plants, and bamboos and rattans.

17.2 Commercial timber trees

Timber is one of the four major raw materials in the world, and China is the world's second largest consumer of timber products (manufactured board, pulp, paper and cardboard) and the third largest consumer of timber itself. Commercial timber trees in China are found mainly in natural forests, with some plantations. Over 800 commercial tree species are grown in China for timber and wood-processing products, and these are concentrated in three main areas: the Changbai Shan, Da Hinggan Ling and Xiao Hinggan Ling of northeastern China, which form the largest natural forest zone; the mountains of southwestern China, an area which is extremely rich in tree diversity; and finally Fujian, Guangdong, Jiangxi and Taiwan in southeastern China, which contain the largest area of secondary forest and plantations.

In this chapter, we describe the significant timber species, selected because they are either important and commonly-used timber trees, locally-used forest and forest-origin timber trees, or species that have a potential value as commercial timber. They include *Cunninghamia lanceolata* (Chinese fir; Taxodiaceae), which is fast-growing with a long, straight trunk and corrosion-resistant and easily-processed wood, making it suitable for construction, telegraph poles, bridges and railway sleepers. However its wood is too soft to be used for furniture. In contrast, the yellow sandalwood (*Dalbergia hupeana*; Fabaceae) has a beautiful luster and superior hardness, making it suitable for such things as musical instruments, carvings and sports racquets, while the wood of *Paulownia* spp. (Scrophulariaceae) is soft, light and does not deform, making it suitable for ceiling decorations.

17.2.1 Important commercial timber trees

Abies.—Abies (fir; leng shan; Pinaceae) comprises 22 evergreen tree species in China. *Abies ferreana* (zhong dian leng shan) is found at altitudes of 3 300–4 000 m in southwestern Sichuan and northwestern Yunnan. Its timber is used for construction and the bark for producing tannins. *Abies delavayi* (cang shan leng shan) is distributed at 3 000–4 300 m in northwestern Yunnan and southeastern Xizang. The timber is used for construction, furniture and wood pulp and the bark for tannins. *Abies forrestii* (chuan dian leng shan), distributed at 2 500–4 200 m in southwestern Sichuan, northwestern Yunnan and eastern Xizang, is used for construction and the bark for extracting tannins.

Cedrus deodara.—Cedrus deodara (Himalayan cedar; xue song; Pinaceae) is an evergreen tree native to the extreme southwest of Xizang, and cultivated extensively as an ornamental in Anhui, Fujian, Guangdong, Guangxi, Hebei, Henan, Hubei, Hunan, Jiangsu, Jiangxi, Liaoning, Shaanxi, Shandong, Taiwan, Yunnan and Zhejiang. The timber is used in ship-building, furniture, bridges and other construction applications.

Keteleeria fortunei.—Keteleeria fortunei (Pinaceae; you shan) is an evergreen tree up to 30 m tall found at altitudes of 200–1 400 m in Fujian, Guangdong, Guangxi and Zhejiang. It is recognized as a vulnerable species in China. The timber is used for construction and furniture, and the species is also cultivated for afforestation and as an ornamental.

Larix potaninii.—Larix potaninii (Chinese larch; hong shan; Pinaceae) is an evergreen tree to 50 m tall found on mountains and in river basins at 2 500–4 600 m elevation in southern Gansu, southern Shaanxi, Sichuan, southeastern Xizang and northern Yunnan. The timber is used for construction, pit props, railway sleepers and furniture, and the bark yields tannins. The species is also used in afforestation.

Picea asperata.—Picea asperata (dragon spruce; yun shan; Pinaceae; Figure 17.1) is an evergreen tree to 45 m tall distributed on mountains and in river basins at 2 400–3 600 m elevation, in eastern and southern Gansu, northern Ningxia, Qinghai, southwestern Shaanxi and Sichuan. The timber is used for construction, aircraft-building, railway sleepers, furniture and fibers. The trunk is used for its resin, and the roots, branches and leaves for their aromatic oils.

Pinus.—There are 39 species of pine (song shu; Pinaceae) in China. *Pinus densiflora* (Japanese red pine; chi song) occurs from near sea level to 900 m in eastern and southern Heilongjiang, northeastern Jiangsu, southeastern

Figure 17.1 *Picea asperata* in forest.

Jilin, Liaoning, eastern and northern Shandong. The timber is used in construction, for poles and furniture, the trunk as a source of resin. *Pinus massoniana* (Chinese red pine; ma wei song) is distributed from near sea level to 2 000 m in Anhui, Fujian, Guangdong, Guangxi, Guizhou, Hainan, western Henan, Hubei, Hunan, southern Jiangsu, Jiangxi, southeastern Shaanxi, Sichuan, Taiwan, eastern Yunnan and Zhejiang, and is an important afforestation species in southern China. The timber is used for construction, railway sleepers, pit props, furniture and wood pulp, and the trunk as a source of resins and tannins and for cultivating fungi. *Pinus tabuliformis* (Chinese red pine; you song) is distributed at 100–2 600 m elevation in Gansu, Hebei, Henan, western Hubei, Hunan, southern Jilin, Liaoning, Nei Mongol, Ningxia, Qinghai, Shaanxi, Shandong, Shanxi and Sichuan. Its timber is used in construction, for poles, pit props, ship-building and furniture. The trunk can be used as a source of resins, the bark for tannins, and the needles in medicine. *Pinus yunnanensis* (yun nan song; Figure 17.2) is found at 400–3 100 m in Guangxi, Guizhou, southwestern Sichuan, southeastern Xizang and Yunnan. The timber is used for construction, railway sleepers, boards, furniture and wood pulp. The trunk can be used as a source of resins, the roots for cultivating fungi, and the bark for producing tannins.

Pseudotsuga sinensis.—Pseudotsuga sinensis (Chinese Douglas fir; huang shan; Pinaceae) is an evergreen tree to 50 m tall found at elevations of 600–3 300 m in southern Anhui, northern Fujian, northern Guizhou, western Hubei, northwestern Hunan, northeastern Jiangxi, southern Shaanxi, southeastern Sichuan, Taiwan, central and northeastern Yunnan, and Zhejiang. The timber is tough, dense and resilient and is used in construction, bridge building, furniture and for fibers.

Tsuga chinensis.—Tsuga chinensis (Chinese hemlock; tie shan; Pinaceae; Figure 17.3) is an evergreen tree to 50 m tall distributed at 1 000–3 500 m in Anhui, Fujian, southern Gansu, Guangdong, Guangxi, northern Guizhou, western Henan, western Hubei, Hunan, Jiangxi, southern Shaanxi, Sichuan, Taiwan, Xizang, Yunnan and Zhejiang. The timber

Figure 17.2 *Pinus yunnanensis* forest.

Figure 17.3 *Tsuga chinensis* in forest.

is used for construction, aircraft-building, furniture and pit props, while the bark is used for producing tannins, the trunk for resins and the roots, trunk and branches for their aromatic oils.

Cryptomeria japonica.—Cryptomeria japonica (Japanese cedar; ri ben liu shan; Taxodiaceae) is an evergreen tree to 40 m tall found up to 2 500 m and native to Fujian, Jiangxi, Sichuan, Yunnan and Zhejiang; introduced in Anhui, Fujian, Gansu, Guangdong, Guangxi, Guizhou, Henan, Hubei, Hunan, Jiangsu, Jiangxi, Shandong, Sichuan, Taiwan, Yunnan and Zhejiang. It is both cultivated as an ornamental and planted for timber. The species grows fast on deep, well-drained soils in montane areas with warm, moist climates, but is intolerant of poor soils and cold, drier climates. The wood is strongly rot-resistant, easily worked, and is used for construction, bridge-building, ship-building, lamp posts, furniture, utensils and paper manufacture.

Cunninghamia lanceolata.—Cunninghamia lanceolata (Chinese fir; shan mu; Taxodiaceae), is an evergreen tree to 50 m tall distributed at 200–2 800 m. Its exact native distribution is uncertain due to widespread planting across Anhui, Fujian, Gansu, Guangdong, Guangxi, Guizhou, Hainan, Henan, Hubei, Hunan, Jiangsu, Jiangxi, Shaanxi, Sichuan, Taiwan, Yunnan and Zhejiang. *C. lanceolata* is the most important fast-growing timber tree in the warm regions south of the Chang Jiang valley, and is propagated by seed, cuttings or suckers. The wood is strongly resistant to rot and termites, and easily worked; it is used in construction, bridges, ship-building, lamp posts, furniture and for fiber.

Taiwania cryptomerioides.—Taiwania cryptomerioides (tai wan shan; Taxodiaceae) is an evergreen tree to 75 m tall distributed at elevations of 500–2 800 m in southeastern Guizhou, southwestern Hubei, southeastern Sichuan, Taiwan, southeastern Xizang and western Yunnan. One of the world's tallest tree species, this is now rare in China, with populations decreasing as a result of cutting for timber. Individuals of *T. cryptomerioides* may attain an age of over 2 000 years. The wood is easily worked and is used in construction, furniture-making and coffins (hence one of its local names, "Chinese coffin tree"), bridges, ship-building and paper manufacture; within its native range it can also be used for afforestation.

Cupressus funebris.—Cupressus funebris (Chinese weeping cypress; bai mu; Cupressaceae) is an evergreen tree found below 2 000 m in Anhui, Fujian, Gansu, northern Guangdong, northern Guangxi, eastern Guizhou, Henan, western Hubei, Hunan, Jiangxi, Shaanxi, Sichuan, Yunnan and Zhejiang; it is also widely cultivated across southern China. The structure of wood is dense and rigid, making it useful for furniture and crossbeams.

Fokienia hodginsii.—Fokienia hodginsii (fu jian bai; Cupressaceae) is an evergreen tree to 30 m tall distributed from 100–1 800 m in Fujian, northern Guangdong, Guangxi, Guizhou, southern Hunan, western Jiangxi, southeastern Sichuan, southeastern Yunnan and southern Zhejiang. A vulnerable species in China, its wood is used for furniture and construction.

Juniperus chinensis L.*—Juniperus chinensis* (Chinese juniper; yuan bai; Cupressaceae) is an evergreen tree to 25 m found at elevations of 1 400–2 300 m in Anhui, Fujian, southern Gansu, Guangdong, northern Guangxi, Guizhou, Hebei, Heilongjiang, Henan, western Hubei, Hunan, Jiangsu, Jiangxi, Nei Mongol, southern Shaanxi, Shandong, Shanxi, Sichuan, eastern Taiwan, Yunnan and Zhejiang. The species is cultivated for afforestation and as an ornamental.

Podocarpus neriifolius.—*Podocarpus neriifolius* (bai ri qing; Podocarpaceae) is an evergreen tree to 25 m tall distributed from 100–1 000 m elevation in Fujian, Guangdong, Guangxi, Guizhou, Hunan, Jiangxi, Sichuan, Xizang, Yunnan and Zhejiang. The wood is used in making furniture, musical instruments, carvings and paper.

Cephalotaxus sinensis.—*Cephalotaxus sinensis* (cu fei; Cephalotaxaceae) is a small evergreen tree or shrub distributed at 600–3 200 m elevation in southern Anhui, Fujian, southern Gansu, southwestern Guangdong, Guangxi, northeastern Guizhou, Henan, Hubei, Hunan, southern Jiangsu, Jiangxi, southern Shaanxi, Sichuan, central and northern Taiwan, southeastern Yunnan and Zhejiang, and cultivated in Shandong. The wood is used to make furniture, farm implements and utensils. The branches, roots, leaves and seeds are sources of many alkaloids, which are used to treat leukemia and lymphosarcoma. The tree is also cultivated as an ornamental.

Torreya.—*Torreya* (fei shu shu; Taxaceae) includes four species in China. *Torreya fargesii* (ba shan fei shu) is distributed from 1 000–3 400 m in southern Anhui, western Hubei, northwestern Hunan, Jiangxi, southern Shaanxi, Sichuan and northwestern Yunnan. It is considered suitable for afforestation and reforestation, while the high quality timber is used in construction, bridges, furniture, tools and utensils. *Torreya grandis* (fei shu) is found at 200–1 400 m in southern Anhui, northern Fujian, northeastern Guizhou, western Hunan, southern Jiangsu, northern Jiangxi and Zhejiang. The wood is used in construction, bridges and furniture. The seeds are edible and also oil-rich; an essential oil is extracted from the aril. *Torreya jackii* (chang ye fei shu) is found at 400–1 000 m in northern Fujian, northeastern Jiangxi and southern Zhejiang. A rare and vulnerable species, its wood is very fragrant and is used to make agricultural implements, utensils and handicrafts. The leaves are also very aromatic when bruised or burned, giving off a fragrance similar to that of sandalwood oil.

Manglietia fordiana.—*Manglietia fordiana* (mu lian; Magnoliaceae) is an evergreen tree to 25 m tall distributed at 300–1 200 in southern Anhui, Fujian, Guangdong, Guangxi, Guizhou, Hainan, Hunan, Jiangxi, Yunnan and Zhejiang. This species is used in herbal medicine and for timber.

Michelia.—*Michelia* (han xiao shu; Magnoliaceae) includes 37–39 species in China and one or two hybrids, of which 18–20 are endemic and one introduced. *Michelia baillonii* (he guo mu) is distributed at 500–1 500 m in southern Yunnan and is used for timber. *Michelia maudiae* (shen shan han xiao), distributed at 600–1 500 m in southern Anhui, Fujian, Guangdong, Guangxi, Guizhou, Jiangxi, Hunan and southern Zhejiang, is grown as an ornamental and is used medicinally and for timber. *Michelia odora* (guan guang mu), distributed at 300–1 100 m in Fujian, Guangdong, Guangxi, Hainan, southern Hunan, southern Jiangxi and southeastern Yunnan, is grown as an ornamental and for timber.

Tetracentron sinense.—*Tetracentron sinense* (shui qing shu; Tetracentraceae; Figure 17.4) is an evergreen tree with vessel-less wood which is pale, structurally delicate, and considered beautiful. It is found at 1 100–3 500 m in southern Gansu, Guizhou, southwestern Henan, western Hubei, northwestern and southwestern Hunan, southern Shaanxi, Sichuan, southern and southeastern Xizang and Yunnan.

Figure 17.4 *Tetracentron sinense*. A, tree in forest. B, infructescence.

Cinnamomum.—*Cinnamomum* (zhang shu; Lauraceae) comprises 49 species of evergreen trees and shrubs in China. *Cinnamomum camphora* (camphor tree; zhang) is native to Taiwan as well as Japan, Korea and Vietnam, and cultivated in valleys and on mountain slopes south of the Chang Jiang in mainland China. This species is the main source of camphor, derived from the chipped wood of the stems and roots and also by steam distillation from branchlets and leaves. It contains a volatile oil which, when fixed, is used medicinally as a stimulant, antispasmodic, antiseptic and rubefacient. It also is used in the manufacture of celluloid. The wood is beautifully-grained, pale brown, easily polished and used in furniture, cabinets and interior finishes in buildings. *Cinnamomum mollifolium* (mao ye zhang) is distributed at 1 100–1 300 m in southern Yunnan. The leafy branchlets contain essential oils used in medicine and light industry. The fruit kernel contains oils and fats with various uses. *Cinnamomum septentrionale* (yin mu) is found at 600–1 000 m in southern Gansu, southern Shaanxi and Sichuan. The root wood, known as "silver wood," is used in sculpture; it also contains a large amount of camphor. The timber of the trunk is yellow-brown, finely grained and used in cabinet- and furniture-making.

Lindera.—*Lindera* (shan hu jiao shu; Lauraceae) includes 38 species of trees and shrubs in China. *Lindera communis* (xiang ye shu) is found in dry sandy places and the broad-leaved evergreen forests of Fujian, Gansu, Guangdong, Guangxi, Guizhou, Hubei, Hunan, Jiangxi, Shaanxi, Sichuan, Taiwan, Yunnan and Zhejiang. The pericarp contains aromatic oil, the seed oil is used in food and making soap and machine oil, and the branchlets and leaves are used medicinally. *Lindera glauca* (shan hu jiao), distributed below 900 m in Anhui, Fujian, Gansu, Guangdong, Guangxi, Guizhou, southwestern Henan, Hubei, Hunan, Jiangxi, southwestern Shaanxi, eastern Shandong, Shanxi, Sichuan, Taiwan and Zhejiang, is used in carpentry. The leaves and pericarp are processed for their aromatic oil, while the seed oil is used in making soap and machine oil. The roots, branchlets, leaves and fruit are used medicinally. *Lindera megaphylla* (hei ke nan) is found at 1 600–2 000 m elevation in Anhui, Fujian, Gansu, Guangdong, Guangxi, Guizhou, Hubei, Hunan, Jiangxi, Shaanxi, Sichuan and Yunnan. The wood is also used in carpentry and as a building material. The leaves and pericarp are processed for their aromatic oil, and the seed oil is used in making soap.

Machilus yunnanensis.—*Machilus yunnanensis* (dian run nan; Lauraceae) is an evergreen tree to 30 m tall distributed in broad-leaved evergreen forests and moist and fertile places on mountain slopes at elevations of 1 500–2 100 m in northwestern Guangxi, western Sichuan, southern Xizang, and central, northwestern and western Yunnan. The wood is used in construction and furniture-making, and the fruit and seed contain oil. This species is also cultivated as an ornamental.

Phoebe zhennan.—*Phoebe zhennan* (nan mu; Lauraceae) is a large evergreen tree up to 30 m or more in height, with a straight trunk. It is usually found in mixed broad-leaved forests or cultivated, below 1 500 m, in northwestern Guizhou, western Hubei and Sichuan. It provides high grade, shrink-resistant, faintly-scented and light-textured wood for making furniture.

Sassafras tzumu.—*Sassafras tzumu* (cha mu; Lauraceae) is a deciduous tree up to 35 m tall, distributed from 100–1 900 m in Anhui, Fujian, Guangdong, Guangxi, Guizhou, Hubei, Hunan, Jiangsu, Sichuan, Yunnan and Zhejiang. The wood is yellowish, finely-grained, durable and used in ship-building and furniture-making. The roots and bark are used in treating traumatic injuries and rheumatic disorders. The fruits, leaves and roots contain essential oils, the main component of which is the phenylpropene safrole, at concentrations of up to 1% in the roots.

Lagerstroemia tomentosa.—*Lagerstroemia tomentosa* (crape myrtle; rong mao zi wei; Lythraceae) is an evergreen tree found at 600–1 200 m in Yunnan. The timber is used in ship-building, bridges, construction and making athletic equipment.

Calophyllum inophyllum.—*Calophyllum inophyllum* (Alexandrian laurel; hong hou ke; Clusiaceae) is a small evergreen tree found wild or cultivated on open sites on hills, seashores and sandy wastelands at around 100 m in Hainan and Taiwan. The seeds yield 20–30% oil, the seed kernels 50–60%. This seed oil is used in industry or as a medicine; it is also edible after refinement and detoxification. The timber is hard and heavy and is used for making furniture. The bark also contains ca. 15% tannins.

Mesua ferrea.—*Mesua ferrea* (Ceylon ironwood; tie li mu; Clusiaceae; Figure 17.5) is an evergreen tree 20–30 m tall, usually cultivated, escaped and locally naturalized in southwestern Yunnan but also at 500–600 m in Guangdong, Guangxi, and south, southwestern and western Yunnan. The seeds yield up to 79% oil, which is inedible but is used industrially. The wood is very hard and the tree is also a handsome ornamental.

Figure 17.5 *Mesua ferrea* in flower.

Eucommia ulmoides.—Eucommia ulmoides (du zhong; Eucommiaceae) is a deciduous tree growing naturally but also widely cultivated and becoming locally naturalized at elevations of 100–2 000 m in Gansu, Guizhou, Henan, Hubei, Hunan, Shaanxi, Sichuan, Yunnan and Zhejiang; it is cultivated in Anhui and Beijing. The timber is used for furniture and fuel; the bark, which contains the glycoside aucubin, is used medicinally as an invigorator, a tonic for arthritis and for reducing blood pressure; the solidified latex is used for lining pipes, insulating electric cables and dental fillings.

Populus.—Populus (poplar; yang shu; Salicaceae) comprises 71 species in China, including *P. adenopoda* (Chinese aspen; xiang ye yang), distributed at 300–2 500 m elevation in Anhui, Fujian, Guangxi, Guizhou, Hebei, Henan, Hubei, Hunan, Jiangsu, Jiangxi, Shaanxi, Sichuan, Yunnan and Zhejiang and used for timber, construction, furniture-making, farm tools and wood pulp. *Populus simonii* (xiao ye yang), distributed from near sea level to 3 000 m in Hebei, Heilongjiang, Jiangsu, Jilin, Liaoning, Nei Mongol, Qinghai, Shanxi, Sichuan, Yunnan and Zhejiang, is used for timber. *Populus tomentosa* (Chinese white poplar; mao bai yang) is found from near sea level to 1 500 m in Anhui, Gansu, Hebei, Henan, Jiangsu, southern Liaoning, Shaanxi, Shandong, Shanxi, Sichuan, Yunnan and Zhejiang, and is used for wood pulp, construction, farm tools, furniture and matchwood; it is a common street tree and ornamental.

Chukrasia tabularis.—Chukrasia tabularis (Indian mahogany; ma lian; Meliaceae) is a deciduous tree to 25 m tall distributed at elevations of 300–1 600 m in Fujian, Guangdong, Guangxi, southern Guizhou, Hainan, Xizang, Yunnan and Zhejiang. The timber is valued for making furniture, plywood, carving planks and for the woodwork in railway carriages. A yellow transparent gum is obtained from the sap; the astringent bark is used medicinally.

Melia azedarach L.*—Melia azedarach* (white cedar; lian; Meliaceae) is a deciduous tree up to 10 m tall found in mixed broad-leaved evergreen and deciduous forests, field margins and roadsides at elevations of 500–2 100 m in Anhui, Fujian, southern Gansu, Guangdong, Guangxi, Guizhou, Hainan, southern Hebei, Henan, Hubei, Hunan, Jiangsu, Jiangxi, southern Shaanxi, Shandong, southern Shanxi, Sichuan, Taiwan, Yunnan, southeastern Xizang and Zhejiang. It is used medicinally, for industrial oil and for timber.

Toona sinensis.—Toona sinensis (Chinese mahogany; xiang chun; Meliaceae; see also Chapter 16 for its use as a vegetable) is a deciduous tree to 40 m tall usually found in primary montane forests, especially on steep hillsides or open slopes between 100 and 2 900 m in Anhui, Fujian, southern Gansu, Guangdong, Guangxi, Guizhou, Hebei, Henan, Hubei, Hunan, Jiangsu, Jiangxi, southern Shaanxi, Sichuan, southeastern Xizang, Yunnan and Zhejiang. The timber is used in making furniture and sieve hoops and in bridge construction. The trees are widely used medicinally, with the bark being used as an astringent and depurative, the powdered root as a refreshment and diuretic, and the tender leaves as a carminative.

Acer.—Acer (maple; qi shu; Aceraceae) includes 99 species in China, of which three are introduced. For example, *Acer davidii* (Père David's maple; qing zha feng), distributed in Anhui, Fujian, southeastern Gansu, Guangdong, Guangxi, Guizhou, Hebei, Henan, Hubei, Hunan, Jiangsu, Jiangxi, Ningxia, southern Shaanxi, Shanxi, Sichuan, Yunnan and Zhejiang, is used as an ornamental for its snakeskin-like bark and autumn color.

Excentrodendron tonkinense.—Excentrodendron tonkinense (xian mu; Tiliaceae) is an evergreen tree to 40 m tall found in evergreen forests on limestone in Guangxi and southeastern Yunnan. The wood is very hard and is prized for making chopping boards. It is classed by the IUCN as Vulnerable (IUCN, 2011).

Tilia tuan.—Tilia tuan (duan shu; Tiliaceae) is a deciduous tree 10–20 m tall found at 1 200–2 400 m in Guangxi, Guizhou, Hubei, Hunan, Jiangsu, Jiangxi, Sichuan, Yunnan and Zhejiang. The wood is light, easily dried and processed, stable and without fragrance. It is used for food vessels, shutters and furniture. The flowers yield aromatic oils and the root is used medicinally.

Alnus hirsuta.—Alnus hirsuta (Manchurian alder; liao dong qi mu; Betulaceae) is a deciduous tree up to 20 m tall distributed in temperate forests and on streamsides at 700–1 500 m in Heilongjiang, Jilin, Liaoning, Nei Mongol and Shandong. The wood is hard and dense and is used for making agricultural tools and furniture.

Betula.—There are 32 species of *Betula* (birch; hua mu shu; Betulaceae) in China, including *B. chinensis* (jian hua), distributed at 700–3 000 m in eastern Gansu, Hebei, Henan, Liaoning, Nei Mongol, Shaanxi, Shandong and Shanxi, one of the most valuable timber trees in northern China. The wood is extremely hard and dense, close-grained, very fine-textured and is used for making pestles and wagon axles. *Betula luminifera* (liang ye hua), distributed at 200–2 900 m in Anhui, Fujian, Gansu, northern Guangdong, Guangxi, Guizhou, Henan, Hubei, Hunan, Jiangsu, Jiangxi, Shaanxi, Sichuan, Yunnan and Zhejiang has wood that is hard, close-grained and fine-textured, and is used in construction and for making agricultural tools and furniture. *Betula platyphylla* (Japanese white birch; bai hua), distributed at 700–4 200 m in Gansu, Hebei, Heilongjiang, Henan, Jiangsu, Jilin, Liaoning, Nei Mongol, Ningxia, Qinghai, Shaanxi, Shanxi, southwestern and western Sichuan, southeastern Xizang and northwestern Yunnan, is another very important timber tree. The wood is soft, light and close-grained, and is used in making agricultural tools, wooden ware and tobacco boxes and, rarely, for construction. The white, papery bark is made into canoes, fancy articles and food baskets and is also used as a substitute for tiles; the inner bark is rich in resin. *Betula utilis* (Himalayan birch; cao pi hua) is found at 2 500–3 800 m in Gansu, Hebei, Ningxia, Qinghai, Shaanxi, western Sichuan, eastern and southern Xizang and northwestern Yunnan. A valuable timber tree of commercial importance, the wood is hard, dense and rather brittle, with pinkish or light reddish-brown heartwood with a silky luster.

Castanopsis fargesii.—*Castanopsis fargesii* (kao; Fagaceae) is an evergreen tree, 10–30 m tall, found in broad-leaved evergreen forests at 200–2 100 m elevation in Anhui, Fujian, Guangdong, Guangxi, Guizhou, Hubei, Hunan, Jiangsu, Jiangxi, Sichuan, Taiwan, Yunnan and Zhejiang. The bark can be used to produce dyes and the skincare ingredient known as "mangrove extract." The wood is very hard and good for making tools, furniture and construction materials. The nuts contain high-quality edible starch.

Cyclobalanopsis glauca.—*Cyclobalanopsis glauca* (qing gang; Fagaceae) is an evergreen tree to 20 m tall found in broad-leaved evergreen and mixed mesophytic forests on mountain slopes and valleys up to 2 600 m in Anhui, Fujian, Gansu, Guangdong, Guangxi, Guizhou, Henan, Hubei, Hunan, Jiangsu, Jiangxi, Shaanxi, Sichuan, Taiwan, Xizang, Yunnan and Zhejiang. The hard timber can be used for furniture, construction, making vehicles, farm tools and bridges.

Fagus longipetiolata.—*Fagus longipetiolata* (shui qing gang; Fagaceae) is a deciduous tree to 25 m tall found in broad-leaved evergreen and mixed mesophytic forests on mountain slopes at 300–2 400 m in Anhui, Fujian, Guangdong, Guangxi, Guizhou, Hubei, Hunan, Jiangxi, Shaanxi, Sichuan, Yunnan and Zhejiang. The wood is used in making agricultural tools and furniture. The seed oil is used in food and oil paints.

Lithocarpus xylocarpus.—*Lithocarpus xylocarpus* (muguo shili; Fagaceae) is an evergreen tree rarely reaching 20 m tall found in mixed mesophytic forests at 1 000–2 800 m in Guizhou, southwestern Sichuan, southeastern Xizang and Yunnan. The hard wood is good for making tools, furniture and construction materials.

Quercus acutissima.—*Quercus acutissima* (sawtooth oak; ma li; Fagaceae) is a deciduous tree up to 30 m tall distributed at 100–2 200 m in Anhui, Fujian, Guangdong, Guangxi, Guizhou, Hainan, Hebei, Henan, Hubei, Hunan, Jiangsu, Jiangxi, Liaoning, Shaanxi, Shandong, Shanxi, Sichuan, southeastern Xizang, Yunnan and Zhejiang. The hard timber can be used for construction and making furniture, vehicles, farm tools and bridges, and is also good for making charcoal. The nut is edible, used for food and animal forage.

Ulmus pumila.—*Ulmus pumila* (Siberian elm; yu shu; Ulmaceae) is a deciduous tree to 25 m tall distributed at 1 000–2 500 m elevation in Gansu, Hebei, Heilongjiang, Henan, Jilin, Liaoning, Nei Mongol, Ningxia, eastern Qinghai, Shaanxi, Shandong, Shanxi, Sichuan, Xinjiang and Xizang. The species is cultivated throughout China. Its wood is hard, strong and beautifully-colored, good for making furniture; the bark is used for fibers; and the fruit, leaves and bark are used to produce medicine and seed oil.

Zelkova.—*Zelkova* (ju shu; Ulmaceae) contains three species in China, of which *Z. schneideriana* (da ye ju shu) is distributed at 200–1 100 m (1 800–2 800 m in Xizang and Yunnan) in Anhui, Fujian, southern Gansu, Guangdong, Guangxi, Guizhou, southern Henan, Hubei, Hunan, Jiangsu, Jiangxi, southern Shaanxi, southeastern Sichuan, southeastern Xizang, Yunnan and Zhejiang. The high-quality timber is resistant to decay, strong and beautifully-colored, renowned for furniture-making; the bark fiber is used for manufacturing ropes and paper. *Zelkova serrata* (ju shu) is distributed at 500–2 000 m in Anhui, Fujian, Gansu, northern Guangdong, Guizhou, Henan, Hubei, Hunan, Jiangsu, Jiangxi, Liaoning, Shaanxi, Shandong, Sichuan, Taiwan and Zhejiang. The bark and leaves are used medicinally.

Nyssa yunnanensis.—*Nyssa yunnanensis* (yun nan lan guo shu; Nyssaceae) is a deciduous tree 25–30 m tall found in dense valley forests at 500–1 100 m in southern Yunnan. The wood is heavy, a commercial timber used in furniture and construction.

Madhuca.—*Madhuca* (zi jing mu shu; Sapotaceae) has hard wood with a dense structure, considered a potential high-quality industrial tropical hardwood. Furthermore the seed kernel oil, which is present at 50% concentrations, is edible. There are two species in China. *Madhuca pasquieri* (zi jing mu) is found below 1 100 m in southwestern Guangdong, southern Guangxi and southeastern Yunnan. The seeds contain 30% edible oil and the trees are a source of valuable timber. *Madhuca hainanensis* (hai nan zi jing mu) is found at around 100 m altitude in Hainan. The seeds contain 55% oil and the timber is used for boats and bridges. Both species are classed by the IUCN as Vulnerable due to over-exploitation (IUCN, 2011).

Gmelina arborea.—*Gmelina arborea* (beechwood; Yunnan shi zi; Verbenaceae) is a deciduous tree ca. 15 m tall, found below 1 500 m in southern Yunnan. The timber is very good for furniture and construction.

Tectona grandis.—*Tectona grandis* (tea you mu; Verbenaceae) is a deciduous tree to 40 m tall with fragrant flowers, found below 900 m in Fujian, Guangdong, Guangxi, Taiwan and Yunnan. It is a medicinal species and provides luxury timber.

Schima wallichii.—*Schima wallichii* (hong mu he; Theaceae) is an evergreen tree 10–20 m tall, found in southern Guangxi, southern Guizhou, southeastern Xizang and southern Yunnan and used for veneers, furniture, construction and interior joinery.

Albizia.—*Albizia* (he huan shu; Fabaceae) has hard wood with a dark texture, and is also the host species for the lac insect, the source of shellac. There are 16 species in China, including *A. lebbeck* (kuo jia he huan), which is planted along roadsides and in gardens in Fujian, Guangdong, Guangxi, Hainan and Taiwan, as well as being used for timber, and *A. lucidior* (guang ye he huan), distributed at 600–1 900 m elevation in Guangxi, Guizhou, Taiwan and Yunnan, grown as an ornamental.

Dalbergia.—There are 29 species of *Dalbergia* (huang tan shu; Fabaceae) in China, including *D. cultrata* (hei huang tan), found at around 1 700 m elevation in Yunnan, which provides valuable timber and shellac as well as red wood for high-quality furniture. *Dalbergia odorifera* (jiang xiang huang tan; Figure 17.6), distributed at 100–500 m in Fujian, Hainan and Zhejiang, provides good quality wood and can also be used medicinally in weight-loss and to lower blood pressure.

Pterocarpus indicus.—*Pterocarpus indicus* (amboine; zi tan; Fabaceae) is an evergreen tree 15–25 m tall, found in open forests on slopes, or cultivated in villages and gardens in Guangdong, Taiwan and Yunnan. It is valued for its timber and renowned for its red wood, used in top quality furniture.

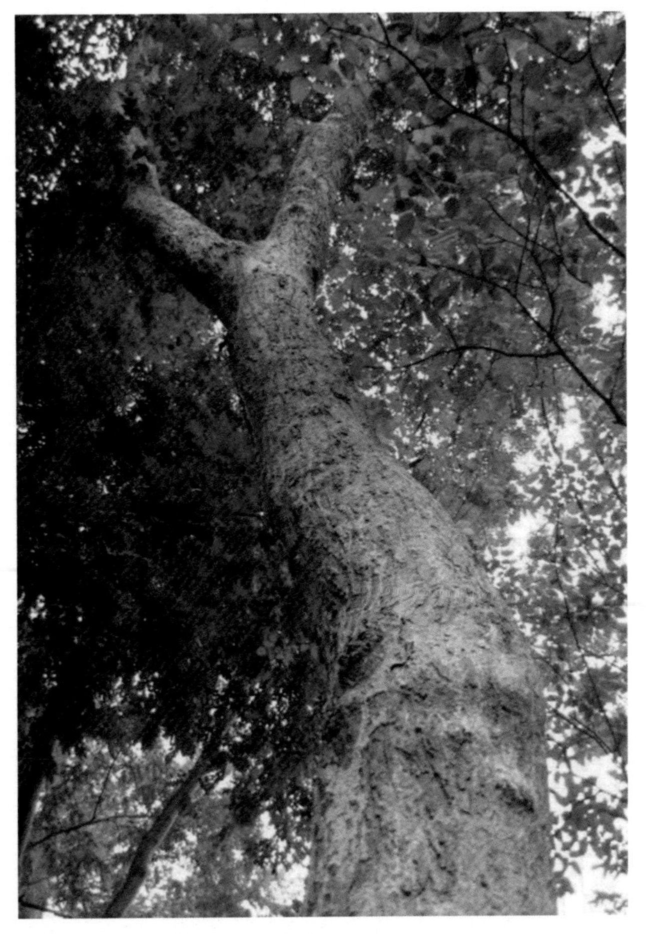

Figure 17.6 *Dalbergia odorifera.*

Senna.—*Senna* spp. (jue ming shu; Fabaceae) are cultivated in southwestern and southern China, where they are common ornamental trees. The bark contains tannins, a source of red dye. The fruit pulp and seeds are used medicinally as a laxative. The wood is hard and durable and is used for props, bridges and farm tools. *Senna siamea* (kassod tree; tie dao mu) is widely cultivated in southern Yunnan, in particular in Xishuangbanna, below 1 200 m, where local people use it for fuel and timber. It is also used for its edible flowers and in herbal medicine. The tree is fast-growing, and coppicing as a means of harvesting fuelwood has potential as a management technique allowing forest restoration and timber production in tropical areas.

Zenia insignis.—*Zenia insignis* (ren dou; Fabaceae) is a deciduous tree 15–20 m tall found at 200–1 000 m in Guangdong, Guangxi and Yunnan. The wood is hard and beautifully-colored, considered good for high-quality furniture and construction.

Neolamarckia cadamba.—*Neolamarckia cadamba* (kadam; tuan hua shu; Rubiaceae) is a large deciduous tree to 30 m tall found in Guangdong, Guangxi and Yunnan. The wood has a beautiful color and venation pattern, and is good for making machinery and for many decorative purposes.

Neonauclea griffithii.—*Neonauclea griffithii* (xin wu tan; Rubiaceae) is a large evergreen tree to 20 m, found in dense valley forests or on humid slopes at 800–1 000 m elevation in Guangxi, Guizhou and Yunnan. The wood is hard and good for tools, musical instruments and furniture.

Terminalia myriocarpa.—*Terminalia myriocarpa* (qian guo lan ren; Combretaceae) is an evergreen tree to 35 m tall found at 600–2 500 m in Guangdong (where it is probably planted), southwestern Guangxi, southeastern Xizang and Yunnan. Its hard wood is used in ship-building and construction.

Pometia pinnata.—*Pometia pinnata* (taun tree; fan long yan; Sapindaceae) is a large evergreen tree ca. 20 m, tall found in Taiwan and Yunnan, used for timber.

Fraxinus.—*Fraxinus* (bai la shu shu; Oleaceae) includes 22 taxa in China, such as *F. chinensis* subsp. *rhynchophylla* (hua qu liu), found below 1 500 m in Gansu, Hebei, Heilongjiang, Henan, Jilin, Liaoning, Shaanxi, Shandong and Shanxi, the wood of which is used for cabinet-making, and *F. mandschurica* (Manchurian ash; shui qu liu), found at 700–2 100 m in Gansu, Hebei, Heilongjiang, Henan, Hubei, Jilin, Liaoning, Shaanxi and Shanxi, which provides an important commercial timber good for furniture, decorative materials and construction.

17.3 Industrial oil- and lacquer-producing plants

China is rich in forest species bearing industrial oils, either under cultivation or growing in the wild. These may be used as sources of edible or industrial oils including biodiesel, and are as important as fossil oils in the world oil industry today. Over 70% of edible oils consumed come from plants—more than twice those provided by animal fats. Furthermore, many oil-bearing plants have multiple uses. For example, the fruit of the Chinese tallow tree, *Triadica sebiferum* (wu jiu shu; Euphorbiaceae), contains over 14% glycerine, a raw material for making epoxy resin, glass fiber, reinforced plastics, nitroglycerin and other explosive materials. Chinese tallow tree seed oil can be used to produce high quality spray-paints, and its natural wax as a raw material for soap, film, plastic membrane, waxed paper, barrier cream and anti-corrosion paints. In general, many industrial oil-producing plants are potential sources of biodiesel, and many are being investigated by the institutions of the Chinese Academy of Sciences (some of these are discussed in Chapter 13). According to historical records, there are more than 800 species of oil-bearing plants which contain at least 10% oil in their seed or fruit (Jia & Zhou, 1987). A recent publication reported 213 species and varieties of biodiesel plants alone in China (Long & Song, 2012).

Camellia.—*Camellia* (shan cha shu; Theaceae) comprises 97 species in China, including *C. oleifera* (tea oil camellia; you cha), distributed at 200–1 800 m in Anhui, Fujian, Guangdong, Guangxi, Guizhou, Hainan, Henan, Hubei, Hunan, Jiangsu, Jiangxi, southern Shaanxi, Sichuan, Yunnan and Zhejiang. This species is extensively cultivated as an oil seed. *Camellia reticulata* (dian shan cha or hong hua you cha) is traditionally cultivated or semi-cultivated in western Yunnan as an edible seed oil crop with great potential for extensive cultivation in the province.

Juglans.—There are three species of *Juglans* (walnut; hu tao shu; Juglandaceae) in China including *J. regia* (Persian walnut; hu tao), which is widely distributed in central, eastern, northwestern and southwestern China, and commonly cultivated from 23°–42° N for its edible, oily nuts and hard, fine-grained wood. *Juglans sigillata* (iron walnut; pao he tao; Figure 17.7) is found at 1 300–3 300 m in Guizhou, Sichuan, southeastern Xizang and Yunnan. It is cultivated in Yunnan for its edible nuts and hard wood.

Figure 17.7 *Juglans sigillata* fruits.

Toxicodendron vernicifluum.—*Toxicodendron vernicifluum* (qi shu; Anacardiaceae) is a deciduous tree to 20 m tall found in the hill and mountain forests between 800 and 2 800 m in Anhui, Fujian, Gansu, Guangdong, Guangxi, Guizhou, Hebei, Henan, Hubei, Hunan, Jiangsu, Jiangxi, Liaoning, Shaanxi, Shandong, Shanxi, Sichuan, Xizang, Yunnan and Zhejiang. This species is a major source of varnish used in the manufacture of lacquerware. In some parts of northwestern Yunnan, local people use the seeds of the tree to make edible oil.

17.4 Edible forest plants

In addition to vegetables and fruits, edible forest plants can provide starches, proteins, oils, sugars, vitamins and other useful components of the diet. The flavor, nutritional value and non-polluting nature of edible forest resources has made them very popular. Many are considered, through their nutritional value and vitamin content, to be preventative and curative of chronic age-related diseases. Although the concentration of proteins in plant material is much lower than in animals, plants contain more protein types, and are generally easier to digest. Furthermore, plants contain many trace elements that play important roles in maintaining a functional metabolism and immune system. Alongside the major (animal-based) protein sources, safe and healthy wild edible plants play an important role in Chinese cuisine. Many wild edible plant resources commonly used in China are also suitable for industrial processing. However, because they are often found in remote areas, lack of preservation techniques hinders their transportation and therefore the development of their industrialization and marketing. Thus, many edible plants are mainly sold locally. Examples of well-known edible forest products include the tubers of *Amorphophallus* (Araceae), *Dioscorea* (Dioscoreaceae) and *Pueraria* (Fabaceae), nuts of *Corylus* (Betulaceae) and *Pinus,* fruits of *Actinidia* (Actinidiaceae), *Amygdalus, Docynia* and *Malus* (all Rosaceae), *Canarium* (Burseraceae), *Myrica* (Myricaceae) and *Phyllanthus* (Euphorbiaceae), tender leaves of *Aralia* (Araliaceae) and *Ficus* (Moraceae), as well as fern shoots. These and more are described in detail below.

Aralia L.—There are 29 species of *Aralia* (cong mu shu) in China, including *A. cordata* (shi yong tu dang gui), found at 1 300–1 600 m in Anhui, Fujian, Guangxi, Hubei, Jiangxi, Taiwan and Zhejiang. This species is used medicinally and as a vegetable. *Aralia fargesii* (long yan du huo) is found at 1 800–2 700 m in Shaanxi, Sichuan and Yunnan. The young leafy shoots of this species are used as a herbal food and vegetable.

Eleutherococcus senticosus.—*Eleutherococcus senticosus* (ci wu jia; Araliaceae) is an evergreen shrub found below 2 000 m in Hebei, Heilongjiang, Henan, Jilin, Liaoning, northern Shaanxi, Shanxi and Sichuan. This species is used medicinally as a herbal food, and the leaves are used all over the country as a popular vegetable.

Sauropus.—There are 15 species of *Sauropus* (shou gong mu shu; Euphorbiaceae) in China, which are used as wild vegetables. *Sauropus spatulifolius* (long li ye) is cultivated for medicine in Fujian, Guangdong and Guangxi. The leaves are used to treat coughs, sore throats and acute bronchitis.

Baccaurea.—*Baccaurea* (mu nai guo shu; Euphorbiaceae) includes two species in China. *Baccaurea ramiflora* (Burmese grape; mu nai guo) is found at 100–1 300 m in Guangdong, Guangxi, Hainan and Yunnan. It has edible fruits; the pulpy, yellow aril is somewhat acid with an agreeable taste. The wood is used for furniture and cabinetwork. *Baccaurea motleyana* (rambai; duo mai mu nai guo) is cultivated in southern Yunnan. The fruits are edible; the aril is sweet to acid and eaten raw, cooked, or preserved. The bark is rich in tannins and is used in dyeing.

Phyllanthus.—There are 32 species of *Phyllanthus* (ye xia zhu shu; Euphorbiaceae) in China, including *P. emblica* (Indian gooseberry; yu gan zi), distributed from 200 to 2 300 m in Fujian, Guangdong, Guangxi, Guizhou, Hainan, Jiangxi, Sichuan, Taiwan and Yunnan. The mature fruits of this species are very sour, containing 1–2% vitamin C. They are eaten raw, sweetened or preserved. The seeds, roots and leaves are also used medicinally, while the dried leaves are sometimes used as pillow-stuffing. *Phyllanthus glaucus* (qing hui ye xia zhu) is distributed from 200 to 1 000 m in Anhui, Fujian, Guangdong, Guangxi, Guizhou, Hainan, Hubei, Hunan, Jiangsu, Jiangxi, Sichuan, Xizang, Yunnan and Zhejiang. The roots are used medicinally to treat infantile malnutrition due to intestinal parasites.

Acacia.—There are 18 species of *Acacia* (jin he huan shu; Fabaceae) in China. *Acacia pennata* (climbing wattle; chou cai), distributed in Fujian, Guangdong, Guangxi, Guizhou, Hainan, Yunnan and Zhejiang, is used as a wild vegetable. *Acacia catechu* (er cha), native to Yunnan and cultivated in Fujian, Guangdong, Guangxi, Hainan, Taiwan, Yunnan and Zhejiang, is used medicinally.

Illicium verum.—*Illicium verum* (star anise; ba jiao; Illiciaceae; Figure 17.8) is an evergreen tree to 15 m tall found in forests at 200–1 600 m in southern and western Guangxi. The species is cultivated for perfume, medicine and as a culinary spice in Fujian, Guangdong, Guangxi, Jiangxi and Yunnan.

Zanthoxylum bungeanum.—*Zanthoxylum bungeanum* (hua jiao; Rutaceae) is a deciduous tree 3–7 m tall, found below 3 200 m in Anhui, Fujian, Gansu, Guangxi, Guizhou, Hebei, Henan, Hubei, Hunan, Jiangsu, Jiangxi, Liaoning, Ningxia, Qinghai, Shaanxi, Shandong, Shanxi, Sichuan, southeastern Xinjiang, southern and southeastern Xizang, Yunnan and Zhejiang. The dried fruit follicles are used as a culinary spice and are particularly popular in Sichuan cuisine. The fruit of *Z. myriacanthum* (common prickly ash; da ye chou hua jiao), is also used locally in southern Yunnan as a spice.

Figure 17.8 *Illicium verum*. A, tree. B, fruits.

Myrica esculenta.—*Myrica esculenta* (box myrtle; mao yang mei; Myricaceae) is a wild evergreen tree distributed at 300–2 500 m elevation in northwestern Guangdong, Guangxi, southern and western Guizhou, western Sichuan and Yunnan, with edible fruits. *Myrica nana* (yun nan yang mei) is widely distributed in forested areas of southwestern China at elevations of 1 200–2 400 m; the wild fruits are eaten and used in wine-making in Yunnan.

Ceratopteris thalictroides.—Ceratopteris thalictroides (shui jue cai; Pteridaceae) is an evergreen herb distributed under humid forests or in paddy fields and wetlands in Anhui, Fujian, Guangdong, Guangxi, Hainan, Hubei, Jiangsu, Jiangxi, Shandong, Sichuan, Taiwan, Yunnan and Zhejiang. The edible leaves are used as wild vegetable.

17.5 Bamboos and rattans

According to archeological evidence, the use of bamboos in Chinese culture has a history as long as 5 000–6 000 years. Since the Han Dynasty (202 BC–220 AD) there has been a body—the "Sizhujian"—to manage bamboo at each governmental level from local to central. In ancient China, bamboo plantations were planted over 2 440 years ago. There are numerous ancient records of research into bamboo and its uses, such as the *Paper Manual* by SU Yi-Jian (976–997 AD), *Shoot Illustration* by ZAN Ning (a monk of the Song Dynasty, 960–1279 AD), and the *Detailed Bamboo Chart* by LI Kan, written in 1312.

China has abundant resources of bamboo, with more than 500 species, about 50% of all woody bamboo species in the world. The country is also particularly rich in economically-important species of bamboo. The uses of bamboo fall into three main types in China: harvesting the culms for timber or fibers, weaving, and use of the shoots. Those used for culm timber or fibers include *Phyllostachys edulis* (Poaceae), the sympodial bamboo *Dendrocalamus giganteus* (Poaceae), and *D. sinicus*. Those used for weaving include *Bambusa intermedia* and *Schizostachyum funghomii* (both Poaceae). Edible-shoot bamboos include *Dendrocalamus asper*, *D. brandisii* and *D. latiflorus*. Bamboos play an important role in rural life and are very important to economic development in China, because of their fast growth, high yield and annual, rationed harvest. China is already known throughout the world for its bamboos, and at present bamboo has reached a level of industrialized production. The Chinese population has a rich experience in the use of wild bamboos, which bodes well for future prospects for sustainable management and development.

17.5.1 Poaceae (bamboos)

Bambusa intermedia.—Bambusa intermedia (mian zhu; Figure 17.9) is widely cultivated in river valleys and hilly areas at elevations of 500–2 300 m in Guizhou, Sichuan, and central and southern Yunnan. The culm bears long fibers with a soft surface, good for weaving and paper-making. The strong culms are also good for furniture-making and construction.

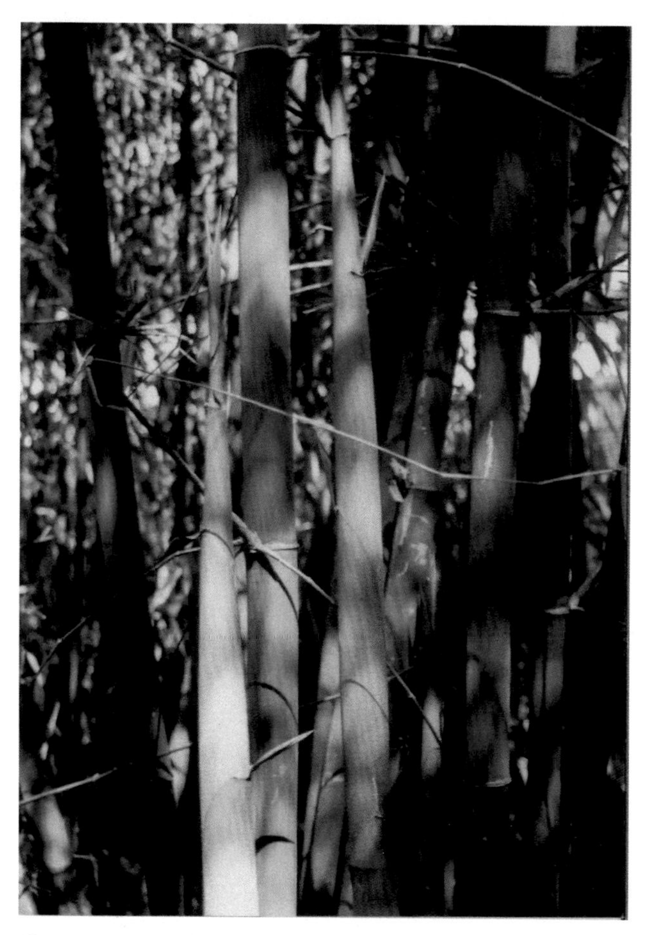

Figure 17.9 *Bambusa intermedia.*

Cephalostachyum pergracile.—Cephalostachyum pergracile (xiang nuo zhu; Figure 17.10) is found only in southern Yunnan, Myanmar and Laos, at around 500–1 200 m, where it is planted around houses and villages. The shoots are of high quality, and the young culms are specifically used by the Dai ethnic minority as receptacles for cooking fragrant sticky rice, giving this bamboo the common name "xiangnuo" ("fragrant and sticky"). Its chestnut-red, shiny sheath also has high ornamental value.

Chimonobambusa tumidissinoda.—Chimonobambusa tumidissinoda (qiong zhu), distributed at 1 500–2 200 m in southwestern Sichuan and northeastern Yunnan, is famed for its culms which have been used for walking sticks since the Han Dynasty; they are also used for umbrella handles and the shoots are eaten as a vegetable.

Chimonocalamus delicatus.—Chimonocalamus delicatus (xiang zhu; Figure 17.11), found at 1 400–2 000 m in southern Yunnan, is used for construction and for its edible shoots.

Dendrocalamus.—Dendrocalamus asper (ma lai tian zhu) is a huge, tropical sympodial bamboo, famed for its high value sweet shoots. Its original distribution was in southwestern Yunnan, but it is now introduced into Southeast Asia and most tropical parts of the world. The large culms are used for furniture and construction and the shoots for food, making it a valuable species of which to develop production in tropical areas. *Dendrocalamus brandisii* (yun nan tian long zhu) is a large, tropical, sympodial bamboo occurring chiefly in southern Yunnan, and widely cultivated in tropical regions of southeastern China and Southeast Asia. It produces high-quality edible

Figure 17.10 *Cephalostachyum pergracile.* A, monodominant stand. B, detail of culms.

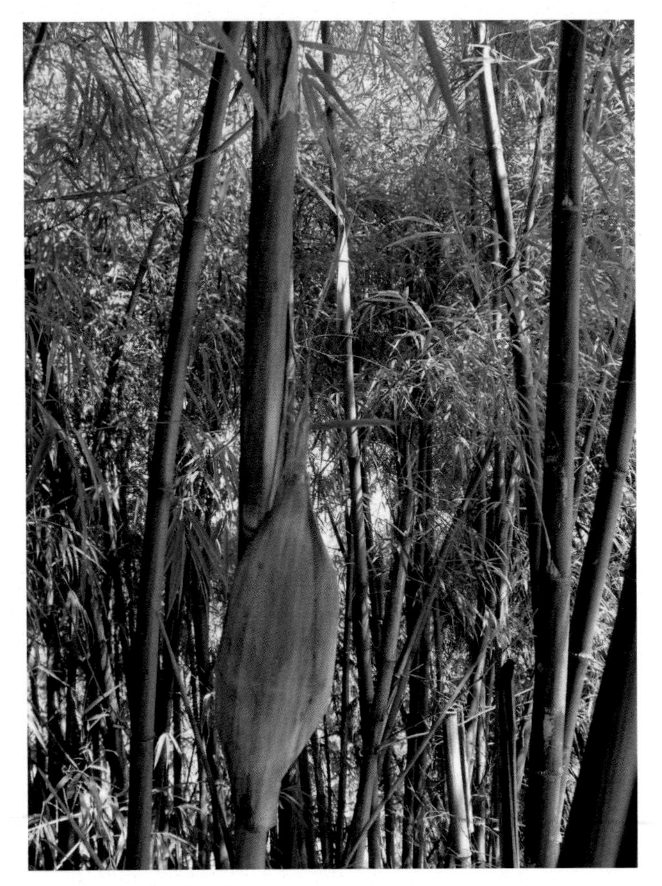

Figure 17.11 *Chimonocalamus delicatus.*

shoots, and very good timber for construction. Its fibers are made into high-quality floorboards, while the culm is also good for weaving material and wood block-making. *Dendrocalamus giganteus* (long zhu; Figure 17.12), like all 27 species of *Dendrocalamus* (mu zhu shu) in China, is a large, arborescent bamboo forming dense clumps. It is tropical and sympodial with gigantic culms 20–35 m tall and 12–18 cm in diameter. Its fibers, which are comparatively thick and strong, are processed into superior-quality floorboards. The shoots, though unsuitable to eat fresh because of their bitter taste, can be made into dried shoots and slivers that are no longer bitter when rinsed and cooked. This is the most widely-distributed bamboo in southern Yunnan and is cultivated in tropical and subtropical areas of Southeast and southern Asia.

Dendrocalamus latiflorus (ma zhu; Figure 17.13) is also a large, tropical, sympodial bamboo, the culm of which is used for making rafts, water pipes and construction materials. It is found in southern and southeastern Yunnan, and is widely cultivated in Fujian, Guangdong, Guanxi, Guizhou, Hainan, Sichuan and Taiwan. The shoots, produced at a high rate over a long period in summer and autumn, are delicious as a vegetables. *Dendrocalamus membranaceus* (huang zhu) is another large, tropical, sympodial bamboo growing naturally in tropical mixed deciduous and monsoon forests in southern Yunnan, and forming pure bamboo forests along the Lancang Jiang. The large culms are straight and strong enough for construction; the fibers are long, good for weaving and pulping. The shoots are edible when fresh and can also be dried. *Dendrocalamus sinicus* (ju long zhu; Figure 17.14) is a huge, tropical sympodial bamboo found only in southwestern Yunnan. It can grow to more than 30 m tall and 35 cm in diameter; the biggest bamboo found in the literature. The culm is widely used as a construction material for houses, bridges and water pipes; the shoots are edible. This species has great potential for extensive development, but is also a rare species in other parts of the world and therefore an important part of China's plant heritage.

Fargesia yunnanensis.—Fargesia yunnanensis (yun nan jian zhu) is distributed at 1 700–2 500 m elevation in southwestern Sichuan and Yunnan. The shoots are delicious and the culms are used for making agricultural tools.

Figure 17.12 *Dendrocalamus giganteus*. A, stand. B, flowers.

Figure 17.13 *Dendrocalamus latiflorus*.

Phyllostachys edulis.—*Phyllostachys edulis* (moso; mao zhu; Figure 17.15) is distributed on mountain slopes in Anhui, Fujian Guangdong, Guangxi, Guizhou, Henan, Hubei, Hunan, Jiangsu, Jiangxi, Shaanxi, Sichuan, Taiwan, Yunnan and Zhejiang. This is the most economically-important bamboo in China, widely cultivated for its versatile culms and delicious shoots. Many cultivars have been named.

Figure 17.14 *Dendrocalamus sinicus* flowers.

Schizostachyum funghomii.—*Schizostachyum funghomii* (sha luo dang zhu) is another huge, tropical sympodial bamboo, distributed in a belt from southern to southwestern China, forming a large patch of natural forest in Hekou, southern Yunnan. It provides fine timber, shoots and fibers for weaving, and is also used for ornaments. The hand-made rocket firecrackers known as "gaosheng," which are set off during the Water-Sprinkling Festival of the Dai ethnic group in Xishuangbanna, are made by filling internodes of *S. funghomii* with gunpowder.

17.5.2 Arecaceae (rattans)

Rattans have been widely used for many millennia in the provinces of southern and southwestern China including Guangdong, Guangxi, Hainan, Xizang and Yunnan, for construction, furniture, ropes, stakes and woven materials. The fruits and young shoots are also eaten locally. Local people have developed technologies for the cultivation of rattans in forest areas. However, due to deforestation and over-harvesting, wild rattans are now

Figure 17.15 Stand of *Phyllostachys edulis.*

very rare and some species are endangered; urgent action is needed to protect rattan species and develop sustainable cultivation systems for them in the forested and agro-forested areas of tropical and sub-tropical China.

Calamus.—Calamus (sheng teng shu) species are used for making canes and for weaving, making them important in handicrafts. The genus has 28 species in China, of which 15 are endemic. *Calamus acanthospathus* (yun nan sheng teng; Figure 17.16), distributed at 800–2 400 m in Xizang and Yunnan, provides canes used in basketry and furniture-making, while *C. erectus* (zhi li sheng teng; Figure 17.17) is found below 1 400 m in Yunnan and provides short, thick, rigid canes used in construction and furniture-making.

Figure 17.16 *Calamus acanthospathus.* A, plants in forest. B, detail of fruits.

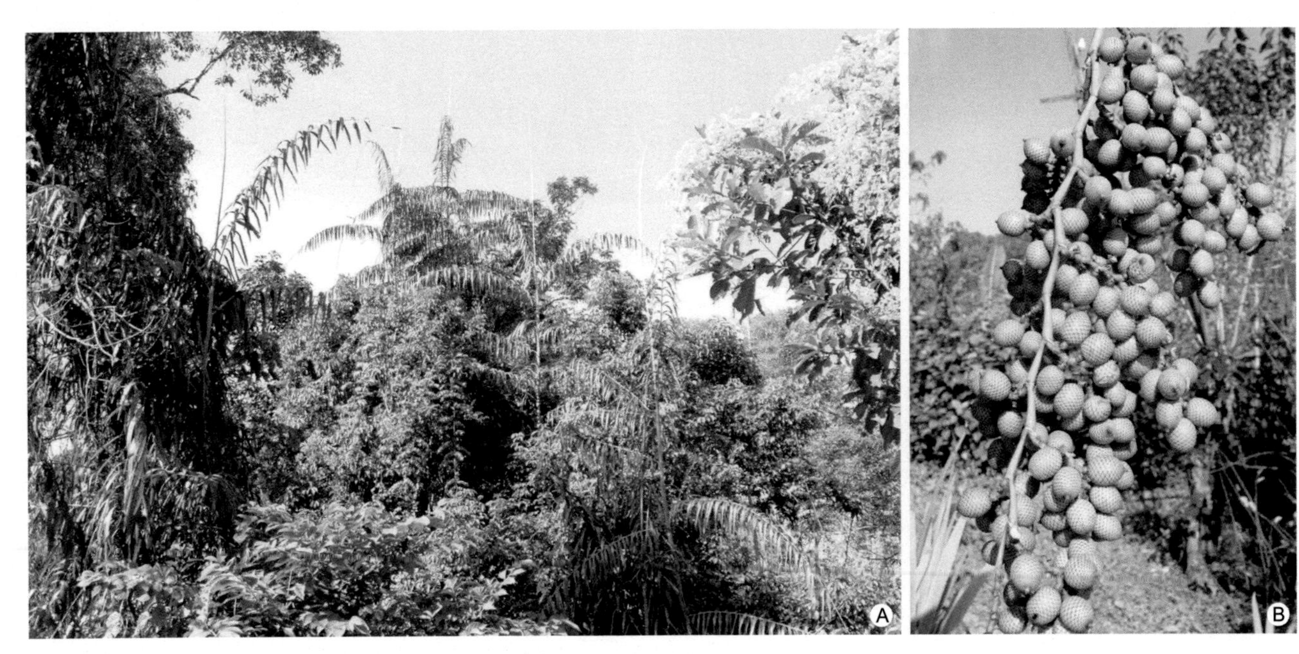

Figure 17.17 *Calamus erectus.* A, plants in natural habitat. B, detail of fruits.

Daemonorops jenkinsiana.—Daemonorops jenkinsiana (huang teng) is distributed in lowland rainforests, often persisting in disturbed areas, below 1 000 m in Guangdong, Guangxi and Hainan. The species provides canes used in furniture-making and the leaves are used for thatching.

Plectocomia himalayana.—Plectocomia himalayana (gou ye teng) is found in mountainous areas of southwestern Yunnan, where the stems are used locally as a large rattan for construction material and furniture.

References

IUCN (2011). IUCN Red List of Threatened Species. Version 2011.2. <http://www.iucnredlist.org/>.

Jia, L. Z. & Zhou, J. (1987). *Fat and Oil Plants of China*. Beijing, China: Science Press.

Jiang, Y. C. & Dang, R. L. (2002). *Plant Resources*. Urumqi, China: Xiangjiang People's Publishing House.

Long, C. L. & Song, H. C. (2012). *Chinese Biodiesel Plants*. Beijing, China: Science Press.

Pei, S. J. & Huai, H. Y. (2007). *Ethnobotany*. Shanghai, China: Shanghai Science and Technology Press.

State Forestry Administration (2009). State Forestry Administration. <http://www.forestry.gov.cn/>.

Wu, D., Wang, H., Li, D. Z. & Blackmore, S. (2005). Pollen morphology of *Parnassia* L. (Parnassiaceae) and its systematic implications. *Journal of Integrative Plant Biology* 47(1): 2-12.

Zheng, W. J. (1983). *Arbororum Sinicorum*. Beijing, China: Chinese Forestry Publishing House.

Zheng, W. J. (1985). *Arbororum Sinicorum*. Beijing, China: Chinese Forestry Publishing House.

Zheng, W. J. (1997). *Arbororum Sinicorum*. Beijing, China: Chinese Forestry Publishing House.

Zheng, W. J. (2004). *Arbororum Sinicorum*. Beijing, China: Chinese Forestry Publishing House.

Chapter 18

Medicinal Plants

PEI Sheng-Ji and HUAI Hu-Yin

18.1 Introduction

China has a long history of using medicinal plants, dating back to the Paleolithic Period, and plants are widely used in all systems of traditional medicine in China, attesting to a long history of interactions between people and plants. One of the earliest documented examples of the use of plants in medicine is SHEN Nong's *Classic Herbal*, dating to around 3000 BC and mentioning 365 types of plants, animals and minerals held to be of medicinal value. Another, major source of reference, still consulted today, is the *Compendium of Materia Medica*, published in 1578 during the Ming Dynasty and referring to 1 095 types of plants. More recently, a *Dictionary of Traditional Chinese Medicine*, by Jiangsu New Medical College (1997), provides an authoritative source of information on Traditional Chinese Medicine (TCM), recording a total of 4 773 species of medicinal plants (altogether used in 82% of the drugs listed).

The total number of plant species used medicinally in China is reported to be 10 000–11 250, about 41% of the native flora (He & Gu, 1997; Xiao & Yong, 1998; Pei, 2000; Pei, 2002; Hamilton, 2004), although this is likely to be a considerable underestimate compared to those used but not documented. Within gymnosperms, species in almost every family have been used as herbal medicines. Among angiosperms, many families and genera have been investigated for medicinal uses. Notable examples include Ranunculaceae, of which 39 genera and ca. 700 species are used medicinally in China, Berberidaceae (11 genera and 260 native species used medicinally), Papaveraceae (the majority of the 300 species in 19 genera used medicinally), Verbenaceae (80 native species in 14 genera), Campanulaceae (50 species in 11 genera), Lamiaceae (250 species in 50 genera), Zingiberaceae (80 species in 12 genera), and Liliaceae, the majority of which are medicinal plants in China. Furthermore, the species of some smaller families such as Aristolochiaceae are all medicinally used. The medicinal uses of more plant species continue to be discovered (Zhu, 2004).

Most medicinal plants used in China are collected from the wild although some, such as *Angelica polymorpha* (Apiaceae), *Ephedra sinensis* (Ephedraceae), *Eucommia ulmoides* (hardy rubber tree; Eucommiaceae), *Gastrodia elata* (Orchidaceae), *Gentiana robusta* (Gentianaceae), *Gingko biloba* (Ginkgoaceae), *Rheum palmatum* (rhubarb; Polygonaceae) and *Aucklandia costus* (snow lotus; Asteraceae), are domesticated and widely cultivated. Cultivation is useful when market demand is high or for taking the pressure off wild populations, and thus contributes to conservation. Guidelines for good agricultural practice are currently being developed in China for a number of medicinal plants (especially those used in TCM), which will assist in ensuring that their production is environmentally sustainable. Unfortunately, some popular medicinal plants have proved very difficult to domesticate or cultivate, such as some species of Orchidaceae.

According to PEI (2000), traditional medicine in China can broadly be classified into three categories: traditional medical systems (formalized through written pharmacopeias and formal apprenticeships), folk medicine (transmitted orally) and shamanistic medicine (spiritually-orientated, associated with certain indigenous groups). Traditional Chinese Medicine is the most widely used of the traditional medicinal systems. Each ethnic minority has its own medical traditions, some of which have written pharmacopeias and are therefore classifiable as traditional medicinal systems. Only traditional medical systems have been studied systematically with regard to the types of plants utilized; relatively little research has been conducted as yet on either folk or shamanistic medicine. In this chapter, therefore, we define species as belonging to either TCM, traditional ethnic minority medicine, and folk medicine (we do not cover shamanistic medicine as so little about this practice is formally documented); we also look at species used by the modern pharmaceutical industry. Note, however, that there is a degree of overlap between these categories; for example, some species are popular both in TCM and folk medicine.

18.1.1 Traditional Chinese Medicine

Traditional Chinese Medicine is the most widely used traditional medicinal system in China. Historically catering especially for the healthcare needs of the Han Chinese, TCM continues to play a vital role in Chinese healthcare, especially in rural areas. Species recognized as TCM plants are those which are used clinically by TCM doctors following the guidelines of TCM theory. The medicines used in TCM come not only from plants but also from animals, fungi and minerals; however plants are the most important sources (Zhang *et al.*, 1995).

Many factors affect the use-value of TCM plants.

Traditional Chinese Medicine emphasizes the influence of environmental factors, such as climate and physical and chemical properties of the soil, as significant determinants of medicinal quality. The quality of TCM products is also considered to be influenced by how the plants are harvested and processed, with time of harvesting in relation to the season of the year an important consideration as it is likely to be related to changes in the abundance of certain physiologically-active chemicals within the plants. Thus, the same species can be considered as yielding excellent drug material from one region, locality or harvest, but not from others. For example, the fruits of *Lycium chinense* (wolfberry or goji berry; Solanaceae) yield a particularly valued tonic when collected from trees planted in Ningxia, where the conditions are held to be especially suitable, but otherwise are not highly regarded. The term "dao di yao cai," indicating genuineness or authenticity, is used for crude drugs originating from plants growing in certain defined environments or geographical areas, varying from species to species. Various traditional technologies are used for processing plant materials, part of a process of improving the final quality of the drugs.

18.1.2 Traditional ethnic minority medicine

In addition to the majority Han people, there are 55 minority ethnic groups in China, each with its own traditional medical knowledge and associated traditional doctors and guidelines of medical theory and practice (see Table 18.1). Ethnic minority populations are mostly concentrated in certain geographical areas and have a high dependency on the wild medicinal plants found in these places. However some species can come from far afield. For example, while many medicinal plants used in Tibetan medicine are collected from the local Qinghai-Xizang Plateau, some of the species used here are imported from the tropics or sub-tropics (Luo, 1985). The medicinal practices of some ethnic minorities, such as the Tibetans, Mongolians and Yi, are well documented; other minorities have only oral medical traditions. Even among minorities with written traditions, much medical practice can be based on oral rather than written knowledge. The distinction between traditional ethnic minority medicine and folk medicine can therefore be blurred.

Table 18.1 Species used in major traditional ethnic minority medicinal systems (from Pei & Huai, 2007).

Ethnic minority	Number of species used medicinally
Dai	at least 800
Hani	ca. 200 (commonly-used species only)
Miao	1 300
Mongolian	1 430
Naxi	at least 800
Qiang	2 310
Tibetan	3 294
Wa	ca. 200
Yi	1 189

It can be difficult to obtain information on the use of medicinal plants by minority groups, which often requires extensive field investigation and can be hampered by reluctance on the part of the holders of such information to disclose it. Local medical knowledge is often regarded as a precious secret, which it would be taboo to reveal to outsiders. One of the best known traditions is Tibetan medicine, with a total of about 3 294 species known to be used in China (Pei & Huai, 2007). Based on nearly 20 years of fieldwork, LUO (1997) has reported 2 085 species of plants used on the eastern and southeastern parts of the Qinghai-Tibetan Plateau, 1 441 of which are described in his book *Tibetan Materia Medica of China* (Luo, 1997). Forty of the 2 085 species originate from as far away as southeastern China. Within the restricted area of Dêqên, Yang and Chuchengijiancuo (1987) found 448 species of medicinal plants used in Tibetan medicine. As with TCM, the number of medicinal plants recorded from Tibetan medicine is increasing as more research is conducted.

Considerable success has been achieved in developing new drugs based on traditional ethnic minority medicine, for example from *Alstonia scholaris* (blackboard tree; Apocynaceae), *Bergenia purpurascens* (Saxifragaceae), *Dracaena cambodiana* (Liliaceae) and *Erigeron breviscapus* (Asteraceae; Pei & Huai, 2007). Conservation of the plants used in traditional ethnic minority medicine has received attention in many parts of China, through cooperation with international organizations. In some places, traditional home gardens have played an important role in conserving and domesticating minority medicinal plants.

18.1.3 Folk medicine

Folk medicine is widely used in China, especially in rural areas, following traditions passed down orally within households, communities or among ethnic minorities, and benefiting from personal experiences accumulating over individual lives. Many types of plants are used, many of which are also found in traditional medical systems. Despite having no basis in medical theory, certain folk remedies are used widely within geographical areas or communities, providing convenient and often effective treatments for common maladies. Undoubtedly much remains to be learnt about folk medicine in China, some of which will likely prove useful for the development of new pharmaceutical products. Community and village folk healers commonly grow medicinal plants in their own gardens, into which they transplant plants from the wild. This practice is useful for providing a convenient source of plant material for treatments and can contribute to the conservation of the species concerned.

18.1.4 The pharmaceutical industry

The total number of species from which active chemical compounds are extracted for use by pharmaceutical industries used in China has been given as between 400 (Zhou, 1999) and 717 (Wang & Liu, 2002). The latter authors provide detailed descriptions of 262 of these species in *The Chinese Plant Raw Material Medicines* (Wang & Liu, 2002). The number of species used can be expected to grow in the future following advances in medical phytochemistry and industrial developments. One example is *Taxus* (yew; Taxaceae), from which the drug taxol (also marketed as paclitaxel) is extracted. Taxol is a patented medicine used to inhibit the growth of cancer cells. *Artemisia annua* (sweet wormwood; Asteraceae), a plant used in TCM, is the source of artemisin, a compound used as an anti-malarial. The antipsychotic and antihypertensive reserpine is extracted from *Rauvolfia verticillata* (Apocynaceae), and the leaves of *Ginkgo biloba* are collected for the extraction of flavonoids, the main chemical constituents of drugs used for improving blood circulation in the brain.

The taxa described below have been chosen as representative medicinal plants which either have a high frequency of use, whether in TCM, ethnic minority medicine or folk medicine, are endemic to China and have important medical qualities, are economically important to the pharmaceutical industry today or likely to become so in the future, or are of high conservation concern. The information is based mainly on the *Dictionary of Traditional Chinese Medicine* (Jiangsu New Medical College, 1997), *Pharmacopeia of the People's Republic of China* (Editor Committee of National Pharmacopeia, 2005), *The Chinese Plant Raw Material Medicines* (Wang & Liu, 2002), *Zhongguo Minzu Yao Zhi Yao* (*Compendium of Ethno-Medical Flora of China*; Jia & Li, 2005) and the *Tibetan Materia Medica of China* (Luo, 1997).

18.2 Important medicinal plant taxa

Selaginella tamariscina.—*Selaginella tamariscina* (juan bai; Selaginaceae; Figure 18.1) is found in southern, eastern, central, northern and northeastern China. The whole plant is used in TCM to treat amenorrhea, dysmenorrhea, fractures, blood stasis and hemostasis. It is also widely used in Bai, Bulang, Dong, Lisu, Miao, Mongolian, Naxi, Pumi, She, Tujia, Wa, Yi, Zang and Zhuang medicine to treat similar maladies.

Figure 18.1 *Selaginella tamariscina.*

Lygodium japonicum.—*Lygodium japonicum* (hai jin sha; Lygodiaceae; Figure 18.2) is found in Fujian, Guangdong, Guangxi, Jiangsu, Jiangxi, Hunan, Hubei, Sichuan and Zhejiang, among others. The dried mature spores are used in TCM to treat urethral pain. They are also widely used in Achang, Bai, Dai, Deang, Dong, Jingpo, Jinuo, Lahu, Lisu, Miao, Tujia, Yi and Zang medicine to treat urethral calculus, nephritis, urinary tract infections and diarrhea. This plant is invasive in many countries.

Figure 18.2 *Lygodium japonicum.*

Ginkgo biloba.—*Ginkgo biloba* (ginkgo; yin xing; Ginkgoaceae) is a deciduous tree that is possibly native to northwestern Zhejiang and widely cultivated at altitudes below 2 000 m in almost all parts of China. The dried seeds are used in TCM to treat congestive coughs, enuresis and frequent urination, as well as in Jingpo, Miao, Mongolian, Tujia and Zhuang medicine. The leaves are used by the pharmaceutical industry for the extraction of flavonoids.

Taxus.—*Taxus* (yew; hong dou shan; Taxaceae) comprises three species in China. *Taxus cuspidata* (Japanese yew; dong bei hong dou shan) is found mainly in northeastern China and Shaanxi, while *T. wallichiana* (Himalayan yew; xu mi hong dou shan) is found at 100–3 500 m elevation in southwestern China, the northern part of southern China, part of eastern China, central China, southern Gansu and southern Shaanxi. The bark from both species is used as a raw material for extracting taxol, an important chemical constituent in antitumor drugs. These two species are also cultivated in some areas of China for commercial extraction. Tissue culture has also successfully been used to produce taxol.

Ephedra.—Many species of *Ephedra* (ma huang shu; Ephedraceae) are used medicinally, including *E. equisetina* (bluestem joint fir; mu zei ma huang), *E. intermedia* (zhong ma huang), *E. likiangensis* (li jiang ma huang), *E. minuta* (ai ma huang), *E. przewalskii* (mo guo ma huang) and *E. sinica* (cao ma huang). The whole plant is used medicinally in TCM, traditional ethnic minority medicine and folk medicine to treat colds, bronchial asthma and night sweats. The pharmaceutical industry uses *Ephedra* for the extraction of alkaloids. Alkaloid content varies between species, from about 1% in *E. sinica* to 2–2.5% in *E. equisetina* (Wang & Liu, 2002).

Magnoliaceae.—The bark and flowers of *Houpoëa officinalis* (hou pu) are used in TCM to treat abdominal distension, shortness of breath and asthma. The dried young flower buds (xin yi), of *Yulania biondii* (wang chun hua; Magnoliaceae), *Y. denudata* (Yulan magnolia; yu lan) and *Y. sprengeri* (wu dang yu lan) are used in TCM to treat headaches and rhinorrhea. Many species of Magnoliaceae, such as *H. officinalis, H. rostrata, Lirianthe delavayi* (Chinese evergreen magnolia), *Y. campbellii, Y. denudata* and *Y. liliiflora* (Mulan magnolia), are used widely in Achang, Hani, Lisu, Miao She, Yi, Wa and Zang medicine to treat dyspepsia, coughs, asthma and rhinitis, among other complaints.

Chloranthus serratus.—*Chloranthus serratus* (ji ji; Chloranthaceae) is a herb found in wet places at 100–1 800 m elevation in southern, most of eastern, and southwestern China. The roots are used in TCM to invigorate the circulation and to treat traumatic injuries and abnormal menstruation. The roots or whole plant are used in Dong and Miao medicine to treat fractures and rheumatoid arthritis (Jia & Li, 2005).

Sarcandra glabra.—*Sarcandra glabra* (cao shan hu; Chloranthaceae) is a subshrub found in forests, thickets, grasslands, swamps, along streamsides and on sandy soils from near sea level to 2 000 m in southern, southwestern, and part of eastern China. The whole plant is used in TCM to treat purpura, rheumatism and fractures. It is also widely used in Achang, Dai, Deang, Dong, Jingpo, Lahu, Miao, She, Shui, Tujia, Yao and Zhuang medicine to treat fractures, rheumatism, skin ulcers and other maladies.

Morus alba.—*Morus alba* (white mulberry; sang; Moraceae) is cultivated widely throughout China. The leaves, roots, bark and branches are all used in TCM to treat edema, coughs, asthma, joint pain, headaches and colds. The plant is also used widely in Achang, Dai, Dong, Hani, Lisu, Miao, Mongolian, Shui, Tujia, Wa, Yao and Zang medicine to treat fractures, colds and rheumatism.

Cannabis sativa.—*Cannabis sativa* (cannabis or hemp; da ma; Cannabaceae; Figure 18.3) is a herb native to, or perhaps naturalized in, Xinjiang, and cultivated throughout China. The leaves are used in TCM to treat malaria and ascariasis; the male inflorescences are used to treat numbness of the limbs and amenorrhea; the roots to treat fractures and to speed up labor when giving birth. The plant is widely used in Achang, Deang, Dong, Jingpo, Lisu, Hani, Korean, Mongolian, Yi and Zang medicine to treat shortness of breath, dysentery, insomnia, headaches and rheumatoid arthritis, among other complaints.

Aristolochia debilis.—*Aristolochia debilis* (Dutchman's pipe; ma dou ling; Aristolochiaceae) is a herb found in thickets, on mountain slopes and in moist valleys at 200–1 500 m in parts of eastern, southern, southwestern and central China. The fruits are used in TCM to treat coughs and hemorrhoids; and the above-ground parts to treat joint pain and edema during pregnancy. At least 17 species of *Aristolochia* are used in Bai, Dai, Miao, Pumi, Wa, Yi and Zang medicine, to treat stomach aches, rheumatism, hypertension and fractures.

Figure 18.3 *Cannabis sativa.*

Asarum sieboldii.—*Asarum sieboldii* (han cheng xi xin; Aristolochiaceae) is a herb found in the forests and moist valleys of Liaoning. The root and rhizome is used in TCM to treat colds, headaches, toothache, rheumatism and coughs (Editor Committee of National Pharmacopeia, 2005). It is also used widely in Mongolian, Tujia and Uyghur medicine to treat fractures, itching and numbness of the limbs (Jia & Li, 2005).

Fallopia multiflora.—*Fallopia multiflora* (Chinese knotweed; he shou wu; Polygonaceae) is a herb found on mountain slopes, in rock crevices and valley thickets at 200–3 000 m elevation over most of China. Both the stems and roots are used in TCM to treat insomnia, rheumatism, itching, weakness, and other complaints (Editor Committee of National Pharmacopeia, 2005). The roots also are used in Dong, Maonan, Miao, Mongolian, Shui, Yao, Yi and Zhuang medicine to treat physical weakness (Jia & Li, 2005).

Rheum.—The roots of herbs of the genus *Rheum* (da huang shu; Polygonaceae) are used commonly in TCM to treat shortness of breath, dyspepsia, jaundice and fractures. Examples include *R. officinale* (rhubarb; yao yong da huang; Figure 18.4), found mainly under forests at 1 200–4 000 m in Fujian, Guizhou, southwestern Henan, western Hubei, Shaanxi, Sichuan and Yunnan, *R. palmatum* (Turkish rhubarb; zhang ye da huang), found on slopes and in valleys at 1 500–4 400 m in Gansu, Hubei, Nei Mongol, Qinghai, Shaanxi, Sichuan, Xizang and Yunnan, and *R. tanguticum* (ji zhua da huang), found in valleys at 1 600–3 000 m in Gansu, Qinghai, Shaanxi and Xizang. *Rheum palmatum*

Figure 18.4 *Rheum officinale.*

and *R. officinale* have long been cultivated commercially in Gansu and Yunnan. Many species of *Rheum* are used in minority medical traditions, such as those of the Achang, Dai, Deang, Jingpo, Miao, Mongolian, Naxi, Nu, Yugu and Zang people, with a particularly large number of species used in Zang medicine.

Achyranthes bidentata.—Achyranthes bidentata (ox knee; niu xi; Amaranthaceae) is a herb found on hillsides at 200–1 800 m in southwestern, northwestern, eastern, part of northern and part of southern China. The roots are collected in winter for TCM and used to treat aches, muscle and bone weakness, amenorrhea and vertigo. In Achang, Bai, Dai, Jingpo, Mongolian, Yi and Zang medicine they are used to treat sore throats, hypertension, rheumatoid arthritis, menstrual disorders and fractures. Henan is the main area producing medicinal material of this species.

Vaccaria hispanica.—Vaccaria hispanica (cowherb; mai lan cai; Caryophyllaceae) is an annual herbaceous weed of wheat fields throughout China. The seeds are used in TCM, as well as Achang and Mongolian medicine, to treat amenorrhea, dysmenorrhea and acute mastitis.

Pseudostellaria heterophylla.—Pseudostellaria heterophylla (false starwort; hai er shen; Caryophyllaceae) is found mainly in mountain valleys and moist shady forests at 800–2 700 m in northern, eastern, central, southwestern, and parts of northwestern and northeastern China. The tubers are used in TCM to treat disorders of the spleen, lack of appetite, weakness and spontaneous perspiration. In Mongolian and Korean medicine, they are used to treat coughs, amnesia and insomnia.

Nelumbo nucifera.—Nelumbo nucifera (lotus; lian; Nelumbonaceae) is cultivated in water-bodies throughout China except in Nei Mongol, Qinghai and Xizang. The seeds are used in TCM to treat spermatorrhea, insomnia and palpitations. The mature receptacle is used to treat hemorrhoids, hematuria and postpartum blood stasis, while the leaves are used to relieve thirst and fatigue, eliminate blood stasis and staunch bleeding. *Nelumbo nucifera* is widely used in Achang, Dai, Deang, Jingpo, Korean, Lisu, Miao, Mongolian, She, Yi, Zang and Zhuang medicine to treat hemoptysis, hematemesis, spermatorrhea, and lack of appetite.

Aconitum.—Many species of *Aconitum* (aconite; wu tou shu; Ranunculaceae) are rich in alkaloids with important medical values. *Aconitum carmichaelii* (wu tou; Figure 18.5), found at forest margins, in scrub, on grassy slopes and mountains at 100–2 200 m in most parts of China, is used in TCM to treat rheumatism, joint pain, as an anesthetic and for pain-relief. The leaves of *A. kusnezoffii* (bei wu tou), found mainly on grassy slopes, forest margins and by streams at 2 000–2 400 m in northeastern China, Hebei, Nei Mongol and Shanxi, are used in TCM to treat fever, diarrhea, stomach aches, headaches and toothache. It is recorded in the *Zhongguo Minzu Yao Zhi Yao* that at least 51 species of *Aconitum* are used across various Chinese medicinal systems.

Figure 18.5 *Aconitum carmichaelii.*

Anemone altaica.—Anemone altaica (a er tai yin lian hua; Ranunculaceae) is a herb found in forests, scrub and beside streams at 1 200–1 800 m in Gansu, western Henan, northwestern Hubei, Shaanxi, southern Shanxi and northern Xinjiang. The rhizome is used in TCM to treat epilepsy, amnesia, rheumatism and skin ulcers (Jiangsu New Medical College, 1997). It is used in Mongolian medicine to treat fever-induced stupors, epilepsy, neurosis, loss of appetite and skin ulcers (Jia & Li, 2005). The entire market demand for this plant is met by wild resources.

Clematis.—Many species of *Clematis* (tie xian lian shu; Ranunculaceae) are used medicinally in China. *Clematis chinensis* (wei ling xian) is found mainly in southwestern, southern and eastern China. Its roots and rhizomes are used in TCM to treat rheumatism and numbness of the limbs, as are those of *Clematis hexapetala* (mian tuan tie xian lian), found mainly in northwestern, northern, parts of northeastern and eastern China. The latter is recorded in the *Zhongguo Minzu Yao Zhi Yao*, which notes that at least 31 species of *Clematis* are used in various minority systems of medicine.

Coptis.—The rhizomes of *C. chinensis* (Chinese goldthread; huang lian; Ranunculaceae), found in forests and shady places at 500–2 000 m in southern Anhui, Fujian, northern Guangdong, northern Guangxi, Guizhou, Hubei, Hunan, southern Shaanxi, Sichuan and Zhejiang, *C. teeta* (yun nan huang lian), from evergreen broad-leaved forests at 1 500–2 300 m in northwestern Yunnan and southeastern Xizang, and *C. deltoidea* (san jiao ye huang lian), restricted to forests at 1 600–2 000 m in western Sichuan, are commonly used in TCM to treat diarrhea, jaundice and toothache, among other complaints (Editor Committee of National Pharmacopeia, 2005). Nearly all species of *Coptis* found in China are used in various minority systems of medicine (Jia & Li, 2005). For example, the rhizomes of *C. quinquesecta* (wu lie huang lian) are used in Zang medicine to treat plague, and *C. chinensis* is used in Deang, Jingpo, Lisu, Miao, Tujia and Uyghur medicine (Jia & Li, 2005). The wild population of *C. teeta* is small and the species is classified as protected in China.

Paeonia.—The roots of *P. lactiflora* (common peony; shao yao; Paeoniaceae; Figure 18.6), found in woods and grasslands at 400–2 300 m in northeastern China, southern Gansu, Hebei, Nei Mongol, southern Ningxia, Shaanxi and Shanxi, are used in TCM to treat light-headedness, stomach aches, menstrual problems and night sweats. *Paeonia suffruticosa* (tree peony; mu dan) is native to Anhui and Henan and widely cultivated in China. The root bark is used in TCM to treat hematemesis, amenorrhea, dysmenorrhea and fractures. Many species of *Paeonia*, such as *P. delavayi* (dian mu dan) and *P. mairei* (mei li shao yao), are used in Bai, Korean, Lisu, Miao, Mongolian and Zang medicine.

Figure 18.6 *Paeonia lactiflora*. A, plants. B, detail of flowers.

Mahonia fortunei.—*Mahonia fortunei* (shi da gong lao; Berberidaceae; Figure 18.7) is found from 300 to 2 000 m in Sichuan, Hubei and Zhejiang, and is widely cultivated in China. The leaves are used in TCM to treat hemoptysis, light-headedness and backache. It is used in Buyi, Dai, Dong, Lahu, Maonan, Miao, Molao, She, Shui, Tujia, Yao, Yi and Zhuang medicine to treat colds, diarrhea, hepatitis and dysentery, among other complaints.

Menispermum dauricum.—*Menispermum dauricum* (moonseed; bian fu ge; Menispermaceae) is a vine found in shrubland, on roadsides and in open forests below 800 m in Anhui, Gansu, Guizhou, Hebei, Heilongjiang, Hubei, Hunan, Jiangsu, Jiangxi, Jilin, Liaoning, Nei Mongol, Ningxia, Shaanxi, Shandong, Shanxi and Zhejiang. The rootstock is used in TCM to treat sore throats, enteritis, dysentery and rheumatism (Editor Committee of National Pharmacopeia, 2005), and in Mongolian and Tibetan medicine to treat diarrhea, skin diseases, cancer and sore throats (Jia & Li, 2005). This species is also in demand by the pharmaceutical industry for the extraction of alkaloids (Wang & Liu, 2002).

Stephania.—Many species of *Stephania* (qian jin teng shu; Meniospermaceae) are found in southwestern China, and between them supply more than 50 known alkaloids. The tuberous root is widely used in TCM and minority medicine. Four species, *S. delavayi* (yi wen qian), *S. epigaea* (di bu rong), *S. japonica* (snake vine; qian jin teng) and *S. tetrandra* (fen fang ji), are widely used in many types of traditional medicine. The roots of *S. tetrandra* are used in TCM to treat urinary difficulties, eczema, rheumatism, numbness and hypertension (Editor Committee of National Pharmacopeia, 2005), and this species is also widely used in Deang, Jingpo, She and Tujia medicine (Jia & Li, 2005). According to the *Zhongguo Minzu Yao Zhi Yao*, 22 species of *Stephania* are used in minority medical traditions in China (Jia & Li, 2005). Many species of *Stephania* are also widely used by the pharmaceutical industry to extract alkaloids (Wang & Liu, 2002).

Cinnamomum cassia.—*Cinnamomum cassia* (Chinese cassia; rou gui; Lauraceae) is a tree found wild in southern China and now widely cultivated in tropical and subtropical parts of Fujian, Guangdong, Guangxi, Guizhou, Hainan,

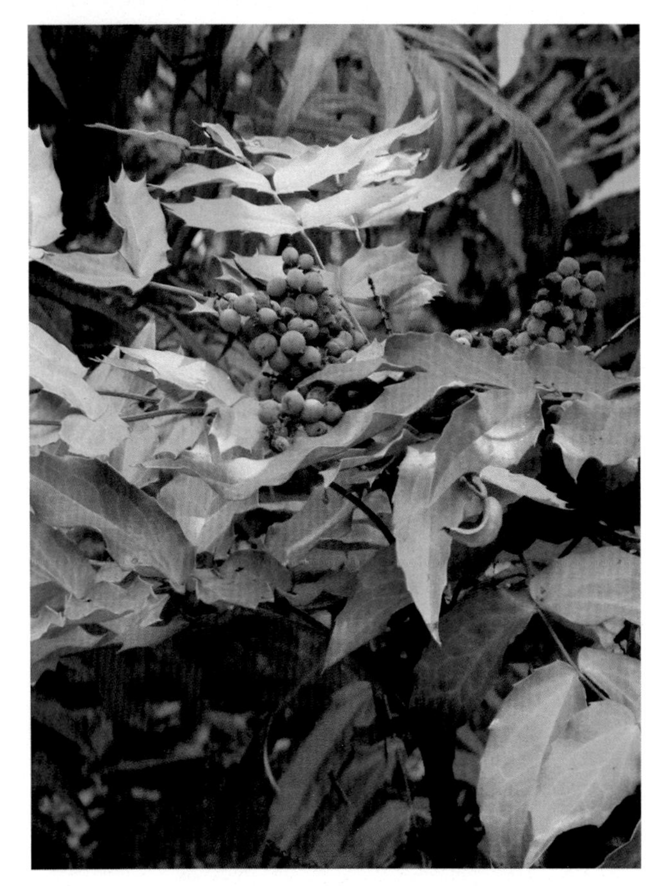

Figure 18.7 *Mahonia fortunei* in fruit.

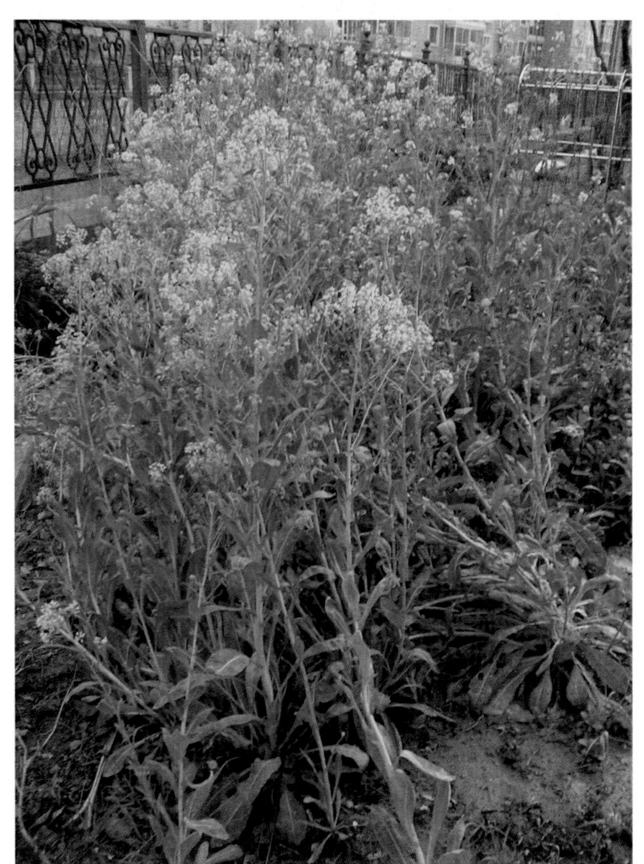

Figure 18.8 *Isatis tinctoria.*

Taiwan and Yunnan. The dried bark is used in TCM to treat erectile dysfunction, amenorrhea and dysmenorrhea. It is used in Achang, Dai, Deang, Done, Jingpo, Jinuo, Miao, Uyghur, Yao, Zang and Zhuang medicine to treat stomach aches, rheumatism and fractures.

Corydalis decumbens.—Corydalis decumbens (xia tian wu; Papaveraceae) is found on mountainsides, roadsides and coastal wastelands at 100–300 m in Anhui, Fujian, Hubei, Hunan, Jiangsu, Jiangxi, Shaanxi, Taiwan and Zhejiang. The tuber or whole plant is used in TCM to treat hypertension, rheumatoid arthritis, ischialgia and poliomyelitis (Jiangsu New Medical College, 1997). It is also used in Miao medicine to treat similar diseases (Jia & Li, 2005).

Meconopsis quintuplinervia.—Meconopsis quintuplinervia (wu mai lü rong hao; Papaveraceae) is a herb found in alpine grassland and meadows at 2 300–4 600 m in southern and southwestern Gansu, western Hubei, northeastern Qinghai, Shaanxi, northwestern Sichuan and northeastern Xizang. The whole plant is used in Tibetan medicine to treat pneumonia, hepatitis, phthisis and gastrohelicosis (Jia & Li, 2005). Many other species of *Meconopsis* are used Zang medicine.

Isatis tinctoria.—Isatis tinctoria (woad; song lan; Brassicaceae; Figure 18.8) is found wild and also widely cultivated in Fujian, Gansu, Guizhou, Hebei, Henan, Hubei, Jiangxi, Liaoning, Nei Mongol, Shaanxi, Shandong, Shanxi, Sichuan, Xinjiang, Xizang, Yunnan and Zhejiang. The roots and leaves are used in TCM to treat flu, cerebrospinal meningitis, Japanese encephalitis B, pneumonia, sore throats and other complaints (Jiangsu New Medical College, 1997).

Rhodiola.—Many species of *Rhodiola* (hong jing tian shu; Crassulaceae) are used medicinally in China, especially in Zang medicine. The root and rhizome of *R. crenulata* (da hua hong jing tian) are used in TCM to treat lack of energy, blood stasis, hemiplegic paralysis caused by apoplexy, and asthma. *Rhodiola tangutica* (tang gu hong jing tian), found in rock crevices in alpine areas and beside water at 2 100–4 700 m in Gansu, Qinghai and Sichuan, is used in Zang medicine to treat lung disorders. Bai, Dai, Jingpo, Tujia and Yi medicine also use various species of *Rhodiola* to treat pneumonia, fever, diarrhea and fractures.

Eucommia ulmoides.—Eucommia ulmoides (hardy rubber tree; du zhong; Eucommiaceae) is a deciduous tree found at 100–2 000 m elevation in Gansu, Guizhou, Henan, Hubei, Hunan, Shaanxi, Sichuan, Yunnan and Zhejiang, and also cultivated in Anhui, Beijing Shi, Gansu and Shaanxi. The bark, which contains aucubin, is used in TCM as an invigorator, a tonic for arthritis and for reducing blood pressure. The solidified latex is used as a dental filling. In recent years, the leaves have been used to make a medicinal tea for reducing blood pressure and improving immunity.

Agrimonia pilosa.—Agrimonia pilosa (agrimony; long ya cao; Rosaceae) is found throughout China. The above-ground parts are used in TCM to treat hemoptysis, malaria, dysentery, carbuncles, and other maladies. In Wa medicine, the whole plant is used against blood clots.

Rosa laevigata.—*Rosa laevigata* (Cherokee rose; jin ying zi; Rosaceae) is found at 200–1 600 m in southern, eastern, central and southwestern China. The roots, leaves and fruit are all used in TCM and minority medical systems. The highly-expensive fruits are used to make a medicinal tea used as a general tonic. In TCM, the fruits are used to treat spermatorrhea and night sweats.

Sanguisorba officinalis.—*Sanguisorba officinalis* (great burnet; di yu; Rosaceae) occurs in almost all of China. The root is used in TCM to treat hematochezia, dysentery and fevers. It is used by the Bai, Dong, Jingpo, Korean, Miao, Mongolian, Yao, Yi, Zang and Zhuang people to treat bleeding of the gastrointestinal tract, hemorrhoids, dyspepsia, stomach aches and fevers.

Astragalus.—Many species of *Astragalus* (huang qi shu; Fabaceae) are important medicinal plants in China. *Astragalus mongholicus* (meng gu huang qi) is found on steppe, in meadows, xerophytic shrubland and coniferous forests at 800–2 000 m in northwestern and northeastern China. The roots are used in TCM to treat chronic nephritis, proteinuria, glycuresis and lack of energy. At least 31 species of *Astragalus* are also widely used in minority medical traditions.

Glycyrrhiza.—The roots and rhizomes of *G. glabra* (liquorice; yang gan cao; Fabaceae), *G. inflata* (zhang guo gan cao) and *G. uralensis* (Chinese liquorice; gan cao; Figure 18.9) are important medicinal ingredients in TCM, widely used to treat exhaustion, palpitations and shortness of breath. These species are also used in Achang, Dai, Deang, Hui, Mongolian, Uyghur and Zang medicine. Many species are used for extraction of saponins by the pharmaceutical industry.

Figure 18.9 *Glycyrrhiza uralensis*. A, foliage. B, fruit.

Sophora flavescens.—*Sophora flavescens* (ku shen; Fabaceae) is found in fields and scrub below 1 500 m throughout China. The roots are used in TCM to treat eczema, hematochezia, itching and trichomoniasis. They are widely used in Achang, Bai, Deang, Done, Jingpo, Korean, Lisu, Miao, Mongolian, She, Tujia, Yi and Zang medicine. The plant is rich in alkaloids and flavonoids, which are extracted by the pharmaceutical industry.

Phellodendron amurense.—*Phellodendron amurense* (Amur cork tree; huang bo; Rutaceae) is a tree of montane and valley forests in northeastern, parts of northern, eastern and central China. It is also widely cultivated. The bark is used in TCM to treat morbid leucorrhoea, jaundice, dermatophytosis, spermatorrhea and infections. The bark is used in Man, Tujia and Zang medicine to treat scabies, ringworm, diabetes and other diseases, while the seeds are used to treat urinary problems and shortness of breath.

Croton tiglium.—*Croton tiglium* (purging croton; ba dou; Euphorbiaceae) is a treelet found in open montane forest and shrubland on limestone at 300–700 m in southern, southwestern and parts of eastern China. The fruits are used in TCM to treat skin ulcers and warts. The roots and bark are used to treat fractures and rheumatism in Achang, Hani, Yao and Zhuang medicine, the fruits to treat shortness of breath by the Dai people, and the fruits and seeds to treat dyspepsia in Mongolian and Zang medicine.

Euphorbia.—Many, perhaps most, species of *Euphorbia* (da ji shu; Euphorbiaceae) have been used to some extent medicinally. Five species are mentioned in the *National Pharmacopoeia of China* (Editor Committee of National Pharmacopeia, 2005), while it is reported in the *Zhongguo Minzu Yao Zhi Yao* (Jia & Li, 2005) that at least 18 species are used widely in many minority medical systems, such as those of the Lisu, Miao, Naxi, Yao and Zang. *Euphorbia kansui* (gan sui) is a herb of fields, roadsides and low mountain slopes in Gansu, Henan, Ningxia, Shaanxi and Shanxi. The root is used in TCM to treat edema, epilepsy, urinary obstruction and constipation. *Euphorbia lathyris* (caper spurge; xu sui zi) is a common weed of cultivated land, found in almost all parts of China. The seeds, known as "qianjinzi," are used in TCM to treat edema, amenorrhea, scabies and other maladies.

Celastrus orbiculatus.—*Celastrus orbiculatus* (oriental bittersweet; nan she teng; Celastraceae) is found in mixed forests at 400–2 200 m in eastern, central, northern, northeastern, parts of northwestern China and Sichuan. The stem is used in TCM to treat joint and muscle pain, numbness of the limbs and dysentery, and in Achang, Deang, Jingpo, Korean, Miao, Mongolian and Yi medicine for rheumatoid arthritis, fractures, amenorrhea, snake bites and icteric hepatitis.

Maytenus hookeri.—*Maytenus hookeri* (mei deng mu; Celastraceae; Figure 18.10), a shrub found at altitudes of 600–1 200 m in Yunnan, like many species of *Maytenus*, contains the anti-cancer constituents maytansine, maytanprine and maytanbutine, which are extracted by the pharmaceutical industry.

Figure 18.10 *Maytenus hookeri*. A, flowering branch. B, fruiting branch.

Tripterygium wilfordii.—*Tripterygium wilfordii* (thunder god vine; lei gong teng; Celastraceae), a toxic, deciduous subshrub, is found in mixed forests up to 3 500 m in southwestern, southern, eastern and parts of northeastern China. The roots are used in TCM as an insecticide, detoxicant and anti-inflammatory. They are used in She medicine to treat rheumatoid arthritis, phthisis, ischialgia and skin problems.

Ziziphus jujuba.—*Ziziphus jujuba* (jujube; zao; Rhamnaceae) is a small deciduous tree found wild and widely cultivated at altitudes below 1 700 m in nearly all parts of China. The fruits (as well as being eaten; see Chapter 16) are used in TCM to improve circulation, as a tranquilizer, and to improve the vital energy known in TCM as "qi," and are also made into medicinal foods. In Miao, Mongolian and Zang medicine they are used to treat hypochondria, insomnia, night sweats and dystrophy.

Althaea officinalis.—*Althaea officinalis* (marshmallow; yao kui; Malvaceae) is a perennial herb native to riverbanks in Xinjiang, and cultivated in Beijing Shi, Jiangsu, Shaanxi and Yunnan. The roots are used in Uyghur medicine to treat colds, coughs, dehydration and boils.

Viola philippica.—*Viola philippica* (zi hua di ding; Violaceae) is a perennial herb found at altitudes below 1 700 m throughout most of China. The whole plant is used in TCM to treat boils, infections, erysipelas and snake bites. It is used in Dong, Lisu, Miao, Mongolian, Pumi, She and Wa medicine to teat icteric hepatitis, mastitis, epipephysitis and other diseases.

Daphne giraldii.—*Daphne giraldii* (huang rui xiang; Thymelaeaceae) is a deciduous shrub found in forest margins and shrubland at 1 600–3 100 m in Gansu, Heilongjiang, Liaoning, Qinghai, Shanxi, Sichuan and Xinjiang. It is also widely cultivated. The velamen and bark (known as "zu shi ma") are used in TCM to treat rheumatism, headaches, stomach aches and fractures. The fruits, flowers, bark and roots are used in Zang medicine to treat syphilis and hydrarthrosis. In Hui medicine, the roots and bark are used to treat rheumatism, rheumatoid arthritis, headaches and stomach aches.

Stellera chamaejasme.—*Stellera chamaejasme* (lang du; Thymelaeaceae) is a toxic perennial herb found on dry, sunny slopes and in sandy places at 2 600–4 200 m in northern and southwestern parts of China. The root is used in TCM to treat abdominal distension, chronic bronchitis, coughs and skin problems. It is also widely used in Lisu, Mongolian, Naxi, Oroqen, Pumi, Yugu, Zang and Zhuang medicine to treat scrofula, fractures, pneumonia and skin problems.

Eleutherococcus senticosus.—*Eleutherococcus senticosus* (ci wu jia; Araliaceae; Figure 18.11) is a shrub found below 2 000 m in northeastern, northern and central China, northern Shaanxi and Sichuan. The roots, rhizomes and stems are used in TCM to treat loss of appetite, soreness of the waist or knee, insomnia and disturbing dreams. They are used in Korean and Mongolian medicine to treat neurasthenia, diabetes mellitus, erectile dysfunction and rheumatoid arthritis.

Panax.—Nearly all species of *Panax* (ginseng; ren shen shu; Araliaceae) in China are used medicinally. *Panax ginseng* (ren shen), found in the forests of northeastern China, is cultivated here and in Hebei and Shanxi. The roots and rhizomes are important in TCM as a tonic. *Panax notoginseng* (Tienchi ginseng; san qi) is found in forests at 1 200–1 800 m in southeastern Yunnan and is cultivated in Fujian, southwestern Guangxi, Jiangxi and Zhejiang. The roots and rhizomes are used in TCM to treat hemoptysis, fractures, hematochezia and stabbing thoraco-abdominal pain. Species of *Panax* (Figure 12) are widely used in minority medical systems across China.

Figure 18.12 *Panax pseudoginseng.*

Figure 18.11 *Eleutherococcus senticosus.*

Angelica.—Many species of *Angelica* (dang gui shu; Apiaceae) are commonly used in China. *Angelica dahurica* (bai zhi) is found at 500–1 000 m in northern and northeastern China, Shaanxi, and northern Taiwan, and is widely cultivated in northern China. The roots are used in TCM to treat colds, headaches, rhinitis and other complaints. In Dong and Molao medicine, *A. dahurica* is used for colds, headaches, dysmenorrhea and chest pain. *Angelica sinensis* (female ginseng; dang gui) is found in forests and shrubby thickets at 2 500–3 000 m as well as cultivated in Gansu, Hubei, Shaanxi, Sichuan and Yunnan. Its roots are used in TCM to treat menstrual problems, amenorrhea, dysmenorrhea, fractures and rheumatism, and similarly in Achang, Jingpo, Miao, Mongolian Yi and Zang medicine.

Bupleurum.—The roots of several species of *Bupleurum* (chai hu shu; Apiaceae), such as *B. chinense* (bei chai hu) and *B. scorzonerifolium* (hong chai hu), both found mainly in northern, eastern, northeastern, northwestern and central China, are famous for their use in TCM (where they are known as "chai hu") for the treatment of coughs, fevers and influenza. The *Zhongguo Minzu Yao Zhi Yao* records that at least 15 species of *Bupleurum* are used in Bai, Dai, Lisu, Miao, Mongolian, She, Yi and Zang medicine (Jia & Li, 2005). Several species are cultivated for commercial use in parts of northwestern China.

Cnidium monnieri.—*Cnidium monnieri* (she chuang; Apiaceae) is an annual herb of riparian grasslands and field margins throughout most of China. The fruits are used in TCM, Mongolian, Tujia and Zang medicine to treat erectile dysfunction, vulval eczema and trichomoniasis.

Ligusticum.—Many species of *Ligusticum* (gao ben shu; Apiaceae) are widely used medicinally in China. *Ligusticum sinense* (Chinese lovage; gao ben) is found in southwestern and northern China, Gansu, Henan, Jiangxi and Shaanxi, and is also cultivated. Its roots and rhizomes are used in TCM to treat colds, headaches and rheumatism. The cultivar "chuanxiong" is cultivated in Gansu, Henan, Hubei, Nei Mongol, Shaanxi and western Sichuan. It roots and rhizomes are used in TCM to treat menstrual problems including amenorrhea and dysmenorrhea, rheumatism and fractures. Both *Ligusticum sinense* and the chuanxiong cultivar are used in Achang, Deang, Dong Jingpo, Lisu, Miao, Mongolian, She and Yi medicine to treat colds, fractures and rheumatism.

Saposhnikovia divaricata.—*Saposhnikovia divaricata* (fang feng; Apiaceae) is a herbaceous perennial of scrub, grasslands and slopes at 400–800 m in Gansu, Hebei, Heilongjiang, Jilin, Liaoning, Nei Mongol, Ningxia, Shaanxi, Shandong and Shanxi. The root is used in TCM and in Dai, Deang, Jingpo and Mongolian medicine to treat colds, headaches, rheumatism and tetanus.

Forsythia suspensa.—*Forsythia suspensa* (lian qiao; Oleaceae) occurs from 300 to 2 200 m in Anhui, Hebei, Henan, Hubei, Shaanxi, Shandong, Shanxi and Sichuan, and is cultivated in Jiangsu. The fruits are used in TCM to treat colds, erysipelas, hyperpyrexia and other complaints, in Achang, Deang and Jingpo medicine to treat colds and acute nephritis, and in Miao and Mongolian medicine against fever, erysipelas and urinary tract infections.

Ligustrum lucidum.—*Ligustrum lucidum* (Chinese privet; nu zhen; Oleaceae) occurs in woods below 2 900 m in eastern, southern, central and part of northwestern China, and is widely cultivated in eastern China as an ornamental tree. The fruit is used in TCM, Achang, Deang, Dong, Jingpo, Jinuo, Lisu, Miao, Mongolian and Shui medicine to treat vertigo, tinnitus, age-related blindness and similar conditions.

Gentiana.—The roots of *Gentiana crassicaulis* (cu jing qin jiao; Gentianaceae), found at 2 100–4 500 m in northwestern Guizhou, western Sichuan, southeastern Xizang and northwestern Yunnan, *G. dahurica* (da wu li qin jiao), *G. macrophylla* (qin jiao), found at 400–3 700 m in Hebei, Nei Mongol, Ningxia, Shaanxi, Shandong and Shanxi, and *G. straminea* (ma hua jiao), found at 2 000–5 000 m in Gansu, Hubei, Ningxia, Qinghai and Xizang, are used in TCM to treat rheumatism, joint pain, malnutrition and childhood indigestion. Many species of *Gentiana* (gentian; long dan shu) are widely used in minority medical systems in China. *Gentiana crassicaulis* is cultivated for commercial use in northwestern Yunnan.

Apocynum venetum.—*Apocynum venetum* (dogbane or Indian hemp; luo bu ma; Apocynaceae) is a perennial herb of saline regions, desert margins and riversides in northwestern, northern, northeastern, part of eastern China and Xizang. The leaves are used in TCM to treat palpitations, insomnia, hypertension, neurasthenia and nephritic edemas. The leaves alone or the whole plant may be used in Mongolian, Uygur and Yugu medicine to treat hypertension, insomnia, neurasthenia and cirrhosis.

Rauvolfia.—The roots of *Rauvolfia serpentina* (snakeroot; she gen mu; Apocynaceae), found in montane forests at 800–1 500 m in southern Yunnan, cultivated in southern Guangdong, southern Guangxi and Hainan, are used as a sedative and for the treatment of hypertension. The bark, leaves and roots are used against snake bites and scorpion stings. *Rauvolfia verticillata* (luo fu mu; Figure 18.13), from southwestern and southern China, is used against snake bite, malaria and typhus, while the roots are used to treat hypertension and as a sedative. The roots of several species of *Rauvolfia* (luo fu mu shu) are used industrially for the extraction of reserpine, and many species are also used in Dai, Jinuo, Lahu, Wa, Yao, Yi and Zhuang medicine to treat fractures, stomach ache, hepatitis, insomnia, malaria and snake bites.

Figure 18.13 *Rauvolfia verticillata.*

Cynanchum.—*Cynanchum atratum* (bai wei; Asclepiadaceae) is a herb of grasslands and riversides up to 2 000 m in southern, southwestern, eastern, northern, northeastern and central China and Shaanxi. Its roots and rhizomes, as well as those of *C. versicolor* (bian southeastern bai qian), are used in TCM to treat fevers, weakness after childbearing, and ulcerous infections, and in Achang, Deang, Dong, Hani, Jingpo, Lisu, Miao, Mongolian, Shui and Zhuang medicine to treat snake bites, hepatitis, malaria, eczema, rheumatoid arthritis and phthisis. *Cynanchum paniculatum* (xu chang qing) is a herbaceous perennial found on sunny slopes throughout China. The roots and rhizomes are used in TCM to treat rheumatism, stomach ache, toothache, fractures and eczema, and in Dong, Miao, Mongolian, Shui, Tujia, Yao, Yi and Zhuang medicine to treat itching, dysmenorrhea, rheumatism, dysentery, snake bites and menstrual problems.

Cuscuta chinensis.—*Cuscuta chinensis* (tu si zi; Convolvulaceae) is a parasitic herb of Asteraceae, Fabaceae and Zygophyllaceae, among others, found up to 3000 m throughout China. The seeds are used in TCM to treat erectile dysfunction, spermatorrhea, frequent urination and tinnitus, and in Dai, Dong, Mongolian, Yi and Zang medicine to treat hematemesis, hepatitis and boils, among other illnesses.

Arnebia euchroma.—*Arnebia euchroma* (ruan zi cao; Boraginaceae) is a herb of rocky slopes and meadows in Xinjiang and western Xizang. The roots, as well as those of *A. guttata* (huang hua ruan zi cao), are known as "zi cao" in TCM and are used to treat eczema, burns and scalds and skin diseases. In Mongolian medicine they are used in cases of acute and chronic hepatitis, frostbite, eczema, burns and scalds.

Verbena officinalis.—*Verbena officinalis* (common vervain; ma bian cao; Verbenaceae) is found in grassy places on mountain slopes up to 1 800 m throughout most of China except the northeast. The above-ground parts are used in TCM to treat amenorrhea, dysmenorrhea and malaria. *Verbena officinalis* is widely used in Achang, Benglong, Bulang, Buyi, Dai, Deang, Dong, Hani, Jingpo, Jinuo, Korean, Lahu, Lisu, Miao, Molao, Naxi, Pumi, Tujia, Wa, Yao, Yi, Zang and Zhuang medicine to treat malaria, hepatitis, dysentery, paristhmitis, fractures and acute gastritis, among other complaints.

Lamiophlomis rotata.—*Lamiophlomis rotata* (du yi wei; Labiatae) is a prostrate, rosettiform herb of weathered alpine alluvial fans, stony alpine meadows and floodplains at 2 700–4 900 m in Gansu, Qinghai, Sichuan, Xizang and Yunnan. The whole plant is widely used in Achang, Deang Mongolian, Naxi, Pumi and Zang medicine to treat fractures, arthritis, osteomyelitis and other conditions. The *National Pharmacopeia of China* (Editor Committee of National Pharmacopeia, 2005) also describes indications including fractures and rheumatism.

Leonurus japonicus.—*Leonurus japonicus* (Chinese motherwort; yi mu cao; Lamiaceae) is found in sunny places up to 3 400 m throughout China. The mature fruits, known as "chong wei zi" in TCM, are used to treat amenorrhea, dysmenorrhea and light-headedness. The above-ground parts are used to treat amenorrhea, dysmenorrhea and acute nephritis. The species is also widely used in Bai, Dai, Dong, Hani, Miao, Mongolian, Tujia, Yi, Zang and Zhuang medicine.

Mentha canadensis.—*Mentha canadensis* (bo he; Lamiacaee) is found wild in wet areas below 3 500 m and cultivated throughout China. The above-ground parts are used in TCM to treat colds, headaches, measles and aphtha. It is also widely used in Dai, Hani, Maonan, Miao, Mongolian, She, Tujia, Wa, Yao, Yugu and Zang medicine to treat typhoid, sore throats, urticaria, itching and other complaints.

Prunella vulgaris.—*Prunella vulgaris* (common self-heal; xia ku cao; Lamiaceae) is a perennial herb found below 3 200 m throughout China. Mature infructescences are used in TCM to treat red, swollen and painful eyes, light-headedness, scrofula, hypertension, and other disorders. In Achang, Bai, Deang, Dong, Jingpo, Lisu, Shui, Tujia and Yi medicine it is used to treat hemostasis, burns and scalds, goiter, dysentery and stomatitis.

Salvia.—Many species of *Salvia* (shu wei cao shu; Lamiaceae) are used medically in China. For example, the roots and rhizomes of *Salvia miltiorrhiza* (red sage; dan shen), found on hillsides, streamsides and in forests up to 1 300 m in eastern, northern and central China and Shaanxi, are used in TCM to treat amenorrhea, dysmenorrhea, insomnia and stenocardia. The *Zhongguo Minzu Yao Zhi Yao* records at least 15 species of *Salvia* used in Bai, Jingpo, Lisu, Miao, Mongolian, Naxi, Yi and Zang medicine.

Scutellaria baicalensis.—One of the most commonly-collected species in China is *Scutellaria baicalensis* (Baikal skullcap; huang qin; Lamiaceae), a perennial herb of sunny grassy slopes, wasteland and cultivated areas up to 2 000 m in northwestern, northeastern, northern and part of eastern China. The roots are used in TCM to relieve fever and inflammation, and in Mongolian and Oroqen medicine to treat respiratory tract infections, hypertension, hepatitis, dysentery and pneumonia.

Anisodus.—Nearly all species of *Anisodus* (shan lang dang shu; Solanaceae) are used medicinally in China. *Anisodus acutangulus* (san fen san) is found in sparse forests, roadsides and fields in Yunnan. Its roots, leaves and seeds are all used medically in folk medicine to treat fractures, rheumatism and stomach aches. The rhizomes of *A. carniolicoides* (sai lang dang), found at altitudes of 3 000–4 500 m in southeastern Qinghai, Sichuan, and northwestern Yunnan, are used medicinally to treat injuries, bleeding and pain associated with rheumatism and arthritis. The roots and seeds of *A. luridus* (ling dang zi) are used for alleviating pain and spasms, while the roots of *A. tanguticus* (shan lang dang) relieve pain. Many species of *Anisodus* are widely used in minority medical systems.

Datura.—The flowers of *Datura metel* (devil's trumpet; yang jin hua; Solanaceae; Figure 18.14), found on grassy and sunny slopes and around human habitation, and also commonly cultivated in many cities of southern and southeastern China and Yunnan, are used in TCM to treat asthma, stomach pain and rheumatism, and as an anesthetic during surgery. In Bulang, Dai, Lisu, Miao and Zhuang medicine they are also used to treat rheumatism, fractures, ischiatitis and other maladies. All parts of *Datura stramonium* (Jimson weed; man tuo luo) are used to treat fractures, rheumatism, chronic bronchitis and dermatophytosis in Bai, Dai, Deang, Dong, Jingpo, Lisu, Miao, Naxi, Yao, Yi and Zang medicine.

Figure 18.14 *Datura metel* flowers.

Hyoscyamus niger.—*Hyoscyamus niger* (henbane; tian xian zi; Solanaceae) is a herb found near habitation, and occasionally cultivated, at 700–3 600 m in northwestern, southwestern, northern, northeastern and eastern China. The seeds are used in TCM to treat gastrointestinal spasms, coughs and dementia. In Mongolian and Zang medicine they are used against stomach aches, rheumatoid arthritis, asthma and other maladies, while the pharmaceutical industry extracts atropine from this species.

Lycium.—The fruits of *L. barbarum* (wolfberry or goji berry; ning xia gou qi; Solanaceae), found in Ningxia, are a well-known tonic in TCM, and can also be used as a tea. The roots, together with the velamen (known as "di gu pi") of *L. chinense* (also known as wolfberry or goji berry; gou qi), are used in TCM to treat night sweats, coughs, hemoptysis and diabetes. Species of *Lycium* (gou qi shu) are used in Deang, Dong, Jingpo, Korean, Mongolian, She, Yi and Zang medicine to treat neurasthenia, diabetes mellitus, insomnia, and other complaints.

Rehmannia glutinosa.—*Rehmannia glutinosa* (di huang; Scrophulariaceae) is a perennial herb of mountain slopes and tracksides up to 1 100 m in Gansu, Hebei, Henan, Hubei, Jiangsu, Liaoning, Nei Mongol, Shaanxi, Shandong and Shanxi. It is widely cultivated for medical use. The roots are used in TCM to treat bad indigestion and hematemesis. In Dai, Miao and Mongolian medicine they are used in the treatment of sore throats, hematochezia and ringworm, among other complaints.

Scrophularia ningpoensis.—*Scrophularia ningpoensis* (Chinese figwort; xuan shen; Scrophulariaceae) is a perennial herb of bamboo forests and tall grass below 1 700 m altitude in Anhui, Fujian, Guangdong, Guizhou, northern Hebei, Henan, Jiangsu, Jiangxi, southern Shaanxi, Shanxi, Sichuan and Zhejiang. The roots are used in TCM to treat shortness of breath, eye problems, pharyngalgia and diphtheria. In Miao and Mongolian medicine they are used to treat scrofula and boils.

Andrographis paniculata.—*Andrographis paniculata* (chuan xin lian; Acanthaceae) is an annual herb found wild in Yunnan and cultivated in Fujian, Guangdong, Hainan, Guangxi and Yunnan. The above-ground parts are used to treat colds, fevers, sore throats, coughs, infections, abscesses and snake bites. In Achang, Jingpo and Miao medicine they are used against mumps, paristhmitis, pharyngitis, tracheitis, otitis media, and other diseases.

Strobilanthes cusia.—*Strobilanthes cusia* (ban lan gen; Figure 18.15; Acanthaceae) is a perennial herb found in southwestern China and widely cultivated in southeastern Yunnan. The whole plant is used for the extraction of a compound known as "qing dai," which is used in TCM to treat hemoptysis, aphtha and pharyngitis. It is widely used in Bai, Bulang, Dai, Dong, Hani, Jinuo, Lisu, Miao, Wa, Yao and Zhuang medicine to treat fractures, flu, fever, tonsillitis and other complaints.

Morinda officinalis.—*Morinda officinalis* (Indian mulberry; ba ji tian; Rubiaceae; Figure 18.16) is a liana

Figure 18.15 *Strobilanthes cusia*.

Figure 18.16 *Morinda officinalis*, flowering branch.

of the montane forests of Fujian, Guangdong, Guangxi and Hainan. The roots are used in TCM to treat erectile dysfunction, spermatorrhea, menstrual problems and rheumatism. In Mongolian medicine they are used to treat joint pain, urinary incontinence and uterine problems.

Uncaria.—The branches of *U. macrophylla* (da ye gou teng; Rubiaceae; Figure 18.17), found in the canopy of secondary forests in Guangdong, Guangxi, Hainan and Yunnan, *U. rhynchophylla* (cat's claw; gou teng), found in open forests, thickets and on streamsides in Guangdong, Guangxi, Yunnan, Guizhou, Fujian, Hunan, Hubei and Jiangxi, *U. sessilifructus* (bai gout eng), found in dense forest or valley thickets in Guangxi and Yunnan, and *U. sinensis* (hua gout eng), found in secondary forests at mid altitudes in Gansu, Guangxi, Guizhou, Hubei, Hunan, Shaanxi, Sichuan and Yunnan, are all used in TCM to treat hypertension, eclampsia gravidarum, headaches and dizziness. The *Zhongguo Minzu Yao Zhi Yao* (Jia & Li, 2005) records at least 11 species of *Uncaria* (gou teng shu) used widely in minority medical systems in China against complaints including rheumatoid arthritis, tracheitis, ischialgia and fractures.

Figure 18.17 Foliage of *Uncaria macrophylla.*

Lonicera.—Many species of *Lonicera* (honeysuckle; ren dong shu; Caprifoliaceae) are used medicinally in China. The flowers ("jin yin hua"), of *L. japonica* (Japanese honeysuckle; ren dong), are used in TCM against colds, erysipelas, fevers and boils, while the stems are used to treat boils, fevers and swollen joints. The flowers, known as "shan yin hua," of *L. confusa* (soft-leaved honeysuckle; hua nan ren dong), *L. hypoglauca* (hong xian ren dong) and *L. macrantha* (da hua ren dong), are used in TCM to treat colds, erysipelas and fevers. Many species of *Lonicera* are widely used in minority medical systems.

Nardostachys jatamansi.—Nardostachys jatamansi (spikenard; gan song; Valerianaceae) is a perennial herb of alpine thickets or meadows at 2 500–5 000 m in southeastern Gansu, southern Qinghai, western Sichuan, Xizang and northern Yunnan. The roots and rhizomes are used in TCM to treat abdominal swelling or distension, loss of appetite, vomiting and toothache. They are used in Mongolian, Uyghur and Zang medicine to treat complaints including flu, fever and hypertension.

Gynostemma pentaphyllum.—Gynostemma pentaphyllum (jiaogulan; jiao gu lan; Cucurbitaceae; Figure 18.18) is found in forests, thickets, on roadsides and mountains in eastern, southern, southwestern and central China. The whole plant or the leaves alone are used in folk medicine to reduce blood pressure, blood fat and blood sugar. In Achang, Bai, Deang, Hani, Jingpo, Jinuo and Yao medicine it is used against hepatitis, pyelonephritis, chronic bronchitis and fractures. The leaves are used as an immune-boosting medicinal tea.

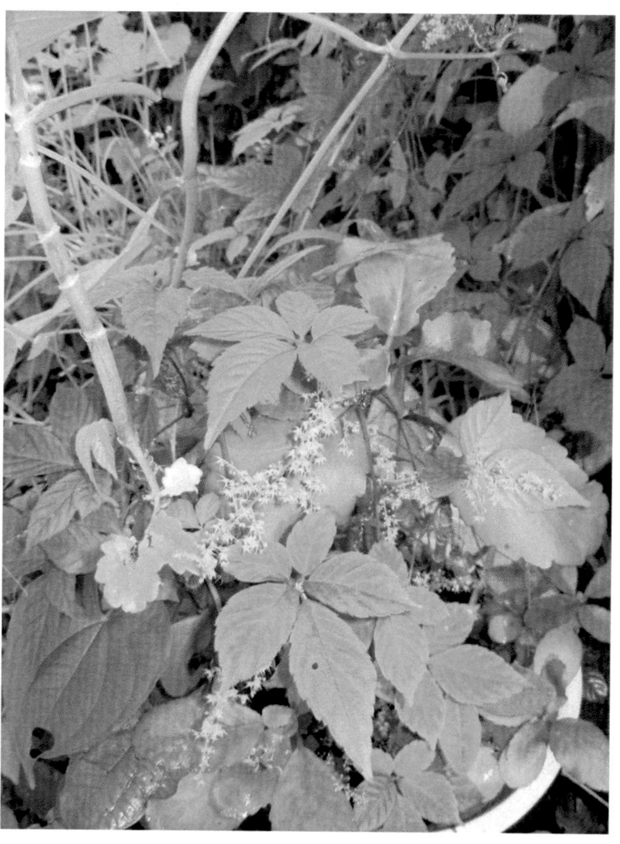

Figure 18.18 *Gynostemma pentaphyllum.*

Siraitia grosvenorii.—Siraitia grosvenorii (luo han guo; Cucurbitaceae) is found in Guangdong, Guangxi, Guizhou, southern Hunan and Jiangxi. The fruits are used in TCM for dry coughs, pharyngalgia and shortness of breath. In Achang, Dong, Miao, Yao and Zhuang medicine they are also used against coughs, tracheitis, phthisis, bronchitis and shortness of breath.

Trichosanthes kirilowii.—Trichosanthes kirilowii (gua lou; Cucurbitaceae) is a climbing herb of open forests, grasslands and fields from 200 to 1 800 m altitude in Gansu, Hebei, Henan, Jiangsu, Jiangxi, Shandong, Shanxi and Zhejiang. The dried roots, known as "tian hua fen" in TCM, are used to treat coughs, diabetes, swelling and skin ulcers, while the seeds, "gua lou zi," are used against dry coughs, intestinal problems and constipation. The peel, "gua lou pi," is also used to treat coughs, chest pain and hypochondria. The species is widely used in Dong, Miao, Mongolian, She and Yi medicine to treat constipation, angina, diabetes and sore throats.

Adenophora.—The roots, known as "nan sha shen," of *Adenophora stricta* (ladybell; sha shen; Campanulaceae), found below 3 300 m in eastern, southwestern, central and parts of northwestern China, and *A. tetraphylla* (fourleaf ladybell; lun ye sha shen), found throughout most of the country, are used in TCM and in Dong, Miao and Tujia medicine to treat coughs, dryness of the mouth and general dysphoria. Many species of *Adenophora* (sha shen shu) are used in Achang, Jingpo, Mongolian and Yi and medicine to treat leprosy, arthritis, psoriasis, and other diseases.

Codonopsis.—Nearly all species of *Codonopsis* (dang shen shu; Campanulaceae) are used medically in China. *Codonopsis pilosula* (poor man's ginseng; dang shen), found in forests, thickets and meadows at 900–3 900 m in northeastern, northwestern, eastern, northern China and parts of southwestern China, and *C. pilosula* subsp. *tangshen* (chuan dang shen), found in scrub and on forest margins at 900–2 300 m in Chongqing Shi, northern Guizhou, western Hubei, northwestern Hunan, southern Shaanxi and northeastern Sichuan and cultivated on a large scale, are important in TCM. Their roots are used as tonics to improve general health. Many species of *Codonopsis* are widely used in minority medical traditions, such as those of the Achang, Deang, Bai, Jingpo, Lahu, Lisu, Miao, Mongol, Yi and Zang, to treat rheumatoid arthritis, skin problems, loss of appetite, and other complaints.

Platycodon grandiflorus.—*Platycodon grandiflorus* (jie geng; Campanulaceae) is a perennial herb of open places below 2 000 m throughout most of China. The roots are used in TCM to treat coughs, sore throats and lung infections. They are used in Dong, Mongolian, Yao and Zhuang medicine to treat respiratory diseases and headaches caused by colds. This plant is cultivated in many parts of China.

Artemisia.—Many species of *Artemisia* (hao shu; Asteraceae) are rich in polyacetylenes, flavonoids, terpenoids and cyanogenic glycosides and are well known medicinal plants. The above-ground parts of *A. annua* (sweet wormwood; huang hua hao), widespread in China except Xizang, are used to treat malaria, fevers and jaundice in TCM. The leaves of *A. argyi* (ai), widespread across the country, are used to treat menstrual problems, infertility and skin problems, while those of *A. capillaris* (yin chen hao) are used for icteric hepatitis and itching. Many species of *Artemisia* are used widely in other ethnomedicinal traditions of China.

Saussurea.—Many species of *Saussurea* (feng mao ju shu; Asteraceae) are found at high altitudes and are in need of conservation measures. The above-ground parts of *S. involucrata* (xue lian hua) are used in TCM to treat rheumatoid arthritis and menstrual problems. *Saussurea medusa* (shui mu xue tu zi), from rocky and scree slopes at 3 000–5 600 m in Gansu, Qinghai, Sichuan, Xizang and Yunnan, is an important medicinal plant in Zang medicine, used in the treatment of gynecological complaints, rheumatism, rheumatoid arthritis and apoplexy. Many species are used widely in Dong, Lisu, Miao, Mongolian, Naxi, Pumi, Uyghur, Yi, Yugu and Zang medicine against hepatitis, menstrual problems and rheumatoid arthritis.

Taraxacum.—Many species of *Taraxacum* (dandelion; pu gong ying shu; Asteraceae) are used medicinally in China. The whole plants of *T. mongolicum* (Chinese dandelion; pu gong ying), found throughout most of the country, *T. sinicum* (hua pu gong ying), found mainly in salt marshes and fields at 300–2 900 m in most of northern China, and other species, are used in TCM to treat boils, infections, acute mastitis, pharyngalgia and jaundice. In Dai, Lisu, Miao, Oroqen, Tujia, Yi and Zang medicine, many species of *Taraxacum* are used to treat upper respiratory tract infections, mastitis, tonsillitis, mumps, ulcers, and other complaints.

Coix lacryma-jobi.—*Coix lacryma-jobi* (Job's tears; yi yi; Poaceae) is an annual herb of wet places near settlements, and often cultivated in eastern, southern, parts of northern, northwestern and northeastern China. The seeds are used in TCM to treat liver and urinary problems, ringworm and pulmonary abscesses. In Dong, Miao, Mongolian, Yi and Zang medicine they are used to treat rheumatism, gonorrhea and cirrhosis, often in the form of a tonic wine.

Arisaema.—Many species of *Arisaema* (tian nan xing shu; Araceae) are of medical importance in China, including the tubers of *A. amurense* (dong bei nan xing), found in forests below 200 m in eastern China, Hebei, Henan, Nei Mongol, Ningxia, Shandong and Shanxi and *A. erubescens* (yi ba san nan xing), found throughout most of the country. *Arisaema heterophyllum* (tian nan xing), found in most areas below 2 700 m except in Xizang, is used in TCM to treat coughs, facial paralysis, hemiplegia, epilepsy, tetanus and snake bites. Many species of *Arisaema* are used widely in other ethnomedicinal traditions to treat complaints including parotiditis, neurodermatitis, fractures and chronic bronchitis.

Pinellia.—*Pinellia ternata* (crow-dipper; ban xia; Araceae) is widespread below 2 500 m in China, excluding Nei Mongol, Qinghai, Xinjiang and Xizang, and is a common remedy in TCM. The tubers are used to reduce phlegm and relieve coughs, and as an antiemetic. *Pinellia cordata* (di shui zhu) is used in Tujia and Yi medicine to treat problems including fractures, vomiting and tuberculosis.

Anemarrhena asphodeloides.—*Anemarrhena asphodeloides* (zhi mu; Liliaceae) is a perennial herb found, sometimes cultivated, below 1 500 m in northeastern, most of northwestern, southwestern, northern and eastern China. It is also cultivated in Taiwan. The rhizomes are used in TCM to treat fevers, shortness of breath and coughs, and in Deang, Jingpo and Mongolian medicine to treat night sweats, glycuresis and shortness of breath.

Lilium. Many species of *Lilium* (bai he shu; Liliaceae) are used medicinally in China. For example, the bulbs of *L. brownii* var. *viridulum* (bai he), found at 300–1 000 m altitude in eastern, southern, central, southwestern and northwestern China, and *L. pumilum* (shan dan), at 400–2 600 m in northeastern, northwestern, some parts of eastern and northern China, are used in TCM to treat coughs, insomnia and absent-mindedness. Many species are used in Bai, Dong, Lisu, Miao, Molao, Mongolian, She, Shui, Tujia, Wa, Yi, Zang and Zhuang medicine as a tonic, cough medicine and sedative. Some species are cultivated in Gansu for commercial use.

Fritillaria.—Several species of *Fritillaria* (bei mu shu; Liliaceae) are cultivated for their bulbs, which are used medicinally. Those of *F. cirrhosa* (chuan bei mu), found from 3 200 to 4 600 m in Gansu, Qinghai, Sichuan, Xizang and Yunnan, *F. delavayi* (suo sha bei mu), *F. przewalskii* (gan su bei mu), found from 2 800 to 4 400 m in southern Gansu, eastern Qinghai and Sichuan, and *F. unibracteata* (an zi bei mu), from 3 200 to 4 700 m in southern Gansu, southeastern Qinghai, and northwestern Sichuan, together supply an important TCM treatment for lung conditions, dry coughs and expectoration. Many species of *Fritillaria* are used in Deang, Jingpo, Kazakh, Korean, Lisu, Miao, Mongolian, Tujia, Uyghur and Zang medicine against diseases of the respiratory system.

Ophiopogon japonicus.—*Ophiopogon japonicus* (mondo grass or fountain plant; mai dong; Liliaceae) is found up to 2 800 m in eastern, southern, southwestern and central China and Shaanxi, and is widely cultivated for medicinal use. The tubers are used in TCM to treat insomnia, shortness of breath and coughs. Many species are used in Bai, Dai, Dong, Lahu, Lisu, Miao, Mongolian, She, Yao, Yi, Zang and Zhuang medicine to treat coughs, shortness of breath, asthma, phthisis and tracheitis.

Paris.—The rhizomes of many species of *Paris* (chong lou shu; Liliaceae; Figure 18.19) are used medicinally in China. *Paris polyphylla* var. *yunnanensis* (dian chong lou) is found in forests at 1 400–3 100 m in southwestern China, and is used in TCM to treat boils, fractures, snake bites and sore throats, as is *P. polyphylla* var. *chinensis* (hua chong lou). Several species are used in Dai, Dong, Jinuo, Miao, Mongolian, Tujia, Wa, Yao and Yi medicine. The market demand for *Paris* is met mainly from cultivated material.

Polygonatum.—Many species of *Polygonatum* (Solomon's seal; huang jing shu; Liliaceae) are used medicinally in China. *Polygonatum odoratum* (scented Solomon's seal; yu zhu) is found in forests from 500 to 3 000 m altitude in eastern, northern, southern, northwestern and parts of northeastern China. The rhizomes are used in TCM to treat coughs, dry throats and thirst. The rhizomes of *P. cyrtonema* (duo hua huang jing), *P. kingianum* (dian huang jing) and *P. sibiricum* (huang jing) are used in TCM to treat spleen and stomach problems, general weakness and blood disorders. The *Zhongguo Minzu Yao Zhi Yao* records ca. ten species used in other ethnomedicinal traditions in China (Jia & Li, 2005).

Figure 18.19 *Paris polyphylla* in fruit.

Dioscorea.—Many species of *Dioscorea* (shu yu shu; Dioscoreaceae) are used medicinally. For example, the rhizomes of *D. futschauensis* (fu zhou shu yu), found in Fujian, northern Guangdong, northeastern Guangxi, Hunan and southern Zhejiang, are used in TCM to treat leukorrhagia, carbuncles and joint pain. The rhizomes, known as "chuan shan long," of *D. nipponica* (chuan long shu yu), found from 100 to 1 800 m in eastern, northern, northeastern and some parts of northwestern China, are used in TCM to treat rheumatoid arthritis, waist pain, leg pain, fractures and chronic bronchitis. The *Zhongguo Minzu Yao Zhi Yao* (Jia & Li, 2005) records at least 15 species of *Dioscorea* used in ethnomedicine in China; many species are also sources of steroids for the pharmaceutical industry.

Belamcanda chinensis.—*Belamcanda chinensis* (leopard lily; she gan; Iridaceae) is found below 2 200 m in nearly all of China, and is also widely cultivated. The rhizomes are used in TCM to treat sore throats, coughs with an abundance of phlegm, and asthma. It is widely used in Bai, Dai, Dong, Miao, Mongolian Mulao, She, Shui, Yao, Zang and Zhuang medicine to treat hepatitis, parotitis and bronchitis, among other complaints.

Alpinia oxyphylla.—*Alpinia oxyphylla* (yi zhi; Zingiberaceae) is a herbaceous perennial of southern China and Yunnan, and cultivated in some parts of this area. The fruits are used in TCM, Mongolian and Zang medicine to treat stomach aches, enuresis, frequent urination and spermatorrhea.

Curcuma phaeocaulis.—*Curcuma phaeocaulis* (e zhu; Zingiberaceae) is found in Yunnan and cultivated in Fujian, Guangdong, Guangxi and Sichuan. The rhizomes are used in TCM to treat amenorrhea, dyspepsia and cervical cancer. It is also widely used in Achang, Dai, Deang, Dong, Hani, Jingpo, Jinuo Lahu, Maonan and Zhuang medicine.

Dendrobium.—Dendrobium (shi hu shu; Orchidaceae) are mostly epiphytic herbs. The dried stems ("shi hu") of *D. catenatum* (huang shi hu) and *D. nobile* (jin chai shi hu) are well-known in TCM for their activity against dry mouths and excessive thirst, weakness and anxiety. According to incomplete statistics, more than ten species of *Dendrobium* are used to supply this medicine in China. Many species are also widely used in Achang, Bai, Dai, Deang, Dong, Jingpo, Jinuo, Lahu, Lisu, Miao, Yao, Yi and Zang medicine to treat fractures, coughs and paralysis. As an important medicinal plant, the relevant species have successfully been cultivated in Yunnan, Hainan, Guizhou and elsewhere.

Gastrodia elata.—Gastrodia elata (tian ma; Orchidaceae) is a terrestrial herb of forests between 400 and 3 200 m across most of China, and is also widely cultivated. The tubers are used in TCM to treat headaches and dizziness, numbness of the limbs, epilepsy and tetanus. It is also widely used in Achang, Bai, Deang, Hani, Jingpo, Lisu, Miao and Mongolian medicine for similar diseases as well as hemiplegia and hypertension. This species has a high economic value due to high market demands, and most of the market demand is supplied from cultivation.

References

Editor Committee of National Pharmacopeia (2005). *Pharmacopeia of the People's Republic of China*. Beijing, China: People's Medical Publishing House.

Hamilton, A. C. (2004). Medicinal plants, conservation and livelihoods. *Biodiversity and Conservation* 13: 1477-1517.

He, S. A. & Gu, Y. (1997). The challenge for the 21st Century for Chinese botanic gardens. In: Touchell, D. H. & Dixon, K. W. (eds.) *Conservation into the 21st Century. Proceedings of the 4th International Botanic Gardens Conservation Congress (Perth, 1995)*. pp. 21-27. Perth, Australia: Kings Park and Botanic Garden.

Jia, M. & Li, X. (2005). *Zhongguo Minzu Yao Zhi Yao*. Beijing, China: China Medical Technology Publishing House.

Jiangsu New Medical College (1997). *Dictionary of Traditional Chinese Medicine*. Shanghai, China: Shanghai People Press.

Luo, D. (1985). Study on the imported Tibetan medicine. *Journal of Chinese Medicinal Materials* 2.

Luo, D. S. (1997). *Tibetan Materia Medica of China*. Beijing, China: Nationality Press.

Pei, S. J. (2000). Application of Ethnobotany in the study of ethno- and folk medicine. *Chinese Journal of Ethnomedicine and Ethnopharmacy* 6: 316-319.

Pei, S. J. (2002). A brief review of ethnobotany and its curriculum development in China. In: Shinwari, Z. K., Hamilton, A. & Khan, A. A. (eds.) *Proceedings of a Workshop on Curriculum Development in Applied Ethnobotany, Nathiagali, 2-4 May 2002*. p. 41. Lahore, Pakistan: WWF-Pakistan.

Pei, S. J. & Huai, H. Y. (2007). *Ethnobotany*. Shanghai, China: Shanghai Science and Technology Press.

Wang, Z. & Liu, L. (2002). *The Chinese Plant Raw Material Medicines*. Kunming, China: Yunnan Science and Technology Press.

Xiao, P. G. & Yong, P. (1998). Ethnopharmacology and research on medicinal plants in China. In: Prendergast, H. D. V., Etkin, N. L., Harris, D. R. & Houghton, P. J. (eds.) *Plants for Food and Medicine. Proceedings from a Joint Conference of the Society for Economic Botany and the International Society for Ethnopharmacology, London, 1-6 July 1996*. pp. 31-39. London, UK: Royal Botanic Gardens, Kew.

Yang, J. S. & Chuchengjiangcuo (1987). *Diqing Tibetan Medicine*. Kunming, China: Yunnan Nationality Press.

Zhang, H., Yuan, C., Sun, C. & Zhang, Z. (1995). Species of traditional Chinese medicine in China. *China Journal of Chinese Material Medica* 20(7): 387-390.

Zhou, H. (1999). Thinking on standardization of plant medicines. *Chinese Pharmaceutical Journal* 34(1): 1-2.

Zhu, T. P. (2004). Plant resource in China. In: Wu, Z. Y. & Chen, X. Q. (eds.) *Flora Reipublicae Popularis Sinicae*. pp. 584-657. Beijing, China: Science Press

Chapter 19

Ornamental Plants

HE Shan-An and XING Fu-Wu

19.1 Introduction

In this chapter we discuss plants cultivated primarily for their ornamental features, whether grown outside or indoors. These include both herbaceous and woody species, selected for their flowers, foliage, fruits, general shape or other characteristics (Editorial Board of Chinese Agriculture Encyclopedia, 1996).

China provides the richest diversity of ornamental and gardening plants in the world, with more than 7 500 popularly-cultivated species native to the country, accounting for 60–70% of the world total (see Table 19.1; Zhu, 2004). An incredible 20 000 species have potential as ornamentals, and 2 000 species have been used for ornamental purposes within China itself. Many popular flowers and ornamentals grown around the world originated or have their natural distribution centers in China. Moreover, rich genetic diversity coupled with a long history of selection and breeding have generated a huge variety of ornamental cultivars, such as *Chrysanthemum* (Asteraceae), with more 3 000 varieties, lotus flower (*Nelumbo nucifera;* Nelumbonaceae), with over 500, azalea (*Rhododendron* spp.; Ericaceae, ca. 500), tree peony (*Paeonia* spp.; Paeoniaceae, more than 460), flowering plum (*Armeniaca mume;* Rosaceae, more 300), *Camellia* (Theaceae, more than 300), herbaceous peony (*Paeonia* spp., more than 200), rose (*Rosa* spp.; Rosaceae, 150), and garden balsam (*Impatiens* spp.; Balsaminaceae, more than 50). The Chinese cultural fascination with the orchid genus *Cymbidium* has resulted in around 100 cultivars of the single species, *Cymbidium ensifolium,* 100 of *Cymbidium goeringii* and 130 of *Cymbidium sinense.* In addition, famous Chinese landscape plants, including *Cathaya argyrophylla* (Pinaceae), *Davidia involucrata* (dove or handkerchief tree; Nyssaceae), *Emmenopterys henryi* (Rubiaceae), *Ginkgo biloba* (Ginkgoaceae), *Metasequoia glyptostroboides* (dawn redwood; Taxodiaceae), *Michelia odora* (Magnoliaceae), *Pseudotaxus chienii* (white-berry yew; Taxaceae) and *Tapiscia sinensis* (Tapisciaceae), are widely grown worldwide.

Plants have been used widely and for a long time to enhance the landscape in China. As long ago as 221 BC, when Emperor YING Zheng of the Qin Dynasty reunited China, he began building the gigantic Epang Palace with its Shanglin Garden. This incorporated many ornamental plants such as *Armeniaca mume, Citrus* spp. (Rutaceae), *Cotinus coggygria* (Anacardiaceae), *Diospyros kaki* (kaki persimmon; Ebenaceae), *Eriobotrya japonica* (loquat; Rosaceae), *Ligustrum lucidum* (Chinese privet; Oleaceae), *Magnolia* spp. (Magnoliaceae) and *Myrica rubra* (Myricaceae). The *Shu Du Fu,* written by YANG Xiong (53 BC–18 AD) records that, "cities grow with cherry, plum and Magnolia," indicating that landscape plants have been used in urban greening in China for more than 2 000 years.

Table 19.1 Significant families of ornamental plants in China.

Family	Number of species/ genera in China	Number of species/ genera worldwide
Aceraceae	101/2	ca. 131/2
Asteraceae	2 336/248	ca. 24 000/1 600–1 700
Caprifoliaceae	66/5	ca. 207/5
Ericaceae	826/22	ca. 4 000/ca. 125
Liliaceae	726/57	ca. 3 500/ca. 250
Magnoliaceae	112/13	ca. 300/17
Nymphaeaceae	8/3	ca. 70/6
Paeoniaceae	15/1	ca. 30/1
Ranunculaceae	921/38	ca. 2 500/ca. 60
Rosaceae	960/55	2 825–3 500/95–125
Theaceae	274/12	600/19

The history of ornamental planting in China can be divided into three periods: the "utility garden," "royal and private garden," and "public garden" periods (Xing *et al.*, 2009). During the utility garden period, lasting from the Neolithic Era to the Warring States periods (475–221 BC), with productivity little developed, most ornamental plants were grown for their economic benefits. The ancient *Book of Songs, Chu Ci, Erya* and *Shan Hai Jing* each recorded a number of economic plants also used as landscape plants, including peach (*Amygdalus persica*; Rosaceae), apricot (*Armeniaca vulgaris*), plum (*Armeniaca mume*), mulberry (*Morus* spp., Moraceae), poplar (*Populus* spp.), willow (*Salix* spp.; both Salicaceae) and elm (*Ulmus* spp., Ulmaceae).

In the royal and private garden period, lasting from the Qin Dynasty beginning in 221 BC to the Qing Dynasty (ending in 1911 AD), due to the sustained growth of productivity, more and more plants were cultivated intentionally for landscaping purposes. For example, prior to the Northern and Southern Dynasties (420–581 AD) the plum (*Armeniaca mume*) was cultivated mainly for its fruit; after this time it became cultivated mostly for its ornamental value. In this period, royal and private gardens, only enjoyed by royalty and noblemen, become popular. Many monographs on ornamental landscape plants were written at this time, such as the *Zhang Wu Zhi* (*Treatise on Superfluous Things*; Wen, 1621), *Hua Jing* (Wu, 1688), *Guang Qun Fang Pu* (Wang, 1708) and *Zhi Wu Ming Shi Tu Kao* (Wu, 1848).

During the public garden period, lasting since the end of the nineteenth century, again due to highly developed productivity, gardens became open not just to the elite but also to the public. Gardens, and the effects of landscape gardening, expanded to include public green spaces. During this period, significant botanical works, such as *Flora Reipublicae Popularis Sinicae* (Editorial Committee of Flora Reipublicae Popularis Sinicae, 1959–2004) and *Iconographia Cormophytorum Sinicorum* (Institute of Botany - Academia Sinica, 1972–1976), studies of landscape plants, such as *Landscape Plants of China* (Xing *et al.*, 2009), and specialized monographs, such as *The Orchids of China* (Chen & Ji, 1998), *Ornamental Palms* (Lin, 2003) and *Magnolias of China* (Law, 2004) were published, introducing more and more beautiful plants to the landscape. In this chapter, we describe a selection of commonly-planted ornamental species selected from the rich natural resource of China.

19.2 Important taxa of ornamental plants

Cycas.—There are about 60 species of *Cycas* (su tie shu; Cycadaceae), of which 16 are found in China, and eight are endemic to the country. These are distributed across eight provinces or autonomous regions, concentrated in southwestern (Guangxi, Guizhou and Yunnan) and southern China (Guangdong and Hainan). *Cycas balansae* (kuan ye su tie), *C. pectinata* (bi chi su tie), *C. revoluta* (king sago; su tie; Figure 19.1) and *C. szechuanensis* (si chuan su tie) are the most commonly cultivated cycads because of their emerald-green leaves, attractive appearance and strong adaptability, making them easy to grow in lowland and mid-altitude areas. *Cycas revoluta* (Figure 19.1) has probably been cultivated for more than 1 800 years. Cycads are sacred to Buddhists and are often planted near temples (Wang & Liang, 1996). A paste of *Cycas* seeds and coconut oil is used for the treatment of skin complaints, wounds, ulcers, sores and boils (Chen & Stevenson, 1999).

Figure 19.1 *Cycas revoluta.*

Ginkgo biloba.—*Ginkgo biloba* (ginkgo; yin xing; Figure 19.2) is widely cultivated in China and also in Europe and North America. A relict species of the Mesozoic Era. *G. biloba* has probably been cultivated for more than 3 000 years. It also is sacred to Buddhists and is often planted near temples (Chen, 1937; Fu *et al.*, 1999).

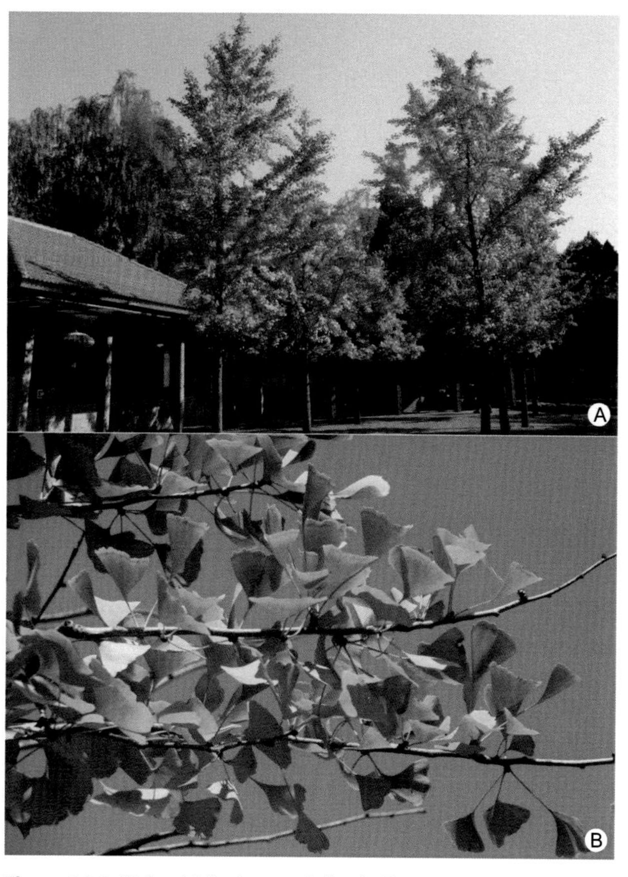

Figure 19.2 *Ginkgo biloba.* A, trees. B, detail of leaves.

Keteleeria fortunei.—*Keteleeria fortunei* (you shan; Pinaceae) is an evergreen coniferous tree to 30 m tall, native to the hills, mountains and broad-leaved forests of Fujian, Guangdong, Guangxi, Guizhou, Hunan, Jiangxi, Yunnan and Zhejiang, as well as Vietnam, and is a vulnerable species in China. It is cultivated as an ornamental but also for afforestation, and the timber is used for construction and furniture (Fu *et al.*, 1999).

Larix gmelinii.—*Larix gmelinii* (Dahurian larch; luo ye song; Pinaceae) is a deciduous conifer native to Hebei, Heilongjiang, Henan, Jilin, Nei Mongol and Shanxi, as well as eastern Siberia, Mongolia and Korea. It is cultivated for afforestation as well as for its ornamental properties. *Larix gmelinii* var. *principis-rupprechtii* (hua bei luo ye song) is cultivated for the same purposes (Fu *et al.*, 1999).

Picea.—*Picea asperata* (dragon spruce; yun shan; Pinaceae) is native to altitudes of 2 400–3 600 m in

Gansu, Ningxia, Qinghai, Shaanxi and Sichuan. As well as being grown as an ornamental in China, Europe and North America, the timber is used for construction, aircraft, railway sleepers, furniture and wood fibers, the trunk produces resin, and the roots, branches and leaves produce aromatic oils. *Picea meyeri* (bai qian), native to the mountains of Gansu, Hebei, Nei Mongol, Shaanxi and Shanxi, and *P. wilsonii* (qing qian), native to mountains and river basins in Gansu, Hebei, Hubei, Nei Mongol, Qinghai, Shaanxi, Shanxi and Sichuan, are also cultivated as ornamentals, for afforestation, and for construction, poles, bridge building, furniture and wood pulp (Fu *et al.*, 1999).

Pinus.—Pinus armandii (hua shan song; Pinaceae) is native to mountains and river basins at elevations of 1 000–3 300 m in Gansu, Guizhou, Hainan, Henan, Hubei, Shaanxi, Shanxi, Sichuan, Taiwan, Xizang and Yunnan, and also Myanmar. It is cultivated as an ornamental, and the timber is used for construction, railway sleepers, furniture and wood fibers. *Pinus bungeana* (lacebark pine; bai pi song), with its very attractive flaky bark, is native to the hills of Gansu, Henan, Hubei, Shaanxi, Shandong, Shanxi and Sichuan. It is used as an ornamental plant in eastern Asian classical gardens, where it symbolizes longevity. *Pinus koraiensis* (Korean pine; hong song) has grayish bark fissured into oblong plates and is native to the mountains of Heilongjiang and Jilin, Japan, Korea and Russia. It is a popular ornamental tree in parks and large gardens with cold climates, such as in eastern Canada and the northeastern USA. *Pinus sylvestris* var. *mongolica* (zhang zi song), native to Heilongjiang, Nei Mongol, Mongolia and Russia, is not only an important forestry species but one of the most popular Christmas trees, while *P. sylvestris* var. *sylvestriformis* (chang bai song), from Jilin, is a vulnerable species cultivated as an ornamental and for afforestation.

Pinus tabuliformis (Chinese red pine; you song), from Gansu, Hebei, Henan, Hubei, Hunan, Jilin, Liaoning, Nei Mongol, Ningxia, Qinghai, Shaanxi, Shandong, Shanxi and Sichuan, and Korea, is cultivated as an ornamental, for afforestation, and for its timber, resin, tannins, and the medicinal properties of the needles. *Pinus taiwanensis* (Taiwan red pine; huang shan song), native to Anhui, Fujian, Guangxi, Guizhou, Henan, Hubei, Hunan, Jiangsu, Jiangxi, Taiwan, Yunnan and Zhejiang, is cultivated as an ornamental, for afforestation, and as an important timber tree in Taiwan, while *P. yunnanensis* (yun nan song), from Guangxi, Guizhou, Sichuan, Xizang and Yunnan, is grown as an ornamental, for afforestation, timber, pulp, resin and tannins (Chen, 1990; Fu *et al.*, 1999; Xing *et al.*, 2009).

Pseudolarix amabilis.—Pseudolarix amabilis (golden larch; jin qian song; Pinaceae) is a deciduous tree to 40 m tall, probably native to the lower Chang Jiang valley and cultivated in Anhui, Hubei, Jiangsu and Sichuan as well as in Asia, Europe and America. An excellent ornamental tree with its attractive crown and golden fall color, it is a attractive addition to parks and large gardens (Chen, 1990; Fu *et al.*, 1999).

Glyptostrobus pensilis.—Glyptostrobus pensilis (Chinese swamp cypress; shui song; Taxodiaceae) is also a deciduous tree, 15–25 m tall, native to the river deltas of Fujian, Guangdong, Guangxi, Hainan, Jiangxi, Sichuan, Yunnan and Zhejiang as well as Vietnam. It may in fact no longer exist in the wild anywhere in China and is rare in all provinces except Guangdong. The species is often planted in wet places for erosion control, as a windbreak, and because it is believed to bring good luck (Chen, 1990; Fu *et al.*, 1999).

Metasequoia glyptostroboides.—Metasequoia glyptostroboides (dawn redwood; shui shan) is a deciduous tree to 50 m tall, native only to Sichuan and Hubei in western-central China. Its persistence here is a relic of a former worldwide distribution. It is now widely cultivated as an ornamental or planted for afforestation in Anhui, Fujian, Guangdong, Guangxi, Guizhou, Hebei, Henan, Hubei, Hunan, Jiangsu, Jiangxi, Liaoning, Shaanxi, Shandong, Shanxi, Sichuan, Yunnan and Zhejiang (Chen, 1990; Fu *et al.*, 1999).

Cupressus funebris.—Cupressus funebris (Chinese weeping cypress; bai mu; Cupressaceae) is native to altitudes below 2 000 m in Anhui, Fujian, Gansu, Guangdong, Guangxi, Guizhou, Henan, Hubei, Hunan, Jiangxi, Shaanxi, Sichuan, Yunnan and Zhejiang, and is widely cultivated in southern China, as an ornamental and for afforestation (Fu *et al.*, 1999).

Fokienia hodginsii.—Fokienia hodginsii (fu jian bai; Cupressaceae; Figure 19.3) is native to the montane forests below 1 800 m of Fujian, Guangdong, Guangxi, Guizhou, Hunan, Jiangxi, Sichuan, Yunnan and Zhejiang, as well as Laos and Vietnam. A vulnerable species in China, it is used as an ornamental tree or shrub in gardens (Fu *et al.*, 1999; Zhang, 2005).

Juniperus.—Juniperus chinensis (Chinese juniper; yuan bai; Cupressaceae) is native to the mountains, at elevations of 1 400–2 300 m, of Anhui, Fujian, Gansu, Guangdong, Guangxi, Guizhou, Hebei, Heilongjiang, Henan, Hubei,

Figure 19.3 *Fokienia hodginsii* fruiting branch.

Hunan, Jiangsu, Jiangxi, Nei Mongol, Shaanxi, Shandong, Shanxi, Sichuan, Taiwan, Yunnan and Zhejiang, and also Japan, Korea, Myanmar and Russia. It is a popular ornamental tree or shrub in gardens and parks, with over 100 named cultivars selected for various characters such as yellow foliage (e.g. "aurea" and "tremonia"), permanently juvenile foliage (e.g. "shoosmith"), a columnar crown ("columnaris") or abundant cones ("kaizuka"). The cultivar "shimpaku" is a very important subject for bonsai (Fu *et al.*, 1999; Zhang, 2005). *Juniperus formosana* (ci bai) is native to the forests of Anhui, Fujian, Gansu, Guizhou, Hubei, Hunan, Jiangsu, Jiangxi, Qinghai, Shaanxi, Sichuan, Taiwan, Xizang, Yunnan and Zhejiang, and is grown as an ornamental tree or shrub in gardens (Chen, 1990; Fu *et al.*, 1999).

Platycladus orientalis.—Natural occurrences of *Platycladus orientalis* (ce bai; Cupressaceae) are difficult to distinguish from local introductions, owing to extensive cultivation and planting in the past. The species is widely naturalized in Asia, as far east as Korea and Japan, south to northern India and west to northern Iran. It is very widely used as an ornamental tree, both in its homeland, where it is associated with long life and vitality, and elsewhere in temperate climates. The wood is used in Buddhist temples; both for construction and, chipped, for incense burning (Chen, 1990; Fu *et al.*, 1999).

Nageia.—There are five to seven species of *Nageia* (zhu bai shu; Podocarpaceae), of which three are found in China. They are distributed in ten provinces or autonomous regions, concentrated in the southwest (Guangxi and Yunnan) and south (Guangdong and Hainan). The majority of species are highly adaptable. *Nageia fleuryi* (zhu bai) and *N. nagi* (Asian bayberry; chang ye zhu bai; Figure 19.4) are the most commonly cultivated because of their emerald-green leaves, handsome shape and strong adaptability, making them easy to grow in mid- and low-altitude areas.

Podocarpus.—*Podocarpus* (lo han song shu; Podocarpaceae) comprises about 100 species, of which seven are found in China, three of them endemic. They are found in 13 provinces or autonomous regions, concentrated in the southwest (Guangxi, Guizhou and Yunnan) and south (Guangdong and Hainan). Like *Nageia*, they are highly adaptable. *Podocarpus macrophyllus* (kusamaki; luo han song; Figure 19.5), *P. neriifolius* (bai ri qing) and *P. wangii* (xiao ye luo han song) are the most commonly cultivated ornamental trees because of their emerald-green leaves, handsome shape, colorful and unusual seeds and ability to be cultivated in both mid- and low-altitude areas.

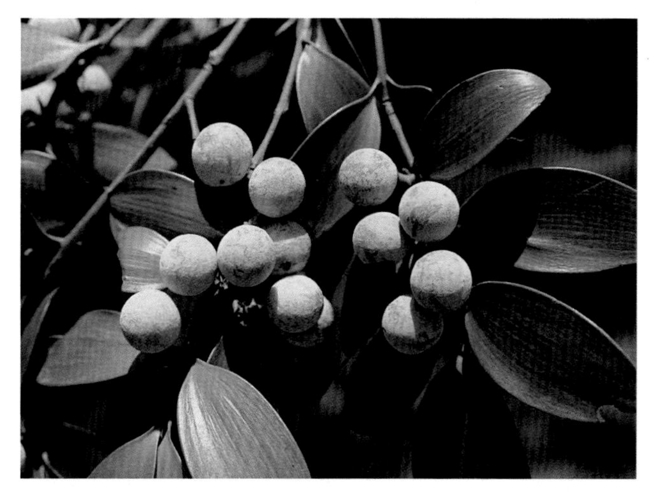

Figure 19.4 *Nageia nagi* fruiting branch.

Figure 19.5 *Podocarpus macrophyllus* fruiting branch.

Cephalotaxus. —There are eight to 11 species of *Cephalotaxus* (san jian shan shu; Cephalotaxaceae), with six in China, of which three are endemic. They are found across 17 provinces or autonomous regions, concentrated in the southwest (Guangxi, Guizhou, Sichuan and Yunnan) and south (Guangdong and Hunan). Again, most species have high adaptability. *Cephalotaxus fortunei* (Chinese plum yew; cu fei) and *C. sinensis* (san jian shan) are the most commonly cultivated ornamental trees because of their emerald-green leaves, handsome shape and colorful seeds, and can easily be cultivated in both mid- and low-altitude areas.

Salix.—*Salix babylonica* (weeping willow; chui liu) is a deciduous tree with striking pendulous branches. Widespread throughout China, it is cultivated in Asia, Europe and America and used for weaving wicker baskets, timber and reforestation. *Salix matsudana* (Chinese willow; han liu) is commonly planted on plains and riverbanks (Xing *et al.*, 2009).

Morus alba.—*Morus alba* (white mulberry; sang; Figure 19.6) was once endemic to central and northern China but is now widely cultivated throughout the country and the world. It is considered very beautiful, with its dark green leaves and densely-clustered fruit. Common cultivated varieties include *M. alba* "pendula," with weeping branches and "tortuosa," with twisted ones (Xing *et al.*, 2009).

Nymphaea.—There are about 50 species of *Nymphaea* (waterlilies; shui lian shu; Nymphaeaceae), widespread in temperate and tropical regions. Five species are found in

Figure 19.6 *Morus alba* fruiting branch.

Figure 19.7 *Nelumbo nucifera.*

China (Fu & John, 2001). Many species are naturalized or cultivated as ornamentals in ponds including, in China, *Nymphaea alba* var. *rubra* and *N. mexicana* in addition to the native species (Fu & John, 2001).

Nelumbo nucifera.—*Nelumbo nucifera* (lotus; he hua; Figure 19.7) is native to tropical Asia and Queensland, Australia, and commonly cultivated in water gardens. The white-and pink-flowered forms are the national flowers of India and Vietnam, respectively (Fu & John, 2001).

Paeonia.—*Paeonia lactiflora* (common garden peony; shao yao; Figure 19.8) is cultivated in northern China as well as in Japan, Korea, Mongolia, the far east of Russia and Siberia, as an ornamental and for medicinal use (Hong *et al.*, 2001; Xing *et al.*, 2009). There are around 400 cultivars in China (Li, 1999; Li, 2006). *Paeonia suffruticosa* (tree peony; mu dan) is native to central Anhui and western Henan, but widely cultivated in China and elsewhere for its large, brightly-colored and richly-patterned flowers which are used *in situ* and as a cut flower (Hong *et al.*, 2001; Xing *et al.*, 2009). *Paeonia suffruticosa* has been grown in China for about 1 600 years. Worldwide, there are seven cultivar groups, four of which are found in China (the "northwestern tree peony," "central plains tree peony," "southern Chang

Figure 19.8 *Paeonia lactiflora* flower.

Jiang tree peony" and "southwestern tree peony" groups), and there are around 1 000 cultivars in China (Li, 1999). Common cultivars include "hua wang," a double-flowered pink variety, "ruan zhi lan," pink but not distinctly layered, "xiu tao hua," in a pinky-purple color, "yu mei ren," with unusual large, flattened outer petals and long narrow inner ones, and "zu yu lan," another pink-flowered cultivar (Xing *et al.*, 2009).

Aconitum coreanum.—*Aconitum coreanum* (huang hua wu tou; Ranunculaceae) is found in northern Hebei, eastern Heilongjiang, Jilin and Liaoning, as well as Korea, Mongolia and the far east of Russia. Its unique and beautiful flowers are often grown in meadows or under a thin forest canopy, as well as in gardens, to give the impression of a natural population.

Aquilegia oxysepala.—*Aquilegia oxysepala* (columbine; jian eastern lou dou cai; Ranunculaceae) is native to Gansu, Guizhou, Hebei, Hubei, Heilongjiang, Jilin, Liaoning, Nei Mongol, southern Ningxia, eastern Qinghai, southern Shaanxi, Sichuan and northern Yunnan as well as Korea and far eastern Russia. Its unique flowers and beautiful habit make it popular for cultivation in meadows, at forest edges and in gardens.

Lirianthe.—There are about 12 species of *Lirianthe* (chang hui mu lan shu; Magnoliaceae) worldwide, of which ten are found in China, with five endemic. They are distributed across 11 provinces or autonomous regions, concentrated in the southwest (Guangxi, Guizhou and Yunnan) and south (Guangdong and Hainan). Most species are very adaptable, being shade-bearing when young, but requiring gradually increasing levels of light with age. *Lirianthe coco* (shan yu lan) and *L. delavayi* (ye xiang mu lan) are the most commonly cultivated ornamental trees, for their handsome habit and large fragrant flowers. The flower and root bark of *L. coco* is also used medicinally. The bark of *L. delavayi* is also sometimes used as a substitute for *Magnolia* bark, known in Chinese medicine as "houpu."

Manglietia.—There are around 40 species *of Manglietia* (mu lian shu), with around 29 in China, where 17 are endemic. They are distributed in 13 provinces and autonomous regions, concentrated in the southwest (Guizhou, Sichuan and Yunnan) and south (Guangdong, Guangxi and Hunan). As *Lirianthe*, most species are highly adaptable, bearing shade when young but requiring more light with age. *Manglietia chevalieri* (mu nan mu lian), *M. fordiana* var. *hainanensis* (hai nan mu lian), *M. insignis* (hong hua mu lian) and *M. lucida* (liang ye mu lian) are the most commonly cultivated as ornamental trees because of their handsome habit, large, fragrant flowers and bright red fruits. Several species are also used medicinally, including the bark and fruit of *M. fordiana* for constipation and coughs and that of *M. fordiana* var. *hainanensis* for swelling and inflammation. *Manglietia lucida* is used in breeding programs due to its large purple flowers (more than 25 cm in diameter) and bright green leaves.

Michelia.—There are around 70 species of *Michelia* (han xiao shu; Magnoliaceae). Around 40 species are found in China, to which 20 are endemic. The genus is found in 14 provinces, autonomous regions or centrally-administered municipalities, concentrated in the southwest (Guizhou, Sichuan and Yunnan) and south (Guangdong, Guangxi and Hunan). As with all Magnoliaceae described here, most species are highly adaptable, bearing shade when young but requiring more light with age. The taxa most commonly cultivated as ornamental trees are *M.* ×*alba* (bai lan; Figure 19.9), *M. champaca* (huang lan; Figure 19.10), *M. chapensis* (le chang han xiao), *M. figo* (han xiao hua), *M. macclurei* (zui xiang han xiao), *M. maudiae* (shen shan han xiao), *M. shiluensis* (shi lu han xiao) and *M. wilsonii* (e mei han xiao), for their handsome habit, very fragrant flowers and bright red fruits.

Figure 19.9 *Michelia* ×*alba* flowering branch.

Figure 19.10 *Michelia champaca* flowering branch.

Yulania.—*Yulania* (yu lan shu; Magnoliaceae) comprises about 25 species, ca. 18 of which are found in China, and 16 of which are endemic. The genus is distributed over 15 provinces, autonomous regions or centrally-administered municipalities, concentrated in the southwest (Guizhou, Sichuan and Yunnan) and the south-central part of the country (Hubei, Hunan and Zhejiang). Most species are highly adaptable, bearing shade when young but requiring more light with age. The majority are used for timber, in landscape planting, and in medicine. The most commonly-cultivated taxa are *Y. denudata* (yu lan), *Y. liliiflora* (zi yu lan) and *Y.* ×*soulangeana* (er qiao yu lan), for their handsome habit, large, fragrant flowers and bright red fruits. Of these, *Y. denudata* has been cultivated for more than 2 500 years (Chen & Cheng, 1999). The flower buds of most species have been used in Chinese medicine, where they are known as "xin yi," for more than 2 000 years, as an anti-inflammatory and analgesic.

Calycanthus chinensis.—*Calycanthus chinensis* (Chinese sweetshrub; xia la mei; Calycanthaceae) is native to Lin'an and Tiantai, in mountainous central and northern Zhejiang, where it is distributed near streams at 600–1 000 m. This species is cultivated in public parks, forests and gardens for its exotic-looking, colorful flowers (Li *et al.*, 2009; Xing *et al.*, 2009).

Chimonanthus praecox.—*Chimonanthus praecox* (wintersweet; la mei; Calycanthaceae) originates from central China, where it is mainly distributed in montane forests, at 500–1 100 m above sea level, in Anhui, Fujian, Guizhou, Henan, Hubei, Hunan, Jiangsu, Jiangxi, Shaanxi, Shandong, Sichuan, Yunnan and Zhejiang. This species is cultivated throughout most of China as well as in other temperate areas of the world, particularly for its sweetly fragrant flowers. The leaves, roots, flowers and seeds are also used medicinally. Common cultivars include "concolor," with small flowers, and "grandiflorus," a larger-flowered form with a lighter fragrance (Li *et al.*, 2009; Xing *et al.*, 2009).

Cinnamomum.—There are about 250 species of *Cinnamomum* (zhang shu; Lauraceae), distributed in tropical and subtropical Asia, Australia and the Pacific islands. Around 50 species, of which 30 are endemic and one introduced, are found in China (Li *et al.*, 2009). The species of *Cinnamomum* are economically important as sources of medicines, timber and nutritious fruits, and the leaves and fruits of some, such as *C. cassia* (Chinese cassia; rou gui) and *C. subavenium* (xiang gui), are perfumed. The fruits of species such as *C. burmanni* (Indonesian cinnamon; yin xiang), *C. camphora* (zhang; camphor tree; Figure 19.11), *C. glanduliferum* (false camphor tree; yun

Figure 19.11 *Cinnamomum camphora* fruiting branch.

nan zhang) and *C. parthenoxylon* (Selasian wood; huang zhang) contain abundant oils and fats; the oil has been used for making perfumes and medicines. The bark of *C. cassia* is a traditional Chinese medicine (Li *et al.*, 2009).

Machilus.—Machilus (run nan shu; Lauranceae) comprises around 100 species distributed in tropical and subtropical southern and Southeast Asia. Around 90 species, of which 63 are endemic, are found in China (Li *et al.*, 2009). Some species of *Machilus*, such as *M. chinensis* (hua run nan), *M. grijsii* (huang rong run nan), *M. nanmu* (run nan), *M. velutina* (rong mao run nan) and *M. yunnanensis* (dian run nan), are economically-important as sources of timber, while others contain mucilage which can be used as a glue. The bark of some species can be used for making a brown dye, and the branches, leaves and fruits for making perfumes. *Machilus salicina* (liu ye run nan; Figure 19.12) is suitable for riverbank stabilization (Li, 1982; Li *et al.*, 2009).

Begonia.—There are around 200 species of *Begonia* (qiu hai tang shu; Begoniaceae) in China, of which around 150 are endemic. They are mostly distributed to the south of the Chang Jiang and particularly concentrated in southeastern

Figure 19.12 *Machilus salicina* fruiting branch.

Yunnan and southwestern Guangxi. Some native species, such as *B. cathayana* (hua ye qiu hai tang), *B. gulinqingensis* (gu lin qing qiu hai tang), *B. leprosa* (lai ye qiu hai tang), *B. masoniana* (tie jia qiu hai tang) and *B. rex* (da wang qiu hai tang), have beautifully colored and patterned leaf blades. Others, such as *B. grandis* (qiu hai tang), have numerous flowers. All can be used as ornamentals themselves, or as good materials for hybridization. However, in China, most *Begonia* cultivars widely used in gardens and sold in markets are introduced from abroad. There are more than 10 000 cultivars of *Begonia* across the world, but less than 100 have been bred in China. Considering the abundant wild resources of *Begonia* in China, there is great potential for exploitation to produce new ornamentals. In addition, some *Begonia* plants are used locally as herbal medicines and vegetables.

Camellia.—Around 100 species of *Camellia* (shan cha shu), ca. 75 of them endemic, have been recorded in China. Many members of the genus are of great ornamental value, with some having been cultivated for more than 1 800 years and recorded in many old documents under the common names "cha hua" and "shan cha hua." In the Qing Dynasty (1644–1911), PU Jing-Zi wrote a monograph on the cultivated *Camellia*, the *Cha Hua Pu*, and recorded many breeds of *Camellia* available in his era. Today, *C. japonica* (Japanese camellia; shan cha), *C. reticulata* (dian shan cha) and *C. sasanqua* (cha mei) are the most popular and have hundreds of traditional cultivars in China. *Camellia japonica*, native to Shandong, Taiwan and Zhejiang as well as Japan and Korea, has been grown as a garden plant throughout southern China and has by far the greatest number of named cultivars. In the eighteenth and early nineteenth centuries, many cultivars of *C. japonica* were introduced into Europe and have become popular garden here plants too.

Camellia reticulata, native to Guizhou, Sichuan and Yunnan, is the largest-flowered *Camellia*. It has been propagated as garden plant for at least 1 300 years in Yunnan, especially in Chuxiong, Dali, Kunming Shi, Lijiang, Tengchong and adjacent mountainous areas. Many cultivars with large, double or semi-double flowers have been selected, and there are now more than 100 named traditional cultivars of *C. reticulata* grown in these areas. Some famous cultivars, such as "shi zi tou," "da li cha," "mu dan cha," "heng tian gao," "zi pao" and "song zi lin" exist as trees hundreds of years old, and are seen as very precious in Chinese garden culture. The form "hong hua you cha," which used to be considered as *C. reticulata* forma *simplex*, has traditionally been used as a source of edible oil in western Yunnan, and was recently identified as an important potential edible oil production species which has become widely popularized in this area.

Many other species of *Camellia* have been used for hybridization, to incorporate desirable characteristics, such as the golden yellow petal color of section *Archecamellia* (gu cha zu), the frost hardiness of *C. oleifera* (you cha; Figure

19.13), or the fragrance of *C. grijsii* (chang ban duan zhu cha) and *C. lutchuensis* (tai wan lian rui cha). The camellia also has also traditional cultural meaning for some indigenous groups, such as the Yi people, who worship *Camellia* trees as sacred and celebrate "Camellia flower festival day" in February every year in the Chuxiong area of Yunnan.

Figure 19.13 *Camellia oleifera* flower and fruit.

Amygdalus.—About 40 species of *Amygdalus* (tao shu) are found in central, eastern and southwestern Asia and southern Europe. Eleven species are found in China, of which four are endemic and one introduced (Lu *et al.*, 2003). Peaches are important, fruit-bearing and ornamental plants cultivated throughout temperate and subtropical zones. *Amygdalus persica* (peach; Figure 19.14) is native to northern China. Its roots, leaves, flowers and seeds are used in Traditional Chinese Medicine and its dried latex can be used as an adhesive (Yu, 1985; Lu *et al.*, 2003). Many other species and cultivars have been grown for their edible fruit or as garden ornamentals and a few species are also grown for their edible seeds.

Armeniaca mume.—*Armeniaca mume* (plum; mei; Figure 19.15) grows in forests and on mountains below 3 100 m in western Sichuan and western Yunnan, and is cultivated throughout most of China particularly in areas south of the Chang Jiang. There are many cultivated varieties of *A. mume* (Yu, 1985; Lu *et al.*, 2003), differing in flower and fruit. *Armeniaca mume* var. *mume* is commonly cultivated throughout most of China both for its edible fruit and as a winter ornamental, while *A. mume* var. *pallescens* grows in western Sichuan and western Yunnan. *Armeniaca mume* var. *cernua* and *A. mume* var. *pubicaulina* are confined to western Yunnan (Yu, 1985; Lu *et al.*, 2003).

Cotoneaster.—There are about 90 species of *Cotoneaster* (xun zi shu; Rosaceae), widespread in temperate northern Africa, Asia (except Japan), Central America (Mexico) and Europe, and most abundant in southwestern China. Around 60 species (with 37 endemics) are found in China (Lu *et al.*, 2003). These shrubs are widely planted as ornamentals for their attractive fruits and flowers and used as borders, hedges and ground cover.

Figure 19.14 *Amygdalus persica* flowering branch.

Figure 19.15 *Armeniaca mume* flowering branch.

Photinia.—*Photinia* (shi nan shu; Rosaceae) comprises about 60 species distributed in eastern, southern and Southeast Asia, and also in Mexico. Around 50 species are found in China, of which 32 are endemic (Lu *et al.*, 2003). Many species of *Photinia* are used as ornamental trees and shrubs, with large lustrous leaves and attractive white flowers in spring followed by red fruits in fall. The wood is hard and heavy, suitable for making furniture and other small articles (Lu *et al.*, 2003).

Prunus.—There are about 30 species of *Prunus* (li shu) in Asia, Europe and North America, with around a dozen species (including two endemic and three introduced) found in China (Lu *et al.*, 2003). Some species of *Prunus* are important fruit trees in temperate zones, both for consumption and wine-making. However their flowers are also considered beautiful and they may be planted in gardens or on roadsides. They are also important honey trees for insects. For example *Prunus salicina* (Chinese plum; li) is an important, temperate fruit tree widely cultivated in China, particularly the variety *P. salicina* var. *pubipes* (mao geng li; Yu, 1985; Lu *et al.*, 2003).

Rosa.—*Rosa* (roses; qiang wei shu) comprises about 200 species widely distributed from subtropical to cold-temperate regions. There are over 100 species in China, of which 65 are endemic (Lu *et al.*, 2003). Species of *Rosa* are famous as an ornamental plant resources and many species and hybrids are cultivated in China, including *R.* ×*alba*, *R. centifolia*, *R. damascena* and *R. gallica* (Yu, 1985; Lu *et al.*, 2003). *Rosa chinensis* (yue ji hua; China rosa), *R. multiflora* (baby rose; ye qiang wei), *R. odorata* (xiang shui yue ji), *R. rugosa* (mei gui) and *R. luciae* var. *luciae* (guang ye qiang wei) are very important for creating new hybrid cultivars of *Rosa*, and their flowers have also been used for making perfumes. The leaves, flowers, roots and fruits of species such as *R. bracteata* (shuo bao qiang wei), *R. chinensis*, *R. cymosa* (xiao guo qiang wei), *R. davidii* (xi bei qiang wei), *R. laevigata* (Cherokee rose; jin ying zi) can be used medicinally. *Rosa chinensis* is widely cultivated throughout China, while *R. rugosa*, native to Jilin, Liaoning and Shandong, has many horticultural forms widely cultivated in China, but is endangered in the wild due to picking and uprooting.

Spiraea.—*Spiraea* (xiu xian ju shu; Rosaceae) includes about 100 species of deciduous shrubs distributed in the northern temperate zone and extending to mountainous subtropical areas. At least 70 species are found in China, of which 47 are endemic (Lu *et al.*, 2003). Many species of *Spiraea*, with their beautiful flowers and nectar, are long established in horticulture and often cultivated in gardens and parks, including *S. chamaedryfolia* (shi can ye xiu xian ju), *S. dasyantha* (mao hua xiu xian ju), *S. fritschiana* (hua bei xiu xian ju), *S. lobulata* (lie ye xiu xian ju) and *S. veitchii* (e xi xiu xian ju; Yu, 1974; Lu *et al.*, 2003).

Bauhinia ×*blakeana*.—*Bauhinia* ×*blakeana* (yang zi jing; Fabaceae) is cultivated as an ornamental and roadside tree because of its red flowers and elegant form (Chen *et al.*, 2005; Xing *et al.*, 2009). It is a hybrid, so usually does not produce fruit. It is found in Fujian, Guangdong, Guangxi and Yunnan, and also in Thailand and India. *Bauhinia variegata* (gong fen yang ti jia) is a deciduous tree 15 m tall and native to southern China. It is cultivated worldwide on sunny slopes or open land, mostly for its beautiful flowers (Chen *et al.*, 2005; Xing *et al.*, 2009).

Callerya speciosa.—*Callerya speciosa* (mei li ji xue teng; Fabaceae) is a liana of relatively open places from 200 to 1 700 m elevation in Fujian, Guangdong, Guangxi, Guizhou, Hainan, Hunan and Yunnan, as well as Vietnam. The roots are rich in starch and used to make a tonic wine, while the pretty, flowering vines are planted in gardens (Xu *et al.*, 2010).

Caragana sinica.—*Caragana sinica* (jin ji er; Fabaceae) is a shrub to 2 m tall native to hills below 1 800 m in Anhui, Fujian, Gansu, Guangxi, Guizhou, Hebei, Henan, Hubei, Hunan, Jiangsu, Jiangxi, Liaoning, Shaanxi, Shandong, Sichuan, Yunnan and Zhejiang; it is also found in Korea. It is used as an ornamental garden plant (Zhang, 2005; Xu *et al.*, 2010).

Codoriocalyx motorius.—*Codoriocalyx motorius* (telegraph plant; wu cao; Fabaceae) is found in Fujian, Guangdong, Guangxi, Guizhou, Jiangxi, Sichuan, Taiwan, Xizang and Yunnan as well as Bhutan, Cambodia, India, Indonesia, Laos, Malaysia, Myanmar, Nepal, the Philippines, Sri Lanka, Thailand and Vietnam. Famous for the movement of its small lateral leaflets at speeds rapid enough to be perceivable with the naked eye, *C. motorius* is widely cultivated.

Lespedeza thunbergii subsp. *formosa*.—*Lespedeza thunbergii* subsp. *formosa* (bushclover; mei li hu zhi zi; Fabaceae) is native to areas below 2 800 m in Fujian, Guangdong, Guangxi, Jiangsu, Jiangxi, Taiwan and Zhejiang. It is used as an ornamental garden shrub (Zhang, 2005; Xu *et al.*, 2010).

Lysidice rhodostegia.—*Lysidice rhodostegia* (yi hua; Fabaceae) is distributed in Guangdong, Guangxi, Yunnan and elsewhere. It is commonly cultivated in southern China, often grown in low-elevation forest. With its elegant habit and flowers, it can be planted alone or in groups as an ornamental, or used as a roadside tree (Chen *et al.*, 2005; Xing *et al.*, 2009).

Saraca dives.—*Saraca dives* (sorrowless tree; zhong guo wu you hua; Fabaceae) is native to forests and riversides at 200–1 000 m altitude in Guangdong, Guangxi and Yunnan as well as Laos and Vietnam. The flowers are large and showy and the trees are often cultivated ornamentally (Zhang, 2005; Xu *et al.*, 2010).

Senna siamea.—*Senna siamea* (kassod tree; tie dao mu; Fabaceae) is grown as a roadside or garden tree (Xing *et al.*, 2009), naturalized in Yunnan and cultivated in most of southern China. It is also found in India, Myanmar and Thailand.

Sophora japonica.—*Sophora japonica* (pagoda tree; huai; Fabaceae; Figure 19.16) is a tree to 25 m native to eastern Asia and widely cultivated throughout China and the world. It is a popular ornamental tree in Europe, North America and South Africa, grown for its white flowers, borne in late summer after most other flowering trees have long finished flowering (Chen *et al.*, 2005; Zhang, 2005; Xu *et al.*, 2010).

Wisteria sinensis.—*Wisteria sinensis* (zi teng; Fabaceae; Figure 19.17) is a woody climber reaching up to 25 m tall, with purple or white flowers, native to mountain forests at 500–1 800 m elevation in Anhui, Fujian, Guangxi, Hebei, Henan, Hubei, Hunan, Jiangsu, Jiangxi, Shaanxi, Shandong,

Figure 19.16 *Sophora japonica* flowering branch.

Figure 19.17 *Wisteria sinensis* flowering branch.

southern Shanxi and Zhejiang, and Japan. It was introduced from China to Europe and North America in 1816, and has become one of the most popular flowering vines in gardens throughout the temperate zone (Chen *et al.*, 2005; Zhang, 2005; Xu *et al.*, 2010).

Acer.—There are about 129 species of *Acer* (maples; qi shu; Aceraceae), widespread in both temperate and tropical regions of northern Africa, Asia, Europe and Central and North America. Around 100 species are found in China, of which 61 are endemic and three introduced (Xu *et al.*, 2008). *Acer palmatum* (Japanese maple; ji zhua feng) is widely cultivated in Chinese gardens, a small ornamental tree with many cultivars. Many Chinese maples have been cultivated widely for their beautiful fall colors, including *A. buergerianum* (trident maple; san jiao feng), *A. davidii* (snakebark maple; qing zha feng), *A. duplicatoserratum* (chong chi feng), *A. griseum* (paperbark maple; xue pi feng), *A. pictum* (painted maple; se mu feng), *A. tataricum* subsp. *ginnala* (Amur maple; cha tiao feng), *A. triflorum* (san hua feng), *A. truncatum* (Shandong maple; yuan bao feng; Figure 19.18) and *A. wilsonii* (san xia feng). In winter, the bark of the snakebark maples (*A. davidii* and relatives) and paperbark maple (*A. griseum*) is also eye-catching. Maple wood is used for furniture, interior beams

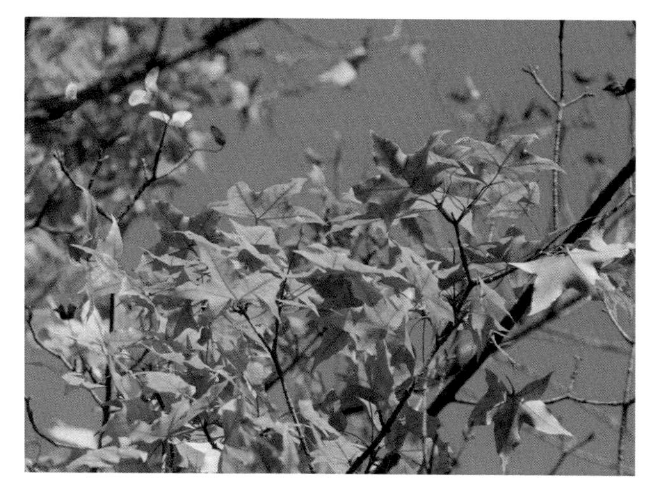

Figure 19.18 *Acer truncatum* leaves in fall color.

and wooden tools. The young leaves of *A. tataricum* subsp. *ginnala* can also be used as a substitute for tea (Fang, 1981; Xu *et al.*, 2008).

Davidia involucrata.—*Davidia involucrata* (dove or handkerchief tree; gong tong) is a deciduous tree to 20 m tall, mainly distributed in montane mixed forest in Guizhou, western Hubei, western Hunan, Sichuan and northern Yunnan. Its common names stem from the large, double, white bracts surrounding its flowers (Qin & Chamlong, 2007; Xing *et al.*, 2009).

Rhododendron.—There are around 600 species of *Rhododendron* (du juan shu) in China, with over 400 of them endemic. Yunnan is one of the distribution centers of the genus, with about 250 species. Many species of *Rhododendron* have beautiful, vividly-colored flowers and are of great ornamental value, including *R. agastum* (mi ren du juan), *R. decorum* (da bai du juan), *R. delavayi* (ma ying du juan), *R. forrestii* (zi bei du juan), *R. fortunei* (yun jin du juan), *R. irroratum* (lu zhu du juan), *R. latoucheae* (xi shi hua), *R. molle* (yang zhi zhu), *R. moulmainense* (mao mian du juan), *R. simsii* (yan shan hong; Figure 19.19), *R. sinogrande* (tu jian du ju), *R. traillianum* (chuan dian du juan), *R. uvariifolium* (zi yu pan du juan), *R. vernicosum* (liang ye du juan) and *R. wardii* (huang bei du juan). There

Figure 19.19 *Rhododendron simsii* flowers.

is a long history of cultivation of *Rhododendron* in China. The earliest recorded transplant of *Rhododendron* was in the Tang Dynasty, about 1 200 years ago. Historically, the most popular cultivated species are *R. molle* and *R. simsii*. In Yunnan, *R. delavayi*, with its large, globose, vermillion inflorescences, is a sacred flower of mountain communities and has been given special cultural and religious significance. For example the Yi people of Chuxiong celebrate "*Rhododendron* flower festival day" ("Ma yin hua jie"), and the Bai community in Dali have established a culture of eating the flowers of *R. decorum*, known as "Da bai hua."

Because most members of *Rhododendron* grow on high mountains and have symbiotic associations with fungi, they are difficult to cultivate. In modern Chinese gardens, only a few native species are used directly or for hybridization, including *R. delavayi, R. fortunei, R. molle* and *R. simsii* (Figure 19.19). The most popular cultivated ornamental *Rhododendron* is *R.* ×*pulchrum* (jin xiu du juan), a horticultural hybrid of uncertain origin. *Rhododendron indicum* (gao yue du juan), native to Japan, is also widely cultivated in China. In recent decades, many botanical gardens in China, such as those at Guizhou, Hangzhou Shi, Kunming Shi, Lushan and Shanghai Shi, have established good collections of *Rhododendron* and have instigated work in the introduction, hybridization and cultivation of *Rhododendron* species. At Kunming Botanical Garden in particular, 141 wild species of *Rhododendron* have successfully been introduced, and tens of new cultivars have been bred, many of them displayed in the new "Yu Xi *Rhododendron* Garden."

Clerodendrum.—There are about 400 species of *Clerodendrum* (glorybower; da qing shu; Verbenaceae), mostly found in tropical and subtropical Asia, Africa and America, with a few in temperate regions. Thirty-four species are found in China (Chen & Michael, 1994). Some species of *Clerodendrum*, such as *C. chinense* var. *simplex* (xiu mo li) and *C. kwangtungense* (guang dong da qing), are used in Chinese medicine. Others, such as *C. japonicum* (cheng tong), *C. trichotomum* (hai zhou chang shan) and *C. wallichii* (bridal veil; chui mo li), are very beautiful ornamentals (Pei & Chen, 1982; Chen & Michael, 1994).

Gardenia.—*Gardenia* (zhi zi shu; Rubiaceae) comprises about 250 species of beautiful-flowered shrubs in tropical and subtropical regions in the eastern hemisphere, including five species in China. The fruits of *G. hainanensis* (hai nan zhi zi), *G. jasminoides* (Cape jasmine; zhi zi) and *G. stenophylla* (xia ye zhi zi) are used in Chinese medicine and for producing yellow dye.

Ixora. There are about 300–400 species of *Ixora* (flame of the woods; long chuan hua shu; Rubiaceae) in tropical Asia, Oceania, tropical Africa and tropical America, including 18 species in China. Many species of *Ixora* are cultivated as ornamentals and there are many cultivated varieties, for example, of *I. chinensis* (long chuan hua; Figure 19.20).

Figure 19.20 *Ixora chinensis* in flower.

Lonicera.—There are about 180 species of *Lonicera* (honeysuckle; reng dong shu; Caprifoliaceae) in northern Africa, Asia, Europe and North America. More than 50 species are found in China, including 23 endemic to the country. Many species and their hybrids are valued garden plants. The shrubby species, such as *L. fragrantissima* (winter honeysuckle; yu xiang ren dong), *L. ligustrina* var. *yunnanensis* (box honeysuckle; liang ye ren dong) and *L. maackii* (Amur honeysuckle; jin yin ren dong), are excellent in borders, while the climbers, such as *L. confusa* (shui ren dong), *L. henryi* (dan hong ren dong) and *L. japonica* (Japanese honeysuckle; ren dong), can be used to cover trellises, walls or fences. The flowers of *L. japonica* and *L. confusa* are also used in Chinese medicine (Page & Olds, 2004; Yang *et al.* 2011).

Viburnum opulus subsp. *calvescens*.—*Viburnum opulus* subsp. *calvescens* (tian mu qiong hua, fu hua tou, ji shu tiao; Adoxaceae) is found in northeastern, northern and eastern China south to the Chang Jiang valley. With beautiful leaves, flowers and fruit, it is cultivated in borders, as a screen, or in mass plantings.

Chrysanthemum indicum.—*Chrysanthemum indicum* (ye ju; Asteraceae; Figure 19.21) is a perennial herb widely distributed in China except Xinjiang. With brightly-colored

Figure 19.21 *Chrysanthemum indicum*.

flowers and a graceful habit, it is suitable for flower beds, potted plants, cut flowers and making garlands (Xing *et al.*, 2009).

Ligularia nelumbifolia.—*Ligularia nelumbifolia* (lian ye tuo wu; Asteraceae) is a perennial herb found in Gansu, Hubei, Sichuan and Yunnan. With its brightly-colored and elegant flowers, it is grown as an outdoor or potted plant (Xing *et al.*, 2009).

Alpinia zerumbet.—*Alpinia zerumbet* (light galangal; yan shan jiang; Zingiberaceae) is a perennial herb distributed in Guangdong, Guangxi, Hainan, Taiwan, Yunnan and other parts of tropical Asia (Wu & Larsen, 2000). The plants have a beautiful habit and large, elegant, white racemes. They are often planted by rocks or flowing water, on corners or under sparse forest. There are two common horticultural cultivars, "spring," with pale yellow stripes on the leaf surface, and "variegata," with golden-yellow striped leaves and red tips to the flower buds (Xing *et al.*, 2009).

Hedychium coccineum.—*Hedychium coccineum* (orange gingerlily; hong jiang hua; Zingiberaceae) is a perennial herb distributed in Guangxi, Xizang and Yunnan. It is cultivated as an ornamental for its red flowers, elegant habit and long flowering time (Wu & Larsen, 2000; Xing *et al.*, 2009). *Hedychium coronarium* (white gingerlily; bai jiang hua) is also cultivated as an ornamental, as well as for medicine and aromatic oils (Wu & Larsen, 2000; Xing *et al.*, 2009).

Zingiber zerumbet.—*Zingiber zerumbet* (shampoo ginger; hong qiu jiang; Zingiberaceae; Figure 19.22) is from Guangdong, Guangxi, Taiwan and Yunnan, as well as Cambodia, India, Laos, Malaysia, Myanmar, Sri Lanka, Thailand and Vietnam. It is cultivated as a garden plant and for cut flowers (Wu & Larsen, 2000).

Costus speciosus.—*Costus speciosus* (crape ginger; bi qiao jiang; Costaceae; Figure 19.23) is an erect perennial herb found in Guangdong, Guangxi, Taiwan and Yunnan as well as tropical Southeast and southern Asia. With its elegant flowers and graceful habit, it is a good shade-tolerant species to plant in clumps in corners, forest margins, or pots (Wu & Larsen, 2000; Xing *et al.*, 2009).

Lilium.—*Lilium* (bai he shu; Liliaceae) comprises about 115 species in temperate and alpine regions of the northern hemisphere, especially eastern Asia. There are over 50 species in China, of which 35 are endemic and one introduced. Many species have been in cultivation for centuries and have acquired many religious and mystical associations. *Lilium davidii* (chuan bai he), from western China, has orange-red flowers in late summer. *Lilium formosanum* (tai wan bai he), native to Taiwan, grows up to two meters high, and in late summer produces highly fragrant flowers, pure white on the inside, pink or purple-brown on the outside.

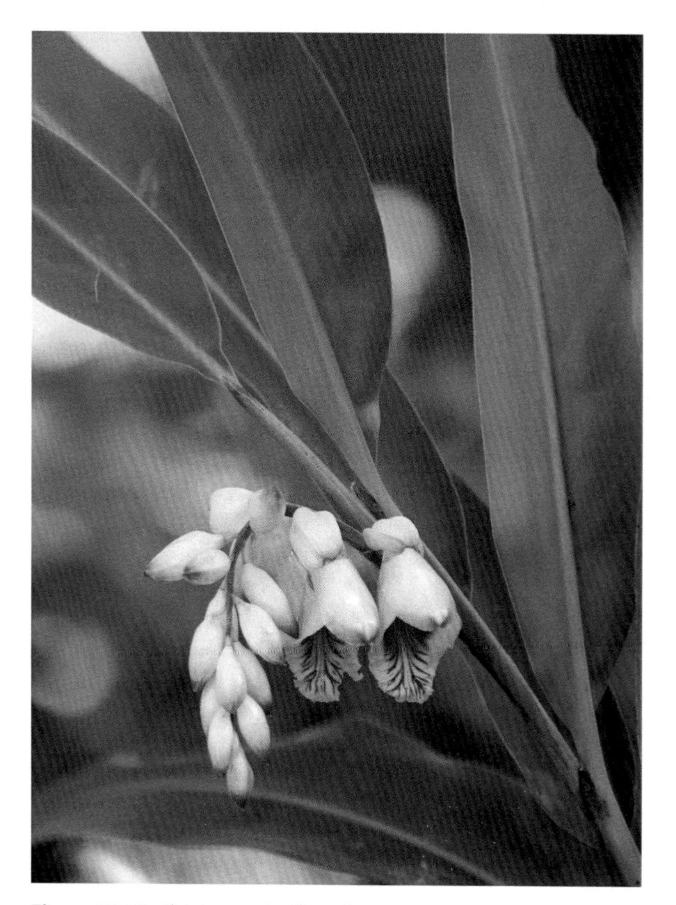

Figure 19.22 *Alpinia zerumber* flowering stem.

Figure 19.23 *Costus speciosus* in flower.

Lilium henryi (tiger lily; hu bei bai he), from central China, grows to three meters in cultivation, but much smaller in the wild. In summer it can have as many as 40 reflexed flowers, bright orange with carmine spots. *Lilium tigrinum* (tiger lily; juan dan) from eastern China, Japan and Korea, is one of the most widely grown and also the oldest in cultivation. The bulbs and flower buds are used for food in China and Japan (Liang & Tamura, 2000; Page & Olds, 2004).

Iris.—There are about 225 species of *Iris* (yuan wei shu; Iridaceae) in temperate regions of the northern hemisphere, including at least 50 species in China, to which 21 are endemic. This genus is valued for its beautiful and distinctive flowers, with many species and their hybrids being valuable garden plants. *Iris japonica* (crested iris; hu die hua), from China and Japan, forms large clumps of almost evergreen leaves and flowers in late winter and spring. *Iris tectorum* (wall iris; yuan wei), native to China, produces fans of bright green leaves up to 30 cm long, and flowers in spring and early summer (Zhao *et al.*, 2000; Page & Olds, 2004).

Lycoris.—There are about 20 species of *Lycoris* (shi suan shu; Amaryllidaceae) across China, India, Japan, Korea, Laos, Myanmar, Pakistan, Thailand and Vietnam. Fifteen species and hybrids are found in China, of which ten are endemic. *Lycoris aurea* (golden spider lily; hu di xiao), native to China and Japan, grows to 60 cm tall and bears a cluster of four or five golden-yellow flowers in summer. *Lycoris radiata* (red spider lily; shi suan), also from China and Japan, is the most commonly-grown species, producing 30—45-cm stems bearing clusters of four or five rose-red flowers in late summer or early fall (Ji & Meerow, 2000; Page & Olds, 2004).

Musella lasiocarpa.—*Musella lasiocarpa* (Chinese dwarf banana; di yong jin lian; Musaceae) is a suckering perennial herb, the only species in its genus, and endemic to Yunnan. It is cultivated or semi-cultivated in Guangxi, Guizhou, Hunan, Sichuan and Yunnan. Traditionally, local people cultivated *M. lasiocarpa* in their private gardens and fields as a hedgerow plant, using its pseudostem as a vegetable and for forage, but more recently it has become popular as a garden plant (Liu, 2001).

References

Chen, D. Z., Zhang, D. X., Hou, D., Larsen, K. & Larsen, S. S. (2005). Caesalpinieae. In: Wu, Z. Y., Raven, P. H. & Hong, D. Y. (eds.) *Flora of CHina*. pp. 7-10, 22. Beijing, China and St. Louis, MO: Science Press and Missouri Botanical Garden Press.

Chen, J. R. & Stevenson, D. W. (1999). Cycadaceae. In: Wu, Z. Y. & Raven, P. H. (eds.) *Flora of China*. pp. 1-7. Beijing, China and St. Louis, MO: Science Press and Missouri Botanical Garden Press.

Chen, J. Y. & Cheng, X. K. (1999). *China Flowers and Horticulture*. Shanghai, China: Shanghai Culture Press.

Chen, S. L. & Michael, G. G. (1994). Verbenaceae. In: Wu, Z. Y., Raven, P. H. & Hong, D. Y. (eds.) *Flora of China*. Beijing, China and St. Louis, MO: Science Press and Missouri Botanical Garden Press.

Chen, X. Q. & Ji, Z. H. (1998). *The Orchids of China*. Beijing, China: China Forestry Press.

Chen, Y. (1937). *Forest-Plant Taxonomy of China*. Nanjing, China: The Agronomy Society of China.

Chen, Y. M. (1990). *Landscape Dendrology*. Beijing, China: China Forestry Press.

Editorial Board of Chinese Agriculture Encyclopedia (1996). *Chinese Agriculture Encyclopedia, Volume of Ornamental Horticulture*. Beijing, China: Beijing: China Agriculture Press.

Editorial Committee of Flora Reipublicae Popularis Sinicae (1959-2004). *Flora Reipublicae Popularis Sinicae*. Beijing, China: Sciences Press.

Fang, W. P. (1981). *Aceraceae*. Beijing, China: Science Press.

Fu, D. Z. & John, H. W. (2001). Nymphaeaceae. In: Wu, Z. Y., Raven, P. H. & Hong, D. Y. (eds.) *Flora of China*. Beijing, China and St. Louis, MO: Science Press and Missouri Botanical Garden Press.

Fu, L. G., Lin, N. N. & Mill, R. R. (1999). Ginkgoaceae. In: Wu, Z. Y. & Raven, P. H. (eds.) *Flora of China*. Beijing, China and St. Louis, MO: Science Press and Missouri Botanical Garden Press.

Hong, D. Y., Pan, K. Y. & Turland, N. J. (2001). Ranunculaceae. In: Wu, Z. Y., Raven, P. H. & Hong, D. Y. (eds.) *Flora of China* pp. 127-132. Beijing, China and St. Louis, MO: Science Press and Missouri Botanical Garden Press.

Institute of Botany - Academia Sinica (1972-1976). *Iconographia Cormophytorum Sinicorum*. Beijing, China: Science Press.

Ji, J. H. & Meerow, A. W. (2000). *Lycoris* Herbert. In: Wu, Z. Y., Raven, P. H. & Hong, D. Y. (eds.) *Flora of China*. Beijing, China and St. Louis, MO: Science Press and Missouri Botanical Garden Press.

Law, Y. W. (2004). *Magnolias of China*. Beijing, China: Beijing Science and Technology Press.

Li, J. J. (1999). *Chinese Peony*. Beijing, China: Chinese Forest Industry Press.

Li, J. J. (2006). *Pictorial Record of Chinese Tree Peony Varieties*. Beijing, China: Chinese Forestry Press.

Li, X. W. (1982). Lauraceae. In: Delectis Florae Reipublicae Popularis Sinicae Agendae Academiae Sinicae Edita (ed.) *Flora Reipublicae Popularis Sinicae*. pp. 1-2. Beijing, China: Science Press.

Li, X. W., Li, J. & Huang, P. H. (2009). Lauraceae. In: Wu, Z. Y., Raven, P. H. & Hong, D. Y. (eds.) *Flora of China*. Beijing, China and St. Louis, MO: Science Press and Missouri Botanical Garden Press.

Liang, S. Y. & Tamura, M. N. (2000). *Lilium* L. In: Wu, Z. Y., Raven, P. H. & Hong, D. Y. (eds.) *Flora of China*. pp. 135-149. Beijing, China and St. Louis, MO: Science Press and Missouri Botanical Garden Press.

Lin, Y. R. (2003). *Ornamental Palms*. Harbin, China: Heilongjiang Science and Technology Press.

Liu, A. Z. (2001). *Phylogeny and Biogeography of Musaceae*. Ph.D. thesis, Kunming, China: Kunming Institute of Botany, Chinese Academy of Sciences.

Lu, L. D., Gu, C. Z. & Li, C. L. (2003). Rosaceae. In: Wu, Z. Y., Raven, P. H. & Hong, D. Y. (eds.) *Flora of China*. Beijing, China and St. Louis, MO: Science Press and Missouri Botanical Garden Press.

Page, S. & Olds, M. (2004). *Botanica*. Germany: Könemann, Verlag GmbH.

Pei, J. & Chen, S. L. (1982). Verbenaceae. In: Delectis Florae Reipublicae Popularis Sinicae Agendae Academiae Sinicae Edita (ed.) *Flora Reipublicae Popularis Sinicae*. p. 150. Beijing, China: Science Press.

Qin, H. N. & Chamlong, P. (2007). Nyssaceae. In: Wu, Z. Y., Raven, P. H. & Hong, D. Y. (eds.) *Flora of China*. p. 301. Beijing, China and St. Louis, MO: Science Press and Missouri Botanical Garden Press.

Wang, F. X. & Liang, H. B. (1996). *Cycads in China*. Guangzhou: Guangdong Science and Technology Press.

Wang, H. (1708). *Guang Qun Fang Pu*.

Wen, Z. H. (1621). *Zhang Wu Zhi*.

Wu, D. L. & Larsen, K. (2000). Zingiberaceae Lindley. In: Wu, Z. Y. & Raven, P. H. (eds.) *Flora of China*. pp. 322-377. Beijing & St. Louis, MO: Science Press & Missouri Botanical Garden Press.

Wu, H. Z. (1688). *Hua Jing*.

Wu, Q. J. (1848). *Zhi Wu Ming Shi Tu Kao*.

Xing, F. W., Zeng, Q. W., Chen, H. F. & Wang, F. G. (2009). *Landscape Plants of China*. Wuhan, China: Huazhong University of Science and Technology Press.

Xu, L. R., Chen, D. Z., Zhu, X. Y., Huang, P. H., Wei, Z., Sa, R., Zhang, D. X., Bao, B. J., Wu, D. L., Sun, H., Gao, X. F., Liu, Y. X., Chang, Z. Y., Li, J. Q., Zhang, M. L., Podlech, D., Ohashi, H., Larsen, K., Welsh, S. L., Vincent, M. A., Gilbert, M. G., Pedley, L., Schrire, B. D., Yakovlev, G. P., Thulin, M., Nielsen, I. C., Choi, B. H., Turland, N. J., Polhill, R. M., Larsen, S. S., Hou, D., Iokawa, Y., Wilmot-Dear, C. M., Kenicer, G. J., Nemoto, T., Lock, J. M., Salinas, A. D., Kramina, T. E., Brach, A. R., Bartholomew, B. & Sokoloff, D. D. (2010). Fabaceae. In: Wu, Z. Y., Raven, P. H. & Hong, D. Y. (eds.) *Flora of China*. Beijing, China and St. Louis, MO: Science Press and Missouri Botanical Garden Press.

Xu, T. Z., Chen, Y. S., Piet, C. D. J., Oterdoom, H. J. & Chang, C. S. (2008). Aceraceae. In: Wu, Z. Y., Raven, P. H. & Hong, D. Y. (eds.) *Flora of China*. pp. 516-517. Beijing, China and St. Louis, MO: Science Press and Missouri Botanical Garden Press.

Yang, Q. E., Landrein, S., Osborn, J. M. & Borosova, R. (2011). *Lonicera* L. In: Wu, Z. Y., Raven, P. H. & Hong, D. Y. (eds.) *Flora of China*. pp. 620-641. Beijing, China and St. Louis, MO: Science Press and Missouri Botanical Garden Press.

Yu, D. J. (1985). Rosaceae. In: Delectis Florae Reipublicae Popularis Sinicae Agendae Academiae Sinicae Edita (ed.) *Flora Reipublicae Popularis Sinicae*. Beijing, China: Science Press.

Yu, T. T. (1974). Rosaceae (1). In: Ling, Y. (ed.) *Flora Reipublicae Popularis Sinicae*. Beijing, China: Science Press.

Zhang, T. L. (2005). *1200 Species of Landscape Trees*. Beijing, China: China Architecture and Building Press.

Zhao, Y. T., Noltie, H. J. & Mathew, B. F. (2000). *Iris* L. In: Wu, Z. Y., Raven, P. H. & Hong, D. Y. (eds.) *Flora of China*. pp. 297-312. Beijing, China and St. Louis, MO: Science Press and Missouri Botanical Garden Press.

Zhu, T. P. (2004). Plant resource in China. In: Wu, Z. Y. & Chen, X. Q. (eds.) *Flora Reipublicae Popularis Sinicae*. pp. 584-657. Beijing, China: Science Press.

Chapter 20

Major Introduced Economic Plants

YI Ting-Shuang
Peter L. Morrell
PEI Sheng-Ji and HE Shan-An

20.1 Introduction

The history of Chinese agriculture involves not just the domestication of native species but the introduction and adoption of many foreign crops. Based on archeological evidence, a number of introduced plants have been cultivated in China for millennia, including bottle gourd (*Lagenaria siceraria*; Cucurbitaceae) from 7 000 years ago (Chen, 2002), sorghum (gao liang; *Sorghum bicolor*; Poaceae) from 5 000 years ago (Chen, 2002), wheat (*Triticum* spp. and hybrids; Poaceae) from at least 4 000 years ago (Chen, 2009) and barley (*Hordeum vulgare*; Poaceae) from 3 500 years ago (Fu *et al.*, 2000). During the Shang Dynasty, from around 1600 BC, the use of wheat was recorded in oracle-bone inscriptions, the earliest written record of an introduced crop. During the Western Zhou Dynasty (ca. 1100–771 BC), introduced crops recorded in the *Book of Songs* include barley, wheat, bottle gourd, radish (*Raphanus sativus*; Brassicaceae) and others.

During periods of the Qin, Han, Wei, Jin, Northern and Southern Dynasties (from 221 BC to 589 AD), especially after the imperial envoy ZHANG Qian was sent on a diplomatic mission to central and southern Asia and established trade routes between the east and west, many foreign crops were introduced into China (An, 1950). Broad bean (*Vicia faba*; Fabaceae), carrot (*Daucus carota* var. *sativa*; Apiaceae), clover (*Trifolium* spp.; Fabaceae), coriander (*Coriandrum sativum*; Apiaceae), cucumber (*Cucumis sativus*; Cucurbitaceae), garlic (*Allium sativum*; Alliaceae), grape (*Vitis vinifera*; Vitaceae), mung bean (*Vigna radiata*; Fabaceae), onion (*Allium cepa*; Alliaceae), pea (*Pisum sativum*; Fabaceae), pepper (*Piper nigrum*; Piperaceae), pomegranate (*Punica granatum*; Lythraceae), safflower (*Carthamus tinctorius*; Asteraceae) and sesame (*Sesamum indicum*; Pedaliaceae). As the territory of China expanded southward during the Qin and Han Dynasties, more tropical crops were introduced into China including sugar cane (*Saccharum officinarum*; Poaceae), while during the Sui and Tang Dynasties (581–907 AD), crops including cauliflower (*Brassica oleracea*; Brassicaceae), chile pepper (*Capsicum annuum*; Solanaceae), maize (*Zea mays*; Poaceae), peanut (*Arachis hypogaea*; Fabaceae), potato (*Solanum tuberosum*; Solanaceae), sunflower (*Helianthus annuus*; Asteraceae), sweet potato (*Ipomoea batatas*; Convolvulaceae), tobacco (*Nicotiana tabacum*; Solanaceae) and tomato (*Lycopersicon esculentum*; Solanaceae), among others, were introduced. At the end of the Ming Dynasty and the beginning of Qing Dynasty, around the middle of the seventeenth century, a further number of economically-important crops were introduced from North and South America including peanut (luo hua sheng; *Arachis hypogaea*; Fabaceae), chili pepper (la jiao shu; *Capsicum* spp., in China mostly *Capsicum annuum*), tomato (fan qie; *Lycopersicon esculentum*), tobacco (yan cao; *Nicotiana tabacum*) and potato (yang yu; *Solanum tuberosum*; all Solanaceae), sunflower (xiang ri kui; *Helianthus annuus*; Asteraceae), sweet potato (fan shu; *Ipomoea batatas*; Convolvulaceae) and maize (yu shu shu; *Zea mays*; Poaceae). Around this time the population of China was increasing rapidly, and there were a series of natural and man-made disasters; introduced crops, especially sweet potato and maize, greatly relieved the consequent food shortages.

Following the founding of the People's Republic of China in 1949, Chinese agriculture developed rapidly, with both the area under cultivation, and productivity, increasing dramatically. Agriculture in China experienced another period of rapid development following reform and opening-up in 1978. Many exotic plant species and varieties were introduced at this time, dramatically affecting the agricultural structure of the country. More than 600 major crop species (excluding forestry species) are now cultivated in China, and more than half of these have been introduced from other countries (Zheng, 2000). Some of the most important crops in China are now introduced species, including the staple foods of maize, sweet potato, wheat; important oil crops such as oil palm (*Elaeis guineensis*; Arecaceae), peanut, rape (*Brassica napus*; Brassicaceae), sesame and sunflower; the most important fiber crop, cotton (*Gossypium hirsutum*; Malvaceae); economically important crops such as rubber tree (*Hevea brasiliensis*; Euphorbiaceae), sugarcane, tobacco; the most important vegetables such as cucumber, potato, radish, tomato; the most important fruits such as apple (*Malus domestica*; Rosaceae), banana (*Musa paradisiaca*, Musaceae) and grape; the primary cultivated flowers such as carnation (*Dianthus caryophyllus*; Caryophyllaceae), lily (*Lilium brownii* var. *viridulum*; Liliaceae), modern roses (*Rosa* hybrids; Rosaceae), and other introduced species. These introductions greatly affected both Chinese agriculture and lifestyle. For example, the introduction of wheat completely changed dietary habits in northern China, and the introduction of tobacco led to widespread smoking.

In this chapter we discuss the many non-native plants that are used in China. The large number of native economic plants are dealt with in Chapters 16–19 and 21. Of the thousands of plant species have been introduced and are now cultivated in China, this chapter focuses on those that are economically important, widely cultivated or essential to the population. The timing of many plant introductions and changes in patterns of crop production coincides with particular dynastic periods in Chinese history. A summary of Chinese dynasties and their timespans may be found in the Appendix (page 473).

20.2 Introduced economic plants in China

20.2.1 Staple food crops

The domestication of staple food crops was essential to the origin of agriculture and the development of modern human society, occupying more than 50% of cultivated land worldwide and more than 75% of the total cultivated area in China, where they are of three main types: cereals, legumes and tubers. In addition to its hugely important native staples, chief among them being rice (*Oryza sativa;* Poaceae), China has a long history of introducing and using staple food crops domesticated in other regions. Of the 40 or so major cultivated staple food crops, 11 cereals, 19 legumes, ten tubers, and quinoa (*Chenopodium quinoa;* Chenopodiaceae; which is used in a manner similar to cereals) are all introduced (see Table 20.1). These introduced staples now form important components of Chinese agriculture and the Chinese lifestyle.

Table 20.1 Introduced staple food crops.

Category	Family	Latin Name	Chinese Name	English Name
grain crops	Poaceae	*Avena sativa* L.	燕麦 (yan mai)	oat
	Poaceae	*Coix lacryma-jobi* L.	薏苡 (yi yi)	job's tears
	Poaceae	*Eleusine coracana* (L.) Gaertn.	䅟 (can)	finger millet
	Poaceae	*Hordeum vulgare* L.	大麦 (da mai)	barley
	Poaceae	*Pennisetum glaucum* (L.) R. Br.	御谷 (yu gu)	pearl millet
	Poaceae	*Secale cereale* L.	黑麦 (hei mai)	ryes
	Poaceae	*Sorghum bicolor* (L.) Moench	高粱 (gao liang)	sorghum
	Poaceae	× *Tricosecale* spp. Wittm. ex A. Camus	小黑麦 (xiao hei mai)	triticale
	Poaceae	*Triticum aestivum* L.	小麦 (xiao mai)	common wheat
	Poaceae	*Triticum turgidum* L.	圆锥小麦 (yuan zhui xiao mai)	durum wheat
	Poaceae	*Zea mays* L.	玉蜀黍 (yu shu shu)	maize
legume crops	Fabaceae	*Cajanus cajan* (L.) Millsp.	木豆 (mu dou)	pigeon pea
	Fabaceae	*Canavalia ensiformis* (L.) DC.	直生刀豆 (zhi sheng dao dou)	jack bean
	Fabaceae	*Canavalia gladiata* (Jacq.) DC.	刀豆 (dao dou)	sword bean
	Fabaceae	*Cicer arietinum* L.	鹰嘴豆 (ying zui dou)	chickpea
	Fabaceae	*Lablab purpureus* (L.) Sweet	扁豆 (bian dou)	hyacinth bean
	Fabaceae	*Lathyrus sativus* L.	家山黧豆 (jia shan li dou)	grass pea
	Fabaceae	*Lens culinaris* Medik.	兵豆 (bing dou)	lentil
	Fabaceae	*Mucuna pruriens* var. *utilis* (Wall. ex Wight) L. H. Bailey	黧豆 (li dou)	velvet bean
	Fabaceae	*Phaseolus coccineus* L.	荷包豆 (he bao dou)	runner bean
	Fabaceae	*Phaseolus lunatus* L.	棉豆 (mian dou)	lima bean
	Fabaceae	*Phaseolus vulgaris* L.	菜豆 (cai dou)	common bean
	Fabaceae	*Pisum sativum* L.	豌豆 (wan dou)	pea
	Fabaceae	*Psophocarpus tetragonolobus* (L.) DC.	四棱豆 (si leng dou)	winged bean
	Fabaceae	*Trigonella foenum-graecum* L.	葫芦巴 (hu lu ba)	fenugreek
	Fabaceae	*Vicia faba* L.	蚕豆 (can dou)	faba bean
	Fabaceae	*Vigna mungo* L. Hepper	黑绿豆 (hei lü dou)	black gram
	Fabaceae	*Vigna radiata* (L.) R. Wilczek	绿豆 (lü dou)	mung bean
	Fabaceae	*Vigna umbellata* (Thunb.) Ohwi & H. Ohashi	赤小豆 (chi xiao dou)	rice bean
	Fabaceae	*Vigna unguiculata* (L.) Walp.	豇豆 (jiang dou)	cowpea
tuber crops	Araceae	*Colocasia esculenta* (L.) Schott	芋头 (yu tou)	taro
	Asteraceae	*Helianthus tuberosus* L.	菊芋 (ju yu)	Jerusalem artichoke
	Cannaceae	*Canna indica* L.	蕉芋 (jiao yu)	achira
	Convolvulaceae	*Ipomoea batatas* (L.) Lam.	番薯 (fan shu)	sweet potato
	Dioscoreaceae	*Dioscorea alata* L.	参薯 (shen shu)	water yam
	Dioscoreaceae	*Dioscorea hispida* Dennst.	白薯莨 (bai shu liang)	white yam
	Euphorbiaceae	*Manihot esculenta* Crantz	木薯 (mu shu)	cassava
	Fabaceae	*Pachyrhizus erosus* (L.) Urb.	豆薯 (dou shu)	yam bean
	Solanaceae	*Solanum tuberosum* L.	马铃薯 (ma ling shu)	potato
	Zingiberaceae	*Zingiber officinale* Roscoe	姜 (jiang)	ginger
others	Amaranthaceae	*Chenopodium quinoa* Willd.	藜谷 (li gu)	quinoa

Table 20.2 Production statistics for some of China's major introduced crops in 2008 (from FAO, 2012).

Crop	Annual production (China; million tons)	Annual production (world; million tons)	China's contribution (% of world production)	Area covered (China; million ha)	Area covered (world; million ha)
Apple (*Malus* spp.)	30	70	43	2	5
Tobacco (*Nicotiana* spp.)	3	7	43	1.25	4
Peanut (*Arachis hypogaea*)	14	38	37	4.6	24.5
Cotton (*Gossypium* spp.)	22	66	33	5.7	31
Walnut (*Juglans* spp.)	0.8	2.4	33	0.3	0.8
Tomato (*Lycopersicon esculentum*)	34	130	26	0.9	5
Rapeseed (*Brassica napus*)	12	58	21	6.6	30
Maize (*Zea mays*)	166	823	20	30	161
Sesame (*Sesamum indicum*)	0.6	3.6	17	0.5	7.5
Wheat (*Triticum* spp.)	112.5	690	16	23.5	224
Castor oil (*Ricinus communis*)	0.2	1.6	13	0.2	1.5
Mango (*Mangifera indica*)	4	34	12	0.5	4.7
Grape (*Vitis vinifera*)	7	67	10	0.4	7
Banana (*Musa* spp.)	8	91	9	0.3	0.3
Pepper (*Piper* spp.)	0.03	0.4	8	0.01	0.5
Sugar cane (*Saccharum* spp.)	125	1700	7	1.7	24
Pineapple (*Ananas comosus*)	1.4	19	7	0.07	0.8
Sunflower (*Helianthus annuus*)	1.8	35.5	5	1	25
Rubber (*Hevea brasiliensis*)	0.5	10.2	5	0.5	9.3
Sorghum (*Sorghum* spp.)	1.8	65.5	3	0.58	45
Barley (*Hordeum* spp.)	3.5	158	2	1	57
Oats (*Avena* spp.)	0.4	26	2	0.23	11
Coconut (*Cocos nucifera*)	0.6	60	1	0.03	11
Coffee (*Coffea* spp.)	0.03	8.3	0.4	0.02	10.4

Cereals.—The following are examples of some of the major introduced cereals cultivated in China today (for the United Nations Food and Agriculture Organisation (FAO) China and worldwide production statistics for cereals from 2008, see Table 20.2).

Triticum spp. —*Triticum* spp. (wheat; xiao mai; Poaceae) are staple foods for more than one-third of the world's human population, the most important food crop in the world by area of cultivation, and the second most important by productivity. It is widely cultivated in Europe, Asia, North and South America and Africa with, in 2007, China, India, the USA, Russia and France, being the the world's largest single wheat producers. Multiple forms of wheat have been domesticated. The diploid einkorn wheat (*Triticum monococcum*; A^mA^m) was domesticated a single time from its wild ancestor *T. boeoticum* around 10 500 years ago, in southeastern Turkey (Harlan & Zohary, 1966; Heun *et al.*, 1997; Salamini *et al.*, 2002). This crop was important for Neolithic agriculture in western Asia and Europe, but was largely abandoned in the Bronze Age (Salamini *et al.*, 2002). The tetraploid emmer wheat (*T. turgidum* ssp. *dicoccon*; BBAA) was domesticated from its tetraploid wild progenitor *T. dicoccoides* approximately 10 000 years ago, also in southeastern Turkey (Özkan *et al.*, 2002; Salamini *et al.*, 2002). Free-threshing forms of wheat were the most recently domesticated and include tetraploid durum wheat (*T. turgidum* ssp. *durum*; BBAA) and hexaploid bread wheat (*T. aestivum* ssp. *aestivum*; BBAADD). Durum wheat is hypothesized to have evolved from emmer wheat (Salamini et al., 2002). Hexaploid bread wheat derived from *Triticum urartu* (supplying the A genome), *Aegilops speltoides* and *Aegilops tauschii* (Petersen et al., 2006). Recent molecular phylogenies have shown that neither *Triticum, Aegilops*, nor *Triticum* plus *Aegilops* are monophyletic (Petersen *et al.*, 2006). Indeed, the generic relationships of the entire Tritiaceae tribe remain incompletely resolved (Petersen *et al.*, 2006). Most wild species in the Tritiaceae are considered to be potential resources for wheat improvement and breeding.

Three wheat species have been introduced and cultivated in China: *Triticum aestivum, T. durum* and *T. turgidum*. Wheat is one of the earliest crops introduced to China, more than 4 000 years ago. Wheat remains dating from the Neolithic Longshan culture which existed 4 000–5 000 years ago have been found at archeological sites in Gansu, Shaanxi, Shandong and Xijiang, suggesting widespread cultivation at that time (Chen, 2002; Chen, 2009). Remains in Anhui, Henan, Qinghai, Shaanxi, Shandong, Xinjiang and Xizang, dating to the Xia, Shang and Zhou dynasties (ca. 2100–500 BC), suggest that wheat was widely distributed in northern China during this period (Zeng, 2005; Chen, 2009). Cultivation of winter wheat appeared during the Shang Dynasty, but was mainly of spring wheat until the Spring and Autumn Period, beginning in 770 BC (Zeng, 2005). Wheat cultivation increased rapidly during the Han Dynasty (202 BC–220 AD), with the appearance of new techniques for making foods from wheat. However, in northern China, wheat remained a secondary crop relative to foxtail millet (*Setaria italica*; Poaceae), soybean (*Glycine max;* Fabaceae) and others until the beginning of the Tang Dynasty (around 620 AD). By the late Tang Dynasty, wheat was roughly as important as foxtail millet in northern China. During the Ming Dynasty (1368–1644 AD), wheat was cultivated across almost all of China, and was second only to rice in importance. After this time, wheat became the most important crop in northern China, comprising almost half of all crop production during the late Ming Dynasty. Cultivation of wheat also expanded somewhat southward during the Warring States Period (475–221 BC), but was still primarily limited to northern China. In southern China, wheat cultivation was initiated at a minimal level during the Western Jin Dynasty (around 300 AD), and expanded during the late Northern Song and early Southern Song Dynasties, around 1100 AD (Zeng, 2005).

Today, wheat is widely cultivated across China, stretching north to Mohe in Heilongjiang (52° N), east to the Pacific coast of northern China, west to Taxkorgan Tajik in Xinjiang (75° E) and south to Hainan (15° N). The lowest altitude at which wheat is cultivated is -150 m in the Turpan Pendi, Xinjiang, and the highest is 4 450 m at Xigazê, Xizang (Zheng & Li, 2000). An extremely successful introduced crop, wheat has not only changed Chinese food production, but also greatly affected Chinese dietary habits. Previously, millet was the staple food of northern China, and rice that of southern China, both of them generally boiled and eaten as grains. During its first 2 000 years of cultivation in China, wheat was prepared in a similar manner. Wheat flour first appeared in China during the Han Dynasty, and was not popularized until the Tang Dynasty, when the Chinese invented their own wheat-based foods: noodles, steamed (and sometimes stuffed) buns, dumplings, deep-fried twisted dough sticks and similar pastries. Baking techniques were introduced into China at the beginning of Qing Dynasty (1644 AD), but wheat is still mainly used in a traditional way (Zeng, 2005).

Zea mays. —*Zea mays* (maize; yu shu shu; Poaceae) is a major food crop, important oil crop, and is widely used as fodder and a source of industrial materials. It is the world's most productive food crop and third most widely-planted. Maize is widely cultivated in North America, South America, Asia, Africa and Europe. In 2007, China and the USA were the world's two greatest maize producers. When maize was initially domesticated, it was primarily for human food. Although it is still an important food source in developing countries, only 5% of total production is now used for human consumption. Approximately 70% of maize is used for animal feed, and 25% as a raw material for the production of alcohol, paper, foodstuffs and pharmaceuticals, among other industrial products. The number of such secondary applications is increasing rapidly.

The genus *Zea* contains four wild species, known collectively as teosinte, including *Z. diploperennis*, a diploid perennial from Jalisco, Mexico, *Z. luxurians*, an annual from Central America, *Z. perennis*, a tetraploid perennial also from Jalisco, and *Z. mays. Zea mays* includes four subspecies, *Z. mays* subsp. *mays, Z. mays* subsp. *huehuetenangensis* (a synonym of *Zea mays* L.) from western Guatemala, *Z. mays* subsp. *mexicana* from southwestern Mexico, and *Z. mays* subsp. *parviglumis* (a synonym of *Zea mays* L.), also from southwestern Mexico (Fukunaga *et al.*, 2005). Other authors have synonymized some of the subspecies into *Z. mays.* Maize appears to have been domesticated only once, from *Z. mays* subspecies *parviglumis* (Matsuoka *et al.*, 2002; Tenaillon, 2005), between 6 250 and 10 000 years ago (Smith, 1998; Piperno & Flannery, 2001). Detailed examinations of phylogenetic relationships among *Zea* species and subspecies, and the demographic history of the species, show that *Z. diploperennis* and *Z. perennis* are sister taxa, and *Z. luxurians* is sister to them. *Zea mays* subsp. *huehuetenangensis* is sister to a clade comprising *Z. mays* subsp. *parviglumis* and *Z. mays* subsp. *mexicana* (Fukunaga *et al.*, 2005; Ross-Ibarra *et al.*, 2009).

Maize was introduced to Spain by Christopher Columbus, and subsequently dispersed to many other areas of the world. It entered China at the beginning of the sixteenth century, and was first recorded in the *Chronicles of Yingzhou* in 1511. Possible routes for the introduction of maize into China from Europe include through western Asia, India and into western China; from Spain through Mecca into central China and onto northwestern China; or through the Philippines and into southeastern China by sea. Maize is cultivated in nearly all provinces of northern China, and the uplands of southern China. Over its 400 years of cultivation in the country, multiple varieties have been produced including a waxy maize, bred in southwestern China.

Hordeum vulgare. —*Hordeum vulgare* (common barley; da mai; Poaceae) belongs to the genus *Hordeum*, which contains 32 species distributed in temperate regions, including five species native to China. Barley is an important crop used as a human food, fodder and primary ingredient in brewing beer and producing Scotch whisky. The crop is widely cultivated on all continents except Antarctica, but primarily in Asia and Europe. Primary producers of barley in 2007 were France, the USA, Australia, Russia and China. Barley is one of the earliest domesticated crops, with active cultivation beginning approximately 10 000 years ago (Zohary & Hopf, 2000). Genetic evidence suggests barley was domesticated at least twice, in the Middle East and central Asia (Azhaguvel & Komatsuda, 2007; Morrell & Clegg, 2007). Consistent with this idea, the genetic composition of cultivated barley in Asia is somewhat distinct from that in Europe and North Africa (Morrell & Clegg, 2007; Saisho & Purugganan, 2007). The crop was introduced into China by 3 500 years ago, and has been a major crop in northern China since the Qin Dynasty (221–206 BC). Barley is widely cultivated across China, and remains a major food crop in the highlands of Qingzang, Qinghai, Sichuan and Yunnan. Peak cultivation of barley in China occurred between 1975 and 1977, when 65 million ha were under cultivation, and annual productivity was almost ten million tons.

Sorghum bicolor —*Sorghum bicolor* (sorghum; gao liang; Poaceae) belongs to the genus *Sorghum* Moench., which contains roughly 30 species distributed across tropical regions. *Sorghum bicolor* was domesticated in Africa (de Alencar Figueiredo *et al.*, 2008). It is an important food crop, and is also used in the industrial production of alcohol, sugar, vinegar, and other products. Today sorghum is widely cultivated in 48 countries on five continents, with India, Nigeria, the USA, Mexico and China being the world's largest producers. Seed remains found at Yangshao archeological site suggest that cultivation of sorghum in China has at least a 5 000 year history (Chen, 2002), and it is now cultivated in all provinces. In the early twentieth century, sorghum was cultivated across 7.4 million ha; at peak cultivation in 1952 it covered 9.4 million ha. Cultivation of sorgum decreased with the expansion of rice, wheat and maize.

Avena ssp. —*Avena* ssp. (oat; yan mai; Poaceae) belong to the genus *Avena* (Poaceae), which includes approximately 25 species distributed from Europe and the Mediterranean region to Ethiopia. Five oat species have been cultivated: common oat (*A. sativa*, 2n = 6x, AACCDD), Abyssinica oat (*A. abyssinica*, 2n = 4x, AABB), black oat (*A. strigosa*, 2n = 2x, AsAs), Mediterranean oat (*A. byzantina*, 2n = 6x, AACCDD), and naked oat (*A. nuda*, 2n = 6x, AACCDD). The black oat was also possibly domesticated from the wild species *A. barbata* in the Mediterranean region, but was later abandoned. The Abyssinian oat was domesticated in Ethiopia, perhaps evolving from wild *A. vaviloviana,* or from the black oat. Common and Mediterranean oats were domesticated in the Mediterranean region; their wild ancestors are not clear. The naked oat originated in China, where it potentially derived from the common oat (Loskutov, 2008). Oats are an important fodder and food crop, especially for the countries of northern Europe and North America, cultivated on five continents. Canada, Finland, the USA, Sweden and Russia were the largest producers in 2007; China was the seventeenth largest producer. In China oats, primarily the naked oat, have been cultivated for more than 2 000 years. They are mainly cultivated in Gansu, Nei Mongol and Yunnan, as well as Chongqing Shi, Guangxi, Hebei, Henan, Hubei, Hunan, Jiangxi, Shaanxi, Shanxi, Sichuan and Zhejiang.

Legume crops.—Legume crops (Fabaceae) are an important source of protein for humans, livestock and poultry, and legumes are widely used as food (including as vegetables), fodder and industrial materials. Many legume plants can also fix nitrogen which makes them important in crop rotation. The domestication and cultivation of legumes has a long history: remains of fenugreek (*Trigonella foenum-graecum*) and sword bean (*Canavalia gladiata*) have been found at sites in Africa dated to 8 000 BC; the broad bean and pea (*Pisum sativum*) were cultivated in the Middle East from 6 000 BC; and species of *Phaseolus* (common bean) were cultivated in Mexico as early as 4 000 BC (Dong & Liu, 2006). Based on cultivation area and productivity, the most important cultivated beans worldwide are *Phaseolus* spp., chickpea (*Cicer arietinum*), cowpea (*Vigna unguiculata*), pigeon pea (*Cajanus cajan*), lentil (*Lens culinaris*) and broad bean. The azuki bean (*Vigna angularis*) was domesticated in China; most other cultivated bean species are introduced. China is the primary country for cultivation of broad bean, garden pea, mung bean (*Vigna radiata*), common bean (*Phaseolus vulgaris*), azuki bean, cowpea and rice bean (*Vigna umbellata*). Here we outline some of the major introduced legume crops grown in China.

Cicer—*Cicer* (ying zui dou shu; Fabaceae) contains 44 species distributed from central Asia to the Mediterranean region. Chickpeas were first domesticated in the Middle East approximately 10 000 years ago (Tanno & Willcox, 2006). When and how this crop was introduced into China is not clear. Today, chickpeas are mainly cultivated in the northwest of the country, including Gansu, Qinghai and Xinjiang.

Lens culinaris—*Lens culinaris* (lentil; bing dou; Fabaceae) belongs to the genus *Lens*, which includes four species distributed in western Asia, the Mediterranean and North Africa. Lentils are one of the earliest-domesticated crops, and have been cultivated for more than 10 000 years in the Middle East (Nene, 2006; Sonnante *et al.*, 2009). It is not clear when lentils were introduced into China, but they are now cultivated mainly in Gansu, Hebei, Henan, Nei Mongol, Ningxia, Shaanxi and Shanxi.

Phaseolus—*Phaseolus* (cai dou shu; Fabaceae) contains 60 species distributed in tropical and temperate North and South America. Four *Phaseolus* species are widely cultivated: the common bean, lima bean (*P. lunatus*), runner bean (*P. coccineus*) and tepary bean (*P. acutifolius*). The common bean, *P. vulgaris,* was domesticated in two distinct regions of the New World, one in Mesoamerica (principally Mexico) and one on the eastern slopes of the Andes in South America (southern Peru, Bolivia and northwestern Argentina; Gepts & Bliss, 1986), and has a 5 500–7 000 year history of cultivation in South America (Kaplan, 1981). The common bean was introduced into China during the fifteenth century and is now widely cultivated, primarily in Gansu, Guizhou, Helongjiang, Jilin, Nei Mongol, Shaanxi, Shanxi and Yunnan. There are two types within the species, the common green bean and the snap bean. Both types are cultivated widely in China, but the country is the center of variation of the snap bean, which has crispy pods. Runner and lima beans are only rarely cultivated in China, and the tepary bean not at all.

Pisum sativum—*Pisum sativum* (Pea; wan dou; Fabaceae).

The taxonomy of *Pisum* is somewhat controversial but it includes two species (*P. fulvum* and *P. sativum*) distributed in the Mediterranean region. *Pisum sativum* is often divided into *P. sativum* subsp. *sativum* and *P. sativum* subsp. *elatius*. *Pisum sativum* subsp. *sativum* itself sometimes includes two domesticated varieties: var. *sativum* and var. *arvense*, while *P. sativum* subsp. *elatius* comprises three wild varieties, var. *elatius*, var. *pumilio* and var. *brevipedunculatum*. The garden pea is one of the earliest-domesticated crops; seed remains from ca. 7000 BC have been found in Turkey (Dong & Liu, 2006). It has been cultivated in China for about 2 000 years, although the precise time of introduction is not clear. The first recorded mention of the crop is in the *Guang Ya*, written by ZHANG Ji during the Wei Dynasty of the Three Kingdoms Period, 220–265 AD. Garden peas are now widely cultivated across the country.

Vicia faba—Vicia faba (Broad bean; can dou; Fabaceae) belongs to the genus *Vicia*, which comprises 160 species, widely distributed in the temperate and tropical Old and New Worlds. The broad bean was domesticated in the Middle East around 10 000 years ago (Tanno & Willcox, 2006) and has been cultivated in China since at least 2 000 years ago. Like *Pisum sativum*, it was first recorded in the *Guang Ya*. Broad beans are now widely cultivated in China, mainly in Gansu, Guangxi, Guizhou, Hubei, Jiangsu, Nei Mongol, Qinghai, Shaanxi, Shanghai Shi, Shanxi, Sichuan, Xinjiang, Xizang, Yunnan and Zhejiang.

Vigna—Vigna (jiang dou shu; Fabaceae) includes over 100 species distributed in the New World, Africa, Asia, Australia and Pacific regions. Multiple species of this genus have been domesticated including the azuki bean, black gram (*V. mugno*), cowpea, moth bean (*V. aconitifolia*), mung bean, rice bean. Mung bean, azuki bean and black gram are the most important worldwide in terms of production (Tomooka *et al.*, 2002), and the azuki bean is widely cultivated across China. The rice bean is cultivated mainly in eastern and southern Asia, and in China is primarily cultivated in Guizhou, Henan, Nei Mongol, Shaanxi, Shandong, Shanxi and Yunnan. Black gram is not widely cultivated, but is grown in Guangdong, Guangxi and Yunnan. The moth bean is cultivated only in Xishuangbanna (Yunnan). The earliest written records of cowpea in China are from the Sui Dynasty (581–618 AD), and this bean is mainly cultivated in Guangxi, Henan, Hubei, Hunan, Liaoning, Shanxi, Sichuan and Yunnan. *Vigna unguiculata* (cowpea), including several varieties, is native to Africa and Asia but widely cultivated in China as a vegetable. *Vigna unguiculata* subsp. *sesquipedalis* was introduced into China in around 1000 BC. The tender pods are used as a vegetable, the seeds also dried for food.

Tuber crops.—Two of the major introduced tuber crops cultivated in China include the potato and sweet potato.

Ipomoea batatas—Ipomoea batatas (sweet potato; fan shu; Convolvulaceae) belongs to the genus *Ipomoea*, a genus of roughly 650 species distributed in tropical to warm-temperate regions (Mabberley, 2008). Sweet potato was possibly first domesticated in Mexico, with *I. trifida* from Central America being its closesr wild relative (Austin, 1988; Zhang *et al.*, 2000). However, the oldest known remains of sweet potato were found in Peru, and date from ca. 2500 BC (Urgent & Peterson, 1988). As an important food, fodder and industrial crop, sweet potato is cultivated in more than 100 countries. At the peak of worldwide cultivation in 1962 it was cultivated over nearly 14 million ha; production has decreased slowly since that time. In 2008, the world cultivation area was eight million ha and total productivity 110 million tons (FAO, 2012). *Ipomoea batatas* was introduced into China during the Wanli Period of the Ming Dynasty (1572–1620). Today, China is the world's largest producer of sweet potato, with 62% of the total world area of cultivation. Sweet potatoes are widely cultivated across China; their peak cultivation area was 11 million ha in 1962, and reduced to 3.7 million ha by 2008.

Solanum tuberosum—Solanum tuberosum (Potato; ma ling shu; Solanaceae) belongs to the genus *Solanum*, which contains approximately 1 500 species distributed in tropical North and South America (Bohs, 2007). Potato (yang yu; *S. tuberosum*) was domesticated once in South America from the *S. brevicaule* complex (Spooner *et al.*, 2005). Archeological data suggest that potato cultivation began in South America about 10 000 years ago (Ames & Spooner, 2008). Potatoes are an important food, fodder and industrial crop, ranked fourth in overall production behind rice, wheat and maize. They are cultivated in 146 countries across 18 million ha, and total productivity in 2008 was 314 million tons (FAO, 2012). Like sweet potatoes, potatoes were introduced into China during the Wanli Period. China is now the world's largest potato cultivator, and the crop is widely cultivated across the country. In 2008, the area under potato cultivation was 4.5 million ha and total productivity was 57 million tons.

20.2.2 Oil crops

After food crops, oil crops are the second most important class of cultivated plants and are widely grown across five continents. Major oil crop producers include the USA, India, China, Brazil, Argentina, Indonesia and Malaysia. The four most productive oil crops are soybean, rapeseed, peanut and sunflower. Sixteen major oil crop species are cultivated in China, five of which were domesticated here: perilla (*Perilla frutescens*; Lamiaceae), soybean, shinyleaf yellowhorn (*Xanthoceras sorbifolium*; Sapindaceae), oiltea camellia (*Camellia oleifera*; Theaceae) and tung oil tree (*Vernicia fordii*). The remaining 11 species are introduced (see Tables 20.2, 20.3); some of them are described below.

Table 20.3 Introduced oil crops.

Family	Latin Name	Chinese Name	English Name
Arecaceae	*Elaeis guineensis* Jacq.	油棕 (you zong)	oil palm
Asteraceae	*Carthamus tinctorius* L.	红花 (hong hua)	safflower
Asteraceae	*Guizotia abyssinica* (L. f.) Cass.	小葵子 (xiao kui zi)	Indian niger
Asteraceae	*Helianthus annuus* L.	向日葵 (xiang ri kui)	sunflower
Brassicaceae	*Brassica napus* L.	欧洲油菜 (ou zhou you cai)	rape
Cyperaceae	*Cyperus esculentus* L.	油莎草 (you sha cao)	chufa
Euphorbiaceae	*Jatropha curcas* L.	麻风树 (ma feng shu)	physic nut
Euphorbiaceae	*Ricinus communis* L.	蓖麻 (bi ma)	castor
Fabaceae	*Arachis hypogaea* L.	落花生 (luo hua sheng)	peanut
Linaceae	*Linum usitatissimum* L.	亚麻 (ya ma)	flax
Pedaliaceae	*Sesamum indicum* L.	芝麻 (zhi ma)	sesame

Brassica ssp. —*Brassica* ssp. (rape; you cai; Brassicaceae). The common name rape (or rapeseed) refers to six species of the genus *Brassica*, a genus of approximately 40 species, distributed primarily in the Mediterranean region but also elsewhere in Europe, western, central and eastern Asia; 14 species are cultivated in China. Rape is the second most important oil crop worldwide, after soybean, with China, Canada, India, Germany and France contributing the greatest production. Rape has a high protein content, and is an important protein source for humans and livestock. It is also an important industrial material used widely in metallurgy, machinery, rubber, chemicals, energy, textiles and other applications.

There are three basic species of rape: black mustard (*B. nigra*, 2n = 16, BB), cabbage (*B. oleracea*, 2n = 18, CC) and colewort (*B. campestis*, 2n = 20, AA) and three hybrid species: cabbage type rape (*B. napus*, 2n = 38, AACC), Ethiopian rape (*B. carinata*, 2n = 34, BBCC), mustard type rape (*B. juncea*, 2n = 36, AABB). The relationships among these six species have been represented in a diagram known as the "Triangle of U" (Nagahara, 1935; Figure 20.1). The triangle shows that *B. napus*, a crop that appeared around 1 600 years ago, is a hybrid of *B. rapa* and *B. oleracea*, but it is not clear where this crop was domesticated (Gómez-Campo, 1999). *Brassica juncea* is a hybrid of *B. nigra* and *B. campestris*, and may have been domesticated in the Middle East or India (for oil production) and China (for vegetable use). *Brassica carinata* is a hybrid of *B. nigra* and *B. oleracea* and may have originated in Ethiopia. The vegetable known as Chinese colewort is hypothesized to have evolved from Asian oil-use colewort (*B. campestris*). However, 6 000–7 000 year old archeological remains of colewort seeds from Banpo, Shaanxi, suggest a possible independent domestication of Chinese colewort (Chen, 2002).

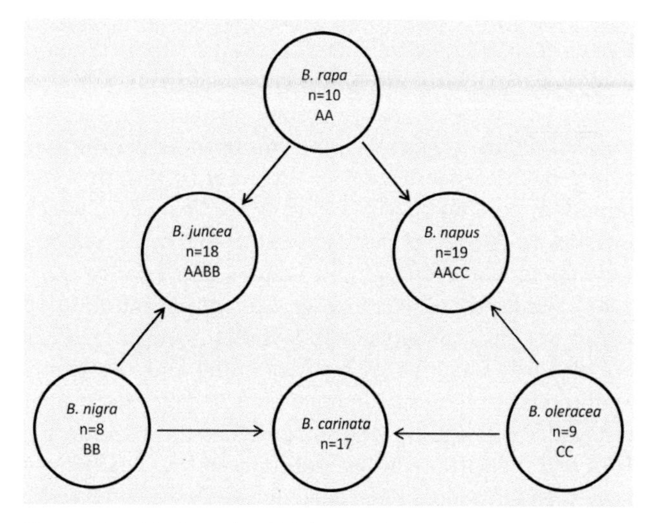

Figure 20.1 The "triangle of U" diagram, showing genetic relationships between six species of *Brassica*, with genome sizes and genome codes (after Nagahara, 1935).

China has a long history of rape cultivation, as evidenced by the Banpo remains. The earliest written record of rape cultivation appears in the *Book of Songs*, written during the Spring and Autumn Period (770 to ca. 476 BC). The cabbage-type rapes (*B. campestris*) and mustard-type rapes (*B. juncea*), that were traditionally cultivated in China, have relatively low oil yields. *Brassica napus* was first introduced into China from Japan in the early 1930s and European *B. napus* varieties were introduced in 1941 from the UK (Fang & Chang, 2007). After 1949, *B. napus* became the main cultivated species in China, with a large proportion of breeding effort focused on the development of locally-adapted varieties. China is now the world's largest rapeseed cultivator.

Other oil crops. —Arachis hypogaea—Arachis hypogaea (Peanut; luo hua sheng; Fabaceae) belongs to the genus *Arachis*, which contains 69 species and 12 unnamed species distributed in South America (Krapovickas & Gregory, 1994). Krapovickas and Gregory (1994) have divided *A. hypogaea* into *A. hypogaea* subsp. *hypogaea* and *A. hypogaea* subsp. *fastigiata*, with *A. hypogaea* subsp. *hypogaea* further divided into two varieties, and *A. hypogaea* subsp. *fastigiata* into four. The cultivated peanut is an allotetraploid, and was domesticated at least 1 000 years ago, possibly in present-day southern Bolivia or northern Argentina (Gregory *et al.*, 1973). There is still debate as to which species is the wild ancestor of the cultivated peanut.

Peanuts are considered to have been introduced to Europe by Columbus in the 1500s, and then dispersed to other regions of Eurasia. However, the introduction of the peanut into China may pre-date this, as the crop is first recorded in the *You Yang Za Zu* written by DUAN Cheng-Hi during the Tang Dynasty, prior to 1000 AD. The peanut emerged as a worldwide crop in the late nineteenth century, with cultivation increasing rapidly during the twentieth century. It is now one of the five most important oil crops worldwide, and is also an important protein source. It is widely cultivated throughout the world, with production concentrated in Asia, Africa, and North and South America. In 2007, China, India, and Nigeria were the largest producers of peanuts, with the cultivation area under peanuts in China increasing rapidly into the late nineteenth century. The country I now placed second in the world for peanut cultivation in terms of area and first in total production. Peanuts are cultivated mainly in Anhui, Guangdong, Guangxi, Hebei, Henan, Hubei, Hunan, Jiangsu, Jiangxi, Liaoning, Shandong and Sichuan.

Helianthus annuus—Helianthus annuus (sunflower; xiang ri kui; Asteraceae) is the world's fourth most important oil crop belongs to the genus *Helianthus*, which includes ca. 50 species native to North America. Sunflowers were domesticated once, in the eastern USA (Harter *et al.*, 2004). Sunflower oil is a high-grade cooking and food oil which also has industrial applications. Sunflower seeds are also eaten as a snack food in China and elsewhere. Sunflowers are cultivated in more than 70 countries around the world, with China, Argentina, Russia, the Ukraine, the USA and India being the largest producers. Sunflowers may have been introduced into China during the Ming Dynasty, although the timing and mechanism of introduction are not clear (Fang & Chang, 2007). Sunflowers are cultivated mainly in northeastern China.

Sesamum indicum—Sesamum indicum (Sesame; zhi ma; Pedaliaceae) belongs to the genus *Sesamum* , which contains 36 species distributed in Africa and India (Kobayashi, 1981). *Sesamum indicum* has been divided into *S. indicum* subsp. *bicarpellatum* and *S. indicum* subsp. *quadricarpellatum.* Sesame may have been first domesticated in Ethiopia. However, the location and number of domestication events has been debated (Bedigian & Harlan, 1986; Thangavelu, 1994). The Indian species *S. orientale*, which has now been synonymized under *S. indicum*, has been considered the wild ancestor of cultivated sesame (John *et al.*, 1950), but *S. triphyllum* var. *grandiflorum* from India and eastern Africa, and *S. latifolium* from eastern Africa, are also possible ancestors (Nayar & Mehra, 1970; Bedigian & Harlan, 1986). Sesame is the fifth most important oil crop in the world, and is also an important protein source and industrial material. It is mainly cultivated in tropical regions of Asia and Africa, as well as North and South America and Australia, with India, China and Myanmar being the largest contributors (FAO, 2012). Sesame has been cultivated in China for more than 2 000 years. Seeds of sesame have been found at the Qianshanyang and Shuitianfan archeological sites in Zhejiang, and dated to about 770–480 BC. Sesame is now widely cultivated across China, mainly in Anhui, Hebei, Henan, Hubei and Jiangxi, and China is now the largest contributor to sesame production worldwide.

Carthamus tinctorius—C. tinctorius (safflower; hong hua; Asteraceae) belongs to the genus *Carthamus*, which contains approximately 20 species distributed in Africa, Europe and Asia (Lopez Gonzalez, 1989). Safflower oil is a high-quality edible and industrial oil. Safflower is also good fodder, and in China is sometimes used as a medicine. Safflower is primarily cultivated in Asia, Europe, North America and Australia, with India the world's largest contributor to safflower production. Safflower may have been domesticated in eastern Africa and the Mediterranean region, from the putative wild ancestor *Carthamus oxyacanthus* (Chapman & Burke, 2007). It was introduced into China during the Han Dynasty (from around 200 BC) for use in dyes and medicines, and varieties optimized for oil production were introduced in the 1960s. Safflower is now mainly cultivated in Xinjiang, as well as Gansu, Guizhou, Hebei, Henan, Ningxia, Shaanxi, Shandong, Shanxi, Sichuan, Yunnan and other provinces, over a total of more than a million hectares (FAO, 2012). The flowers are also used in Traditional Chinese Medicine to treat amenorrhea, dysmenorrhea and fractures, and in Achang, Dai, Deang, Jingpo, Korean, Miao, Mongolian, Uyghur, Yi, Zang and Zhuang medicine for hypertension, heart trouble, tracheitis, coronary disease and other ailments. The flowers are also used to extract a red pigment.

Elaeis guineensis—Elaeis guineensis (oil palm; you zong; Arecaceae) belongs to the genus *Elaeis*, which includes two species (*E. guineensis* and *E. oleifera*) distributed in tropical America and Africa. Palm oil is used as both an edible and industrial oil, with world palm oil production in 2011 being over 48 million tons (FAO, 2012), and increasing rapidly. Oil palm was introduced into Indonesia in 1848 as an ornamental, and taken into cultivation in 1911. It was introduced into China in 1926, and since 1960 has principally been cultivated in Hainan and Yunnan (Fang & Chang, 2007). It is now primarily cultivated in Indonesia and Malaysia; China is the world's seventh largest producer.

Ricinus communis—Ricinus communis (castor; bi ma; Euphorbiaceae) belongs to the genus *Ricinus*, *R. communis* , with four subspecies, *R. communis* ssp. *sanguineus*, *R. communis* ssp. *manspuricus*, *R. communis* ssp. *zanzibariensis* and *R. communis* ssp. *Persicus.* Castor oil is an important industrial oil, primarily cultivated in Africa, Asia, South America and Europe. Major castor oil-producing countries include India, China, and Brazil, of which China is the second most significant. Castor oil appears to have been domesticated in eastern Africa, and was introduced into China from India about 1 400 years ago. It is widely cultivated across the country, including Jilin, Liaoning, Nei Mongol, Shaanxi and Shanxi. In 2008, the area under cultivation was approximately 225 000 ha, and productivity 220 000 tons.

20.2.3 Fiber Crops

Fiber crops are cultivated for the production of cloth, paper and rope, and include cotton (*Gossypium* spp.), the seed fibers of which primarily provide material for textiles. The fibers of flax (*Linum usitatissimum*; Linaceae) are used to make linen, hemp (*Cannabis sativa*; Cannabaceae) for twine and canvas, and jute (*Corchorus capsularis*; Malvaceae) and kenaf (*Hibiscus cannabinus*; Malvaceae) for burlap. The leaf fibers of sisal (*Agave sisalana*; Asparagaceae) and manila hemp (*Musa textilis*; Musaceae) are used to make rope. A number of fiber crops were first domesticated in China including kapok (*Bombax ceiba*; Bombacaceae), ramie (*Boehmeria nivea*; Urticaceae), velvetleaf (*Abutilon theophrasti*; Malvaceae), kendir (*Apocynum venetum*; Apocynaceae) and possibly hemp. The major introduced fiber crops are listed in Table 20.4, and selected examples are discussed below.

Table 20.4 Introduced fiber crops.

Family	Latin Name	Chinese Name	English Name
Asparagaceae	*Agave angustifolia* Haw.	狭叶龙舌兰 (xia ye long she lan)	Caribbean agave
Asparagaceae	*Agave cantula* Boxb.	龙舌兰麻 (long she lan ma)	cantala
Asparagaceae	*Agave sisalana* Perrine	剑麻 (jian ma)	sisal
Linaceae	*Linum usitatissimum* L.	亚麻 (ya ma)	flax
Malvaceae	*Gossypium arboreum* L.	树棉 (shu mian)	tree cotton
Malvaceae	*Gossypium barbadense* L.	海岛棉 (hai dao mian)	island cotton
Malvaceae	*Gossypium hirsutum* L.	陆地棉 (lu di mian)	cotton
Malvaceae	*Hibiscus cannabinus* L.	大麻槿 (da ma jin)	kenaf
Musaceae	*Musa textilis* Née	蕉麻 (jiao ma)	Manilla hemp
Tiliaceae	*Corchorus capsularis* L.	黄麻 (huang ma)	jute

Gossypium spp. —*Gossypium* spp. (cotton; mian; Malvaceae) is the world's most important fiber crop, and the only one in which the fiber is produced by the seeds. Wild *Gossypium* species are mainly distributed in Africa, North America, South America, Hawaii and Australia. Fryxell's (1992) taxonomic system for the genus includes 51 species in four subgenera, eight sections and ten subsections, with four cultivated species: Asian cotton (*G. arboreum*), grass cotton (*G. herbaceum*), island cotton (*G. barbadense*) and upland cotton (*G. hirsutum*). Grass cotton and Asian cotton are diploid. Grass cotton evolved from the southern African wild species *G. herbaceum* subsp. *Africanum*, was introduced into India and there evolved into Asian cotton. Upland cotton and island cotton are hetero-tetraploid hybrids of African grass cotton and North American wild cotton species (Wendel & Cronn, 2003). There is still debate as to whether upland cotton and island cotton were domesticated through one or two separate domestication events (Westengen *et al.*, 2004). The North American wild species involved in the formation of upland cotton and island cotton is also not confirmed: possible candidates include *G. gossypioides*, *G. klotzschianum*, *G. raimondii* and *G. trilobum*.

Upland cotton is the most widely cultivated cotton, and contributes 90% of total cotton production. Island cotton provides 5–8% of total productivity, and Asian cotton 2–5%; grass cotton is now rarely cultivated. Cotton is cultivated in more than 150 countries, with China followed by India and the USA being the three largest contributors. Grass cotton was introduced into Asia not later than 1000 BC, and into China, most likely by ZHANG Qian, 2 200 years ago during the Han Dynasty. Grass cotton was cultivated mainly in Gansu and Xinjiang; Asian cotton was introduced from India to Fujian, Guangdong, Guangxi, Hainan and Yunnan, a story which is recorded in the *Yu Gong* from the Warring States Period (475–221 BC). Upland cotton and island cotton were introduced to China during the 1950s, and gradually replaced Asian cotton. Cotton is now cultivated in 28 provinces or municipalities of China, with production concentrated in Anhui, Hebei, Henan, Hubei, Jiangsu, Shandong, and Xinjiang.

Other major introduced fiber crops.—Corchorus capsularis—Corchorus capsularis (Jute; huang ma; Tiliaceae) is an important fiber crop, second only to cotton, and used to make coarse cloth; however the cultivation of jute has reduced greatly with the development and use of synthetic fibers. Jute belongs to the genus *Corchorus*, which contains approximately 70 species distributed across tropical regions; two of them—*C. capsularis* and *C. olitorius*—cultivated, and six found in China. *Corchorus capsularis* in southern China, India or Myanmar, and *C. olitorius* was domesticated in eastern Africa (Fang & Chang, 2007). The major jute cultivation countries are India and Bangladesh, with China the third largest contributor. China has a long history of jute cultivation, first recorded in the *Tu Jing Ben Cao* (1061 AD). Jute is primarily cultivated in Fujian, Guangdong, Jiangxi and Zhejiang, and also in Anhui, Guangxi, Hubei, Hunan, Jiangsu, Sichuan and Taiwan.

Hibiscus cannabinus—Hibiscus cannabinus (kenaf; da ma jin; Malvaceae) is an important fiber crop, domesticated in Africa 4 000 years ago. *Hibiscus mechowii* is the likely ancestral species of kenaf (Fang & Chang, 2007). It is primarily cultivated in Asia, including China, India,

Thailand, Indonesia, Vietnam and neighboring countries. China is one of the largest producers of kenaf, which is cultivated in more than 20 provinces, principally Anhui, Henan and Sichuan.

Linum usitatissimum—Linum usitatissimum (flax; ya ma; Linaceae) is one of the oldest cultivated fiber crops, also used as an edible oil, medicine and vegetable jelly. *Linum* (ya ma shu) includes more than 200 species distributed in temperate and subtropical regions, of which nine are native to China. Flax was cultivated in Iran and Egypt by 7 000 years ago and in China by 5 000 years ago, but the history of its domestication is unclear. Flax is today cultivated in over 20 countries, with China, France, Russia and Belarus the four largest contributors, and in at least ten provinces of China.

20.2.4 Fruits

Fruits contain carbohydrates, proteins, multiple vitamins, organic acids and other essential nutrients, and are important in human diets. Fruit crops are major components of world agriculture, being cultivated in more than 170 countries, in some of which they are a key industry. In addition to the great number of native cultivated fruits, China has a long history of cultivating introduced fruits. For instance, grapes (*Vitis vinifera*) were introduced into China more than 2 000 years ago. China is now the largest world fruit producer; fruit farming provided an annual revenue of 90 billion Yuan at end of the twentieth century, and is a key industry for some regions of China. The major introduced fruit species are listed in Table 20.5, and some examples are discussed below.

Malus domestica Borkh. —*Malus domestica* (Apple; ping guo; Rosaceae) is widely cultivated throughout the world, with the majority of cultivation in Asia and Europe. It was one of the earliest fruits to be domesticated: apple remains dating to 7 000 years ago have been found in Europe. However, its early domestication history remains unclear. The central Asian *M. sieversii* has been hypothesized as a wild ancestor based on morphological and molecular data (Robinson *et al.*, 2001; Forte *et al.*, 2002; Harris *et al.*, 2002), while other studies have reported *M. sylvestris* as a closer wild relative of *M. domestica* (Coart *et al.*, 2006). In China, a few local *Malus* species were domesticated and cultivated as fruit crops, including *M. asiatica*, *M. prunifolia*. and *M. domestica* subsp. *chinensis*. The cultivation of *M. domestica* subsp. *chinensis* has a more than 2 000 year history and is recorded in the *Shang Lin Fu*, written by SI-MA Xiang-Ru from 126–118 BC. *Malus domestica* subsp. *chinensis* was most likely domesticated from *M. sieversii*, which is still distributed in some areas of Xinjiang. *Malus domestica* was introduced into Shandong by the American missionary John Livingstone Nevius in 1871. Cultivation of *M. domestica* developed slowly until the 1950s, but rapidly thereafter, to replace local domesticated species. It is now widely cultivated in more than 25 provinces of northern China and China is the world's largest contributor to apple production.

Other fruits.—Ananas comosus (pineapple; feng li; Bromeliaceae) is native to, and was domesticated in, South America, although the exact location is not clear. Pineapples are today cultivated in 95 countries of central and North America, Europe, Africa, Oceania and Asia. Following its introduction to Europe by Columbus, *A. comosus* was introduced to China in the late 1550s, where it is primarily cultivated in Fujian, Guangdong, Guangxi, Hainan, Taiwan and Yunnan. *Annona squamosa* (fan li zhi; custard or sugar apple; Annonaceae) is an evergreen shrub with berry-like, edible fruits, native to tropical America. It was introduced to China in the seventeenth century and is cultivated throughout southern China. The jackfruit, *Artocarpus heterophyllus* (bo luo mi), is a large evergreen tree introduced to China from its native India in the ninth century and now cultivated at low elevations in Guangdong, Guangxi, Hainan and southern Yunnan. The fruits are used fresh, dried and made into wine. A yellow dye is also extracted from the wood.

Carica papaya—Carica papaya (papaya; fan mu gua; Caricaceae) is a small evergreen tree native to southern Mexico, introduced to China in seventeenth century and cultivated in the southern provinces. The fruits are used fresh and canned, as juice or in jams.

Cocos nucifera—Cocos nucifera (coconut; ye zi; Arecaceae) was introduced to China about 2 000 years ago. In the early twentieth century only small plantations existed in Wenchang, Hainan. By 1950, the commercial growing area was 5 000 ha, with a production of 58 400 fruits per year, and by 2000, this had increased to 40 000 ha producing 190 million fruits. Today it is cultivated in Guangdong, Hainan, Taiwan and Yunnan. In Hainan three types are grown: tall, dwarf and hybrid. Coconuts are widely cultivated in more than 80 countries of Asia, Africa, Oceania and South America, and it is not clear where they were first domesticated. The fruits are used for drinks and to produce coconut powder, butter, oil, and other products.

Fragaria × ananassa—Fragaria × ananassa (strawberry; cao mei; Rosaceae) belongs to the genus *Fragaria*, which contains nine species distributed in Asia, Europe, and North and South America. Multiple local wild species of *Fragaria* including the European *F. vesca*, the eastern North American *F. virginiana* and the South American *F. chiloensis* were consumed by humans before the domestication of *F. × ananassa*. *Fragaria × ananassa* appeared in Europe around 1750 as a hybrid between *F. vesca* and *F. chiloensis*. Strawberries are mainly cultivated in Europe, North America and Asia, with countries that contributed most to cultivation in 2007 including the USA, Spain, Turkey, Russia and Korea. China has a long history of using local wild strawberries; *Fragaria × ananassa* was introduced at the beginning of the twentieth century and is now cultivated across the country.

Table 20.5 Introduced fruits.

Family	Latin Name	Chinese Name	English Name
Anacardiaceae	*Anacardium occidentale* L.	腰果 (yao guo)	cashew
Anacardiaceae	*Mangifera indica* L.	芒果 (mang guo)	mango
Annonaceae	*Annona squamosa* L.	番荔枝 (fan li zhi)	sugar apple
Arecaceae	*Cocos nucifera* L.	椰子 (ye zi)	coconut
Arecaceae	*Phoenix dactylifera* L.	海枣 (hai zao)	date palm
Bromeliaceae	*Ananas comosus* (L.) Merr.	凤梨 (feng li)	pineapple
Caricaceae	*Carica papaya* L.	番木瓜 (fan mu gua)	papaya
Clusiaceae	Garcinia × mangostana L.	莽吉柿 (mang ji shi)	mangosteen
Elaeagnaceae	*Elaeagnus angustifolia* L.	沙枣 (sha zao)	Russian olive
Ericaceae	*Vaccinium vitis-idaea* L.	越桔 (yue ju)	cowberry
Grossulariaceae	*Ribes nigrum* L.	黑茶藨子 (hei cha biao zi)	black currant
Grossulariaceae	*Ribes rubrum* L.	红茶藨子 (hong cha biao zi)	red currant
Lauraceae	*Persea americana* Mill.	鳄梨 (e li)	avocado
Lythraceae	*Punica granatum* L.	石榴 (shi liu)	pomegranate
Malvaceae	*Durio zibethinus* L.	榴莲 (liu lian)	durian
Moraceae	*Artocarpus altilis* (Parkinson ex F. A. Zorn) Fosberg	面包果 (mian bao guo)	breadfruit
Moraceae	*Artocarpus heterophyllus* Lam.	波罗蜜 (bo luo mi)	jackfruit
Moraceae	*Ficus carica* L.	无花果 (wu hua guo)	fig
Musaceae	*Musa paradisiaca* L.	香蕉 (xiang jiao)	banana
Musaceae	*Musa basjoo* Siebold & Zucc. ex Iinuma	芭蕉 (ba jiao)	Japanese banana
Musaceae	*Musa× paradisiaca* L.	大蕉 (da jiao)	plantain
Myrtaceae	*Psidium guajava* L.	番石榴 (fan shi liu)	guava
Oxalidaceae	*Averrhoa carambola* L.	阳桃 (yang tao)	star fruit
Passifloraceae	*Passiflora caerulea* L.	西番莲 (xi fan lian)	granadilla
Rosaceae	*Amygdalus communis* L.	扁桃 (bian tao)	almond
Rosaceae	*Cerasus mahaleb* (L.) Mill.	圆叶樱桃 (yuan ye ying tao)	mahaleb cherry
Rosaceae	*Cerasus vulgaris* Mill.	欧洲酸樱桃 (ou zhou suan ying tao)	sour cherry
Rosaceae	*Cydonia oblonga* Mill.	榅桲 (wen po)	quince
Rosaceae	*Fragaria × ananassa* (Weston) Duchesne	草莓 (cao mei)	strawberry
Rosaceae	*Malus domestica* Borkh.	苹果 (ping guo)	apple
Rosaceae	*Prunus americana* Marshall	美洲李 (mei zhou li)	American plum
Rosaceae	*Prunus avium* (L.) L.	欧洲甜樱桃 (ou zhou tian ying tao)	sweet cherry
Rosaceae	*Prunus cerasifera* Ehrh.	樱桃李 (ying tao li)	cherry plum
Rosaceae	*Prunus domestica* L.	欧洲李 (ou zhou li)	plum
Rosaceae	*Prunus insititia* L.	乌荆子李 (wu jing zi li)	damson plum
Rosaceae	*Prunus serotina* Ehrh.	黑樱桃 (hei ying tao)	black cherry
Rosaceae	*Prunus spinosa* L.	黑刺李 (hei ci li)	blackthorn
Rosaceae	*Prunus ussuriensis* Kovalev & Kostina	东北李 (dong bei li)	Ussuri plum
Rosaceae	*Pyrus communis* L.	西洋梨 (xi yang li)	common pear
Rutaceae	*Citrus bergamia* Risso & Poit.	巴柑檬 (ba gan meng)	bergamot
Rutaceae	*Citrus limon* (L.) Burm. f.	柠檬 (ning meng)	lemon
Rutaceae	*Citrusparadisi* Macfad.	葡萄柚 (pu tao you)	grapefruits
Sapindaceae	*Nephelium lappaceum* L.	红毛丹 (hong mao dan)	rambutan
Sapotaceae	*Manilkara zapota* (L.) P. Royen	人心果 (ren xin guo)	sapodilla
Vitaceae	*Vitis vinifera* L.	葡萄 (pu tao)	grape

Mangifera indica—Mangifera indica (mango; mang guo; Anacardiaceae) belongs to the genus *Mangifera*, which contains 40–60 species distributed in southern to Southeast Asia. Mangoes were domesticated in Southeast Asia, and introduced from there to China as early as the Ming Dynasty (1368–1644). Mangoes are primarily cultivated in tropical Asia, Africa and North America, particularly in India and China, the world's second largest producer. Chinese mango cultivation centers on Fujian, Guangdong, Guangxi, Hainan, Sichuan, Taiwan and Yunnan. Since the middle of the twentieth century, a series of new cultivars has been released in Taiwan. The fruits are used fresh and in juice, jam, wine, vinegar and pickles.

Musa paradisiaca—Musa paradisiaca (banana; xiang jiao; Musaceae) belongs to the genus *Musa*, which contains 37 species distributed in tropical Asia, with ca. ten in China. Banana is one of the oldest domesticated fruits, with a 3 000—4 000-year cultivation history, a hybrid by *M. acuminata* and *M. balbisiana*. Banana is one of the most important fruits, behind only grape and orange in terms of volume of production worldwide, and is widely cultivated in 130 countries across Africa, Asia, North and South America. In China bananas have been cultivated for 2 000 years, primarily in Fujian, Guangdong, Guangxi, Taiwan and Yunnan, as well as Chongqing Shi, Guizhou, Sichuan and other provinces.

Vitis vinifera—Vitis vinifera (grape; pu tao; Vitaceae) belongs to the genus *Vitis* L., which contains 65 species distributed in Asia, Europe and North America and centered on China, with 37 species. Grapes is one of the world's most important fruits, consumed directly as table grapes and used in the production of wine (Vivier & Pretorius, 2002). The seed oil is also becoming increasingly popular. Grapes are widely cultivated in Europe, Asia, North and South America, Oceania and Africa. In 2007, Italy, China, the USA, France and Spain were the world's largest grape producers. Grapes were domesticated at least twice from their wild ancestor, *V. vinifera* subsp. *sylvestris*: once in the Middle East and once in the western Mediterranean (Arroyo-Garcia *et al.*, 2006). Grapes have been cultivated in China for more than 2 000 years, with a number of centers of production developing including Hetian and Turpan in Xinjiang, Lichang in Hebei, and Pingdu in Shangdong, among others. Grape cultivation in China increased dramatically in the late nineteenth century with the introduction of new cultivars and wine-making techniques. Now, grapes are widely cultivated across Anhui, Hebei, Henan, Hubei, Jiangsu, Jilin, Liaoning, Shaanxi, Shandong, Shanxi, Sichuan, Taiwan, Tianjin, Xinjiang and Zhejiang.

Persea americana—Persea americana (avocado; e li; Lauraceae) is an evergreen broad-leaved tree bearing pear-shaped, oil-rich fruits used both fresh and processed. Native to tropical America and introduced to China in the 1920s, it is cultivated in Taiwan and southern China. *Synsepalum dulcificum* (miracle fruit; shen mi guo; Sapotaceae) is an evergreen tree native to tropical western Africa. Introduced to China in the early 1960s, it is now cultivated in Fujian, Guangdong, Guangxi and Yunnan. The fruits contain the compound miraculin, which alters the sense of taste, making sour foods taste sweet. *Syzygium samarangensis* (champoo or wax apple; yang pu tao; Myrtaceae), from Indonesia, Malaysia, Papua New Guinea and Thailand, is introduced and cultivated in Fujian, Guangdong, Guangxi, Sichuan, Taiwan and Yunnan. *Syzygium malaccensis* (Malay rose apple; ma liu jia pu tao), probably native to Malaysia and commonly cultivated for its fruit in wet-tropical areas around the world, is introduced, cultivated and naturalized in Taiwan and Yunnan.

20.2.5 Nuts

Nuts, which are defined as fruits with a dry pericarp when mature, contain substantial quantities of essential nutrients in balanced proportions. Thus they are an excellent source of minerals, vitamins and enzymes and many are widely cultivated throughout the world, including pecan (*Carya illinoensis*; Juglandaceae), chestnut (li; *Castanea mollissima*; Fagaceae), walnut (hu tao shu; *Juglans* spp.) and pistachio (kai xin guo; *Pistacia vera;* Anacardiaceae). China was an important center of domestication for nut trees including chestnut, gingkgo (*Ginkgo biloba*; Ginkgoaceae), and possibly walnut. Multiple nut species have also been introduced into China (see Table 20.6).

Juglans ssp. —*Juglans* ssp. (walnut; hu tao; Juglandaceae). Walnut may or may not be exotic to China. Two species, the iron walnut (*Juglans regia*) and walnut (*J. sigillata* Dode),

Table 20.6 Introduced nuts.

Family	Latin Name	Chinese Name	English Name
Anacardiaceae	*Pistacia verna* L.	开心果 (kai xin guo)	pistachio
Arecaceae	*Areca catechu* L.	槟榔 (bin lang)	betel palm
Betulaceae	*Corylus avellana* L.	欧洲榛子 (ou zhou zhen zi)	Hazelnut
Combretaceae	*Terminalia catappa* L.	榄仁树 (lan ren shu)	tropical almond
Fagaceae	*Castanea crenata* Siebold & Zucc.	日本栗 (ri ben li)	Japanese chestnut
Fagaceae	*Castanea dentata* (Marshall) Brokb.	美洲栗 (mei zhou li)	American chestnut
Fagaceae	*Castanea sativa* Mill.	欧洲栗 (ou zhou li)	European chestnut
Juglandaceae	*Carya illinoinensis* (Wangenh.) K. Koch	美国山核桃 (mei guo shan he tao)	pecan
Oleaceae	*Olea europaea* L.	油橄榄 (you gan lan)	olive

have been domesticated. Wild iron walnut occurs in Sichuan, Yunnan and Xizang, and was domesticated in southwestern China where records of its cultivation date to the Ming Dynasty. The domestication history of the true walnut, *J. sigillata,* however, remains unclear: it is possible that domestication occurred in either Xinjiang or central Asia; wild populations are still found in both regions. The walnut was certainly introduced into central China by envoy ZHANG Qian during the Han Dynasty (202 BC–220 AD). Walnuts are used both as edible nuts and for oil; the wood of the walnut tree is also valued in furniture-making and charcoal production, among other purposes. Walnuts are widely cultivated in Asia, Europe, North and South America and Africa, with China, followed by the USA, Turkey, and Iran, having the largest walnut production. Walnuts are the most important nut crop in China, produced primarily in 21 provinces and administrative districts: Anhui, Beijing Shi, Gansu, Guangxi, Guizhou, Heibei, Henan, Hubei, Hunan, Jiangsu, Liaoning, Ningxia, Qinghai, Shaanxi, Shandong, Shanxi, Sichuan, Tianjin, Yunnan, Xinjiang and Xizang, but also cultivated in Fujian, Nei Mongol and Zhejiang.

Corylus avellana—Corylus avellana (hazelnut; ou zhou zhen zi; Betulaceae) is the major cultivated species of *Corylus*, which includes 16 species distributed in Asia, Europe and North America. *Corylus avellana* is found in Europe and western Asia, and may have been domesticated in the Mediterranean region. Hybrids between *C. avellana* and *C. maxima* are also cultivated. Hazelnuts are primarily grown in Europe, western Asia and North America. Wild hazelnuts have been used in China for more than 6 000 years but were not formally cultivated until domesticated *C. avellana* was introduced in the 1940s. Using a Chinese wild species (*C. heterophylla*) as another parent, hybrids were successfully bred, and are now cultivated in more than ten provinces in northern and southwestern China.

20.2.6 Sugar crops

Sugar is a primary resource of energy for the human body, providing a carbon skeleton for metabolism. The main crops grown for sugar production worldwide are sugar cane (*Saccharum officinarum*; Poaceae) and sugar beet (*Beta vulgaris*; Amaranthaceae). All major sugar crops cultivated in China were introduced from other countries (see Table 20.7).

Table 20.7 Introduced sugar crops.

Family	Latin Name	Chinese Name	English Name
Amaranthaceae	*Beta vulgaris* L.	甜菜 (tian cai)	sugar beet
Arecaceae	*Borassus flabellifer* L.	糖棕 (tang zong)	sugar palm
Asteraceae	*Stevia rebaudiana* (Bertoni) Bertoni	甜叶菊 (tian ye ju)	stevia
Poaceae	*Saccharum sinense* Roxb.	竹蔗 (zhu zhe)	Uba cane
Poaceae	*Saccharum officinarum* L.	甘蔗 (gan zhe)	sugarcane
Poaceae	*Sorghum bicolor* (L.) Moench	高粱 (gao liang)	sorghum

Saccharum officinarum—Saccharum officinarum (sugarcane; gan zhe; Poaceae) is the primary source of sugar, and is also used to produce alcohol. Cultivated sugar cane includes three species Indian *S. barberi,* tropical *S. officinarum* and Chinese *S. sinense*. The domestication history of sugar cane is not clear: *S. officinarum* is hypothesized to have evolved from *S. robustum* in New Guinea, with *M. sinensis* and *M. sacchariflorus* possibly involved in this process (Grassl, 1967). The Chinese species *S. sinense* is a possible hybrid between *S. officinarum* and either wild *S. spontaneum* (Brandes, 1958) or *M. sacchariflorus* (Grassl, 1964). *Saccharum barberi* was derived from the hybridization of *S. officinarum* and *S. spontaneum* or other species of the *Saccharum* complex (Grassl, 1977).

Sugar cane is cultivated in more than 130 countries of South and Central America, Oceania, Asia and Africa. The ten biggest sugar cane producing countries in 2007 were Brazil, India, China, Thailand, Mexico, Pakistan, Australia, Colombia, USA and Guatemala. China has a long history of sugar cane cultivation. The earliest records appeared in the fourth century BC, during the Warring States Period. Sugar-refining techniques were invented in India and introduced into China around 645 AD, which dramatically promoted the cultivation of sugar cane. Sugar cane is cultivated mainly in 11 provinces: Fujian, Guangdong, Guangxi, Guizhou, Hainan, Hunan, Jiangxi, Sichuan, Taiwan, Yunnan and Zhejiang. In 2008, the area cultivated for sugar cane was 1.7 million ha and total production was 125 million tons.

Beta vulgaris—Beta vulgaris (sugar beet; tian cai; Chenopodiaceae) belongs to the genus *Beta*, which contains 12 species distributed in Europe, the Mediterranean region and Asia. *Beta vulgaris* (tian cai) includes four varieties, with the most widely-cultivated being common beet, *Beta vulgaris* var. *vulgaris*. This is further divided into four cultivar groups: sugar beet used for making sugar, garden beet used for vegetables, fodder beet used for animal feed and leaf beet (also known as chard), used as a green vegetable. Sugar beet was first domesticated in Iraq and nearby countries. From 2 500 years ago, the crop dispersed eastward through Iran, Turkmenistan, Uzbekistan and Kazakhstan to China, and westward through Syria and Egypt to countries around the Mediterranean and into Europe. It is now cultivated in more than 50 countries, with primary cultivation in Europe, Asia and the Americas. Sugar beet was introduced into China during the fifth century, and is known by a host of local names including "tian cai," "jun da cai" and "huo yan cai." It was used mainly as a vegetable until the Ming Dynasty, since when it has seen increased use as fodder. Its use in sugar production has only a 100-year history in China, and beet sugar occupies about only 10% of total sugar production in the country. Sugar beet is cultivated in around ten provinces of northern China.

20.2.7 Vegetable crops

Vegetables are a primary source of vitamins, minerals, micronutrients, carbohydrates and cellulose, and are indispensable as human food. Around 200 major vegetable species are cultivated around the world, with only 50–60 species being commonly cultivated (Mbagwu *et al.*, 2008). In 2008, the total world cultivation area under vegetables was 54 million ha, and total production 916 million tons (FAO, 2012). China is the world's largest contributor to vegetable cultivation: in 2008, China's vegetable cultivation area was 24 million ha and production was 460 million tons (FAO, 2012). About 158 species of vegetable in 45 families are commonly cultivated in China, around half of which are introduced (Zhou *et al.*, 2008; Table 20.8). China is one of the world's major centers of vegetable domestication, with important species including soybean (*Glycine max*), bamboo shoot (Poaceae spp.) and Chinese cabbage (*Brassica rapa* var. *chinensis*) domesticated here.

China has a long history of introduction of exotic vegetables: gourds (*Lagenaria siceraria*) and radishes (luo bo; *Raphanus sativus*) were introduced more than 2 500 years ago. Through commerce on the "Silk Road," vegetables domesticated in Central Asia, the Middle East, Africa and the Mediterranean region including alfalfa (*Medicago sativa*; Fabaceae), coriander (*Coriandrum sativum*), cucumber (*Cucumis sativus*), garlic (*Allium sativum*) and sweet melon (*Cucumis* spp.) were introduced during the Qin Dynasty, spinach (*Spinacia oleracea*) during the Tang Dynasty, and lettuce (*Lactuca sativa*) before or during the Song Dynasty. Through sea or overland trade, vegetables domesticated from Southeast Asia including eggplant (*Solanum melongena*; Solanaceae), loofah (*Luffa* spp.; Cucurbitaceae) and wax gourd (*Benincasa hispida*; Cucurbitaceae) were introduced into China between the Han and Qing Dynasties. During the Ming and Qing Dynasties, many North and South American vegetables were also introduced including squashes and pumpkins (*Cucurbita* spp.; Cucurbitaceae), chili pepper (*Capsicum* spp.; Cucurbitaceae), tomato (*Lycopersicon esculentum*) and potato (*Solanum tuberosum*; see Table 20.8). As in previous chapters, many foods that are technically fruits but commonly used in China as foods (such as most Cucurbitaceae) are treated here as vegetables.

Table 20.8 Introduced vegetables.

Family	Latin Name	Chinese Name	English Name
Aizoaceae	*Tetragonia tetragonioides* (Pall.) Kuntze	番杏 (fan xing)	New Zealand spinach
Alliaceae	*Allium ampeloprasum* L.	南欧蒜 (nan ou suan)	leek
Alliaceae	*Allium ascalonicum* L.	胡葱 (hu cong)	shallot
Alliaceae	*Allium ascalonicum* L.	细香葱 (xi xiang cong)	chive
Alliaceae	*Allium cepa* L.	洋葱 (yang cong)	onion
Alliaceae	Allium × proliferum (Moench) Schrad. ex Willd.	楼子葱 (lou zi cong)	bunching onion
Alliaceae	*Allium fistulosum* L.	大葱 (da cong)	welsh onion
Alliaceae	*Allium fistulosum* L.	葱 (cong)	welsh onion
Alliaceae	*Allium porrum* L.	韭葱 (jiu cong)	leek
Alliaceae	*Allium sativum* L.	蒜 (suan)	garlic
Amaranthaceae	*Beta vulgaris* L.	甜菜 (tian cai)	chard
Amaranthaceae	*Spinacia oleracea* L.	菠菜 (bo cai)	spinach
Apiaceae	*Anethum graveolens* L.	莳萝 (shi luo)	dill
Apiaceae	*Apium graveolens* L.	旱芹 (han qin)	celery
Apiaceae	*Coriandrum sativum* L.	芫荽 (yan sui)	coriander
Apiaceae	*Daucus carota* var. *sativa* Hoffm.	胡萝卜 (hu luo bo)	carrot
Apiaceae	*Foeniculum vulgare* (L.) Mill.	茴香 (hui xiang)	fennel
Apiaceae	*Pastinaca sativa* L.	欧防风 (ou fang feng)	parsnip
Apiaceae	*Petroselinum crispum* (Mill.) Nyman ex A. W. Hill	欧芹 (ou qin)	parsley
Asparagaceae	*Asparagus officinalis* L.	石刁柏 (shi diao bai)	asparagus
Asteraceae	*Chrysanthemum morifolium* Ramat.	茼蒿 (tong hao)	Florist's daisy
Asteraceae	*Cichorium endivia* L.	苦苣 (ku ju)	endive
Asteraceae	*Cichorium intybus* L.	菊苣 (ju ju)	common chicory
Asteraceae	*Cynaracardunculus* L.	朝鲜蓟 (chao xian ji)	Cardoon
Asteraceae	*Lactuca sativa* L.	莴苣 (wo ju)	lettuce

Continued

Family	Latin Name	Chinese Name	English Name
Asteraceae	*Sonchus oleraceus* L.	苦苣菜 (ku ju cai)	common sowthistle
Asteraceae	*Tragopogon porrifolius* L.	蒜叶婆罗门参 (suan ye po luo men shen)	salsify
Basellaceae	*Anredera cordifolia* (Ten.) Steenis	落葵薯 (luo kui shu)	Madeira vine
Brassicaceae	*Armoracia rusticana* P. Gaert., B. Mey. & Scherb.	辣根 (la gen)	horseradish
Brassicaceae	*Brassica napus* var. *napobrassica* (L.) Rchb.	蔓菁甘蓝 (man jing gan lan)	rutabaga
Brassicaceae	*Brassica oleracea* var. *botrytis* L.	花椰菜 (hua ye cai)	cauliflower
Brassicaceae	*Brassica oleracea* var. *capitata* L.	甘蓝 (gan lan)	cabbage
Brassicaceae	*Brassica rapa* L.	蔓菁 (man jing)	turnip rape
Brassicaceae	*Eruca vesicaria* (L.) Cav.	芝麻菜 (zhi ma cai)	arugula
Brassicaceaet	*Nasturtium officinale* R. Br.	豆瓣菜 (dou ban cai)	watercress
Brassicaceae	*Raphanus sativus* L.	萝卜 (luo bo)	radish
Convolvulaceae	*Ipomoea aquatica* Forssk.	蕹菜 (weng cai)	water spinach
Cucurbitaceae	*Benincasa hispida* (Thunb.) Cogn.	冬瓜 (dong gua)	wax gourd
Cucurbitaceae	*Citrullus lanatus* (Thunb.) Matsum. & Nakai	西瓜 (xi gua)	watermelon
Cucurbitaceae	*Cucumis melo* L.	甜瓜 (tian gua)	muskmelon
Cucurbitaceae	*Cucumis melo* L.	越瓜（梢瓜）[yue gua (shao gua)]	snake melon
Cucurbitaceae	*Cucumis melo* L.	菜瓜 (cai gua)	serpent melon
Cucurbitaceae	*Cucumis sativus* L.	黄瓜 (huang gua)	cucumber
Cucurbitaceae	*Cucurbita argyrosperma* Huber	墨西哥南瓜 (mo xi ge nan gua)	Mexican pumpkin
Cucurbitaceae	*Cucurbita ficifolia* Bouché	黑籽南瓜 (hei zi nan gua)	blackseed squash
Cucurbitaceae	*Cucurbita maxima* Duchesne	笋瓜 (sun gua)	winter squash
Cucurbitaceae	*Cucurbita moschata* Duchesne	南瓜 (nan gua)	pumpkin
Cucurbitaceae	*Cucurbita pepo* L.	西葫芦 (xi hu lu)	marrow
Cucurbitaceae	*Lagenaria siceraria* (Molina) Standl.	葫芦 (hu lu)	bottle gourd
Cucurbitaceae	*Luffaacutangula* (L.) Roxb.	丝瓜 (si gua)	loofah
Cucurbitaceae	*Momordica charantia* L.	苦瓜 (ku gua)	bitter gourd
Cucurbitaceae	*Sechium edule* (Jacq.) Sw.	佛手瓜 (fo shou gua)	chayote
Cucurbitaceae	*Trichosanthes anguina* L.	蛇瓜 (she gua)	snake gourd
Fabaceae	*Cajanus cajan* (L.) Millsp.	木豆 (mu dou)	pigeon pea
Fabaceae	*Canavalia ensiformis* (L.) DC.	直生刀豆 (zhi sheng dao dou)	jack bean
Fabaceae	*Canavalia gladiata* (Jacq.) DC.	刀豆 (dao dou)	sword bean
Fabaceae	*Cicer arietinum* L.	鹰嘴豆 (ying zui dou)	chickpea
Fabaceae	*Lablab purpureus* (L.) Sweet	扁豆 (bian dou)	hyacinth bean
Fabaceae	*Lens culinaris* Medik.	兵豆 (bing dou)	lentil
Fabaceae	*Medicago polymorpha* L.	南苜蓿 (nan mu xu)	burr medic
Fabaceae	*Phaseolus coccineus* L.	荷包豆 (he bao dou)	runner bean
Fabaceae	*Phaseolus vulgaris* L.	菜豆 (cai dou)	common bean
Fabaceae	*Pisum sativum* L.	豌豆 (wan dou)	pea
Fabaceae	*Psophocarpus tetragonolobus* (L.) DC.	四棱豆 (si leng dou)	winged bean
Fabaceae	*Trigonella foenum-graecum* L.	葫芦巴 (hu lu ba)	fenugreek
Fabaceae	*Vicia faba* L.	蚕豆 (can dou)	broad bean
Fabaceae	*Vigna unguiculata* (L.) Walp.	豇豆 (jiang dou)	cowpea
Lamiaceae.	*Mentha longifolia* (L.) L.	欧薄荷 (ou bo he)	habek mint
Lamiaceae	*Ocimum basilicum* L.	罗勒 (luo le)	basil
Malvaceae	*Abelmoschus esculentus* (L.) Moench	咖啡黄葵 (ka fei huang kui)	okra
Portulacaceae	*Talinum paniculatum* (Jacq.) Gaertn.	土人参 (tu ren shen)	fameflower
Solanaceae	*Capsicum annuum* L.	辣椒 (la jiao)	chili pepper
Solanaceae	*Capsicum annuum* L.	小米辣 (xiao mi la)	hot chili
Solanaceae	*Cyphomandra betacea* (Cav.) Miers	树番茄 (shu fan qie)	tree tomato
Solanaceae	*Lycopersicon esculentum* Mill.	番茄 (fan qie)	tomato
Solanaceae	*Solanum melongena* L.	茄 (qie)	eggplant

Spinacia oleracea—Spinacia oleracea (spinach; bo cai; Amaranthaceae) is native to Iran but introduced to China in the seventh century. Cultivated mainly in Guangdong, Guangxi, Henan, Hebei, Hubei, Jiangsu, Shandong and Sichuan, it is a popular vegetable, the tender stems and leaves rich in vitamins, phosphate and iron. It is also considered to have some medicinal value.

Cucurbita ssp. —*Cucurbita* (nan gua shu; Cucurbitaceae) includes about 15 species from warm regions of the Americas, which are widely cultivated in tropical, subtropical and temperate regions. Three species were introduced to China in the fourteenth century. *Cucurbita moschata* is one of the most popular vegetables in China, cultivated all over the country. The young vine tips, young flowers, immature and mature fruits are used as a highly nutritious vegetable which can be fried, steamed, baked or prepared in many other ways. The seeds are also edible and popularly used as a nut in China. *Cucurbita pepo* is also commonly cultivated throughout China for its fruits, used as a vegetable, as is *C. maxima* (winter squash; sun gua), the seeds of which are also rich in oil.

Luffa acutangula—Luffa acutangula (loofah; si gua; Cucurbitaceae) originated in tropical Asia and was introduced to China during the Song Dynasty, between 960 and 1279 AD. It is now commonly cultivated in China, while naturalized plants are found in Yunnan. The young fruits are used as a vegetable, and the fibrous endocarp for making items such as loofahs, sandals, insulation, slipper-soles, saddle-padding and pillow-stuffing, as well as being used in traditional Chinese medicine.

Lagenaria siceraria—Lagenaria siceraria (bottle gourd; hu lu; Cucurbitaceae) originated in tropical Africa and is one of the oldest crops cultivated across the warmer areas of the world. In China it has been cultivated for over 7 000 years. China is rich in germplasm resources, with many local varieties found in Fujian, Guangdong, Guangxi, Guizhou, Yunnan and Zhejiang. The tender fruits, shoots and young leaves are used as a vegetable, while the woody mature fruits are used to make vessels and other handicrafts.

Brassica oleracea—Brassica oleracea (cabbage; gan lan; Brassicaceae) includes the cultivated variety *B. oleracea* var. *capitata,* introduced in the sixteenth century and cultivated mainly in Fujian, Guangdong, Hebei, Henan, Hubei and Shandong for its dense leafy heads. One-third of total world production occurs in China. *Brassica oleracea* var. *botrytis* originated in the eastern Mediterranean and was introduced to China in the mid nineteenth century. Cultivated mainly in southeastern China, Gansu and Henan, *B. oleracea* var. *gongylodes* is also widely cultivated as a vegetable in China; *B. oleracea* var. *italica* is only recently introduced; and *B. oleracea* var. *alboglabra* (a synonym of *B. oleracea*) has been cultivated since the eighth century in Guangdong. *Brassica nigra* distributed in Gansu, Jiangsu, Qinghai, Xinjiang and Xizang, has seeds used in seasoning and pickling.

Raphanus sativus—Raphanus sativus (radish; luo bo; Brassicaceae) contains just three species, native to the Mediterranean region, two of which were introduced, cultivated and naturalized in China in ancient times. *Raphanus sativus* (radish; luo bo) has been cultivated since the first century BC and is grown in most provinces of China, from the east coast to the Xizang Plateau, and from Mohe in Heilongjiang to Hainan. There are as many as 300 cultivars, which between them cover 12 million ha, with an annual output of more than 30 million tons. The fleshy roots are eaten raw, fried, cooked or made into pickles. *Raphanus raphanistrum* (wild radish or jointed charlock; ye luo bo) does not have fleshy roots, and is found in Qinghai, Sichuan and Taiwan.

Apium graveolens—Apium graveolens (celery; han qin; Apiaceae) originated in the Mediterranean region, was introduced into China 2 000 years ago, and is now cultivated over almost all of the country, with Guangdong, Hebei, Henan, Jiangsu, Shandong and Sichuan the main growing regions. The stems and leaves are used as a vegetable with high nutritional and medicinal value, especially for reducing blood pressure.

Coriandrum sativum—Coriandrum sativum (coriander; yan sui; Apiaceae) was also originated in the Mediterranean. It is introduced and widely cultivated, sometimes naturalized, in China, where the young stems and leaves are used as a herb, and the fruits for medicine and as a spice.

Daucus carota var. *sativa—Daucus carota* var. *sativa* (carrot; hu luo bo; Apiaceae) is cultivated worldwide as a root vegetable. It was introduced to China in the twelfth century from central and western Asia and is now widely cultivated, especially in Fujian, Guangdong, Heilongjiang, Jiangsu, Nei Mongol and Shandong. The escaped wild relative *D. carota* (ye hu luo bo) is found in Anhui, Guizhou, Hubei, Jiangsu, Jiangxi, Sichuan and Zhejiang, where its fruits are used as a medicine and for oil.

Capsicum annuum.—Capsicum annuum (chili pepper; la jiao) is native to Mexico and South America, but cultivated and naturalized throughout the world. Introduced to China in the seventeenth century, it is now widely cultivated as an important vegetable, covering 1.3 million ha; it is also used as a diaphoretic medicine, and sometimes as an ornamental. Many cultivars have been selected, differing in fruit appearance and taste (from sweet to pungent).

Capsicum annuum—Capsicum annuum (chili pepper; la jiao; Solanaceae) is one of the most important and popular vegetables in the world. It originated in the South American Andes, somewhere in modern-day Chile, Bolivia, Ecuador, Colombia or Peru, but the precise timing and place of domestication are not clear (Peralta & Spooner, 2007). Tomatoes were introduced into Europe in the fifteenth century and greatly improved during the eighteenth and nineteenth centuries (Sims, 1980). They were introduced from Europe into China during the late Ming Dynasty (1368–1644) and are now widely cultivated. The delicious, nutritious fruits are eaten raw or cooked, or processed in many different ways.

Solanum melongen L. —*Solanum melongena* (eggplant; qie; Solanaceae) originated in tropical areas of Southeast Asia and has been cultivated in China for more than 2 000 years, where it is widely distributed in Guangdong, Guangxi, Hainan, Yunnan and other regions. The young fruits are used as a vegetable, an important part of the diet in China and other parts of Asia. China produces about 50% of the world output and the south of the country houses many related wild species, providing important germplasm for breeding, including *S. undatum* (ye qie), distributed in Guangdong, Guangxi, Guizhou, Hainan, Taiwan and Yunnan. *Solanum pseudocapsicum* (Jerusalem cherry; shan hu yin), native to South America, is also cultivated in China as an ornamental.

Lactuca sativa—*Lactuca sativa* (lettuce; wo ju; Asteraceae) originated in the Mediterranean region, was introduced to China in the fifth century, and has been cultivated here for more than a millennium. There are three cultivated types in China: those with a thickened stem, those cultivated for their leaves, and those cultivated for their dense leafy heads. Mainly distributed in Chongqing Shi, Fujian, Jiangsu, Shandong, Shanghai Shi and Zhejiang.

Allium sativum—*Allium sativum* (garlic; suan; Alliaceae) is from central Asia and was introduced into China in the second century BC, where it is widely cultivated. The bulbs, scapes and leaves are commonly used, fresh or cooked, as a vegetable or seasoning. They are believed to have great health benefits, being antibacterial and antifungal, especially in processed foods.

20.2.8 Animal feed and green manure crops

Animal-feed crops are cultivated on arable land and grassland, producing forage and fodder for livestock. Green manure crops are grown primarily to add nutrients and organic matter to the soil. Most green manure crops are also used as forage, thus we discuss these two types of crops together. Together, cultivated fodder, forage and green manure crops in China include approximately 210 species belonging to 17 families and 96 genera. This includes 91 species in 37 genera of Poaceae, 76 species (33 genera) of Fabaceae, and 43 species (26 genera) of other families (Jiang, 2007). Around 100 of these are introduced. The most important species are listed in Table 20.9 (based on Jiang, 2007).

Table 20.9 Introduced animal feed and green manure crops (based on Jiang, 2007).

Family	Latin Name	Chinese Name	English Name
Amaranthaceae	*Alternanthera philoxeroides* (Mart.) Griseb.	喜旱莲子草 (xi han lian zi cao)	alligator weed
Amaranthaceae	*Amaranthus cruentus* L.	老鸦谷 (lao ya gu)	paniculate amaranth
Amaranthaceae	*Amaranthus hybridus* L.	绿穗苋 (lü sui xian)	green amaranth
Amaranthaceae	*Amaranthus hypochondriacus* L.	千穗谷 (qian sui gu)	amaranth
Amaranthaceae	*Beta vulgaris* var. *lutea* CD.	饲用甜菜 (si yong tian cai)	fodder beet
Amaranthaceae	*Kochia prostrata* (L.) Schrad.	木地肤 (mu di fu)	forage kochia
Apiaceae	*Daucus carota* var. *sativa* Hoffm.	饲用胡萝卜 (si yong hu luo bo)	forage carrot
Asteraceae	*Cichorium intybus* L.	菊苣 (ju ju)	chicory
Asteraceae	*Silphium perfoliatum* L.	松香草 (song xiang cao)	cup plant
Boraginaceae	*Symphytum peregrinum* Ledeb.	聚合草 (ju he cao)	common comfrey
Brassicaceae	*Brassica oleracea* L.	甘蓝 (gan lan)	cabbage
Brassicaceae	*Raphanus sativus* L.	饲用萝卜 (si yong luo bo)	forage radish
Convolvulaceae	*Ipomoea batatas* (L.) Lam.	番薯 (fan shu)	sweet potato
Cucurbitaceae	*Citrullus lanatus* (Thunb.) Matsum. & Nakai	饲用西瓜 (si yong xi gua)	forage watermelon
Cucurbitaceae	*Cucurbita moschata* Duchesne	饲用南瓜 (si yong nan gua)	crookneck
Fabaceae	*Aeschynomene americana* L.	敏感合萌 (min gan he meng)	American jointvetch
Fabaceae	*Amorpha fruticosa* L.	紫穗槐 (zi sui huai)	bastard indigobush
Fabaceae	*Astragalus cicer* L.	鹰嘴紫云英 (ying zui zi yun ying)	cicer milkvetch
Fabaceae	*Calopogonium caeruleum* (Benth.) C. Wright	蓝花毛蔓豆 (lan hua mao man dou)	blue flower calopo
Fabaceae	*Calopogonium mucunoides* Desv.	毛蔓豆 (mao man dou)	calopo
Fabaceae	*Centrosema pubescens* Benth.	距瓣豆 (ju ban dou)	buttefly pea
Fabaceae	*Crotalaria juncea* L.	菽麻 (shu ma)	sunn hemp
Fabaceae	*Desmodium intortum* (Mill.) Urb.	扭曲山蚂蝗 (niu qu shan ma huang)	greenleaf ticktrefoil
Fabaceae	*Lathyrus sativus* L.	家山黧豆 (jia shan li dou)	grass pea

Continued

Family	Latin Name	Chinese Name	English Name
Fabaceae	*Lotus corniculatus* L.	百脉根 (bai mai gen)	bird's-foot trefoil
Fabaceae	*Lotus tenuis* Waldst. & Kit.	细叶百脉根 (xi ye bai mai gen)	narrow leaf trefoil
Fabaceae	*Lupinus luteus* L.	黄花羽扇豆 (huang hua yu shan dou)	yellow lupine
Fabaceae	*Macroptilium atropurpureum* (DC.) Urb.	紫花大翼豆 (zi hua da yi dou)	purple bushbean
Fabaceae	*Macroptilium lathyroides* (L.) Urb.	大翼豆 (da yi dou)	phasey bean
Fabaceae	*Medicago falcata* L.	野苜蓿 (ye mu xu)	yellow alfalfa
Fabaceae	*Medicago lupulina* L.	天蓝苜蓿 (tian lan mu xu)	black medic
Fabaceae	*Medicago polymorpha* L.	金花菜 (jin hua cai)	burclover
Fabaceae	*Medicago sativa* L.	南苜蓿 (nan mu xu)	alfalfa
Fabaceae	*Onobrychis viciifolia* Scop.	驴食草 (lü shi cao)	sainfoin
Fabaceae	*Pisum sativum* L.	豌豆 (wan dou)	pea
Fabaceae	*Stylosanthes guianensis* (Aubl.) Sw.	圭亚那笔花豆 (gui ya na bi hua dou)	Guyana stylo
Fabaceae	*Stylosanthes hamata* (L.) Taub.	有钩柱花草 (you gou zhu hua cao)	Caribbean stylo
Fabaceae	*Stylosanthes scabra* Vogel	粗糙柱花草 (cu cao zhu hua cao)	shrubby stylo
Fabaceae	*Stylosanthes sympodialis* Taub.	合轴柱花草 (he zhou zhu hua cao)	sypodium stylo
Fabaceae	*Trifolium fragiferum* L.	草莓车轴草 (cao mei che zhou cao)	strawberry clover
Fabaceae	*Trifolium hybridum* L.	杂种车轴草 (za zhong che zhou cao)	alsike clover
Fabaceae	*Trifolium incarnatum* L.	绛车轴草 (jiang che zhou cao)	crimson clover
Fabaceae	*Trifolium pratense* L.	红车轴草 (hong che zhou cao)	red clover
Fabaceae	*Trifolium repens* L.	白车轴草 (bai che zhou cao)	white clover
Fabaceae	*Vicia sativa* L.	救荒野豌豆 (jiu huang ye wan dou)	common vetch
Poaceae	*Agropyron desertorum* (Fisch. ex Link) Schult.	沙生冰草 (sha sheng bing cao)	desert wheatgrass
Poaceae	*Agropyron fragile* (Roth) P. Candargy	西伯利亚冰草 (xi bo li ya bing cao)	Siberian wheatgrass
Poaceae	*Agrostis gigantea* Roth	巨序剪股颖 (ju xu jian gu ying)	redtop
Poaceae	*Alopecurus arundinaceus* Poir.	苇状看麦娘 (wei zhuang kan mai niang)	creeping foxtail
Poaceae	*Alopecurus pratensis* L.	大看麦娘 (da kan mai niang)	meadow foxtail
Poaceae	*Avena sativa* L.	燕麦 (yan mai)	oat
Poaceae	*Axonopus compressus* (Sw.) P. Beauv.	地毯草 (di tan cao)	broadleaf carpet grass
Poaceae	*Axonopus fissifolius* (Raddi) Kuhlm.	类地毯草 (lei di tan cao)	carpet grass
Poaceae	*Brachiaria brizantha* (A. Rich.) Stapf	珊瑚臂形草 (shan hu bi xing cao)	marandu grass
Poaceae	*Brachiaria decumbens* Stapf	俯仰臂形草 (fu yang bi xing cao)	signal grass
Poaceae	*Brachiaria dictyoneura* (Fig. & De Not.) Stapf	网脉臂形草 (wang mai bi xing cao)	Keoronivia grass
Poaceae	*Brachiaria mutica* (Forssk.) Stapf	巴拉草 (ba la cao)	paragrass
Poaceae	*Brachiaria ruziziensis* Germ. & C. M. Evrard	刚果旗草 (gang guo qi cao)	Congo grass
Poaceae	*Bromus carinatus* Hook. & Arn.	显脊雀麦 (xian ji que mai)	California brome
Poaceae	*Bromus catharticus* Vahl	扁穗雀麦 (bian sui que mai)	rescue grass
Poaceae	*Bromus inermis* Leyss.	无芒雀草 (wu mang que cao)	smooth brome
Poaceae	*Bromus japonicus* Thunb.	雀麦 (que mai)	brome
Poaceae	*Bromus marginatus* Nees ex Steud.	山地雀麦 (shan di que mai)	mountain brome
Poaceae	*Bromus pumpellianus* Scribn.	紧穗雀麦 (jin sui que mai)	Pumpelly's brome
Poaceae	*Bromus tectorum* L.	旱雀麦 (han que mai)	downy brome
Poaceae	*Chloris gayana* Kunth	非洲虎尾草 (fei zhou hu wei cao)	rhodes grass
Poaceae	*Cynodon dactylon* (L.) Pers.	狗牙根 (gou ya gen)	bermuda grass
Poaceae	*Dactylis glomerata* L.	鸭茅 (ya mao)	orchard grass
Poaceae	*Digitaria eriantha* Steud.	南非马唐 (nan fei ma tang)	digit grass
Poaceae	*Digitaria eriantha* subsp. *pentzii* (Stent) Kok	俯仰马唐 (fu yang ma tang)	pangola grass
Poaceae	*Elymus elongatus* (Host) Runemark.	长穗偃麦草 (chang sui yan mai cao)	tall wheatgrass
Poaceae	*Elymus hispidus* (Opiz) Melderis	中间偃麦草 (zhong jian yan mai cao)	intermediate wheatgrass

Continued

Family	Latin Name	Chinese Name	English Name
Poaceae	*Elymus sibiricus* L.	老芒麦 (lao mang cao)	Siberian wildrye
Poaceae	*Elytrigia repens* Desv.	偃麦草 (yan mai cao)	quackgrass
Poaceae	*Festuca arundinacea* Schreb.	苇状羊茅 (wei zhuang yang mao)	tall fescue
Poaceaet	*Festuca pratensis* Huds.	草甸羊茅 (cao dian yang mao)	meadow fescue
Poaceae	*Hemarthria altissima* (Poir.) Stapf & C. E. Hubb.	大牛鞭草 (da niu bian cao)	limpograss
Poaceae	*Hordeum vulgare* L.	大麦 (da mai)	common barley
Poaceae	*Lolium perenne* L.	黑麦草 (hei mai cao)	perennial ryegrass
Poaceae	*Melinis minutiflora* P. Beauv.	糖蜜草 (tang mi cao)	molasses grass
Poaceae	*Panicum maximum* Jacq.	大黍 (da shu)	guinea grass
Poaceae	*Paspalum dilatatum* Poir.	毛花雀稗 (mao hua que bai)	dallis grass
Poaceae	*Paspalum notatum* Flüggé	百喜草 (bai xi cao)	bahia grass
Poaceae	*Paspalum virgatum* L.	粗秆雀稗 (cu gan que bai)	broad-leaf paspalum
Poaceae	*Pennisetum clandestinum* Hochst. ex Chiov.	铺地狼尾草 (pu di lang wei cao)	Kikuyu grass
Poaceae	*Pennisetum glaucum* (L.) R. Br.	御谷 (yu gu)	pearl millet
Poaceae	*Pennisetum purpureum* Schumach.	象草 (xiang cao)	elephant grass
Poaceae	*Phalaris aquatica* L.	水虉草 (shui ge cao)	bulbous canary grass
Poaceae	*Phalaris arundinacea* L.	虉草 (ge cao)	reed canary grass
Poaceae	*Phalaris canariensis* L.	加那利虉草 (jia na li ge cao)	canary grass
Poaceae	*Phleum pratense* L.	梯牧草 (ti mu cao)	timothy
Poaceaet	*Poa compressa* L.	加拿大早熟禾 (jia na da zao shu he)	Canada bluegrass
Poaceae	*Poa pratensis* L.	草地早熟禾 (cao di zao shu he)	Kentucky bluegrass
Poaceae	*Poa trivialis* L.	普通早熟禾 (pu tong zao shu he)	rough bluegrass
Poaceae	*Psathyrostachys juncea* (Fisch.) Nevski	新麦草 (xin mai cao)	Russian wildrye
Poaceae	*Sorghum bicolor* (L.) Moench	高粱 (gao liang)	sorghum
Poaceae	*Sorghum sudanense* (Piper) Stapf	苏丹草 (su dan cao)	Sudan grass
Poaceae	*Spartina anglica* C. E. Hubb.	大米草 (da mi cao)	common cordgrass
Poaceae	*Zea mays* L.	玉蜀黍 (yu shu shu)	maize
Poaceae	*Zoysia matrella* (L.) Merr.	沟叶结缕草 (gou ye jie lü cao)	Manila grass
Polygonaceae	*Calligonum arborescens* Litv.	乔木沙拐枣 (qiao mu sha guai zao)	arborescent calligonum
Polygonaceae	*Calligonum caput-medusae* Schrenk	头状沙拐枣 (tou zhuang sha guai zao)	capitate calligonum
Polygonaceae	*Rumex patientia* L.	巴天酸模 (ba tian suan mo)	garden patience
Pontederiaceae	*Eichhornia crassipes* (Mart.) Solms	凤眼莲 (feng yan lian)	water-hyacinth

20.2.9 Ornamental plants

Here we use the term ornamentals to refer to plants that have a viewing value in the broad sense. In 2003, the area under cultivation of commercial ornamental plants was 235 000 ha, with an output value of ca. 26 billion US Dollars. China has a 2 000-year history of cultivation of ornamentals and many popular species were domesticated in China. However, many species have also been introduced into China, with more than 1 000 species and varieties introduced since 1978 (Zhang & Pan, 2001), giving a total of around 2 500–3 000 ornamental species recognized in China (Fei *et al.*, 2008). Cultivation of ornamental plants developed rapidly after the establishment of the People's Republic of China in 1949, and especially from the 1980s onwards. In 2005, commercial ornamental plants were cultivated over an area of 810 000 ha, and had an output value of 50 billion Yuan. Table 20.10 lists the most common and widely cultivated introduced ornamental species.

As an example, Indian siris (*Albizia lebbeck*; Fabaceae) is a deciduous tree cultivated in the south of China but native to tropical Africa. It is grown as a roadside tree, shade tree and garden ornamental for its open crown, feathery leaves, clustered flowers and long inflorescence (Wu & Nielsen, 2005; Xing *et al.*, 2009). Four O'Clock (*Mirabilis jalapa*; Nyctaginaceae) was also originally introduced as an ornamental and is now a ruderal weed in some areas of China. The purple flowers, commonly known as "yin-zhi-hua" ("rouge flower"), are used as rouge.

Table 20.10 Introduced ornamental plants.

Family	Latin Name	Chinese Name	English Name
Agavaceae	*Agave americana* L.	龙舌兰 (long she lan)	mescal
Amaranthaceae	*Celosia cristata* L.	鸡冠花 (ji guan hua)	cockscomb
Amaranthaceae	*Gomphrena globosa* L.	千日红 (qian ri hong)	globe amaranth
Amaryllidaceae	*Clivia miniata* (Lindl.) Bosse	君子兰 (jun zi lan)	bush lily
Amaryllidaceae	*Crinum asiaticum* L.	文殊兰 (wen shu lan)	crinum lily
Amaryllidaceae	*Hippeastrum striatum* (Lam.) H. E. Moore	朱顶红 (zhu ding hong)	amaryllis
Amaryllidaceae	*Polianthes tuberosa* L.	晚香玉 (wan xiang yu)	tuberose
Amaryllidaceae	*Zephyranthes carinata* Herb.	韭莲 (jiu lian)	pink rain lily
Apocynaceae Juss.	*Nerium oleander* L.	夹竹桃 (jia zhu tao)	oleander
Araceae	*Zantedeschia aethiopica* (L.) Spreng.	马蹄莲 (ma ti lian)	calla lily
Asparagaceae	*Asparagus setaceus* (Kunth) Jessop	文竹 (wen zhu)	asparagus fern
Asparagaceae	*Chlorophytum comosum* (Thunb.) Jacques	吊兰 (diao lan)	spider plant
Asparagaceae	*Hyacinthus orientalis* L.	风信子 (feng xin zi)	common hyacinth
Asteraceae	*Ageratum conyzoides* L.	藿香蓟 (huo xiang ji)	billygoat-weed
Asteraceae	*Aster novi-belgii* L.	荷兰菊 (he lan ju)	New York aster
Asteraceae	*Bellis perennis* L.	雏菊 (chu ju)	daisy
Asteraceae	*Calendula officinalis* L.	金盏菊 (jin zhan ju)	marigold
Asteraceae	*Centaurea cyanus* L.	矢车菊 (shi che ju)	cornflower
Asteraceae	*Cosmos bipinnatus* Cav.	秋英 (qiu ying)	garden cosmos
Asteraceae	*Dahlia × pinnata* Cav.	大丽花 (da li hua)	dahlia
Asteraceae	*Gerbera jamesonii* Bolus ex Hook. f	非洲菊 (fei zhou cao)	barberton daisy
Asteraceae	*Pericallis hybrida* (Regel) B. Nord.	瓜叶菊 (gua ye ju)	florist's cineraria
Asteraceae	*Rudbeckia laciniata* L.	金光菊 (jin guang ju)	coneflower
Asteraceae	*Tagetes erecta* L.	万寿菊 (wan shou ju)	marigold
Asteraceae	*Tagetes patula* L.	孔雀草 (kong que cao)	French marigold
Asteraceae	*Zinnia elegans* Jacq.	百日草 (bai ri cao)	common zinnia
Balsaminaceae	*Impatiens balsamina* L.	凤仙花 (feng xian hua)	balsam
Begoniaceae	*Begonia × tuberhybrida* Voss.	球根海棠 (qiu gen hai tang)	tuberous begonia
Bignoniaceae	*Fuchsia × hybrida* Voss.	倒挂金钟 (dao gua jin zhong)	hybrid fuchsia
Brassicaceae	*Lobularia maritima* (L.) Desv.	香雪球 (xiang xue qiu)	sweet alyssum
Cactaceae	*Disocactus × hybidus* (Geel) Barthlott	令箭荷花 (ling jian he hua)	nopalxochia
Cactaceae	*Echinopsis tubiflora* (Pfeiff.) Zucc. ex A. Dietr.	仙人球 (xian ren qiu)	prickly pear
Cactaceae	*Epiphyllum oxypetalum* (DC.) Haw.	昙花 (tan hua)	Dutchman's pipe
Cactaceae	*Opuntia dillenii* (Ker Gawl.) Haw.	仙人掌 (xian ren zhang)	cactus
Cannaceae	*Canna indica* L.	美人蕉 (mei ren jiao)	canna
Caryophyllaceae	*Dianthus caryophyllus* L.	康乃馨 (kang nai xin)	carnation
Convolvulaceae	*Ipomoea quamoclit* L.	茑萝 (niao luo)	cypressvine
Ericaceae	*Rhododendron × hybridum* Ker Gawl.	西洋杜鹃 (du juan)	western rhododendron
Euphorbiaceae	*Euphorbia pulcherrima* Willd. ex Klotzsch	一品红 (yi pin hong)	poinsettia
Fabaceae	*Albizia lebbeck* (L.) Benth.	阔荚合欢 (kuo jia he huan)	Indian siris
Fabaceae	*Lathyrus odoratus* L.	香豌豆 (xiang wan dou)	sweet pea
Geraniaceae	*Pelargonium × hortorum* L. H. Bailey	天竺葵 (tian zhu kui)	geranium
Gesneriaceae	*Sinningia speciosa* Hiern	大岩桐 (da yan tong)	gloxinia
Iridaceae	*Gladiolus × gandavensis* Van Houtte	唐菖蒲 (tang chang pu)	gladiolus
Iridaceae	*Freesia refracta* (Jacq.) Klatt.	小苍兰 (xiao cang lan)	common freesia
Lamiaceae	*Salvia splendens* Sellow ex Roem. & Schult.	一串红 (yi chuan hong)	scarlet sage
Liliaceae	*Lilium brownii* var. *viridulum* Baker	百合 (bai he)	lily
Liliaceae	*Tulipa gesneriana* L.	郁金香 (yu jin xiang)	tulip
Meliaceae	*Aglaia odorata* Lour.	米仔兰 (mi zi lan)	Chinese perfume Plant
Moraceae	*Ficus elastica* Roxb. ex Hornem.	印度榕 (yin du rong)	rubber fig

Continued

Family	Latin Name	Chinese Name	English Name
Nyctaginaceae	*Bougainvillea spectabilis* Willd.	叶子花 (ye zi hua)	bougainvillea
Nyctaginaceae	*Mirabilis jalapa* L.	紫茉莉 (zi mo li)	Four O'Clock
Oleaceae	*Jasminum sambac* (L.) Aiton	茉莉花 (mo li hua)	jasmine
Papaveraceae	*Papaver rhoeas* L.	虞美人 (yu mei ren)	corn poppy
Polemoniaceae	*Phlox drummondii* Hook.	福禄考 (fu lu kao)	annual phlox
Pontederiacea	*Eichhornia crassipes* (Mart.) Solms	凤眼莲 (feng yan lian)	common waterhyacinth
Portulacaceae	*Portulaca grandiflora* Hook.	半支莲 (ban zhi lian)	rose moss
Primulaceae	*Primula* ssp.	报春 (bao chun)	primrose
Punicaceae	*Punica granatum* L.	石榴 (shi liu)	pomegranate
Ranunculaceae	*Delphinium grandiflorum* L.	翠雀 (cui que)	larkspur
Rosaceae	*Rosa* hybrids	玫瑰 (mei gui)	rose
Rutaceae	*Ruta graveolens* L.	芸香 (yun xiang)	rue
Scrophulariaceae	*Antirrhinum majus* L.	金鱼草 (jin yu cao)	snapdragon
Solanaceae	*Petunia hybrida* Vilm.	碧冬茄 (bi dong qie)	common garden petunia
Verbenaceae	*Verbena × hybrida* Voss.	美女樱 (mei nv ying)	verbena
Violaceae	*Viola tricolor* L.	三色堇 (san se jin)	pansy
Xanthorrhoeaceae	*Kniphofia uvaria* (L.) Oken	火炬花 (huo ju hua)	torch lily
Zingiberaceae	*Hedychium coronarium* J. König	姜花 (jiang hua)	ginger flower

20.2.10 Medicinal herbs

China is the world's largest contributor to the cultivation of medicinal herbs, with about 500 species used in the country. More than 100 species have been domesticated and cultivated for medicinal use in China, and only a few important medicinal herbs are introduced (see Table 20.11).

Table 20.11 Introduced medicinal plants.

Family	Latin Name	Chinese Name	English Name
Apiaceae	*Levisticum officinale* W. D. J. Koch	欧当归 (ou dang gui)	lovage
Apocynaceae	*Catharanthusroseus* (L.) G. Don	长春花 (chang chun hua)	Madagascar periwinkle
Araliaceae	*Panax quinquefolius* L.	西洋参 (xi yang shen)	American ginseng
Arecaceae	*Areca catechu* L.	槟榔 (bin lang)	betel nut palm
Asteraceae	*Carthamus tinctorius* L.	红花 (hong hua)	safflower
Asteraceae	*Pyrethrum cinerariifolium* Trevir.	除虫菊 (chu chong ju)	pyrethrum
Euphorbiaceae	*Sauropusandrogynus* (L.) Merr.	守宫木 (shou gong mu)	katuk
Fabaceae	*Amorpha fruticosa* L.	紫穗槐 (zi sui huai)	desert false Indigo

Katuk (*Sauropus androgynus*; Euphorbiaceae) introduced from South America, is widely cultivated in tropical Guangdong, Guangxi, Hainan and Yunnan; the leaves are used as a vegetable and herbal medicine for coughs, to soothe the lungs, as a tonic and to relieve fever. Madagascar periwinkle (*Catharanthus roseus*; Apocynaceae), native to Madagascar, is widely cultivated in eastern and southwestern China. The whole plant is rich in alkaloids, flavones and organic acids. Decoctions made from all parts are used in the treatment of malaria, skin diseases, Hodgkin's disease, diarrhea, hypertension and diabetes. *Catharanthus roseus* is also used for the extraction of vincaleukoblastinum, an important anti-cancer compound. Betel nut palm (*Areca catechu*; Arecaceae) is an evergreen tree probably native to central Malesia and commonly cultivated in the lowland settlements of Guangxi, Guangdong, Hainan, Taiwan and Yunnan, and throughout tropical Asia. In Traditional Chinese Medicine the seeds provide a remedy for stomach and intestinal parasites and the fibrous seed-coat, known as "da fu pi," is used in other gastrointestinal disorders. The flowers are used in local medicine and the seeds have been chewed for many centuries as a mild narcotic in Guangdong, Guangxi, Hunan, Taiwan and Yunnan. *Areca catechu* is still seen as an important local product today.

20.2.11 Forestry plants

Plants used in artificial forest planting include some that are also ornamental or horticultural plants. Many important afforestation plants are Chinese native species, but introduced plants also play an important role. The common introduced afforestation plants are listed in Table 20.12. As one example, southern magnolia (*Magnolia grandiflora*; Magnoliaceae), an evergreen tree to 30 m tall native to southeastern North America, is grown in provinces south of the Chang Jiang. The timber is hard, heavy and dense, used, for example, in furniture and stationery. With its large white flowers it is also considered a good landscape tree.

Table 20.12 Introduced afforestation plants.

Family	Latin Name	Chinese Name	English Name
Araucariaceae	*Araucaria cunninghamii* Aiton ex D. Don	南洋杉 (nan yang shan)	hoop pine
Araucariaceae	*Araucaria heterophylla* (Salisb.) Franco	异叶南洋杉 (yi ye nan yang shan)	Norfolk Island pine
Casuarinaceae	*Casuarina equisetifolia* L.	木麻黄 (mu ma huang)	coast she-oak
Cupressaceae	*Chamaecyparis obtusa* (Siebold & Zucc.) Endl.	日本扁柏 (ri ben bian bai)	Japanese cypress
Cupressaceae	*Chamaecyparis pisifera* (Siebold & Zucc.) Endl.	日本花柏 (ri ben hua bai)	sawara cypress
Cupressaceae	*Cryptomeria japonica* (Thunb. ex L.f.) D. Don	日本柳杉 (ri ben liu shan)	Japanese cedar
Cupressaceae	*Juniperus virginiana* L.	北美圆柏 (bei mei yuan bai)	red cedar
Cupressaceae	*Sciadopitys verticillata* (Thunb.) Siebold & Zucc.	金松 (jin song)	Japanese umbrella pine
Cupressaceae	*Sequoia sempervirens* (D. Don) Endl.	北美红杉 (bei mei hong shan)	coast redwood
Cupressaceae	*Sequoiadendrongiyanteum* (Lindl.) J. Buchholz.	巨杉 (ju shan)	giant sequoia
Cupressaceae	*Taxodium distichum* (L.) Rich.	落羽杉 (luo yu shan)	bald cypress
Cupressaceae	*Taxodium distichum* var. *imbricatum* (Nutt.) Croom	池杉 (chi shan)	pond cypress
Cupressaceae	*Taxodium mucronatum* Ten.	墨西哥落羽杉 (mo xi ge luo yu shan)	Montezuma cypress
Cupressaceae	*Thuja occidentalis* L.	北美香柏 (bei mei xiang bai)	white cedar
Cupressaceae	*Thuja standishii* (Gordon) Carrière	日本香柏 (ri ben xiang bai)	Japanses cedar
Cupressaceae	*Thujopsis dolabrata* (L. f.) Siebold & Zucc.	罗汉柏 (luo han bai)	Japanese thuja
Fabaceae	*Amorpha fruticosa* L.	紫刺槐 (zi ci huai)	false indigo
Fabaceae	*Robinia pseudoacacia* L.	刺槐 (ci huai)	black locust
Juglandaceae	*Carya illinoinensis* (Wangenh.) K. Koch	美国山核桃 (mei guo shan he tao)	pecan
Magnoliaceae	*Magnolia grandiflora* L.	荷花木兰 (he hua mu lan)	southern magnolia
Myrtaceae	*Eucalyptus camaldulensis* Dehnh.	赤桉 (chi an)	red gum
Oleaceae	*Fraxinus americana* L.	大叶白蜡 (da ye bai la)	white ash
Pinaceae	*Abies firma* Siebold & Zucc.	日本冷杉 (ri ben leng shan)	Japanese fir
Pinaceae	*Cedrus deodara* (Roxb. ex Lamb.) G. Don	雪松 (xue song)	Himalayan cedar
Pinaceae	*Pinus elliottii* Engelm.	湿地松 (shi di song)	slash pine
Pinaceae	*Pinus parviflora* Siebold & Zucc.	日本五针松 (ri ben wu zhen song)	Japanese white pine
Pinaceae	*Pinus taeda* L.	火炬松 (huo ju song)	loblolly pine
Pinaceae	*Pinus thunbergii* Parl.	黑松 (hei song)	black pine
Platanaceae	*Platanus orientalis* L.	悬铃木 (xuan ling mu)	oriental plane
Salicaceae	*Populus × canadensis* Moench cv. 'robusta'	健杨 (jian yang)	robusta populus
Salicaceae	*Populus × canadensis* Moench cv. 'Sacrau 79'	沙兰杨 (sha lan yang)	sacrau populus

20.2.12 Other economically-important introduced crops

Nicotiana tabacum—Nicotiana tabacum (tobacco; yan cao; Solanaceae) belongs to the genus *Nicotiana*, which includes approximately 60 species, divided by Goodspeed (1954) into three subgenera and 14 sections. There are two primary cultivated species, red-flowered tobacco (*Nicotiana tabacum*) and yellow-flowered tobacco (*N. rustica*). Tobacco is primarily used for making cigarettes but also has medicinal applications (Charlton, 2004). It is widely cultivated in more than 120 countries, most importantly China, Brazil, India and the USA. *Nicotiana tabacum* and *N. rustica* both originated in the South American Andes. *Nicotiana tabacum* is a hybrid between *N. sylvestris* and *N. tomentosiformis*; *N. rustica* between *N. paniculata* and *N. Undulata* (Chase *et al.*, 2003). There is still debate regarding how tobacco dispersed out of South America: it was introduced into Europe during the fifteenth century and China in the Jiajing (1521–1566) or Wanli Period (1573–1619) of the Ming Dynasty. Four main routes of introduction are hypothesized: from the Philippines to Fujian between 1563 and 1640; from Indonesia or Vietnam to Guangdong between 1620 and 1627; from Japan into Manchuria in 1616 or 1617; or from Portugal into Guangxi between 1522 and 1560. For 300 years after introduction, only sun-cured tobacco was cultivated. Cured tobacco was introduced into China during the early twentieth century and became the primary form of cultivated tobacco. Tobacco is cultivated in approximately 30 provinces or municipalities, especially in southwestern China. China is the largest contributor to world tobacco production, and tobacco cultivation is the primary source of income for 30 million growers.

Hevea brasiliensi—*Hevea brasiliensis* (rubber tree; xiang jiao shu; Euphorbiaceae) is one of the four most important industrial materials together with iron, petroleum and coal. Rubber trees are widely cultivated in tropical countries including Thailand, Indonesia, Malaysia, India, China, Vietnam, Côte d'Ivoire, Sri Lanka, Liberia, Brazil, the Philippines, Cameroon, Nigeria, Cambodia, Guatemala, Myanmar, Ghana, Congo, Papua New Guinea, Bengal and Gabon. *Hevea brasiliensis* is native to Brazil from where, in 1876, Henry Wickham took seeds to the Royal Botanical Gardens, Kew, UK, and cultivated seedlings in Sri Lanka and Singapore. Rubber tree cultivation developed rapidly following the invention of tyres and the beginnings of the automotive industry in the late nineteenth century, and China began to cultivate rubber trees on a large scale from 1952. Rubber trees are primarily cultivated in the tropical regions of Guangdong, Guangxi, Hainan and Yunnan.

Piper nigrum—*Piper nigrum* (pepper; hu jiao; Piperaceae) is an important spice crop also used in medicine. The genus *Piper* contains ca. 2 000 species distributed in tropical regions, of which approximately 60 are native to China. *Piper nigrum* was domesticated in India and has a 2 000-year history of cultivation, today cultivated in more than 20 countries of Asia, Africa and Latin America; primarily Vietnam, Brazil, Indonesia, India and China. It was introduced into China in the 1950s, and is mainly cultivated in Hainan and Guangdong, as well as Fujian, Guangxi and Yunnan.

Coffea ssp. —*Coffea* ssp. (coffee; ka fei; Rubiaceae) are important beverage crop also used medicinally. The genus *Coffea* (ka fei shu; Rubiaceae) includes approximately 100 species distributed from tropical Africa to Madagascar. Five species of *Coffea* have been introduced into southern and southwestern China, but two main species are cultivated: *Coffea arabica* (xiao li ka fei) and *C. canephora* (zhong li ka fei), known as arabica and robusta coffee respectively. *Coffea liberica* (da li ka fei) is rarely cultivated due to its poor quality. Coffee is cultivated in more than 70 countries across Latin America, Asia and Africa. It was first introduced into Taiwan in 1884 and in China is mainly cultivated in Hainan and Yunnan.

Other species.—*Santalum album*—*Santalum album* (Indian sandalwood; tan xiang; Santalaceae), native to India and the islands of tropical Asia and the Pacific, has been imported into China for 1 500 years for use as incense. The yellow-brown heartwood, which has a strong aroma, is considered a precious medicine and spice, as well as being one of the finest woods for carving. *Jatropha curcas* (physic nut; xiao tong zi, ma feng shu; Euphorbiaceae), native to tropical America, has been cultivated in China for 200 years for medicinal use and has now become a very important bio-diesel energy resource. *Pelargonium graveolens* (Geraniaceae) is a perennial herb native to South Africa and cultivated all over China for geranium oil, which is sold for aromatherapy and massage therapy applications, and known locally as "mo mo xiang." It is cultivated in large numbers, especially on the central plateau of Yunnan, and the production of geranium oil has become a very successful and important industry. *Trigonella foenum-graecum* (Cucurbitaceae), although primarily known as a vegetable or spice, is also exploited for fenugreek gum, a kind of galactomannan, which can be used for oil exploitation. Finally, *Vanilla planifolia* (Orchidaceae) is a climbing, herbaceous vine native to eastern Mexico and cultivated in Fujian, Hainan and Yunnan. The long, slender black fruits are the vanilla "beans" of commerce from which vanilla extract is made. They are cured to ferment and dry the pods while minimizing the loss of essential oils. A list of other economically important introduced plants not discussed here is provided in Table 20.13.

Table 20.13 Other economically-important introduced plants.

Family	Latin Name	Chinese Name	English Name
Annonaceae	*Cananga odorata* (Lam.) Hook. f. & Thomson	依兰 (yi lan)	ylang-ylang
Asteraceae	*Parthenium argentatum* A. Gray	银胶菊 (yin jiao ju)	guayule
Cannabaceae	*Humulus lupulus* L.	啤酒花 (pi jiu hua)	hop
Cucurbitaceae	*Hodgsonia heteroclita* (Roxb.) Hook. f. & Thomson	油渣果 (you zha guo)	hodgsonia
Euphorbiaceae	*Hevea brasiliensis* (Willd. ex A. Juss.) Müll. Arg.	橡胶树 (xiang jiao shu)	rubber tree
Geraniaceae	*Pelargonium graveolens* L'Hér.	香叶天竺葵 (xiang ye tian zhu kui)	Scented Geranium
Lamiaceae	*Lavandula angustifolia* Mill.	薰衣草 (xun yi cao)	lavender
Malvaceae	*Cola acuminata* (P. Beauv.) Schott & Endl.	可拉 (ke la)	cola
Malvaceae	*Theobroma cacao* L.	可可 (ke ke)	cocao
Orchidaceae	*Vanilla planifolia* Jacks. ex Andrews	香荚兰 (xiang jia lan)	flat-leaved vanilla
Piperaceae	*Piper nigrum* L.	胡椒 (hu jiao)	pepper
Poaceae	*Cymbopogon nardus* (L.) Rendle	亚香茅 (ya xiang mao)	citronella grass
Rubiaceae	*Cinchona calisaya* Wedd.	金鸡纳树 (jin ji na shu)	cinchona tree
Rubiaceae	*Coffea* L. ssp.	咖啡 (ka fei)	coffee
Santalaceae	*Santalum album* L.	檀香 (tan xiang)	Indian sandalwood
Solanaceae	*Nicotiana tabacum* L.	烟草 (yan cao)	tabacco

References

Ames, M. S. & Spooner, D. M. (2008). DNA from herbarium specimens settles a controversy about origins of the European potato. *American Journal of Botany* 95: 252-257.

An, Z. S. (1950). *The Relationship History between West Han Dynasty and the Western Regions*. Shandong, China: Shandong People's Publishing House.

Arroyo-Garcia, R., Ruiz-Garcia, L., Bolling, L. & Ocete, R. (2006). Multiple origins of cultivated grapevine (*Vitis vinifera* L. ssp. *sativa*) based on chloroplast DNA polymorphism. *Molecular Ecology* 15: 3707-3714.

Austin, D. F. (1988). The taxonomy, evolution and genetic diversity of sweet potatoes and related wild species. In: Gregory, P. (ed.) *Exploration, Maintenance, and Utilization of Sweet Potato Genetic Resources*. pp. 27-60. Lima, Peru: International Potato Center.

Azhaguvel, O. & Komatsuda, T. (2007). A phylogenetic analysis based on nucleotide sequence of a marker linked to the brittle rachis locus indicates a diphyletic origin of barley. *Annals of Botany* 100: 1009-1015.

Bedigian, D. & Harlan, J. R. (1986). Evidence for cultivation of sesame in the ancient world. *Economic Botany* 40: 137-154.

Bohs, L. (2007). Phylogeny of the *Cyphomandra* clade of the genus *Solanum* (Solanaceae) based on ITS sequence data. *Taxon* 56: 1012-1026.

Brandes, E. W. (1958). Origin, classification and characteristics of sugar cane. In: Artschwager, E. & Brandes, E. W. (eds.) *Sugar Cane (Saccharum officinarum L.)*. pp. 1-35, 260-262. Washington, DC: US Government Print Office.

Chapman, M. A. & Burke, J. M. (2007). DNA sequence diversity and the origin of cultivated safflower *Carthamus tinctorius* L.; Asteraceae). *BMC Plant Biology* 7: 60.

Charlton, A. (2004). Medicinal uses of tobacco in history. *Journal of the Royal Society of Medicine* 97: 292-296.

Chase, M. W., Knapp, S., Cox, A. V. & Klarkson, J. J. (2003). Molecular systematics, GISH and the origin of hybrid taxa in *Nicotiana*. *Annals of Botany* 92: 107-127.

Chen, W. (2002). *Nong Ye Kao Gu*. Beijing, China: Cultural Relics Publishing House.

Chen, X. (2009). Discussing wheat in ancient China. *Dong Yue Tribune* 30: 102-105.

Coart, E., Van Glabeke, S., De Loose, M., Larsen, A. S. & Roldán-Ruiz, I. (2006). Chloroplast diversity in the genus *Malus*: new insights into the relationship between the European wild apple (*Malus sylvestris* (L.) Mill.) and the domesticated apple (*Malus domestica* Borkh.). *Molecular Ecology* 15: 2171-2182.

de Alencar Figueiredo, L. F., Calatayud, C., Dupuits, C. & Billot, C. (2008). Phylogeographic evidence of crop neodiversity in *Sorghum*. *Genetics* 179: 997-1008.

Dong, Y. & Liu, D. (2006). Food crops. In: Dong, Y. & Liu, X. (eds.) *Crops and Their Wild Relatives in China*. Beijing, China: China Agriculture Press.

Fang, J. & Chang, R. (2007). Economic crops. In: Dong, Y. & Liu, X. (eds.) *Crops and Their Wild Relatives in China*. Beijing, China: China Agriculture Press.

FAO (2012). FAOSTAT. <http://faostat3.fao.org/home/index.html>.

Fei, Y., Liu, Q. & Ge, H. (2008). Flowers. In: Dong, Y. & Liu, X. (eds.) *Crops and Their Wild Relatives in China*. pp. 31-48. Beijing, China: China Agriculture Press.

Forte, A. V., Ifnatov, A. N., Ponomarenko, V. V., Dorokhov, D. B. & Savelyev, N. I. (2002). Phylogeny of the *Malus* (apple tree) species, inferred from the morphological traits and molecular DNA analysis. *Russian Journal of Genetics* 38: 1150-1160.

Fryxell, P. A. (1992). A revised taxonomic interpretation of *Gossypium* L. (Malvaceae). *Rheedea* 2: 108-165.

Fu, D. X., Xu, T. W. & Feng, Z. Y. (2000). The ancient carbonized barley (*Hordeum vulgare* L. var. *nudum*) kernel discovered in the middle Yalu Tsangpo river basin in Tibet. *Southwest China Journal of Agricultural Sciences* 13: 38-41.

Fukunaga, K., Hill, J., Vigouroux, Y., Matsuoka, Y., Sanchez, G. J., Liu, K., Buckler, E. S. & Doebley, J. (2005). Genetic diversity and population structure of teosinte. *Genetics* 169: 2241-2254.

Gepts, P. & Bliss, F. A. (1986). Phaseolin variability among wild and cultivated common beans (*Phaseolus vulgaris*) from Colombia. *Economic Botany* 40: 469-478.

Gómez-Campo, C. (1999). *Biology of Brassica Coenospecies (Developments in Plant Genetics and Breeding)*. Amsterdam, Netherlands: Elsevier Science B.V.

Goodspeed, T. H. (1954). The genus *Nicotiana*. *Chronica Botanica* 16: 1-489.

Grassl, C. O. (1964). Problems relative to the origin and evolution of wild and cultivated *Saccharum*. *Indian Journal of Sugar Cane Research and Development* 8: 106-116.

Grassl, C. O. (1967). Introgression between *Saccharum* and *Miscanthus* in New Guinea and the Pacific area. *Proceedings of the International Society of Sugar-Cane*

Technologists 12: 995-1003.

Grassl, C. O. (1977). The origin of the sugar producing cultivars of *Saccharum*. *Sugar Cane Breeders Newsletter* 39: 8-33.

Gregory, W. C., Gregory, M. P., Krapovickas, A., Smith, B. W. & Yarbrough, J. A. (1973). Structures and genetic resources of peanuts. In: Wilson, C. T. (ed.) *Peanuts: Culture and Uses*. pp. 47-133. Oklahoma: American Peanut Research and Education Society. Inc.

Harlan, J. R. & Zohary, D. (1966). Distribution of wild wheats and barley. *Science* 153: 1074-1080.

Harris, S. A., Robinson, J. P. & Juniper, B. E. (2002). Genetic clues to the origin of the apple. *Trends in Genetics* 18: 426-430.

Harter, A. V., Gardner, K. A., Falush, D. & Lentz, D. L. (2004). Origin of extant domesticated sunflowers in eastern North America. *Nature* 430: 201-205.

Heun, M., R., S. P., Klawan, D. & Castagna, R. (1997). Site of einkorn wheat domestication identified by DNA fingerprinting. *Science* 278: 1312-1314.

Jiang, Y. (2007). Forage and manure crops. In: Dong, Y. & Liu, X. (eds.) *Crops and Their Wild Relatives in China*. pp. 31-49. Beijing, China: China Agriculture Press.

John, C. M., Narayana, G. V. & Seshadri, C. R. (1950). The wild gingelly of Malabar (*Sesamum orientate* var. *malabaricum* var. nov.). *Madras Agricultural Journal* 37: 47-50.

Kaplan, L. (1981). What is the origin of the common bean? *Economic Botany* 35: 240-254.

Kobayashi, T. (1981). The wild and cultivated species in genus *Sesamum*. Sesame: status and improvement. In: FAO (ed.) *Proceedings of Expert Conclusion*. pp. 157-163. Rome, Italy: FAO,.

Krapovickas, A. & Gregory, W. C. (1994). Taxonomy of the genus *Arachis* (Leguminosae). *Bonplandia* 8: 1-186.

Lopez Gonzalez, G. (1989). Observations about the natural classification of the genus *Carthamus* L., s.l. *Anales del Jardín Botánico de Madrid* 47: 11-34.

Loskutov, I. G. (2008). On evolutionary pathways of *Avena* species. *Genetic Resources and Crop Evolution* 55: 187-202.

Mabberley, D. J. (2008). *Mabberley's Plant Book. A Portable Dictionary of Plants, their Classification and Uses*. Cambridge, UK: Cambridge University Press.

Matsuoka, Y., Vigouroux, Y., Goodman, M. M. & Sanchez, G. J. (2002). A single domestication for maize shown by multilocus microsatellite genotyping. *Proceedings of the National Academy of Sciences of the United States of America* 99: 6080-6084.

Mbagwu, F. N., Chime, E. G. & Unamba, C. I. N. (2008). Palynological studies on five species of Asteraceae. *Life Science Journal-Acta Zhengzhou University Overseas Edition* 5(1): 73-76.

Morrell, P. L. & Clegg, M. T. (2007). Genetic evidence for a second domestication of barley (*Hordeum vulgare*) east of the Fertile Crescent. *Proceedings of the National Academy of Sciences of the United States of America* 104: 3289-3294.

Nagahara, U. (1935). Genome analysis in *Brassica* with special reference to the experimental formation of *B. napus* and peculiar mode of fertilisation. *Journal of Japanese Botany* 7: 389-452.

Nayar, N. M. & Mehra, K. L. (1970). Sesame: its uses, botany, cytogenetics, and origin. *Economic Botany* 24: 20-31.

Nene, Y. L. (2006). Indian pulses through the millennia. *Asian Agri-History* 10: 179-202.

Özkan, H., Brandolini, A., Schäfer-Pregl, R. & Salamini, F. (2002). AFLP analysis of a collection of tetraploid wheats indicates the origin of emmer and hard wheat domestication in Southeast Turkey. *Molecular Biology and Evolution* 19: 1797-1801.

Peralta, I. E. & Spooner, D. M. (2007). History, origin and early cultivation of tomato (Solanaceae). In: Razdan, M. K. & Mattoo, A. K. (eds.) *Genetic Improvement of Solanaceous Crops, Volume 2: Tomato*. pp. 1-27. Enfield, USA: Science Publishers.

Petersen, G., Seberg, O., Yde, M. & Berthelsen, K. (2006). Phylogenetic relationships of *Triticum* and *Aegilops* and evidence for the origin of the A, B, and D genomes of common wheat (*Triticum aestivum*). *Molecular Phylogenetics and Evolution* 39: 70-82.

Piperno, D. R. & Flannery, K. V. (2001). The earliest archaeological maize (*Zea mays* L.) from highland Mexico: new accelerator mass spectrometry dates and their implications. *Proceedings of the National Academy of Sciences of the United States of America* 98: 2101-2103.

Robinson, J. P., Harris, S. A. & Juniper, B. E. (2001). Taxonomy of the genus *Malus* Mill. (Rosaceae) with emphasis on the cultivated apple, *Malus × domestica* Borkh. *Plant Systematics and Evolution* 226: 35-58.

Ross-Ibarra, J., Tenaillon, M. & Gaut, B. S. (2009). Historical divergence and gene flow in the genus *Zea*. *Genetics* 181: 1399-1413.

Saisho, D. & Purugganan, M. D. (2007). Molecular phylogeography of domesticated barley traces expansion of agriculture in the Old World. *Genetics* 177: 1765-1776.

Salamini, F., Özkan, H., Brandolini, A., Schäfer-Pregl, R. & Martin, W. (2002). Genetics and geography of wild cereal domestication in the Near East. *Nature Reviews Genetics* 3: 429-441.

Sims, W. L. (1980). History of tomato production for industry around the world. *Acta Horticulturae* 100: 25-26.

Smith, B. D. (1998). *The Emergence of Agriculture*. New York: W.H. Freeman.

Sonnante, G., Hammer, K. & Pignone, D. (2009). From the cradle of agriculture a handful of lentils: history of domestication. *Rendiconti Lincei* 20: 21-37.

Spooner, D. M., McLean, K., Ramsay, G., Waugh, R. & Bryan, G. J. (2005). A single domestication for potato based on multilocus amplified fragment length polymorphism genotyping. *Proceedings of the National Academy of Sciences of the United States of America* 102: 14694-14699.

Tanno, K. I. & Willcox, G. (2006). The origins of cultivation of *Cicer arietinum* L. and *Vicia faba* L.: early finds from Tell el-Kerkh, north-west Syria, late 10th millennium B.P. *Vegetation History and Archaeobotany* 15: 197-204.

Tenaillon, M. (2005). Selection versus demography: a multilocus investigation of the domesticated process in maize. *Molecular Biology and Evolution* 21: 1214-1225.

Thangavelu, S. (1994). Diversity in wild and cultivated species of *Sesamum* and its use. In: Arora, R. K. & Riley, K. W. (eds.) *Sesame Biodiversity in Asia: Conservation, Evaluation and Improvement*. pp. 13-23. New Delhi, India: International Plant Genetic Resources Institute, Office for South Asia.

Tomooka, N., Vaughan, D. A., Moss, H. & Maxted, N. (2002). *The Asian Vigna: Genus Vigna Subgenus Ceratotropis Genetic Resources*. Dordrecht, Netherlands: Kluwer Academic Publishers.

Urgent, D. & Peterson, L. (1988). Archeological remains of potato and sweet potato in Peru. *CIP Circular* 16: 1-10.

Vivier, M. A. & Pretorius, I. S. (2002). Genetically tailored grapevines for wine industry. *Trends in Biotechnology* 20: 472-478.

Wendel, J. F. & Cronn, R. C. (2003). Polyploidy and the evolutionary history of cotton. *Advances in Agronomy* 78: 139-186.

Westengen, O. T., Huamán, Z. & Heun, M. (2004). Genetic diversity and geographic pattern in early South American cotton domestication. *Theoretical and Applied Genetics* 110: 392-402.

Wu, D. L. & Nielsen, I. C. (2005). Mimosaceae. In: Wu, Z. Y., Raven, P. H. & Hong, D. Y. (eds.) *Flora of China*. pp. 62-64. Beijing, China and St. Louis, MO: Science Press and Missouri Botanical Garden Press.

Xing, F. W., Zeng, Q. W., Chen, H. F. & Wang, F. G. (2009). *Landscape Plants of China*. Wuhan, China: Huazhong University of Science and Technology Press.

Zeng, X. (2005). On the expansion of wheat in ancient China. *Chinese Food Culture* 1: 1.

Zhang, D., Cervantes, J., Huamán, Z., Carey, E. & Ghislain, M. (2000). Assessing genetic diversity of sweet potato (*Ipomoea batatas* (L.) Lam.) cultivars from tropical America using AFLP. *Genetic Resources and Crop Evolution* 47: 659-665.

Zhang, Q. & Pan, H. (2001). The introducing status and measures of foreign ornamental plant in China. In: Gao, J. & Jiang, W. (eds.) *Advances in Science and Technology of Chinese Flowers*. pp. 133-142. Beijing, China: China Agriculture Press.

Zheng, D. (2000). Diversity of crop genetic resources in China. *Review of China Agricultural Science and Technology* 2: 45-49.

Zheng, D. & Li, L. (2000). Analysis of varieties and evolution of major cultivars in *Triticum aestivum* in China. *Zhiwu Yichuan Ziyuan Kexue* 1: 14-20.

Zhou, D., Wang, D. & Li, X. (2008). Vegetable crops. In: Dong, Y. & Liu, X. (eds.) *Crops and Their Wild Relatives in China*. pp. 31-67. Beijing, China: China Agriculture Press.

Zohary, D. & Hopf, M. (2000). *Domestication of Plants in the Old World*. Oxford, UK: Oxford University Press.

Chapter 21

Other Important Economic Plants

PEI Sheng-Ji and HU Guang-Wan

21.1 Introduction

In addition to the medicinal plants, forest economic plants, ornamental plants and crop plants and their wild relatives discussed in previous chapters, there are Chinese plants that do not fit into these categories and are also widely used and economically valued. Some have been extensively cultivated and some are collected largely from wild resources. During the long history of their interaction with plants, the Chinese people have accumulated large amounts of experience and knowledge in using plant resources. For instance, from as many as 5 000 years ago, they made use of tea (cha; *Camellia sinensis*; Theaceae) as a detoxifying medicine, as recorded in SHEN Nong's *Herbal Classic*. More than 2 000 years ago, tea also became a popular beverage (Zhu *et al.*, 2007). In the first century AD, CAI Lun of the Eastern Han dynasty developed the world's first paper-making technology, using plant fibers. This development had far-reaching impacts in promoting the development of world culture. The Chinese were also the first, at least 3 000 years ago, to make use of indigo from plants, including *Indigofera suffruticosa* (anil; ye qing shu; Fabaceae), *Indigofera tinctoria* (true indigo; mu lan), *Isatis tinctoria* (woad; song lan; Brassicaceae), *Polygonum tinctorium* (Chinese indigo; liao lan; Polygonaceae) and *Strobilanthes cusia* (ban lan; Acanthaceae), to dye cloth. Even today, the technology of tie-dyeing cloth with natural plant indigo is widely used in China by the Bai, Buyi, Han, Miao, Tujia and Zhuang peoples (Zhang *et al.*, 2003; Pei & Huai, 2007).

In this chapter, we describe over 100 native Chinese plant taxa of economic importance, sorted into seven groups: beverage plants, dye plants, aromatic plants, rubber and gum plants, fiber plants, forage plants and bio-energy plants. These were selected from the huge number of economically-important species on the basis of their high economic value and popularity of usage in China.

21.2 Beverage plants

Beverage plants include all plants that are used by people for making drinks. The Chinese have a very long history of using beverage plants, and the most famous and popular is tea (cha; *Camellia sinensis* var. *sinensis* and *C. sinensis* var. *assamica* pu er cha; Theaceae). SHEN Nong's *Herbal Classic*, one of the earliest books about plant utilization in China, describes how the Chinese people began using tea from the period of SHEN Nong (3000 BC), firstly as a herbal medicine for detoxification. Now, tea is one of the three most popular plant-based beverages in the world, alongside coffee and cocoa (the former has also been cultivated on large-scale plantations in Hainan and Yunnan since the 1980s). As well as tea, many other beverage plants have a long history of local utilization, some of which have become important economic products in their own right.

Some beverage plants contain stimulants such as the caffeine present in tea and coffee, while others can benefit human health and prevent disease. For example, "ku ding cha" (a tea made from the leaves of *Ilex* (dong qing shu; Aquifoliaceae), *Ligustrum* (nv zhen shu; Oleaceae), or other species) can reduce blood pressure and cholesterol, and Chinese herbal teas are used to treat sunstroke. Herbal teas, traditionally supplied via local markets and brewed in homes only in a restricted area or culture, have recently developed into important products with a large market, which are becoming increasingly popular. These include ku ding cha, "wang lao ji" from Guangdong, southern China, and mixed herbal teas.

According to the origin of the material that is used to make the drink, beverage plants fall into four main types. Firstly, those whose leaves, especially the young leaves or buds, provide the beverage, such as *Camellia* and *Ilex*. Secondly, those in which the flowers are used: *Chrysanthemum* (ju shu; Asteraceae), *Jasminum* (mo li shu; Oleaceae) and *Rosa* (mei gui shu; Rosaceae). Thirdly, fruit-beverage plants, based on taxa such as *Coffea* (Rubiaceae), *Hippophae* (sha ji shu; Elaeagnaceae), *Nitraria* (bai ci shu; Nitrariaceae), *Passiflora* (ji dan guo shu; Passifloraceae) and, again, *Rosa*. Finally, there are those plants whose roots or stems are used in beverages, such as *Imperata* (bai mao shu; Poaceae) and *Pueraria* (ge gen shu; Fabaceae). Beverage plants also differ in their geographical distribution: tea is mainly cultivated in southern, southwestern and central China. Specialist varieties are even more restricted, such as "Pu-er" tea (*Camellia sinensis* var. *assamica*), which is cultivated only in Yunnan. *Humulus lupulus* (hop; pi jiu hua; Cannabaceae), an important raw material for the brewing industry, is mainly cultivated in eastern Shandong and Xinjiang. Many traditional local beverage plants have very narrow distributions or are used only by a restricted population.

21.2.1 Examples of important beverage plants

Lindera fragrans.—*Lindera fragrans* (xiang ye zi; Lauraceae) is a small evergreen tree found in ditch sides or thickets on mountain slopes from 700 to 2 100 m in

Guangxi, Guizhou, Hubei, Shaanxi and Sichuan. The young leaves are used as a substitute for tea known as "lao ying cha," which has become an important local commodity in its place of origin. In the area of the Wuling Shan, in particular, local people like drinking it to prevent sunstroke.

Humulus lupulus.—*Humulus lupulus* (hop; pi jiu hua; Cannabaceae) is a climbing perennial herb widely cultivated throughout China, especially in eastern Shandong and Xinjiang. The flowers and infructescences are important ingredients for brewing, while the female flowers and bracts are used medicinally.

Lithocarpus litseifolius.—*Lithocarpus litseifolius* (mu jiang ye ke; Fagaceae) is an evergreen tree found in the broad-leaved evergreen forests of Fujian, Guangdong, Guangxi, Guizhou, Hainan, Hubei, Hunan, Jiangxi, Sichuan, Yunnan and Zhejiang. The young leaves are sweet-tasting and traditionally used, in the mountains south of the Chang Jiang, as a substitute for tea known as "tian cha" or "sweet tea." Tian cha has become a popular beverage in China and is also exported to Japan and the USA today for producing sweeteners. There are two varieties of *L. litseifolius* in China, *L. litseifolius* var. *litseifolius* and *L. litseifolius* var. *pubescens*.

Adinandra nitida.—*Adinandra nitida* (liang ye yang tong; Theaceae) is an evergreen shrub found in montane forests or on streamsides at 500–1 000 m elevation in Guangdong, eastern and southern Guangxi and southeastern Guizhou. The young leaves have a long history of use locally as a substitute for tea known as "shi ya cha." Shi ya cha has become an important local product in this area today.

Figure 21.1 *Camellia sinensis* var. *assamica*. A, flowering branch. B, fruiting branch.

Camellia sinensis.—*Camellia sinensis* (tea; cha; Theaceae; Figure 21.1) is an shrub grown in the evergreen broad-leaved forests of Anhui, Fujian, Guangdong, Guangxi, Guizhou, Hainan, Henan, Hubei, Hunan, Jiangsu, Jiangxi, southern Shaanxi, Sichuan, Taiwan, southeastern Xizang, Yunnan and Zhejiang. The vegetative buds and young leaves of *C. sinensis* var. *sinensis* and *C. sinensis* var. *assamica* (pu-er cha; Theaceae; Figure 21.1) are both used to make tea. There is a long history of the use and cultivation of tea in China. Chinese people usually distinguish between small-leaved (*C. sinensis* var. *sinensis*) and large-leaved tea (*C. sinensis* var. *assamica*). Unless it is to be used as green tea, large-leaved tea is usually made into black tea by wilting, fermenting and drying the leaves. Sometimes the buds and young leaves of other varieties of *C. sinensis* and other species of wild *Camellia* are used locally as tea.

Cratoxylum.—There are two species of *Cratoxylum* (huang niu mu shu; Clusiaceae) in China, *C. cochinchinense* (huang niu mu), distributed in southern Guangdong, southern Guangxi and southern Yunnan, and *C. formosum* (yue nan huang niu mu). The latter comprises two subspecies, *C. formosum* subsp. *formosum* and *C. formosum* subsp. *pruniflorum* (hong ya mu), distributed in southern Guangxi, Hainan and southern Yunnan. The young leaves are used locally as the tea-substitute known as ku ding cha.

Passiflora edulis.—*Passiflora edulis* (passion fruit; ji dan guo; Passifloraceae) is a herbaceous vine cultivated—and sometimes escaped—in forested mountain valleys below 1 900 m in Fujian, Guangdong, Hainan, Taiwan and Yunnan. The fruit is used to make passion fruit juice, a very tasty and popular beverage in China today.

Gynostemma pentaphyllum.—*Gynostemma pentaphyllum* (jiao gu lan; Cucurbitaceae) is a climbing herb distributed in Anhui, Fujian, Guangdong, Guangxi, Guizhou, Hainan, southern Henan, Hubei, Hunan, Jiangsu, Jiangxi, Shandong, Sichuan, Taiwan, Yunnan and Zhejiang. There are two varieties in China, *G. pentaphyllum* var. *dasycarpum* (mao guo jiao gu lan) and *G. pentaphyllum* var. *pentaphyllum* (jiao gu lan). The whole plant of both contains ginsenosides—a type of saponins—and is used dried as a substitute for tea. This is widely considered to be a health-giving drink, important as a local beverage commodity.

Rosa roxburghii.—*Rosa roxburghii* (sao si hua; Rosaceae) is found at 500–1 400 m elevation in Anhui, Fujian, southern Gansu, Guangxi, Guizhou, Hubei, Hunan, Jiangxi, southern Shaanxi, Sichuan, Xizang, Yunnan and Zhejiang. The sweet and sour edible fruits, known locally as "ci li," are very rich in vitamins and the enzyme superoxide dismutase, and are used to make a health-giving drink and also frequently in fermenting wine.

Rubus chingii var. *suavissimus*.—*Rubus chingii* var. *suavissimus* (tian cha; Rosaceae) is a lianescent shrub found in the hillside forests of eastern-central Guangxi. The leaves are sweet, mainly because of their high levels (4–6 %) of rubusoside, which has a sweetness 300 times that of sucrose (Zhang & Ye, 2007). The leaves are used in southern China to make a sweet tea, which is an important local beverage commodity and is also exported to Europe.

Pueraria montana.—*Pueraria montana* (ge; Fabaceae) is a herbaceous vine with tuberous roots found widely in montane forests, thickets and open places throughout China with the exceptions of Qinghai, Xinjiang and Xizang. The species includes three varieties, *P. montana* var. *lobata* (ge ma mu), *P. montana* var. *montana* (shan ge) and *P. montana* var. *thomsonii* (fen ge), all of which are commonly known as the kudzu vine. The tuberous roots are full of starch, which is extracted to make kudzu tea ("ge cha"), a very popular beverage in China.

Hippophae rhamnoides subsp. *sinensis*.—*Hippophae rhamnoides* subsp. *sinensis* (Chinese sea buckthorn; zhong guo sha ji; Rhamnaceae; Figure 21.2) is found from 800 to 3 600 m altitude in Gansu, Hebei, Nei Mongol, Qinghai, Shaanxi, Shanxi and western Sichuan. The fruits are juicy, sweet and sour in taste, and rich in vitamin C. Their juice is used as a drink and to make wine, especially in Qinghai.

Figure 21.2 *Hippophae rhamnoides.*

Ilex.—There are over 200 species of *Ilex* (dong qing shu; Aquifoliaceae) in China, ca. 150 of which are endemic, distributed in regions south of the Chang Jiang and Qin Ling, mainly in the south and southwest of the country. The leaves of four species of *Ilex* are used locally as the tea-substitute ku ding cha, a very popular healthy beverage. The species used are *I. dunniana* (long li dong qing), from eastern Guizhou, western Hubei, southwestern Sichuan and northeastern Yunnan, *I. kaushue* (kou shu), from Guangdong, Guangxi, Hainan, Hubei, Hunan, Sichuan and southeastern Yunnan, *I. latifolia* (da ye dong qing), from Anhui, Fujian, Guangdong, Guangxi, Henan, Hubei, Hunan, Jiangsu, Jiangxi, Yunnan and Zhejiang, and *I. pentagona* (wu leng ku ding cha) found on the forested limestone hills of western Guangxi, Guizhou, Hunan and southeastern Yunnan.

Ampelopsis grossedentata.—*Ampelopsis grossedentata* (snake wine vine; xian chi she pu tao; Vitaceae) is a woody vine of forests and shrublands at 200–1 500 m elevation in Fujian, Guangdong, Guangxi, Guizhou, Hubei, Hunan, Jiangxi and Yunnan. The young branches and leaves are used locally as substitute for tea known as "teng cha." In its production area, especially southern Hunan, it has become a popular healthy beverage and important local commodity.

Nitraria tangutorum.—*Nitraria tangutorum* (bai ci; Nitrariaceae) is a shrub found at 1 900–3 500 m in Gansu, Hebei, western Nei Mongol, Ningxia, Qinghai, northern Shaanxi, Xinjiang and northeastern Xizang. The fruits are edible, sweet and sour in taste, and rich in vitamin C. Their juice is used as a drink and in making wine, especially in Qinghai.

Elsholtzia rugulosa.—*Elsholtzia rugulosa* (ye ba zi; Lamiaceae) grows wild at 1 300–2 800 m elevation in Guangxi, Guizhou, Sichuan and Yunnan. It is a very important substitute for tea in its area of production, especially popular with the Yi community in Yunnan, and is also a very important nectar plant.

Prunella vulgaris.—*Prunella vulgaris* (common self-heal; xia ku cao; Lamiaceae) is a herb found up to 3 200 m in Fujian, Gansu, Guangdong, Guangxi, Guizhou, Henan, Hubei, Hunan, Jiangxi, Shaanxi, Sichuan, Taiwan, Xinjiang, Xizang, Yunnan and Zhejiang. It is a very important ingredient in traditional herbal teas used to prevent heatstroke. There are two varieties, *P. vulgaris* var. *lanceolata* (xia ye xia ku cao) and *P. vulgaris* var. *vulgaris*.

Ligustrum.—There are 27 species of *Ligustrum* (nü zhen shu; Oleaceae) in China. The young leaves of four species and one subspecies are another source of the tea substitute and healthy beverage ku ding cha: *L. henryi* (li ye nü zhen), found in Gansu, Guangxi, Guizhou, Hubei, western Hunan, Shaanxi, Sichuan and Yunnan, *L. amamianum* (tai wan nü zhen), cultivated widely, *L. lucidum* (nü zhen), from Anhui, Fujian, Gansu, Guangdong, Guangxi, Guizhou, Hainan, Henan, Hubei, Hunan, Jiangsu, Jiangxi, Shaanxi, Sichuan, Xizang, Yunnan and Zhejiang, *L. pedunculare* (zong gen nü zhen), from Guizhou, western Hubei, Hunan, Shaanxi and Sichuan, and *L. robustum* subsp. *chinense* (chu zhuang nü zhen), distributed in southern Anhui, northwestern Fujian, Guangdong, western Guangxi, Guizhou, eastern Hubei, western Hunan, Jiangxi, Sichuan and Yunnan.

Osmanthus matsumuranus.—*Osmanthus matsumuranus* (niu shi guo; Oleaceae) grows on densely-wooded slopes and in valley thickets at 800–1 500 m in Anhui, Guangdong, Guangxi, Guizhou, Jiangxi, Taiwan, Yunnan and Zhejiang. The young leaves are used locally as another source of ku ding cha.

Imperata cylindrica.—*Imperata cylindrica* (bai mao;

Poaceae) is a perennial herb growing wild on sandy riverbanks and coasts and disturbed grasslands in Anhui, Fujian, Guangdong, Guangxi, Guizhou, Hainan, Hebei, Heilongjiang, Henan, Hubei, Hunan, Jiangsu, Jiangxi, Liaoning, Nei Mongol, Shaanxi, Shandong, Shanxi, Sichuan, Taiwan, Xinjiang, Xizang, Yunnan and Zhejiang. The sweet-tasting rhizomes are used widely as a substitute for tea. It flourishes in grasslands that are frequently burned and the young shoots provide good fodder. It is also used as a medicine and fiber. There are two varieties, *I. cylindrica* var. *cylindrica* (bai mao) and *I. cylindrica* var. *majar* (da bai mao).

Lophatherum.—There are two species of the perennial herb *Lophatherum* (dan zhu ye shu; Poaceae) in China: *L. gracile* (dan zhu ye), distributed in Anhui, Fujian, Guangdong, Guangxi, Guizhou, Hainan, Hubei, Hunan, Jiangsu, Jiangxi, Sichuan, Taiwan, Yunnan and Zhejiang, and *L. sinense* (zhong hua dan zhu ye), in Hunan, Jiangsu, Jiangxi and Zhejiang. The whole plant of both species is an important ingredient in traditional herbal teas used to prevent heatstroke in summer.

21.3 Dye plants

Many plants contain natural coloring matter which can be extracted for dyeing fibers. For example, alizarin, a red tincture, is extracted from *Rubia cordifolia* (qian cao; Rubiaceae), indigo, a blue tincture, from multiple species as described above, and gardenin, a yellow tincture, from *Gardenia jasminoides* (zhi zi; Rubiaceae). The most common uses of dye plants are in cloth dyeing and food dyeing. They may also be used to color handicrafts and wooden furniture, fishing nets and body tattoos in some cultures in China.

The Chinese people have a very long history of using dye plants. SHI Ji recorded that from the period of HUANG Di (3000 BC), plants were used to dye cloth (Zhu *et al.*, 2007). Even today, local people among many ethnic minorities still extract coloring matter from plants to dye their cloth using traditional technology. *Strobilanthes cusia* ban lan; Acanthaceae and *Polygonum tinctorium* liao lan; Polygonaceae are widely used to dye cloth by the Bai, Buyi, Miao, Tujia and Yi people. The Yi people in Liangshan, Sichuan, cultivate this plant in their fields and gardens. The leaves are collected to extract the indigo-containing material, which is compressed and dried into disc-shaped pieces like cakes which can be stored for several years. *Buddleja officinalis* (mi meng hua; Loganiaceae) and *Iresine herbstii* (xue xian; Amaranthaceae) are used by the Buyi people in Yunnan to dye rice. In southern Yunnan, the Dai people still use *Indigofera suffruticosa* to dye blue clothing, the tubers of *Curcuma longa* (turmeric; jiang huang; Zingiberaceae) to dye yellow clothing, and the tubers of *Dioscorea cirrhosa* (shu liang; Dioscoreaceae) to dye fishing nets and make them durable. Dye plants are traditionally used all over China, with different species being used in different areas. Those described below are drawn from a variety of parts of China.

21.3.1 Examples of important dye plants

Polygonum tinctorium.—*Polygonum tinctorium* (liao lan; Polygonaceae Figure 21.3) is an annual herb 50–80 cm tall found widely in moist valleys and on stream banks and also extensively cultivated in China below 1 000 m. The leaves have a long history of use for making indigo, and the species is also used medicinally.

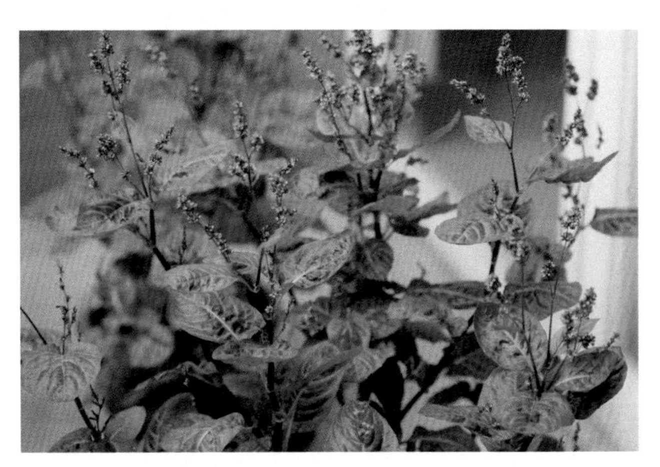

Figure 21.3 *Polygonum tinctorium.*

Isatis tinctoria.—*Isatis tinctoria* (song lan; Cruciferae) is a biennial herb of fields, pastures, roadsides and waste places in Fujian, Gansu, Guizhou, Hebei, Henan, Hubei, Jiangxi, Liaoning, Nei Mongol, Shaanxi, Shandong, Shanxi, Sichuan, Xinjiang, Xizang, Yunnan and Zhejiang. It is a very important dye plant. The dye is extracted from the roots and leaves, which are also used for medicinal purposes, and the seed oil is used in industry.

Vaccinium bracteatum.—*Vaccinium bracteatum* (nan zhu; Ericaceae) is a small evergreen tree or shrub from Anhui, Fujian, Guangdong, Guangxi, Guizhou, Hainan, Hunan, Jiangsu, Jiangxi, Sichuan, Taiwan, Yunnan and Zhejiang, commonly known as "wu-fan-shu." The ripe fruits are used widely in the region south of the Chang Jiang to stain cooked rice. There are four varieties, *V. bracteatum* var. *bracteatum*, *V. bracteatum* var. *chinense* (xiao ye nan zhu), *V. bracteatum* var. *obovatum* (dao nuan ye nan zhu), and *V. bracteatum* var. *rubellum* (dan hong nan zhu).

Caesalpinia sappan.—*Caesalpinia sappan* (su mu; Fabaceae; Figure 21.4) is a small tree widely cultivated in Fujian, Guangdong, Guangxi, Guizhou, Hainan, Sichuan, Taiwan and Yunnan. The heartwood is used to extract a red dye used to color cotton, hemp, wool and paper. It can also be used as a mordant for pigments in painting the background colors of woodenware, and is used medicinally to relieve pain.

Indigofera tinctoria.—*Indigofera tinctoria* (mu lan; Fabaceae) is a shrublet cultivated in southwestern Anhui, Guangdong, Guangxi, Guizhou, Hainan, Taiwan and Yunnan. The leaves of this species are widely used as a source of the dye indigo. *Indigofera suffruticosa* (ye qing

Figure 21.4 *Caesalpinia sappan* in fruit.

shu; Figure 21.5) is cultivated as an alternative to *I. tinctoria* and is used by the Dai people of Xishuangbanna to dye cloth and fishing nets to make them more compact and stand friction, thus extending their service time.

Rhamnus utilis.—Rhamnus utilis (Chinese buckthorn; dong lü; Rhamnaceae) is a small tree found below 3 300 m in Anhui, Fujian, Gansu, Guangdong, Guangxi, Guizhou, Hebei, Henan, Hubei, Hunan, Jiangsu, Jiangxi, Shaanxi, Shanxi, Sichuan and Zhejiang. The fruit, bark and leaves contain green and yellow dyes. The *Compendium of Materia Medica* records it as a popular dye plant for producing green cloth. An oil extracted from the seeds is used for making lubricating oil. The species includes three varieties, *R. utilis* var. *hypochrysa* (mao dong lü), *R. utilis* var. *szechuanensis* Chou (gao shan dong lü) and *R. utilis* var. *utilis*.

Figure 21.5 *Indigofera suffruticosa.*

Rhus.—There are six species of *Rhus* (yan fu mu shu; Anarcardiaceae), with four endemic, in China. Many species in this genus are the hosts of gall-producing insects such as aphids, the galls of which are full of tannins and traditionally used to dye cloth. All six species are usually cultivated to host the gall-producing insects, with *R. chinensis* (yan fu mu; Chinese sumac) being the most popular, distributed in Anhui, Fujian, Gansu, Guangdong, Guangxi, Guizhou, Hainan, Hebei, Henan, Hubei, Hunan, Jiangsu, Jiangxi, Ningxia, Qinghai, Shaanxi, Shandong, Shanxi, Sichuan, Taiwan, Xizang, Yunnan and Zhejiang. The other five species are *R. hypoleuca* (bai bei fu yang), from Fujian, Guangdong, southern Hunan and Taiwan, *R. potaninii* (qing fu yang), from southern Gansu, Henan, southern Shaanxi, southern Shanxi, Sichuan and northwestern Yunnan, *R. punjabensis* (pang zhe pu fu yang) in Gansu, Guizhou, western Hubei, western Hunan, southern Shaanxi, Sichuan, southern and southeastern Xizang and northern Yunnan, *R. teniana* (dian fu yang), in northern Yunnan, and *R. wilsonii* (chuan fu yang), distributed in southwestern Sichuan and northeastern Yunnan.

Buddleja officinalis.—Buddleja officinalis (mi meng hua; Loganiaceae) is a shrub found on montane forest edges and riverbanks at 200–2 800 m in Anhui, Fujian, Gansu, Guangdong, Guangxi, Guizhou, Henan, Hubei, Hunan, Jiangsu, Shaanxi, Shanxi, Sichuan, Xizang and Yunnan. The flowers are a source of food-coloring and are also used as a medicine and perfume.

Peristrophe bivalvis.—Peristrophe bivalvis (guan yin cao; Acanthaceae; Figure 21.6) is a perennial herb found in the forests of Fujian, Guangdong, Guangxi, Guizhou, Hainan, Hunan, Jiangxi, Shanghai and Yunnan between 500 and 1 000 m. This species is an important dye plant, from which is extracted orange-yellow to orange-red dye.

Figure 21.6 *Peristrophe bivalvis* in flower.

Strobilanthes cusia.—*Strobilanthes cusia* (ban lan; Acanthaceae) is a herb of moist wooded places below 2 000 m, and sometimes cultivated, in Fujian, Guangdong, Guangxi, The leaves contain indigo, and the species has been widely cultivated as a dye for over 3 000 years in central, southern and western China. The roots and leaves are also used medicinally.

Gardenia jasminioides.—*Gardenia jasminioides* (zhi zi; Rubiaceae; Figure 21.7) is a shrub found below 1 500 m in Anhui, Fujian, Guangdong, Guangxi, Guizhou, Hainan, Hong Kong, Hubei, Hunan, Jiangsu, Jiangxi, Shangdong, Sichua, Taiwan and Yunnan. It is cultivated in Hebei, Gansu and Shanxi. The fruits are a very important source of yellow dye used widely for coloring food and fibers; they are also used in traditional medicine.

Figure 21.7 Flower of *Gardenia jasminioides*.

Rubia cordifolia.—*Rubia cordifolia* (qian cao; Rubiaceae) is a lianescent herb found in northeastern, northern and northwestern China, Sichuan and Xizang. The roots contain alizarin, an important red dye which is widely used to color fibers and foods.

Curcuma longa.—*Curcuma longa* (turmeric; jiang huang; Zingeberaceae; Figure 21.8) is a herb cultivated in Fujian, Guangdong, Guangxi, Sichuan, Taiwan, Xizang and Yunnan. The rhizomes contain curcumin, a good yellow dye used for all kinds of animal- and plant-based fibers. The dye can be used directly or with only a little denaturing using alum, acid or acid salts. The rhizomes are also used medicinally and as a spice.

Figure 21.8 *Curcuma longa*.

Dioscorea cirrhosa.—*Dioscorea cirrhosa* (shu liang; Dioscoreaceae) is a herbaceous vine found in forests below 1 500 m in Fujian, Guangdong, Guangxi, Guizhou, Hainan, Hunan, southern Jiangxi, southern and western Sichuan, Taiwan, southeastern Xizang, Yunnan and southern Zhejiang. The tubers have a very high tannin content and have long been used to dye and preserve fabrics such as silk, cotton and fishing nets. There are two varieties, *D. cirrhosa* (shu liang) var. *cirrhosa* and *D. cirrhosa* var. *cylindrica* (kuai jing shu liang).

21.4 Aromatic plants

Plants that contain a fragrant component, especially volatile oils, are used for making perfumes, flavorings, food additives, cigarette additives and incense, among others. Some of these plants have very high economic value and high market price. In China, there is a very long history of using aromatic plants: the use of *Ocimum basilicum* (sweet basil; luo le; Lamiaceae) as an incense was recorded more than 2 000 years ago, in the poem *Li Sa* by the great poet QU Yuan (340–278 BC), who lived in the State of Chu during the Warring States Period (Zhang *et al.*, 2007). SHEN Nong's *Herbal Classic* also records several spice plants, such as *Cinnamomum* (zhang shu; Lauraceae) and *Zanthoxylum* (hua jiao shu; Rutaceae; Zhang *et al.*, 2007). Aromatic plants are particularly concentrated in southern and southwestern parts of China: *Cinnamomum* spp., *Zanthoxylum* spp. and *Piper nigrum* (pepper; hu jiao; Piperaceae) are cultivated only in the south of the country.

21.4.1 Examples of important aromatic plants

Illicium verum.—*Illicium verum* (star anise; ba jiao; Illiciaceae; Figure 21.9) is a tree to 15 m tall found in the forests below 1 600 m of southern and western Guangxi. It is also cultivated in Fujian, Guangdong, Guangxi, Jiangxi and Yunnan for perfume, medicine, and as a culinary spice.

Cinnamomum pauciflorum.—*Cinnamomum pauciflorum* (shao hua gui; Lauraceae) is a tree of hillside forests on calcareous or sandy soils in Chongqing Shi, Guangdong, Guangxi, Guizhou, Hubei, western Hunan, Jiangxi, eastern Sichuan and northeastern Yunnan. The bark and roots are used as a medicine for treating abdominal pain, while the leafy branchlets contain a volatile oil (ca. 35%), the main chemical constituent of which is safrole, also known as shikimol (80–95%). In Chongqing Shi and Sichuan, it is known as "xiang gui" or "yan gui" and is planted in large numbers for the production of safrole.

Lindera.—*Lindera* (shan hu jiao shu; Lauraceae) includes around 40 species of evergreen or deciduous trees in China, to which 23 are endemic; one additional species is of uncertain placement. Three species are widely used as spices: *L. angustifolia* (xia ye shan hu jiao), *L. communis*

Figure 21.9 *Illicium verum*. A, tree. B, fruits.

(xiang ye shu) and *L. glauca* (shan hu jiao). *Lindera angustifolia* (xia ye shan hu jiao) is found in montane thickets and sparse forests in Anhui, Fujian, Guangdong, Guangxi, Henan, Hubei, Jiangsu, Jiangxi, Shaanxi, Shandong and Zhejiang. The leaves are processed for their aromatic oils, and the seed oil is used for making soap and machine oil. *Lindera communis* (xiang ye shu) is found in dry sandy places and evergreen broad-leaved forests in Fujian, Gansu, Guangdong, Guangxi, Guizhou, Hubei, Hunan, Jiangxi, Shaanxi, Sichuan, Taiwan, Yunnan and Zhejiang. The pericarp contains aromatic oil, the seed oil is used in food and for making soap and machine oil, and the branchlets and leaves are used medicinally. *Lindera glauca* (shan xiang zhi) is found in montane forests and on roadsides below 900 m in Anhui, Fujian, Gansu, Guangdong, Guangxi, Guizhou, southwestern Henan, Hubei, Hunan, Jiangxi, southwestern Shaanxi, eastern Shandong, Shanxi, Sichuan, Taiwan and Zhejiang. The wood is used in woodwork, the leaves and pericarp are processed for their aromatic oil, the seed oil is used for making soap and machine oil, and the roots, branchlets, leaves and fruit are used medicinally.

Litsea.—There are over 70 species of *Litsea* (mu jiang zi shu; Lauraceae) in China, of which 47 are endemic to the country. The flowers, leaves and fruits of most species contain aromatic oils, which may be processed to produce the compound citral, and are used for their fragrance and medicinal properties. Three species are most widely used: *L. cubeba* (may chang; shan ji jiao), *L. pungens* (mu jiang zi) and *L. mollis* (mao ye mu jiang zi). *Litsea cubeba* is found in Anhui, Fujian, Guangdong, Guangxi, Guizhou, Hainan, Hubei, Hunan, Jiangsu, Jiangxi, Sichuan, Taiwan, Xizang, Yunnan and Zhejiang; its flowers, leaves and fruit walls are processed for production of citral and used for their fragrance and medicinal properties. In Kunming Shi, Shanghai and Sichuan, the fruits are used as a medicine known as "bi cheng qie." *Litsea pungens* is found in Gansu, northern Guangdong, Guangxi, Guizhou, Henan, Hubei, Hunan, Shaanxi, southern Shanxi, Sichuan, Xizang, Yunnan and southern Zhejiang. The aromatic oil content of the fruit is 3–4% when fresh, 2–6% when dried, and its main ingredients are citral (60–90%) and geraniol (5–19%). The fruit is used in food and cosmetics and also as a component of high-grade spices, vitamin A and ionones. The seeds contain ca. 48% oil, which may be used in industry and for making soap. *Litsea mollis* is found in montane thickets and broad-leaved forests in Guangdong, Guangxi, Guizhou, Hubei, Hunan, Sichuan, eastern Xizang and Yunnan. The fruits are processed for their aromatic oil (3–5%). The seeds contain ca. 25% oil and are used as a main ingredient in soaps. The roots and fruits are also used medicinally.

Lysimachia foenum-graecum.—*Lysimachia foenum-graecum* (ling xiang cao; Primulaceae) is a perennial herb of wet mixed forests, streams and humus-rich soils in northern Guangdong, Guangxi, southwestern Hunan and southeastern Yunnan. The whole plant has a strong aroma when dried, and is used as a flavor additive in tobacco.

Murraya paniculata.—*Murraya paniculata* (orange jessamine; qian li xiang; Rutaceae) is a shrub or tree of montane forests and thickets below 1 300 m in Fujian, Guangdong, Guangxi, southern Guizhou, Hainan, southern Hunan, Taiwan and Yunnan. The flowers have a strong aroma and are refined for their essence, while the plant is also used as an ornamental.

Zanthoxylum bungeanum.—*Zanthoxylum bungeanum* (hua jiao; Rutaceae) is a small deciduous shrub or tree found in many habitats below 3 200 m in Anhui, Fujian, Gansu, Guangxi, Guizhou, Hebei, Henan, Hubei, Hunan, Jiangsu, Jiangxi, Liaoning, Ningxia, Qinghai, Shaanxi, Shandong, Shanxi, Sichuan, southeastern Xinjiang, southern and southeastern Xizang, Yunnan and Zhejiang. The dried fruit follicles are used as a culinary spice and are particularly popular in Sichuan cuisine (hence the common name "Sichuan pepper"). In Yunnan, the fruit follicles of *Z. acanthopodium* (ci hua jiao) are sometimes similarly used.

Cuminum cyminum.—Cuminum cyminum (cumin; zi ran qin; Apiaceae) is an annual or biennial herb cultivated in Xinjiang. The aromatic fruits are used as a flavoring, to aid digestion, and are reputed to have medicinal value. This species is widely cultivated in favorable climates outside its presumed native range.

Elsholtzia ciliata.—Elsholtzia ciliata (Vietnamese balm; xiang ru; Lamiaceae) is a herb found below 3 400 m in all provinces except Qinghai and Xinjiang. The whole plant has a strong aroma when dried and is used to produce an essence. It is also used traditionally as a medicine. *Elsholtzia communis* (ji long cao) is cultivated and occasionally naturalized in Yunnan. The whole plant has strong aroma and is used to produce an essence.

Mentha canadensis.—Mentha canadensis (bo he; Lamiaceae) is a perennial rhizomatous herb found in wet places below 3 500 m throughout China. It is a source of mint oil which is used in cigarettes, pharmaceuticals and oral preparations. In Yunnan, local people use it as spice and vegetable.

Ocimum.—There are five species of *Ocimum* (luo le shu; Lamiaceae) in China, all of which are aromatic. Two species are cultivated as aromatic herbs, sources of essential oil, condiments, and for use in cosmetics and perfumes: *O. basilicum* (sweet basil; luo le; Figure 21.10), is distributed in Anhui, Fujian, Guangdong, Guangxi, Guizhou, Hebei, Henan, Hubei, Hunan, Jiangsu, Jiangxi, Jilin, Sichuan, Taiwan, Xinjiang, Yunnan and Zhejiang, mostly cultivated but sometimes wild, and *O. sanctum* (holy basil; sheng luo le), from Hainan, Sichuan and Taiwan, is used as a condiment in salads and other foods and as a substitute for tea.

Perilla frutescens.—Perilla frutescens (zi su; Lamiaceae) is an aromatic herb found on wasteland and in gardens in Fujian, Guangdong, Guangxi, Guizhou, Hebei, Hubei, Jiangsu, Jiangxi, Shanxi, Sichuan, Taiwan, Xizang, Yunnan and Zhejiang. The whole plant is very aromatic; the leaves are widely used as a flavoring and are also distilled to produce an essential oil. Both leaves and flower clusters are used as a condiment or salted and eaten; the seedling cotyledons are also used as a condiment. Today it is also used for production of industrial oil (see Chapter 16).

Figure 21.10 *Ocimum basilicum.*

Cyperus rotundus.—Cyperus rotundus (coco grass; xiang fu zi; Cyperaceae) is a perennial herb found below 2 100 m in northeastern, northern, northwestern, eastern, central, southern and southwestern China. The tubers are used in Chinese medicine and contain a fragrant oil which may be extracted.

Cymbopogon.—Cymbopogon (xiang mao shu; Poaceae) includes ca. 25 species in China, of which seven are endemic and up to five introduced. Six species are cultivated commercially for the aromatic oils that are distilled from their leaves. The oils are often lemon-scented and are used in perfumes and as an insect repellent. The six cultivated species are *C. citratus* (lemongrass; xiang mao), grown in Fujian, Guangdong, Guizhou, Hainan, Hubei, Taiwan, Yunnan and Zhejiang, *C. distans* (yun xiang cao), of southern Gansu, Guizhou, Shaanxi, Sichuan, Xizang and Yunnan, *C. flexuosus* (Cochin grass; qu xi xiang mao), found on grassy slopes below 1 000 m in southwestern Yunnan, *C. jwarancusa* (la bo he cao), of mountain slopes and dry valleys in southwestern Sichuan, Xizang and Yunnan, *C. nardus* (ya xiang mao), cultivated in Fujian, Guangdong, Hainan, Taiwan and Yunnan, and *C. winterianus* (feng mao), grown in Guangdong, Hainan, Sichuan and Yunnan, known only in cultivation, and a source of high quality citronella oil.

Amomum tsaoko.—Amomum tsaoko (cao guo; Zingiberaceae) is an aromatic, perennial herb growing or cultivated in clumps in sparse forests at 1 100–1 800 m elevation in southern Yunnan. The dried fruits are used as a flavoring, to aid digestion, and as a medicine.

21.5 Rubber and gums

As well as the introduced rubber tree, *Hevea brasiliensis* (Euphorbiaceae), two introduced species of *Parthenium* (Asteraceae) are used to produce rubber in northwestern China. A wider range of plants produce natural gum, rich in the macromolecule amylose, some as resins, some in fruits and some in the seeds. Plant gums are very important raw materials for industry. For example, guar gum from *Cyamopsis tetragonoloba* (cluster bean; gua er dou; Fabaceae) used to be necessary for oil exploitation. In China, guar gum has more recently been supplanted by sesbania gum from the native species *Sesbania cannabina* (tian jing; Fabaceae; Zhang, 2008).

21.5.1 Examples of important rubber and gum plants

Amygdalus persica.—Amygdalus persica (peach; tao; Rosaceae) is a tree 3–8 m tall cultivated throughout China and escaped from cultivation on wastelands and disturbed slopes, at least in Gansu, Hebei and Shanxi. The trunk produces peach gum which is harvested by tapping. As described in Chapter 16, *A. persica* is also an important source of fruit in China.

Armeniaca vulgaris.—*Armeniaca vulgaris* (apricot; xing; Rosaceae) is a tree found wild or cultivated in sparse forests on mountain slopes at 700–3 000 m in Gansu, Hebei, Henan, Jiangsu, Liaoning, Nei Mongol, Ningxia, Qinghai, Shaanxi, Shandong, Shanxi, Sichuan and Xinjiang. While apricots are cultivated for their edible fruit and seed throughout China (see Chapter 16), the trunk can also produce gum for industrial uses, harvested by tapping.

Prunus salicina.—*Prunus salicina* (Chinese plum; li; Rosaceae) is a tree found wild or cultivated in Anhui, Fujian, Gansu, Guangdong, Guangxi, Guizhou, Hebei, Heilongjiang, Henan, Hubei, Hunan, Jiangsu, Jiangxi, Jilin, Liaoning, Ningxia, Shaanxi, Shandong, Shanxi, Sichuan, Taiwan, Yunnan and Zhejiang. It is an important, temperate fruit tree widely cultivated in China (see Chapter 16), the trunk of which also produces gum for industrial use.

Dalbergia obtusifolia.—*Dalbergia obtusifolia* (dun ye huang tan; Fabaceae) is a tree 13–17 m tall found in open forests, on mountainsides or in riverside vegetation at 800–1 300 m in Guizhou and southern Yunnan. It is also known as "niu lei ba" and is a good host for the shellac insect which produces lac, a resinous secretion that can be used in making shellac, used as a colorant, glaze, varnish or polish for anything from food to furniture, as is *Dalbergia assamica* (yang qing), a tree 7–10 m tall found in Fujian, Guangdong, Guangxi, Guizhou, Hainan, Sichuan and Yunnan.

Flemingia macrophylla.—*Flemingia macrophylla* (da ye qian jin ba; Fabaceae) is a shrub of meadows, roadsides and forest margins at 200–1 800 m in Fujian, Guangdong, Guangxi, Guizhou, Hainan, Jiangxi, Sichuan, Taiwan, Yunnan, and is another very good host for the shellac insect, cultivated largely for its shellac yield.

Gleditsia.—There are six species of *Gleditsia* (locust trees; zao jia shu; Fabaceae) in China, of which three are endemic and one introduced. All of them are good hosts of the shellac insect. The six species of *Gleditsia* are *G. australis* (xiao guo zao jia), in Guangdong and Guangxi, *G. fera* (hua nan zao jia), in Fujian, Guangdong, Guangxi, Guizhou, Hainan, Hunan, Jiangxi, Taiwan and Yunnan, *G. japonica* (shan zao jia), in Anhui, Guizhou, Hebei, Henan, Hunan, Jiangsu, Jiangxi, Liaoning, Shandong, Yunnan and Zhejiang, *G. microphylla* (ye zao jia), in Anhui, Guizhou, Hebei, Henan, Jiangsu, Shaanxi, Shandong and Shanxi, *G. sinensis* (zao jia; Figure 21.11), in Anhui, Fujian, Gansu, Guangdong, Guangxi, Guizhou, Hebei, Henan, Hubei, Hunan, Jiangsu, Jiangxi, Qinghai, Shaanxi, Shanxi, Sichuan, Yunnan and Zhejiang, and *G. triacanthos* (honey locust; meu guo zao jia), cultivated in Hong Kong and Shanghai. The introduced pigeon pea (mu dou; *Cajanus cajan*; Fabaceae), is also a very good host for the shellac insect and is particularly valued in areas of low or erratic rainfall.

Figure 21.11 *Gleditsia sinensis*, fruiting branch.

Sesbania.—There are four species of *Sesbania* (tian jing shu; Fabaceae) in China, one of which may be native (ci tian jing *S. bispinosa*), although its precise natural distribution is unknown. The seeds of *Sesbania* contain sesbania gum, a kind of galactomannan, which can be used as a substitute for guar gum in oil exploitation.

21.6 Fiber plants

Many plants are rich in fine fibers, which can be used for making cloth, paper, ropes, sacks or woven materials. Plant fibers such as flax and cotton have been very important through human history. The Chinese first made use of flax to weave cloth in 3000–4000 BC (Zhu *et al.*, 2007). SHEN Nong's *Herbal Classic* records the use of fibers from *Cannabis sativa* (hemp; da ma; Cannabaceae) and *Pueraria* spp. By the Han dynasty (202 BC–220 AD), cotton cultivation and cloth-weaving had became popular. HUANG Dao-Po (1245–1330 AD), a famous weaver living a the end of Song and beginning of the Ming dynasty developed a technology for weaving cotton cloth, which had very far-reaching impacts on cloth-weaving in China. Plant fibers are also widely used in China to make paper. Around 200 BC, in the Western Han dynasty, the Chinese began to make paper with plant fibers, even before CAI Lun developed paper-making technology. Plant fibers can also be used to make other commodities. For example, those of *Caragana* (jin ji er shu; Fabaceae) are used to make ropes, and those of *Carex meyeriana* (wu la cao; Cyperaceae) to make insoles for shoes. Recently, the extraction of fibers from plants to make artificial cotton has become an important industry.

Fiber plants are distributed and cultivated widely in China, with different plants having different distribution areas and being popular in different parts of the country. *Cannabis sativa* and *Linum usitatissimum* (common flax; ya ma; Linaceae) are cultivated widely across the whole country, as is *Gossypium herbaceum* (grass cotton; cao mian; Malvaceae; Figure 21.12), which is particularly prevalent in the northwestern province of Xinjiang. *Bombax* spp. (mu mian shu; Bombacaceae) and *Boehmeria nivea* (ramie; zhu ma; Urticaceae) are grown in southern China, the latter especially Hubei, Hunan, Jiangxi and Sichuan, and *Carex meyeriana* is native to northeastern China. *Pteroceltis tatarinowii* (qing tan; Ulmaceae) and *Broussonetia* spp. (gou shu; Moraceae) used to be important raw materials for making traditional paper types, such as "xuan zhi." The former is distributed widely across China, but historically was only used for making xuan zhi in Jinxian, Anhui.

Figure 21.12 *Gossypium herbaceum* with flower and young fruit.

21.6.1 Examples of important fiber plants

Gnetum.—Gnetum (mai ma teng shu; Gnetaceae) is an evergreen woody vine with nine species in China, six of which are endemic. The bark of many species is used for making gun bags, ropes and fishing nets. The species usually used for fibers are *G. montanum* (mai ma teng), which is found in Guangdong, Guangxi, Hainan and southern Yunnan, and *G. parvifolium* (xiao ye mai ma teng), from Fujian, Guangdong, Guangxi, Guizhou, Hainan, Hunan and Jiangxi.

Pteroceltis tatarinowii.—Pteroceltis tatarinowii (qing tan; Ulmaceae) is a tree to 20 m tall found on limestone mountains and alongside rivers and streams below 1 500 m in Anhui, Fujian, southern Gansu, Guangdong, Guangxi, Guizhou, Hebei, Henan, Hubei, Hunan, Jiangsu, Jiangxi, Liaoning, southeastern Qinghai, Shaanxi, Shandong, Shanxi, Sichuan and Zhejiang. The bark fiber is used in the manufacture of traditional paper (xuan zhi).

Cannabis sativa.—Cannabis sativa (hemp; da ma; Cannabaceae) is cultivated throughout China and possibly native to, or naturalized in, Xinjiang. Its long, strong fibers are used in the paper-making industry and for weaving cloth. The female inflorescences (particularly the glandular leafy bracts and bracteoles) are also used as a drug.

Antiaris toxicaria.—Antiaris toxicaria (jian xue feng hou; Moraceae) is a tree 25–40 m tall found in rainforests below 1 500 m in Guangdong, Guangxi, Hainan and southern Yunnan. The latex contains varying amounts of cardiac glycosides and can be very poisonous; the bark fibers are used for cordage. In southern Yunnan, the whole raw bark fibers are also used directly to make clothes.

Broussonetia.—Broussonetia (gou shu; Moraceae) comprises four taxa of deciduous trees in China. The bark fibers of three taxa have traditionally been used in paper-making for many years: *B. kaempferi* var. *australis* (teng gou), from Anhui, Chongqing, Fujian, Guangdong, Guangxi, Guizhou, Hubei, Hunan, Jiangxi, Taiwan, southeastern Yunnan and Zhejiang, *B. kazinoki* (chu), from Anhui, Fujian, Guangdong, Guangxi, Guizhou, Hainan, Henan, Hubei, Hunan, Jiangsu, Jiangxi, Sichuan, Taiwan, Yunnan and Zhejiang, and *B. papyrifera* (paper mulberry; gou shu; Figure 21.13), found in Anhui, Fujian, Gansu, Guangdong, Guangxi, Guizhou, Hainan, Hebei, Henan, Hubei, Hunan, Jiangsu, Jiangxi, Shaanxi, Shandong, Shanxi, Sichuan, Taiwan, southeastern Xizang, Yunnan and Zhejiang.

Figure 21.13 *Broussonetia papyrifera*, flowering branch.

Boehmeria nivea.—Boehmeria nivea (ramie; zhu ma; Urticaceae) is a monoecious shrub found at 200–1 700 m elevation and often cultivated in southern Anhui, Fujian, Guangdong, Guangxi, Guizhou, Hainan, Hubei, Hunan, Jiangxi, southern Shaanxi, Sichuan, Taiwan, Yunnan and Zhejiang, separated into two varieties, *B. nivea* var. *nivea* and *B. nivea* var. *tenacissima* (qing ye zhu ma). It provides a high-quality fiber, ramie, used to make ropes, cloth and some industrial materials; the young leaves are also used as fodder for silkworms. The history of cultivation of *B. nivea* in China can be traced back at least 3 000 years, and ramie was introduced to Europe and North and South America in the early eighteenth century.

Bombax.—Bombax (mu mian shu; Bonbacaceae) are large deciduous trees with three taxa in China: *B. ceiba* (cotton tree; mu mian), found in Fujian, Guangdong, Guangxi, Guizhou, Hainan, Jiangxi, Sichuan, Taiwan and Yunnan, and *B. cambodiense* (lan cang mu mian) and *B.*

insigne var. *tenebrosum* (chang guo mu mian), both from southern Yunnan.

Caragana jubata.—*Caragana jubata* (gui jian jin ji er; Fabaceae) is a shrub found on slopes and at forest margins from 2 400 to 4 700 m elevation in Gansu, Hebei, Nei Mongol, Ningxia, Qinghai, northern Shaanxi, Shanxi, Sichuan, Xinjiang, Xizang and northern Yunnan. The bark is rich in fine fibers, used for making ropes and sacks.

Edgeworthia.—There are four species of *Edgeworthia* (jie xiang shu; Thymelaeaceae) in China, three of which are endemic. The inner bark (bast fiber) is very pliable and strong, a good raw material for making high grade artificial cotton and paper. *Edgeworthia albiflora* (bai jie xiang) is found in forests and valleys at 1 000–1 200 m in southwestern Sichuan, *E. chrysantha* (jie xiang) in forests and also cultivated in Fujian, Guangdong, Guangxi, Guizhou, Henan, Hunan, Jiangxi, Yunnan and Zhejiang, *E. eriosolenoides* (xi chou jie xiang) in southeastern Yunnan, and *E. gardneri* (dian jie xiang) in eastern Xizang and northwestern Yunnan.

Stellera chamaejasme.—*Stellera chamaejasme* (lang du; Thymelaeaceae) is a perennial herb found on dry, sunny slopes and in sandy places at 2 600 to 4 200 m elevation in Gansu, Hebei, Heilongjiang, Henan, Jilin, Liaoning, Nei Mongol, Ningxia, Qinghai, Shaanxi, Shanxi, Sichuan, Xinjiang, Xizang and Yunnan. The bast fiber of the rhizomes and stems is used in making paper.

Apocynum.—There are two species of *Apocynum* (luo bu ma shu; Apocynaceae) in China, the bark fiber of which is very fine, strong and shiny, a very good material for making cloth, strings, sails, fishing nests and high-quality paper. *Apocynum venetum* (dogbane; luo bu ma), known as "hong ma," is found on barren saline substrates, desert margins, alluvial flats and riversides in Gansu, Hebei, Henan, Jiangsu, Liaoning, Nei Mongol, Qinghai, Shaanxi, Shandong, Shanxi, Xinjiang and Xizang. *Apocynum pictum* (bai ma) is found in similar habitats in Gansu, Qinghai and Xinjiang.

Carex meyeriana.—*Carex meyeriana* (wu la cao; Cyperaceae) is a perennial herb of marshes in Heilongjiang, Jilin, Liaoning, Nei Mongol and Sichuan. The leaves are full of fine soft, fibers, considered very good for making insoles to keep the feet warm, as well as artificial cottons, fiberboard, paper and woven handicrafts.

Achnatherum splendens.—*Achnatherum splendens* (ji ji cao; Poaceae) is a perennial herb forming dense tussocks on dry mountain slopes and grassy places on slightly alkaline or sandy soils at 900–4 500 m elevation in Gansu, Heilongjiang, Henan, Nei Mongol, Ningxia, Qinghai, Shanxi, Sichuan, Xinjiang, Xizang and Yunnan. The culms and tough, long, smooth leaves, are good sources of fibers, providing the raw materials for paper-making pulp, grass baskets, shades and brooms. *A. splendens* can tolerate high levels of salinity and is also used for improving saline and alkaline land.

Eulaliopsis binata.—*Eulaliopsis binata* (ni jin mao; Poaceae) is a perennial herb of dry mountain slopes in Guangdong, Guangxi, Guizhou, Henan, Hubei, Shaanxi, Sichuan, Taiwan and Yunnan. The species is a good source of fibers for making paper, artificial cotton and artificial silk.

Phragmites australis.—*Phragmites australis* (lu wei; Poaceae) is a robust perennial herb forming large colonies in damp places throughout China. The culms are a very important material for paper-making, harvested in large quantities each year in China.

21.7 Forage plants

Forage plants are plants used for feeding livestock, poultry or even silkworms. Most forage plants are highly nutritious, particularly in terms of protein content. As soon as the Chinese began keeping livestock, they also began to utilize forage plants. Thus, chickens were being fed as long ago as the Neolithic Era (7 000–8 000 years ago), pigs from around 6 000 years ago, and cattle and horses about 4 000 years ago (Wang, 1991; Li & Li, 2008). From an early stage, wild forage was the main resource for feeding these animals. From about 2 000 years ago, cultivation of *Medicago sativa* (alfalfa; zi mu xu; Fabaceae) for forage became popular in northwestern China, and from at least 1 000 years ago *Astragalus sinicus* (Chinese milkvetch; zi yun ying; Fabaceae) was cultivated widely as a forage and green manure in southern China. The use of white mulberry (*Morus alba*; sang; Moraceae) trees to feed silkworms and produce silk dates back as far as 7 000 years. Inscribed "oracle bones" dating from the Shang Dynasty (ca. 1600--1100 BC) mention the mulberry, silkworm and silk. Mulberry and silkworm cultivation are now popular in eastern and southern China, especially in Jiangsu, Sichuan and Zhejiang.

Common forage plants today include, in northern and northwestern China, *Blysmus sinocompressus* (hua bian sui cao) and *Kobresia* spp. (bog sedges; song cao shu; both Cyperaceae), *Leymus chinensis* (yang cao) and *Lolium* spp. (hei mai cao shu; ryegrasses; both Poaceae); in southern China, *Astragalus sinicus* (Fabaceae) remains the most popular forage plant. Other forage species are distributed or cultivated all over the country, including *Melilotus* spp. (sweet clover; cao mu xi shu), *Trifolium* spp. (clover; che zhou cao shu) and *Vicia* spp. (vetches; ye wan dou shu; all Fabaceae), and species of *Morus* (Moraceae).

21.7.1 Examples of important forage plants

Morus alba.—*Morus alba* (white mulberry; sang; Moraceae) is originally endemic to central and northern China, and now cultivated throughout China. The leaves provide food for silkworms, the bark fibers are used for textiles and paper, and the bark is also used medicinally. There are two varieties, *M. alba* var. *alba* and *M. alba* var.

multicaulis (lu sang).

Polygonum viviparum.—*Polygonum viviparum* (alpine bistort; zhu ya quan shen; Polygonaceae) is a perennial herb of forest margins, grassy slopes and alpine steppe at 1 200–5 100 m altitude in Gansu, Guizhou, Hebei, Heilongjiang, Henan, Hubei, Jilin, Liaoning, Nei Mongol, Ningxia, Qinghai, Shaanxi, Shanxi, Sichuan, Xinjiang, Xizang and Yunnan. The young leaves provide good forage for goats and sheep. There are two varieties, *P. viviparum* var. *tenuifolium* (xi ye zhu ya quan shen) and *P. viviparum* var. *viviparum*.

Astragalus sinicus.—*Astragalus sinicus* (zi yun ying; Fabaceae) is found as a weed in wet places and cultivated in rice fields in Fujian, Gansu, Guangdong, Guangxi, Guizhou, Hebei, Hunan, Jiangsu, Jiangxi, Shaanxi, Sichuan, Taiwan, Yunnan and Zhejiang. It is an important forage plant and green manure, popularly cultivated in the areas south of the Chang Jiang.

Medicago sativa.—*Medicago sativa* (alfalfa; zi mu xu; Fabaceae) is a perennial herb cultivated throughout China, often escaped on roadsides, streamsides, in fields and grasslands. This species is rich in protein, providing high-quality forage and pasture.

Melilotus.—*Melilotus* (sweet clover; cao mu xi shu; Fabaceae) comprises four species in China, all rich in protein. The root nodules fix nitrogen and the plants are adapted to a wide range of habitats. They are usually grown as green manure, forage crops or bee plants. *Melilotus albus* (Bokhara clover; bai hua cao mu xi) is found in Anhui, Fujian, Gansu, Guizhou, Hebei, Heilongjiang, Henan, Hubei, Jiangsu, Liaoning, Nei Mongol, Ningxia, Shaanxi, Shandong, Shanxi, Sichuan, Xinjiang and Yunnan, *M. dentatus* (xi chi cao mu xi) in Gansu, Hebei, Heilongjiang, Liaoning, Nei Mongol, Shaanxi, Shandong, Shanxi and Xinjiang, *M. indicus* (yin du cao mu xi) in Anhui, Fujian, Guangdong, Guangxi, Guizhou, Hainan, Jiangsu, Sichuan, Taiwan, Yunnan and Zhejiang, and *M. officinalis* (yellow sweet clover; cao mu xi) throughout China.

Trifolium repens.—*Trifolium repens* (white clover; bai che zhou cao; Fabaceae) is a perennial herb cultivated, escaped and naturalized in grasslands, ditches and roadsides throughout China. It provides very good forage and feedstuff. *Trifolium pratense* (red clover; hong che zhou cao) is cultivated and naturalized on woodland margins, wet meadows and roadsides in Anhui, Fujian, Guangdong, Hebei, Heilongjiang, Henan, Hubei, Jiangsu, Jiangxi, Jilin, Liaoning, Shandong, Taiwan, Xinjiang and Yunnan, also providing good forage and feed.

Vicia.—There are ca. 40 species of *Vicia* (vetches; ye wan dou shu; Fabaceae) in China. Many are used from the wild or cultivated as a green mature and forage, with five taxa being the most popularly cultivated: *V. cracca* (tufted vetch; guang bu ye wan dou), distributed and cultivated in Anhui, Chongqing, Fujian, Gansu, Guangdong, Guangxi, Guizhou, Hebei, Heilongjiang, Henan, Hubei, Jiangxi, Jilin, Liaoning, Nei Mongol, Shaanxi, Shanghai, Shanxi, Sichuan, Taiwan, Xinjiang, Xizang, Yunnan and Zhejiang, *V. hirsuta* (hairy tare; xiao chao cai), found and cultivated in Anhui, Fujian, Gansu, Guangdong, Guangxi, Guizhou, Jiangsu, Qinghai, Shaanxi, Sichuan, Taiwan, Xinjiang, Yunnan and Zhejiang, *V. sepium* (bush vetch; ye wan dou), found and cultivated in Gansu, Guizhou, Shaanxi, Sichuan, Xinjiang and Yunnan, *V. tetrasperma* (smooth tare; si zi ye wan dou), found and cultivated in Anhui, Fujian, Gansu, Guizhou, Hebei, Henan, Hubei, Hunan, Jiangsu, Jiangxi, Shaanxi, Sichuan, Taiwan, Xinjiang, Yunnan and Zhejiang, and *V. villosa* subsp. *varia* (ou zhou tiao zi), cultivated and possibly naturalized in Guangdong, Shandong and Taiwan.

Blysmus sinocompressus.—*Blysmus sinocompressus* (hua bian sui cao; Cyperaceae) is a perennial herb of wet places in Gansu, Hebei, Liaoning, Nei Mongol, Ningxia, Qinghai, Shaanxi, Shanxi, Sichuan, Xinjiang, Xizang and northwestern Yunnan, particularly widely distributed on plateaus. This soft sedge, 13–17% protein, 4–5% fat and 20–27% fiber, provides high-quality and pretty grazing. There are three varieties, *B. sinocompressus* var. *nodosus* (jie gan bian sui cao), *B. sinocompressus* var. *sinocompressus*, and *B. sinocompressus* var. *tenuifolius* (xi ye bian sui cao).

Kobresia pygmaea.—*Kobresia pygmaea* (gao shan song cao; Cyperaceae) is a cushion-forming perennial herb found at 3 100–5 600 m in Gansu, Hebei, Nei Mongol, Qinghai, Shanxi, Sichuan, Xinjiang, Xizang and Yunnan. This low-growing, soft sedge is 11–17% protein, 4–6% fat and 24–27% fiber, providing high-quality grazing. *Kobresia royleana* (xi ma la ya song cao) is found at 700–5 200 m in Gansu, Qinghai, Sichuan, Xinjiang, Xizang and Yunnan and contains 12–16% protein, 4–5% fat and 25–29% fiber. There are two subspecies, *K. royleana* subsp. *minshanica* and *K. royleana* subsp. *royleana*.

Bromus inermis.—*Bromus inermis* (smooth brome; wu mang que mai; Poaceae) is a perennial herb, often the dominant species of mountain meadows at 1 000–3 500 m elevation, in Gansu, Guizhou, Hebei, Heilongjiang, Jiangsu, Jilin, Liaoning, Nei Mongol, Qinghai, Shaanxi, Shandong, Shanxi, Sichuan, Xinjiang, Xizang and Yunnan. It provides high-quality grazing with a high yield and is adaptable to many sites. The species has been spread worldwide through seed distribution for pasture and fodder.

Lolium perenne.—*Lolium perenne* (perennial ryegrass; hei mai cao; Poaceae) is a perennial herb commonly cultivated in China and extensively cultivated in temperate regions of the world as an excellent forage and lawn grass. *Lolium multiflorum* (Italian ryegrass; duo hua hei mai cao) has been introduced to grasslands in Anhui, Fujian, Guizhou, Hebei, Henan, Hunan, Jiangxi, Nei Mongol, Shaanxi, Sichuan, Taiwan, Xinjiang and Yunnan, and is also widely grown in temperate regions of the world for pasture and forage.

Leymus chinensis.—Leymus chinensis (yang cao; Poaceae) is a perennial herb of grassy places in Gansu, Hebei, Heilongjiang, Henan, Jilin, Liaoning, Nei Mongol, Qinghai, Shaanxi, Shandong, Shanxi and Xinjiang. It is tolerant of cold, drought, alkaline soils and trampling, and is an important grazing plant in the pastures of northeastern and northern China.

21.8 Bio-energy plants

Plants that convert solar energy quickly into chemical energy can provide an energy resource for industrial development. In this chapter, we focus on species that can be used to produce biological diesel (bio-diesel) oil, a substitute for fossil oil, which is very important in a rapidly-developing and energy-requiring country such as China, in a situation of finite, limited fossil fuels. Not all green plants provide an easily usable energy resource for industry; whereas previously the biomass of plants has traditionally been used as firewood, only two main forms of plant energy are usable in modern industries: vegetable oil and bio-ethanol. It is estimated that there are at least 100 species of plants whose seed can be used as a source of bio-diesel oil (i.e. those with an oil content greater than 25%), but until recently only a few have been exploited. Examples include the introduced *Jatropha curcas* (xiao tong zi; Euphorbiaceae; Figure 21.14; see Chapter 20), cultivated on a large scale in Yunnan and Sichuan for producing bio-diesel, and *Vernicia fordii* (tung oil tree; you tong; Euphorbiaceae) and *V. montana* (mu you tong), which used to be cultivated in large numbers for the production of tung oil (used as a coating to protect wooden and bamboo furniture), and are now increasingly used to produce bio-diesel (Long & Song, 2012). In China, policies on developing bio-energy have clearly stated a need to avoid using farmland and food crops to produce bio-energy. Based on recent research, we here describe the native taxa that have been highlighted by scientists as having greatest potential value for green energy exploitation (for introduced examples see Chapter 20).

Figure 21.14 *Jatropha curcas* in fruit.

21.8.1 Examples of important energy plants

Pongamia pinnata.—Pongamia pinnata (shui huang pi; Fabaceae) is a tree 8–15 m tall of coastal areas and riverbanks in Fujian, Guangdong, Hainan and Taiwan. The seed oil can be used as a fuel and it is a potential bio-diesel plant.

Triadica.—There are three species of *Triadica* (wu jiu shu; Euphorbiaceae) in China. The seeds have long been used as a source of fat for making candles and soap, and are now considered to have great potential as a source of bio-diesel. *Triadica cochinchinensis* (shan wu jiu) is found in Anhui, Fujian, Guangdong, Guangxi, Guizhou, Hainan, Hubei, Hunan, Jiangxi, Sichuan, Taiwan, Yunnan and Zhejiang, *T. rotundifolia* (yuan ye wu jiu) in Guangdong, Guangxi, Guizhou, Hunan and Yunnan, and *T. sebifera* (Chinese tallow tree; wu jiu) in Anhui, Fujian, Gansu, Guangdong, Guangxi, Guizhou, Hainan, Hubei, Jiangxi, Jiangsu, Shaanxi, Shandong, Sichuan, Taiwan, Yunnan and Zhejiang.

Vernicia.—Vernicia (you tong shu; Euphorbiaceae) comprises two species in China. The seeds are a source of drying oils, used in paints and varnishes traditionally used on wooden and bamboo furniture, now considered to have great potential as a source of bio-diesel. *Vernicia fordii* (tung oil tree; you tong) is found or cultivated in Anhui, Fujian, Guangdong, Guangxi, Guizhou, Hainan, Henan, Hubei, Hunan, Jiangsu, Jiangxi, Shaanxi, Sichuan, Yunnan and Zhejiang, and *V. montana* (mu you tong) is cultivated in Anhui, Fujian, Guangdong, Guangxi, Guizhou, Hainan, Hubei, Hunan, Jiangxi, Taiwan, Yunnan and Zhejiang.

Xanthoceras sorbifolium.—Xanthoceras sorbifolium (shiny-leaved yellowhorn; wen guan guo; Sapindaceae) is a small deciduous tree found on hills and slopes in Gansu, Hebei, Henan, Nei Mongol, Ningxia, Shaanxi, Shandong and Shanxi. The seeds are edible with a very high oil content and potential to produce bio-diesel in northern China.

Pistacia.—There are two species of *Pistacia* (huang lian mu shu; Anacardiaceae) in China, one of which is endemic. The oil content of their seeds is very high, and can be used as a bio-diesel. *Pistacia chinensis* (huang lian mu) is found in Anhui, Fujian, Gansu, Guangdong, Guangxi, Guizhou, Hainan, Hebei, Henan, Hubei, Hunan, Jiangsu, Jiangxi, Shaanxi, Shandong, Shanxi, Sichuan, Taiwan, southeastern Xizang, Yunnan and Zhejiang, and *P. weinmanniifolia* (qing xiang mu) is found in southwestern Guangxi, southwestern Guizhou, southwestern Sichuan, southeastern Xizang and Yunnan.

References

Li, Q. & Li, X. (2008). Poultry breeding history in China. *China Poultry* 30(23): 5-8.

Long, C. L. & Song, H. C. (2012). *Bio-diesel Plants of China*. Beijing, China: Sciences Press.

Pei, S. J. & Huai, H. Y. (2007). *Ethnobotany*. Shanghai, China: Shanghai Science and Technology Press.

Wang, M. N. (1991). Origination and transmission of fous in China. *Agricultural History of China* 4: 43-49.

Zhang, B. B. (2008). RAPD analysis on the relation of 17 wild *Malus* species. *China Fruits* 2: 24-26.

Zhang, G. X., Pei, S. J., Yang, Y. M. & Li, B. J. (2003). Application of ethnobotanical methods in the study of traditional dye plants. *Acta Botanica Yunnanica* Supplement XIV: 115-122.

Zhang, J. & Ye, L. Q. (2007). Determination of rubusoside in *Rubus suavissimus* S.Lee by HPLC. *Food Industry* 1: 55-57.

Zhang, W. M., Xiao, Z. C. & et al. (2007). *Exploiture and Utilization of Spicy Plants of China*. Nanjing, China: Southeast University Press.

Zhu, T. P., Liu, L. & Zhu, M. (2007). *Plant Resources of China*. Beijing, China: Science Press.

Chapter 22

Naturalized and Invasive Plants in China

LI Zhen-Yu
FAN Xiao-Hong and David E. Boufford

22.1 Introduction

One of the less-recognized effects of recent global climatic change is the increasing worldwide ecological and economic issue of invasive species (D'Antonio & Vitousek, 1992; Vitousek *et al.*, 1994; Baskin, 1998), one of the knottiest environmental problems to solve (Perrings *et al.*, 2000; Perrings *et al.*, 2002). Alien species, subspecies, or lower taxa are defined as those occurring outside their natural range and beyond that into which they have natural potential to disperse (IUCN Species Survival Commission, 2000). The varied topography, climate and ecosystems of China, as described in previous chapters, means that most plant species from other parts of the world are likely to find a suitable habitat somewhere in the country (Xie *et al.*, 2001). Among those alien species introduced into China are some, including important grains and economic crops, which play a positive role in agricultural development, economics, construction and the improvement of living standards (see Chapter 20). However, a large number of less useful alien species have become naturalized in China, among which are several invasive species, i.e. those that change or threaten native biological diversity. Invasive species may also bring enormous losses to the economy and human health. China, with its heavy dependence on agriculture, large population and rich biodiversity, is particularly susceptible to invasive alien species, which may have serious deleterious consequences. In this chapter we illustrate the history, pathways of introduction, and impact of invasive species in China, using representative examples and statistical data based upon many years of field surveys and observation by the authors.

22.2 The threat of invasive species

Invasive species have been reported from all 34 provinces, autonomous regions and municipalities of China. Except for a few cold, alpine and remote areas such as Hoh Xil in Qinghai and Qiantang in Xizang, at least some invasive species are also found in nearly all of China's nature reserves. These invasive species may have impacts upon biodiversity, the economy, human health and cultural heritage.

22.2.1 Impacts on biodiversity

Impacts on ecosystems.—Invasive species tend to be from tropical and subtropical areas rather than temperate ones, from low elevations rather than high, and to occur more widely and with more impact on islands and in wet areas than on mainland, dry areas. Examples include *Ageratina adenophora* (Asteraceae), a native of central America: in just one province of China, Yunnan, *A. adenophora* occupies an area of 247 000 km^2 and is an aggressive invader of agricultural lands and pastures, accessing valleys along roadsides and stream banks. It causes not only damage to agricultural production and animal husbandry, but also difficulties in restoring natural vegetation (Wang *et al.*, 2004; Lu & Ma, 2006). *Mikania micrantha* (Asteraceae), a native of tropical America, was introduced into southern Guangdong in the 1980s, where it has severely damaged local ecosystems by rapidly forming dense ground cover and by climbing into the crowns of trees and strangling them. Almost all native species are inhibited by *M. micrantha*, with the exception of *Cuscuta campestris* (Convolvulaceae), which parasitizes it, and could even be considered a biological control agent (Wang *et al.*, 2004). In southern China, the worldwide invasive water hyacinth, *Eichhornia crassipes* (Pontederiaceae) is a major problem, covering more than 1 000 km^2 of Dian Chi in Kunming Shi, more than 2 000 km^2 in Wenzhou Shi, Zhejiang, and more than 61 000 ha of Taiwan. *Eichhornia crassipes* has caused great damage to local fisheries and to tourism by destroying aquatic ecosystems and blocking water traffic (Ding *et al.*, 1995; Ding & Wang, 1998; Wang *et al.*, 2004).

Impacts on species diversity.—Invasive species compete with local plants for nutrients and space, and may also reduce population sizes of local plants to sometimes dangerous levels through allelopathy. The negative impact of invasive species on endemic species is most severe in aquatic environments and on islands. For instance, *Eichhornia crassipes* and *Pistia stratiotes* (water lettuce; Araceae) form dense mats on the surface of water, destroying native aquatic plant populations and endangering the survival of certain species. The spread of *E. crassipes* in Dian Chi has led to the local extinction of at least ten species of aquatic plant (Ding *et al.*, 1995). This has consequently impacted on the survival of aquatic animals, cutting the number of species by half since 1960, and reducing the number of species of fish from 15 to five (Ding *et al.*, 1995). On Nei Lingding Dao, southern Guangdong, *Mikania micrantha* has caused a large number of trees to die, which in turn has endangered a population of over 600 Rhesus macaques (*Macaca mulatta*), plus birds and other animals and plants.

Impacts on genetic diversity.—Invasive plants, particularly those with ruderal characteristics, can encroach upon the ecological niches of native species, leading to population fragmentation and reduction, directly resulting in the loss of genetic diversity. Impacts on the genetic diversity of native plants can also be indirect, through changing patterns of natural selection, changing gene flow between local populations, hybridization and introgression (Fang, 2000). The phenomenon of hybridization is particularly significant. For example, natural hybrids have been reported from Nanjing between *Pinus massoniana* (Pinaceae), a native species, and *P. thunbergii*, introduced from Japan (Chen, 1994). *Pinus thunbergii* is also reported to have hybridized with native *P. densiflora* in Dalian (Wu, 1983).

Elsewhere, in the UK, interspecific crosses have been noted between native *Rumex conglomeratus* (Polygonaceae) and *R. frutescens* from South America, forming *R.* ×*wrightii*; and between native *R. obtusifolius* and *R. frutescens* (*R.* ×*cornubiensis*; Stace, 1997). *Spartina alterniflora* (Poaceae, with 31 pairs of chromosomes), brought in a ship's ballast from North America to the UK, hybridized with native *S. maritima* to produce the sterile hybrid *S.* ×*townsendii*. This then gave rise to the fertile *S. anglica*, which has become an invasive weed, by doubling its chromosomes to produce an allopolyploid with 60–62 pairs (Thompson, 1991; Williamson, 1996; Ayres & Strong, 2001). In China, species of *Spartina* were deliberated introduced in the 1960s and 1970s to prevent coastal erosion, but *S. alterniflora* and *S. anglica* have now spread throughout coastal areas of China. The genera *Aster, Senecio* and *Solidago* (all Asteraceae), *Pennisetum* and *Sorghum* (both Poaceae), and *Rumex*, among others, are rich in both native species and non-native relatives, resulting in great competition for ecological niches and potential genetic impacts by their interaction.

Transgenic plants—artificial cultivars bred for stress resistance, high yields or insect resistance formed from recombining genes from different organisms—are also of concern, particularly that insect- and herbicide-resistant, highly fertile "super weeds" could be formed if genetically modified plants escaped from cultivation or crossed with their wild relatives.

22.2.2 Impacts on the economy

On Nei Lingding Dao, the area infested with *Mikania micrantha* has reached 300 ha, 35% of the area of the island. This has resulted in the loss of 50 ha of forest. Over 60% of the 468 ha of shrub forest are also densely covered, causing seven ha to have deteriorated to grassland (Wang *et al.*, 2004). The resultant decrease in the island's ecosystem services, such as water conservation, carbon fixation, oxygen production and environmental purification, have caused estimated economic losses of 3 800 000–8 600 000 Yuan per year. The cost of losses of biological diversity is estimated at 670 000–1 500 000 Yuan per year, giving a total annual economic impact of 4 500 000–10 000 000 Yuan (Wang *et al.*, 2004). Economic losses caused by *Eichhornia crassipes* are estimated at ten billion Yuan per year in China, with the major proportion being the cost of prevention and control (Wang *et al.*, 2004).

Alien species such as *Ageratum conyzoides* and *Erigeron* spp. (both Asteraceae), *Alternanthera philoxeroides* and *Amaranthus retroflexus* (both Amaranthaceae), and *Lolium temulentum* (Poaceae) have now become China's most noxious agricultural weeds (Qiang & Cao, 2000; Wan *et al.*, 2002). According to the United Nations' Food and Agriculture Organization (FAO) statistics, weeds cause an average annual decline in worldwide crop output of 0.7%, equivalent to a loss of 20.4 billion US dollars (China Import and Export Corporation for Farms and Land Reclamation, 1992). More than 50 of the 90 most noxious weeds in the world occur in China, causing the output of 40 million ha of farmland to decrease by as much as 10% per year (Zhang *et al.*, 1997). It has been estimated that, in 2004, the Chinese economic losses caused by alien invasive species were 14.5 billion US dollars, 1.4% of the country's GDP (Xu *et al.*, 2004).

22.2.3 Impacts on human health

The pollen of alien *Ambrosia artemisiifolia* (Asteraceae) and *A. trifida* is a leading cause of allergic rhinitis (hayfever), which is now reported to affect 1.5% of the population of China (Xia, 1983). *Ambrosia artemisiifolia* occurs in 17 provinces and municipalities of China. *Ambrosia* is wind-pollinated and its pollen can be carried by wind to remote regions (Raynor *et al.*, 1974). In tropical areas, *Parthenium hysterophorus* (Asteraceae) also causes pollen allergies. The seeds of *Agrostemma githago* (Caryophyllaceae) contain toxins which can poison both humans and livestock. Seeds of the poison darnel (*Lolium temulentum*) are often infected by the ergot fungus (*Stromatinia temulenta*) which produces highly toxic compounds causing serious poisoning in humans at even low concentrations (Chen & Zheng, 1987). Because *L. temulentum* grows intermixed with wheat, it poses a serious threat to human health. Both the nectar and pollen of toxic plants can also have an adverse impact on human health through contamination of honey.

Toxic alien species of *Chamaecrista, Clitoria* and *Crotalaria* (all Fabaceae), *Datura* and *Solanum* (both Solanaceae), *Digitalis* (Scrophulariaceae), *Euphorbia* (Euphorbiaceae) and *Linum* (Linaceae) are usually avoided by animals, causing the plants to increase greatly in abundance under intensive grazing. As more and more animals are grazed on less and less land, opportunities for the survival and spread of toxic invasive plants increases.

22.2.4 Impacts on cultural heritage

Invasive species can impose negative impacts on local societies and cultures by changing the original landscape around important cultural sites. For example, Beijing Shi has been the capital of six ancient dynasties, with many monuments and cultural sites reflecting its historical importance. Twenty-nine alien plant species are now known to have naturalized around Zhoukoudian, the southwestern gate of Beijing Shi, close to the site of the *Homo erectus* fossil "Peking Man" and more than a dozen other cultural sites and early human landscapes. The Zhoukoudian river is densely overgrown with alien *Alternanthera philoxeroides*, while terrestrial habitats have been invaded and occupied by *Amaranthus hybridus*, *Amaranthus retroflexus*, *Bidens frondosa* and *Symphyotrichum subulatum* (both Asteraceae), *Ipomoea purpurea* (Convolvulaceae), and *Panicum dichotomiflorum* (Poaceae), which together have seriously reduced Zhoukoudian's biodiversity, greatly changed its ecology and altered the scenic views and harmony of the environment.

22.3 Patterns of introduction of alien species

Introduction refers to the movement of species within or between countries, mainly through human means, and is the principal way by which alien species arrive in regions outside their natural range. In this chapter we focus upon movement between countries. Of the estimated ca. 300 000 higher plant species in the world, 80 000 have been introduced into areas outside their native range, accounting for approximately 27% of the total flora of any given area. Of these, only a few have become invasive (Zhao & Xia, 2002). According to data published by the IUCN Invasive Species Specialist Group (IUCN, 2001), 36 of the world's 100 worst invasive alien species are plants (IUCN, 2001). Randall (2002) recorded 13 702 escaped and naturalized plants, environmental weeds and invasive plants worldwide, while 277 invasive alien species have been identified in the Global Invasive Species Database (http://www.issg.org/database/welcome/), as posing a threat to biological diversity. Of those where the pathway of introduction is known, 64% of introductions were intentional.

China has been introducing and making use of alien plants since ancient times, especially along the "Silk Road," "Tea-Horse Road," maritime trade routes and inland tributary systems, but many unintentional plant introductions have occurred since 1860, when China began to open its ports to the world. Many of these have become problematic weeds.

22.3.1 Intentional introductions

Initially, most introduced species were useful crops. Many intentionally-introduced crops, such as *Ipomoea batatas* (sweet potato; Convolvulaceae) and *Zea mays* (maize; Poaceae), did not establish wild populations, but other exotics introduced as useful later became weeds. In the fourth century BC, the road between Chengdu Shi in Sichuan and ancient India was formed from several connected paths, passing through Tengchong in Yunnan and northern Myanmar. Along this road, *Amaranthus tricolor* (Joseph's coat) and *Basella alba* (Malabar spinach; Basellaceae) were introduced as vegetables from India. The Silk Road, formed in the Han Dynasty (around 210 BC), connected the Wei He basin with eastern Turkey via central Asia and was the route of introduction for *Punica granatum* (pomegranate; Lythraceae) from central Asia, *Linum usitatissimum* (flax; Linaceae), *Raphanus sativus* (radish; Brassicaceae) and *Vitis vinifera* (grape; Vitaceae), from the Mediterranean, and *Sesamum indicum* (sesame; Pedaliaceae) from Africa (Xie *et al.*, 2001).

From the Tang Dynasty (618–907 AD) onwards, increasing maritime trade with neighboring countries and the Middle East led to the introductions of *Euphorbia lathyris* (caper spurge), *Piper nigrum* (pepper; Piperaceae), *Quisqualis indica* (Chinese honeysuckle; Combretaceae) and *Ricinus communis* (castor oil plant; Euphorbiaceae), as recorded in Tang Dynasty pharmacopeias (Hu, 1984).

From 1405 to 1433, the grand fleets of the famous Chinese navigator ZHENG He (1371–1433) conducted seven voyages as far as eastern Africa, promoting massive exchange of culture, personnel and plant and animal species between China and countries along the Nan Hai (South China Sea) and the Indian Ocean. During his voyages, ZHENG paid special attention to local natural resources and geographical conditions, often collecting species, including large mammals such as the giraffe (*Giraffa camelopardalis*), lion (*Panthera leo*) and ostrich (*Struthio camelus*), recorded as the "superstars" of his acquisitions by the fleet's interpreter MA Huan in his *Tiang Ya Sheng Lan*, published in 1451. ZHENG also brought back a large number of medicinal plants, many never before seen in China (Vesugi, 2003). Following this period of exploration, many east African succulent plants, such as *Aloe vera* (Liliaceae) and *Bryophyllum pinnatum* (Crassulaceae), probably related to those intercontinental voyages, were found on uninhabited islands in southern China.

After Vasco da Gama's successful voyage to India around the southern tip of Africa in the late 1490s, Europeans began to bring plants of the Americas to southern and Southeast Asia. Overseas Chinese residing in those areas introduced *Capsicum annuum* (chili pepper) and *Nicotiana tabacum* (tobacco; both Solanaceae), *Ipomoea batatas*, *Mirabilis jalapa* (marvel of Peru; Nyctaginaceae) and other economic plants to China. In 1553 the Portuguese occupied Macao and in the seventeenth century the Spanish and the Dutch annexed parts of Taiwan. During this period, the Dutch East India Company introduced *Acacia farnesiana* (needle bush) and *Leucaena leucocephala* (both Fabaceae), *Hylocereus undatus* (dragonfruit; Cactaceae), *Psidium guajava* (guava; Myrtaceae) and other American plants into Taiwan. After the end of the Opium War in 1842, Dalian, Guangzhou, Hong Kong, Qingdao Shi, Shanghai Shi, Xiamen, Yantai Shi and other cities were forced to open for trade and quickly became entry ports for exotic weeds. *Erigeron annuus*, *E. bonariensis*, *E. canadensis* and others landed in rapid succession (Xie *et al.*, 2001).

Fodder plants.—Alternanthera philoxeroides (alligator weed), *Eichhornia crassipes* (water hyacinth) and *Pistia stratiotes* (water lettuce) were commonly used as fodder plants from the 1950s to the 1970s, but were replaced by more nutritious species in the 1980s. Once they were no longer harvested for animal feed, they spread rapidly and rampantly. Other fodder plants, such as *Macroptilium lathyroides* (Fabaceae), *Medicago* spp., *Melilotus* spp., *Panicum maximum, Pennisetum polystachion, Pennisetum purpureum, Tithonia diversifolia* (Asteraceae) and *Trifolium* spp., escaped from cultivation to become invasive (Li, 1998; Qiang & Cao, 2000).

*Ornamentals.—*The *Solidago canadensis* complex (including *S. altissima* and *S. canadensis*), introduced as ornamental plants prior to 1926, has became a serious problem in eastern China and Hubei, where it is now widespread and difficult to control. *Ipomoea cairica* (mile-a-minute vine), *I. purpurea* (morning glory) and *Merremia tuberosa* (woodrose; all Convolvulaceae), *Mimosa pudica* (sensitive plant or touch-me-not; han xiu cao), *Mirabilis jalapa* (marvel of Peru), *Oxalis corymbosa* (Oxalidaceae), *Phytolacca americana* (American pokeweed; Phytolaccaceae), *Tagetes erecta* (French marigold; Asteraceae) and others arrived, pre-adapted to local ecosystems, and have become thoroughly established. A large number of ornamental aquarium plants entered China in the 1980s: *Cabomba caroliniana* (Cabombaceae) was first reported in Jinxian, Zhejiang, in 1993, then appeared in Tai Hu in Wuxian, Jiangsu, in 1998 (Ding, 2000). *Gymnocoronis spilanthoides* (Asteraceae), from South America, became established beside the Li Jiang in Yangshuo, Guangxi, in 2006 and has continued to spread along the river since (Gao & Liu, 2007).

Textile plants.—Abutilon theophrasti (velvetleaf; Malvaceae), *Agave americana* (century plant; Amaryllidaceae), *A. sisalana* (sisal), *Cannabis sativa* and *Linum usitatissimum* were cultivated as important textile plants in the past, but declined to the status of weeds in many areas after being replaced by more desirable cotton and chemical fibers.

Green manure species.—Azolla caroliniana (Azollaceae) and species of *Clitoria* (Fabaceae), *Crotalaria* and a number of other plants were introduced as green manure crops in the 1950s, before the widespread use of chemical fertilizers. Now that chemical fertilizers are widely available and popular, the ranges of many green manure crops have expanded and have spread into the wild.

*Medicinal plants.—*The introduction of medicinal plants has a long history (Hu, 1984). Ancient medicinal introductions, such as *Anredera cordifolia* (Basellaceae), *Ricinus communis, Senna occidentalis* (Fabaceae), *Talinum paniculatum* (Portulacaceae), *Vaccaria hispanica* (Caryophyllaceae), and species of *Datura*, are now rarely used medicinally and have become weeds.

*Fruit trees.—*Seeds of the ancient introductions *Clausena anisum-olens* (Rubiaceae), *Passiflora edulis* (passionfruit; Passifloraceae), *Psidium guajava* (guava) and *Tamarindus indica* (Fabaceae), as well as *Muntingia calabura* (Muntingiaceae), introduced in recent times, have been spread by birds and mammals, causing the range of these aliens to extend in China.

*Vegetables.—*Early vegetable introductions, such as *Basella alba* (Malabar spinach), *Cichorium intybus* (chicory) and *Helianthus tuberosus* (Jerusalem artichoke; both Asteraceae), *Dioscorea alata* (purple yam; Dioscoreaceae), *Eryngium foetidum* (cilantro, Mexican coriander; Apiaceae), *Nasturtium officinale* (Brassicaceae) and *Tetragonia tetragonioides* (New Zealand spinach; Aizoaceae) have escaped from cultivation to a limited extent.

Aromatic plants.—Chrysopogon zizanioides (Poaceae), *Ocimum americanum* (limehairy, hoary basil; Lamiaceae), *O. basilicum* (sweet basil) and *O. sanctum* (holy basil) entered China as aromatic plants, then became naturalized after cultivation. A naturalized population of *Chrysopogon zizanioides*, which is cultivated in southern China, was found in 1936 on Hainan.

Landscaping plants.—Leucaena leucocephala, a fast growing species, was brought to Taiwan in 1645 by the Dutch, then used widely as landscape tree in southern and southwestern China over the following 300 years. When villagers used the branches for firewood, the population balance was maintained, but *L. leucocephala* became invasive after the 1980s when energy resources in developed areas shifted to coal, electricity and natural gas. *Acacia farnesiana* came to China at around the same time and followed the same pattern. *Robinia pseudoacacia* (Fabaceae) has become naturalized in Anhui, Hebei, Shandong and several other temperate provinces, while *Pithecellobium dulce* (Fabaceae) has invaded tropical rainforests on Hainan. *Hylocereus undatus* and four species of *Opuntia* (Cactaceae) were introduced into southern China during the late Ming Dynasty for use as hedge plants. Their seeds were spread by animals to natural ecosystems and pastures, causing damage to the environment and to animal husbandry.

*Cover plants.—*As described above, *Spartina alterniflora* and *S. anglica* were introduced for coastal protection. *Brachiaria mutica* (Poaceae) and *Sphagneticola trilobata* (Asteraceae) have been widely used for water and soil conservation. *Axonopus compressus* and *Buchloe dactyloides* (both Poaceae) have been used as lawns, and *Calopogonium mucunoides* (Fabaceae) and *Crotalaria trichotoma* as ground-cover in rubber plantations. All of these have become invasive.

Experimental material.—Flaveria bidentis (Asteraceae) was introduced for experiments into the evolution of photosynthetic pathways. A few scattered individuals of *F. bidentis* were noted near Nankai University in Tianjin Shi in 2001, but in the years since it has spread to 45 counties and towns around Tianjin Shi, and to Hebei and Shandong, where it occurs over an area of 20 000 ha. *Helenium autumnale*, *H. flexuosum* and *Mikania micrantha* have all escaped from botanic garden collections.

22.3.2 Unintentional introductions

Many exotic plant disseminules or propagules (mainly weed seeds) have been inadvertently transmitted by human activities, usually alongside goods, in transport vehicles, or by passengers. Disposal of plant propagules sometimes also leads to transmission.

Human-carried plants.—Mimetic weeds of wheat, such as *Aegilops tauschii*, *Avena fatua* and *Lolium temulentum* (all Poaceae), *Agrostemma githago* and *Vaccaria hispanica* (both Caryophyllaceae), grow alongside wheat and their seeds are spread with those of the crop. *Camelina sativa* (Brassicaceae) and *Lolium remotum*, weeds commonly found in fields of *Linum usitatissimum*, and *Cuscuta epilinum*, a parasite of *Linum*, often have their seeds spread with the crop species. In crops farmed by manpower alone, mimetic weeds have a greater chance of spreading with the crop plants than other weeds, which may explain why these were the major trans-boundary weeds of ancient times.

More recently, *Plantago virginica* (Plantaginaceae), the seeds of which become sticky when moist, was first found in Liantang, Nanchang Shi, Jiangxi, in 1951, where the largest airport of the area was located. *Sideritis lanata* (Lamiaceae), which first appeared in downtown Taipei, is estimated, based on the time and place of discovery, to have been brought into Taiwan by African students studying at the China-Africa Technical Cooperation Centre (Liu, 1972). The seeds of *Bidens* spp., *Synedrella nodiflora* and *Xanthium* spp. (all Asteraceae) and *Cenchrus* (Poaceae) are all dispersed by adhering to animal fur, feathers, clothes and luggage. Mechanized harvesting also extends the capacity of all weed species to be spread with agricultural products. Seeds of *Peperomia pellucida* (Piperaceae) and *Pilea microphylla* (Urticaceae) are commonly spread through seed banks in the soil accompanying nursery stock and seedlings.

Seeds carried in imported goods.—Over 550 species and varieties of weed seeds, originating in ca. 30 countries, have been intercepted at 12 Chinese ports including Dalian Shi, Guangzhou Shi, Nanjing Shi, Qingdao Shi, Shanghai Shi and Zhangjiagang. Aliens have been found in imported grains, oil-bearing crops, feedstocks, cotton, wool and among seeds of economic plants (Li, 1998). After the reforms and opening to the world that were initiated in the late 1970s, China's international trade expanded from encompassing only Soviet-era socialist countries to most countries and regions of the world. Consequently, pathways for weed introduction have become more diverse, and the number and variety of weeds introduced have steadily increased. For example, several species of *Amaranthus* from North America—which, even as recently as 2003, were not recorded in the flora of China—were found and destroyed near oil pressing mills (Bao *et al.*, 2003).

Plants accidentally carried by ship.—Prior to the 1880s, cargo ships carrying goods in only one direction often used locally-gathered sand or earth as ballast on the return journey. Weeds such as *Amaranthus spinosus*, *Euphorbia hirta*, *Gamochaeta purpurea* (Asteraceae), *Hyptis rhomboidea* (Lamiaceae), *Herissantia crispa* and *Malvastrum coromandelianum* (both Malvaceae), *Scoparia dulcis* (Scrophulariaceae), *Sesuvium portulacastrum* (Aizoaceae) and *Waltheria indica* (Sterculiaceae) are all port-related introductions to China that are unlikely to have been intentional. For instance, the first documentation of *E. hirta* in China comprises specimens collected in Macao in 1820. From 1553 to 1999, Macao was a colony of Portugal, a major seafaring nation; *E. hirta* clearly became naturalized through solid ballast. After the 1880s, solid ballast was replaced by water. The International Maritime Organization estimates that more than ten billion tones of water is transferred as ballast in ocean-going vessels worldwide per year. Released water ballast has carried a number of aquatic animals, plants—such as *Myriophyllum heterophyllum* (Haloragaceae), collected in the Zhu Jiang estuary, Guangzhou, in 1917 (Yu *et al.*, 2002)—and algae, including about 16 organisms known to cause "red tide" algal blooms, into littoral regions of China. Red tides are highly problematic, resulting in the death of marine life, coastal pollution, harm to marine aquaculture and tourism and even poisoning to humans (Liang & Wang, 2001; Zhang, 2002).

In addition to the issue of ballast, when plant propagules carried by ship are discarded or lost during a shipwreck, they may be carried by currents to nearby coasts and islands. *Aloe vera*, *Kalanchoe integra* (Crassulaceae), *K. pinnata* and *Narcissus tazetta* (Amaryllidaceae) all very probably arrived on uninhabited islands in the Nan Hai (South China Sea) and Dong Hai (East China Sea) in garbage thrown from ships (Li & Xie, 2002). Seeds in bottles or plastic bags, in particular, can travel great distances on ocean currents.

22.3.3 Cross-border transmission from neighboring countries

Some alien species were not introduced into China directly from their country of origin, but were first introduced and naturalized in neighboring countries before spreading to China by natural forces (wind, currents, waves or animals) and sometimes by humans or human transport. *Crassocephalum crepidioides* (Asteraceae), which is known as "Annam cao" (Annam weed) in Guangxi, came to China from Indochina. Since Vietnam was only known as Annam from 679 to 1803, the entry of *C. crepidioides* probably occurred before 1803. *Chromolaena odorata* (Asteraceae) was introduced to Thailand as an aromatic plant in the early 1920s (Gagnepain, 1924), then spread throughout Indochina where it became known as "Siam weed." It was first reported in southern Yunnan in 1934. *Galinsoga parviflora* (Asteraceae), first reported on the Tea-Horse Road in Ninglang, Yunnan, and Muli, Sichuan, was probably carried into China by horse caravans.

22.4 Naturalized and invasive plants in China

22.4.1 Statistical details

Statistical data regarding naturalized plants in China differ according to the research methods and definitions of the word "naturalized" that are used. In floras and articles published before September 2008 it was estimated that China had 420 naturalized species in 273 genera and 84 families (Wu *et al.*, 2010), while JIANG and colleagues (2011) reported 861 species in 465 genera and 110 families. However, some Chinese authors may have treated some native species as exotic; for instance *Coix lacryma-jobi* (Poaceae) and *Polygonum aviculare* (Polygonaceae) are both widely distributed in China and have been used as medicines for centuries, perhaps since before 100 AD. *Sonchus oleraceus* (Asteraceae), which appears to be native to Eurasia, also has a long history in China. *Alternanthera sessilis*, of the Old World tropics, is also found in southern China where it may be native. *Juglans regia* (Juglandaceae) in Xinjiang, *Plantago lanceolata* in northwestern China, and *Schima crenata* (Theaceae) in Hainan may also be native. Further inflation of the figures may occur due to misidentification: for example, *Passiflora gracilis* has been misapplied to another exotic species, *Passiflora suberosa*.

Based on observations during field surveys and on specimens identified nationwide over the past ten years, as well as a survey of the literature, we calculate that that there are 864 alien species and infraspecific taxa naturalized in China. These belong to 102 families and 426 genera (see Table 22.1). Our figures are roughly in line with those of JIANG and colleagues (2011). Table 22.1 shows that Asteraceae (with 140 species), Fabaceae (116 species) and Poaceae (81 species) account for the most naturalized species in China. Other families with more than ten naturalized species are Solanaceae (47 species), Brassicaceae (32), Euphorbiaceae (32), Amaranthaceae (31), Convolvulaceae (28), Lamiaceae (21), Scrophulariaceae (19), Malvaceae (17), Rubiaceae (15), Onagraccac (13), Caryophyllaceae (11), Myrtaceae (11) and Verbenaceae (11).

22.4.2 Newly invasive species

Among the 864 naturalized plants, 369 (42%), in 223 genera and 70 families are considered to be invasive. Some newly-invasive species, such as *Amaranthus palmeri*, *Azolla filiculoides*, *Chenopodium pumilio* (Chenopodiaceae), *Crassocephalum rubens*, *Cyclachaena xanthiifolia*, *Erigeron philadelphicus*, *Flaveria bidentis*, *Gymnocoronis spilanthoides*, *Pluchea carolinensis*, *Pluchea sagittalis* and *Symphyotrichum subulatum* var. *parviflorum* (all Asteraceae), *Euphorbia hyssopifolia*, *E. peplus*, *Jacquemontia tamnifolia* (Convolvulaceae), *Lindernia rotundifolia* and *Mecardonia procumbens* (both Scrophulariaceae), *Panicum dichotomiflorum*, *Pennisetum polystachion*, *Salvinia molesta* (Salviniaceae), *Sicyos angulatus* (Cucurbitaceae), *Solanum rostratum*, *Solanum sisymbriifolium* and *Sporobolus pulvinatus* (Poaceae), have started to show aggressiveness in several areas. Others, such as *Croton capitatus* (Euphorbiaceae), *Croton hirtus*, *Cyperus eragrostis* (Cyperaceae), *Sesbania herbacea* (Fabaceae), *Spermacoce tenuior* (Rubiaceae), *Stachytarpheta dichotoma*, *Verbena bonariensis* and *V. bracteata* (all Verbenaceae) and *Stemodia verticillata* (Scrophulariaceae), have established populations in small areas where they need to be controlled and contained.

The status of some invasive species has undergone a significant change in part of their naturalized range since introduction. For example, *Chromolaena odorata*, once rampant in the Zhu Jiang delta, has been almost totally replaced by a more-recently invasive species, *Praxelis clematidea* (Asteraceae). *Alternanthera pungens*, a common weed in Xiamen in the 1970s, is becoming rare as the city is transformed into a "garden city." Over the last 20 years, the total number of naturalized plant species, and the extent of their coverage, has increased, but due to a lack of comprehensive and systematic monitoring the number of naturalized and invasive plants is still not accurately known. Nevertheless, these data show some clear general features. Around half of the invasive species in China have come from America; the remainder are from Europe's Mediterranean region, Asia and Africa, with only a few from Oceania. More than half of all invasive species entered China since 1949, most of the remainder between 1842 and 1949, the rest before that. This pattern is closely related to the expansion of international exchange and trade.

22.5 Prospects

Patterns of biodiversity and geographic distributions in China have formed over long periods of time through the processes of evolution, dispersal and extinction. China's heterogeneous environments can act as constraints on the natural spread of species: oceans, high mountains, deserts and ice caps often pose insurmountable barriers to dispersal; lakes, rivers and low mountains less so. Under the continuous impact of human activities, the effect of these natural barriers has been weakened. The ease with which goods and people are now able to move around the globe has provided new means of long-distance dispersal and has changed natural distribution patterns of many species to a large extent. As international trade and travel activities increase and many countries relax restrictions on trade and tourism, the potential for species to disperse beyond their native range is increased. In sum, species can now easily cross what were once insurmountable barriers and spread rapidly to every corner of the planet (Xie, 2008).

Table 22.1 Numbers of naturalized plants in China.

Family	Naturalized taxa		Of which invasive taxa	
	Number of species	Number of genera	Number of species	Number of genera
Asteraceae	140	71	61	43
Fabaceae	116	45	37	20
Poaceae	81	35	37	23
Solanaceae	47	14	15	4
Brassicaceae	32	15	25	13
Euphorbiaceae	32	8	15	4
Amaranthaceae	31	6	21	5
Convolvulaceae	28	5	10	4
Lamiaceae	21	9	11	6
Scrophulariaceae	19	10	12	7
Malvaceae	17	9	8	4
Rubiaceae	15	8	8	5
Onagraceae	13	3	8	2
Caryophyllaceae	11	10	8	8
Verbenaceae	11	5	5	4
Myrtaceae	11	4	1	1
Chenopodiaceae	9	4	4	2
Cucurbitaceae	8	7	2	2
Amaryllidaceae	8	7	1	1
Cactaceae	8	5	4	2
Apiaceae	7	7	4	4
Araceae	7	6	1	1
Lythraceae	7	4	3	2
Polygonaceae	7	3	4	2
Crassulaceae	7	3	2	1
Cyperaceae	7	2	3	2
Acanthaceae	6	5	0	0
Liliaceae	6	4	1	1
Ranunculaceae	6	2	1	1
Oxalidaceae	6	1	4	1
Passifloraceae	6	1	3	1
Nyctaginaceae	5	3	1	1
Portulacaceae	5	2	2	2
Aizoaceae	4	4	3	3
Apocynaceae	4	4	1	1
Papaveraceae	4	4	1	1
Campanulaceae	4	3	2	1
Commelinaceae	4	3	1	1
Iridaceae	4	3	1	1
Plantaginaceae	4	1	2	1
Phytolaccaceae	3	3	2	2
Bignoniaceae	3	3	1	1
Boraginaceae	3	3	1	1
Hydrocharitaceae	3	2	3	2
Rutaceae	3	2	1	1
Basellaceae	3	2	1	1
Capparidaceae	3	1	1	1
Tiliaceae	3	1	1	1
Violaceae	3	1	0	0
Geraniaceae	2	2	2	2
Resedaceae	2	2	2	2
Lemnaceae	2	2	1	1
Moraceae	2	2	1	1
Piperaceae	2	2	1	1
Primulaceae	2	2	1	1
Urticaceae	2	2	1	1

Continued

Family	Naturalized taxa		Of which invasive taxa	
	Number of species	Number of genera	Number of species	Number of genera
Anacardiaceae	2	2	0	0
Pinaceae	2	2	0	0
Balsaminaceae	2	1	1	1
Haloragaceae	2	1	1	1
Annonaceae	2	1	0	0
Casuarinaceae	2	1	0	0
Dioscoreaceae	2	1	0	0
Linaceae	2	1	0	0
Asclepiadaceae	1	1	1	1
Azollaceae	1	1	1	1
Butomaceae	1	1	1	1
Cabombaceae	1	1	1	1
Marantaceae	1	1	1	1
Martyniaceae	1	1	1	1
Melastomataceae	1	1	1	1
Pontederiaceae	1	1	1	1
Pteridaceae	1	1	1	1
Salviniaceae	1	1	1	1
Sapindaceae	1	1	1	1
Sterculiaceae	1	1	1	1
Vitaceae	1	1	1	1
Aceraceae	1	1	0	0
Aristolochiaceae	1	1	0	0
Begoniaceae	1	1	0	0
Berberidaceae	1	1	0	0
Bixaceae	1	1	0	0
Bombacaceae	1	1	0	0
Burseraceae	1	1	0	0
Callitrichaceae	1	1	0	0
Cannaceae	1	1	0	0
Caprifoliaceae	1	1	0	0
Caricaceae	1	1	0	0
Celastraceae	1	1	0	0
Cyatheaceae	1	1	0	0
Moringaceae	1	1	0	0
Muntingiaceae	1	1	0	0
Musaceae	1	1	0	0
Platanaceae	1	1	0	0
Plumbaginaceae	1	1	0	0
Polemoniaceae	1	1	0	0
Polygalaceae	1	1	0	0
Sonneratiaceae	1	1	0	0
Tropaeolaceae	1	1	0	0
Ulmaceae	1	1	0	0
Valerianaceae	1	1	0	0
Totals	**864**	**426**	**369**	**223**

Many alien species can bring tremendous benefits to mankind, but they can also cause ecological and economic disaster, warranting a great deal of interest. Once introduced, eradication of invasive species is often almost impossible to achieve (Mack *et al.*, 2000). To prevent the unintentional introduction and spread of exotic weeds, quarantine, monitoring and early prevention and control should be strengthened. Besides limiting the introduction of new pests, such monitoring will also minimize economic costs (Chen & Kang, 2003). Where exotic plants are introduced intentionally, risk assessment is essential. For invasive species that have already spread, well-studied, integrated control measures should be undertaken. The movement of species beyond their native range is of worldwide concern, and international cooperation is therefore crucial in raising public awareness of the threat and dangers of biological invasions. The effective utilization of land and plant resources may serve as a potential driving force to eliminate invasive plants.

References

AYRES, D. R. & STRONG, D. R. (2001). Origin and genetic diversity of *Spartina anglica* (Poaceae) using nuclear DNA markers. *American Journal of Botany* 88: 1863-1867.

BAO, B. J., CLEMANTS, S. E. & BORSCH, T. (2003). Amaranthaceae. In: Wu, Z. Y. & Raven, P. H. (eds.) *Flora of China*. pp. 415-429. Beijing, China and St. Louis, MO: Science Press and Missouri Botanical Garden Press.

BASKIN, Y. (1998). Winners and losers in changing world: global changes may promote invasion and alter the fate of invasive species. *BioScience* 48(10): 788-792.

CHEN, B. & KANG, L. (2003). Biological invasion and its relation with global changes. *Chinese Journal of Ecology* 22(1): 31-34.

CHEN, J. S. & ZHENG, S. (1987). *Poisonous Plants of China*. Beijing, China: Science Press.

CHEN, T. H. (1994). A study on natural hybridization between *Pinus thunbergii* Parl. and *P. massoniana* in Nanjing Region, Jiangsu. *Journal of Nanjing Forestry University* 18(4): 6-12.

CHINA IMPORT AND EXPORT CORPORATION FOR FARMS AND LAND RECLAMATION (1992). *Complete Book of Weed Chemical Control*. Shanghai, China: Shanghai Science & Technology Press.

D'ANTONIO, C. M. & VITOUSEK, P. M. (1992). Biological invasions by exotic grasses, the grass fire cycle, and global change. *Annual Review of Ecology and Systematics* 23: 63-87.

DING, B. Y. (2000). *Cabomba* Aublet (Cabombaceae), a newly naturalized genus of China. *Acta Phytotaxonomica Sinica* 38(2): 198-200.

DING, J. Q., WANG, R. & FAN, Z. N. (1995). Occurrence, harms and control of *Eichhornia crassipes* (Mart.) Solms in China. *Chinese Journal of Weed Science* 9(2): 49-52.

DING, J. Q. & WANG, R. (1998). *Alien Harmful Plants*. Beijing, China.

FANG, W. (2000). Biological invasion and global climate changes. In: Fang, J. Y. (ed.) *Global Ecology, Climate Change and Ecological Responses*. pp. 43-70. Beijing, China and Heidelberg, Germany: China Higher Education Press and Springer-Verlag.

GAGNEPAIN, F. (1924). Composées. In: Lecomte, H. (ed.) *Flore Générale L'Indo-Chine*. pp. 448-663. Paris, France: Masson & Cie.

GAO, T. G. & LIU, Y. (2007). *Gymnocoronis*, a new naturalized genus of the tribe Eupatorieae, Asteraceae in China. *Acta Phytotaxonomica Sinica* 45(3): 329-332.

HU, S. Y. (1984). *The Investigation of Exotic Elements in Traditional Chinese Medicine*. Beijing, China: China Agriculture Press.

IUCN (2001). *100 of the World's Worst Invasive Alien Species*. Auckland, New Zealand: Invasive Species Specialist Group.

IUCN SPECIES SURVIVAL COMMISSION (2000). *IUCN Guidelines for the Prevention of Biodiversity Loss Caused by Alien Invasive Species*. Gland, Switzerland: IUCN.

JIANG, H., FAN, Q., LI, J. T., SHI, S., LI, S. P., LIAO, W. B. & SHU, W. S. (2011). Naturalization of alien plants in China. *Biodiversity and Conservation* 20: 1545-1556.

LI, Y. H. (1998). *The Weeds in China*. Beijing, China: China Agriculture Press.

LI, Z. Y. & XIE, Y. (2002). *Invasive Alien Species in China*. Beijing, China: China Forestry Publishing House.

LIANG, Y. B. & WANG, B. (2001). Alien marine species and their impacts in China. *Biodiversity Science* 9(4): 458-465.

LIU, T. R. (1972). *Zhongshan Natural Science Dictionary*. Taipei, Taiwan: The Commercial Press.

LU, Z. J. & MA, K. P. (2006). Spread of the exotic croftonweed (*Eupatorium adenophorum*) across southwest China along roads and streams. *Weed Science* 54: 1068-1072.

MACK, R. N., SIMBERLOFF, D., LONSDALE, W. M., EVANS, H., CLOUT, M. & BAZZAZ, F. (2000). Biotic invasions: causes, epidemiology, global consequences, and control. *Ecological Applications* 10(3): 689-710.

PERRINGS, C., WILLIAMSON, M. & DALMAZZONE, S. (2000). *The Economics of Biological Invasions*. Cheltenham, UK: Edward Elgar Publishing Ltd.

PERRINGS, C., WILLIAMSON, M., BARBIER, E. B., DELFINO, D., DALMAZZONE, S., SHOGREN, J., SIMMONS, P. & WATKINSON, A. (2002). Biological invasion risks and the public good: an economic perspective. *Conservation Ecology* 6(1): 1.

QIANG, S. & CAO, X. Z. (2000). Survey and analysis of exotic weeds in China. *Journal of Plant Resources and Environment* 9(4): 34-38.

RANDALL, J. M. (2002). A Global Compendium of Weeds. <http://www.hear.org/gcw/>.

RAYNOR, G. S., HAYES, J. V. & OGDEN, E. C. (1974). Mesoscale transport and dispersion of airborne pollens. *Journal of Applied Meteorology* 13: 87-95.

STACE, C. A. (1997). *New Flora of the British Isles*. Cambridge, UK: Cambridge University Press.

THOMPSON, J. D. (1991). The biology of an invasive plant:

What makes *Spartina anglica* so successful? *BioScience* 41(6): 393-401.

Vesugi, C. (2003). *Great Voyages of Zhenghe*. Shanghai, China: Shanghai Academy of Social Sciences Press.

Vitousek, P. M., Turner, D. R., Parton, W. J. & Sanford, R. L. (1994). Litter decomposition on the Mauna Loa environment matrix Hawaii: I. Patterns, mechanisms, and models. *Ecology* 75(2): 418-429.

Wan, F. H., Guo, J. Y. & Wang, D. H. (2002). Alien invasive species in China: their damages and management strategies. *Biodiversity Science* 10(1): 119-125.

Wang, B. S., Wang, Y. J., Liao, W. B., Zan, Q. J., Li, M. G., Peng, S. L., Hang, S. C., ZHang, W. Y. & Chen, R. P. (2004). *The Invasion Ecology and Management of Alien Weed Mikania micrantha H.B.K.* Beijing, China: Science Press.

Williamson, M. (1996). *Biological Invasions*. London, UK: Chapman & Hall.

Wu, C. I. (1983). *Conspectus of the Introduced Foreign Trees*. Beijing, China: Science Press.

Wu, S. H., Sun, H. T., Teng, Y. C., Rejmánek, M., Chaw, S. M., Yang, T. Y. A. & Hsieh, C. F. (2010). Patterns of plant invasions in China: taxonomic, biogeographic, climatic approaches and anthropogenic effects. *Biological Invasions* 12: 2179-2206.

Xia, F. Y. (1983). Ragweed pollen allergy. *Plant Quarantine* 6: 16-19.

Xie, Y., Li, Z. Y., Gregg, W. P. & Li, D. M. (2001). Invasive species in China - an overview. *Biodiversity and Conservation* 10(8): 1317-1341.

Xie, Y. (2008). *Bioinvasion and Ecological Security in China*. Shijiazhuang, China: Hebei Science and Technology Press.

Xu, H. G., Mao, J. M., Qiang, S. & Wang, Y. J. (2004). *Hotspot Study on Convention on Biological Diversity: Alien Invasive Species, Biosecurity, Genetic Resources*. Beijing, China: Science Press.

Yu, D., Wang, D., Li, Z. Y. & Funston, A. M. (2002). Taxonomic revision of the genus *Myriophyllum* (Haloragaceae) in China. *Rhodora* 104: 396-421.

Zhang, C. X., Hu, X. S. & Qian, Y. X. (1997). Trend of herbicide use in developed countries and current research and future directions in weed science research in China. *Journal of Plant Protections* 24(3): 278-282.

Zhang, S. H. (2002). Discussion on mechanisms of non-indigenous species transfer via ship's ballast water. In: Wang, D. H. & McNeely, J. A. (eds.) *Prevention and Control of Alien Invasive Species - Collected Papers of International Symposium on Biodiversity and Management of Alien Invasive Species*. pp. 181-187. Beijing, China: China Environmental Science Press.

Zhao, N. X. & Xia, N. H. (2002). Effect of alien plant species on biological diversity. In: Wang, D. H. & McNeely, J. A. (eds.) *Prevention and Control of Alien Invasive Species - Collected Papers of International Symposium on Biodiversity and Management of Alien Invasive Species*. pp. 103-111. Beijing, China: China Environmental Science Press.

Chapter 23

The Extinction Crisis

HUANG Hong-Wen and Sara Oldfield

23.1 Introduction

The flora of China, as discussed throughout this book, is of immense importance but is facing a potential crisis of extinction. Over 5 000 plant species in China are threatened but these have not been fully documented and the areas of distribution for many species are still poorly known. Up to 200 plant species (Zhang *et al.*, 2000) may have become extinct since the 1950s, with examples including *Adiantum lianxianense* (Adiantaceae), *Lepisanthes unilocularis* (Sapindaceae) and *Ormosia howii* (Fabaceae).

Adiantum lianxianense (Figure 23.1) was endemic to Guangdong, China, with its type specimen collected in Lian, Guangdong. It became extinct due to loss of its natural habitat—subtropical and tropical moist lowland forests in limestone areas. *Lepisanthes unilocularis* (Figure 23.2) was first found in Dong, Hainan, in 1935 when the type specimen was collected, but has not been seen since and is currently considered extinct in the wild due to habitat loss. The location of the type specimen – Fuluo – comprises a very fragile coastal habitat. *Ormosia howii* (Hainan ormosia; Figure 23.3) was first found on Diaoluo Shan, Hainan, in 1954, and in Yangchun, Guangdong, in 1957, where it had extremely small populations of several individuals. It has not been seen in recent years and is considered extinct in the wild due to a sharp decline in coverage of natural forests and environmental change which has severely hampered natural reproduction and regeneration.

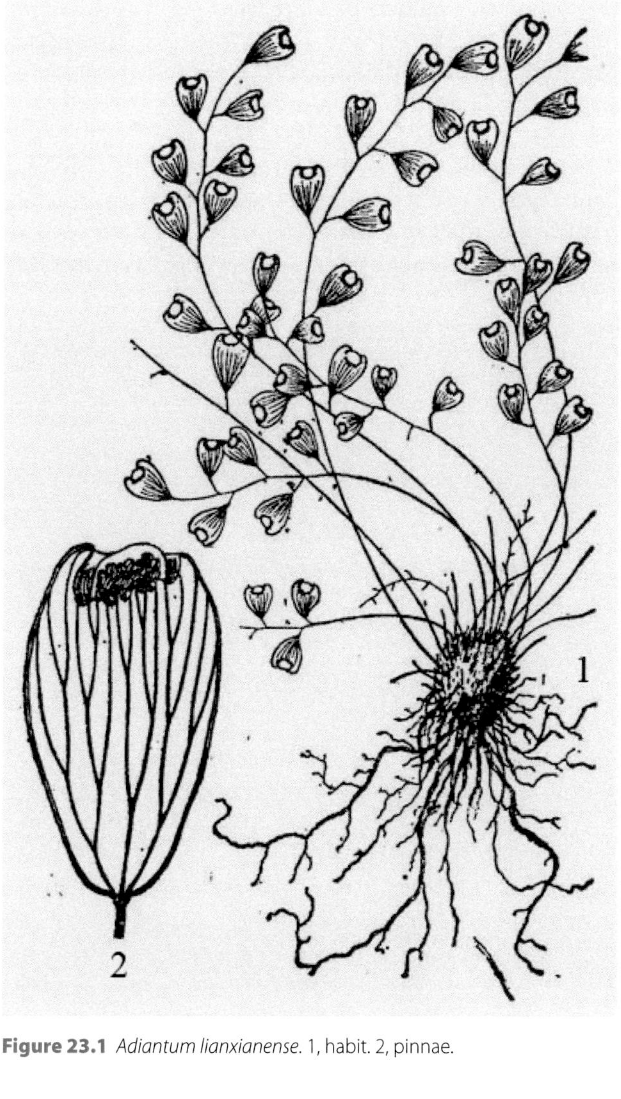

Figure 23.1 *Adiantum lianxianense.* 1, habit. 2, pinnae.

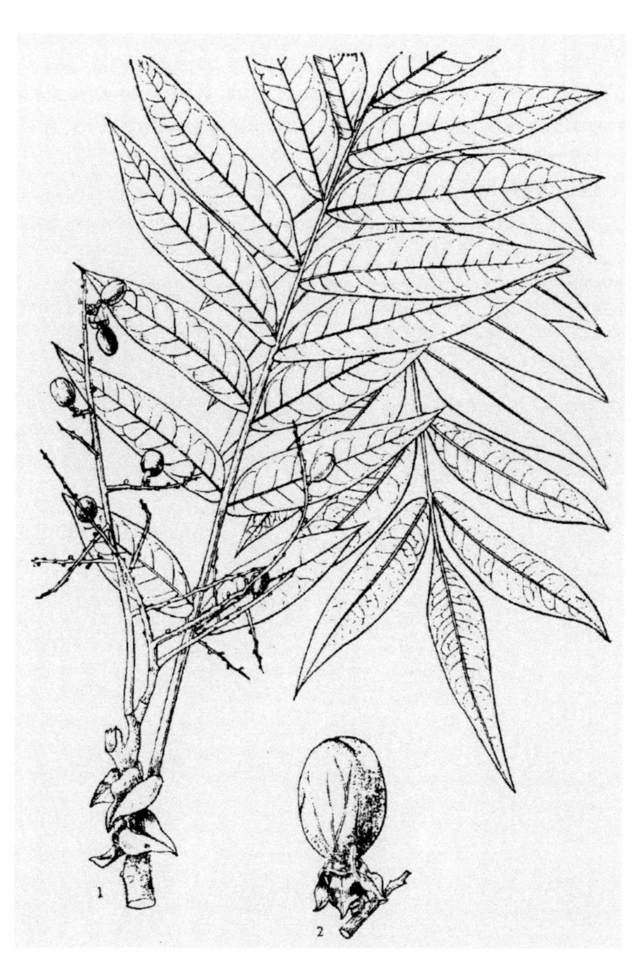

Figure 23.2 *Lepisanthes unilocularis.* 1, fruiting branch. 2, drupe showing persistent sepals.

Figure 23.3 *Ormosia howii*, fruiting branch.

Many more species have been reduced to a critical threshold for extinction. For example, *Acer yangbiense* (Aceraceae) has only four individuals found in a valley on the western slopes of Cangshan, Yangbi, in northwestern Yunnan, *Carpinus putoensis* (Betulaceae) comprises a single plant discovered in 1930 on Putuo Shan in the Zhoushan Qundao, Zhejiang, *Firmiana danxiaensis* is limited to Renhua, Guangdong, *F. kwangsiensis* has only three plants found in Jingxi, Guangxi, and *Gleditsia japonica* var. *velutina* (Fabaceae) comprises only two trees located in Nanyue Hengshan National Nature Reserve (Hunan). Many populations of the fern *Cystoathyrium chinense* (Cystopteridaceae) existed in 1963 when it was first discovered on Erlang Shan, Sichuan, but in 1984 only one individual was found following destruction of the forest habitat and loss of the moist conditions needed for the species' survival. It is likely that these and many other taxa, such as *Abies beshanzuensis* var. *ziyuanensis* (Pinaceae), *Camellia mairei* var. *mairei* (Theaceae), *Neolitsea sericea* (Lauraceae), *Picea neoveitchii* (Pinaceae), *Sinojackia xylocarpa* (Styracaceae) and *Thymus mandschuricus* (Lamiaceae), will go extinct if effective conservation measures are not taken immediately.

23.2 General threat status of plant taxa

23.2.1 Pteridophytes

Deforestation, industrialization, over-harvesting of medicinal plants and tourism are major factors contributing to the endangerment of ferns in China. Damage and deforestation of native forests has caused significant reductions in environmental humidity and a lowering of the underground water table, which has greatly altered the original habitat of many fern species, severely impacting on their survival and reproduction, as seen for *Cystoathyrium chinense*, above. Industrial and agricultural disturbance from economical and social development may also damage the local niche habitats of some ferns, such as *Adiantum reniforme* var. *sinense* and *Isoetes sinensis* (Isoetaceae; Ministry of Environmental Protection of the People's Republic of China, 1998; Figure 23.4).

Over-harvesting for medicinal uses or for food has contributed greatly to losses in ferns, severely reducing natural populations, depleting natural resources and taking some taxa to the brink of extinction. Many ferns are

Figure 23.4 A, Natural habitat of *Adiantum reniforme* var. *sinense*. B, C, *Adiantum reniforme* var. *sinense* (images courtesy KANG Ming, South China Botanical Garden).

highly valued as medicinal and vegetable plants in China, accounting for an amazing 10% of plants used in Chinese medicine. For example, *Cyrtomium fortunei* and *Dryopteris crassirhizoma* (both Dryopteridaceae), *Lygodium japonicum* (Lygodiaceae) and *Pyrrosia* spp. (Polypodiaceae) are widely used as Chinese herbal medicine for the physiological enhancement of bones and kidneys. Even worse, medicinal ferns are highly demanded in international trade: several hundred tons of *Cibotium barometz* (Dicksoniaceae), used as an alternative treatment for kidney, liver and joint complaints, have been exported to Asian markets, although the species occurs widely and abundantly. Many fern species are also used as vegetables or ornamental plants. For example, *Callipteris esculenta* (Athyriaceae), *Matteuccia struthiopteris* (Onocleaceae) and *Pteridium aquilinum* var. *latiusculum* (Dennstaedtiaceae) have been over-harvested for production of canned vegetables. Urgent measures need to be taken for both protection and public awareness (Ministry of Environmental Protection of the People's Republic of China, 1998).

23.2.2 Gymnosperms

Unsustainable logging during past the 50 years has severely decreased China's conifer forests, and many gymnosperms are currently threatened or endangered, the majority being regarded as a timber resource. The large natural conifer forests of the Da Xingan Ling, Xiao Xingan Ling and Changbai Shan in northeastern China have been over-harvested, greatly reducing this natural resource since the 1950s. Similarly, the natural conifer forests of the Hengduan Shan (Figure 23.5) in southwestern China have been over-harvested since the 1960s, with only patches of forest remaining on inaccessible montane cliffs. The situation in central, eastern and southern China is even worse because of high levels of economic development and dense population, resulting in almost all natural conifer forests being logged and replaced by artificial plantings of Chinese fir (*Cunninghamia lanceolata*; Taxodiaceae), Chinese weeping cypress (*Cupressus funebris*; Cupressaceae) and mason pine (*Pinus massoniana*; Pinaceae). Deforestation and damage to conifer forests

Figure 23.5 The Hengduan Shan (image courtesy SUN Hang, Kunming Institute of Botany, Chinese Academy of Sciences).

have changed the forest ecosystem and natural habitats, accelerating the rate of disappearance of many understory gymnosperms. Approximately 63 species (28% of the total in China) are threatened or endangered, including *Cycas taiwaniana* (Cycadaceae) and *Thuja sutchuenensis* (Cupressaceae), which are extinct in the wild, and many others that are close to extinction; the critically endangered *Abies beshanzuensis*, *A. yuanbaoshanensis*, *Amentotaxus formosana* (Taxaceae), *Cephalotaxus lanceolata* (Cephalotaxaceae), *Cycas multipinnata*, *Keteleeria fortunei* var. *oblonga* (Pinaceae), *K. hainanensis*, *K. pubescens*, *Picea likiangensis* var. *montigena* and *Pinus squamata* (Ministry of Environmental Protection of the People's Republic of China, 1998).

23.2.3 Angiosperms

The large Chinese population and long history of exploitation of plant resources in the country have imposed severe pressure on the diversity of angiosperm plants, because many angiosperm species are targeted as high-quality timbers, valuable medicinal herbs and economic crops. Angiosperms have suffered irremediable damage in the past, with many species depleted to the brink of extinction by over-harvesting in the wild for high-value products such as herbal medicines, ornamentals and timbers, including *Cinnamomum* spp. and *Phoebe* spp. (both Lauraceae), *Coptis* spp. (Ranunculaceae), *Ferula* spp. (Apiaceae), *Fritillaria* spp. (Liliaceae), *Paeonia* spp. (Paeoniaceae), and many species of Orchidaceae. Currently, there are estimated to be more than 4 000 threatened or endangered species in China (Wang & Xie, 2004), 10–20% of which are critically endangered with minimal population sizes in the wild. Angiosperms make up a large proportion of the world's plant resources and are thus crucially important to safeguard economical and social sustainability, which fundamentally depend on natural resources for food, medicine and the raw materials of industry. Effective conservation measures are urgently needed to cope with increasing demand.

The underlying drivers of ecosystem change in China include factors of population growth, economic development, migration, trade, governance, land tenure and public participation and environmental awareness (Anon, 2008), and other issues that also impact on related aspects of poverty. In many parts of the country, population density exceeds the productive capacity of local ecosystems, many of which are now degraded. The clearance of vast tracts of land for rural development, followed over the past two decades by rapid economic development, with China showing an average 9.4% rate of increase in GDP from 1979–2002 (World Bank, 2003), have placed sustained pressure on the biodiversity of China. Unfortunately, despite the resulting huge changes within the rural landscape of the country, the concept of allocating wild areas for conservation was realized relatively late in China. While the first nature reserves were established in 1956 amid a surge of interest that followed the establishment of the People's Republic of China, but both research and conservation activity slowed to a standstill during the Cultural Revolution of the late 1960s and early 1970s. An emerging middle class, stronger education programs and increasing understanding of the links between biodiversity and poverty do, however, increase the potential for biodiversity conservation in the future.

As elsewhere in the world, the main direct human threats to Chinese plant diversity include habitat modification and destruction, environmental contamination, overexploitation of species for human use, introduction of exotic species and the impacts of climate change. Destruction and fragmentation of natural habitats are perhaps the most important causes of species extinction: habitats in all biomes of China have been subject to modification and loss.

23.3 Ecosystem degradation

Forests cover 1.75 billion ha in China, 18% of the country's landmass, with the total land area covered by forest and shrubland together amounting to 2.6 billion ha (Ren *et al.*, 2007). Although forest cover is increasing as a result of government policies promoting afforestation, natural forests have sustained particular pressures, and primary forest accounts for less than 5% of China's forest today. The forests of China include lowland rainforest, occupying a very small area in southern Hainan, Guangxi on the border with Vietnam, and southern Yunnan, montane rainforest in Yunnan, monsoon forest, which occurs on limestone in southern Guangxi and more extensively in western and southern Hainan, and small areas of stunted northern mangrove forests in southern China and Hainan.

According to Collins and colleagues (1991) the main cause of loss of rainforest in China has been agriculture, both shifting and settled. During the country's long period of closure to the outside world there was a need to be self-sufficient in tropical commodities for which there was very little suitable land. This was one of the key factors leading to widespread deforestation in the tropical forest region. Collection of fuelwood has been an additional threat together, more recently, with illegal logging. Xishuangbanna has suffered greatly from deforestation, with forest cover decreasing from more than 60% to less than 30% due to fuelwood collection, shifting cultivation (included repeated slash and burn) and massive rubber plantations (Zhang & Cao, 1995). As a consequence, about 340 plant species have been recognized as threatened in Xishuangbanna (Zhang & Cao, 1995), and about 600 have probably been lost (Liu *et al.*, 2002). Elsewhere in the country, two historical events contributed greatly to the loss of China's forests: during the Second World War, massive cutting occurred to provide fuelwood for factories, while in the 1950s the campaign ordered the establishment of thousands of village iron-smelting works. The forests of Taiwan were heavily cut in the period from the Second World War to the 1960s (Yang & Pan, 1996).

Grasslands form the most extensive land cover in China, accounting for 42% of the country's land surface. Of this total area, half is used for pasture, 15% for grazing and livestock, 31% for agriculture and forest stands, and the rest occurs in lake, river and coastal belts (Ren *et al.*, 2007). Degradation of grasslands has been caused by overgrazing, overexploitation of medicinal and woody plants, conversion to crop land, expansion of mining and industry, and the impact of rodent and insect pests (Anon, 2008). There is approximately 600 000 km^2 of grassland in northern China, mainly in Nei Mongol. The risk of degradation in this area of low precipitation is considered to be very high. In western China, soil erosion by wind is a major problem in degraded grassland areas, reducing the biodiversity and productivity of the land and leading to sand storms.

Deserts in China cover about ca. 1.3 billion ha, accounting for ca. 13% of the country. The Gobi, stretching from the Tian Shan in the west, across southeastern Mongolia and northern China, covers over one million km^2. The Taklimakan Shamo, in southern Xinjiang, is the largest desert in the country. Overall the area of desert in China is increasing as a result of degradation of grasslands. The biodiversity of China's arid areas is sensitive and increasingly under threat.

Wetlands in China, including swamps, lakes and rivers, cover ca. 4% of the country's land surface and account for ca. 10% of the world's total wetlands. Over half the area of wetlands has been converted into agricultural land since the 1950s. In coastal areas, wetlands have been lost as a result of urbanization, industrial development and agriculture (Anon, 2008).

23.4 Exploitation of wild plant resources

In addition to habitat loss and modification, direct exploitation of species is also a serious threat to many Chinese plants. Groups of species that have been subjected to particularly intense exploitation pressures include trees which yield valuable timber, species harvested for gums, resins and fibers, ornamental species such as cycads and orchids, and plants harvested for their medicinal products.

23.4.1 Timber

Exploitation of timber resources in China has contributed to both the loss of forest habitats and pressure on the targeted species themselves. Conifers are one of the groups that have been highly valued for timber and sought after as a consequence. Chinese gymnosperms comprise 207 species, representing 36 genera of ten families(eflora of China), ca. 80% of which are threatened due to a variety of factors (Wang & Xie, 2004), and 63 of which are listed as state protected plants in China (Fu & Jin, 1992). For example, *Taiwania cryptomerioides* (Taxodiaceae, Figure 23.6) is one species that has decreased drastically due to logging for the quality of its straight-trunked timber. Poor regeneration is another factor: *Picea neoveitchii* was extensively logged from the 1850s to the

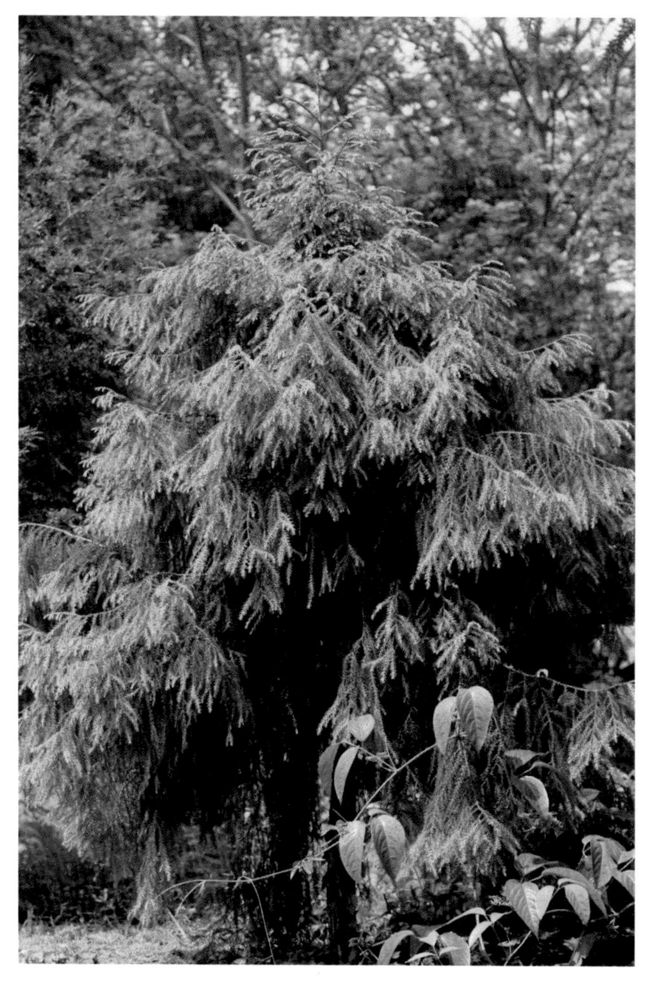

Figure 23.6 *Taiwania cryptomerioides* (image courtesy YANG Ke-Ming, South China Botanical Garden).

1990s for its straight trunks, resulting in its endangered status and destruction of associated vegetation, due to poor regeneration the four remaining, widely-separated subpopulations of this species cover less than four hectares in total. *Picea neoveitchii* is assigned an IUCN Red List assessment of Critically Endangered (IUCN, 2011).

Fokienia hodginsii (Cupressaceae), a conifer species that occurs in China, Laos and Vietnam, yields a fragrant timber that is favored in the manufacture of Chinese coffins. It is now considered to be a vulnerable species, as is *Pseudotsuga forrestii* (Pinaceae), the timber of which is used for construction, bridge-building, vehicles and furniture. The species is also grown for afforestation. The genus *Dalbergia* (Fabaceae), which includes species commonly known as "rosewoods," has around 30 species in China. *Dalbergia odorifera*, which occurs in generally open forests and forest margins in Fujian, Hainan and Zhejiang, has been heavily exploited for its fragrant wood.

There are about a dozen species of Dipterocarpaceae in China, in five genera, distributed in Guangxi, Hainan, Xizang and Yunnan (Zhang & Wang, 1985). Various species are important as a source of timber and also for the production of balsam. *Dipterocarpus retusus* (Figure 23.7) is now rare in Yunnan, its distribution affected by habitat loss. The wood is heavy and is used for building, and the tree is also a source of a balsam used for caulking boats. Of the four species of *Hopea* in China, two are considered under threat. *Hopea hainanensis*, which is likely a Chinese endemic, is considered to be an endangered species of scattered occurrence. Its durable wood is used for making boats, bridges and houses. *Hopea chinensis* (Figure 23.8), a species of Guangxi, Yunnan and also northern Vietnam is considered to be an endangered species threatened by timber exploitation, again for making boats, furniture and bridges. Dipterocarpaceae threatened by a combination of habitat loss and felling for timber include *Vatica guangxiensis*, a species occurring in Guangxi and Yunnan together with Vietnam, with wood used for making boats, houses and furniture, and *Vatica mangachapoi*, used for similar purposes.

Figure 23.7 *Dipterocarpus retusus*. A, habitat. B, detail of leaves and fruit (images courtesy LI Er-Liu).

Figure 23.8 *Hopea chinensis*. A, habit. B, flowers. C, fruit (images courtesy LUO Mei-Zhen, YANG Ke-Ming and ZENG Qin-Wen, South China Botanical Garden).

23.4.2 Ornamental plants

With China's long tradition of gardening and appreciation of ornamental plants, harvesting from the wild has contributed to the decline of many species. This process was accelerated with the opening of China to external horticultural exploration in the nineteenth century, and again with the rapid urbanization and increasing wealth of the late twentieth century. The Orchidaceae in particular have been heavily exploited worldwide by collectors. China has ca. 1 400 orchid species in 194 genera, with 11 endemic genera and ca. 490 endemic species. The country has a long tradition, spanning 2 000 years, of using orchid species in herbal medicine, and a 1 000-year tradition of orchid cultivation for ornamental horticulture. Wild populations have been subject to destructive collection and local extinction both by farmers gaining a supplementary source of income and breeders looking to develop new orchid forms (Liu *et al.*, 2009). Orchids used medicinally include *Gastrodia elata* and various species of *Bletilla* and *Dendrobium*.

Ornamental Orchidaceae that have been subject to intense collecting pressures include many species of *Paphiopedilum*. For instance, the discovery of a new species, *P. armeniacum*, in Yunnan in the 1980s led to the intense exploitation of wild populations for the international market. With its bright yellow flowers it was immediately highly sought-after by orchid collectors in Japan, Europe and the USA. It is claimed that 35 000 *P. armeniacum* plants were exported via Hong Kong in the six months to April 1986. Another species, *P. micranthum*, with large pink and white flowers, occurs in a remote area of southeastern Yunnan, close to the Vietnamese border, and has also been heavily collected for the international market (Cribb, 1998).

Species of *Cymbidium* (Orchidaceae) have also been subject to collecting pressures in China (Du Puy & Cribb, 2007). There is a long tradition of cultivation of *Cymbidium* hybrids in China, originating over 2 500 years ago. Nevertheless, collection of species from the wild is still commonplace. Species commonly collected for sale include widespread species such as *C. ensifolium, C. faberi, C. goeringii* and *C. sinense*, which are becoming of conservation concern in parts of their range. Recently, species with larger flowers, such as the endemic and vulnerable *C. wilsonii*, have become fashionable and are collected for sale in the cities of Yunnan. As with *Paphiopedilum*, newly described species are highly sought-after by collectors. The description of *C. wenshanense* in 1990 and *C. nanulam* in 1991, both in Yunnan, immediately resulted in exploitation of the natural populations. *Cymbidium wenshanense* is now considered to be endangered. In total, Du Puy & Cribb (2007) consider three Chinese endemic species of *Cymbidium* to be threatened, together with *C. wenshanense* which may also occur in Vietnam. They also consider 12 non-endemic species of *Cymbidium* that occur in China to be under threat (see Figure 23.9).

It has been estimated that around 60% of China's native Cycadaceae have been lost over the past century, mainly as a result of collection for horticulture, particularly in the 1980s, and clearance of land for agriculture (Chen & Liu, 2004). Species that are close to extinction include *Cycas revoluta* and *C. szechuanensis* in Fujian, and *C. hongheensis* in Yunnan. A small population of *C. micholitzii* in Xinlong, Hainan, has probably disappeared. In the past five years, the populations of *C. debaoensis, C. hainanensis* and *C. multipinnata* have significantly reduced in number, by ca. 20–30% (Chen & Liu, 2004).

Wang and Liang (1996) have described the factors leading to the loss of Cycadaceae in China. These include deforestation in mountain areas that has caused considerable reduction in the area of cycad habitat and reduced habitat continuity. This has in turn significantly influenced regeneration and gene flow within cycad populations. Deforestation within areas of cycad habitat has also caused soil erosion. This has been a particular problem in the limestone mountain areas of Guangxi, Guizhou and Yunnan, where large areas of barren land and exposed rock remain. An additional threat during the period of economic depression and famine in China (1958–1962), was direct harvesting of cycad trunks to produce a form of starch which became known as "immortal rice" in Guangxi. It is also customary in southwestern Guizhou to harvest cycad trunks to obtain starch from the pith to make wine.

In the 1970s, the high ornamental value of cycads as garden plants became recognized. Many plants were harvested from the wild and sold throughout the country. In the early 1980s, horticultural dealers from Guangdong,

Figure 23.9 Examples of *Cymbidium* spp. threatened due to collecting pressure as ornamental orchids. A, *C. faberi*. B, *C. wenshanense*. C, *C. ensifolium*. D, *C. sinense*. E, *C. goeringii* (images courtesy LUO Yi-Bo, ZENG Qing-Wen, XU Ke-Xue and LI Min).

Figure 23.10 *Cycas revoluta*. A, habit. B, cone (images courtesy YANG Ke-Ming and LIAO Li-Fang, South China Botanical Garden).

Jiangsu, Shanghai Shi and Zhejiang travelled to Lianjiang, Luoyuan and Xiapu in Fujian to buy wild cycads. Fuzhou Shi and Fuqing became the main trade centers for cycads in China. As a result, cycad populations in coastal areas of eastern Fujian became extinct by 1983–1984 through over-harvesting. The most sought after species was *Cycas revoluta* (Figure 23.10). The price of cycad stems rose dramatically, causing attention to be drawn to other areas of cycad distribution within China. Cycads of other species were harvested and their roots and leaves removed so that they could be sold as *C. revoluta*. In the late 1980s and early 1990s, for example, *C. hongheensis* was still present at most of its original population sites. However it has been targeted because of its similarity to *C. revoluta*, with many whole trunks being removed and those that remain damaged by removal of branches and basal offsets, leading to death by stump decay or a hindering of reproduction. This "cycad craze" gradually reduced in intensity but harvesting of wild cycads still continues and has caused other Cycadaceae to become threatened. *Cycas szechuanensis*, for example, has a very wide distribution area but has nevertheless become threatened in the wild. This species has become extinct in certain counties, with populations in Guangxi and Yunnan being severely damaged. The species has only survived in mountainous areas where there are poor transport links. Other cycads that have become threatened as a result of illegal harvesting are listed in Table 23.1.

Table 23.1 Some examples of threatened cycads in China (Donaldson, 2003; Hill, 2008). IUCN *Red List* status is taken from IUCN (2011). Asterisks indicate taxa described since the publication of Cycadaceae in *Flora of China*.

Species	IUCN Red List status	Distribution (number of populations)	Conservation issues	Reserves in which conserved
Cycas balansae	Near threatened	Guangxi (1)	Dioecious, low seed set, niche habitat, illegal poaching	Shiqan Dashan
C. changjiangensis	Endangered	Hainan (1)	Dioecious, low seed set, niche habitat, illegal poaching	Bawangling
C. debaoensis	Critically endangered	Guangxi (3)	Dioecious, low seed set, niche habitat, illegal poaching	Huanglianshan
*C. dolichophylla**	Vulnerable	Guangxi, Yunnan (5)	Dioecious, low seed set, niche habitat, illegal poaching	Daweishan
C. hongheensis	Critically endangered	Yunnan (2)	Collection for horticulture, reproductive barriers	-
C. micholitzii	Vulnerable	Yunnan (2)	Dioecious, low seed set, niche habitat, illegal poaching	Daweishan
C. multipinnata	Endangered	Yunnan (4)	Dioecious, low seed set, niche habitat, illegal poaching	Daweishan
C. segmentifida	Vulnerable	Guangxi, Guizhou, Yunnan (7)	Dioecious, low seed set, niche habitat, illegal poaching	Jinzhongshan
C. szechuanensis	Critically endangered	Fujian (3)	Dioecious, low seed set, niche habitat, illegal poaching	-
C. taiwaniana	Endangered	Guangdong (4, 6)	Dioecious, low seed set, niche habitat, illegal poaching	Local reserve

In 1987, the Environment Protection Bureau of China classified *Cycas micholitzii* and *C. panzhihuaensis* as Grade 2 and *C. pectinata* and *C. taiwaniana* as Grade 3 protected species. In a revised list of rare and endangered plants drafted by the Ministry of Forestry in 1996, all species of Cycadaceae were classified with Grade 1 protection. Various nature reserves have been set up in China, specifically to protect cycad habitats. These include the Natural Reserves of *C. panzhihuaensis* in Panzhihua, Sichuan, and Luquan, Yunnan, the Natural Reserve of *C. taitungensis* in Taidong, Taiwan, and the Honghe Cycad Nature Reserve established to protect *C. hongheensis*, *C. multipinnata* and other species in southeastern Yunnan in 1996. The Debao Cycad Nature Reserve was established in Guangxi in 1999 to protect *Cycas debaoensis* (Figure 23.11) which was discovered in 1984 and described as a new species in 1997. The species is very rare, consisting of only one population with ca. 1 500 individuals in 1997. By January 2000, the population had been reduced to ca. 500 individuals (including 200 reproductively-mature plants) because of illegal harvesting for commerce. Among the adult individuals, 20% are female and 80% are male. Only two or three female plants in the population produce seeds in any given year.

Botanic gardens are also helping to conserve China's cycad diversity. The first botanic garden to carry out *ex situ* protection of Cycadaceae was the South China Botanical Garden, in which a special collection of cycads was set up in the early 1970s. Species included *C. micholitzii, C. revoluta, C. rumphii* (not a Chinese native), *C. segmentifida, C. szechuanensis* and *C. taiwaniana* etc. The garden was also the first to introduce *C. rumphii* to cultivation and holds the largest *ex situ* collection of individuals in China. In 1989, a special collection of cycads was also established at Shenzhen Fairy Lake Botanical Garden, to which *C. micholitzii, C. miquelii, C. revoluta, C. segmentifida, C. szechuanensis* and *C. taitungensis* were introduced. Since 1993, an international centre for the *ex situ* protection of cycads has been built in the garden and more than 100 species of cycads introduced, making it one of the largest collections of cycads in China.

23.4.3 Medicinal plants

Around 11 000 species are used in Traditional Chinese Medicine (see Chapter 18). Of the 600 plant species that are regularly used, about one-third are well-established in

Figure 23.11 *Cycas debaoensis*. A, habit. B, male cone. C, female cone (Images courtesy ZENG Qing-Wen, LIAO Li-Fang and YANG Ke-Ming, South China Botanical Garden).

cultivation which helps to prevent unsustainable harvesting from the wild. Nevertheless, wild harvesting of medicinal plants for local consumption and international trade continues to be a major threat to plant diversity in China. Threatened medicinal plants of the Himalaya Shan region are, for example, listed by Hamilton and Radford (2007). China is a major and growing manufacturing centre for medicinal plant extracts and finished products. These rely increasingly on cultivated material, but were initially based on exploitation of plants from the wild. One medicinal species that has been heavily exploited on an industrial level is *Taxus wallichiana* (Taxaceae), a species of yew native to southwestern China. This tree was traditionally used in Tibetan medicine and is also a source of new anticancer drugs known as taxanes, such as paclitaxel. Until recently, China's manufacture of such drugs was based on *T. wallichiana* and a few other native yew species, harvested from the wild. However, all wild harvest of native yew species for taxane extraction in China was banned in 2003 and there has been increased emphasis on cultivation, including of non-native species and hybrids such as *T.* ×*media* (a hybrid of *T. baccata* and *T. cuspidata*) from Canada. Wild-collected material is also imported from other countries in the region (TRAFFIC, 2004).

Cistanche deserticola (Orobanchaceae) is a medicinal species used to treat kidney disorders and to enhance fertility, with dried stems being the main product utilized. The distribution of this desert species covers China, Mongolia and eastern Asia. Within China, it occurs in Gansu, Nei Mongol, Ningxia, Qinghai, Shaanxi and Xinjiang, a range which is determined by the distribution of the host plant *Haloxylon ammodendron* (Chenopodiaceae). The species is not included in the IUCN *Red List* but it is classified as "critically endangered" in the *China Species Red List* (Wang & Xie, 2004), with an estimated 80% decline owing primarily to harvest for medicinal use, cutting of the host plant for fuelwood, overgrazing by camels and pressures from expanding agriculture. Although *C. deserticola* occurs over a wide area, its distribution area has declined and in Shanxi and Shaanxi, for example, the species' populations are considered to have been reduced to the point that reproduction no longer takes place (Tan *et al.*, 2004). The greatest losses occur near centers of human population. Although it was until recently considered difficult to cultivate due to its parasitic life history strategy, an estimated 30–40% of *C. deserticola* is now said to come from cultivation, a figure which is soon expected to overtake wild collection (Mulliken & Croften, 2008).

The bulbs of *Fritillaria*, known locally as "bei mu," are one of the most popular herbal medicines in China and are collected in the wild throughout the country. There is a huge annual demand for manufactured products, produced mainly from *F. cirrhosa,* which were worth around 400 million US dollars per year in 2002–2003 (Leon *et al.*, 2009). The bulbs are dried, powdered and used as an infusion or made into pills to treat coughs and asthma. Several species of *Fritillaria* are cultivated for Traditional Chinese Medicine in different parts of China (Leon *et al.*, 2009), but commercial cultivation of *F. cirrhosa* has not yet proven successful, so a major reintroduction program of *F. cirrhosa* bulbs has been underway for the last 20 years in an attempt to re-populate the natural low-growing alpine scrub habitat of the species in Ganzi, Sichuan. Using bulbs grown in a nearby nursery from wild-collected seed, some 10 000 ha of habitat have been set aside for the reintroduction, and the future potential to establish a sustainable harvesting program is considered to be promising (Leon *et al.*, 2009).

23.4.4 Other types of exploitation

The plants of China are subjected to a wide range of threats. For example, the palm *Chuniophoenix hainanensis*

(Arecaceae) is considered to be highly endangered as a result of over-exploitation of the durable stems to make arts and crafts. Although the species occurs within a protected area, control of cutting remains important (Johnson & IUCN/SSC Palm Specialist Group, 1996).

23.5 Invasive species

China has a long history of introduction of non-native species (see Chapter 20), due to their potential economic value or other benefits as medicinal, ornamental, soil-improving, erosion-controlling or landscaping plants, for example. Fifty percent of invasive alien plants were intentionally introduced for reasons of this nature (Ding & Wang, 1998). Twenty-five percent of alien animals were also intentionally introduced. In a survey carried out from 2001–2003, 188 alien plant taxa were recorded in China (Xu *et al.*, 2004). Rapid growth in trade and transportation systems within China is increasing the introduction and spread of invasive species leading to transformations in the structure of ecosystems (Ding *et al.*, 2008). One example of an invasive plant is *Tithonia diversifolia* (Asteraceae), an alien ornamental introduced from central America, which has become widely naturalized in Yunnan where it is currently distributed in at least 64 counties, covering about 60% of the province (Wang *et al.*, 2008). *Tithonia diversifolia* probably first escaped to become naturalized in the 1930s (Wang *et al.*, 2008) and now forms naturalized populations in which mature plants can produce a large number of light-winged fertile seeds that are readily dispersed by air, water, vehicles, human activity or livestock. Human introductions for green manure and ornamentals, alongside extensive transportation, have all played an important role in the spread of the species.

References

ANON (2008). Contributing to One World, One Health, A Strategic Framework for Reducing Risks of Infectious Diseases at the Animal-Human-Ecosystems Interface. <http://www.fao.org/docrep/011/aj137e/aj137e00.htm>.

CHEN, C. J. & LIU, N. (2004). New discoveries of cycads and advancement of conservation of cycads in China. *The Botanical Review* 70(1): 93-100.

COLLINS, M., SAYER, J. A. & WHITMORE, T. C. (eds.) (1991). *The Conservation Atlas of Tropical Forests: Asia and the Pacific*. London, UK: Macmillan Press.

CRIBB, P. J. (1998). *The Genus Paphiopedilum*. Kota Kinabalu, Borneo and Richmond, UK: Natural History Publications.

DONALDSON, J. S. (ed.) (2003). *Cycads. Status Survey and Conservation Action Plan*. Gland, Switzerland and Cambridge, UK: IUCN.

DING, J. & WANG, R. (1998). Invasive alien species and their impact on biodiversity in China. In: Compilation Group (ed.) *Biodiversity: A Country Study*. pp. 58-63. Beijing, China: China Environmental Science Press.

DING, J., MACK, R. N., LU, P., REN, M. & HUANG, H. (2008). China's booming economy is parking and accelerating biological invasions. *BioScience* 58: 317-324.

DU PUY, D. & CRIBB, P. J. (2007). *The Genus Cymbidium*. Richmond, UK: Kew Publishing.

FU, L. G. & JIN, J. M. (1992). *China Plant Red Book*. Beijing, China: Science Press.

HAMILTON, A. C. & RADFORD, E. A. (2007). *Identification and Conservation of Important Plant Areas for Medicinal Plants in the Himalaya*. Salisbury, UK and Kathmandu, Nepal: Plantlife International and Ethnobotanical Society of Nepal.

HILL, K. D. (2008). The genus *Cycas* (Cycadaceae) in China. *Telopea* 12(1): 71-118.

IUCN (2011). IUCN Red List of Threatened Species. Version 2011.2. <http://www.iucnredlist.org/>.

JOHNSON, D. & IUCN/SSC PALM SPECIALIST GROUP (eds.) (1996). *Palms: their Conservation and Sustained Utilization. Status Survey and Conservation Action Plan*. Gland, Switzerland and Cambridge, UK: IUCN.

LEON, C., FAY, M. F. & RIX, M. (2009). *Fritillaria yuminensis*. *Curtis's Botanical Magazine* 26(1-2): 21-32.

LIU, H., XU, Z., XU, Y. & WANG, J. (2002). Practice of conserving plant diversity through traditional beliefs: a case study in Xishuangbanna, southwest China. *Biodiversity and Conservation* 11: 705-713.

LIU, Z. J., CHEN, X. Q. & CRIBB, P. J. (2009). *Paphiopedilum*. In: Wu, Z. Y., Raven, P. H. & Hong, D. Y. (eds.) *Flora of China*. pp. 33-34. Beijing, China and St. Louis, MO: Science Press and Missouri Botanical Garden Press.

MINISTRY OF ENVIRONMENTAL PROTECTION OF THE PEOPLE'S REPUBLIC OF CHINA (1998). *Country Status Report of China's Biodiversity*. Beijing, China: Environmental Science Press.

MULLIKEN, T. & CROFTEN, P. (2008). Review of the status, harvest, trade and management of seven Asian CITES-listed medicinal and aromatic plant species. *BfN Scripten* 277.

REN, H., LI, Z. A., SHEN, W. J., YU, Z. Y., PENG, S. L., LIAO, C. H., DING, M. M. & WU, J. G. (2007). Changes in biodiversity and ecosystem function during the restoration of a tropical forest in south China. *Science in China Series C: Life Sciences* 50(2): 277-284.

TAN, D., GUO, Q., WANG, C. L. & MA, C. (2004). Effects of the parasite plant (*Cistanche deserticola*) on growth and biomass of the host plant (*Haloxylon ammodendron*). *Forest Research* 17: 24-29.

TRAFFIC (2004). *TRAFFIC recommendations on the proposals to amend the CITES appendices*. Bangkok, Thailand: TRAFFIC.

WANG, F. X. & LIANG, H. B. (1996). *Cycads in China*. Guangzhou: Guangdong Science and Technology Press.

WANG, S. & XIE, Y. (eds.) (2004). *China Species Red List 1*. Beijing, China: Higher Education Press.

WANG, S. H., SUN, W. B., CHEN, X. & YANG, Y. M. (2008). Reproductive characteristics of *Tithonia diversifolia* and its geographical spread in Yunnan Province of South-West China. *Acta Ecologica Sinica* 28: 21-29.

WORLD BANK (2003). *China Promoting Growth with Equity*. World Bank.

XU, H., QIANG, S., HAN, Z., GUO, J., HUANG, Z., SUN, H., HE, S., DING, H., WU, H. & WAN, F. (2004). The distribution and introduction pathway of alien invasive species in China. *Biodiversity Science* 12: 626-638.

YANG, F. J. & PAN, C. C. (1996). The current status of native woody vegetation in Taiwan. *In*: Hunt, D. (ed.) *Temperate Trees Under Threat. Proceedings of an IDS Symposium on the Conservation Status of Temperate Trees*. pp. 108-110. University of Bonn: International Dendrology Society.

ZHANG, J. & CAO, M. (1995). Tropical forest vegetation of Xishuangbanna, SW China and its secondary changes, with special reference to some problems in local nature conservation. *Biological Conservation* 73: 229-238.

ZHANG, J. Q. & WANG, L. Z. (1985). Geographical distribution of Dipterocarpaceae. *Chinese Bulletin of Botany* 8: 1-8.

Chapter 24

Conservation Strategies

HUANG Hong-Wen
Peter S. Wyse Jackson and CHEN Ling-Zhi

24.1 Introduction: the international context for conservation of plant diversity in China

Over the last two decades there has been a growing international dimension to biodiversity conservation, which has had a significant impact on the development of conservation policies and their implementation at national levels. The development of this international focus has been, in part, a response to the need for the conservation of biodiversity to be a shared international concern, and in part a result of greater international understanding of the growing environmental crisis and the failure or inability of individual countries to address biodiversity issues sufficiently to ensure that biodiversity is safeguarded. The international dimension has also sought to address inequality in available resources for biodiversity conservation, including funding and financial mechanisms, technical know-how and human and institutional resources.

In this chapter we outline the wide range of international instruments, policies and other initiatives for biodiversity conservation, and specifically plant conservation, in the context of which the conservation of the plants of China must lie. It is important that plant conservation at all levels is linked to such initiatives because governmental priorities and programs today are often directed towards meeting international obligations, and funding is most readily available to programs that fulfill such obligations. In addition, aligning local and national programs with international priorities provides added value by links to other initiatives that may share the same broad objectives.

Prior to 1992 there were three major international conventions of direct importance for biodiversity conservation: the World Heritage Convention, the Convention on International Trade in Endangered Species of Fauna and Flora (CITES), and the Convention on Wetlands (Ramsar Convention).

24.1.1 The World Heritage Convention

The World Heritage Convention is an intergovernmental agreement concerning the protection of the world's cultural and natural heritage. It came into being after a 1965 conference in Washington DC, USA, which called for a "world heritage trust" to stimulate international cooperation to protect "superb natural and scenic areas and historic sites." The Convention was adopted by the United Nations Educational, Scientific and Cultural Organization (UNESCO) in 1972. At its heart is a "World Heritage List" that currently includes 891 properties (759 cultural, 193 natural and 29 mixed), recognized as having international heritage importance. The Convention lists sites under threat on the "World Heritage in Danger" list. One hundred and ninety countries are parties to the Convention, China ratified it in 1985. China's nature-related world heritage sites are listed in Table 24.1.

Table 24.1 Natural World Heritage Sites of China included on the World Heritage List. Note that geographic names follow the World Heritage Site listing and not the geographic standards of the *Flora of China*.

Site	Year listed
Natural	
Chengjiang Fossil Site	2012
China Danxia	2010
Huanglong Scenic and Historic Interest Area	1992
Jiuzhaigou Valley Scenic and Historic Interest Area	1992
Mount Sanqingshan National Park	2008
Sichuan Giant Panda Sanctuaries – Wolong, Mt Siguniang and Jiajin Mountains	2006
South China Karst	2007
Three Parallel Rivers of Yunnan Protected Areas	2003
Wulingyuan Scenic and Historic Interest Area	1992
Xinjiang Tianshan	2013
Mixed cultural and natural	
Mount Emei Scenic Area, including Leshan Giant Buddha Scenic Area	1996
Mount Huangshan	1990
Mount Taishan	1987
Mount Wuyl	1999

24.1.2 The Convention on Wetlands

The Convention on Wetlands was signed in 1971 in Ramsar, Iran, from where it gains its name. It is an intergovernmental treaty, providing a framework for national action and international cooperation for the conservation and wise use of wetlands and their resources. To date there are 168 parties to the Convention, China became a party in 1992. Worldwide, 2 143 sites (covering a total of 185 million ha) are designated for inclusion in the "Ramsar List of Wetlands of International Importance." China has 41 Ramsar sites, covering ca. 3.7 million ha (see Table 24.2). The Convention was established primarily to protect habitats for waterfowl but has been broadened to include a much wider range of considerations, including ecology, botany, zoology, limnology and hydrology, and many of the sites are of national and international importance for their flora.

Table 24.2 The 37 Ramsar Sites in China. Note that geographic names follow the Ramsar Site listing and not the geographic standards of the *Flora of China*.

Ramsar Site	Year designated	Province	Approximate area covered (ha)
Bitahai Wetland	2004	Yunnan	2 000
Chongming Dongtan Nature Reserve	2002	Shanghai Shi	33 000
Dafeng (*Elaphurus davidianus*) National Nature Reserve	2002	Jiangsu	78 000
Dalai Lake National Nature Reserve,	2002	Nei Mongol	740 000
Dalian National Spotted Seal (*Phoca vitulina*) Nature Reserve	2002	Liaoning	12 000
Dashanbao	2004	Yunnan	6 000
Dongdongtinghu	1992	Hunan	190 000
Dongzhaigang	1992	Hainan	5 400
Eerduosi National Nature Reserve	2002	Nei Mongol	7 70
Eling Lake	2004	Qinghai	66 000
Fujian Zhangjiangkou National Mangrove Nature Reserve	2008	Fujian	2 400
Gansu Gahai Wetlands Nature Reserve	2011	Gansu	250 000
Guangdong Haifeng Wetlands	2008	Guangdong	12 000
Guangxi Beilun Estuary National Nature Reserve	2009	Guangxi	3 000
Hangzhou Xixi Wetlands	2009	Zhejiang	325
Heilongjiang Nanweng River National Nature Reserve	2011	Heilongjiang	230 000
Heilongjiang Qixing River National Nature Reserve	2011	Heilongjiang	20 000
Heilongjiang Zhenbaodao Wetland National Nature Reserve	2011	Heilongjiang	44 000
Honghe National Nature Reserve	2002	Heilongjiang	22 000
Hubei Honghu Wetlands	2008	Hubei	43 000
Huidong Harbor Sea Turtle National Nature Reserve	2002	Guangdong	400
Lashihai Wetland	2004	Yunnan	3 600
Mai Po Marshes and Inner Deep Bay	1995	Hong Kong	1 500
Maidika	2004	Xizang	43 000
Mapangyong Cuo	2004	Xizang	74 000
Nan Dongting Wetland and Waterfowl Reserve	2002	Hunan	170 000
Napahai Wetland	2004	Yunnan	2 000
Niaodao ("Bird Island")	1992	Qinghai	54 000
Poyanghu	1992	Jiangxi	22 000
San Jiang National Nature Reserve	2002	Heilongjiang	160 000
Shanghai Yangtze Estuarine Wetland Nature Reserve	2008	Shanghai Shi	3 800
Shankou Mangrove Nature Reserve	2002	Guangxi	4 000
Shuangtai Estuary	2004	Liaoning	130 000
Sichuan Ruoergai Wetland National Nature Reserve	2008	Sichuan	167 000
Xi Dongting Lake (Mupinghu) Nature Reserve	2002	Hunan	35 000
Xianghai	1992	Jilin	110 000
Xingkai Lake National Nature Reserve	2002	Heilongjiang	220 000
Yancheng National Nature Reserve	2002	Jiangsu	450 000
Zhaling Lake	2004	Qinghai	65 000
Zhalong	1992	Heilongjiang	210 000
Zhanjiang Mangrove National Nature Reserve	2002	Guangdong	20 000

24.1.3 The Convention on International Trade in Endangered Species of Fauna and Flora

The international wildlife trade is estimated to be worth billions of US dollars per year, and to include hundreds of millions of plant and animal specimens (Convention on International Trade in Endangered Species of Wild Fauna and Flora, 2012). The trade is diverse, ranging from live animals and plants to a vast array of derived products including foods, musical instruments, timber, curios and medicines. The Convention on International Trade in Endangered Species of Wild Fauna and Flora (CITES) aims to ensure that international trade in specimens of wild animals and plants does not threaten their survival, promoting sustainability in trade, even for non-endangered species. Efforts to regulate such cross-border trade clearly require international cooperation. China signed and became a party to the Convention in 1981. At present, over 30 000 species are protected by CITES, under three categories: "CITES Appendix I" lists species that are the most endangered, "Appendix II" those that are not necessarily threatened with extinction but may become so unless trade is closely controlled (trade in these species may be authorized with an export permit), and "Appendix III" species already regulated by one country which need the cooperation of other countries to prevent unsustainable or illegal exploitation. China has 13 plant species listed in CITES Appendix I, 167 in Appendix II and three in Appendix III.

24.1.4 International conservation agreements developed after 1992

Beginning in 1992, a remarkable range of new international initiatives relating to biodiversity conservation came into being. Many of them have had a very significant impact on plant conservation priorities and actions throughout the world. They include the United Nations (UN) Convention on Biological Diversity (CBD), UN Convention to Combat Desertification, UN Framework Convention on Climate Change, Food and Agriculture Organization (FAO) International Treaty on Plant Genetic Resources for Food and Agriculture, World Summit on Sustainable Development, and various National Biodiversity Strategies and Action Plans.

The Convention on Biological Diversity.—In relation to plant conservation, by far the most important of the post-1992 agreements is the CBD, which has had a profound impact on the development and implementation of biodiversity conservation strategies throughout the world. The rationale for the development of the CBD was that the Earth's biological resources are vital to humanity's economic and social development; thus biological diversity is a global asset of tremendous value to both present and future generations, and one under greater threat than ever before. The United Nations Environment Program (UNEP) convened a high-level group in November 1988 to explore the need for an international convention on biological diversity, followed by a working group of technical and legal experts who were to take into account the need to share costs and benefits between developed and developing countries and ways and means to support innovation by local people. Intergovernmental negotiations continued until May 1992 when the text of the CBD was agreed in Nairobi, Kenya. China was one of the first countries to sign this agreement, and formally ratified the Convention in 1993.

The CBD is a legally-binding agreement ratified by almost all of the world's governments, setting out commitments for maintaining the world's biological diversity while balancing economic development. In accordance with the UN Charter and international law, states have the authority to exploit their own resources in accordance with their own environmental policies and the responsibility to ensure that activities within their jurisdiction or control do not cause damage to the environment elsewhere. The Convention has three main objectives: conservation of biological diversity; sustainable use of its components; and fair and equitable sharing of the benefits from the use of genetic resources. It is comprehensive in its goals and stands as a landmark in international law, recognizing that conservation of biological diversity is a common concern of humankind and an integral part of development. The agreement covers all ecosystems, species, genetic resources and the variability between and within species, as well as interactions between species at all levels of biological organization. It links traditional conservation efforts to the economic goal of sustainable resource use, sets principles for fair and equitable sharing of the benefits arising from the use of genetic resources, in particular those destined for commercial use, and addresses the rapidly-expanding fields of biotechnology and biosafety.

The CBD has had a major impact on the way in which biological institutions and organizations act worldwide, in particular because it affirms the sovereign rights of states over their genetic resources. For example botanic gardens, which have traditionally enjoyed virtually open access to plant material for their collections, have needed to review collection policies and procedures *in lieu* of national legislation regarding the collection and export of wild plants. The CBD also provides new opportunities for scientific institutions to become involved in national biodiversity conservation issues and sustainable development, with the potential to gain a greater influence, profile and recognition of their role in plant conservation and the implementation of the CBD both nationally and internationally.

As part of its CBD commitments, China published a "Biodiversity Conservation Action Plan for China" in 1994—a comprehensive document adopted by the government. Its overall objectives are to undertake urgent and effective measures to avoid further damage to the natural environment and resources of China, and to mitigate or reverse existing detrimental impacts to China's natural environment that impact on biodiversity. Specifically, the Action Plan identified seven major objectives: strengthening fundamental studies on biodiversity to generate the knowledge necessary to guide its conservation; improving networks of protected areas; protecting wild species of biodiversity importance, including species that are rare or endangered; protecting genetic resources of crops and domesticated animals; *in situ* conservation of biodiversity outside protected areas; establishing national networks for biodiversity monitoring and information; and ensuring coordination between measures undertaken to ensure biodiversity conservation and sustainable development.

Under this Action Plan, China set a target of expanding its protected area coverage to 18% of its total land area by 2050, and establishing a comprehensive legal and management system for protected areas by 2010. In addition, by 2020, China intends to have set up 200 protected areas to safeguard the biodiversity of wild agricultural crops originating in China. Other initiatives of importance to plant conservation include the development of regulations related to the protection of genetic resources, protection of new plant varieties and trade in seeds; surveys to document traditional knowledge about biodiversity in parts of Yunnan, Guizhou and Qinghai and its relationship to the conservation and sustainable use of biodiversity; and the conservation of Chinese medicinal herbs and traditional knowledge related to their use.

24.2 China's National Strategy for Plant Conservation

China's flora is coming under increasing threat from its explosive economic growth, rapid industrialization, large development schemes, intensive agricultural projects and the over-harvesting of timber and medicinal plants. The cumulative effect of these factors has been a ten-fold increase in the number of threatened plant species since 1992, with up to 20% of China's native plants now considered at risk. Despite good progress in several areas, such as the establishment of nature reserves, which have increased in number from just 45 in the late 1970s to over 3 000 today, the recent "National Strategy for Plant Conservation" is the first time a coordinated, nationwide plan has been developed to address China's loss of plant diversity.

The launch of China's National Strategy aims to tackle the plant conservation issue head-on, demonstrating its firm commitment to the environment as well as its international obligations as a signatory to the CBD, and setting out specific and measurable conservation targets for measuring the success of existing and future conservation efforts. Indeed China is one of the first of the CBD's signatory countries to fulfill its commitment by publishing a national strategy following the agreed targets of the Global Strategy for Plant Conservation (GSPC). The catalyst for the creation of the strategy was a meeting of Chinese state officials, initiated by Botanic Gardens Conservation International (BGCI) in Beijing Shi in November 2006. Coordinated by the Chinese Academy of Sciences (CAS), the State Environmental Protection Administration (SEPA) and the State Forestry Administration (SFA) in combination with other related departments. The far-reaching strategy covers 16 targets focusing on four key plant diversity themes: understanding and documenting existing plant diversity; conservation of plant diversity; sustainable use of wild plant resources; and generating public awareness and education of plant diversity.

24.2.1 Understanding and documenting existing plant diversity (Theme 1)

In order efficiently to conserve a nation's flora, it is vital to gain a thorough understanding of the current state of wild populations, present botanical collections and conservation efforts. As described in previous chapters, China's flora has been documented in the *Flora Reipublicae Popularis Sinicae* (Editorial Committee of Flora Reipublicae Popularis Sinicae, 1959–2004; Ma & Clemants, 2006) and the English-language *Flora of China* (Wu *et al.*, 1994–2013). Additionally, most of China's provinces and municipalities have published a local flora. An innovative "virtual herbarium" of online scanned images of pressed plants and associated information, is under development. The first *Red Data Book*, a catalogue that lists rare species and those in danger of extinction, for China's plants was published in 1992 (Fu, 1992). In 1999 this was complemented by the publishing of China's *Catalogue of Wild Plants Under Special State Control* (State Council, 1999) and the *Red List* of native threatened and endangered plants was updated in early 2013. A number of models for the conservation and sustainable use of plants have also been developed, tailored specifically to the local situation in China.

Planned actions under Theme 1 include a new national survey of plant species and habitats, starting with key areas and groups of plants. This will include ensuring that recently-discovered species are added to the flora and that old records are verified. Significantly, this will incorporate conservation assessments, focusing initially on known threatened plants, plants of economic importance and endemic species. There are also plans to accelerate the construction of China's National Herbarium, develop a national network of herbaria, complete the development of the virtual herbarium and make digitized specimens available via the internet. A survey will be undertaken to ascertain the population status of plants *in situ* and *ex situ* conservation programs, and models for the further conservation and sustainable use of plants in China will be scaled-up.

24.2.2 Conservation of plant diversity (Theme 2)

The National Strategy for Plant Conservation details a comprehensive plan for conserving native plants, both *in situ* and *ex situ*. This dual method is considered the most robust approach to wildlife conservation, mitigating against uncertain prospects. Since 1956 China has established more than 3 000 nature reserves, covering 15% of its land area. In addition, a large number of forest parks and small reserves provide a basis for the development of a network of *in situ* conservation areas for wild plants. Sixty-five percent of China's higher plant communities, including most plants under national key protection, are currently conserved in these state-managed nature reserves. To date more than 180 botanic gardens and arboreta have been established, cultivating 60% of its regionally-important plants, and a national seed bank (containing 340 000 accessions) and network of regional seed banks also ensure the long-term conservation of important crop genetic resources, such as rice (*Oryza sativa*; Poaceae) and soybean (*Glycine max;* Fabaceae). A national medicinal plant genetic resource bank is also currently being established.

Actions planned under Theme 2 include a review of China's current network of nature reserves to ensure that all major ecological zones are represented, with particular emphasis on desert, grassland and wetland ecosystems, assessments of the impact of proposed development projects on plant diversity, and prohibition of development projects in areas where pollution is deemed likely to impact on key biodiversity. There are also plans to establish a wild plant monitoring system, to include protecting 90% of key conserved wild plants, by 2015, and a national network of *ex situ* conservation centers based in botanic gardens. Proposals aim to ensure the conservation of 70% of the genetic diversity of China's major economic plants.

24.2.3 Sustainable use of wild plant resources (Theme 3)

Ensuring the sustainable use of wild plant resources is of key importance to conservation, as unsustainable collection is often a key driver in the extinction of economically-valuable species. China, as described above,, is a signatory to CITES and has enacted laws to prevent illegal trade in endangered plants. Inspection stations have been established at major ports such as Shenzhen Shi and Dalian Shi to enforce these laws. Sustainable cultivation systems have been developed for 200 species widely used in Traditional Chinese Medicine, thereby preventing their unsustainable harvesting from the wild. China's first organic and "green" product certification schemes have been introduced and are being actively promoted—currently over 12 868 products have been certified; forest certification schemes for sustainably-sourced timber are also under development. Finally, several volumes of the *China Ethno-medicine Encyclopedia,* officially cataloguing the uses and preparation methods of a wide range of medicinal plants, have been published.

Future actions include the active state promotion of a wide range of sustainable agricultural practices, including organic growing, building ecological criteria into new development projects, including the construction of 10 000 new "eco-villages," commissioning a study into the scale of the international trade in wild Chinese plants and an aggressive crack-down on illegal logging and over-exploitation of medicinal plants and timber. There are also plans to strengthen support for developing forest certification schemes, actively to promote the cultivation of useful wild plant species to relieve pressure of harvesting on wild plant populations, including close monitoring of the populations of plant resources in the wild, and systematically to collect and document traditional knowledge, particularly that of minority ethnic groups, as part of a broader effort to preserve traditional cultures.

24.2.4 Generating public awareness and education of plant diversity (Theme 4)

Grass roots support is vital to the success of any major conservation project, and generating public awareness for plant conservation is particularly important in China, where popular knowledge and support for environmental issues were previously low. Thus, plant diversity conservation, as part of a broader environmental education, has been incorporated into Chinese primary and secondary school curricula. A number of large-scale public environmental awareness campaigns are now run each year, tied to specific themes such as "Tree Planting Day," "Earth Day," or "National Laws and Regulations Publicity Day," and typically including news conferences, international conventions and art exhibitions. Forest parks, botanic gardens and nature reserves are overwhelmingly popular attractions in China, and play a significant role in plant diversity conservation and education. In 2005 these facilities together attracted over 440 million visitors (China's Strategy for Plant Conservation Editorial Committee, 2008).

Further plans to generate public awareness include equipping every administrative level of the Chinese state government with a wild plant conservation campaign branch, establishing a specialized plant diversity campaign team in each regional and local nature conservation institution, designating a national "Wild Plant Conservation Day" as part of the existing calendar of large-scale state public environmental awareness campaigns, investing heavily in improving environmental campaigns and education websites as well as visual and audio databases, in order to modernize education and training approaches, founding "green nursery schools," "green schools" and "green universities," as well as ensuring that environmental courses and lectures are incorporated into non-environmental fields, and opening China's botanic gardens to the outside world in order to share scientific knowledge and expertise with partners around the globe.

24.3 *In situ* conservation and natural reserves

A philosophical and conceptual ideology of nature conservation has a long history in China, and may date back as early as the Zhou Dynasty (ca. 1100–221 BC). The ancient tradition of protecting holy places around Buddhist and Taoist temples and sacred mountains continues today. The historical relationships between nature conservation and minority cultures have evolved into a deeply-rooted cultural desire for nature protection, ranging from the exemplary conservation of holy lands, holy mountains and holy trees to various styles of worshipping nature by many ethnic minorities in China (Yang *et al.*, 2004). For example, in Xishuangbanna (Yunnan) some 400 "spirit mountains" (Figure 24.1), with a total area of nearly 500 km^2, have historically been protected by the Dai people (Walters & Hu, 2003).

Figure 24.1 Spirit Mountains in Yunnan (image courtesy WANG Xue-Hai, South China Botanical Garden).

A modern nature reserve network has also been developing rapidly to ensure effective *in situ* conservation of Chinese native plants (López-Pujol & Zhao, 2004). This process began a little more than 50 years ago, with the flagship Dinghushan National Nature Reserve, the first in China, being established by the South China Botanical Garden (SCBG), CAS in 1956. The reserve is found in Zhaoqing Shi, Guangdong, and covers an area of 1 155 ha, selected for the uniqueness of its flora—including more than 1 900 native vascular plant species—and habitats including the regional climax vegetation, monsoon evergreen broadleaved forest. In 1979 Dinghushan, with two other National Nature Reserves, was designated part of the UNESCO "Man and Biosphere Program" network.

The establishment of Dinghushan National Nature Reserve was followed by the creation of the Wanmu Forest Nature Reserve in Fujian in 1957 and Xishuangbanna Nature Reserve in Yunnan in 1958 (Gu, 1998), making up the three nature reserves established during the 1950s. During the 1960s and 1970s only about 20–30 reserves were established, covering less than 1% of China's land area; most Chinese nature reserves have been developed in the past 30 years, a time of tremendous progress. By 2007, 2 531 natural reserves had been established, covering 152 million ha—15% of mainland China (Figure 24.2). These natural reserves play important roles in plant conservation, covering 85% of the types of terrestrial natural ecosystems, 40% of the types of natural wetlands, 20% of native forests and 65% of natural communities of vascular plants. The number of natural reserves and the area covered already exceed the world average and this achievement has been well received by the international conservation community. In addition, more than 50 wetland parks, 2 151 forest parks, 187 national parks and approximately 480 provincial scenic spots and historic sites have been established across

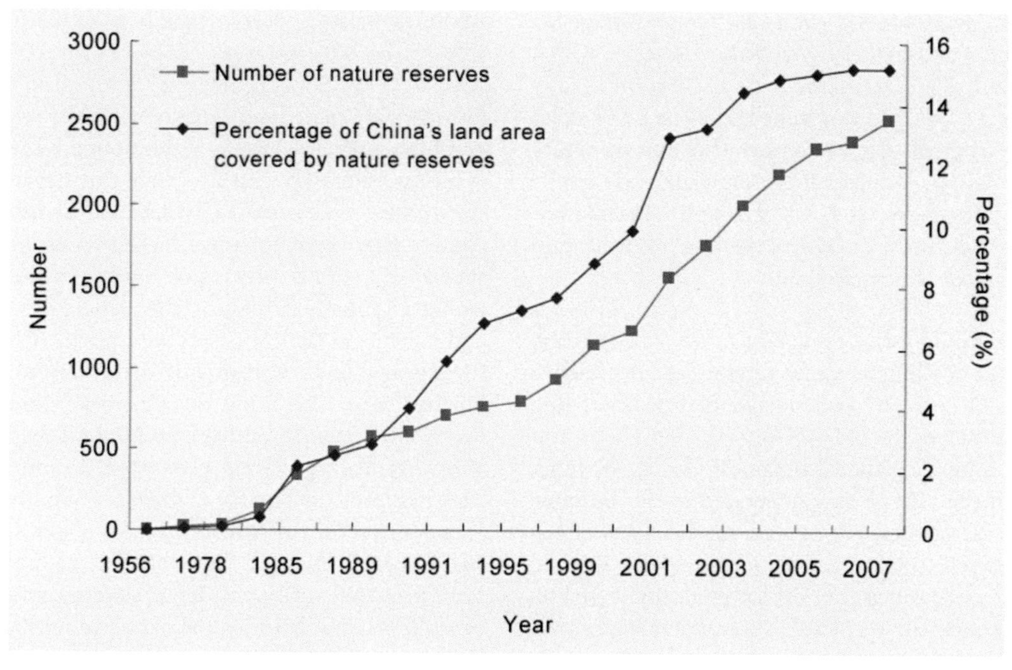

Figure 24.2 Increases in the number of nature reserves in China since 1956.

the country (Ministry of Environmental Protection of the People's Republic of China, 2003).

A statistical analysis of nature reserves in China conducted in 2003 (Ministry of Environmental Protection of the People's Republic of China, 2003) revealed that the number of nature reserves established in each province or autonomous region was led by Guangdong, with more than 200, Yunnan (more than 180), Nei Mongol (more than 180), Heilongjiang (more than 150), Jiangxi (more than 120), Sichuan (more than 120) and Guizhou (more than 100), with these seven regions accounting for about half of the total nature reserves in China. Fourteen provinces or autonomous regions were found to have nature reserves covering more than one million ha, including Xizang with more than 40 million ha, Xinjiang (more than 21 million ha) and Nei Mongol (more than 15 million ha).

Of the more than 3 000 nature reserves established to date, more than 60% are designated for natural ecosystem protection, approximately 30% for wildlife and 1% for natural relics. Thus, high priority has been given to protection of the ecosystems, natural habitats and wildlife species in China. Chinese nature reserves are subdivided into nine categories: forest ecosystems, grassland and meadow ecosystems, desert ecosystems, wetland and river ecosystems, marine and coastal ecosystems, wild animals, wild plants, geological relics and paleontological relics. Forest ecosystems account for more than 1 000 nature reserves and more than 20% of the land area covered by Chinese nature reserves; wetland and river ecosystems account for more than 200 nature reserves and at least 10% of area coverage; and wild plants for ca. 140 nature reserves with more than 6% of the land area.

Each of these nature reserves has a flora ranging from several hundreds to more than 3 000 species. In general, checklists are available for each reserve showing, for instance, that Dinghushan National Nature Reserve, with a total area of only 1 155 ha, harbors ca. 1 900 plant species of which 100 are rare and threatened. However, the exact total number of plant species conserved *in situ* in nature reserves is unknown because of the rapid increase in the number of nature reserves and lack of complete inventories. In particular, it is difficult to assess the level of overlap in the floras of China's nature reserves.

One of most impressive results for *in situ* conservation of plant species in Chinese nature reserves is the effective conservation of critically endangered plants, according to surveys conducted by the SFA during 1997–2003 and recent statistics by the National Botanical Garden Working Committee of the CAS. These generated useful baseline data that provide valuable information on the status of 200 critically endangered plants with minimal population sizes, of top national or provincial priority for protection based on four criteria: species with very few individuals and high risk of extinction in the wild, those in niche habitats and extremely narrow distribution, those with potential genetic value unknown but wild extinction resulting in loss of gene flow and reduction of local diversity, and those of potential ecological and social-economical value. More than 182 plant species with minimal population sizes are at high risk of extinction, including *Abies beshanzuensis* (Pinaceae), *Bhesa robusta* (Celastraceae), *Carpinus putoensis* (Betulaceae), *Diploknema yunnanensis* (Sapotaceae), *Gleditsia japonica* (Fabaceae) and *Sinojackia xylocarpa* (Styracaceae). Of the species assessed, 12 have minimal population sizes (fewer than ten individuals in the wild), 45 species have 10–100 individuals, 68 have 100–1 000, 57 have 1 000–5 000 and 13 have 5 000–10 000. Most of the species with minimal population sizes have one to three distribution sites, usually in very narrow distributions and niche habitats, and are highly vulnerable to any anthropogenic and environmental disturbance. Approximately 70% of these critically endangered and highly prioritized plants have been protected in nature reserves categorized as either national, provincial, city or county-designated reserves. The remaining 30% are probably protected in either natural forest parks or small designated protection spots. However, an inventory and monitoring system for *in situ* conservation is not well developed and is urgently needed.

To strengthen further Chinese *in situ* plant conservation, priorities have been set to improve the national system of nature reserves, to carry out pilot projects for establishing national nature reserves to enhance *in situ* management, and to improve *in situ* conservation outside nature reserves and establish a series of mini-nature reserves and conservation sites (Ministry of Environmental Protection of the People's Republic of China, 2008). Equally emphasized in term of *in situ* conservation is the need to protect key areas of plant species richness and high endemism. The species richness and high number of endemic plants in China have been identified in 17 recognized key biodiversity zones with globally high conservation significance. These sites include 14 land ecosystems (a total area of 230 million ha). Within these 14 sites, 418 nature reserves have been established conserving 34% of the hotspot areas (Table 24.3). According to its "Outlines of National Ecological Environment Conservation" (Ministry of Science and Technology of the People's Republic of China, 2001), China has developed a few new nature reserves where species are rich and natural biological systems are representative, typical and undisturbed. *In situ* conservation also emphasizes niche habitats harboring plant species of major conservation value and their habitats in western China, particularly the critical desert ecosystem and habitats of typical desert wild plants (China's Strategy for Plant Conservation Editorial Committee, 2008). Furthermore, China has developed a "China Ecosystem Research Network" and set up 36 field research stations and a number of forest ecosystem monitoring sample sites in places such as the Changbai Shan, Dongling Shan, Shennongjia, Gutianshan, Dinghu Shan and Xishuangbanna (see Figures 24.3–24.7). Some national nature reserves have been provided with considerable biodiversity monitoring capabilities and facilities for conservation-based research (Ministry of Environmental Protection of the People's Republic of China, 2008).

Table 24.3 Key Chinese areas of terrestrial ecosystem biodiversity with globally significant protection (China's Strategy for Plant Conservation Editorial Committee, 2008). Note that geographic names do not necessarily follow the geographic standards of the *Flora of China*.

Key zone of biodiversity	Approximate area (ha)	Number of nature reserves	Approximate area under nature reserves (ha)	Proportion of area protected (%)
Southern ridge of the Hengduan Shan	25 000 000	47	6 200 000	24
Min Shan to northern ridge of the Hengduan Shan	11 000 000	29	3 000 000	27
Junction of Xinjiang, Qinghai and Xizang	101 000 000	9	60 000 000	59
Xishuangbanna, southern Yunnan	1 700 000	3	320 000	19
Mountains of Hunan, Guizhou, Sichuan and Hubei borders	10 000 000	61	1 100 000	11
Mountains of central and southern Hainan	1 000 000	21	72 000	5
Limestone area of southwestern Guangxi	3 800 000	22	530 000	14
Mountains of Zhejiang, Fujian and Jiangxi borders	4 900 000	18	170 000	3
Qin Ling	4 800 000	19	370 000	8
Ili to the western ridge of TianShan	12 000 000	11	1 400 000	12
Changbai Shan	14 000 000	42	1 100 000	7
Huitu coastal wetlands (Liaohe Kou, Huang He delta, Yancheng beach, eastern beach of Chongming Dao, Shanghai Shi)	12 000 000	56	2 600 000	21
Songneng Pingyuan and Sanjiang Pingyuan in the northeast	23 000 000	71	2 300 000	10
The lakes of the downstream Chang Jiang	4 100 000	9	85 000	2
Totals	**230 000 000**	**418**	**80 000 000**	**34**

Figure 24.3 Images of the mountains of the Hunan, Guizhou, Sichuan and Hubei borders (images courtesy HUANG Hong-Wen, South China Botanical Garden).

Figure 24.4 The Hengduan Shan (image courtesy China's Strategy for Plant Conservation Editorial Committee).

Figure 24.5 The mountains of central and southern Hainan (image courtesy XU Ke-Xue).

24.4 *Ex situ* conservation and sustainable use of plant resources

Ex situ conservation plays a key role in securing the conservation of plant diversity, not only as an insurance policy for the future, but also as a basis for restoration and reintroduction of threatened and endangered plants into natural habitats. The most effective *ex situ* conservation measures are the conservation of living plants in botanic

Figure 24.6 Images of the mountains of Xinjiang and Xizang (images courtesy WANG Xue-Hai, South China Botanical Garden).

Figure 24.7 Images of the limestone mountains of southwestern Guangxi (images courtesy WANG Xue-Hai, South China Botanical Garden).

gardens (and arboreta), and field germplasm repositories and seed banks (Hawkins *et al.*, 2008). Currently, increasing awareness of the impacts of climate change on plant distributions *in situ* has made the appropriate application of *ex situ* techniques more crucial. At present, more than 25 000 plant species are estimated to be conserved in the living collections of 160 botanic gardens across China, and the recently-completed China Southwest Wildlife Germplasm Genebank at the Kunming Institute of Botany has collected over 54 000 accessions of seeds of more than 7 000 species. In addition, 412 000 accessions of crop germplasm belonging to 1 890 species are managed in seed banks by the Ministry of Agriculture, and 40 germplasm repositories, mostly for fruit and nut crops, have been established (Huang & Zhang, 2012). These enhancements in *ex situ* conservation capacity over the past ten years provide great potential rapidly to increase the accessions representing species, genera and families in the future.

24.4.1 *Ex situ* conservation

Historically, botanic gardens have had a major role in plant introduction and conservation, botanical research and sustainable utilization of plant resources. Ancient Chinese gardens can be dated back to the Xia Dynasty (ca. 2100–1600 BC), when unimproved wild species were used as ornamentals. Even earlier, in ca. 2800 BC, the legendary SHEN Nong established a medicinal garden, currently regarded as the earliest botanic garden in the world (Xu, 1997). Plant introductions also originated in very early civilizations in China. The earliest recorded plant hunter was ZHANG Qian, who introduced alfalfa (*Medicago sativa;* Fabaceae) and grapes (*Vitis vinifera;* Vitaceae) in 126 BC during his tenure as ambassador of the Han Dynasty to the Western Territory (Sheng, 1980; Heywood, 2006).

Several botanic gardens were established around the turn of the nineteenth century by western or colonial gardeners, such as the Hong Kong Zoological and Botanic Garden in 1861, Hengchun Tropical Botanical Garden (Taiwan) in 1901, Xiongyue Arboretum (Liaoning) in 1915 and Taipei Botanical Garden (Taiwan) in 1921 (Xu, 1997; Chapman & Wang, 2002; Lopez-Pujol *et al.*, 2006). However, the western concept of botanic gardens designated for plant introduction and botanical research was established in China during the 1920s and 1930s. The first modern botanic gardens (botanical institutes) were set up in Jiangsu (Nanjing Botanical Garden, in 1929) and Guangdong (SCBG, formerly South China Institute of Botany, 1929). In general, botanic gardens in China have developed in four phases: an initiation phase from 1950 to 1965, when a total of 43 botanic gardens were established, a slow development phase from 1966 to 1975, when only ca. nine gardens were developed due to the social and economical instability of the "Cultural Revolution," a booming development phase (1976–1990), when the number of botanic gardens increased rapidly to 123, and, from 1991 to the present day, a period of continuously rapid growth up to a total of 160 botanic gardens (He, 2002; Figure 24.8). Most botanic gardens (145 of the total of 160) are located in the temperate and subtropical regions of China (Table 24.4). A significant feature of the distribution of Chinese botanic gardens is their high concentration in eastern (40%) and south-central China (43%), with only 10% in western China. In the past ten years, more efforts have put into the establishment of botanic gardens in southern and southwestern China where most centers of plant diversity and high plant endemism occur.

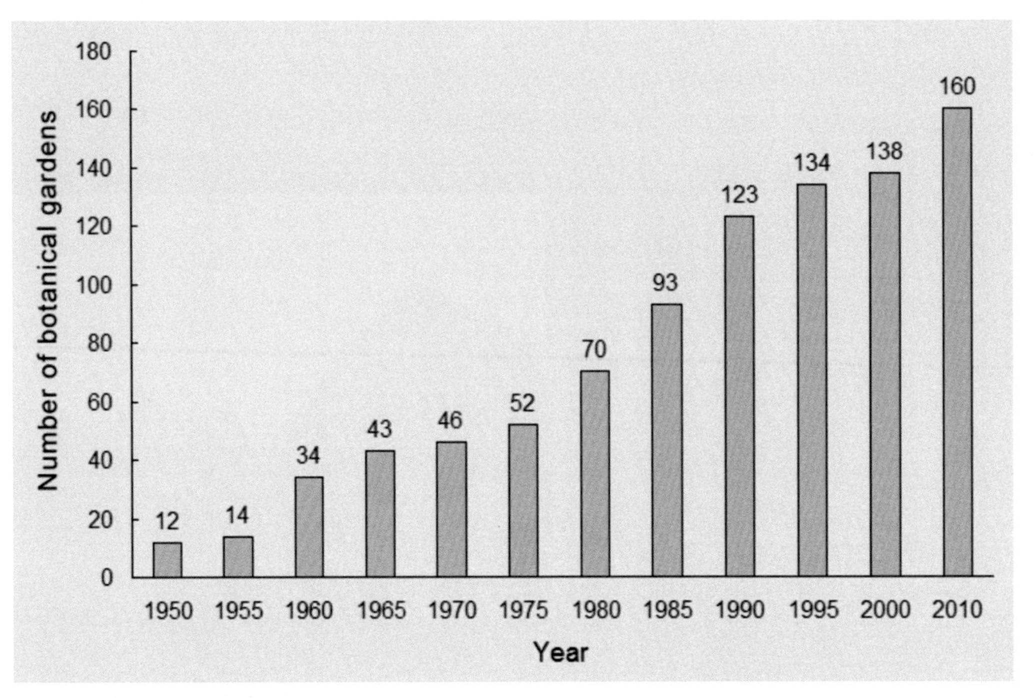

Figure 24.8 Increases in the number of Chinese botanic gardens since 1950 (updated from He, 2002).

Table 24.4 Distribution of Chinese botanic gardens across geographical and climatic zones.

Zone	Vegetation type	Approximate land area (1 000 km^2)	Number of botanic gardens
Plateau and cold	Qinghai-Xizang Plateau alpine	1 800	0
Temperate	Boreal-temperate coniferous forest, temperate coniferous and broad-leaved mixed forest, temperate grassland, warm temperate deciduous broad-leaved forest, temperate desert	5 500	61
Subtropical	Subtropical evergreen broad-leaved forest	2 100	67
Marginal tropical	Tropical rainforest and monsoon forest	200	15
Totals		**9 600**	**160**

The management of China's botanic gardens varies considerably, with differently-prioritized roles given to each garden by different government agencies. Most *ex situ* conservation efforts and programs are conducted by botanic gardens under the administration of the CAS. Other gardens, under the administration of the Ministry of Housing and Urban-Rural Development, SFA, Ministry of Health and state universities are also devoted to *ex situ* conservation to a certain extent, but their prioritized roles are more in terms of landscaping, tourist attraction and public education. Overall national plant conservation strategy and implementation is generally coordinated by the CAS Botanical Garden Working Committee and Chinese Botanical Garden Society. In 2013 a Chinese Union of Botanical Gardens was established and officially launched in Beijing Shi, linking 48 botanic gardens or arboreta jointly to construct a platform for technical exchange and information-sharing. The Union is sponsored jointly by the CAS, SFA and the Ministry of Housing and Urban-Rural Development.

The CAS, as the principal national academic organization, manages its affiliated botanic gardens for the exploration, utilization and conservation of strategic plant resources. The CAS botanic gardens are designated as essential organizations for *ex situ* conservation and utilization of plant resources. A total of 16 botanic gardens are under the management of CAS: Beijing Botanical Garden, Chenshan Botanical Garden, Dinghushan Arboretum, Fairy Lake Botanical Garden, Guilin Botanical Garden, Kunming Botanical Garden, Lushan Botanical Garden, Nanjing Botanical Garden, Qinling Botanical Garden, Shenyang Arboretum, SCBG, Three Gorges

Botanical Garden, Turpan Botanical Garden, West China Subalpine Botanical Garden, Wuhan Botanical Garden (WBG) and Xishuangbanna Tropical Botanical Garden (XTBG). Of these, seven (at Chenshan, Fairy Lake, Guilin, Lushan, Nanjing, Qinling and Three Gorges) are under joint management with the local government and one is a National Nature Reserve (Dinghushan Arboretum). These 16 botanic gardens are distributed over ten provinces (or autonomous regions) in China, occupying a total area of 14 000 ha.

The CAS Botanical Garden Working Committee was established in 1963 as a research and conservation-oriented management organization promoting the development of Chinese botanic gardens and coordinating national plans and strategies for *ex situ* conservation and sustainable use of plant resources under the guidance of the CAS Life Science and Biotechnology Bureau. The committee has played a key role in botanic garden-based research and conservation in China. Its main tasks include the formulation of a master plan for the CAS botanic gardens and verification and approval of development plans for each garden, organization of projects for botanic garden-based research, conservation, public education, research and development for the sustainable use of plant resources, promotion and coordination of scientific and management exchange and technical training, and enhancement of international exchange and collaboration among worldwide botanic gardens and botanical institutes.

The CAS botanic gardens are the largest group of botanic gardens devoted to *ex situ* conservation and the implementation of China's National Strategy for Plant Conservation, as well as botanical research, public education and tourist services. Enhancements in infrastructure and capacity over the past ten years have seen CAS botanic gardens well developed as national centers for *ex situ* conservation and the sustainable utilization of plant resources. The ten main botanic gardens have maintained living collections of ca. 43 500 plant species, with over 24 500 species and infraspecific taxa conserved *ex situ* in total across the gardens (see Table 24.5). This accounts for approximately over 90% of the total *ex situ* conservation living collections of Chinese botanic gardens. Of the species conserved in these *ex situ* living collections, an estimated 90% are native to China and 10% are introduced species. This provides an important reserve of plant resources for sustainable economic and social development in China. Different regional native flora are represented across China, as most botanic gardens have emphasized the collection and conservation of local or regional floras. Duplication of *ex situ* conserved plants across different gardens has been shown to provide an insurance policy against unpredictable losses. Of the *ex situ* conservation collections, 8 215 species (33%) are duplicated in at least one other botanic garden (see Table 24.5). Duplication of rare and threatened species in the ten main botanic gardens ranges from 58% at Turpan Botanical Garden to 95% at Nanjing Botanical Garden, with

Table 24.5 Contribution of the ten main Chinese botanic gardens to *ex situ* living collections for plant conservation in China, 2012 (Huang, 2011; Huang & Zhang, 2012). * Includes infraspecific varieties. ** Species noted on China's *Red List* for plants as of national conservation concern, most not yet evaluated by the IUCN.

Botanic garden	Number of verified plant taxa*	Number of verified Chinese Red List plant taxa
Beijing Botanical Garden	3 463	177
Fairy Lake Botanical Garden	4 956	441
Guilin Botanical Garden	3 843	445
Kunming Botanical Garden	3 330	423
Lushan Botanical Garden	4 378	229
Nanjing Botanical Garden	2 701	263
South China Botanical Garden	7 898	710
Turpan Botanical Garden	490	26
Wuhan Botanical Garden	5 023	652
Xishuangbanna Tropical Botanical Garden	7 420	571
Total number of verified taxa *in ex situ* collections	**43 502**	**3 973**
Summary statistics		
Measure	Number of verified species	Number of verified Chinese Red List species**
Total number of plant species conserved *ex situ* in China	24 667	1 633
Total number of native plant species conserved *ex situ* in China	ca. 19 000	1 633
Total number of plant species recorded in China	33 000	4 408
Proportion of total native plant species conserved *ex situ* in China	58%	37%
Proportion of conserved species unduplicated across gardens	67%	46%
Proportion of conserved species duplicated across gardens	33%	54%

Table 24.6 Number of plant taxa (species and infraspecific varieties, including introduced taxa) held in Chinese botanic gardens, and duplication of *ex situ* living collections between gardens within regions, by 2012.

Region	Number of taxa held (regional total)	Level of duplication within region	China Red List taxa held (regional total)	Level of duplication of Red List taxa within region
Central China	7 740	65%	740	80%
Eastern China	2 701	72%	263	95%
Northern China	3 463	42%	177	84%
Northwestern China	490	51%	26	58%
Southern China	12 074	69%	996	85%
Southwestern China	9 779	57%	839	73%
Average duplication		**59%**		**79%**

an average value of 80%. Duplication within regions ranges from 51% to 72% (Table 24.6), a highly-focused conservation strategy for the local flora surrounding each botanic garden. Regional duplication is also the most practical way to provide such insurance, because similarity of climates and environments within regions provide favorable conditions for *ex situ* conservation of living collections. When individual species are considered, 874 rare or threatened species are backed-up in at least two botanic gardens, 54% of the total number of rare or threatened plants conserved in *ex situ* collections by the ten main botanic gardens.

Since 1998, with a re-evaluation of the global view of China's role in plant conservation and sustainable use, the CAS has developed a new strategy for botanic garden research and formulated a five-year master plan for CAS botanic garden development into the twenty-first century. Subsequently, the CAS re-organized its botanic gardens and designated three National Core Botanic Gardens (South China, Wuhan and Xishuangbanna Botanical Gardens) and also promoted the concept of Scientific Botanic Gardens. The CAS encapsulates its efforts for enhancement of its botanic gardens as a "supporting platform for innovation research in life science," in line with leading international botanic gardens. It has initiated a "Knowledge Innovation Program " at the three core gardens, with the idea of establishing a National Scientific Botanic Garden of China. As these projects continue successfully, capacity for scientific and technological innovation has shown great improvement in these core botanic gardens. In terms of *ex situ* conservation, they have reinforced their efforts and extensively enhanced collection and conservation of resource-plant species. For example, XTBG has increased the number of species under *ex situ* conservation from ca. 4 000 in 2002 to ca. 10 000 in 2005, in which 7 420 have been verified, becoming the first garden in China to conserve 10 000 plant taxa and one of the few botanic gardens in the world with such a large number of plant species in living collections. South China Botanical Garden also increased its number of species from ca. 4 500 in 2002 to more than 10 000 in 2006, and WBG from ca. 4 000 in 2002 to ca. 8 000. In the meantime, a large number of specialized collections have been renovated and enhanced for botanical research and germplasm discovery (Huang & Zhang, 2012). Today, SCBG holds the world's largest collections of Magnoliaceae (more than 130 species), Zingiberaceae (more than 120 species), Arecaceae (382 species), and China's largest collection of bamboos (over 200 species), while WBG has the world's largest collection of *Actinidia* (Actinidiaceae), with more than 52 species, and aquatic plants, with over 800 taxa. Thus the core botanic gardens have become important centers in China, with standards of species conservation aligned to those of the top international botanic gardens, and provide an important resource platform to serve social and economic sustainable development in China.

South China Botanical Garden.—Originally established as the Institute of Agriculture and Forestry affiliated with the National Sun Yat-Sen University in 1929, this institute, given its present name in 2003, soon became one of the two most important botanical organizations in China. As a national research institution and one of the core botanic gardens of China, SCBG comprises three components: a research institute for plant sciences and ecology science, a garden for *ex situ* plant conservation, garden-based research and horticulture-related services, and an arboretum (Dinghushan Arboretum or Dinghushan National Nature Reserve) for *in situ* conservation, nature reserve-based research and management. The research institute occupies an area of 37 ha, with a herbarium holding over one million plant specimens together with state-of-the-art research laboratories and facilities. Research undertaken here includes global change and ecosystem services, environmental degradation and ecological restoration, plant systematic and evolutionary biology, biodiversity conservation and sustainable utilization, agricultural food quality and safety and phytochemical resources, and sustainable use of plant and gene resources. At present, there are 30 research groups associated with these six research fields.

The garden itself, with an area of 283 ha, integrates scenic landscaping, scientific elements and cultural heritage and has been recognized nationally and internationally as one of the world's best classical gardens. Around 13 000 plant taxa are held as living collections in 30 themed gardens plus specialist collections such as Arecaceae,

Cycadaceae, Magnoliaceae, Orchidaceae, Zingiberaceae, bamboo, ferns, shade-loving plants, medicinal plants, energy plants and Australian plants. The conservatory complex includes a tropical house, desert house, alpine and polar house, rare flower and fruit house and a plant aquarium. The well known landscape of Longdong Qi-Lin ("jade forest by dragon cave") integrates tropical, subtropical and temperate plant communities with unusual seasonal colors, and it has been elected as one of the eight "Fantastic Scenes in Guangzhou," recognized as a classical masterpiece of modern landscape-gardening.

Wuhan Botanical Garden.—Founded in 1956 as a national reserve of plant resources and comprehensive research-base for *ex situ* plant conservation, more than 8 000 species are conserved in WBG at present. The kiwifruit (*Actinidia*) garden holds the largest number of species in the world and the garden also has the largest aquatic plant garden in eastern Asia and the biggest garden of genetic resources of special wild fruit trees in central China, as well as theme gardens such as the central China relict, rare and endangered plant garden and central China medicinal herb garden. WBG maintains strong research programs in plant conservation genetics and sustainable use of genetic resources, aquatic plant biology and environmental health of inland water, watershed ecology and ecological security of large projects, and resource botany and natural products.

Xishuangbanna Tropical Botanical Garden.—Founded in 1959, XTBG covers an area of 900 ha. Over 10 000 species of tropical plants are preserved in its 34 living collections. Xishuangbanna Tropical Botanical Garden is a comprehensive research institution engaged in biodiversity conservation and sustainable use of plant resources, focusing on forest ecosystem ecology, conservation biology and resource plant development. Facilities available for scientific research include two national field research stations (Xishuangbanna Tropical Rainforest Ecosystem Station, and Ailao Shan Station for Forest Ecosystem Studies); laboratories for biogeochemistry, molecular biology and biotechnology, plant phylogenetics and conservation biology, physiological ecology, a germplasm bank for rare and endangered plants, and a herbarium of tropical plants.

Prospects for ex situ conservation.—The conservation of rare and threatened plants has been prioritized in botanic gardens. *In situ* conservation cannot save all threatened or rare plants, so *ex situ* conservation offers an effective supplementary activity (Hawkins *et al.*, 2008). Work towards ensuring the *ex situ* conservation of threatened and rare plants is guided by GSPC Target 8, and major progress has been made to rescue threatened and rare plants by botanic gardens: it is estimated that all 388 species categorized as nationally protected are included in *ex situ* living collections in China (Fu, 1992). However, species included in the more recently published *Red List* of 4 404 threatened or endangered species (Wang & Xie, 2004) have not been extensively collected because the list is not yet widely accepted nor finalized by Chinese government conservation agencies. Each of the ten main botanic gardens currently holds between 26 and 710 rare and threatened plants (Table 24.7), accounting for an average of 9% of their total *ex situ* living collections. Thus, a total of 1 633 rare and endangered plants are maintained in living collections, 37% of those on the recent *Red List* (Wang & Xie, 2004). Further gap analysis and re-evaluation of *ex situ* conservation strategies are urgently needed to ensure a true "safety net" for rare and endangered species, and an in-depth assessment of all plant species presently known to be threatened at national level should be an urgent priority. Gap analyses are also critical for selected groups of *ex situ* collections as a basis for strengthening these and planning for the restoration of key species (BGCI, 2007).

Table 24.7 Number of rare and threatened species conserved *ex situ* in the ten largest botanic gardens in China, by 2012.

Botanic garden	Total verified species held	China Red List species	Proportion of species Red Listed (%)
Beijing Botanical Garden	3 463	177	5%
Fairy Lake Botanical Garden	4 956	441	9%
Guilin Botanical Garden	3 843	445	11%
Kunming Botanical Garden	3 330	423	13%
Lushan Botanical Garden	4 378	229	5%
Nanjing Botanical Garden	2 701	263	10%
South China Botanical Garden	7 898	710	9%
Turpan Botanical Garden	490	26	5%
Wuhan Botanical Garden	5 023	652	13%
Xishuangbanna Tropical Botanical Garden	7 420	571	8%
Total		3 937	9%
Total unduplicated		1 633	37%

24.5 Sustainable use of plant resources

One area of progress complementing *ex situ* living plant collections is the construction of national seed banks. While prioritization of *ex situ* living collections continues to be important, wide-scale seed banking efforts have been developing rapidly, particularly in response to growing concern about the impacts of the changing climate which puts virtually all species at risk. These efforts were initiated when the first National Seed Bank was established in Beijing Shi in 1983, with capacity for 230 000 accessions of seeds. Now, for example, the China Southwest Wildlife Germplasm Genebank project, operated by the Kunming Institute of Botany, CAS, included 54 292 accessions from 7 271 plant species in 1 638 genera and 188 families by 2012. The goal of the gene bank is to target for seed banking a total of 190 000 accessions and 19 000 plant species by 2020.

The need for *ex situ* conservation of crop material, in particular, has been recognized as increasingly urgent—with a particular need to focus on the wild relatives and local varieties of crops as a rich source of diversity and adaptive traits for extreme abiotic conditions. The Ministry of Agriculture has established cold storage seed banks safely to conserve crop genetic resources. These include long-term, medium-term and duplicate cold storage facilities for crop seeds, and 390 000 accessions of crop germplasm belonging to 450 species, 160 genera and 28 families have been made. The second National Seed Bank, established in 1988 with a capacity of more than 450 000 accessions, is committed to the global germplasm conservation of *Brassica rapa* (turnip rape; Brassicaceae), *B. rapa* var. *glabra* (Chinese cabbage), *Raphanus sativus* (radish; Brassicaceae) and *Triticum aestivum* (wheat; Poaceae; Ministry of Agriculture, 1995; Gu, 1998). To safeguard the security of crop germplasm in future, a national duplication seed bank was developed in far western China (Xining, Qinghai) in 1995, with a total capacity for 400 000 accessions. Together these form a basic national network for long-term seed banking of crop seeds including about 208 000 accessions of cereals, 61 000 of rice, more than 1 000 for apple, pear and peach, 25 000 of vegetables, and 3 000 of grass and green manure crops (Ministry of Agriculture, 1995). In addition, 40 national field germplasm repositories designated for perennial and vegetatively-propagated crops, mostly fruit and nut crops and their wild relatives, have collected more than 1 300 species of rare and endangered plants (Ministry of Environmental Protection of the People's Republic of China, 2008). These efforts have greatly enhanced *ex situ* conservation capacity of crop diversity for food security.

The CAS-affiliated botanic gardens have also established 90 specialized gardens or collections to create a solid foundation for botanic garden research into the conservation and utilization of plant resources, as well as for public education. The plant resources collected (or introduced), conserved and evaluated by botanic gardens have made a significant contribution to developing new industries, fostering new economic growth and enhancing economic development. Historically, research conducted in CAS botanic gardens has made many significant innovations and transfers of technology to agricultural and industrial developments in China. Many new cultivars in current agricultural use are derived from early research conducted by CAS botanic gardens, based on the model known as "3R"—resource, research, and resolution—for the sustainable use of plant resources formulated within botanic gardens. This model is based on germplasm collection, *ex situ* conservation and evaluation of genetic resources, while conducting multi-disciplinary on plant communities, individuals, cells and genomes and identification of functional genes. The model integrates evaluation of *ex situ* valuable genotypes with novel approaches to gene discovery and new breeding technologies to develop new bio-products with intellectual property rights, including new genes, new crop varieties, natural medicines, and other high-value plant-based products.

For example, Kunming Botanic Garden successfully introduced the "golden dollar" tobacco cultivar from the USA; subsequent cultivar improvement substantially changed the cultivar structure of tobacco production in Yunnan, and made a significant contribution to the tobacco industry in the province. South China Botanical Garden's 30-year research program on the introduction and evaluation of sandalwood (*Santalum album*; Santalaceae) has filled the gap in sandalwood production in China.

The Chengdu Diao company, a CAS-founded enterprise, and subsequently Chengdu Diao Medicine Cooperation, developed a cardio-vascular drug originally initiated from a survey of peltate yam (*Dioscorea zingiberensis*; Dioscoreaceae) resources in China by CAS botanic gardens. Beijing Botanical Garden has developed about ten high-quality grape (*Vitis vinifera*) varieties which are now widely cultivated commercially over hundreds of thousands of hectares, providing great economic benefits to the Chinese grape industry and its growers. Xishuangbanna Tropical Botanic Garden's inter-cropping system for rubber (*Hevea brasiliensis*; Euphorbiaceae) and tea (*Camellia sinensis;* Theaceae) has now been adopted over a total plantation area of about 133 000 ha, contributing significant economic, ecological and social benefits. A new resource of dragon-blood plants (*Dracaena cochinchinensis*; Liliaceae), also discovered by XTBG, ended the importation of such herbal medicines into China, with a dozen Chinese pharmaceutical factories now manufacturing this traditional medicine for circulatory problems.

A showcase of the 3R model is the kiwifruit (*Actinidia*) project conducted by WBG, which holds the biggest germplasm repository of kiwifruit resources in the world. This included intensive research into the genetic diversity of wild species, identifying superior and special genotypes from *ex situ* genotypes, plus research at the molecular level focused on species' origins and systematics, polyploidy evolution, and color regulation of the fruit flesh and skin. Wuhan Botanic Garden scientists then integrated traditional breeding with molecular-aided techniques to develop new varieties. Thus, WBG has developed more than ten new cultivars or varieties for fresh markets and the fruit-processing industry, including the widely-cultivated commercial cultivars "jintao," "jinxia," "jinzao," "jinyan" and "wuzhi 3," as well other new cultivars of the male pollenizer "moshan 4" and ornamental varieties "jianshanjiao" and "mantianxin." The novel cultivar "jintao," with golden-colored flesh, is the first example of a Chinese-owned patent of a new fruit cultivar made available to the international kiwifruit market by granting propagation rights for 28 years. This success showcases a sustainable development model for plant genetic research, innovating new cultivars and generating economic profit, which in return funds further genetic resource conservation, research and development.

Since the initiation of the Knowledge Innovation Program, the core botanic gardens at SCBG, WBG and XTBG have enhanced their research activities and attracted more prestigious research funds from national programs. In recent years, the scientific publications of the three core botanic gardens have increased substantially, including in international journals such as *Science* and *Nature*. In 2001, XTBG scientists published an article on systematic research and the name of a new pollination mechanism, "flexistyly," in *Nature* (Li *et al.*, 2001), which aroused attention from international biologists. In 2002, WBG scientists and others published an article in *Science* presenting strategies and action plans for Chinese native plant conservation (Huang *et al.*, 2002), which raised substantial attention from related international experts and academics. In 2004, SCBG scientists published an article, also in *Nature*, reporting the discovery of a new self-pollination mechanism (Wang *et al.*, 2004), which also aroused considerable attention from international biologists. These three core botanic gardens have applied for over 100, and have been granted 50, new patents. In recent years, they have developed and released more than 40 new cultivars or varieties into the agriculture, forestry and horticulture industries. The Knowledge Innovation Program has not only stimulated science and technological innovation within the CAS botanic gardens, but also provided guidance to other Chinese botanic gardens. Furthermore, the three core botanic gardens have played an increasingly important role in the international botanic garden community, developing a leading position amongst Asian botanic gardens.

24.6 Recovery and restoration of plant diversity: progress and case studies

The recovery and restoration of threatened and endangered plant species in the wild has so far been relatively limited in China. However, recovery of wild populations of depleted species may be increasingly important in repairing damaged ecosystems and restoring connectivity at a landscape scale. Genetically diverse populations of species therefore need to be maintained in *ex situ* collections, and the capacity to propagate and cultivate endangered species needs to be enhanced, in support of such habitat restoration work. In this respect, the use of locally-appropriate species in restoration work should be encouraged, with the aim of creating diverse, resilient native communities (Hawkins *et al.*, 2008). The following case studies provide examples of the recovery and restoration of plant diversity in China.

24.6.1 Primulina tabacum

Primulina tabacum (Gesneriaceae; Figure 24.9) is a calciphilous perennial herb first documented by Hance in 1883, based on a specimen collected by Andrew Henry in 1881 from Lianzhou, Guangdong. *Primulina tabacum* was subsequently considered to be extinct in the wild for more than 100 years until a few plants were rediscovered in Huangchuan and Lianzhou in the 1990s. There are now only eight small populations known in the wild. The species is a slow-growing, shade-tolerant plant, pollinated by insects and dispersed by wind. Rapid reductions in distribution, and the disappearance of populations due to increasing anthropogenic activity has resulted in the listing of *P. tabacum* in the first category of the national *Red List* (Peng & Chen, 2002), and the species has attracted great attention from conservationists.

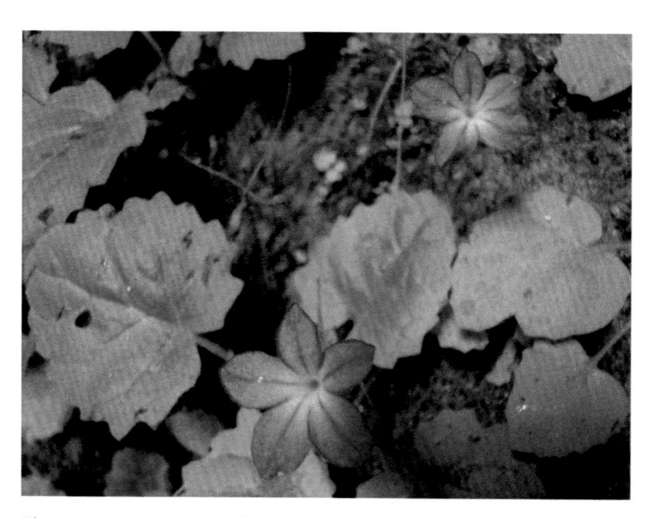

Figure 24.9 *Primulina tabacum* in natural habitat (image courtesy REN Hai, South China Botanical Garden).

Natural distribution and habitat.—Primulina tabacum is extremely narrowly-distributed and restricted to a small boundary locality between Guangdong and Hunan in southern China, where the climate is dominated by the central subtropical monsoon and the vegetation by evergreen broad-leaved forest. The species naturally occurs in karst landscapes with other calciphilous and shade-tolerant plants including bryophytes, mostly near the entrances to deep karst caves, occasionally on cave walls where light-scattering is apparent. Its population density is greatest at cave entrances, gradually becoming sparser, and the individuals smaller, with distance from the entrance. Individuals form a uniform distribution at cave entrances, but are clustered inside. In this niche habitat, the soil is thin, alkaline and nutrient-poor and both the concentration of carbon dioxide and relative humidity are higher than in the surrounding area. Of some 19 plant species coexisting with *P. tabacum*, the most frequently occurring are *Pilea notata* (Urticaceae) and *Pteris cretica* var. *nervosa* (Pteridaceae). Several moss species are found, including *Gymnostomiella longinervis* (Splachnaceae) and *Heteroscyphus coalitus* (Lophocoleaceae). The number of species declines with depth into the cave, while *P. tabacum* increase relative to the number of flowering plants in the community. The three most important factors associated with the habitat environment of *P. tabacum* are thought to be relative humidity, atmospheric carbon dioxide concentration and soil potassium content (Ren *et al.*, 2010).

Conservation and recovery programs.—A five-year project conducted by SCBG in cooperation with local conservation agencies included population surveys and taxonomic verification, characterization of reproductive biology, investigation of habitats, population genetic evaluation, propagule increase by *in vitro* culture, and reintroduction and recovery of wild populations (Ren *et al.*, 2010). The project has also identified threat factors, mostly anthropogenic disturbance resulting from tourism and environmental changes—karst cave plants are highly sensitive to any changes in their environment. By the end of 2009, five trials of reintroduction of tissue-cultured propagules into native habitats had been carried out, with a 40% survival rate. Two key techniques for restoration of this critically endangered plant have been developed. Firstly, a specific tissue culture protocol providing rapid increase of propagules was developed for restoration, with a scientific basis in population genetic evaluation, and secondly a "nursing" planting protocol was developed based on field survey and detailed investigation of the habitat; companion mosses *Gymnostomiella longinervis* and *Heteroscyphus coalitus* were successfully used as nurse plants for *P. tabacum* reintroduction (Ren *et al.*, 2010).

24.6.2 Tigridiopalma magnifica

Tigridiopalma magnifica (Melastomataceae; Figure 24.10) is a rare and endangered species endemic to Guangdong and is the only species in the monotypic genus *Tigridiopalma*. It has an extremely narrow and isolated distribution, found only in Ehuangzhang nature reserve (Yangchun), Xinyi and Dianbai in southwestern Guangdong. The plant is a succulent, shade-tolerant, perennial evergreen herb, with short creeping stems and large leaves, and small and exquisite flowers. Since it was first documented and published in 1974, *T. magnifica* has been valued as an excellent indoor and outdoor ornamental.

Figure 24.10 *Tigridiopalma magnifica*. A, plant in natural habitat. B, young plant. C, flowers (images courtesy WANG Shao-Ping and WEI Qiang, South China Botanical Garden).

Factors contributing to endangered status.—*Tigridiopalma magnifica* forms a patchy distribution in which populations have continually declined, now only existing in a few places in southwestern Guangdong. Only a small number of populations have been found in the wild and all are of minimal population size. The largest population, in Xinyi nature reserve, has 100–300 plants. This species usually occurs on riversides or damp cliffs alongside small rivers or creeks in montane forest. It is highly habitat-specific, dependent on appropriate humidity and light-intensity, adapted to a cool and humid climate and very sensitive to high temperatures and drought. Regeneration in the wild is weak and it appears that seedling success is the major bottleneck to survival in the wild because seed production and seed germination are normal but seedling success is limited. Deforestation has been a major threat, with many populations in riverside habitats lost due to logging or orchard-planting. A field survey has observed rapid population decline or population extinction in at least three localities in Yangchun (Li *et al.*, 2009).

Conservation and recovery programs.—In situ conservation to mitigate population decline of *Tigridiopalma magnifica* has been undertaken in Ehuangzhang and Dahuling nature reserves, with collection of mature seeds, support for seed germination and seedling establishment. *Ex situ* conservation to rescue critically-endangered populations near villages and disturbed habitats has been conducted, and several botanic gardens have conserved the introduced plants. At the same time, domestication and evaluation of germplasm for commercial growth have been successful in reducing harvesting in the wild. Micropropagation protocols have been developed for commercial production. Reintroduction and recovery of populations to original habitats in local nature reserves has also been successful, with new populations established by translocating micropropagated seedlings. Extensive field surveys have been conducted to obtain details of the natural distribution of *T. magnifica* today, and have verified 11 localities including seven sites in Gaozhou and four in Yangchun. However, extinction was noted at two previously-recorded sites in Yangxi. In an application of an ecosystem approach, more than 1 000 seedlings have been reintroduced into three carefully-selected sites at the species' original habitats, and on average 50% success rates for seedling establishment have been achieved. Several local populations, previously severely impacted, have shown progressive recovery within nature reserves with artificial support.

24.6.3 Pachylarnax sinica

Pachylarnax sinica (Magnoliaceae; Figure 24.11) is critically endangered and has been categorized under first grade protection in China since 1999. It is also listed as Critically Endangered in the IUCN *Red List* (IUCN, 2011). *Pachylarnax sinica* is a canopy tree in evergreen broad-leaved forests, endemic to Yunnan. The largest tree observed is 40 m high and 1.2 m in diameter at breast height. The species is of great value for studies of Magnoliaceae taxonomy and the paleo-floristic origins of the family. It is also a high quality timber and ornamental tree.

Factors contributing to endangered status.—Pachylarnax sinica occurs in highly fragmented and isolated habitats with five minimal-sized populations of five, six, seven, ten and 20 individuals in four counties of southeastern Yunnan. There are estimated to be less than 50 plants in total. A few factors have been identified contributing to the current endangered status of *P. sinica*. Firstly, habitat damage by deforestation and agricultural activities including planting of fruit crops has resulted in highly fragmented and isolated habitats and caused a great deterioration in the environmental parameters required for regeneration of natural populations, such as humidity, temperature and soil fertility in the habitat microenvironment. Secondly,

Figure 24.11 *Pachylarnax sinica*. A, tree. B, fruit. C, flowers (images courtesy YANG Ke-Ming and ZENG Qing-Wen, South China Botanical Garden).

the species' weak reproductive fitness is related to low seed set, difficulties in seed germination and slow population regeneration. Insect damage to flowers and a long juvenile stage also contribute to reduced productivity and seedling regeneration.

Conservation and recovery programs.—Ex situ conservation has been conducted by staff at Kunming Botanical Garden, who have obtained over 6 000 seedlings from seeds collected from different wild populations. Key maternal trees are well protected with the assistance of local government agencies and nature reserves. A trial reintroduction into original habitats was conducted using 200 seeds collected from the wild, but unfortunately no seed germination occurred, indicating that *Pachylarnax sinica* will not regenerate in the wild without artificial aid. A further 400 seedlings propagated at Kunming Botanical Garden were reintroduced into two selected sites in Xiaoqiaogou National Nature Reserve in 2007 and are being monitored. Ongoing efforts for the reintroduction of garden-propagated seedlings into original habitats and translocation for the recovery of small natural populations has made significant progress: monitoring revealed an annual average 30–33 cm growth in height and 1.3–1.9 mm increase in trunk diameter at two experimental sites. The reproductive biology of *P. sinica* has been investigated for the appropriate formulation of a conservation plan. However, it is too early to conclude that reintroduction has been a success because normal maturity and flowering of *P. sinica* requires about 30 years. Long-term monitoring, reintroduction and recovery enhancement are required.

24.6.4 Formanodendron doichangensis

Formanodendron doichangensis (Fagaceae; Figure 24.12) is most closely related to the genus *Quercus* (Fagaceae) and is an important species in the study of Fagaceae phylogeny and floristic distribution patterns including intercontinental disjunctions between Southeast Asian and South America. *Formanodendron doichangensis* is an evergreen timber tree with excellent form for shade tree landscaping. Due to heavy exploitation for fuelwood, habitat destruction and its restricted distribution, *F. doichangensis* is seriously threatened and is being pushed to the verge of extinction. The species is currently categorized at the second grade for protection in China and is also listed as Endangered in the IUCN *Red List* (IUCN, 2011).

Figure 24.12 *Formanodendron doichangensis.* A, tree. B, male inflorescences. C, female inflorescences (images courtesy SUN Wei-Bang, Kunming Institute of Botany).

Factors contributing to endangered status.—Formanodendron doichangensis is confined to four localities in southwestern Yunnan and is also known from northern Thailand. Most extant individuals occur in a highly scattered distribution in tropical and subtropical montane evergreen broad-leaved forest at 1 000–1 700 m altitude. Anthropogenic habitat fragmentation is a major threat to *F. doichangensis,* with habitat destruction and disappearance through expansion of agricultural land, and logging, causing rapid population declines. Reproductive barriers have probably also contributed to the current endangered status. Although *F. doichangensis* does not show any abnormal phenomena in anther wall development, microspore genesis or male gamete formation, it has been shown to display serious ovule developmental abortion, with only 17% of ovules developing into embryos. Examination of pollen germination revealed a 38% germination rate. Furthermore, its inverse flowering and fruiting period (October to May) might be detrimental to seed development due to drought. Only 10% of nuts are viable, suggesting that *F. doichangensis* exhibits a reproductive barrier in seed development. Seedling establishment needs a moist environment and the establishment of young trees from seedlings needs comparatively sunny conditions. Evidently, the environment in its severely fragmented habitat is not suitable for seed germination and plant development into the reproductive phase. Population genetic analyses have revealed that *F. doichangensis* has high genetic differentiation, but low levels of genetic diversity and a poor gene flow compared to the other Fagaceae. It therefore seems that habitat degradation and reproductive barriers are most likely to be the main factors threatening the species. Nevertheless, the current endangered status could also result from a combination of catastrophic geographical events, habitat deterioration, declining genetic diversity and physiological stress. Therefore, a practical conservation strategy for *F. doichangensis* is urgently needed.

Conservation and recovery programs.—In situ conservation and enhancement of natural populations through translocation have been conducted to enhance one population, at Canyuan, the only population in China which is protected inside a national nature reserve. This population is a vital resource for further conservation actions, as the three other populations are still exposed to high habitat destruction and cutting for fuelwood. It is essential that new *in situ* conservation sites in these areas are urgently provided, and will play an important role in habitat recovery and population restoration. Although transplanting is controversial in plant conservation measures, in the case of *Formanodendron doichangensis* seedling transplantation between populations should be considered and at least studied, to determine if the low gene flow could be enhanced and overall diversity increased. *Ex situ* conservation has been successfully implemented to preserve existing genetic diversity, taking account of seed sources from all extant populations. Seeds have been collected from all four extant populations in China and the one known site in Thailand. More than 1 000 seedlings were obtained from these seed sources, and so far a total of about 2 000 seedlings established from all sources has successfully been cultivated at Kunming Botanical Garden. A first trial reintroduction of 240 seedlings into habitats at the species' Langcan site has been successful, and a long term monitoring program is underway (Sun *et al.*, 2006).

24.6.5 Paphiopedilum armeniacum

Paphiopedilum armeniacum (Orchidaceae; Figure 24.13) is rare, endangered and endemic to China. It is a highly-valued slipper orchid species on the ornamental

Figure 24.13 *Paphiopedilum armeniacum*. A, leaf rosette. B, flower (images courtesy WEI Qiang and YANG Ke-Ming, South China Botanical Garden).

market, very closely related to *P. micranthum* but with significantly apricot-yellow flowers rather than pink-white flowers of *P. micranthum*. These two slipper orchids are together marketed as a "golden couple" in the orchid trade. Due to illegal over-harvesting, *P. armeniacum* is listed as Endangered in the IUCN *Red List* (IUCN, 2011).

Factors contributing to endangered status.—Paphiopedilum armeniacum has a very narrow distribution, confined to two localities in western Yunnan. It frequents niche habitats, normally at 1 400–2 100 m on limestone cliffs and well-drained rocky slopes. The few populations are found in a relatively restricted area, usually with small population sizes. *Paphiopedilum armeniacum* reproduces well in the wild, and is tolerant of poor soil conditions; thus its current endangered status is mostly caused by over-harvesting, encouraged by high prices in the illegal orchid trade.

Conservation and recovery programs.—Considering the extremely narrow distribution and niche habitat of *Paphiopedilum armeniacum*, protection of the original populations and habitats are high priorities to allow species recovery. So far, law enforcement for protection has been tightened and wild harvesting has been banned. Many monitoring and *in situ* conservation sites have been established within the species' range of distribution. Research and conservation have been conducted and an effective protocol for tissue culture of seedlings using the tiny seeds has been successful, providing sufficient propagules for *ex situ* conservation and the restoration of natural populations. *Ex situ* conservation and increase of propagated populations have been established at the National Orchid Germplasm Conservation Center, Shenzhen Shi, Guangdong. The mycorrhizal associations of *P. armeniacum* have also been investigated at Kunming Botanical Garden, providing for development of techniques for transplanting when reintroduction is conducted. Thus, an integrated approach to the restoration of endangered *P. armeniacum* populations has been established by a combination of original wild population monitoring, *ex situ* conservation, and reintroduction to the wild. The conservation ecology of *P. armeniacum* has been investigated at Luoshanpo, one of the original population sites on Lu Shan. Since 2000, conservationists have studied the reproductive biology, life cycle, ecological parameters of the niche habitats, and population structure and features of *P. armeniacum*. Subsequently, based on solid research results, a number of *ex situ* propagated plants at the Shenzhen Center were reintroduced into the original site. This five-year program has so far been successful, with monitoring showing that reintroduced plants grow well at the restoration site (Liu *et al.*, 2006b).

24.6.6 Pinus squamata

Pinus squamata (Pinaceae; Figure 24.14) was discovered and taxonomically documented in 1992. It is a locally-

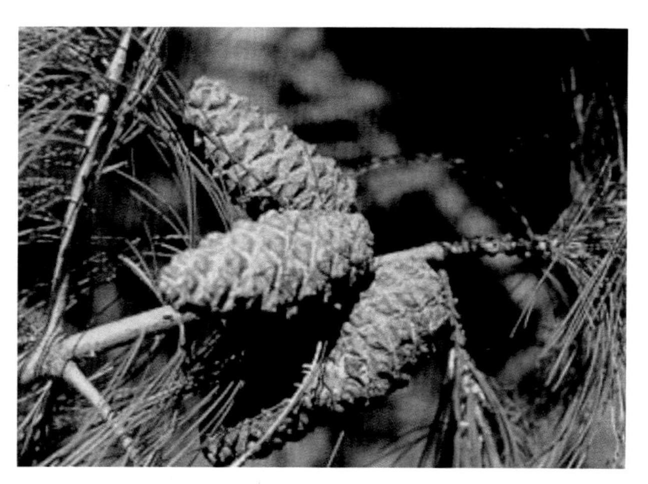

Figure 24.14 Cones of *Pinus squamata* (image courtesy LI Xiang-Wang, http://www.conifers.org).

endemic species with extremely narrow distribution and minimal population size of only 34 plants. However, *P. squamata* is of high value for academic research, ecosystem services and the ornamental and timber industries. It has been nationally categorized at the first grade for protection in China since 1999.

Factors contributing to endangered status.—This species has only a small remnant population of 34 individual plants scattered on a mountain range over 5 km^2 at an altitude of 2 000–2 300 meters in Qiaojia, northeastern Yunnan. The local vegetation comprises transition forests, between warm-temperate coniferous and broad-leaved forests. Of the 34 plants, 32 are more than 50 years old and only 18 mature trees produce fruit. Regeneration is severely limited, with only two plants younger than eight years old. According to the *IUCN Red List Categories and Criteria* (IUCN, 2001), *P. squamata* should be listed as Critically Endangered.

Conservation and recovery programs.—*In situ* conservation programs have been conducted to designate the whole remnant population within the management of the Yaoshan National Nature Reserve. On-site monitoring of each individual tree and any regenerated seedlings has continued since 2005. A near-*in situ* conservation program has been conducted by collecting seeds and propagating seedlings for reinforcement of the remnant population. Measures to prevent and reduce catastrophic environmental events, such as snow storms, rodent or pest damage have been taken. These programs have been successful in increasing population size and a number of planted populations have been increased using seeds from original trees. *Ex situ* conservation programs have been undertaken by several local research institutes and universities. A protocol for seedling propagation was developed, with a total of about 3 000 seedlings obtained for reintroduction and population recovery. Based on the results of these programs and experiments, an approach for the long-term survival of *Pinus squamata* has been formulated, including technical development of seed germination, seedling propagation, population reinforcement, seedling establishment and increase of population size. Preliminary progress has been demonstrated, with a total of 1 500 four-year-old seedlings reintroduced into original habitats in 2008. The reintroduction sites have been under rigorous monitoring to establish survival rates and population change, while the original population shows promising sign of regeneration. However, even greater efforts are needed to remove *P. squamata* from its critically endangered status (Zhang *et al.*, 2004).

24.6.7 Cyclobalanopsis sichourensis

Cyclobalanopsis sichourensis (Fagaceae) is an important species in evergreen broad-leaved forests on karst limestone mountains in southeastern Yunnan. *Cyclobalanopsis sichourensis* has the largest nuts and most distinct leaf pubescence in the genus *Cyclobalanopsis*, which makes it of great value for studying the phylogeny of Fagaceae and floristic patterns in the limestone mountain ranges of Guangxi, Guizhou and Yunnan. According to notes on the type specimen collected in 1947, it occurred frequently as a large timber tree in subtropical humid evergreen broad-leaved forests at altitudes of 760–1 500 m. However, *C. sichourensis* has hardly been seen for more than 50 years due to severe damage to the local forests.

Factors contributing to endangered status.—There are only three known sites where five mature *Cyclobalanopsis sichourensis* trees have recently been found, two in Yunnan and one in Guizhou. It should be included as Critically Endangered on the IUCN *Red List*. The main factors contributing to its endangerment are deforestation, habitat modification and poor regeneration of seedlings in the wild. The recalcitrant seeds of *C. sichourensis* are easily dehydrated, resulting in low seed germination (Chen *et al.*, 2007).

Conservation and recovery programs.—*In situ* conservation, *ex situ* conservation and reintroduction measures have been taken to rescue *Cyclobalanopsis sichourensis*. Seed germination and reproductive biology have been investigated for the appropriate formulation of conservation and restoration strategies. All known natural sites have been designated for *in situ* conservation, with remnant trees now protected and monitored. In terms of *ex situ* conservation, more than 300 seedlings have been propagated at Kunming Botanical Garden. A first reintroduction trial was conducted in 2009, with 35 seedlings translocated into the native habitat. All transplanted seedlings survived for the first six months but more than half died in the following year due to extreme drought.

24.6.8 Cycas debaoensis

Cycas debaoensis (Cycadaceae; Figure 24.15) is a narrowly-distributed endemic, of great value in the study of remnant cycad occurrence patterns, and is of high ornamental value. Over-harvesting has severely reduced the natural population bringing the species to the brink of extinction. *C. debaoensis* is now listed at the first grade of national protection in China. There has been an enthusiastic response from the public and government agencies for

Figure 24.15 *Cycas debaoensis* (image courtesy YANG Ke-Ming, South China Botanical Garden).

conservation of this species.

Factors contributing to endangered status.—*Cycas debaoensis* has an extremely narrow distribution in a niche habitat, across eight known sites in Baise, southwestern Guangxi. *Cycas debaoensis* usually occurs on karst limestone mountain slopes in small isolated populations. The main threat is illegal harvesting for the ornamental trade. There were more than 2 000 plants in 1997, but over-harvesting has caused a sharp decline to about 600 plants over past ten years, and two natural sites have been depleted completely. The illegal harvest for mature plants has made the situation even worse in that the sex ratio has been altered from ten males per female, to five, which greatly hampers the regeneration of wild populations.

Conservation and recovery programs.—The original distribution range of *Cycas debaoensis* has been designated for protection, and the prohibition of illegal harvesting is enforced by the local community. *Ex situ* conservation has been implemented at SCBG and Shenzhen Fairylake Botanical Garden, which has established a specialized cycad collection and conservation center. More than 500 plants are propagated and conserved in the center. Research programs for reproductive biology have been conducted and a protocol for propagation developed. A state program for restoration was initiated in 2007 and the first group of 500 garden-propagated plants was reintroduced into native habitats in Huanglianshan National Nature Reserve in 2008. Efforts have been made to nurse the reintroduced plants. When evaluated in 2009, 472 plants were growing well, with established root systems and new growth of leaves. This preliminary 94% survival rate is promising and also provides evidence that *C. debaoensis* should recover when illegal harvesting is banned and transplanting enhancement is provided (Chia & Liu, 2004).

24.6.9 Myricaria laxiflora

Myricaria laxiflora (Tamaricaceae; Figure 24.16) is an evergreen shrub, the only species in its genus that occurs entirely in the low-altitude region of the Chang Jiang valley,

Figure 24.16 Restoration of *Myricaria laxiflora*. A, habitat: the Chang Jiang valley. B, population in natural habitat. C, individual in natural habitat. D, flowering individual. E, detail of flowers. F, team evaluating recovery program (images courtesy WU Jing-Qing and colleagues, Wuhan Botanical Garden).

with a reversed seasonal growth habit of summer dormancy and winter growth due to long evolutionary adaptation to the natural dynamics of seasonal fluctuations in water level, particularly summer flooding. *Myricaria laxiflora* normally undergoes summer dormancy when the water level rises from May to October, but resumes rapid growth when the water level falls in October, until April the following year. It is highly tolerant to water submergence, with characteristics of mixed sexual and asexual reproduction, excessive seed production, and highly clonal propagation, probably as a result of long adaptive evolution in the riverbank habitat within the water-level fluctuating zone. *Myricaria laxiflora* has become endangered in the wild owing to the construction of the Three Gorges Dam, evoking great concern within the Chinese conservation community and attracting enthusiastic attention among the public and government agencies for rescuing this species. It has been listed as endangered on the *China Species Red List* (Wang & Xie, 2004).

Factors contributing to endangered status.—Myricaria laxiflora is a narrowly-distributed species restricted to the water-level fluctuation zone in twelve counties from Chongqing to Yichang in the Chang Jiang valley. A previously exhaustive survey identified 31 natural populations and approximately 90 000 plants over the natural range of the species. Recent reports of three additional populations found by local people and tourists have been confirmed by conservation researchers from WBG. The habitats of *M. laxiflora* are typical riverbanks with sandy or gravelly, alkaline, poorly-fertile soils with low nitrogen and phosphorous contents but high salt contents. The community structure of *M. laxiflora* natural populations is relatively simple: most coexisting plants are shrubs or grasses including *Arundinella fluviatilis, Beckmannia syzigachne, Capillipedium assimile, Cynodon dactylon* and *Miscanthus sinensis* (all Poaceae), *Distylium chinense* (Hamamelidaceae), *Equisetum palustre* (Equisetaceae) and *Salix variegata* (Salicaceae), with the latter being predominant. *Cynodon dactylon* acts as a nurse plant covering the ground of many *M. laxiflora* habitats.

Myricaria laxiflora can produce many flowering branches, inflorescences and seeds. Seeds are very small and output is high although biomass allocation to reproduction is low (ca. 4%). The seeds are dispersed by both wind and river currents: wind-dispersed seeds usually settle within 25 m of the parent plants leading to a clumped distribution of individuals in populations; long-distance seed dispersals are mostly driven by water flow. The life-span of *M. laxiflora* seeds is about seven days and seeds can germinate within 24 hours when adequate amounts of water are absorbed. Soil water content is a key factor limiting the establishment ability of *M. laxiflora.*

An investigation of genetic diversity and variation of the nine main natural populations across the whole geographical distribution of *Myricaria laxiflora* using amplified fragment length polymorphisms found a moderate level of gene diversity among natural populations, with a high level of genetic differentiation among populations. Estimates of gene flow indicate restricted recurrent gene exchange among populations. Migration patterns in *M. laxiflora* are best explained by a classical metapopulation model, but with a unique unidirectional attribute contributed by the hydrochoric force driving dispersal of seeds and propagules from upstream to downstream. Allozyme variation in *M. laxiflora* populations shows a similar pattern of population genetic diversity and differentiation: a moderate level of allozyme variation with high population genetic divergence (Liu *et al.*, 2006a). Phylogeographical analysis of the chloroplast DNA haplotypes in natural populations of the genus *Myricaria* has revealed a relatively ancient origin but recent demographic expansion for *M. laxiflora*, around 15 000 years ago. The phylogeographic and demographic history of *M. laxiflora* has been strongly driven by geological events and climatic effects within the Three Gorges region (Liu *et al.*, 2009).

Conservation and recovery programs.—Most natural populations of *Myricaria laxiflora* were submerged when the Three Gorges Dam was completed in 2009. Before that, conservation efforts focused on *ex situ* conservation of the natural populations of this species. During the past ten years, many field surveys and collections have been conducted, covering most individual plants in the main natural populations of *M. laxiflora*, and an extensive collection of natural populations has been assembled into a specialized *ex situ* conservation section at WBG. Two satellite *ex situ* conservation sites have also been established locally on branches of the Chang Jiang, close to the natural range of *M. laxiflora.*

While reintroduction programs are underway, recovery of this species in the wild is challenging. Because the natural dispersal dynamics relying on seasonal fluctuations in the water level of the Chang Jiang are now regulated by the Three Gorges Dam, natural habitats for *M. laxiflora* no longer exist. Moreover, the high level of genetic differentiation between populations should be taken into account carefully to preventing possible genetic risks such as inbreeding or out-crossing depression when mixing different individuals in a single reintroduced population. Several sites downstream of the dam have been selected and rigorously evaluated as reintroduction sites, and a preliminary investigation on reintroduced *M. laxiflora* plants at one site has shown a ca. 80% survival rate for reintroduced individuals. Furthering recovery experiments and programs are underway (Liu *et al.*, 2006a).

24.6.10 Prospects

Although similar examples of the restoration of threatened plant species in the wild to reach Target 8 of the GSPC have so far been relatively restricted worldwide, the ongoing efforts of the BGCI China office, coordinated with many Chinese botanic gardens, have initiated the recovery of wild populations of depleted species, repair of damaged ecosystems and restoration of connectivity at a landscape scale. Genetic materials propagated in *ex situ* collections are crucial to these attempts, and detailed research providing the capacity to propagate and cultivate endangered species has supported such habitat restoration work. Several projects are implementing the use of locally-appropriate species in restoration work such as tree planting schemes, with the aim of creating diverse, resilient native communities.

Since the establishment of the BGCI China office in June 2007, BGCI has initiated more than ten threatened plant conservation and recovery projects in China, with target species including *Acer leipoense* (Aceraceae), *A. yangbiense*, *Bretschneidera sinensis* (Bretschneideraceae), *Davidia involucrata* (Nyssaceae), *Dipteronia dyeriana* (Aceraceae), *Euryodendron excelsum* (Theaceae), *Lirianthe fistulosa* (Magnoliaceae), *L. odoratissima*, *Manglietia aromatica* (Magnoliaceae), *Manglietia longipedunculata*, the critically endangered *Manglietia ventii*, endangered *Michelia angustioblonga* (Magnoliaceae), *Michelia coriacea* and endangered *Michelia fulva*. These projects focus mainly on the conservation and restoration of highly threatened and endangered plants in China. All are integrating *ex situ* and *in situ* conservation and restoration approaches.

24.7 Conservation of habitats and ecosystems

Much of the previous part of this chapter has focused on the conservation of individual species through targeted approaches, or protected areas designated to protect the greatest number of individual species. However, biodiversity may also be defined by its ecosystems: biological communities and their related physical and chemical environment. The characteristics of ecosystems are determined by ecosystem processes, such as cycles of water and matter, and the flow of energy, and the plant communities within them are crucial for the fixation of solar energy, absorption of water and nutrients, and as the base of the food chain, providing energy and habitats for other biological communities. Thus ecosystems, and the vegetation within them, provide not only food, water and fuel for human beings, but also play important roles in soil and water conservation, climate regulation, and providing fresh air and clean water. Therefore, the conservation, restoration and sustainable use of ecosystems are of immense significance to humankind.

As noted above, rapid development can have serious influences upon ecosystems and the natural environment. In recent years, the Chinese government has paid great attention to the ecological and environmental problems that are developing and has instigated six ecological construction projects at the ecosystem level: Natural Forest Conservation, Shelter Forest System Construction for the Three Northern Regions and the Middle and Lower Reaches of the Chang Jiang; Conversion of Land to Forestry and Pasture; Sand Defense and Control for the Beijing Shi Region; Wildlife Conservation and Nature Reserve Construction; and Forestry Estate Base Construction (for fast-growing and high-yielding timber forests). The core of these important ecological projects is the improvement of the environment through restoration and reconstruction of damaged vegetation, especially in forest and steppe communities.

24.7.1 Conservation, restoration and rehabilitation of forest

Forests are very widely distributed in China, and play an important role in economic development. Their multi-layered vertical structure performs many functions such as intercepting rainfall, controlling runoff and preventing erosion. Underground, the multi-layered root system fixes the soil, reduces soil and water erosion, decreases overground flow and prevents landslides. And, of course, there are the wildlife and other resources that dwell in the forest. Therefore, the unsustainable cutting and use of forests is of long-term concern, and the conservation, restoration and reconstruction of forests are now considered important measures for improving the environment and, ultimately, for human survival. Two principles have been emphasized regarding the processes of restoration and reconstruction of vegetation: firstly, it should consider the natural geographical distribution of vegetation, using local species in restoration programs, and secondly degraded vegetation communities may be restored through the promotion and utilization of natural successional processes (Chen & Chen, 1997).

Conservation and management of primary forest.—Many forest protected areas have now been established, including some primary forests to which an ecosystem management model could be suitable. Selective removal of over-mature or unhealthy mature wood results in gap formation, making space available for the growth of saplings and thereby improving natural regeneration. Under this management model, a variety of biological niches are formed in the forest for colonization by animals, fungi and microorganisms.

Restoration of secondary forest.—Secondary forests are formed naturally after the destruction of primary forests. Although their ecological function is generally not as effective as primary forests, ecosystem services are still present. In China, the main secondary forests are coniferous and broad-leaved forests, the former resulting from large areas of clear-cutting deciduous broad-leaved forests. For example, in the northeast of the country, forest management comprised a cycle of several years in which secondary broad-leaved forests were cut in a ribbon shape, and seedlings of *Pinus koraiensis* planted in the gaps, which grew vigorously, gradually to restore a mixed coniferous-broad-leaved forest (Li, 1997). In subtropical regions, evergreen broad-leaved saplings were planted among secondary *Pinus massoniana* forest, resulting in the development of mixed coniferous-broad-leaved forest after a few years, followed by a gradual shift to zonal evergreen broad-leaved forest (Peng & Fang, 1995). In this way, natural forest regeneration is promoted towards primary forest by human management, and is consistent with both economic and ecological benefits.

Closure of hillsides to facilitate afforestation.—This is an important method of ecological engineering. Closure of hillsides results first in the development of scrub and coppice on clear-cut forest stands. Coppice is similar in physiognomy to scrub with, in general, five to six shoots per coppice and a height of 1–2 m. Artificial management is necessary to foster coppices, with the retention of two to three shoots per plant to ensure nutrients and space are maintained. After a few years, forest is restored. In temperate and warm-temperate forest regions, coniferous and broad-leaved saplings planted amongst scrub will regenerate to form mixed coniferous and broad-leaved forest within around 15 years (Xu *et al.*, 2005). In subtropical regions, scrub can develop through deciduous broad-leaved forest to mixed evergreen and deciduous broad-leaved forest after 20 years of hillside closure, and the development towards evergreen broad-leaved forest is coincident with an improvement in the general environment (Liu, 1983).

Plantations.—The success of plantations for conservation is determined by the selection of tree species and the land used. Perhaps the most important plantations for conservation are those designed for water conservation, a crucial forest function. Planting of forests for water conservation can help in the conservation of primary forests, artificial afforestation, promoting natural progressive succession, and forest expansion. Research has shown that broad-leaved forests and mixed forests have higher water conservation properties than coniferous forest, with multi-species mixed forests being the best (He & Chen, 1995; Chen & Chen, 1997).

The planting of forest for soil conservation is another important measure for improving the environment. Leguminous shrubs such as *Caragana korshinskii, Lespedeza bicolor* and *Sophora davidii* (all Fabaceae), and herbs, can greatly improve soil fertility and are therefore often planted first in areas where the vegetation has been severely damaged such as the Huangtu Gaoyuan (Loess Plateau). Once soil fertility has improved, trees can be planted, often making use of planting techniques such as terracing and fish-scale-pattern pits. Clear-cutting of any grassy vegetation is avoided when planting shrubs and trees, as this would intensify water and soil loss. In subtropical regions, some coniferous and broad-leaved trees such as *Liquidambar formosana* (Hamamelidaceae), *Pinus massoniana, Quercus* spp., *Populus adenopoda* (Salicaceae) and some Fabaceae can be planted as pioneers. In some cases the ultimate aim is the development of timber forest, such as the important *Abies* forests of southern China, which require fertile and humid conditions. Repeated planting of *Abies* does not maintain fertility, but success has been achieved by planting mixed *Abies* and evergreen broad-leaved forest.

In rural China energy is at a premium, and firewood is one of the most important living sources of energy. We recommend the planting of fuelwood forests in fuel-deficient regions, including trees and shrubs that are easy to harvest. Deciduous *Quercus* and evergreen *Castanopsis* (Fagaceae) species are suited to different climatic zones and habitats, and can be cut in rotation, allowing others to grow. In flood-prone areas, shrubby *Salix* may be more appropriate. *Salix* clumps grow fast, can be harvested for fuel and cut again after only two or three years. We consider that, with the popularization of marsh gas and solar energy as sources of power, the demand of fuelwood will gradually decrease in China.

China has many bare hillsides and wastelands suitable for afforestation. Economically-productive forests featuring oil-producing and crop trees such as *Camellia oleifera, Castanea mollissima* (Fagaceae), *Juglans regia* (Juglandaceae), *Triadica sebifera* and *Vernicia fordii* (both Euphorbiaceae), tea gardens, orchards and bamboo forests can all be developed on such land under suitable climate and soil conditions. Agroforestry systems may also be extended onto low mountains, hills and plateaus: because the required spacing between some economic trees, such as species of *Castanea, Crataegus* (Rosaceae), *Juglans* and *Ziziphus* (Rhamnaceae), is wide, and the duration of fruiting is long, inter-planting with crops can increase yields from the land, giving both economic and ecological improvements. Planted forests may serve other functions including as shelter belts and road protection.

24.7.2 Restoration and reconstruction of steppe and desert vegetation

Northwestern China is home to a vast steppe, supporting widespread animal husbandry. One-third of this natural steppe is in serious decline, partly due to overgrazing and unsustainable utilization. Much of the desertification of steppe occurs due to "blind reclamation," in which the steppe is taken for cultivation, farmed intensively, and soil fertility declines rapidly over several years, leading to the land being abandoned. Continuous cutting of steppe vegetation year on year may also cause productivity to decline and vegetation to alter, as grasses and legumes decrease and forbs increase, with an overall decrease in height and vegetation biomass. In some parts of Nei Mongol, scrub plants and *Artemisia* (Asteraceae) are cut for fuel. Elsewhere, medicinal plants such as *Anemarrhena* spp. (Liliaceae), *Astragalus mongholicus* and *Glycyrrhiza uralensis* (both Fabaceae) and *Ephedra sinica* (Ephedraceae) as well as edible fungi and the cyanobacterium *Nostoc flagelliforme* are harvested from the steppe on a large scale, with effects upon the vegetation, soil structure, and ecological function of the steppe ecosystem, ultimately causing desertification (Li & Chen, 1995). Other activities such as the exploitation of mineral resources and unsustainable development for tourism can have further negative impacts. The situation is similar in desert vegetation communities.

The main modes of restoration of steppe are natural restoration, artificial promotion of natural restoration, and planting of grassland. Closing grassland to grazing by fencing to promote natural restoration can be useful, and is popular due to its low cost; however the natural restoration process is slow. For this reason, artificial means of promoting regeneration have been adopted, including shallow plowing to loosen the soil. This causes the soil to become more porous, increasing the activity of soil animals and microorganisms. Plowing also cuts the underground stem of rhizomatous plants, encouraging the tiller to sprout more densely, accelerating the rate of restoration. For severely-damaged steppe, replanting with high quality vegetation can be important for promoting restoration. For example, replanting with *Leymus chinensis* and Fabaceae on degraded steppe in Nei Mongol has clearly shortened the duration of restoration and increased productivity of the community (Li *et al.*, 1995).

To minimize degradation of grasslands, an estimated "one-tenth grassland conversion" paradigm has been developed, based on long-term field surveys and experimental research (Xu *et al.*, 2010). Fertile, moist, lowland sites with well-preserved grassland can be cultivated for forage pasture, while sites with intermediate conditions can support moderate summer grazing, rotational and delayed grazing systems. For seriously-degraded grassland, three to five years of protection from grazing by fencing should be supplemented with re-seeding and shallow plowing in wet years. Empirical data suggest that each hectare of forage pasture can yield the same mass of dry matter as ten hectares of natural grassland or 100 ha of heavily-degraded grassland. Therefore the formation of a single hectare of forage pasture could protect ten of natural grassland, or 100 ha if heavily degraded, from grazing for one year, helping to restore and improve natural grasslands. This action should strictly be confined to the areas with over 300 mm of rainfall per year and soils not prone to erosion (Ren *et al.*, 2008).

The main method of restoration for desertified farmland in the agro-pastoral areas of northern China is conversion of the land to artificial pasture by planting perennial legumes, which improves both soil structure and fertility. Experiments similar to those above have shown that for every hectare of artificial pasture planted in this way, livestock can be kept off five hectares of natural grassland (Ren *et al.*, 2008). Research on Hunshandake Shadi, Nei Mongol, showed that some vegetation remains on two-thirds of this desertified steppe, and a certain amount of seeds remain in the soil seed-bank, such that this kind of steppe could be restored naturally by prohibiting grazing for several years. This area also suffers from sand and wind erosion, which could be prevented by the use of biological meshes. Biological meshes can be formed by planting wheat (*Triticum* spp.), oats (*Avena* spp.; Poaceae) or the alien *Salix floridana*, followed by sand-fixing pioneer plants such as *Caragana korshinskii* and *Corethrodendron lignosum* var. *lignosum* (Fabaceae), enabling gradual restoration of degraded steppe (Ren *et al.*, 2008).

In desert vegetation, the main measures of restoration are prohibition or rotation of grazing and reduction of human activities, to promote the growth of herbs and shrubs. After exclusion of grazing livestock, the surface of mobile sand dunes can develop a biological crust, which both fixes the sand and traps rainfall in a shallow layer. Multiple experiments have shown that preventing livestock grazing and fuel gathering can effectively promote natural regeneration and restoration of desert vegetation, serving an important role in sustainable development (Ren *et al.*, 2008). Studies on degraded desert on the northern piedmont slopes of Tian Shan, where sand dunes have migrated to low altitudes due to drought, and desert vegetation has regressed due to overgrazing and reclaiming of land, have shown that excluding grazing and conversion of land to pasture were necessary. Seeding with desert plants such as *Tamarix* spp. (Tamaricaceae) accelerated the restoration of vegetation on the diluvial fan and plain areas. Unsustainable utilization and irrigation have reduced soil fertility and caused secondary salinization on the alluvial plain, creating a highly adverse environment for agriculture. Sustainable utilization of water resources is very important, because planting of feeder plants can cause further desertification where overgrazing is also occurring. Grazing levels in desert areas should also be carefully managed to match the available grass biomass. With careful treatment, regression of surrounding desert vegetation can be reduced (Li *et al.*, 1995; Li & Chen, 1995).

These examples demonstrate the importance of learning from the experiences of past years. Desert and steppe vegetation in arid and semi-arid regions are the result of long-term adaptation to the environment, through continuous evolution and development. It is not feasible to increase vegetation coverage very much in desert and steppe regions, for instance by trying to increase forest coverage or to grow irrigated trees and shrubs along roadsides—such actions result not only in failure but in destruction of the original, fragile environment. Similarly, construction of shelter forest on a large scale is not suitable given water shortages in arid and semi-arid regions. As with vegetation management throughout China, measures need to be suited to local conditions (Zheng, 2006).

References

BGCI (2007). Thorny challenge clears the way for conservation. <http://www.bgci.org/resources/news/0435/>.

Chapman, G. P. & Wang, Y. Z. (2002). *The Plant Life of China: Diversity and Distribution*. Berlin and Heidelberg, Germany: Springer-Verlag.

Chen, L. Z. & Chen, Q. L. (1997). *Forest Diversity and its Geographical Distribution in China*. Beijing, China: Science Press.

Chen, W. J., Ji, Y. H., Deng, M., Chen, S. T. & Zhou, Z. K. (2007). *Cyclobalanopsis sichourensis,* a new record species of Fagaceae in Guizhou. *Acta Botanica Yunnanica* 29(4): 395-396.

Chia, J. C. & Liu, N. (2004). New discovery of cycads and advancement of conservation of cycads in China. *The Botanical Review* 70: 93-100.

China's Strategy for Plant Conservation Editorial Committee (2008). *China's Strategy for Plant Conservation*. Guangzhou, China: Guangdong Press Group.

Convention on International Trade in Endangered Species of Wild Fauna and Flora (2012). CITES Trade Database. <http://www.unep-wcmc-apps.org/citestrade/trade.cfm>.

Editorial Committee of Flora Reipublicae Popularis Sinicae (1959-2004). *Flora Reipublicae Popularis Sinicae*. Beijing, China: Sciences Press.

Fu, L. K. (ed.) (1992). *China Plant Red Data Book. Rare and Endangered Plants 1*. Beijing, China and New York: Science Press.

Gu, J. (1998). Conservation of plant diversity in China: achievements, prospects and concerns. *Biological Conservation* 85: 321-327.

Hawkins, B., Sharrock, S. & Havens, K. (2008). *Plants and Climate Change: Which Future?* Richmond, UK: BGCI.

He, J. S. & Chen, W. L. (1995). The degraded ecdosystem in subtropical China: their types, distribution structure characteristics and approaches to their recovery. In: Chen, L. Z. & Chen, W. L. (eds.) *The Studies on Degraded Ecosystem in China*. pp. 61-93. Beijing, China: Science and Technology Press.

He, S. A. (2002). Fifty years of botanical gardens in China. *Acta Botanica Sinica* 44(9): 1123-1133.

Heywood, V. H. (2006). Human use of plant resources - the knowledge base and conservation needs. In: Ertug, Z. F. (ed.) *IVth International Congress of Ethnobotany (ICEB 2005)*. pp. 365-372. Istanbul, Turkey: Ege Yaynllar1.

Huang, H. (2011). Plant diversity and conservation in China: planning a strategic bioresource for a sustainable future. *Botanical Journal of the Linnean Society* 166: 282-300.

Huang, H. W., Han, X. G., Kang, L., Raven, P. H., Jackson, P. W. & Chen, Y. Y. (2002). Conserving native plants in China. *Science* (935-936).

Huang, H. W. & Zhang, Z. (2012). Current status and prospects of *ex situ* cultivation and conservation of plants in China. *Biodiversity Science* 20(5): 559-571.

IUCN (2001). *Red List Categories and Criteria. Version 3.1*. Gland, Switzerland: IUCN Species Survival Commission.

IUCN (2011). IUCN Red List of Threatened Species. Version 2011.2. <http://www.iucnredlist.org/>.

Li, J. W. (1997). *Ecology and management of Pinus koraiensis mixed forest*. Harbin, China: NE Forestry University Press.

Li, L. N., Chen, Y. J., Zeng, S. J., Huang, Z. F., Xu, Y., Ye, H. G., Wang, S. P. & Duan, J. (2009). Study on habitat and endangered causes of *Tigridiopalma magnifica*. *Guangdong Landscape Architecture* 31: 12-15.

Li, Q. J., Xu, Z. F., Kress, W. J., Xia, Y. M., Zhang, L., Deng, X. B., Gao, J. Y. & Bai, Z. L. (2001). Flexible style that encourages outcrossing. *Nature* 410: 432.

Li, S. G., Chen, X. J. & Liu, S. (1995). The degraded desert and desert grassland ecosystems. In: Chen, L. Z. & Chen, W. L. (eds.) *The Studies on Degraded Ecosystem in China*. pp. 195-220. Beijing, China: Beijing Science and Technology Press.

Li, Y. H. & Chen, Z. Z. (1995). The degradation of grassland ecosystem in the temperate of China and their restoration and rehabilitation. In: Chen, L. Z. & Chen, W. L. (eds.) *Studies on Degraded System in China*. pp. 186-194. Beijing, China: Beijing, Science and Technology Press.

Liu, J. R. (1983). Phytosociological analysis during restorated process of secondary vegetation on Wuchaoshan, Zhejiang. *Acta Phytoecologica et Geobotanica Sinica* 7(1): 8-9.

Liu, Y. F., Wang, Y. & Huang, H. W. (2009). Species-level phylogeographical history of *Myricaria* plants in the mountain ranges of western China and the origin of *M. laxiflora* in the Three Gorges mountain region. *Molecular Ecology* 18: 2700-2712.

Liu, Y. F., Wang, Y. & Huang, H. W. (2006a). High interpopulation genetic differentiation and unidirectional linear migration patterns in *Myricaria laxiflora* (Tamaricaceae), an endemic riparian plant in the Three Gorges Valley of the Yangtze River. *American Journal of Botany* 93: 206-215.

Liu, Z. J., Liu, K. W., Chen, L. J., Lei, S. P., Li, L. Q., Shi, X. C. & Huang, L. Q. (2006b). Conservation ecology of endangered species *Paphiopedilum armeniacum* (Orchidaceae). *Acta Ecologica Sinica* 26(9): 2791-2800.

Lopez-Pujol, J., Zhang, F. M. & Ge, S. (2006). Plant biodiversity in China: richly varied, endangered, and in need of conservation. *Biodiversity and Conservation* 15: 3983-4026.

López-Pujol, J. & Zhao, A. M. (2004). China: a rich flora needed of urgent attention. *Orsis* 19: 49-89.

Ma, J. S. & Clemants, S. (2006). A history and overview of the *Flora Reipublicae Popularis Sinicae* (FRPS, Flora of China, Chinese Edition, 1959-2004). *Taxon* 55(2): 451-460.

Ministry of Agriculture (1995). *China: Country Report to the FAO International Technical Conference on Plant Genetic Resources (Leipzig, 1996)*. Beijing, China: Ministry of Agriculture.

Ministry of Environmental Protection of the People's Republic of China (2003). *List of Nature Reserves in China 2003*. Beijing, China: China Environmental Science Press.

Ministry of Environmental Protection of the People's Republic of China (2008). China's Fourth National Report on Implementation of the Convention on Biological Diversity. <http://www.cbd.int/countries/profile.shtml?country=cn>.

Ministry of Science and Technology of the People's Republic of China (2001). China Science and Technology Newsletter. <http://www.most.gov.cn/eng/newsletters>.

Peng, S. L. & Fang, W. (1995). Studies on dynamics of *Castanopsis chinenses* and *Schima superba* population in forest succession of Dinghushan Mountain. *Chinese Journal of Plant Ecology* 19(4): 311-318.

Peng, S. L. & Chen, W. C. (2002). *Rare and Endangered Plants in Guangdong*. Beijing, China: Science Press.

Ren, H., Liu, Q. & Li, L. H. (2008). *An Introduction to Restoration Ecology*. Beijing, China: Science Press.

Ren, H., Zhang, Q. M., Wang, Z. F., Guo, Q. F., Wang, J., Liu, N. & LIang, K. M. (2010). Conservation and possible reintroduction of an endangered plant based on analysis of community ecology: a case study of *Primulina tabacum* Hance in China. *Plant Species Biology* 25: 43-50.

Sheng, C. K. (1980). Plant introduction and botanic gardens in China. *Plant Science Bulletin* 26: 25-28.

State Council (1999). Catalogue of national key conserved wild plants (first list). <http://www.forestry.gov.cn/portal/main/s/3094/minglu1.htm>.

Sun, W. B., Zhou, Y., Han, C. Y., Zeng, C. X., Shi, X. D., Xiang, Q. B. & Coombes, A. (2006). Status and conservation of *Trigonobalanus doichangensis* (Fagaceae). *Biodiversity and Conservation* 15: 1303-1318.

Walters, M. & Hu, H. (2003). Yunnan. Ethnobotany in the service of conservation. *Plant Talk* 34: 19-23.

Wang, S. & Xie, Y. (eds.) (2004). *China Species Red List 1*. Beijing, China: Higher Education Press.

Wang, Y. Q., Zhang, D. X., Renner, S. S. & Chen, Z. Y. (2004). A new self-pollination mechanism. *Nature* 431: 39-40.

Wu, Z. Y., Raven, P. H. & Hong, D. Y. (eds.) (1994-2013). *Flora of China*. Beijing, China: Science Press.

Xu, W. H., Liu, C. H. & Lian, Z. M. (2005). The restoration of degraded ecosystem on Loess Plateau, N Shanxi. *Journal of NW Forestry College* 20(3): 23-25.

Xu, Z. W., Wan, S. Q., Zhu, G. L., Ren, H. Y. & Han, X. G. (2010). The influence of historical land us and water availability on grassland restoration. *Restoration Ecology* 18: 217-225.

Xu, Z. F. (1997). The status and strategy for ex situ conservation of plant diversity in Chinese botanic gardens - discussion of principles and methodologies of ex situ conservation for plant diversity. In: Schei, P. J. & Wang, S. (eds.) *Conserving China's Biodiversity*. pp. 79-95. Beijing, China: China Environmental Science Press.

Yang, Y., Tan, K., Hao, J., Pei, S. & Yang, Y. (2004). Biodiversity and biodiversity conservation in Yunnan, China. *Biodiversity and Conservation* 13: 813-826.

Zhang, Z. Y., Chen, Y. Y. & Li, D. Z. (2004). Detection of low genetic variation in a critically endangered Chinese pine, *Pinus squamata*, using RAPD and ISSR markers. *Biochemical Genetics* 43: 239-249.

Zheng, D. (2006). The environment and ecological construction in dry region of NW China. *Journal of Hebei Normal University (Nature Science)* 30(3): 349-352.

Chapter 25

Prospects for the Future

HUANG Hong-Wen

25.1 Introduction

Plants are essential components of biodiversity, the primary source of the materials for sustaining human life and economic productivity, as well as the ecological support on which the survival of humankind depends. They are the basic source of human necessities including food, timber, oil and medicine. So far we have only explored and utilized a small proportion of all plant species; the potential of the vast majority of wild plants has yet to be understood and is still beyond human knowledge. Although we depend on plants for our livelihood and socio-economic development, damage to plant communities by human activities has reached an unprecedented level, such that human survival is gravely threatened. Currently, about one-third of the world's 300 000–450 000 vascular plant species face the threat of extinction (CBD Secretariat, 2003) due to a variety of activities, including over-harvesting, over-exploitation by destructive agricultural and forestry practices, urbanization, environmental pollution, land-use changes, exotic invasive species and global climate change. Even more alarming estimates suggest that up to two-thirds of higher plants will have disappeared by the end of this century (BGCI, 2000; Pitman & Jørgensen, 2002; CBD Secretariat, 2003). The current rate of extinction for plant species is 2 000–5 000 times faster than would normally occur in nature (Brooks *et al.*, 2006).

One of the major challenges facing humanity in the twenty-first century is how to resolve the paradox of increasing demands on plant resources while providing sustainability for the future. Yet biological resources themselves may be one of the most crucial resources supporting global, social and economic sustainability. Developing biotechnologies provide some hope, through exploration of useful genes widely existing in wild plant resources (e.g. for cold-hardiness, drought-tolerance, disease-resistance, fast-growth, high-yield or quality) in order to develop new "super plant" varieties to keep pace with social and economic development and population growth. This is encapsulated in LI's (1996) statement that "a special gene can affect a country's prosperity, and one important species can affect a country's economy." Thus, botanical research, knowledge, technological innovation and sustainable utilization of plants in a country can be a reflection of that country's overall development and strength. Countries, such as China, that are rich in plant resources as well as innovative knowledge and new technologies for plant conservation and sustainable utilization, will have a distinct advantage in this respect.

China's plant diversity, in terms of species richness, high endemism, paleo-floristic origins and a rich germplasm reservoir of cultivated plants, has been explored throughout this book. However, the conservation and sustainable utilization of Chinese plant resources and sustainable social and economic development in China face increasing challenges due to depleting plant resources and a deteriorating ecological environment caused by rapid economic development and population growth over the past 30 years. As previous chapters have indicated, these challenges are many-fold. The number of endangered plants in China is increasing rapidly: the *China Species Red List* (Wang & Xie, 2004) lists 4 400 gymnosperm and angiosperm species as endangered or threatened, accounting for about 14% of the total plants of China. Of particular concern is the accelerating extinction of the wild ancestors and relatives of crop plants. At the same time, research and development into the use of high-value crops from native plant resources has lagged behind: many developed countries are developing commercial varieties based on China's endemic plant resources such as fruits, vegetables and ornamentals, which are then widely marketed in China. For example, Yunnan supplies 80% of the cut flower market of China, but over 95% of the flowers grown are cultivars imported from other countries. Crop plant genetic resources have also been increasingly eroded, with a loss of genetic diversity and narrowing of the genetic pool of traditional cultivars and landraces. In 1949 China had 10 000 wheat (*Triticum* spp.; Poaceae) cultivars and varieties, but by the 1970s only 1 000 remained in cultivation, as figure which may be even lower today (FAO, 1998). Finally, invasive alien species and ecological degradation have also increasingly impacted on China's native flora and ecosystem health.

25.1.1 China's National Centennial Survey of Plant Diversity and Resources

The currently available information on China's plant resources—including the published *Flora of China* itself—is based mainly on specimens taken from the nineteenth to the middle of the twentieth century. Modern-day information on the current status of plant diversity and resources in China is somewhat inadequate, inaccurate, and out-of-date for practical use consistent with today's rapid economic and social development. The initiation of a "National Centennial Survey of Plant Diversity and Resources" is therefore one of the most urgent tasks in China's botanical research. This survey will be published in a *Country Report on the Current Status of China's Plant Resources,* which should provide a solid scientific basis for the formulation of future social and economic development strategies. The goals of the Centennial Survey are to obtain accurate information on the status of plant diversity and resources in China following the rapid economic development of the past 30 years and, subsequently, to compile ongoing advisory reports monitoring changes in China's plant diversity and resources every three to five years. The survey also aims to increase the specimen collections in China's national herbaria, to enhance capacity for a young generation of experienced field botanists and taxonomists, and to provide baseline data for new editions of Floras and vegetation studies in China (Huang, 2008; Zhang & Huang, 2010).

25.2 China's strategy for plant conservation beyond 2010

Conserving the native plants and flora of China should be further prioritized and urgently enacted. The richness of future plant diversity depends on how we act and what we conserve today (Hawkins *et al.*, 2008). China has developed a national strategy for plant conservation along the lines of the "Global Strategy for Plant Conservation" (China's Strategy for Plant Conservation Editorial Committee, 2008). As China houses some 10% of the world's flora, the implementation of the Chinese national strategy is of global significance. Although China's Plant Conservation Strategy has set up 16 targets designated for completion by 2010, the Strategy needs to be updated and reformulated with a broader scope and achievable framework into a strategic plan beyond this date, taking into consideration increasing pressures on plant diversity and the underlying causes of plant diversity losses. To provide for China's economic and social sustainability in the twenty-first century, the updating and implementation of China's new strategy beyond 2010 should assure that China will be able to continue to rely upon plants for ecosystem goods and services, including food, medicines, clean water, climate amelioration, rich, productive landscapes, energy sources, and a healthy atmosphere, that China can secure the ability to fully utilize the potential of plants to mitigate and adapt to climate change, recognizing the role of plant diversity in maintaining the resilience of ecosystems, that the risk of plant extinctions caused by human activities will be greatly diminished and the genetic diversity of plants safeguarded, and that the rich evolutionary legacy of plant diversity will be used sustainably and derived benefits enhanced to solve pressing problems, support livelihoods and improve social-economic development (Hawkins *et al.*, 2008). Below, I outline how I believe conservation strategies should be prioritized.

25.2.1 Protected areas and *in situ* conservation

Defining priority areas is crucial to formulating a sound national *in situ* conservation plan, because habitat protection is of great importance to saving plant diversity and individual plant species, particularly for plants in niche habitats. Based on extensive research and information, further *in situ* conservation plans and implementation should emphasize the hotspots, ecoregions and centers of plant diversity and take into consideration sound, scientifically-based ecosystem, phylogeographic, refugium and biological corridor approaches. As noted in Chapter 2, a total of 41 sites have been identified as centers of plant diversity and endemism in China and should be a focus of attention when the national conservation strategy is reformulated in terms of protected areas and *in situ* conservation of plants in the future. These centers of plant diversity cover a wide range of regions and span all temperate zones in China, as well as various habitats of significant importance for *in situ* conservation including the Changbai Shan, Taibai Shan region of the Qin Ling, the Nanling Shan, Gaoligong Shan, tropical forests of Hainan, Xishuangbanna region, the limestone region of Guangxi, the Hengduan Shan and the Min Jiang basin.

The genus *Myricaria* (Tamaricaceae) provides a case study of the phylogeographic approach to conservation. Phylogeography examines the spatial arrangements of genetic lineages and provides a powerful tool to understand biogeographic patterns and processes that determine the genetic composition of species or species groups and to allow coherent basic biological units to be defined and better conserved. In this sense, a phylogeographic conservation approach is of critical importance to the conservation of diversity in specific regions. *Myricaria* occurs mainly at high altitudes in the mountain ranges of eastern Asia, centering on the Qinghai-Xizang Plateau and extending to central Asia and Europe. Of the 13 *Myricaria* species described, ten occur in China: *M. bracteata, M. paniculata* and *M. squamosa* are widespread, while the other seven species are narrow endemics. Overall, the geographical distribution of the genus covers 14 provinces of China.

Species-level phylogeographic investigations of *Myricaria* species in China (Liu *et al.*, 2009) have revealed that most chloroplast haplotypes are species-specific, except for one which is shared by the three widespread species. Higher haplotype diversity within the Qinghai-Xizang Plateau region supports the hypothesis that the Himalaya Shan are the center of diversity for *Myricaria*. The phylogeny of *Myricaria* is geographically structured, and an Bayesian-estimated chronology suggests that the main divergence events occurred during the Late Pliocene and Early Pleistocene Epochs (ca. 1.5–2.5 million years ago). The overall phylogeographic pattern is characterized by vicariance events and regional demographical expansion, reflecting the major influence of geological and climatic events on the evolution of *Myricaria* species. Phylogeographic analysis of chloroplast DNA has also suggested an overall correspondence between the current geographic distribution of *Myricaria* and the effects of geology and climate on its evolution in the mountain ranges of western China. It is notable that the main genetic divergence between *Myricaria* species occurred between the Neogene and Early Quaternary Periods, and was predominantly driven by the recent rapid uplift of mountains in the Qinghai-Xizang Plateau region. Local survival in refugia and subsequent expansion in response to Quaternary glaciations has also significantly contributed to genetic isolation and endemic species formation over different regional scales (Liu *et al.*, 2009). This information provides a for formulating a conservation strategy on a phylogeographic basis for *Myricaria*.

A second example is provided by *Taxus wallichiana* (Taxaceae) in China and northern Vietnam. The phylogeography of *Taxus wallichiana*, a species distributed over most of southern China and adjacent regions, has been studied by Gao and colleagues (2007). A total of 1 235 individuals from 50 populations in China and northern Vietnam were analyzed for chloroplast DNA variation using polymerase chain reaction and restriction fragment length polymorphism of the *trn*L-F intron-spacer region. A total of 19 different haplotypes were distinguished and a very high level of population differentiation and a strong phylogeographic pattern was found, suggesting low levels of recurrent gene flow among populations. Haplotype differentiation was most marked along the boundary between the Sino-Himalayan and Sino-Japanese Forest Floristic Subkingdoms, with only one haplotype being shared among these two subkingdoms. The Malesian and Sino-Himalayan Forest Subkingdoms had five and ten haplotypes, respectively, while the relatively large Sino-Japanese Forest Subkingdom had only eight. This strong geographic-haplotype correlation persisted at a regional floristic level with most regions, except for central China, possessing a unique set of haplotypes. Strong landscape effects were observed in the Hengduan Shan and Daba Shan, where steep mountains and valleys might have acted as natural dispersal barriers. The molecular-phylogenetic data, together with the geographic distribution of haplotypes, suggests the existence of several localized refugia during the last glaciation, from which the present-day distribution may be derived. The pattern of haplotype distributions across China and northern Vietnam corresponds well with the current taxonomic delineation of the three intraspecific varieties of *T. wallichiana* (Gao *et al.*, 2007). This study provides useful insights into how conservation plans can be more specifically formulated to address regional diversity with better understanding of evolutionary lineages and floristic similarities and differences in China and eastern Asia.

A further case study of a canopy tree species—*Eurycorymbus cavaleriei* (Sapindaceae)—provides an example of the importance of refugia. *Eurycorymbus cavaleriei* is a relict tree species endemic to subtropical China, a center for speciation and evolution within the East Asia biome, and one of its most important refugial locations. WANG and colleagues (2009) conducted a study aimed at elucidating phylogeographic patterning in *E. cavaleriei*, in order to identify the locations of the species' main refugia and the predominant patterns of migration that have led to the contemporary spatial genetic structure of chloroplast variation. Eighteen populations of *E. cavaleriei*, from throughout its geographical range, were sampled, and chloroplast DNA sequence data were obtained from 170 individuals for phylogeographic analysis. The results revealed that strong phylogeographic structure is exhibited by this species and there are five diverging groups of related haplotypes, which coincide with major landscape features of the region (Wang *et al.*, 2009). This information contributes to our growing understanding of the phylogeography of tree species in subtropical China. In the absence of fossil pollen data for *E. cavaleriei*, population-genetic analysis of chloroplast DNA sequences has provided the first attempt to identify glacial refugia and to infer the population history of the species in southern China. The high differentiation observed among populations is probably a combined effect of contemporary and historical processes, such as the low effective population size of the chloroplast genome in a dioecious species, long-term range fragmentation and limited seed dispersal. Furthermore, clear-cut geographical distributions of ancestral haplotypes of the species suggest multiple potential refugia across subtropical China.

A contrasting level and pattern in genetic variation between eastern and western regions indicates different histories for *Eurycorymbus cavaleriei* populations in these two regions. A higher genetic diversity and a stronger phylogeographic structure detected in the west may be explained by the more complex geological and ecological conditions in southwestern compared to southeastern China. This large-scale phylogeographic investigation demonstrates that *E. cavaleriei* is structured into several distinct groups, mainly corresponding to geographical location. Three potential refugia, one located in each of the Daba Shan, the eastern Yungui Gaoyuan and the Nanling Shan, have been identified for *E. cavaleriei* (Wang *et al.*, 2009). It appears that populations from each of these refugia have contributed to the recent post-glacial expansion of the species. These refugial regions have long been recognized as centers of plant diversity and endemism in China and have also been suggested as glacial or interglacial refugia for many other plant species based on population-genetic data. WANG and colleagues' (2009) study demonstrates that these regions should be of the highest importance for consideration when developing sound conservation policies and sustainable forest management strategies for subtropical China.

25.2.2 *Ex situ* conservation and garden floras

As explained in Chapter 24, *ex situ* conservation plays a key role in securing plant diversity, not only as an insurance policy for the future, but also as a basis for restoration and reintroduction of endangered plants into the wild. The increasing awareness of climate change impacting on plant distributions *in situ* has made *ex situ* conservation more crucial. Moreover, endemic and naturally rare species are especially vulnerable to extinction, since they are narrowly distributed with few populations and small numbers of individuals, which make them particularly vulnerable to changes in habitats. While *ex situ* conservation priorities have continuously focused on threatened and endangered plants, species which are most vulnerable to environmental and habitat changes in the overall background of climate changes deserve more conservation attention. These include taxa with nowhere to go, such as those restricted to mountain summits, low-lying islands, high latitudes and the edges of continents, plants with restricted ranges such as rare and endemic species, taxa with poor dispersal capacity or long generation times, species that are susceptible to extreme conditions such as flood or drought, plants with extreme habitat or niche specialization such as narrow tolerance to climate-sensitive variables, taxa with co-evolved or synchronous relationships with other species, and species with inflexible physiological responses to climate variables (Gran Canaria Group, 2006).

Although the 160 botanical gardens in China hold *ex situ* living collections of almost 25 000 plant species in total, this is still inadequate in relation to overall representation of the Chinese native flora, genetic integrity and population diversity, assessment of genetic risks designated for recovery and restoration, and coverage of threatened or potentially threatened plants (currently 38% of the newly-updated *Red List* species are conserved in this way). Future priorities for China's national *ex situ* conservation strategy need to include enhancement of *ex situ* living collections, increasing coverage of the Chinese native flora, backing-up of important taxa by replication in at least two botanical gardens, increasing the proportion of taxa directly collected from the wild, and maintaining the species integrity of existing *ex situ* living collections. Immediate measures should be taken to collect the taxa most vulnerable to climate change to ensure against future losses (Huang, 2010; Huang, 2011).

A nationwide inventory of *ex situ* living collections is needed to focus future developments towards a goal of a well-documented Chinese garden flora. A ongoing five-year national project of this inventory focus on providing current *ex situ* conserved plant species with adequate collecting data, comprehensive evaluation of *ex situ* living collections, development of standardized information systems for data collection and management of *ex situ* living collections in Chinese botanical gardens, establishment of a database of *ex situ* living collections for 90% of species currently cultivated in botanical gardens (i.e. an *ex situ* Flora of China), enhanced conservation management and information service provision to the public, and provision of sound, scientifically-based measures for conservation of plant diversity and sustainable use of plant resources (Huang, 2010).

Complementing the *ex situ* living plant collections is the rapidly increasing capacity of seed banks. For example, the China Southwest Wildlife Germplasm Genebank now includes over 54 000 accessions of seeds from almost 7 300 plant species. The goal of plant seed collection has been set for the next ten years: to hold 190 000 accessions and 19 000 plant species by 2020. In particular, the need for *ex situ* conservation of diverse crop material has been recognized as increasingly urgent, with a particular need to focus on the wild relatives and local varieties of crops as a rich source of diversity and adaptive traits for extreme abiotic conditions. The national seed bank system operated by the Ministry of Agriculture has the capacity to hold more than one million accessions. However, the extent of genetic diversity of crop species presently in cold storage facilities varies considerably from crop to crop. More emphasis should be put on the many minor and local crop plants and wild relatives of major crops, and on increasing representation of the genetic diversity of each crop.

Many positive case studies of *ex situ* conservation are presented in the previous chapter. To provide just one note of caution here, *Sinojackia xylocarpa* (Styraceae) provides an example of genetic risk in *ex situ* conservation. From a conservation perspective, there is a growing awareness of the negative impact of hybridization on the conservation of species through genetic processes, such as out-breeding depression, genetic assimilation, and hybridization swarming. The harmful effects of hybridization leading to the extinction of threatened and endangered populations and species have been reported for many plants in the wild. However, little attention has been paid to the threat of hybridization to endangered or rare species in *ex situ* living collections in botanic gardens, despite the fact that a large number of congeneric species are cultivated in close proximity in botanic gardens. This proximity may raise potential hybridization risks to species that are extinct in the wild, when garden-produced offspring of the wild extinct species are reintroduced.

Sinojackia xylocarpa is a well-documented Chinese endemic species that is extinct in the wild and has been investigated as a model to address this concern. Paternity analyses have been used to assess spontaneous hybridization and patterns of pollen flow between *S. xylocarpa* and *S. rehderiana* (Ye *et al.*, 2006; Zhang *et al.*, 2010), in conserved populations in Wuhan Botanic Garden. Interspecific hybridization events were detected in seven out of eight maternal trees of *S. xylocarpa*, and an average of 33% seeds collected from maternal trees of *S. xylocarpa* were hybrids. The paternity of 93 out of 249 seedlings from *S. xylocarpa* assigned to *S. rehderiana* provided convincing evidence that spontaneous interspecific hybridization occurred extensively in the living garden collection. Different patterns of pollen dispersal (predominantly short-distance compared to long-distance pollination) were observed between intraspecific and interspecific hybridization events in the garden. Pollen dispersal within the *ex situ* populations was not restricted by distance between maternal and paternal trees, ranging from ten to 620 m. Such extensive hybridization in *ex situ* collections could jeopardize the genetic integrity of endangered species and irrevocably contaminate the gene pool if such hybrids are used for reintroduction and restoration. This case study recommends strongly that measures be taken to minimize the genetic risks of this kind of hybridization, including establishing buffer zones in *ex situ* collections, manipulating flowering phenology, testing seeds before use in reintroduction programs, and controlling pollination for seed purity. An increasing number of rare and endangered plant species are being conserved in botanic gardens. To achieve successful reintroduction to the wild of these endangered species, the design and management of such *ex situ* populations before reintroduction are important tasks. This study provides convincing evidence challenging ongoing seed-exchange practices among botanical gardens worldwide (Ye *et al.*, 2006; Zhang *et al.*, 2010).

25.3 Research and sustainable utilization of plants

Although first Chinese garden may date back more than 3 000 years, the modern concept of a botanic garden originated in Europe, with a history of about 470 years (Italy's Orto Botanico di Padova was built in 1545). In general, scientific botanic gardens in a modern sense started with the collection, evaluation and display of medicinal plants. These were expanded as western nations built colonies across the world, from which plants were collected, introduced, domesticated, utilized and studied on an unprecedentedly large scale. About a century ago, botanic gardens gradually settled into a role as integrated scientific institutions conducting botanical research. There is now a strategic need to think deeply about the research focus and direction of Chinese botanic gardens in the twenty-first century from a global perspective of plant resource conservation and sustainable utilization. As Wagner (1972) has stated, "a botanic garden without scientific research is just a park. A botanic garden with scientific research plays an important role in social development." Moreover, the modern life sciences have entered the era of molecular biology, so that traditional scientific research in botanic gardens now faces great opportunities as well as a serious challenges, particularly in macro-morphological research.

25.3.1 Research scope and targets

Botanical institutions and botanic gardens should take into account the current status of China's native plant resources and nationwide needs in shaping their position. I believe that, while they should focus on the safety of plant resources and the sustainable utilization of plant resources in China, they should also support the sustainable development of the national economy and the establishment of a harmonious society in China. Botanical institutions and botanic gardens should reinforce the construction of specialized gardens of endemic and native plant resources (especially plants with economic and environmental potential). Focus should be placed on the selection, evaluation, sustainable utilization and research into critical plant resources that are required by China's various biological industries. The scope and targets of this research should further be focused on the basis of current plant resource collections, conservation and research, shifting from germplasm collection and conservation to the discovery and sustainable utilization of useful resources. Major principles should be securing quantity, improving quality, enhancing discovery and encouraging utilization.

In order to enhance scientific exchange and data-sharing nationally and internationally, a national scientific criterion system for the strategic conservation of plant resources and sustainable utilization practices in the Chinese botanic garden network (including establishment of a digitalized botanic garden system) should be formulated and developed. The national conservation network should be strengthened to cover at least 80% of all native plant resources in China (about 30 000 higher plant species). This could make China a major component of the international system of biodiversity conservation and a leader in the sustainable utilization of biological resources. Botanical institutions and botanic gardens should take advantage of their accumulated research experience and the fact that China accounts for about 10% of the total number of plant species in the world. Chinese botanic gardens should take a leading role in the eastern Asian botanic garden network, and strengthen their role in guiding neighboring countries which are dedicated to Asian plant resource research (Huang, 2008; Huang, 2010).

25.3.2 Strengthening capacity for scientific research

Specialized gardens and related scientific research.— To keep pace with social and economic developments in China, botanical institutions and botanic gardens should further reinforce the construction of practical specialized gardens on the basis of current plant collections, and enhance plant germplasm evaluation and discovery of useful plant genetic resources to meet China's social and economic development.

Considering the growing demand for healthcare by the Chinese population, and on the basis of the geographic distribution of medicinal plants and the traditional status of Chinese medicines, the construction of medicinal plant gardens and discovery and research on medicinal plant resources should be reinforced through provision of medicinal plant gardens. Botanic gardens should focus on research into medicinal plant resources that can serve as a health supplements, tonics or special regulatory medicines and provide such information and products to the industry and society.

Considering also the demand for energy sources in China, the construction of gardens for industrial bio-energy sources and research into bio-diesel plants should be enhanced. Botanical institutions and botanic gardens should focus on research into bio-diesel plant resources that are rich in carbon and hydrogen. Screening, evaluation and cultivation of bio-diesel plants, and their application for industrial uses should be encouraged.

With increasing demand for urban landscape

development and environmental improvement in China, and the country's abundant endemic ornamental plant resources, the construction of gardens for landscaping and ornamental plants should also be reinforced. Botanic gardens should conduct research into new cultivars of flowers, landscaping plants and turf that will be suitable for different climatic zones in China and for commercial production. Botanic gardens should also focus on the discovery of native plant resources so to reinforce supply of fruit and vegetables, through specialized edible plant gardens, to develop and release new varieties of fruits and vegetables (such as special aquatic vegetables and healthy, functional fruit and vegetables).

Finally, considering the rate of construction in China, resulting in damage to the environment especially caused by mining, large-scale construction projects and urbanization, botanical institutions and botanic gardens should set up specialized gardens to conduct research into useful tree and shrub species and improved varieties for the restoration of degraded habitats and environmental amelioration (Huang, 2008).

Enhancing a research and development platform for molecular biology.—For more than 400 years, botanic gardens have made significant contributions to comparative biological research through traditional comparative morphology, anatomy and physiology. Modern life science and biotechnology research has now entered a molecular era, in which the substantial resource materials held by botanic gardens will provide valuable material for studies such as comparative functional genomics, phytochemistry and proteomics. Also, research into model genes and model plant species should be considered in botanic gardens, taking account of the enriched plant resources in *ex situ* collections. In this way botanic gardens can continue to support research in the life sciences and modern comparative biology.

Construction of digitalized botanic gardens.—A germplasm collection management system and an information sharing system for a National Botanic Garden System of China should be set up, in order to enhance resource management within the botanic garden network, improve the efficiency of research, development and information exchange, and raise the level of popular science education in the public media. This could be achieved by setting up, firstly, a National Botanic Gardens information management protocol including policies and technical standards, secondly a digitized platform for information management, based on geographic information systems and covering all Chinese botanic gardens, thirdly a National Botanic Gardens popular science education website for the public, and finally an intranet covering all National Botanic Gardens in China, using wireless technology to cover all areas of all botanical gardens (Huang, 2008; Huang, 2010; Huang, 2011).

25.3.3 Priority research targets

The extent of conservation of existing biological and genetic resources is strong proof of the sustainable development of China's economy and society. Certain key research targets should be greatly enhanced and prioritized in China, such as the implementation of *ex situ* conservation for plant community structure, and the scientific basis for sustainable reproduction in garden-based *ex situ* conservation. Other targets include the genetic mechanisms and population eco-genetic equilibrium of in-breeding and out-breeding depression resulting from species co-existence in the process of genetic enhancement in garden cultivation, and the genetic mechanisms underlying community reproduction, structural stability and ecological adaptability of endangered species when returned to their natural habitats. Such science-based research should enhance botanic gardens in China, as a reserve for strategically important plants to accomplish the conservation goal of covering over 80% of native plant species and 500 000 special genetic accessions in *ex situ* conservation. This in turn guarantees the safety of China's biological and genetic resources.

There is also a need to set up a phytochemical resource bank for China's medicinal plants, a DNA bank for rare, endangered and endemic plants, and related data information management systems. All these will be in support of knowledge innovation and valuable raw materials for the development of relevant biological industries in China. It is necessary to conduct research and development for the sustainable utilization and effective models for economic plants which are urgently needed by biological industries in China. Such research and development should reach the standard of industrial testing. Important research topics and directions that we believe should be developed with priority in the near future include: the theory and practice of plant genetic enrichment in gardens *ex situ* conservation, the population-genetic basis of plant resource conservation and sustainable utilization of garden-conserved genetic resources, plant resource recovery and its ecological function, the impact of botanical bio-safety and plant resources on global environment changes, and the theory and practice of integrated conservation for plant genetic resources (Huang, 2008).

Botanical institutions and botanic gardens are undertaking to secure the safety of China's plant resources and ensure the sustainable use of these resources. They are irreplaceable in achieving these two missions. Meanwhile, further refining the research scope of botanical institutions and botanic gardens as well as strengthening both the networking and research facilities for botanic gardens in China will provide a solid foundation for the sustainable utilization of plant resources. This will in turn support the safety of plant germplasm resources in China, especially endemic plants, critically endangered plants, species with high economic or scientific research value, and keystone species in major ecosystems in China, allowing the sustainable utilization of China's rich plant germplasm resources to be realized (Zhang & Huang, 2010; Huang, 2011).

References

BGCI (2000). *The Gran Canaria Declaration Calling for a Global Program for Plant Conservation.* London, UK: BGCI.

Brooks, T. M., Mittermeier, R. A., da Fonseca, G. A. B., Gerlach, J., Hoffmann, M., Lamoreux, J. F., Mittermeier, C. G., Pilgrim, J. D. & Rodrigues, A. S. L. (2006). Global biodiversity conservation priorities. *Science* 313: 58-61.

CBD Secretariat (2003). *Global Strategy for Plant Conservation.* Montreal, Canada: CBD Secretariat.

China's Strategy for Plant Conservation Editorial Committee (2008). *China's Strategy for Plant Conservation.* Guangzhou, China: Guangdong Press Group.

FAO (1998). *The State of the World's Plant Genetic Resources for Food and Agriculture.* Rome, Italy: FAO.

Gao, L. M., Möller, M., Zhang, X. M., Hollingsworth, M. L., Liu, J., Mill, R. R., Gibby, M. & Li, D. Z. (2007). High variation and strong phylogeographic pattern among cpDNA haplotypes in *Taxus wallichiana* (Taxaceae) in China and North Vietnam. *Molecular Ecology* 16: 4684-4698.

Gran Canaria Group (2006). *The Gran Canaria Declaration II on Climate Change and Plants.* Gran Canaria, Spain: Jardin Botanico Canario "Viera y Clavijo" and BGCI.

Hawkins, B., Sharrock, S. & Havens, K. (2008). *Plants and Climate Change: Which Future?* Richmond, UK: BGCI.

Huang, H. W. (2008). Future scientific research in botanic gardens. *In:* Huang, H. W. & Leadlay, E. (eds.) *Third Global Botanical Gardens Congress.* pp. 11-18. Wuhan, China: BGCI.

Huang, H. (2010). *Ex situ* plant conservation-a key role of Chinese botanical gardens in implementing China's strategy for plant conservation. *Botanic Gardens Journal* 7(2): 14-19.

Huang, H. (2011). Plant diversity and conservation in China: planning a strategic bioresource for a sustainable future. *Botanical Journal of the Linnean Society* 166: 282-300.

Li, Z. S. (1996). Speech. *In: CAS Biodiversity Symposium.* Beijing, China.

Liu, Y. F., Wang, Y. & Huang, H. W. (2009). Species-level phylogeographical history of *Myricaria* plants in the mountain ranges of western China and the origin of *M. laxiflora* in the Three Gorges mountain region. *Molecular Ecology* 18: 2700–2712.

Pitman, N. C. A. & Jørgensen, P. M. (2002). Estimating the size of the world's threatened flora. *Science* 298: 989.

Wagner, W. H. (1972). Botanical research at botanical gardens (Proceedings of the symposium in a national garden system for Canada). *Technical Bulletin of the Royal Botanical Gardens, Hamilton* 6: 1-6.

Wang, J., Gao, P. X., Kang, M., Lowe, J. A. & Huang, H. W. (2009). Refugia within refugia: the case study of a canopy tree (*Eurycorymbus cavaleriei*) in subtropical China. *Journal of Biogeography* 36: 2156-2164.

Wang, S. & Xie, Y. (eds.) (2004). *China Species Red List 1.* Beijing, China: Higher Education Press.

Ye, Q. G., Yao, X. H., Zhang, S. J., Kang, M. & Huang, H. H. (2006). Potential risk of hybridization in *ex situ* collections of two endangered species of *Sinojackia* Hu (Styracaceae). *Journal of Integrative Plant Biology* 48(7): 867-872.

Zhang, J. J., Ye, Q. G., Yao, X. H. & Huang, H. H. (2010). Spontaneous interspecific hybridization and patterns of pollen dispersal in *ex situ* populations of a tree species (*Sinojackia xylocarpa*) that is extinct in the wild. *Conservation Biology* 24: 246-255.

Zhang, X. S. & Huang, H. H. (2010). *Science and Technology on Bio-hylic and Biomass Resources in China: a Roadmap to 2050.* Beijing, China: Science Press.

Index to Scientific Names

I

J

K

M

U

V

W

X

Y

Z

APPENDIX Table of Chinese Dynasties

YI Ting-Shuang and Peter L. Morrell

Dynasty	Dates
Xia	ca. 2100–1600 BC
Shang	ca. 1600–1100 BC
Zhou:	
Western Zhou	ca. 1100–771 BC
Eastern Zhou – Spring and Autumn Period	770–ca. 476 BC
Eastern Zhou – Warring States Period	475–221 BC
Qin	221–206 BC
Han:	
Western Han	202 BC–8 AD
Eastern Han	25–220 AD
Three Kingdoms:	
Wei	220–265 AD
Shu	221–263 AD
Wu	229–280 AD
Jin:	
Western Jin	265–316 AD
Eastern Jin	317–420 AD
Northern and Southern Dynasties	420–581 AD
Sui	581–618 AD
Tang	618–907 AD
Five Dynasties Period	907–960 AD
Song:	
Northern Song	960–1127 AD
Southern Song	1127–1279 AD
Yuan	1271–1368 AD
Ming	1368–1644 AD
Qing	1644–1911 AD

Figure Credits

Some of the illustrations and tables in this book are reproduced from sources noted in the individual captions. The authors and publishers acknowledge the sources of copyright material below and are grateful for the permission granted. While every effort has been made, it has not always been possible to identify the sources of all material used, or to trace all copyright holders. If any omissions are brought to our notice, we will be happy to include the appropriate acknowledgements on reprinting.

Figure 1.1 is courtesy of HONG De-Yuan, Institute of Botany, Chinese Academy of Sciences. Reprinted with permission.

Figures 2.1, 2.2, 2.3, 2.4, 2.5, 2.6, 2.7, 2.8, 2.9, 2.10, 2.11, 2.12, 2.13, 2.14, 2.15, 2.16, 2.17, 2.18, 2.19, 2.20, 2.21, 2.22, 2.23, 2.24, 2.25, 2.26, 2.27, 2.28, 2.29, 2.30, 2.31, 2.32, 2.33, 23.1, 23.2, 23.3, 23.4, 23.5, 23.6, 23.7, 23.8, 23.9, 23.10, 23.11, 24.1, 24.2, 24.3, 24.4, 24.5, 24.6, 24.7, 24.8, 24.9, 24.10, 24.11, 24.12, 24.13, 24.14, 24.15 and 24.16 are courtesy of the South China Botanical Garden, Chinese Academy of Sciences. Reprinted with permission.

Figures 3.1, 3.2, 3.3, 3.4, 3.5, 3.6 and 3.7 are courtesy of Zheng Du, Dai Erfu, Zhang Xue-Qin and Wang Xiuhong. Reprinted with permission.

Figures 4.1, 4.2, 4.3, 4.4, 4.5, 4.6, 4.7, 4.8, 4.9, 4.10, 4.12 and 4.13 are courtesy of Zhang Xue-Qin. Reprinted with permission.

Figures 4.11 and 4.14 are courtesy of Zheng Du, Dai Erfu, Zhang Xue-Qin and Wang Xiuhong. Reprinted with permission.

Figures 5.1 and 5.2 are courtesy of GONG Zi-Tong. Reprinted with permission.

Figures 6.1, 6.2, 6.3, 6.4, 6.5, 7.1 and 7.2 are courtesy of Zhou Zhe-Kun (Z. K. Zhou). Reprinted with permission.

Figures 8.1, 8.11 and 8.12 are courtesy of SANG Wei-Guo. Reprinted with permission.

Figures 8.2, 8.3, 8.4, 8.5, 8.6, 8.7, 8.9, 8.10, 8.13, 8.14, 8.15, 8.16, 8.19 and 8.25 are courtesy of CHEN Ling-Zhi. Reprinted with permission.

Figures 8.8, 8.29, 8.30, 8.32 and 8.34 are courtesy of GUO Ke. Reprinted with permission.

Figures 8.17, 8.18, 8.20, 8.22 and 8.23 are courtesy of CHEN Wei-Lie. Reprinted with permission.

Figures 8.24, 8.26 and 8.27 are courtesy of ZHU Hua. Reprinted with permission.

Figures 8.21 and 8.28 are courtesy of CHEN Bin. Reprinted with permission.

Figures 8.31, 8.33, 8.35, 8.36, 8.37, 8.38 and 8.39 are courtesy of BAI Yong-Fei. Reprinted with permission.

Figures 9.1, 9.2, 9.3, 9.4, 9.5, 9.6, 9.7, 9.8, 9.9 and 9.10 are courtesy of Peng Hua. Reprinted with permission.

Figure 10.1 courtesy of Hang Sun

Figures 12.1, 12.2, 12.3 and 12.4 are courtesy of the Royal Botanic Garden Edinburgh

Figures 12.6, 12.7, 12.8, 12.9, 12.10, 12.11 and 12.12 are courtesy of Hu Qi Ming. Reprinted with permission.

Figure 12.13 is courtesy JIN Xiao-Hua

Figures 12.14 and 12.15 are courtesy of James L. Luteyn

Figures 12.16 and 12.17 are courtesy of Bruce Bartholomew from an original by Linsley Gressitt

Figure 12.18 is courtesy Robert J. Mitchell from a photograph by GUAN Kai-Yun

Figure 12.19 is courtesy of Robert J. Mitchell

Figure 12.20 is courtesy of David G. Long

Figures 12.21, 12.22 and 12.23 are courtesy of Mark F. Watson

Figure 12.24 is courtesy of Susan Kelley

Figure 14.1 is courtesy of Kunming Institute of Botany archive

Figure 14.2 is courtesy of YANG Yun-Shan

Figures 16.1 and 16.2 are courtesy of WANG, L. S. Reprinted with permission.

Figure 16.3 is courtesy HE, S. A. Reprinted with permission.

Figures 17.1, 17.2, 17.7, 17.8, 17.9, 17.10, 17.11, 17.12, 17.13, 17.14 and 17.15 are courtesy of Yang Yuming. Reprinted with permission.

Figures 17.3, 17.4 and 17.6 are courtesy of Wang Juan. Reprinted with permission.

Figures 17.16, 17.17, 18.4, 18.5, 18.6, 18.7, 18.8, 18.9, 18.10, 18.11, 18.12, 18.13, 18.14, 18.15, 18.16, 18.17, 18.18, 18.19, 21.1, 21.2, 21.3, 21.4, 21.5, 21.6, 21.7, 21.8, 21.9, 21.10, 21.11, 21.12, 21.13 and 21.14 are courtesy of Pei Shengji. Reprinted with permission.

Figures 19.1, 19.2, 19.3, 19.4, 19.5, 19.6, 19.7, 19.8, 19.9, 19.10, 19.11, 19.12, 19.13, 19.14, 19.15, 19.16, 19.17, 19.18, 19.19, 19.20, 19.21, 19.22 and 19.23 are courtesy of He Shanan. Reprinted with permission.

Figure 20.1 is courtesy of Tingshuang Yi. Reprinted with permission.

Printed in the United States
by Baker & Taylor Publisher Services